The University of Chicago School Mathematics Project

Algebra

Second Edition

Teacher's Edition

Part 2, Chapters 7-13

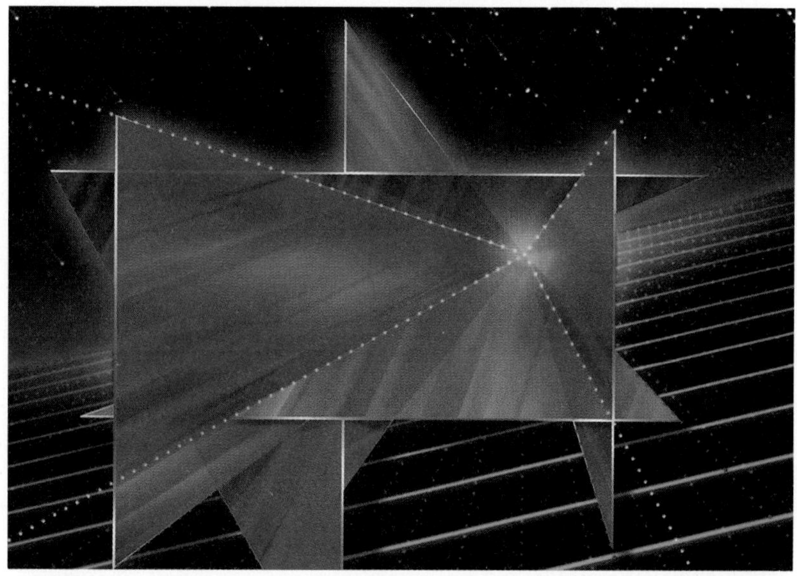

About the Cover The art on the cover was generated by a computer. The planes, grid, and intersecting lines suggest the integrated approach of *UCSMP Algebra.* This course uses geometry and statistics as a setting for work with linear expressions and sentences, and much work is done with graphing.

Authors

John W. McConnell   Susan Brown

Zalman Usiskin   Sharon L. Senk   Ted Widerski   Scott Anderson

Susan Eddins   Cathy Hynes Feldman   James Flanders   Margaret Hackworth

Daniel Hirschhorn   Lydia Polonsky   Leroy Sachs   Ernest Woodward

ScottForesman

A Division of HarperCollins*Publishers*

Editorial Offices: Glenview, Illinois

Regional Offices: Sunnyvale, California • Tucker, Georgia

Glenview, Illinois • Oakland, New Jersey • Dallas, Texas

Contents

of Teacher's Edition

The complete Table of Contents for the Student Edition begins on page *vi*.

Your UCSMP Professional Sourcebook is found at the back of Part 1, starting on page T20.

ISBN: 0-673-45767-2

3 4 5 6 7 8 9—DR—0 1 0 0 9 9 9 8 9 7 9 6

CONTENTS

USES OF VARIABLES

MULTIPLICATION IN ALGEBRA

Adapting to Individual Needs

The student text is written for the vast majority of students. The chart at the right suggests two pacing plans to accommodate the needs of your students. Students in the Full Course should complete the entire text by the end of the year. Students in the Minimal Course will spend more time when there are quizzes and more time on the Chapter Review. Therefore, these students may not complete all of the chapters in the text.

Options are also presented to meet the needs of a variety of teaching and learning styles. For each lesson, the Teacher's Edition provides sections entitled: *Video* which describes video segments and related questions that can be used for motivation or extension; *Optional Activities* which suggests activities that employ materials, physical models, technology, and cooperative learning; and, *Adapting to Individual Needs* which regularly includes **Challenge** problems, **English Language Development** suggestions, and suggestions for providing **Extra Help.** The Teacher's Edition also frequently includes an **Error Alert,** an **Extension,** and an **Assessment** alternative. The options available in Chapter 7 are summarized in the chart below.

Chapter 7 Pacing Chart

Day	Full Course	Minimal Course
1	7-1	7-1
2	7-2	7-2
3	7-3	7-3
4	Quiz*; 7-4	Quiz*; begin 7-4.
5	7-5	Finish 7-4.
6	7-6	7-5
7	Quiz*; 7-7	7-6
8	7-8	Quiz*; begin 7-7.
9	7-9	Finish 7-7.
10	Self-Test	7-8
11	Review	7-9
12	Test*	Self-Test
13		Review
14		Review
15		Test*

*in the Teacher's Resource File

In the Teacher's Edition...

Lesson	Optional Activities	Extra Help	Challenge	English Language Development	Error Alert	Extension	Cooperative Learning	Ongoing Assessment
7-1	●	●	●	●	●	●	●	Oral
7-2	●	●	●	●	●	●	●	Written
7-3	●	●	●	●	●	●	●	Group
7-4	●	●	●			●	●	Written
7-5	●	●	●	●	●	●	●	Group
7-6	●	●	●	●	●	●	●	Oral
7-7	●	●	●	●		●	●	Written
7-8	●	●	●	●	●	●	●	Written
7-9	●	●	●	●		●		Oral

In the Additional Resources...

	In the Teacher's Resource File								
Lesson	Lesson Masters, A and B	Teaching Aids*	Activity Kit*	Answer Masters	Technology Sourcebook	Assessment Sourcebook	Visual Aids**	Technology	Video Segments
Opener		74					74		
7-1	7-1	28, 71, 75–77		7-1			28, 71, 75–77, AM		
7-2	7-2	26, 71		7-2			26, 71, AM		
7-3	7-3	28, 71, 78–80		7-3		Quiz	28, 71, 78–80, AM		7-3
7-4	7-4	28, 72	16	7-4	Comp 16		28, 72, AM	GraphExplorer	
7-5	7-5	28, 72		7-5			28, 72, AM		
7-6	7-6	28, 72		7-6		Quiz	28, 72, AM		
In-class Activity		81		7-7			81, AM		
7-7	7-7	28, 73, 82–84	17	7-7	Comp 17		28, 73, 82–84, AM	GraphExplorer	
7-8	7-8	28, 73		7-8	Calc 5		28, 73, AM		
7-9	7-9	28, 73, 85, 86		7-9	Comp 18		23, 73, 85, 86, AM	GraphExplorer	
End of chapter				Review		Tests			

*Teaching Aids, except Warm-ups, are pictured on pages 416C and 416D. The activities in the Activity Kit are pictured on page 416C. **Teaching Aid 81** which accompanies the In-class Activity preceding **Lesson 7-7** is pictured with the lesson notes on page 456.

**Visual Aids provide transparencies for all Teaching Aids and all Answer Masters.

Also available is the Study Skills Handbook which includes study-skill tips related to reading, note-taking, and comprehension.

Integrating Strands and Applications

	7-1	7-2	7-3	7-4	7-5	7-6	7-7	7-8	7-9
Mathematical Connections									
Algebra	●	●	●	●	●	●	●	●	●
Geometry		●	●	●		●			●
Measurement			●	●		●			●
Statistics/Data Analysis	●	●	●		●	●	●	●	●
Patterns and Functions	●	●	●	●	●	●	●		●
Discrete Mathematics		●					●	●	
Interdisciplinary and Other Connections									
Literature									●
Science	●	●		●	●	●	●		●
Social Studies	●		●	●	●	●	●	●	●
Multicultural	●		●		●		●		
Technology		●	●	●			●	●	●
Consumer		●	●	●	●	●	●	●	●
Sports		●	●					●	●

Teaching and Assessing the Chapter Objectives

				In the Teacher's Resource File		
Chapter 7 Objectives (Organized into the SPUR categories—Skills, Properties, Uses, and Representations)	**Lessons**	**Progress Self-Test Questions**	**Chapter Review Questions**	**Chapter Test, Forms A and B**	**Chapter Test, Forms C**	**D**
Skills						
A: Find the slope of the line through two given points.	7-2	3	1–6	3	1	
B: Find an equation for a line given two points on it, or its slope and one point on it.	7-4, 7-5, 7-6	7, 14	7–16	7, 13	2	
C: Write an equation for a line in standard form or slope-intercept form, and from either form find its slope and y-intercept.	7-4, 7-8	5, 6, 8	17–24	4, 5, 8	2	
Properties						
D: Use the definition and properties of slope.	7-2, 7-3	4, 9, 10, 20	25–32	6, 9, 16, 18	1	
Uses						
E: Calculate rates of change from real data.	7-1, 7-3	11, 12	33–40	14, 15	4	✓
F: Use equations for lines to describe real situations.	7-4, 7-5, 7-6, 7-8	13, 15	41–48	10, 20	5	
G: Given data whose graph is approximately linear, find a linear equation to fit the graph.	7-7	21	49, 50	17		✓
Representations						
H: Graph a straight line given its equation, or given a point and the slope.	7-3, 7-4, 7-8	14, 16, 17, 19	51–58	11	3	
I: Graph linear inequalities.	7-9	18	59–68	12, 19	6	

Multidimensional Assessment
Quiz for Lessons 7-1 through 7-3
Quiz for Lessons 7-4 through 7-6
Chapter 7 Test, Forms A–D
Chapter 7 Test, Cumulative Form

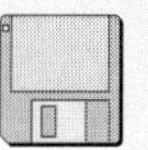

Quiz and Test Writer
Multiple forms of chapter tests and quizzes; Challenges

Activity Kit

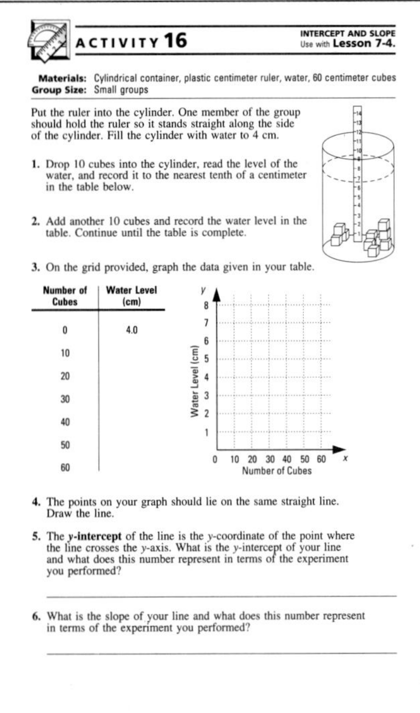

ACTIVITY 16 — INTERCEPT AND SLOPE. Use with **Lesson 7-4.**

Materials: Cylindrical container, plastic centimeter ruler, water, 60 centimeter cubes
Group Size: Small groups

Put the ruler into the cylinder. One member of the group should hold the ruler so it stands straight along the side of the cylinder. Fill the cylinder with water to 4 cm.

1. Drop 10 cubes into the cylinder, read the level of the water, and record it to the nearest tenth of a centimeter in the table below.
2. Add another 10 cubes and record the water level in the table. Continue until the table is complete.
3. On the grid provided, graph the data given in your table.

Number of Cubes	Water Level (cm)
0	4.0
10	
20	
30	
40	
50	
60	

4. The points on your graph should lie on the same straight line. Draw the line.
5. The ***y*-intercept** of the line is the *y*-coordinate of the point where the line crosses the *y*-axis. What is the *y*-intercept of your line and what does this number represent in terms of the experiment you performed?
6. What is the slope of your line and what does this number represent in terms of the experiment you performed?

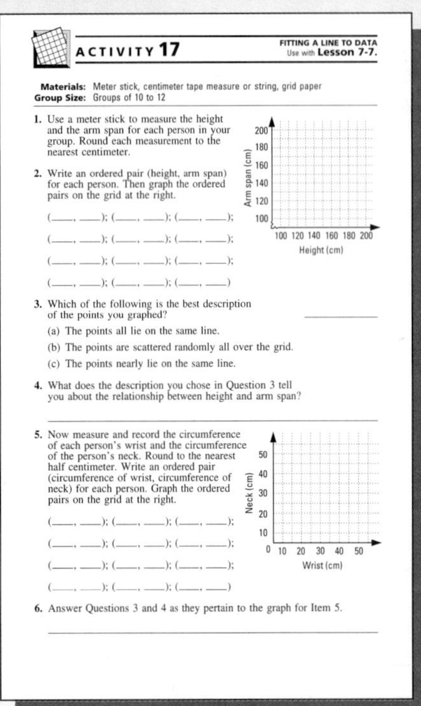

ACTIVITY 17 — FITTING A LINE TO DATA. Use with **Lesson 7-7.**

Materials: Meter stick, centimeter tape measure or string, grid paper
Group Size: Groups of 10 to 12

1. Use a meter stick to measure the height and the arm span for each person in your group. Round each measurement to the nearest centimeter.
2. Write an ordered pair (height, arm span) for each person. Then graph the ordered pairs on the grid at the right.

(___, ___); (___, ___); (___, ___);
(___, ___); (___, ___); (___, ___);
(___, ___); (___, ___); (___, ___);
(___, ___); (___, ___); (___, ___)

3. Which of the following is the best description of the points you graphed?
   (a) The points all lie on the same line.
   (b) The points are scattered randomly all over the grid.
   (c) The points nearly lie on the same line.
4. What does the description you chose in Question 3 tell you about the relationship between height and arm span?
5. Now measure and record the circumference of each person's wrist and the circumference of the person's neck. Round to the nearest half centimeter. Write an ordered pair (circumference of wrist, circumference of neck) for each person. Graph the ordered pairs on the grid at the right.

(___, ___); (___, ___); (___, ___);
(___, ___); (___, ___); (___, ___);
(___, ___); (___, ___); (___, ___);
(___, ___); (___, ___); (___, ___)

6. Answer Questions 3 and 4 as they pertain to the graph for Item 5.

Teaching Aids

Teaching Aid 26, Graph Paper, (shown on page 140D) can be used with **Lesson 7-2. Teaching Aid 28, Four-Quadrant Graph Paper,** (shown on page 140D) can be used with **Lessons 7-1** and **7-3 through 7-9.**

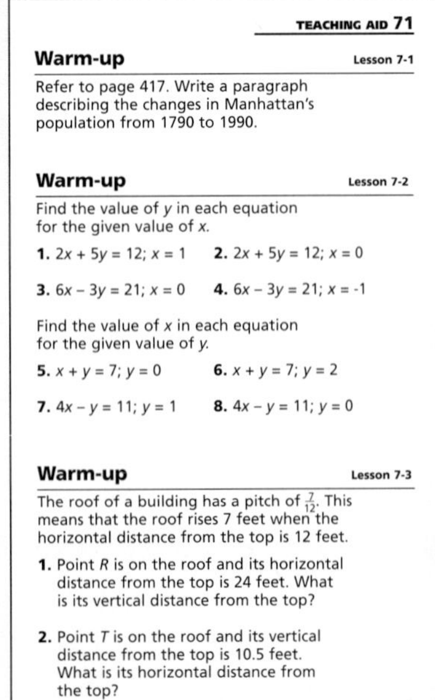

TEACHING AID 71

Warm-up — Lesson 7-1

Refer to page 417. Write a paragraph describing the changes in Manhattan's population from 1790 to 1990.

Warm-up — Lesson 7-2

Find the value of *y* in each equation for the given value of *x*.

1. $2x + 5y = 12$; $x = 1$
2. $2x + 5y = 12$; $x = 0$
3. $6x - 3y = 21$; $x = 0$
4. $6x - 3y = 21$; $x = -1$

Find the value of *x* in each equation for the given value of *y*.

5. $x + y = 7$; $y = 0$
6. $x + y = 7$; $y = 2$
7. $4x - y = 11$; $y = 1$
8. $4x - y = 11$; $y = 0$

Warm-up — Lesson 7-3

The roof of a building has a pitch of $\frac{7}{12}$. This means that the roof rises 7 feet when the horizontal distance from the top is 12 feet.

1. Point *R* is on the roof and its horizontal distance from the top is 24 feet. What is its vertical distance from the top?
2. Point *T* is on the roof and its vertical distance from the top is 10.5 feet. What is its horizontal distance from the top?

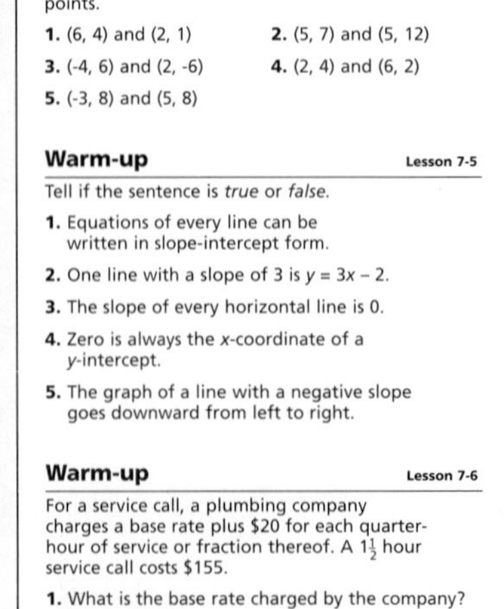

TEACHING AID 72

Warm-up — Lesson 7-4

Find the slope of a line through the given pair of points.

1. (6, 4) and (2, 1)
2. (5, 7) and (5, 12)
3. (-4, 6) and (2, -6)
4. (2, 4) and (6, 2)
5. (-3, 8) and (5, 8)

Warm-up — Lesson 7-5

Tell if the sentence is *true* or *false*.

1. Equations of every line can be written in slope-intercept form.
2. One line with a slope of 3 is $y = 3x - 2$.
3. The slope of every horizontal line is 0.
4. Zero is always the *x*-coordinate of a *y*-intercept.
5. The graph of a line with a negative slope goes downward from left to right.

Warm-up — Lesson 7-6

For a service call, a plumbing company charges a base rate plus $20 for each quarter-hour of service or fraction thereof. A $1\frac{1}{2}$ hour service call costs $155.

1. What is the base rate charged by the company?
2. Write an equation relating the cost of a service call *y* to the number of quarter-hours of service *x*.

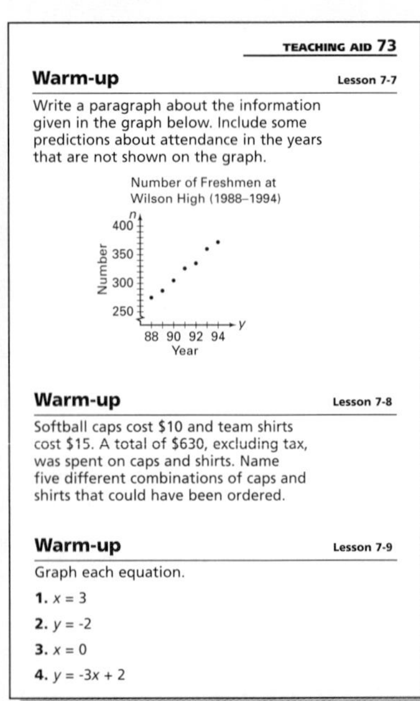

TEACHING AID 73

Warm-up — Lesson 7-7

Write a paragraph about the information given in the graph below. Include some predictions about attendance in the years that are not shown on the graph.

Warm-up — Lesson 7-8

Softball caps cost $10 and team shirts cost $15. A total of $630, excluding tax, was spent on caps and shirts. Name five different combinations of caps and shirts that could have been ordered.

Warm-up — Lesson 7-9

Graph each equation.

1. $x = 3$
2. $y = -2$
3. $x = 0$
4. $y = -3x + 2$

Population of Manhattan Island

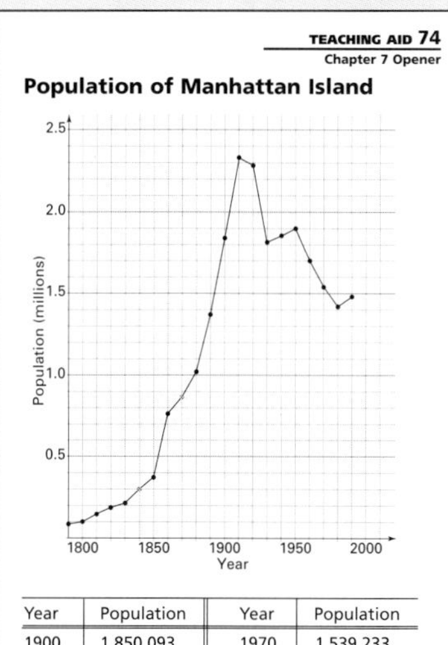

Year	Population	Year	Population
1900	1,850,093	1970	1,539,233
1910	2,331,542	1980	1,428,285
1920	2,284,103	1990	1,487,536
1960	1,698,281		

Manhattan Spreadsheet

	A	B	C
1			Rate of Change
2	Year	Population	for Previous Decade
3	1900	1850093	
4	1910	2331542	48144.9
5	1920	2284103	-4743.9
6	1930	1876412	
7	1940	1889924	
8	1950	1960101	
9	1960	1698281	
10	1970	1539233	
11	1980	1428285	
12	1990	1487536	

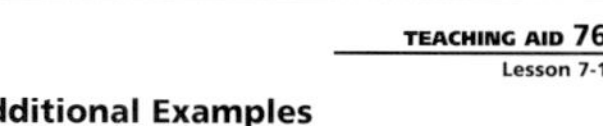

Additional Examples

1. The chart below shows Tom's income for each week of a six-week period.

Week	1	2	3	4	5	6
Dollars	11	15	0	21	25	25

a. Graph these points and connect them.

b. Find the rate of change of Tom's income between weeks 1 and 6.

c. During what time period was the rate of change of income negative? How can you tell?

d. When was the rate of change 0? How can you tell?

2. The graph shows the weight of Li-Li, the first giant panda born in captivity, at age 1 month, 3 months, and 6 months.

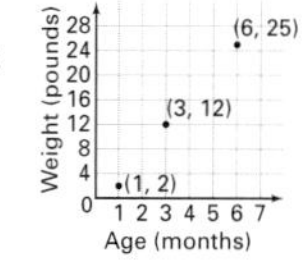

a. Find the rate of change in weight per month from 1 month to 3 months and from 3 months to 6 months.

b. During which time period did Li-Li's weight increase faster?

Questions 14–17

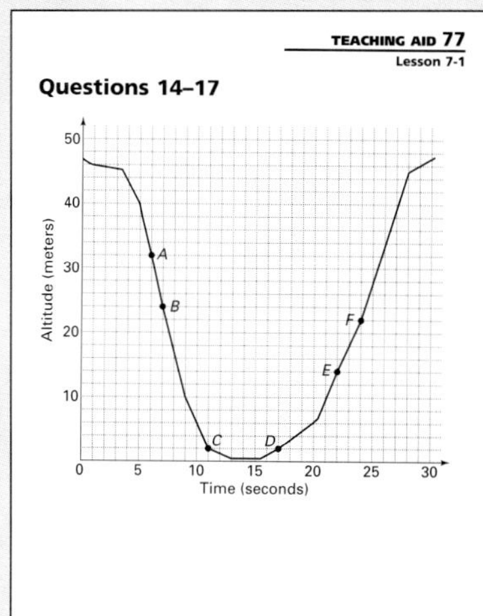

Example 2

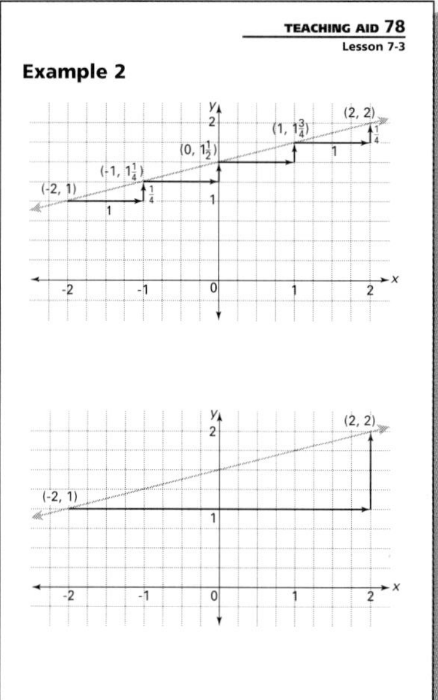

Questions 5–6, 16–18

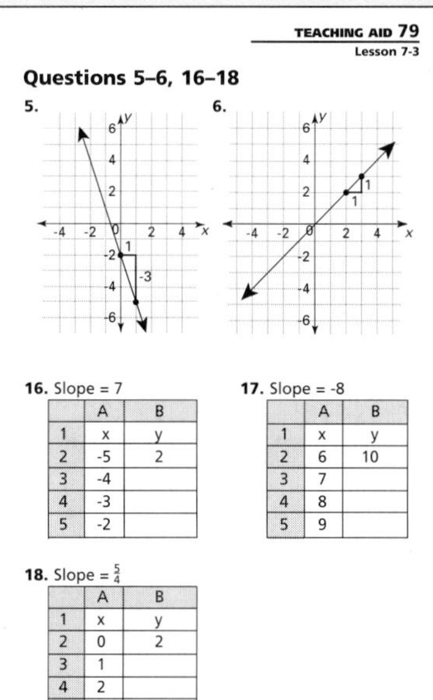

16. Slope = 7

	A	B
1	x	y
2	-5	2
3	-4	
4	-3	
5	-2	

17. Slope = -8

	A	B
1	x	y
2	6	10
3	7	
4	8	
5	9	

18. Slope = $\frac{5}{4}$

	A	B
1	x	y
2	0	2
3	1	
4	2	
5	3	

Questions 26–28

Average Home Runs per Game

Year	American League	National League
1981	1.42	1.12
1982	1.83	1.34
1983	1.68	1.44
1984	1.75	1.32
1985	1.93	1.47
1986	2.02	1.57
1987	2.32	1.88
1988	1.68	1.32
1989	1.51	1.41
1990	1.59	1.56
1991	1.72	1.47
1992	1.57	1.30
1993	1.83	1.72

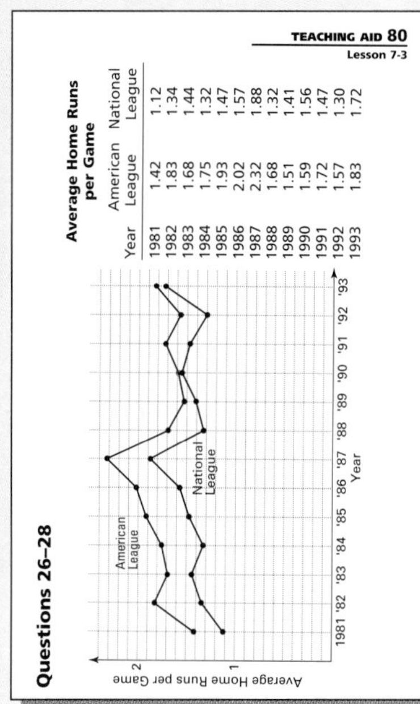

Additional Examples

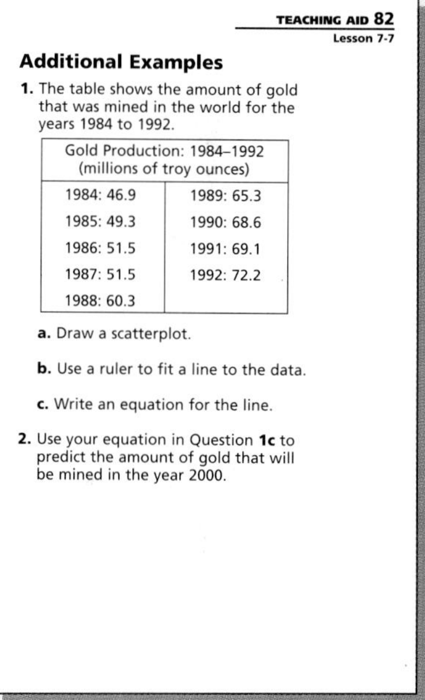

1. The table shows the amount of gold that was mined in the world for the years 1984 to 1992.

Gold Production: 1984–1992 (millions of troy ounces)	
1984: 46.9	1989: 65.3
1985: 49.3	1990: 68.6
1986: 51.5	1991: 69.1
1987: 51.5	1992: 72.2
1988: 60.3	

a. Draw a scatterplot.

b. Use a ruler to fit a line to the data.

c. Write an equation for the line.

2. Use your equation in Question **1c** to predict the amount of gold that will be mined in the year 2000.

Questions 12–15

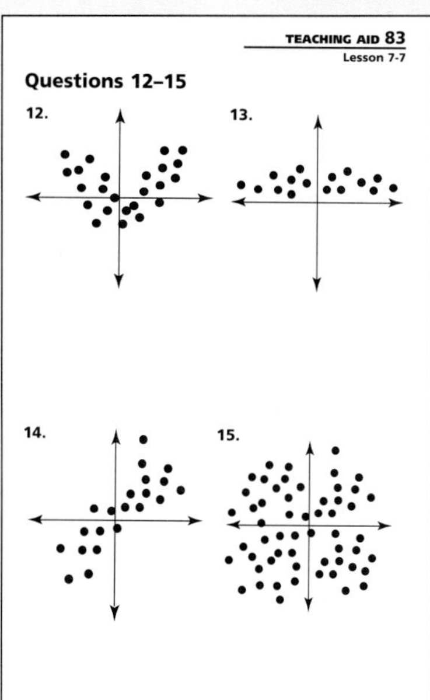

Question 16

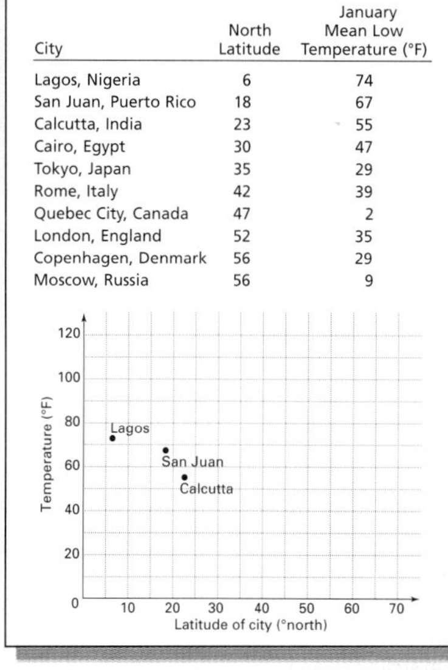

City	North Latitude	January Mean Low Temperature (°F)
Lagos, Nigeria	6	74
San Juan, Puerto Rico	18	67
Calcutta, India	23	55
Cairo, Egypt	30	47
Tokyo, Japan	35	29
Rome, Italy	42	39
Quebec City, Canada	47	2
London, England	52	35
Copenhagen, Denmark	56	29
Moscow, Russia	56	9

Additional Examples

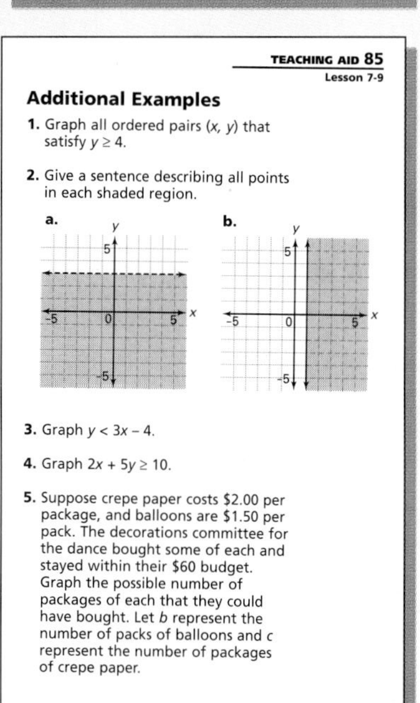

1. Graph all ordered pairs (x, y) that satisfy $y \geq 4$.

2. Give a sentence describing all points in each shaded region.

a. **b.**

3. Graph $y < 3x - 4$.

4. Graph $2x + 5y \geq 10$.

5. Suppose crepe paper costs $2.00 per package, and balloons are $1.50 per pack. The decorations committee for the dance bought some of each and stayed within their $60 budget. Graph the possible number of packages of each that they could have bought. Let b represent the number of packs of balloons and c represent the number of packages of crepe paper.

Question 24

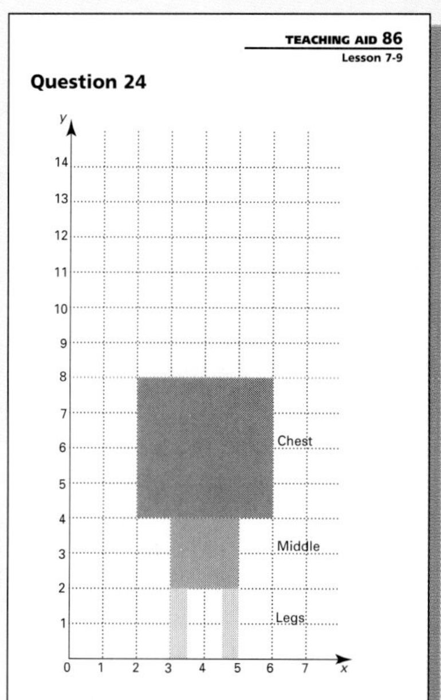

Chapter Opener

Pacing

All lessons in this chapter are designed to be covered in 1 day. At the end of the chapter, you should plan to spend 1 day to review the Progress Self-Test, 1 to 2 days for the Chapter Review, and 1 day for a test. You may wish to spend a day on projects, and possibly a day is needed for quizzes. This chapter should therefore take 12 to 15 days. We strongly advise you not to spend more than 16 days on this chapter.

Using Pages 416–417

When discussing the graph you might want to use **Teaching Aid 74**. The graph may surprise those students who think that cities are growing in population or who think that populations are either increasing or decreasing in some reasonably consistent manner. Point out that by connecting the points, we get a visual picture of the changes in population. Explain that the graph should be read from left to right. Discuss the upward and downward slant of the line segments as representing a growth or a decline in the population. The steeper the line, the faster the rate of increase or decrease.

Photo Connections

The photo collage makes real-world connections to the content of the chapter: slopes and lines.

Manhattan: Because of new bridges and subways built around 1910, the population of New York City spread out and the population of Manhattan decreased. The most recent population increase is due to a surge of immigration and to higher birth rates.

CHAPTER 7

416

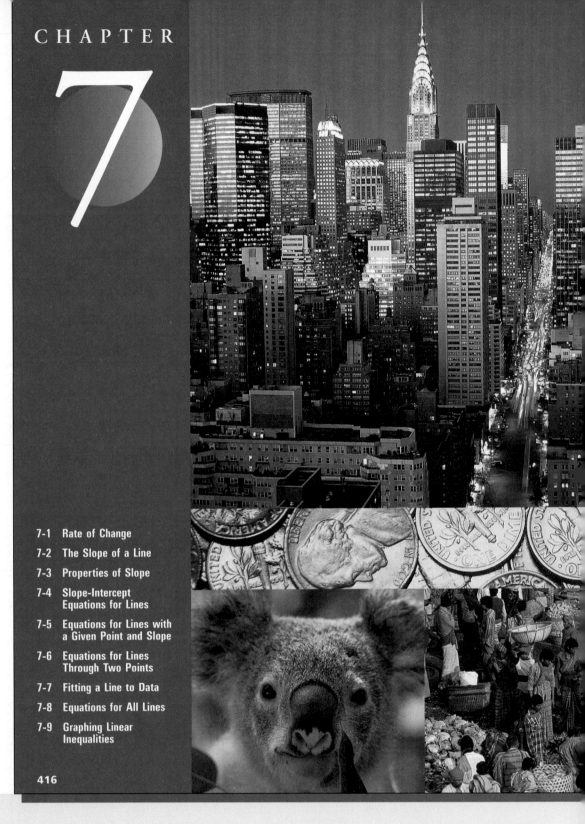

Chapter 7 Overview

Now, more than ever before, it is important that students have a good understanding of the relationships between linear equations and their graphs. The ability of computers to display graphs quickly has made graphs far more important than they were a generation ago. Graphing lines by using a table of values was covered in Chapters 4 and 5. The emphasis in this chapter is on finding and analyzing equations of lines, and relating the equations to the applications that give rise to them and with the geometry of the plane.

The chapter begins with rates. The goal is to have students interpret the slope of a line as a rate. In Lesson 7-2, a line emerges as a graph of a situation involving a constant rate of change, and the slope is defined in the traditional fashion. The results should remind students of their work with rates in Chapter 2 and their work with change in Chapter 4. This preparation leads students to represent slope as a rate of change and to use the slope to draw a line in Lesson 7-3.

The remainder of the chapter covers five categories of applications. One, which leads to $y = mx + b$, involves an initial quantity and the rate at which it increases or decreases. These settings are covered in Lesson 7-4. In Lesson 7-5, there are

SLOPES AND LINES

Below is a graph of the population of Manhattan Island (part of New York City) every ten years from 1790 to 1990. Coordinates of some of the points are shown.

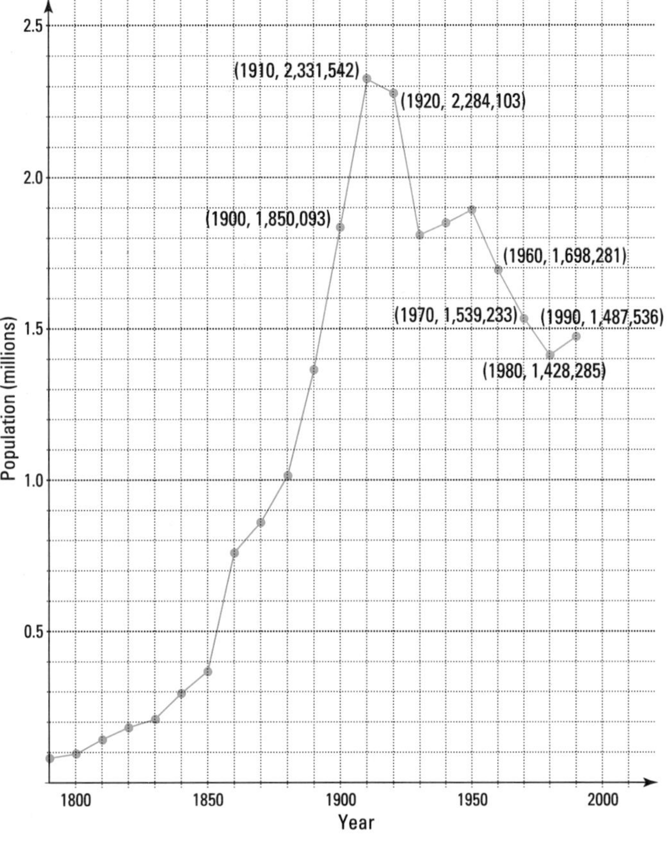

The slopes of the lines connecting the points indicate how fast the population increased or decreased. In this chapter, you will study many examples of lines and slopes.

Flags: The scholarly study of all aspects of flags is known as vexillology (from the Latin *vexillum* which means banner). Lesson 7-2 considers the slope of a line. A flagpole suggests a vertical line which has no slope.

Nickels and Dimes: The United States has two coinage mints, one in Philadelphia and one in Denver. Proof sets of coins for collectors are minted in San Francisco. Students graph linear inequalities when they find ordered pairs of dimes and nickels that total a given amount of money in Lesson 7-9.

Koala: A newborn koala remains in its mother's pouch for as long as seven months and then clings to its mother's back until it is about a year old. Koalas mature at about 4 years. A koala is approximately 2 cm long at birth and until maturity grows at an average rate of 1.5 cm per month. These numerical data can be used to write an equation in slope-intercept form.

Dhaka, Bangladesh: In 1990, the population of Bangladesh was estimated to be 116,000,000 people. In 1992, its population was around 119,000,000. The slope of the line connecting the points indicates how fast the population increased.

Chapter 7 Projects

At this time you might want to have students look over the projects on pages 476–477.

problems in which a data point and the rate of change are known.

In Lesson 7-6, two data points are known. Lesson 7-7 discusses situations in which more than two data points are known and students approximate the data with a line. The last type of application involves linear-combination situations which lead to sentences of the form $Ax + By = C$. These sentences are covered in Lesson 7-8 as the standard form for equations of lines. The last lesson is on graphing linear inequalities.

Encourage students to use graph paper or **Teaching Aids 26 or 28** to save time, increase accuracy, and observe patterns more easily. Time is wasted when students repeatedly have to draw axes and tick marks on notebook paper when graph paper has such marks already prepared. Graph paper also aids understanding. For instance, slope is illustrated better when students can accurately count squares to move between two points on the graph. The geometry of lines is also easier for students to assimilate from accurate drawings.

Although an automatic grapher is not assumed in this course, on several occasions we point out how the use of automatic graphers can aid understanding.

Lesson 7-1

Objectives

E Calculate rates of change from real data.

Resources

From the ***Teacher's Resource File***

- Lesson Master 7-1A or 7-1B
- Answer Master 7-1
- Teaching Aids
  28 Four-Quadrant Graph Paper
  71 Warm-up
  75 Manhattan Spreadsheet
  76 Additional Examples
  77 Questions 14–17

Additional Resources

- Visuals for Teaching Aids 28, 71, 75–77

Teaching Lesson 7-1

Warm-up

Writing Refer to page 417. Write a paragraph describing the changes in Manhattan's population from 1790 to 1990. **Answers will vary.**

Notes on Reading

Reading Mathematics This entire chapter provides many opportunities to concentrate on careful reading of data presented graphically. For example, when you discuss the first graph in this lesson, be careful to distinguish *rate* of change from *amount* of change. Point out that the faster the rate of growth, the steeper the incline of the segment. Karen's rate of growth was faster between ages 9 and 11. Between ages 11

LESSON

7-1 *Rate of Change*

Different rates. *The fastest rate of growth for girls usually begins between the ages of 9 and 11. Boys usually begin their growth spurt about 2 years later than girls.*

What Is a Rate of Change?

At age 9 Karen was 4′3″ (4 feet 3 inches) tall. At age 11 she was 4′9″ tall. How fast did she grow from age 9 to age 11? To answer this question, we calculate the *rate of change* of Karen's height, that is, how much she grew per year.

The rate of change in her height per year from age 9 to age 11 is

$$\frac{\text{change in height}}{\text{change in age}} = \frac{4'9'' - 4'3''}{(11 - 9)\text{ years}} = \frac{6 \text{ inches}}{2 \text{ years}} = 3\,\frac{\text{inches}}{\text{year}}.$$

At age 14 Karen was 5′4″ tall. How fast did she grow from age 11 to age 14? Use the same method. The rate of change in her height per year from age 11 to age 14 is

❶ $$\frac{\text{change in height}}{\text{change in age}} = \frac{5'4'' - 4'9''}{(14 - 11)\text{ years}} = \frac{64'' - 57''}{3 \text{ years}} = \frac{7 \text{ inches}}{3 \text{ years}} = 2.\overline{3}\,\frac{\text{inches}}{\text{year}}.$$

Notice that Karen grew at a faster rate from age 9 to age 11 than from age 11 to 14.

At the right, the data points are plotted and connected. The rate of change measures how fast the segment goes up as you read from left to right along the x-axis.

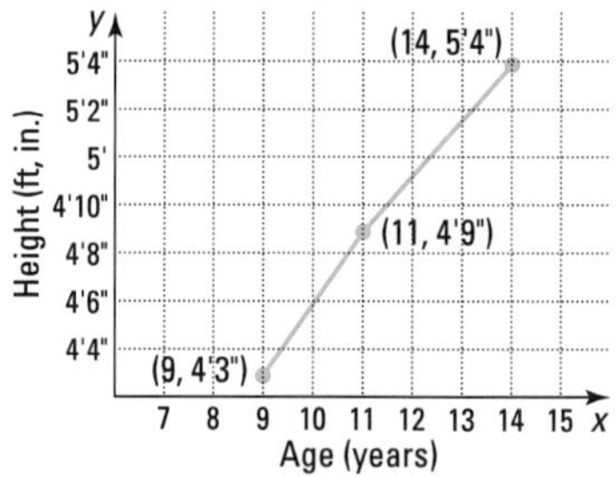

Since the rate of change of Karen's height is greater from age 9 to age 11 than from age 11 to age 14, the segment connecting (9, 4′3″) to (11, 4′9″) is steeper than the one connecting (11, 4′9″) and (14, 5′4″).

Lesson 7-1 Overview

Broad Goals This lesson uses applications to introduce the formula for rate of change between two points and shows this idea graphically.

Perspective In this lesson, we use situations and graphs in which quantities change over time. The problems in this lesson result in graphs that are broken lines, and we talk about rate of change between pairs of points. In the next lesson we look at lines and introduce the term *slope*.

One reason to start with rate of change rather than slope is that no lines are needed, just two points. This is intuitively easier. Also, the subtraction and division in the formula come from change and rate, so the name *rate of change* is easily associated with the operations in the formula. This makes the formula easier to understand. This application approach helps students make sense of positive, negative, and zero rates of change. It is easy to picture population or height increasing, decreasing, or remaining constant.

and 14, there is greater growth, but not faster growth, because the time period is 3 years, not 2 years.

❶ Because we emphasize the rate of change, students should label their answers with units of rate. Here the rate unit is *inches per year*. You may have to remind students that the bar above the 3 in $2.\overline{3}$ means that the 3 repeats forever.

❷ **Error Alert** Point out that x_1 stands for the "first *x*-value" and x_2 stands for the "second *x*-value," and a similar correspondence holds for y_1 and y_2. Some students may still want to think of the 2 as being multiplied by *x* or as being an exponent. It may be useful to point out that in some computer programs x_1 and x_2 would be typed as X1 and X2, and in that sense they are like locations in a column of a spreadsheet.

As illustrated in the following example, rates of change may be positive or negative.

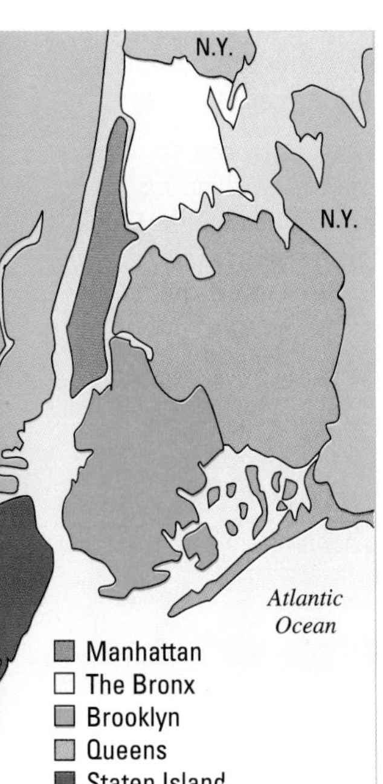

New York City. *New York City consists of five boroughs: The Bronx, Brooklyn, Manhattan, Queens, and Staten Island.*

Example 1

The graph on page 417 shows the population of Manhattan. Find the rate of change of the population of Manhattan (in people per year) during the given time, and tell how that rate is pictured on the graph

a. between 1900 and 1910. **b.** between 1910 and 1920.

Solution

To find the rate of change in people per year, calculate $\frac{\text{change in population}}{\text{change in years}}$.

a. Between 1900 and 1910, the rate of change is

$$\frac{2{,}331{,}542 - 1{,}850{,}093}{1910 - 1900} = \frac{481{,}449 \text{ people}}{10 \text{ years}}$$
$$= 48{,}144.9 \text{ people per year.}$$

Between 1900 and 1910, the population increased. So the rate of change is positive. As you read from left to right, the graph slants upward.

b. Between 1910 and 1920 the rate of change is

$$\frac{2{,}284{,}103 - 2{,}331{,}542}{1920 - 1910} = \frac{-47{,}439 \text{ people}}{10 \text{ years}}$$
$$= -4{,}743.9 \text{ people per year.}$$

Between 1910 and 1920, the population decreased. So the rate of change is negative. Between those dates, as you read from left to right, the graph slants downward.

When you read graphs, read from left to right just as you read prose. A positive rate of change indicates that the graph slants upward. A negative rate of change indicates that the graph slants downward.

There is a general formula for rate of change, in terms of coordinates. In Example 1, the year is the *x*-coordinate on the graph on page 417 and the population is the *y*-coordinate. The rate of change between two points is calculated by dividing the difference in the *y*-coordinates by the difference in the *x*-coordinates. We use the subscripts $_1$ and $_2$ to identify the coordinates of the two points. x_1 is read "*x* one" or "*x* sub one" or "the first *x*."

❷ **Definition**
The **rate of change** between points (x_1, y_1) and (x_2, y_2) is $\frac{y_2 - y_1}{x_2 - x_1}$.

Because every rate of change comes from a division, the unit of a rate of change is a rate unit. For instance, in Example 1, since a number of people is divided by a number of years, the unit of the rate of change is

$$\frac{\text{number of people}}{\text{year}}.$$

Optional Activities

If students seem to have a solid understanding of rate of change and percent, you may want to compare rate of change with percent of change. When populations are small, the rate of change will generally be small, but the percent of change will be large. When populations are large, the rate of change may be large, but the percent of change will be lower. Have students calculate the percent of change of Manhattan Island's population for each decade.

The percent of change is the quotient

$$\frac{(\text{new population} - \text{old population})}{\text{old population}}.$$

See page 420 for a table showing the percent of change in population for Manhattan Island.

Note that because the population tends to increase, the percents of change have tended to be less in recent years, even though the amount of change may be greater now than in the early part of the 19th century. When people talk about change, it is important to distinguish whether they are speaking of *amount* of change or *percent* of change. This is particularly confusing when people speak of changes in the percent of change, because then both the amount of change and the percent of change are percents.

(Optional Activities continue on page 420.)

❸ This table is reproduced on **Teaching Aid 75.**

Additional Examples

These examples are also given on **Teaching Aid 76.**

1. The chart below shows Tom's income for each week of a six-week period.

Week	1	2	3	4	5	6
Dollars	11	15	0	21	25	25

   a. Graph these points and connect them.

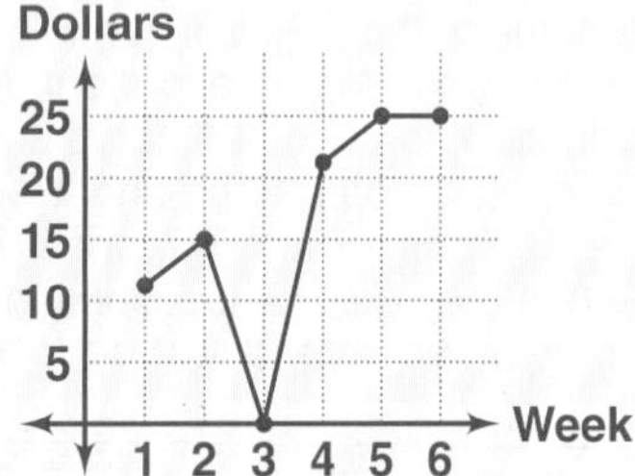

   b. Find the rate of change of Tom's income between weeks 1 and 6.
   $\frac{(25-11) \text{ dollars}}{(6-1) \text{ week}} = \$2.80/\text{wk}$

   c. During what time period was the rate of change of income negative? How can you tell? **Between weeks 2 and 3; the graph slants down.**

   d. When was the rate of change 0? How can you tell? **Between weeks 5 and 6; the graph is horizontal.**

Rates of Change in Tables

Spreadsheets and other table generators can be used to calculate rates of change. For instance, the spreadsheet below shows the years from 1900 to 1990 in column A, the population of Manhattan in column B, and the rate of change of population for the decade ending that year in column C.

Each rate of change is found using years and populations from two rows of the spreadsheet. Each formula in column C does two subtractions, one to find the change in population and one to find the change in years. Then the population change is divided by the change in years. For instance, the formula for C4 is = (B4−B3)/(A4−A3). You should check this. Notice that cell C3 cannot be filled because the population previous to 1900 has not been entered.

	A	B	C
1			Rate of Change
2	Year	Population	for Previous Decade
3	1900	1850093	
4	1910	2331542	48144.9
5	1920	2284103	-4743.9
6	1930	1876412	
7	1940	1889924	
8	1950	1960101	
9	1960	1698281	
10	1970	1539233	
11	1980	1428285	
12	1990	1487536	

A deer problem. *In some places, the deer population has grown too large for its natural habitat. So deer often search outside their area for food.*

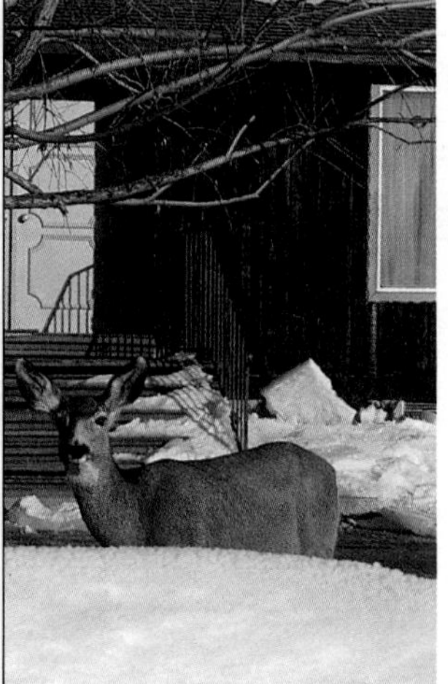

Constant Rates of Change

Sometimes quantities do not change over a certain interval.

Example 2

The graph at the right shows the number of deer in a park between 1975 and 1990.

a. During what time period did the deer population not change?

b. Find the rate of change of population during this time.

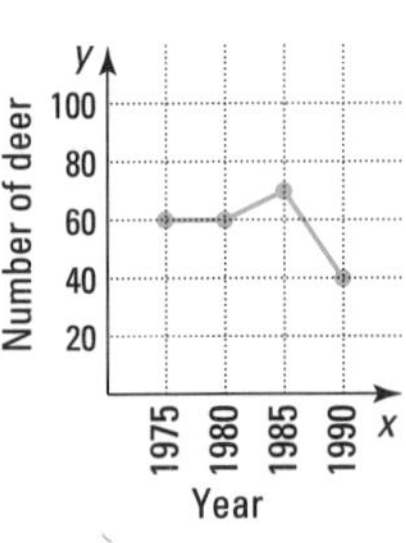

▶

Optional Activities, continued

Percent of Change of Manhattan Island's Population

Year	Population (thousands)	Percent Change	Year	Population (thousands)	Percent Change	Year	Population (thousands)	Percent Change
1790	33		1860	814	[58%]	1930	1867	[-18%]
1800	61	[85%]	1870	942	[16%]	1940	1890	[1%]
1810	96	[57%]	1880	1165	[24%]	1950	1960	[4%]
1820	124	[29%]	1890	1441	[24%]	1960	1698	[-13%]
1830	203	[64%]	1900	1850	[28%]	1970	1539	[-9%]
1840	313	[54%]	1910	2332	[26%]	1980	1428	[-7%]
1850	516	[65%]	1920	2284	[-2%]	1990	1488	[4%]

Solution

a. When the number of deer has been constant, there is zero change in the y-coordinate. This means that the line segment is horizontal. The *deer population did not change between 1975 and 1980.*

b. Use the coordinates of the endpoints, (1975, 60) and (1980, 60), of the horizontal segment.

$$\text{rate of change} = \frac{60 - 60}{1980 - 1975} = \frac{0}{5} = 0\ \frac{\text{deer}}{\text{year}}$$

In general, a rate of change of 0 corresponds to a horizontal segment on a graph.

The following table summarizes the relationship between rates of change and their graphs.

Situation	Rate of Change	Slant (from left to right)	Sketch of Graph
increase	positive	upward	
no change	zero	horizontal	
decrease	negative	downward	

QUESTIONS

Covering the Reading

1. The rate of change in height per year is the change in __?__ divided by the change in __?__. **height, years**

In 2–4, refer to the situation at the beginning of the lesson about Karen's height. Suppose Karen is 5′5″ tall at age 18 and 5′5″ tall at age 19.

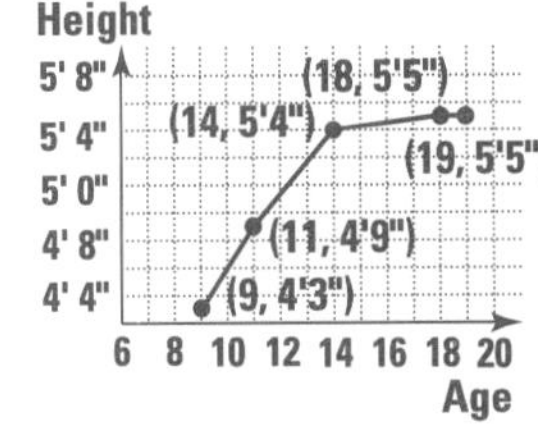

2. a. Copy the graph of her height on page 418. Extend the axes and plot her height at ages 18 and 19. **See left.**
   b. Which is steeper, the segment connecting (9, 4′3″) to (11, 4′9″) or the segment connecting (11, 4′9″) to (14, 5′4″)? **segment connecting (9, 4′3″) to (11, 4′9″)**
3. a. What is the rate of change of Karen's height from age 14 to age 18?
   b. Was the rate of change of her height greater from age 9 to age 11 or from age 14 to age 18? **a) 0.25 in. per year; b) from age 9 to 11**
4. What is the rate of change in her height from age 18 to age 19? **0 in. per year**

In 5–10, use the graph of the population of Manhattan on page 417.

5. If the rate of change in population in a given time period is positive, has the population increased or decreased? **increased**
6. Identify all decades in which the rate of change of Manhattan's population was negative. **1910 to 1930; 1950 to 1980**

2. The graph shows the weight of Li-Li, the first giant panda born in captivity, at age 1 month, 3 months, and 6 months.

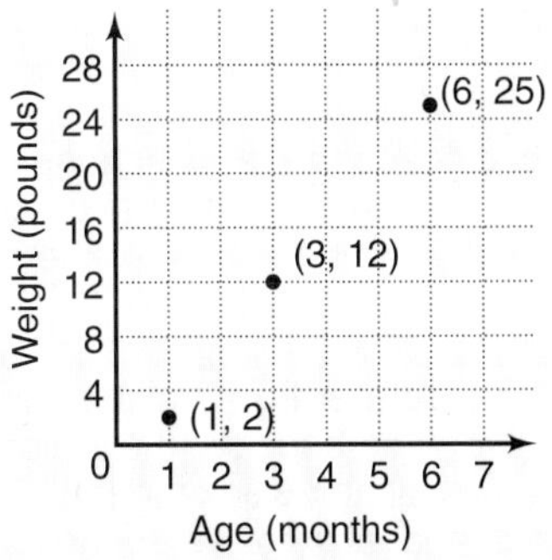

   a. Find the rate of change in weight per month from 1 month to 3 months and from 3 months to 6 months. **$5\ \frac{\text{lb}}{\text{mo}}$; about $4.3\ \frac{\text{lb}}{\text{mo}}$**
   b. During which time period did Li-Li's weight increase faster? **From 1 to 3 months**

Notes on Questions

Questions 2b and 3b Emphasize the connection between the steepness of the line and the rate of change.

LESSON MASTER 7-1 A

Questions on SPUR Objectives
See pages 480-483 for objectives.

Uses Objective E

In 1–5, use the chart below. It shows the number to the nearest million of U.S. households with cable TV every five years from 1970 to 1990.

Year	1970	1975	1980	1985	1990
Households	4 million	9 million	15 million	36 million	55 million

Households (millions)
50
40
30
20
10
'70 '80 '90
Year

1. Graph the information on the grid at the left. Connect the points.
2. Which segment is steeper, the one connecting (1975, 9 million) to (1980, 15 million) or the one connecting (1980, 15 million) to (1985, 36 million)? **from (1980, 15) to (1985, 36)**
3. What was the rate of change of number of households with cable from 1970 to 1975? **1 million households/year**
4. What was the rate of change of the number of households with cable from 1985 to 1990? **3.8 million households/year**
5. For the years shown, was the rate of change ever negative? **no**

In 6 and 7, use the table at the right. It shows the prices a vacation resort charges for stays of the given durations.

Days	Cost
3	$289
4	$359
7	$449

6. a. What is the rate of change from a 4-day vacation to a 7-day vacation? **$30/day**
   b. What is the unit of the rate of change? **dollars per day**
7. Is the rate of change from 3 to 4 days the same as the rate of change from 4 days to 7 days? **no**

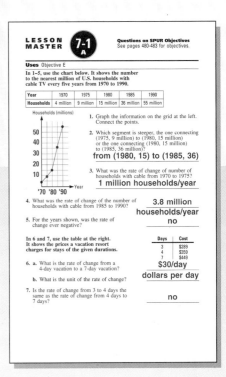

Adapting to Individual Needs

Extra Help

Some students may have difficulty deciding which way to subtract when calculating the rate of change. You might want to have these students name each of the ordered pairs and each of the coordinates. For the opening situation, students might write $(x_2, y_2) = (11, 4'9'')$ and $(x_1, y_1) = (9, 4'3'')$ to help them get $y_2 = 4'9''$, $y_1 = 4'3''$, $x_2 = 11$, and $x_1 = 9$. Then they can substitute these values in the rate-of-change formula.

Notes on Questions

Questions 14–17 You might want to use **Teaching Aid 77** when discussing this question. The graph provides many opportunities to compare the different rates of change. During the swoop, the height is decreasing, so the rate of change is negative. Then the vulture takes off and the height increases, so the rate of change becomes positive.

Science Connection Vultures do not kill for food because their talons are too small and weak for that. They are scavengers who live on the leftovers of another animal's kill. Vultures serve a purpose by getting rid of waste that would otherwise decay and spread disease. All continents except Australia and Antarctica have vultures.

Question 20 Multicultural Connection Students might be interested in knowing that *bonsai* is the art of growing miniature trees in pots. *Bonsai* originated in the homes of Chinese and Japanese aristocrats as early as 1000 A.D. The Japanese used *bonsai* in an attempt to bring nature into the house. A single bonsai tree brings an image of a large tree in the middle of a field. Several trees in a single pot suggest a forest.

Question 21 Even though no scale is given on the horizontal axis, we assume that the axis is marked in equal intervals.

(Notes on Questions continue on page 424.)

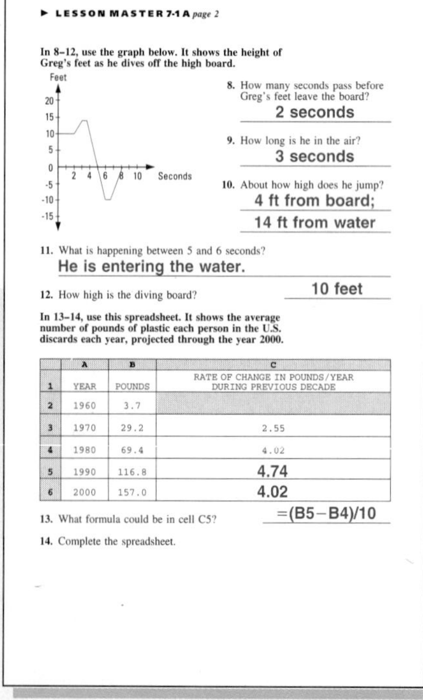

▸ **LESSON MASTER 7-1 A** *page 2*

In 8–12, use the graph below. It shows the height of Greg's feet as he dives off the high board.

8. How many seconds pass before Greg's feet leave the board? **2 seconds**

9. How long is he in the air? **3 seconds**

10. About how high does he jump? **4 ft from board; 14 ft from water**

11. What is happening between 5 and 6 seconds? **He is entering the water.**

12. How high is the diving board? **10 feet**

In 13–14, use this spreadsheet. It shows the average number of pounds of plastic each person in the U.S. discards each year, projected through the year 2000.

	A	B	C
1	YEAR	POUNDS	RATE OF CHANGE IN POUNDS/YEAR DURING PREVIOUS DECADE
2	1960	3.7	
3	1970	29.2	2.55
4	1980	69.4	4.02
5	1990	116.8	**4.74**
6	2000	157.0	**4.02**

13. What formula could be in cell C5? **=(B5−B4)/10**

14. Complete the spreadsheet.

10) Sample: The slope of the line connecting these points is steeper than the slope of the line connecting 1860 to 1870. Therefore, the population increased faster between 1850 to 1860.

7. Between 1950 and 1960, did the population increase or decrease? **decreased**

8. Find the rate of change in population between 1980 and 1990. **5925.1 people per year**

9. Find the rate of change in population from 1900 to 1990. **-4028.4 people per year**

10. During which 10-year period—1850 to 1860 or 1860 to 1870—did the population change more rapidly? Justify your answer. **1850 to 1860; See left for reasons.**

11. In terms of coordinates, the rate of change between two points is the _?_ of the y-coordinates divided by the difference of the _?_. **difference; x-coordinates**

12. a. Copy the spreadsheet below and fill in the information for the deer populations graphed in Example 2.

	A	B	C
1	Year	Deer	Rate of change in Previous 5-Yr. Period
2	1975	60	
3	1980	60	0
4	1985	70	2
5	1990	40	-6

b. What is the formula in cell C4? **= (B4−B3)/(A4−A3)**

c. What is the value of C4? **2 deer per year**

d. Which cell cannot be filled in? **C2**

13. Describe the graph of the segment connecting two points when the rate of change between them is **a.** positive; **b.** negative; **c.** zero. **a) slants upward from left to right; b) slants downward from left to right; c) horizontal**

Scavenger hunt. *This Rüppell's griffon vulture lives in the dry bush and desert areas of Kenya.*

Applying the Mathematics

In 14–17, use this graph showing the altitude of a vulture over a period of time.

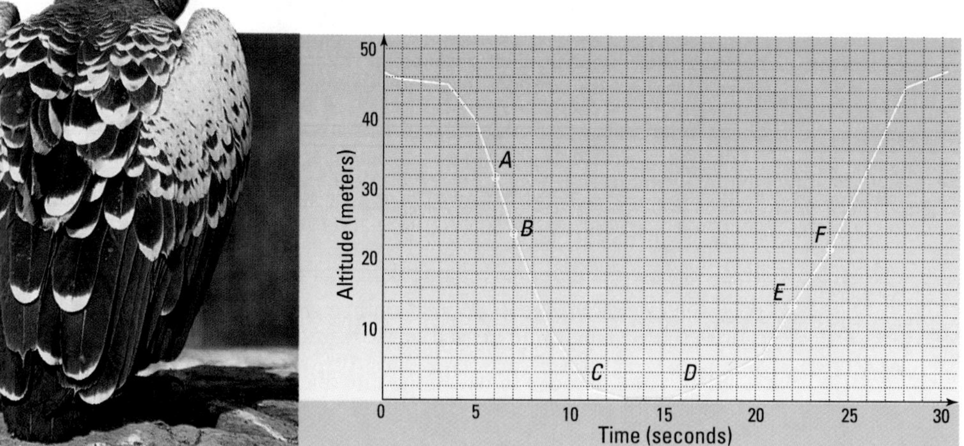

Adapting to Individual Needs

English Language Development

Note that the word *steep* refers to slant or incline. It is a quality of a line; it describes how quickly a line goes up or down. Give a visual example by drawing three "mountains."

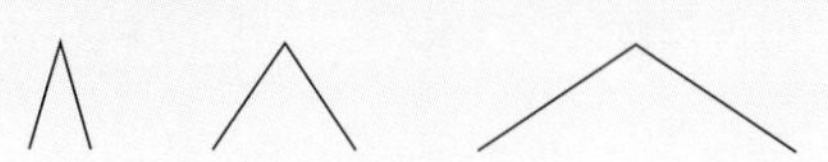

Point out that the first mountain is considered steeper or more slanted than the next one. Discuss the incline of each mountain with respect to the others.

Have students copy the table on page 421, which summarizes the relationships between rates of change and slant of graphs.

In 14–17, refer to the graph on page 422.

14. **a.** Give the coordinates of point A. **(6, 32)**
b. Give the coordinates of point B. **(7, 24)**
c. What is the rate of change of altitude (in meters per second) between points A and B? **−8 meters per second**

15. What is the rate of change of altitude between points C and D?
0 meters per second

16. After about how many seconds does the altitude of the vulture begin to increase? **after about 15.5 seconds**

17. Is the rate of change between E and F positive, negative, or zero?
positive

18. Below are heights (in inches) for a boy from age 9 to age 15.

Age	9	11	13	15
Height	51″	58″	63″	65″

a. Accurately graph these data and connect the points. **See left.**
b. In which two-year period did the boy grow the fastest? How can you tell? **from 9 to 11; The slope of the line is steepest.**
c. Calculate his rate of growth in that two-year period.
3.5 inches per year

19. Older people tend to lose height. Tim reached his full height at age 20, when he was 74″ tall. He stayed at that height for 35 years, and then started losing height. At age 65 his height was 73″. What was the rate of change of his height from age 55 to age 65?
−0.1 inch per year

20. When the bonsai tree was bought, it was 12 years old and h cm tall. When the tree was 15 years old, it was H cm tall.
a. Write an expression for the rate of change in the height of the tree from ages 12 to 15. **$\frac{H-h}{3}$ cm per yr**
b. If the bonsai did not change during these years, what must be true about the rate of change in its height?
The rate of change is 0 cm per year.

21. In the graph at the right, find the rate of change in meters per minute.
$\frac{y}{4}$ meters per minute

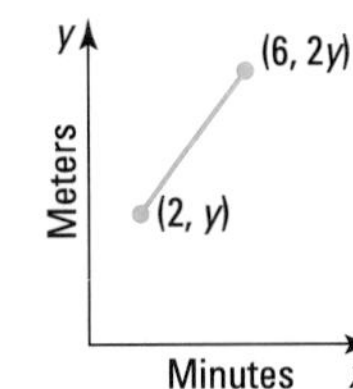

22. The graph below shows the population of Hong Kong at seven different times. Which line segment—$\overline{AB}$, $\overline{BC}$, $\overline{CD}$, $\overline{DE}$, $\overline{EF}$, or $\overline{FG}$—matches the following description?
a. The population decreased. **$\overline{CD}$**
b. The rate of change is 0. **$\overline{AB}$**
c. The rate of change was greatest. **$\overline{DE}$**

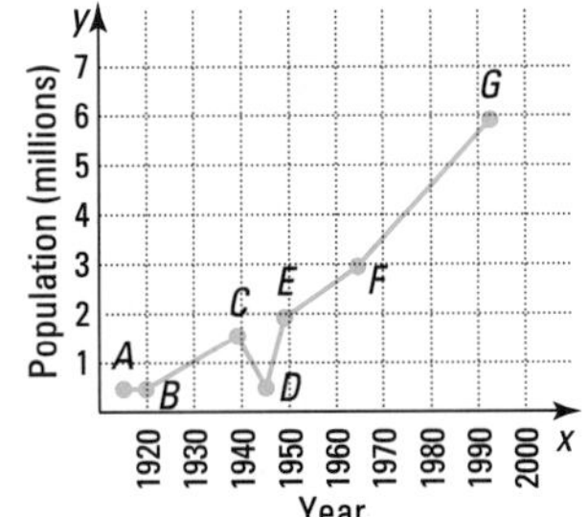

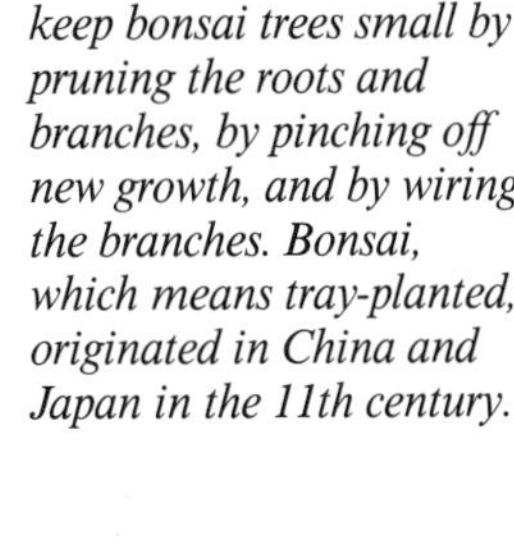

Miniature trees. *Bonsai is the art of growing miniature trees. Growers keep bonsai trees small by pruning the roots and branches, by pinching off new growth, and by wiring the branches. Bonsai, which means tray-planted, originated in China and Japan in the 11th century.*

18a)

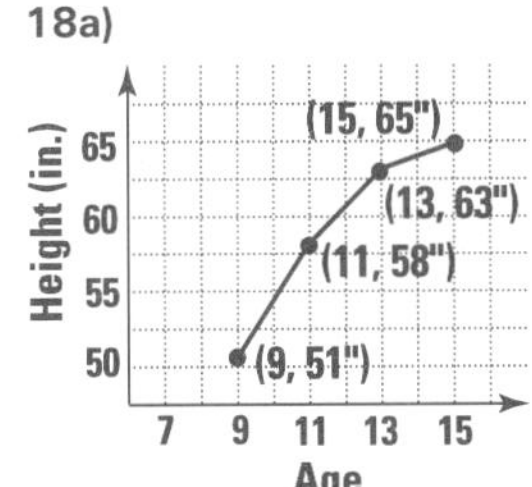

Follow-up for Lesson 7-1

Practice

For more questions on SPUR Objectives, use **Lesson Master 7-1A** (shown on pages 421–422) or **Lesson Master 7-1B** (shown on pages 423–424).

Assessment

Oral Communication Have students explain how they can tell by looking at a graph if the rate of change between two points is positive, negative, or zero. [Students describe an upward slant from left to right as positive, a downward slant from left to right as negative, and a horizontal slant as zero.]

Extension

Give students the population of their school for each decade of its history (or for each year, if the school is not very old) and have them find the rate of change over each period of time. You might discuss if the student population seems to be increasing, decreasing, or staying about the same. You might also discuss how this information might be used.

Project Update Project 5, *Populations of Cities*, on page 477 relates to the content of this lesson.

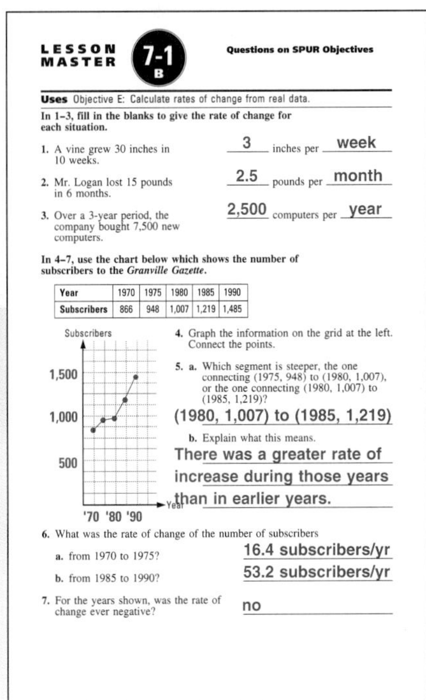

LESSON MASTER 7-1 B **Questions on SPUR Objectives**

Uses Objective E: Calculate rates of change from real data.

In 1–3, fill in the blanks to give the rate of change for each situation.

1. A vine grew 30 inches in 10 weeks. **3** inches per **week**
2. Mr. Logan lost 15 pounds in 6 months. **2.5** pounds per **month**
3. Over a 3-year period, the company bought 7,500 new computers. **2,500** computers per **year**

In 4–7, use the chart below which shows the number of subscribers to the *Granville Gazette*.

Year	1970	1975	1980	1985	1990
Subscribers	866	948	1,007	1,219	1,485

4. Graph the information on the grid at the left. Connect the points.
5. a. Which segment is steeper, the one connecting (1975, 948) to (1980, 1,007), or the one connecting (1980, 1,007) to (1985, 1,219)? **(1980, 1,007) to (1985, 1,219)**
b. Explain what this means. **There was a greater rate of increase during those years than in earlier years.**
6. What was the rate of change of the number of subscribers
a. from 1970 to 1975? **16.4 subscribers/yr**
b. from 1985 to 1990? **53.2 subscribers/yr**
7. For the years shown, was the rate of change ever negative? **no**

Adapting to Individual Needs

Challenge

Have students solve the following problem: Human hair grows at an average rate of 1.5 to 3.0 millimeters per week. Suppose you had not cut your hair since you were born and your hair grew at the maximum rate. Estimate how long your hair would be now. [Answers will vary according to the ages of the students. For instance, if a student is 14 years 2 weeks old, the student's hair length would be about 2.19 meters.]

Notes on Questions

Question 25 This *Skill sequence* emphasizes manipulation skills that will be needed in later lessons of the chapter.

Question 27 This application has a starting quantity that is increased (or decreased) at a constant rate. It previews an idea found in Lesson 7-4.

Question 31 Cooperative Learning You may want to have students **work in small groups** while researching these data. If the population of their city or town is difficult for students to find, have them select a city from an almanac for which the data are given.

Review

23. *Skill sequence.* *(Lessons 3-9, 5-9, 6-8)*
   a. Simplify $\frac{2}{a} - \frac{1}{a}$. $\frac{1}{a}$
   b. Simplify $\frac{2}{3+4x} - \frac{1}{3+4x}$. $\frac{1}{3+4x}$
   c. Solve $\frac{2}{3+4x} - \frac{1}{3+4x} = \frac{4}{x}$. $x = -\frac{4}{5}$

24. Simplify $-\frac{\frac{2}{5}}{4}$. *(Lesson 6-1)* $-\frac{1}{10}$

25. *Skill sequence.* Solve for y. *(Lesson 5-7)*
   a. $33 - 4y = 12$ $y = 5.25$
   b. $3x - 4y = 12$ $y = \frac{3}{4}x - 3$
   c. $ax - 4y = 12$ $y = \frac{a}{4}x - 3$

26. a. Graph $y = 3x + 2$ and $y = -3x + 2$ on the same axes. **See left.**
   b. Where do the two lines intersect? *(Lesson 4-9)* **(0, 2)**

26a)

27. Suppose a stamp collection now contains 10,000 stamps. If it grows at 1000 stamps per year, how many stamps will there be in x years? *(Lesson 3-8)* **10,000 + 1,000x**

28. *True or false.* The rate "3 pizzas for 4 people" is equal to the rate "$\frac{3}{4}$ pizza for 1 person." Explain your answer. *(Lesson 6-2)* **See left.**

28) True; by the Rate Model for Division, if 3 pizzas are divided among 4 people, the result is $\frac{3}{4}$ pizza per person.

In 29 and 30, evaluate $\frac{y_2 - y_1}{x_2 - x_1}$ at the given values. *(Lesson 1-5)*

29. $y_2 = 5$, $y_1 = 6$, $x_2 = -2$, and $x_1 = -4$ $-\frac{1}{2}$

30. $y_2 = 8$, $y_1 = 4$, $x_2 = 2$, and $x_1 = \frac{1}{3}$ $\frac{12}{5}$ **or 2.4**

Exploration

31. a. Find the population of the town or city where you live (or near where you live) in 1940, 1950, 1960, 1970, 1980, and 1990. (You may want to call a local historical society or public library.)
   b. In which 10-year period did the population change the most? Was the rate of change positive or negative? **a, b) Answers will vary.**

Shown are postcards from some cities and towns in the U.S.A.

Setting Up Lesson 7-2

Discussing **Questions 24 and 29** of this lesson will help students when they read and answer the questions for Lesson 7-2.

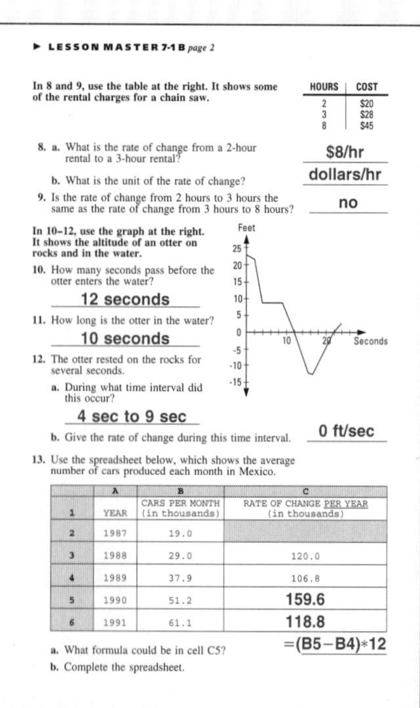

▸ **LESSON MASTER 7-1B** *page 2*

In 8 and 9, use the table at the right. It shows some of the rental charges for a chain saw.

HOURS	COST
2	$20
3	$28
8	$45

8. a. What is the rate of change from a 2-hour rental to a 3-hour rental? **$8/hr**
   b. What is the unit of the rate of change? **dollars/hr**
9. Is the rate of change from 2 hours to 3 hours the same as the rate of change from 3 hours to 8 hours? **no**

In 10–12, use the graph at the right. It shows the altitude of an otter on rocks and in the water.

10. How many seconds pass before the otter enters the water? **12 seconds**
11. How long is the otter in the water? **10 seconds**
12. The otter rested on the rocks for several seconds.
   a. During what time interval did this occur? **4 sec to 9 sec**
   b. Give the rate of change during this time interval. **0 ft/sec**
13. Use the spreadsheet below, which shows the average number of cars produced each month in Mexico.

	A	B	C
1	YEAR	CARS PER MONTH (in thousands)	RATE OF CHANGE PER YEAR (in thousands)
2	1987	19.0	
3	1988	29.0	120.0
4	1989	37.9	106.8
5	1990	51.2	**159.6**
6	1991	61.1	**118.8**

   a. What formula could be in cell C5? **=(B5−B4)∗12**
   b. Complete the spreadsheet.

LESSON

7-2

The Slope of a Line

Scuba-diving pressure. *As scuba divers descend, pressure increases at a constant rate. For example, in salt water, pressure increases by about 15 pounds per square inch for every 33 feet of descent.*

Consider the following situation. An ant, 12 feet high on a flagpole, walks down the flagpole at a rate of 8 inches (which is $\frac{2}{3}$ foot) per minute.

This is a *constant-decrease* situation. Each minute the height of the ant decreases by $\frac{2}{3}$ foot. You can see the constant decrease by graphing the height of the ant after 0, 1, 2, 3, 4, . . . minutes of walking. Below are the ordered pairs (time, height) charting the ant's progress.

Time (min)	Height (ft)
0	12
1	$11\frac{1}{3}$
2	$10\frac{2}{3}$
3	10
4	$9\frac{1}{3}$
5	$8\frac{2}{3}$
6	8

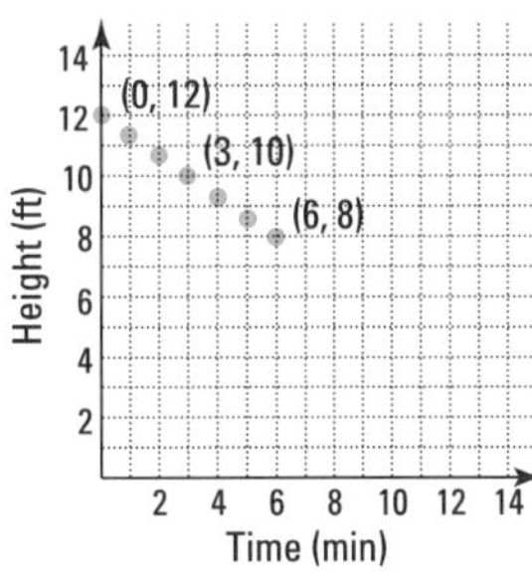

Notice that the rate of change between the points (0, 12) and (3, 10) is

$$\frac{\text{change in height}}{\text{change in time}} = \frac{10 \text{ feet} - 12 \text{ feet}}{3 \text{ minutes} - 0 \text{ minutes}} = \frac{-2 \text{ feet}}{3 \text{ minutes}}.$$

The rate of change is $-\frac{2}{3}$ foot per minute, the same as the ant's rate.

Similarly, after 1 minute the ant is $11\frac{1}{3}$ feet high. After 5 minutes the ant is $8\frac{2}{3}$ feet high. The rate of change of height in $\frac{\text{ft}}{\text{min}}$ between these points is

$$\frac{\text{change in height}}{\text{change in time}} = \frac{8\frac{2}{3} - 11\frac{1}{3}}{5 - 1} = \frac{-2\frac{2}{3}}{4} = \frac{-\frac{8}{3}}{4} = \frac{-8}{3} \cdot \frac{1}{4} = \frac{-2}{3}.$$

Lesson 7-2

Objectives

A Find the slope of the line through two given points.

D Use the definition of slope.

Resources

From the ***Teacher's Resource File***
- Lesson Master 7-2A or 7-2B
- Answer Master 7-2
- Teaching Aids
  26 Graph Paper
  71 Warm-up

Additional Resources
- Visuals for Teaching Aids 26, 71

Teaching Lesson 7-2

Warm-up

Find the value of y in each equation for the given value of x.

1. $2x + 5y = 12; x = 1$ $y = 2$
2. $2x + 5y = 12; x = 0$ $y = \frac{12}{5}$
3. $6x - 3y = 21; x = 0$ $y = -7$
4. $6x - 3y = 21; x = -1$ $y = -9$

Find the value of x in each equation for the given value of y.

5. $x + y = 7; y = 0$ $x = 7$
6. $x + y = 7; y = 2$ $x = 5$
7. $4x - y = 11; y = 1$ $x = 3$
8. $4x - y = 11; y = 0$ $x = \frac{11}{4}$

Notes on Reading

There are many important ideas presented in this lesson. You might choose to divide the lesson into six parts: the introductory situation about an ant, the definition of slope, and the four examples.

Lesson 7-2 Overview

Broad Goals The fact that the slope between any two points on a line is constant may not be obvious to students. In this lesson, students consider situations in which the rate of change is constant and thus the points lie on a line. That rate of change is then defined to be the slope of the line.

Perspective The lesson begins with a constant-decrease situation, and the rate of change between different points is calculated and seen to be constant. The points also lie on the same line. You may wish to introduce the term *collinear,* though we do not use it.

Two methods are given for finding the slope of a line through two points. The first is to use the familiar formula $\frac{y_2 - y_1}{x_2 - x_1}$. The second is to examine the graph and find the rate of change from that.

If the algebra of slope is divorced from its applications, students often wonder why slope is not defined as $\frac{x_2 - x_1}{y_2 - y_1}$. The approach in the lesson should make this clear. Many applications deal with things that change over time. Since time is usually graphed on the horizontal axis, the slope measures the rate of change in the quantity measured along the vertical axis.

(Overview continues on page 426.)

Science Connection When discussing the introductory situation, you might want to include some of the following information about ants. Ants are known as social insects because they live in organized communities called colonies. An ant colony may have millions of members or it might consist of only a dozen ants. Ants vary in size from the largest, which are more than 1 inch long, to the smallest, which are about $\frac{1}{25}$ inch long. Most of the nearly 20,000 species of ants are varying shades of black, brown, or rust. However, some species are yellow, blue, green, or purple.

The ant's speed is given in two ways: $-\frac{2 \text{ feet}}{3 \text{ minutes}}$ and $-\frac{2}{3}\frac{\text{foot}}{\text{minute}}$. Show each of these on the graph. The first is the change over three-minute intervals. The second is the change from one minute to the next. Stress that the *mathematical* slope of a line can be visually determined only from the *apparent* slope when the scales on the axes are the same, which is the case here.

As long as the ant has a constant rate of walking, the rate of change in its height will always equal that constant rate.

Notice that all the points on the graph of the ant's height lie on the same line. In *any* situation in which there is a constant rate of change between points, the points lie on the same line.

Contrast this situation to the Manhattan population situation graphed on page 417. There, no three consecutive points lie on the same line. The rates of change of Manhattan's population are different for each decade.

This leads to an important property of lines. On a line the rate of change between any two points is always the same. The constant rate of change is called the *slope* of that line.

Definition
The slope of the line through (x_1, y_1) and (x_2, y_2) is
$$\frac{y_2 - y_1}{x_2 - x_1}.$$

Example 1

Find the slope of line ℓ at the right.

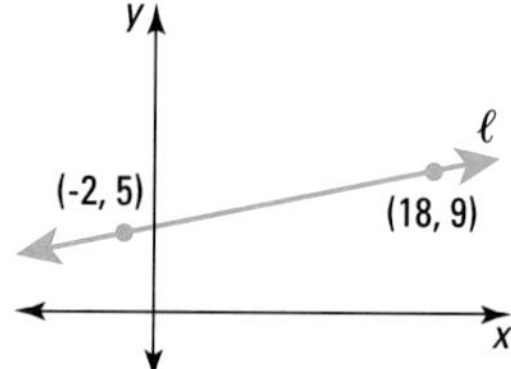

Solution 1

Note that the coordinates of two points are given. Let $(x_1, y_1) = (-2, 5)$ and $(x_2, y_2) = (18, 9)$. Apply the formula for slope.

$$\text{slope} = \frac{9 - 5}{18 - -2} = \frac{4}{20} = \frac{1}{5}$$

Solution 2

Let $(x_1, y_1) = (18, 9)$ and $(x_2, y_2) = (-2, 5)$. Apply the formula for slope.

$$\text{slope} = \frac{5 - 9}{-2 - 18} = \frac{-4}{-20} = \frac{1}{5}$$

Check

As you look from left to right, the graph slants upward. This checks that the slope is positive.

Notice that it does not matter which point is chosen as (x_1, y_1) and which one is (x_2, y_2). The resulting slope is the same.

When lines are drawn on a coordinate grid where each tick mark represents one unit, you can find the slope by counting or by measuring.

Lesson 7-2 Overview, continued

Expect that some students will count units on graphs in preference to using the formula. They need to convince themselves that the line has constant slope. There are questions in this lesson and the next in which some points have fraction coordinates. Students will probably find the formula easier for those problems.

Optional Activities

Activity 1 Writing After students complete the reading, you might ask them to think of another constant-increase or constant-decrease situation, and explain how that situation leads to a line and what the slope of that line is.

Activity 2 After students complete the reading, you might have them draw any line in a coordinate plane and then choose at least three different pairs of points on the line to determine its slope. Be sure students understand and show that the slope will be the same no matter what pair of points they choose.

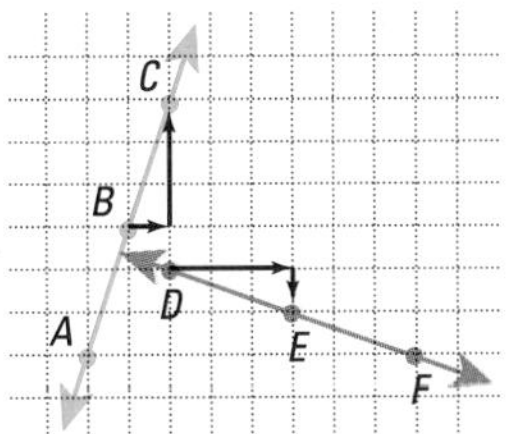

Example 2

In the coordinate grid at the left, the side of each square is one unit. Find the slope of the line.

a. $\overleftrightarrow{AC}$ b. $\overleftrightarrow{DF}$

Solution

a. Pick two points on $\overleftrightarrow{AC}$. We pick points B and C. The change in the x-coordinate as we read from left to right (B to C) is 1 unit. The change in the y-coordinate is 3 units. The slope of $\overleftrightarrow{AC} = \frac{3}{1} = 3$.

b. We pick points D and E. The change in the x-coordinate as we read from left to right (D to E) is 3 units. The change in the y-coordinate is -1 unit. The slope of $\overleftrightarrow{DF} = \frac{-1}{3}$.

Lines with negative slope, such as $\overleftrightarrow{DF}$ in Example 2, go downward as you read from left to right. Lines with positive slope, such as $\overleftrightarrow{AB}$ in Example 2, go upward as you read from left to right.

If the rate of change or slope is the same for a set of points, then all the points lie on the same line. If the rate of change is different for different parts of a graph, the graph is *not* a line.

Example 3

a. Show that (0, 3), (4, 1), and (-8, 7) lie on the same line.
b. Give the slope of that line.

Solution

a. Pick pairs of points and calculate the rate of change between them. The rate of change between (0, 3) and (4, 1) is

$$\frac{1-3}{4-0} = \frac{-2}{4} = -\frac{1}{2}.$$

The rate of change between (-8, 7) and (4, 1) is

$$\frac{1-7}{4--8} = \frac{-6}{12} = -\frac{1}{2}.$$

The rate of change between (-8, 7) and (0, 3) is

$$\frac{3-7}{0--8} = \frac{-4}{8} = -\frac{1}{2}.$$

Since the rate of change between any pair of the given points is $-\frac{1}{2}$, the points lie on the same line.

b. The slope of the line is the constant rate of change, $-\frac{1}{2}$.

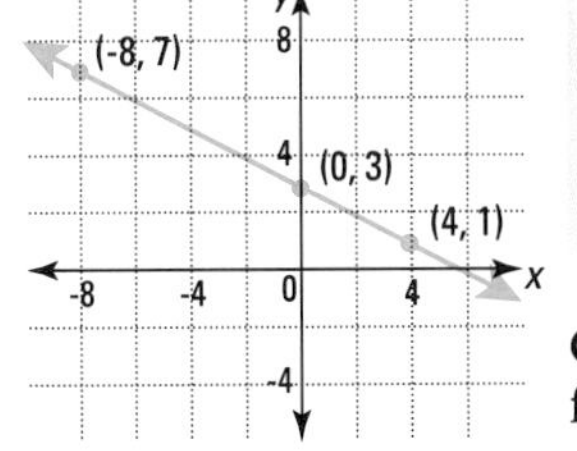

Check

Graph the points. They do lie on the same line. The line goes down as you read from left to right, so a negative slope is correct. Counting the change in the units from (0, 3) to (4, 1), we see that the slope is $\frac{-2}{4}$ or $\frac{-1}{2}$.

Given an equation for a line, it is easy to find the slope of the line. Just find two points on it and calculate the rate of change between them.

In the previous lesson, the issue of which point was called (x_1, y_1) and which (x_2, y_2) was not discussed because the assignment was naturally based on order of occurrence. However, many of the problems in this lesson are abstract, and there is no natural order. **Example 1** shows how it is possible to interchange the points when calculating slope. You might have students determine the slope of the line in **Example 3** by interchanging points. Then show students the general pattern below.

$$\frac{y_2-y_1}{x_2-x_1} = \frac{y_1-y_2}{x_1-x_2}$$

Example 2 shows how to calculate the slope given a graph of a line. Emphasize that the scale must be the same on the two axes in order for this method to work without knowing the coordinates. This is particularly important to emphasize when working with automatic graphers.

When three or more points are on the same line, the slope determined by any two of them is constant. Thus, we can define *the* slope of a line to be that constant. The converse of the statement, which is also true, is used in **Example 3:** *If the slopes determined by three or more points are the same, then the points lie on the same line.*

Adapting to Individual Needs

English Language Development

To help students understand the term *slope,* have them participate in the following activity. Tell students to imagine that the classroom floor is a coordinate plane. Mark two points, A and B, on the floor and connect them with a string. The points should be at least 6 feet apart and the string should not be parallel to either axis. Have a volunteer walk from A to B by walking only horizontally and vertically as shown in the diagram. Have another student record the number of horizontal steps and the number of vertical steps as $\frac{\text{vertical steps}}{\text{horizontal steps}}$. (Note: if your classroom floor has square tiles, select tile corners for A and B; then students can count the number of vertical and horizontal tiles rather than footsteps.) Next extend the line to point C, and repeat the procedure. Compare the fractions for the two situations and connect the ideas of constant rate of change and slope to the activity.

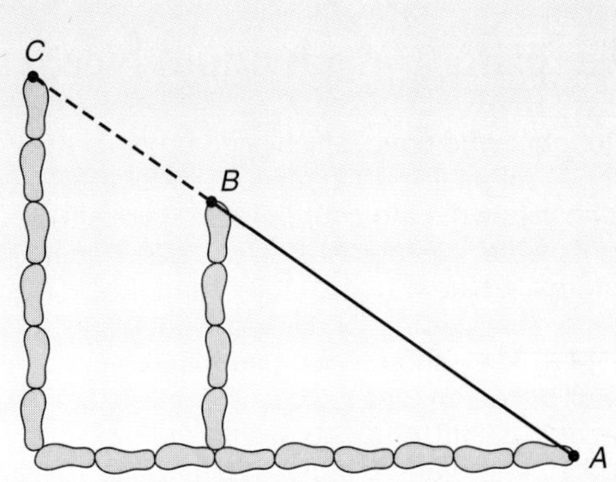

Example 4 uses the standard form of the equation of a line. Students should be reminded of why the sentences, "If $x = 2$, then $y = 0$" and "If $x = 0$, then $y = \frac{3}{2}$," are true. Stress that any two values for x can be chosen to find points to use in determining the slope.

Additional Examples

Students will need graph paper or **Teaching Aid 26** for Additional Example 2b.

1. Find the slope of the line through (3, 1) and (-5, 7). $-\frac{3}{4}$
2. **a.** If a line goes up 5 units for each 2 units that it goes to the right, what is the slope of the line? **$\frac{5}{2}$ or 2.5**

   b. Sketch such a line.

3. **a.** Show that (1, 5), (3, 11), and (-3, -7) lie on the same line. **$\frac{11-5}{3-1} = 3$; $\frac{-7-11}{-3-3} = 3$; $\frac{-7-5}{-3-1} = 3$**

   b. Find the slope of that line. **3**
4. Find the slope of the line with equation $x - 4y = 8$. **$\frac{1}{4}$**

Example 4

Find the slope of the line with equation $3x + 4y = 6$.

Solution

First find two points that satisfy the equation. Pick values for x or y, then substitute into $3x + 4y = 6$. To make the calculations easier, let $x = 0$ for the first point and then $y = 0$ for the second point.

Let x = 0:

$$3 \cdot 0 + 4y = 6$$
$$0 + 4y = 6$$
$$y = \frac{6}{4} = \frac{3}{2}$$

The point $(0, \frac{3}{2})$ is on the line.

Let y = 0:

$$3x + 4 \cdot 0 = 6$$
$$3x + 0 = 6$$
$$x = 2$$

The point (2, 0) is on the line.

Substitute $\left(0, \frac{3}{2}\right)$ and (2, 0) into the slope formula.

$$\text{Slope} = \frac{\frac{3}{2} - 0}{0 - 2} = \frac{\frac{3}{2}}{-2} = \frac{3}{2} \cdot -\frac{1}{2} = -\frac{3}{4}$$

Check 1

Graph the line. Does its slope seem to be $-\frac{3}{4}$? Yes, the graph slants downward, so the slope must be negative.

Check 2

Find another ordered pair that satisfies $3x + 4y = 6$. We substitute $x = 1$ in the equation.

$$3(1) + 4y = 6$$
$$4y = 3$$
$$y = \frac{3}{4}$$

So $(1, \frac{3}{4})$ is on the line. Now calculate the slope determined by it and one of the points in the solution, say (2, 0). This gives $\frac{\frac{3}{4} - 0}{1 - 2}$, which equals $-\frac{3}{4}$. This checks.

QUESTIONS

Covering the Reading

1. In a constant-increase or constant-decrease situation, all points lie on the same __?__. **line**
2. What is the constant rate of change between any two points on a line called? **slope**

Adapting to Individual Needs

Extra Help

Students who have difficulty computing slopes might find it helpful to use colored pencils to write the coordinates. Have students write y coordinates in red and x coordinates in blue. Suggest they keep the same color coding for all their work with slopes. They should also use the same color codes on their graphs. For the example at the right, students would write 27 and 12 in red and 8 and 3 in blue. This coding should help students to use the correct values for computing the slope:

$$\frac{27 - 12}{8 - 3} = \frac{15}{5} = 3$$

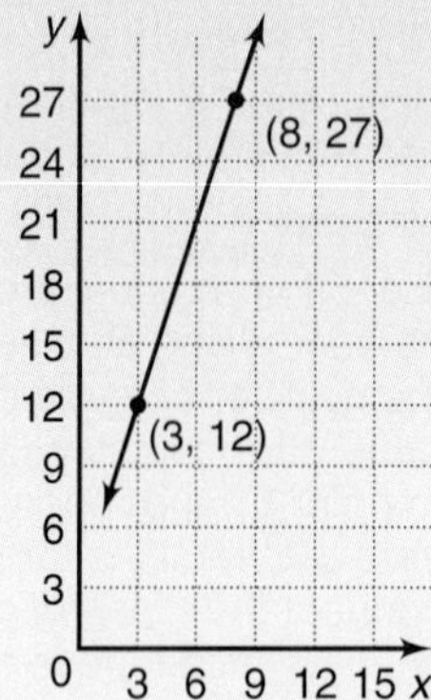

3a) Sample: $(7, 7\frac{1}{3})$; (9, 6)

4a)

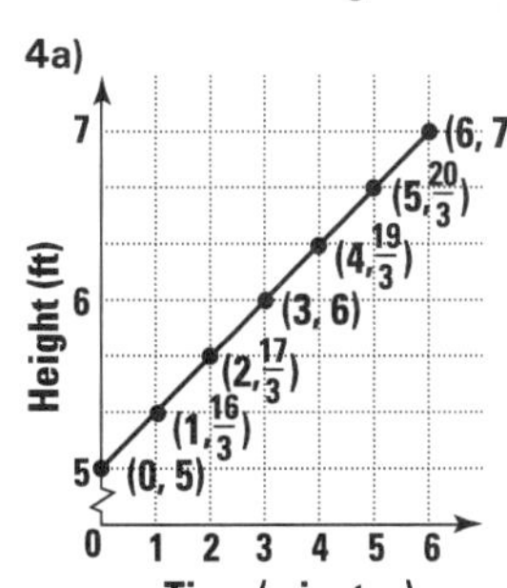

3. An equation for the height y of the ant in this lesson after x minutes is $y = -\frac{2}{3}x + 12$.
   a. Find two points on this line not graphed in this lesson. **See left.**
   b. Find the rate of change between those points. $-\frac{2}{3}$

4. An ant starts 5 feet from the base of a flagpole and climbs $\frac{1}{3}$ foot up the pole each minute.
   a. Graph the ant's progress for the first 6 minutes using ordered pairs (time, height). **See left.**
   b. Find the rate of change between any two points on the graph. **1/3 foot per minute**

In 5 and 6, calculate the slope of the line through the given pair of points.

5. (0, 1) and (2, 7) **3**
6. (4, 1) and (-2, 4) $-\frac{1}{2}$

7. a. Calculate the slope of the line through (1, 2) and (6, 11). $\frac{9}{5}$
   b. Calculate the slope of the line through (6, 11) and (-10, -16). $\frac{27}{16}$
   c. Do the points (1, 2), (6, 11), and (-10, -16) lie on the same line? How can you tell? **No; the slope through (1, 2) and (6, 11) is different from the slope through (6, 11) and (-10, -16).**

In 8 and 9, refer to the graphs at the left. Find the slope of the line.

8. line ℓ **2**
9. line m **-4**

In 10 and 11, an equation is given. **a.** Find two points on the line. **b.** Find the slope of the line. **c.** Check your work by graphing the line.

10. $5x - 2y = 10$ **a) Sample: (0, -5); (2, 0) b) $\frac{5}{2}$ or 2.5**
11. $x + y = 6$ **a) Sample: (0, 6); (6, 0) b) -1**

See margin for 10c and 11c.

Applying the Mathematics

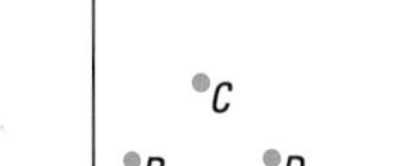

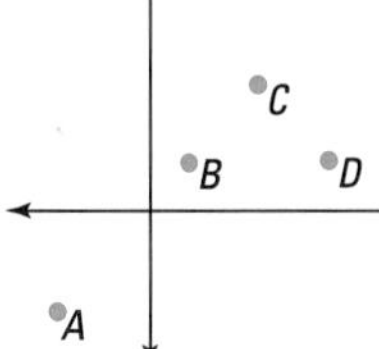

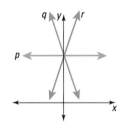

In 12–14, use the figure at the left.

12. The slope determined by points A and B seems to equal the slope determined by B and what other point? **C**
13. The slope determined by __?__ and __?__ is negative. **C; D**
14. The slope determined by __?__ and __?__ is zero. **B; D**
15. Consider lines p, q, and r graphed at the left.
    a. Which line has slope $\frac{5}{2}$? ***r***
    b. Which line has slope $-\frac{5}{2}$? ***q***
    c. Which line has slope zero? ***p***
16. Consider the following set of points. High-rise apartment monthly rents: (14th floor, \$535), (17th floor, \$565), (19th floor, \$585).
    a. Calculate the slope of the line through these points. **10**
    b. What does the slope represent in this situation? **an increase in rent of \$10 per floor**
17. The points (2, 5) and (3, y) are on the same line. Find y when the slope of the line is -2. **$y = 3$**

Notes on Questions

Questions 5–6 Encourage students to plot the points and to check the visual image of the line (slants up, slants down, or is horizontal) against the sign of their numerical responses (positive, negative, zero).

Questions 12–15 Error Alert Some students have difficulty with the concept of slope when no coordinates are given. Here, students need to think about positive, negative, and zero slopes without resorting to calculations. If students cannot answer these questions, ask them to refer to Lesson 7-1 and to decide whether each case reflects a constant increase (positive slope), a constant decrease (negative slope), or "no change" (zero slope) situation.

Question 16 This is an application in which a quantity is a function of height rather than time. The slope equals a change in rent price divided by a change in the floor level.

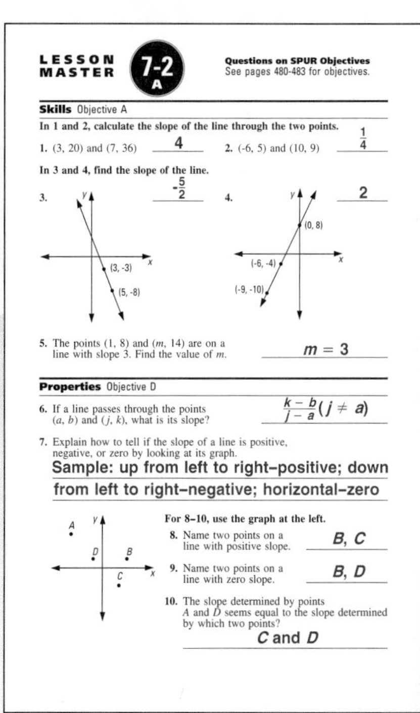
LESSON MASTER 7-2 A

Questions on SPUR Objectives
See pages 480-483 for objectives.

Skills Objective A

In 1 and 2, calculate the slope of the line through the two points.

1. (3, 20) and (7, 36) 4
2. (-6, 5) and (10, 9) $\frac{1}{4}$

In 3 and 4, find the slope of the line.

3. $-\frac{5}{2}$
4. 2

5. The points (1, 8) and (m, 14) are on a line with slope 3. Find the value of m. $m = 3$

Properties Objective D

6. If a line passes through the points (a, b) and (j, k), what is its slope? $\frac{k-b}{j-a} (j \neq a)$
7. Explain how to tell if the slope of a line is positive, negative, or zero by looking at its graph. Sample: up from left to right–positive; down from left to right–negative; horizontal–zero

For 8–10, use the graph at the left.

8. Name two points on a line with positive slope. B, C
9. Name two points on a line with zero slope. B, D
10. The slope determined by points A and D seems equal to the slope determined by which two points? C and D

Additional Answers

10c.

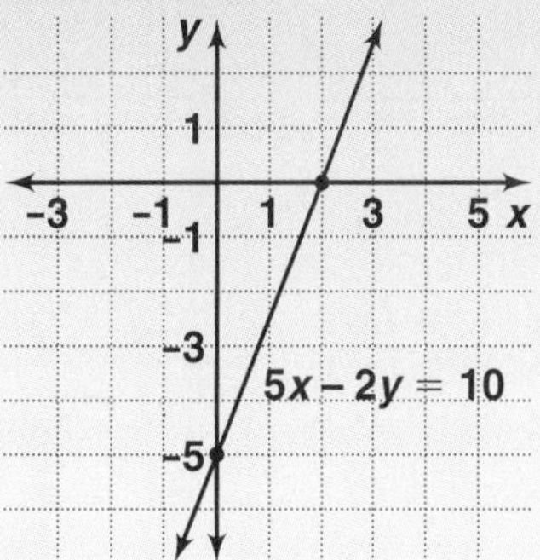

11c.

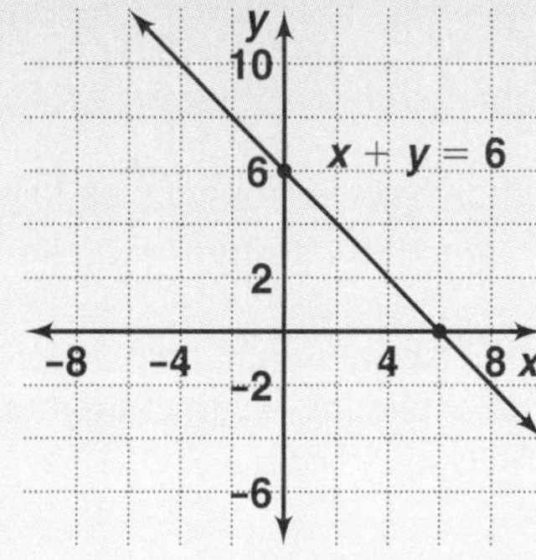

Notes on Questions

Question 22 If you run the computer program in class, you might want to have students enter (2, 3) and (–1, 3), or (4, –1) and (4, 8). The first gives a slope of zero (horizontal line) and the second gives an error message about division by zero (vertical line).

Question 23d Writing The distinction between discrete and continuous situations is reviewed. You might have students critique the graph of the population of Manhattan Island from this perspective. [Sample answer: This situation is a discrete situation. By connecting the points, you get a visual picture of the changes in population. Points on the line segments do not describe the population between the decades.]

Question 30 This question provides an opportunity to combine individual statistics and to examine the average change in height for the whole class.

Follow-up for Lesson 7-2

Practice

For more questions on SPUR Objectives, use **Lesson Master 7-2A** (shown on page 429) or **Lesson Master 7-2B** (shown on pages 430–431).

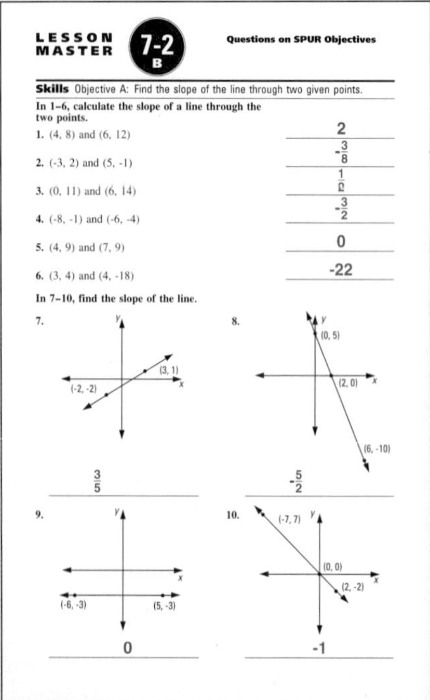

LESSON MASTER 7-2 B Questions on SPUR Objectives

Skills Objective A: Find the slope of the line through two given points.

In 1–6, calculate the slope of a line through the two points.

1. (4, 8) and (6, 12) 2
2. (-3, 2) and (5, -1) $-\frac{3}{8}$
3. (0, 11) and (6, 14) $\frac{1}{2}$
4. (-8, -1) and (-6, -4) $-\frac{3}{2}$
5. (4, 9) and (7, 9) 0
6. (3, 4) and (4, -18) -22

In 7–10, find the slope of the line.

7. (-2, -2), (3, 1) $\frac{3}{5}$
8. (0, 5), (2, 0), (6, -10) $-\frac{5}{2}$
9. (-6, -3), (5, -3) 0
10. (-7, 7), (0, 0), (2, -2) -1

In 18 and 19, coordinates of points are given in columns A and B of a spreadsheet.

a. Fill in the rate-of-change column, leaving C2 blank.
b. Do the points lie on a line? How can you tell?

18.

	A	B	C
1	x	y	rate of change
2	-5	19	
3	0	9	-2
4	7	-5	-2
5	10	-11	-2

b) Yes; the rate of change is constant.

19.

	A	B	C
1	x	y	rate of change
2	0	1	
3	4	6	1.25
4	8	7	0.25
5	12	11	1

b) No; the rate of change is not constant.

20. Refer to the graph below at the left.
a. Find the slope of the horizontal line $y = 2$. 0
b. Find the slope of the horizontal line $y = -3$. 0
c. From parts **a** and **b**, what do you think can be said about the slope of all horizontal lines? The slope is zero.

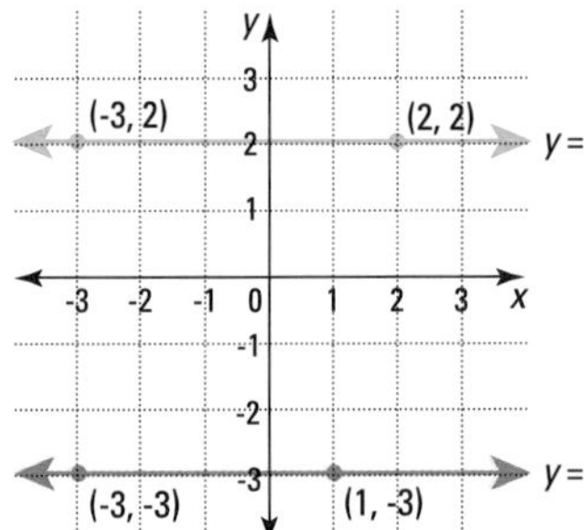

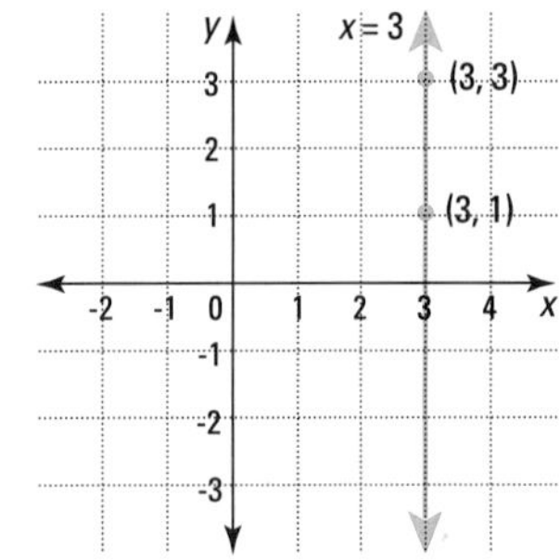

21. Consider the vertical line $x = 3$ graphed above at the right.
a. What happens when you try to find its slope?
b. From your answer to part **a**, do you think that vertical lines have slope? a) Zero is in the denominator; division by zero is impossible. b) No

22. The program below determines the slope of a line given two points.

```
10 PRINT "DETERMINE SLOPE FROM TWO POINTS"
20 PRINT "GIVE COORDINATES OF FIRST POINT"
30 INPUT X1, Y1
40 PRINT "GIVE COORDINATES OF SECOND POINT"
50 INPUT X2, Y2
60 LET M = (Y2 – Y1)/(X2 – X1)
70 PRINT "THE SLOPE IS "; M
80 END
```

What would you input at lines 30 and 50 to answer Question 5? (0, 1); (2, 7)

Adapting to Individual Needs

Challenge

Have students solve the following problem:

The recommended maximum slope for a wheelchair ramp is $\frac{1}{10}$. Suppose a store owner needs to build a ramp that rises 24 inches (the height of 3 stairs). What would be the minimum length of the ramp? [Approximately 20.1 feet]

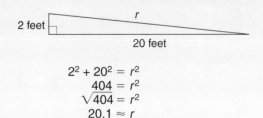

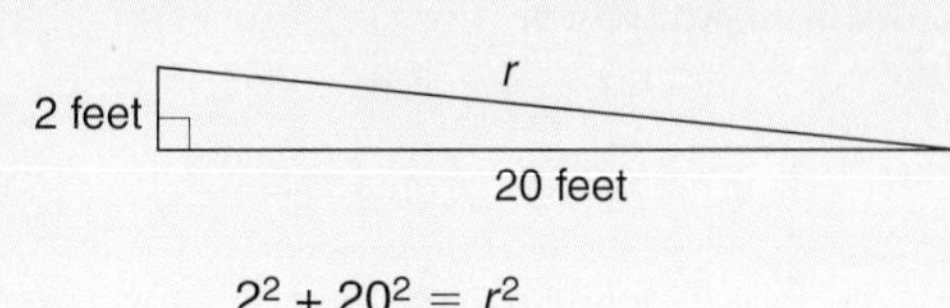

$$2^2 + 20^2 = r^2$$
$$404 = r^2$$
$$\sqrt{404} = r^2$$
$$20.1 \approx r$$

23d) The cost of stamps did not gradually rise over each 5-year period. For example, the cost was never 4.3 cents.

Review

23. Below is a graph showing the cost (in cents) of sending a one-ounce first-class letter. The graph is shown in 5-year intervals beginning in 1960. *(Lesson 7-1)*

a. During which 5-year period did the cost of postage increase the most? **1980–1985**

b. During which 5-year periods was the increase in postage the same? **1960–1965 and 1965–1970**

c. Calculate the average increase per year from 1965 to 1975.

d. Give a reason why the points on the graph are not connected.

c) $\frac{1}{2}$ cent per year; d) See left.

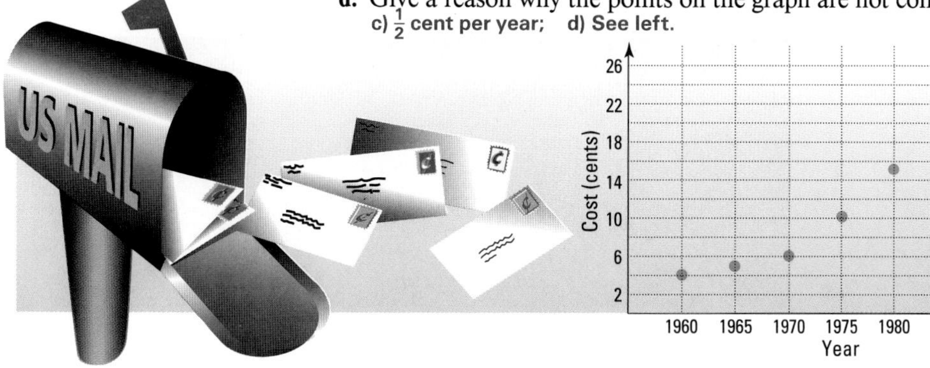

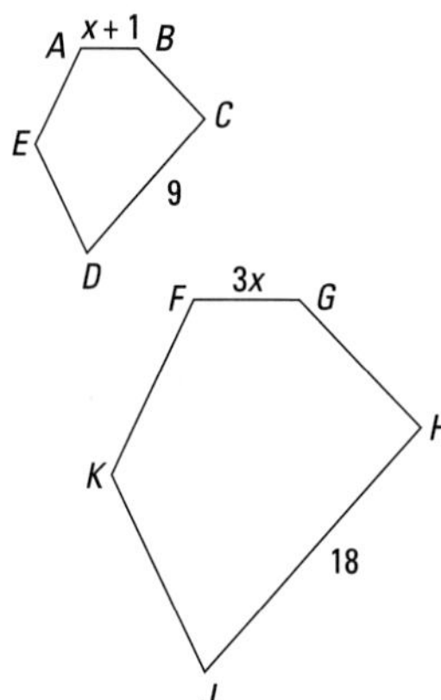

24. The two figures at the left are similar. Corresponding sides are parallel. Find AB and FG. *(Lessons 6-8, 6-9)* **$AB = 3$; $FG = 6$**

In 25 and 26, graph the line. *(Lessons 4-9, 5-1)*

25. $y = 5x$ **See left.**

26. $y = 5$ **See below.**

27. Is the point (7, 6) on the line $y = 2x - 5$? Explain your reasoning. *(Lessons 4-6, 4-9)* **No. If you substitute 7 for x in the equation $y = 2x - 5$, then $y = 9$. $y \neq 6$**

In 28 and 29, write an expression for the amount of money the person has after x weeks. *(Lesson 3-8)*

28. Eddie is given \$100 and spends \$4 a week. **$100 - 4x$**

29. Gretchen owes \$350 on a stereo and is paying it off at \$5 a week. **$-350 + 5x$**

Exploration

30. Find a record of your height at some time over a year ago. Compare it with your height now. How fast has your height been changing from then until now? **Answers will vary.**

25)

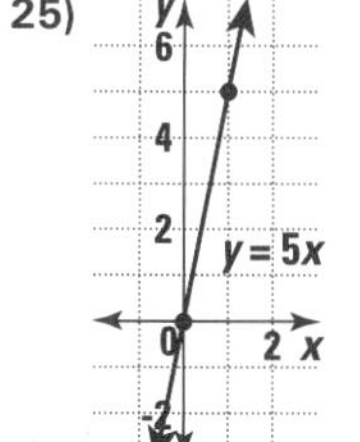

26)

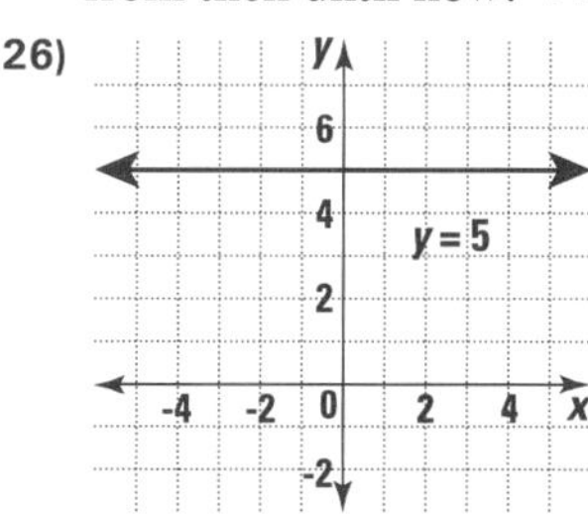

Assessment

Written Communication Refer students to the introductory situation and to **Question 4.** Then have them describe a situation about an ant climbing up or down a flagpole. Ask students to find the constant rate of increase or decrease for their situation. [Students use ordered pairs (time, height) to find rate of change in terms of height per time.]

Extension

Cooperative Learning Give each group of students graph paper or **Teaching Aid 26,** and a long piece of thin pasta which represents a line. Have students place the pasta on the x-axis. Ask students to write about what happens to the slope of a line as the line moves counterclockwise about the origin through the four quadrants. [The slope starts at zero and gets greater as it approaches the y-axis. When the line is on the y-axis, the slope is undefined. Crossing over into the second quadrant, the slope is negative and approaches zero. Going into the third quadrant, the slope is positive and becomes greater as it approaches the y-axis. As the line crosses the y-axis into the fourth quadrant, the slope is negative and becomes greater as it approaches the x-axis.]

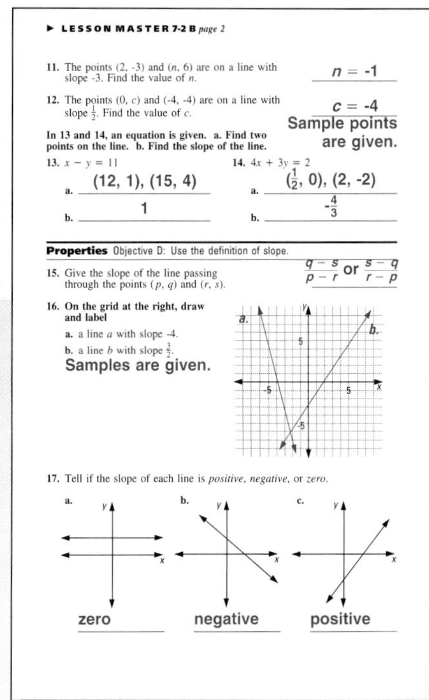

▶ **LESSON MASTER 7-2 B** *page 2*

11. The points (2, -3) and (n, 6) are on a line with slope -3. Find the value of n. **$n = -1$**

12. The points (0, c) and (-4, -4) are on a line with slope $\frac{1}{2}$. Find the value of c. **$c = -4$**

In 13 and 14, an equation is given. a. Find two points on the line. b. Find the slope of the line. **Sample points are given.**

13. $x - y = 11$
a. **(12, 1), (15, 4)**
b. **1**

14. $4x + 3y = 2$
a. **($\frac{1}{2}$, 0), (2, -2)**
b. **$-\frac{4}{3}$**

Properties Objective D: Use the definition of slope.

15. Give the slope of the line passing through the points (p, q) and (r, s). **$\frac{q-s}{p-r}$ or $\frac{s-q}{r-p}$**

16. On the grid at the right, draw and label
a. a line a with slope -4.
b. a line b with slope $\frac{3}{2}$.
Samples are given.

17. Tell if the slope of each line is *positive, negative,* or *zero.*
a. **zero**
b. **negative**
c. **positive**

Setting Up Lesson 7-3

Discussing **Question 19** in this lesson helps prepare students for the characterization of slope given in Lesson 7-3.

Lesson 7-3

Objectives

D Use the definition and properties of slope.
E Calculate rates of change from real data.
H Graph a straight line given a point and the slope.

Resources

From the ***Teacher's Resource File***
- Lesson Master 7-3A or 7-3B
- Answer Master 7-3
- Assessment Sourcebook: Quiz for Lessons 7-1 through 7-3
- Teaching Aids
  28 Four-Quadrant Graph Paper
  71 Warm-up
  78 Example 2
  79 Questions 5–6, 16–18
  80 Questions 26–28

Additional Resources
- Visuals for Teaching Aids 28, 71, 78–80

LESSON

7-3 *Properties of Slope*

Land movers. *Bulldozers are machines used in grading and excavating land for construction projects such as highways. In road construction, the slant or angle of the roadway is carefully calibrated to meet certain specifications.*

A test ramp for bulldozers goes down 0.6 foot for each foot it goes across, as illustrated below at the left.

The part of the ramp with the triangle is enlarged and graphed below at the right. Two points are shown on the graph. The rate of change between these points is the slope of the line.

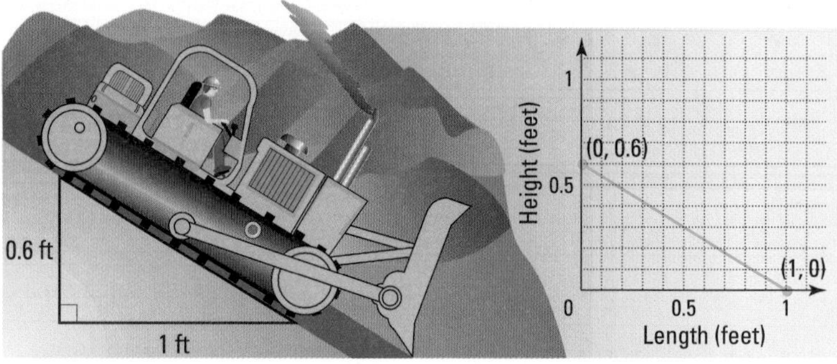

$$\text{slope} = \frac{\text{vertical change}}{\text{horizontal change}} = \frac{\text{change in height}}{\text{change in length}} = \frac{0 - 0.6}{1 - 0} = -0.6$$

This verifies an important property of slopes of lines that is given at the top of page 433.

Lesson 7-3 Overview

Broad Goals This lesson gives various applications of the property that the slope of a line is the amount of change in the height of the line as the line goes one unit to the right.

Perspective All of the examples deal with graphing a line given a point and the slope of the line. The first two examples are abstract, and the third is from a real situation.

At first, it may seem strange to graph a line with slope $\frac{1}{4}$ (as in **Example 2**) by moving up $\frac{1}{4}$ unit when going right 1 unit instead of going up 1 and over 4. However, the use of this approach in earlier editions of this book has shown that it works very well. There are several advantages to this interpretation. First, by stating that slope *always* measures the change in *height*, we eliminate one of the most common errors in calculating slope—reversing the horizontal and vertical changes.

The second advantage of this view of slope is that it fits naturally with the idea of rate that has been developed since Chapter 2. In this lesson, exercises deal with change in dollars/week, magazines/week, and dollars/day.

The slope of a line is the amount of change in the height of the line for every change of one unit to the right.

For instance, if a line has slope 3, as you move one unit to the right, the line goes up three units.

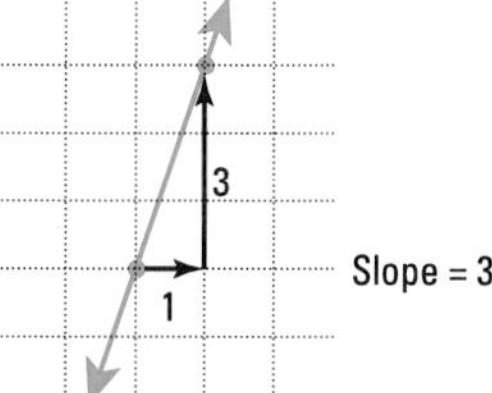

If the slope is $-\frac{3}{2}$, for every change of one unit to the right, the line goes down $\frac{3}{2}$ units.

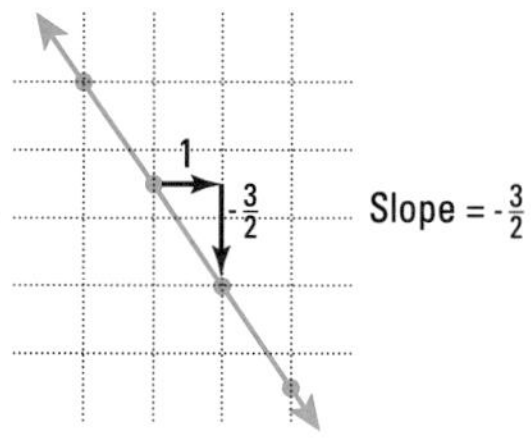

Graphing Lines by Using Properties of Slope

Example 1

Graph the line which passes through (3, -1) and has a slope of -2.

Solution

Plot the point (3, -1). Since the line has slope -2, start at (3, -1), and then move one unit to the right and two units down. Plot that point and draw the line.

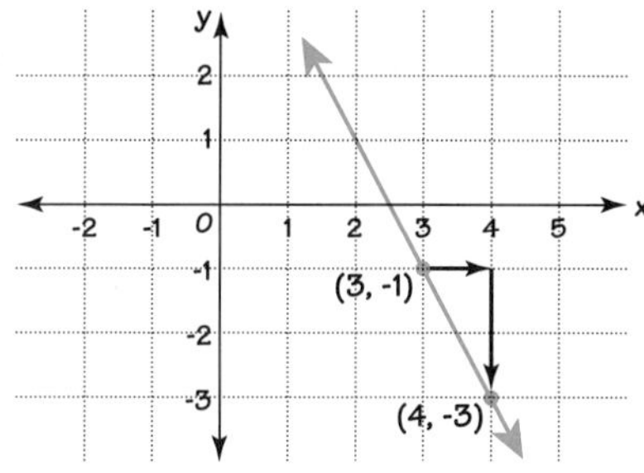

To graph lines having slopes that are given as fractions, it may be helpful to split unit intervals.

Warm-up

The roof of a building has a pitch of $\frac{7}{12}$. This means that the roof rises 7 feet when the horizontal distance from the top is 12 feet.

1. Point *R* is on the roof and its horizontal distance from the top is 24 feet. What is its vertical distance from the top? **14 feet**
2. Point *T* is on the roof and its vertical distance from the top is 10.5 feet. What is its horizontal distance from the top? **18 feet**

If students are having difficulty with the *Warm-up,* suggest they draw a diagram, or you might want to give them this diagram:

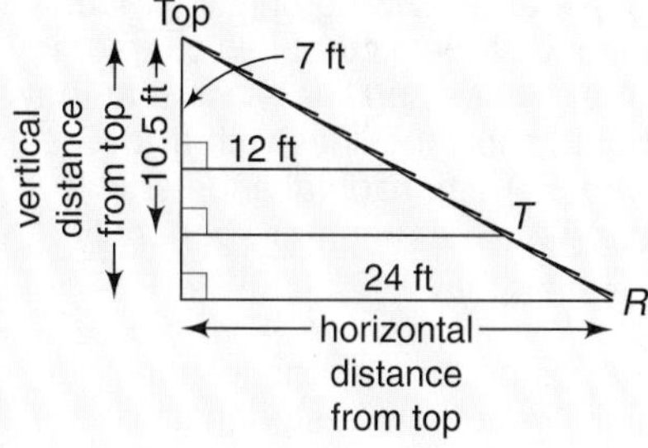

A third advantage is that it is easier to compare slopes and to see that a greater slope means a faster increase (if both are positive).

Fourth, it is easier to make sense of positive, negative, and zero slopes. Since we always move to the right, negative slope means to go down. Zero slope means that you don't change the height. Also, the fact that a vertical line has no slope is almost obvious. You cannot measure change as you go to the right if you can't move to the right on the line.

Fifth, in the slope-intercept equation for a line, an increase of 1 in x means an increase of m in y. Thus thinking of increasing x by 1 offers a firm interpretation of this important parameter in the slope-intercept equation.

Optional Activities

You might have students read through this lesson and in every case in which an example or instance is given, have them change numbers and tell what would happen in the drawing. For instance, what if the bulldozer goes down $\frac{1}{2}$ foot for each foot it goes across? [Its slope is $-\frac{1}{2}$.] You might note that the absolute value of the slope of the ramp, converted to a percent, is called the *grade* of the ramp.

Notes on Reading

When discussing **Example 2,** you might want to use **Teaching Aid 78** which contains the graphs for the two solutions.

History Connection When discussing **Example 3,** you might mention that our modern postal system was greatly affected by extensive road building in the 18th century and the introduction of railroads and steamships in the 19th century. The development of reliable airmail service, the introduction of automated mail handling, and the initiation of worldwide overnight mail are three of the most significant advancements for the postal system in the 20th century.

Error Alert Students often draw a line only long enough to connect the two points they plot. You can compare this to taking a snapshot of people and showing only their feet. Remind students that this chapter is about slopes and lines, and that they should show the lines with arrowheads. The horizontal and vertical changes on a graph are generally represented as segments or rays.

After going through the lesson, you might want to have students do *Optional Activities* below.

Example 2

a. Graph the line through (-2, 1) with slope $\frac{1}{4}$.

b. Name another point on the line with integer coordinates.

Solution 1

Draw the axes with the unit intervals split into fourths. Plot (-2, 1), then move right 1 unit and up $\frac{1}{4}$ unit. Plot the resulting point, $\left(-1, 1\frac{1}{4}\right)$, and draw the line through the two points.

a.

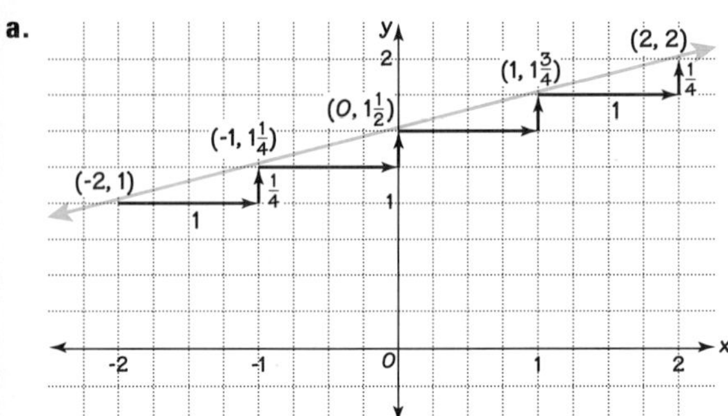

b. One point with integer coordinates (-2, 1) is given. Continue plotting points by moving right 1 unit and up $\frac{1}{4}$ unit, until you reach another point with integral coordinates. As shown in the graph, the point (2, 2) is also on the line.

Solution 2

a. Plot (-2, 1) as in Solution 1. However, instead of going across 1 and up $\frac{1}{4}$, go across $4 \cdot 1$ and up $4 \cdot \frac{1}{4}$. That is, go across 4 and up 1 to the point (2, 2). The graph is shown below.

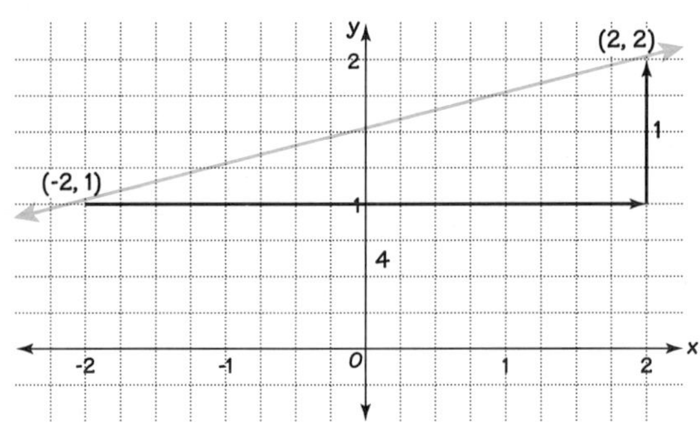

b. The solution to part a shows that (2, 2) is also on the line.

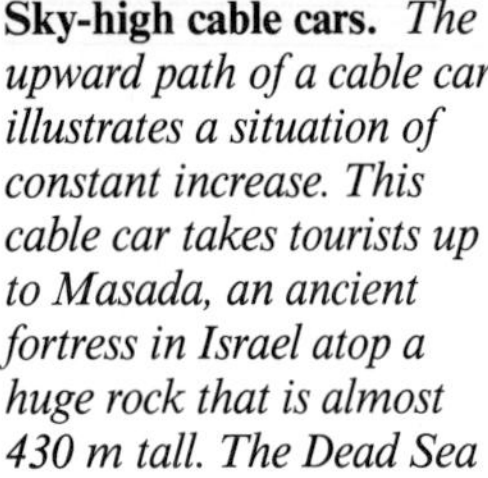

Sky-high cable cars. *The upward path of a cable car illustrates a situation of constant increase. This cable car takes tourists up to Masada, an ancient fortress in Israel atop a huge rock that is almost 430 m tall. The Dead Sea is seen in the background.*

The idea in Examples 1 and 2 helps in drawing graphs of situations of constant increase.

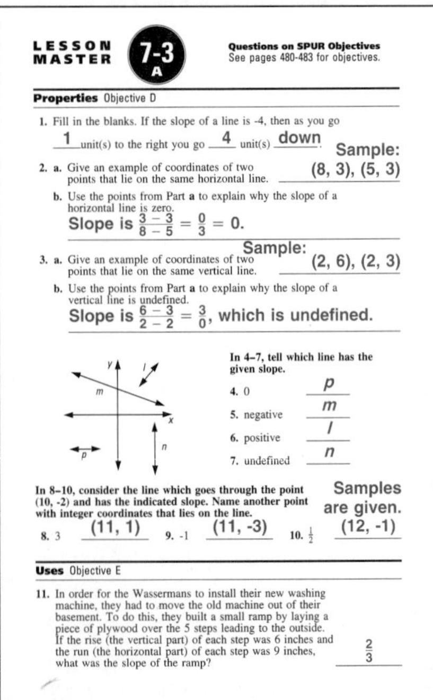

LESSON MASTER 7-3 A

Questions on SPUR Objectives
See pages 480-483 for objectives.

Properties Objective D

1. Fill in the blanks. If the slope of a line is -4, then as you go __1__ unit(s) to the right you go __4__ unit(s) __down__.

2. a. Give an example of coordinates of two points that lie on the same horizontal line. Sample: __(8, 3), (5, 3)__

   b. Use the points from Part **a** to explain why the slope of a horizontal line is zero.
   Slope is $\frac{3-3}{8-5} = \frac{0}{3} = 0$.

3. a. Give an example of coordinates of two points that lie on the same vertical line. Sample: __(2, 6), (2, 3)__

   b. Use the points from Part **a** to explain why the slope of a vertical line is undefined.
   Slope is $\frac{6-3}{2-2} = \frac{3}{0}$, which is undefined.

In 4–7, tell which line has the given slope.

4. 0 __p__
5. negative __m__
6. positive __l__
7. undefined __n__

In 8–10, consider the line which goes through the point (10, -2) and has the indicated slope. Name another point with integer coordinates that lies on the line. Samples are given.

8. 3 __(11, 1)__ 9. -1 __(11, -3)__ 10. $\frac{1}{2}$ __(12, -1)__

Uses Objective E

11. In order for the Wassermans to install their new washing machine, they had to move the old machine out of their basement. To do this, they built a small ramp by laying a piece of plywood over the 5 steps leading to the outside. If the rise (the vertical part) of each step was 6 inches and the run (the horizontal part) of each step was 9 inches, what was the slope of the ramp? __$\frac{2}{3}$__

Video

Wide World of Mathematics The segment, *Ski Flying,* provides dramatic footage of a championship 200-meter jump competition. The segment, which includes supporting diagrams, provides an exciting way to introduce a lesson on slope. Related questions and an investigation are provided in videodisc stills and in the Video Guide. A related CD-ROM activity is also available.

Videodisc Bar Codes

Search Chapter 34

Play

Example 3

Beginning in 1991, postage rates for first-class mail were $.29 for the first ounce and $.23 for each additional ounce up to 10 ounces. Graph the relation between the weight of a letter and its mailing cost for whole-number weights.

Solution

The starting point is (1, 0.29). Because the rate goes up $.23 for each ounce, the slope is $.23/oz. The points are on a line because the situation is a constant-increase situation.

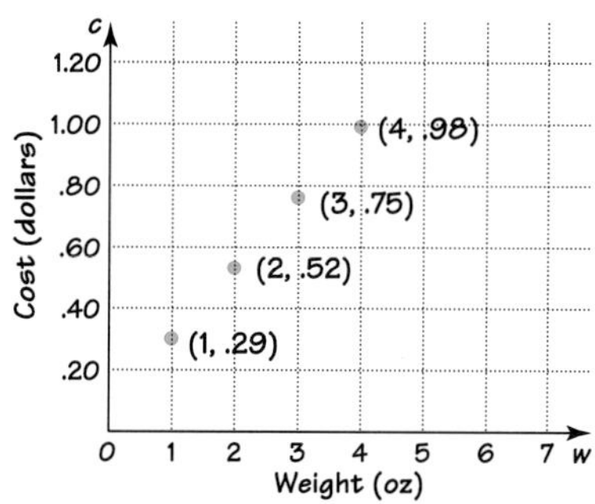

An equation for the cost C to mail a first-class letter weighing w whole ounces is $C = 0.29 + 0.23(w - 1)$. This is an equation for the line that contains the points graphed in Example 3. Notice that the given numbers 0.29 and 0.23 both appear in this equation. In the next lesson, you will learn how the slope can be easily determined from the equation.

Zero Slope and Undefined Slope

Lines with slope 0 may be drawn using the method of Examples 2 and 3.

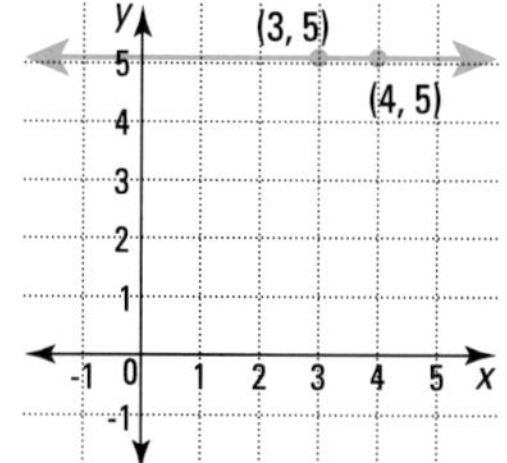

Example 4

Draw a line through the point (3, 5) with slope 0.

Solution

Plot the point (3, 5). From this point move one unit to the right, and 0 units up. Draw the line through these two points as shown at the left.

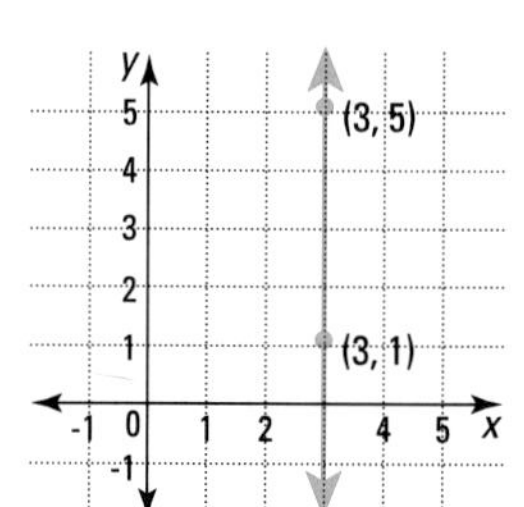

There is one kind of line for which the methods of Examples 2 and 3 does not work. You cannot go to the right on a vertical line. Furthermore, if you try to calculate the slope from two points on a vertical line, the denominator will be zero. For instance, to find the slope of the vertical line through (3, 5) and (3, 1), you would be calculating $\frac{5-1}{3-3} = \frac{4}{0}$, which is not defined. Thus the slope of a vertical line is *not defined.*

The slope of every horizontal line is 0.
The slope of every vertical line is undefined.

Additional Examples

1. Graph the line with slope 3 which contains (-4, 0).

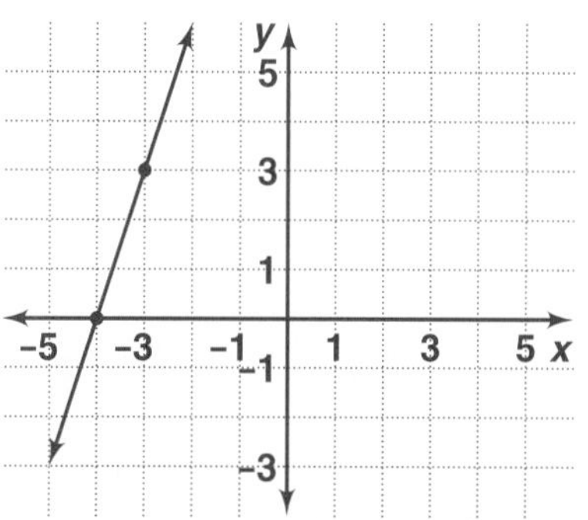

2. a. A line passes through the point (3, -5) and has slope $-\frac{1}{3}$. Graph the line.

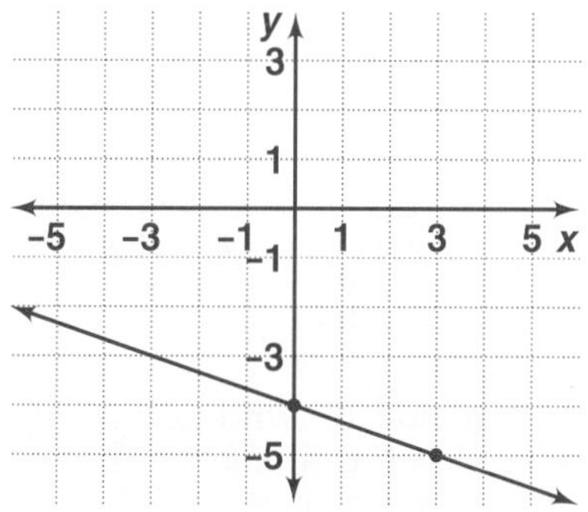

   b. Name another point on the line with integer coordinates. **Sample answer: (0, -4)**

3. Mrs. Wagner drained her childrens' wading pool. The pool contained 10 inches of water and

(Additional Examples continue on page 436.)

Adapting to Individual Needs

Extra Help

When graphing a line with a fractional slope, it may be helpful to some students to adjust the scale on the grid so they do not have to estimate a fractional part of a unit. In **Example 2,** point out to students that the slope is $\frac{1}{4}$, so the units are scaled with 4 squares per unit. Thus, a move of one unit to the right is represented by moving 4 squares to the right. A move of $\frac{1}{4}$ unit up is represented by moving one square up.

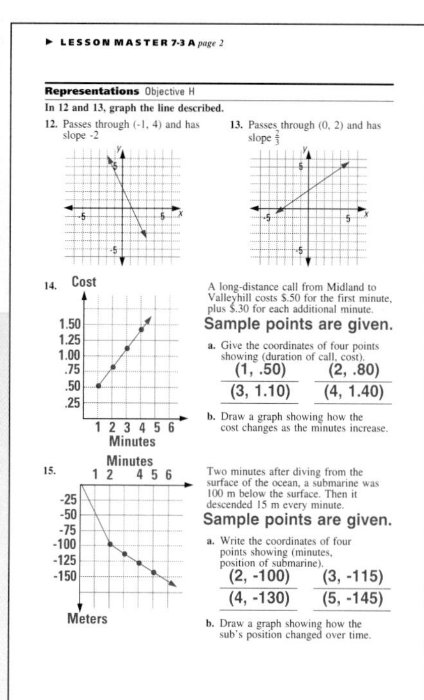

► **LESSON MASTER 7-3 A** *page 2*

Representations Objective H

In 12 and 13, graph the line described.

12. Passes through (-1, 4) and has slope -2

13. Passes through (0, 2) and has slope $\frac{2}{3}$

14. A long-distance call from Midland to Valleyhill costs $.50 for the first minute, plus $.30 for each additional minute.

Sample points are given.

a. Give the coordinates of four points showing (duration of call, cost).
(1, .50) **(2, .80)** **(3, 1.10)** **(4, 1.40)**

b. Draw a graph showing how the cost changes as the minutes increase.

15. Two minutes after diving from the surface of the ocean, a submarine was 100 m below the surface. Then it descended 15 m every minute.

Sample points are given.

a. Write the coordinates of four points showing (minutes, position of submarine).
(2, -100) **(3, -115)** **(4, -130)** **(5, -145)**

b. Draw a graph showing how the sub's position changed over time.

drained at a rate of $\frac{1}{2}$ inch per minute. Graph the relationship between time and height of the water.

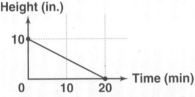

Height (in.)
10
0 10 20 Time (min)

4. Draw a line that passes through the point (3, –1) and for which the slope of the line is undefined. **The graph is the vertical line through (3, –1).**

Notes on Questions

Questions 5–6 The graphs for these questions are given on **Teaching Aid 79.**

Questions 7–11 Cooperative Learning These are good questions for small group work. Have students record at least four points for each line, and plot the points.

Question 15 Writing You may use "no slope" as a synonym for "undefined slope." If you do so, make sure students understand the distinction between "0 slope" and "no slope." You may want to have students make up their own examples of each of these ideas and to write about their understanding of the distinction.

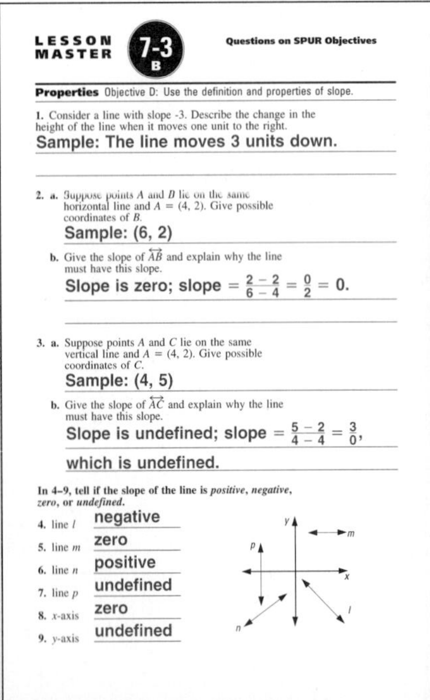

LESSON MASTER 7-3 B — Questions on SPUR Objectives

Properties Objective D: Use the definition and properties of slope.

1. Consider a line with slope -3. Describe the change in the height of the line when it moves one unit to the right.
Sample: The line moves 3 units down.

2. a. Suppose points A and B lie on the same horizontal line and A = (4, 2). Give possible coordinates of B.
Sample: (6, 2)
b. Give the slope of $\overleftrightarrow{AB}$ and explain why the line must have this slope.
Slope is zero; slope = $\frac{2-2}{6-4} = \frac{0}{2} = 0$.

3. a. Suppose points A and C lie on the same vertical line and A = (4, 2). Give possible coordinates of C.
Sample: (4, 5)
b. Give the slope of $\overleftrightarrow{AC}$ and explain why the line must have this slope.
Slope is undefined; slope = $\frac{5-2}{4-4} = \frac{3}{0}$, which is undefined.

In 4–9, tell if the slope of the line is *positive, negative, zero,* or *undefined.*

4. line l: negative
5. line m: zero
6. line n: positive
7. line p: undefined
8. x-axis: zero
9. y-axis: undefined

2)
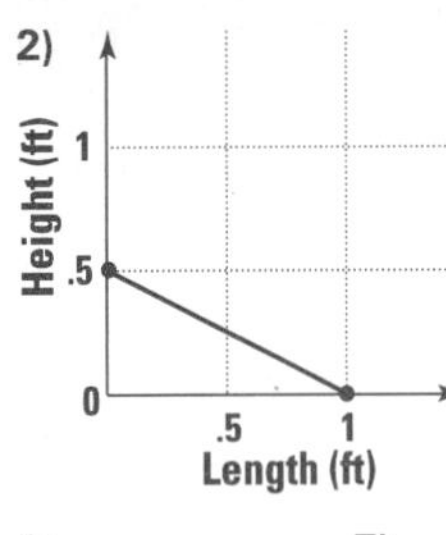

8) 7)
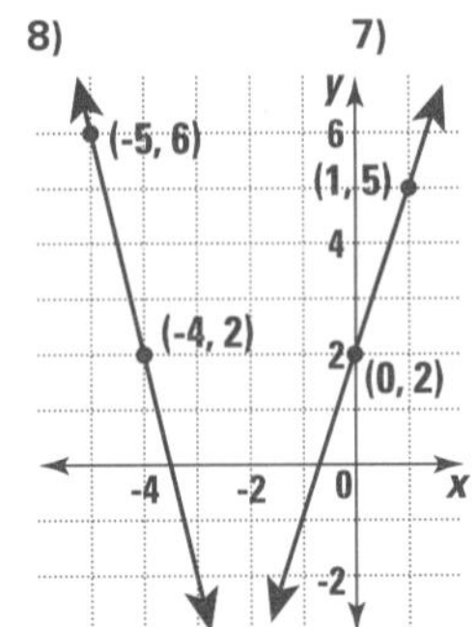

9)
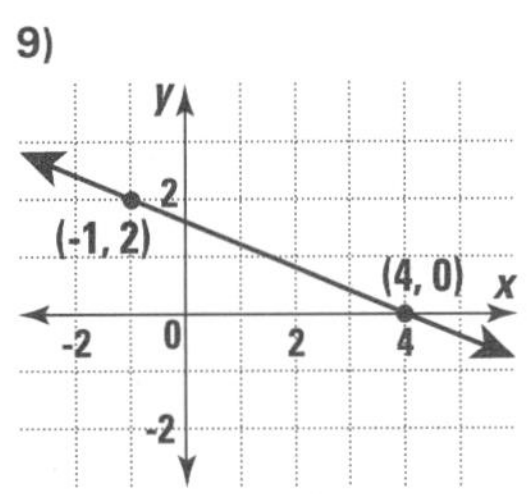

10)
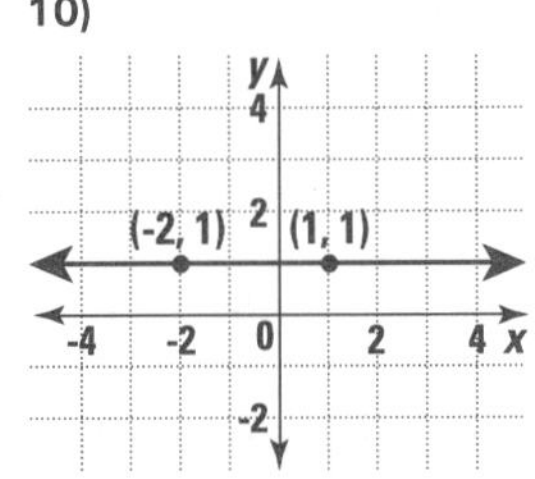

13)
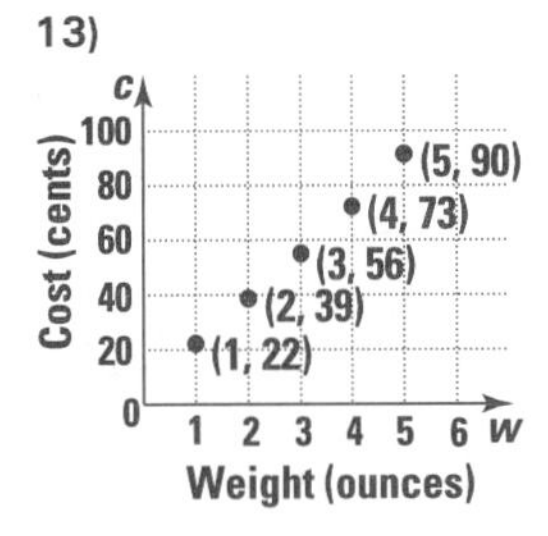

QUESTIONS

Covering the Reading

In 1 and 2, a test ramp for bulldozers goes down 0.5 unit for each unit across.

1. What is its slope? **-0.5**
2. Draw a picture of the test ramp. **See left for sample.**
3. Slope is the amount of change in the __?__ of a graph for every change of one unit to the right. **height**
4. If the slope of a line is 8, then you move __?__ as you move one unit to the right. **up 8 units**

In 5 and 6, give the slope of the line.

5. **-3**

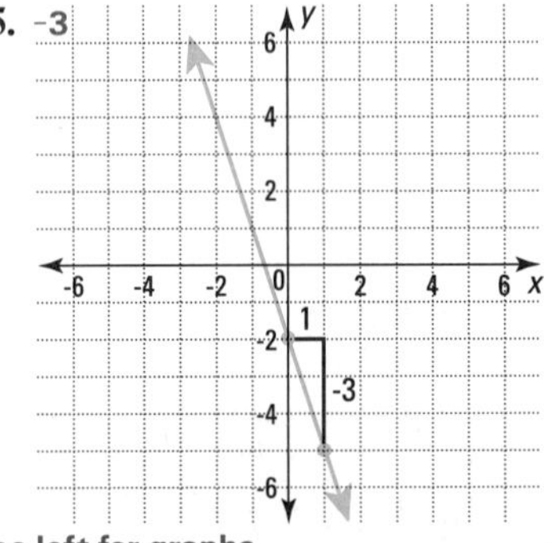

6. **1**

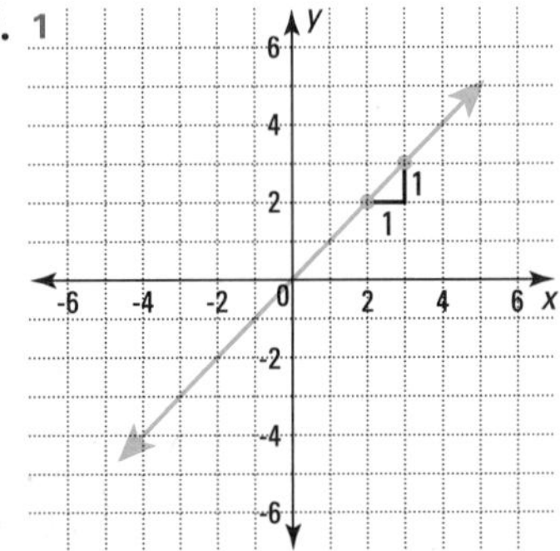

See left for graphs

In 7–10, a point and slope are given. **a.** Graph the line with the given slope that goes through the given point. **b.** Find one other point on this line.

7. point (0, 2), slope 3 **Sample: (1, 5)**
8. point (-5, 6), slope -4 **Sample: (-4, 2)**
9. point (-1, 2), slope $-\frac{2}{5}$ **Sample: (4, 0)**
10. point (-2, 1), slope 0 **Sample: (1, 1)**
11. A line has slope $\frac{1}{3}$ and passes through the point (-4, -2). Name one other point with integer coordinates on this line. **Sample: (-1, -1)**
12. In 1991, what was the cost to mail a letter weighing 9 oz? **$2.13**
13. In 1987, postal rates were 22¢ for the first ounce and 17¢ for each additional ounce up to 10 ounces. Graph at least five points showing the relation between weight and cost for whole-number weights. **See left.**
14. Explain why the slope of a vertical line is not defined. Include a specific example not in the text. **See below.**
15. Graph the line through (-2, 1) with an undefined slope. **See margin.**

14) **Sample: Slope is defined to be the amount of change in the height of a line for every change of one unit to the right; vertical lines do not change to the left or right. The slope of the line through (-5, 2) and (-5, 12) is $\frac{10}{0}$, which is not defined.**

Adapting to Individual Needs

English Language Development
Non-English-speaking students might benefit by being paired with English-speaking students. Encourage the pairs of students to discuss the graphs in the lesson and to describe the relationships the graphs show.

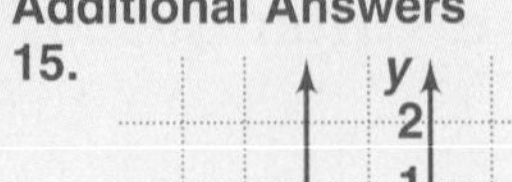

Additional Answers
15.

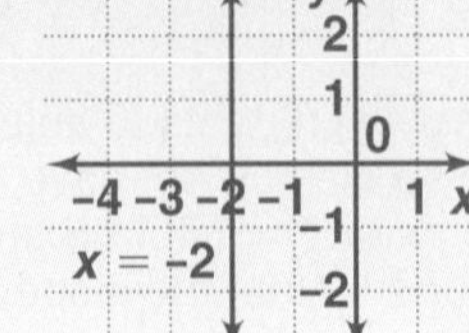

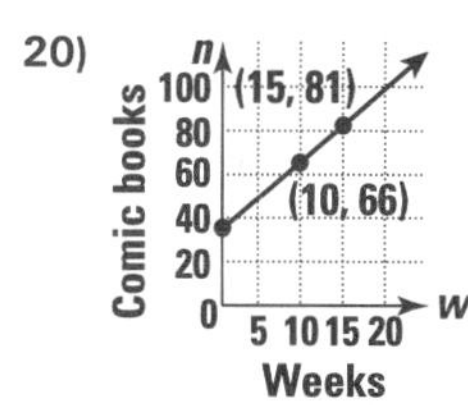

Hills of Dunedin. *In addition to having the world's steepest street, Dunedin is also known for its cathedrals, chocolate factories, and scenic railway station.*

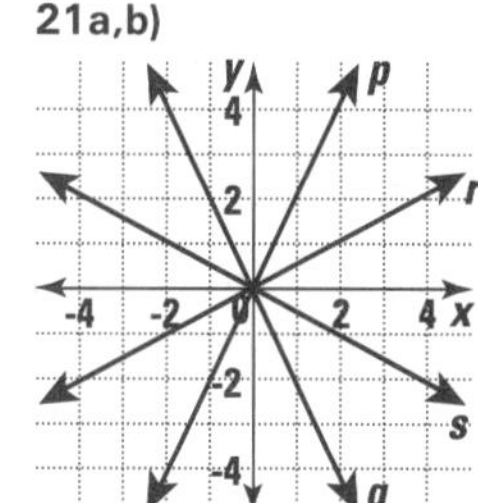

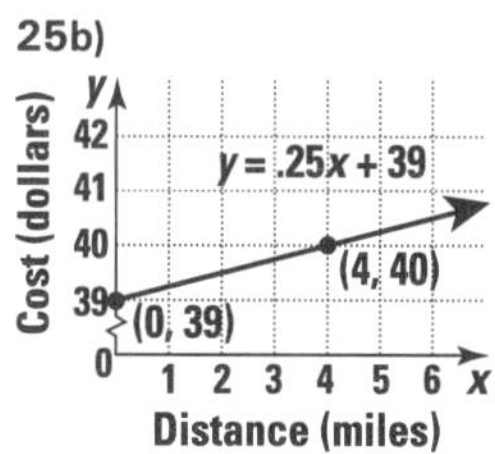

Applying the Mathematics

In 16–18, the slope of a line is given. Fill in the cells of the spreadsheets so that the coordinates are points on the line.

16. slope = 7

	A	B
1	x	y
2	-5	2
3	-4	9
4	-3	16
5	-2	23

17. slope = -8

	A	B
1	x	y
2	6	10
3	7	2
4	8	-6
5	9	-14

18. slope = $\frac{5}{4}$

	A	B
1	x	y
2	0	0
3	1	1.25
4	2	2.5
5	3	3.75

19. The steepest street in the world is Baldwin Street in Dunedin, New Zealand, where the maximum *gradient* is 1 in 1.266. This means that for every horizontal change of 1.266 units, the road goes 1 unit up. Assume that $\overleftrightarrow{UP}$ represents Baldwin Street. What is the slope of $\overleftrightarrow{UP}$? **≈0.79**

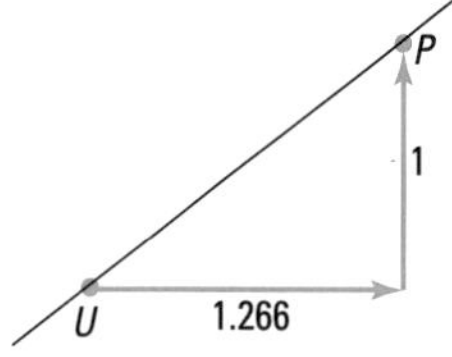

20. Kareem collects comic books. His budget allows him to buy 3 comic books a week. This year, after 10 weeks he has 66 comic books in his collection. Draw a graph to represent the growth of Kareem's collection over the weeks. **See left.**

21. a. Draw a set of coordinate axes and plot the point (0, 0).

b. Draw the lines through (0, 0) with the following slopes:

line *p:* slope 2; line *q:* slope -2; **See left for 21 a,b.**

line *r:* slope $\frac{1}{2}$; line *s:* slope $-\frac{1}{2}$.

c. Describe how the slopes of the lines are related to the appearance of their graphs. **Sample: Lines with positive slopes slant upward from left to right. Lines with negative slopes slant downward from left to right.**

Review

In 22 and 23, calculate the slope determined by the two points. *(Lesson 7-2)*

22. (5, 3) and (8, -2) $-\frac{5}{3}$

23. (-6, -9) and (-13, 5) **-2**

24. Find the slope of the line $3x - y = 15$. *(Lesson 7-2)* **3**

25. A rental truck costs $39 for a day plus $.25 a mile. After *x* miles, the total cost will be *y* dollars.

a. Write an equation relating *x* and *y*. **y = .25x + 39**

b. Graph the line. **See left.**

c. Find the rate of change between any two points on the graph. *(Lessons 3-8, 4-9, 7-2)* $\frac{1}{4}$

Questions 16–18 The tables for these questions are given on **Teaching Aid 79.**

Question 19 Multicultural Connection Students might be interested in knowing that New Zealand is a small island nation in the South Pacific Ocean located midway between the equator and the South Pole and 1200 miles southeast of Australia. The country's first people, the Maori, arrived from an island in Polynesia about 1200 years ago. British settlers began moving to New Zealand in the early 1800s and became the majority population by 1860. New Zealand was under Britain's authority until 1947 when it became completely independent.

Question 20 A point and the slope of the line are given and another point must be found. In graphing these lines, students should be aware that the size of the interval between tick marks on the axes is an important decision.

Question 21 This question previews work with slopes of perpendicular lines which will be done in later courses.

Question 24 Students should use the method from the previous lesson, but you can use this question to lead into the next lesson, which discusses the slope-intercept form of a line, $y = mx + b$.

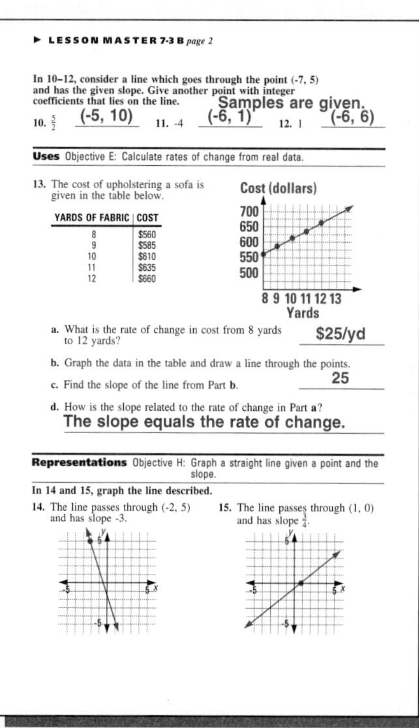

► **LESSON MASTER 7-3 B** *page 2*

In 10–12, consider a line which goes through the point (-7, 5) and has the given slope. Give another point with integer coefficients that lies on the line. Samples are given.

10. $\frac{5}{2}$ (-5, 10) 11. -4 (-6, 1) 12. 1 (-6, 6)

Uses Objective E: Calculate rates of change from real data.

13. The cost of upholstering a sofa is given in the table below.

YARDS OF FABRIC	COST
8	$560
9	$585
10	$610
11	$635
12	$660

a. What is the rate of change in cost from 8 yards to 12 yards? $25/yd

b. Graph the data in the table and draw a line through the points.

c. Find the slope of the line from Part **b**. 25

d. How is the slope related to the rate of change in Part **a**? The slope equals the rate of change.

Representations Objective H: Graph a straight line given a point and the slope.

In 14 and 15, graph the line described.

14. The line passes through (-2, 5) and has slope -3.

15. The line passes through (1, 0) and has slope $\frac{3}{4}$.

Notes on Questions

Questions 26–28 The graph and table are given on **Teaching Aid 80.**

Question 33 This puzzle relates to the subject matter of this lesson because Henri is climbing at a rate of (3 feet – 2 feet)/day, or 1 foot/day. However, that rate changes to 3 feet/day on the last day, which is why the question seems to be a trick.

Follow-up for Lesson 7-3

Practice

For more questions on SPUR Objectives, use **Lesson Master 7-3A** (shown on pages 434–435) or **Lesson Master 7-3B** (shown on pages 436–437).

Assessment

Quiz A quiz covering Lessons 7-1 through 7-3 is provided in the *Assessment Sourcebook.*

Group Assessment
Materials: Four-Quadrant Graph Paper or **Teaching Aid 28**

Have each student **work with a partner.** Ask each student to draw four lines on a coordinate grid and label them *a, b, c,* and *d.* Then have them write the slope of each line, in any order, at the bottom of the paper. Have partners exchange papers and match each line with its slope. [Students can recognize lines with positive or negative slopes and can find the slope of a line.]

Extension

Have students update the information in the graph for **Questions 26–28** by finding the average number of home runs per game for both the American League and the National League for any seasons completed after 1993.

In 26–28, refer to the table and graph below. *(Lesson 7-1)*

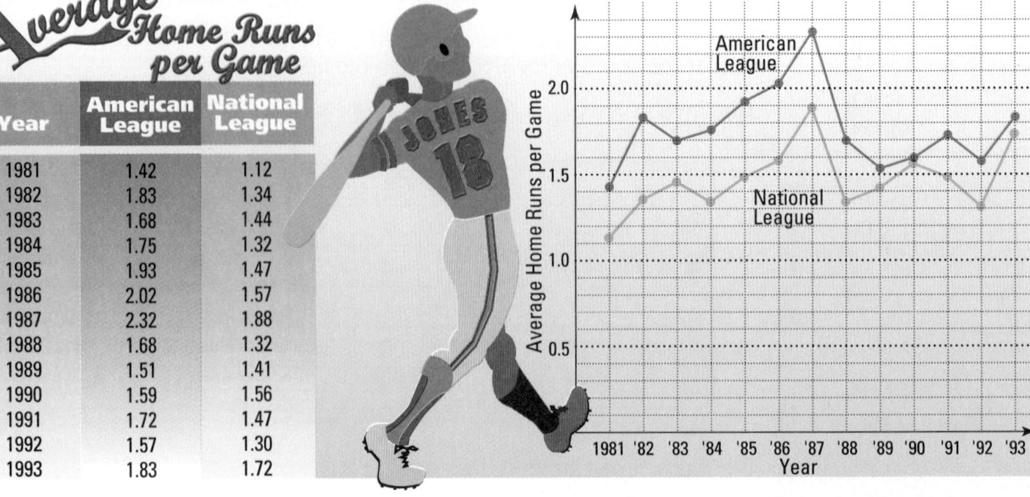

Year	American League	National League
1981	1.42	1.12
1982	1.83	1.34
1983	1.68	1.44
1984	1.75	1.32
1985	1.93	1.47
1986	2.02	1.57
1987	2.32	1.88
1988	1.68	1.32
1989	1.51	1.41
1990	1.59	1.56
1991	1.72	1.47
1992	1.57	1.30
1993	1.83	1.72

26. Between which two consecutive years was there the biggest increase in average home runs per game for the
a. American League? **1981–1982** **b.** National League? **1992–1993**

27. Between which two consecutive years was there the biggest decrease in average home runs per game for the
a. American League? **1987–1988** **b.** National League? **1987–1988**

28. What is the rate of change from 1981 to 1990 in average home runs per game for the
a. American League? **≈0.019** **b.** National League? **≈0.049**

29. If $4a = 15$, find $12a - 5$. *(Lesson 5-8)* **40**

30. If the triangle below has perimeter 71, what is z? *(Lessons 3-5, 4-6)*
$z = 9$

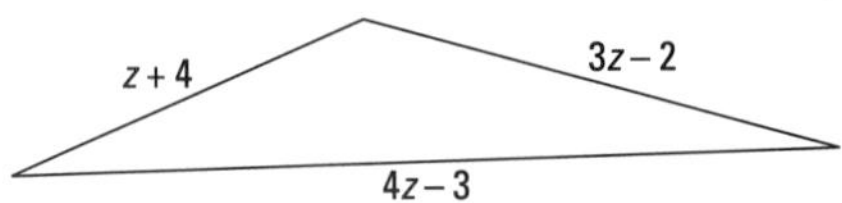

31. Calculate in your head. *(Lesson 3-7)*
a. 15% of \$8.00 **\$1.20** **b.** 200% of x dollars **2x dollars**

32. Find b if $y = mx + b$, $y = -4$, $m = 3$, and $x = 2$. *(Lessons 1-5, 3-7)*
$b = -10$

Exploration

33. Here is a famous puzzle. Beware of the trick. Henri Escargot, a snail, is $14\frac{1}{2}$ feet deep in a well. Every day he climbs 3 feet up the walls of the well. At night the walls are damper and he slips down 2 feet. How many days will it take him to climb out of the well? Explain how you arrived at your answer. **13 days; On the 12th day, he has gone 12 feet. On the 13th day, he starts at 12 feet, goes up 3 feet and is out of the well before he slides back down.**

Adapting to Individual Needs

Challenge
Have students solve the following problem: The Great Pyramid of Cheops in Egypt originally had a height of 482 feet. The base is a square with each side having a length of 755 feet. Find the slope of each sloping edge of the pyramid. Note that this is not the same as the slope of each triangular face. [Approximately 0.90]

Setting Up Lesson 7-4

Materials If automatic graphers are available, we recommend that students use them for **Question 21** in Lesson 7-4.

LESSON

7-4

Slope-Intercept Equations for Lines

Saving for the future. *Many people, such as this painter, save a fixed amount of their weekly earnings in a savings account. When a bank balance grows at a constant rate, the savings can be represented by a graph of a linear equation.*

Shawn has \$300 saved for a used car and saves an additional \$25 a week. After 3 weeks, Shawn will have $300 + 3 \cdot 25$ dollars, or \$375. After x weeks, Shawn will have y dollars, where $y = 300 + 25x$.

The line $y = 300 + 25x$ is graphed below. There are two key numbers in the equation for this line. The number 25 is the slope of the line. The number 300 indicates where the line crosses the y-axis. That number is the y-intercept of the line. In general, when a graph intersects the y-axis at the point $(0, b)$, the number b is a ***y*-intercept** for the graph.

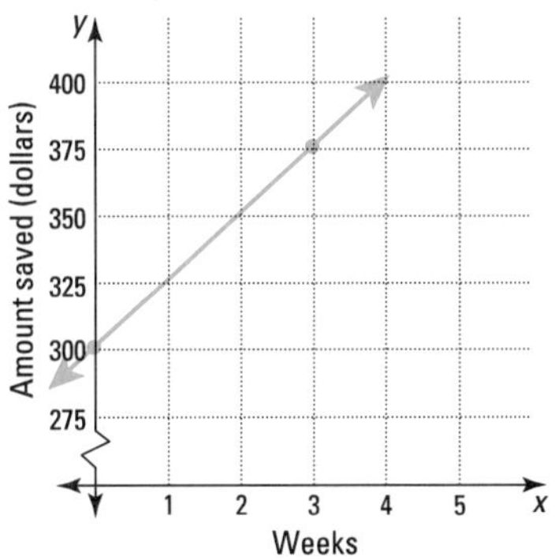

The graph of this line is completely determined by its slope, 25, and y-intercept, 300. When the equation is rewritten in the form $y = 25x + 300$, it is said to be in *slope-intercept* form.

Lesson 7-4

Objectives

B Find an equation for a line given its slope and one point on it.
C Write an equation for a line in slope-intercept form, and find its slope and y-intercept.
F Use equations for lines to describe real situations.
H Graph a straight line given its equation, or given a point and the slope.

Resources

From the ***Teacher's Resource File***

- Lesson Master 7-4A or 7-4B
- Answer Master 7-4
- Teaching Aids
  28 Four-Quadrant Graph Paper
  72 Warm-up
- Activity Kit, Activity 16
- Technology Sourcebook
  Computer Master 16

Additional Resources

- Visuals for Teaching Aids 28, 72
- Automatic grapher (Question 21)

Teaching Lesson 7-4

Warm-up

Find the slope of a line through the given pair of points.

1. (6, 4) and (2, 1) $\frac{3}{4}$
2. (5, 7) and (5, 12) **No slope**
3. (−4, 6) and (2, −6) **−2**
4. (2, 4) and (6, 2) $-\frac{1}{2}$
5. (−3, 8) and (5, 8) **0**

Lesson 7-4 Overview

Broad Goals This lesson culminates the discussion of slope by developing and applying the wonderful statement that there is a simple way to determine the slope of a line from its equation, and the y-intercept is obtained as a bonus.

Perspective Equations for lines are generally found in either of two common forms: the slope-intercept form $y = mx + b$ or the standard form $Ax + By = C$. The slope-intercept form has two major advantages: it is unique, in that a line has only one such equation, and it is akin to the function notation $f(x) = mx + b$. The standard form also has advantages: every line, including each vertical line, has an equation in this form, and systems of linear equations in this form are often easier to analyze. There are applications for which each form is natural. In this text, we call the *value* of b in the equation $y = mx + b$ the y-intercept. This fits well with the term *slope-intercept form.*
In later courses, it extends easily to x-intercepts being solutions to equations of the form $f(x) = 0$. Some people call the point $(0, b)$, where the line crosses the y-axis, the y-intercept. Our definition is more common.

Notes on Reading

Pacing There are many activities that can be used with this lesson, and since a quiz covering Lessons 7-1 through 7-3 is suitable with the lesson, you might consider spending two days on this material.

You might want to use Activity 1 in *Optional Activities* below to introduce this lesson.

The zig-zag in the vertical axis in the graph on page 439 indicates that there is a gap. Leaving out a section of the graph makes it possible to fit the graph onto the page. This practice is not used consistently in graphs outside the classroom, so we do not use it consistently in this book. Yet, students should know how to interpret the zig-zag symbol.

Reading Mathematics If the slope and y-intercept can be found without computation, some people speak of *reading* the slope and y-intercept from the equation.

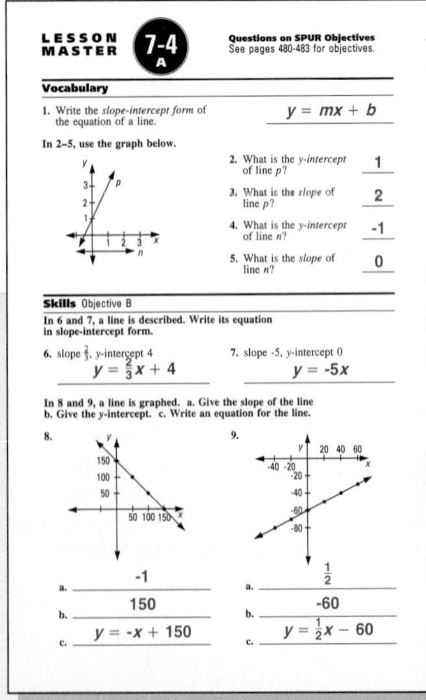

LESSON MASTER 7-4 A

Questions on SPUR Objectives
See pages 480-483 for objectives.

Vocabulary

1. Write the *slope-intercept form* of the equation of a line. $y = mx + b$

In 2–5, use the graph below.

2. What is the *y-intercept* of line *p*? 1
3. What is the *slope* of line *p*? 2
4. What is the *y-intercept* of line *n*? -1
5. What is the *slope* of line *n*? 0

Skills Objective B

In 6 and 7, a line is described. Write its equation in slope-intercept form.

6. slope $\frac{2}{3}$, y-intercept 4 $y = \frac{2}{3}x + 4$
7. slope -5, y-intercept 0 $y = -5x$

In 8 and 9, a line is graphed. a. Give the slope of the line b. Give the y-intercept. c. Write an equation for the line.

8. a. -1 b. 150 c. $y = -x + 150$
9. a. $\frac{1}{2}$ b. -60 c. $y = \frac{1}{2}x - 60$

Slope-Intercept Property
The line with equation $y = mx + b$ has slope m and y-intercept b.

The form $y = mx + b$ is called the **slope-intercept form** for the line.

Finding the Slope and y-intercept from the Equation of a Line

Example 1

Give the slope and the y-intercept of $y = \frac{1}{2}x + 4$.

Solution

The equation is already in slope-intercept form. Compare $y = \frac{1}{2}x + 4$ to $y = mx + b$. $m = \frac{1}{2}$ and $b = 4$. The slope is $\frac{1}{2}$ and the y-intercept is 4.

Check

A y-intercept of 4 means the line must contain the point (0, 4). Does (0, 4) satisfy $y = \frac{1}{2}x + 4$? Yes, because $4 = \frac{1}{2} \cdot 0 + 4$. A slope of $\frac{1}{2}$ implies that the point (2, 5) is also on the line. Does $5 = \frac{1}{2}(2) + 4$? Yes; it checks.

The advantage of slope-intercept form is that it tells you much about the line. So it is often useful to convert other equations for lines into slope-intercept form.

Example 2

Give the slope and y-intercept of $y = -8 - 3x$.

Solution

Rewrite $y = -8 - 3x$ in the form $y = mx + b$ using the definition of subtraction and the Commutative Property of Addition.

$$y = -3x - 8$$

The slope is -3 and the y-intercept is -8.

Check

Draw a graph of $y = -8 - 3x$. Verify that (0, -8) is on the graph. Note that the point (1, -11) is also on the graph. This verifies that the slope is -3.

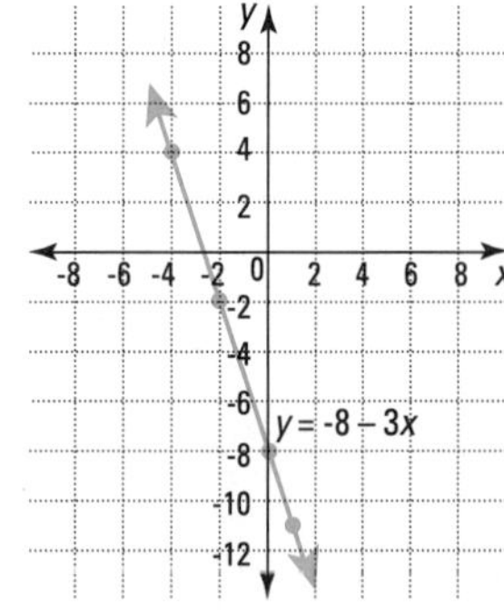

440

Optional Activities

Activity 1 You can use *Activity Kit,* Activity 16, to introduce this lesson. In this activity, students drop centimeter cubes into a cylinder of water to generate their own data, make a linear graph, and analyze the y-intercept and slope of the line.

Activity 2 Consumer Connection
Materials: Graph paper or **Teaching Aid 28**

You might give students this problem after they have completed the lesson. The Internal Revenue Service allows a method called *linear depreciation* for depreciating business property. If the original cost of the property is C dollars and it is depreciated at a constant rate over N years, its value at the end of n years is given by $V = C(1 - \frac{n}{N})$. Have

Being able to change the form of an equation is an important skill. Many automatic graphers will not accept the equation $3x + 4y = 9$ for graphing. It needs to be solved for y. On such graphers, the slope-intercept form of a linear equation is acceptable.

Example 3

Write the equation $3x + 4y = 9$ in slope-intercept form. Give the slope and y-intercept of the line.

Solution

❶ Solve $3x + 4y = 9$ for y.

$$3x + 4y = 9$$
$$4y = -3x + 9$$
$$\frac{4y}{4} = \frac{-3x + 9}{4}$$
$$y = -\frac{3}{4}x + \frac{9}{4}$$

The slope is $-\frac{3}{4}$. The y-intercept is $\frac{9}{4}$ or $2\frac{1}{4}$.

Writing an Equation for a Line with Given Slope and y-intercept

Equations for all nonvertical lines can be written in slope-intercept form.

❷ ### Example 4

a. Write an equation for the line with slope 5 and y-intercept -1.
b. Graph the line.

Solution

a. Use $y = mx + b$. Substitute 5 for m and -1 for b.
$y = 5x + -1$ or $y = 5x - 1$ is an equation for the line.

b. Since the y-intercept is -1, the graph contains (0, -1). Plot this point. Since the slope is 5, the graph goes up 5 units for each unit it goes to the right. Plot another point 1 unit over and 5 units up from (0, -1). That point is (1, 4).

Check

Both points (0, -1) and (1, 4) satisfy the equation $y = 5x - 1$; and the slope determined by the points is 5. This agrees with the given information.

Every constant-increase or constant-decrease situation can be described by an equation whose graph is a line. The y-intercept of that line can be interpreted as the starting amount. The slope of that line is the amount of increase or decrease per unit.

❶ Note that in solving $3x + 4y = 9$ for y, rather than dividing both sides by 4, it is just as easy to multiply both sides by $\frac{1}{4}$. In converting equations of the form $by = ax + c$ to slope-intercept form, multiplying by the reciprocal of b usually results in fewer errors. If students divide both sides by b, they must remember to divide *both* terms of $ax + c$ by b.

❷ **Part a** is new to this lesson, but **part b**, graphing a line knowing its slope and one point on it, is review. In **part b,** instead of giving the coordinates of a point, we give the y-intercept. This situation is equivalent to knowing one point.

students write an equation for determining the value of a machine after n years if its original value was \$5000 and it is being depreciated over 20 years. You might also have them draw the graph of the equation. [$V = 5000(1 - \frac{n}{20})$; a sample graph is shown at the right.]

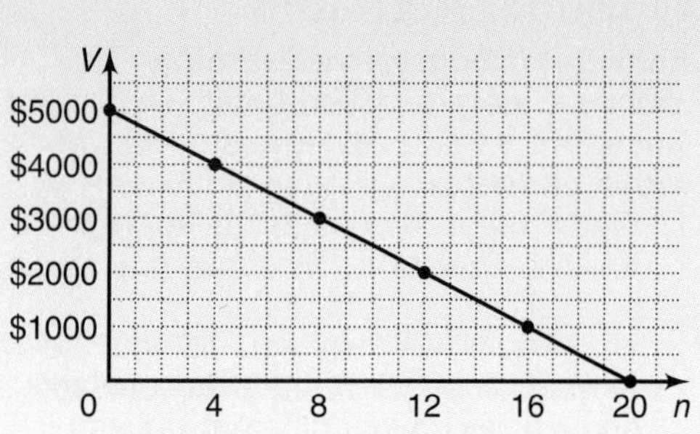

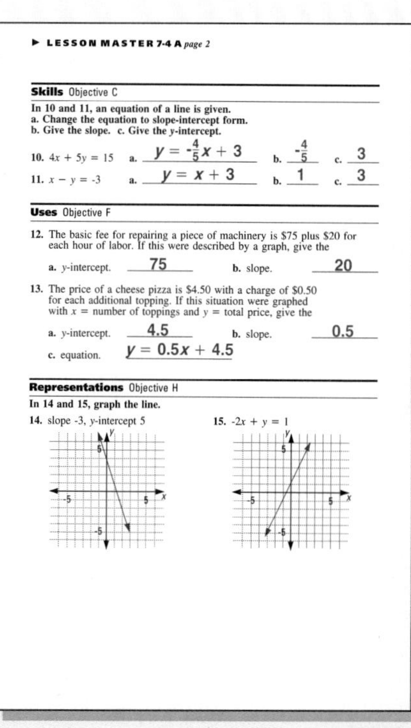

▶ **LESSON MASTER 7-4 A** *page 2*

Skills Objective C

In 10 and 11, an equation of a line is given.
a. Change the equation to slope-intercept form.
b. Give the slope. c. Give the y-intercept.

10. $4x + 5y = 15$ a. $y = -\frac{4}{5}x + 3$ b. $-\frac{4}{5}$ c. 3

11. $x - y = -3$ a. $y = x + 3$ b. 1 c. 3

Uses Objective F

12. The basic fee for repairing a piece of machinery is \$75 plus \$20 for each hour of labor. If this were described by a graph, give the
a. y-intercept. 75 b. slope. 20

13. The price of a cheese pizza is \$4.50 with a charge of \$0.50 for each additional topping. If this situation were graphed with x = number of toppings and y = total price, give the
a. y-intercept. 4.5 b. slope. 0.5
c. equation. $y = 0.5x + 4.5$

Representations Objective H

In 14 and 15, graph the line.

14. slope -3, y-intercept 5

15. $-2x + y = 1$

❸ You can use an algebraic argument to show that a line with equation $x = h$ cannot be written in slope-intercept form.

$$x = h$$
$$x + 0y = h$$
$$0y = -x + h$$

It is impossible to modify the equation for y to have a coefficient of 1. Students may try dividing through by 0, but, of course, that can't be done.

Students will need graph paper or **Teaching Aid 28** for the *Additional Examples* and for the *Questions.*

Additional Examples

In 1 and 2, give the slope and the y-intercept of the line.

1. $y = -2x + 6$ **-2; 6**
2. $y = 16 + 4x$ **4; 16**
3. Write the equation $5x + 3y = 10$ in slope-intercept form. Then give the slope and y-intercept. **$y = -\frac{5}{3}x + \frac{10}{3}$; $-\frac{5}{3}$; $\frac{10}{3}$**
4. a. Write an equation of the line with y-intercept 8 and slope -6. **$y = -6x + 8$**
   b. Graph the line. **Graph should pass through (0, 8) and ($1\frac{1}{3}$, 0).**
5. Jeremy had $42 in his savings account when he decided to start adding $5 a week.
   a. Write an equation for Jeremy's total savings y after x weeks. **$y = 42 + 5x$**
   b. Give the slope and the y-intercept. **5; 42**

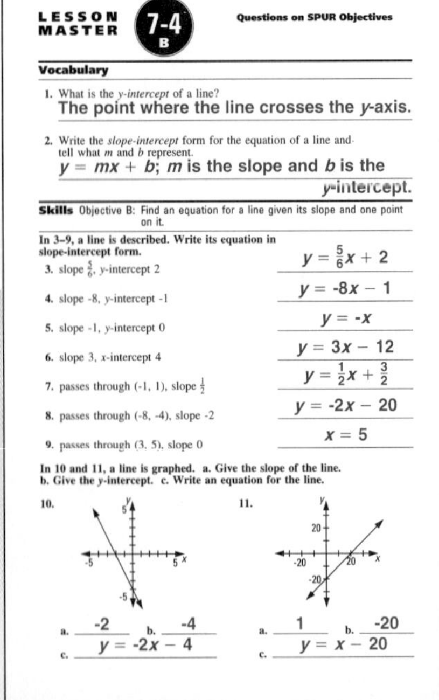

LESSON MASTER 7-4 B Questions on SPUR Objectives

Vocabulary

1. What is the *y-intercept* of a line? The point where the line crosses the y-axis.
2. Write the *slope-intercept* form for the equation of a line and tell what m and b represent. $y = mx + b$; m is the slope and b is the y-intercept.

Skills Objective B: Find an equation for a line given its slope and one point on it.

In 3–9, a line is described. Write its equation in slope-intercept form.

3. slope $\frac{5}{6}$, y-intercept 2 — $y = \frac{5}{6}x + 2$
4. slope -8, y-intercept -1 — $y = -8x - 1$
5. slope -1, y-intercept 0 — $y = -x$
6. slope 3, x-intercept 4 — $y = 3x - 12$
7. passes through (-1, 1), slope $\frac{1}{2}$ — $y = \frac{1}{2}x + \frac{3}{2}$
8. passes through (-8, -4), slope -2 — $y = -2x - 20$
9. passes through (3, 5), slope 0 — $x = 5$

In 10 and 11, a line is graphed. a. Give the slope of the line. b. Give the y-intercept. c. Write an equation for the line.

10. a. -2 b. -4 c. $y = -2x - 4$
11. a. 1 b. -20 c. $y = x - 20$

2c)

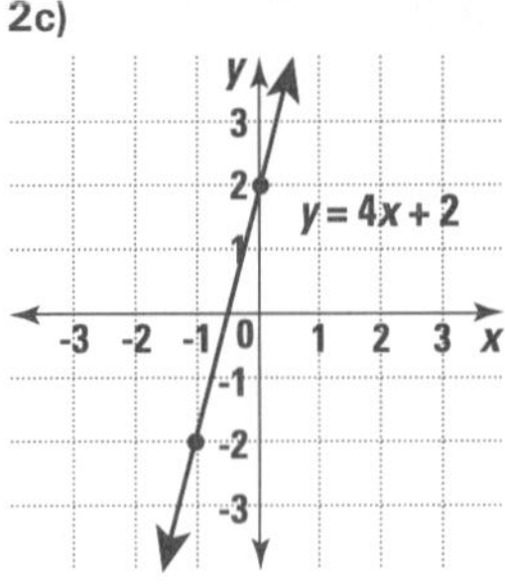

Example 5

Pam received $100 for her birthday and spends $4 of it a week.
a. Find an equation for the amount y she has after x weeks.
b. What are the slope and the y-intercept of the graph?

Solution

a. The equation is found by methods you have learned in previous chapters.

$$y = 100 - 4x$$

b. Rewrite in slope-intercept form.

$$y = -4x + 100.$$

The slope is -4 and the y-intercept is 100.

❸ Recall that every vertical line has an equation of the form $x = h$ where h is a fixed number. Equations of this form clearly cannot be solved for y. Thus, equations of vertical lines cannot be written in slope-intercept form. (They cannot be graphed on many automatic graphers.) This confirms that the slope of vertical lines cannot be defined.

Rising graduation rates. *In 1950, only about 59% of teenagers in the U.S. graduated from high school with their age group. By 1993, that figure had risen to about 74%.*

QUESTIONS

Covering the Reading

1. In what form is the equation of the line $y = mx + b$? **slope-intercept form**

In 2 and 3, an equation of a line is given. **a.** What is its slope? **b.** What is its y-intercept? **c.** Graph the line.

2. $y = 4x + 2$ **a) 4; b) 2; c) See left.**
3. $y = -\frac{1}{3}x + 6$ **a) $-\frac{1}{3}$; b) 6; c) See page 443.**

In 4–7, an equation of a line is given. **a.** Rewrite the equation in slope-intercept form. **b.** Give its slope. **c.** Give its y-intercept. **See margin.**

4. $y = 7.3 - 1.2x$
5. $x + 6y = 7$
6. $3x + 2y = 10$
7. $y = x$

In 8 and 9, write an equation of the line with the given characteristics.

8. slope -3, y-intercept 5 **$y = -3x + 5$**
9. slope $\frac{2}{3}$, y-intercept -1 **$y = \frac{2}{3}x - 1$**
10. Suppose you receive $100 for a graduation present, and you deposit it in a savings account. Then each week thereafter, you add $5 to the account but no interest is earned.
    a. Find an equation for the amount y you have after x weeks.
    b. Draw a graph of the equation in part **a.**
    c. What are the slope and the y-intercept of the graph in part **b**?
    a) $y = 5x + 100$; b) See page 443. c) 5; 100
11. When a constant-increase situation is graphed, the y-intercept can be interpreted as the __?__. **starting point**
12. Equations of __?__ lines cannot be written in slope-intercept form. **vertical**

Optional Activities

Activity 3 Cooperative Learning You might use these questions after you discuss **Questions 19 and 20**. Have students **work with a partner**.

1. Find a set of 4 equations (different from those in **Question 19**) whose graphs are parallel lines. [There are many possible answers. All will have equations of the form $y = mx + b$, where m is constant and b takes on four different values.]
2. Find a set of 4 equations whose graphs contain the same point, and intersect forming eight 45° angles. [Sample: $y = x + 3$, $y = -x + 3$, $y = 3$, $x = 0$]

Activity 4 Technology Connection With *Technology Sourcebook, Computer Master 16,* students use *GraphExplorer* or similar software to discover how the values of m and b affect the graph of $y = mx + b$.

3c)

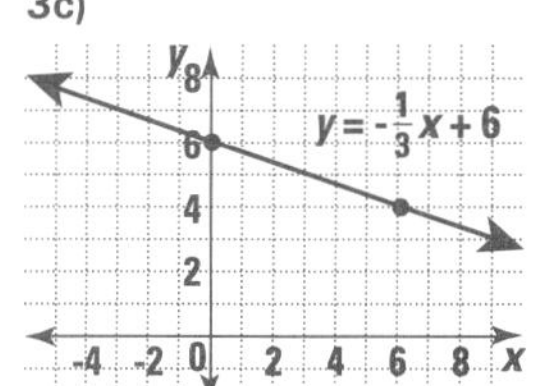

10b)

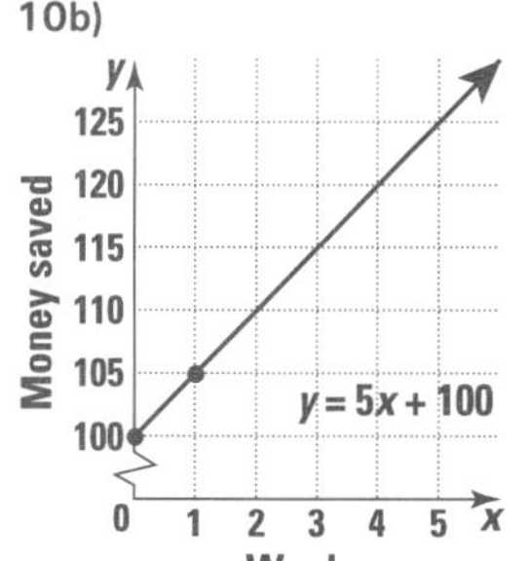

17b)

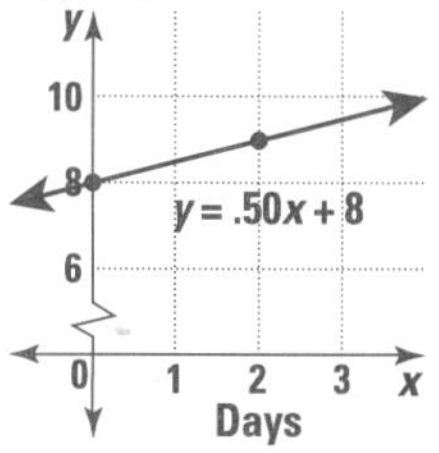

Dollars

y
10
8
6
$y = .50x + 8$
0 1 2 3 x
Days

18b)

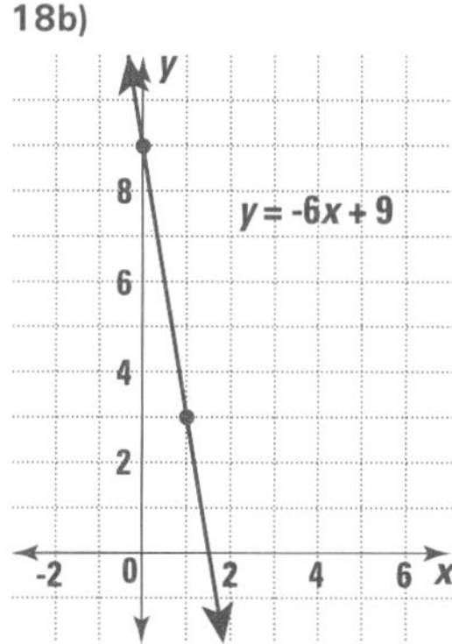

Applying the Mathematics

In 13–15, consider these three graphs. **a.** Match the situation with its graph. **b.** Give the slope of the line. **c.** Give the y-intercept.

(i)

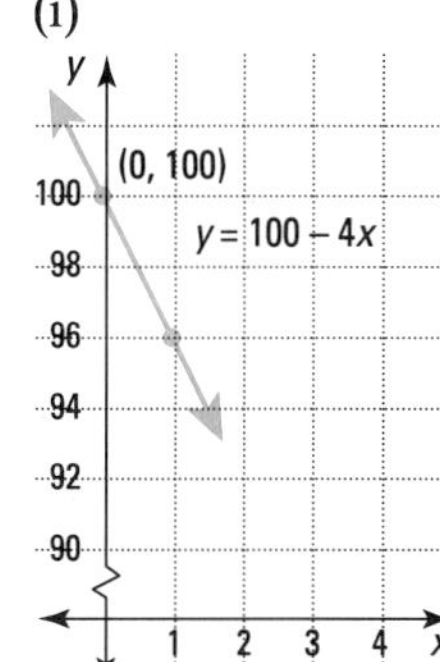

(ii)

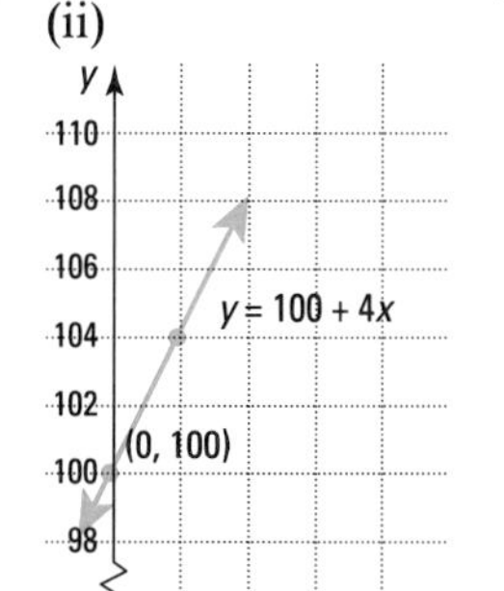

(iii)

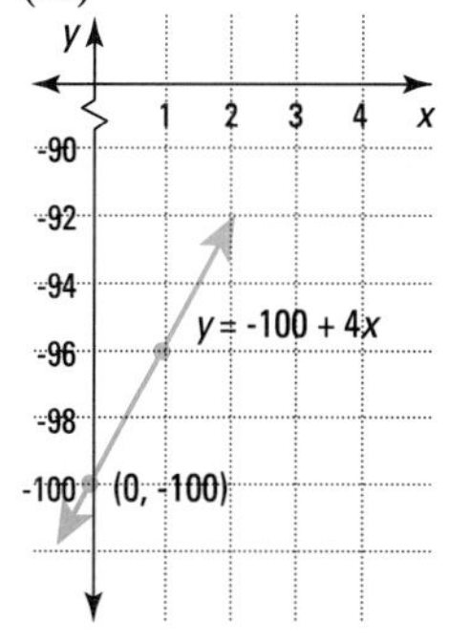

13. Shawn began with \$100 and saves \$4 a week. a) ii; b) 4; c) 100

14. Pam received \$100 for her birthday and spends \$4 a week. a) i; b) -4; c) 100

15. The Carter family owes \$100 on luggage and is paying it off at \$4 a week. a) iii; b) 4; c) -100

16. Write an equation of the horizontal line with y-intercept -4. $y = -4$

In 17 and 18, each situation leads to an equation of the form $y = mx + b$. **a.** Give the equation. **b.** Graph the equation. See left for 17b and 18b.

17. Begin with \$8.00. Collect \$.50 per day. a) $y = .50x + 8$

18. A boat is 9 miles from you. It travels towards you at a rate of 6 mph. a) $y = -6x + 9$

19. Match each line n, p, q, and r with its equation.

a. $y = x$ q **b.** $y = x + 1$ p

c. $y = x + 3$ n **d.** $y = x - 3$ r

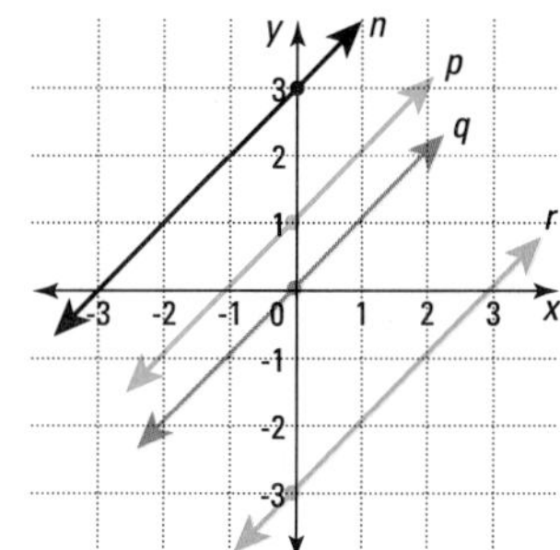

6. Mr. Campbell's students asked him how much he weighs. He wouldn't say, but he gave them a hint. First, Mr. Campbell (with an empty bucket) stepped on the scale which showed 195 lb. Then he had students fill soup cans with water and pour the water into the bucket. With 33 cans full of water in the bucket he weighed 217 lb.

a. Find the weight of the water in one soup can. $\frac{2}{3}$ lb

b. Find Mr. Campbell's weight. **It cannot be determined, but it is less than 195 pounds.**

Notes on Questions

Questions 19–20 Students need to know how changes in the values of m and b affect the graphs of equations. Begin by asking what is the same in the graphs and equations for each problem. [In **Question 19,** the lines are parallel; the equations have the same slope. In **Question 20,** the lines all intersect the y-axis at (0, -2); the equations have the same y-intercept.] Activity 3 in *Optional Activities* on page 442 extends these questions.

Additional Answers

4. a. $y = -1.2x + 7.3$; b. -1.2; c. 7.3

5. a. $y = -\frac{1}{6}x + \frac{7}{6}$; b. $-\frac{1}{6}$; c. $\frac{7}{6}$

6. a. $y = -\frac{3}{2}x + 5$; b. $-\frac{3}{2}$; c. 5

7. a. $y = x + 0$; b. 1; c. 0

► LESSON MASTER 7-4 B *page 2*

Skills Objective C: Write an equation for a line in slope-intercept form, and find its slope and y-intercept.

In 12–14, an equation of a line is given. a. Write it in slope-intercept form. b. Give the slope. c. Give the y-intercept.

12. $3x + y = 36$
a. $y = -3x + 12$
b. -3
c. 12

13. $-5x + 2y = -8$
a. $y = \frac{5}{2}x - 4$
b. $\frac{5}{2}$
c. -4

14. $x - y = 6$
a. $y = x - 6$
b. 1
c. -6

Uses Objective F: Use equations for lines to describe real situations.

In 15 and 16, a situation is given that can be described by a line. a. Give the y-intercept. b. Find the slope. c. Write an equation for the line.

15. Margie has 88 CDs. Through a club, she will purchase 3 new CDs per month.
a. 88 b. 3 c. $y = 3x + 88$

16. Engraving costs \$18 plus .75 per letter.
a. 18 b. .75 c. $y = .75x + 18$

Representations Objective H: Graph a straight line given its equation, or given a point and the slope.

In 17 and 18, graph the equation.

17. slope $-\frac{1}{3}$, y-intercept -4

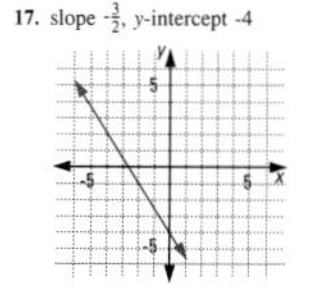

18. $-3x + y = 1$

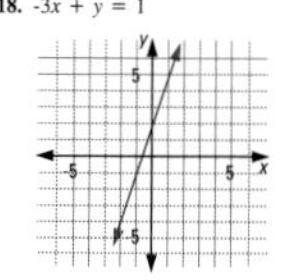

Adapting to Individual Needs

Extra Help

You might want to show students how the properties are used to rewrite the equation in **Example 2.**

$y = -8 - 3x$	
$y = -8 + -3x$	Definition of Subtraction
$y = -3x + -8$	Commutative Property of Addition
$y = -3x - 8$	Definition of Subtraction

Point out that, while $y = -3x - 8$ is a simpler form of the equation, $y = -3x + -8$ shows the values of m and b more clearly. Caution students that, if the subtraction form is used, they need to be careful to read the y-intercept as -8, not 8. **Example 4** also points out that either form is acceptable. Write the following equations on the board and ask students to give the y-intercept: $y = 2x + 4$ [4]; $y = x + -5$ [-5]; $y = 3x - 7$ [-7]; $y = 2x - 3$ [-3]

Notes on Questions

Question 21 If automatic graphers are not available, this question set will take a rather long time.

Question 22 Science Connection The Blue Mountains feature deep gorges, escarpments, and peaks. A bluish haze produced by exuding oils of the eucalyptus trees gives the mountains their name.

Question 29 This question leads to the generalization that two lines are perpendicular if and only if the product of their slopes is -1. This generalization is considered in the *Extension* below.

Follow-up for Lesson 7-4

Practice

For more questions on SPUR Objectives, use **Lesson Master 7-4A** (shown on pages 440–441) or **Lesson Master 7-4B** (shown on pages 442–443).

Assessment

Written Communication Have each student write a paragraph explaining what he or she has learned about slope in this chapter. [Students' paragraphs demonstrate understanding of slope and contain explanations of how to find the slope of a line.]

Extension

Have students **work in groups.** Give them these equations: $y = \frac{2}{3}x + 1$ and $y = -\frac{3}{2}x - 2$. Ask them how the graphs of the lines are related. [The lines are perpendicular.] Next ask them to compare the slopes of the lines. [Their product is -1.] Finally ask them to give other pairs of equations for perpendicular lines. [Samples: $y = 4x$ and $y = -\frac{1}{4}x$; $y = -2x + 3$ and $y = \frac{1}{2}x + 5$]

Project Update Project 1, *The Intercept Form of an Equation of a Line*, on page 476, relates to the content of this lesson.

Shown is the Katoomba Scenic Railway located in the Blue Mountains of New South Wales, Australia. The railway is a series of cable cars that drops at an angle close to 52° into the Jamison Valley.

21a)

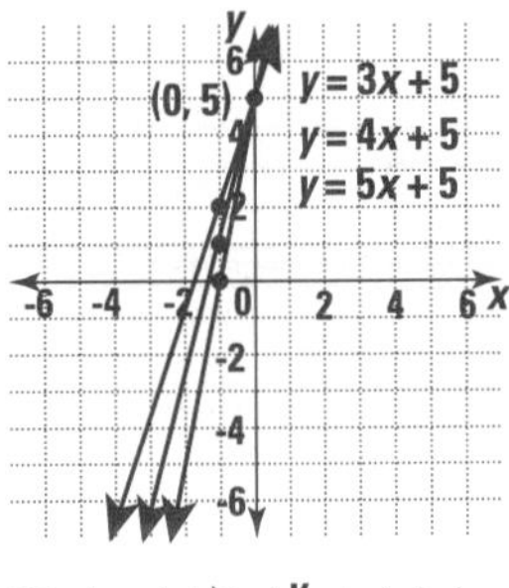

21c)

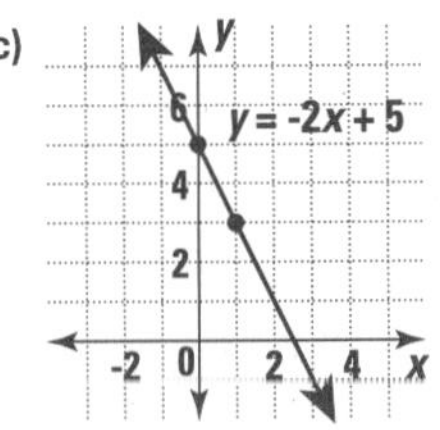

20. Match each line *s, t, u,* and *v* with its equation.

a. $y = 2x - 2$ *u* **b.** $y = \frac{2}{3}x - 2$ *v*

c. $y = -2x - 2$ *t* **d.** $y = -\frac{2}{3}x - 2$ *s*

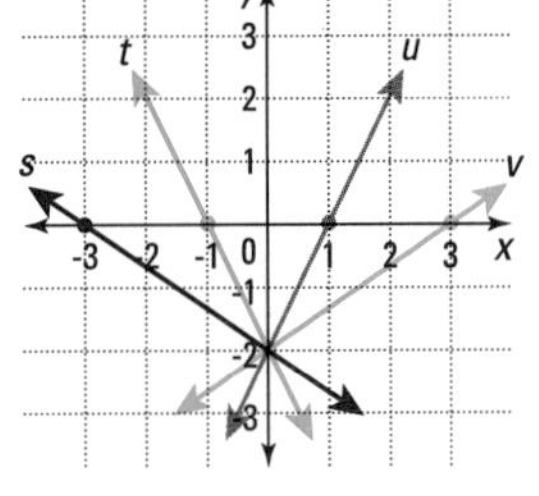

21. An automatic grapher may be helpful in this question.

a. Graph the lines with equations $y = 3x + 5$, $y = 4x + 5$, and $y = 5x + 5$. See left.

b. Describe the graph of $y = mx + 5$. a line that passes through (0, 5) with slope *m*

c. Graph $y = -2x + 5$. Is your description in part **b** true when *m* is negative? Yes; See left.

Review

In 22 and 23, the gradient of a railway is given. Find the slope of the line made as the train goes up the incline. *(Lesson 7-3)*

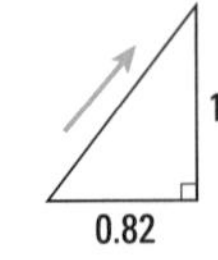

22. The Katoomba Scenic Railway in the Blue Mountains of New South Wales, Australia, is the steepest railway in the world with gradient of 1 in 0.82. ≈1.22

23. The steepest standard gauge railway in the world is between Chedde and Seovoz, in France, with a gradient of 1:11. ≈ .09

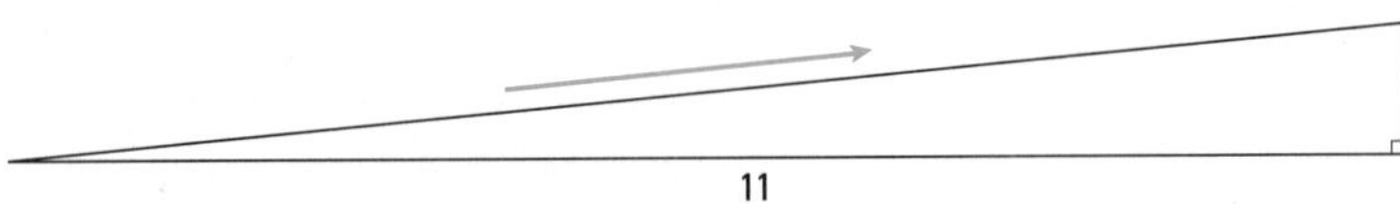

24. Which lines have no slope? Explain your reasoning. *(Lesson 7-3)* Sample: Vertical lines have no slope, since you cannot divide by 0.

25. Find the slope of the line through (5, -3) and (2, 7). *(Lesson 7-2)* $-\frac{10}{3}$

In 26 and 27, solve. *(Lesson 6-8)*

26. $\frac{3x + 5}{2 - 4x} = \frac{2}{3}$ $x = \frac{-11}{17}$ **27.** $\frac{z + 2}{2} = 2z - 2$ $z = 2$

28. A rectangle is 5 cm longer than twice its width. Its perimeter is 58 cm.

a. Write an expression for the length.

b. Find its width and length.

c. What is its area? *(Lessons 2-1, 3-5)*

a) $l = 2w + 5$; b) 8 cm; 21 cm; c) 168 cm^2

w

Exploration

29. Find equations for four lines, none of them horizontal or vertical, which intersect at the vertices of a rectangle. What is true about the slopes of these lines? Sample: $y = \frac{1}{2}x + 2$; $y = \frac{1}{2}x - 4$; $y = -2x - 4$; $y = -2x + 2$. The slopes of the lines that form parallel sides of the rectangle are equal. The slopes of the lines that form perpendicular sides are negative reciprocals of one another (their product is -1).

Adapting to Individual Needs

Challenge

Have students solve the following problems.

1. A line contains points with the coordinates (-1, *a*) and (3, 9). The slope of the line is 3. Find *a*. [-3]
2. What is the equation of the line in Question 1? [$y = 3x$]

Setting Up Lesson 7-5

Materials If automatic graphers are available, we recommend that students use them for Lesson 7-5. Students doing the *Challenge* on page 448 will need an almanac.

LESSON

7-5

Equations for Lines with a Given Point and Slope

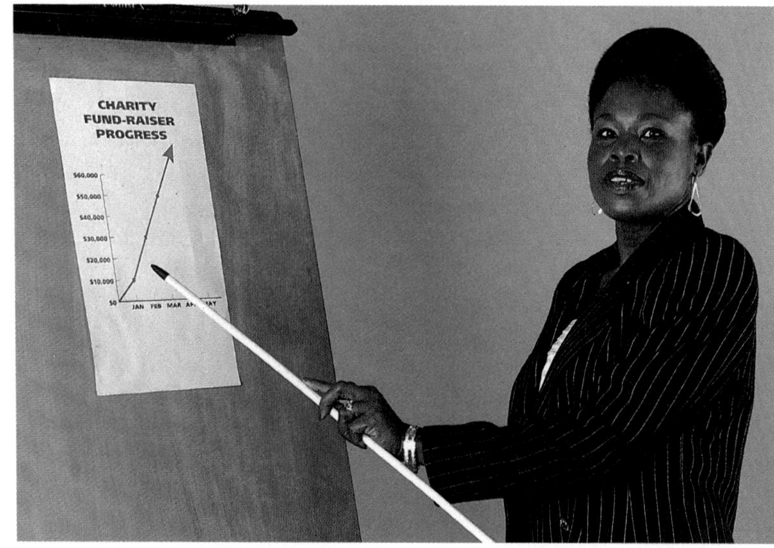

Reaching the goal. *This businesswoman is showing the progress made in the company's charity fund-raiser. The steep slope of the line indicates that large donations have increased over time.*

In the last lesson you found an equation of a line given its slope and a particular point on the line, its y-intercept. You can, in fact, find an equation in slope-intercept form of a line given its slope and any point on the line, not necessarily the y-intercept.

Example 1

Find an equation in slope-intercept form for the line through (-3, 5) with a slope of 2.

Solution

You know that $y = mx + b$ is the slope-intercept equation of a line. In this case you are given $m = 2$. All that is needed is b. Follow these three steps.

1. Substitute 2 for m and the coordinates (-3, 5) for (x, y) into $y = mx + b$. This gives
$$5 = 2 \cdot -3 + b.$$
2. Solve this equation for b.
$$5 = -6 + b$$
$$11 = b$$
3. Substitute the slope for m and the value you found for b in $y = mx + b$.
The equation is $y = 2x + 11$.

Lesson 7-5

Objectives

B Find an equation for a line given its slope and one point on it.
F Use equations for lines to describe real situations.

Resources

From the ***Teacher's Resource File***

- Lesson Master 7-5A or 7-5B
- Answer Master 7-5
- Teaching Aids
  28 Four-Quadrant Graph Paper
  72 Warm-up

Additional Resources

- Visuals for Teaching Aids 28, 72
- Automatic grapher

Teaching Lesson 7-5

Warm-up

Tell if the sentence is *true* or *false*.

1. Equations of every line can be written in slope-intercept form. False
2. One line with a slope of 3 is $y = 3x - 2$. True
3. The slope of every horizontal line is 0. True
4. Zero is always the x-coordinate of a y-intercept. True
5. The graph of a line with a negative slope goes downward from left to right. True

Lesson 7-5 Overview

Broad Goals This lesson discusses the standard procedure for finding an equation for a line given its slope and a point on it.

Perspective The situations in this chapter are gradually becoming more general. In Lesson 7-4, an equation for a line was found given the slope and the y-intercept. That is equivalent to being given the slope and the particular point $(0, b)$. In this lesson, the procedure is generalized, and an equation is found for the line with a given slope and containing any given point. Thus, in the applications in this lesson, the rate of change is described along with information that corresponds to some point on the line other than the y-intercept or the initial value.

Notice that we ask for "an equation" for a line rather than "the equation" for a line. We do so because every line has many equations. At this point, be flexible in what you consider an acceptable answer. Distinguish between getting *an* equation for a line (which is sufficient in many situations) and the extra step that is often necessary to find *the* equation in slope-intercept form.

Notes on Reading

The procedure for finding an equation for a line given one point and the slope is difficult for some students to follow because the uses of the variables change during the problem. In **Example 1,** the slope m is a constant throughout the problem. In Step 1, x and y begin as variables in the equation $y = mx + b$ but end as constants (because values are substituted for them), and b becomes an unknown. In Step 2, b becomes a constant. In Step 3, x and y return to their position as variables. It is helpful to demonstrate the procedure step by step on the board.

Error Alert In Step 1, some students put the y-coordinates in place of x because y comes first in the equation. It may help if students set the coordinates of the given point equal to (x, y) before substituting into the equation.

❶ Students who have automatic graphers can use them to check their answers.

❷ The procedure used in **Example 1** is summarized in this table. It is important to emphasize the steps because an additional step is added to the procedure in the next lesson.

❸ These data assume that the population of Ontario illustrates a linear relationship over time, which is not true in the long term. However, this model can be used to describe short-term events. Some lively discussion can arise from this issue, paving the way for Lesson 7-7, which discusses fitting a line to data.

Multicultural Connection The majority of Canada's population is made up of people of British or French descent. English and French are the official languages in Canada.

Check 1

Plot the point (-3, 5). Draw the line through it having slope 2. Notice that the graph as shown at the right has y-intercept 11.

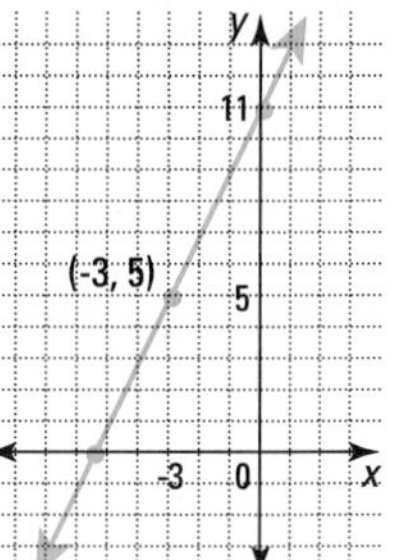

❶ **Check 2**

Use an automatic grapher to graph the equation $y = 2x + 11$. Use the trace feature to check that the line appears to pass through the point (-3, 5), and that a slope of 2 seems to fit the line.

The method of Example 1 may be used to find an equation for any nonvertical line, if you know its slope and the coordinates of one point on the line.

❷ To find an equation of a nonvertical line given one point and the slope:
Step 1: Substitute the slope for m and the coordinates of the given point for x and y in the equation $y = mx + b$.
Step 2: Solve the equation from Step 1 for b.
Step 3: Substitute the slope for m and the value you found for b in $y = mx + b$.

In particular, you can use the algorithm above to find an equation for a line whose slope and x-intercept are known. The ***x*-intercept** of a graph is the x-coordinate of a point where the graph intersects the x-axis.

Example 2

A line has slope -4, and its x-intercept is 6. Find an equation for the line in slope-intercept form.

Solution

Because the x-intercept is 6, we know that (6, 0) is a point on this line. Follow the steps above with $m = -4$, $x = 6$, and $y = 0$.

1. Substitute for m, x, and y. $\quad 0 = -4 \cdot 6 + b$
2. Solve for b. $\quad 24 = b$
3. Substitute for m and b. $\quad y = -4x + 24$

Check

A check is left for you to do.

This procedure is often useful for finding an equation to describe a real-life situation when you know a rate of change and information that gives coordinates for one point.

Optional Activities

Cooperative Learning After students finish reading the lesson, have them **work in pairs.** Students should write the slope of a line and the coordinates of a point on the line at the top of a sheet of graph paper or on **Teaching Aid 28,** and then exchange papers. The partner draws the line on a coordinate plane, writes the equation of the line, and returns the paper. Students check their partner's graphs by locating another point on the line and use the definition of slope to verify that the line is correct. Students should also check that the equation of the line is correct.

Adapting to Individual Needs

Extra Help

When substituting the slope and the coordinates of a given point into $y = mx + b$, some students might transpose the x and y values. This can happen because x is stated first in the ordered pair form (x, y), but y is first in the equation. To avoid this mistake, have students write the values of m, x, and y before substituting into the equation. Point out that this is done in **Example 2** with $m = -4$, $x = 6$, and $y = 0$.

Example 3

The population of the province of Ontario in Canada was 10,085,000 in 1991. At that time, the population was increasing at a rate of about 80,000 people per year. Assume this rate continues.

a. Find an equation relating the population y of Ontario to the year x.

b. Use the equation to predict the population of Ontario in the year 2001.

Solution

a. This is a constant-increase situation, so it can be described by a line with equation $y = mx + b$. The rate 80,000 people per year is the slope, so $m = 80{,}000$. The population of 10,085,000 in 1991 is described by the point (1991, 10,085,000). Now follow the steps listed on page 446.

1. Substitute for m, x, and y. $\quad 10{,}085{,}000 = 80{,}000 \cdot 1991 + b$
2. Solve for b. $\quad 10{,}085{,}000 = 159{,}280{,}000 + b$

$$-149{,}195{,}000 = b$$

3. Substitute for m and b. $\quad y = 80{,}000x - 149{,}195{,}000.$

b. Evaluate the formula $y = 80{,}000x - 149{,}195{,}000$ when $x = 2001$.

$$y = 80{,}000(2001) - 149{,}195{,}000$$
$$= 160{,}080{,}000 - 149{,}195{,}000$$
$$= 10{,}885{,}000.$$

A prediction for the population in the year 2001 is about 10,885,000.

Check

a. When $x = 1991$, does $y = 10{,}085{,}000$?
Does $80{,}000(1991) - 149{,}195{,}000 = 10{,}085{,}000$? Yes, so it checks.

b. Does the rate of change in population between 1991 and 2001 equal 80,000 people per year?

$$\frac{10{,}885{,}000 - 10{,}085{,}000}{2001 - 1991} = \frac{800{,}000}{10} = 80{,}000.$$ Yes, it checks.

A scenic square. *Nathan Phillips Square is located in Toronto in Ontario. Ontario has the largest population of the Canadian provinces. More than 20 million tourists visit Ontario each year.*

QUESTIONS

Covering the Reading

1. Describe the steps for finding an equation for a line, given the slope and one point on the line. **See left.**

1) Step 1: In the equation $y = mx + b$, substitute the slope m and the coordinates of the point (x, y).
Step 2: Solve the equation from step 1 for b.
Step 3: Substitute the values of m and b into the equation $y = mx + b$.

2. Check the answer to Example 2.
$0 = -4 \cdot 6 + 24$? $\;0 = -24 + 24$? $\;0 = 0$? Yes; it checks.

In 3–6, find an equation for the line given its slope and one point.

3. point (2, 3); slope 4
$y = 4x - 5$

4. point (-10, 3); slope -2
$y = -2x - 17$

5. point (-6, 0); slope $\frac{1}{3}$
$y = \frac{1}{3}x + 2$

6. point $(4, -\frac{1}{2})$; slope 0
$y = -\frac{1}{2}$

7. Find an equation for the line with slope -4 and x-intercept 7.
$y = -4x + 28$

8. Refer to Example 3. If the rate of population increase stays constant, what will be the population of Ontario in the year 2005? **11,205,000**

Additional Examples

1. Write the equation in slope-intercept form of the line that has slope $\frac{1}{2}$ and contains (6, –5).
$y = \frac{1}{2}x - 8$

2. Write the equation in slope-intercept form of the line that contains (–4, 11) and has slope –1.
$y = -x + 7$

3. When Larry became president of a service organization, he decided to start a membership campaign with the goal of signing up 4 new members each week. He hoped that after 10 weeks the club's total membership would be 375.
   a. Write an equation to relate the desired number of members y to the number of weeks x that Larry has been president.
   $y = 335 + 4x$
   b. Assume the trend of adding 4 new members each week continues. Use the equation to find the total membership after one year (52 weeks).
   543 members

Adapting to Individual Needs

English Language Development

You might ask students to put the following helpful hints for matching lines with equations on index cards.

a. Check the slope. If it is positive, the line goes upward from left to right. If it is negative, the line goes downward from left to right.

b. Check the y-intercept. This will give you a point on the line.

c. Equations for horizontal lines are of the form y = a number.

d. Equations for vertical lines are of the form x = a number.

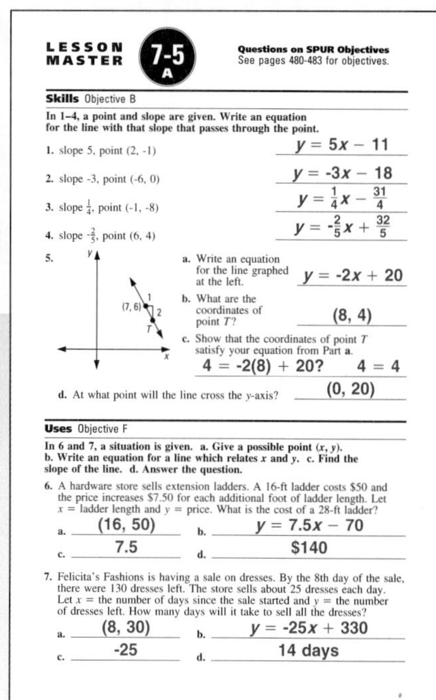

LESSON MASTER 7-5 A

Questions on SPUR Objectives
See pages 480-483 for objectives.

Skills Objective B

In 1–4, a point and slope are given. Write an equation for the line with that slope that passes through the point.

1. slope 5, point (2, -1) $\quad y = 5x - 11$
2. slope -3, point (-6, 0) $\quad y = -3x - 18$
3. slope $\frac{1}{4}$, point (-1, -8) $\quad y = \frac{1}{4}x - \frac{31}{4}$
4. slope $-\frac{2}{5}$, point (6, 4) $\quad y = -\frac{2}{5}x + \frac{32}{5}$

5.

a. Write an equation for the line graphed at the left. $\quad y = -2x + 20$

b. What are the coordinates of point T? $\quad$ (8, 4)

c. Show that the coordinates of point T satisfy your equation from Part **a**.
$4 = -2(8) + 20$? $\quad 4 = 4$

d. At what point will the line cross the y-axis? $\quad$ (0, 20)

Uses Objective F

In 6 and 7, a situation is given. a. Give a possible point (x, y). b. Write an equation for a line which relates x and y. c. Find the slope of the line. d. Answer the question.

6. A hardware store sells extension ladders. A 16-ft ladder costs $50 and the price increases $7.50 for each additional foot of ladder length. Let x = ladder length and y = price. What is the cost of a 28-ft ladder?
a. (16, 50) $\quad$ b. $y = 7.5x - 70$
c. 7.5 $\quad$ d. $140

7. Felicita's Fashions is having a sale on dresses. By the 8th day of the sale, there were 130 dresses left. The store sells about 25 dresses each day. Let x = the number of days since the sale started and y = the number of dresses left. How many days will it take to sell all the dresses?
a. (8, 30) $\quad$ b. $y = -25x + 330$
c. -25 $\quad$ d. 14 days

Notes on Questions

Question 12 Error Alert If students do not understand how to proceed, you many have to remind them of the meaning of *reciprocal.*

Questions 13–14 It is often the case that applications are easier than theoretical problems because intuition can be used to find alternative strategies. For instance, in **Questions 13 and 14**, the y-intercept can also be found by adding the amount of change to, or subtracting it from, the current value to find the initial value. This is equivalent to solving $y = mx + b$ for b. In **Question 13**, Marty has spent $3 \cdot 14 = \$42$, so he began with $68 + 42 = \$110$. In **Question 14**, at 10¢/min, the first three minutes cost 30¢, so the initial cost is (in theory) $25 - 30 = -5$¢.

Questions 15–16 These questions are designed to continue the discussion of how changing the values of m and b affects the graph of $y = mx + b$.

Question 24 Part b generalizes **part a**, and **part c** further generalizes the first two parts.

Question 25c Students should solve the equation by trial and error.

LESSON MASTER 7-5 B Questions on SPUR Objectives

Skills Objective B: Find an equation for a line given its slope and one point on it.

In 1–6, a point and slope are given. Write an equation of the line given its slope and one point.

1. slope $\frac{2}{3}$, point (3, 5) $y = \frac{2}{3}x + 3$
2. slope -6, point (1, 0) $y = -6x + 16$
3. slope -1, point (0, 3) $y = -x + 3$
4. slope 2, point (0, 0) $y = 2x$
5. slope 0, point (-3, 8) $y = 8$
6. slope -3.5, point (-4, 5) $y = -3.5x - 9$
7. a. Write an equation for the line graphed at the right. $y = -\frac{1}{2}x + 3$
   b. What are the coordinates of point P? (2, 2)
   c. Show that the coordinates of point P satisfy your equation from Part **a**. $2 = -\frac{1}{2}(2) + 3$; $2 = -1 + 3$; $2 = 2$
   d. At what point will the line cross the y-axis? (0, 3)
8. Match each line r, s, t, and u with its equation below.
   a. $y = -x$ s
   b. $y = -x + 3$ r
   c. $y = 3x - 6$ u
   d. $y = 3x$ t

Applying the Mathematics

9. The population of Anchorage, Alaska, was 226,300 in 1990. Suppose the population is increasing at a rate of 5,200 people per year, and the rate stays constant indefinitely. **a)** $y = 5200x - 10{,}121{,}700$
  a. Find an equation relating the population of Anchorage to the year.
  b. Predict the population of Anchorage in the year 2002. **288,700**
  c. At this rate of growth, in what year will the population of Anchorage first reach 500,000? **2042**

Are koalas bears? *No. Koalas are marsupials. These Australian mammals spend the first seven months of their life in their mother's pouch and the next six months riding on her back.*

10. A newborn koala (age 0 months) is 2 cm long. Until maturity, it grows at an average rate of 1.5 cm per month. Koalas mature at about 4 years.
  a. How long are mature koalas? **74 cm**
  b. The given information represents the slope m and the (x, y) coordinates of a point. Give the values of m, x, and y. **1.5; (0, 2)**
  c. Find an equation estimating the Koala's length y at age x months. $y = 1.5x + 2$
  d. Estimate the koala's length at age 6 months. **11 cm**

11. a. Write an equation for the line graphed at the right. $y = 3x - 7$
  b. At what point does the line cross the y-axis? **(0, -7)**
  c. What are the coordinates of point P? **(4, 5)**
  d. Check your answer to part **a** by showing that the coordinates of P make the equation in part **a** true. $5 = 3(4) - 7$? $5 = 5$? **Yes; it checks.**

12. The slopes of two lines are reciprocals. The equation of one of the lines is $y = 2x + 1$.
  a. Find the slope of the second line. $\frac{1}{2}$
  b. Find an equation for the second line if it passes through the point (2, 3). $y = \frac{1}{2}x + 2$

In 13 and 14, some information is given. **a.** Write the slope and an ordered pair described by the information. **b.** Write an equation relating x and y.

13. Marty is spending money at the average rate of \$3 a day. After 14 days he has \$68 left. Let y be the amount left after x days. **a)** -3; (14, 68) **b)** $y = -3x + 110$

14. Diane knows a phone call to a friend costs 25¢ for the first 3 minutes and 10¢ for each additional minute. Let y be the cost of a call that lasts x minutes. **a)** 0.10; (3, 0.25) **b)** $y = 0.10(x - 3) + .25$ or $y = .10x - .05$

Review

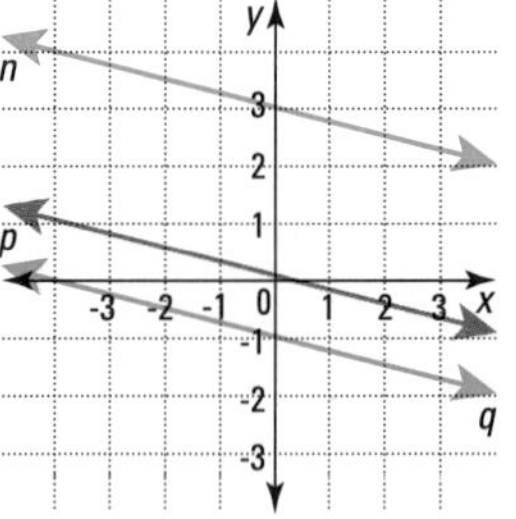

15. Match each of lines n, p, and q at the left with its equation. *(Lesson 7-4)*
  a. $y = -\frac{1}{4}x$ p **b.** $y = -\frac{1}{4}x + 3$ n **c.** $y = -1 - \frac{1}{4}x$ q

Adapting to Individual Needs

Challenge

Materials: Almanacs

Have students use almanacs to find information about consumer products that were produced in two different years. For example, they might find the number of automobiles or television sets produced in a recent year compared to the previous year. (Almanacs are full of such information, so allow students some freedom in choosing a topic.) From this information, have them find a linear equation relating the number of products y to the year x. Then have them use this equation to predict the number of products that will be produced in some year in the future.

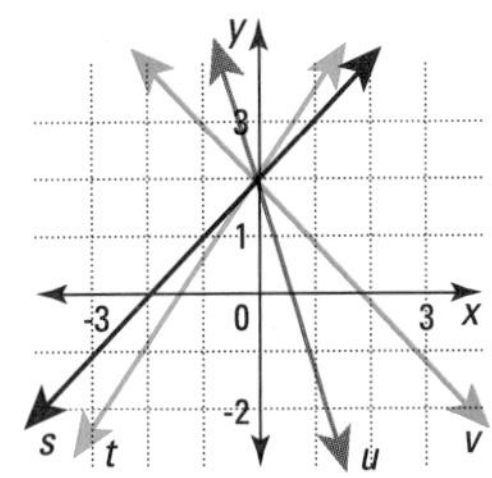

17)

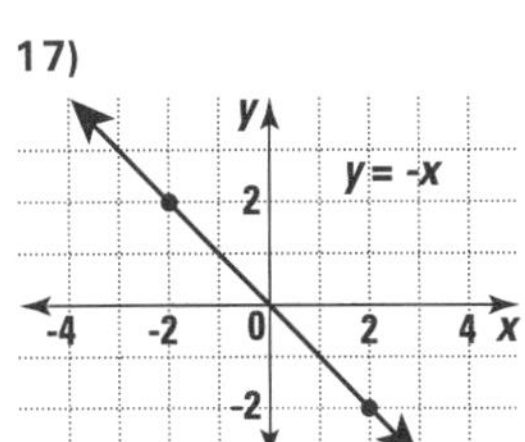

22) Elevation (feet)

11,570 (2, 11,569) (5, 11,566)
11,565 (6, 11,566)
11,560 (0, 11,565)
(7, 11,561)
0 1 2 3 4 5 6 7 8
Time (min.)

Mexican mountain. *Popocatépetl, a volcanic mountain in Cholula, Mexico, is 5,452 m high. It is a popular mountain-climbing site.*

16. Match each of lines *s, t, u,* and *v* with its equation. *(Lesson 7-4)*
- **a.** $y = x + 2$ s
- **b.** $y = -x + 2$ v
- **c.** $y = 2 - 3x$ u
- **d.** $y = \frac{3}{2}x + 2$ t

17. Graph the line with equation $y = -x$. *(Lesson 7-4)* **See left.**

18. Describe a real-world situation that could fit the equation $y = 15x + 45$. *(Lesson 7-4)* **Sample: Adam has $45 in a savings account and adds $15 to it each week. After x weeks, he has y total dollars.**

19. Do the points (1, 3), (-3, -5), and (3, 6) lie on the same line? Justify your answer. *(Lesson 7-2)* **No; (1, 3) and (-3, -5) lie on a line with slope 2, while (1, 3) and (3, 6) lie on a line with slope 3/2.**

20. The following two points give information about an overseas telephone call: (5 minutes, $5.91), (10 minutes, $10.86). Calculate the slope and describe what it stands for. *(Lesson 7-2)* **0.99; average cost of long-distance calls between 5 and 10 minutes long is $0.99 per minute**

21. Which section, or sections, of this graph shows:
a. the fastest increase? **b.** the slowest decrease? *(Lesson 7-1)* **a) A b) C**

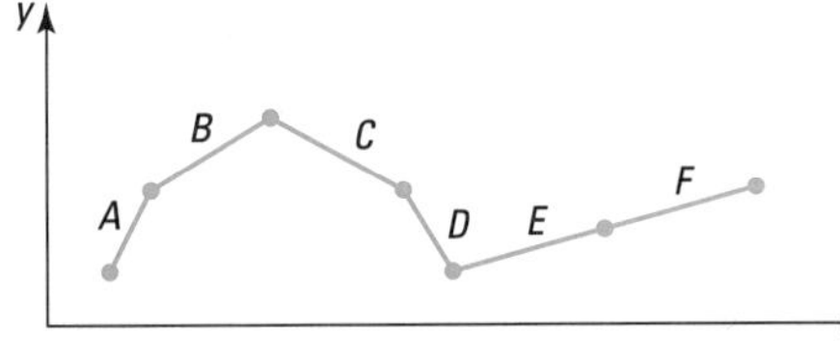

22. A climber starts at an elevation of 11,565 feet on the volcano Popocatépetl in Mexico, the 5th-highest mountain in North America. The climber goes up 2 ft/min for 2 minutes, then down 1 ft/min for 3 minutes. The climber rests for one minute and then descends 5 feet in one minute. Graph the climber's elevation during the seven minutes. *(Lesson 7-1)* **See left.**

23. A survey showed that 7 out of 10 adults drink coffee in the morning. If 650 adults live in an apartment complex, how many of them would you expect to drink coffee in the morning? *(Lesson 6-8)* **455 adults**

24. *Skill sequence.* Simplify. *(Lesson 6-1)*
- **a.** $\frac{15}{2} \div \frac{15}{3}$ $\frac{3}{2}$
- **b.** $\frac{a}{2} \div \frac{a}{3}$ $\frac{3}{2}$
- **c.** $\frac{a}{b} \div \frac{a}{c}$ $\frac{c}{b}$

25. Solve for *n*. *(Lessons 1-6, 2-10)*
- **a.** $n^2 = 24$ $n = \sqrt{24}$ or $n = -\sqrt{24}$
- **b.** $\sqrt{n} = 24$ $n = 576$
- **c.** $n! = 24$ $n = 4$

Exploration

26. Examine Example 3. **Answers will vary.**
- **a.** Find the population of the state where you live at the last census.
- **b.** Estimate or find out how fast your state is growing per year.
- **c.** Using your estimate, find a linear equation relating the population *y* to the year *x*.
- **d.** Use this equation to estimate the population of your state when you will be 50 years old.

Follow-up for Lesson 7-5

Practice

For more questions on SPUR Objectives, use **Lesson Master 7-5A** (shown on page 447) or **Lesson Master 7-B** (shown on pages 448–449).

Assessment

Group Assessment Have students **work in small groups.** Have each group decide on a slope for a line and a point on the line. Then have each student in the group name another point. Have the rest of the group decide whether or not that point is on the line. [Students use the slope-intercept form of an equation to check if a point is on a line.]

Extension

Have students find the equation for a line that has a *y*-intercept of 5 and is parallel to a line whose equation is $3x - 4y = 6$. [$y = \frac{3}{4}x + 5$]

Project Update Project 1, *The Intercept Form of an Equation of a Line,* on page 476, relates to the content of this lesson.

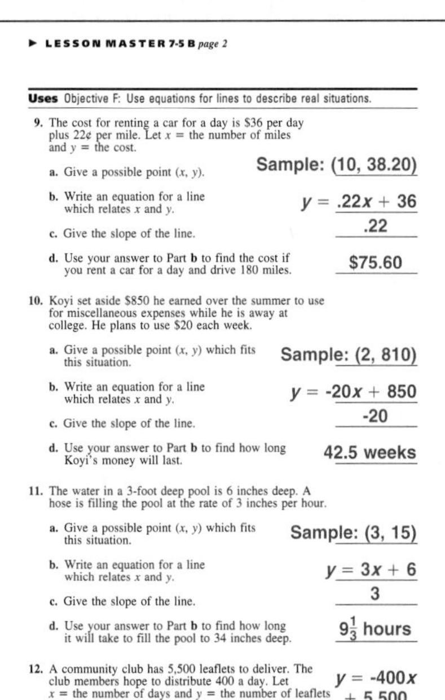

▸ LESSON MASTER 7-5 B *page 2*

Uses Objective F: Use equations for lines to describe real situations.

9. The cost for renting a car for a day is $36 per day plus 22¢ per mile. Let x = the number of miles and y = the cost.
   a. Give a possible point (x, y). Sample: (10, 38.20)
   b. Write an equation for a line which relates x and y. $y = .22x + 36$
   c. Give the slope of the line. .22
   d. Use your answer to Part **b** to find the cost if you rent a car for a day and drive 180 miles. $75.60

10. Koyi set aside $850 he earned over the summer to use for miscellaneous expenses while he is away at college. He plans to use $20 each week.
   a. Give a possible point (x, y) which fits this situation. Sample: (2, 810)
   b. Write an equation for a line which relates x and y. $y = -20x + 850$
   c. Give the slope of the line. -20
   d. Use your answer to Part **b** to find how long Koyi's money will last. 42.5 weeks

11. The water in a 3-foot deep pool is 6 inches deep. A hose is filling the pool at the rate of 3 inches per hour.
   a. Give a possible point (x, y) which fits this situation. Sample: (3, 15)
   b. Write an equation for a line which relates x and y. $y = 3x + 6$
   c. Give the slope of the line. 3
   d. Use your answer to Part **b** to find how long it will take to fill the pool to 34 inches deep. $9\frac{1}{3}$ hours

12. A community club has 5,500 leaflets to deliver. The club members hope to distribute 400 a day. Let x = the number of days and y = the number of leaflets still available. Write an equation relating x and y. $y = -400x + 5,500$

Setting Up Lesson 7-6

Materials If automatic graphers are available, we recommend that students use them for Lesson 7-6. Students doing the *Challenge* on page 455 will need telephone books.

Lesson 7-6

Objectives

B Find an equation for a line given two points on it.
F Use equations for lines to describe real situations.

Resources

From the ***Teacher's Resource File***
- Lesson Master 7-6A or 7-6B
- Answer Master 7-6
- Assessment Sourcebook: Quiz for Lessons 7-4 through 7-6
- Teaching Aids
  28 Four-Quadrant Graph Paper
  72 Warm-up

Additional Resources
- Visuals for Teaching Aids 28, 72
- Automatic grapher

Teaching Lesson 7-6

Warm-up

For a service call, a plumbing company charges a base rate plus \$20 for each quarter-hour of service or fraction thereof. A $1\frac{1}{2}$-hour service call costs \$155.

1. What is the base rate charged by the company? **\$35**
2. Write an equation relating the cost of a service call y to the number of quarter-hours of service x. $\mathbf{y = 20x + 35}$

LESSON

7-6

Equations for Lines Through Two Points

Cellular phone costs. *The average cost of a cellular phone call is greater than the cost of a regular telephone call. One reason cellular costs are higher is that cellular telephones use a combination of radio, telephone, and computer technology.*

A phone book might tell you that a call to a friend costs 25¢ for the first 3 minutes and 10¢ each additional minute. This is like being given the point (3, 25) and the slope 10. If you know an equation of the line through this point with this slope, then you have a formula which can help you determine the cost of any phone call to that friend. In Lesson 7-5 you learned how to find such an equation.

Sometimes in a constant-increase situation, the given information does not include the slope, but includes two points. For instance, you might be told that a 5-minute call overseas costs \$5.91 and a 10-minute call costs \$10.86. This means that you have two data points, (5, 5.91) and (10, 10.86). To obtain an equation for the line through these points, first calculate the slope. Then work as before.

Before finding this equation for the cost of a phone call, we work through an example with simpler numbers.

Example 1

Find an equation for the line through (5, -1) and (-3, 3).

1 **Solution**

1. First find the slope m.

$$m = \frac{3 - -1}{-3 - 5} = \frac{4}{-8} = -\frac{1}{2}$$

Now you have the slope and two points. This is more information than is needed. Pick one of these points. We pick (5, -1). Now follow the steps given in Lesson 7-5 to find an equation for the line through (5, -1) and with slope $-\frac{1}{2}$. ▶

Lesson 7-6 Overview

Broad Goals As its title indicates, the goal of this lesson is to show how to find an equation for the line through two points.

Perspective This lesson discusses the last in a three-problem sequence on finding equations of lines. Each type of problem requires one more step for its solution than the previous one.

1. Given slope and y-intercept (Just substitute into $y = mx + b$.)
2. Given slope and one point (Reduce to type 1 by calculating b.)
3. Given two points (Reduce to type 2 by calculating slope.)

When working with applications, there is always a question about the naming of variables. If cost depends on time, should we name the variables c and t so that the referent is clear, or should we name them x and y? In solving sentences in previous lessons and chapters, we have usually opted for the first letters of the quantity; in this case that would be c and t. In this lesson, we usually use x and y. Though the resulting equations are not quite as easy to relate to the application, these variables make it easier to think about intercepts and slopes. It is one of the few times that we encourage the use of generic names.

2. Substitute $-\frac{1}{2}$ and the coordinates of (5, -1) into $y = mx + b$.

$$-1 = -\frac{1}{2}(5) + b$$

3. Solve for b.

$$-1 = -\frac{5}{2} + b$$

$$\frac{3}{2} = b$$

4. Substitute the values of m and b into the equation.

$$y = -\frac{1}{2}x + \frac{3}{2}$$

Check 1

Substitute the coordinates of the point not used to see if they work.

Point (-3, 3): Does $3 = -\frac{1}{2}(-3) + \frac{3}{2}$?

Does $3 = \frac{3}{2} + \frac{3}{2}$? Yes.

Check 2

Use an automatic grapher to draw a graph of $y = -.5x + 1.5$. Use the trace feature to verify that (5, -1) and (-3, 3) are on the graph.

Notice that the procedure in Example 1 involves just one more step than the procedure of Lesson 7-5.

To find an equation for a nonvertical line given two points on it:
Step 1: Find the slope determined by the two points.
Step 2: Substitute the slope for m and the coordinates of one of the points (x, y) in the equation $y = mx + b$.
Step 3: Solve for b.
Step 4: Substitute the values you found for m and b in $y = mx + b$.

❷ This method of finding the equation of a line was developed by René Descartes in the early 1600s. We now apply it to a problem about the cost of a phone call.

Example 2

❸ Suppose a 5-minute overseas call costs $5.91 and a 10-minute call costs $10.86.

a. What is the cost y of a call of x minutes duration? (Assume that this is a constant-increase situation and x is a positive integer.)

b. How long could you talk for $20.00?

Solution

a. You need an equation for the line through (5, 5.91) and (10, 10.86). Find the slope.

$$m = \frac{10.86 - 5.91}{10 - 5} = \frac{4.95}{5} = 0.99$$

Substitute 0.99 and the coordinates of one of the points into $y = mx + b$.

We pick (5, 5.91). $5.91 = 0.99(5) + b$

Solve for b. $5.91 = 4.95 + b$

$b = 0.96$

Notes on Reading

❶ Some students may be uncomfortable with the number of choices they must make when solving these problems, while others may enjoy the challenge. Students can reverse their choice of points in computing slope and still get the right answer. They can also choose a different point for substitution in step 2 and still get the right answer. To emphasize this, you could show another solution for **Example 1**. The slope can be computed in step 1 as $m = \frac{-1 - 3}{5 - -3} = -\frac{4}{8} = -\frac{1}{2}$. In step 2, you can use (−3, 3) instead of (5, −1); then $3 = -\frac{1}{2}(-3) + b$, but b is still $\frac{3}{2}$. In either case, you should have students check their equation with the "other" point. This reiterates the idea that a point on the line must have coordinates that satisfy the equation.

If automatic graphers are available, have students use them to check **Example 1.**

❷ **History Connection** You might remind students that our coordinate system is sometimes called the *Cartesian* coordinate system in honor of the French mathematician and philosopher René Descartes. (The word *Cartesian* is from Descartes' Latin name *Cartesius.*) In the time of Descartes, most science and mathematics texts were written in Latin—a situation analogous to that of today, when many scientists and mathematicians write in English even if English is not their native language.

❸ In an application problem, there is often a choice about which quantity will be represented by the independent variable x and which by the dependent variable y. Sometimes a problem will naturally lend itself toward one choice. In **Example 2,** it makes sense to view the cost of the call as a function of its duration.

Optional Activities

Activity 1 Writing
After discussing the lesson, you might have students **work in small groups** and do this activity. Ask each group to make up three questions; each question should involve finding an equation for a line given one of the following conditions:
a. slope and y-intercept
b. slope and one point
c. two points.

Each group should provide answers to their own questions. Encourage students to put at least one of their questions in some context. If there is time, have groups trade questions. If there is no time for this exchange, use some of the students' questions as additional *Warm-up* questions over the next several days.

Activity 2 Consumer Connection
You might want to extend *Additional Example 2* by noting that the unit for the slope 0.10 is $\frac{\text{cost}}{\text{copy}}$. This means that it costs 10¢ to print more than one copy. In business, that cost is called the *marginal cost.* Students might interview business people, and ask how marginal costs are related to their businesses.

Some problems, like **Example 3,** work well with the variables assigned in either order. Our solution is stated with chirps per minute as the independent variable *x* and temperature as the dependent variable *y*. That is, we think of knowing the chirps and determining the temperature from them.

Additional Examples

1. Find an equation for the line that contains (8, 7) and (−4, −2).
   $y = \frac{3}{4}x + 1$
2. A printer charges $65 for 100 copies of a booklet and $105 for 500 copies. Assume the relationship between the number of copies and the cost is linear.
   a. Write an equation relating the cost *y* and the number of copies *x*. $y = .1x + 55$
   b. How may copies can be printed for $200? **1450 copies**
3. Blood pressure is given by two numbers: the systolic pressure over the diastolic pressure. Both numbers are usually measured in millimeters. Suppose the blood pressure of a 20-year-old is 110 over 70 and that of a 50-year-old is 125 over 80. Assume the increase is constant over time.
   a. Write an equation relating age *x* and systolic (the greater number) blood pressure *y*.
   $y = \frac{1}{2}x + 100$
   b. What might the systolic blood pressure of a 40-year-old be? **120 millimeters**

Notes on Questions

Students will need graph paper or **Teaching Aid 28**.

Questions 1–4 Cooperative Learning Reassure students who used different choices of points that the intermediate work shown in class might be different from theirs, but that the slope-intercept equation of the line will be the same. It may be informative to have students show different strategies for solving at least one of these problems on the board. We suggest **Question 2,** since no answer is given in the text and because the negative numbers may cause some trouble.

(Notes on Questions continue on page 454.)

Substitute the values for *m* and *b* into $y = mx + b$.

$$y = 0.99x + 0.96$$

The equation $y = 0.99x + 0.96$ tells you that a call costs $.96 to make plus $.99 for each minute you talk.

b. Substitute $y = 20$ into the equation in part **a,** and solve for *x*.

$$0.99x + 0.96 = 20.00$$
$$0.99x = 19.04$$
$$x \approx 19.23$$

You can talk for about 19 minutes for $20.00.

Check

a. Does this equation give the correct cost for a 10-minute call? Substitute 10 for *x* in the equation $y = 0.99x + 0.96$. Then $y = 0.99(10) + 0.96 = 9.9 + 0.96 = 10.86$, as it should.

b. The cost of a 19-minute call is $0.99(19) + 0.96 = 19.77$. The cost of a 20-minute call is $0.99(20) + 0.96 = 20.76$. It checks.

Relationships that can be described by an equation of a line are called *linear relationships.* They occur in many places. Here is how a relationship mentioned in Chapter 1 was found.

Example 3

Biologists have found that the number of chirps some crickets make per minute is related to the temperature. The relationship is very close to being linear. When crickets chirp 124 times a minute, it is about 68°F. When they chirp 172 times a minute, it is about 80°F. Below is a graph of this information.

a. Find an equation for the line through the two points.

b. About how warm is it if you hear 100 chirps in a minute?

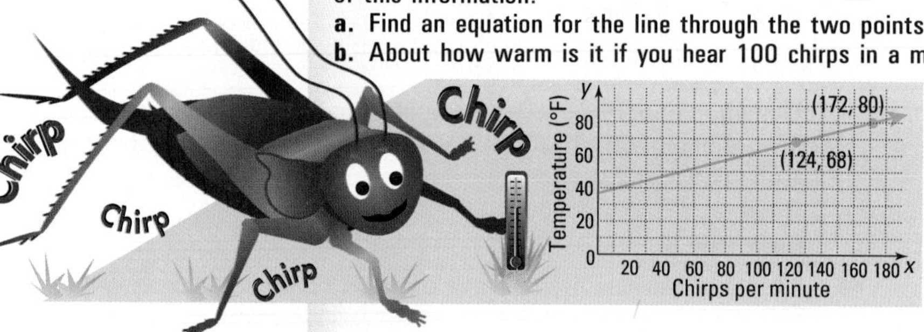

Solution

a. First, find the slope.

$$m = \frac{80 - 68}{172 - 124} = \frac{12}{48} = \frac{1}{4}$$

Substitute $\frac{1}{4}$ and the coordinates of (124, 68) into $y = mx + b$.

$$68 = \frac{1}{4}(124) + b$$

Solve for *b*.

$$68 = 31 + b$$
$$37 = b$$

Substitute for *m* and *b*. An equation is $y = \frac{1}{4}x + 37$.

Optional Activities

Activity 3 Health Connection

Materials: Graph paper or **Teaching Aid 28**

As an extension of *Additional Example 3,* you might have students graph the equations they found in **part a** and check their answer to **part b.**

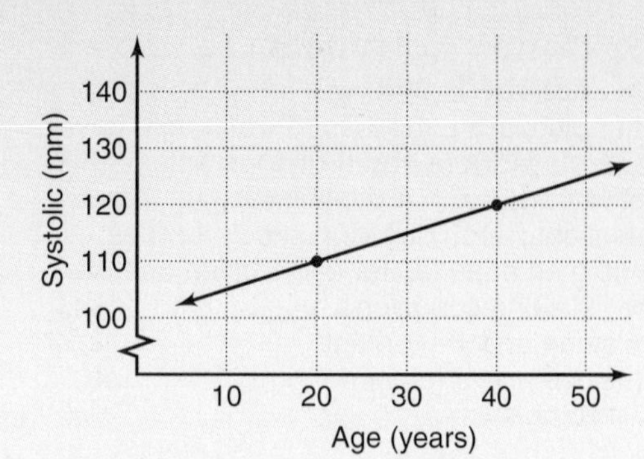

b. Substitute 100 for x in the equation $y = \frac{1}{4}x + 37$.

$$y = \frac{1}{4}(100) + 37$$
$$y = 25 + 37$$
$$y = 62$$

It is about 62°F when you hear 100 chirps in a minute.

Check

a. Substitute the coordinates of the point (172, 80) which were not used in finding the equation.

Does $80 = \frac{1}{4}(172) + 37$?

Does $80 = 43 + 37$? Yes.

b. Look at the graph. Is the point (100, 62) on the line? Yes, it checks.

The equation in Example 3 enables the temperature to be estimated for any number of chirps. By solving for x in terms of y, you could get a formula for the number of chirps to expect at a given temperature. Formulas like these seldom work for values far from the given data points. Crickets tend not to chirp at all below 50°F, yet the formula $y = \frac{1}{4}x + 37$ predicts about 50 chirps a minute at 50°F.

Notice that in Example 2 you were told to let x equal the number of minutes and y equal the cost. In Example 3, the graph told you to use x for the number of chirps and y for the temperature. If you must decide which variable is x, and which is y, use a question about rate of change to suggest which makes more sense. For instance, in a situation about distances and miles, ask yourself, "Does it make more sense to think about miles per hour or hours per mile in this situation?" If you need miles per hour, let y = the number of miles and x = the number of hours. If you need hours per mile, let y = the number of hours and x = the number of miles. If you can solve the problem using either rate, then x and y can represent either quantity.

QUESTIONS

Covering the Reading

In 1–4, find an equation for the line through the two given points. Check your answer. See left for check.

1. (1, 0), (4, 15) $y = 5x - 5$
2. (1, 9), (7, 3) $y = -x + 10$
3. (6, -3), (-8, -10) $y = \frac{1}{2}x - 6$
4. (0, 11), (13, 0) $y = \frac{-11}{13}x + 11$
5. Who developed the method used in this lesson for finding the equation of a line? René Descartes
6. In Example 2, what would an 8-minute call cost? $8.88

1) $0 = 5(1) - 5$? $0 = 0$? Yes; it checks. $15 = 5(4) - 5$? $15 = 15$? Yes; it checks.

2) $9 = -1 + 10$? $9 = 9$? Yes; it checks. $3 = -7 + 10$? $3 = 3$? Yes; it checks.

3) $-3 = \frac{1}{2} \cdot 6 - 6$? $-3 = 3 - 6$? $-3 = -3$? Yes; it checks. $-10 = \frac{1}{2} \cdot (-8) - 6$? $-10 = -4 - 6$? $-10 = -10$? Yes; it checks.

4) $11 = \frac{-11}{13} \cdot 0 + 11$? $11 = 0 + 11$? $11 = 11$? Yes; it checks. $0 = -\frac{11}{13} \cdot 13 + 11$? $0 = -11 + 11$? $0 = 0$? Yes; it checks.

Follow-up for Lesson 7-6

Practice

For more questions on SPUR Objectives, use **Lesson Master 7-6A** (shown on page 453) or **Lesson Master 7-6B** (shown on pages 454–455).

Assessment

Quiz A quiz covering Lessons 7-4 through 7-6 is provided in the *Assessment Sourcebook.*

Oral Communication Ask students if an equation of a line can be found if only the x-intercept and y-intercept are known. Have them explain their answers. [Students understand that an equation for a line can be found given any two points on the line.]

Extension

If the variables are reversed in **Example 3**, we think of the temperature as determining the number of chirps. Ask students to determine the relationship between x and y in this case. [The slope is 4 instead of $\frac{1}{4}$, and the y-intercept is –148. The slope-intercept form of the equation is $y = 4x - 148$. This is an equation for the inverse function of the one that is graphed in **Example 3.**]

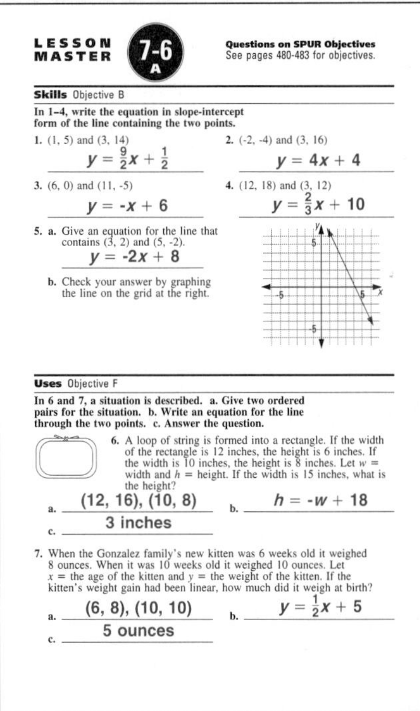

LESSON MASTER 7-6 A

Questions on SPUR Objectives
See pages 480-483 for objectives.

Skills Objective B

In 1–4, write the equation in slope-intercept form of the line containing the two points.

1. (1, 5) and (3, 14) $y = \frac{9}{2}x + \frac{1}{2}$
2. (-2, -4) and (3, 16) $y = 4x + 4$
3. (6, 0) and (11, -5) $y = -x + 6$
4. (12, 18) and (3, 12) $y = \frac{2}{3}x + 10$
5. a. Give an equation for the line that contains (3, 2) and (5, -2). $y = -2x + 8$
   b. Check your answer by graphing the line on the grid at the right.

Uses Objective F

In 6 and 7, a situation is described. a. Give two ordered pairs for the situation. b. Write an equation for the line through the two points. c. Answer the question.

6. A loop of string is formed into a rectangle. If the width of the rectangle is 12 inches, the height is 6 inches. If the width is 10 inches, the height is 8 inches. Let w = width and h = height. If the width is 15 inches, what is the height?
   a. (12, 16), (10, 8) b. $h = -w + 18$
   c. 3 inches
7. When the Gonzalez family's new kitten was 6 weeks old it weighed 8 ounces. When it was 10 weeks old it weighed 10 ounces. Let x = the age of the kitten and y = the weight of the kitten. If the kitten's weight gain had been linear, how much did it weigh at birth?
   a. (6, 8), (10, 10) b. $y = \frac{1}{2}x + 5$
   c. 5 ounces

Adapting to Individual Needs

Extra Help

Be sure students realize that in $y = mx + b$, letters are used in two ways—x and y represent variables, and the letters m and b represent constants. That is, for a given equation, m and b each represent fixed values even though we have to find them. The variables x and y take on different values for different points on a line. Explain that the constants m and b describe the position of a line, and the variables x and y describe different points on the line.

Notes on Questions

Question 13 Science Connection This application is a classic in that virtually every textbook uses it, and virtually every student sees it. The given is information they should know; the result is an equation that may come in handy. For this reason, it should be discussed in some detail. Here we have arbitrarily selected the Fahrenheit temperature as x and the Celsius temperature as y. However, outside the classroom, students might see the formula solved for either variable.

Question 15 Error Alert Some students may have difficulty with this question because of the choice of letters for the variables. If so, suggest that they use x and y even though the question identifies the variables as n and S.

Students can check their results by noting that the sum of the measures of the interior angles of a pentagon is 540° and of a hexagon is 720°. Pictures, such as those on page 455, show the figures separated into triangles, and they will help students see how the sum of the interior angle measures can be obtained.

LESSON MASTER 7-6 B Questions on SPUR Objectives

Skills Objective B: Find an equation for a line given two points on it.

In 1–10, find the equation in slope-intercept form for the line containing the two points.

1. (1, 3) and (-2, 1) $y = \frac{2}{3}x + \frac{7}{3}$
2. (-6, -6) and (0, -4) $y = \frac{1}{3}x - 4$
3. (1, 4) and (-2, -5) $y = 3x + 1$
4. (-3, 13) and (-1, -1) $y = -7x - 8$
5. (3, 12) and (-4, -16) $y = 4x$
6. (-3, 7) and (6, -8) $y = -\frac{5}{3}x + 2$
7. (9, 2) and (18, 11) $y = x - 7$
8. (1, 1) and (5, 4) $y = \frac{3}{4}x + \frac{1}{4}$
9. (8, 6) and (-3, 6) $y = 6$
10. (100, -300) and (101, -299) $y = x - 400$

In 11 and 12, two points are given. a. Write an equation in slope-intercept form for the line containing the two points. b. Graph the line from Part a and check that both points are on the line.

11. (-3, 7) and (5, -1)
 a. $y = -x + 4$
 b.
12. (6, -3) and (0, -3)
 a. $y = -3$
 b.

7. If a 5-minute overseas call to Bonn costs \$4.50 and a 10-minute call costs \$8.50, find a formula relating time and cost. $y = 0.80x + 0.50$

In 8–10, refer to Example 3.

8. The number of times a cricket chirps in a minute and the temperature is very close to what kind of relationship? linear
9. By substituting in the equation $y = \frac{1}{4}x + 37$, about how many chirps per minute would you expect if the temperature is 70°F? about 132 chirps
10. Suppose the cricket chirps 90 times per minute.
 a. Use the graph to estimate the temperature. about 60°F
 b. Use the equation to estimate the temperature. 59.5°F

Applying the Mathematics

11. Find an equation for the line with x-intercept 6 and y-intercept 5. $y = -5/6x + 5$
12. a. Show that $A = (-4, 7)$, $B = (1, 5)$, and $C = (16, -1)$ lie on the same line. See left.
 b. Find an equation for that line. $y = -\frac{2}{5}x + \frac{27}{5}$

12a) $\frac{5 - 7}{1 - -4} = \frac{-1 - 5}{16 - 1} = \frac{-2}{5}$

13. The graph below shows the linear relationship between Fahrenheit and Celsius temperatures. The freezing point of water is 32°F and 0°C. The boiling point of water is 212°F and 100°C.

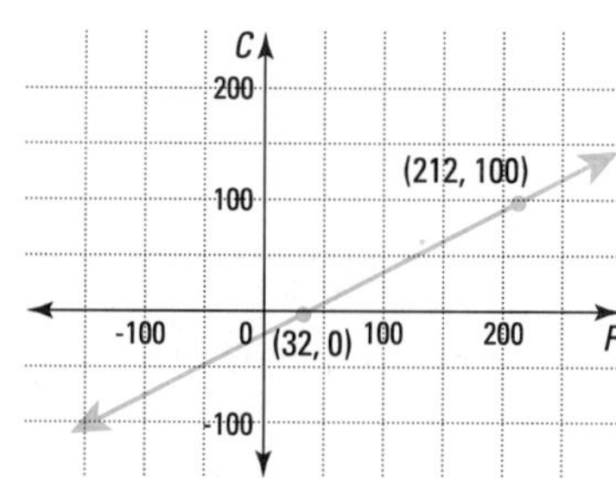

 a. Find an equation that relates Celsius and Fahrenheit temperatures. (Hint: $C = mF + b$.) $C = \frac{5}{9}F - \frac{160}{9}$
 b. When it is 150°F, what is the temperature in degrees Celsius?
 c. When it is 150°C, what is the temperature in degrees Fahrenheit?
 b) ≈ 65.6°C; c) 302°F

It's all relative. *Shown is a milk snake, a member of the* Lampropeltis *family.*

14. The total length y and the tail length x of females of the snake species *Lampropeltis polyzona* have close to a linear relationship. When $x = 60$ mm, $y = 455$ mm. When $x = 140$ mm, $y = 1050$ mm.
 a. Find the slope of this linear relationship. (Approximate your answer to the nearest tenth.) ≈ 7.4
 b. Find an equation for y in terms of x using (60, 455) and your slope from part **a.** $y = 7.4x + 11$
 c. Check your equation using (140, 1050).
 d. If a female of this species has a tail of length 100 mm, how long is the snake? ≈ 751 mm
 c) 7.4(140) + 11 = 1047 which is close enough to 1050.

Adapting to Individual Needs

English Language Development
Have students summarize the different ways they can write an equation depending on what is given.

1. **Given**: Slope and y-intercept
 Example: Given $m = 5$ and $b = -1$
 What to do: Substitute m and b in $y = mx + b$: $y = 5x + -1$ or $y = 5x - 1$.
2. **Given**: Slope and any point on the line
 Example: Given $m = 2$ and point (–3, 5)
 What to do: Substitute 2 for m, –3 for x, and 5 for y in $y = mx + b$ to find $b = 11$. Substitute the values of m and b in $y = mx + b$: $y = 2x + 11$.
3. **Given:** Two points on the line
 Example: Given (–4, 3) and (–2, –1)
 What to do: Find the slope: $\frac{-1 - 3}{-2 - -4} = -2$. Use the slope and one point to find $b = -5$. Substitute the slope and value for b in $y = mx + b$: $y = -2x + -5$ or $y = -2x - 5$.

15. The sum of the measures of the angles in a triangle is 180°. In a quadrilateral, the sum is 360°. Find an equation relating *n*, the number of sides in a polygon, and the sum of its angles, *S*.
$S = 180n - 360$

Review

16b) Sample: How much would a 10-mile cab ride cost if one uses this cab company? Answer: $13.50

16. A cab company charges a base rate plus $1.20 per mile. A 12-mile cab ride costs $15.90.
 a. Write an equation relating the number of miles driven to the cost of the cab ride. $C = 1.20m + 1.50$
 b. Make up a question about this cab company. Use your equation from part **a** to answer your question. *(Lesson 7-5)* See left.

17. What is an equation for the line with slope 7 which passes through the point (7, 7)? *(Lesson 7-5)* $y = 7x - 42$

18)

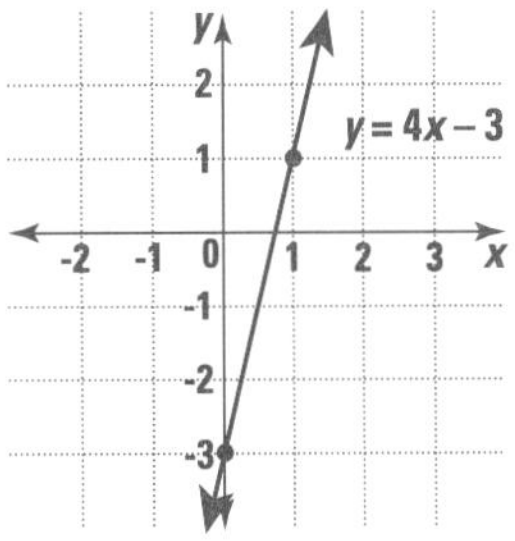

18. Graph $y = 4x - 3$ by using its slope and *y*-intercept. *(Lesson 7-4)* See left.

19. The points (*a*, 5) and (-2, -4) lie on a line with slope $\frac{3}{4}$. Find *a*. *(Lesson 7-2)* $a = 10$

20. A road leads from the town of Salida. There are signs every 5 miles that tell the elevation. *(Lesson 7-1)*

Miles from Salida	Elevation (feet)
0	1744
5	1749
10	1749
15	1759
20	1757

 a. Between which two signs is the rate of change of elevation negative? between 15 and 20
 b. Calculate the rate of change of elevation for the entire distance. .65 foot per mile

21. *Skill sequence.* Solve. *(Lesson 5-9)*
 a. $x^2 = 64$ 8 or -8 **b.** $(y - 7)^2 = 64$ -1 or 15 **c.** $(3y - 7)^2 = 64$ $-\frac{1}{3}$ or 5

22. *Skill sequence.* Simplify. *(Lesson 2-3)*
 a. $\frac{12}{144}$ $\frac{1}{12}$ **b.** $\frac{3}{3^2}$ $\frac{1}{3}$ **c.** $\frac{d}{d^2}$ $\frac{1}{d}$

Exploration

23. In many places, a taxicab ride costs a fixed number of dollars plus a constant charge per mile. Answers will vary.
 a. Find a rate for taxi rides in your community or in a nearby city.
 b. Graph your findings on coordinate axes like the ones at the right.
 c. Find an equation relating distance traveled and cost.

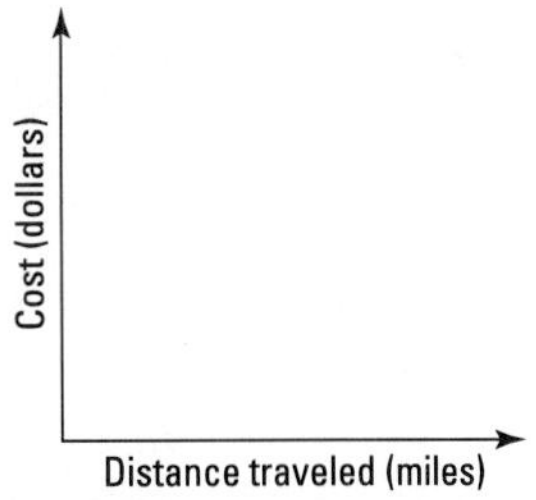

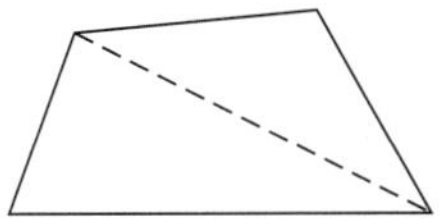

Quadrilateral: 2 · 180° = 360°

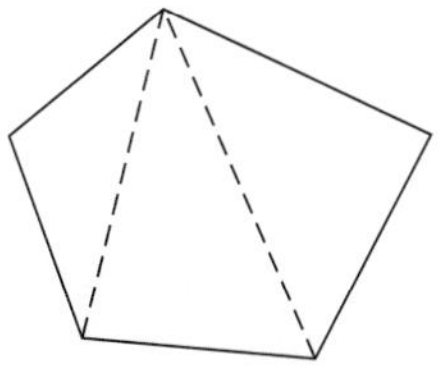

Pentagon: 3 · 180° = 540°

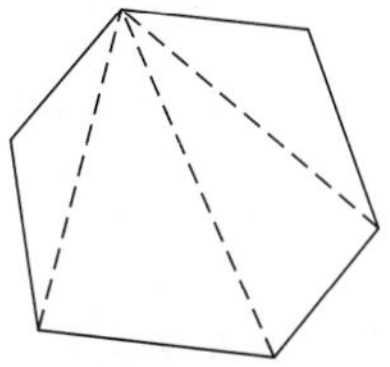

Hexagon: 4 · 180° = 720°

Question 20 You may also want to ask, "Between which consecutive mile markers does the rate of change exceed the average?" [Between 0 and 5; between 10 and 15]

Question 23 To be most accurate, the graph of cost versus distance is a step function. However, the steps are so small that we have ignored them. The step function will pass through the points on the linear function, so the approximation seems reasonable.

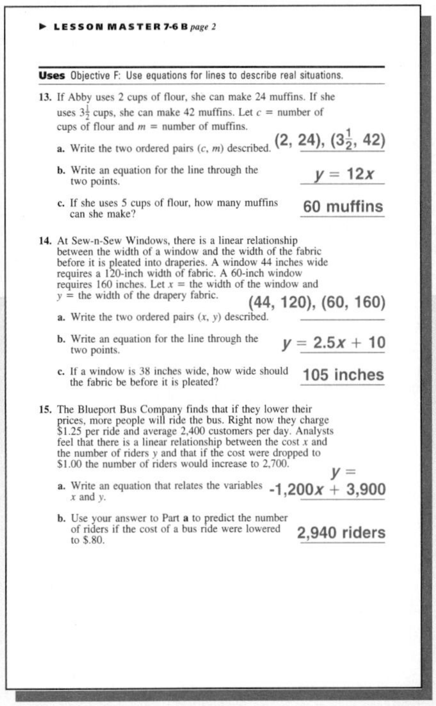

▸ LESSON MASTER 7-6 B *page 2*

Uses Objective F: Use equations for lines to describe real situations.

13. If Abby uses 2 cups of flour, she can make 24 muffins. If she uses $3\frac{1}{2}$ cups, she can make 42 muffins. Let *c* = number of cups of flour and *m* = number of muffins.
 a. Write the two ordered pairs (*c*, *m*) described. (2, 24), ($3\frac{1}{2}$, 42)
 b. Write an equation for the line through the two points. $y = 12x$
 c. If she uses 5 cups of flour, how many muffins can she make? 60 muffins

14. At Sew-n-Sew Windows, there is a linear relationship between the width of a window and the width of the fabric before it is pleated into draperies. A window 44 inches wide requires a 120-inch width of fabric. A 60-inch window requires 160 inches. Let *x* = the width of the window and *y* = the width of the drapery fabric.
 a. Write the two ordered pairs (*x*, *y*) described. (44, 120), (60, 160)
 b. Write an equation for the line through the two points. $y = 2.5x + 10$
 c. If a window is 38 inches wide, how wide should the fabric be before it is pleated? 105 inches

15. The Blueport Bus Company finds that if they lower their prices, more people will ride the bus. Right now they charge $1.25 per ride and average 2,400 customers per day. Analysts feel that there is a linear relationship between the cost *x* and the number of riders *y* and that if the cost were dropped to $1.00 the number of riders would increase to 2,700.
 a. Write an equation that relates the variables *x* and *y*. $y = -1{,}200x + 3{,}900$
 b. Use your answer to Part **a** to predict the number of riders if the cost of a bus ride were lowered to $.80. 2,940 riders

Adapting to Individual Needs

Challenge
Materials: Telephone book

Refer students to the opening paragraph of the lesson on page 450. Then have them use the rate data in a telephone book to write an equation to determine the cost of making a long distance call to a place of their choice.

Setting Up Lesson 7-7

Materials Students will need their calculators for the computations of slope and intercept for the questions in Lesson 7-7. They will need atlases or globes for **Question 27.**

In-class Activity

Resources

From the ***Teacher's Resource File***

- Answer Master 7-7
- Teaching Aid 81: Latitude and Temperature

Additional Resources

- Visual for Teaching Aid 81

This activity is designed to introduce students to the example that appears in Lesson 7-7. It should take ten to fifteen minutes of class time. **Teaching Aid 81** contains the table and coordinate grid for this activity.

It may seem obvious to us that a line of best fit is the best way to arrive at the answer in step 3 of the activity. However, this idea is unlikely to occur to students unless they have read ahead. Allow students to give virtually any reason for choosing the estimate they did. Use the variety of responses to each of the parts as evidence for the need to have some systematic way of making an estimate. Point out that a systematic way is described in Lesson 7-7.

TEACHING AID 81
In-class Activity, Lesson 7-7

Latitude and Temperature

City	North Latitude	April Mean High Temperature (°F)
Lagos, Nigeria	6	89
San Juan, Puerto Rico	18	84
Calcutta, India	23	97
Cairo, Egypt	30	83
Tokyo, Japan	35	63
Rome, Italy	42	68
Quebec City, Canada	47	46
London, England	52	56
Copenhagen, Denmark	56	50
Moscow, Russia	56	47

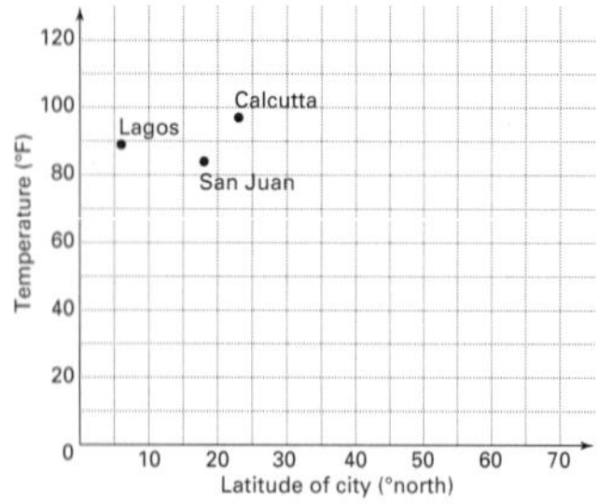

Introducing Lesson 7-7

Latitude and Temperature

Materials: Graph paper
Work in small groups.

When a situation involves a relation between two variables, a graph of the relation can be drawn. If the points all lie on a line, an equation for the line can be found. When the points do not lie on a line, but are approximately on a line, an equation can be found that closely describes the relation. In this activity, you will explore how to do this.

1 The *latitude* of a place on Earth tells how far the place is from the equator. Latitudes in the Northern Hemisphere range from 0° at the equator to 90° at the North Pole. In the table below are the latitudes and mean high temperatures in April for selected cities in the Northern Hemisphere. (The mean high temperature is the mean of all the daily high temperatures for the month.) Although in all of these cities temperature is measured in degrees Celsius, we have converted the temperatures to Fahrenheit for you.

Latitude and Temperature in Selected Cities

City	North Latitude	April Mean High Temperature (°F)
Lagos, Nigeria	6	89
San Juan, Puerto Rico	18	84
Calcutta, India	23	97
Cairo, Egypt	30	83
Tokyo, Japan	35	63
Rome, Italy	42	68
Quebec City, Canada	47	46
London, England	52	56
Copenhagen, Denmark	56	50
Moscow, Russia	56	47

North Pole: latitude 90° north
Minneapolis, MN: latitude 45° north
Equator: 0° latitude

456

Optional Activities

Materials: Globe

Geography Connection Because national studies have shown that the typical U.S. student's knowledge of geography is not particularly strong, you might want to bring a globe to class to explain *latitude* and *longitude*. Ask students to point out where it is generally coldest (near the poles, latitude 90° N and latitude 90° S) and where it is usually hottest (near the equator, latitude 0°). Then have them locate the cities listed in the table.

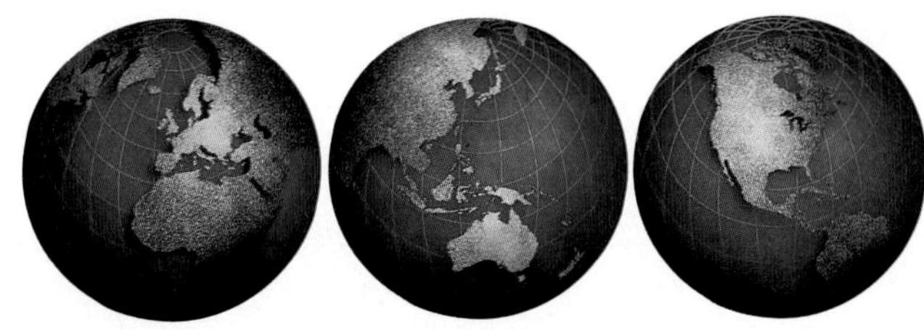

Graph the data points in the table. Let the latitude be the x-value and the temperature be the y-value in degrees. Label each point with the city it represents. We have plotted the first three entries from the table.

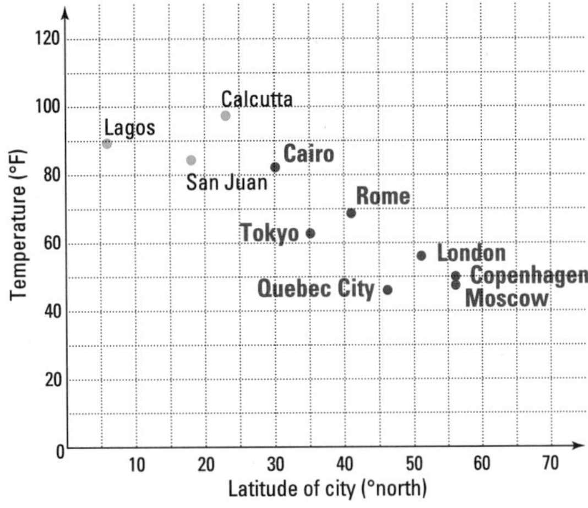

2 What pattern(s) do you notice in this scatterplot? In particular, as the latitude increases, what generally happens to the temperature in April? **See below.**

3 Based on the data above, what do you expect the mean high temperature to be in April in cities at the following north latitudes?

a. Singapore at 1° north latitude
b. a city at 19° north latitude, like Mexico City or Bombay, India
c. Madrid, Spain, at 40° north latitude
d. Helsinki, Finland, at 60° north latitude

Explain how you arrived at your answers. **Answers will vary.**

Samples: a) 90°; Singapore would probably be like Lagos. b) 85°; Bombay is at about the same latitude as San Juan. c) 70°; Madrid is near Rome. d) 40°; Helsinki is farther north than Moscow and should be colder.

2) Answers will vary. Sample: As the latitude increases, the temperature decreases.

457

Lesson 7-7

Objectives

G Given data whose graph is approximately linear, find a linear equation to fit the graph.

Resources

From the ***Teacher's Resource File***
- Lesson Master 7-7A or 7-7B
- Answer Master 7-7
- Teaching Aids
  28 Four-Quadrant Graph Paper
  73 Warm-up
  82 Additional Examples
  83 Questions 12–15
  84 Question 16
- Activity Kit, Activity 17
- Technology Sourcebook
  Computer Master 17

Additional Resources
- Visuals for Teaching Aids 28, 73, 82–84
- Rulers or **Geometry Templates**
- Atlas or globe

Teaching Lesson 7-7

Warm-up

Writing Write a paragraph about the information given in the graph below. Include some predictions about attendance in the years that are not shown on the graph.
Answers will vary.

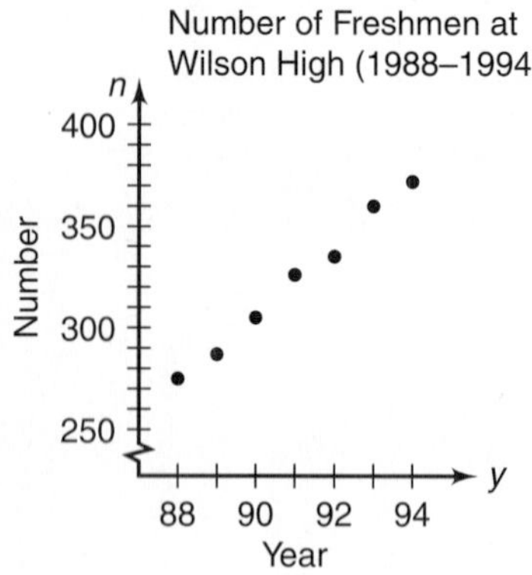

LESSON

7-7 Fitting a Line to Data

Denmark's charm. *Shown is Copenhagen's Nyhavn Canal with its ships, old buildings, and cafes. Copenhagen is the capital of Denmark. Because the country is almost entirely surrounded by water, the climate is mild and damp.*

❶ If the data points of a set do not all lie on one line, but are close to being linear, you can often use an equation for a line to describe trends in the data. For instance, below is a graph of the latitude and temperature data you used in the In-class Activity on page 456.

Fitting a Line to Data

Even though no line passes through all the data points, you can find a line which describes the trend of higher latitude, lower temperature. The simplest way is to take a ruler and draw a line that seems "close" to all the points. This is called "fitting a line by eye," and one such line is graphed at the right. Notice that the line we drew does not pass through any of the original data points.

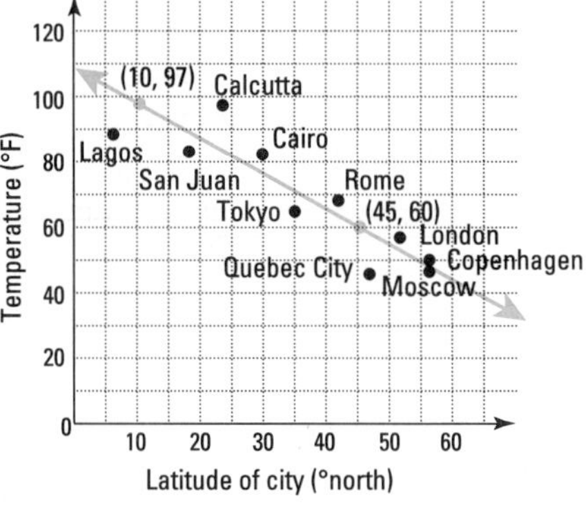

❷ Once a line has been fitted to the data, you can find an equation for the line, and then use the equation to predict temperatures for different latitudes.

458

Lesson 7-7 Overview

Broad Goals The major goal of this lesson is to find an equation for a line which roughly approximates a set of points.

Perspective Lines are the simplest descriptors for trends in data that arise in the real world. Many students don't encounter situations in which data fail to lie in a perfect line until later courses. Fitting a line to data, however, is a technique that is appropriate for algebra students.

This lesson is centered on one set of data—the April mean high temperatures of cities at various latitudes in the Northern Hemisphere—but the questions and activities provide additional data sets. In the lesson, we do not give a precise way to determine a line of good fit or the least squares "line of best fit" that is found in statistics. However, students may have access to an automatic grapher or a statistical package that will identify that line.

Example 1

Find an equation for the line drawn to fit the latitude and temperature data points on page 457.

Solution

Notice that the line contains the points (10, 97) and (45, 60). Use these points to find the slope of the line.

$$\text{slope} = \frac{60 - 97}{45 - 10} = \frac{-37}{35} \approx -1.06$$

Now substitute this slope and the coordinates of one of the points into $y = mx + b$ and solve. We use (10, 97).

$$97 = -1.06 \cdot 10 + b$$
$$97 = -10.6 + b$$
$$107.6 = b$$

An equation for the line is $y = -1.06x + 107.6$.

The negative slope -1.06 means as you move 1° north in latitude, the April mean high temperature is lower by about 1°F. The y-intercept 107.6 means that the high temperature at the equator (0° latitude) should be about 108°F. Notice that this agrees with the graph drawn on the previous page. The line crosses the vertical axis at about 108°F.

Activity 1

Use the scatterplot of latitude and temperature data you made in the In-class Activity on page 457.

a. Draw another line that seems to fit the data. Graphs will vary.
b. Give the coordinates of two points on your line. Sample: (5, 110); (60, 35)
c. Find an equation for your line. Sample: $y = -1.36x + 116.8$

Famous Spanish writer. *Miguel de Cervantes (1547–1616) ranks as one of the world's greatest writers. Statues of two of his most famous characters, Don Quixote and Sancho Panza, stand in Madrid, Spain.*

Using an Equation for a Fitted Line to Make Predictions

Using the equation $y = -1.06x + 107.6$ found in Example 1, you can estimate the April mean high temperature for any city in the Northern Hemisphere.

Example 2

Use the equation for the fitted line to predict the April mean high temperature for Madrid, Spain, which is at 40° north latitude.

Solution

Substitute 40° for x in the equation.

$$y = -1.06 \cdot x + 107.6$$
$$y = -1.06 \cdot 40 + 107.6$$
$$y = 65.2$$

You can predict that Madrid would have an April mean high temperature of about 65°F.

The mean high temperature in April is actually 64°, so the predicted temperature is quite close.

Notes on Reading

❶ You might want to have a globe or world map for students to examine. Have them locate the cities that are mentioned in the lesson.

Point out that there are no new mathematical techniques in this lesson—merely the application of the idea from the previous lesson to situations in which there are more than two points that do not lie on the same line.

❷ Students will naturally wonder how to determine if a line is a line of good fit. Because we are using the line to predict values, the critical idea in how well the line fits the point is the vertical distance from the point to the line. A line of good fit should have about as many points above it as below it, and the distances from the points to the line should balance. (In statistics, the line for which the sum of the squares of the vertical distances from the points to the line is minimized is defined as having the "best fit.")

Multicultural Connection All or part of Cervante's *Don Quixote* has been translated into more than 60 languages. Due to the widespread representation of Don Quixote and Sancho Panza in art, drama, and film, Cervante's figures are probably familiar to more people than any other fictitious characters in world literature.

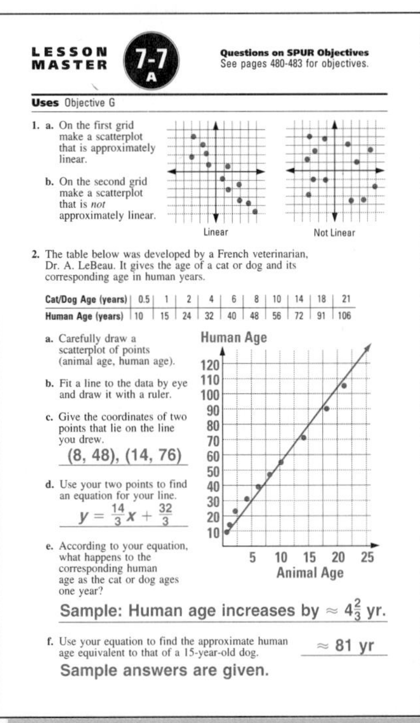

LESSON MASTER 7-7 A

Questions on SPUR Objectives
See pages 480-483 for objectives.

Uses Objective G

1. a. On the first grid make a scatterplot that is approximately linear.
   b. On the second grid make a scatterplot that is *not* approximately linear.

Linear   Not Linear

2. The table below was developed by a French veterinarian, Dr. A. LeBeau. It gives the age of a cat or dog and its corresponding age in human years.

Cat/Dog Age (years)	0.5	1	2	4	6	8	10	14	18	21
Human Age (years)	10	15	24	32	40	48	56	72	91	106

a. Carefully draw a scatterplot of points (animal age, human age).

b. Fit a line to the data by eye and draw it with a ruler.

c. Give the coordinates of two points that lie on the line you drew. (8, 48), (14, 76)

d. Use your two points to find an equation for your line. $y = \frac{14}{3}x + \frac{32}{3}$

e. According to your equation, what happens to the corresponding human age as the cat or dog ages one year?
Sample: Human age increases by $\approx 4\frac{2}{3}$ yr.

f. Use your equation to find the approximate human age equivalent to that of a 15-year-old dog. ≈ 81 yr

Sample answers are given.

Optional Activities

Activity 1 You might want to use *Activity Kit, Activity 17,* to introduce the lesson. In this activity, students gather data and draw graphs. Their graphs should show points clustering in such a way that a line can be drawn to "fit" the data.

Activity 2 Technology Connection You might want to assign *Technology Sourcebook, Computer Master 17.* Students use *GraphExplorer* or similar software to plot data. Students then graph lines through pairs of data points and choose the line they think best fits the data.

Additional Examples

These examples are also given on **Teaching Aid 82.**

1. The table shows the amount of gold that was mined in the world for the years 1984 to 1992.

Gold Production: 1984–1992 (millions of troy ounces)	
1984: 46.9	1989: 65.3
1985: 49.3	1990: 68.6
1986: 51.5	1991: 69.1
1987: 51.5	1992: 72.2
1988: 60.3	

   a. Draw a scatterplot. **Sample graph is shown below.**
   b. Use a ruler to fit a line to the data. **One line is shown on the graph.**
   c. Write an equation for the line. **An equation for the line through (88, 60) and (91, 69) is $y = 3x - 204$.**

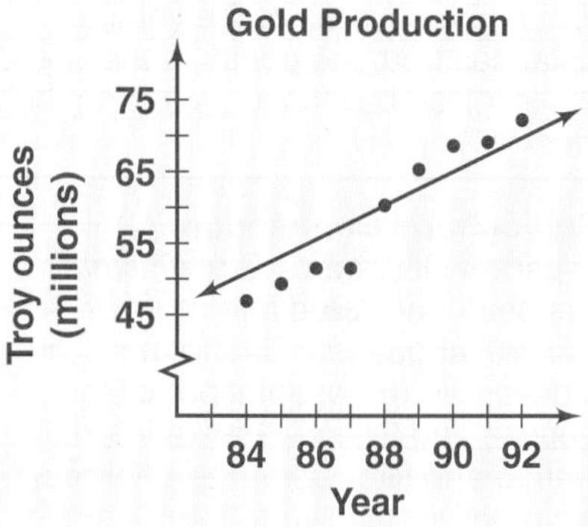

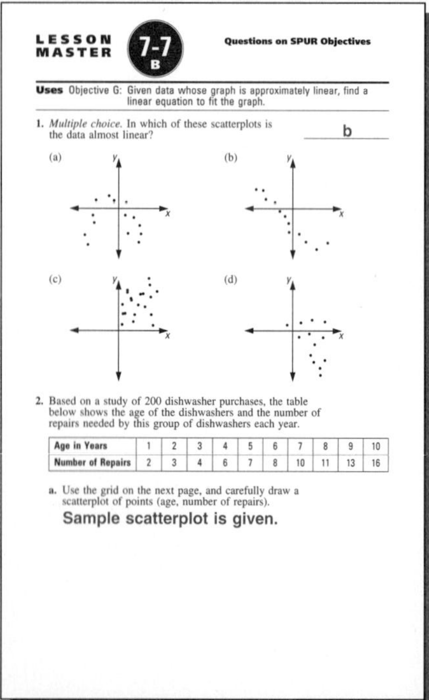

LESSON MASTER 7-7 B Questions on SPUR Objectives

Uses Objective G: Given data whose graph is approximately linear, find a linear equation to fit the graph.

1. *Multiple choice.* In which of these scatterplots is the data almost linear? b

(a) (b) (c) (d)

2. Based on a study of 200 dishwasher purchases, the table below shows the age of the dishwashers and the number of repairs needed by this group of dishwashers each year.

Age in Years	1	2	3	4	5	6	7	8	9	10
Number of Repairs	2	3	4	6	7	8	10	11	13	16

a. Use the grid on the next page, and carefully draw a scatterplot of points (age, number of repairs).
Sample scatterplot is given.

Located in the National Palace of Mexico City are the Diego Rivera murals. This one shows Hidalgo, Morelos, and Juárez—men who made important political contributions.

Shown is an outdoor marketplace in Calcutta, India. The city lies along the east bank of the Hooghly river and serves as India's chief port for trade with Southeast Asia.

Activity 2

Use the equation you found in Activity 1 to predict the mean high temperature in April for Madrid. By how much does your prediction differ from the actual value? Sample: $y = -1.36(40) + 116.8 = 62.4$. The prediction was off by 1.6°F.

Sometimes the fitted line does not predict temperature accurately. For instance, for a city at a latitude of 19° north, the line predicts a temperature of 87°. Both Bombay, India, and Mexico City are at this latitude. For Bombay the predicted temperature is accurate. But in Mexico City, the actual April mean temperature is 78°. The prediction is too high because Mexico City is at an altitude of about one mile, and temperatures at high altitudes are lower than those at sea level.

Being able to fit a line to data allows you to obtain a general formula from a few cases. This is such an important skill that some computer software and graphing calculators will find the line that *best fits* data. If you have access to such technology, you might want to enter these latitude and temperature data.

QUESTIONS

Covering the Reading

1. What does the latitude of a place on Earth signify? **the distance a place is from the equator**
2. What is the latitude of the equator? **0°**
3. Which city is farther north, Calcutta or Cairo? **Cairo**
4. What does it mean to "fit a line by eye" to a scatterplot? **Draw a line that seems closest to all of the points in the graph.**
5. Once a line is fitted, what is a good first step toward getting an equation for the line? **Estimate the coordinates of two points on the line.**
6. a. Use the graph of the fitted line in this lesson to predict the mean high temperature in April in a city at 25° north latitude. **about 81°F**
   b. Use the equation in Example 1 to predict this temperature. **81.1°F**
   c. Use the equation you found in Activity 1 to predict this temperature.
   d. What is true about your answers to parts **a, b,** and **c?**
   c) Answers will vary. Sample: 82.8°F; d) They are all close to 82°F
7. Refer to Activity 2. By how much does your prediction differ from the actual mean high temperature in Madrid? **Answers may vary. Sample: My prediction differed from the actual mean temperature by 1.6°F.**
8. a. Acapulco, Mexico, is at 17° north latitude. Use the equation in Example 1 to estimate its average April high temperature. **≈ 90°F**
   b. The actual mean high temperature in April for Acapulco is 87°F. Give a reason why the answer in part **a** is closer to the actual value for Acapulco than the prediction was for Mexico City. **Sample: Acapulco is at sea level.**
9. a. What is the latitude of the North Pole? **90° North**
   b. What is the predicted mean high temperature in April at the North Pole? **≈ 12.2°F**

Adapting to Individual Needs

Challenge Geography Connection

Have students look up the meanings of the *prime meridian of longitude* and the *International Date Line* and explain how they are related to international time zones. [Students' answers might include: The prime meridian is an imaginary north-south line that passes through both the North and South poles and Greenwich, England. Longitude is measured 180° both east and west of the prime meridian. In 1884, worldwide time zones were established with the Greenwich meridian as the starting point. The zones are theoretically 15° longitude wide. The Greenwich meridian is in the middle of the first zone. Twelve zones are to the east of Greenwich, and twelve zones are to the west. The 12th zones east and west are each a half zone wide and are separated by the International Date Line. The time east of the Date Line is one day earlier than the time to the west of the line.]

Applying the Mathematics

16a, b)

Temperature F°

y

100 80 60 40 20 0

10 20 30 40 50 60 x

Degrees north latitude

c) Sample: (6, 74); (30, 47)

d) $y = -1.125x + 80.75$

e) be about 1° colder.

10. If a city in the Northern Hemisphere has a mean high temperature in April of about 80°, at what latitude would you expect it to be? Explain how you got your answer. **26° north; by substituting 80 for *y* in the equation in Example 1, you get *x* = 26.**

11. Below are the latitudes and April mean high temperatures for two cities in the Southern Hemisphere.

Rio de Janeiro, Brazil (23° south, 69°F)
Cape Town, South Africa (34° south, 58°F)

Could you predict temperatures for cities south of the equator by using negative values of x in the equation in Example 1? **No**
Explain your reasoning. **Negative values of *x* would result in values of *y* that are always greater than 107.6°F.**

In 12–15, tell whether fitting a line to the data points would be appropriate.

12. 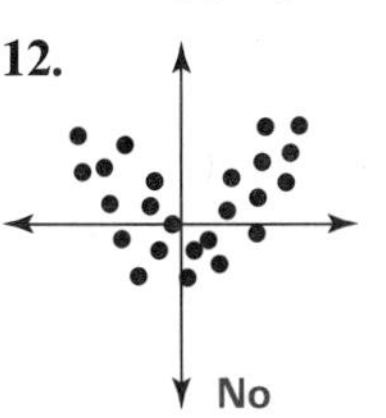 **No**

13. 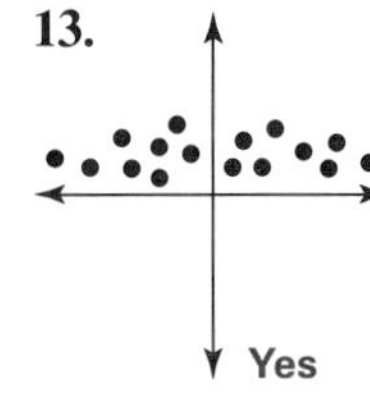 **Yes**

14. 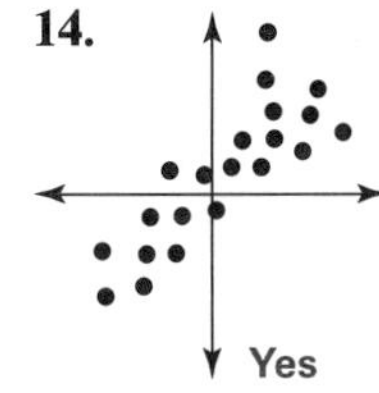 **Yes**

15. 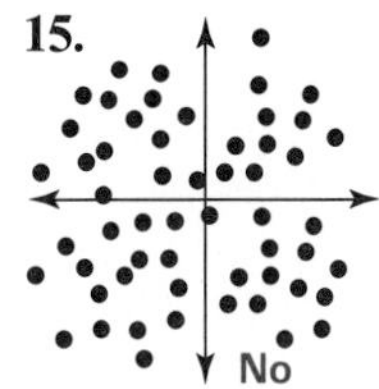 **No**

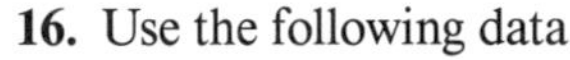

16. Use the following data.

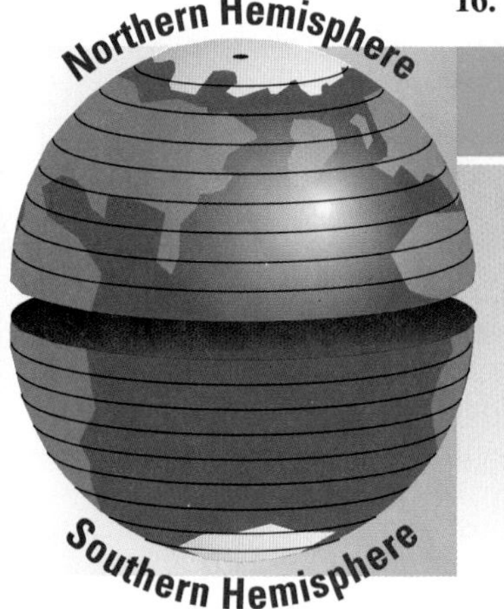

City	North Latitude	January Mean Low Temperature (°F)
Lagos, Nigeria	6	74
San Juan, Puerto Rico	18	67
Calcutta, India	23	55
Cairo, Egypt	30	47
Tokyo, Japan	35	29
Rome, Italy	42	39
Quebec City, Canada	47	2
London, England	52	35
Copenhagen, Denmark	56	29
Moscow, Russia	56	9

See left for a, b, c, d, e.

a. Carefully draw a scatterplot showing a point for each city.
b. Fit a line to the data by eye and draw the line with a ruler.
c. Estimate the coordinates of two points on the line you drew.
d. Find an equation for the line through the points in part **c.**
e. Complete the following sentence: "As you go one degree north, the January mean low temperature tends to __?__."
f. What does the equation predict for a January mean low temperature at the equator? **about 81°F; g) about −21°F**
g. Predict the January mean low temperature for the North Pole.
h. Use your equation to predict the January mean low temperature for Acapulco, which is at 17° north latitude. (Note: the actual mean low is 70°F.) **about 62°F**

2. Use your equation in Question 1c to predict the amount of gold that will be mined in the year 2000. **For $y = 3x - 204$, about 96 million troy ounces; note that for the year 2000, *x* is 100.**

Notes on Questions

Students will need their calculators for the computations of slope and intercept. Even so, because students are estimating the best-fitting line, they will get different slopes and intercepts as answers to some of the questions. They will also need graph paper or **Teaching Aid 28.**

Questions 12–15 Teaching Aid 83 shows these scatterplots. These questions are important conceptually because they demonstrate that not all sets of points are best fitted by a line. For **Question 12**, a curve might be appropriate. For **Question 15,** there seems to be no particular pattern except that the points are rather evenly spaced within a circle.

Question 16 Cooperative Learning The table and a grid for showing the scatterplot is given on **Teaching Aid 84.** After students have drawn the scatterplot, you might have them use a thin piece of uncooked spaghetti as a line which can be moved easily around on the graph to help them find a line of best fit. Students can discuss the best fit and establish the points in class before they do **parts d–g** on their own.

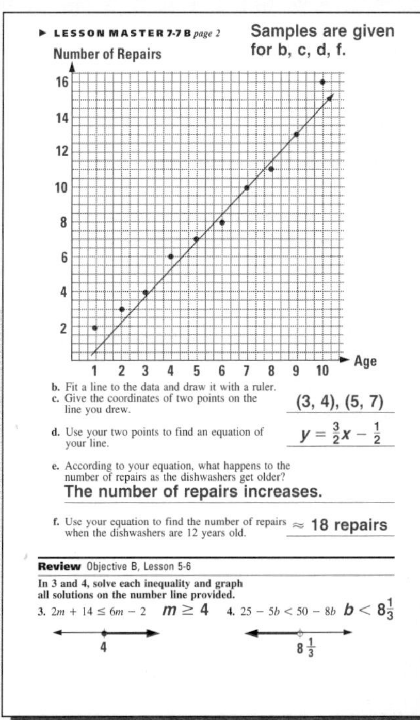
► LESSON MASTER 7-7 B *page 2* — Samples are given for b, c, d, f.

Number of Repairs — Age

b. Fit a line to the data and draw it with a ruler.
c. Give the coordinates of two points on the line you drew. (3, 4), (5, 7)
d. Use your two points to find an equation of your line. $y = \frac{3}{2}x - \frac{1}{2}$
e. According to your equation, what happens to the number of repairs as the dishwashers get older? The number of repairs increases.
f. Use your equation to find the number of repairs when the dishwashers are 12 years old. ≈ 18 repairs

Review Objective B, Lesson 5-6

In 3 and 4, solve each inequality and graph all solutions on the number line provided.

3. $2m + 14 \le 6m - 2$ $m \ge 4$
4. $25 - 5b < 50 - 8b$ $b < 8\frac{1}{3}$

Adapting to Individual Needs

Extra Help

Some students are frustrated by problems that do not have one correct answer. When trying to draw a line to fit data, they might attempt to connect all of the data points, forgetting that they have to find a straight line. Emphasize that the goal is to find a line that describes *trends* in the data and that the line may or may not pass through any of the actual data points. Help students to think of examples of data where trends might be important. One example would be the growth rates in children; another would be the winning times in certain Olympic events over a period of many years. Point out that some data will not suggest a linear trend, such as annual snowfall in a particular city.

Notes on Questions

Question 25 Multicultural Connection In ancient Rome the term *circus* was used to denote an enclosure for chariot racing. The modern day circus was founded in 1768 when Philip Astley, an English rider, discovered that centrifugal force made it fairly easy to stand on his horse's back while it galloped in circles around a ring. He built a roof and viewing stands to attract spectators and other riders. One of the performers, Charles Hughes, established his own ring nearby in 1782 and called it the Royal Circus, the first modern use of the name. Ask students to tell about circuses they have seen or read about.

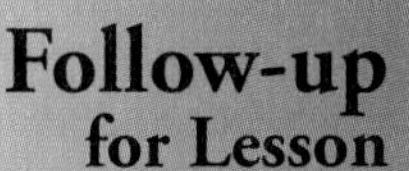

Follow-up for Lesson 7-7

Practice

For more questions on SPUR Objectives, use **Lesson Master 7-7A** (shown on page 459) or **Lesson Master 7-7B** (shown on pages 460–461).

Assessment

Written Communication Ask students to write a summary of this lesson. Tell them to include the main objective and a question that relates to the lesson. [Students understand that equations for lines can be written to describe data that is approximately linear.]

Extension

There is said to be a *positive correlation* when two variables are related in such a way that one increases as the other increases. Have students look at **Questions 12–15** and determine which one illustrates a positive correlation. [Question 14]

Project Update Projects 2, 3, 4, and 6 on pages 476–477 relate to the content of this lesson. Students choosing Projects 2, 3, or 4 might begin reseaching their projects in almanacs and encyclopedias. Students who have chosen Project 6 might want to **work with a partner or as a group** since this project has students perform several experiments. These students can begin gathering the materials for the experiment.

Ringling Brothers, 1907; Circus of Seville, 1993.

Review

17. Find an equation for the line with y-intercept equal to 7 and x-intercept equal to 4. *(Lesson 7-6)* $y = -\frac{7}{4}x + 7$

18. a. Find an equation for the line through (3, 2) with a slope of $\frac{3}{5}$.
b. Give the coordinates of one other point on this line. *(Lesson 7-5)*
a) $y = (3/5)x + 1/5$; b) Sample: $(0, 1/5)$

19. a. Give the slope and y-intercept of the line $3x + 5y = 2$. $-\frac{3}{5}$; $\frac{2}{5}$
b. Graph the line. *(Lesson 7-4)* See left below.

In 20 and 21, give **a.** the slope; and **b.** the y-intercept of the line. *(Lesson 7-4)*

20. $y = -\frac{x}{2}$ a) $-\frac{1}{2}$; b) 0

21. $y = -2$ a) 0; b) -2

22. a. Draw a line which has no y-intercept. See below.
b. What is the slope of your line? *(Lesson 7-4)* no slope or undefined

23. Seattle scored 17 points in the first nine minutes of a game. At that rate about how many points would they score in a 48-minute game? *(Lesson 6-8)* about 91 points

24. Solve for y. *(Lesson 5-7)*
a. $8y = 2 + y$ $y = \frac{2}{7}$
b. $By = C + y$ $y = \frac{C}{B-1}$

25. If a is the cost of an adult's ticket to the circus and c is the cost of a child's ticket, what is the cost of tickets for 2 adults and 4 children? *(Lessons 2-4, 3-1)* $2a + 4c$

26. Graph the solution set to $-3x \ge 15$ using the given domain. *(Lessons 1-2, 2-8)* See below.
a. set of integers
b. set of real numbers

Exploration

27. a. Find the latitude of your school, to the nearest degree.
b. Predict the April mean high temperature in °F for your latitude using the graph in Example 1.
c. Check your prediction in part **b** against some other source of the April mean high temperature. (Newspapers or TV weather records are possible sources.)
(a–c) Answers will vary.

28. What is meant by *longitude?* How far a place is, east or west, from the great circle through the North Pole and the South Pole passing through Greenwich, England.

19b)

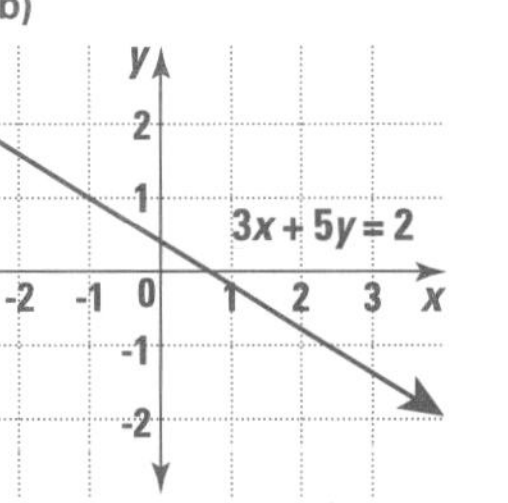

22a) Sample:

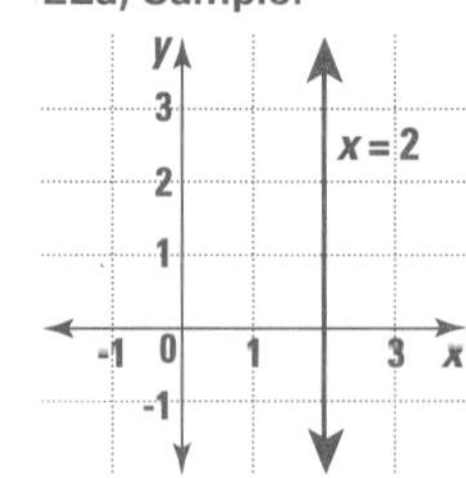
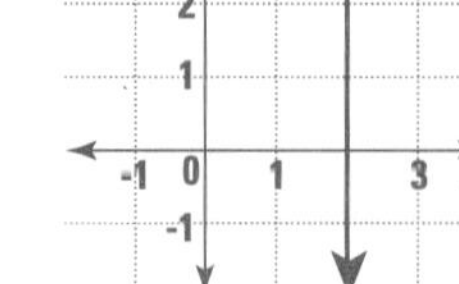

Adapting to Individual Needs

English Language Development
Materials: Globe

You might want to use a globe and review the following vocabulary: *latitude, longitude, equator, Northern Hemisphere,* and *Southern Hemisphere.* If some students were born in other communities or countries, ask them to find the latitude of their place of birth.

Setting Up Lesson 7-8

Materials If automatic graphers are available, we recommend that students use them for Lesson 7-8.

Use **Question 25** on page 462 to lead into Lesson 7-8. Have students imagine that \$20 is spent. Ask, "What could the values of a and c be?" Have students write the equation $2a + 4c = 20$ and graph it. Point out that this equation is in standard form—the form they will study in Lesson 7-8.

LESSON

7-8

Equations for All Lines

In this chapter, lines have been used to describe situations involving constant increase or decrease. We have also fitted lines to data. The slope-intercept form $y = mx + b$ arises naturally from these applications. All lines except vertical lines can have equations in slope-intercept form.

Some situations, such as the one below, naturally lead to equations of lines in a different form.

The Ramirez family bought 4 sandwiches and 3 salads. They spent \$24.00. If x is the cost of a sandwich and y the cost of a salad, then

$$4x + 3y = 24.$$

The pairs of values of x and y that make this equation true are the possible costs of the items. For instance, because

$$4(4.50) + 3 \cdot 2 = 18 + 6 = 24$$

each sandwich could have cost \$4.50 and each salad \$2. This yields the point (4.5, 2). Other possible costs of sandwiches and salads can be found by rewriting this equation in slope-intercept form.

$$4x + 3y = 24$$
$$3y = -4x + 24$$
$$y = -\frac{4}{3}x + 8$$

The graph of this equation is a line with slope $-\frac{4}{3}$ and y-intercept 8. Since the graph shows the possible costs of the sandwiches and salads, we use only the first quadrant where both x and y are positive numbers.

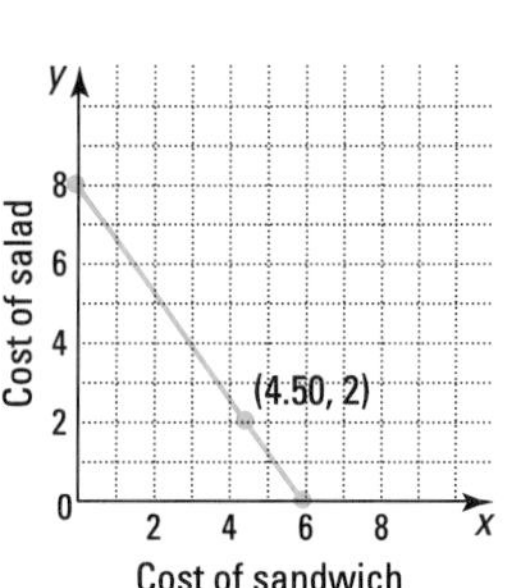

Lesson 7-8

Objectives

C Write an equation for a line in standard form, and from that form find its slope and y-intercept.
F Use equations for lines to describe real situations.
H Graph a straight line given its equation.

Resources

From the ***Teacher's Resource File***
- Lesson Master 7-8A or 7-8B
- Answer Master 7-8
- Teaching Aids
  28 Four-Quadrant Graph Paper
  73 Warm-up
- Technology Sourcebook
  Calculator Master 5

Additional Resources
- Visuals for Teaching Aids 28, 73
- Automatic grapher

Teaching Lesson 7-8

Warm-up

Softball caps cost \$10 and team shirts cost \$15. A total of \$630, excluding tax, was spent on caps and shirts. Name five different combinations of caps and shirts that could have been ordered. **Any whole-number solutions to $10x + 15y = 630$, such as (0, 42), (6, 38), (12, 34), (20, 22), and (63, 0), are solutions.**

Lesson 7-8 Overview

Broad Goals This lesson introduces the standard form for the equation of a line, $Ax + By = C$, and presents situations that lead to that form.

Perspective Linear-combination situations naturally lead to equations of lines in standard form. This is one of the reasons for studying this form. Another reason for studying the standard form is that all lines can be written in standard form, whereas vertical lines do not have equations in slope-intercept form.

The lesson begins with a situation that leads to the standard form, and the resulting line is graphed in **Example 1**. **Example 2** presents a second situation. The lesson ends with a discussion of the conversion of equations for lines into standard form. This idea has applications beyond the study of lines: there are standard forms for quadratics, polynomials, and virtually every other type of expression. It is common to prefer that the coefficients in $Ax + By = C$ be integers and that A be positive, but you should not be overly rigid. The context of the problem should determine the kinds of numbers used. Percent problems might best use decimals, and a problem about a recipe might use fractions. If the situation involves first a loss and then a gain, the value of A might be negative.

Notes on Reading

Reading Mathematics This is another lesson that you may wish to read in class. As you read, you might have students consider similar situations in which the numbers are different.

For instance, on page 463, you could ask what the equation would be if the Ramirez family had spent \$18 or if they had bought tickets for 3 adults and 5 children.

❶ This example is important because it shows students an "easy" way to graph lines in standard form: substitute 0 for x to get the y-intercept 8; substitute 0 for y to get the x-intercept 6. This is a particularly effective way to graph $Ax + By = C$ when A and B are factors of C; that is, when the intercepts are integers.

❷ Emphasize the idea of equivalent formulas that was first discussed in Lesson 5-7. In **Example 2**, the given answer is equivalent to $x + 2y = 16$, and either equation could be used from that point on.

Error Alert Some students may have trouble understanding **Examples 3 and 4** which apply the idea of clearing fractions. Refer these students to Lesson 5-8 where the idea was first discussed. You might also remind them of the Distributive Property.

The Standard Form of an Equation for a Line

The equation $4x + 3y = 24$ has the form

$$Ax + By = C,$$

where A, B, and C are constants. The variables x and y are on one side of the equation. The constant term C is on the other. The equation $Ax + By = C$, where A, B, and C are constants, is the **standard form** for an equation of a line.

To graph a line whose equation is in standard form, you do not need to rewrite the equation in slope-intercept form. Instead, you can find the intercepts and draw the line that contains the intercepts.

Example 1

Graph $4x + 3y = 24$.

❶ **Solution**

Find the x- and y-intercepts.

Let $x = 0$. $4 \cdot 0 + 3y = 24$

$3y = 24$

$y = 8$

The y-intercept is 8, so one point on the line is (0, 8).

Let $y = 0$. $4x + 3 \cdot 0 = 24$

$4x = 24$

$x = 6$

The x-intercept is 6, so (6, 0) is also on the line.

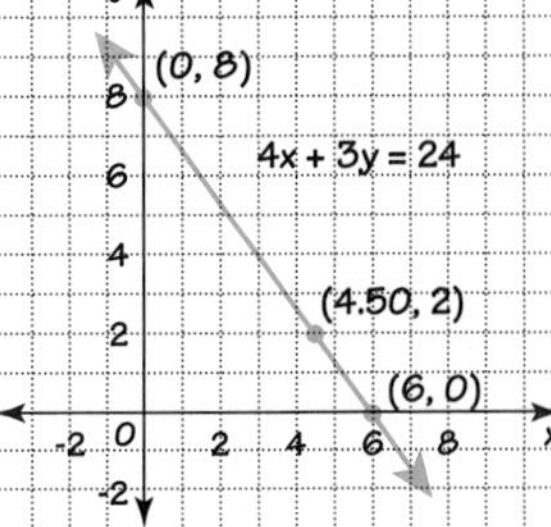

Now plot (6, 0) and (0, 8) and draw the line through them, as shown.

Check

Find a third point satisfying the equation. Earlier we noted that (4.50, 2) satisfies the equation. Is this point on the graph? Yes, it is.

You should be able to recognize real situations that lead to equations of lines.

Example 2

Roast beef sells for \$6 a pound. Shrimp costs \$12 a pound. Andy has \$96 to buy beef and shrimp for a party. Write an equation in standard form to describe the different possible combinations Andy could buy.

❷ **Solution**

Let x = pounds of roast beef bought. So 6x = cost of roast beef.

Let y = pounds of shrimp bought. So 12y = cost of shrimp.

Cost of roast beef	+	cost of shrimp	=	total cost
6x	+	12y	=	96

▶

Optional Activities

Activity 1

Materials: Graph paper or **Teaching Aid 28**

In **Question 27**, students work with equations of the form $Ax + By = C$ in which A and B remain constant and C changes. After answering the question, you might have students **work in groups** and determine what happens if B and C remain constant and A varies. They might begin by graphing $x + y = 12$, $2x + y = 12$, $3x + y = 12$, and $-3x + y = 12$ on the same coordinate plane. [See the graphs at the right. In general, all the equations will yield graphs having the same y-intercept, but the slopes will vary.] Next ask students to speculate what will happen if A and C remain constant and B varies. [In general, the equations will yield graphs having different y-intercepts and different slopes.]

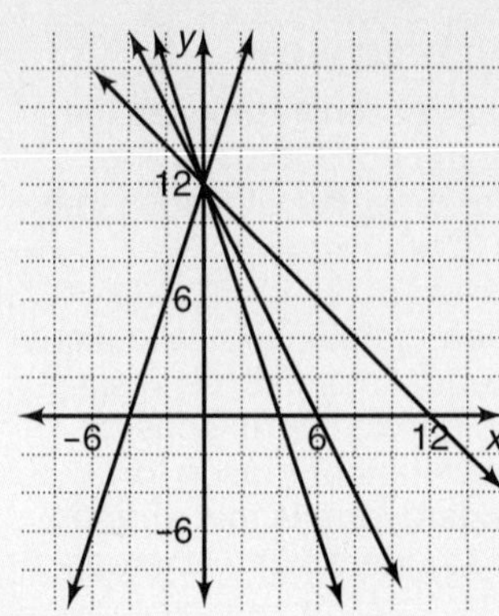

Check

To check the equation, suppose Andy buys 6 pounds of roast beef. He will have spent $6 \cdot \$6 = \$36$. He will have $\$96 - \$36 = \$60$ left to spend on shrimp. At \$12 per pound he can buy $\frac{60}{12} = 5$ pounds of shrimp. Does (6, 5) work in $6x + 12y = 96$? Yes, $6 \cdot 6 + 12 \cdot 5 = 96$.

Notice that the equation $6x + 12y = 96$ can be simplified by dividing each term by 6.

$$\frac{6x}{6} + \frac{12y}{6} = \frac{96}{6}$$
$$x + 2y = 16$$

Notice that this equation is also in standard form, so it also is an answer to Example 2.

Rewriting Equations in Slope-Intercept and Standard Form

The lines in Examples 1 and 2 are **oblique,** meaning they are neither horizontal nor vertical. Recall from Lesson 5-1, if a line is vertical, then its equation has the form $x = h$. If a line is horizontal, it has an equation $y = k$. Notice that these are already in standard form. For example,

$x = 3$ is equivalent to $x + 0y = 3$, where $A = 1, B = 0, C = 3$.
$y = -2$ is equivalent to $0x + y = -2$, where $A = 0, B = 1, C = -2$.

Thus *every line* has an equation in the standard form $Ax + By = C$.

You now have seen the two most common forms of equations of lines.

Slope-Intercept Form: $y = mx + b$
Standard Form: $Ax + By = C$

You should be able to quickly change an equation from one form into the other. In standard form, the equation is usually written with *A, B,* and *C* as integers.

Example 3

Rewrite $y = \frac{3}{5}x + \frac{8}{5}$ in standard form with integer coefficients. Find the values of *A, B,* and *C.*

Solution

First, multiply each side of the equation by 5 to clear the fractions.

$$5y = 5 \cdot \frac{3}{5}x + 5 \cdot \frac{8}{5}$$
$$5y = 3x + 8$$

Now add $-3x$ to both sides so the x and y terms are both on the left. Write the x term first.

$$-3x + 5y = 8$$

This is in standard form with $A = -3$, $B = 5$, and $C = 8$.

Additional Examples

Students will need graph paper or **Teaching Aid 28.**

1. Graph $5x - 2y = -20$.

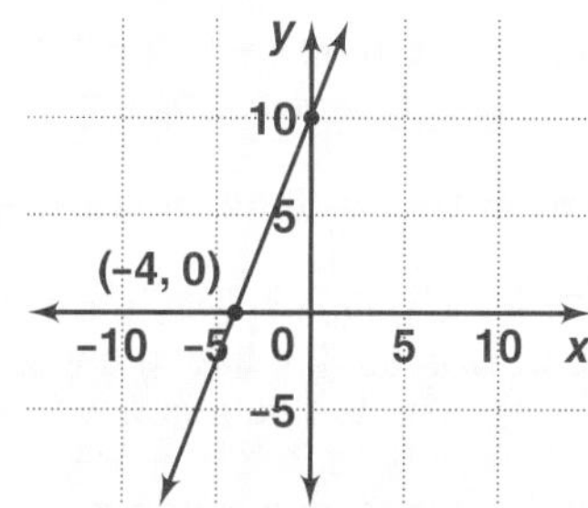

2. During the day, kitchen workers at a restaurant earn \$10 for each lunch meal and \$15 for each dinner meal they work. If a worker earns \$160, write an equation in standard form to describe the possible combinations of lunches and dinners he or she has worked. **$10x + 15y = 160$**
3. Rewrite $y = \frac{3}{8}x + \frac{1}{2}$ in standard form using integer values of *A, B,* and *C.* **$-3x + 8y = 4$; if you want *A* to be positive, then $3x - 8y = -4$.**
4. Rewrite each equation in standard form with integer values for *A, B,* and *C.*
   a. $y = -7x - 2$ **$7x + y = -2$**
   b. $x = \frac{3}{4}y$ **$4x - 3y = 0$**
   c. $.4x + .75y = 5$ **$40x + 75y = 500$, or $8x + 15y = 100$**

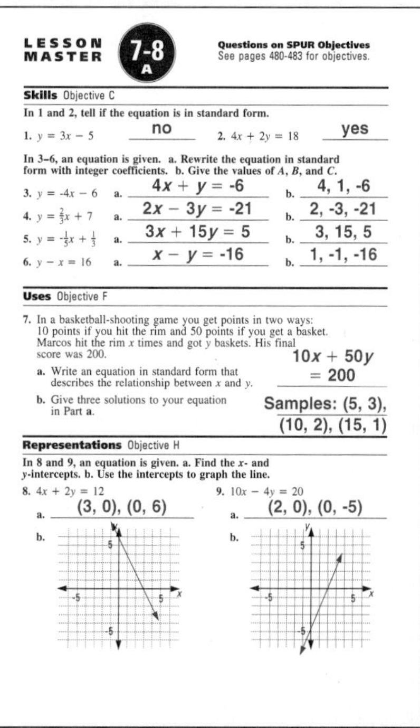

LESSON MASTER 7-8 A

Questions on SPUR Objectives
See pages 480-483 for objectives.

Skills Objective C

In 1 and 2, tell if the equation is in standard form.

1. $y = 3x - 5$ no
2. $4x + 2y = 18$ yes

In 3–6, an equation is given. a. Rewrite the equation in standard form with integer coefficients. b. Give the values of *A*, *B*, and *C*.

3. $y = -4x - 6$ a. $4x + y = -6$ b. 4, 1, -6
4. $y = \frac{2}{3}x + 7$ a. $2x - 3y = -21$ b. 2, -3, -21
5. $y = -\frac{1}{5}x + \frac{1}{3}$ a. $3x + 15y = 5$ b. 3, 15, 5
6. $y - x = 16$ a. $x - y = -16$ b. 1, -1, -16

Uses Objective F

7. In a basketball-shooting game you get points in two ways: 10 points if you hit the rim and 50 points if you get a basket. Marcos hit the rim x times and got y baskets. His final score was 200.
   a. Write an equation in standard form that describes the relationship between x and y. $10x + 50y = 200$
   b. Give three solutions to your equation in Part **a**.   Samples: (5, 3), (10, 2), (15, 1)

Representations Objective H

In 8 and 9, an equation is given. a. Find the *x*- and *y*-intercepts. b. Use the intercepts to graph the line.

8. $4x + 2y = 12$
   a. (3, 0), (0, 6)
   b.
9. $10x - 4y = 20$
   a. (2, 0), (0, -5)
   b.

Activity 2 Technology Connection
You may wish to assign *Technology Sourcebook, Calculator Master 5.* Students solve linear equations for *y*, and then graph these equations with their graphics calculators.

Notes on Questions

Questions 4–7 The equations in **Questions 4 and 5** do not have convenient coefficients for the intercept method of graphing. The equations in **Questions 6 and 7** *do* have convenient coefficients.

Question 16 Since no variables are specified, many equations are possible. Ordered pairs in **parts b and c** can be reversed, but they must be consistent with the equation written.

(Notes on Questions continue on page 468.)

Follow-up for Lesson 7-8

Practice

For more questions on SPUR Objectives, use **Lesson Master 7-8A,** (shown on page 465) or **Lesson Master 7-8B** (shown on pages 466–467).

Assessment

Written Communication Have students write paragraphs summarizing what they have learned about standard forms and slope-intercept forms of equations. Ask them to include an example of each type. [Examples are correct and slope, y-intercept, and x-intercept are identified for each form.]

LESSON MASTER 7-8 B **Questions on SPUR Objectives**

Skills Objective C: Write an equation for a line in standard form, and from that form find its slope and y-intercept.

1. Write the standard form for an equation of a line. Then identify the constants.
   $Ax + By = C$; A, B, and C are constants.

2. Are there any lines that cannot be described by an equation in standard form? If so, which ones?
   No; all lines can be described by an equation in standard form.

In 3–12, tell if the equation is in standard form.

3. $7y + 2x = 18$	no	4. $x + y = -8$	yes
5. $4x - 2y = 5$	yes	6. $y = 3x + 8$	no
7. $-x + 3y = -8$	yes	8. $4x + 2y + 9 = 0$	no
9. $0x - 2y = 7$	yes	10. $22x + 0y = 44$	yes
11. $y = -x + 0$	no	12. $15y - x = 19$	no

In 13–20, a. rewrite the equation in standard form with integer coefficients, and b. give the values of A, B, and C. Samples are given.

	a.	b.
13. $y = -2x + 5$	$2x + y = 5$	2, 1, 5
14. $y = \frac{4}{3}x + 1$	$-4x + 3y = 3$	-4, 3, 3
15. $y = -\frac{7}{8}x - \frac{1}{3}$	$21x + 24y = -8$	21, 24, -8
16. $2y + 3x = 9$	$3x + 2y = 9$	3, 2, 9
17. $y = 4x$	$-4x + y = 0$	-4, 1, 0
18. $-\frac{1}{2}x = 14$	$-x + 0y = 28$	-1, 0, 28
19. $2.1x - .8y = .6$	$21x - 8y = 6$	21, -8, 6
20. $\frac{5}{6}y = \frac{3}{4}x$	$-9x + 10y = 0$	-9, 10, 0

Some people prefer that the coefficient A be positive in standard form. To rewrite the equation in Example 3 with A positive, you could multiply each side by -1.

$$-1(-3x + 5y) = -1(8)$$
$$3x - 5y = -8$$

Every line has many equations in standard form. The equation $3x - 5y = -8$ is usually considered simplest because 3, -5, and -8 have no common factors.

Example 4

Rewrite $y = .25x$ in standard form with integer values of A, B, and C.

Solution

Multiply each side by 100 to clear the decimal.

$$y = .25x$$
$$100y = 25x$$

Now, add $-25x$ to each side to get both variable terms on the left. Write the x-term first.

$$-25x + 100y = 0$$

Here $A = -25$, $B = 100$, and $C = 0$.

QUESTIONS

Covering the Reading

In 1 and 2, refer to the situation about the Ramirez family buying food.

1. **a.** If each sandwich cost \$3.75, how much did each salad cost? \$3
   b. Give the coordinates of the point on the graph in Example 1 corresponding to your answer to part **a.** (3.75, 3)
2. Give another pair of possible costs for the sandwich and salad. Sample: (3, 4)
3. What is the form $Ax + By = C$ called? standard form

In 4 and 5, the equation is in the form $Ax + By = C$. Give the values of A, B, and C.

4. $4x + 2y = 5$
   $A = 4$; $B = 2$; $C = 5$
5. $x - 8y = 2$
   $A = 1$; $B = -8$; $C = 2$

In 6 and 7, an equation for a line is given.
a. Find the x- and y-intercepts.
b. Graph. See left for graphs.

6. $3x + 5y = 30$
   a) x-intercept is 10; y-intercept is 6
7. $2x - 3y = 12$
   a) x-intercept is 6; y-intercept is -4

6b)

(0, 6)
$3x + 5y = 30$
(10, 0)

7b)

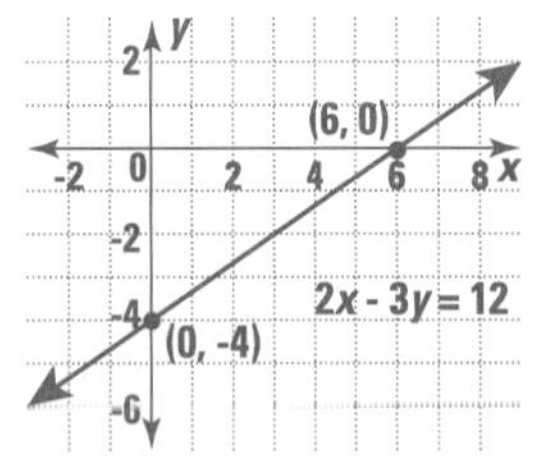

Adapting to Individual Needs

Extra Help

Be sure students understand that, for every linear equation, the x-intercept has a y-coordinate of 0, and the y-intercept has an x-coordinate of 0. Use a coordinate grid (**Teaching Aid 28**). Point to the origin and ask a volunteer to name the coordinates. [(0, 0)] Then mark and label various points on the x-axis, and have students name the coordinates. Do the same for various points on the y-axis. Students should see that going from the origin to any point on the x-axis involves a horizontal move only and that moving from the origin to any point on the y-axis involves a vertical move only.

8. What lines do not have equations in slope-intercept form? vertical lines

9. What lines do not have equations in standard form? Every line has an equation in the standard form.

In 10 and 11, refer to Example 2.

10. Find three different combinations of pounds of roast beef and shrimp Andy could buy. Sample: (4, 6); (6, 5); (8, 4)

11. The store at which Andy usually shops is having a sale. Roast beef costs \$4 a pound and shrimp costs \$10 a pound.

a. Write an equation in standard form to describe the different possible combinations of roast beef and shrimp he can buy for \$96.

b. What is the greatest amount of roast beef he can buy? 24 pounds

c. What is the greatest amount of shrimp he can buy? 9.6 pounds

d. Graph the solutions to the equation in part **a.** See left.

a) $4x + 10y = 96$

11d)

$4x + 10y = 96$; (4, 8); Shrimp (lb); Roast beef (lb)

In 12–14, an equation in slope-intercept form is given.

a. Rewrite the equation in standard form with integer coefficients.

b. Give the values of *A*, *B*, and *C*.

12. $y = \frac{2}{3}x + 12$ $-2x + 3y = 36$; $A = -2$; $B = 3$; $C = 36$

13. $y = 4x$ $-4x + y = 0$; $A = -4$; $B = 1$; $C = 0$

14. $y = -8x - 3$ $8x + y = -3$; $A = 8$; $B = 1$; $C = -3$

Applying the Mathematics

15. A 100-point test has *x* questions worth 2 points apiece and *y* questions worth 4 points apiece.

a. Write an equation in standard form that describes all possible numbers of questions that might be on the test. $2x + 4y = 100$

b. Give three solutions to the equation in part **a.** Samples: (2, 24); (10, 20); (30, 10)

16. Louise has \$36 in five-dollar bills and singles. How many of each kind of bill does she have? a) $5f + s = 36$, where f = number of fives and s = number of singles.

a. Write an equation to describe this situation.

b. Give three solutions. Samples: (0, 36); (1, 31); (2, 26)

c. Graph all possible solutions. (The graph will be discrete.) See left.

16c)

(0, 36); (1, 31); (2, 26); (3, 21); (4, 16); (5, 11); (6, 6); (7, 1)

17. Suppose $Ax + By = C$, with $B \neq 0$.

a. Solve this equation for *y*. $y = -\frac{A}{B}x + \frac{C}{B}$

b. Identify the slope and the *y*-intercept of this line. slope = $-A/B$; *y*-intercept = C/B

18. a. What are the *x*- and *y*-intercepts of the line graphed at the right? 2; 4

b. Write an equation in standard form for the line. $2x + y = 4$

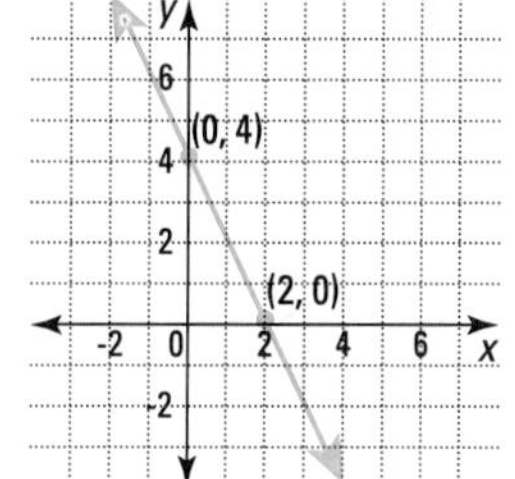

Extension

You might want to discuss the idea of Project 1 on page 476 with the entire class. Explain that any oblique line that does not contain the origin has a unique equation of the form $\frac{x}{a} + \frac{y}{b} = 1$ where *a* and *b* are its *x*-intercept and *y*-intercept, respectively. This is called the *intercept form* for the equation of a line. Ask students to rewrite the equations in **Questions 4–7** and **Questions 12–14** in intercept form.

[**4.** $\frac{x}{\frac{5}{4}} + \frac{y}{\frac{5}{2}} = 1$; **5.** $\frac{x}{2} + \frac{y}{-\frac{1}{4}} = 1$; **6.** $\frac{x}{10} + \frac{y}{6} = 1$; **7.** $\frac{x}{6} + \frac{y}{-4} = 1$; **12.** $\frac{x}{-18} + \frac{y}{12} = 1$; **13.** Cannot be written in this form; **14.** $\frac{x}{-\frac{3}{8}} + \frac{y}{-3} = 1$]

Project Update Project 1, *The Intercept Form of an Equation of a Line*, on page 476, relates to the content of this lesson.

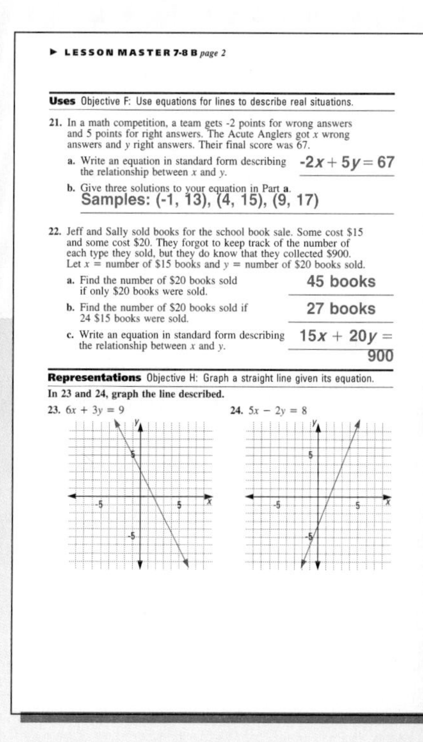

▶ **LESSON MASTER 7-8 B** *page 2*

Uses Objective F: Use equations for lines to describe real situations.

21. In a math competition, a team gets -2 points for wrong answers and 5 points for right answers. The Acute Anglers got *x* wrong answers and *y* right answers. Their final score was 67.

a. Write an equation in standard form describing the relationship between *x* and *y*. $-2x + 5y = 67$

b. Give three solutions to your equation in Part **a**. Samples: (-1, 13), (4, 15), (9, 17)

22. Jeff and Sally sold books for the school book sale. Some cost \$15 and some cost \$20. They forgot to keep track of the number of each type they sold, but they do know that they collected \$900. Let x = number of \$15 books and y = number of \$20 books sold.

a. Find the number of \$20 books sold if only \$20 books were sold. 45 books

b. Find the number of \$20 books sold if 24 \$15 books were sold. 27 books

c. Write an equation in standard form describing the relationship between *x* and *y*. $15x + 20y = 900$

Representations Objective H: Graph a straight line given its equation.

In 23 and 24, graph the line described.

23. $6x + 3y = 9$

24. $5x - 2y = 8$

Adapting to Individual Needs

English Language Development

Materials: Index cards with equations of lines (some in the slope-intercept form and others in standard form)

To check students' understanding of the material in this lesson, you might pair non-English-speaking students with English-speaking students. Pass out the index cards, and have each pair of students study their equation. Then ask one of them to read the equation aloud, tell what form the equation is in, give the slope, and give the *y*-intercept.

Notes on Questions

Question 19 Social Studies Connection Ask students why there are no data for 1916 [World War I] and 1940 and 1944 [World War II]. Some students may know that the United States did not participate in the 1980 Moscow Olympics and that the Soviet Union did not participate in the 1984 Los Angeles Olympics. If all nations had participated, the discus throw might have produced new records that would have followed the linear trend represented by most of the data. Here is a situation in which politics may have affected the data.

Sports Connection The first Olympiad was held in ancient Greece in 776 B.C. It took place in one day and had a single event—a footrace that was the length of the stadium. Eventually, additional races and other events were added, and the games took place over a number of days.

Questions 24–25 These questions prepare students for the next lesson.

Question 27 An automatic grapher is helpful for this question. This question helps to prepare students for the case of parallel lines in the discussion of systems, which they will encounter in a later chapter. A similar problem is given in *Optional Activities* on page 464.

19b) $y = .38x - 686$
d) Sample: The line is not exact, or there may be a threshold distance beyond which it is physically impossible to throw.

21a) No; $\frac{0 - (-2)}{10 - 8} = \frac{2}{2} = 1 \neq 2$

b) Yes; $\frac{18 - (-2)}{18 - 8} = \frac{20}{10} = 2$

22a)

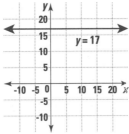

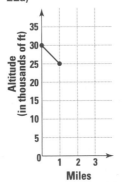

Altitude (in thousands of ft)
35, 30, 25, 20, 15, 10, 5, 0
1 2 3
Miles

23)

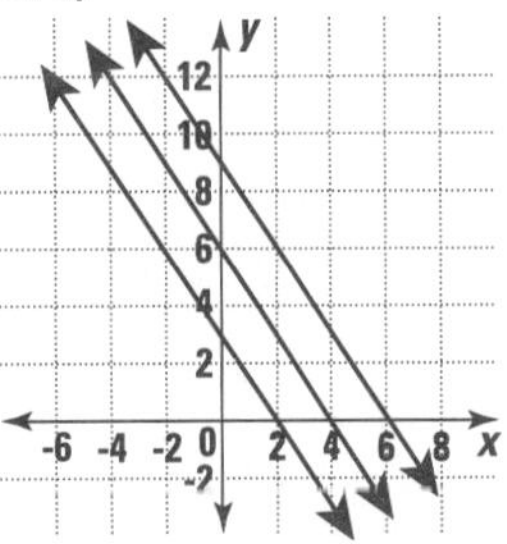

27a)

y
12
10
8
6
4
2
-6 -4 -2 0 2 4 6 8 x
-2

Review

19. The length of the winning toss in the men's discus throw has increased since the first Olympics in 1896. The scatterplot below shows the length of the winning tosses in meters in each Olympic year through 1992. *(Lesson 7-7)*

a. Trace the graph and fit a line to the data.
b. Find an equation for the line in part **a.** See left.
c. Use your equation to predict the length of the winning toss in the year 2000. ≈ 74 meters
d. Why might the prediction for 2000 be incorrect? See left.

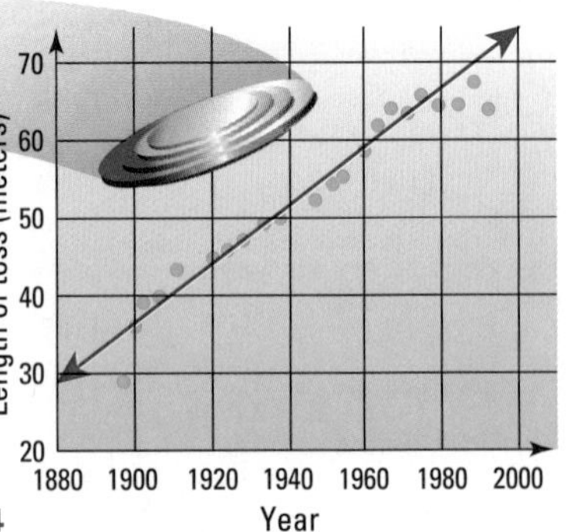

20. Find an equation for the line through the points (-1, 7) and (1, 1). *(Lesson 7-6)* $y = -3x + 4$

21. A robot is moving along a floor which has a coordinate grid. The robot starts at the point (8, -2) and moves along a line with slope 2. Does the robot pass through the following points? Justify your answer.
a. (10, 0) **b.** (18, 18) *(Lessons 7-2, 7-5)* See left.

22. An airplane is flying at an altitude of 30,000 ft. The pilot is instructed to drop to an altitude of 25,000 ft in the next mile relative to the ground. Assume the pilot flies in a straight line.
a. Sketch the path of the plane. See left.
b. What is the rate of change of altitude over horizontal distance in the path? *(Lesson 7-1)* $\frac{-5000 \text{ ft}}{1 \text{ mile}}$

23. Graph on a coordinate plane: $y = 17$. *(Lesson 5-1)* See left.

In 24 and 25, tell whether (0, 0) is a solution to the sentence. *(Lesson 3-10)*

24. $x + y < -4$ No **25.** $6x - y > -6$ Yes

26. Write an inequality to describe this graph. *(Lesson 1-1)* $n > 23$

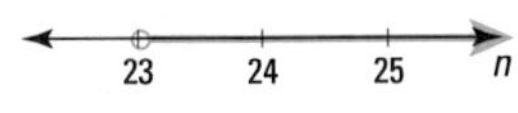

Exploration

27. An automatic grapher may be useful in this question.
a. Graph the lines with equations: See left.

$$3x + 2y = 6$$
$$3x + 2y = 12$$
$$3x + 2y = 18.$$

b. What happens to the graph of $3x + 2y = C$ as C gets larger?
c. Try values of C that are negative. What can you say about the graphs of $3x + 2y = C$ then?

b) The graph remains parallel to the first line, and slides to the upper-right.
c) Sample: Try $3x + 2y = -6$. The graph is still parallel to the first line, but slides to the lower-left.

Adapting to Individual Needs

Challenge

Have students find the slope *m* and the *y*-intercept for the line with equation $Ax + By = C$. Tell them to assume $B \neq 0$ and to give each in terms of *A*, *B*, and *C*. [$m = \frac{A}{B}$ and *y*-intercept $= \frac{C}{B}$]

Setting Up Lesson 7-9

Questions 24 and 25, on page 468, review a simple skill that is utilized in Lesson 7-9. Point out that because (0, 0) does not satisfy the sentence in **Question 24,** it is not on the graph. Since (0, 0) satisfies the sentence in **Question 25**, it is on the graph of $6x - y > -6$.

LESSON

7-9

Graphing Linear Inequalities

Surveying the land. *These two surveyors are using a transit to determine the boundary line between two areas of land in a subdivision. Boundary lines are critical in graphs of linear inequalities.*

You have graphed solutions to inequalities since the very first lesson of Chapter 1. Recall how to graph $n < 3$ on a number line. First find the point where $n = 3$.

-2 -1 0 1 2 3 4 5 n

Next, decide which part of the line contains the solutions to the inequality, that is, which part to shade. The sentence $n < 3$ states that we want values smaller than 3. Shade the points which are to the left of 3.

-2 -1 0 1 2 3 4 5 n

Recall that the 3 is marked with an open circle because 3 doesn't actually make the sentence $n < 3$ true, but it is important. It is the *boundary point* between the values that satisfy $n < 3$ and those that don't. It separates the line into two parts.

These ideas can be extended to graphs of inequalities in two dimensions. In this case, the boundary is a line instead of a point.

Lesson 7-9

Objectives

I Graph linear inequalities.

Resources

From the ***Teacher's Resource File***

- Lesson Master 7-9A or 7-9B
- Answer Master 7-9
- Teaching Aids
  28 Four-Quadrant Graph Paper
  73 Warm-up
  85 Additional Examples
  86 Question 24
- Technology Sourcebook
  Computer Master 18

Additional Resources

- Visuals for Teaching Aids 28, 73, 85, 86

Teaching Lesson 7-9

Warm-up

Graph each equation.

1. $x = 3$
2. $y = -2$
3. $x = 0$
4. $y = -3x + 2$

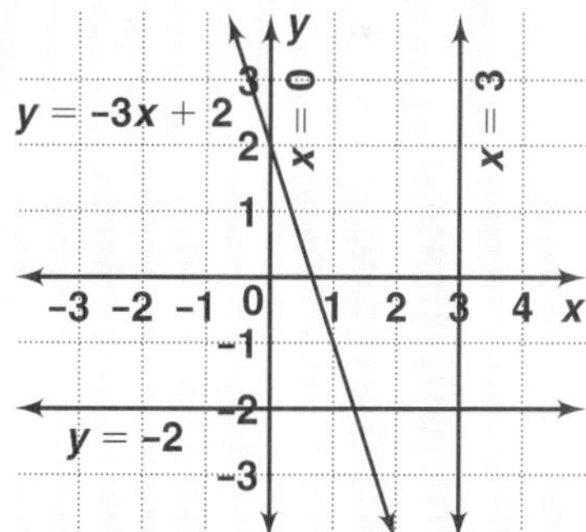

Lesson 7-9 Overview

Broad Goals The lesson presents the graphing of two-dimensional inequalities $Ax + By < C$, where the $<$ sign may be replaced by $\leq$, $>$, or $\geq$. Special cases are the forms $x < h$, $y < k$, and $y < mx + b$.

Perspective Linear inequalities help with the understanding of lines. A student cannot fully understand what it means for a point to be on a line without understanding which points are not on the line.

Geometrically, linear inequalities allow one to easily describe half-planes. Intersections of half-planes are interiors of angles, strips between parallel lines, or interiors of triangles or other polygons. Thus, these sentences are quite useful in describing other figures. Graphing linear inequalities is important in business applications, and it is the foundation of the technique of linear programming.

If an inequality is in slope-intercept form, it is easy to decide which half-plane to shade. If $y > mx + b$, then the y-coordinates of the points shaded should be greater than the y-coordinates of corresponding points on the line $y = mx + b$; thus, you shade the upper half-plane. (**Example 3** illustrates this idea.) If $y < mx + b$, shade the lower half-plane. (**Example 1** could be said to illustrate this idea.)

(Overview continues on page 470.)

Notes on Reading

Note that $y < 3$ could be graphed on a number line as a ray without its endpoint. But in **Example 1,** the graph is a half-plane. There is no way to tell what dimensions the graph has without some indication; for that reason, the direction in **Example 1** indicates "coordinate plane."

Also note that a half-plane does not contain its edge.

As you go through examples and questions, you might switch the verb in a sentence and ask students how that affects the graph. For instance, with **Example 4,** consider $3x - 4y \geq 12$, $3x - 4y < 12$, $3x - 4y \leq 12$, or $3x - 4y = 12$. Once a student has graphed one of these sentences, graphing any of the others should be relatively easy.

Reading Mathematics Emphasize the paragraph before **Example 3** and the paragraphs preceding and following **Example 4**. These are the general principles that students should learn in this lesson. You might want to have students read this material aloud in class.

Inequalities Involving Horizontal or Vertical Lines

Example 1

Graph $y < 3$ on a coordinate plane.

Solution

Graph the line $y = 3$. This horizontal line is the boundary of the solution set. The line is dashed to show that the points having a y-coordinate of 3 are not included in the solution set. The solution set consists of all points having a y-coordinate less than 3. This is the region below the boundary line.

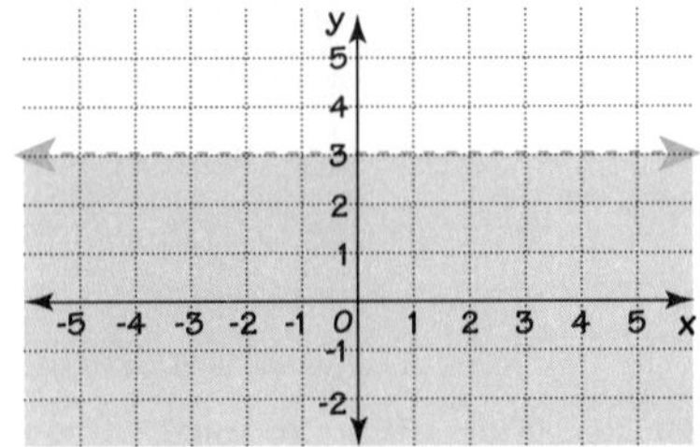

Check

Pick a point in the shaded region. We choose (0, 0). Do the coordinates of this point satisfy the inequality? Yes. $y = 0$ and $0 < 3$.

The regions on either side of a line in a plane are called **half-planes.** The boundary line is the edge of the half-plane. In Example 1, the line $y = 3$ is the edge of the half-plane $y < 3$. If Example 1 had asked you to graph $y \leq 3$, then the edge $y = 3$ would be included and shown as a solid line.

Example 2

Give a sentence describing all points in each region.

a.

b. 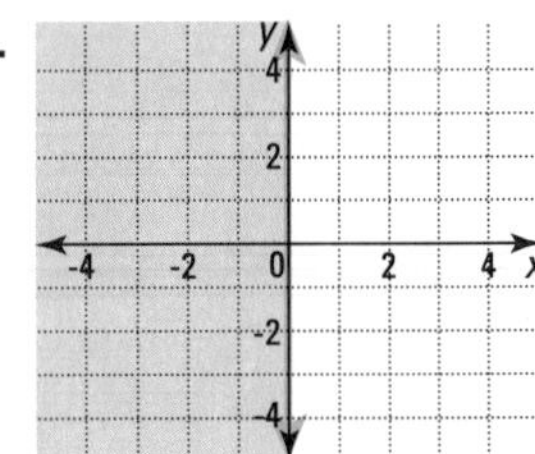

Solution

a. Every point in the half-plane has a y-coordinate greater than -2. The dotted line shows that points with a y-coordinate equal to -2 are not included. So a sentence describing this half-plane is $y > -2$.

b. Every point to the left of the y-axis has a negative x-coordinate. The solid line indicates that all points with x-coordinate equal to 0 are also to be included. So a sentence describing the region is $x \leq 0$. This region is the union of a half plane and its edge.

Lesson 7-9 Overview, continued

The reasoning in the previous paragraph seems natural to adults, but the concept is a new one to students; encourage your students to check points by substituting. Checking is certainly necessary if the line is in standard form, and is good verification of the fundamental idea of a solution to a sentence: an ordered pair is in the solution set of the sentence if and only if the corresponding point lies on the graph.

Inequalities in standard form are a little trickier to graph, and students should definitely check using the idea in **Example 4.**

Example 5 shows some of the utility of being able to graph linear inequalities and is an important precursor to the linear programming that students should encounter in their second course in algebra.

Optional Activities

Activity 1 Technology Connection
With *Technology Sourcebook, Computer Master 18,* students use *GraphExplorer* or similar software to graph linear inequalities.

Inequalities Involving Oblique Lines

Every oblique line has an equation of the form $y = mx + b$. It is the boundary of two half-planes, described by $y < mx + b$ and $y > mx + b$.

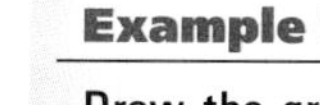

Example 3

Draw the graph of $y \geq -3x + 2$.

Solution

Start by graphing the boundary line $y = -3x + 2$ using techniques you have learned.

The $\geq$ sign in the inequality $y \geq -3x + 2$ indicates that points are desired whose y-coordinate is greater than or equal to the y values which satisfy $y = -3x + 2$. Since y values get larger as one goes higher, shade the entire region above the line.

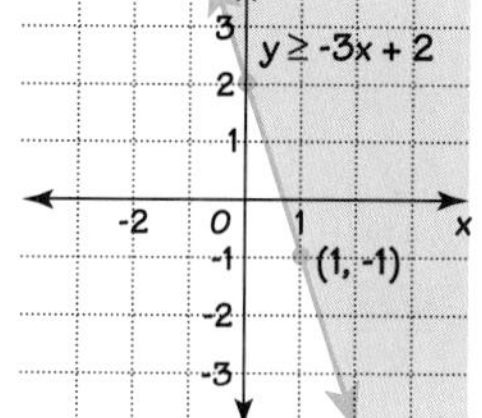

Check

Try the point (0, 3), which is in the shaded region. Does it satisfy the inequality? Is $3 > -3 \cdot 0 + 2$? Yes, 3 is greater than 2. So the correct side of the line has been shaded.

When an inequality is written in slope-intercept form, it is easy to tell which half-plane of the boundary line to shade. For $y < mx + b$ shade below; for $y > mx + b$ shade above. But when an inequality is in the standard form $Ax + By > C$, you cannot use the inequality sign to determine which side of the line to shade. A more direct method, the testing of a point, is usually used. The point (0, 0) is usually chosen if it is not on the boundary line. If (0, 0) is a solution to the inequality, then the half-plane that contains it is shaded. If (0, 0) does not satisfy the inequality, shade the other side of the boundary line.

Example 4

Graph $3x - 4y > 12$.

Solution

First graph the boundary line $3x - 4y = 12$. This line is dashed because it is not part of the solution. To determine which side of the line is the solution set, substitute (0, 0) into the original equation. Is $3 \cdot 0 - 4 \cdot 0 > 12$? No. Since (0, 0) is in the upper half-plane and is *not* a solution, shade the lower half-plane. The graph of $3x - 4y > 12$ is shown at the right.

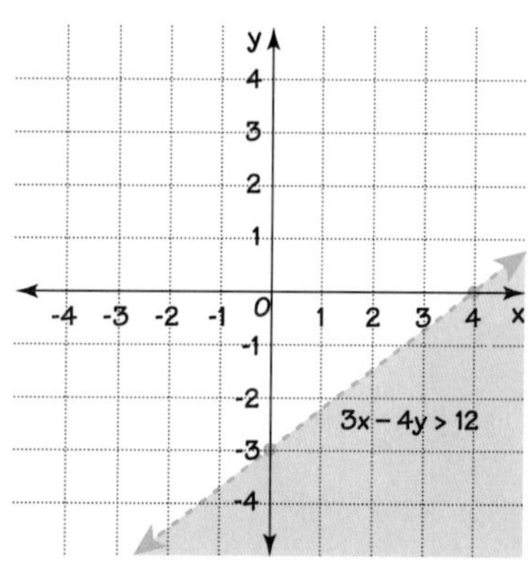

Additional Examples

These examples are also given on **Teaching Aid 85.** Students will need graph paper or **Teaching Aid 28** for these examples.

1. Graph all ordered pairs (x, y) that satisfy $y \geq 4$.

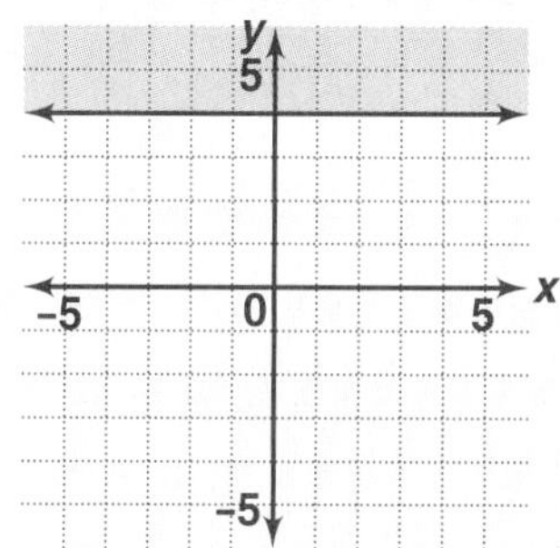

2. Give a sentence describing all points in each shaded region.
   a. $y < 3$; **b.** $x \geq 1$

a.

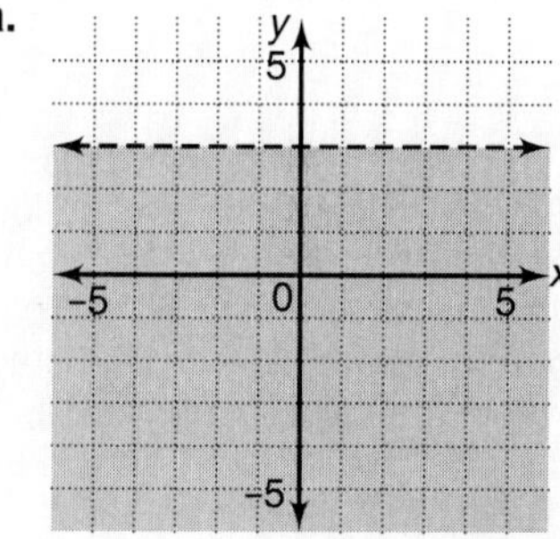

b.

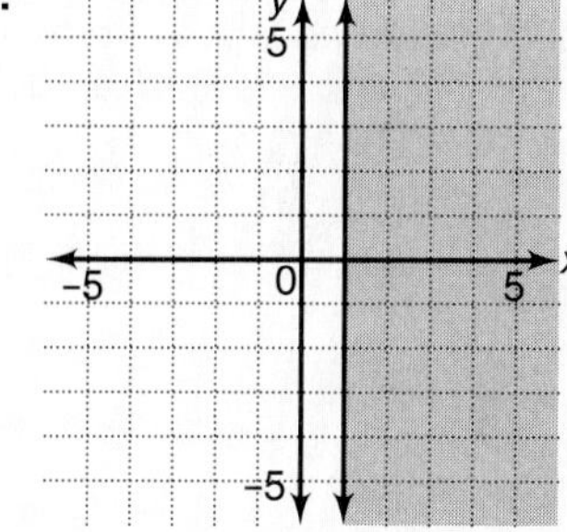

(Additional Examples continue on page 472.)

Optional Activities

Activity 2

Materials: Graph paper or **Teaching Aid 28**

As an extension of the lesson, have students solve the following problem by drawing a graph. You have \$60 to buy tickets to a ball game. Adult tickets cost \$6 each, and student tickets cost \$4 each.

1. If you spend all the money on tickets, how many combinations of adult and student tickets are possible? [6]
2. If you spend less than \$60, how many combinations are possible, assuming that you buy at least one ticket? [84]

In the graph at the right, the solid dots show combinations of adult and student tickets that cost exactly \$60. The open dots show combinations that include at least one ticket and cost less than \$60.

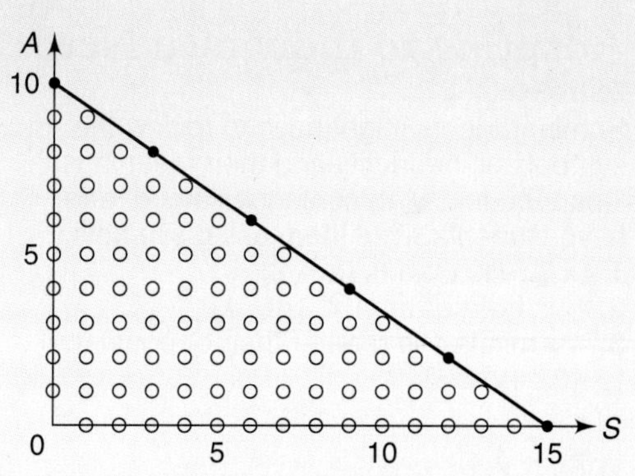

3. Graph $y < 3x - 4$.

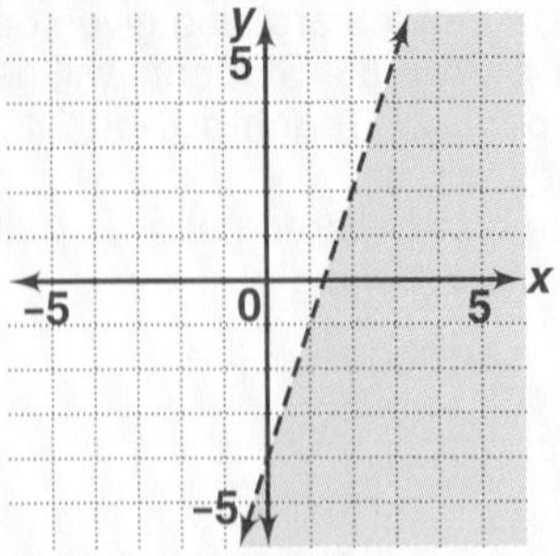

4. Graph $2x + 5y \geq 10$.

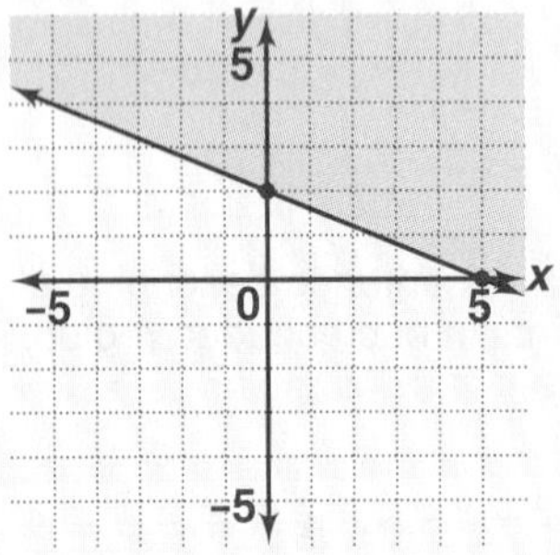

5. Suppose crepe paper costs \$2.00 per package, and balloons are \$1.50 per pack. The decorations committee for the dance bought some of each and stayed within their \$60 budget. Graph the possible number of packages of each that they could have bought. Let b be the number of packs of balloons and c be the number of packages of crepe paper. **$1.50b + 2.00c \leq 60.00$, or $1.5b + 2c \leq 60$**

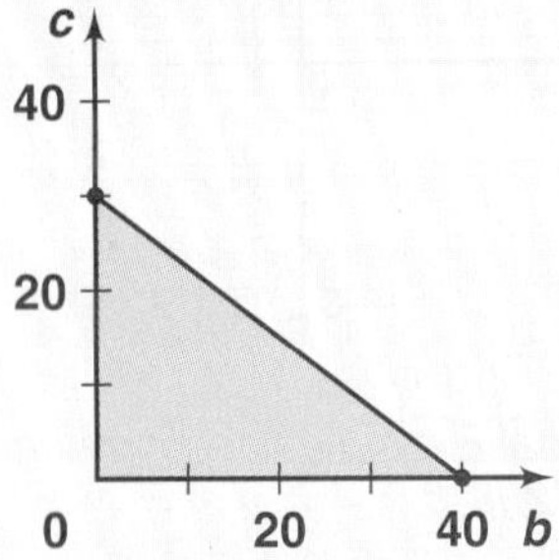

Watching your money grow. *Some "piggy banks" are made of transparent plastic. That makes it easy to see how much money you have.*

1b)

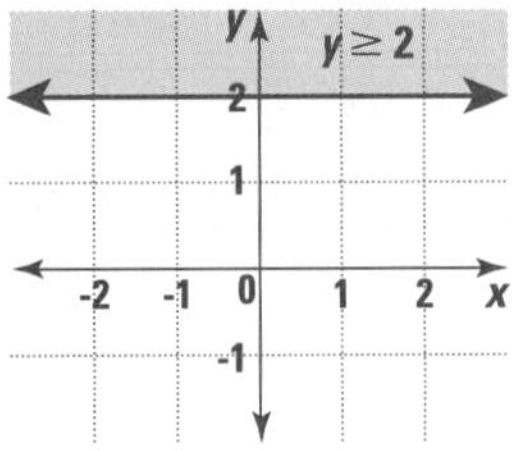

2b)

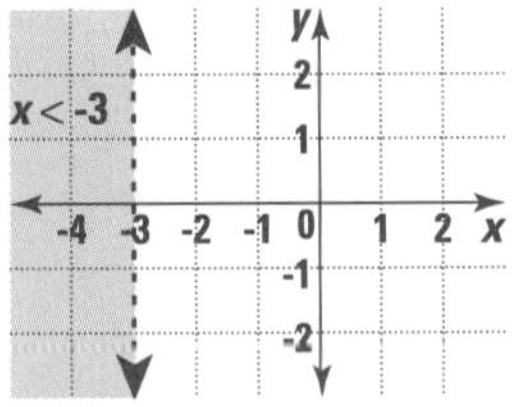

Sentences equivalent to $Ax + By < C$ or $Ax + By \leq C$ are called **linear inequalities.** The preceding examples show that there are two steps to graphing linear inequalities.

Step 1: Graph a dashed or solid line as the solution to the corresponding linear equation.

Step 2: Shade the half-plane that makes the inequality true. (You may have to test a point. If possible, use (0, 0).)

Sometimes the graph of all solutions to a linear inequality is not an entire half-plane.

Example 5

Suppose you have less than \$5.00 in nickels and dimes. Find an inequality and sketch a graph to describe how many of each coin you might have.

Solution

Let n = the number of nickels you have.
Let d = the number of dimes you have.
Since each nickel is worth .05 and each dime is worth .10,

$$0.05n + 0.10d < 5.00.$$

Multiply both sides by 100 to clear the decimals.

$$5n + 10d < 500$$

This inequality is graphed at the left.

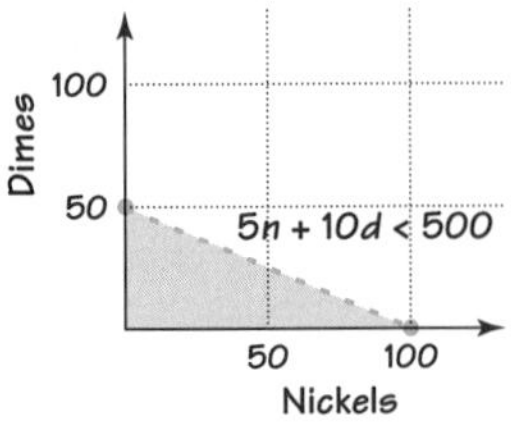

Since n and d cannot be negative, only the points in the first quadrant or on the axes are shaded. The graph is actually discrete since n and d must be integers, but there are so many points that shading might be easier. The graph thus shows that there are very many solutions.

Check

The point (0, 0) is in the shaded region. If you have 0 nickels and 0 dimes, do you have less than \$5.00? Yes.

QUESTIONS

Covering the Reading

In 1 and 2, a sentence is given. **1b, 2b) see left.**

a. Graph the set of points satisfying the sentence on a number line.

b. Graph the points satisfying the sentence on a coordinate plane.

1. $y \geq 2$

1a) 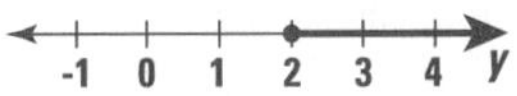

2. $x < -3$

2a) 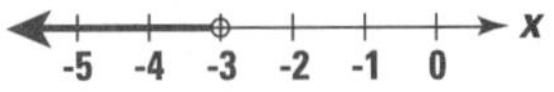

Adapting to Individual Needs

Extra Help

Some students might have to review the symbols of inequality and their meanings. Read the following sentences aloud, and have students write them using symbols.

1. x is less than 8. [$x < 8$]
2. y is greater than x. [$y > x$]
3. $2x$ minus y is greater than or equal to 4. [$2x - y \geq 4$]
4. y is less than or equal to $4x$ plus 2. [$y \leq 4x + 2$]

Some students reverse the < and > symbols. Encourage these students to check their work by reminding them that the symbol points to the smaller quantity and opens to the larger quantity.

English Language Development

Before students read this lesson, you might want to make sure they understand these terms: *boundary, boundary point, plane, half-plane, edge, horizontal line, vertical line, oblique line, linear equation,* and *linear inequality.*

8) 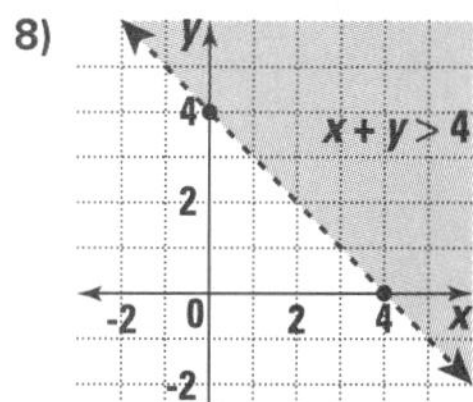

9) 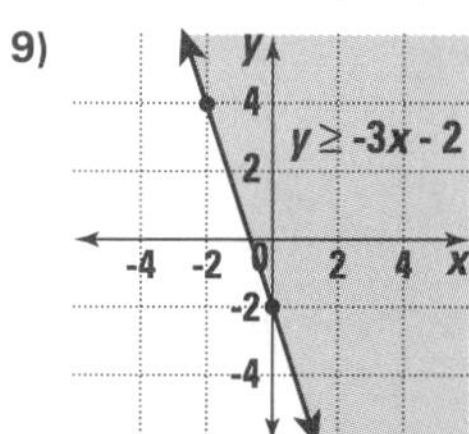

10) 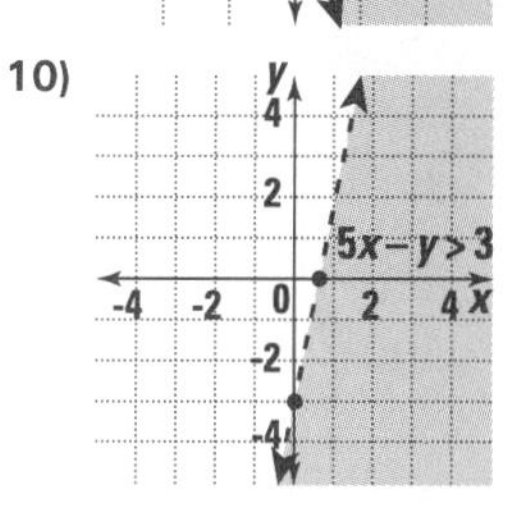

In 3 and 4, write an inequality describing each graph.

3. 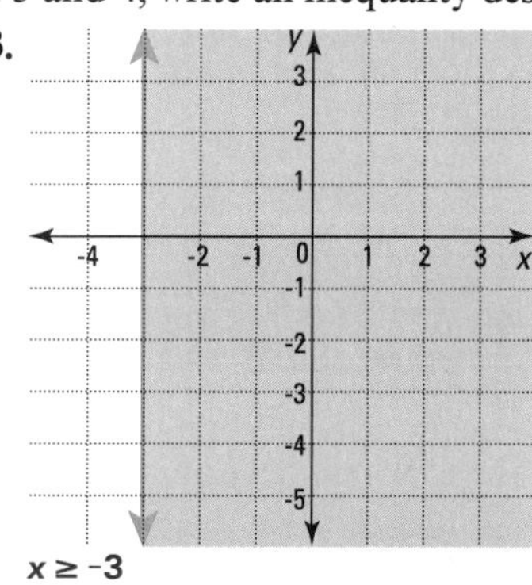
$x \geq -3$

4. 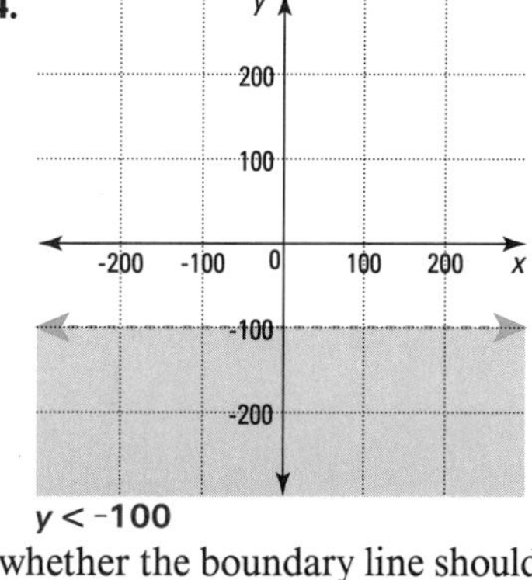
$y < -100$

In 5 and 6, an inequality is given. Tell whether the boundary line should be drawn solid or dashed. (You do not need to draw the graph.)

5. $y > -4x - 7$ **dashed**

6. $4x + y < -7$ **dashed**

7. a. What point is usually chosen to test which half-plane is the solution set? **(0, 0)**

b. When would you not choose that point? **when it lies on the boundary line**

In 8–11, graph all points (x, y) that satisfy the inequality.

8. $x + y > 4$

9. $y \geq -3x - 2$

10. $5x - y > 3$ **8–10) See left.**

11. $5x - y \leq 3$ **See margin.**

12. Suppose a person has less than \$4.00 all in quarters or dimes. Let x = the number of quarters and y = the number of dimes the person has. **See margin.**

a. Write an inequality to describe this situation.

b. Graph the possible numbers of quarters and dimes the person might have.

Applying the Mathematics

13. "It will take at least 20 points to make the playoffs," the hockey team coach told the players. "We get 2 points for a win and 1 for a tie." Let W be the number of wins and T the number of ties.

a. Write an inequality to describe the values of W and T that will enable the team to make the playoffs. $2W + T \geq 20$

b. Graph these values. **See margin.**

Stanley Cup champs. *Shown is Mark Messier of the New York Rangers holding the Stanley Cup trophy after his team won the championship in 1994. This was the Rangers' first championship in 54 years.*

14. Suppose CDs cost \$10 and tapes cost \$8. Miguel has \$40 to spend. Let x = the number of CDs, and y = the number of tapes Miguel buys.

a. Write an inequality to describe the number of CDs and tapes Miguel can buy. $10x + 8y \leq 40$

b. Graph the solution set to the inequality. **See margin.**

c. Which points on your graph represent ways Miguel can buy CDs and still have money left?
(1, 0); (1, 1); (1, 2); (1, 3); (2, 0); (2, 1); (2, 2); (3, 0); (3, 1)

Notes on Questions

Students will need graph paper or **Teaching Aid 28** for many of the questions.

Question 13 The situation has a limited domain, but there are enough solutions to make graphing worthwhile.

Question 14 In this instance, it may be easier to list the solutions than to graph them.

Additional Answers, page 473

11. 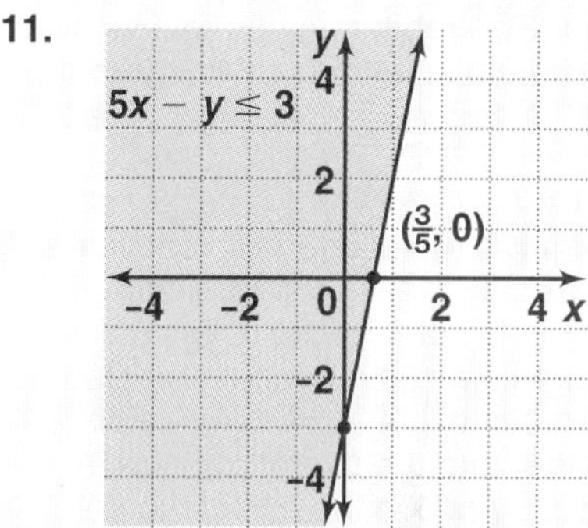

12a. $.25x + .1y < 4$

b. 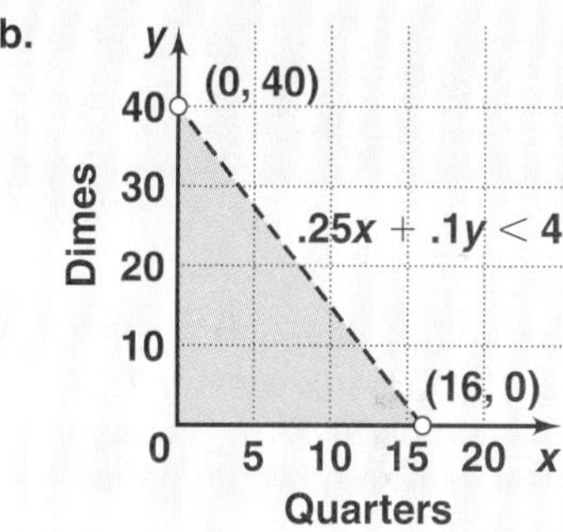

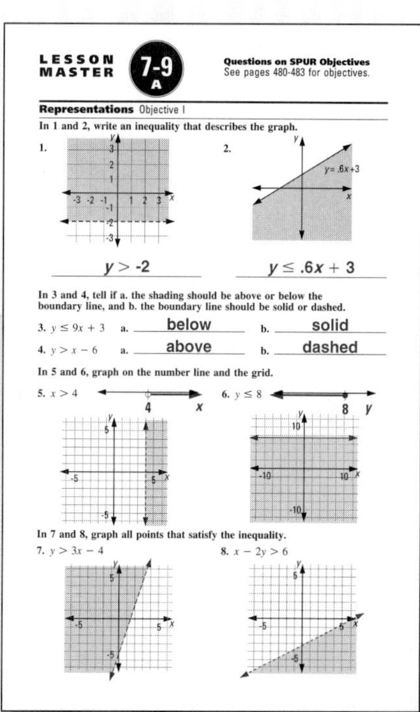

LESSON MASTER 7-9 A

Questions on SPUR Objectives
See pages 480-483 for objectives.

Representations Objective I

In 1 and 2, write an inequality that describes the graph.

1. $y > -2$

2. $y \leq .6x + 3$

In 3 and 4, tell if a. the shading should be above or below the boundary line, and b. the boundary line should be solid or dashed.

3. $y \leq 9x + 3$ a. below b. solid

4. $y > x - 6$ a. above b. dashed

In 5 and 6, graph on the number line and the grid.

5. $x > 4$

6. $y \leq 8$

In 7 and 8, graph all points that satisfy the inequality.

7. $y > 3x - 4$

8. $x - 2y > 6$

Additional Answers, continued

13a. $2W + T \geq 20$

b. 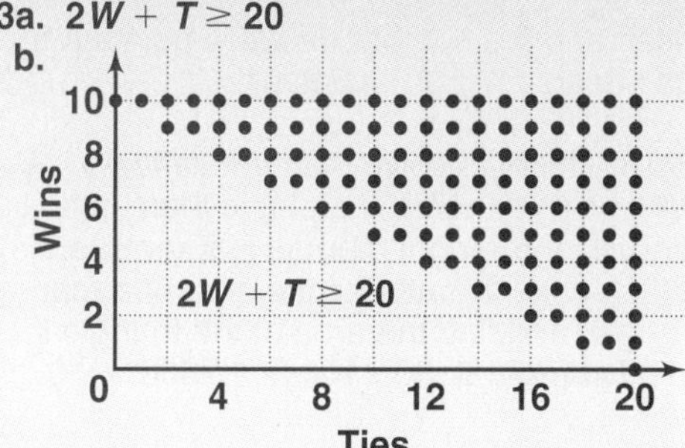

14b. 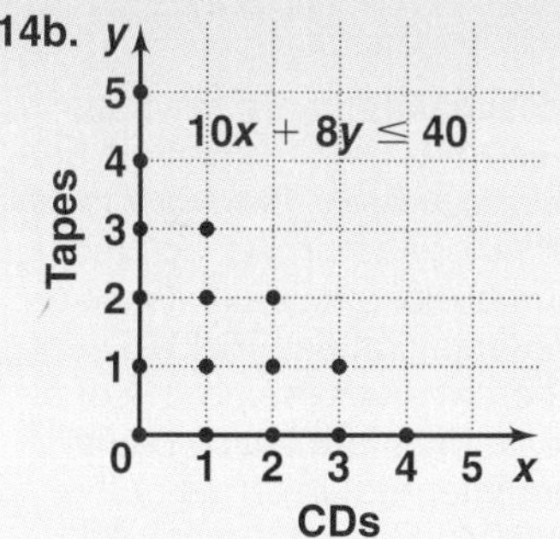

Notes on Questions

Question 24 This diagram is given on **Teaching Aid 86.**

Literature Connection Students who like science fiction are familiar with the many portrayals of robots. These students are probably familiar with the American author Isaac Asimov who introduced the "Three Laws of Robotics": (1) A robot may not injure a human being, or through inaction allow a human to be harmed; (2) A robot must obey orders given by humans except when it conflicts with the first law; and (3) A robot must protect its own existence unless it conflicts with the first or second law. George Lucas's robots, C3PO and R2D2, in the three *Star Wars* movies exemplify Asimov's moral code for robots. Ask students to tell the class about the robots in some of the science fiction stories they have read. Interested students might want to research the history and the use of robotics in industry and in the exploration of space and the ocean floor.

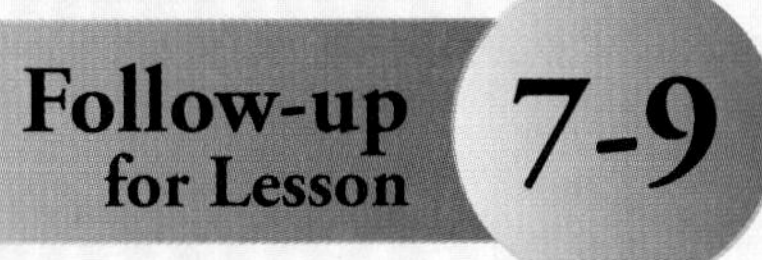

Practice

For more questions on SPUR Objectives, use **Lesson Master 7-9A** (shown on page 473) or **Lesson Master 7-9B** (shown on pages 474–475).

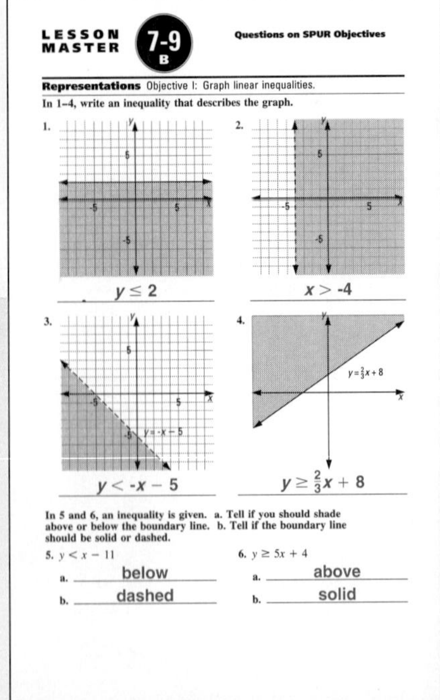

LESSON MASTER 7-9 B Questions on SPUR Objectives

Representations Objective I: Graph linear inequalities.

In 1–4, write an inequality that describes the graph.

1. $y \le 2$
2. $x > -4$
3. $y < -x - 5$
4. $y \ge \frac{2}{3}x + 8$

In 5 and 6, an inequality is given. a. Tell if you should shade above or below the boundary line. b. Tell if the boundary line should be solid or dashed.

5. $y < x - 11$
   a. below
   b. dashed
6. $y \ge 5x + 4$
   a. above
   b. solid

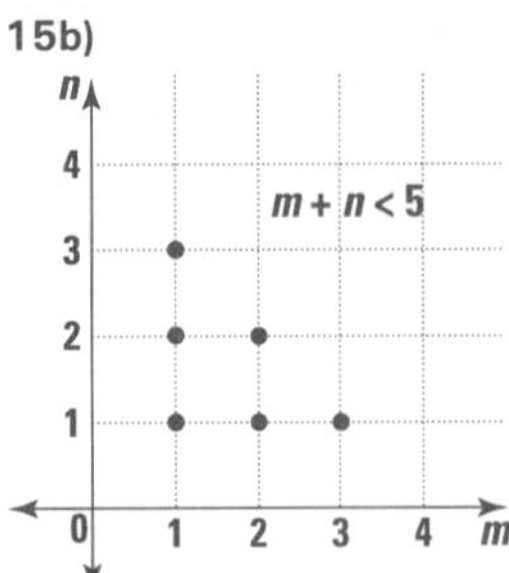

15. Suppose m and n are positive integers.
 a. How many points (m, n) satisfy $m + n < 5$? **6**
 b. Graph them all. Plot m on the x-axis and n on the y-axis. **See left.**

Review

16. a. Rewrite $4x - 28 = 3y$ in standard form. $4x - 3y = 28$
 b. Give the values of A, B, and C. *(Lesson 7-8)* $A = 4$; $B = -3$; $C = 28$

17. The table below gives latitude and the average annual snowfall of some cities in the U.S. that have an average of at least 1" of snow per year.

City	Latitude (°North)	Average Annual Snowfall (in.)
New York, NY	41	29
Chicago, IL	42	40
Philadelphia, PA	40	20
Boston, MA	42	42
Louisville, KY	38	17
Charlotte, NC	35	6
Dallas, TX	33	3

a, b) See left.

 a. In general, how is the average annual snowfall of a city related to its latitude?
 b. Find an equation of a line that seems to fit these data.
 c. Use the equation to predict the average annual snowfall of a U.S. city at 36° north latitude. **Sample: 8 inches**
 d. Nashville, Tennessee, and Las Vegas, Nevada, are both at 36° north latitude. Nashville has about 11" of snow per year and Las Vegas only 1". What factors other than latitude affect the amount of snowfall a city receives? *(Lesson 7-7)* **Sample: altitude, geographic position**

17a) Sample: Average annual snowfall is linearly related to latitude.
b) Sample: $y = 5x - 172$

In 18 and 19, consider the rectangular field pictured below.

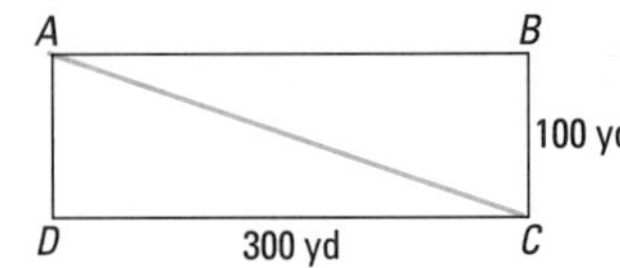

18. How much shorter is the distance from A to C if you walk diagonally across the field instead of along its sides? *(Lesson 1-8)* **≈ 83.8 yds**

19. If D is the origin of a coordinate system with y-axis on $\overleftrightarrow{AD}$ and the x-axis on $\overleftrightarrow{DC}$, what is the slope of $\overleftrightarrow{AC}$? *(Lesson 7-2)* $-\frac{1}{3}$

Adapting to Individual Needs

Challenge

Example 5 notes that there are too many possibilities to list. You might explain to students that, more accurately, there are too many possibilities for a *short* list. To calculate the total number of lattice points inside the triangle, students can use Pick's Theorem. Pick's Theorem gives the area A of a polygon whose vertices are lattice points (coordinates are integers) in terms of the number B of lattice points on the polygon itself and the number of lattice points N in its interior: $A = N + B/2 - 1$.

Have students solve the formula for N. [$N = A + 1 - B/2$] Then have them find B and N for the right triangle in **Example 5**.

1. Use the formula for the area of a triangle, and find the area of the triangle in **Example 5**: $50 \cdot 100/2 = 2500$.

In 20 and 21, use the graph below.

20. Of the 10,886,000 males aged 25–29 in 1990, how many were single? *(Lesson 5-4)* **4,920,472 men**

21. a. What was the average yearly increase in the percentage of females aged 20–24 staying single from 1970 to 1980? **1.44%**

b. Answer part **a** for the decade from 1980 to 1990. **1.26%**

c. Is the rate of increase constant from 1970 to 1990? *(Lesson 7-1)* **No**

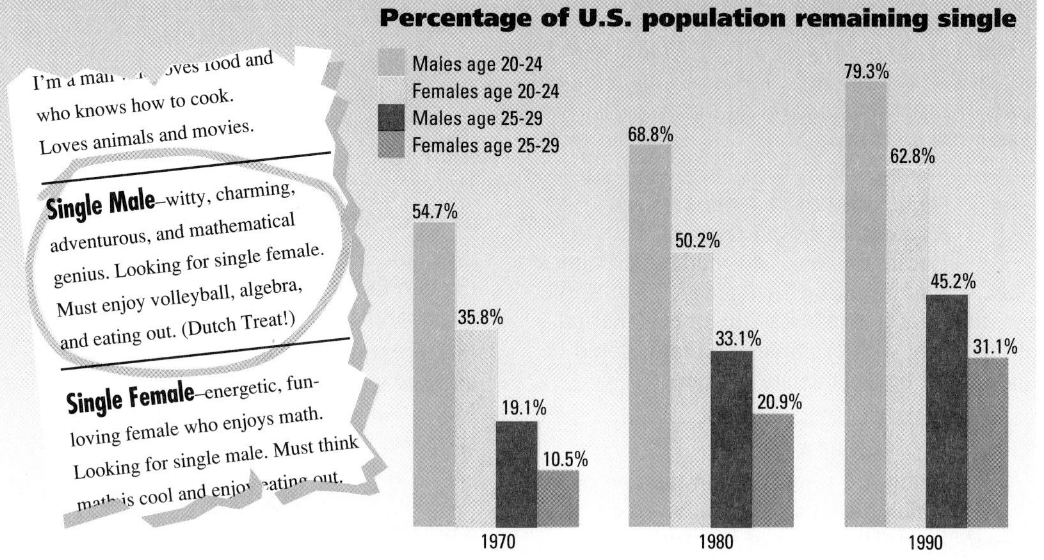

22. Evaluate the following. *(Lesson 2-5)*

a. 5^3 **125** **b.** $(-3)^5$ **-243** **c.** $(-2)^{10}$ **1024**

23. Rewrite in decimal form. *(Previous course, Appendix B)*

a. 10^{-1} **0.1** **b.** $3 \cdot 10^{-2}$ **.03** **c.** $2 \cdot 5 \cdot 10^{-6}$ **.00001**

Exploration

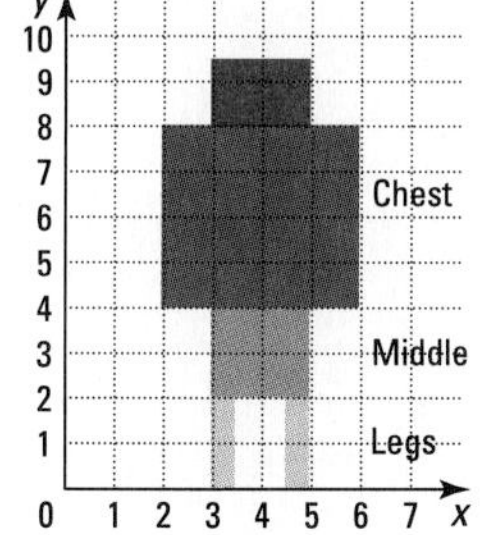

24. The robot at the left is described by the following inequalities:

Chest:	$2 \le x \le 6$	and	$4 \le y \le 8$.
Middle:	$3 \le x \le 5$	and	$2 \le y \le 4$.
Legs:	$3 \le x \le 3.5$	and	$0 \le y \le 2$.
	$4.5 \le x \le 5$	and	$0 \le y \le 2$.

Copy the diagram and put a head on the robot. Describe the head with inequalities. **Sample: $3 \le x \le 5$ and $8 \le y \le 9.5$**

Assessment

Oral Communication Ask students why they think the point (0, 0) is often used as the point that is tested to decide which half-plane to shade when graphing an inequality. [Students recognize that it is easy to tell whether the inequality is true or false when (0, 0) is used.]

Extension

In **Example 2** on page 470, the graph of $y > -2$ is shown at the left, and the graph of $x \le 0$ is shown at the right. Have students use one coordinate plane and show the graph of $y > -2$ *and* $x \le 0$. [The region above the graph of the line $y = -2$ and to the left of and including the graph of the line $x = 0$.] You might give students other similar problems, such as having them graph the points (x, y) for which $x + y > 5$ *and* $x - y < 8$. [The region between the graphs of the lines $x + y = 5$ and $x - y = 8$]

Project Update Project 4, *Describing Figures with Coordinates*, on page 477, relates to the content of this lesson.

2. Find B, the number of points on the boundary of the triangle: there are 51 points on the y-axis, 100 more on the x-axis, and 49 more on the hypotenuse. Thus $B = 200$.

3. Now find N: $N = 2500 + 1 - 200/2 = 2401$.

You might also have students use Pick's Theorem to find the answer for Question 2 in *Optional Activities* on page 471. [The area of the triangle is 75 and B is 30 (11 points on the y-axis, 15 more on the x-axis, and 4 more on the hypotenuse). Now N can be computed from Pick's Theorem: $N = 75 + 1 - 30/2 = 61$. Note that 61 does not include the 14 points on the x-axis and the 9 points on the y-axis that are part of the solution. Thus the answer to Question 2 is $61 + 14 + 9 = 84$.]

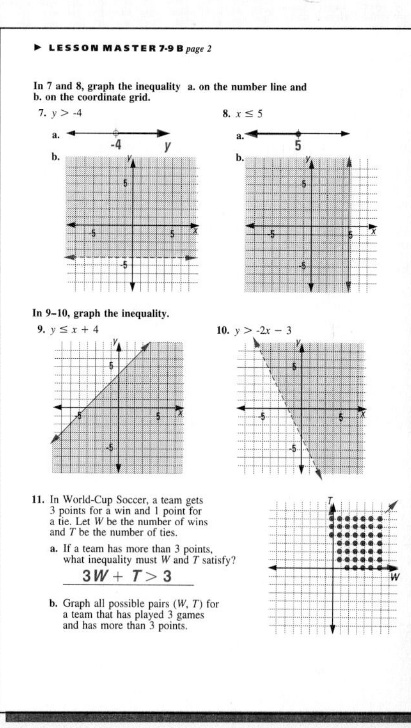
▸ LESSON MASTER 7-9 B *page 2*

In 7 and 8, graph the inequality **a.** on the number line and **b.** on the coordinate grid.

7. $y > -4$

8. $x \le 5$

In 9–10, graph the inequality.

9. $y \le x + 4$

10. $y > -2x - 3$

11. In World-Cup Soccer, a team gets 3 points for a win and 1 point for a tie. Let W be the number of wins and T be the number of ties.

a. If a team has more than 3 points, what inequality must W and T satisfy? $3W + T > 3$

b. Graph all possible pairs (W, T) for a team that has played 3 games and has more than 3 points.

Chapter 7 Projects

The projects relate to the content of the lessons of this chapter as follows:

Project	Lesson(s)
1	7-4, 7-5, 7-6, 7-8
2	7-7
3	7-7
4	7-7, 7-9
5	7-1
6	7-7

1 The Intercept Form of an Equation of a Line Once students have found the equation $\frac{x}{a} + \frac{y}{b} = 1$, make sure they understand that $(a, 0)$ is the x-intercept and that $(0, b)$ is the y-intercept. The line can easily be graphed using this information.

2 Marriage Age One source of information for this project can be found in the *Information Please Almanac.* This project is important in that it provides students with the opportunity to see how data can be used to support a point of view, prediction, or conclusion. When students use the marriage data from 1950 to 1990, it seems clear that the median age at marriage is increasing. However, when earlier data are used, the picture becomes much less clear.

3 Foot Length and Shoe Size Stress the importance of taking accurate measurements in order to start with reliable data. Students may be surprised that the difference in shoe size is determined by a rather small change in foot length. In fact, one sixth of an inch is the actual difference between two consecutive half-sizes for both men and women. Students may be interested in knowing that metric shoe sizes are based on increments of $\frac{2}{3}$ of a centimeter

A project presents an opportunity for you to extend your knowledge of a topic related to the material of this chapter. You should allow more time for a project than you do for typical homework questions.

1 The Intercept Form of an Equation of a Line

Locate an advanced mathematics book that discusses the *intercept form* of the equation of a line. Then write an explanation of this form, with examples, so that it could be understood by a classmate of yours.

2 Marriage Age

Below are some data on the ages at which men and women have first married in the past 40 years.

Median Age (in years) at First Marriage

Year	Men	Women
1950	22.8	20.3
1960	22.8	20.3
1970	23.2	20.8
1980	24.7	22.0
1990	26.1	23.9

a. Plot all the data on one graph. Describe some trends in your graph.

b. Use an automatic grapher to find the lines that best fit the data. Make some predictions based on these.

c. Find some earlier data (for the years 1900, 1910, . . . , 1940) on age of first marriage. Do the earlier data suggest any ways you might modify the predictions you made?

3 Foot Length and Shoe Size

Find ten males and ten females who agree to have you measure their feet.

a. For each person measure each foot (without shoes) to the nearest $\frac{1}{8}$ of an inch, and record the person's shoe size. If the two feet are different lengths, take the larger of the two measurements. For instance, one woman's feet are $9\frac{5}{8}''$ long, and she wears a size 7 shoe.

b. Make two graphs, one showing how men's shoe size varies with foot length, the other showing the relation between women's shoe size and foot length. In each graph, plot foot length on the x-axis.

c. Are the graphs approximately linear? If so, find equations for them. Compare them to the equations given in Lesson 4-4.

Possible responses

1. Students should explain that if a and b are not zero, the line with x-intercept a and y-intercept b has equation $\frac{x}{a} + \frac{y}{b} = 1$. Samples:
   a. For x-intercept 3 and y-intercept -1, the equation is $\frac{x}{3} + \frac{y}{-1} = 1$. Multiplying both sides by 3 gives the equation in standard form: $x - 3y = 3$.
   b. For x-intercept $\frac{1}{2}$ and y-intercept $\frac{3}{4}$, the equation is $\frac{x}{\frac{1}{2}} + \frac{y}{\frac{3}{4}} = 1$. To put this equation in standard form, clear fractions in the denominators by multiplying both sides by the least common denominator, $\frac{3}{8}$: $\frac{3}{4}x + \frac{1}{2}y = \frac{3}{8}$. Then, to get integer coefficients, multiply both sides of the equation by the least common denominator, 8: $6x + 4y = 3$.

   Students might note that the intercept form of an equation provides a convenient way to the find the equation of a line, given the graph of the line.

4 Describing Figures with Coordinates

In the Exploration for Lesson 7-9, you were given a drawing of a robot in a coordinate plane. Design your own figure in the coordinate plane. Write a set of clear instructions using equations and inequalities to graph that figure. Ask a friend to use your instructions to reproduce the figure you designed.

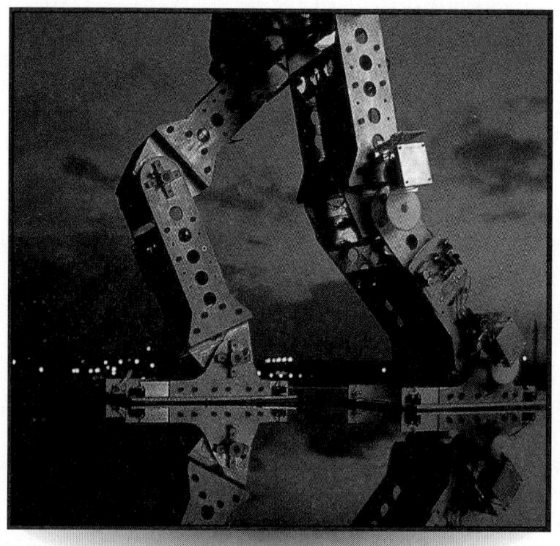

5 Populations of Cities

The population of Manhattan grew at a great pace from 1790 to 1910. Since that time, it has tended to decline. In 1990, its population was only about $\frac{2}{3}$ what it was in 1910. Has this period of great growth followed by a period of sizeable decline happened in other large cities in the U.S.? Examine the census data for the largest cities to see whether Manhattan is unusual, or whether this is a common pattern. Write a summary of what you find.

6 Paper Towels: Price vs. Absorbency

Are more expensive paper towels more absorbent than less expensive ones? Get samples of about six different kinds of paper towels, record their prices, and calculate the price paid per towel. Perform the following experiment to measure absorbency. Fold one piece of towel in half vertically, and then in half again horizontally.

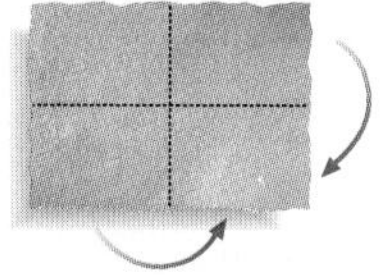

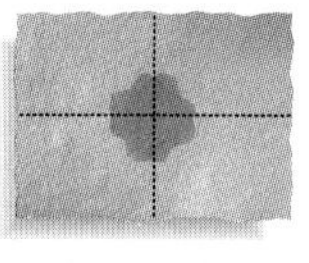

Fill an eyedropper with a fixed amount of water and drop it on the corner with the folds. Open the towel and record and measure the diameter of the circular area that is wet. Repeat for each type of towel you have, making sure you use the same amount of water each time.

a. Plot your data with unit price on the horizontal axis and diameter on the vertical axis. Is the relation linear? If so, find a line to describe your data.

b. Plot your data with unit price on the horizontal axis and the area of the wet region on the vertical axis. If the relation is linear, describe it with a line.

c. What advice would you give someone shopping for paper towels?

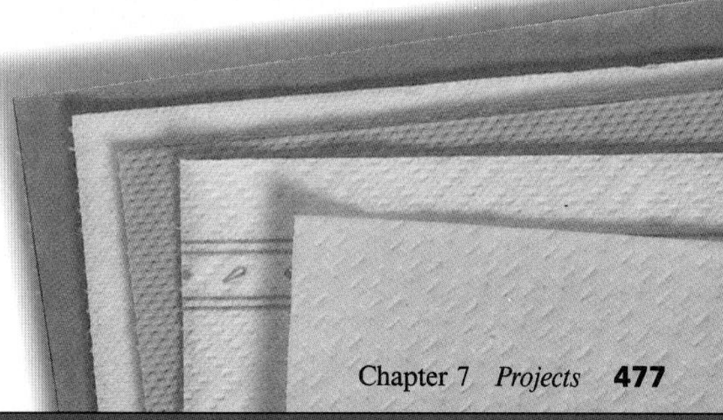

and this applies to both men's and women's shoes.

4 Describing Figures with Coordinates Encourage students to be creative when they design their figures. This project leads to a discussion about similar figures if students graph the same figure on coordinate grids with proportional scales.

5 Populations of Cities A good source for information needed for this project is the *World Almanac and Book of Facts.* Suggest that students find census data for cities in various areas of the United States. From 1980 to 1990, of the ten most populous cities, Chicago, Philadelphia, and Detroit showed a decrease in population and New York, Los Angeles, Houston, San Diego, Dallas, Phoenix, and San Antonio showed an increase. Students may wish to compare these trends with the population trends in their own geographic area.

6 Paper Towels: Price vs. Absorbency Explain to students that the paper towel with the greatest absorbency will have the smallest area that is wet. Have students put the same amount of water on each towel and then outline the wet area. Have students use the regular price of paper towels for this project. However, sale prices should be considered in **part c** when students are asked to advise someone buying paper towels. Factors other than cost, such as the use of recycled paper or brand loyalty, might be included in students' advice, as well.

2a. See the part of the sample graph from 1950 to 1990. The mean age at which men and women marry has increased from 1970 to 1990. The difference between the ages at which men and women marry seems to have stayed relatively constant, at about 2.5 years.

b. Estimate of lines of best fit:
Men: $y = .09x - 143.53$
Women: $y = .09x - 153.87$
Sample prediction: The marriage age will continue to increase, but at a slower rate. The gap will remain the same.

c. Data for median age (in years) at time of first marriage for 1900–1940 is given in the table. The data for 1900–1990 is graphed on page 478.

Year	Men	Women
1900	25.9	21.9
1910	25.1	21.6
1920	24.6	21.2
1930	24.3	21.3
1940	24.3	21.5

Summary

The Summary gives an overview of the entire chapter and provides an opportunity for students to consider the material as a whole. Thus, the Summary can be used to help students relate and unify the concepts presented in the chapter.

Vocabulary

Terms, symbols, and properties are listed by lesson to provide a checklist of concepts that students must know. Emphasize that students should read the vocabulary list carefully before starting the Progress Self-Test. If students do not understand the meaning of a term, they should refer back to the indicated lesson.

Additional Answers, Progress Self-Test, page 479

14a.

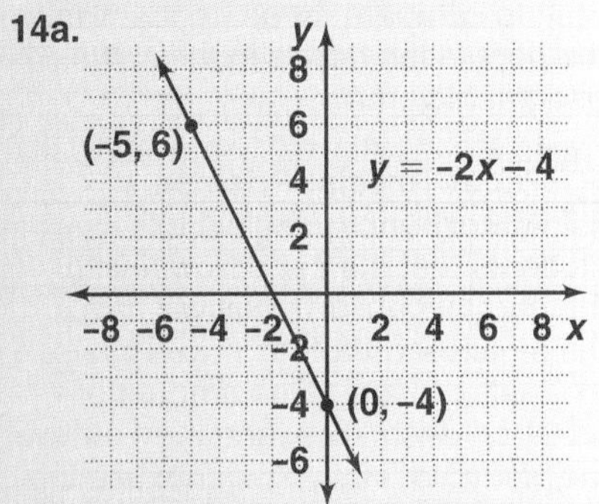

16. 17.

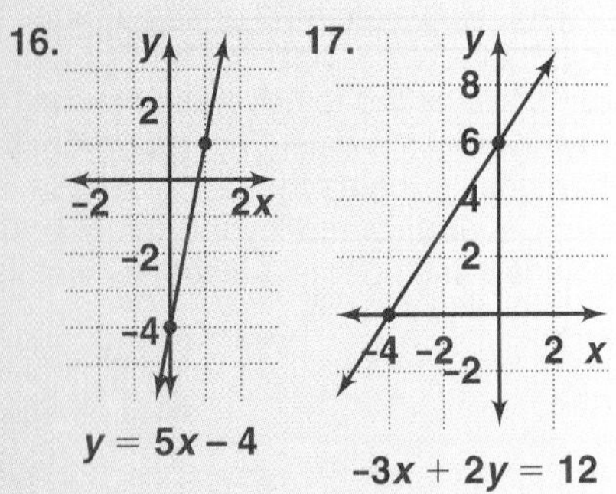

SUMMARY

The rate of change between two points (x_1, y_1) and (x_2, y_2) is $\frac{y_2 - y_1}{x_2 - x_1}$. When points all lie on the same line, the rate of change between them is constant and is called the slope of the line. The slope tells how much the line rises or falls for every move of one unit to the right. When the slope is positive, the line goes up to the right. When the slope is negative, the line falls to the right. When the slope is 0, the line is horizontal. The slope of vertical lines is not defined.

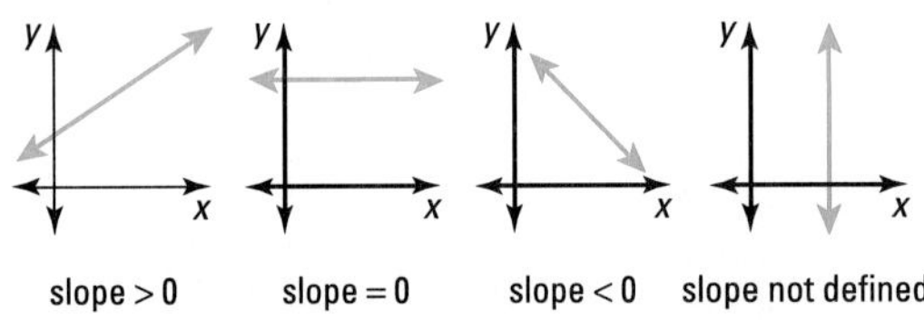

Constant-increase or constant-decrease situations lead naturally to linear equations of the form $y = mx + b$. The graph of the set of points (x, y) satisfying this equation is a line with slope m and y-intercept b.

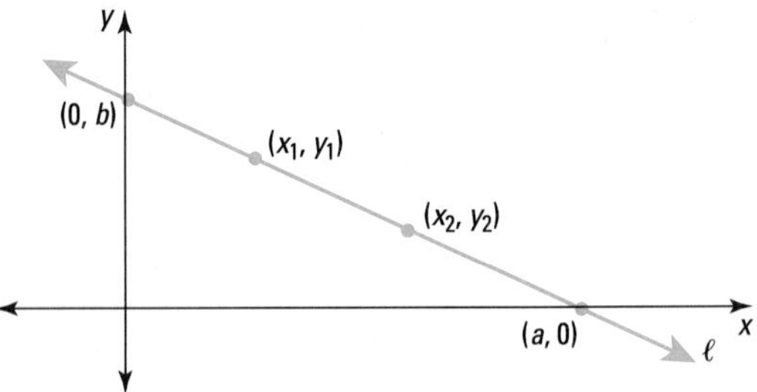

Line ℓ has slope $m = \frac{y_2 - y_1}{x_2 - x_1}$, y-intercept b, and x-intercept a.

Other situations lead to linear equations in the standard form $Ax + By = C$. When the = sign in equations of either form is replaced by < or >, the graph of the resulting linear inequality is a half-plane, the set of points on one side of a line.

A line is determined by any point on it and its slope, and its equation can be found from this information. Likewise, an equation can be found for the line containing two given points. If more than two points are given, then there may be more than one line determined. You can then draw a line that comes close to all the points, and use these points to determine an equation for the line.

VOCABULARY

Below are the most important terms and phrases for this chapter. You should be able to give a definition or general description and a specific example of each.

Lesson 7-1
rate of change

Lesson 7-2
constant decrease
slope

Lesson 7-4
slope-intercept form for an equation of a line
Slope-Intercept Property
y-intercept

Lesson 7-5
x-intercept

Lesson 7-7
latitude
fitting a line to data

Lesson 7-8
standard form for an equation of a line
oblique line

Lesson 7-9
boundary point
half-plane
linear inequality

Additional responses, pages 476–477

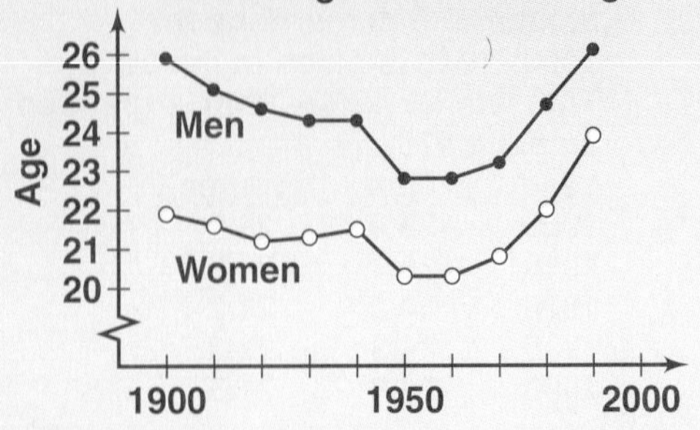

The earlier data suggest that the ages decreased from 1900 to 1930, leveled off from 1930 to 1940, and then decreased again from 1940 to 1950. Including these data, students might predict that the marriage age will level off or begin to decrease.

3a. Responses will vary.
b,c. Students' graphs will vary, but, since shoe size increases linearly, the data should lie very close to a line. The formulas in Lesson 4-5 are as follows, with S = size and L = length of foot in inches:
Man's size: $S = 3L - 26$
Woman's size: $S = 3L - 22$.
4. Responses will vary.

PROGRESS SELF-TEST

In 1–3, refer to the line graphed at the right.

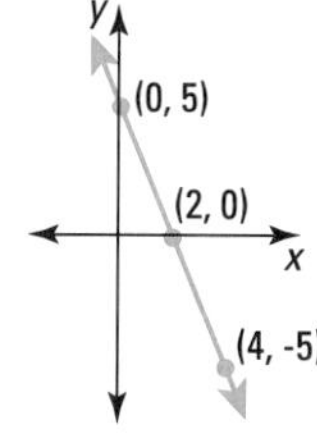

1. Give its y-intercept. 5
2. Give its x-intercept. 2
3. Calculate its slope. $\frac{-5}{2}$

4. Do the points (4, -5), (-2, 1) and (20, -20) all lie on the same line? Explain your answer. See below.

In 5 and 6, find the slope and y-intercept.

5. $y = 8 - 4x$ -4; 8
6. $5x + 2y = 1$ $\frac{-5}{2}$; $\frac{1}{2}$
7. Find an equation of the line with slope $\frac{3}{4}$ and y-intercept 13. $y = \frac{3}{4}x + 13$
8. Rewrite the equation $y = 5x - 2$ in standard form $Ax + By = C$ and give the values of A, B, and C. $-5x + y = -2$; $A = -5$; $B = 1$; $C = -2$
9. What is the slope of every horizontal line? 0
10. If a line's slope is $\frac{3}{5}$, how will the y-coordinate change as you go one unit to the right? up $\frac{3}{5}$ unit

In 11 and 12, use the following data of total U.S. Army personnel during and after World War II.

Year	Total Personnel
1942	3,074,184
1943	6,993,102
1944	7,992,868
1945	8,266,373
1946	1,889,690

11. Between which two consecutive years was there the greatest increase in U.S. Army personnel? 1942 and 1943
12. What is the rate of change in Army personnel per year from 1942 to 1946? -296,123.5
13. A basketball team scored 67 points, from x baskets worth 2 points each and y free throws worth 1 point each. Write an equation that describes all possible values of x and y. $2x + y = 67$

4) No; the slope of the line between the first two points (-1) is not the same as the slope of the line between the last two points $\left(\frac{-21}{22}\right)$.

14a) See margin. b) $y = -2x - 4$

14. a. Graph the line with slope -2 containing the point (-5, 6).
    b. Find an equation for this line.
15. At age 12 Patrick weighed 43 kg; at 14 he weighed 50 kg. Find a linear equation relating Patrick's weight y to his age x. $y = \frac{7}{2}x + 1$

In 16–18, graph. 16-18) See margin.

16. $y = 5x - 4$
17. $-3x + 2y = 12$
18. $y \leq x + 1$
19. *Multiple choice.* Which of the four lines below left could be the graph of $y = -x + 3$? c

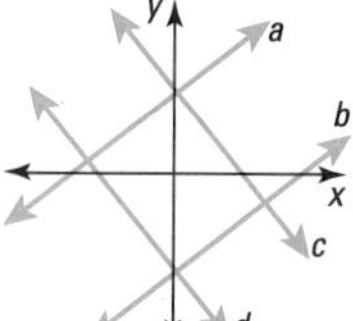

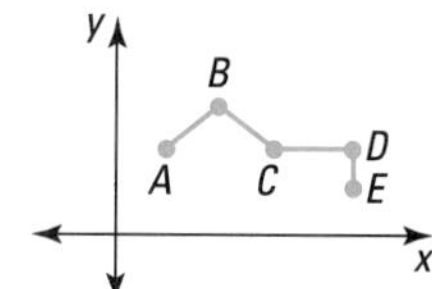

20. Using the graph above at the right, name the segment that
    a. has a positive slope; $\overline{AB}$
    b. does not have a slope. $\overline{DE}$
21. The scatterplot below illustrates the wingspan and length of 2-engine and 3-engine jet planes. 21a-d are sample answers.
    a. Estimate the coordinates of two points on a line that would fit the data. (60, 70); (120, 140)
    b. Find the slope of your line. $\frac{7}{6}$
    c. Write an equation for your line. $y = \frac{7}{6}x$
    d. Use the equation to estimate the length of a jet with a wingspan of 100 feet. ≈ 117 ft Show your work. $y = \frac{7}{6} \cdot 100 \approx 117$

Length and wingspan of 2- and 3-engine jets

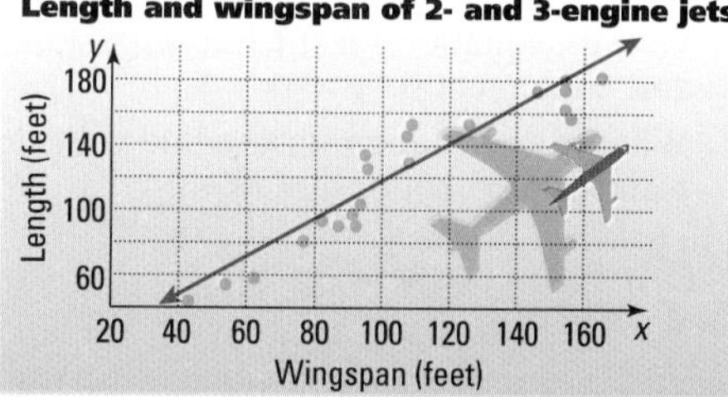

Progress Self-Test

For the development of mathematical competence, feedback and correction, along with the opportunity to practice, are necessary. The Progress Self-Test provides the opportunity for feedback and correction; the Chapter Review provides additional opportunities and practice. We cannot overemphasize the importance of these end-of-chapter materials. It is at this point that the material "gels" for many students, allowing them to solidify skills and understanding. In general, student performance should be markedly improved after these pages.

Assign the Progress Self-Test as a one-night assignment. Worked-out *solutions* for all questions are in the Selected Answers section of the student book. Encourage students to take the Progress Self-Test honestly, grade themselves, and then be prepared to discuss the test in class.

Advise students to pay special attention to those Chapter Review questions (pages 480–483) which correspond to the questions that they missed on the Progress Self-Test.

Additional Answers, continued

18.

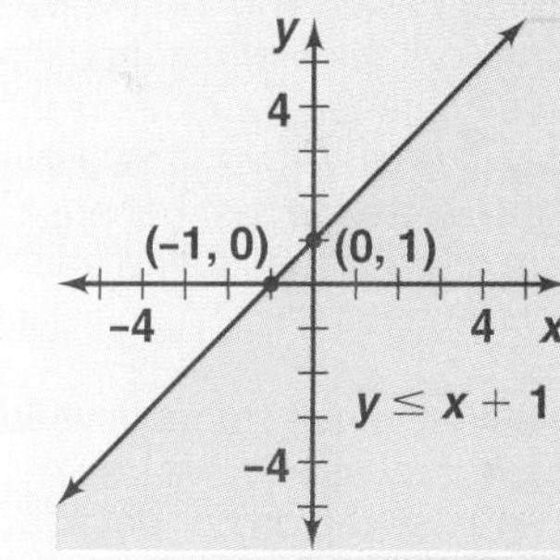

5. The following information is a sample of what students might include in their projects. The populations of cities increase or decrease due to a variety of reasons, such as availability of space to grow, availability of jobs, quality of education, cost of living, taxes, climate, and recreational opportunities. Currently, older cities in the eastern United States, such as Baltimore, Detroit, Cleveland, and Chicago, are declining in population. New York City, however, increased in population from 7,071,639 in 1980 to 7,322,564 in 1990. Newer cities in the western United States, such as Seattle, Los Angeles, Phoenix, and Dallas, are still growing. Other cities with room for expansion, such as Jacksonville, San Jose, and Indianapolis, are also still growing.

6a,b. Results may be influenced by experimental procedures and vary widely. In general, students will find that the graphs are linear.

c. Responses will vary. Students might mention that the price of the towels should not be the only criterion used when purchasing them; absorbency, the number of towels, and size should also be considered.

Chapter 7 Review

Resources

From the ***Teacher's Resource File***
- Answer Master for Chapter 7 Review
- Assessment Sourcebook: Chapter 7 Test, Forms A–D Chapter 7 Test, Cumulative Form

Additional Resources
- Quiz and Test Writer

The main objectives for the chapter are organized in the Chapter Review under the four types of understanding this book promotes—Skills, Properties, Uses, and Representations.

Whereas end-of-chapter material may be considered optional in some texts, in *UCSMP Algebra* we have selected these objectives and questions with the expectation that they will be covered. Students should be able to answer these questions with about 85% accuracy after studying the chapter.

You may assign these questions over a single night to help students prepare for a test the next day, or you may assign the questions over a two-day period. If you work the questions over two days, then we recommend assigning the *evens* for homework the first night so that students get feedback in class the next day, then assigning the *odds* the night before the test because answers are provided to the odd-numbered questions.

It is effective to ask students which questions they still do not understand and use the day or days as a total class discussion of the material which the class finds most difficult.

CHAPTER REVIEW

Questions on SPUR Objectives

SPUR stands for **S**kills, **P**roperties, **U**ses, and **R**epresentations. The Chapter Review questions are grouped according to the SPUR Objectives for this chapter.

SKILLS DEAL WITH THE PROCEDURES USED TO GET ANSWERS.

Objective A: *Find the slope of the line through two given points.* *(Lesson 7-2)*

1. Calculate the slope of the line containing (2, 4) and (6, 2). $-\frac{1}{2}$
2. Calculate the slope of the line through (1, 5) and (-2, -2). $\frac{7}{3}$
3. Find the slope of line ℓ at the left below. $.\overline{54}$ or $\frac{6}{11}$

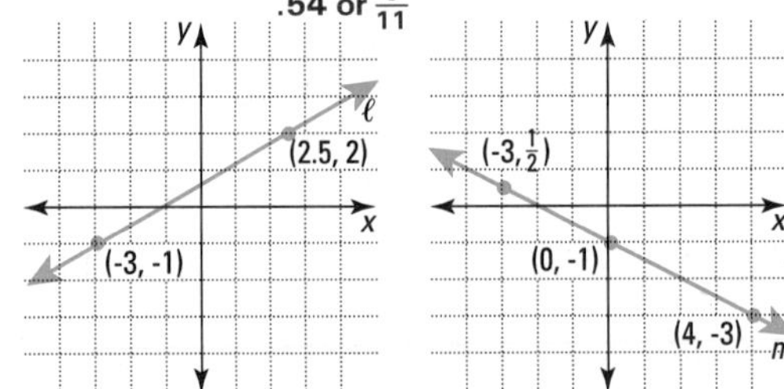

4. Use two different pairs of points to confirm that the slope of line m above is $-\frac{1}{2}$. See below.

In 5 and 6, the points (4, 10) and (8, y) are on the same line. Find y when the slope of the line is the number given.

5. -2 $y = 2$
6. $\frac{5}{4}$ $y = 15$

Objective B: *Find an equation for a line given two points on it, or its slope and one point on it.* *(Lessons 7-4, 7-5, 7-6)*

7. Give an equation for the line with slope 4 and y-intercept 3. $y = 4x + 3$
8. What is an equation for the line with slope p and y-intercept q? $y = px + q$

4) $\frac{\frac{1}{2} - (-1)}{-3 - 0} = -\frac{1}{2}$; $\frac{-1 - (-3)}{0 - 4} = -\frac{1}{2}$

In 9–11, find an equation for the line through the given point with slope m.

9. (-4, 1), $m = -2$ $y = -2x - 7$
10. (6, 10), $m = 0$ $y = 10$
11. $\left(3, \frac{1}{4}\right)$, $m = 30$ $y = 30x - \frac{359}{4}$
12. What is an equation for the line through (3, -1) with undefined slope? $x = 3$

On 13–16, find an equation for the line through the two given points. 13) $y = \frac{1}{2}x - \frac{9}{2}$ 14) $y = 4x + 4$

13. (5, -2), (-7, -8)
14. (0.5, 6), (0, 4)
15. (6, 9), (6, 0)
16. (-5, 2), (3, 2)

15) $x = 6$ 16) $y = 2$

Objective C: *Write an equation for a line in standard form or slope-intercept form, and from either form find its slope and y-intercept.* *(Lessons 7-4, 7-8)*

In 17 and 18, write in the form $Ax + By = C$. Then give the values of A, B, and C. See below.

17. $x - 22 = 5y$
18. $y = \frac{2}{5}x + 7$

In 19 and 20, rewrite the equation in slope-intercept form. 19) $y = -2x + 4$ $y = -\frac{1}{3}x + \frac{11}{3}$

19. $2x + y = 4$
20. $x + 3y = 11$

In 21–24, find the slope and the y-intercept of the line.

21. $y = 7x - 3$ 7; -3
22. $4x + 5y = 1$ $-\frac{4}{5}$; $\frac{1}{5}$
23. $y = -x - 1$; 0
24. $48x - 3y = 30$ 16; -10

17) $x - 5y = 22$; $A = 1, B = -5, C = 22$ 18) $-2x + 5y = 35$; $A = -2, B = 5, C = 35$

PROPERTIES DEAL WITH THE PRINCIPLES BEHIND THE MATHEMATICS.

Objective D: *Use the definition and properties of slope.* *(Lessons 7-2, 7-3)*

25. Find the slope of the line through (a, b) and (c, d). $\frac{d-b}{c-a}$ or $\frac{b-d}{a-c}$

26. The slope determined by two points is the change in the _?_ coordinates divided by the _?_ in the x-coordinates. **y; change**

27. Slope is the amount of change in the _?_ of the graph for every change of one unit to the _?_. **height; right**

In 28 and 29, refer to the graph at the right.

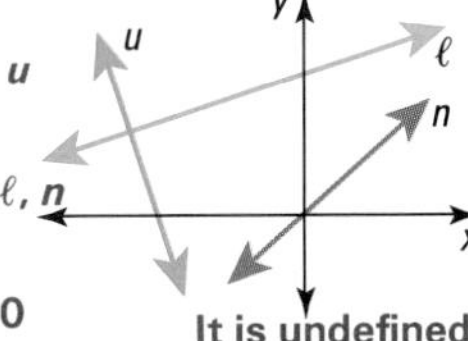

28. Which line or lines have negative slope? **u**

29. Which line or lines have positive slope? **ℓ, n**

30. What is the slope of any horizontal line? **0**

31. What is true about the slope of a vertical line? **It is undefined.**

32. Explain how you can use slope to determine whether three points all lie on the same line. **Sample: Check to see if the slope between points *A* and *B* is the same as the slope between points *B* and *C*.**

USES DEAL WITH APPLICATIONS OF MATHEMATICS IN REAL SITUATIONS.

Objective E: *Calculate rates of change from real data.* *(Lessons 7-1, 7-3)*

33. Assume an ascending jet plane climbs 0.46 km for each kilometer it travels away from its starting point.

a. Draw a picture of this situation. **See margin.**

b. What is the slope of the ascent? **0.46**

34. The picture at the right represents part of a ski slope. What is the slope of this part? $-\frac{1}{4}$

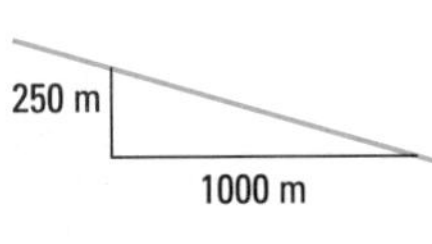

In 35–37, use the average height (in cm) for girls between birth and age 14 given below.

Age (yr)	Height (cm)
birth	50.8
2	83.8
4	99.0
6	111.7
8	127.0
10	137.1
12	147.3
14	157.5

35. Find the rate of change of height from age 10 to 12. **5.1 cm per year**

36. Find the rate of change of height between age 2 and age 14. **6.14 cm per year**

37. a. According to these data, in which two-year period do girls gain height fastest? **birth to 2 years**

b. What is this rate of change? **16.5 cm per year**

In 38–40, use the chart below of temperatures recorded at Capitol City Airport in Lansing, Michigan, starting at 8 P.M. on January 15, 1994.

	January 15 P.M.					January 16 A.M.											P.M.
time	8	9	10	11	12	1	2	3	4	5	6	7	8	9	10	11	12
temp. (°F)	-3	-8	-9	-6	-3	-3	-3	-4	-4	-9	-11	-11	-11	-8	-8	-5	-5

38. Find the rate of change in temperature per hour from midnight to 8 A.M. **-1° per hour**

39. What was the rate of change of temperature per hour over the period of time reported?

40. a. During which two-hour period did the temperature decrease the most? **4 A.M.-6 A.M.**

b. In that period, what was the average rate of change of the temperature per hour? **-3.5° per hour**

39) about -0.1° per hour

Objective F: *Use equations for lines to describe real situations.* *(Lessons 7-4, 7-5, 7-6, 7-8)*

In 41 and 42, each situation can be represented by a straight line. Give the slope and the y-intercept of the line describing this situation.

41. Julie rents a truck. She pays an initial fee of \$15 and then \$0.25 per mile. Let y be the cost of driving x miles. **0.25; 15**

42. Nick is given \$50 to spend on a vacation. He decides to spend \$5 a day. Let y be the amount Nick has left after x days. **-5; 50**

Assessment

Evaluation The *Assessment Sourcebook* provides five forms of the Chapter 7 Test. Forms A and B present parallel versions in a short-answer format. Forms C and D offer performance assessment. The fifth test is Chapter 7 Test, Cumulative Form. About 50% of this test covers Chapter 7; 25% covers Chapter 6, and 25% covers earlier chapters.

For information on grading see *General Teaching Suggestions: Grading* in the *Professional Sourcebook* which begins on page T20 in Part 1 of the Teacher's Edition.

Feedback After students have taken the test for Chapter 7 and you have scored the results, return the tests to students for discussion. Class discussion on the questions that caused trouble for most students can be very effective in identifying and clarifying misunderstandings. You might want to have them write down the items they missed and work either in groups or at home to correct them. It is important for students to receive feedback on every chapter test, and we recommend that students see and correct their mistakes before proceeding too far into the next chapter.

Additional Answers

33a.

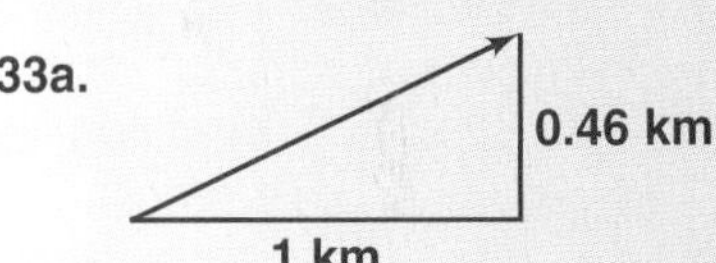

Additional Answers, page 483

50a.

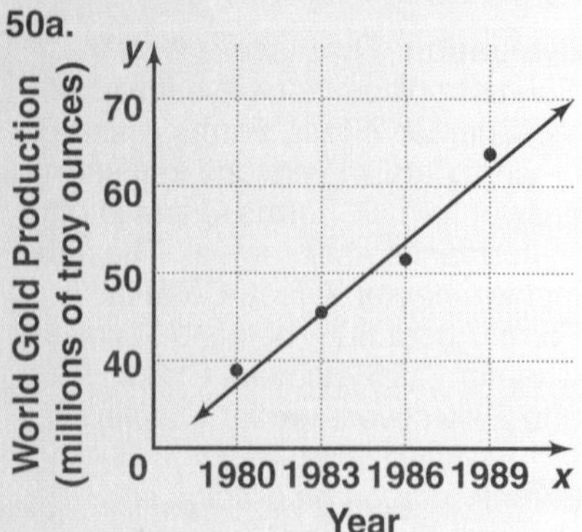

51.

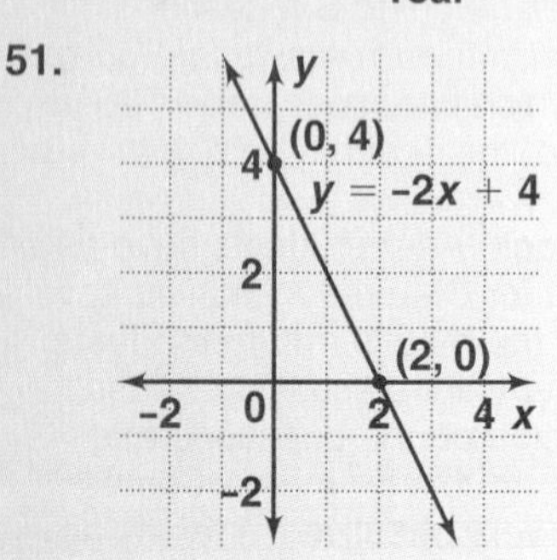

52.

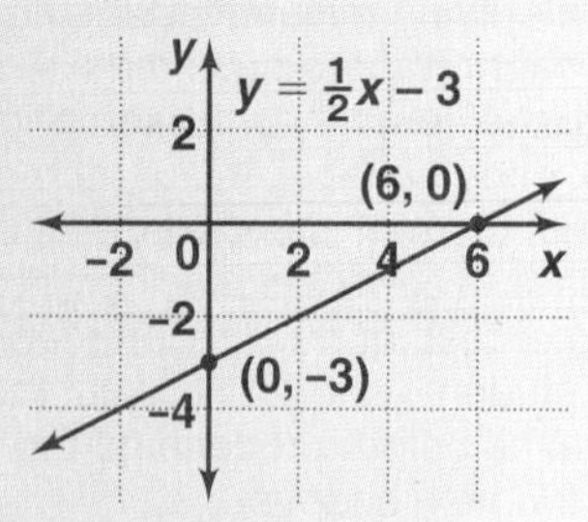

53.

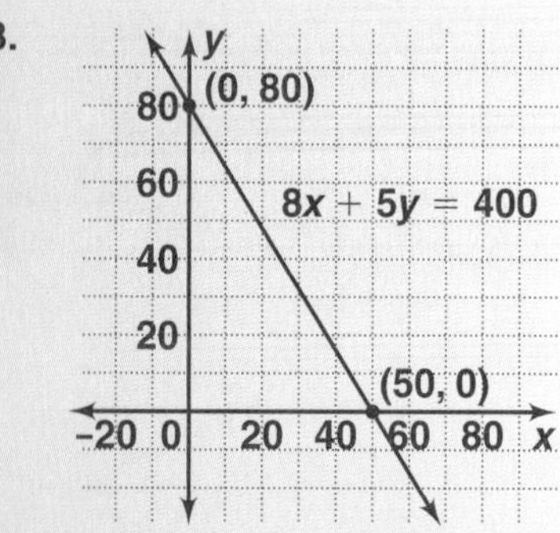

54.

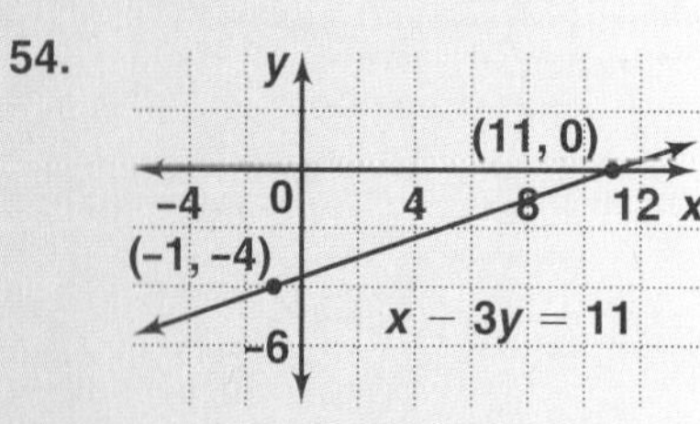

55.

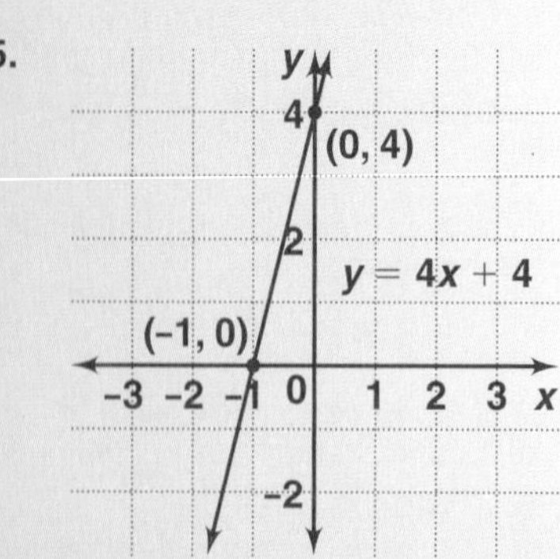

56.

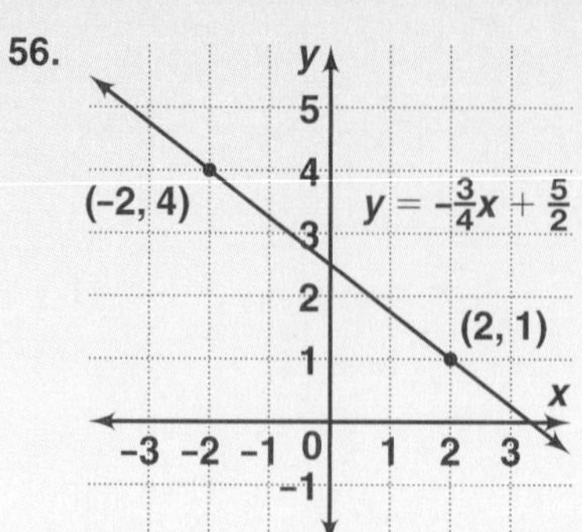

57.

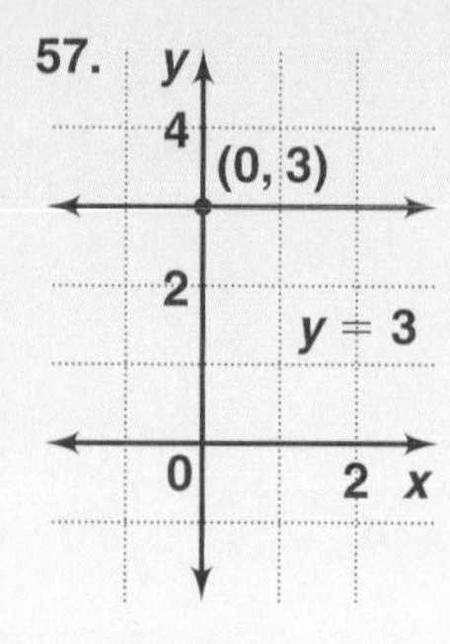

58.

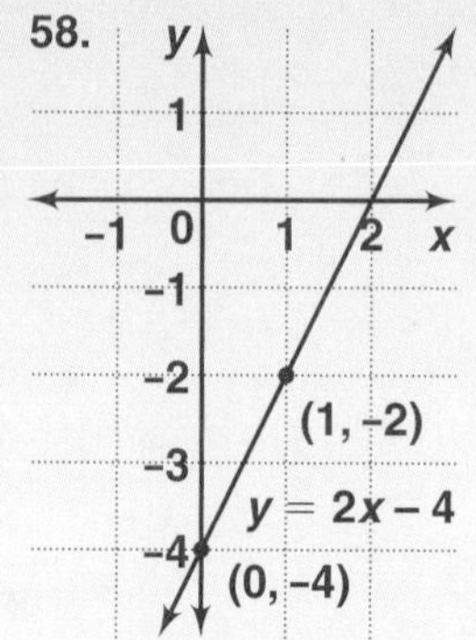

In 43 and 44, each situation leads to an equation of the form $y = mx + b$. Find that equation. **See below.**

43. A student takes a test and gets a score of 50. He gets a chance to take the test again. He estimates that every hour of studying will increase his score by 3 points. Let x be the number of hours studied and y be his score.

44. A plane loses altitude at the rate of 5 meters per second. It begins at an altitude of 8500 meters. Let y be its altitude after x seconds.

45. Julio plans a diet to gain 0.2 kg a day. After 2 weeks he weighs 40 kg. Write an equation relating Julio's weight w to the number of days d on his diet. **$w = 0.2d + 37.2$**

46. Each month about 50 new people come to live in a town. After 3 months the town has 25,500 people. Write an equation relating the number of months m to the number of people p in the town. **$p = 50m + 25{,}350$**

47. Robert babysat for \$2.50 an hour and mowed lawns for \$5 an hour. He earned a total of \$25. Write an equation that describes the possible babysitting hours B and lawn mowing hours L he could have spent at these jobs. **$2.5B + 5L = 25$**

48. The games of the 21st Modern Olympiad were in 1976. The games of the 20th Olympiad were 4 years earlier. Let y be the year of the nth summer Olympic games. Give a linear equation which relates n and y. **$y = 4n + 1892$**

43) $y = 3x + 50$
44) $y = -5x + 8{,}500$

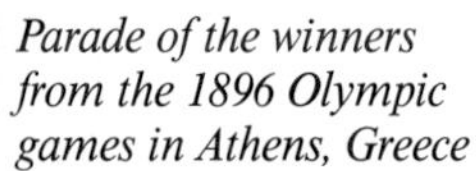

Parade of the winners from the 1896 Olympic games in Athens, Greece

49c) The Olympic winning times for this event drop (improve) by about .03 min/year.
d) $y = -0.029x + 61.7$

Objective G: *Given data whose graph is approximately linear, find a linear equation to fit the graph.* (*Lesson 7-7*) **49a-e are sample answers.**

49. Olympic swimmers tend to get faster over time. The scatterplot below shows the times for the winners of the Women's 400-meter freestyle for the years 1924 to 1992.

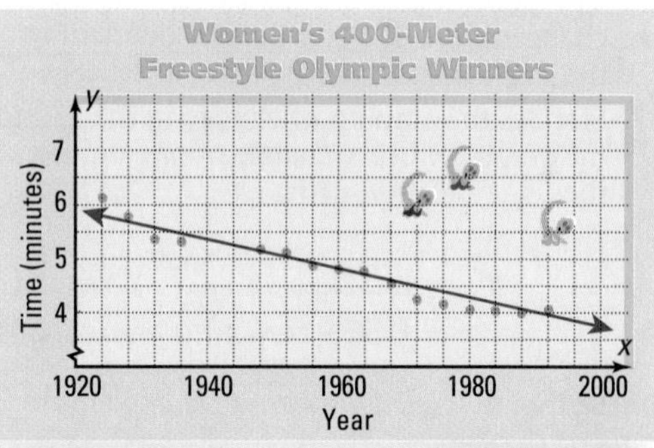

Year	Winner	Time	Decimal Time
1924	**Martha Norelius,** U.S.	6:02.2	6.04
1928	**Martha Norelius,** U.S.	5:42.8	5.71
1932	**Helene Madison,** U.S.	5:28.5	5.48
1936	**H. Mastenbroek,** Netherlands	5:26.4	5.44
1948	**Anne Curtis,** U.S.	5:17.8	5.30
1952	**Valerie Gyenge,** Hungary	5:12.1	5.20
1956	**Lorraine Crapp,** Australia	4:54.6	4.91
1960	**Chris von Saltza,** U.S.	4:40.6	4.68
1964	**Virginia Duenkel,** U.S.	4:43.3	4.72
1968	**Debbie Meyer,** U.S.	4:31.8	4.53
1972	**Shane Gould,** Australia	4:19.44	4.32
1976	**Petra Thümer,** E. Germany	4:09.89	4.17
1980	**Ines Diers,** E. Germany	4:08.76	4.15
1984	**Tiffany Cohen,** U.S.	4:07.10	4.12
1988	**Janet Evans,** U.S.	4:03.85	4.06
1992	**Dagmar Hase,** Germany	4:07.18	4.12

The times have been converted to decimal parts of a minute: 6:02.2 is graphed as 6.04 minutes. Use the converted times to complete the instructions below. **c, d) See left.**

a. Draw a line to fit the data. **See chart above.**

b. Find the slope of the fitted line. **≈ -.029**

c. Explain what the slope tells you about the trend in these data.

d. Find an equation for the line.

e. Predict the winning time in 1996. **3.82 min or 3:49.2 min**

50. The amount of gold mined in the world increased throughout the 1980s, as shown in the table below.

Year	World Gold Production (millions troy ounces)
1980	39.2
1983	45.2
1986	51.5
1989	65.3
1992	72.3

(Source: Bureau of Mines, U.S. Department of the Interior)

a) See margin.

a. Graph the data and a line to fit it.

b. Find an equation for the line.

c. Use the equation to predict the amount of gold produced in the world in 1995.

b) Sample: $y = 2.96x - 5824.5$

c) Sample: 80.7 million troy ounces

Shown are two men who are "gold-washing" at a mine in Brazil.

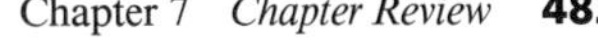

REPRESENTATIONS DEAL WITH PICTURES, GRAPHS, OR OBJECTS THAT ILLUSTRATE CONCEPTS.

Objective H: *Graph a straight line given its equation, or given a point and the slope.* *(Lessons 7-3, 7-4, 7-8)* **51-58) See margin.**

In 51–54, graph the line with the given equation.

51. $y = -2x + 4$ **52.** $y = \frac{1}{2}x - 3$

53. $8x + 5y = 400$ **54.** $x - 3y = 11$

In 55–58, graph the line satisfying the given condition.

55. passes through (0, 4) with a slope of 4

56. passes through (-2, 4) with a slope of $-\frac{3}{4}$

57. slope 0 and y-intercept 3

58. slope 2 and y-intercept -4

Objective I: *Graph linear inequalities.* *(Lesson 7-9)* **60–68) See margin.**

59. What are the regions on either side of a line in a plane called? half-planes

60. If you have only x nickels and y quarters and a total of less than $2.00, graph all possible values of x and y.

In 61–68, graph.

61. $x \geq 5$ **62.** $y \leq 4$

63. $y < -3$ **64.** $x > 0$

65. $y \geq x + 1$ **66.** $y < -3x + 2$

67. $3x + 2y > 5$ **68.** $x - 8y \leq 0$

60. Quarters

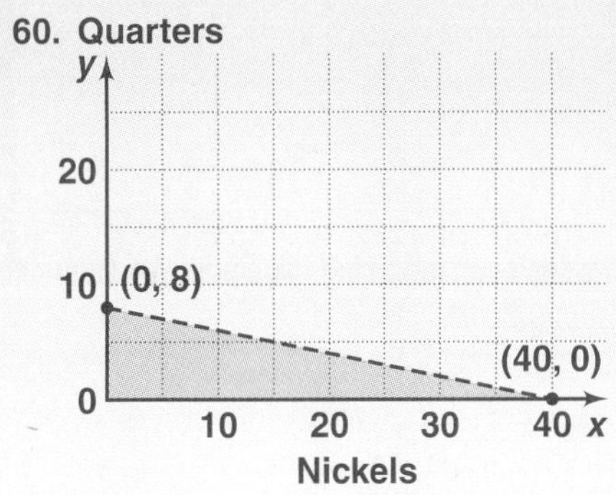

61.

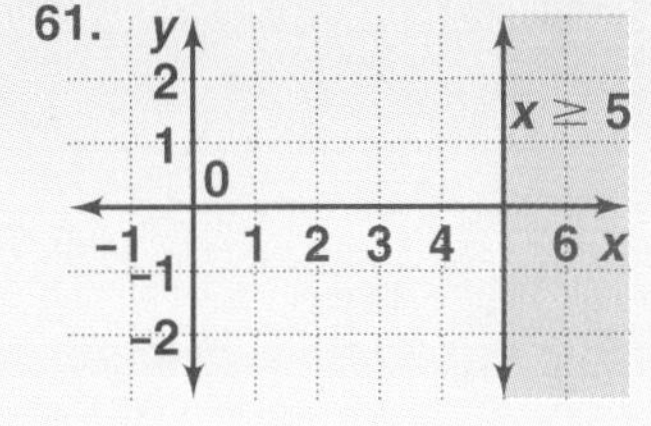

62.

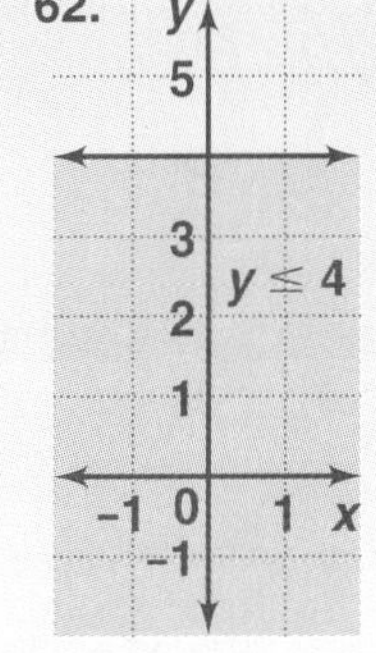

Additional Answers, continued

63.

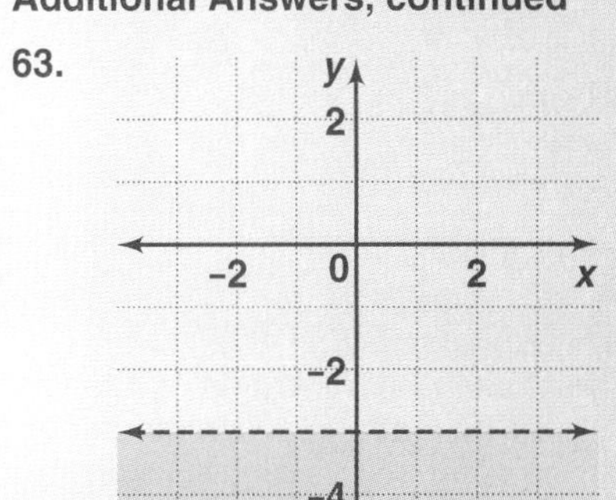

64.

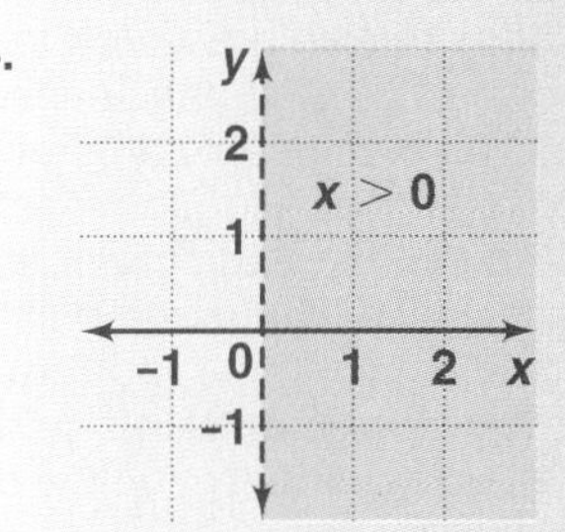

65.

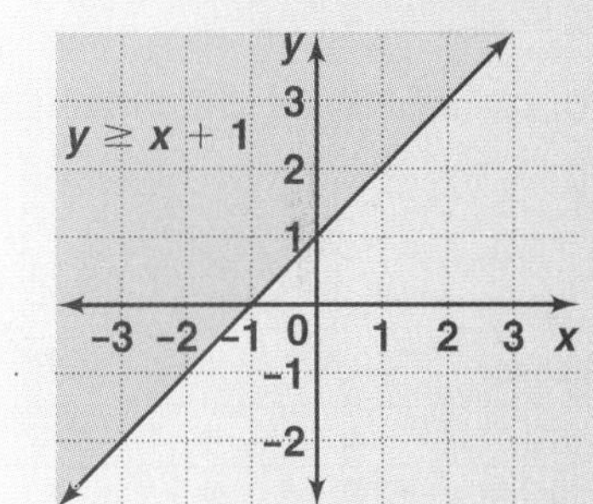

66.

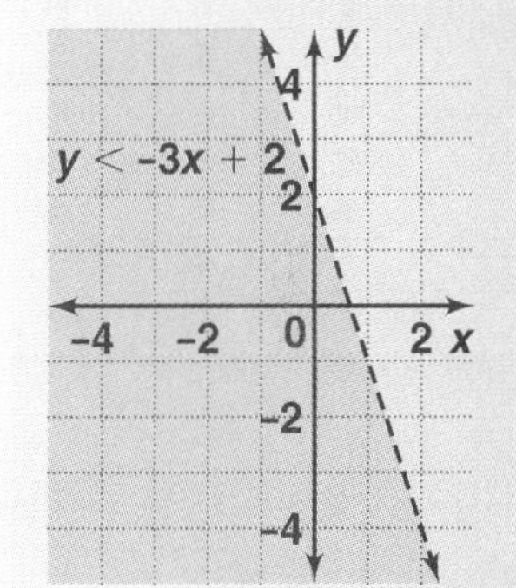

67.

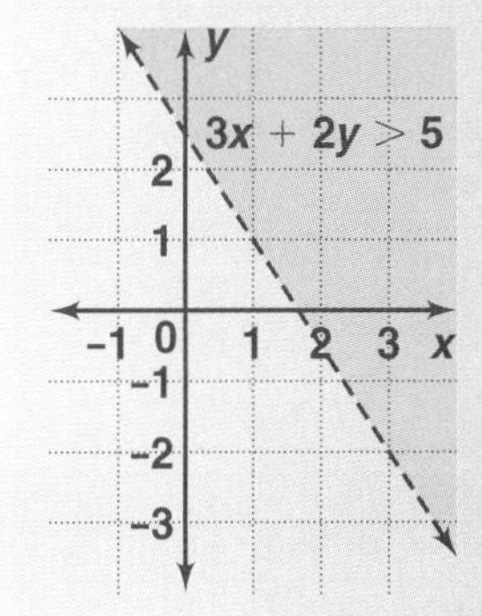

68.

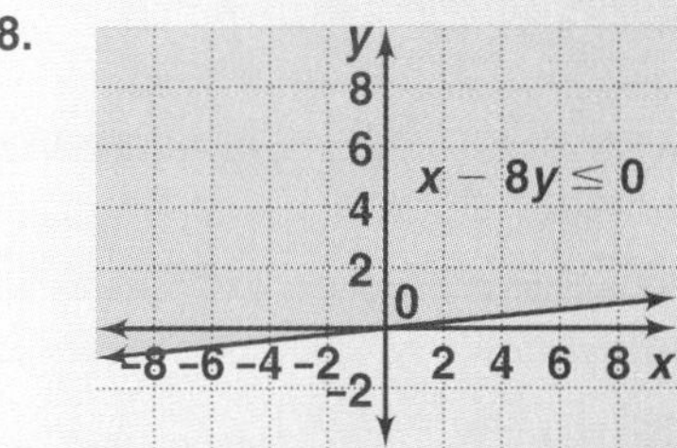

Chapter 8 Planner

Adapting to Individual Needs

The student text is written for the vast majority of students. The chart at the right suggests two pacing plans to accommodate the needs of your students. Students in the Full Course should complete the entire text by the end of the year. Students in the Minimal Course will spend more time when there are quizzes and more time on the Chapter Review. Therefore, these students may not complete all of the chapters in the text.

Options are also presented to meet the needs of a variety of teaching and learning styles. For each lesson, the Teacher's Edition provides sections entitled: *Video* which describes video segments and related questions that can be used for motivation or extension; *Optional Activities* which suggests activities that employ materials, physical models, technology, and cooperative learning; and, *Adapting to Individual Needs* which regularly includes **Challenge** problems, **English Language Development** suggestions, and suggestions for providing **Extra Help.** The Teacher's Edition also frequently includes an **Error Alert,** an **Extension,** and an **Assessment** alternative. The options available in Chapter 8 are summarized in the chart below.

Chapter 8 Pacing Chart

Day	Full Course	Minimal Course
1	8-1	8-1
2	8-2	8-2
3	8-3	8-3
4	Quiz*; 8-4	Quiz*; begin 8-4.
5	8-5	Finish 8-4.
6	8-6	8-5
7	Quiz*; 8-7	8-6
8	8-8	Quiz*; begin 8-7.
9	8-9	Finish 8-7.
10	Self-Test	8-8
11	Review	8-9
12	Test*	Self-Test
13		Review
14		Review
15		Test*

*in the Teacher's Resource File

In the Teacher's Edition...

Lesson	Optional Activities	Extra Help	Challenge	English Language Development	Error Alert	Extension	Cooperative Learning	Ongoing Assessment
8-1	●	●	●	●	●	●	●	Written
8-2	●	●	●	●	●	●	●	Written
8-3	●	●	●	●		●	●	Oral
8-4	●	●	●	●		●	●	Written
8-5	●	●	●			●		Oral/Written
8-6	●	●	●			●	●	Written
8-7	●	●	●	●	●	●	●	Group
8-8	●	●	●	●	●	●		Oral
8-9	●	●	●	●		●	●	Written

In the Additional Resources...

	In the Teacher's Resource File								
Lesson	Lesson Masters, A and B	Teaching Aids*	Activity Kit*	Answer Masters	Technology Sourcebook	Assessment Sourcebook	Visual Aids**	Technology	Video Segments
Opener		90					90		
8-1	8-1	26, 87		8-1	Comp 19		26, 87, AM	Spreadsheet	
8-2	8-2	26, 87		8-2			26, 87, AM		
8-3	8-3	26, 37, 87, 91, 92		8-3	Comp 20	Quiz	26, 37, 87, 91, 92, AM	GraphExplorer	
8-4	8-4	26, 87, 93	18	8-4	Comp 21		26, 87, 93, AM	GraphExplorer	8-4
8-5	8-5	88		8-5			88, AM		
8-6	8-6	88		8-6		Quiz	88, AM		
8-7	8-7	88	19	8-7			88, AM		
8-8	8-8	89		8-8			89, AM		
8-9	8-9	89, 94		8-9			89, 94, AM		
End of chapter				Review		Tests			

*Teaching Aids, except Warm-ups, are pictured on pages 484C and 484D. The activities in the Activity Kit are pictured on page 484C.

**Visual Aids provide transparencies for all Teaching Aids and all Answer Masters.

Also available is the Study Skills Handbook which includes study-skill tips related to reading, note-taking, and comprehension.

Integrating Strands and Applications

	8-1	8-2	8-3	8-4	8-5	8-6	8-7	8-8	8-9
Mathematical Connections									
Algebra	●	●	●	●	●	●	●	●	●
Geometry	●			●			●	●	●
Measurement	●						●	●	
Probability	●					●		●	
Statistics/Data Analysis		●	●						
Patterns and Functions	●	●	●	●	●	●	●	●	●
Interdisciplinary and Other Connections									
Science		●		●	●	●	●	●	
Social Studies	●	●	●	●	●	●	●	●	●
Multicultural		●		●	●			●	
Technology	●	●	●	●		●	●		●
Consumer	●	●	●	●		●	●	●	●

Teaching and Assessing the Chapter Objectives

Chapter 8 Objectives (Organized into the SPUR categories—Skills, Properties, Uses, and Representations)	**Lessons**	**Progress Self-Test Questions**	**Chapter Review Questions**	**In the Teacher's Resource File:** Chapter Test, Forms A and B	Chapter Test, Forms C	Chapter Test, Forms D
Skills						
A: Evaluate integer powers of real numbers.	8-1, 8-2, 8-6, 8-7, 8-8, 8-9	1–3, 10	1–12	5, 7	1	✓
B: Simplify products, quotients, and powers of powers.	8-5, 8-6, 8-7	4, 6, 7, 9	13–28	1–3, 20, 21	3	
C: Rewrite powers of products and quotients.	8-8	5, 8	29–40	4, 6		
Properties						
D: Test a special case to determine whether a pattern is true.	8-9	4, 11	41–44	10		✓
E: Identify properties of exponents and use them to explain operations with powers.	8-2, 8-5, 8-6, 8-7, 8-8	12	45–54	8, 19	2	
Uses						
F: Calculate compound interest.	8-1	13, 14	55–58	9	4	
G: Solve problems involving exponential growth and decay.	8-2, 8-3, 8-4, 8-6	15, 16, 19	59–67	11–13, 15, 16	5	✓
H: Use and simplify expressions with powers in real situations.	8-6, 8-7, 8-8	17, 20	68–73	18		✓
Representations						
I: Graph exponential relationships.	8-2, 8-3, 8-4	18	74–82	14, 17	5	

Multidimensional Assessment
Quiz for Lessons 8-1 through 8-3
Quiz for Lessons 8-4 through 8-6
Chapter 8 Test, Forms A–D
Chapter 8 Test, Cumulative Form

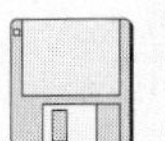

Quiz and Test Writer
Multiple forms of chapter tests and quizzes; Challenges

Activity Kit

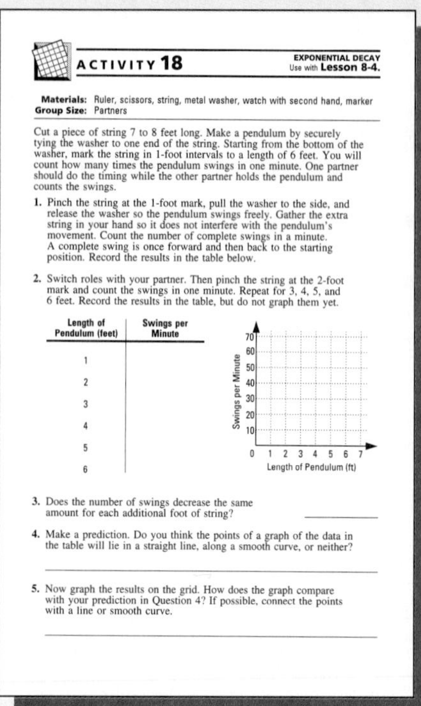

ACTIVITY 18 — **EXPONENTIAL DECAY** Use with **Lesson 8-4.**

Materials: Ruler, scissors, string, metal washer, watch with second hand, marker
Group Size: Partners

Cut a piece of string 7 to 8 feet long. Make a pendulum by securely tying the washer to one end of the string. Starting from the bottom of the washer, mark the string in 1-foot intervals to a length of 6 feet. You will count how many times the pendulum swings in one minute. One partner should do the timing while the other partner holds the pendulum and counts the swings.

1. Pinch the string at the 1-foot mark, pull the washer to the side, and release the washer so the pendulum swings freely. Gather the extra string in your hand so it does not interfere with the pendulum's movement. Count the number of complete swings in a minute. A complete swing is once forward and then back to the starting position. Record the results in the table below.

2. Switch roles with your partner. Then pinch the string at the 2-foot mark and count the swings in one minute. Repeat for 3, 4, 5, and 6 feet. Record the results in the table, but do not graph them yet.

Length of Pendulum (feet)	Swings per Minute
1	
2	
3	
4	
5	
6	

3. Does the number of swings decrease the same amount for each additional foot of string? ______

4. Make a prediction. Do you think the points of a graph of the data in the table will lie in a straight line, along a smooth curve, or neither?

______

5. Now graph the results on the grid. How does the graph compare with your prediction in Question 4? If possible, connect the points with a line or smooth curve.

______

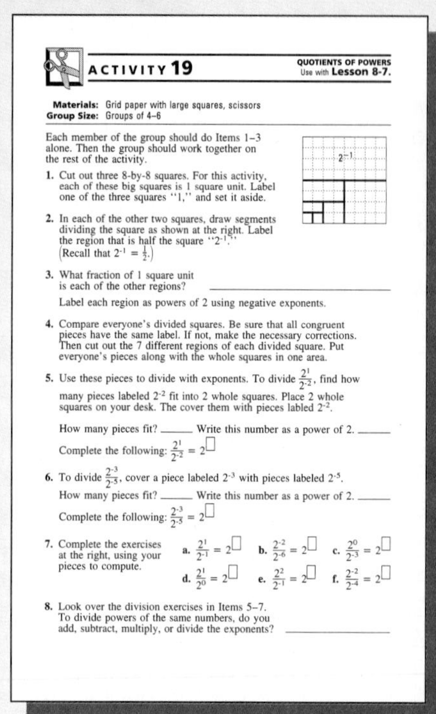

ACTIVITY 19 — **QUOTIENTS OF POWERS** Use with **Lesson 8-7.**

Materials: Grid paper with large squares, scissors
Group Size: Groups of 4–6

Each member of the group should do Items 1–3 alone. Then the group should work together on the rest of the activity.

1. Cut out three 8-by-8 squares. For this activity, each of these big squares is 1 square unit. Label one of the three squares "1," and set it aside.

2. In each of the other two squares, draw segments dividing the square as shown at the right. Label the region that is half the square "2^{-1}." (Recall that $2^{-1} = \frac{1}{2}$.)

3. What fraction of 1 square unit is each of the other regions? ______

   Label each region as powers of 2 using negative exponents.

4. Compare everyone's divided squares. Be sure that all congruent pieces have the same label. If not, make the necessary corrections. Then cut out the 7 different regions of each divided square. Put everyone's pieces along with the whole squares in one area.

5. Use these pieces to divide with exponents. To divide $\frac{2^1}{2^{-2}}$, find how many pieces labeled 2^{-2} fit into 2 whole squares. Place 2 whole squares on your desk. The cover them with pieces labled 2^{-2}.

   How many pieces fit? ______ Write this number as a power of 2. ______

   Complete the following: $\frac{2^1}{2^{-2}} = 2^{\square}$

6. To divide $\frac{2^{-3}}{2^{-5}}$, cover a piece labeled 2^{-3} with pieces labeled 2^{-5}.

   How many pieces fit? ______ Write this number as a power of 2. ______

   Complete the following: $\frac{2^{-3}}{2^{-5}} = 2^{\square}$

7. Complete the exercises at the right, using your pieces to compute.

   a. $\frac{2^1}{2^{-1}} = 2^{\square}$ b. $\frac{2^{-2}}{2^{-6}} = 2^{\square}$ c. $\frac{2^0}{2^{-3}} = 2^{\square}$

   d. $\frac{2^1}{2^0} = 2^{\square}$ e. $\frac{2^2}{2^{-1}} = 2^{\square}$ f. $\frac{2^{-2}}{2^{-4}} = 2^{\square}$

8. Look over the division exercises in Items 5–7. To divide powers of the same numbers, do you add, subtract, multiply, or divide the exponents? ______

Teaching Aids

Teaching Aid 26, Graph Paper, (shown on page 140D) can be used with **Lessons 8-1 through 8-4.**

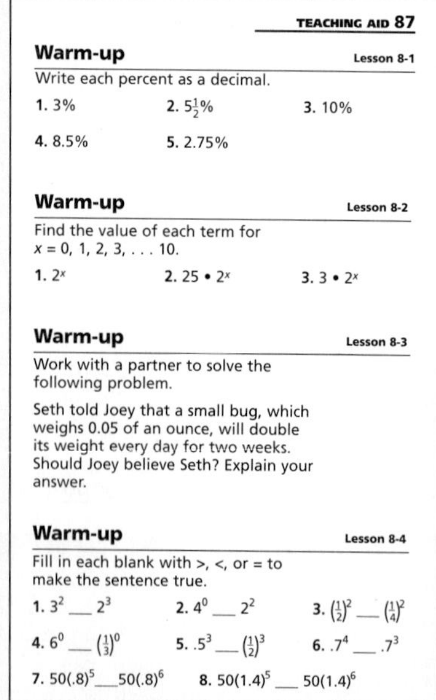

TEACHING AID 87

Warm-up — Lesson 8-1

Write each percent as a decimal.

1. 3% 2. $5\frac{1}{2}$% 3. 10%
4. 8.5% 5. 2.75%

Warm-up — Lesson 8-2

Find the value of each term for x = 0, 1, 2, 3, . . . 10.

1. 2^x 2. $25 \bullet 2^x$ 3. $3 \bullet 2^x$

Warm-up — Lesson 8-3

Work with a partner to solve the following problem.

Seth told Joey that a small bug, which weighs 0.05 of an ounce, will double its weight every day for two weeks. Should Joey believe Seth? Explain your answer.

Warm-up — Lesson 8-4

Fill in each blank with >, <, or = to make the sentence true.

1. 3^2 ___ 2^3 2. 4^0 ___ 2^2 3. $(\frac{1}{2})^2$ ___ $(\frac{1}{4})^2$
4. 6^0 ___ $(\frac{1}{3})^0$ 5. $.5^3$ ___ $(\frac{1}{2})^3$ 6. $.7^4$ ___ $.7^3$
7. $50(.8)^5$ ___ $50(.8)^6$ 8. $50(1.4)^5$ ___ $50(1.4)^6$

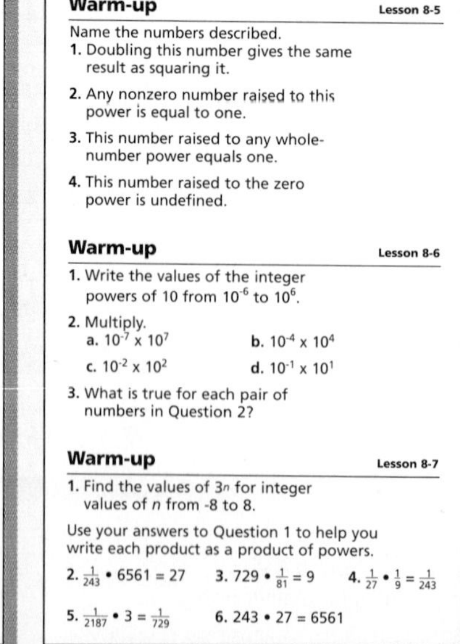

TEACHING AID 88

Warm-up — Lesson 8-5

Name the numbers described.

1. Doubling this number gives the same result as squaring it.
2. Any nonzero number raised to this power is equal to one.
3. This number raised to any whole-number power equals one.
4. This number raised to the zero power is undefined.

Warm-up — Lesson 8-6

1. Write the values of the integer powers of 10 from 10^{-6} to 10^6.
2. Multiply.
   a. $10^{-7} \times 10^7$ b. $10^{-4} \times 10^4$
   c. $10^{-2} \times 10^2$ d. $10^{-1} \times 10^1$
3. What is true for each pair of numbers in Question 2?

Warm-up — Lesson 8-7

1. Find the values of 3^n for integer values of n from -8 to 8.

Use your answers to Question 1 to help you write each product as a product of powers.

2. $\frac{1}{243} \bullet 6561 = 27$ 3. $729 \bullet \frac{1}{81} = 9$ 4. $\frac{1}{27} \bullet \frac{1}{9} = \frac{1}{243}$
5. $\frac{1}{2187} \bullet 3 = \frac{1}{729}$ 6. $243 \bullet 27 = 6561$

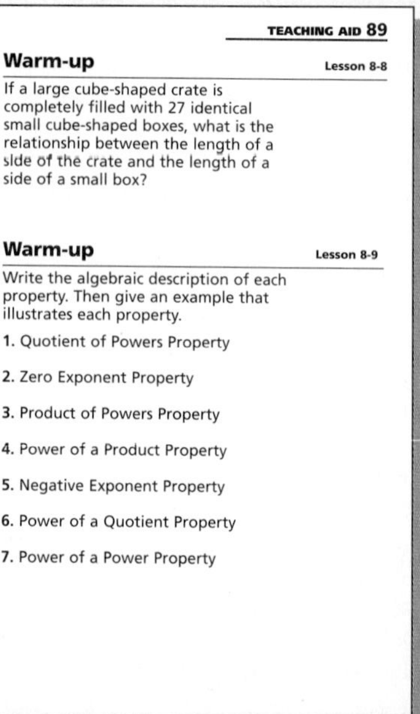

TEACHING AID 89

Warm-up — Lesson 8-8

If a large cube-shaped crate is completely filled with 27 identical small cube-shaped boxes, what is the relationship between the length of a side of the crate and the length of a side of a small box?

Warm-up — Lesson 8-9

Write the algebraic description of each property. Then give an example that illustrates each property.

1. Quotient of Powers Property
2. Zero Exponent Property
3. Product of Powers Property
4. Power of a Product Property
5. Negative Exponent Property
6. Power of a Quotient Property
7. Power of a Power Property

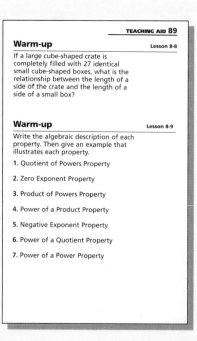

TEACHING AID 90
Chapter 8 Opener

Population Growth

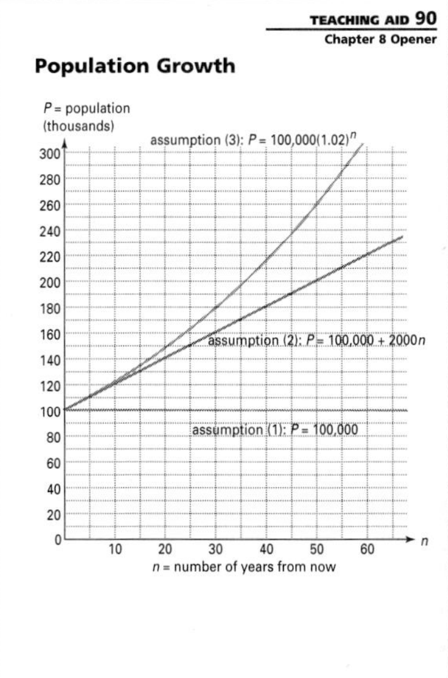

TEACHING AID 91
Lesson 8-3

Example 1

	Option (1): add 50	Option (2): multiply by 1.5
1st day	$10 + 50 \cdot 1 = \$60$	$10 \cdot 1.5^1 = \$15.00$
2nd day	$10 + 50 \cdot 2 = \$110$	$10 \cdot 1.5^2 = \$22.50$
3rd day	$10 + 50 \cdot 3 = \$160$	$10 \cdot 1.5^3 = \$33.75$
4th day	$10 + 50 \cdot 4 = \$210$	$10 \cdot 1.5^4 \approx \$50.63$
5th day	$10 + 50 \cdot 5 = \$260$	$10 \cdot 1.5^5 \approx \$75.94$
6th day	$10 + 50 \cdot 6 = \$310$	$10 \cdot 1.5^6 \approx \$113.91$
7th day	$10 + 50 \cdot 7 = \$360$	$10 \cdot 1.5^7 \approx \$170.86$
•	•	•
•	•	•
•	•	•
*n*th day	$10 + 50n$	$10 \cdot 1.5^n$

TEACHING AID 92
Lesson 8-3

Constant Increase and Exponential Growth

Constant Increase

1. Begin with an amount *b*.
2. *Add m* (the slope) in each of *x* time periods.
3. After the *x* time periods, there will be $b + mx$.

Exponential Growth

1. Begin with an amount *b*.
2. *Multiply* by *g* (the growth factor) in each of *x* time periods.
3. After the *x* time periods, the amount will be $b \cdot g^x$.

Constant Increase
$y = mx + b,\ m > 0$
(0, b)

Exponential Growth
$y = b \cdot g^x,\ g > 1$
(0, b)

TEACHING AID 93
Lesson 8-4

Linear and Exponential Increase and Decrease

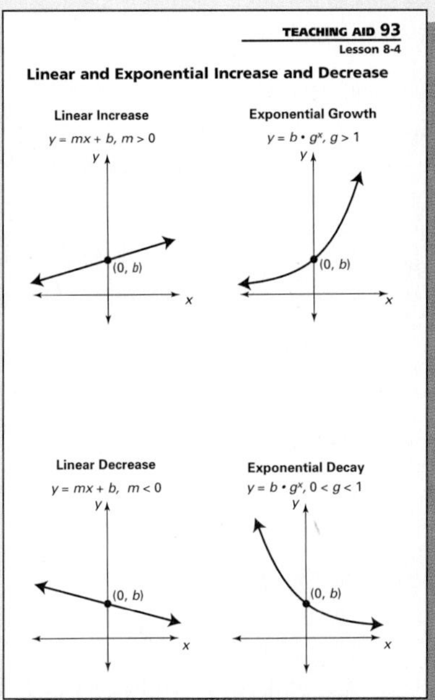

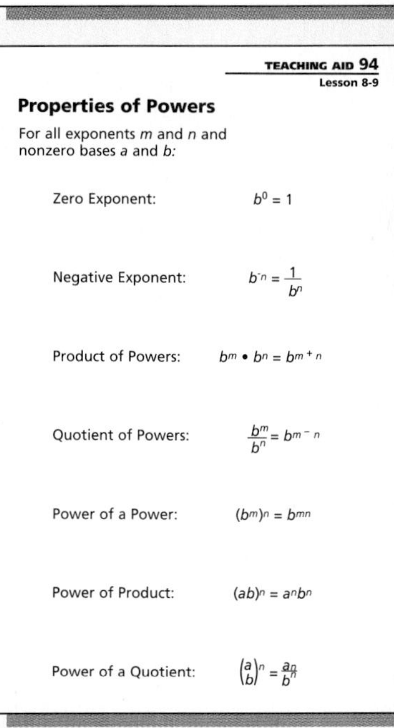

TEACHING AID 94
Lesson 8-9

Properties of Powers

For all exponents *m* and *n* and nonzero bases *a* and *b:*

Zero Exponent: $b^0 = 1$

Negative Exponent: $b^{-n} = \frac{1}{b^n}$

Product of Powers: $b^m \cdot b^n = b^{m+n}$

Quotient of Powers: $\frac{b^m}{b^n} = b^{m-n}$

Power of a Power: $(b^m)^n = b^{mn}$

Power of Product: $(ab)^n = a^n b^n$

Power of a Quotient: $\left(\frac{a}{b}\right)^n = \frac{a^n}{b^n}$

Chapter Opener

Pacing

All lessons in this chapter are designed to be covered in one day. At the end of the chapter, you should plan to spend 1 day to review the Progress Self-Test, 1 to 2 days for the Chapter Review, and 1 day for a test. You may wish to spend a day on projects, and possibly a day is needed for quizzes. This chapter should therefore take 12 to 15 days. We strongly advise that you not spend more than 16 days on this chapter; there is ample opportunity to review ideas in later chapters.

Notes on Reading

The graphs of the three equations introduce a theme of the chapter—the difference between exponential growth and linear increase or decrease. When discussing the graph, you might want to use **Teaching Aid 90.**

Have students consider ten-year periods. With assumption 1, for any ten-year period, the corresponding change in the population *P* is always 0. With assumption 2, for any ten-year period, say 30 – 20 = 10, the corresponding change in *P* is always 20,000. However, for assumption 3, the change in *P* is different. It increases with time as follows:

Population Growth under Assumption 3

Change in time (years)	Change in *P* (closest 1000)
10 – 0 = 10	122 – 100 = 22
20 – 10 = 10	149 – 122 = 27
30 – 20 = 10	181 – 149 = 32
40 – 30 = 10	221 – 181 = 40
50 – 40 = 10	269 – 221 = 48
60 – 50 = 10	328 – 269 = 59

CHAPTER 8

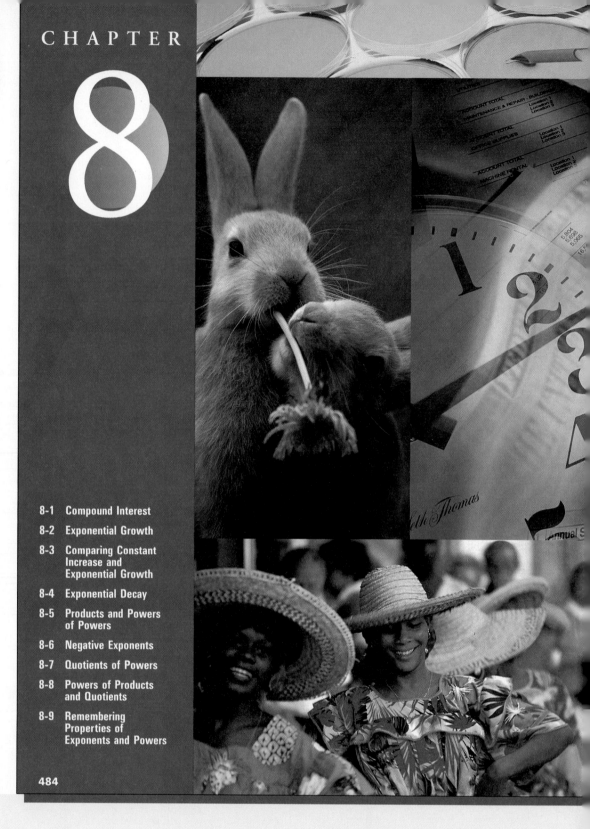

Chapter 8 Overview

The first nonlinear graph in some algebra books is the parabola, which arises from a quadratic equation. However, since exponential growth and exponential decay are among the most-used mathematical concepts in modern business, finance, biology, physics, and sociology, we discuss these concepts (and their corresponding nonlinear graphs) first. Parabolas appear in the next chapter.

Because exponential growth problems give different values of *y* for different values of *x*, it is easier for students to study the problem "what *x* will produce a given value of *y*" in this context than it is with quadratics, where multiple and extraneous solutions may arise.

Lesson 8-1 introduces a topic—compound interest—that all high school students should study, whether they take algebra or not. Problems involving compound interest can be solved by using a scientific calculator. Historically, compound interest was not included in first-year algebra books because the computations required logarithms.

Lessons 8-2 through 8-4 stress the uses and representations that arise from exponential growth, and contrast them with situations of constant increase or decrease. This work builds on the strong ties between

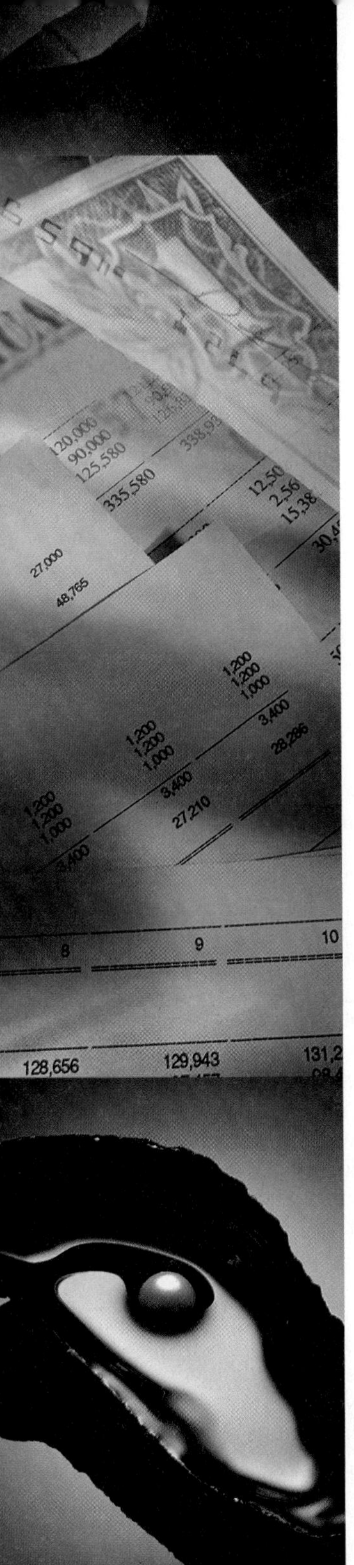

EXPONENTS AND POWERS

A city of 100,000 people is planning for the future. The planners want to know how many schools the town will need during the next 50 years. They consider three possibilities.

(1) The population will stay the same.
(2) The population will increase by 2,000 people per year. (Increases by a constant amount.)
(3) The population will grow by 2% a year. (Increases at a constant growth rate.)

Here is a graph of what would happen under the three possibilities. P is the population n years from now.

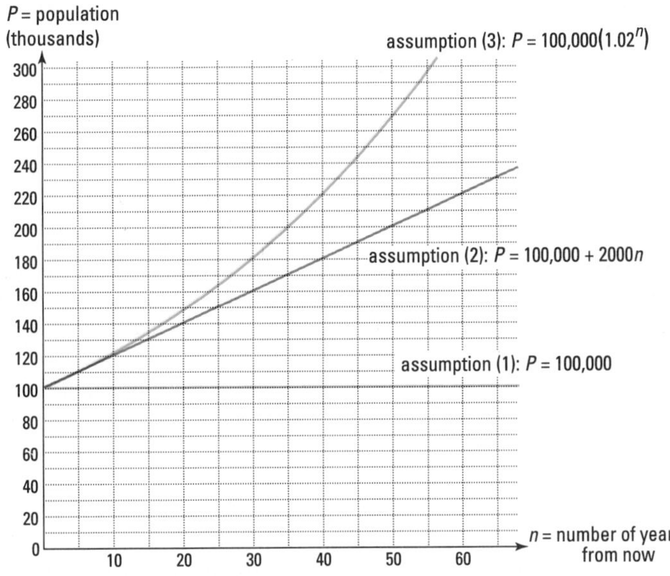

Possibility (3) is often considered the most reasonable. Under this assumption, $P = 100{,}000(1.02)^n$. Because the variable n is an exponent, this equation is said to represent *exponential growth.* Exponential growth is among the many applications of exponents and powers you will study in this chapter.

powers and repeated multiplication. This leads to the development of negative exponents and properties of exponents, which are discussed in Lessons 8-5 through 8-8, and the skills associated with those properties. Lesson 8-9 is a problem-solving lesson that discusses strategies which students can use to help them remember the properties.

The result of each assumption can be seen in the graph. For assumption 1, the graph is horizontal, so P is not changing. For assumption 2, the slope of the graph does not change and the graph is a line. That is, for each change of 1 in n, the change in P is always 2000. For assumption 3, the graph gets steeper as you move to the right. So, for each succeeding move of one unit to the right, the change in P increases and the graph is a curve.

Photo Connections

The photo collage makes real-world connections to the content of the chapter: exponents and powers.

Petri Dishes: Bacteria are often grown in laboratories for experimental purposes. Growth of bacteria refers to the growth of population rather than the growth of an individual microorganism. An equation with exponents can be used to describe the fast growth of bacteria.

Rabbits: Rabbit populations grow at a rapid rate with females bearing several litters of two to eight young each year. The rabbit population is said to grow exponentially.

Clock, Statements, and Money: An important application of exponents and powers involves savings accounts.

Jamaican Women: Jamaica's population of 2,466,000 in 1990 was expected to grow exponentially by 1.1% each year for the rest of the twentieth century. This is a relatively low growth rate for a developing country.

Pearl: The richest harvests of pearls come from the Persian Gulf where mollusks are found at depths of 8 to 20 fathoms (48 to 120 feet). The probability of finding pearls with perfect shapes, sizes, and color is very low.

Chapter 8 Projects

At this time you might want to have students look over the projects on pages 539–540.

Lesson 8-1

Objectives

A Evaluate integer powers of real numbers.
F Calculate compound interest.

Resources

From the ***Teacher's Resource File***
- Lesson Master 8-1A or 8-1B
- Answer Master 8-1
- Teaching Aids
  26 Graph Paper
  87 Warm-up
- Technology Sourcebook
  Computer Master 19

Additional Resources
- Visuals for Teaching Aids 26, 87
- Spreadsheet software

Teaching Lesson 8-1

Warm-up

Write each percent as a decimal.
1. 3% .03
2. $5\frac{1}{2}$% .055
3. 10% .1
4. 8.5% .085
5. 2.75% .0275

Notes on Reading

You might want to use *Optional Activities* on page 487 to introduce this lesson.

Reading Mathematics This lesson is appropriate for reading in class. If you do so, make sure that every student has a calculator, and stop to give students time to do the calculations.

LESSON

8-1 *Compound Interest*

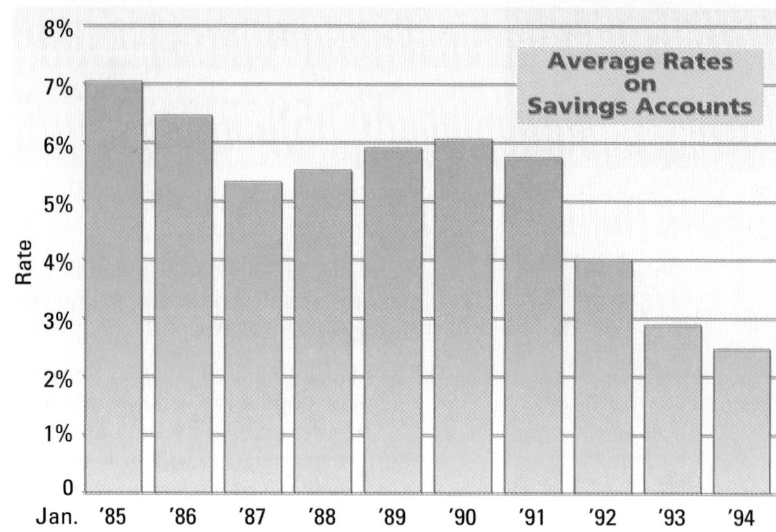

Changing bank interest rates. *Interest rates used by U.S. banks are based on many economic factors. Between 1985 and 1994, the trend was toward lower interest rates on savings accounts.*

Powers and Repeated Multiplication

A number having the form x^n is called a **power.** Certain powers can be changed to decimals by repeated multiplication. For instance, $3^4 = 3 \cdot 3 \cdot 3 \cdot 3 = 81$ and $10^7 = 10 \cdot 10 \cdot 10 \cdot 10 \cdot 10 \cdot 10 \cdot 10 = 10{,}000{,}000$. These are instances of the following model.

Repeated Multiplication Model for Powering
When n is a positive integer, $x^n = \underbrace{x \cdot x \cdot \ldots \cdot x}_{n \text{ factors}}$.

The number x^n is called the ***n*th power** of x and is read "x to the nth power" or just "x to the n." In the expression x^n, x is the **base** and n is the **exponent.** Thus 3^4 is read "3 to the 4th power," or "3 to the 4th"; 3 is the base and 4 is the exponent. In the expression $100{,}000(1.02)^n$ found on the previous page, 1.02 is the base and n is the exponent. The number 100,000 is the **coefficient** of the power 1.02^n.

How Is Interest Calculated?

An important application of exponents and powers involves savings accounts. Banks, savings and loan associations, and credit unions will pay you to give them your money to keep for you. The amount you give them is called the **principal.** The amount they pay you is called **interest.**

Interest is always a percent of the principal. The percent the money earns per year is called the **annual yield.**

Lesson 8-1 Overview

Broad Goals The purpose of this lesson is to review the meaning of powers as repeated multiplication and to introduce students to one of the most important applications of powers—compound interest.

Perspective Why does this chapter begin with the topic of compound interest? One reason is that understanding compound interest is very important for students. Savings accounts, loans, and investments all involve compound interest. It is fair to say that it is more important for students to know the Compound Interest Formula, $I = P(1 + i)^n$, than it is for them to know the Simple Interest Formula, $I = prt$. Few situations today involve $I = prt$ when $t > 1$.

A second reason is pedagogical: compound interest deals with powering as repeated multiplication, and thus naturally introduces students to exponents greater than 2 or 3. If you invest money at 6% for 10 years, your investment is multiplied by 1.06^{10}.

A third reason is that our studies show that students who have not studied compound interest score less than chance on multiple-choice items involving this topic. Apparently, they have more misconceptions than correct conceptions about compound interest. For such an important consumer topic, it is important to correct these misconceptions.

❶ **Example 1**

Suppose you deposit P dollars in a savings account upon which the bank pays an annual yield of 3%. If you make no other deposits or withdrawals, how much money will be in the account at the end of a year?

Solution 1

$$\begin{aligned}\text{Total} &= \text{principal} + 3\% \text{ of principal}\\ &= P + 0.03P\\ &= (1 + 0.03)P\\ &= 1.03P\end{aligned}$$

Solution 2

$$\begin{aligned}\text{Total} &= 100\% \text{ of principal} + 3\% \text{ of principal}\\ &= 103\% \text{ of principal}\\ &= 1.03P\end{aligned}$$

So, if you deposited \$1000 in a savings account with an annual yield of 3%, a bank would pay you 0.03(\$1000), or \$30, interest. You would have 1.03(\$1000), or \$1030, after one year.

Compound Interest, and How It Is Calculated

Savings accounts pay **compound interest,** which means that the interest earns interest.

❷ **Example 2**

Suppose you deposit \$100 in a savings account upon which the bank pays an annual yield of 3%. If the account is left alone, how much money will be in it at the end of four years?

Solution

Refer to Example 1. Each year the amount in the bank is multiplied by 1 + 0.03, or 1.03.

End of first year:

$$100(1.03) = 100(1.03)^1 = 103.00$$

End of second year:

$$100(1.03)(1.03) = 100(1.03)^2 = 106.09$$

End of third year:

$$100(1.03)(1.03)(1.03) = 100(1.03)^3 \approx 109.2727 \approx 109.27$$

End of fourth year:

$$100(1.03)(1.03)(1.03)(1.03) = 100(1.03)^4 \approx 112.5509 \approx 112.55$$

At the end of four years there will be \$112.55 in the account.

❶ **History Connection** You might want to clarify for students that interest is *given* when money is deposited, but interest is *charged* when money is borrowed. Historically, when someone borrowed money from another person, it was considered wrong or sinful to charge the borrower interest. Some thought it was a duty of the rich to loan money (interest free) to the poor. In Europe, during the Middle Ages, people who were found guilty of *usury* (charging an excessive interest rate for borrowing money) could be banished or whipped, or have their possessions taken away from them. By 1545, English law allowed for some forms of interest, and since the 1700s, charging interest on a loan has been accepted as a fair business practice. Now most disagreements about interest concern the maximum rates that lenders should be allowed to charge.

❷ **Consumer Connection** The difference between annual *rate* and annual *yield* is as follows: the interest rate determines the multiplier used when interest is calculated for each compounding period. The annual yield is the result of doing this over a year's time. For instance, suppose an investment has an 8.25% *annual rate* compounded daily. We would expect the *annual yield* to be slightly higher than 8.25%. Each day the account would earn $\frac{1}{365}$ of 8.25%, so the amount is multiplied by $1 + \frac{.0825}{365}$ each day. When compounded 365 times, the multiplier is $\left(1 + \frac{.0825}{365}\right)^{365} \approx 1.086 =$ 108.6%. In effect, the account has grown by 8.6%. Thus the rate of 8.25% compounded daily equals a rate of about 8.6% compounded yearly, and 8.6% is the annual yield.

Optional Activities

Activity 1 Cooperative Learning
As an introduction to the lesson, you might want to use the following activity. Have students **work in small groups** to complete this table of values of P under each of the assumptions on page 485. Call P_1 the value of P under assumption 1, P_2 the value of P under assumption 2, and P_3 the value of P under assumption 3, each rounded to the nearest whole number.

n	P_1	P_2	P_3
0	[100,000]	[100,000]	[100,000]
1	[100,000]	[102,000]	[102,000]
2	[100,000]	[104,000]	[104,040]
3	[100,000]	[106,000]	[106,121]
4	[100,000]	[108,000]	[108,243]
5	[100,000]	[110,000]	[110,408]
10	[100,000]	[120,000]	[121,899]
20	[100,000]	[140,000]	[148,595]
30	[100,000]	[160,000]	[181,136]

When a group of students has completed the table of values, ask a student in the group to describe orally or in writing the trends that the group members observe in their data.

Technology Connection If students have spreadsheet software available to them, have them use it to complete the table.

In UCSMP *Algebra*, students study only interest that is compounded annually, so annual rate and annual yield can be used interchangeably. In UCSMP *Advanced Algebra*, students study interest that is compounded more often during the year.

❸ The Compound Interest Formula need not be learned by rote if compounding is thought of as a scale change of $1 + i$ repeated n times. Two ideas are fundamental to having students understand this concept. One is the fact that the Distributive Property converts an addition to a multiplication, as shown in **Example 1.** Even though we have used this concept before in conjunction with discounts, the idea may have to be re-emphasized here. The other idea is expressing products involving the same factors as powers, as in **Example 2.**

❹ **Calculator** Even though students have used the y^x key before, some may need additional instruction and practice in its use. Since we often use long time periods and sometimes use noninteger values for y and for x, the calculator is indispensable here.

Instead of using the key sequence shown in **Example 3**, students may find it easier to evaluate $P(1 + i)^n$ by keying in $(1 + i)^n$ first and then multiplying by P. Parentheses are needed either way, unless the student mentally adds 1 to the rate i before entering it into the calculator.

Home improvements. *Many banks offer home equity loans to customers who own homes. People often use the money from these loans to make home improvements, such as adding a deck.*

Examine the pattern in the solution to Example 2. At the end of n years there will be

$$100(1.03)^n$$

dollars in the account. The general formula for compound interest uses this expression, but variables replace the principal and annual yield.

❸ **Compound Interest Formula**
If a principal P earns an annual yield of i, then after n years there will be a total T, where $T = P(1 + i)^n$.

The compound interest formula is read "T equals P times the quantity 1 plus i, that quantity to the nth power."

Example 3

$1500 is deposited in a savings account. What will be the total amount of money in the account after 10 years at an annual yield of 6%?

Solution

Here $P = \$1500$, $i = 6\%$, and $n = 10$.
Substitute the values in the Compound Interest Formula. Use 6% = 0.06.

$$T = P(1 + i)^n$$
$$= 1500(1 + 0.06)^{10}$$

❹ Use a calculator key sequence such as one of the following:

1500 [×] 1.06 [y^x] 10 [=],
or 1500 [×] 1.06 [^] 10 [ENTER].

Displayed will be 2686.2715, which is approximately $2686.27. **After 10 years, the account will contain $2686.27.**

In Example 3, because the $1500 will increase to $2686.27, the saver will earn $1186.27. Since 6% of $1500 is $90, if the interest is taken out each year for 10 years, the saver will earn $900. The compounding will increase the interest by $286.27.

Ten years may seem like a long time, but it is not an unusual amount of time for money to be in retirement accounts or in accounts parents set up to save for their children's college expenses.

Why Do You Receive Interest on Savings?

Banks and other savings institutions pay you interest because they want money to lend to other people. Banks earn money by charging a higher rate of interest on the money they lend than the rate they pay customers who deposit money. Thus, if the bank loans your $1000 (perhaps to someone buying a car or a house) at 12%, the bank receives 0.12($1000), or $120, from that person. If the bank pays you 3% interest, or $30, the bank earns $120 − $30, or $90 on your money.

Optional Activities

Activity 2 Technology Connection
You may wish to assign *Technology Sourcebook, Computer Master 19.* Students use a spreadsheet program or a BASIC program to produce tables that show the value of investments over several years.

Adapting to Individual Needs

Extra Help
Be sure students understand what each variable represents in the Compound Interest Formula. Also, help students see why the principal is multiplied by a power of $1 + i$ rather than by a power of i. Remind them that each year the interest is added to the beginning principal to get the new principal a year later. If P is the beginning principal, then Pi is the interest. Therefore, $P + Pi$ is the principal a year later. Note that $P + Pi = P(1 + i)$.

QUESTIONS

Covering the Reading

1b) 6 [×] 6 [×] 6 [=] 216; 6 [y^x] 3 [=] 216

1. a. Calculate 6^3 without a calculator. $6 \cdot 6 \cdot 6 = 216$
b. Calculate 6^3 with a calculator (show your key sequence).

2. Refer to assumptions (1), (2), and (3) on page 485. What is the predicted population in 50 years under each assumption? **100,000; 200,000; ≈ 269,159**

3. Consider the expression $50x^{10}$. Name each.
a. base x **b.** power x^{10}
c. exponent 10 **d.** coefficient 50

4. Match each term with its description.
a. money you deposit iii (i) annual yield
b. interest paid on interest ii (ii) compound interest
c. yearly percentage paid i (iii) principal

In 5 and 6, write an expression for the amount in the bank after one year if *P* dollars are in an account with an annual yield as given.

5. 6% $P(1.06)$ **6.** 2.75% $P(1.0275)$

7. Consider the situation of Example 2.
a. How much money will you have in your savings account at the end of 5 years? **$115.93**
b. How much interest will you have earned? **$15.93**

8. Follow Example 2 to explain why $100(1.054)^3$ is the total value of $100.00 invested for 3 years at an annual yield of 5.4%. **See below.**

9. a. Write the Compound Interest Formula. $T = P(1 + i)^n$
b. What does *T* represent? **total amount in account**
c. What does *P* stand for? **principal; amount invested**
d. What is *i?* **annual yield**
e. What does *n* represent? **number of years invested**

10. Write a calculator key sequence for your calculator to evaluate $573(1.063)^{24}$. **Sample: 573 [×] 1.063 [y^x] 24 [=]**

11. Suppose you deposit $150 in a new savings account paying an annual yield of 4%. If the account is left alone, how much money will be in the account at the end of 5 years? **$182.50**

12. A bank advertises an annual yield of 3.81% on a 6-year CD (certificate of deposit). If the CD's original amount was $2,000, how much will it be worth after 6 years? **$2503.02**

13. How much interest will be earned in 7 years on a principal of $500 at an annual yield of 5.6%? **$232.18**

8) Each year, the amount in the bank is multiplied by 1 + 0.054 = 1.054. So, at the end of 1 year, there will be 100(1.054); 2 years, there will be $100(1.054)(1.054) = 100(1.054)^2$; 3 years, there will be $100 \cdot (1.054) \cdot (1.054) \cdot (1.054) = 100(1.054)^3$.

Additional Examples

1. If *P* dollars are invested in an account at 5.2% annual yield, what will the value of the account be at the end of a year? Assume no other deposits or withdrawals are made. **1.052*P***

2. Suppose $500 is deposited in an annuity with a 7% annual yield. If there are no deposits or withdrawals, how much will be in the account after 8 years? **$500(1.07)^8 \approx \$859.09$**

3. A baby's grandparents invest $1000 on the day their grandchild is born.
a. How much is the investment worth on the grandchild's 18th birthday if it earns 6.3% annual yield? **$1000(1.063)^{18} \approx \$3,003.30$**
b. How much interest was earned? **$2,003.30**

Notes on Questions

For many questions in this chapter, students will derive an expression and then evaluate it on the calculator. For **Questions 7a and 11–13**, we suggest that you require students to write both the expression and the result, as shown by the answer in *Additional Example 2.* In this way, they can determine whether an error is due to having an incorrect expression or to incorrect use of the calculator.

Questions 3, 4, and 9 These questions emphasize the symbols and vocabulary of this lesson.

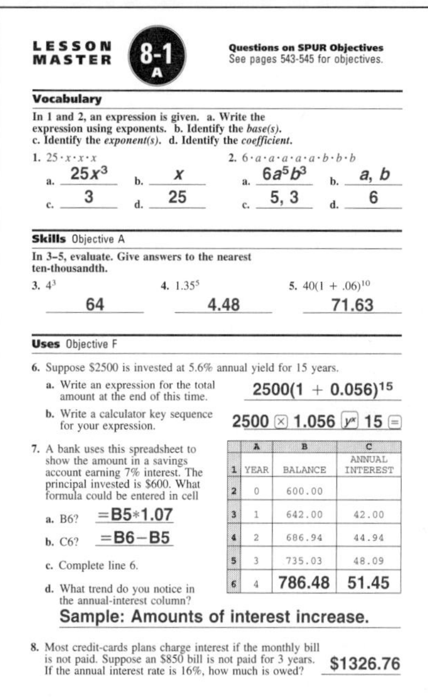

LESSON MASTER 8-1 A
Questions on SPUR Objectives
See pages 543-545 for objectives.

Vocabulary

In 1 and 2, an expression is given. a. Write the expression using exponents. b. Identify the *base(s)*. c. Identify the *exponent(s)*. d. Identify the *coefficient*.

1. $25 \cdot x \cdot x \cdot x$
a. $25x^3$ b. x
c. 3 d. 25

2. $6 \cdot a \cdot a \cdot a \cdot a \cdot a \cdot b \cdot b \cdot b$
a. $6a^5b^3$ b. a, b
c. 5, 3 d. 6

Skills Objective A

In 3–5, evaluate. Give answers to the nearest ten-thousandth.

3. 4^3 64 4. 1.35^5 4.48 5. $40(1 + .06)^{10}$ 71.63

Uses Objective F

6. Suppose $2500 is invested at 5.6% annual yield for 15 years.
a. Write an expression for the total amount at the end of this time. $2500(1 + 0.056)^{15}$
b. Write a calculator key sequence for your expression. 2500 [×] 1.056 [y^x] 15 [=]

7. A bank uses this spreadsheet to show the amount in a savings account earning 7% interest. The principal invested is $600. What formula could be entered in cell
a. B6? =B5*1.07
b. C6? =B6−B5
c. Complete line 6.
d. What trend do you notice in the annual-interest column? Sample: Amounts of interest increase.

	A	B	C
1	YEAR	BALANCE	ANNUAL INTEREST
2	0	600.00	
3	1	642.00	42.00
4	2	686.94	44.94
5	3	735.03	48.09
6	4	786.48	51.45

8. Most credit-cards plans charge interest if the monthly bill is not paid. Suppose an $850 bill is not paid for 3 years. If the annual interest rate is 16%, how much is owed? $1326.76

Adapting to Individual Needs

English Language Development
You might want to go over the vocabulary in this lesson: *principal, interest, annual yield,* and *compound interest.* The word *annual* means yearly, as in the annual school play. In Spanish, the word for year is *año;* in French it is *année.* The word *yield* means to produce or earn. Have students add the Compound Interest Formula to their index cards. On the front of the card, they can write the formula, and on the back they can write what each term means.

Notes on Questions

Question 14 Some students are surprised that after several years the interest from 10% annual yield is more than double the interest earned from 5% annual yield. Encourage students to explain in their own words why this happens. You might extend this situation to investigate what happens to each investment after 6 years. [$134.01 versus $177.16]

Question 18 Students can solve this problem by using *trial and error.*

Question 19 Students will need graph paper or **Teaching Aid 26.**

Question 22 Students are expected to use chunking to solve this problem—first taking the square roots of both sides, and then solving for x.

Question 25 Error Alert Employing the correct order of operations in expressions that involve exponents is critical. Many students will multiply by coefficients before taking the power, whether they use calculators or not. Review order of operations with these students.

Question 27 Problems such as this appear throughout the chapter. At this point, *trial and error,* with or without a calculator, is an appropriate method of solution.

LESSON MASTER 8-1 B — Questions on SPUR Objectives

Vocabulary

1. Consider a number of the form x^n.
   a. x^n is called a power.
   b. x is the base.
   c. n is the exponent.

In 2 and 3, an expression is given. a. Write the expression using exponents. b. Identify the *base(s)*. c. Identify the *exponent(s)*. d. Identify the *coefficient*.

2. $38 \cdot c \cdot c \cdot c \cdot c$
   a. $38c^4$
   b. c
   c. 4
   d. 38

3. $8 \cdot x \cdot x \cdot y \cdot y \cdot y \cdot y \cdot y$
   a. $8x^2y^5$
   b. x, y
   c. 2, 5
   d. 8

4. The data below show the amount of money in a bank account.

Deposit	$500.00
End of first year	500(1.07) = $535.00
End of second year	500(1.07)(1.07) = $572.45

   a. What is the annual yield? 7%
   b. What is the principal? $500
   c. What is the interest paid during the two-year period? $72.45

Skills Objective A: Evaluate integer powers of real numbers.

In 5–10, evaluate the expression. Give answers rounded to the nearest ten-thousandth.

5. 5^3 125
6. 19^2 361
7. 3.4^2 11.56
8. 1.06^6 1.4185
9. $(1 + .04)^7$ 1.3159
10. $200(1.085)^{12}$ 532.3372

Applying the Mathematics

14. Susan invests $100 at an annual yield of 5%. Jake invests $100 at an annual yield of 10%. They leave the money in the bank for 2 years.
a. How much interest does each person earn?
b. Does Jake earn exactly twice the interest that Susan does? **No**
a) Susan, $10.25; Jake, $21.00

15. Which results in more interest, (a) an amount invested for 5 years at an annual yield of 6%, or (b) the same amount invested for 3 years at an annual yield of 10%? Justify your answer. **a;** $P(1.06)^5 > P(1.1)^3$

In 16 and 17, use the spreadsheet below. It was produced to show the interest and total owed on a loan account of $1000 with a monthly interest rate of 1.5%, assuming none was paid back.

	A	B	C
1	Month	Total Owed	Monthly Interest
2	1	1015.00	15.00
3	2	1030.23	15.23
4	3	1045.68	
5	4	1061.36	15.69
6	5	1077.28	15.92
7	6	1093.44	16.16
8	7		16.40
9	8	1126.49	16.65
10	9	1143.39	16.90
11	10	1160.54	17.15
12	11	1177.95	17.41
13	12	1195.62	17.67

16. a. What formula should be typed into cell B8? **= 1000 * 1.015 ^ A8**
b. What value should be in B8? **$1109.84**
c. What is the *yearly* interest rate? **18%**

17. a. What formula should be typed into cell C4?
b. What value should be in C4? **$15.45**
a) = 1000 * 1.015 ^ A4 – 1000 * 1.015 ^ A3

18. Use your calculator. If a principal of $1000 is saved at an annual yield of 8% and the interest is kept in the account, in how many years will it double in value? **in about 9 years**

490

Adapting to Individual Needs

Challenge

In this book, all the examples involve interest that is compounded annually. As a challenge, you might give students the following formula to compute total interest T for 1 year when the annual rate of interest is i and the interest is compounded n times during the year. P represents the initial principal: $T = P\left(1 + \frac{i}{n}\right)^n$. Have students complete the following table to investigate the results of computing interest for $1000 earning an annual yield of 5% over shorter and shorter time periods.

Compounded	Total after 1 year
yearly	[$1050.00]
monthly	[$1051.16]
weekly	[$1051.25]
daily	[$1051.27]
hourly	[$1051.27]

Review

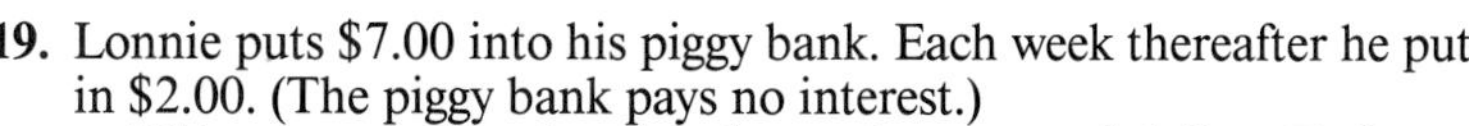

19. Lonnie puts $7.00 into his piggy bank. Each week thereafter he puts in $2.00. (The piggy bank pays no interest.)
- **a.** Write an equation showing the total amount of dollars T after W weeks. $T = 2W + 7$
- **b.** Graph the equation. *(Lesson 7-2)* See left.

19b)

$T = 2W + 7$

In 20 and 21, find the probability of the event. *(Lesson 6-4)*

20. getting a prime number in one toss of a fair die $\frac{1}{2}$

21. getting a face card (jack, queen, or king) when one card is pulled at random from a standard deck $\frac{3}{13}$

22. Solve $(1 + x)^2 = 1.1664$. *(Lesson 5-9)* $x = 0.08$ or $x = -2.08$

23. *Skill sequence.* Simplify. *(Lessons 2-1, 3-9)*
- **a.** $12(3n)$ $36n$
- **b.** $12(3n - 7)$ $36n - 84$
- **c.** $-12(3n - 7)$ $-36n + 84$
- **d.** $n(3n - 7)$ $3n^2 - 7n$

24. *Multiple choice.* Which formula describes the numbers in this table? *(Lesson 1-9)* d

x	1	2	3	4	5
y	2	4	8	16	32

(a) $y = 2x$ (b) $y = x + (x + 1)$ (c) $y = x^2$ (d) $y = 2^x$

25. Evaluate $-4x^5$ when $x = \frac{1}{2}$. *(Lesson 1-3)* $-\frac{1}{8}$

26. The volume V of a sphere with diameter d is given by the formula $V = \frac{\pi}{6}d^3$. Find, to the nearest cubic millimeter, the volume of a ball bearing with a diameter of 8 mm. *(Lesson 1-3)* 268 mm^3

27. Find t if $2^t = 64$. *(Previous course)* 6

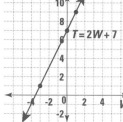

Exploration

28. Find out the yield for a savings account in a bank or other savings institution near where you live. (Often these yields are in newspaper ads.) Answers will vary.

29. Explore the meaning of 0 as an exponent on your calculator.
- **a.** Enter 25 [y^x] 0 [=] on your calculator. (You may have to use [x^y] or [^] instead of [y^x].) What is displayed? 1
- **b.** Enter 0 [y^x] 0 [=] on your calculator. What is displayed? Error
- **c.** Enter [(] 1 [+/−] [)] [y^x] 0 [=] on your calculator. What is displayed? Error or 1 depending on the calculator.
- **d.** Try other numbers and generalize what happens. Answers may vary depending on the calculator.

Follow-up for Lesson 8-1

Practice

For more questions on SPUR Objectives, use **Lesson Master 8-1A** (shown on page 489) or **Lesson Master 8-1B** (shown on pages 490–491).

Assessment

Written Communication Have students write a note to their parents explaining the Compound Interest Formula and how to find compound interest using a calculator. Ask them to give an example in their notes. [Notes are clearly written, and the terms *principal* and *interest* are defined. Students use an appropriate key sequence to find compound interest.]

Extension

Another way to answer **Question 18** is to use a rule of thumb called the *Rule of 72*. If money is invested at x% annual yield, it will double in about $\frac{72}{x}$ years. You might have students test this rule with several interest rates.

Project Update Project 5, *Interest Rates*, on page 540, relates to the content of this lesson.

► **LESSON MASTER 8-1 B** *page 2*

Uses Objective F: Calculate compound interest.

11. a. Write an expression for the amount in an account after 18 years if $3,500 is invested at 7.2% annual yield. $3{,}500(1.072)^{18}$

b. Write a key sequence to enter this on your calculator. Sample: 3500 [×] 1.07 [y^x] 18 [=]

c. To the nearest dollar, how much is in the account after 18 years? $12,234

12. A bank uses the spreadsheet below to show the amount in a savings account earning 6% interest. The principal invested is $800.

	A	B	C
1	YEAR	BALANCE	YEARLY INTEREST
2	0	800.00	
3	1	848.00	48.00
4	2	898.88	50.88
5	3	952.81	53.93
6	4		

a. What formula can be entered in cell B6? Sample: =B5*1.06

b. What number should appear in cell B6? 1009.98

c. What formula can be entered in cell C6? Sample: =B6−B5

d. What number should appear in cell C6? 57.17

e. What trend do you notice in Column C? What do you think accounts for this? Sample: The amount of interest increases because the balance each year is greater.

13. Barb won $5,000 in a lottery and decided to put it in the bank for an emergency. An emergency never arose. How much was in the account after 15 years if the account had an annual yield of 9%? $18,212.41

14. A department store charges 19.6% interest per year on unpaid monthly bills. How much would you owe if you did not pay a bill of $272 for 2 years? $389.07

Setting Up Lesson 8-2

Question 29 deals with the Zero Exponent Property, which is discussed in Lesson 8-2.

Lesson 8-2

Objectives

A Evaluate zero and positive integer powers of real numbers.
E Identify and apply the Zero Exponent Property.
G Solve problems involving exponential growth.
I Graph situations involving exponential growth.

Resources

From the ***Teacher's Resource File***
- Lesson Master 8-2A or 8-2B
- Answer Master 8-2
- Teaching Aids
  26 Graph Paper
  87 Warm-up

Additional Resources
- Visuals for Teaching Aids 26, 87

Teaching Lesson 8-2

Warm-up

Find the value of each term for $x = 0, 1, 2, 3, \dots 10$.

1. 2^x **1, 2, 4, 8, 16, 32, 64, 128, 256, 512, 1024**
2. $25 \cdot 2^x$ **25, 50, 100, 200, 400, 800, 1600, 3200, 6400, 12,800, 25,600**
3. $3 \cdot 2^x$ **3, 6, 12, 24, 48, 96, 192, 384, 768, 1536, 3072**

LESSON

8-2 *Exponential Growth*

A vegetarian diet. *Wild rabbits, such as these from Australia, feed on grass, weeds, bushes, and trees. Crops are sometimes damaged because rabbits eat vegetable sprouts and the bark of fruit trees.*

Powering and Population Growth

An important application of powers is in population growth situations. As an example, consider rabbit populations, which can grow quickly. In 1859 in Australia, 22 rabbits were imported from Europe as a new source of food. Rabbits are not native to Australia, but conditions there were ideal for rabbits, and they flourished. Soon, there were so many rabbits that they damaged grazing land. By 1889, the government was offering a reward for a way to control the rabbit population.

Example 1

Twenty-five rabbits are introduced to an area. Assume that the rabbit population doubles every six months. How many rabbits will there be after 5 years?

Solution

Since the population doubles twice each year, in 5 years it will double 10 times. The number of rabbits will be

$$25 \cdot \underbrace{2 \cdot 2 \cdot 2 \cdot 2 \cdot 2 \cdot 2 \cdot 2 \cdot 2 \cdot 2 \cdot 2}_{\text{10 factors}},$$

To evaluate this expression on a calculator rewrite it as

$$25 \cdot 2^{10}.$$

Use the y^x or $\wedge$ key. After 5 years there will be 25,600 rabbits.

Lesson 8-2 Overview

Broad Goals The purpose of this lesson is to help students understand how some situations lead to exponential growth and to introduce them to graphs associated with such growth.

Perspective The Growth Model for Powering is similar to the models that students have seen for addition, subtraction, multiplication, and division. That is, it indicates a class of applications of the operation; in so doing, it gives us a way of thinking about what the operation of powering means. We identify the base with the letter g to emphasize that it is the growth factor. In the Compound Interest Formula, $g = 1 + i$.

Even though we use integer values for g and x in the examples, the growth model holds for any positive real values of g and any real values of x. However, in this lesson g is always greater than 1 because we are dealing with growth. In Lesson 8-4 we introduce values of g such that $0 < g < 1$. These lead to exponential decay.

What Is Exponential Growth?

The rabbit population in Example 1 is said to grow exponentially. In **exponential growth,** the original amount is repeatedly *multiplied* by a positive number called the *growth factor.*

Growth Model for Powering
If a quantity is multiplied by a positive number g (the **growth factor**) in each of x time periods, then after the x periods, the quantity will be multiplied by g^x.

In Example 1, the population doubles (is multiplied by 2) every six months, so $g = 2$. There are 10 time periods (think 5 years = 10 half-years), so $x = 10$. The original number of rabbits, 25, is multiplied by g^x, or 2^{10}.

Compound interest is another example of the growth model. Suppose an annual yield is 5%. Then the growth factor $g = 1.05$. If the money is kept in for 7 years, then $x = 7$. The original principal is multiplied by g^x, which is 1.05^7.

What Happens if the Exponent Is Zero?

In the growth model, x can be any real number. Consider the statement of the growth model when $x = 0$. It reads:

❶ If a quantity is multiplied by a positive number g in each of 0 time periods, then after the 0 time periods, the quantity will be multiplied by g^0.

In 0 time periods, no time can elapse. Thus the quantity remains the same. It can remain the same only if it is multiplied by 1. This means that $g^0 = 1$, regardless of the value of the growth factor g. This property applies also when g is a negative number.

Zero Exponent Property
If g is any nonzero real number, then $g^0 = 1$.

In words, the zero power of any nonzero number equals 1. For example, $4^0 = 1$, $(-2)^0 = 1$, and $\left(\frac{5}{7}\right)^0 = 1$. The zero power of 0, which would be written 0^0, is undefined.

What Does a Graph of Exponential Growth Look Like?

❷ An equation of the form $y = b \cdot g^x$, where g is a number greater than 1, can describe exponential growth. Graphs of such equations are not lines. They are *exponential growth curves.*

Notes on Reading

You might want to use Activity 1 in *Optional Activities* below to introduce this lesson.

Reading Mathematics This lesson is appropriate for reading in class. As students read, you might interrupt them at the end of some of the sections to go over the corresponding questions in *Covering the Reading.*

If your students have read the lesson on their own, ask them for the main ideas. In our opinion, there are four key concepts: the growth model and its application to rabbits; the use of the growth model to explain why $g^0 = 1$ when $g \neq 0$; the graphing of equations of the form $y = b \cdot g^x$ and why such equations are not linear; the fact that such equations model real population growth in some situations. Each of these ideas is covered again in later lessons in the chapter; even if students are unsure about them, it is still reasonable to proceed to the next lesson.

❶ When the only meaning that students have for powering is repeated multiplication, a zero exponent is not meaningful because one cannot repeat something 0 times. However, the idea of "zero time periods from now," namely the present, is meaningful. Thus, when $x = 0$, it is natural that $b \cdot g^x = b$, which implies that $g^x = 1$ for all nonzero g.

❷ We use the letter b for the coefficient in $y = b \cdot g^x$ because we will compare $y = b \cdot g^x$ with $y = b + mx$ in the next lesson. In both cases, b is the y-intercept, which is the starting point for the situation.

Optional Activities

Activity 1 Technology Connection
To introduce the lesson, you might want to demonstrate repeated doubling with the following activity. Ask students to **work in pairs** and enter a number less than 20 into a calculator. Have them press [×] 2 ten times, writing down the product each time. Then have students repeat the process, pressing [×] 3 ten times and writing down each product.

Activity 2 Science Connection
To extend the topic of animal populations, you might want to have students use a science book or an encyclopedia to research the introduction of some of the following plants and animals into the United States, as well as the unforeseen results: Argentine fire ant, English sparrow, European starling, gypsy moth, Japanese beetle, kudzu, sea lamprey, and water hyacinth.

❸ Students are not expected to determine growth factors, but they may ask you how California's growth factor was calculated. The average growth factor for California over the six decades from 1930 to 1990 is the geometric mean of the growth factors for the decades. It is approximately equal to

$$\sqrt[6]{1.22 \cdot 1.53 \cdot 1.48 \cdot 1.27 \cdot 1.19 \cdot 1.26}.$$

Additional Examples

Students will need graph paper or **Teaching Aid 26.**

1. If you save 1 cent in January and double the amount of savings each month, how much would you save in a year? **$40.95**
2. Graph the equation $y = 2 \cdot 3^x$, when x is 0, 1, 2, 3, and 4.

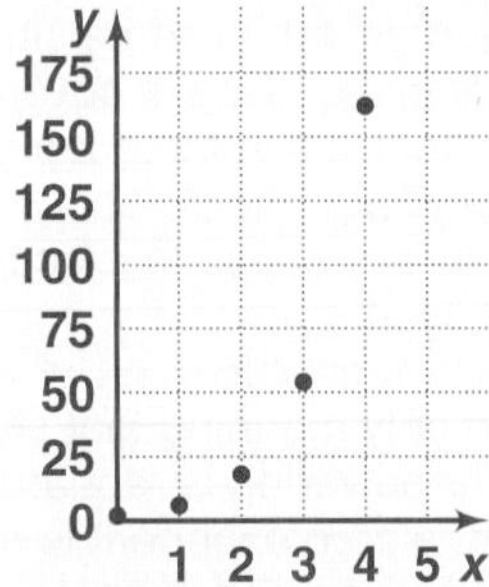

3. Through the 1980s, the population of Central and South America grew at a rate of about 2.1% per year. In 1991 the population was 458 million people. If this growth continues, what will the population of Central and South America be in the year 2000? **$458(1.021)^9 \approx 552$ million**

LESSON MASTER 8-2 A

Questions on SPUR Objectives
See pages 543-545 for objectives.

Skills Objective A

In 1–7, evaluate the expression.

1. 5^0 **1**
2. $8^0 \cdot 8^2$ **64**
3. $5^6 \cdot 5^0$ **15,625**
4. $(a + 15)^0$ when $a = 6$ **1**
5. $\left(\frac{3}{8}\right)^2 + \left(\frac{7}{8}\right)^0$ $\frac{73}{64}$
6. $3 \cdot 10^2 + 4 \cdot 10^1 + 9 \cdot 10^0$ **349**
7. $400 \cdot 3^n$ when $n = 0$ **400**

Properties Objective E

8. Describe a situation that can be modeled with the expression $100 \cdot 2^0$.
Sample: A colony of 100 organisms doubles its population every 15 minutes.

9. A colony of bacteria doubles every 6 hours. How many times will it double in
   a. 24 hours? **4 times**
   b. 5 days? **20 times**
   c. 2 weeks? **56 times**

Uses Objective G

10. An orange grower introduced a new type of ladybug to his orchards to control pests. He began with 500 ladybugs and feels that the number will naturally increase 5% each month. Based on this information, answer the following questions: In Parts **b** and **c**, round values down to the preceding integer.
    a. By how much is the number of ladybugs multiplied each month? **1.05**
    b. How many ladybugs will there be after 6 months? **≈670 ladybugs**
    c. How many ladybugs will there be after 2 years? **≈1,612 ladybugs**

Example 2

Graph the equation $y = 3 \cdot 2^x$, when x is 0, 1, 2, 3, 4.

Solution

Substitute $x = 0, 1, 2, 3$, and 4 into the formula $y = 3 \cdot 2^x$. Below we show the computation and the results listed as (x, y) pairs.

$3 \cdot 2^0 = 3 \cdot 1 = 3$
$3 \cdot 2^1 = 3 \cdot 2 = 6$
$3 \cdot 2^2 = 3 \cdot 4 = 12$
$3 \cdot 2^3 = 3 \cdot 8 = 24$
$3 \cdot 2^4 = 3 \cdot 16 = 48$

x	$y = 3 \cdot 2^x$
0	3
1	6
2	12
3	24
4	48

Plot these points. The graph is shown below.

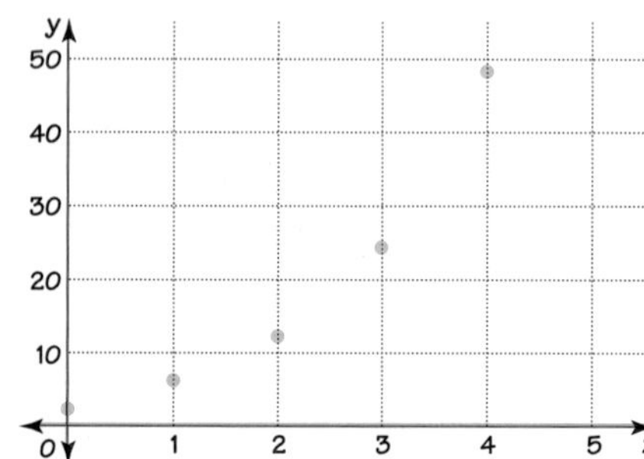

Notice that the graph in Example 2 does not have a constant rate of change. When something grows exponentially, its rate of change is continually increasing.

An Example of Exponential Population Growth

Other than money calculated using compound interest, few things in the real world grow exactly exponentially. Over the short term, however, exponential growth approximates the changes that have been observed in many different populations. For instance, consider the population of California from 1930 to 1990.

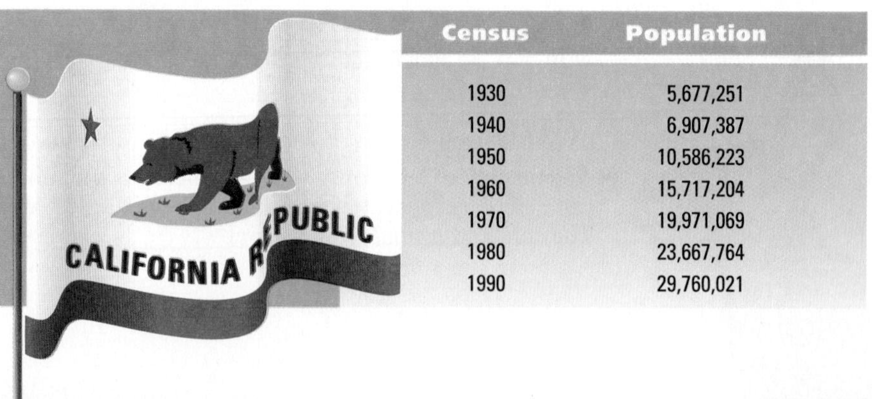

Census	Population
1930	5,677,251
1940	6,907,387
1950	10,586,223
1960	15,717,204
1970	19,971,069
1980	23,667,764
1990	29,760,021

494

Adapting to Individual Needs

Extra Help

If students need further convincing that $g^0 = 1$, show them the following pattern.

$2^4 = 2 \cdot 2 \cdot 2 \cdot 2 = 16$
$2^3 = 2 \cdot 2 \cdot 2 = 8$
$2^2 = 2 \cdot 2 = 4$
$2^1 = 2$

Point out that the exponents decrease by 1, so the next power in the sequence would be 2^0. Ask students how the product changes as the exponent decreases by 1. [It is divided by 2.] Ask what the next product would be. [It is $2 \div 2$, or 1.] Then write $2^0 = 1$ as the fifth step in the pattern.

Streets of San Francisco. *Shown is a trolley car on one of San Francisco's busy streets. Like other parts of California, San Francisco experienced rapid growth from 1930 to 1945. In 1993, its population was 723,959.*

3 To see how close California's population growth is to exponential growth, compare the two graphs below. On the left is a graph of California's population. On the right we have plotted points on the graph of $y = 5.68(1.32)^x$. We picked 1.32 because it is an "average" growth factor for California for the six decades. Except for the differences in the scale on the horizontal axis, they look quite a bit alike.

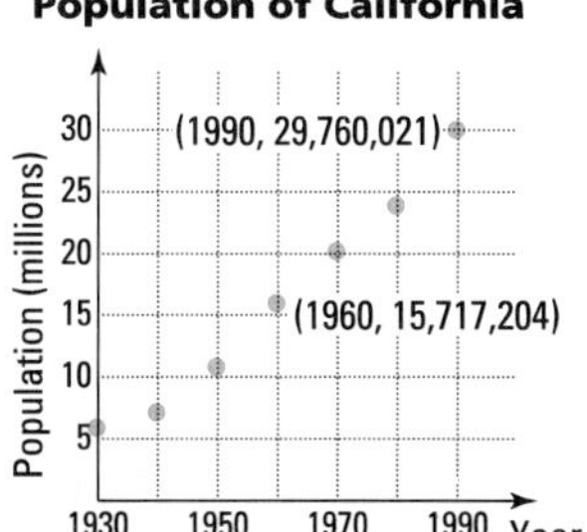

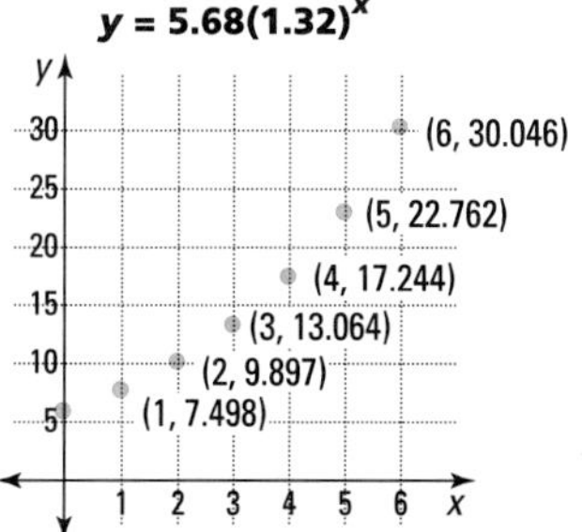

QUESTIONS

Covering the Reading

In 1–3, refer to Example 1.

1. a. After 7 years, how many times will the rabbit population have doubled? **14**
   b. How many rabbits will there be? **409,600**
2. How many rabbits will there be after 10 years? **26,214,400**
3. If the rabbit population triples in 6 months, rather than doubles, how many rabbits would there be after 5 years? **1,476,225**
4. Assume the rabbit population in Australia doubled every *year* from 1859 to 1889. How many rabbits were in Australia in 1889? **about 23,622,000,000 rabbits**
5. State the Growth Model for Powering. **See left.**

5) If a quantity is multiplied by a positive number g in each of x time periods, then after the x periods, the quantity will be multiplied by g^x.

6. Suppose money is put into a savings account with an annual yield of 3.5%. It is left in the account for 2 years.
   a. What is the growth factor? **1.035**
   b. What is the unit period? **one year**
   c. By how much will the money be multiplied in the 2-year period? **$(1.035)^2 = 1.071225$**
7. *True or false.* An amount is multiplied by 10 in each of 7 time periods. Then after the 7 periods, the original amount will be multiplied by 70. **False**
8. State the Zero Exponent Property. **If g is any nonzero real number, then $g^0 = 1$.**
9. Calculate 17^0. **1**
10. Explain how $g^0 = 1$ fits the Growth Model for Powering. **A period of length 0 means no time has elapsed, so a quantity multiplied by g^0 doesn't change. This means $g^0 = 1$.**

Notes on Questions

Questions 1–4 Science Connection Because of the lack of natural predators, the rabbit population in Australia grew at an alarming rate between 1859 and 1889. Similar but less dramatic situations have occurred throughout the history of the United States. Non-native plants and animals have been introduced and in some cases this has led to unfortunate results. Sometimes the species were introduced accidentally, and in other cases the introduction was intentional. See Activity 2 in *Optional Activities* on page 493.

Questions 1–3 Error Alert A common error is not realizing that the number of time periods is twice the number of years. For example, if students got 3200 as the answer to **Question 1,** they probably used 7 as the number of time periods. Have these students reread the information in **Example 1**. Ask them what time period is mentioned for doubling the rabbit population [6 months], and how many such periods there are in 7 years. [14 periods.]

Question 4 Error Alert Students might mistakenly use the value of 25 (from **Example 1**) as the number of rabbits introduced into Australia. The correct value, 22, is given in the paragraph above **Example 1.**

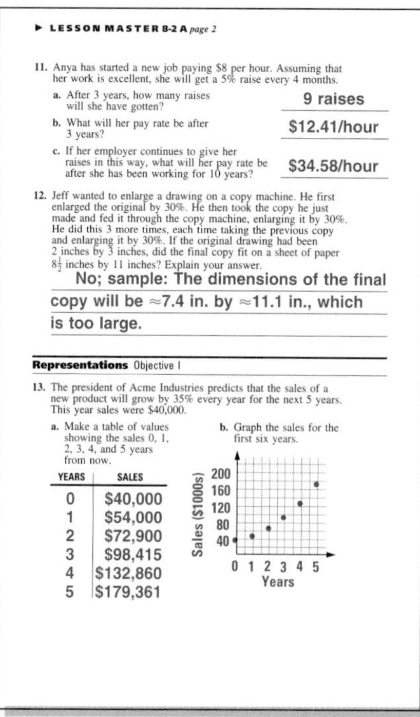

▸ **LESSON MASTER 8-2 A** *page 2*

11. Anya has started a new job paying \$8 per hour. Assuming that her work is excellent, she will get a 5% raise every 4 months.
    a. After 3 years, how many raises will she have gotten? **9 raises**
    b. What will her pay rate be after 3 years? **\$12.41/hour**
    c. If her employer continues to give her raises in this way, what will her pay rate be after she has been working for 10 years? **\$34.58/hour**
12. Jeff wanted to enlarge a drawing on a copy machine. He first enlarged the original by 30%. He then took the copy he just made and fed it through the copy machine, enlarging it by 30%. He did this 3 more times, each time taking the previous copy and enlarging it by 30%. If the original drawing had been 2 inches by 3 inches, did the final copy fit on a sheet of paper $8\frac{1}{2}$ inches by 11 inches? Explain your answer.
    No; sample: The dimensions of the final copy will be ≈7.4 in. by ≈11.1 in., which is too large.

Representations Objective I

13. The president of Acme Industries predicts that the sales of a new product will grow by 35% every year for the next 5 years. This year sales were \$40,000.
    a. Make a table of values showing the sales 0, 1, 2, 3, 4, and 5 years from now.

YEARS	SALES
0	\$40,000
1	\$54,000
2	\$72,900
3	\$98,415
4	\$132,860
5	\$179,361

    b. Graph the sales for the first six years.

Adapting to Individual Needs

English Language Development

Ask students what they think of when they hear the word *power* in an everyday setting. Most students will think of strength, force, or influence. Relate this to the mathematical meaning of power by reminding students how quickly powers increase; illustrate by reviewing powers of 10 through 10^6.

Notes on Questions

Question 15 Ask students to explain how they did **parts c and d.** Some students will continue the tripling pattern set up in **parts a and b.** Others may notice a formula, $b = 2000 \cdot 3^{3t}$, where t = the number of hours and b = the number of bacteria.

Question 16 Geography Connection *Xaymaca,* which means *land of wood and water,* was the name given to Jamaica by the Arawak Indians who once lived there. Jamaica is the third largest of the Caribbean islands. Its year-round warm climate, sandy beaches, and mountains make it a popular vacation area.

Question 19 This problem should be done by using the Repeated Multiplication Model for Powering. You can use this question to anticipate the Product of Powers Property, which is covered in Lesson 8-5.

Question 22 Students can use the pattern $b \cdot g^x$ or the Compound Interest Formula to find the answer; either way is reasonable.

Question 28 Students can use calculators to multiply $11^2 \cdot 11^3$. You might ask if the answer, 161,051, is a power of 11 and if so, which power it is. [Yes; $11^5 = 161{,}051$] The discovery that $11^2 \cdot 11^3 = 11^5$ previews the study of properties of powers that begins in Lesson 8-5.

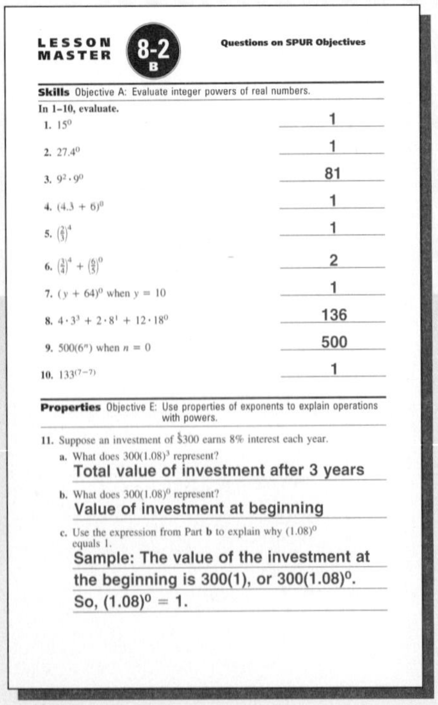

LESSON MASTER 8-2 B Questions on SPUR Objectives

Skills Objective A: Evaluate integer powers of real numbers.

In 1–10, evaluate.

1. 15^0 — **1**
2. 27.4^0 — **1**
3. $9^2 \cdot 9^0$ — **81**
4. $(4.3 + 6)^0$ — **1**
5. $\left(\frac{2}{3}\right)^4$ — **1**
6. $\left(\frac{3}{4}\right)^4 + \left(\frac{6}{5}\right)^0$ — **2**
7. $(y + 64)^0$ when $y = 10$ — **1**
8. $4 \cdot 3^3 + 2 \cdot 8^1 + 12 \cdot 18^0$ — **136**
9. $500(6^n)$ when $n = 0$ — **500**
10. $133^{(7-7)}$ — **1**

Properties Objective E: Use properties of exponents to explain operations with powers.

11. Suppose an investment of \$300 earns 8% interest each year.
    a. What does $300(1.08)^3$ represent? **Total value of investment after 3 years**
    b. What does $300(1.08)^0$ represent? **Value of investment at beginning**
    c. Use the expression from Part **b** to explain why $(1.08)^0$ equals 1. **Sample: The value of the investment at the beginning is 300(1), or $300(1.08)^0$. So, $(1.08)^0 = 1$.**

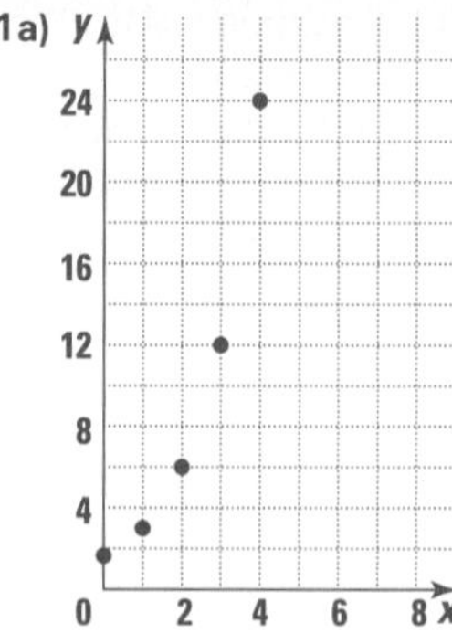
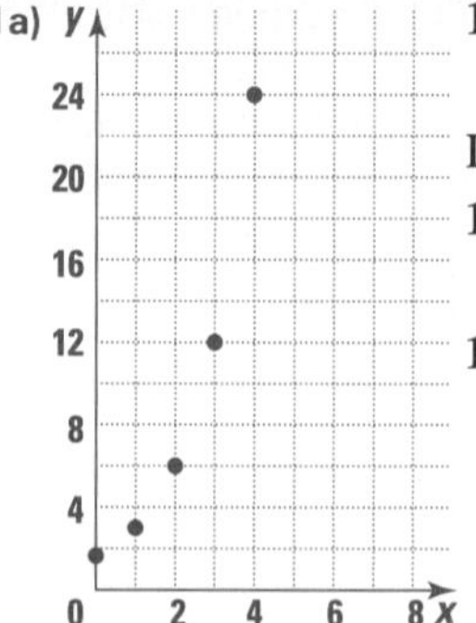

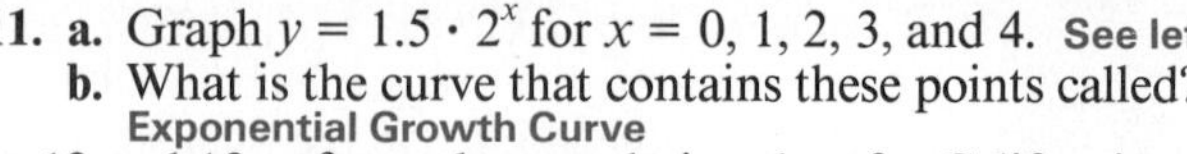

11. a. Graph $y = 1.5 \cdot 2^x$ for $x = 0, 1, 2, 3,$ and 4. **See left.**
    b. What is the curve that contains these points called? **Exponential Growth Curve**

In 12 and 13, refer to the population data for California.

12. What has been an "average" growth factor for California for the 6 decades from 1930 to 1990? **1.32**

13. Suppose California's population grows by 20% in the decade from 1990 to 2000.
    a. What would be the growth factor for that decade? **1.2**
    b. What would the population be in the year 2000? **about 35,712,025**

Applying the Mathematics

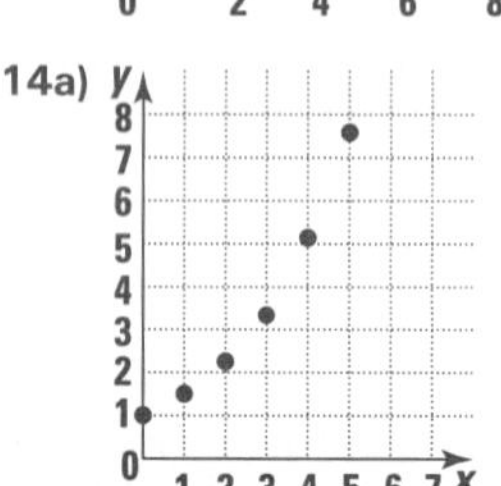

14. a. Graph $y = 1.5^x$ for $x = 0, 1, 2, 3, 4,$ and 5. **See left.**
    b. Calculate the rate of change on the graph from $x = 0$ to $x = 1$. **0.5**
    c. Calculate the rate of change on the graph from $x = 4$ to $x = 5$.
    d. What do the answers to parts **b** and **c** tell you about this graph? **Sample: The rate of change is increasing as x increases. c) 2.53125**

15. The following chart describes the exponential growth of a colony of bacteria. You can see that this strain of bacteria grows very fast. In only one hour it grows from 2,000 to 54,000 bacteria.

Time Intervals from Now	Time (min)	Number of Bacteria
0	0	2000
1	20	$6000 = 2000 \cdot 3^1$
2	40	$18{,}000 = 2000 \cdot 3^2$
3	60	$54{,}000 = 2000 \cdot 3^3$

a. How long does it take the population to triple? **20 minutes**
b. How many times will the population triple in two hours? **6**
c. How many bacteria will be in the colony after two hours?
d. How many bacteria will be in the colony after four hours? **1,062,882,000; c) 1,458,000**

The marketplace. *This is an outdoor market in Jamaica, an island in the West Indies. Sugar cane is Jamaica's most important crop. Other farm products include bananas, cacao, coconuts, coffee, and citrus fruits.*

16. Jamaica's population of 2,466,000 in 1990 was expected to grow exponentially by 1.1% each year for the rest of the twentieth century.
    a. With this growth, what will the population be in 1995? **2,604,647**
    b. With this growth, what will the population be in 2010? **3,069,136**

17. On September 30, 1992, the U.S. national debt was about 4.065 trillion dollars and was growing at a rate of about 12.5% per year. At this rate, estimate the national debt on September 30, 1994. **about 5.14 trillion dollars**

In 18–21, simplify.

18. $(4y)^0$ when $y = \frac{1}{2}$ **1**

19. $7^0 \cdot 7^1 \cdot 7^2$ **343**

20. $6(x + y)^0$ when $x = 3$ and $y = -8$ **6**

21. $\frac{2}{3}\left(\frac{1}{2}\right)^0 + \frac{1}{2}\left(\frac{2}{3}\right)^2$ $\frac{8}{9}$

Adapting to Individual Needs

Challenge

Have students solve this problem. Suppose you begin with one sheet of notebook paper which is about 0.0037 inches thick. How many times would you have to cut the paper in half and put the halves on top of each other to get a stack at least as tall as you are? [Answers will vary. Sample: For students 5 feet tall through 5 feet 5 inches tall, there would have to be 14 cuts.]

Review

22. \$2200 is deposited in a savings account. *(Lesson 8-1)*
   a. What will be the total amount of money in the account after 6 years at an annual yield of 6%? **\$3,120.74**
   b. How much interest will have been earned in those 6 years? **\$920.74**

23. Suppose a letter from the alphabet is chosen randomly. What is the probability that it is a vowel from the first half of the alphabet? *(Lesson 6-4)* $\frac{3}{26}$

24. *Skill sequence.* Solve. *(Lesson 5-9)*
   a. $y^2 = 144$ **12; -12**
   b. $(4y)^2 = 144$ **3; -3**
   c. $(4y - 20)^2 = 144$ **8; 2**
   d. $y^2 + 80 = 144$ **8; -8**

In 25 and 26, simplify. *(Lessons 3-9, 4-6)*

25. $6(n + 8) + 4(2n - 1)$ **$14n + 44$**
26. $13 - (2 - x)$ **$11 + x$**

In 27 and 28, evaluate the expression. *(Lesson 1-3)*

27. $4s^9$ when $s = \frac{1}{2}$ $\frac{1}{128}$
28. $t^2 \cdot t^3$ when $t = 11$ **161,051**

Exploration

29. This exploration helps to explain why 0^0 is undefined.
   a. Use your calculator to give values of x^0 for $x = 1, 0.1, 0.01, 0.001$, and so on. What does this suggest for the value of 0^0? **See below.**
   b. Use your calculator to give values of 0^x for $x = 1, 0.1, 0.01, 0.001$, and so on. What does this suggest for the value of 0^0? **See below.**
   c. What does your calculator display when you try to evaluate 0^0? Why do you think it gives that display? **Error message. 0^0 cannot equal both 0 and 1, so 0^0 is undefined.**

30. There is an old story about a man who did a favor for a king. The king wished to reward the man and asked how he could do so. The man asked for a chessboard with one kernel of wheat on the first square of the chessboard, two kernels on the second square, four on the third square, eight on the fourth square, and so on for the entire sixty-four squares of the board. Find how many grains of wheat would be on the last square. (The answer may amaze you. It is about 250 times the total present yearly wheat production of the world.) **about 9.22×10^{18}**

29a) 1, 1, 1, 1, . . . ; $0^0 = 1$ b) 0, 0, 0, 0, . . . ; $0^0 = 0$

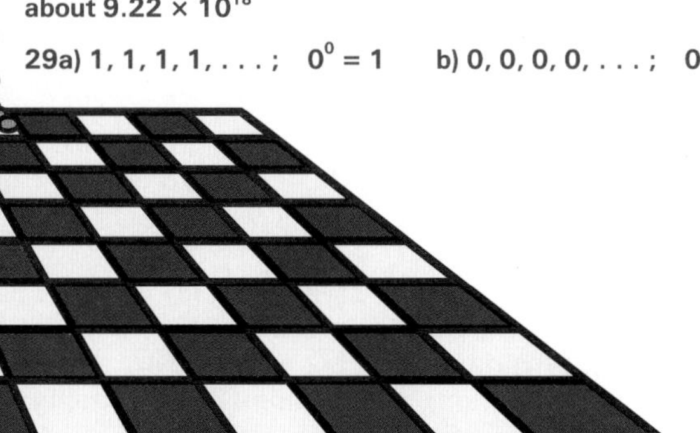

Follow-up for Lesson 8-2

Practice

For more questions on SPUR Objectives, use **Lesson Master 8-2A** (shown on pages 494–495) or **Lesson Master 8-2B** (shown on pages 496–497).

Assessment

Written Communication Have students **work in pairs**. Ask each student to write four expressions that are similar to those in **Questions 18–21**. Then have partners exchange papers, simplify the expressions, and check each other's work. [Students demonstrate the ability to evaluate zero and positive integer powers of real numbers.]

Extension

Project Update Project 1, *Population Growth over Long Periods*, on page 539, relates to the content of this lesson.

Setting Up Lesson 8-3

Materials If automatic graphers are available, we recommend that students have them for **Questions 16–19** in Lesson 8-3.

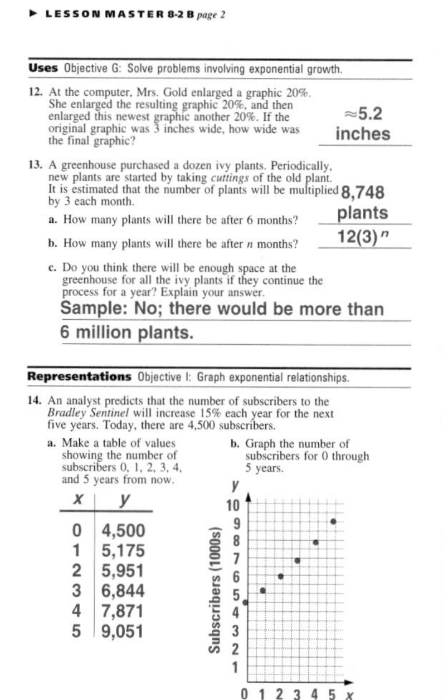

▶ **LESSON MASTER 8-2 B** *page 2*

Uses Objective G: Solve problems involving exponential growth.

12. At the computer, Mrs. Gold enlarged a graphic 20%. She enlarged the resulting graphic 20%, and then enlarged this newest graphic another 20%. If the original graphic was 3 inches wide, how wide was the final graphic? **≈5.2 inches**

13. A greenhouse purchased a dozen ivy plants. Periodically, new plants are started by taking *cuttings* of the old plant. It is estimated that the number of plants will be multiplied by 3 each month.
   a. How many plants will there be after 6 months? **8,748 plants**
   b. How many plants will there be after n months? **$12(3)^n$**
   c. Do you think there will be enough space at the greenhouse for all the ivy plants if they continue the process for a year? Explain your answer. **Sample: No; there would be more than 6 million plants.**

Representations Objective I: Graph exponential relationships.

14. An analyst predicts that the number of subscribers to the *Bradley Sentinel* will increase 15% each year for the next five years. Today, there are 4,500 subscribers.
   a. Make a table of values showing the number of subscribers 0, 1, 2, 3, 4, and 5 years from now.

x	y
0	4,500
1	5,175
2	5,951
3	6,844
4	7,871
5	9,051

   b. Graph the number of subscribers for 0 through 5 years.

Lesson 8-3

Objectives

G Solve problems involving exponential growth and decay.
I Graph exponential relationships.

Resources

From the ***Teacher's Resource File***
- Lesson Master 8-3A or 8-3B
- Answer Master 8-3
- Assessment Sourcebook: Quiz for Lessons 8-1 through 8-3
- Teaching Aids
  26 Graph Paper
  37 Spreadsheet
  87 Warm-up
  91 Example 1
  92 Constant Increase and Exponential Growth
- Technology Sourcebook Computer Master 20

Additional Resources
- Visuals for Teaching Aids 26, 37, 87, 91, 92
- GraphExplorer or other automatic graphers
- Spreadsheet software

Teaching Lesson 8-3

Warm-up

Writing Have students **work with a partner** to solve the following problem.

Seth told Joey that a small bug, which weighs 0.05 of an ounce, will double its weight every day for two weeks. Should Joey believe Seth? Explain your answer. **No; after 14 days the bug would weigh 819.2 oz or about 51 lb.**

LESSON

8-3 Comparing Constant Increase and Exponential Growth

Uncle Scrooge. *Walt Disney's Uncle Scrooge generously doled out money to Donald Duck's nephews, Huey, Dewey, and Louie. If they had invested this money wisely, their savings might have grown exponentially.*

What Is the Difference Between Constant Increase and Exponential Growth?

In constant-increase situations, a number is repeatedly *added.* In exponential-growth situations, a number is repeatedly *multiplied.* If the growth factor g is greater than 1, exponential growth always overtakes constant increase. This is illustrated in the following example.

Example 1

Suppose you have \$10. Your rich uncle agrees each day either to (1) increase what you had the previous day by \$50, or (2) increase what you had the previous day by 50%. Which is the better option?

Solution

Make a table to compare the two options during the first week. The exponential growth factor is $1 + 50\% = 1 + 0.5 = 1.5$.

❶

	option (1): add 50	option (2): multiply by 1.5
1st day	$10 + 50 \cdot 1 =$ \$ 60.00	$10 \cdot 1.5^1 =$ \$ 15.00
2nd day	$10 + 50 \cdot 2 =$ \$110.00	$10 \cdot 1.5^2 =$ \$ 22.50
3rd day	$10 + 50 \cdot 3 =$ \$160.00	$10 \cdot 1.5^3 =$ \$ 33.75
4th day	$10 + 50 \cdot 4 =$ \$210.00	$10 \cdot 1.5^4 \approx$ \$ 50.63
5th day	$10 + 50 \cdot 5 =$ \$260.00	$10 \cdot 1.5^5 \approx$ \$ 75.94
6th day	$10 + 50 \cdot 6 =$ \$310.00	$10 \cdot 1.5^6 \approx$ \$113.91
7th day	$10 + 50 \cdot 7 =$ \$360.00	$10 \cdot 1.5^7 \approx$ \$170.86
nth day	$10 + 50n$	$10 \cdot 1.5^n$

In all the first 7 days, option (1) is better. But examine the 14th day. Substitute 14 for n in the bottom row.

14th day $10 + 50 \cdot 14 =$ \$710 $\quad 10 \cdot 1.5^{14} =$ \$2919.29

In the long run, option (2), increasing by 50% each day, is the better choice.

Lesson 8-3 Overview

Broad Goals This lesson affords a second day to examine exponential growth and to contrast exponential-growth situations with the constant-increase situations that students encountered in Chapter 3.

Perspective The word *rate* is often used in discussing both constant increase and exponential growth. In the former, it is a rate of change; in the latter it is a growth rate. This dual use of *rate* can confuse students.

We try to avoid confusion by using the word *growth* (and *decay*) when referring to change under *exponential growth* (and *decay*) and the word *increase* (and *decrease*) when referring to *linear increase* (and *decrease).* Yet students must realize that the language is not always so clear outside the classroom.

It may take a few days before students see that quantities which grow exponentially will eventually increase *very quickly.* Examining the patterns $b + mx$ and $b \cdot g^x$ will help them understand this concept.

We use the complete equations $y = mx + b$ and $y = b \cdot g^x$ for descriptions and graphs; they are convenient forms for automatic graphing. The graphs enable students to estimate solutions to equations. Spreadsheets and automatic graphers will simplify the job of calculating and graphing.

To see what has happened in Example 1, examine the spreadsheet below, which lists the amount received on each of the first 14 days.

	A	B	C
1	day	constant increase	exponential growth
2	x	10 + 50x	10 * 1.5^x
3	0	10	10.00
4	1	60	15.00
5	2	110	22.50
6	3	160	33.75
7	4	210	50.63
8	5	260	75.94
9	6	310	113.91
10	7	360	170.86
11	8	410	256.29
12	9	460	384.44
13	10	510	576.65
14	11	560	864.98
15	12	610	1297.46
16	13	660	1946.20
17	14	710	2919.29

How Do the Graphs of Constant Increase and Exponential Growth Compare?

Another way to compare the two options is to graph the two equations

$y = 10 + 50x$ and $y = 10 \cdot 1.5^x$.

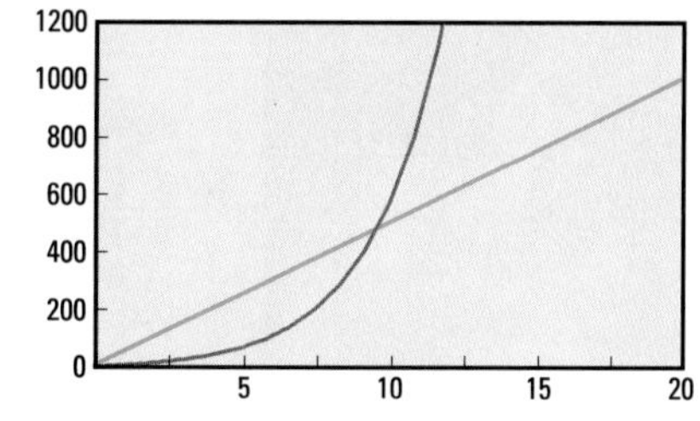

The graphs at the right were drawn with an automatic grapher. Each graph has y-intercept of 10. As you learned in Chapter 7, a line has a constant rate of change. But the graph of $y = 10 \cdot 1.5^x$ is not a line. It is a curve that gets steeper and steeper as you move to the right. Notice that at first the exponential curve is below the line. But toward the middle of the window, it intersects the line and passes above it. On later days, the graph of the curve rises farther and farther above the line. The longer your uncle gives you money, the better option 2 is when compared to option 1.

Notes on Reading

Both **Examples 1 and 2** help students to recognize the differences between constant increase and exponential growth. So, although the lesson may seem long, it has only one major idea.

In this lesson, all situations involve growth. Situations in the next lesson involve decay. All of the graphs use $x \geq 0$ as the domain because we do not discuss negative exponents until Lesson 8-6.

❶ **Teaching Aid 91** contains the facts for **Example 1.**

Technology Connection If you have access to either a spreadsheet or an automatic grapher for demonstration, you may want to demonstrate **Examples 1 and 2.**

You might want to use **Teaching Aid 92** when you discuss the summary of Constant Increase and Exponential Growth on page 500.

Optional Activities

Activity 1 After students have completed the lesson, you might want to have them invent a situation in which there might be both exponential growth and constant increase, and then have them describe the differences. For instance, if you start with 10,000 people, does a 2% yearly growth rate ever overcome a yearly increase of 10,000? [Yes, at about 286 years]

Technology Connection It students have spreadsheet software available to them, have them use it for their situations.

Activity 2 Technology Connection With *Technology Sourcebook, Computer Master 20,* students use *GraphExplorer* or similar software to make graphs comparing constant increase and exponential growth.

Additional Examples

1. Two villages each have 50 inhabitants. Suppose the first village grows by 20 people each year and the second village grows by 20% each year.
   a. Make a table of values to show the growth of the two villages over the next 10 years.
   1st village: $y = 50 + 20x$; 2nd village: $y = 50 \cdot 1.2^x$

1st village: Add 20	2nd village: Multiply by 1.2
$50 + 20 = 70$	$50 \cdot 1.2 = 60$
$70 + 20 = 90$	$50 \cdot (1.2)^2 = 72$
$90 + 20 = 110$	$50 \cdot (1.2)^3 \approx 86$
$110 + 20 = 130$	$50 \cdot (1.2)^4 \approx 104$
$130 + 20 = 150$	$50 \cdot (1.2)^5 \approx 124$
$150 + 20 = 170$	$50 \cdot (1.2)^6 \approx 149$
$170 + 20 = 190$	$50 \cdot (1.2)^7 \approx 179$
$190 + 20 = 210$	$50 \cdot (1.2)^8 \approx 215$
$210 + 20 = 230$	$50 \cdot (1.2)^9 \approx 258$
$230 + 20 = 250$	$50 \cdot (1.2)^{10} \approx 310$

   For the first five years, the first village grows faster; after that the second village grows faster. The rate of growth is the change in population over the change in time between one year and the previous year, not the amount of the population change.

   b. Graph the populations of the two villages over the next 10 years.

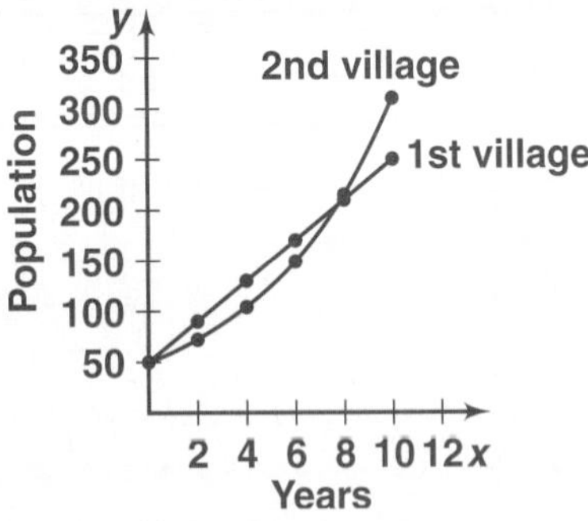

A Summary of Constant Increase and Exponential Growth

Here is how constant increase and exponential growth compare, in general.

Constant Increase	Exponential Growth
1. Begin with an amount b.	1. Begin with an amount b.
2. *Add* m (the slope) in each of x time periods.	2. *Multiply* by g (the growth factor) in each of x time periods.
3. After the x time periods, the amount will be $b + mx$.	3. After the x time periods, the amount will be $b \cdot g^x$.

The difference can be seen in the graph of each type.

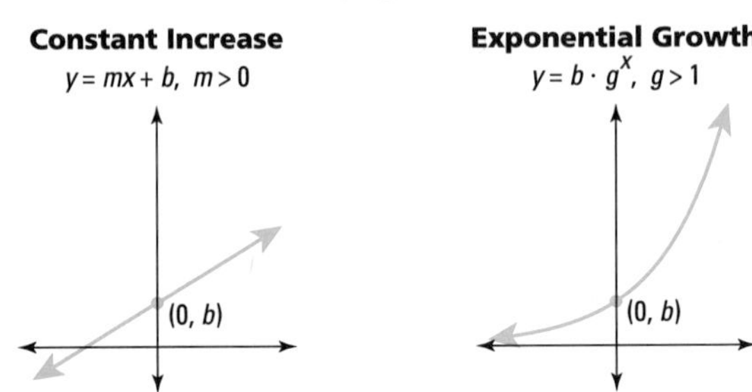

Mind over mattress. *Comedian Jack Benny played a character who stored his money in his home. However, in this scene from his TV show, even he realized it is wiser to save at a bank.*

Example 2

Suppose you have saved \$100 and are able to earn an annual yield of 6%.

a. Graph your savings if you leave your \$100 in the bank, but take the interest out and put it in a piggy bank each year. (This is *not* compound interest.)

b. On the same set of axes, graph your savings if you leave the interest in the bank account. Round values to the nearest dollar.

c. After 30 years, how much more will you have if you let your interest compound in the bank account?

Solution

a. Each year 6% of the original \$100 (or \$6) is earned in interest. If y is the total savings you have after x years, then $y = 100 + 6x$. Make a table.

number of years	0	5	10	15	20	25	30
value	\$100	\$130	\$160	\$190	\$220	\$250	\$280

b. Make a table for the amount saved at 6% compound interest. Use the formula $y = 100(1.06)^x$.

number of years	0	5	10	15	20	25	30
value	\$100	\$134	\$179	\$240	\$321	\$429	\$574

Adapting to Individual Needs

Extra Help

Some students might not understand how repeated multiplication by a given percent translates into exponential growth. Showing them the following steps might help. Use the numbers from **Example 1**. First show that the total for the first day is 10×1.5, or 15. Then point out that the second day's total is found by multiplying the first day's total by 1.5. Then use the Associative Property.

$$\begin{aligned} \text{2nd day} &= \text{1st day} \cdot 1.5 \\ &= (10 \cdot 1.5) \cdot 1.5 \\ &= 10 \cdot (1.5 \cdot 1.5) \\ &= 10 \cdot 1.5^2 \end{aligned}$$

Plot the points and connect them for each graph. In the graph below, the piggy bank graph (blue) is one of constant increase. The compound interest graph (orange) is of exponential growth.

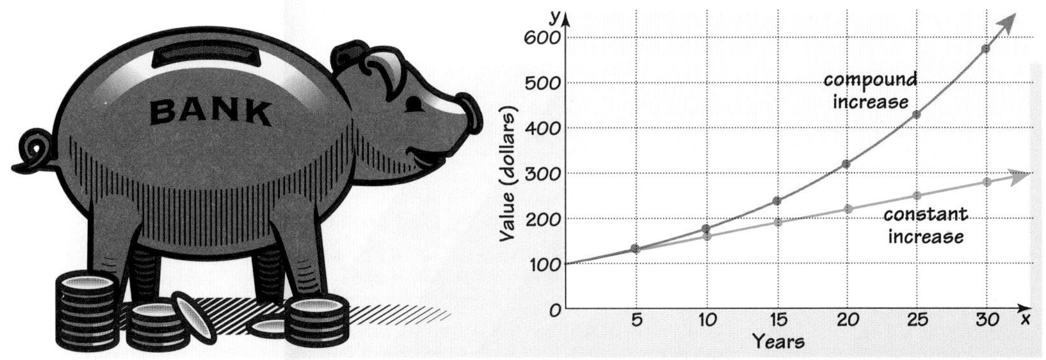

c. You can use either the table or the graph. From the table you see that after 30 years the first plan yields \$280, while the second plan yields \$574. *You will have \$574 − \$280 = \$294 more with compound interest.*

QUESTIONS

Covering the Reading

1) Sample: In constant-increase situations, a number is repeatedly added; in exponential-growth situations, a number is repeatedly multiplied.

1. What is the difference between a constant-increase situation and an exponential-growth situation? **See left.**

2. In Example 2, as x increases, which increases more rapidly, $y = 100 + 6x$ or $y = 100(1.06)^x$? **$y = 100(1.06)^x$**

In 3–5, use Example 1 or the spreadsheet that follows it.

3. After one month (30 days), how much money would you have under option (1)? under option (2)? **\$1510; \$1,917,510.59**

4. Which is the first day the amount received under exponential growth is greater than the amount received under constant increase? **the tenth day**

5. Give the numbers that go in row 18 of the spreadsheet. **15; 760; 4378.94**

6. How does the graph of $y = 10 + 5x$ differ from the graph of $y = 10 \cdot 1.5^x$? **The graph of $y = 10 + 5x$ is a line. The graph of $y = 10 \cdot 1.5^x$ is a curve.**

7. Give the general formula for the indicated type of situation.
   a. constant increase **$y = mx + b,\ m > 0$**
   b. exponential growth **$y = b \cdot g^x,\ g > 1$**

2. Brian put \$1000 into a certificate of deposit (CD) with an annual yield of 7%.
   a. Graph the value of the CD and interest Brian receives after 10 years if Brian takes the interest out each year.
   b. On the same grid, graph the value of the CD if Brian leaves the interest in the certificate.
   c. After 10 years how much more will Brian earn if he leaves the interest in the bank? **About \$267**

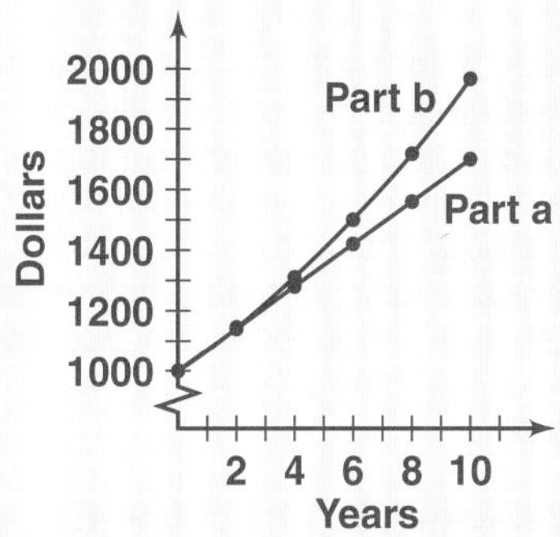

3. Company A earned a \$1000 profit this month; its profits are increasing by \$40 each month. Company B also earned a \$1000 profit this month; its profits are growing by 2% per month. If these trends continue, after 3 years, which company will have the greater monthly profit?
Co. A: $1000 + 40 \cdot 36 = \$2440$;
Co. B: $1000(1.02)^{36} = \$2039.89$;
Company A will have the greater monthly profit.

LESSON MASTER 8-3 A

Questions on SPUR Objectives
See pages 543-545 for objectives.

Uses Objective G
Representations Objective I

1. A biologist is studying the effect of a new medicine on the production of antibodies. The antibodies may grow either at a constant rate or exponentially. The biologist studies two cases beginning with 100 antibodies. Case 1: There are 30 more each day. Case 2: There are 20% more each day. Let x = the number of days the antibodies have been growing.

Day	Increase By 30	Increase By 20%
1	130	120
2	160	144
3	190	172
4	220	207
5	250	248
6	280	298
7	310	358
8	340	429
9	370	515
10	400	619

   a. Write an expression for Case 1. **$100 + 3x$**
   b. Write an expression for Case 2. **$100(1.2)^x$**
   c. Complete the table at the right. Round values down to the preceding integer.
   d. Which case yields more antibodies after 4 days? **Case 1**
   e. Which case yields more antibodies after 10 days? **Case 2**
   f. How many antibodies will there be after 100 days in each case?
   Case 1 **3100 antibodies**
   Case 2 **$\approx 8.28(10)^9$**
   g. Graph the information in the table on the grid at the right.

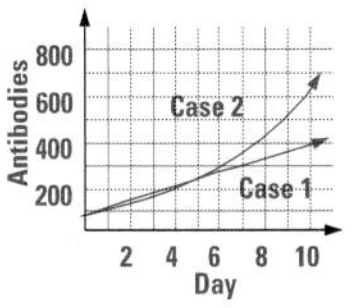

In 2–5, tell which is being described:
(a) *constant increase* or
(b) *exponential growth*.

2.
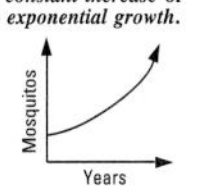

(b)

3. Each day the water level rose $\frac{1}{2}$ inch. **(a)**
4. $2000 + 30x$ **(a)**
5. $450 \cdot 1.8^x$ **(b)**

Adapting to Individual Needs

English Language Development
The first two sentences of this lesson explain that constant increase involves a number being *repeatedly added* while exponential growth involves a number being *repeatedly multiplied.* Be sure students understand the meaning of *repeatedly* as it is used here. To demonstrate, write the two columns of the chart in **Example 1** in the following way.

10 + 50	10 · 1.5
10 + 50 + 50	10 · 1.5 · 1.5
10 + 50 + 50 + 50	10 · 1.5 · 1.5 · 1.5

Notes on Questions

Question 15 Students can do this question without a spreadsheet. However, if you have a spreadsheet program and a computer in your classroom, use them to demonstrate the complete table of values. If there is time, you may wish to explore the time required for the enrollment to double.

Social Studies Connection There are about 15,700 public school districts in the United States. Each district has a school board (or board of education) with members that are either elected by voters in the district or appointed by the government of the district. Board members determine such things as school policy, finances, hiring of staff, and courses of study. Ask students if they know any of the members of your school board.

Questions 16–19 Cooperative Learning If students have automatic graphers, have them **work in groups** of 2 or 4 and check their answers to these questions.

LESSON MASTER 8-3 B Questions on SPUR Objectives

Uses Objective G: Solve problems involving exponential growth.
Representations Objective I: Graph exponential relationships.

1. The transportation department is studying two reports related to the predicted increase in the number of passengers at the airport. Report *A* predicts an increase of 1 million passengers each year, while Report *B* predicts that each year there will be a 6% increase. This year there are 14 million passengers at the airport. Let x = the number of years.

a. Write an expression for the number of passengers if that number increases by 1 million per year. $14{,}000{,}000 + 1{,}000{,}000x$

b. Write an expression for the number of passengers if that number increases by 6% per year. $14{,}000{,}000(1.06)^x$

c. Fill in the table below.

YEAR	INCREASE BY 1,000,000	INCREASE BY 6%
0	14,000,000	14,000,000
1	15,000,000	14,840,000
2	16,000,000	15,730,400
3	17,000,000	16,674,224
4	18,000,000	17,674,677
5	19,000,000	18,735,158
6	20,000,000	19,859,267
7	21,000,000	21,050,823
8	22,000,000	22,313,873
9	23,000,000	23,652,705
10	24,000,000	25,071,867

In 8 and 9, refer to Example 2.

8. How much more will the amount be in 15 years at compound interest than with the piggy bank? **$50.00**

9. Use the graph to estimate when the investment will be worth $500 with compound interest. **about 27.5 years**

Applying the Mathematics

In 10–13, an equation is given. **a.** Tell whether its graph is linear or exponential. **b.** Tell whether its graph is a curve or a line.

10. $y = 3x - 2$ a) linear b) line
11. $y = 3x$ a) linear b) line
12. $y = 3^x$ a) exponential b) curve
13. $y = 3(1.05)^x$ a) exponential b) curve

14. Suppose you have 60¢ and a 2 cm by 3 cm picture of a building that you want to enlarge as much as possible. Copies on a photocopy machine cost 10¢ and the machine can enlarge any image put through it to 120% of its original size. How large a picture of the building can you obtain? **about 6 cm by 9 cm**

Teen trends. *In 1960, 10.2 million teenagers were enrolled in U.S. high schools. By 1991, 13.1 million teenagers attended high school—about a 28% increase in student enrollment.*

15. A school board is making long-range budget plans. This year Central High went from 2400 to 2520 students. The number of students may be increasing at a constant rate of 120 students per year or exponentially by 5% per year.

a. Copy and complete the spreadsheet below. Show the future enrollments in the two possible situations.

	A	B	C
1	years from now	constant increase	exponential
2	0	2520	2520
3	1	2640	2646
4	2	2760	2778
5	3	2880	2917
6	4	3000	3063
7	5	3120	3216

b) There are 96 more students if the growth is exponential.

b. Describe how the predicted enrollments differ after 5 years.

c. Find the projected enrollment 15 years from now if the number of students increases at a constant rate. **4,320 students**

d. Find the projected enrollment 15 years from now if the number of students grows exponentially. **5,239 students**

Adapting to Individual Needs

Challenge

Have students find the least positive integer other than 1 that can be written as a square, a cube, and a fourth power. [It must be a 12th power of an integer. Since the integer cannot be 1, it must be 2. $2^{12} = 4096$ and $4096 = 64^2 = 16^3 = 8^4$.]

In 16–19, *multiple choice.* Each graph is drawn on the window $0 \le x \le 10$, $0 \le y \le 4000$. Match the graph with its equation.

(a) $y = 200 + 400x$ (b) $y = 200 + 40x$
(c) $y = 200 \cdot 1.4^x$ (d) $y = 200 \cdot 1.24^x$

16. d

17. c

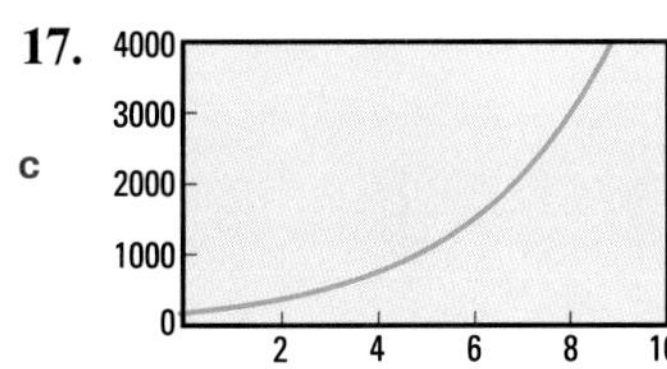

18. a

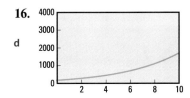

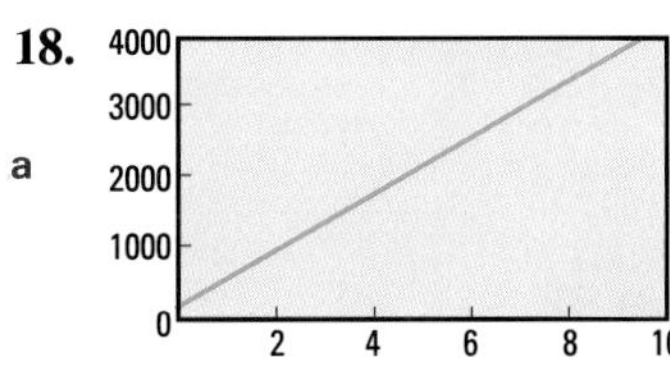

19. b

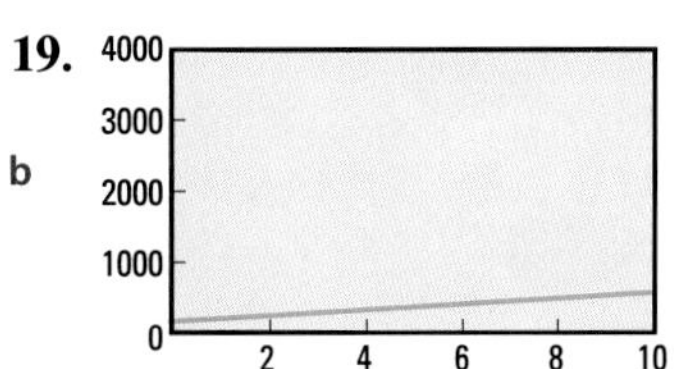

Review

20. a. Evaluate $-4x^n$ when $x = 1.2$ and $n = 3$; **-6.912**
b. Evaluate $4x^n$ when $x = -1.2$ and $n = 0$. *(Lessons 1-4, 8-2)* **4**

21. a. Copy and complete the computer program below to find the total T in the bank when \$100 is invested at an annual yield of 6% throughout the next 20 years. **100 * 1.06 ^ year**

```
10 PRINT "$100 AT 6 PER CENT"
20 PRINT "YEARS", "TOTAL"
30 FOR YEAR = 0 TO 20
40 LET TOTAL = _?_
50 PRINT YEAR, TOTAL
60 NEXT YEAR
```

b. Write the last line that will be printed when this program is run. **20 320.7135**
c. How would you modify line 30 to print the table for YEAR = 1, 2, 3, . . . , 100? **30 FOR YEAR = 1 TO 100**
d. After the change in part **c** is made, what will be the last line printed when the program is run? *(Lesson 8-1)* **100 33930.2084**

22. According to the Repeated Multiplication Model of Powering, $x^5 = \underline{\ ?\ }$. *(Lesson 8-1)* **x · x · x · x · x**

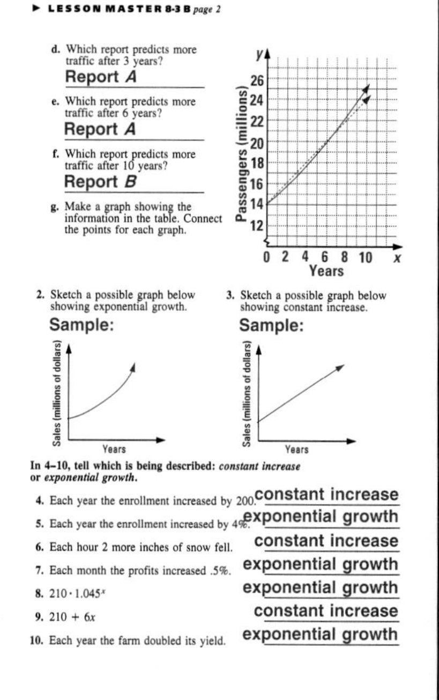
► LESSON MASTER 8-3 B *page 2*

d. Which report predicts more traffic after 3 years? **Report A**
e. Which report predicts more traffic after 6 years? **Report A**
f. Which report predicts more traffic after 10 years? **Report B**
g. Make a graph showing the information in the table. Connect the points for each graph.

2. Sketch a possible graph below showing exponential growth. **Sample:**

3. Sketch a possible graph below showing constant increase. **Sample:**

In 4–10, tell which is being described: ***constant increase or exponential growth.***

4. Each year the enrollment increased by 200. **constant increase**
5. Each year the enrollment increased by 4%. **exponential growth**
6. Each hour 2 more inches of snow fell. **constant increase**
7. Each month the profits increased .5%. **exponential growth**
8. $210 \cdot 1.045^x$ **exponential growth**
9. $210 + 6x$ **constant increase**
10. Each year the farm doubled its yield. **exponential growth**

Notes on Questions

Questions 25–26 Aspects of the Distributive Property are reviewed in preparation for the study of quadratics and other polynomials in the next chapters.

Question 29 Stress the order of operations. You may want to anticipate later lessons in the chapter by using repeated multiplication and showing that this expression equals $108x^2y^3$ for the values in question.

Question 30 This question anticipates the study of negative exponents in Lesson 8-6. To demonstrate, you might want to extend the graph in **Example 2** to negative values of x. In this case, $x = -1$ means "1 year earlier"; $100(1.06)^{-1}$ gives the amount you would have had to have in the account one year ago at a rate of 6% in order to now have $100.

Follow-up for Lesson 8-3

Practice

For more questions on SPUR Objectives, use **Lesson Master 8-3A** (shown on page 501) or **Lesson Master 8-3B** (shown on pages 502–503).

Assessment

Quiz A quiz covering Lessons 8-1 through 8-3 is provided in the *Assessment Sourcebook.*

Oral Communication Ask students to explain the difference between the rate of change in a constant increase situation and in an exponential growth situation. [Students understand that the rate of change is constant in a constant increase situation while exponential growth has a constantly increasing rate of change.]

Extension

Have students look up the meaning and origin of the word *googol.* [A googol is 1 followed by 100 zeros, or 10^{100}.]

23. The graph below shows the average diameter of rocks in a stream at half-mile intervals from the stream's source.

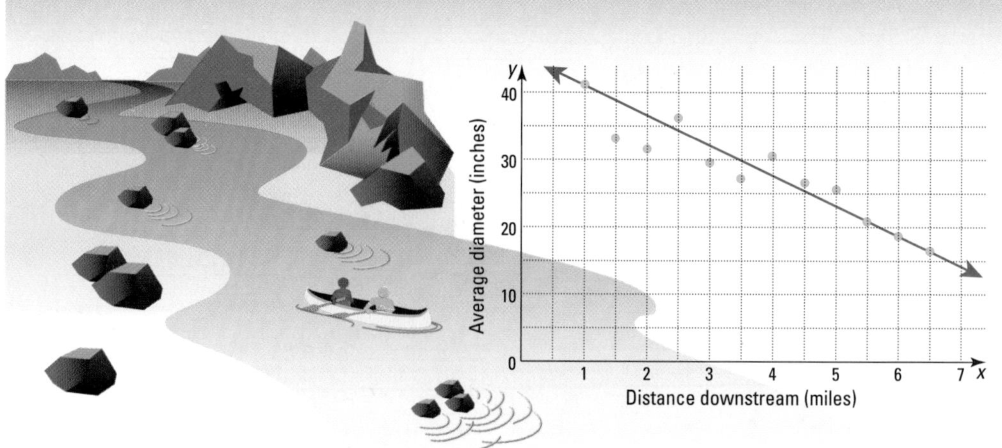

a. Fit a line to the data. See graph above for sample.
b. Find the coordinates of two points on your line in part **a.**
c. Determine an equation for your line using the points in part **b.**
d. Predict the average diameter of rocks that are 7.5 miles from the stream's source. *(Lesson 7-7)* Sample: about 12.4 in.

b) Sample: (1, 41); (6, 19) c) Sample: $y = -4.4x + 45.4$

24. Solve $1.3d + 5.2 = 9.4d - 9.0$. *(Lessons 5-3, 5-6)* 1.75

In 25 and 26, simplify. *(Lesson 3-9)*

25. $4a^2(a + 3) + 2a(a^2 - 1)$ $6a^3 + 12a^2 - 2a$

26. $b(b + 3) - b(b + 1)$ $2b$

27. *Skill sequence.* Add and simplify. *(Lesson 3-9)*

a. $\frac{4}{x} + \frac{5}{x}$ $\frac{9}{x}$

b. $\frac{4}{y} + \frac{5}{2y}$ $\frac{13}{2y}$

c. $4 + \frac{5}{2z}$ $\frac{8z + 5}{2z}$

28. Recall that if an item is discounted x%, you pay $(100 - x)$% of the original price. Calculate in your head the amount you pay for a jacket originally costing $200 and discounted the indicated amount.

a. 10% $180.00
b. 20% $160.00
c. 33% *(Lesson 3-3)* $134.00

29. Evaluate $(2x)^2(3y)^3$ when $x = 4$ and $y = -2$. *(Lesson 1-4)* -13,824

Exploration

30. Use your calculator to evaluate $y = 100(1.06)^x$ when $x = -1$ and $x = -2$. What do the answers mean for the situation of Example 2?
When $x = -1$, $y = 94.34$; when $x = -2$, $y = 89$. The answers give the amount in the account 1 and 2 years ago respectively.

504

LESSON

8-4

Exponential Decay

French-speaking regions. *This is a street scene in Quebec, where French is the official language spoken. French is also the official language of Belgium, France, Haiti, Switzerland, parts of Africa, and the French West Indies.*

An Example of Exponential Decay

A student crams for a Friday French test, learning 100 vocabulary words Thursday night. Each day the student expects to forget 20% of the words known the day before. If the test is delayed from Friday to Monday, what will happen if the student does not review?

To answer this question, referring to a table is convenient. Since 20% of the words are forgotten each day, 80% are remembered.

Day	Day Number	Words Known
Thursday	0	100
Friday	1	$100(.80) = 80$
Saturday	2	$100(.80)(.80) = 100(.80)^2 = 64$
Sunday	3	$100(.80)(.80)(.80) = 100(.80)^3 = 51.2 \approx 51$
Monday	4	$100(.80)(.80)(.80)(.80) = 100(.80)^4 = 40.96 \approx 41$

The pattern is like that of the growth model or compound interest. After d days, this student will know about $100(0.80)^d$ words. Because the growth factor 0.80 is less than 1, the number of words remembered decreases. This type of situation is called **exponential decay.**

What Are the Characteristics of a Graph of Exponential Decay?

A table of values and a graph of this situation is shown at the top of the next page. As with exponential growth, the points lie on a curve rather than a straight line.

Lesson 8-4

Objectives

G Solve problems involving exponential decay.

I Graph exponential relationships with the base between 0 and 1.

Resources

From the ***Teacher's Resource File***

- Lesson Master 8-4A or 8-4B
- Answer Master 8-4
- Teaching Aid
  26 Graph Paper
  87 Warm-up
  93 Linear and Exponential Increase and Decrease
- Activity Kit, Activity 18
- Technology Sourcebook
  Computer Master 21

Additional Resources

- Visuals for Teaching Aids 26, 87, 93
- GraphExplorer or other automatic graphers

Teaching Lesson 8-4

Warm-up

Fill in each blank with $>$, $<$, or $=$ to make the sentence true.

1. 3^2 ____ 2^3 $>$
2. 4^0 ____ 2^2 $<$
3. $(\frac{1}{2})^2$ ____ $(\frac{1}{4})^2$ $>$
4. 6^0 ____ $(\frac{1}{3})^0$ $=$
5. $.5^3$ ____ $(\frac{1}{2})^3$ $=$
6. $.7^4$ ____ $.7^3$ $<$
7. $50(.8)^5$ ____ $50(.8)^6$ $>$
8. $50(1.4)^5$ ____ $50(1.4)^6$ $<$

Lesson 8-4 Overview

Broad Goals This lesson extends the Growth Model of Powering to situations where the growth factor is less than 1, but always positive. They are defined as situations of decay.

Perspective In the preceding lessons the growth factor g has been greater than 1. This lesson deals with the exponential growth model $b \cdot g^x$ when g is a positive number less than 1. The result is decay.

Students should realize that it is the value of the growth factor and not the starting quantity that affects whether there is growth or decay, just as the slope of a line and not the y-intercept determines increase or decrease.

Constant increase/decrease and exponential growth/decay are mathematically *isomorphic;* that is, they have the same structure. The former is related to addition as the latter is related to multiplication. For instance, the additive identity 0 is the pivot in constant increase/decrease; a slope greater than 0 means increase whereas a slope less than 0 means decrease. Similarly, the multiplicative identity 1 is the pivot in exponential growth/decay; a growth factor greater than 1 means growth whereas a growth factor less than 1 signals decay. **Teaching Aid 93** shows the characteristic graphs of these kinds of change.

Notes on Reading

Multicultural Connection In the opening situation, a student is studying for a French test. You might tell students that French ranks second in non-English languages spoken in American homes. The first is Spanish, the third is German, followed by Italian, Chinese, Tagalog, Polish, Korean, Vietnamese, Portuguese, and Japanese.

Reading Mathematics This lesson is shorter than previous lessons. If you decide to have students read it in class, be sure that they have calculators so that they can perform the calculations on page 506. You might also discuss the questions in *Covering the Reading* as you proceed.

❶ The formula for exponential decay is the same as the formula for exponential growth, namely $y = b \cdot g^x$. In the **Example,** the starting amount is 67,000 and the growth factor g is $1 - .05$, or .95. Since $0 < .95 < 1$, there is decay. You might graph the population of the town over time to provide a picture of the decrease.

Students are often intrigued by the fact that the graph of an exponential decrease situation approaches the x-axis (as an asymptote) but never touches it. This can be shown by extending the graph on page 506. In the French word memory situation, this means that you would never forget all the words.

LESSON MASTER 8-4 A **Questions on SPUR Objectives** See pages 543-545 for objectives.

Uses Objective G

1. Investors in the Get-Rich-Quick Investment company lose about 15% of the value of their money each year.
   a. Fill in the table to show the value of a \$2,000 investment for each of the first seven years.

YEAR	VALUE
0	\$2,000
1	\$1,700
2	\$1,445
3	\$1,228.25
4	\$1,044.01
5	\$887.41
6	\$754.30

   b. How much was lost the first year? \$300
   c. How much was lost from the fifth to the sixth year? \$133.11
   d. Write an expression for the value of the investment in year x. $2000(0.85)^x$
   e. Use trial and error to find the first year in which the value is less than \$100. year 19
2. Amalgamated Industries receives hundreds of applications for each job opening it has. Their selection process is to review the applications and discard 50% of them. Then this is repeated until only one applicant is left. Let n = the number of times that half the applications are discarded.
   a. Write an expression of the form $b \cdot g^n$ to describe the number of people left after the applications have been reviewed n times. $b(0.5)^n$
   b. If 512 people apply for a job, how many are left when $n = 4$? 32 people
   c. If 512 people apply, what is the value of n when one person is left? 9
3. In 1993, the national debt was estimated at 4.4 trillion dollars. Suppose this were cut 10% each year. Let x be the number of years since 1993.
   a. Write an equation to describe this situation. $d = 4.4 \cdot 10^{12}(0.9)^x$
   b. What would the deficit be after 20 years? $\approx 5.35 \cdot 10^{11}$ dollars

Day	Words
0	100
1	80
2	64
3	≈ 51
4	≈ 41
5	≈ 33
6	≈ 26
7	≈ 21
8	≈ 17
9	≈ 13
10	≈ 11

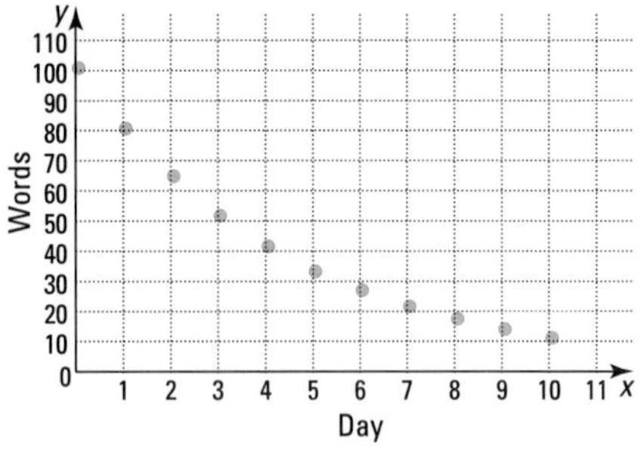

Exponential-decay situations, like constant-decrease situations, have graphs that go downward as you go to the right. However, in a constant decrease graph the slope is constant. Here the rate of change is not constant, as the chart shows. From day 0 to day 1, the rate of change is $\frac{80 - 100}{1 - 0}$, or -20 $\frac{\text{words}}{\text{day}}$. The student forgot 20 words. But from day 9 to day 10, the rate of change is $\frac{11 - 13}{10 - 9} = -2$. The student forgot only 2 words that day.

Exponential Decay with Populations

Exponential decay can occur with populations if the growth factor is less than 1.

Population: zero. *This is an abandoned ghost town in Wyoming that was once inhabited by early settlers and fur traders.*

Example

A town with population 67,000 is losing 5% of its population each year. At this rate, how many people will be left in the town after 10 years?

Solution

If 5% of the population is leaving, 95% is staying. Thus, every year, the population is multiplied by 0.95.

$$\text{population} = 67{,}000 \cdot (0.95)^{10} \approx 40{,}115$$

After ten years, the population will be about 40,115.

Determining Growth or Decay from the Value of the Growth Factor

❶ Exponential growth and exponential decay are both described by an equation of the same form,

$$y = b \cdot g^x.$$

The value of g determines whether the equation describes growth or decay. For exponential growth, $g > 1$. For instance, compound interest at a 3% rate means $g = 1.03$. Doubling means $g = 2$. When there is exponential decay, $0 < g < 1$. For instance, in the vocabulary situation on page 505, the growth factor is 0.8. In the population decline example, the growth factor is 0.95.

Optional Activities

Activity 1

You can use *Activity Kit, Activity 18* to introduce Lesson 8-4 or you might use it as a follow-up for the lesson. In this activity, students experiment with pendulums to gather their own data and then they draw graphs that exhibit non-linear decay.

Activity 2

Materials: Automatic graphers

Question 16c asks, "Can $(\frac{1}{2})^x$ ever be zero?" in a graphical context. You might want to have students graph the equation on an automatic grapher and to use the trace feature to examine values of $y = .5^x$ for large values of x.

506

❷ Between exponential growth and exponential decay is the situation where $g = 1$. When $g = 1$, then $y = b \cdot g^x = b \cdot 1^x = b$.

So y remains equal to b, its initial value, regardless of the value of x. So if 1 is the growth factor, then over time there is neither growth nor decay. The original value remains constant.

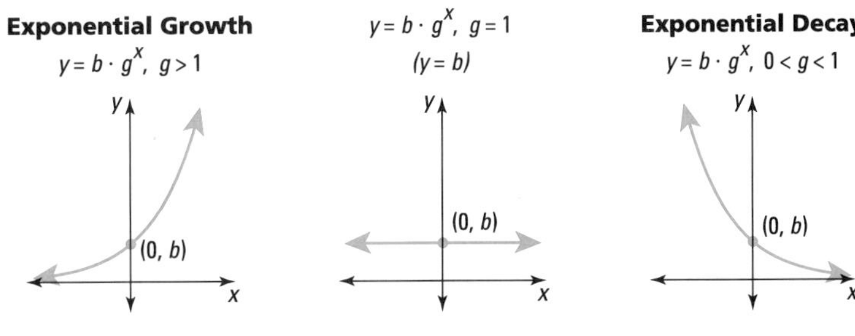

QUESTIONS

Covering the Reading

In 1–4, refer to the French test situation of this lesson. Assume the student does not review.

1. If the test is delayed a week until the next Friday, about how many words will the student know? **17**
2. About how many words did the student forget the first 3 days? **49**
3. About how many words did the student forget the second 3 days? **25**
4. Evaluate $100x^5$ when $x = 0.80$. What does this answer mean? **See left.**
5. State the general form of an equation for exponential decay. **$y = b \cdot g^x$, $0 < g < 1$**
6. What is a difference between the shape of the graphs of constant decrease and exponential decay? **The shape of the constant decrease graph is linear, while the shape of the exponential decay graph is curved.**
7. *Multiple choice.* Which formula could describe an exponential decay situation? **c**
   (a) $T = 5(3)^n$ (b) $T = \frac{1}{5}(3)^n$ (c) $T = 5\left(\frac{1}{3}\right)^n$ (d) $T = 5 + \frac{1}{3}n$
8. What possible values can the growth factor have in an exponential decay equation? **$0 < g < 1$**

In 9 and 10, refer to the Example.

9. What is the population of the town after 1 year? **63,650**
10. What is the population of the town after n years? **$67{,}000\ (0.95)^n$**
11. If the population of a town declines by 6.3% each year, by what number would you multiply to find the population each year? **0.937**

4) ≈ 33. About 33 words will be remembered on the 5th day if 20% of the words are forgotten each day.

❷ Make sure students understand that exponential growth occurs when the growth factor g is greater than 1, and that exponential decay occurs when $0 < g < 1$. **Teaching Aid 93** provides a good summary of this relationship.

Additional Examples

1. For a certain type of calculator that cost $350 in 1973, the price dropped about 19% each year. What was the price 15 years later? **$350(.81)^{15} \approx$ $14.84**
2. Scrooge is such a cruel boss that each year 40% of his employees quit. If 300 people work for Scrooge now, graph the number of these employees that will still be working for him n years from now. Let $n = 1, 2, 3, 4, 5$.

Years	Employees
0	300
1	180
2	108
3	65
4	39
5	23

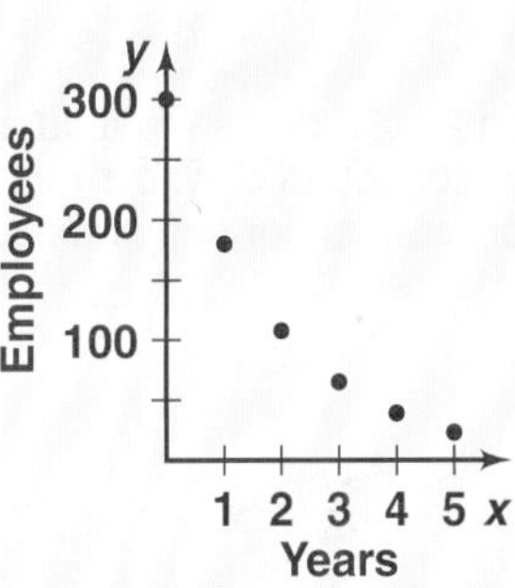

Activity 3
After completing the lesson, you might ask some students to research *Zeno's Paradox* and tell the class about how it relates to the lesson. [The paradox is an exponential decay situation whose equation might be written as $y = b\left(\frac{1}{2}\right)^x$ where b is the total distance, $\frac{1}{2}$ is the growth factor, and x is the number of increments.]

Activity 4 Technology Connection
You may wish to assign *Technology Sourcebook, Computer Master 21.* Students use *GraphExplorer* or similar software to graph situations that follow an exponential decay model. Students then make predictions based on their graphs.

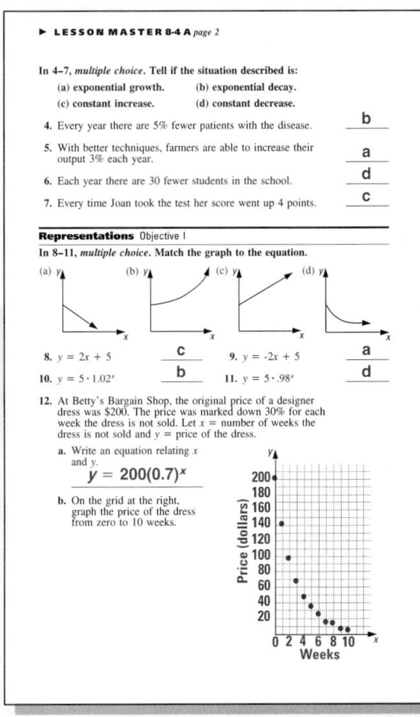

► **LESSON MASTER 8-4 A** *page 2*

In 4–7, *multiple choice*. Tell if the situation described is:
(a) exponential growth. (b) exponential decay.
(c) constant increase. (d) constant decrease.

4. Every year there are 5% fewer patients with the disease. **b**
5. With better techniques, farmers are able to increase their output 3% each year. **a**
6. Each year there are 30 fewer students in the school. **d**
7. Every time Joan took the test her score went up 4 points. **c**

Representations Objective I

In 8–11, *multiple choice*. Match the graph to the equation.

(a) (b) (c) (d)

8. $y = 2x + 5$ **c**
9. $y = -2x + 5$ **a**
10. $y = 5 \cdot 1.02^x$ **b**
11. $y = 5 \cdot .98^x$ **d**

12. At Betty's Bargain Shop, the original price of a designer dress was $200. The price was marked down 30% for each week the dress is not sold. Let x = number of weeks the dress is not sold and y = price of the dress.
   a. Write an equation relating x and y. **$y = 200(0.7)^x$**
   b. On the grid at the right, graph the price of the dress from zero to 10 weeks.

Price (dollars): 200, 180, 160, 140, 120, 100, 80, 60, 40, 20
Weeks: 0 2 4 6 8 10

Notes on Questions

Question 14 The same idea that is used in this situation can be used to calculate the amount of light that reaches various depths under water. The growth (in this instance, decay) factor depends on the murkiness of the water.

Question 17 Technology Connection Note that one would write this program only if there were a reasonable number of powers to compute. Otherwise, it is much faster to use a calculator. However, you or a student could build in a FOR/NEXT loop to calculate several values during one run of the program. For example:

```
80  PRINT "TIME
    PERIODS," "TOTAL AMOUNT"
90  FOR I = 1 TO N
100 LET AMT = P*X^N
110 PRINT I, AMT
120 NEXT I
130 END
```

Question 25b Cooperative Learning You might have students **work in groups** to find the dimensions of at least five boxes with the given volume.

Question 26 Students must be able to recognize factors of whole numbers in order to factor polynomials in later chapters.

LESSON MASTER 8-4 B Questions on SPUR Objectives

Uses Objective G: Solve problems involving exponential decay.

1. National Cables has launched a program to reduce its debt by 12% per year. The debt is currently $9,000,000.
   a. Fill in the table to show the amount of the company's debt for 0 through 6 years.

YEAR	AMOUNT OF DEBT
0	$9,000,000
1	$7,920,000
2	$6,969,600
3	$6,133,248
4	≈$5,397,258
5	≈$4,749,587
6	≈$4,179,637

   b. How much did the company reduce its debt during the first year? $1,080,000
   c. How much did the company reduce its debt during the fifth year? ≈$647,671
   d. Write an expression for the amount of debt remaining after x years. $9{,}000{,}000(.88)^x$
   e. Use trial and error to find how long it will take to reduce the debt to less than $1,000,000. 18 years

2. A single thickness of one type of glass cuts down the emission of light by 10%. Consider a 60-watt light bulb which has an intensity of about 800 *lumens*.
   a. By what number would you multiply to find the number of the light bulb's lumens emitted through a single thickness of the glass? .9
   b. How many lumens are emitted through a single thickness of the glass? 720 lumens
   c. How many lumens are emitted through two thicknesses of the glass? 648 lumens
   d. Let t = the number of thicknesses of glass. Write an expression to describe the number of lumens that are emitted through t thicknesses of the glass. $800(.9)^t$
   e. How many lumens are emitted through five thicknesses of the glass? ≈472 lumens

Applying the Mathematics

12. Suppose a school has 2,500 students and the number of students is decreasing by 3% each year.
  a. By what number would you multiply to find the number of students after each year? **0.97**
  b. If this rate continues, write an expression for the number of students after n years. **$2500\,(0.97)^n$**
  c. If this rate continues, how many students will the school have 10 years from now? **about 1844**

13. *True or false.* $\frac{1}{2}$ can be the growth factor in an exponential decay situation. **True**

14. When one plate of tinted glass allows 60% of the light through, the amount of light y passing through x panes of glass is described by the formula $y = (0.6)^x$. Plot 5 points on the graph of this equation. **See left.**

14)
y: Amount of light (0 to 1); x: Panes of glass (1 to 8); $y = (0.6)^x$

15. You fold a sheet of paper in half, then fold it again in half, and so on. Let f = the number of folds; let A = the area of what is folded; and let t = the total thickness of the sheets.
  a. As f increases, does A grow or decay exponentially? What is the growth factor? ***A* decays exponentially; the growth factor is $\frac{1}{2}$.**
  b. As f increases, does t grow or decay exponentially? What is the growth factor? ***t* grows exponentially; the growth factor is 2.**

16. Consider the equation $y = \left(\frac{1}{2}\right)^x$. **See left.**
  a. Find the value of y as both a fraction and a decimal when x takes on each value in this domain: {0, 1, 2, 3, 10, 20}.
  b. Draw the graph using the points from part **a.**
  c. Does this graph ever touch the x-axis? Justify your answer.

16a) $x = 0,\ y = 1$
$x = 1,\ y = \frac{1}{2} = 0.5$
$x = 2,\ y = \frac{1}{4} = 0.25$
$x = 3,\ y = \frac{1}{8} = 0.125$
$x = 10,\ y = \frac{1}{1024} \approx 0.001$
$x = 20,\ y = \frac{1}{1{,}048{,}576} \approx 0.000001$

b) $y = \left(\frac{1}{2}\right)^x$

c) No; there is no power of $\frac{1}{2}$ that equals 0.

17. The following program finds the result when an amount grows or decays exponentially. The BASIC expression for x^n is X ^ N or X ↑ N.

```
10 PRINT "EXPONENTIAL GROWTH"
20 PRINT "WHAT IS THE AMOUNT AT BEGINNING?"
30 INPUT P
40 PRINT "WHAT ARE YOU MULTIPLYING BY?"
50 INPUT X
60 PRINT "HOW MANY TIME PERIODS?"
70 INPUT N
80 LET T = P * X ^ N
90 PRINT "TOTAL AMOUNT IS"
100 PRINT T
110 END
```

  a. What will the total be if 8, 3, and 6 are entered for P, X, and N?
  b. What will the total be if 8, 0.33, and 6 are entered for P, X, and N?
  c. What should be entered to find the population to which a city of 100,000 will grow in 10 years if the growth rate is 2% a year?

a) 5832 b) 1.033174×10^{-2}
c) P = 100,000; X = 1.02; N = 10

Adapting to Individual Needs

Extra Help
Be sure students understand that, in an exponential decay situation, the growth factor is found by subtracting the growth rate from 1. Point out that this corresponds to the growth factor in an exponential growth situation where the growth factor is found by adding the growth rate to 1.

English Language Development
Students who are not fluent in English might need help with the use of the word *decay* in this lesson. If they are familiar with the word in the sense of tooth decay, point out that it can also mean any kind of decrease or decline in quantity as well as in quality.

Tropical fruit. *This pineapple is from a plantation in Honduras, a small Central American country.*

18. Penny Wise had 12 children. In her will she left her first child \$300,000. The second child got $\frac{1}{4}$ of what the first child did. The third child got $\frac{1}{4}$ of what the second child did, and so on. How much did the last child get? about 7¢

Review

In 19 and 20, consider the equations $k = 30 + 1.05n$ and $k = 30(1.05)^n$. *(Lesson 8-3)*

19. In which equation does k increase more rapidly as n increases? $k = 30(1.05)^n$

20. Which equation could represent the value of \$30 invested at 5% compound interest for n years? $k = 30(1.05)^n$

21. In the 1980s, the growth rate of the population of Honduras was about 2.7% per year (a high rate). The 1993 population of Honduras was estimated at 5,170,000. Assuming the 1980s growth rate holds throughout the 1990s, what will the population be (to the nearest thousand) in the year 2000? *(Lesson 8-2)* about 6,230,000

22. Robert buys six guppies. Every month his guppy population doubles. Assume the population continues to grow at this rate.
    a. How many guppies will there be after 4 months? 96
    b. How many will there be after a year? *(Lesson 8-2)* 24,576

23. Give the value of 372.114^0. *(Lesson 8-2)* 1

24. Calculate the interest paid on \$500 at a 6.1% annual yield for 5 years. *(Lesson 8-1)* \$172.27

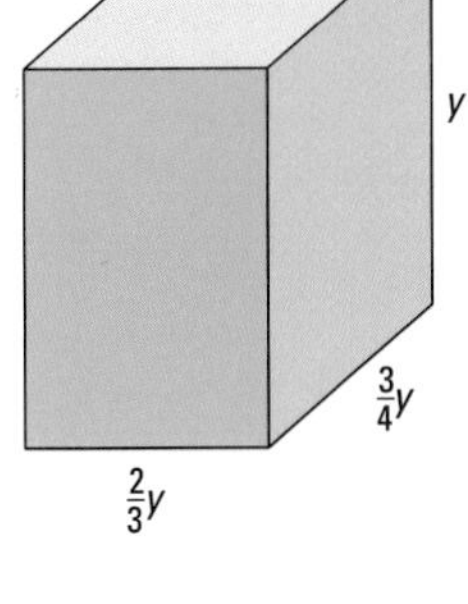

25. Refer to the box pictured at the left.
    a. Write an expression for its volume. $\frac{1}{2}y^3$
    b. Give the dimensions of a box with the same volume but different shape. *(Lessons 2-1, 2-3)* Sample: $\frac{1}{2}y$, y, y

26. Write as the product of prime numbers. *(Previous course)*
    a. 14 $2 \cdot 7$
    b. 40 $2 \cdot 2 \cdot 2 \cdot 5$
    c. 81 $3 \cdot 3 \cdot 3 \cdot 3$

Exploration

27. Exponential decay gets its name from the natural decay of elements, called *radioactive decay*. Radioactive decay is used to determine the approximate age of archaeological objects that were once alive. This can be done because all living things contain radioactive carbon 14, which has a half-life of 5600 years. When an organism dies, the amount of carbon 14 in it begins to diminish as the carbon 14 decays.
    a. What is meant by the *half-life* of an element? See left.
    b. A bone is found to have $\frac{1}{16}$ of the carbon 14 that it had when the animal it came from was alive. How old is the bone? 22,400 years

27a) The half-life of an element is the time it takes for one half the amount of the element to decay.

Follow-up for Lesson 8-4

Practice

For more questions on SPUR Objectives, use **Lesson Master 8-4A** (shown on pages 506–507) or **Lesson Master 8-4B** (shown on pages 508–509).

Assessment

Written Communication Ask each student to write two equations for exponential growth and two equations for exponential decay. [Students understand that the growth factor is greater than 1 in equations for exponential growth and between zero and 1 in equations for exponential decay.]

Extension

Cooperative Learning Have students **work in pairs** to solve the following problem. The nuclei of some elements decay because they are unstable. Suppose a certain element loses 10% of its mass per week. After how many weeks will the element have less than half of its original mass? [7 weeks]

Project Update Project 3, *Fitting an Exponential Curve,* and Project 6, *Cooling Water,* on pages 539–540, relate to the content of this lesson.

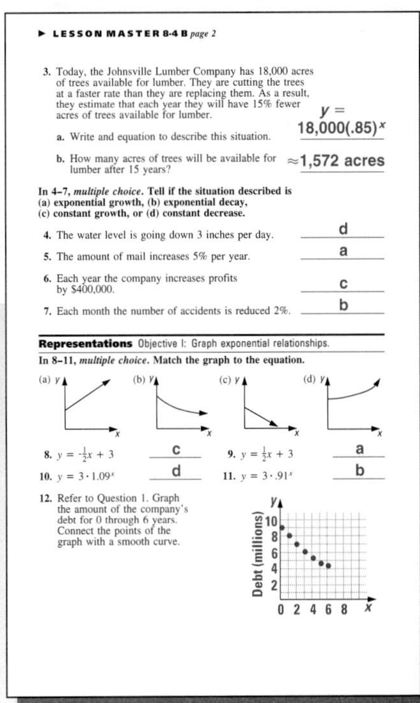

▸ LESSON MASTER 8-4 B *page 2*

3. Today, the Johnsville Lumber Company has 18,000 acres of trees available for lumber. They are cutting the trees at a faster rate than they are replacing them. As a result, they estimate that each year they will have 15% fewer acres of trees available for lumber.
    a. Write and equation to describe this situation. $y = 18{,}000(.85)^x$
    b. How many acres of trees will be available for lumber after 15 years? ≈1,572 acres

In 4–7, *multiple choice*. Tell if the situation described is (a) exponential growth, (b) exponential decay, (c) constant growth, or (d) constant decrease.

4. The water level is going down 3 inches per day. d
5. The amount of mail increases 5% per year. a
6. Each year the company increases profits by \$400,000. c
7. Each month the number of accidents is reduced 2%. b

Representations Objective I: Graph exponential relationships.

In 8–11, *multiple choice*. Match the graph to the equation.

(a) (b) (c) (d)

8. $y = -\frac{1}{2}x + 3$ c
9. $y = \frac{1}{2}x + 3$ a
10. $y = 3 \cdot 1.09^x$ d
11. $y = 3 \cdot .91^x$ b

12. Refer to Question 1. Graph the amount of the company's debt for 0 through 6 years. Connect the points of the graph with a smooth curve.

Video

Wide World of Mathematics The segment, *Endangered Species,* focuses on the decline of various animal and plant species in Hawaii and in other places, as well as on the recovery of some of those species. The segment may be used to extend lessons on exponential growth and decay. Related questions and an investigation are provided in videodisc stills and in the Video Guide. A related CD-ROM activity is also available.

Videodisc Bar Codes

Search Chapter 39

Play

Lesson 8-5

Objectives

B Simplify products of powers and powers of powers.

E Identify and use the Product of Powers Property and the Power of a Power Property.

Resources

From the ***Teacher's Resource File***

- Lesson Master 8-5A or 8-5B
- Answer Master 8-5
- Teaching Aid 88: Warm-up

Additional Resources

- Visual for Teaching Aid 88

Teaching Lesson 8-5

Warm-up

Name the numbers described.

1. Doubling this number gives the same result as squaring it. **2 or 0**
2. Any nonzero number raised to this power is equal to one. **0**
3. This number raised to any whole-number power equals one. **1**
4. This number raised to the zero power is undefined. **0**

Notes on Reading

Reading Mathematics To review this reading, you might ask students what they think the purpose of each example is. [**Example 1** practices the Product of Powers Property; **Example 2** shows the growth of something n years after it has grown for m years with the same growth factor g; **Example 3** utilizes the Product of Powers Property in a situation with more than one base;

LESSON 8-5 *Products and Powers of Powers*

Multiplying Powers with the Same Bases

There is no general way to simplify the sum of two powers, even when the powers have the same base. For instance, $2^5 + 2^3 = 32 + 8 = 40$, and 40 is not an integer power of 2. But some products of powers can be simplified using the repeated multiplication model of x^n. Notice the patterns in the products below.

$$2^4 \cdot 2^3 = \underbrace{(2 \cdot 2 \cdot 2 \cdot 2)}_{\text{4 factors}} \cdot \underbrace{(2 \cdot 2 \cdot 2)}_{\text{3 factors}} = \underbrace{2 \cdot 2 \cdot 2 \cdot 2 \cdot 2 \cdot 2 \cdot 2}_{\text{7 factors}} = 2^7$$

$$10^2 \cdot 10^3 = \underbrace{(10 \cdot 10)}_{\text{2 factors}} \cdot \underbrace{(10 \cdot 10 \cdot 10)}_{\text{3 factors}} = \underbrace{10 \cdot 10 \cdot 10 \cdot 10 \cdot 10}_{\text{5 factors}} = 10^5$$

$$x^2 \cdot x^5 = \underbrace{(x \cdot x)}_{\text{2 factors}} \cdot \underbrace{(x \cdot x \cdot x \cdot x \cdot x)}_{\text{5 factors}} = \underbrace{x \cdot x \cdot x \cdot x \cdot x \cdot x \cdot x}_{\text{7 factors}} = x^7$$

In each case, when we multiplied two powers with the same base the product was also a power of that base. The exponent of the product is the sum of the exponents of the factors.

$$2^4 \cdot 2^3 = 2^{4+3} = 2^7$$
$$10^2 \cdot 10^3 = 10^{2+3} = 10^5$$
$$x^2 \cdot x^5 = x^{2+5} = x^7$$

These examples are instances of the *Product of Powers Property.*

Product of Powers Property
For all m and n, and all nonzero b, $b^m \cdot b^n = b^{m+n}$.

Example 1

Multiply $x^7 \cdot x^5$.

Solution

Use the Product of Powers Property.

$$x^7 \cdot x^5 = x^{7+5} = x^{12}$$

Check

Test a special case. We let $x = 4$.

Does $(4)^7 \cdot (4)^5 = (4)^{12}$? Check with a calculator.

The key sequences 4 [y^x] 7 [×] 4 [y^x] 5 [=] and 4 [y^x] 12 [=] both give 16,777,216. (On some calculators, you need to use [^] rather than [y^x] to calculate a power.) It checks.

Lesson 8-5 Overview

Broad Goals This lesson discusses the most basic law of exponents: For $b \neq 0$ and all m and n, $b^m \cdot b^n = b^{m+n}$. The law is derived from repeated multiplication and applied in growth situations. The related property $(b^m)^n = b^{mn}$ is also examined.

Perspective Starting with this lesson, the rest of this chapter deals with the formal properties of powers, sometimes called the *laws of exponents.* When expressing these properties, we use the letter b for the base. Using x can be confusing to some students who have calculators because some calculators label their powering key [y^x] while others use [x^y] or [^].

The power b^n gets meaning both from repeated multiplication and from growth. From repeated multiplication we derive the Product of Powers Property. This is verified in **Example 2** by using a growth situation. The Product of Powers Property is used to explain the Power of a Power Property. If a property is forgotten, students should learn to go back to more primitive properties or to one of these models.

Recall that b^n is the growth factor in n time periods if there is growth by a factor b in each of the periods. Example 2 shows that $2^5 \cdot 2^3 = 2^8$ using this model.

Example 2

Suppose a colony of bacteria doubles in number every hour. Then, if there were 2000 bacteria in the colony at the start, after h hours there will be T bacteria, where

$$T = 2000 \cdot 2^h.$$

How many bacteria will there be after the 5th hour? How many bacteria will there be 3 hours after the 5th hour?

Solution 1

There will be $2000 \cdot 2^5$ bacteria at the end of the 5th hour. Three hours later the bacteria will have doubled three more times. So there will be $(2000 \cdot 2^5) \cdot 2^3$ bacteria. This equals $2000 \cdot (2^5 \cdot 2^3)$ bacteria.

Solution 2

Three hours after the 5th hour is the 8th hour. From the given formula, there will be $2000 \cdot 2^8$ bacteria.

Since the solutions are equal, $2^5 \cdot 2^3$ must equal 2^8.

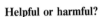

Helpful or harmful? *Certain kinds of bacteria, such as those that aid in digestion, are beneficial to humans. Other kinds of bacteria are harmful, such as this one that causes diphtheria, an infection of the upper respiratory system.*

Multiplying Powers When the Bases Are Not the Same

The Product of Powers Property tells how to simplify the product of two powers with the same base. A product with different bases, such as $a^3 \cdot b^4$, usually cannot be simplified.

Example 3

Simplify $r^4 \cdot s^3 \cdot r^5 \cdot s^8$.

Solution

Use the properties of multiplication to group factors with the same base.

$$\begin{aligned} r^4 \cdot s^3 \cdot r^5 \cdot s^8 &= r^4 \cdot r^5 \cdot s^3 \cdot s^8 \\ &= r^{4+5} \cdot s^{3+8} \quad \text{Apply the Product of Powers Property.} \\ &= r^9 \cdot s^{11} \end{aligned}$$

$r^9 \cdot s^{11}$ cannot be simplified further because the bases are different.

Check

Look at the special case when $r = 2$, $s = 3$.

Does $2^4 \cdot 3^3 \cdot 2^5 \cdot 3^8 = 2^9 \cdot 3^{11}$?

Does $16 \cdot 27 \cdot 32 \cdot 6561 = 512 \cdot 177{,}147$?

Yes, they both equal 90,699,264.

Example 4 applies the Power of a Power Property with variables; **Example 5** combines both of the properties of the lesson.]

Example 2 is a specific case of the following general idea: If a starting amount S grows with a growth factor g, then after m periods, the amount is $S \cdot g^m$. If the amount continues to grow with that same factor for n more periods, the amount will be $(S \cdot g^m) \cdot g^n$. We can also think of this growth taking place for $m + n$ periods, after which the amount is $S \cdot g^{m+n}$. Equating the two expressions and dividing by S, we have the Product of Powers Property. Although this may seem like a distant argument, the advantage is that it applies when m and n are not integers, so it is more general than the use of repeated multiplication.

Not all algebra students are ready to use the properties of powers. Some must revert back to the meaning of powers as repeated multiplication. If some of your students are more comfortable with the latter approach, it is probably best to not discourage them. They are using their understanding of the operations that are involved.

The problem-solving strategy *testing a special case* is used in checking **Examples 1 and 3.** It is very useful and helps students avoid errors with properties of exponents. It will be used throughout the rest of the chapter, especially in Lesson 8-9. You might discuss choosing the best

LESSON MASTER 8-5 A

Questions on SPUR Objectives
See pages 543-545 for objectives.

Uses Objective B

In 1–12, simplify the expression.

1. $n^2 \cdot n^6$	n^8	2. $y^8 \cdot y$	y^9
3. $(x^6)^2$	x^{12}	4. $(5^2)^3$	5^6, or 15,625
5. $(k^0)^2$	1	6. $x^2 \cdot x^4 \cdot x$	x^7
7. $a^2 \cdot x^3 \cdot a^7$	a^9x^3	8. $n(n^5)^4$	n^{21}
9. $6^2 \cdot 6^2$	6^4, or 1296	10. $2x^3 \cdot 8x^4$	$16x^7$
11. $(n \cdot n^4)^3$	n^{15}	12. $x^2(x + x^6)$	$x^3 + x^8$

Properties Objective E

13. Show how to simplify $(n^2)^4$
    a. by treating n^2 as a chunk.
    $(n^2)^4 = n^2 \cdot n^2 \cdot n^2 \cdot n^2 = n^8$
    b. by using the Power of a Power Property.
    $(n^2)^4 = n^{2 \cdot 4} = n^8$

14. Show how to simplify $a^5 \cdot a$
    a. by using repeated multiplication.
    $a^5 \cdot a = (a \cdot a \cdot a \cdot a \cdot a) \cdot a = a^6$
    b. by using the Product of Powers Property.
    $a^5 \cdot a = a^5 \cdot a^1 = a^{5+1} = a^6$

Samples are given.

15. Write a multiplication expression that uses the Product of Powers Property and the value of which is x^{10}. $x^4 \cdot x^6$

16. Write an expression that uses the Power of a Power Property and the value of which is y^{24}. $(y^3)^8$

Optional Activities

After students have answered **Question 19**, you might want to use this activity to lead to a consideration of negative exponents. List pairs of powers whose product is x^7 on the board. Ask students to describe the patterns they notice and to predict what expression might be multiplied by x^8 to give x^7 if the patterns were continued. [The sum of the exponents is 7. So, if the pattern continues, $x^{-1} \cdot x^8 = x^7$.]

$$\begin{aligned} x^4 \cdot x^3 &= x^7 \\ x^3 \cdot x^4 &= x^7 \\ x^2 \cdot x^5 &= x^7 \\ x^1 \cdot x^6 &= x^7 \\ x^0 \cdot x^7 &= x^7 \end{aligned}$$

numbers to test. The numbers 0 and 1 should be avoided because they have special properties. The number 2 can also cause trouble because squaring it gives the same result as doubling it.

Periodically ask students how they can check problems like those in this lesson. Stress the use of special cases and of going back to repeated multiplication or growth situations.

Additional Examples

1. Multiply $k^4 \cdot k^2$. k^6
2. A population of guinea pigs is tripling every year. If there were 6 guinea pigs at the beginning of the year, how many will there be
   a. after 5 years? $6 \cdot 3^5 = 1458$
   b. four years after that? $6 \cdot 3^5 \cdot 3^4 = 6 \cdot 3^9 = 118{,}098$
3. Simplify $a^4 \cdot x^3 \cdot a \cdot x^{10}$. a^5x^{13}
4. Solve $(3^5)^3 = 3^n$. $n = 15$
5. a. Simplify $(x^5)^2$. x^{10}
   b. Verify your answer by testing a special case. **Sample: Let $x = 3$. $(3^5)^2 = 243^2 = 59{,}049$ and $3^{10} = 59{,}049$.**
6. Simplify $3p^2 \cdot (p^3)^4$. $3p^{14}$

What Happens if We Take a Power of Powers?

When powers of powers are calculated, there are also some interesting patterns. Notice how we use chunking in the following examples.

Example 4

Solve $(5^2)^4 = 5^n$.

Solution

Think of 5^2 as a chunk that is raised to the 4th power.
$(5^2)^4 = 5^2 \cdot 5^2 \cdot 5^2 \cdot 5^2 = 5^{2+2+2+2} = 5^8$. *So n = 8.*

Check

Use a calculator, following order of operations. $(5^2)^4 = (25)^4 = 390{,}625$ and $5^8 = 390{,}625$. It checks.

The idea of Example 4 works with variables. For instance,

$$\begin{aligned}(x^3)^4 &= x^3 \cdot x^3 \cdot x^3 \cdot x^3 \\ &= x^{3+3+3+3} \\ &= x^{12}\end{aligned}$$

The general pattern is called the *Power of a Power Property*.

Power of a Power Property
For all m and n, and all nonzero b,
$$(b^m)^n = b^{mn}.$$

In both properties in this lesson b cannot be 0, because 0^0 is not defined. However, m and n may be 0. For instance, $(5^0)^3 = 5^{0 \cdot 3} = 5^0 = 1$. This checks, because $(5^0)^3 = 1^3 = 1$.

Some expressions involve both powers and multiplication.

Example 5

Simplify $2x(x^5)^3$.

Solution 1

Recall that $x = x^1$, and rewrite $(x^5)^3$ as repeated multiplication.

$2x(x^5)^3 = 2 \cdot x^1 \cdot x^5 \cdot x^5 \cdot x^5$	Repeated Multiplication Model of Powering
$= 2 \cdot x^{16}$	Product of Powers Property

Solution 2

Use the Power of a Power Property to simplify $(x^5)^3$.
$2x(x^5)^3 = 2 \cdot x^1 \cdot x^{5 \cdot 3} = 2 \cdot x^1 \cdot x^{15} = 2x^{16}$

Adapting to Individual Needs

Extra Help

Be sure students understand and recall that $x = x^1$, as shown in **Example 5**. Point out that often the exponent 1 is not written but that the meaning is the same without it.

You might give students these numerical examples first.
$5^1 \cdot 5^1 = 5 \cdot 5 = 5^2$
$2^1 \cdot 2^1 \cdot 2^1 = 2 \cdot 2 \cdot 2 = 2^3$
$3^1 \cdot 3^1 \cdot 3^1 \cdot 3^1 = 3 \cdot 3 \cdot 3 \cdot 3 = 3^4$

Then use variables and compare the examples.
$m^1 \cdot m^1 = m \cdot m = m^2$
$n^1 \cdot n^1 \cdot n^1 = n \cdot n \cdot n = n^3$
$r^1 \cdot r^1 \cdot r^1 \cdot r^1 = r \cdot r \cdot r \cdot r = r^4$

LESSON MASTER 8-5 B — Questions on SPUR Objectives

Skills Objective B: Simplify products and powers of powers.

In 1–6, write each expression as a single power.

1. $8^4 \cdot 8^3$ — 8^7
2. $10^2 \cdot 10^2$ — 10^4
3. $5^6 \cdot 5^5 \cdot 5^2$ — 5^{13}
4. $6 \cdot 6^3$ — 6^4
5. $(3^4)^2$ — 3^8
6. $(14^3)^5$ — 14^{15}
7. a. Simplify $m^4 \cdot m^2$. — m^6
   b. Check your answer by letting $m = 3$. $3^4 \cdot 3^2 = 81 \cdot 9 = 729$; $3^6 = 729$

In 8–27, simplify each expression.

8. $m^2 \cdot m^3$ — m^5
9. $x^4 \cdot x^4$ — x^8
10. $(g^4)^3$ — g^{12}
11. $(a^5)^5$ — a^{25}
12. $r^3 \cdot r^4 \cdot r^2$ — r^9
13. $e \cdot e^3$ — e^4
14. $w^4 \cdot w \cdot w^5$ — w^{10}
15. $k^7 \cdot y^3 \cdot k \cdot y^3$ — k^8y^6
16. $(n^3 \cdot n^5)^2$ — n^{16}
17. $(d \cdot d^2)^4$ — d^{12}
18. $4p^3 \cdot 5p^6$ — $20p^9$
19. $-6x^2 \cdot 2x$ — $-12x^3$
20. $x^2 \cdot y \cdot x^4 \cdot y^3 \cdot y$ — x^6y^5
21. $(h^0)^3$ — 1
22. $5n(n^6)^3$ — $5n^{19}$
23. $x^0 \cdot 2x^3$ — $2x^3$
24. $a(a^6 - a^3)$ — $a^7 - a^4$
25. $5x^2(x + 2x^4)$ — $5x^3 + 10x^6$
26. $m^2(m^3 - y^3)$ — $m^5 - m^2y^3$
27. $(b^4)^3 - (b^3)^4$ — 0

512

QUESTIONS

Covering the Reading

1b) b^6 is the growth factor after 6 unit periods, if there is growth by a factor b in each of 6 unit periods.

1. **a.** Explain the meaning of b^6 using the Repeated Multiplication Model of Powering. **$b^6 = b \cdot b \cdot b \cdot b \cdot b \cdot b$**
b. Explain the meaning of b^6 using the Growth Model of Powering. **See left.**

In 2 and 3, write the product as a power.

2. $3^2 \cdot 3^4$ **3^6** **3.** $10^5 \cdot 10^2$ **10^7**

4. **a.** Simplify $a^2 \cdot a^3$. **a^5**
b. Check your answer by letting $a = -2$. **Does $(-2)^2 \cdot (-2)^3 = (-2)^5$? $4 \cdot -8 = -32$? $-32 = -32$? Yes, it checks.**

5. State the Product of Powers Property.
For all m and n, and all nonzero b, $b^m \cdot b^n = b^{m+n}$.

6. Refer to Example 2. Give answers as powers.
a. How many bacteria will there be after 11 hours? **$2000 \cdot 2^{11}$**
b. How many bacteria will there be 6 hours after that? **$2000 \cdot 2^{17}$**
c. Part **b** verifies that x^{17} equals what product? **$x^{11} \cdot x^6 = x^{17}$**

7. State the Power of a Power Property.
For all m and n, and all nonzero b, $(b^m)^n = b^{mn}$.

In 8 and 9, write the result as a single power. Then evaluate the power.

8. $(2^3)^4$ **$2^{12} = 4096$** **9.** $(7^2)^3$ **$7^6 = 117,649$**

10. Find x when $(10^3)^5 = 10^x$. **$x = 15$**

In 11–16, simplify.

11. $x^5 \cdot x^{50}$ **x^{55}** **12.** $(k^{10})^3$ **k^{30}** **13.** $(n^2)^6$ **n^{12}**

14. $a^3 \cdot a^5 \cdot b^0 \cdot a^2 \cdot b^9$ **$a^{10}b^9$** **15.** $b^3(a^3b^5)$ **a^3b^8** **16.** $4n^3(n^{10})^2$ **$4n^{23}$**

A family's keepsake. *Many families maintain records of their ancestors and living relatives by recording the information in a family tree.*

Applying the Mathematics

In 17 and 18, solve.

17. $2^5 \cdot 2^a = 2^{12}$ **$a = 7$** **18.** $(3^2)^x = 3^8$ **$x = 4$**

19. In Example 1 you saw that $x^2 \cdot x^5 = x^7$. Find three other pairs of expressions whose product is x^7. **Sample: x^1, x^6; x^3, x^4; x^0, x^7**

20. Write your answers as powers.
a. If you trace your family tree back through ten generations of natural parents, at most how many ancestors could you have in that generation? **2^{10}**
b. How many times more ancestors could you have five more generations back? **2^5**
c. At most how many ancestors do you have 15 generations back? **2^{15}**
d. How are parts **a**, **b**, and **c** related to one of the properties in this lesson? **$2^{10} \cdot 2^5 = 2^{15}$ uses the Product of Powers Property.**

Notes on Questions

Question 6b Ask for two answers: $2000 \cdot 2^{11} \cdot 2^6$ and $2000 \cdot 2^{17}$. Then ask why the two answers are equal. [Product of Powers Property]

Question 14 Point out that $b^0 \cdot b^9$ can be simplified by adding exponents or by substituting 1 for b^0.

Question 20 Multicultural Connection The study of genealogy has interested people since early times. The most famous ancient genealogical records are contained in the Book of Genesis, which traces the lineage of the Hebrew people. Ancient Arabs, Egyptians, Romans, and Greeks also traced the genealogies of their people. Perhaps the world's largest modern-day collection of genealogical records has been and is continuing to be assembled by the Mormons in Salt Lake City, Utah. Tracing one's genealogy is an important aspect of the Mormon religion.

► **LESSON MASTER 8-5 B** *page 2*

Properties Objective E: Identify properties of exponents and use them to explain operations with powers.

28. Show how to simplify $r^2 \cdot r^4$
a. by using repeated multiplication.
$r^2 \cdot r^4 = r \cdot r(r \cdot r \cdot r \cdot r) = r^6$
b. by using the Product of Powers Property.
$r^2 \cdot r^4 = r^{2+4} = r^6$

29. Show how to simplify $(x^3)^4$
a. by treating x^3 as a chunk.
$(x^3)^4 = x^3 \cdot x^3 \cdot x^3 \cdot x^3 = x^{12}$
b. by using the Power of a Power Property.
$(x^3)^4 = x^{3 \cdot 4} = x^{12}$

30. Write a multiplication expression that uses the Product of Powers Property and has the value w^9. **Sample: $w^3 \cdot w^6$**

31. Write an expression that uses the Power of a Power Property and has the value m^{18}. **Sample: $(m^2)^9$**

In 32–37, solve for x.

32. $8^x \cdot 8^3 = 8^7$ **$x = 4$** **33.** $4^4 \cdot 4^x = 4^5$ **$x = 1$**
34. $n^x \cdot n^3 = n^3$ **$x = 0$** **35.** $(5^4)^x = 5^{12}$ **$x = 3$**
36. $(9^x)^5 = 9^{10}$ **$x = 2$** **37.** $e(e^8)^x = e^{17}$ **$x = 2$**

Review Objective A, Lesson 3-1

38. In a magic square, each row, column, and diagonal has the same sum. Fill in the boxes at the right so each of these sums is $-6x$.
Sample is given.

x	$-6x$	**$-x$**
$-4x$	**$-2x$**	**0**
$-3x$	**$2x$**	$-5x$

Adapting to Individual Needs

Challenge
Have students solve the following problem. Which is greater, 2^{100} or 3^{75}? If students need a hint, tell them to rewrite each expression as a power of a power. [$2^{100} = (2^4)^{25}$ and $3^{75} = (3^3)^{25}$. Since $27 > 16$, then $27^{25} > 16^{25}$. So 3^{75} is greater.]

Notes on Questions

Questions 23 and 25 These expressions show the difference between multiplying by 2 and squaring. Writing $2 \cdot x^3 \cdot x^4$ and $x^3 \cdot x^4 \cdot x^3 \cdot x^4$ may help students to make the distinction.

Question 34 Relate this question to the lesson by asking how many ways the first 5 questions could be answered [2^5], and how many ways the last 10 questions can be answered. [2^{10}] The product is the answer to this question.

Question 35 If students evaluate $(-1)^{100}$ on their calculators, different answers may result. Some calculators can handle integer powers of negative bases; some cannot, and will give an error message.

Follow-up for Lesson 8-5

Practice

For more questions on SPUR Objectives, use **Lesson Master 8-5A** (shown on page 511) or **Lesson Master 8-5B** (shown on pages 512–513).

Assessment

Oral/Written Communication Have students explain the Product of Powers Property and the Power of a Power Property in their own words. Ask them to write an example of each property in their explanations. [Students' explanations are clear and examples are accurate.]

Extension

Ask students to give at least two examples that illustrate this statement: For all m and n, when $b \neq 0$, $(b^m)^n = (b^n)^m$. [Sample: $(3^4)^2 = (81)^2 = 6561$; $(3^2)^4 = (9)^4 = 6561$] Then have them attempt to show how $(b^m)^n = (b^n)^m$ follows from earlier properties. [Sample:

$(b^m)^n = b^{mn}$ Power of a Power Property

$= b^{nm}$ Commutative Property of Multiplication

$= (b^n)^m$ Power of a Power Property.]

Project Update Project 4, *Numbers with a Fixed Number of Factors,* on page 540, relates to the content of this lesson.

21. Suppose a population P of bacteria triples each day.
 a. Write an expression for the number of bacteria after 5 days. **$P \cdot 3^5$**
 b. How many days after the fifth day will the bacteria population be $P \cdot 3^{17}$? **12**

In 22–27, simplify.

22. $(x^2)^3 - (x^3)^2$ **0** **23.** $2(x^3 \cdot x^4)$ **$2x^7$** **24.** $3m^4 \cdot 5m^2$ **$15m^6$**

25. $(x^3 \cdot x^4)^2$ **x^{14}** **26.** $a^2(a^3 + 4a^4)$ **$a^5 + 4a^6$** **27.** $y(y^7 - y^2)$ **$y^8 - y^3$**

28. *True or false.* The tenth power of x^2 equals the square of x^{10}. Explain your answer. **True; the tenth power of x^2 is $(x^2)^{10} = x^{20}$ and the square of x^{10} is $(x^{10})^2 = x^{20}$.**

Review

29. Graph $y = 5\left(\frac{1}{2}\right)^x$ for integer values of x from 0 to 5. *(Lesson 8-4)* **See left.**

29)

30. A city's current population is 2,500,000. Write an expression for the population y years from now under each assumption.
 a. The population is growing 4.5% per year. **$2{,}500{,}000\ (1.045)^y$**
 b. The population is decreasing 3% per year. **$2{,}500{,}000\ (0.97)^y$**
 c. The population is decreasing by 2,500 people each year. *(Lessons 8-2, 8-4)* **$2{,}500{,}000 - 2500y$**

31. Give a situation which is represented by $T = 5000(1.035)^n$. *(Lesson 8-1)* **See left.**

31) Sample: the population T of a city of 5000 that is growing at a rate of 3.5% per year, n years from now

32. Rewrite $7y - 2x - 7 = 19 - 9x$
 a. in standard form. *(Lesson 7-8)* **$7x + 7y = 26$ or $-7x - 7y = -26$**
 b. in slope-intercept form. *(Lesson 7-4)* **$y = -x + \frac{26}{7}$**

33. For a fund-raiser, a club mixes 50 pounds of cashews costing \$3.49/lb and 20 pounds of pecans costing \$6.99/lb. To break even on their costs, how much should they charge for a pound of the cashew-pecan mixtures? *(Lesson 6-2)* **\$4.49/lb**

34. Tell how many ways you could answer a 15-item true-false test by guessing. *(Lesson 2-9)* **$2^{15} = 32{,}768$**

35. **a.** Calculate $(-1)^n$ for $n = 1, 2, 3, 4, 5, 6, 7,$ and 8. **-1; 1; -1; 1; -1; 1; -1; 1**
 b. What is $(-1)^{100}$? *(Lesson 2-5)* **1**

In 36 and 37, write in scientific notation. *(Appendix B)*

36. 4,000,000,000 **4×10^9** **37.** 0.00036 **3.6×10^{-4}**

Exploration

38. There are metric prefixes for many powers of 10^3. For instance, 10^3 meters is one kilometer. Give the metric prefix for each power.
 a. 10^{-3} **milli-** **b.** 10^6 **mega-** **c.** 10^{-6} **micro-** **d.** 10^9 **giga-**
 e. 10^{-9} **nano-** **f.** 10^{12} **tera-** **g.** 10^{-12} **pico-** **h.** 10^{15} **peta-**
 i. 10^{-15} **femto-** **j.** 10^{18} **exa-** **k.** 10^{-18} **atto-**

514

Setting Up Lesson 8-6

Materials If automatic graphers are available, we recommend that students use them to check **Question 19** in Lesson 8-6.

Questions 37 and 38 involve negative exponents with 10 as a base, which students have seen in their work with scientific notation. These questions help prepare the students for the next lesson.

LESSON

8-6

Negative Exponents

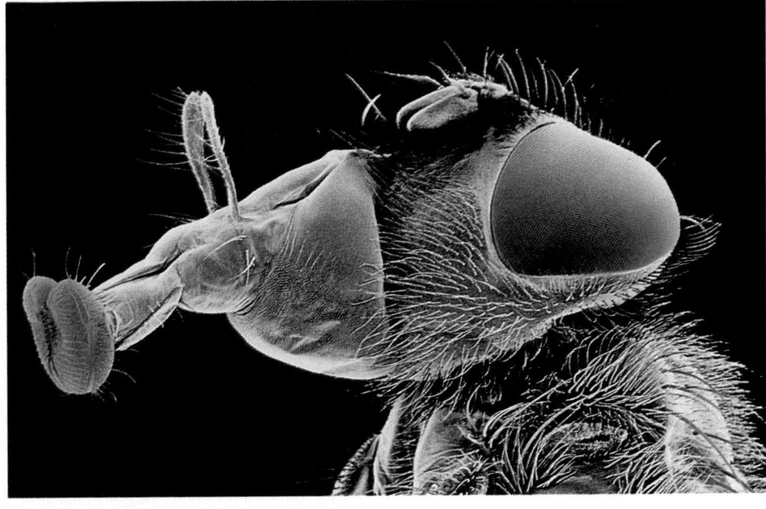

A closer look. *This is the head of a fly as seen under an electron microscope. Some electron microscopes can view objects that are less than .0001, or 10^{-4} meter in size.*

What Is the Value of a Power When Its Exponent Is Negative?

You have used the base 10 with a negative exponent to represent small numbers in scientific notation. For instance,

$$10^{-1} = 0.1 = \frac{1}{10}, \quad 10^{-2} = 0.01 = \frac{1}{10^2}, \quad 10^{-3} = 0.001 = \frac{1}{10^3}, \quad \text{and so on.}$$

Now we consider other powers with negative exponents. That is, we want to know the meaning of b^n when n is negative.

Consider this pattern of the powers of 2.

$$2^4 = 16$$
$$2^3 = 8$$
$$2^2 = 4$$
$$2^1 = 2$$
$$2^0 = 1$$

Each exponent is one less than the one above it. The value of each power is half that of the number above. Continuing the pattern suggests that the following are true.

$$2^{-1} = \frac{1}{2}$$
$$2^{-2} = \frac{1}{4} = \frac{1}{2^2}$$
$$2^{-3} = \frac{1}{8} = \frac{1}{2^3}$$
$$2^{-4} = \frac{1}{16} = \frac{1}{2^4}$$

A general description of the pattern is simple: $2^{-n} = \frac{1}{2^n}$. That is, 2^{-n} is the reciprocal of 2^n.

Lesson 8-6

Objectives

- **A** Evaluate negative integer powers of real numbers.
- **B** Simplify products of powers and powers of powers involving negative exponents.
- **E** Identify the Negative Exponent Property and use it to explain operations with powers.
- **G** Solve problems involving exponential growth and decay.
- **H** Use and simplify expressions with powers involving negative exponents in real situations.

Resources

From the ***Teacher's Resource File***

- Lesson Master 8-6A or 8-6B
- Answer Master 8-6
- Assessment Sourcebook: Quiz for Lessons 8-4 through 8-6
- Teaching Aid 88: Warm-up

Additional Resources

- Visual for Teaching Aid 88
- Automatic graphers

Teaching Lesson 8-6

Warm-up

1. Write the values of the integer powers of 10 from 10^{-6} to 10^6.
$10^{-6} = .000001$; $10^{-5} = .00001$; $10^{-4} = .0001$; $10^{-3} = .001$; $10^{-2} = .01$; $10^{-1} = .1$; $10^0 = 1$; $10^1 = 10$; $10^2 = 100$; $10^3 = 1000$; $10^4 = 10{,}000$; $10^5 = 100{,}000$; $10^6 = 1{,}000{,}000$

(Warm-up continues on page 516.)

Lesson 8-6 Overview

Broad Goals This lesson develops the meaning of powers with negative exponents by extending the properties of powers that were presented in previous lessons.

Perspective Notice that we speak of "negative exponents" rather than "negative powers" even though the latter terminology is commonly used. In 2^{-5}, the value of the exponent is negative, but the value of the power is positive.

Often, the Negative Exponents Property $b^{-n} = \frac{1}{b^n}$ is given as a definition. Here it can be deduced because we have assumed the Product of Powers Property for all exponents. The deductive argument is given. However, the first explanation is via an inductive argument because students at this level are more often convinced by patterns than by deduction.

The Growth Model explains that negative powers of a positive number should be positive, because going back in time does not mean that negative values are introduced. This is confirmed by evaluating powers that have negative exponents.

2. Multiply.
   a. $10^{-7} \times 10^{7}$ **1**
   b. $10^{-4} \times 10^{4}$ **1**
   c. $10^{-2} \times 10^{2}$ **1**
   d. $10^{-1} \times 10^{1}$ **1**
3. What is true for each pair of numbers in Question 2? **They are reciprocals.**

Notes on Reading

Reading Mathematics There is quite a lot of reading in this lesson, and it will be helpful for students to focus on the main ideas.

1. The pattern on page 515 suggests the Negative Exponent Property. **Example 1** applies the property.
2. The argument at the bottom of page 516 shows that the Negative Exponent Property must be true.
3. **Example 2** applies the Negative Exponent Property to a real situation. The paragraph below the example shows how to calculate powers with negative exponents using a calculator.
4. The Product of Powers and the Power of a Power Property both hold when the exponents are negative; **Example 3** shows an instance.
5. Negative exponents can refer to going back in time; therefore, they fit into the Growth Model for Powering. This is the idea behind **Example 4**.

We call the general property the *Negative Exponent Property.*

Negative Exponent Property
For any nonzero b and all n, $b^{-n} = \frac{1}{b^n}$, the reciprocal of b^n.

Example 1

a. Write 4^{-3} as a simple fraction without a negative exponent.
b. Write $\frac{1}{32}$ as a negative power of an integer.
c. Rewrite $q^5 \cdot t^{-3}$ without negative exponents.

Solution

a. Use the Negative Exponent Property.

$$4^{-3} = \frac{1}{4^3} = \frac{1}{64}$$

b. First write the denominator as a power. You must recognize 32 as a power of 2.

$$\frac{1}{32} = \frac{1}{2^5} = 2^{-5}$$

c. Substitute $\frac{1}{t^3}$ for t^{-3}.

$$q^5 \cdot t^{-3} = q^5 \cdot \frac{1}{t^3} = \frac{q^5}{t^3}$$

Notice that even though the exponent in 4^{-3} is negative, the number 4^{-3} is still positive. All negative integer powers of positive numbers are positive.

Because the Product of Powers Property applies to all exponents, it applies to negative exponents.

$$b^n \cdot b^{-n} = b^{n+-n} \quad \text{Product of Powers Property}$$
$$= b^0$$
$$= 1 \quad \text{Zero Exponent Property}$$

Also,
$$b^n \cdot b^{-n} = b^n \cdot \frac{1}{b^n} = \frac{b^n}{b^n} = 1$$

In this way, the Product of Powers Property verifies that b^{-n} must be the reciprocal of b^n. In particular, $b^{-1} = \frac{1}{b}$. That is, the -1 power (read "negative one" or "negative 1st" power) of a number is its reciprocal.

Optional Activities

You might want to use this activity after discussing **Example 1**. Have students graph $y = 2^x$ for integer values of x from -4 to 3. Students are asked to make this graph in **Question 19**. Use a large-scale graph on the board or overhead so that students will be able to distinguish y values as small as $\frac{1}{16}$. Referring to the graph of $y = 2^x$ can help students avoid the most common pitfall of negative exponents. The entire graph of $y = 2^x$ lies above the x-axis because 2^x is always positive, even for negative values of x.

The graph of $y = 2^x$ can also be related to exponential growth situations. Suppose you are doubling your money every year. If you now have $1 million, moving to the right on the graph shows what will happen in the future; moving to the left shows what happened in the past. Each time you move one unit to the left, you multiply by $2^{-1} = \frac{1}{2}$, which is equivalent to dividing by 2.

Cool choices. *Choosing ice cream flavors is usually easier than choosing answers on a multiple-choice test.*

Example 2

Suppose each question on a 10-item multiple choice test has four options. Write the probability of guessing all the correct answers with a negative exponent.

Solution

The Multiplication Counting Principle gives

$$4 \cdot 4 \cdot 4 \cdot 4 \cdot 4 \cdot 4 \cdot 4 \cdot 4 \cdot 4 \cdot 4 = 4^{10}$$

different ways of doing the test. Only one of these ways is correct.

$$P(\text{perfect paper}) = \frac{1}{4^{10}}$$

Use the Negative Exponent Property. $\frac{1}{4^{10}} = 4^{-10}$, so $P = 4^{-10}$.

Evaluating Powers with Negative Exponents

You can evaluate negative exponents on your calculator. Remember to use the [+/–] or [(–)] key, not the subtraction key. Here are sample key sequences and displays for evaluating the answers to Example 2. For 4^{-10}: Key in 4 [y^x] 10 [+/–] [=]. Display: [0.000000954] or [9.5367 -7]

All the properties of powers you have learned are usable with negative exponents. They can help translate an expression with a negative power into one with only positive powers.

Example 3

Rewrite $(x^3)^{-2}$ without negative exponents.

Solution 1

Use the Power of a Power Property, which says to multiply the exponents.

$$(x^3)^{-2} = x^{3 \cdot -2} \quad \text{Power of a Power Property}$$
$$= x^{-6}$$
$$= \frac{1}{x^6} \quad \text{Negative Exponent Property}$$

Solution 2

Use the Negative Exponent Property first instead of last. Treat x^3 as a chunk. Find the reciprocal of the square of this chunk.

$$(x^3)^{-2} = \frac{1}{(x^3)^2} \quad \text{Negative Exponent Property}$$
$$= \frac{1}{x^6} \quad \text{Power of a Power Property}$$

Adapting to Individual Needs

Extra Help

If students are having difficulty understanding the meaning of negative exponents, you might have them choose a positive value of b and any real value of m. Ask them to evaluate b^m and b^{-m} with their calculators. Then have them find the reciprocal of one of these expressions to verify that it equals the other expression.

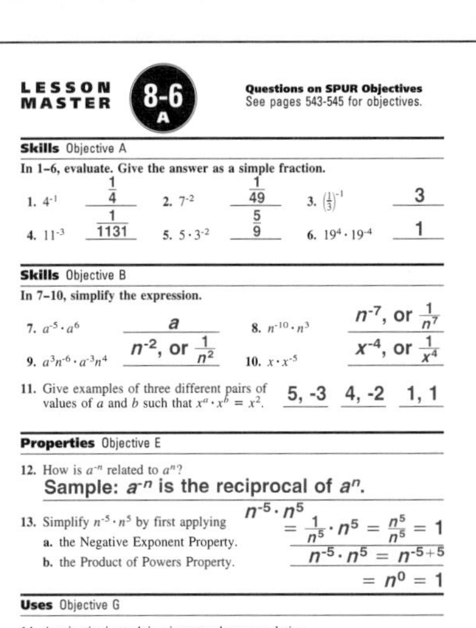

LESSON MASTER 8-6 A

Questions on SPUR Objectives
See pages 543-545 for objectives.

Skills Objective A

In 1–6, evaluate. Give the answer as a simple fraction.

1. 4^{-1} $\frac{1}{4}$
2. 7^{-2} $\frac{1}{49}$
3. $\left(\frac{1}{3}\right)^{-1}$ 3
4. 11^{-3} $\frac{1}{1131}$
5. $5 \cdot 3^{-2}$ $\frac{5}{9}$
6. $19^4 \cdot 19^{-4}$ 1

Skills Objective B

In 7–10, simplify the expression.

7. $a^{-5} \cdot a^6$ a
8. $n^{-10} \cdot n^3$ n^{-7}, or $\frac{1}{n^7}$
9. $a^3n^{-6} \cdot a^{-3}n^4$ n^{-2}, or $\frac{1}{n^2}$
10. $x \cdot x^{-5}$ x^{-4}, or $\frac{1}{x^4}$
11. Give examples of three different pairs of values of a and b such that $x^a \cdot x^b = x^2$. 5, -3 4, -2 1, 1

Properties Objective E

12. How is a^{-n} related to a^n?
Sample: a^{-n} is the reciprocal of a^n.
13. Simplify $n^{-5} \cdot n^5$ by first applying
a. the Negative Exponent Property. $n^{-5} \cdot n^5 = \frac{1}{n^5} \cdot n^5 = \frac{n^5}{n^5} = 1$
b. the Product of Powers Property. $n^{-5} \cdot n^5 = n^{-5+5} = n^0 = 1$

Uses Objective G

14. A scientist is studying insects whose population doubles every 5 days. There are 4800 insects today. How many were there 30 days ago? 75 insects
15. Adria has $4,700 in a bank account that earns 6% interest.
a. How much did she have in the account 5 years ago? $3512.11
b. How much will she have 5 years from now? $6289.66

Uses Objective H

16. A test has 15 multiple-choice questions, each with 5 options. Write the probability, using a negative exponent, of guessing
a. all correct answers. 5^{-15} b. all wrong answers. $4^{15} \cdot 5^{-15}$

The Growth Model can confirm the Negative Exponents Property: Suppose there is a quantity Q that has been growing with a growth factor g for n years. Then n years ago, there was $Q \cdot g^{-n}$. In the n years, it grew by a factor of g^n, so there is now $(Q \cdot g^{-n}) \cdot g^n$. But there is now Q. So $Q = (Q \cdot g^{-n}) \cdot g^n$. Divide both sides by Q to get $1 = g^{-n} \cdot g^n$.

Additional Examples

1. a. Write 7^{-3} as a simple fraction without a negative exponent. $\frac{1}{343}$

   b. Write $\frac{1}{81}$ as the power of an integer. **9^{-2} or 3^{-4} or 81^{-1}**

   c. Simplify $n^5 \cdot n^{-6}$, and write your answer without negative exponents. $\frac{1}{n}$

2. A true-false test has 25 questions. Give the probability that you will guess all 25 questions correctly. $(\frac{1}{2})^{25} = \frac{1}{2^{25}} = 2^{-25}$

3. Rewrite $(m^{-5})^4$ without negative exponents. $\frac{1}{m^{20}}$

4. The viruses in a culture are quadrupling in number each day. Right now, the culture contains about 1,000,000 viruses. About how many viruses did the culture have 6 days ago?
   $1{,}000{,}000 \cdot 4^{-6} \approx 244$

LESSON MASTER 8-6 B Questions on SPUR Objectives

Skills Objective A: Evaluate integer powers of real numbers.

In 1–8, evaluate the expression. Give your answers as a simple fraction.

1. 7^{-1}	$\frac{1}{7}$	2. 11^{-2}	$\frac{1}{121}$
3. $\left(\frac{1}{5}\right)^{-1}$	5	4. 6^{-3}	$\frac{1}{216}$
5. $2 \cdot 3^{-4}$	$\frac{2}{81}$	6. $20^{-8} \cdot 20^8$	1
7. $\left(\frac{2}{3}\right)^{-2}$	$\frac{9}{4}$	8. $3^{-2} \cdot 3^{-3}$	$\frac{1}{243}$

In 9–16, evaluate the expression. Give your answer as a decimal.

9. 10^{-1}	0.1	10. 10^{-4}	0.0001
11. 10^{-2}	0.01	12. 10^{-10}	0.0000000001
13. $6 \cdot 10^{-1}$	0.6	14. $5 \cdot 10^{-3}$	0.005

15. $2 \cdot 10^3 + 8 \cdot 10^2 + 6 \cdot 10^1 + 6 \cdot 10^0 + 3 \cdot 10^{-1} + 4 \cdot 10^{-2} + 5 \cdot 10^{-3}$ 2866.345

16. $9 \cdot 10^5 + 4 \cdot 10^4 + 3 \cdot 10^2 + 8 \cdot 10^0 + 2 \cdot 10^{-1} + 7 \cdot 10^{-4} + 1 \cdot 10^{-5}$ 940,308.20071

Skills Objective B: Simplify products of and powers of powers.

17. $m^{-3} \cdot m^5$	m^2	18. $x^2 \cdot x^{-8}$	x^{-6}
19. $a^{-2} \cdot a^{-6}$	a^{-8}	20. $(a^{-3})^4$	a^{-12}
21. $x^3 \cdot x^{-9} \cdot x^2$	x^{-4}	22. $n \cdot n^{-6}$	n^{-5}
23. $uw^4 \cdot u^{-2}w^{-8}$	$u^{-1}w^{-4}$	24. $(g^{-1})^{-5}$	g^5

In 25–28, write each expression without negative exponents.

25. 4^{-x}	$\frac{1}{4^x}$	26. x^{-3}	$\frac{1}{x^3}$
27. $m^{-2}n^5$	$\frac{n^5}{m^2}$	28. $7a^{-1}b^{-5}$	$\frac{7}{ab^5}$

An Application with Negative Exponents

In the Growth Model for Powering, negative exponents stand for unit periods going back in time.

Example 4

Recall the compound interest formula

$$T = P(1 + i)^n.$$

Three years ago, Mr. Cabot put money in a CD at an annual yield of 7%. If the CD is worth \$3675 now, what was it worth then?

Solution

Here $P = 3675$, $i = 0.07$, and $n = -3$ (for three years ago). So $T = 3675(1.07)^{-3}$. The calculator key sequence

3675 [×] 1.07 [y^x] 3 [±] [=]

gives 2999.8947.

So Mr. Cabot probably started with \$3000. The 11¢ difference is probably due to rounding.

QUESTIONS

Covering the Reading

1. a. Complete the last three equations in this pattern. Then write the next equation in the pattern.

$$4^3 = 64$$
$$4^2 = 16$$
$$4^1 = 4$$
$$4^0 = 1$$
$$4^{-1} = \underline{\ ?\ } \quad \frac{1}{4}$$
$$4^{-2} = \underline{\ ?\ } \quad \frac{1}{16}$$
$$4^{-3} = \underline{\ ?\ } \quad \frac{1}{64}; \quad 4^{-4} = \frac{1}{256}$$

   b. *True or false.* When x is negative, 4^x is negative. **False**

2. *Multiple choice.* When $x \neq 0$, x^{-n} equals which of the following? **d**
   (a) $-x^n$ (b) $(-x)^n$ (c) $\frac{1}{x^{-n}}$ (d) $\frac{1}{x^n}$

3. a. Write 10^{-9} as a decimal. **0.000000001**
   b. Write 10^{-9} as a simple fraction. $\frac{1}{1{,}000{,}000{,}000}$

4. b^{-1} is the _?_ of b. **reciprocal**

In 5–7, write as a simple fraction.

5. 5^{-2} $\frac{1}{25}$ 6. 3^{-6} $\frac{1}{729}$ 7. $\left(\frac{1}{2}\right)^{-1}$ 2

518

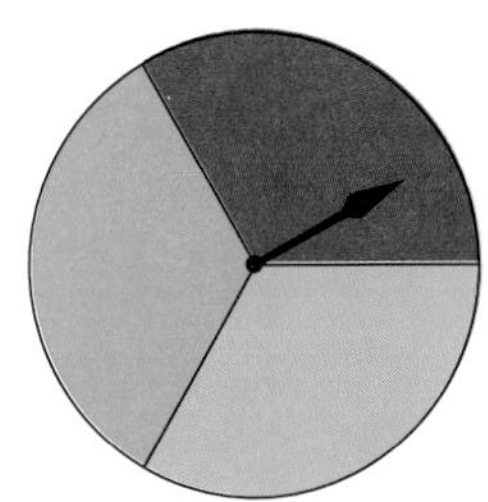

In 8–10, write without negative exponents.

8. x^5y^{-2} $\frac{x^5}{y^2}$ **9.** $3a^{-2}b^{-4}$ $\frac{3}{a^2b^4}$ **10.** 2^{-n} $\frac{1}{2^n}$

11. Suppose the three regions on the spinner at the left are equally likely. What is the probability that 5 spins give five reds? $\frac{1}{243}$ or about .004

12. *True or false.* If x is positive, then x^{-4} is positive. True

In 13 and 14, simplify.

13. $5^3 \cdot 5^{-3}$ 1 **14.** $c^j \cdot c^{-j}$ 1

In 15 and 16, simplify and write without negative exponents.

15. $(x^4)^{-2}$ $\frac{1}{x^8}$ **16.** $(a^{-3})^4$ $\frac{1}{a^{12}}$

17. Refer to Example 4. Find the value of Mr. Cabot's CD one year ago. $3434.58

18. Theresa has $1236.47 in a savings account that has had an annual yield of 5.25% since she opened the account. Assuming no withdrawals or deposits were made, how much was in the account 8 years ago? $821.12

Applying the Mathematics

19a)

19. a. Graph $y = 2^x$ when the domain of x is {-4, -3, -2, -1, 0, 1, 2, 3}.

b. Describe what happens to the graph as x becomes smaller and smaller. The y-coordinate approaches zero.

20. If the reciprocal of $(1.06)^{11}$ is $(1.06)^n$, what is n? $n = -11$

In 21 and 22, solve and check. Check answers using a calculator.

21. $3^{10} \cdot 3^{-12} = 3^z$ $z = -2$ **22.** $5^6 \cdot 5^y = 5^{-9}$ $y = -15$

In 23 and 24, simplify.

23. $t^{-2} \cdot t^{-4}$ t^{-6} **24.** $(x^5y^3) \cdot (x^{-3}y^{-5})$ x^2y^{-2}

25. The human population P (in billions) of the earth x years from 1985 can be estimated by the formula $P = 5 \cdot (1.017)^x$. Use this formula to estimate the earth's population in the given year.

a. 1994 ≈ 5.82 billion **b.** 1980 ≈ 4.60 billion

Predicting the future. *Shown is a scene from the 1968 film,* 2001: A Space Odyssey. *In this science-fiction classic, spaceships rendezvous with space stations, and crew members play chess with a soft-voiced computer, Hal.*

Notes on Questions

Question 6 If students use calculators to evaluate 3^{-6} and then find the reciprocal of the result, some calculators will display 729. This indicates that $3^{-6} = \frac{1}{729}$.

Question 19 An automatic grapher can be used to display the graph.

Question 25 Negative exponents must be used for the years prior to 1985.

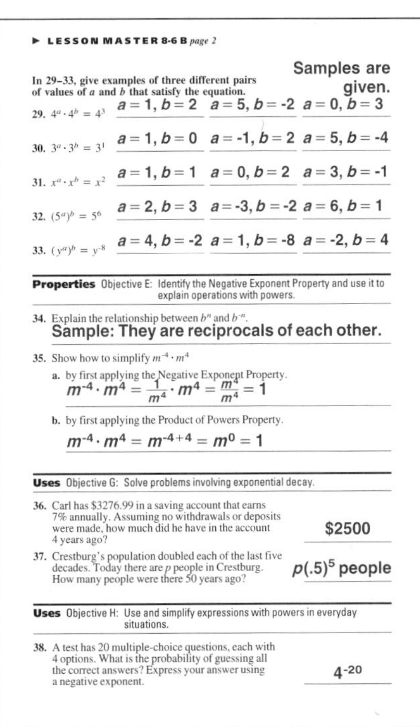

▸ LESSON MASTER 8-6 B *page 2*

In 29–33, give examples of three different pairs of values of a and b that satisfy the equation. Samples are given.

29. $4^a \cdot 4^b = 4^3$ $a = 1, b = 2$ $a = 5, b = -2$ $a = 0, b = 3$

30. $3^a \cdot 3^b = 3^1$ $a = 1, b = 0$ $a = -1, b = 2$ $a = 5, b = -4$

31. $x^a \cdot x^b = x^2$ $a = 1, b = 1$ $a = 0, b = 2$ $a = 3, b = -1$

32. $(5^a)^b = 5^6$ $a = 2, b = 3$ $a = -3, b = -2$ $a = 6, b = 1$

33. $(y^a)^b = y^{-8}$ $a = 4, b = -2$ $a = 1, b = -8$ $a = -2, b = 4$

Properties Objective E: Identify the Negative Exponent Property and use it to explain operations with powers.

34. Explain the relationship between b^n and b^{-n}.
Sample: They are reciprocals of each other.

35. Show how to simplify $m^{-4} \cdot m^4$

a. by first applying the Negative Exponent Property.
$m^{-4} \cdot m^4 = \frac{1}{m^4} \cdot m^4 = \frac{m^4}{m^4} = 1$

b. by first applying the Product of Powers Property.
$m^{-4} \cdot m^4 = m^{-4+4} = m^0 = 1$

Uses Objective G: Solve problems involving exponential decay.

36. Carl has $3276.99 in a saving account that earns 7% annually. Assuming no withdrawals or deposits were made, how much did he have in the account 4 years ago? $2500

37. Crestburg's population doubled each of the last five decades. Today there are p people in Crestburg. How many people were there 50 years ago? $p(.5)^5$ people

Uses Objective H: Use and simplify expressions with powers in everyday situations.

38. A test has 20 multiple-choice questions, each with 4 options. What is the probability of guessing all the correct answers? Express your answer using a negative exponent. 4^{-20}

Adapting to Individual Needs

Challenge

Have students simplify the expression below. If students need a hint, tell them to first multiply both the numerator and denominator by x^{10}. The numerator of the resulting fraction can be simplified by applying the distributive property. [x^{10}]

$$\frac{x + x^2 + x^3 + x^4 + x^5 + x^6 + x^7}{x^{-3} + x^{-4} + x^{-5} + x^{-6} + x^{-7} + x^{-8} + x^{-9}}$$

Notes on Questions

Question 26 A pattern with base 10 is probably the pattern that, more than any other, convinces students of the reasonableness of the properties of negative exponents.

Questions 31, 33, and 34 An automatic grapher can be used to check the answers.

Question 32 Health Connection A virus is an infectious agent that can multiply only in the living cells of animals, plants, or bacteria. Viruses can vary in size from about 20 to 250 nanometers in diameter. (1 nanometer = 10^{-9} meter) Infectious mononucleosis (mono) is an example of a disease that is caused by a virus.

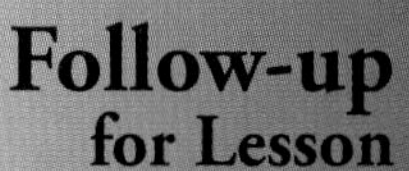

Follow-up for Lesson 8-6

Practice

For more questions on SPUR Objectives, use **Lesson Master 8-6A** (shown on page 517) or **Lesson Master 8-6B** (shown on pages 518–519).

Assessment

Quiz A quiz covering Lessons 8-4 through 8-6 is provided in the *Assessment Sourcebook.*

Written Communication Have students use the information given in **Question 25** to find the population of the earth in the year they were born and in the year that they will graduate from high school. [Students use positive and negative exponents correctly.]

Extension

Writing Have each student write a problem that involves the use of negative exponents. Then have **pairs of students** exchange problems and solve them.

Project Update Project 2, *Applying the Properties,* on page 539, relates to the content of this lesson.

26. Evaluate each expression.

a. $3 \cdot 10^4 + 5 \cdot 10^2 + 6 \cdot 10^1 + 2 \cdot 10^0 + 4 \cdot 10^{-1} + 7 \cdot 10^{-3}$

b. $9 \cdot 10^3 + 8 \cdot 10^2 + 7 \cdot 10^1 + 6 \cdot 10^0 + 5 \cdot 10^{-1} + 4 \cdot 10^{-2} + 9 \cdot 10^{-4}$

a) 30,562.407 b) 9876.5409

Review

In 27–29, simplify. *(Lesson 8-5)*

27. $4x \cdot 3x^2$ $12x^3$

28. $a^2 \cdot a^5 \cdot a$ a^8

29. $3c^3 \cdot 4c^4 + 5c^2 \cdot 2c^5$ $22c^7$

30. $y^m \cdot y^n \cdot y^p$ y^{m+n+p}

31. *Multiple choice.* Choose the equation whose graph is shown at the left. c

(a) $y = 0.3^x$
(b) $y = -(0.3^x) + 1$
(c) $y = 0.8^x$
(d) $y = -(0.8^x) + 1$ *(Lesson 8-3)*

32. A certain kind of virus doubles its population every 3 hours.

a. In two days, how many times does the population double? 16

b. If a culture begins with 25 virus organisms of this kind, how many are there after two days? *(Lesson 8-2)* 1,638,400

33. Write an equation for the line with slope -5 that passes through the point (2, -1). *(Lesson 7-5)* $y = -5x + 9$

34. Consider the line $y = 2x - 3$.

a. Give its slope. 2

b. Give its y-intercept. -3

c. Graph the line. *(Lesson 7-4)* See left.

34c)

35. *Skill sequence.* *(Lesson 3-9)*

a. Simplify $4(7a - 2)$. $28a - 8$

b. Simplify $4(7a - 2) - 3(5a + 1)$. $13a - 11$

c. Solve $4(7a - 2) - 3(5a + 1) = 15$. $a = 2$

36. Simplify $\frac{3x + 6x}{12x}$. *Lessons 2-3, 3-2)* $\frac{3}{4}$

Exploration

37. a. You know that $\left(\frac{1}{2}\right)^2 = \frac{1}{2} \cdot \frac{1}{2} = \frac{1}{4}$. Find other positive and negative integer powers of the number $\frac{1}{2}$.

b. How do the powers of $\frac{1}{2}$ compare with the powers of 2?

c. Generalize to other pairs of reciprocal bases.

a) Samples: $\left(\frac{1}{2}\right)^3 = \frac{1}{2} \cdot \frac{1}{2} \cdot \frac{1}{2} = \frac{1}{8}$

$\left(\frac{1}{2}\right)^4 = \frac{1}{2} \cdot \frac{1}{2} \cdot \frac{1}{2} \cdot \frac{1}{2} = \frac{1}{16}$

$\left(\frac{1}{2}\right)^{-3} = \frac{1}{\left(\frac{1}{2}\right)^3} = \frac{1}{\frac{1}{8}} = 8$

b) Powers of $\frac{1}{2}$ are reciprocals of the corresponding powers of 2.

c) $\left(\frac{1}{x}\right)^n = x^{-n}$

LESSON

8-7

Quotients of Powers

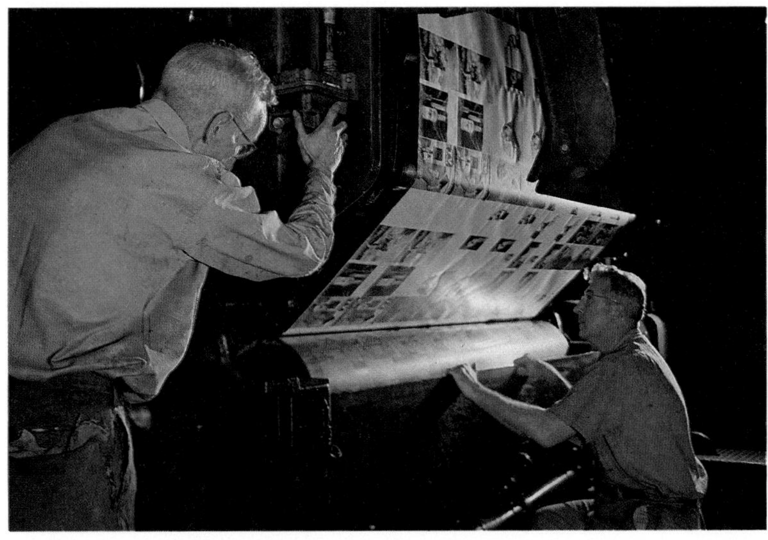

How books are made. *For many books, 32 pages are printed on each side of large sheets of paper. A machine folds each sheet several times. How many folds per sheet are needed to put the pages in order?*

Dividing Powers with the Same Base

Here is part of a list of the integer powers of 2, those from 2^{-5} through 2^{11}.

$$\ldots, \frac{1}{32}, \frac{1}{16}, \frac{1}{8}, \frac{1}{4}, \frac{1}{2}, 1, 2, 4, 8, 16, 32, 64, 128, 256, 512, 1024, 2048, \ldots$$

Multiply any two of these numbers and you will find that the product is on the list. For instance,

$$\frac{1}{8} \cdot 256 = 32.$$

When you write each number in this equation as a power of 2, you can see an instance of the Product of Powers Property.

$$2^{-3} \cdot 2^{8} = 2^{5}.$$

It may surprise you that if you divide any two numbers on the list, their quotient is on the list as well. For instance, consider 512 and 16.

Dividing larger by smaller	*Dividing smaller by larger*
$\frac{512}{16} = 32$	$\frac{16}{512} = \frac{1}{32}$
As powers: $\frac{2^9}{2^4} = 2^5$	As powers: $\frac{2^4}{2^9} = \frac{1}{2^5} = 2^{-5}$

These examples illustrate the Quotient of Powers Property.

Quotient of Powers Property
For all m and n, and all nonzero b,
$$\frac{b^m}{b^n} = b^{m-n}.$$

Lesson 8-7

Objectives

A Evaluate quotients of integer powers of real numbers.
B Simplify quotients of powers.
E Identify the Quotient of Powers Property and use it to explain operations with powers.
H Use and simplify expressions with quotients of powers in real situations.

Resources

From the ***Teacher's Resource File***
- Lesson Master 8-7A or 8-7B
- Answer Master 8-7
- Teaching Aid 88: Warm-up
- Activity Kit, Activity 19

Additional Resources
- Visual for Teaching Aid 88

Teaching Lesson 8-7

Warm-up

1. Find the values of 3^n for integer values of n from -8 to 8.
**$3^{-8} = \frac{1}{6561}$ $3^{-7} = \frac{1}{2187}$ $3^{-6} = \frac{1}{729}$
$3^{-5} = \frac{1}{243}$ $3^{-4} = \frac{1}{81}$ $3^{-3} = \frac{1}{27}$
$3^{-2} = \frac{1}{9}$ $3^{-1} = \frac{1}{3}$ $3^0 = 1$
$3^1 = 3$ $3^2 = 9$ $3^3 = 27$
$3^4 = 81$ $3^5 = 243$ $3^6 = 729$;
$3^7 = 2187$ $3^8 = 6561$**

(Warm-up continues on page 522.)

Answer to caption: 5 folds

Lesson 8-7 Overview

Broad Goals This lesson develops and applies the Quotient of Powers Property: If $b \neq 0$, for all m and n, $\frac{b^m}{b^n} = b^{m-n}$.

Perspective In this lesson, quotients of powers of the same number are simplified. Since negative exponents were covered in the previous lesson, we can say that for $b \neq 0$, $\frac{b^m}{b^n} = b^{m-n}$ for any values of m and n. This is a more efficient approach than the one taken when negative exponents are avoided. That approach requires two rules, one rule for the case $m < n$ and the other for the case $m > n$.

Because students have already discussed negative exponents, there are two reasonable answers to be given when the exponent in the denominator is greater than the exponent in the numerator. Each answer has its advantages. For example, $\frac{2^5}{2^8} = 2^{-3}$ avoids a fraction. On the other hand, $\frac{2^5}{2^8} = \frac{1}{2^3}$ avoids a negative exponent.

Use your answers to Question 1 to help you write each product as a product of powers.

2. $\frac{1}{243} \cdot 6561 = 27$ $3^{-5} \cdot 3^8 = 3^3$
3. $729 \cdot \frac{1}{81} = 9$ $3^6 \cdot 3^{-4} = 3^2$
4. $\frac{1}{27} \cdot \frac{1}{9} = \frac{1}{243}$ $3^{-3} \cdot 3^{-2} = 3^{-5}$
5. $\frac{1}{2187} \cdot 3 = \frac{1}{729}$ $3^{-7} \cdot 3^1 = 3^{-6}$
6. $243 \cdot 27 = 6561$ $3^5 \cdot 3^3 = 3^8$

Notes on Reading

There is only one new idea in this lesson, the Quotient of Powers Property, but it is applied in a variety of ways. In **Example 1**, the property is used to divide powers of the same real number; then it is applied to division of powers of the same variable; in **Example 2**, it is applied to division of powers of numbers in scientific notation; in **Example 3**, it is applied to the division of monomials.

Point out, as stated in the overview, that when an answer involves a negative exponent, it can be rewritten as a fraction. Indicate to your students which type of answer you prefer, or if you want both. Our preference is that students be asked to give one correct answer, but that they know how to convert from fraction to negative exponent and back when requested to do so.

Some students prefer to do the problems in this lesson by rewriting them as repeated multiplication, as is done in Solution 2 to **Example 1**. This method works well since most of the exponents in the problems are small. Such rewriting can also be used to explain why the properties hold. If the exponents are large, it can be cumbersome and time consuming, so try to wean those students away from this method.

When simplifying $\frac{b^m}{b^n}$, if the larger power is in the numerator, the result is a positive power of b. If the larger power is in the denominator, then the result is a negative power of b.

Example 1

Evaluate the following.

a. $\frac{6^{10}}{6^7}$ **b.** $\frac{6^7}{6^{10}}$

Solution 1

Use the Quotient of Powers Property.

a. $\frac{6^{10}}{6^7} = 6^{10-7} = 6^3 = 216$

b. $\frac{6^7}{6^{10}} = 6^{7-10} = 6^{-3} = \frac{1}{6^3} = \frac{1}{216}$

Solution 2

Use the Repeated Multiplication Model for Powering.

a. $\frac{6^{10}}{6^7} = \frac{6 \cdot 6 \cdot 6 \cdot 6 \cdot 6 \cdot 6 \cdot 6 \cdot 6 \cdot 6 \cdot 6}{6 \cdot 6 \cdot 6 \cdot 6 \cdot 6 \cdot 6 \cdot 6} = 6^3 = 216$

b. $\frac{6^7}{6^{10}} = \frac{6 \cdot 6 \cdot 6 \cdot 6 \cdot 6 \cdot 6 \cdot 6}{6 \cdot 6 \cdot 6 \cdot 6 \cdot 6 \cdot 6 \cdot 6 \cdot 6 \cdot 6 \cdot 6} = \frac{1}{6^3} = \frac{1}{216}$

Check

Use a calculator.

a. One possible key sequence is

6 [y^x] 10 [÷] 6 [y^x] 7 [=].

The display shows [216].

b. A possible key sequence is

6 [y^x] 7 [÷] 6 [y^x] 10 [=].

The display shows [0.0046296].

1 [÷] 216 [=] also gives [0.0046296].

It checks.

Optional Activities

Activity 1

You can use *Activity Kit, Activity 19* to introduce Lesson 8-7. In this activity, students use tile-like pieces made from grid paper to model dividing powers.

Activity 2

After students have read the lesson, have them derive the Quotient of Powers Property from other properties. Here is one way.

$\frac{b^m}{b^n} = b^m \cdot \frac{1}{b^n}$	Algebraic Definition of Division
$= b^m \cdot b^{-n}$	Negative Exponent Property
$= b^{m+-n}$	Product of Powers Property
$= b^{m-n}$	Algebraic Definition of Subtraction

Using the Quotient of Powers Property

The Quotient of Powers Property may be used to divide algebraic expressions. For instance, $\frac{y^{12}}{y^5} = y^{12-5} = y^7$.

Now consider the fraction $\frac{b^m}{b^m}$. By the Quotient of Powers Property, $\frac{b^m}{b^m} = b^{m-m} = b^0$. But we know $\frac{b^m}{b^m} = 1$. This is another way of showing why $b^0 = 1$.

The Quotient of Powers Property can be applied to divide numbers written in scientific notation.

A note on notes. *These \$1 and \$3 bills were issued by the Continental Congress to help finance the Revolutionary War (1775–1783). So many of these notes were printed that they became almost worthless. The \$2 bill was first issued in 1862.*

Example 2

In 1993, there were approximately 5.3 billion one-dollar bills in circulation and about 256 million people in the United States. How many dollar bills was this per person?

Solution

Since dollars per person is a rate unit, the answer is found by division.

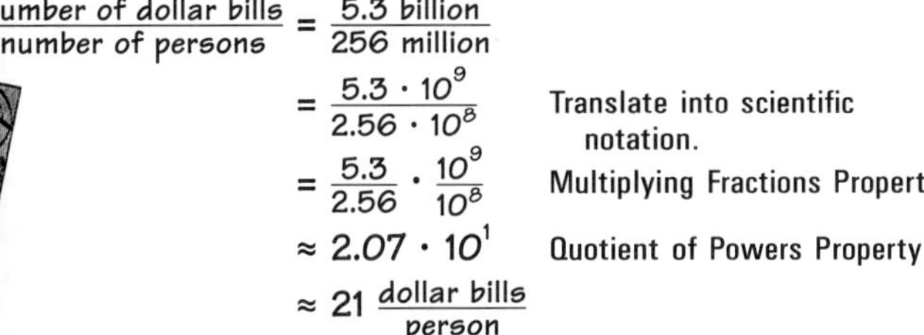

$\frac{\text{number of dollar bills}}{\text{number of persons}} = \frac{5.3 \text{ billion}}{256 \text{ million}}$

$= \frac{5.3 \cdot 10^9}{2.56 \cdot 10^8}$ Translate into scientific notation.

$= \frac{5.3}{2.56} \cdot \frac{10^9}{10^8}$ Multiplying Fractions Property

$\approx 2.07 \cdot 10^1$ Quotient of Powers Property

$\approx 21 \frac{\text{dollar bills}}{\text{person}}$

Check

Change the numbers to decimal notation and simplify the fraction.

$\frac{5{,}300{,}000{,}000}{256{,}000{,}000} = \frac{5300}{256} \approx 21$

Dividing Powers When the Bases Are Not the Same

To use the Quotient of Powers Property, the bases must be the same. For instance, $\frac{a^9}{b^4}$ cannot be simplified. To divide two algebraic expressions that involve different bases, group powers of the same base together and use the Quotient of Powers Property to simplify each fraction.

Consumer Connection When discussing **Example 2,** you might tell students that paper money in the United States is printed in \$1, \$2, \$5, \$10, \$20, \$50, and \$100 denominations. One-dollar bills stay in circulation for an average of 18 months, but larger denominations last for several years. Worn-out bills are sent to the Federal Reserve Banks where they are shredded.

Additional Examples

1. Evaluate the following.
   a. $\frac{4^{12}}{4^3}$ **$4^9 = 262{,}144$**
   b. $\frac{4^3}{4^{12}}$ **$4^{-9} = \frac{1}{262144}$**
2. In March, 1992, there was a total of 283.9 billion dollars in U.S. currency in circulation. The U.S. population was about 252.7 million. How much currency per person was in circulation? **About \$1123**
3. Simplify.
   a. $\frac{30x^2y^{10}}{25xy^{15}}$ **$\frac{6x}{5y^5}$**
   b. $\frac{12x^7}{3x^2}$ **$4x^5$**
   c. $\frac{a^5}{a^8}$ **$\frac{1}{a^3}$ or a^{-3}**
   d. $\frac{28 \cdot 10^{14}}{7 \cdot 10^{17}}$ **$4 \cdot 10^{-3}$ or .004**

Adapting to Individual Needs

Extra Help

Remind students who use the Repeated Multiplication Model for Powering (as in Solution 2 of **Examples 1 and 3)** that they can divide through by common factors.

$$\frac{6^{10}}{6^7} = \frac{\cancel{6} \cdot \cancel{6} \cdot \cancel{6} \cdot \cancel{6} \cdot \cancel{6} \cdot \cancel{6} \cdot \cancel{6} \cdot 6 \cdot 6 \cdot 6}{\cancel{6} \cdot \cancel{6} \cdot \cancel{6} \cdot \cancel{6} \cdot \cancel{6} \cdot \cancel{6} \cdot \cancel{6}}$$

$$= \frac{6 \cdot 6 \cdot 6}{1} = 6^3$$

Often, after repeated use of this method, students will easily make the transition to using the Quotient of Powers Property.

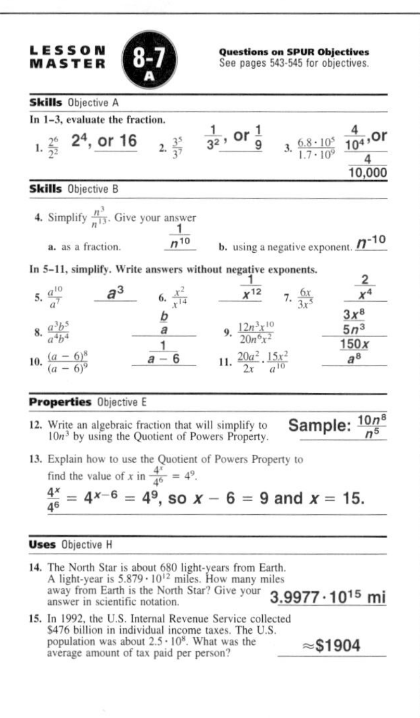

LESSON MASTER 8-7 A

Questions on SPUR Objectives
See pages 543-545 for objectives.

Skills Objective A

In 1–3, evaluate the fraction.

1. $\frac{2^6}{2^2}$ **2^4, or 16**
2. $\frac{3^5}{3^7}$ **$\frac{1}{3^2}$, or $\frac{1}{9}$**
3. $\frac{6.8 \cdot 10^5}{1.7 \cdot 10^9}$ **$\frac{4}{10^4}$, or $\frac{4}{10{,}000}$**

Skills Objective B

4. Simplify $\frac{n^3}{n^{13}}$. Give your answer
   a. as a fraction. **$\frac{1}{n^{10}}$**
   b. using a negative exponent. **n^{-10}**

In 5–11, simplify. Write answers without negative exponents.

5. $\frac{a^{10}}{a^7}$ **a^3**
6. $\frac{x^2}{x^{14}}$ **$\frac{1}{x^{12}}$**
7. $\frac{6x}{3x^5}$ **$\frac{2}{x^4}$**
8. $\frac{a^3b^5}{a^4b^4}$ **$\frac{b}{a}$**
9. $\frac{12n^3x^{10}}{20n^6x^2}$ **$\frac{3x^8}{5n^3}$**
10. $\frac{(a-6)^8}{(a-6)^9}$ **$\frac{1}{a-6}$**
11. $\frac{20a^2}{2x} \cdot \frac{15x^2}{a^{10}}$ **$\frac{150x}{a^8}$**

Properties Objective E

12. Write an algebraic fraction that will simplify to $10n^3$ by using the Quotient of Powers Property. **Sample: $\frac{10n^8}{n^5}$**
13. Explain how to use the Quotient of Powers Property to find the value of x in $\frac{4^x}{4^6} = 4^9$.
    $\frac{4^x}{4^6} = 4^{x-6} = 4^9$, so $x - 6 = 9$ and $x = 15$.

Uses Objective H

14. The North Star is about 680 light-years from Earth. A light-year is $5.879 \cdot 10^{12}$ miles. How many miles away from Earth is the North Star? Give your answer in scientific notation. **$3.9977 \cdot 10^{15}$ mi**
15. In 1992, the U.S. Internal Revenue Service collected \$476 billion in individual income taxes. The U.S. population was about $2.5 \cdot 10^8$. What was the average amount of tax paid per person? **≈\$1904**

Notes on Questions

Question 14 This is an important question to discuss. Show students that the Quotient of Powers Property is consistent with the Zero Exponent Property: $\frac{b^n}{b^n} = b^{n-n} = b^0$, but, obviously, $\frac{b^n}{b^n}$ also equals 1.

Question 15 Ecology Connection The word *plastic* comes from the Greek word *plastics,* which means *able to be shaped.* Plastic can be as rigid as steel or as soft as cotton, and it is shaped into objects ranging from automobile fenders to squeezable bottles. More and more packaging materials are made from plastic and used by consumers; consequently, more plastic waste is being generated. Because most plastics do not break down readily, disposing of plastic waste is a major environmental concern. Scientists are trying to design plastics that can be broken down by nature and time. Meanwhile, recycling plastic materials is the most efficient method to combat the problem of plastic waste.

Questions 16–23 Cooperative Learning These questions can each be done in more than one way. Encourage students to share their strategies, and point out where the various properties of exponents are applied.

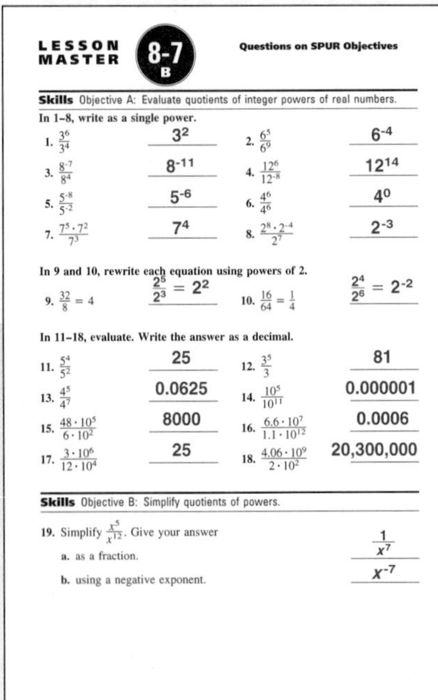

LESSON MASTER 8-7 B Questions on SPUR Objectives

Skills Objective A: Evaluate quotients of integer powers of real numbers.

In 1–8, write as a single power.

1. $\frac{3^6}{3^4}$ 3^2
2. $\frac{6^5}{6^9}$ 6^{-4}
3. $\frac{8^{-7}}{8^4}$ 8^{-11}
4. $\frac{12^6}{12^{-8}}$ 12^{14}
5. $\frac{5^{-8}}{5^{-2}}$ 5^{-6}
6. $\frac{4^6}{4^6}$ 4^0
7. $\frac{7^5 \cdot 7^2}{7^3}$ 7^4
8. $\frac{2^8 \cdot 2^{-4}}{2^7}$ 2^{-3}

In 9 and 10, rewrite each equation using powers of 2.

9. $\frac{32}{8} = 4$ $\frac{2^5}{2^3} = 2^2$
10. $\frac{16}{64} = \frac{1}{4}$ $\frac{2^4}{2^6} = 2^{-2}$

In 11–18, evaluate. Write the answer as a decimal.

11. $\frac{5^4}{5^2}$ 25
12. $\frac{3^5}{3}$ 81
13. $\frac{4^5}{4^7}$ 0.0625
14. $\frac{10^5}{10^{11}}$ 0.000001
15. $\frac{48 \cdot 10^5}{6 \cdot 10^2}$ 8000
16. $\frac{6.6 \cdot 10^7}{1.1 \cdot 10^{12}}$ 0.0006
17. $\frac{3 \cdot 10^6}{12 \cdot 10^4}$ 25
18. $\frac{4.06 \cdot 10^9}{2 \cdot 10^2}$ 20,300,000

Skills Objective B: Simplify quotients of powers.

19. Simplify $\frac{x^5}{x^{12}}$. Give your answer
   a. as a fraction. $\frac{1}{x^7}$
   b. using a negative exponent. x^{-7}

Example 3

Simplify $\frac{7a^3b^2c^6}{28a^2b^5c}$. Write the result as a fraction.

Solution 1

$$\frac{7a^3b^2c^6}{28a^2b^5c} = \frac{7}{28} \cdot \frac{a^3}{a^2} \cdot \frac{b^2}{b^5} \cdot \frac{c^6}{c}$$
$$= \frac{1}{4} \cdot a^{3-2} \cdot b^{2-5} \cdot c^{6-1} \quad \text{(Remember that } c = c^1\text{.)}$$
$$= \frac{1}{4} \cdot a^1 \cdot b^{-3} \cdot c^5$$
$$= \frac{1}{4} \cdot a \cdot \frac{1}{b^3} \cdot c^5$$
$$= \frac{ac^5}{4b^3}$$

Solution 2

$$\frac{7a^3b^2c^6}{28a^2b^5c} = \frac{7 \cdot a \cdot a \cdot a \cdot b \cdot b \cdot c \cdot c \cdot c \cdot c \cdot c \cdot c}{7 \cdot 4 \cdot a \cdot a \cdot b \cdot b \cdot b \cdot b \cdot b \cdot c}$$
$$= \frac{a \cdot c \cdot c \cdot c \cdot c \cdot c}{4 \cdot b \cdot b \cdot b}$$
$$= \frac{ac^5}{4b^3}$$

Experts do all of this work in one step.

QUESTIONS

Covering the Reading

1. Rewrite the multiplication problem $64 \cdot 256 = 16{,}384$ using powers of 2. $2^6 \cdot 2^8 = 2^{14}$
2. Rewrite the division problem $1024 \div 16 = 64$ using powers of 2. $2^{10} \div 2^4 = 2^6$
3. State the Quotient of Powers Property. **For all *m* and *n*, and all nonzero *b*, $b^m/b^n = b^{m-n}$**

In 4 and 5, a fraction is given. **a.** Write the quotient as a power of 3. **b.** Check your work. **See left.**

4b) By calculator $\frac{3^8}{3^2} = 729$ and $3^6 = 729$. It checks.

5b) By calculator $\frac{3^2}{3^8} = 0.0013717$ and $3^{-6} = 0.0013717$. It checks.

4. $\frac{3^8}{3^2}$ 3^6
5. $\frac{3^2}{3^8}$ 3^{-6}

In 6–11, use the Quotient of Powers Property to simplify the fraction.

6. $\frac{x^{12}}{x^6}$ x^6
7. $\frac{y^5}{y^5}$ 1
8. $\frac{19.2^4}{19.2^6}$ $19.2^{-2} = \frac{1}{368.64}$
9. $\frac{9.5 \cdot 10^{12}}{1.9 \cdot 10^4}$ $5 \cdot 10^8$
10. $\frac{3w^2z^6}{42w^2z^3}$ $\frac{z^3}{14}$
11. $\frac{4abc^{10}}{28a^2b^5c}$ $\frac{c^9}{7ab^4}$
12. Why can't $\frac{r^{10}}{s^7}$ be simplified? **The bases are different.**
13. In 1993, there were approximately 1.11 billion five-dollar bills in circulation and 256 million people in the United States. Convert these numbers to scientific notation and find the number of five-dollar bills per person. $\frac{1.11 \cdot 10^9}{2.56 \cdot 10^8} \approx 0.436 \cdot 10^1 \approx 4.4 \frac{\text{five-dollar bills}}{\text{person}}$

Adapting to Individual Needs

English Language Development
Be sure that students who are just learning English can distinguish the *numerator* from the *denominator* of an expression containing powers. Refer to the paragraph at the top of page 522. Read the following descriptions aloud and ask students to write the expression. Then ask what the expressions equal as decimals.

1. The numerator is 2^3 and the denominator is 2^4. [$\frac{2^3}{2^4}$; 0.5]
2. The numerator is 5^2 and the denominator is 5^0. [$\frac{5^2}{5^0}$; 25]
3. The denominator is 4^2 and the numerator is 1. [$\frac{1}{4^2}$; 0.0625]

15) $\frac{6 \cdot 10^9}{2.56 \cdot 10^8} \approx 2.34 \cdot 10^1$
$\approx 23 \frac{\text{pounds}}{\text{person}}$

Applying the Mathematics

14. If $m = n$, then $\frac{b^m}{b^n} = \underline{\ ?\ }$. 1

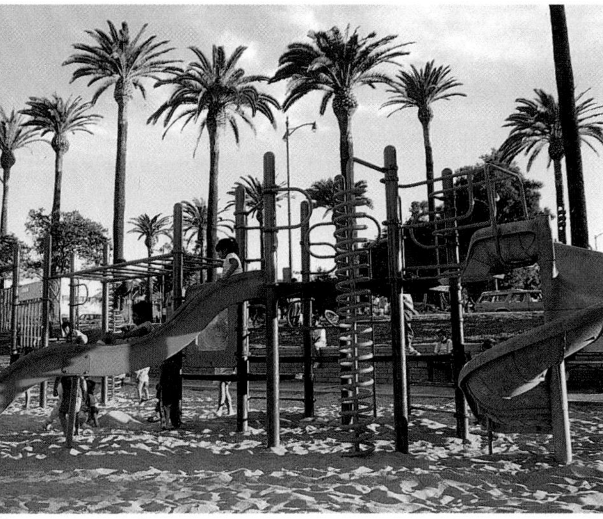

15. About $6 \cdot 10^9$ pounds of plastic were used in 1993 in the United States to produce trash bags, film, and other low density plastic goods. About how many pounds of plastic were used per person? See left.

In 16–19, rewrite to eliminate the fraction.

16. $\frac{3^{-8}}{3^{-2}}$ $\frac{1}{3^6} = 3^{-6}$ **17.** $\frac{3^{-2}}{3^{-8}}$ 3^6

18. $\frac{x^{-2}}{x^{-8}}$ x^6 **19.** $\frac{x^{-2n}}{x^{-3n}}$ x^n

In 20–23, rewrite the expression so that it has no fraction. You may need negative exponents. 22) $3x^{-7}y^2$

20. $\frac{(7m)^2}{(7m)^3}$ $(7m)^{-1}$ **21.** $\frac{(x+3)^6}{(x+3)^6}$ 1

22. $\frac{3x^2}{y^3} \cdot \frac{y^5}{x^9}$ **23.** $\frac{3p^5 + 2p^5}{p^4}$ $5p$

Playing with plastics. *Plastic products are often lighter, longer lasting, and less expensive to make. They can be made in any color and can be recycled and shaped into a variety of objects, such as this playground equipment.*

24. Write an algebraic fraction that can be simplified to $8x^5$ using the Quotient of Powers Property. Sample: $\frac{16x^7}{2x^2}$

25. In 1992, Alaska had a population of about $5.87 \cdot 10^5$ and an area of about $1.48 \cdot 10^6$ square kilometers. Find Alaska's population per square kilometer. about 0.40 $\frac{\text{people}}{\text{km}^2}$

Review

26. a. Write 2^{-6} as a fraction. $\frac{1}{64}$ **b.** Write 2^{-6} as a decimal. *(Lesson 8-6)* 0.015625

In 27–29, simplify. *(Lessons 8-5, 8-6)*

27. $4^x \cdot 4^y$ 4^{x+y} **28.** $-2(y^4)^3$ $-2y^{12}$ **29.** $-2(y^4)^3 \cdot y^4$ $-2y^{16}$

30. a. Find the probability of throwing 5 sixes on five consecutive throws of a fair die. $\left(\frac{1}{6}\right)^5 = \frac{1}{7776}$

b. Write your answer using negative exponents. *(Lessons 2-9, 8-6)* 6^{-5}

31. Simplify $x^5 \cdot x^5 \cdot x^5$. Check by letting $x = 3$. *(Lesson 8-5)*
x^{15}; $3^5 \cdot 3^5 \cdot 3^5 = 243 \cdot 243 \cdot 243 = 14{,}348{,}907$; $3^{15} = 14{,}348{,}907$

32. In 1989, Milo invested \$3000 for 10 years at an annual yield of 10%. In 1994, Sylvia invested \$6000 for 5 years at 5%. By the end of 1999, who would have more money? Justify your answer. *(Lesson 8-1)*
Milo: he would have \$7781.23, while Sylvia would have \$7657.69.

Question 16 Many students will evaluate this as $3^{-8 - -2} = 3^{-6}$. Others may write $\frac{\frac{1}{3^8}}{\frac{1}{3^2}} = \frac{1}{3^8} \cdot \frac{3^2}{1} = \frac{1}{3^6}$.

Question 17 Note that this question involves the reciprocal of the expression in **Question 16.**

Question 19 We expect most students to use the Quotient of Powers Property: $\frac{x^{-2n}}{x^{-3n}} = x^{-2n - -3n} = x^n$. Another approach is to multiply both numerator and denominator by x^{3n}. Some students may rewrite x^{-3n} as $\frac{1}{x^{3n}}$, and x^{-2n} as $\frac{1}{x^{2n}}$, and divide the fractions.

Question 21 Point out to students how chunking simplifies this question.

Question 26b Students can convert to $\frac{1}{64}$ and then divide. They can also use the powering key on a calculator.

Question 27 Error Alert Some students might say that $4^x \cdot 4^y = 16^{x+y}$ rather than 4^{x+y}. Have them check by testing a specific case. For instance, $4^2 \cdot 4^3 = 16 \cdot 64 = 1024$; $1024 = 4^5$, not 16^5.

Adapting to Individual Needs

Challenge

Have students **work in pairs** to solve the following problem.

Find the units digit of $3^{1999} - 2^{1999}$. [9]

Students can try simpler cases and look for a pattern. That is, find $3^1 - 2^1$, $3^2 - 2^2$, $3^3 - 2^3$, and so on. Evaluating the first nine differences establishes the pattern of the units digits to be 1, 5, 9, 5. Dividing 1999 by 4 gives 499 R3, which tells you that 499 complete patterns are used plus the first three of the pattern. Therefore the units digit in the third difference is 9.

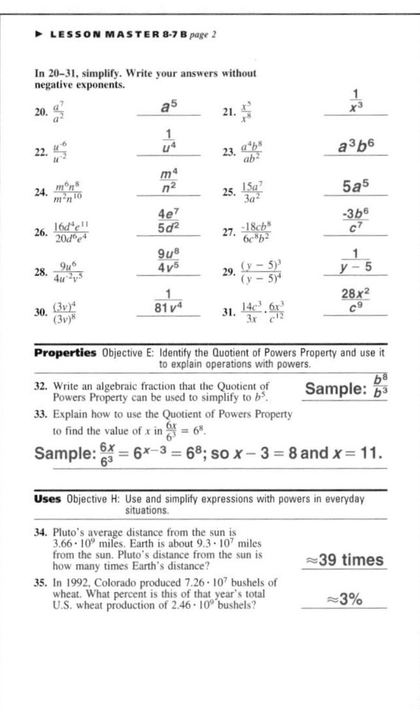

▸ LESSON MASTER 8-7 B *page 2*

In 20–31, simplify. Write your answers without negative exponents.

20. $\frac{a^7}{a^2}$	a^5	21. $\frac{x^5}{x^8}$	$\frac{1}{x^3}$
22. $\frac{u^{-6}}{u^{-2}}$	$\frac{1}{u^4}$	23. $\frac{a^4b^8}{ab^2}$	a^3b^6
24. $\frac{m^6n^8}{m^2n^{10}}$	$\frac{m^4}{n^2}$	25. $\frac{15a^7}{3a^2}$	$5a^5$
26. $\frac{16d^4e^{11}}{20d^6e^4}$	$\frac{4e^7}{5d^2}$	27. $\frac{-18cb^8}{6c^8b^2}$	$\frac{-3b^6}{c^7}$
28. $\frac{9u^6}{4u^{-2}v^5}$	$\frac{9u^8}{4v^5}$	29. $\frac{(y-5)^3}{(y-5)^4}$	$\frac{1}{y-5}$
30. $\frac{(3v)^4}{(3v)^8}$	$\frac{1}{81v^4}$	31. $\frac{14c^3}{3x} \cdot \frac{6x^3}{c^{12}}$	$\frac{28x^2}{c^9}$

Properties Objective E: Identify the Quotient of Powers Property and use it to explain operations with powers.

32. Write an algebraic fraction that the Quotient of Powers Property can be used to simplify to b^5. Sample: $\frac{b^8}{b^3}$

33. Explain how to use the Quotient of Powers Property to find the value of x in $\frac{6x}{6^3} = 6^8$.
Sample: $\frac{6x}{6^3} = 6^{x-3} = 6^8$; so $x - 3 = 8$ and $x = 11$.

Uses Objective H: Use and simplify expressions with powers in everyday situations.

34. Pluto's average distance from the sun is $3.66 \cdot 10^9$ miles. Earth is about $9.3 \cdot 10^7$ miles from the sun. Pluto's distance from the sun is how many times Earth's distance? $\approx$39 times

35. In 1992, Colorado produced $7.26 \cdot 10^7$ bushels of wheat. What percent is this of that year's total U.S. wheat production of $2.46 \cdot 10^9$ bushels? $\approx$3%

Notes on Questions

Question 33b In many buildings the main floor is called *lobby* and the next floor up is 1. In other buildings, the main floor is the first floor. Either interpretation is permissible.

Question 35 Students could find the volume of each box and then form the ratio. Alternately, the students could visualize how many small boxes will form one layer in the big box, and how many layers will completely fill the big box.

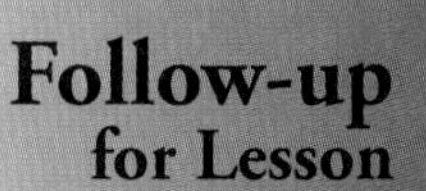

Practice

For more questions on SPUR Objectives, use **Lesson Master 8-7A** (shown on page 523) or **Lesson Master 8-7B** (shown on pages 524–525).

Assessment

Group Assessment Have students **work in small groups**. Tell each group to write as many algebraic expressions as they can that simplify to $3x^2$. [Students work together and use the properties of exponents to write expressions.]

Extension

Have students solve this problem. Light travels at about 3.0×10^5 km/sec. The distance from the earth to the sun is about 1.5×10^8 km. About how many seconds does it take light to travel from the sun to the earth? [500 seconds]

Project Update Project 2, *Applying the Properties,* on page 539, relates to the content of this lesson.

Going up? *This glass elevator gives people a panoramic view of the interior of a hotel in Atlanta, Georgia. Elisha G. Otis invented the first elevator with an automatic safety device in 1854.*

33. After x seconds, an elevator is on floor y, where $y = 54 - 2x$.

a. Give the slope and y-intercept of $y = 54 - 2x$, and describe what they mean in this situation.

b. Graph the line. *(Lesson 7-4)*
See below.

34. In your head, find the ratio of the areas of two squares, the first with a side of length 2 and the second with a side of length 3. *(Lesson 6-3)* **4/9**

35. A box with dimensions 40 cm by 60 cm by 80 cm will hold how many times as much as one with dimensions 4 cm by 6 cm by 8 cm? *(Lesson 6-6)* **1000 times as much**

36. a. Find the perimeter of the rectangle at the right. *(Lesson 3-1)* **4x + 32 units**

b. Write a simplified expression for the area of this rectangle. *(Lessons 3-3, 3-9)* **24x + 48 square units**

c. If the area is 312, find the value of x. *(Lesson 3-7)* **11**

12

$2x + 4$

Exploration

37. The average 14-year-old has a volume of about 3 cubic feet.

a. If you took all the students in your school, would their volume be more or less than the volume of one classroom that is 10 feet high, 30 feet long, and 30 feet wide? **Answers will vary.**

b. Assume the population of the world to be 5.5 billion people. Assume the average volume of a person to be 4 cubic feet. Is the volume of all the people more or less than the volume of a cubic mile? How much more or less? (There are 5280^3 cubic feet in a cubic mile.)
The volume of all the people is less by about $1.25 \cdot 10^{11}$ ft^3.

33a) slope = −2;
y intercept = 54;
The elevator descends at a rate of 2 floors per second and started on the 54th floor.

b)

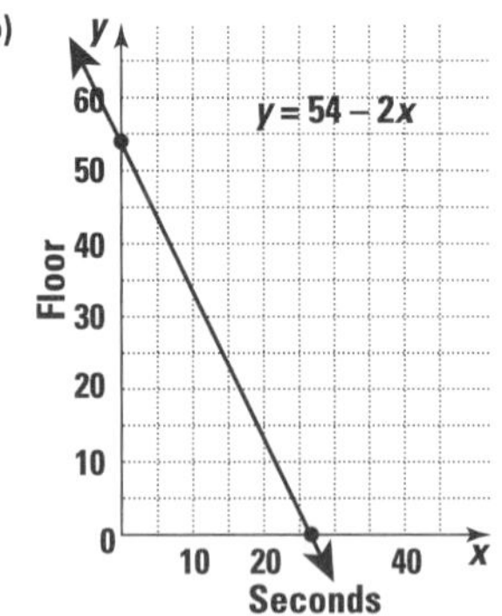

LESSON

8-8

Powers of Products and Quotients

Almost a sphere. *The distances from the center of Earth to points on Earth range from 3950 mi to 3963 mi. These are close enough that Earth is often considered a sphere. See Example 2.*

The Power of a Product

$(3x)^4$ is an example of a power of a product. It can be rewritten using repeated multiplication.

$$\begin{aligned} (3x)^4 &= (3x) \cdot (3x) \cdot (3x) \cdot (3x) \\ &= 3 \cdot 3 \cdot 3 \cdot 3 \cdot x \cdot x \cdot x \cdot x && \text{associative and commutative properties} \\ &= 3^4 \cdot x^4 && \text{Repeated Multiplication Model for Powering} \\ &= 81x^4 \end{aligned}$$

You can check this answer. Consider the special case when $x = 2$. Then $(3x)^4 = 6^4 = 1296$ and $81x^4 = 81 \cdot 2^4 = 81 \cdot 16 = 1296$.

In general, any positive integer power of a product can be rewritten using repeated multiplication.

$$\begin{aligned} (ab)^n &= \underbrace{(ab) \cdot (ab) \cdot \ldots \cdot (ab)}_{n \text{ factors}} \\ &= \underbrace{a \cdot a \cdot \ldots \cdot a}_{n \text{ factors}} \cdot \underbrace{b \cdot b \cdot \ldots \cdot b}_{n \text{ factors}} \\ &= a^n \cdot b^n \end{aligned}$$

This results holds when n is any number, positive, negative, or zero.

Power of a Product Property
For all nonzero a and b, and for all n, $(ab)^n = a^n \cdot b^n$.

In the specific case that began the lesson, $a = 3$, $b = x$, and $n = 4$. So $(3x)^4 = 3^4x^4$, as we found then.

Lesson 8-8

Objectives

A Evaluate integer powers of real-number products and quotients.
C Rewrite powers of products and quotients.
E Identify the Power of a Product and Power of a Quotient Properties and use them to explain operations with powers.
H Use and simplify expressions with powers in real situations.

Resources

From the ***Teacher's Resource File***

- Lesson Master 8-8A or 8-8B
- Answer Master 8-8
- Teaching Aid 89: Warm-up

Additional Resources

- Visual for Teaching Aid 89

Teaching Lesson 8-8

Warm-up

If a large cube-shaped crate is completely filled with 27 identical small cube-shaped boxes, what is the relationship between the length of a side of the crate and the length of a side of a small box? **The length of a side of the crate is three times the length of a side of a small box.**

Lesson 8-8 Overview

Broad Goals This lesson introduces and applies the last two of the basic properties of powers: the Power of a Product Property $(ab)^n = a^nb^n$, and the Power of a Quotient Property $\left(\frac{a}{b}\right)^n = \frac{a^n}{b^n}$, both true for all real numbers a, b, and n for which the expressions are defined.

Perspective The properties of this lesson and those of Lessons 8-6 and 8-7 have similar names. These names come from the order of operations. For example, $(ab)^n$ is viewed as a power of a product because the product is done first and then the power is taken. In contrast, $2^x \cdot 2^y$ is a product of powers. These distinctions may help students remember the names of properties. At this point, however, knowing the actual names of the properties is not important; being able to do the appropriate computation or simplification is the goal.

The two properties of this lesson are sometimes called the Distributive Properties of Powering over Multiplication and Division. There are no similar properties over addition or subtraction.

Notes on Reading

Reading Mathematics A quick review of the reading can be done by asking students for the specific application of the properties of this lesson that are used in each example. This helps to clarify exactly what has been done.

Example 1: $(2e)^3 = 2^3e^3$

Example 2: $(6.36 \cdot 10^3)^3 = 6.36^3 \cdot (10^3)^3$

Example 3: $(-xy)^6 = (-1)^6x^6y^6$

Example 4: $\left(\frac{2}{3}\right)^5 = \frac{2^5}{3^5}$

Example 5: $\left(\frac{2x}{y}\right)^4 = \frac{(2x)^4}{y^4}$

❶ **Error Alert** Some students may have difficulty following the solutions to **Examples 1 and 2**. Go through the solutions step by step, highlighting where the properties of exponents are applied. Or you might ask students to tell the changes from step to step and name the properties that are used.

Example 3 is included because opposites in powering expressions present problems for many students. The special case $-b^2 \neq (-b)^2$ is discussed in the paragraph following **Example 3** because students often fail to interpret $-b^2$ correctly.

LESSON MASTER 8-8 A — **Questions on SPUR Objectives** See pages 543-545 for objectives.

Skills Objective A

In 1–3, tell if the number is *positive* or *negative*.

1. -18^2 negative
2. $(-10)^3$ negative
3. $(-42)^4$ positive

In 4–9, evaluate the expression.

4. -7^2 -49
5. $(2 \cdot 10)^3$ 8,000
6. $\left(\frac{3}{5}\right)^3$ $\frac{27}{125}$
7. $(3^{-2})^2$ $\frac{1}{81}$
8. $(1.4 \cdot 10^8)^5$ $5.37824 \cdot 10^{40}$
9. $(8.95 \cdot 10^6)^3$ $7.1691738 \cdot 10^{20}$

Skills Objective C

In 10–15, simplify the expression.

10. $(3x)^4$ $81x^4$
11. $(2a^4)^3$ $8a^{12}$
12. $\left(\frac{8}{9}ny^{10}\right)^2$ $\frac{64}{81}n^2y^{20}$
13. $10x \cdot (-5x^3)^3$ $-1250x^{10}$
14. $\left(\frac{6n^4}{x}\right)^2$ $\frac{36n^8}{x^2}$
15. $\left(\frac{1}{3}x\right)^2 \cdot (6^2)^3$ $5184x^2$

Properties Objective E

16. Simplify $(ab^5)^3$ by rewriting using
   a. repeated multiplication. $(ab^5)^3 = ab^5 \cdot ab^5 \cdot ab^5 = a^3b^{15}$
   b. the Power of a Product Property. $(ab^5)^3 = a^3 \cdot (b^5)^3 = a^3b^{15}$

Uses of the Power of a Product

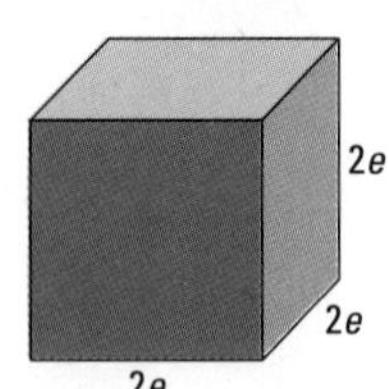

Example 1

Suppose one cube has edges twice the length of another. The volume of the larger cube is how many times the volume of the smaller?

❶ **Solution**

The volume of a cube is the cube of its edge. The volume of the small cube is e^3.

$$\begin{aligned}\text{Volume of larger cube} &= (2e)^3 \\ &= 2^3e^3 \\ &= 8e^3 \\ &= 8 \cdot \text{volume of the smaller cube}\end{aligned}$$

So the larger cube has eight times the volume of the smaller.

The Power of a Product Property can be used to take powers of numbers written in scientific notation.

Example 2

Earth is approximately the shape of a sphere with radius $6.36 \cdot 10^3$ km. The volume of a sphere of radius r is given by the formula

$$V = \frac{4}{3}\pi r^3.$$

Calculate the approximate volume of Earth.

Solution

Substitute $6.36 \cdot 10^3$ for r.

$$V = \frac{4}{3}\pi(6.36 \cdot 10^3)^3$$

Apply the Power of a Product Property to $(6.36 \cdot 10^3)^3$. Think of 10^3 as a single chunk.

$$\begin{aligned}V &= \frac{4}{3}\pi(6.36 \cdot 10^3)^3 \\ &\approx \frac{4}{3}\pi(257.26) \cdot 10^9 \\ &\approx 1077.61 \cdot 10^9 \\ &\approx 1.08 \cdot 10^{12} \text{ km}^3\text{, in scientific notation} \\ &\approx 1{,}080{,}000{,}000{,}000 \text{ km}^3\text{, written as a decimal}\end{aligned}$$

Earth's volume is about 1 trillion cubic kilometers.

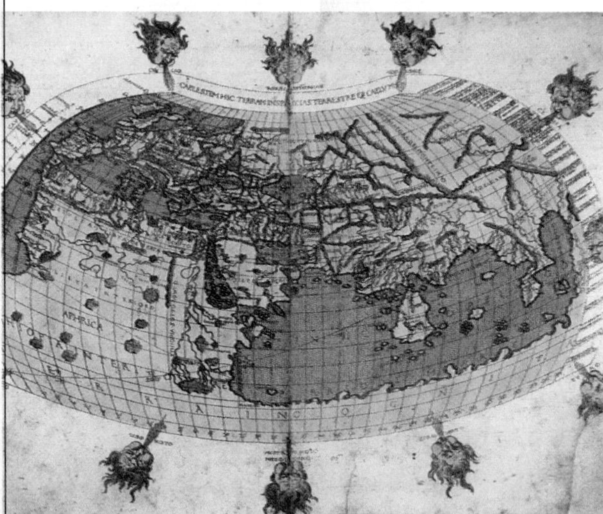

Ptolemy's ancient map. *Ptolemy was a Greek geographer and astronomer who lived about 150 A.D. His map was one of the first to list places with their latitude and longitude, and to depict Earth as spherical. Few people knew about his maps until they were published in an atlas in the late 1400s.*

Optional Activities

After discussing **Example 4**, you might want to use this activity to help students see that the Power of a Quotient Property is just the Power of a Product Property in disguise. The Negative Exponent Property is the "mask." Here is the "unmasking":

$\left(\frac{a}{b}\right)^n = (ab^{-1})^n$	Negative Exponent Property
$= a^n(b^{-1})^n$	Power of a Product Property
$= a^nb^{-n}$	Power of a Power Property
$= a^n \cdot \frac{1}{b^n}$	Negative Exponent Property
$= \frac{a^n}{b^n}$	Rule for Multiplication of Fractions

Recall that an odd power of a negative number is negative, while an even power of a negative number is positive.

Example 3

Simplify $(-xy)^6$.

Solution

First rewrite $-xy$ using $-b = -1 \cdot b$.

$$\begin{aligned}(-xy)^6 &= (-1 \cdot x \cdot y)^6 \\ &= (-1)^6 \cdot x^6 \cdot y^6 \\ &= 1 \cdot x^6 \cdot y^6 \\ &= x^6y^6\end{aligned}$$

Caution: In order of operations, powers take precedence over opposites. In $-b^2$, the power is done first, so the number $-b^2$ is negative. On the other hand, $(-b)^2$ is never negative because

$$(-b)^2 = (-b)(-b) = (-1)(b)(-1)(b) = (-1)(-1)(b)(b) = b^2.$$

So, when $b \neq 0$, $-b^2 \neq (-b)^2$. For instance, $-5^2 = -25$, while $(-5)^2 = 25$.

The Power of a Quotient

The Power of a Quotient Property is very similar to the Power of a Product Property. It enables you to easily find powers of fractions.

Power of a Quotient Property
For all nonzero a and b, and for all n,

$$\left(\frac{a}{b}\right)^n = \frac{a^n}{b^n}.$$

❷ **Example 4**

Write $\left(\frac{2}{3}\right)^5$ as a simple fraction.

Solution

Use the Power of a Quotient Property.

$$\begin{aligned}\left(\frac{2}{3}\right)^5 &= \frac{2^5}{3^5} \\ &= \frac{32}{243}\end{aligned}$$

Check

Change the fractions to decimals. $\left(\frac{2}{3}\right)^5 = (0.\overline{6})^5 \approx 0.1316872\ldots$

$\frac{32}{243} \approx 0.1316872\ldots$ also.

❷ Students often need practice with questions like **Examples 4 and 5**. It is easy to compose questions by using bases that can be written in different forms. For instance, $(1.2)^9 = \left(\frac{6}{5}\right)^9$. Now write $\left(\frac{6}{5}\right)^9$ without parentheses. The answer is $\frac{6^9}{5^9}$. Does it equal $(1.2)^9$? (Yes, since we simply wrote the same number in different forms.)

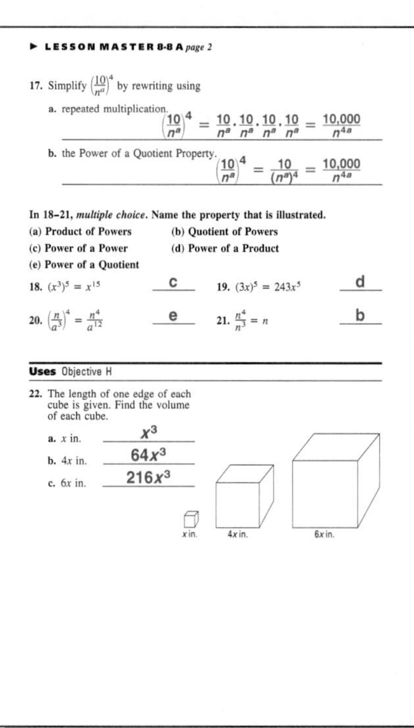

▸ **LESSON MASTER 8-8 A** *page 2*

17. Simplify $\left(\frac{10}{n^a}\right)^4$ by rewriting using
    a. repeated multiplication. $\left(\frac{10}{n^a}\right)^4 = \frac{10}{n^a}\cdot\frac{10}{n^a}\cdot\frac{10}{n^a}\cdot\frac{10}{n^a} = \frac{10{,}000}{n^{4a}}$
    b. the Power of a Quotient Property. $\left(\frac{10}{n^a}\right)^4 = \frac{10}{(n^a)^4} = \frac{10{,}000}{n^{4a}}$

In 18–21, *multiple choice*. Name the property that is illustrated.
(a) Product of Powers **(b) Quotient of Powers**
(c) Power of a Power **(d) Power of a Product**
(e) Power of a Quotient

18. $(x^3)^5 = x^{15}$ c
19. $(3x)^5 = 243x^5$ d
20. $\left(\frac{n}{a^3}\right)^4 = \frac{n^4}{a^{12}}$ e
21. $\frac{n^4}{n^3} = n$ b

Uses Objective H

22. The length of one edge of each cube is given. Find the volume of each cube.
    a. x in. x^3
    b. $4x$ in. $64x^3$
    c. $6x$ in. $216x^3$

Adapting to Individual Needs

Extra Help

If students seem unsure of the order of operations as presented in Lesson 1-4, review it with them now. Remind students of the importance of parentheses as grouping symbols by using the introductory expressions on page 527. Write $3x^4$ (without parentheses) on the board. Ask students to rewrite this expression using repeated multiplication. $[3 \cdot x \cdot x \cdot x \cdot x]$ Then write $(3x)^4$ on the board and have students rewrite it using repeated multiplication. $[3x \cdot 3x \cdot 3x \cdot 3x]$ Then have students compare the two forms. Point out that the parentheses in $(3x)^4$ indicate that the product $3x$ is considered first. Without parentheses, the exponent is considered first.

❸ In doing **Example 5** note that when you follow the order of operations, the power is calculated before the multiplication is done.

Although all the examples deal with positive integer exponents, point out that the properties work with all nonzero values for a and b and with all values for n.

Additional Examples

1. The length of an edge of one cube is four times the length of an edge of another cube. The volume of the larger cube is how many times the volume of the smaller cube? **64**
2. The sun's radius is about $6.96 \cdot 10^5$ km. Estimate the volume of the sun. $\frac{4}{3}\pi \cdot (6.96 \cdot 10^5)^3 \approx 1.41 \cdot 10^{18}$ **km³**
3. Simplify $-(-a^2b^3)^4$. $-a^8b^{12}$
4. Write $\left(\frac{4}{5}\right)^4$ as a simple fraction. $\frac{256}{625}$
5. Rewrite $\frac{2x}{5y} \cdot \left(\frac{7}{xy}\right)^3$ as a single fraction. $\frac{686}{5x^2y^4}$

❸ **Example 5**

Rewrite $3 \cdot \left(\frac{2x}{y}\right)^4$ as a single fraction.

Solution

First rewrite the power using the Power of a Quotient Property.

$$3 \cdot \left(\frac{2x}{y}\right)^4 = 3 \cdot \frac{(2x)^4}{y^4}$$

$$= 3 \cdot \frac{2^4x^4}{y^4} \quad \text{Power of a Product Property}$$

$$= 3 \cdot \frac{16x^4}{y^4}$$

$$= \frac{48x^4}{y^4} \quad \text{Think of 3 as } \frac{3}{1}.$$

Check

By repeated multiplication,

$$3 \cdot \left(\frac{2x}{y}\right)^4 = 3 \cdot \frac{2x}{y} \cdot \frac{2x}{y} \cdot \frac{2x}{y} \cdot \frac{2x}{y} = \frac{48x^4}{y^4}.$$

First people on the moon. *Shown is astronaut Edwin "Buzz" Aldrin, the second person to stand on the surface of the moon, on July 20, 1969. The first was Neil Armstrong, who stepped on the moon moments earlier.*

QUESTIONS

Covering the Reading

1. **a.** Rewrite $(5x)^3$ without parentheses. $125x^3$
   b. Check your answer by letting $x = 2$. $(5 \cdot 2)^3 = 10^3 = 1000$; $125 \cdot 2^3 = 125 \cdot 8 = 1000$
2. In Example 1, suppose the length of each side of the smaller cube is 12.5 feet.
   a. Find the volume of the smaller cube. **1953.125 ft³**
   b. Find the volume of the larger cube. **15,625 ft³**
3. The edge of one cube is k inches. The edge of a second cube is 5 times as long.
   a. Write an expression for the volume of the first cube. k^3 **in³**
   b. Write a simplified expression for the volume of the second cube. $125k^3$ **in³**
4. Calculate $(1.3 \cdot 10^4)^5$. $\approx 3.7 \cdot 10^{20}$
5. The radius of Earth's moon is approximately $1.738 \cdot 10^3$ km. Calculate its approximate volume. Write your answer **a.** in scientific notation. **b.** as a decimal.
   a) $\approx 2.1991 \cdot 10^{10}$ km³ b) $\approx 21{,}991{,}000{,}000$ km³

LESSON MASTER 8-8 B Questions on SPUR Objectives

Skills Objective A: Evaluate integer powers of real number products and quotients.

In 1–6, tell if the number is *positive* or *negative*.

1. -13^5	Negative	2. -11^6	Negative
3. $(-8)^4$	Positive	4. $\left(-\frac{4}{5}\right)^7$	Negative
5. 17^3	Positive	6. $(-2.6)^{87}$	Negative

In 7–20, evaluate each expression.

7. -8^2	-64	8. $(-9)^2$	81
9. -3^2	-9	10. $(-6)^3$	-216
11. -7^3	-343	12. $(-4)^4$	256
13. $\left(\frac{1}{2}\right)^5$	$\frac{1}{32}$	14. $\left(-\frac{3}{10}\right)^4$	$\frac{81}{10{,}000}$
15. $-\left(\frac{4}{3}\right)^2$	$-\frac{16}{9}$	16. $(-5.1)^2$	26.01
17. $(3 \cdot 4)^2$	144	18. $(3^{-2})^3$	$\frac{1}{729}$
19. $(1.8 \cdot 10^4)^2$	324,000,000	20. $(-7.66 \cdot 10^3)^2$	58,675,600

Skills Objective C: Rewrite powers of products and quotient.

In 21–28, simplify each expression.

21. $(4x)^2$	$16x^2$	22. $(3a^5)^3$	$27a^{15}$
23. $(3xyz)^0$	1	24. $\left(\frac{1}{6}a^4b\right)^4$	$\frac{1}{1296}a^{16}b^4$
25. $4n(8n^5)^3$	$2048n^{16}$	26. $(5r^2)^4(-2r^3)^5$	$-20{,}000r^{23}$
27. $-\left(\frac{3a^2}{b}\right)^4$	$-\frac{81x^8}{b^4}$	28. $\left(\frac{3}{4}m\right)^2 \cdot (8m)^2$	$36m^4$

Adapting to Individual Needs

English Language Development

Have students read expressions aloud, paying close attention to how expressions such as $5(x^2)^3$ and $(5x^2)^3$ are read and how they are interpreted. Remind students that the exponent 2 can be read as "squared" or "the second power" and that the exponent 3 can be read as "cubed" or "the third power."

530

In 6–7, write as a simple fraction.

6. $\left(\frac{1}{2}\right)^4$ $\frac{1}{16}$

7. $\left(\frac{7}{10}\right)^3$ $\frac{343}{1000}$

In 8 and 9, answer *true or false.*

8. -5^2 is negative. True

9. $(-7)^2 = 7^2$ True

10. Simplify.

a. -3^2 -9
b. $(-3)^2$ 9
c. -5^3 -125
d. $(-5)^3$ -125
e. $(-5)^4$ 625
f. -5^4 -625

In 11–16, rewrite without parentheses.

11. $(ab)^3$ a^3b^3

12. $(3x^3)^2$ $9x^6$

13. $\left(\frac{I}{S}\right)^3$ $\frac{I^3}{S^3}$

14. $(8y)^3$ $512y^3$

15. $(-ab)^9$ $-a^9b^9$

16. $\left(\frac{a}{b^5}\right)^3$ $\frac{a^3}{b^{15}}$

Applying the Mathematics

In 17–22, rewrite without parentheses and simplify.

17. $\frac{1}{2}(6x)^2$ $18x^2$

18. $(pqr)^0$ 1

19. $\left(\frac{u}{3}\right)^t$ $\frac{u^t}{3^t}$

20. $4L \cdot \left(\frac{5k}{L}\right)^2$ $\frac{100k^2}{L}$

21. $\left(\frac{2}{7}z\right)^4 \cdot z$ $\frac{16z^5}{2401}$

22. $(2q)^5(3q^4)^2$ $288q^{13}$

23. Suppose that about $\frac{1}{3}$ of the time, a pearl found by a pearl fisher is good enough to sell.

a. What is the probability that 5 pearls in a row will not be good enough to sell? $(2/3)^5 = 32/243 \approx 0.13$

b. To what example of this lesson is the answer to part **a** related? Example 4

In 24–26, multiple choice. Name the property that is illustrated.

(a) Product of Powers
(b) Quotient of Powers
(c) Power of a Power
(d) Power of a Product
(e) Power of a Quotient

24. $\left(\frac{a}{2n}\right)^3 = \frac{a^3}{(2n)^3}$ e

25. $8^5 \cdot 8^{10} = 8^{15}$ a

26. $(5x^2)^3 = 5^3 \cdot (x^2)^3$ d

27. If $x = 3$, what is the value of $\frac{(4x)^8}{(4x)^5}$? 1728

28. Ms. Taix incorrectly simplified $3(5x^4)^2$ as $15x^6$.

a. Find a counterexample to show that this is not true for all values of *x*.

b. Write out an explanation for Ms. Taix showing how to get the correct answer.

a) Sample: Let $x = 1$. Then $3(5(1)^4)^2 = 3(5 \cdot 1)^2 = 3(5)^2 = 3 \cdot 25 = 75$; however, $15(1)^6 = 15 \cdot 1 = 15$.

b) Sample: To simplify correctly, use the Power of a Product Property and the Power of a Power Property as follows:

$$3(5x^4)^2 = 3 \cdot 5^2 \cdot (x^4)^2 = 3 \cdot 25x^{4 \cdot 2} = 75x^8$$

Where do pearls come from? *Pearls are found inside oyster shells. This oyster shell is from Thailand; the pearl is in the upper part of the shell.*

Notes on Questions

Questions 4–5 These questions illustrate that knowing how to calculate the power of a product is quite useful when dealing with numbers in scientific notation.

Questions 10–16 If students have trouble with these questions, encourage them to either go back to the definition of powers as repeated multiplication or to test a special case.

Question 23 Science Connection Today, most of the pearls used in jewelry are cultured pearls. To make cultured pearls, young oysters are planted in oyster beds. After three years, trained people open the oysters' shells and insert tiny pieces of shell or other foreign matter. Then the oysters are placed in a cage and kept in calm waters. After another one to three years the shells are opened and any pearls that have grown are removed, washed, graded, and polished. Although not all of these oysters produce pearls, far more do so than oysters in natural settings. Interested students might like to find out what characteristics determine the value of a pearl.

Question 27 Students can substitute 3 for *x* and then evaluate, or they can simplify the fraction first and then substitute. It is interesting to ask students which they think is easier.

Question 28 Writing Share the explanations that students write.

► **LESSON MASTER 8-8 B** *page 2*

Properties Objective E: Identify the Product of a Power and Power of a Quotient Properties and use them to explain operations with powers.

29. Show how to simplify $(mx^{-3})^4$

a. by rewriting using repeated multiplication.
$(mx^{-3})^4 = mx^{-3} \cdot mx^{-3} \cdot mx^{-3} \cdot mx^{-3} = m^4x^{-12}$

b. by rewriting using the Power of a Product Property.
$(mx^{-3})^4 = m^4(x^{-3})^4 = m^4 \cdot x^{-3 \cdot 4} = m^4x^{-12}$

30. Show how to simplify $\left(\frac{3}{u^e}\right)^4$

a. by rewriting using repeated multiplication.
$\left(\frac{3}{u^e}\right)^4 = \left(\frac{3}{u^e}\right)\left(\frac{3}{u^e}\right)\left(\frac{3}{u^e}\right)\left(\frac{3}{u^e}\right) = \frac{81}{u^{4e}}$

b. by rewriting using the Power of a Quotient Property.
$\left(\frac{3}{u^e}\right)^4 = \frac{3^4}{(u^e)^4} = \frac{81}{u^{4e}}$

In 31–34, *multiple choice*. Identify the property that is illustrated.

(a) Power of a Power
(b) Power of a Product
(c) Product of Powers
(d) Power of a Quotient
(e) Quotient of Powers

31. $\left(\frac{a}{3}\right)^4 = \frac{a^4}{81}$ d

33. $m^3 \cdot m^6 = m^9$ c

33. $(e^5)^2 = e^{10}$ a

34. $\frac{x^3}{x^8} = x^{-5}$ e

Uses Objective H: Use and simplify expressions with powers in everyday situations.

35. Find the volume of each cube.

a. n^3 cm^3
b. $27n^3$ cm^3
c. k^3n^3 cm^3

n cm 3*n* cm *kn* cm

Adapting to Individual Needs

Challenge

Point out to students that the decimal (base ten) number system is the one we most commonly use. However, numbers other than ten can be used as the base of a number system. For example, the *binary system* uses base two. In the binary system the only digits are 0 and 1, and the values of the places from right to left are 2^0, 2^1, 2^2, 2^3, 2^4, and so on.

A numeral such as $(1011)_{two}$ means $(1 \times 2^3) + (0 \times 2^2) + (1 \times 2^1) + (1 \times 2^0)$. So in base ten the value is 8 + 0 + 2 + 1, or 11. Have students determine the value of each of these numbers in base ten.

1. $(1100)_{two}$ [12]
2. $(1110)_{two}$ [14]
3. $(11111)_{two}$ [31]
4. $(100000)_{two}$ [32]

Follow-up for Lesson 8-8

Practice

For more questions on SPUR Objectives, use **Lesson Master 8-8A** (shown on pages 528–529) or **Lesson Master 8-8B** (shown on pages 530–531).

Assessment

Oral Communication Ask students to explain how they would check their work to see if they have simplified an expression correctly. [Students understand they can use special cases or repeated multiplication to check their work.]

Extension

Multicultural Connection There is a legend associated with the situation in **Example 1**. The Greeks called this legend the Problem of Duplicating (or Doubling) the Cube. When his son died, the king of the island of Delos had a tomb built in the shape of a cube. When it was finished, he thought it was too small and ordered that it be rebuilt twice as large. His engineers interpreted this to mean twice as long in each dimension. But the king meant twice the volume. The new tomb had 8 times the volume of the old tomb and had to be rebuilt again.

The Greeks searched in vain for a ruler and compass construction that would show how to duplicate the cube; it became known as the Delian problem. More than 2000 years later such a construction was shown to be impossible. Ask students to explore with their calculators to find the length of a cube that would give a volume close to 2 cubic units. [$\sqrt[3]{2} \approx 1.26$; thus, a cube with edge $1.26e$ has *about* twice the volume of a cube with edge e.]

Project Update Project 2, *Applying the Properties,* on page 539, relates to the content of this lesson.

Review

29. Simplify $\frac{5n^2 - 3n^2}{10n^2}$. *(Lessons 3-3, 8-7)* $\frac{1}{5}$

30. Other than the sun, the star nearest to us, Alpha Centauri, is about $4 \cdot 10^{13}$ km away. Earth's moon is about $3.8 \cdot 10^5$ km from us. If it took astronauts about 3 days to get to the moon in 1969, at that speed how long would it take them to get to Alpha Centauri? *(Lesson 8-7)* $3.16 \cdot 10^8$ days

In 31–33, simplify. *(Lessons 8-5, 8-6, 8-7)*

31. $\frac{k^{12}}{k^9}$ k^3 **32.** $y \cdot y^3$ y^4 **33.** $(v^{-2})^3$ v^{-6}

34. Which is larger, $(5^4)^3$ or $5^4 \cdot 5^3$? *(Lesson 8-7)* $(5^4)^3$

35. *Skill sequence.* Simplify. *(Lesson 8-5)*
a. $2(x \cdot x^4)$ $2x^5$ **b.** $x \cdot (x^4)^2$ x^9 **c.** $(x \cdot x^4)^2$ x^{10}

36. a. Graph $y = 3^x$ using $x = -3, -2, -1, 0, 1,$ and 2. See left.
b. What name is given to this curve? *(Lessons 8-2, 8-5)* exponential growth curve

36a)

37. Calculate in your head. *(Lesson 6-2)*
a. the total cost of 6 cans of beans at \$.98 per can
b. the total cost of 4 tickets at \$15.05 per ticket \$60.00 + \$0.20 = \$60.20
c. a 15% tip for a \$40.00 dinner bill \$6.00
a) \$6.00 – \$0.12 = \$5.88

Exploration

38. The number 64 can be written as 8^2, or as 4^3, or as 2^6. Likewise, each of the numbers given here can be written in more than one way in the form a^n, where a and n are positive integers from 2 to 20. For each, find two pairs of values of a and n.
a. 81 **b.** 256 **c.** 32,768 **d.** 43,046,721
Samples: a) 3^4 or 9^2 b) 16^2 or 2^8 c) 32^3 or 2^{15} d) 3^{16} or 9^8

LESSON

8-9

Remembering Properties of Exponents and Powers

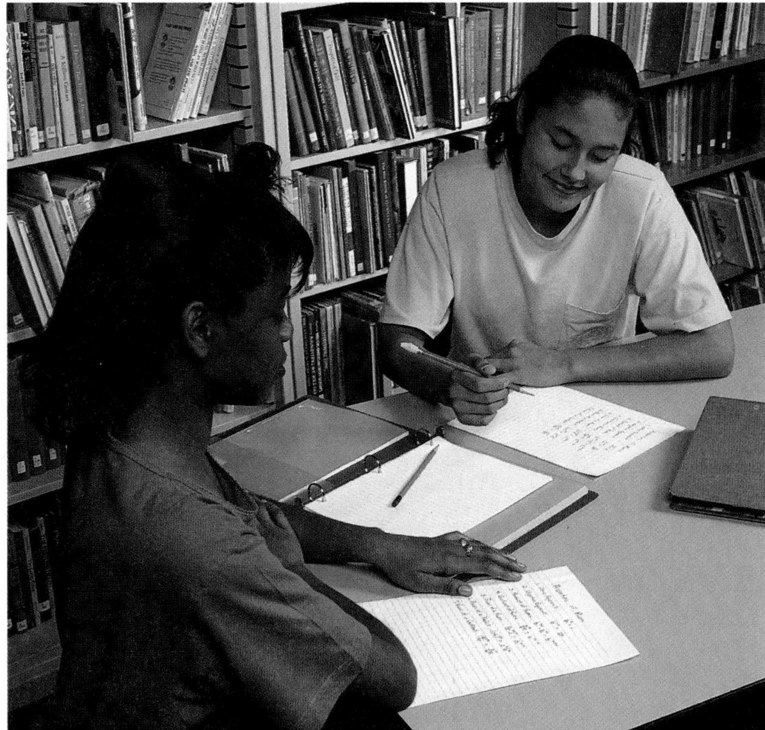

The power of studying. *These students are studying the properties of exponents and powers. By testing a special case of a property, or by showing that a pattern is not always true, these properties can easily be applied correctly.*

Seven properties of powers have been studied in this chapter. They apply to all exponents m and n and nonzero bases a and b.

Zero Exponent	$b^0 = 1$
Negative Exponent	$b^{-n} = \frac{1}{b^n}$
Product of Powers	$b^m \cdot b^n = b^{m+n}$
Quotient of Powers	$\frac{b^m}{b^n} = b^{m-n}$
Power of a Power	$(b^m)^n = b^{mn}$
Power of a Product	$(ab)^n = a^n b^n$
Power of a Quotient	$\left(\frac{a}{b}\right)^n = \frac{a^n}{b^n}$

Some people confuse these properties. Fortunately, mathematics is *consistent.* As long as you apply properties correctly, the results you get using some properties will not disagree with the results you get using other properties. We begin with what may look like a new problem: a negative power of a fraction.

Lesson 8-9

Objectives

A Evaluate integer powers of real numbers.

D Test a special case to determine whether a pattern is true.

Resources

From the ***Teacher's Resource File***

- Lesson Master 8-9A or 8-9B
- Answer Master 8-9
- Teaching Aids
  89 Warm-up
  94 Properties of Powers

Additional Resources

- Visuals for Teaching Aids 89, 94

Teaching Lesson 8-9

Warm-up

Write the algebraic description of each property. Then give an example that illustrates each property.

Examples will vary.

1. Quotient of Powers Property $\frac{b^m}{b^n} = b^{m-n}$
2. Zero Exponent Property $b^0 = 1$
3. Product of Powers Property $b^m \cdot b^n = b^{m+n}$
4. Power of a Product Property $(ab)^n = a^n b^n$
5. Negative Exponent Property $b^{-n} = \frac{1}{b^n}$
6. Power of a Quotient Property $\left(\frac{a}{b}\right)^n = \frac{a^n}{b^n}$
7. Power of a Power Property $(b^m)^n = b^{mn}$

Lesson 8-9 Overview

Broad Goals This lesson shows how students can use general problem-solving strategies to remember the properties of this chapter. In doing so, it offers an extra day of work on the content of the chapter.

Perspective Of all the concepts that students need to learn in mathematics, there is none more important than *consistency* of mathematics. Consistency allows us to *test special cases,* the problem-solving strategy in this lesson. Consistency enables us to generalize, to prove, and to disprove. Yet it is the case that students often see properties of powers as distinct independent facts to be memorized. The purpose of this lesson is to dissuade students from memorizing without understanding. We want students to be able to test a simple case to verify their reasoning. As with the strategy of "chunking," *testing a special case* is used throughout the rest of the book. We take advantage of this powerful tool in many of the checks of problems involving quadratics and other polynomials. Another place that *testing a special case* is important is in calculator use—to test a key sequence. How do you know when you are pressing the keys in the right order, especially in problems involving the y^x key? If you remember that $2^3 = 8$, you can test your key sequence with the special case of 2^3.

Notes on Reading

When discussing this lesson, you might want to use **Teaching Aid 94,** which contains the seven properties of powers given on page 533.

❶ **Example 1** gives three methods for simplifying a negative power of a fraction. Students may have preferences for using one method over the others. But each student should realize that most problems can be approached in more than one way. Emphasize again that powers with negative exponents refer to reciprocals of powers with positive exponents—not opposites. If students forget this or any other property, they can test a special case.

❷ **History Connection** A digital arithmetic machine was first devised by Blaise Pascal in 1642. In the early 20th century, desktop adding machines were developed, and electronic data-processing systems were introduced. By 1975, miniature solid-state electronic devices for pocket or desk were available to perform simple mathematical functions in addition to basic arithmetical operations.

Testing a special case is a strategy that is quite useful on multiple-choice tests. In **Example 2**, the three choices Norm has might be choices on such a test. Testing a special case can be used both to verify properties and to find counterexamples. **Example 3** uses the strategy to find a counterexample.

Optional Activities

Writing You might want to use the following activity after discussing **Example 2**. Have students **work in pairs**. Ask each student to write two multiple-choice test items similar to the question in Example 2. Each item should have three choices for answers, one of which is correct. Then have partners exchange papers and solve the test items by testing a special case. Each answer choice should be tested and found correct or incorrect.

❶ **Example 1**

Write $\left(\frac{2}{3}\right)^{-4}$ as a simple fraction.

Solution 1

Think: The problem asks for the power of a quotient. So use that property.

$$\left(\frac{2}{3}\right)^{-4} = \frac{2^{-4}}{3^{-4}}$$

Now evaluate the numerator and denominator, using the Negative Exponent Property.

$$= \frac{\frac{1}{2^4}}{\frac{1}{3^4}} = \frac{\frac{1}{16}}{\frac{1}{81}} = \frac{1}{16} \div \frac{1}{81} = \frac{1}{16} \cdot 81 = \frac{81}{16}$$

Solution 2

Think: The problem asks for a negative exponent. So use the Negative Exponent Property first.

$$\left(\frac{2}{3}\right)^{-4} = \frac{1}{\left(\frac{2}{3}\right)^4}$$

Now use the Power of a Quotient Property.

$$= \frac{1}{\frac{2^4}{3^4}} = 1 \div \frac{2^4}{3^4} = 1 \cdot \frac{3^4}{2^4} = \frac{81}{16}$$

Solution 3

Think: -4 = -1 · 4. Use the Power of a Power Property.

$$\left(\frac{2}{3}\right)^{-4} = \left(\left(\frac{2}{3}\right)^{-1}\right)^4$$

Now use the Negative Exponent Property to evaluate $\left(\frac{2}{3}\right)^{-1}$.

$$\left(\frac{2}{3}\right)^{-1} = \frac{1}{\frac{2}{3}}$$

$$= \frac{3}{2}$$

So $$\left(\frac{2}{3}\right)^{-4} = \left(\frac{3}{2}\right)^4.$$

Now apply the Power of a Quotient Property.

$$\left(\frac{3}{2}\right)^4 = \frac{3^4}{2^4} = \frac{81}{16}$$

Check

Use a calculator. (2 ÷ 3) y^x 4 ± = gives 5.0625, which is $\frac{81}{16}$.

Testing a Special Case

❷ Before the days of hand-held calculators (before the early 1970s), expressions with large exponents could not be calculated in a first-year algebra course. So it would be difficult to check some answers with arithmetic. However, with a calculator, a strategy called testing a special case is often possible.

Example 2

Norm was asked to simplify $x^8 \cdot x^6$. He wasn't sure of the answer, but knew it should be $2x^{14}$, or x^{48}, or x^{14}. Which is the correct response?

Solution 1

Use a special case. Let x = 3. Now calculate $3^8 \cdot 3^6$ (with a calculator) and see if it equals 3^{14} or 3^{48} or $2 \cdot 3^{14}$. A calculator shows

$$3^8 \cdot 3^6 = 4{,}782{,}969$$
$$2 \cdot 3^{14} = 9{,}565{,}938$$
$$3^{48} = 7.9766 \cdot 10^{22}$$
$$3^{14} = 4{,}782{,}969$$

The answer is 3^{14}. So $x^8 \cdot x^6 = x^{14}$.

Solution 2

Use the Repeated Multiplication Model for Powering to rewrite x^8 and x^6.

$$x^8 \cdot x^6 = (x \cdot x \cdot x \cdot x \cdot x \cdot x \cdot x \cdot x) \cdot (x \cdot x \cdot x \cdot x \cdot x \cdot x)$$

Notice that there are 14 factors of x in the product. So $x^8 \cdot x^6 = x^{14}$.

Showing that a Pattern Is Not Always True

In the test of a special case, the number used should not be too special. A pattern may work for a few numbers but not all. Recall that a counterexample is a special case for which a pattern is false. To show that a pattern is not always true, it is enough to find *one* counterexample.

Example 3

Ali noticed that $2^3 = 2^2 + 2^2$ since 8 = 4 + 4. She guessed that, in general, there is a property

$$x^3 = x^2 + x^2.$$

She tested a second case by letting $x = 0$. She found that $0^3 = 0^2 + 0^2$. She concluded that her property is always true. Is Ali right?

Solution

Try a different number. Let x = 5.
Does $5^3 = 5^2 + 5^2$? 125 = 25 + 25? No.
x = 5 is a counterexample which shows that Ali's property is not always true.

There is a subtle point in the discussion at the top of page 536 that should be mentioned. Although the numbers 0, 1, and 2 are not good for checking answers to problems involving powers, they may be just the numbers you would want to use in some special situations. For instance, Ali could have seen that x^3 does not always equal $x^2 + x^2$ by noting that the expressions are not equal when $x = 1$.

Additional Examples

1. Write $\left(\frac{7}{8}\right)^{-2}$ as a simple fraction.
   $\frac{64}{49}$
2. Simplify $x^{-4} \cdot x^{10}$. Test a special case to check your answer. **x^6; Sample check: if $x = 2$, then $2^{-4} \cdot 2^{10} = \frac{1}{16} \cdot 1024 = 64 = 2^6$.**
3. Use a counterexample to show that $\sqrt{a+b} \neq \sqrt{a} + \sqrt{b}$. **Sample: Let $a = 4$ and $b = 9$. Then $\sqrt{4+9} = \sqrt{13} \approx 3.6$ but $\sqrt{4} + \sqrt{9} = 2 + 3 = 5$.**
4. Ann works at a dress shop and gets 20% off anything she buys. Winter merchandise is on sale for 30% off. If Ann buys something, does it make any difference if the 20% discount is figured before or after the employee discount? **No. This can be answered by testing a special case or examining .8(.7*x*) and .7(.8*x*).**

LESSON MASTER 8-9 A

Questions on SPUR Objectives
See pages 543-545 for objectives.

Uses Objective A

In 1–3, simplify, giving the answer as a simple fraction.

1. $\left(\frac{9}{13}\right)^{-1}$ **$\frac{13}{9}$**
2. $\left(\frac{7}{8}\right)^{-2}$ **$\frac{64}{49}$**
3. $\left(\frac{5}{3}\right)^{-3}$ **$\frac{27}{125}$**

4. *Multiple choice.* Which of the following equals $(4x^{-5})^{-3}$?
   (a) $64x^{15}$ (b) $\frac{64}{x^{15}}$ (c) $\frac{x^{15}}{64}$ (d) $\frac{1}{64x^{15}}$
   Explain your reasoning.
   c; sample: By the Power of a Product and Power of Powers Properties, $(4x^{-5})^{-3} = 4^{-3} \cdot x^{-5 \cdot -3} = \frac{x^{15}}{64}$

5. Find a counterexample for the pattern $x^2 \cdot x^4 = x^8$.
   Sample: $x = 2$, since $2^2 \cdot 2^4 = 4 \cdot 16 = 64$ and $2^8 = 256$.

Properties Objective D

6. Alicia can't remember how to simplify $\frac{x^6}{x^2}$ but she thinks the answer is x^3. For each value of x, tell if $\frac{x^6}{x^2} = x^3$ is *true* or *false*.
   a. $x = 1$ **True** b. $x = 2$ **False**
   c. $x = 0$ **False** d. $x = -1$ **False**

In 7–9, choose the simplified form of the given expression. Check your answer by testing a special case. **Sample checks are given.**

7. $2x^3 \cdot 8x^{10}$ (a) $16x^{30}$ (b) $16x^{13}$ (c) $10x^{30}$ (d) $10x^{13}$ **b**
   $x = 2$: $2 \cdot 2^3 \cdot 8 \cdot 2^{10} = 16 \cdot 2^{13} = 131{,}072$; $16 \cdot 2^{30} = 131{,}072$
8. $(5x^2)^3$ (a) $5x^5$ (b) $5x^6$ (c) $125x^5$ (d) $125x^6$ **d**
   $x = 2$: $(5 \cdot 2^2)^3 = 20^{-3} = 8000$; $125 \cdot 2^6 = 8000$
9. $\frac{x^{10}}{x^2}$ (a) x^8 (b) x^5 (c) $\frac{1}{x^8}$ (d) $\frac{1}{x^5}$ **a**
   $x = 3$: $\frac{3^{10}}{3^2} = \frac{59{,}049}{9} = 6561$; $3^8 = 6561$

Adapting to Individual Needs

Extra Help

Some students might need help with the complex fractions used in Solutions 1 and 2 of **Example 1**. Remind students that the fraction bar is a grouping symbol. So,

$$\frac{\frac{1}{2^2}}{\frac{1}{3^4}} = \frac{\left(\frac{1}{2^2}\right)}{\left(\frac{1}{3^4}\right)}.$$ The numerator and denominator can be simplified separately.

Notes on Questions

Questions 11, 16, and 17 You might discuss algebraic methods of determining if the patterns are true or false. Mathematicians often use special cases to suggest properties that may be true; however, they must use deductive reasoning to prove that these properties are true.

Question 24 Social Studies Connection There are about 1.25 billion beef and dairy cattle in the world and over 30% of them are in Asia. India has the most cattle of any country, but cattle are of little economic value because the cow is considered a sacred animal and consequently it is not slaughtered for food.

(Notes on Questions continue on page 538.)

Follow-up for Lesson 8-9

Practice

For more questions on SPUR Objectives, use **Lesson Master 8-9A** (shown on page 535) or **Lesson Master 8-9B** (shown on pages 536–537).

LESSON MASTER 8-9 B Questions on SPUR Objectives

Skills Objective A: Evaluate integer powers of real numbers.

In 1–6, simplify, giving the answer as a simple fraction.

1. $\left(\frac{3}{14}\right)^{-1}$ $\frac{14}{3}$
2. $\left(\frac{2}{7}\right)^{-2}$ $\frac{49}{4}$
3. $\left(\frac{4}{3}\right)^{-3}$ $\frac{27}{64}$
4. $(-3)^{-4}$ $\frac{1}{81}$
5. -3^{-4} $-\frac{1}{81}$
6. $\left(-\frac{1}{10}\right)^{-4}$ 10,000
7. *Multiple choice.* Which of the following equals $(3m^{-3})^{-4}$? Explain your reasoning.

   (a) $81m^{12}$ (b) $\frac{m^{12}}{81}$ (c) $\frac{81}{m^{12}}$ (d) $\frac{1}{81m^{12}}$

   b; $(3m^{-3})^{-4} = 3^{-4}(m^{-3})^{-4} = \frac{1}{81}m^{12} = \frac{m^{12}}{81}$

Properties Objective D: Test a special case to determine whether a pattern is true.

8. Give a counterexample to the pattern $a^3 \cdot a^2 = a^6$.
   Sample: $3^3 \cdot 3^2 = 27 \cdot 9 = 243$; $3^6 = 729$
9. For each value of a, tell if $\frac{a^3}{a^2} = a^4$ is *true* or *false*.
   a. $a = 2$ false b. $a = 1$ true
   c. $a = .1$ true d. $a = 0$ false
10. For each value of x, tell if $(-x)^5 = -(x)^5$ is *true* or *false*.
    a. $x = 1$ true b. $x = 0$ true
    c. $x = -1$ true d. $x = 2$ true

1b) $\frac{1}{\left(\frac{5}{2}\right)^3} = \frac{1}{\frac{5}{2} \cdot \frac{5}{2} \cdot \frac{5}{2}} = \frac{1}{\frac{125}{8}} = \frac{8}{125}$

2b) $\frac{1}{\left(\frac{3}{4}\right)^3} = \frac{1}{\frac{3}{4} \cdot \frac{3}{4} \cdot \frac{3}{4}} = \frac{1}{\frac{27}{64}} = \frac{64}{27}$

5) c; $\left(\frac{3x}{y}\right)^{-2} = \frac{1}{\left(\frac{3x}{y}\right)^2} = \frac{1}{\frac{9x^2}{y^2}} = \frac{y^2}{9x^2}$

In the questions, if you have trouble remembering a property or are not certain that you have simplified an expression correctly, try going back to the Repeated Multiplication Model for Powering or to testing a special case.

However, remember that some numbers are very special. For instance, the number 2 has properties that other numbers do not have. Squaring it gives the same result as doubling it. So beware of using 2 as a special case. Beware also of using 0 and 1.

QUESTIONS

Covering the Reading

In 1 and 2, **a.** write the number as a fraction. **b.** Check by using the Repeated Multiplication Model for Powering. **See left for check.**

1. $\left(\frac{5}{2}\right)^{-3}$ $\frac{8}{125}$
2. $\left(\frac{3}{4}\right)^{-3}$ $\frac{64}{27}$

In 3 and 4, select the correct choice and check by testing a special case.

3. $\frac{x^6}{x^3} =$ b; Sample: Let $x = 3$. Then $\frac{x^6}{x^3} = \frac{3^6}{3^3} = \frac{729}{27} = 27$ and $x^3 = 3^3 = 27$.
   (a) x^2 (b) x^3 (c) 2 (d) 1
4. $\left(\frac{m}{n}\right)^2 =$ d; Sample: Let $m = 3$ and $n = 2$. Then $\left(\frac{m}{n}\right)^2 = \left(\frac{3}{2}\right)^2 = \frac{9}{4}$ and $\frac{m^2}{n^2} = \frac{3^2}{2^2} = \frac{9}{4}$.
   (a) $\frac{2m}{n}$ (b) $\frac{m}{n}$ (c) $\frac{2m}{2n}$ (d) $\frac{m^2}{n^2}$
5. *Multiple choice.* Which of the following equals $\left(\frac{3x}{y}\right)^{-2}$? Justify your answer. **See left.**
   (a) $\frac{y^2}{3x^2}$ (b) $-6x^2y^2$ (c) $\frac{y^2}{9x^2}$ (d) $\frac{-6x^2}{y^2}$
6. Consider the equation $x^4 = 4x^2$. Tell if the equation is true for the special case indicated.
   a. $x = 0$ True **b.** $x = 2$ True **c.** $x = -2$ True **d.** $x = 3$ False
7. *True or false.* If more than two special cases of a pattern are true, then the pattern is true. False
8. *True or false.* If one special case of a pattern is not true, then the general pattern is not true. True
9. What is a *counterexample?* A counterexample is a special case for which a pattern is false.

Adapting to Individual Needs

English Language Development

Have students make a set of index cards defining each property of exponents. Have them write the property and an example on the front of the index card, and then show the worked-out example on the back of the card.

10) Samples:

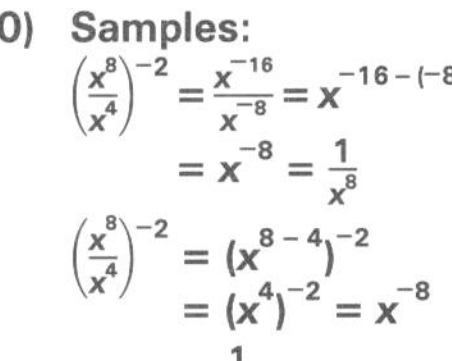

$\left(\frac{x^8}{x^4}\right)^{-2} = \frac{x^{-16}}{x^{-8}} = x^{-16-(-8)} = x^{-8} = \frac{1}{x^8}$

$\left(\frac{x^8}{x^4}\right)^{-2} = (x^{8-4})^{-2} = (x^4)^{-2} = x^{-8} = \frac{1}{x^8}$

16b) Sample: Let $z = y = 3$. Then $\frac{1}{z} + \frac{1}{y} = \frac{1}{3} + \frac{1}{3} = \frac{2}{3}$, and $\frac{y+z}{yz} = \frac{3+3}{3 \cdot 3} = \frac{6}{9} = \frac{2}{3}$.

c) $\frac{1}{z} + \frac{1}{y} = \frac{1}{2} + \frac{1}{5} = .5 + .2 = .7$ and $\frac{y+z}{yz} = \frac{5+2}{5 \cdot 2} = \frac{7}{10} = .7$

17) Sample: Let $a = 2$ and $b = 3$. Then $2^2 + 3^2 = 13$ and $(2 + 3)^2 = 5^2 = 25$. The counterexample shows that the pattern is not true.

Applying the Mathematics

10. Describe at least two different ways to simplify $\left(\frac{x^8}{x^4}\right)^{-2}$. **See left.**

11. Find a counterexample to show that it is not always true that $(2x)^3 = 2x^3$. **Sample: Let $x = 3$. Then $(2x)^3 = (2 \cdot 3)^3 = 6^3 = 216$ and $2x^3 = 2 \cdot 3^3 = 2 \cdot 27 = 54$.**

In 12–15, an instance of a property is given. Describe the general property.

12. $(x + 1)(x + 1)^3 = (x + 1)^4$ **Product of Powers Property**

13. $(4v)^3 = 64v^3$ **Power of a Product Property**

14. $\left(\frac{1}{p}\right)^{10} = \frac{1}{p^{10}}$ **Power of a Quotient Property**

15. $\left(\frac{4}{9}\right)^{-2} = \frac{1}{\left(\frac{4}{9}\right)^2}$ **Negative Exponent Property**

16. Consider the pattern $\frac{1}{z} + \frac{1}{y} = \frac{y + z}{yz}$.

a. Is the pattern true when $y = 3$ and $z = 4$? **Yes**

b. Test a special case when $y = z$. **See left.**

c. Test another special case. Let $y = 5$ and $z = 2$. Convert the fractions to decimals to check. **See left.**

d. Do you think this pattern is true for all nonzero y and z? **Yes**

17. Consider the pattern $a^2 + b^2 = (a + b)^2$. Test special cases to decide whether this pattern is always true. Explain how you arrived at your conclusion. **See left.**

18. Use special cases to answer this question: If a price is discounted 30% and then the sale price is discounted 10%, what percent of the original price is the sale price? **Sample: Let the price be $100. A 30% discount gives .7(100) = $70. Another 10% discount gives .9(70) = $63. So you are paying 63% of the original price.**

Review

In 19–21, simplify. *(Lessons 8-7, 8-8)*

19. $\left(\frac{3}{5x}\right)^4 \cdot \left(\frac{2}{3}\right)^2$ $\frac{36}{625x^4}$

20. $x^5 \cdot \left(\frac{3}{x}\right)^2$ $9x^3$

21. $100\left(\frac{a^3}{2b}\right)^3$ $\frac{25a^9}{2b^3}$

22. *Skill sequence.* Evaluate $\frac{4}{3}\pi r^3$ for the given values of r. Leave your answers in terms of π. *(Lessons 8-5, 8-8)*

a. $r = 3$ 36π

b. $r = 3k$ $36\pi k^3$

23. If $\frac{6n^5}{x} = 3n$, what is x? *(Lesson 8-7)* $2n^4$

24. In 1993, the U.S. Department of Agriculture estimated that on farms in the United States there were 100,892 *thousand* cattle worth an average of 649 dollars each. What was the estimated total value of all these cattle? *(Lesson 8-7)*
about 65.5 billion dollars

Texas cattle. *Beef cattle, like this Texas Longhorn herd, provide about half of Texas' farm income.*

Assessment

Written Communication Ask students to reread the three solutions given for **Example 1**. Then have them write a short paragraph in which they describe which solution they would most likely use and explain their choice. [Students display ability to use properties of exponents.]

Extension

Cooperative Learning Have students **work in small groups** to give counterexamples of the following generalizations.

1. In every ten consecutive integers, there is at least one prime number. [Sample: The integers from 117 to 126 provide a counterexample.]
2. The fourth power of a number is greater than the number itself. [If $0 \le x \le 1$, then $x^4 \le x$.]
3. A polygon has more diagonals than sides. [A quadrilateral has two diagonals, a pentagon has five; an n-gon has $\frac{1}{2}n(n - 3)$ diagonals, and $\frac{1}{2}n(n - 3) > n$ for $n > 5$.]

Project Update Project 2, *Applying the Properties,* on page 539, relates to the content of this lesson.

▶ **LESSON MASTER 8-9 B** *page 2*

In 11–14, *multiple choice*. Choose the simplified form of the given expression. Check your answer by testing a special case. **Sample checks are given.**

11. $(4x^3)^2$ (a) $16x^5$ (b) $16x^6$ (c) $4x^5$ (d) $16x^5$ **b**
$x = 2: (4 \cdot 2^3)^2 = (4 \cdot 8)^2 = 32^2 = 1024; 16 \cdot 2^6 = 16 \cdot 64 = 1024$
check

12. $5x^4 \cdot 3x^3$ (a) $15x^7$ (b) $8x^7$ (c) $15x^{12}$ (d) $16{,}875^{14}x^7$ **a**
$x = 2: (5 \cdot 2^4) \cdot (3 \cdot 2^3) = 80 \cdot 24 = 1920; 15 \cdot 2^7 = 15 \cdot 128 = 1920$
check

13. $\frac{x^{12}}{x^6}$ (a) x^2 (b) $\frac{1}{x^6}$ (c) x^6 (d) $\frac{1}{x^2}$ **c**
$x = 3: \frac{3^{12}}{3^6} = \frac{531{,}441}{729} = 729; 3^6 = 729$
check

14. $\left(\frac{2}{x^2}\right)^{-3}$ (a) $\frac{x^6}{8}$ (b) $\frac{x^6}{6}$ (c) $-\frac{x^6}{8}$ (d) $\frac{1}{8x}$ **a**
$x = 3: \left(\frac{2}{3^2}\right)^{-3} = \frac{2^{-3}}{(3^2)^{-3}} = \frac{2^{-3}}{3^{-6}} = \frac{3^6}{2^3} = \frac{729}{8}; \frac{3^6}{8} = \frac{729}{8}$
check

15. Show two different ways to simplify $\left(\frac{x}{y}\right)^3\left(\frac{x}{y}\right)^{-1}$.
Sample: $\left(\frac{x}{y}\right)^3\left(\frac{x}{y}\right)^{-1} = \left(\frac{x}{y}\right)^{3-1} = \left(\frac{x}{y}\right)^2 = \frac{x^2}{y^2}$;
$\left(\frac{x}{y}\right)^3\left(\frac{x}{y}\right)^{-1} = \frac{x^3}{y^3} \cdot \frac{x^{-1}}{y^{-1}} = \frac{x^{3-1}}{y^{3-1}} = \frac{x^2}{y^2}$

16. Consider the pattern $\sqrt{x^2 + y^2} = x + y$.
a. Test the case with four different pairs of numbers for x and y.
Sample: $\sqrt{1^2 + 2^2} = \sqrt{5} \ne 1 + 2$; $\sqrt{1^2 + 3^2} = \sqrt{10} \ne 1 + 3$; $\sqrt{0^2 + 4^2} = \sqrt{16} = 4 = 0 + 4$; $\sqrt{2^2 + 4^2} = \sqrt{20} \ne 2 + 4$
b. Do you think the pattern is true? Why or why not?
No; there exists at least one counterexample.

Adapting to Individual Needs

Challenge

Refer students to the *Challenge* for Lesson 8-8. Remind them that a numeral in base two stands for a sum of the powers of two in descending order. Then have them copy and complete the table for changing a base-ten numeral to base-two numeral. Two examples are shown.

Base Ten	2^5 32	2^4 16	2^3 8	2^2 4	2^1 2	2^0 1	Base Two
19	0	1	0	0	1	1	$(10011)_{two}$
38	1	0	0	1	1	0	$(100110)_{two}$
16	[0	1	0	0	0	0]	[$(10000)_{two}$]
21	[0	1	0	1	0	1]	[$(10101)_{two}$]
28	[0	1	1	1	0	0]	[$(11100)_{two}$]
45	[1	0	1	1	0	1]	[$(101101)_{two}$]
63	[1	1	1	1	1	1]	[$(111111)_{two}$]

Notes on Questions

Question 27 The numbers are too large to evaluate on most calculators. However, if students realize that the power of 2 with the greatest exponent is the largest, they can use calculators to examine the exponents.

Question 30d Ask why the restriction $a \neq 0$ is included. [It allows you to divide both sides by a.] Would the answer to the problem be different if $a = 0$? [Yes. All ordered pairs would solve the equation. The problem-solving technique *try a simpler case* may help to explain this. Compare the solutions of $5ax = 10$ when $a = 0$ and when $a \neq 0$.]

Question 32 Students should be expected to memorize the small powers in the table.

In 25 and 26, solve. *(Lessons 8-5, 8-6)*

25. $3^a \cdot 3^{10} = 3^{30}$ **a = 20**

26. $\frac{1}{128} = 2^k$ **k = -7**

27. Which is largest: 2^{1492}, $(2^{14})^{92}$, or $((2^{14})^9)^2$? *(Lesson 8-5)* **2^{1492}**

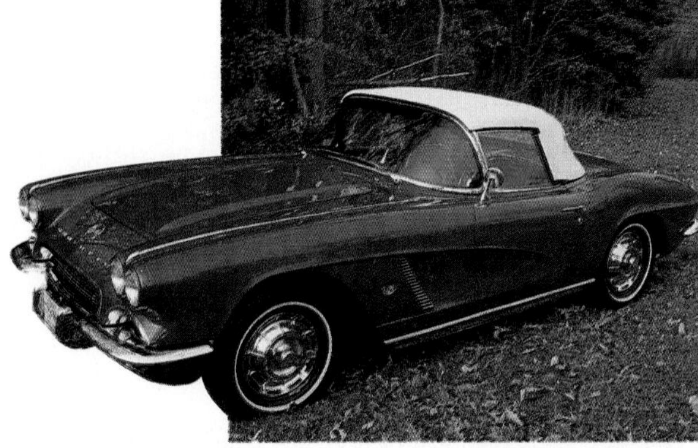

28. Suppose John bought a new car 36 years ago for \$4000. It lost 10% of its value each year for the first 15 years. Then its value stayed the same for 5 years. When it was 20 years old, the car became a collector's item and its value increased 23% each year. Find the value of the car now. *(Lessons 8-2, 8-4)* **about \$23,000**

Classic sports car. *This is a 1962 Corvette convertible. Chevrolet Motor Company first began manufacturing Corvettes in 1953.*

29a)

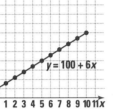

29b)

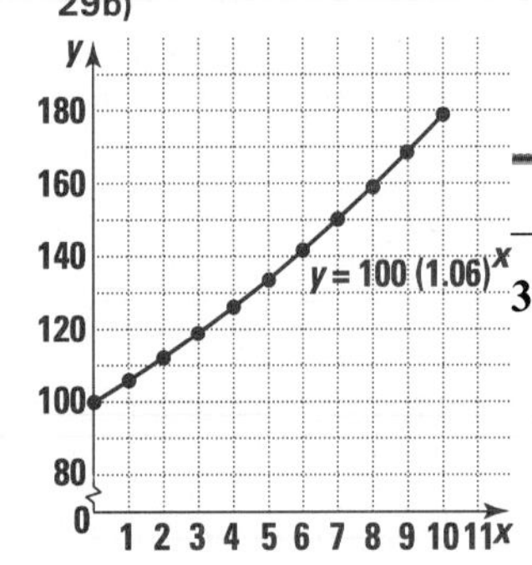

29c) Sample: Of two accounts, each starting with \$100, which has more money in it at the end of 10 years, one that earns 6% compound interest or one that has \$6 deposited in it each year? Answer: the one that earns 6% compound interest.

29. a. Graph $y = 100 + 6x$ for $x = 0$ to $x = 10$. **See left.**

b. Graph $y = 100(1.06)^x$ for $x = 0$ to $x = 10$. **See left.**

c. Make up a question about investments that can be answered by using the graphs in parts **a** and **b**. Answer your question. *(Lesson 8-3)* **See left.**

30. *Skill sequence.* $a \neq 0$. Solve for y. *(Lessons 3-8, 5-7)*

a. $2x + 3y = 4$ **$y = -\frac{2}{3}x + \frac{4}{3}$**

b. $4x + 6y = 8$ **$y = -\frac{2}{3}x + \frac{4}{3}$**

c. $6x + 9y = 12$ **$y = -\frac{2}{3}x + \frac{4}{3}$**

d. $2ax + 3ay = 4a$ **$y = -\frac{2}{3}x + \frac{4}{3}$**

31. *Multiple choice.* Choose the expression that can be simplified and simplify it. *(Lesson 3-6)* **a; $5m^2$**

(a) $3m^2 + 2m^2$ (b) $m^2 + m^3$ (c) $3m^2 + 2m^3$

Exploration

32. If you do not have a calculator, testing a special case can be difficult. So it helps to know the small positive integer powers of 2, 3, 4, 5, and 6. Fill in this table of values of x^n.

$n =$	2	3	4	5	6	7	8	9	10
$x = 2$	4	8	16	32	64	128	256	512	1024
3	9	27	81	243	729				
4	16	64	256	1024					
5	25	125	625						
6	36	216							

How can you use the properties of powers to make it easier to remember these powers? **Samples: $2^{10} = 2^9 \cdot 2^1 = 512 \cdot 2 = 1024$; $4^4 = (2^2)^4 = 2^8 = 256$**

PROJECTS 8 CHAPTER EIGHT

A project presents an opportunity for you to extend your knowledge of a topic related to the material of this chapter. You should allow more time for a project than you do for typical homework questions.

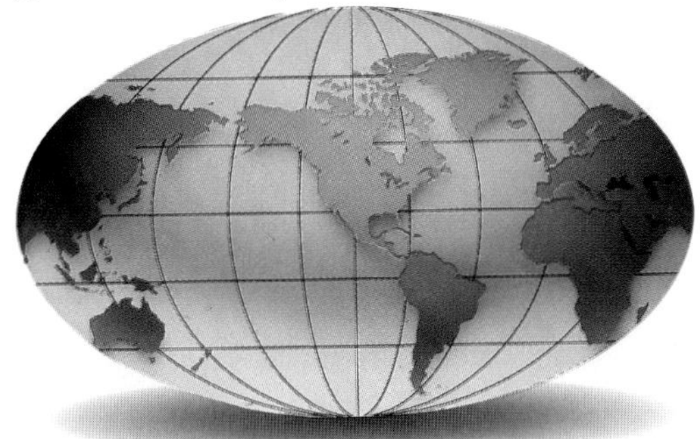

1 Population Growth over Long Periods

Not all populations grow exponentially. You learned in Chapter 7 that the population of Manhattan has not grown exponentially over the past 100 years. From an almanac or other source, examine the population of one of the following areas over the given time period. Graph the data to determine if the growth is approximately exponential. Summarize what you have found.

(a) the United States population each decade from 1790 to the present
(b) the world population since 1900
(c) the world population over the past two thousand years
(d) the population of a continent other than North America over the past 100 years

2 Applying the Properties

You have simplified some expressions like $\left(\frac{5x}{y}\right)^{-2}$ in this chapter. Devise a very complicated expression which would require you to use every property of powers in this chapter. Show how each property would be used to simplify the expression.

3 Fitting an Exponential Curve

The table below gives the speed of several microprocessors introduced since 1970. (Microprocessors function as the "brains" of computers.) The speed is described as the number of seconds it takes to scan the entire *Encyclopaedia Britannica.* Use an automatic grapher which has a "best-fitting exponential curve" option to do the following.

a. Graph the data, using the number of years after 1970 as x-values, and the time to scan as y-values.

b. Have the program determine a and b in $y = a \cdot b^x$.

Time to scan Encyclopaedia Britannica

Microprocessor	Year introduced	Speed (in seconds)
4004	1971	2250
8088	1979	400
80286	1982	45
Intel 386™	1985	13
Intel 486™	1989	4

c. According to your equation, how much time will it take a microprocessor to scan the *Encyclopaedia Britannica* in the year 2000? Do you think this prediction is a good one?

The Intel Pentium™ and Motorola PowerPC™ microprocessors are significantly faster than older processors.

Chapter 8 Projects

The projects relate to the content of the lessons of this chapter as follows:

Project	Lesson(s)
1	8-2
2	8-6, 8-7, 8-8, 8-9
3	8-4
4	8-5
5	8-1
6	8-4

1 Population Growth over Long Periods This project offers students an opportunity to explore population trends in a variety of situations. If students choose to examine the population in **part a,** they should be able to find United States census data for each decade beginning with 1790. You might want to remind students that the growth factor for each decade is found by dividing the population of a given decade by the population of the previous decade. Students may wish to use a spreadsheet to generate the growth factor for each decade. The geometric mean growth factor from 1790 to 1990 is approximately 1.23. If students choose to work on **parts b, c, or d,** explain that they may not be able to find population data for each decade or century. However, if they graph the data they can find, they should be able to determine that the growth is approximately exponential.

2 Applying the Properties Suggest that students refer to the table of properties on page 533 of this chapter. Students may need to work backward or revise their expression as they work in order to use all of the properties at least once. Students should show each step of the simplification and identify the property used. Remind students to use special cases as a check that they are using properties correctly.

3 Fitting an Exponential Curve You might want to point out that the powerful microprocessors are shown close to actual size. Students will recognize the graph of the data as one of exponential decay. They should, therefore, expect the value of b to be between 0 and 1. Remind students of the importance of checking that data found by using calculators or computers are sensible.

Possible responses

1. **The populations in parts a–d grew approximately exponentially. The graphs for parts a–d are based on data from the *1994 World Almanac* and the *1994 Information Please Almanac.***

a.

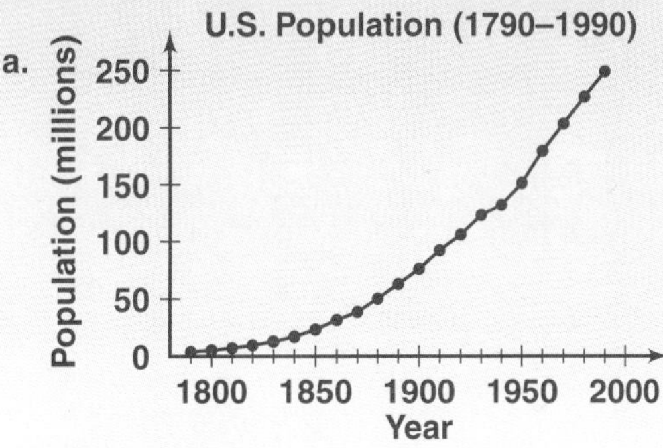

(Responses continue on page 540.)

4 Numbers with a Fixed Number of Factors Suggest that students begin this project by writing the factors of all integers from 2 to 100. This will give examples for each of *a*, *b*, and *c*. In general, if *p* is a prime, then p^{n-1} has exactly *n* factors. Once students have found numbers with 3 and 4 factors, suggest that they write the prime factorization of these numbers and describe any patterns they notice. Students should quickly see that numbers with 3 factors can be written as a prime number to a power. Students may have more difficulty describing numbers with four factors because they are either a prime number to the third power or a product of two different prime numbers. You may wish to suggest that the description of numbers with four factors has two parts.

5 Interest Rates This is a straightforward project. Many banks, savings and loans, and credit unions provide an information sheet listing interest rates for the accounts and loans they offer. Students will notice that annual interest rates as well as annual yields are listed. Students should understand the difference between these terms. The interest rate determines the multiplier used when interest is calculated for each compounding period. The annual yield is the result of doing this over a year. Many students may be surprised to learn that, in some cases, the amount of interest paid exceeds the amount of money actually borrowed.

6 Cooling Water Variations in factors such as the size of the ice cubes, shape of the bowl, and temperature of the water should not have a significant effect on the results of this project. Students' graphs should demonstrate exponential decay. Suggest that students actually repeat the experiment in **part b** to determine if their guess was accurate. Your science department may have computer probes that simultaneously take temperatures and graph the data on a screen.

(continued)

4 Numbers with a Fixed Number of Factors

In this project, consider only whole numbers. Some whole numbers, called *prime numbers,* have exactly two factors, the number itself and 1. For example, 17 has only the factors 1 and 17. In this project, explore whole numbers that have the same number of factors. Do at least the following.

a. The number 49 has exactly three factors: 1, 7, and 49. Find some other numbers that have exactly three factors. Then describe *all* numbers that have exactly three factors. Explain how you determined your answer.

b. The number 8 has exactly four factors: 1, 2, 4, and 8. So does the number 10: 1, 2, 5, and 10. Find some other numbers that have exactly four factors. Then describe *all* numbers that have exactly four factors. Explain how you determine your answer.

c. Use what you have learned to find a number that has exactly 11 factors.

5 Interest Rates

There are many different kinds of interest rates, including mortgage rates, rates on car loans, rates on credit cards, and savings rates of various kinds. Use newspapers and other sources to find at least three examples of current rates of each of these kinds. Give examples of how much interest there would be on various amounts of money at each of these rates.

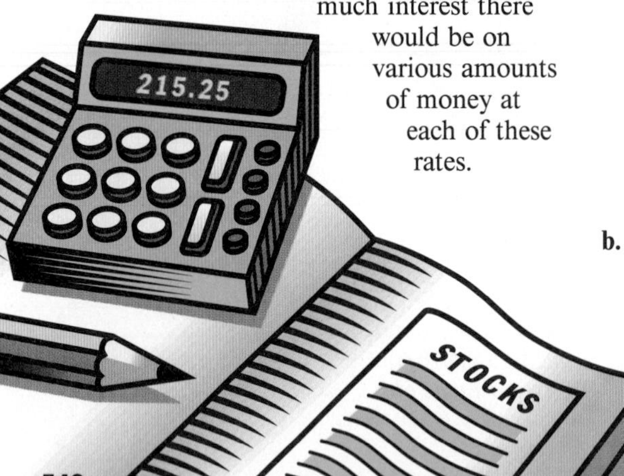

6 Cooling Water

How fast does water cool? What kind of formula describes the cooling? For this project you will need a bowl, hot water, ice cubes, a thermometer, and a watch that shows seconds.

a. As you run the tap to let the water get hot, hold the thermometer under the tap to heat it also. (Caution: Do not use a thermometer designed for body temperature. The high heat of the water may break the thermometer.) Put 1 to 2 cups of water into the bowl. Record the temperature of the water. Add 3 or 4 ice cubes. Record the temperature of the water every minute until it appears to stabilize at room temperature. Make a graph of your data. Plot time in seconds on the horizontal axis and temperature on the vertical axis. What type of equation might fit these points: linear, exponential, or neither? Justify your choice.

b. Using the same quantity of water, heated to the original temperature, add twice as many ice cubes. How does this affect the relation between time and temperature?

Additional responses, page 539

1b.

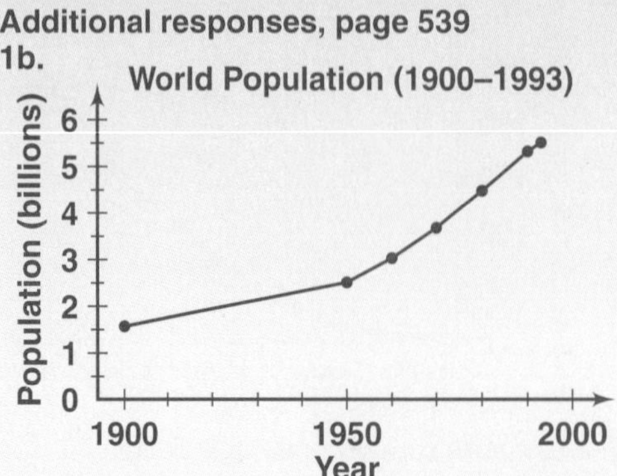

c.

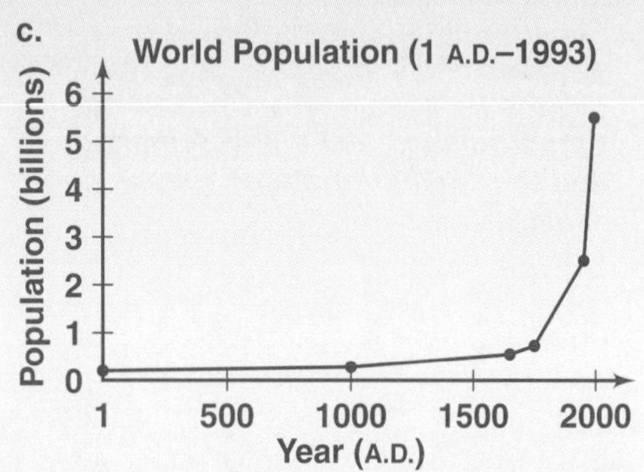

SUMMARY

The nth power of x is written x^n. The number n is called the exponent and x is called the base. Thus, whenever there is an exponent, there is a power. When n is a positive integer, x^n means $x \cdot x \cdot \ldots \cdot x$, where there are n factors. Because powers are related to multiplication, the basic properties of powers involve multiplication and division. For all m and n and all nonzero x and y, the following are true:

$$x^m \cdot x^n = x^{m+n} \qquad \frac{x^m}{x^n} = x^{m-n} \qquad (x^m)^n = x^{mn}$$

$$(xy)^n = x^n y^n \qquad \left(\frac{x}{y}\right)^n = \frac{x^n}{y^n}$$

The expression x^n can also be the growth model in a period of length n, when the growth factor in each unit period is x. Important applications of exponential growth and decay are population growth and compound interest. In compound interest, the growth factor is the quantity $1 + i$, where i is the annual yield. So at an annual yield of i, after n years an amount P grows to $P(1 + i)^n$.

When the growth factor is between 0 and 1, the amount gets smaller, and exponential decay occurs. Graphs of exponential growth or decay are curves.

Exponential Growth

$y = b \cdot g^x,\ g > 1$

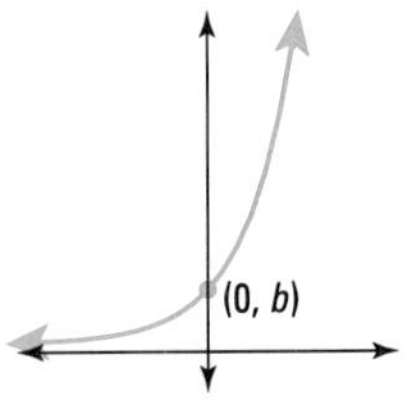

Exponential Decay

$y = b \cdot g^x,\ 0 < g < 1$

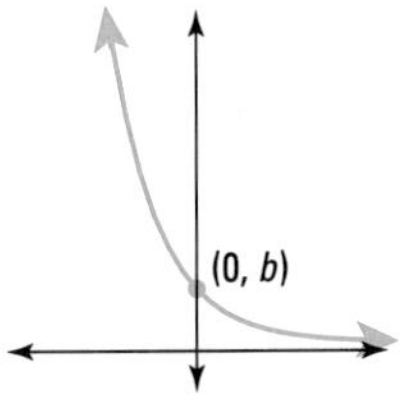

The growth model allows x^n to be interpreted when n is not a positive integer. The number x^0 is the growth factor for a period of length 0, so x^0 is the identity under multiplication. Thus $x^0 = 1$. The number x^{-n} is a growth factor going back in time, and $x^{-n} = \frac{1}{x^n}$.

VOCABULARY

Below are the most important terms and phrases for this chapter. You should be able to give a general description and a specific example of each.

Lesson 8-1
Repeated Multiplication Model for Powering
*n*th power, base, exponent
principal
annual yield
interest, compound interest
Compound Interest Formula

Lesson 8-2
exponential growth
growth factor
Growth Model for Powering
Zero Exponent Property

Lesson 8-4
exponential decay

Lesson 8-5
Product of Powers Property
Power of a Power Property

Lesson 8-6
Negative Exponent Property

Lesson 8-7
Quotient of Powers Property

Lesson 8-8
Power of a Product Property
Power of a Quotient Property

Lesson 8-9
testing a special case

Summary

The Summary gives an overview of the entire chapter and provides an opportunity for students to consider the material as a whole. Thus, the Summary can be used to help students relate and unify the concepts presented in the chapter.

Vocabulary

Terms, symbols, and properties are listed by lesson to provide a checklist of concepts that students must know. Emphasize that students should read the vocabulary list carefully before starting the Progress Self-Test. If students do not understand the meaning of a term, they should refer back to the indicated lesson.

d.

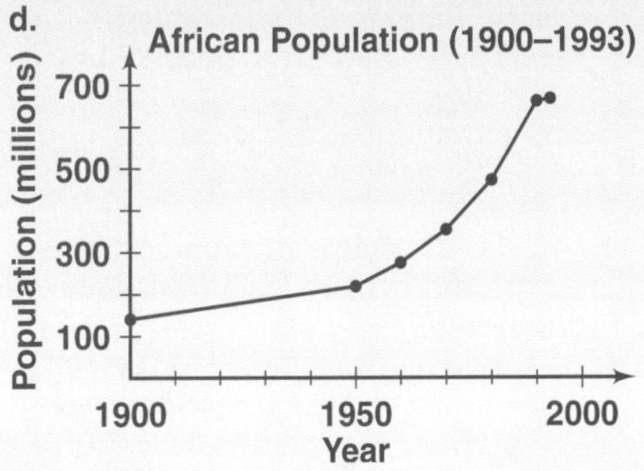

2. Students should use each of the following properties to simplify their expressions: Zero Exponent, Negative Exponent, Product of Powers, Quotient of Powers, Product of a Power, Power of a Product, and Power of a Quotient.
Sample expression:

$$\left(\frac{(2x)^3 \cdot x^0 \cdot x^{-2}}{(x^4)^2 \cdot x^5}\right)^6$$

$$= \left(\frac{(2x)^3 \cdot 1 \cdot x^{-2}}{(x^4)^2 \cdot x^5}\right)^6 \quad \text{Zero Exponent}$$

$$= \left(\frac{(2x)^3 \cdot x^{-2}}{x^8 \cdot x^5}\right)^6 \quad \text{Power of a Power}$$

$$= \left(\frac{8x^3 \cdot x^{-2}}{x^8 \cdot x^5}\right)^6 \quad \text{Power of a Product}$$

$$= \left(\frac{8x}{x^{13}}\right)^6 \quad \text{Product of Powers}$$

(Responses continue on page 542.)

Progress Self-Test

For the development of mathematical competence, feedback and correction, along with the opportunity to practice, are necessary. The Progress Self-Test provides the opportunity for feedback and correction; the Chapter Review provides additional opportunities and practice. We cannot overemphasize the importance of these end-of-chapter materials. It is at this point that the material "gels" for many students, allowing them to solidify skills and understanding. In general, student performance should be markedly improved after these pages.

Assign the Progress Self-Test as a one-night assignment. Worked-out *solutions* for all questions are in the Selected Answers section of the student book. Encourage students to take the Progress Self-Test honestly, grade themselves, and then be prepared to discuss the test in class.

Advise students to pay special attention to those Chapter Review questions (pages 543–545) which correspond to the questions that they missed on the Progress Self-Test.

PROGRESS SELF-TEST

Take this test as you would take a test in class. You will need graph paper and a calculator. Then check your work with the solutions in the Selected Answers section in the back of the book.

1. a. Evaluate $\frac{4^{12}}{4^6}$, and explain how you got your answer. **4096**
   b. Check your answer using another method. **See below.**
2. Evaluate $\frac{5 \cdot 10^{20}}{5 \cdot 10^{10}}$. **$10^{10} = 10{,}000{,}000{,}000$**
3. Write $(8)^{-5}$ as a fraction without negative exponents. **$\frac{1}{32768}$**

In 4–7, simplify.

4. $b^7 \cdot b^{11}$ **b^{18}**
5. $(5y^4)^3$ **$125y^{12}$**
6. $\frac{3z^6}{12z^4}$ **$\frac{z^2}{4}$**
7. $(y^{10})^4$ **y^{40}**
8. Rewrite $\left(\frac{3}{x}\right)^2 \cdot \left(\frac{x}{3}\right)^4$ as a single fraction. **$\frac{x^2}{9}$**
9. Simplify and rewrite $\frac{48a^3b^7}{12a^4b}$ without fractions. **$4a^{-1}b^6$**
10. If $q = 11$, what is the value of $6q^0$? **6**
11. Find a counterexample to the pattern $3x^2 = (3x)^2$. **See below.**
12. Name the general property that justifies $2^{10} \cdot 2^3 = 2^{13}$. **Product of Powers Property**
13. Felipe invests \$6500 in an account with an annual yield of 5%. Without any withdrawals or additional deposits, how much will be in the account after 5 years? **\$8295.83**

1a) $\frac{4^{12}}{4^6} = 4^{12-6} = 4^6 = 4096$;

b) $\frac{4^{12}}{4^6} = \frac{16777216}{4096} = 4096$

11) Sample: $3 \cdot 2^2 = 3 \cdot 4 = 12$; $(3 \cdot 2)^2 = 6^2 = 36$

14) $1900\,(1.058)^3 \approx 2250.15$; $2250.15 - 1900 = \$350.15$

16) $\approx 127{,}000$

17) a and d exponential; b and c not exponential

14. Darlene invests \$1900 for three years at an annual yield of 5.8%. At the end of the three years, how much interest will she earn? Show your work. **See below.**

In 15 and 16, use this information. The population of a city has been growing exponentially at 3% a year. The city currently has a population of 135,000. Assume this growth rate continues.

15. What will the population be five years from now? **$\approx 157{,}000$** **16, 17, 18) See below.**
16. What was the population 2 years ago?
17. For each of the following equations, tell whether or not it can describe exponential growth or decay.
    a. $y = \left(\frac{1}{3}\right)^x$
    b. $y = 27 + 14x$
    c. $y = \frac{1}{3}x$
    d. $y = 27 \cdot 14^x$
18. Graph $y = 3^x$ for $x = -2, -1, 0, 1, 2,$ and 3.
19. A duplicating machine enlarges a picture 30%. If that enlarger is used 3 times, how many times as large as the original picture will the final picture be? **$(1.30)^3 = 2.197$**
20. Recall that the volume V of a sphere with radius r is $V = \frac{4}{3}\pi r^3$. The radius of the sun (roughly a sphere of gas) is about $6.96 \cdot 10^6$ km. Estimate the volume of the sun. **$\approx 1.41 \cdot 10^{21}$ km^3**

18)

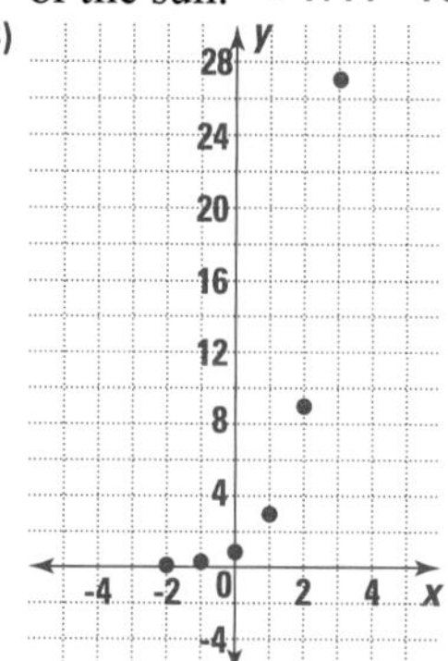

Additional responses, pages 539–540

$= \frac{262{,}144x^6}{x^{78}}$ **Power of a Quotient**

$= 262{,}144x^{-72}$ **Quotient of Powers**

$= \frac{262{,}144}{x^{72}}$ **Negative Exponents**

3. Sample answers for the TI–81 Graphics Calculator are given.
a. Enter the data given in the table. Use the STAT mode. Set the range $0 \le x < 40$; $0 < y < 2300$. Graph the *xy* line. The graph should look like an exponential decay graph. **(1, 2250), (9, 400), (12, 45), (15, 13), and (19, 4) are on the graph.**
b. The calculator will determine *a* and *b* by using the STAT mode and expReg. According to the calculator:
$a = 4598.125535$
$b = 0.6909938515$
$y = 4598.126(.69)^x$.
c. Using the equation in part b, it would take a microprocessor about 0.07 second to scan the *Encyclopedia Britannica* in the year 2000. Students' predictions will vary.

4. Sample responses are given.
a. 4, 9, 25; perfect squares of prime numbers
b. 14, 15, 27; cubes of prime numbers or products of two different prime numbers
c. A prime number raised to the 10th power; $2^{10} = 1024$ and 1024 has 11 factors.

CHAPTER REVIEW

Questions on SPUR Objectives

SPUR stands for **S**kills, **P**roperties, **U**ses, and **R**epresentations. The Chapter Review questions are grouped according to the SPUR Objectives for this chapter.

SKILLS DEAL WITH THE PROCEDURES USED TO GET ANSWERS.

Objective A: *Evaluate integer powers of real numbers.* *(Lessons 8-1, 8-2, 8-6, 8-7, 8-8, 8-9)*

1. Evaluate. **a.** 3^4 81 **b.** -3^4 -81 **c.** $(-3)^4$ 81
2. Simplify $-2^5 \cdot (-2)^5$. 1024
3. If $y = 7$, then $4y^0 = \underline{\ ?\ }$. 4
4. If $x = 2$, then $3x^3 - x^2 = \underline{\ ?\ }$. 20

In 5 and 6, simplify.

5. $(2^3)^3 \div 2^6$ 8
6. $\frac{9 \cdot 10^6}{3 \cdot 10^8}$ .03

In 7 and 8, rewrite as a fraction without an exponent.

7. 5^{-3} $\frac{1}{125}$
8. 2^{-5} $\frac{1}{32}$

In 9–12, write as a simple fraction.

9. $\left(\frac{2}{7}\right)^3$ $\frac{8}{343}$
10. $\left(-\frac{4}{3}\right)^4$ $\frac{256}{81}$
11. $\left(\frac{1}{3}\right)^{-4}$ 81
12. $10 \cdot \left(\frac{2}{5}\right)^{-3}$ $\frac{625}{4}$

Objective B: *Simplify products, quotients, and powers of powers.* *(Lessons 8-5, 8-6, 8-7)*

In 13–22, simplify.

13. $x^4 \cdot x^7$ x^{11}
14. $r^3 \cdot t^5 \cdot r^8 \cdot t^2$ $r^{11}t^7$
15. $y^2(x^3y^{10})$ x^3y^{12}
16. $p^4(pq^2)$ p^5q^2
17. $\frac{n^{15}}{n^2}$ n^{13}
18. $\frac{a^{12}}{a^4} \cdot a^6$ a^{14}
19. $\frac{3a^4c}{3a^5}$ $\frac{c}{a}$
20. $\frac{15x^2y^5}{12xy^6}$ $\frac{5x}{4y}$
21. $(3x^5)^3 + (x^3)^5$ $28x^{15}$
22. $(3m^4)^4 + (9m^2)^2$ $81m^{16} + 81m^4$

23. Rewrite $\frac{4m^6}{20m^2}$ without fractions. $5^{-1}m^4$
24. Rewrite $\frac{60w^8}{15t^2w}$ without fractions. $4t^{-2}w^7$
25. Simplify $\frac{(2+8)^5}{(2+8)^2}$. 1000
26. Describe two different ways to evaluate $\frac{(2t-1)^{11}}{(2t-1)^4}$ when $t = 6$. See below.
27. Rewrite xy^{-2} without a negative exponent. $\frac{x}{y^2}$
28. Rewrite $2m^{-1}n^4p^2$ without a negative exponent. $\frac{2n^4p^2}{m}$

Objective C: *Rewrite powers of products and quotients.* *(Lesson 8-8)*

In 29–40, rewrite without parentheses.

29. $\left(\frac{y}{x}\right)^{-3}$ $\frac{x^3}{y^3}$
30. $\left(\frac{a}{b}\right)^{-5}$ $\frac{b^5}{a^5}$
31. $(4x)^5$ $1024x^5$
32. $(5y)^4$ $625y^4$
33. $\left(\frac{2}{n}\right)^5$ $\frac{32}{n^5}$
34. $\left(\frac{t^7}{2}\right)^4$ $\frac{t^{28}}{16}$
35. $(-3n)^3$ $-27n^3$
36. $-(2y)^3$ $-8y^3$
37. $4 \cdot \left(\frac{k}{3}\right)^3$ $\frac{4k^3}{27}$
38. $45\left(\frac{t}{3}\right)^2$ $5t^2$
39. $2(4x)^2$ $32x^2$
40. $11(10k)^3$ $11{,}000\,k^3$

26) $\frac{(2 \cdot 6 - 1)^{11}}{(2 \cdot 6 - 1)^4} = \frac{11^{11}}{11^4} = 19{,}487{,}171$;
$\frac{(2 \cdot 6 - 1)^{11}}{(2 \cdot 6 - 1)^4} = (2 \cdot 6 - 1)^{11-4} = 11^7 = 19{,}487{,}171$

Chapter 8 Review

Resources

From the ***Teacher's Resource File***

- Answer Master for Chapter 8 Review
- *Assessment Sourcebook:* Chapter 8 Test, Forms A–D Chapter 8 Test, Cumulative Form

Additional Resources

- Quiz and Test Writer

The main objectives for the chapter are organized in the Chapter Review under the four types of understanding this book promotes – Skills, Properties, Uses, and Representations.

Whereas end-of-chapter material may be considered optional in some texts, in *UCSMP Algebra* we have selected these objectives and questions with the expectation that they will be covered. Students should be able to answer these questions with about 85% accuracy after studying the chapter.

You may assign these questions over a single night to help students prepare for a test the next day, or you may assign the questions over a two-day period. If you work the questions over two days, then we recommend assigning the *evens* for homework the first night so that students get feedback in class the next day, then assigning the *odds* the night before the test because answers are provided to the odd-numbered questions.

It is effective to ask students which questions they still do not understand and use the day or days as a total class discussion of the material which the class finds most difficult.

5. Responses will vary.
6. a. The sample graph at the right includes these points: (0, 130), (1, 97), (2, 92), (3, 91), (4, 90), (5, 90), (6, 90), (7, 89), (8, 89),(9, 88), (10, 88). An exponential equation best fits the graph of the data.
   b. The temperature of the water decreases faster when twice as many ice cubes are added.

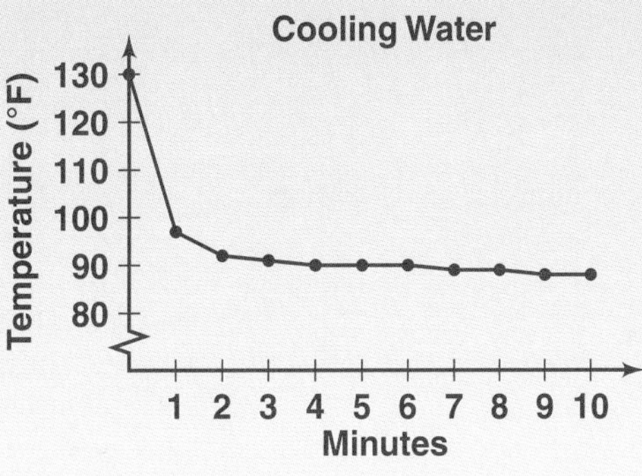

 CHAPTER EIGHT

Assessment

Evaluation The *Assessment Sourcebook* provides five forms of the Chapter 8 Test. Forms A and B present parallel versions in a short-answer format. Forms C and D offer performance assessment. The fifth test is Chapter 8 Test, Cumulative Form. About 50% of this test covers Chapter 8; 25% covers Chapter 7, and 25% covers earlier chapters.

For information on grading see *General Teaching Suggestions; Grading* in the *Professional Sourcebook* which begins on page T20 in Part 1 of the Teacher's Edition.

Feedback After students have taken the test for Chapter 8 and you have scored the results, return the tests to students for discussion. Class discussion on the questions that caused trouble for most students can be very effective in identifying and clarifying misunderstandings. You might want to have them write down the items they missed and work either in groups or at home to correct them. It is important for students to receive feedback on every chapter test, and we recommend that students see and correct their mistakes before proceeding too far into the next chapter.

Additional Answers

45. Power of a Product Property
46. Quotient of Powers Property
47. Zero Exponent Property
48. Product of Powers Property
49. Power of a Quotient Property
50. Zero Exponent Property and Multiplicative Identity Property
51. Negative Exponent Property
52. Negative Exponent Property and Power of a Quotient Property

PROPERTIES DEAL WITH THE PRINCIPLES BEHIND THE MATHEMATICS.

Objective D: *Test a special case to determine whether a pattern is true.* *(Lesson 8-9)*

41. For each case tell whether the pattern $x = x^2$ is true.

a. $x = 0$ **True** **b.** $x = 1$ **True**
c. $x = 2$ **False** **d.** $x = -1$ **False**

42. Consider the pattern $(x^2)^y = x^{2y}$. **See below.**

a. Is the pattern true when $x = 3$ and $y = 4$?
b. Is the pattern true when $x = 5$ and $y = 2$?
c. Based on your answers to parts **a** and **b,** do you have evidence that the pattern is true, or are you sure it is not always true?

In 43 and 44, find a counterexample to the pattern. **Samples:**

43. $(a + b)^3 = a^3 + b^3$ **$(1 + 1)^3 = 2^3 = 8$; $1^3 + 1^3 = 1 + 1 = 2$**

44. $(x^3)^2 = x^{(3^2)}$ **$(2^3)^2 = 8^2 = 64$; $2^{(3^2)} = 2^9 = 512$**

Objective E: *Identify properties of exponents and use them to explain operations with powers.* *(Lessons 8-2, 8-5, 8-6, 8-7, 8-8)*

Here is a list of the power properties in this chapter. For all m and n, and nonzero a and b:

42a) Yes b) Yes c) yes, there is evidence that it is true

Zero Exponent Property: $b^0 = 1$
Product of Powers Property: $b^m \cdot b^n = b^{m+n}$
Power of a Product Property: $(ab)^n = a^n \cdot b^n$
Negative Exponent Property: $b^{-n} = \frac{1}{b^n}$
Quotient of Powers Property: $\frac{b^m}{b^n} = b^{m-n}$
Power of a Power Property: $(b^m)^n = b^{mn}$
Power of a Quotient Property: $\left(\frac{a}{b}\right)^n = \frac{a^n}{b^n}$

45-52) See margin.

In 45–52, describe the general property or properties which justify the simplification.

45. $a^7 \cdot b^7 = (ab)^7$ **46.** $a^7 \div a^2 = a^5$

47. $(4.36)^0 = 1$ **48.** $4^6 \cdot 4^9 = 4^{15}$

49. $\left(\frac{7}{g}\right)^y = \frac{7^y}{g^y}$ **50.** $6^3 \cdot 2^0 = 6^3$

51. $14^{-2} = \frac{1}{14^2}$ **52.** $\left(\frac{x}{y}\right)^{-2} = \frac{y^2}{x^2}$

53. Show two different ways to simplify $\left(\frac{x^3}{x}\right)^8$.

54. Describe two different ways to simplify $(ab^2)^4$. **repeated multiplication, Power of a Product Property**

53) Samples: $\left(\frac{x^3}{x}\right)^8 = (x^{3-1})^8 = (x^2)^8 = x^{16}$
$= \frac{x^{3 \cdot 8}}{x^{1 \cdot 8}} = \frac{x^{24}}{x^8} = x^{24-8} = x^{16}$

USES DEAL WITH APPLICATIONS OF MATHEMATICS IN REAL SITUATIONS.

Objective F: *Calculate compound interest.* *(Lesson 8-1)*

In 55 and 56, use the advertisement at the right.

GUARANTEED
5.7% YIELD
$2,500 MINIMUM

55. Using the annual yield, how much money will there be in an account if $2,500 is deposited for 3 years? **$2952.33**

56. Using the annual yield, calculate how much interest $3000 will earn if deposited for 4 years. **$744.74**

57. Susan has $1200 in a CD with an annual yield of 6%. Without any withdrawals, how much money would she have in the account after 2 years? **$1348.32**

58. Which investment yields more money: (a) x dollars for 2 years at an annual yield of 10%, or (b) the same amount for 10 years at an annual yield of 2%? Explain your reasoning. **(b); $x \cdot (1.10)^2 = 1.21x$; $x \cdot (1.02)^{10} = 1.219x$; $1.219x > 1.21x$**

Objective G: *Solve problems involving exponential growth and decay.* *(Lessons 8-2, 8-3, 8-4, 8-6)*

59. Jennifer earns $7.25 an hour. If she gets a 5.6% raise each year, how much will she earn per hour after 4 years on the job?

60. From 1991 to 1992, the United States had an inflation rate of about 3.0% per year. Thus, an article costing $100 in 1991 cost $103.00 in 1992. Consider a book that sold for $16.95 in 1991. If the rate of inflation remains at 3.0%, how much would the book cost in 1999? **$21.47**

59) $9.02 per hour

544

CHAPTER EIGHT

61b) After 4 hours there will be 128,000 bacteria.

In 61 and 62, suppose that after a few hours a colony of bacteria that doubles every hour has 8000 bacteria. After n more hours there will be T bacteria where $T = 8000 \cdot 2^n$.

61. **a.** Find the value of T when $n = 4$. 128,000

b. In words describe the meaning of your answer to part **a.** See above.

62. **a.** Find T when $n = -3$. 1000

b. Describe the meaning of your answer to part **a.** The bacteria count was 1000 three hours earlier.

In 63–65, suppose that the population in a city of 1,500,000 people is decreasing exponentially at a rate of 3% per year. Let P = the population after n years. 63) $P = 1{,}500{,}000 \cdot (0.97)^n$ 64) 1,249,458

63. Write an equation of the form $y = b \cdot g^x$ to describe the population after n years.

64. What will the population be in 6 years' time?

65. What is the population when $n = 0$? What does your answer mean? 1,500,000; the population now

In 66 and 67, use this information. The death rate from heart disease is decreasing each year. The formula $y = 436.4(0.983)^x$ approximates the number of deaths per 100,000 people x years after 1980.

66. Estimate the death rate per 100,000 population in 1992. ≈ 355

67. **a.** Find y when $x = -3$. ≈ 459

b. Describe the meaning of your answer to part **a.** The death rate 3 years earlier (1977) was about 459 per 100,000 people.

Objective H: *Use and simplify expressions with powers in real situations.* *(Lessons 8-6, 8-7, 8-8)*

68. A certain photographic enlarger can enlarge any picture $\frac{3}{2}$ times. By how many times will a picture be enlarged if the enlarger is used

a. twice? $\frac{9}{4}$ **b.** 5 times? $\frac{243}{32}$

69. Water blocks out light. (At a depth of 10 meters it is not as bright as on the surface.) Suppose 1 meter of water lets in $\frac{9}{10}$ of the light. How much light will get through x meters of water? $(0.9)^x$

70. In 1993 there were about $5.6 \cdot 10^9$ people on Earth. The land area is about $1.48 \cdot 10^8$ km^2. How many people are there per km^2? ≈ 38

71. The moon is nearly a sphere with radius of $1.08 \cdot 10^3$ miles. The volume of a sphere of radius r is $\frac{4}{3}\pi r^3$. To the nearest billion cubic miles, what is the volume of the moon?

In 72 and 73, a spinner has six equal-sized sections: two red, three blue, and one yellow. Spins are random. To win a game, you must spin yellow 3 times in a row.

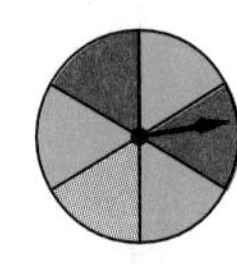

72. **a.** Find the probability of getting yellow on each of three spins. $\left(\frac{1}{6}\right)^3 = \frac{1}{216}$

b. Write your answer with negative exponents. 6^{-3}

73. Find the probability of getting red on four consecutive spins. $\left(\frac{1}{3}\right)^4 = \frac{1}{81}$

71) 5 billion cubic miles

REPRESENTATIONS DEAL WITH PICTURES, GRAPHS, OR OBJECTS THAT ILLUSTRATE CONCEPTS.

Objective I: *Graph exponential relationships.* *(Lessons 8-2, 8-3, 8-4)* 78–81) See margin.

In 74–77, tell whether the graph of the equation is linear or exponential.

74. $y = 4x$ linear **75.** $y = \left(\frac{1}{2}\right)^x$ exponential

76. $y = 100 \cdot (3.4)^x$ exponential **77.** $y = \frac{3}{4}x + 100$ linear

78. Graph $y = 2^x$ for $x = -3, -2, -1, 0, 1, 2,$ and 3.

79. Graph $y = .4^x$ for $x = -3, -2, -1, 0, 1, 2,$ and 3.

80. Suppose an investment worth \$200 is invested in a bank at a 5% annual yield. Graph the amount after 0 to 15 years if the interest is kept in the bank.

81. When x is large, which equation's graph rises faster, $y = 56 + 0.04x$ or $y = 5 \cdot (1.04)^x$? Explain your reasoning.

82. Match each graph below with its description.

a. constant increase **b.** constant decrease

c. exponential growth **d.** exponential decay

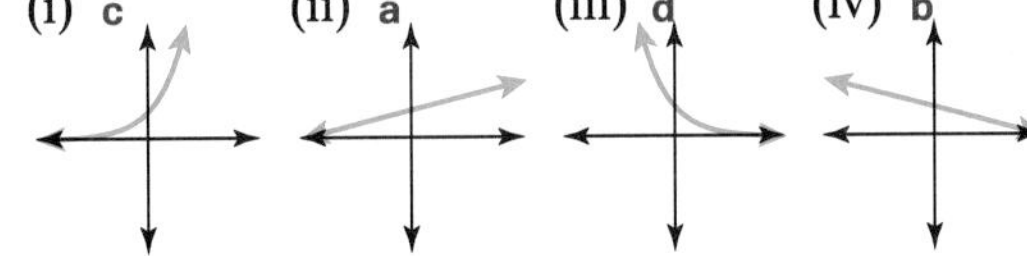

Setting Up Lesson 9-1

We recommend that you assign the Chapter 9 Opener and Lesson 9-1, both reading and some questions, for homework the evening of the test.

Additional Answers

78.

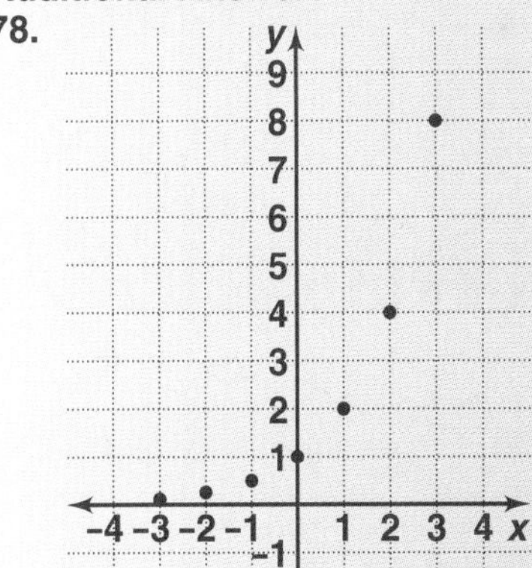

79.

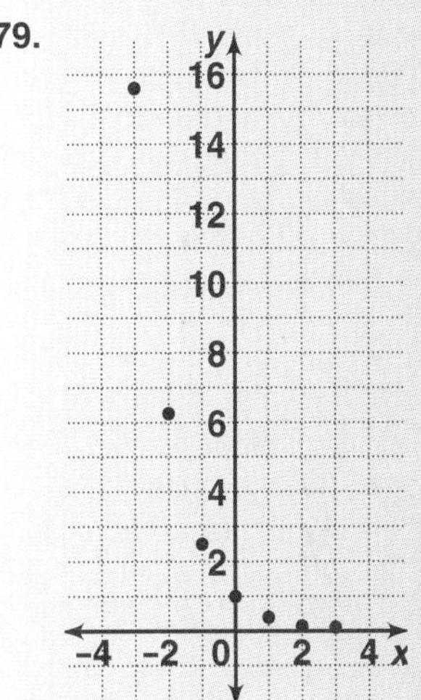

80.

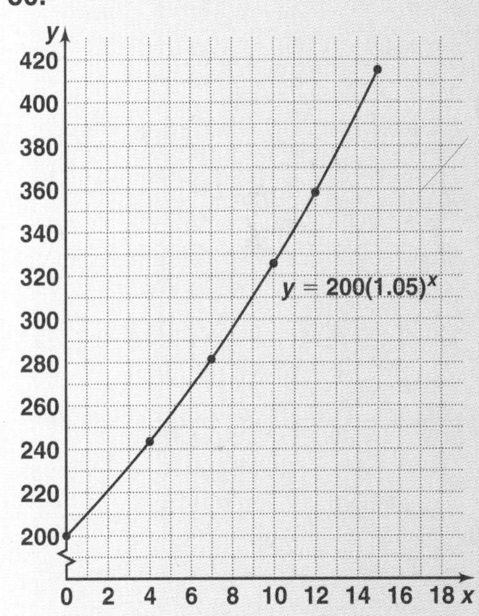

81. $y = 5 \cdot (1.04)^x$; if the growth factor g is greater than 1, exponential growth always overtakes constant increase.

Chapter 9 Planner

Adapting to Individual Needs

The student text is written for the vast majority of students. The chart at the right suggests two pacing plans to accommodate the needs of your students. Students in the Full Course should complete the entire text by the end of the year. Students in the Minimal Course will spend more time when there are quizzes and more time on the Chapter Review. Therefore, these students may not complete all of the chapters in the text.

Options are also presented to meet the needs of a variety of teaching and learning styles. For each lesson, the Teacher's Edition provides sections entitled: *Video* which describes video segments and related questions that can be used for motivation or extension; *Optional Activities* which suggests activities that employ materials, physical models, technology, and cooperative learning; and, *Adapting to Individual Needs* which regularly includes **Challenge** problems, **English Language Development** suggestions, and suggestions for providing **Extra Help.** The Teacher's Edition also frequently includes an **Error Alert,** an **Extension,** and an **Assessment** alternative. The options available in Chapter 9 are summarized in the chart below.

Chapter 9 Pacing Chart

Day	Full Course	Minimal Course
1	9-1	9-1
2	9-2	9-2
3	9-3	9-3
4	Quiz*; 9-4	Quiz*; begin 9-4.
5	9-5	Finish 9-4.
6	9-6	9-5
7	Quiz*; 9-7	9-6
8	9-8	Quiz*; begin 9-7.
9	9-9	Finish 9-7.
10	Self-Test	9-8
11	Review	9-9
12	Test*	Self-Test
13	Comprehensive Test*	Review
14		Review
15		Test*
16		Comprehensive Test*

*in the Teacher's Resource File

In the Teacher's Edition...

Lesson	Optional Activities	Extra Help	Challenge	English Language Development	Error Alert	Extension	Cooperative Learning	Ongoing Assessment
9-1	●	●	●	●	●	●	●	Oral/Written
9-2	●	●	●	●	●	●	●	Group
9-3	●	●	●			●	●	Oral
9-4	●	●	●	●		●		Oral/Written
9-5	●	●	●	●	●	●		Written
9-6	●	●	●	●	●	●	●	Oral
9-7	●	●	●	●	●	●	●	Group
9-8	●	●	●			●	●	Written
9-9	●	●	●	●	●		●	Group

In the Additional Resources...

	In the Teacher's Resource File								
Lesson	Lesson Masters, A and B	Teaching Aids*	Activity Kit*	Answer Masters	Technology Sourcebook	Assessment Sourcebook	Visual Aids**	Technology	Video Segments
9-1	9-1	28, 95, 98	20	9-1	Comp 22		28, 95, 98, AM	Spreadsheet	
9-2	9-2	28, 37, 95, 99–102		9-2			28, 37, 95, 99–102, AM		
In-class Activity				9-3					
9-3	9-3	55, 95		9-3	Comp 23	Quiz	55, 95, AM	GraphExplorer	
9-4	9-4	96, 103		9-4	Comp 24		96, 103, AM	GraphExplorer	
9-5	9-5	96, 104		9-5			96, 104, AM		9-5
9-6	9-6	96, 104		9-6		Quiz	96, 104, AM		
9-7	9-7	97, 105		9-7			97, 105, AM		
9-8	9-8	97		9-8			97, AM		
9-9	9-9	26, 28, 97, 106	21	9-9			26, 28, 97, 106, AM		
End of chapter				Review		Tests			

*Teaching Aids are pictured on pages 546C and 546D. The activities in the Activity Kit are pictured on page 546C.

**Visual Aids provide transparencies for all Teaching Aids and all Answer Masters.

Also available is the Study Skills Handbook which includes study-skill tips related to reading, note-taking, and comprehension.

Integrating Strands and Applications

	9-1	9-2	9-3	9-4	9-5	9-6	9-7	9-8	9-9
Mathematical Connections									
Number Sense	●								
Algebra	●	●	●	●	●	●	●	●	●
Geometry	●	●	●	●	●	●	●	●	●
Measurement	●				●		●	●	●
Logic and Reasoning							●		
Probability					●				
Patterns and Functions	●	●	●	●		●	●		
Interdisciplinary and Other Connections									
Science	●	●		●	●	●	●	●	
Social Studies			●		●		●	●	●
Multicultural		●	●				●		●
Technology	●	●	●	●	●	●		●	●
Career							●	●	
Consumer		●	●		●				●
Sports	●			●	●	●			

Teaching and Assessing the Chapter Objectives

Chapter 9 Objectives (Organized into the SPUR categories—Skills, Properties, Uses, and Representations)	**Lessons**	**Progress Self-Test Questions**	**Chapter Review Questions**	**In the Teacher's Resource File** **Chapter Test, Forms A and B**	**Chapter Test, Forms C**	**D**
Skills						
A: Solve quadratic equations.	9-1, 9-5	1–4, 25	1–14	2–5	1	✓
B: Simplify square roots.	9-7	13–15	15–28	13–15	2	
C: Evaluate expressions and solve equations using absolute value.	9-8	26, 27	29–40	16, 17, 22	3	
Properties						
D: Identify and use properties of quadratic equations.	9-6	5	41–48	6	1	
Uses						
E: Use quadratic equations to solve problems about paths of projectiles.	9-1, 9-4, 9-5	21–24	49–54	11, 12	5	✓
Representations						
F: Graph equations of the form $y = ax^2 + bx + c$ and interpret these graphs.	9-1, 9-2, 9-3	6–12, 16	55–70	1, 7–10, 23	5	✓
G: Calculate and represent distances on the number line or in the plane.	9-8, 9-9	17–20	71–84	18–21	4	

Multidimensional Assessment
Quiz for Lessons 9-1 through 9-3
Quiz for Lessons 9-4 through 9-6
Chapter 9 Test, Forms A–D
Chapter 9 Test, Cumulative Form
Comprehensive Test, Chapters 1–9

Quiz and Test Writer
Multiple forms of chapter tests and quizzes; Challenges

Activity Kit

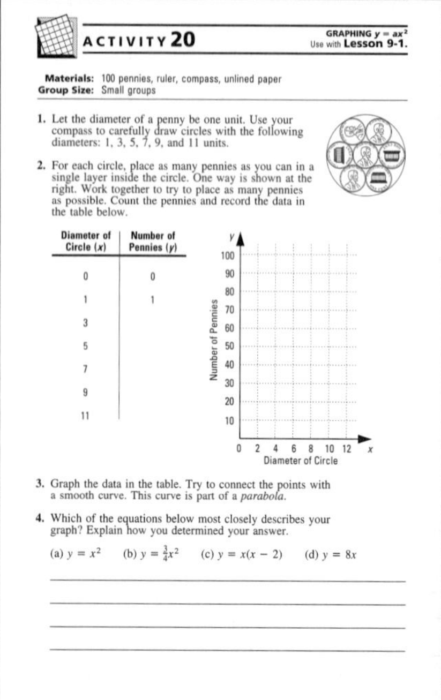

ACTIVITY 20

GRAPHING $y = ax^2$
Use with Lesson 9-1.

Materials: 100 pennies, ruler, compass, unlined paper
Group Size: Small groups

1. Let the diameter of a penny be one unit. Use your compass to carefully draw circles with the following diameters: 1, 3, 5, 7, 9, and 11 units.
2. For each circle, place as many pennies as you can in a single layer inside the circle. One way is shown at the right. Work together to try to place as many pennies as possible. Count the pennies and record the data in the table below.

Diameter of Circle (x)	Number of Pennies (y)
0	0
1	1
3	
5	
7	
9	
11	

3. Graph the data in the table. Try to connect the points with a smooth curve. This curve is part of a *parabola*.
4. Which of the equations below most closely describes your graph? Explain how you determined your answer.

(a) $y = x^2$ (b) $y = \frac{3}{4}x^2$ (c) $y = x(x - 2)$ (d) $y = 8x$

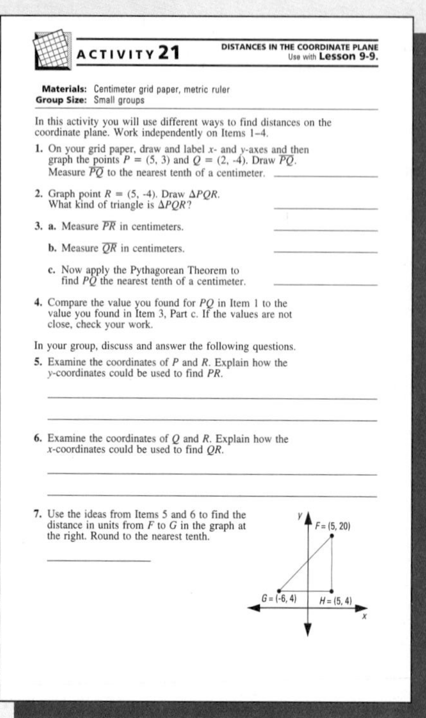

ACTIVITY 21

DISTANCES IN THE COORDINATE PLANE
Use with Lesson 9-9.

Materials: Centimeter grid paper, metric ruler
Group Size: Small groups

In this activity you will use different ways to find distances on the coordinate plane. Work independently on Items 1–4.

1. On your grid paper, draw and label x- and y-axes and then graph the points $P = (5, 3)$ and $Q = (2, -4)$. Draw $\overline{PQ}$. Measure $\overline{PQ}$ to the nearest tenth of a centimeter.
2. Graph point $R = (5, -4)$. Draw ΔPQR. What kind of triangle is ΔPQR?
3. a. Measure $\overline{PR}$ in centimeters.
   b. Measure $\overline{QR}$ in centimeters.
   c. Now apply the Pythagorean Theorem to find PQ the nearest tenth of a centimeter.
4. Compare the value you found for PQ in Item 1 to the value you found in Item 3, Part c. If the values are not close, check your work.

In your group, discuss and answer the following questions.

5. Examine the coordinates of P and R. Explain how the y-coordinates could be used to find PR.
6. Examine the coordinates of Q and R. Explain how the x-coordinates could be used to find QR.
7. Use the ideas from Items 5 and 6 to find the distance in units from F to G in the graph at the right. Round to the nearest tenth.

Teaching Aids

Teaching Aid 26, Graph Paper, (shown on page 140D) can be used with **Lessons 9-8. Teaching Aid 28, Four-Quadrant Graph Paper,** (shown on page 140D) can be used with **Lessons 9-1, 9-2, and 9-8. Teaching Aid 37, Spreadsheet,** (shown on page 214D) can be used with **Lesson 9-2. Teaching Aid 55, Automatic Grapher Grids,** (shown on page 282D) can be used with **Lesson 9-3.**

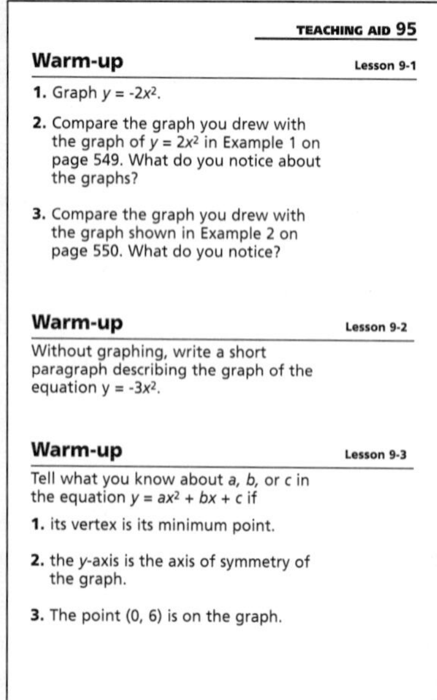

TEACHING AID 95

Warm-up Lesson 9-1

1. Graph $y = -2x^2$.
2. Compare the graph you drew with the graph of $y = 2x^2$ in Example 1 on page 549. What do you notice about the graphs?
3. Compare the graph you drew with the graph shown in Example 2 on page 550. What do you notice?

Warm-up Lesson 9-2

Without graphing, write a short paragraph describing the graph of the equation $y = -3x^2$.

Warm-up Lesson 9-3

Tell what you know about a, b, or c in the equation $y = ax^2 + bx + c$ if

1. its vertex is its minimum point.
2. the y-axis is the axis of symmetry of the graph.
3. The point (0, 6) is on the graph.

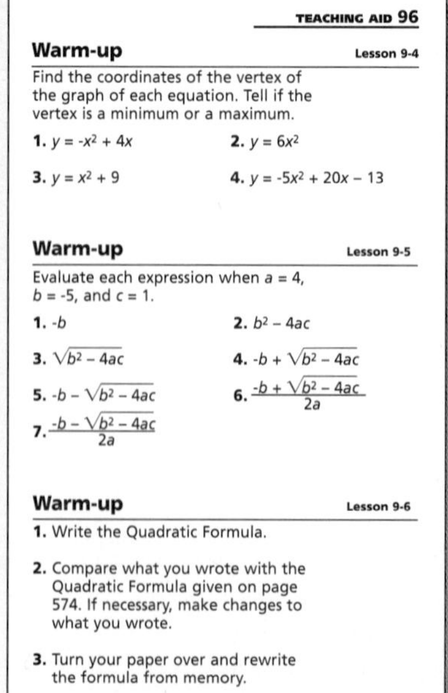

TEACHING AID 96

Warm-up Lesson 9-4

Find the coordinates of the vertex of the graph of each equation. Tell if the vertex is a minimum or a maximum.

1. $y = -x^2 + 4x$
2. $y = 6x^2$
3. $y = x^2 + 9$
4. $y = -5x^2 + 20x - 13$

Warm-up Lesson 9-5

Evaluate each expression when $a = 4$, $b = -5$, and $c = 1$.

1. $-b$
2. $b^2 - 4ac$
3. $\sqrt{b^2 - 4ac}$
4. $-b + \sqrt{b^2 - 4ac}$
5. $-b - \sqrt{b^2 - 4ac}$
6. $\frac{-b + \sqrt{b^2 - 4ac}}{2a}$
7. $\frac{-b - \sqrt{b^2 - 4ac}}{2a}$

Warm-up Lesson 9-6

1. Write the Quadratic Formula.
2. Compare what you wrote with the Quadratic Formula given on page 574. If necessary, make changes to what you wrote.
3. Turn your paper over and rewrite the formula from memory.

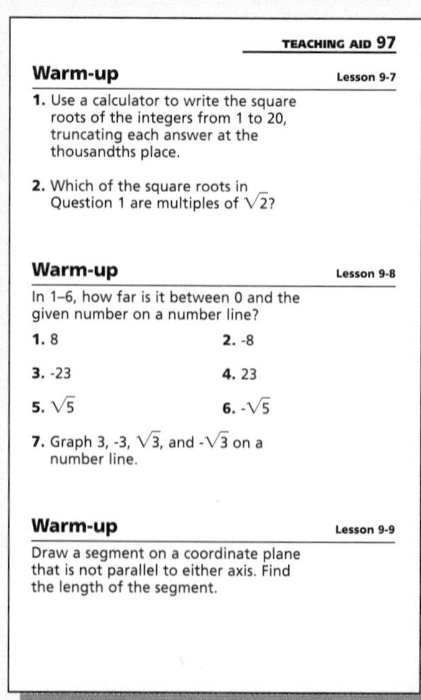

TEACHING AID 97

Warm-up Lesson 9-7

1. Use a calculator to write the square roots of the integers from 1 to 20, truncating each answer at the thousandths place.
2. Which of the square roots in Question 1 are multiples of $\sqrt{2}$?

Warm-up Lesson 9-8

In 1–6, how far is it between 0 and the given number on a number line?

1. 8
2. -8
3. -23
4. 23
5. $\sqrt{5}$
6. $-\sqrt{5}$
7. Graph 3, -3, $\sqrt{3}$, and $-\sqrt{3}$ on a number line.

Warm-up Lesson 9-9

Draw a segment on a coordinate plane that is not parallel to either axis. Find the length of the segment.

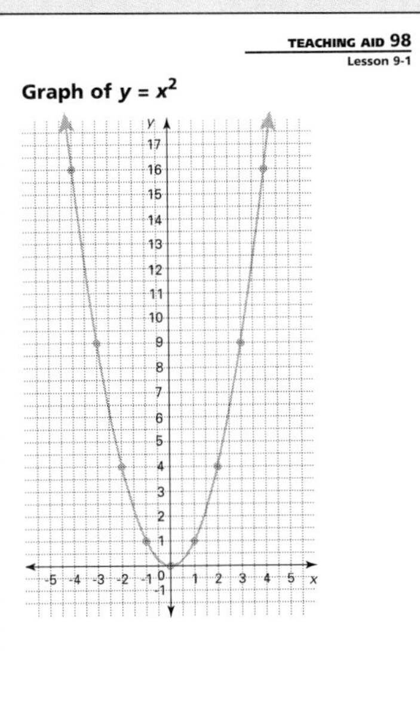

TEACHING AID 98
Lesson 9-1

Graph of $y = x^2$

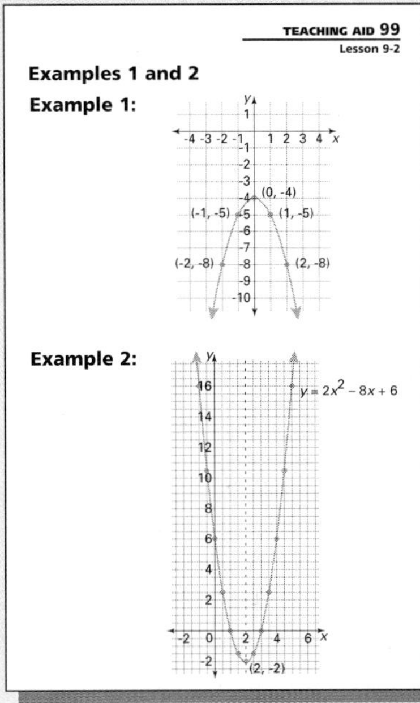

TEACHING AID 99
Lesson 9-2

Examples 1 and 2

Example 1:

Example 2:

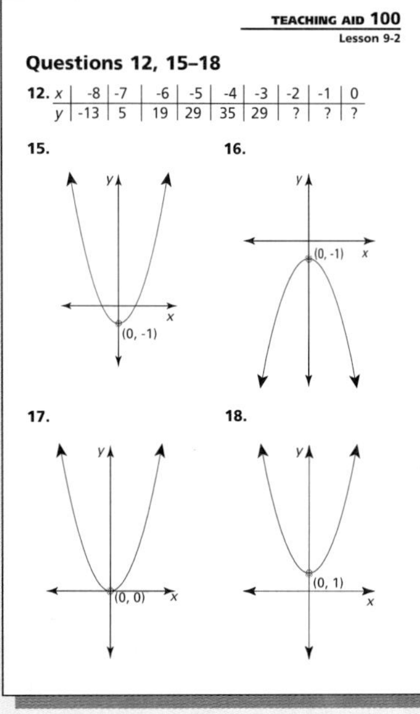

TEACHING AID 100
Lesson 9-2

Questions 12, 15–18

12.

x	-8	-7	-6	-5	-4	-3	-2	-1	0
y	-13	5	19	29	35	29	?	?	?

15.

16.

17.

18.

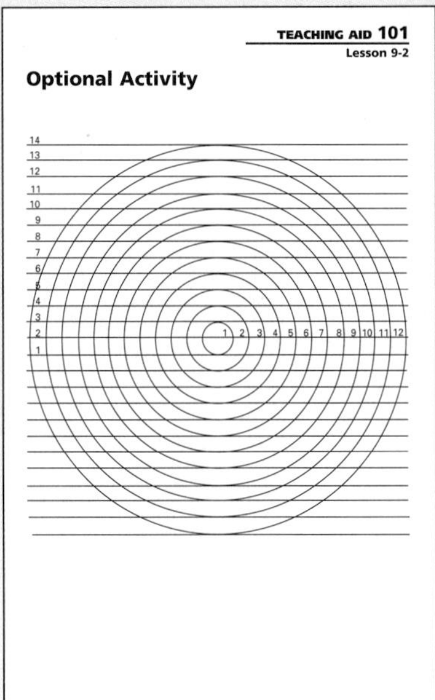

TEACHING AID 101
Lesson 9-2

Optional Activity

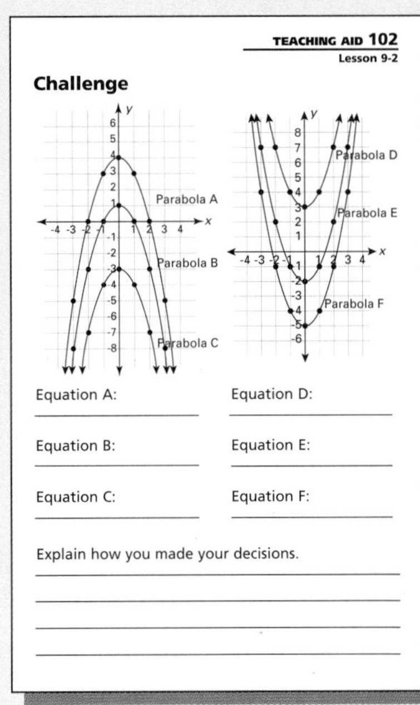

TEACHING AID 102
Lesson 9-2

Challenge

Equation A: ______

Equation B: ______

Equation C: ______

Equation D: ______

Equation E: ______

Equation F: ______

Explain how you made your decisions.

______

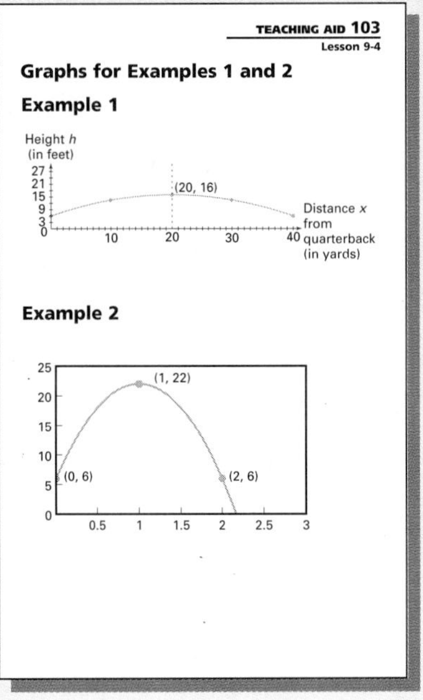

TEACHING AID 103
Lesson 9-4

Graphs for Examples 1 and 2

Example 1

Example 2

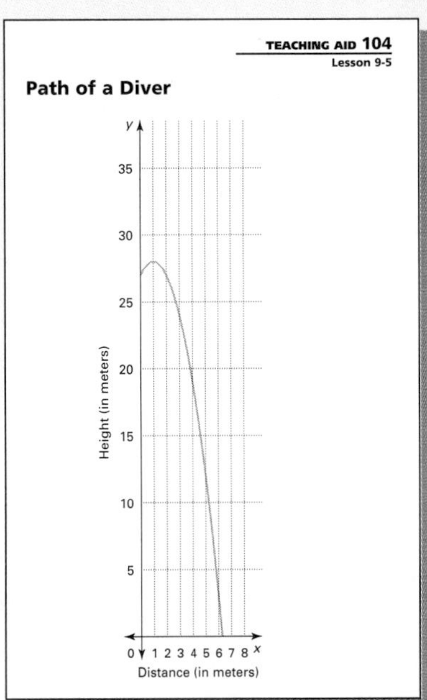

TEACHING AID 104
Lesson 9-5

Path of a Diver

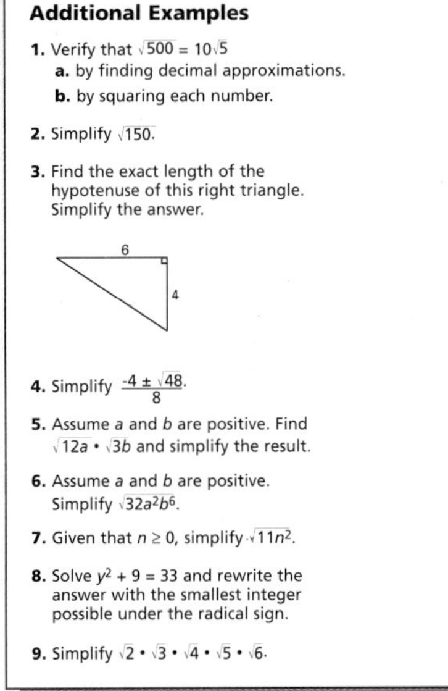

TEACHING AID 105
Lesson 9-7

Additional Examples

1. Verify that $\sqrt{500} = 10\sqrt{5}$
   a. by finding decimal approximations.
   b. by squaring each number.
2. Simplify $\sqrt{150}$.
3. Find the exact length of the hypotenuse of this right triangle. Simplify the answer.

4. Simplify $\frac{-4 \pm \sqrt{48}}{8}$.
5. Assume a and b are positive. Find $\sqrt{12a} \cdot \sqrt{3b}$ and simplify the result.
6. Assume a and b are positive. Simplify $\sqrt{32a^2b^6}$.
7. Given that $n \geq 0$, simplify $\sqrt{11n^2}$.
8. Solve $y^2 + 9 = 33$ and rewrite the answer with the smallest integer possible under the radical sign.
9. Simplify $\sqrt{2} \cdot \sqrt{3} \cdot \sqrt{4} \cdot \sqrt{5} \cdot \sqrt{6}$.

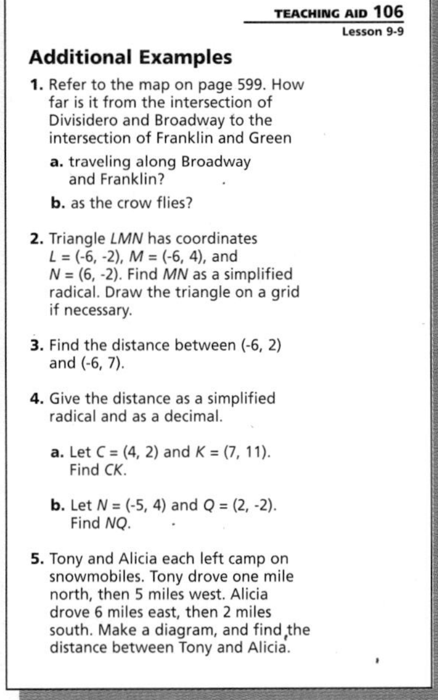

TEACHING AID 106
Lesson 9-9

Additional Examples

1. Refer to the map on page 599. How far is it from the intersection of Divisidero and Broadway to the intersection of Franklin and Green
   a. traveling along Broadway and Franklin?
   b. as the crow flies?
2. Triangle LMN has coordinates $L = (-6, -2)$, $M = (-6, 4)$, and $N = (6, -2)$. Find MN as a simplified radical. Draw the triangle on a grid if necessary.
3. Find the distance between (-6, 2) and (-6, 7).
4. Give the distance as a simplified radical and as a decimal.
   a. Let $C = (4, 2)$ and $K = (7, 11)$. Find CK.
   b. Let $N = (-5, 4)$ and $Q = (2, -2)$. Find NQ.
5. Tony and Alicia each left camp on snowmobiles. Tony drove one mile north, then 5 miles west. Alicia drove 6 miles east, then 2 miles south. Make a diagram, and find the distance between Tony and Alicia.

Chapter Opener

Pacing

With the possible exception of Lesson 9-3, all lessons in this chapter are designed to be covered in one day. At the end of the chapter, you should plan to spend 1 day to review the Progress Self-Test, 1 to 2 days for the Chapter Review, and 1 day for a test. You may want to spend a day on projects, and possibly a day is needed for quizzes. Therefore, this chapter should take 12 to 16 days. We strongly recommend that you not spend more than 17 days on the chapter; there is ample opportunity to review ideas in later chapters.

Using Pages 546–547

Science Connection Galileo's equation $d = 16t^2$ is a special case of the more general equation $d = \frac{1}{2}gt^2 - vt + h$, where the acceleration due to gravity is $g = 32$ feet per second per second (because it is near the surface of the earth), the initial velocity is $v = 0$ (because the object is dropped), and the initial height of the object $h = 0$ (because we are measuring from the point at which the object is dropped).

The unit *feet per second per second* may be easier for students to understand if it is written as (feet per second) per second, or as $\frac{\frac{\text{feet}}{\text{second}}}{\text{second}}$. Either way emphasizes that this acceleration measures the change in velocity per second.

Some students may remember the often-told story of Galileo's dropping objects from the Leaning Tower of Pisa. The story is now thought to be fiction; but suppose it is true. The balcony at the top of the Leaning

CHAPTER 9

546

Chapter 9 Overview

In the United States, the study of quadratic expressions and equations is a traditional topic for first-year algebra courses. Until recently, that study concentrated almost exclusively on solving quadratic equations—first by factoring, then by completing the square, and then by using the quadratic formula. Graphing, the accuracy of which depended on a person's being able to complete the square and solve quadratic equations, was delayed until a later course. The technology of calculators and computers has reversed the order; now graphing is the easiest thing to do, and graphs can be used to picture solutions to quadratic equations.

This chapter has three themes, each of which is covered in three lessons. The first theme, the graphing of parabolas, is presented in Lessons 9-1 through 9-3. The second theme, the solving of quadratic equations, is the subject of Lessons 9-4 through 9-6. The last theme is the study of square roots and their relation to absolute value and distance.

The order in which we have selected the concepts has pedagogical as well as mathematical significance. By covering parabolas first, it is natural to examine the classic application of the path of a projectile. To determine when a projectile reaches a

QUADRATIC EQUATIONS AND SQUARE ROOTS

When an object is dropped from a high place, such as the roof of a building or an airplane, it does not fall at a constant speed. The longer it is in the air, the faster it falls. Furthermore, the distance d that a heavier-than-air object falls in time t does not depend on its weight.

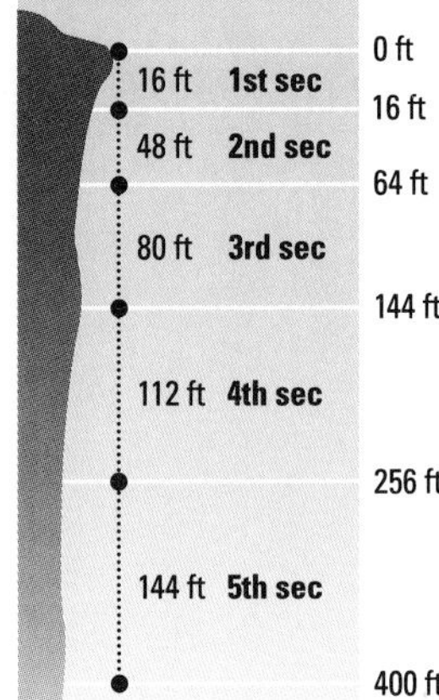

About 400 years ago, the Italian scientist Galileo described the relationship between d and t mathematically. In present-day units, if d is measured in feet and t is in seconds, then

$$d = 16t^2.$$

A table of values and a graph of this equation are shown below.

t (sec)	d (ft)
0	0
1	16
2	64
3	144
4	256
5	400

Total Distance Fallen (d)

Time in Air (t)

The equation $d = 16t^2$ is an example of a *quadratic equation.* The word *quadratic* comes from the Latin word for square. (Notice the t^2.) The points on the graph lie on a curve called a *parabola.* Parabolas and quadratic equations occur often, both in nature and in manufactured objects. A parabola is the shape of the path of a basketball tossed into a hoop, the stream from a water fountain, and the shape of a cross-section of a satellite dish.

In this chapter you will study quadratic equations. Solutions to quadratic equations often involve square roots, so you will also learn more about square roots.

Tower is about 150 feet high. Ask students how long it would take an object dropped from the top of the tower to reach ground level. [A little more than 3 seconds] Ask where it would be after one second [16 feet below the balcony], 2 seconds [64 feet below the balcony], and 3 seconds [only about 6 feet above the ground].

Photo Connections

The photo collage makes real-world connections to the content of the chapter: quadratic equations and square roots.

Fireworks: The path of a projectile, such as a rocket firework, can be graphed as part of a parabola and represented by a quadratic equation.

Solar Collectors: Solar collectors use parabolic reflectors to focus the sun's rays on a target which helps to generate electricity.

Diver: In competitive diving, the diver's parabolic path from the board to the water is scored with attention to take-off, bearing of the body in the air, execution of the prescribed movements, and entry into the water.

Skydivers: The distance of a skydiver's free-fall from the time of the jump until the parachute is opened can be calculated using a quadratic equation.

Automobile: A formula relating the speed x (in mph) of some cars and the stopping distance d (in feet) is the quadratic equation $d = .05x^2 + x$.

Chapter 9 Projects

At this time you might want to have students look over the projects on pages 605–606.

certain height (or when a graph has a certain value), it is necessary to solve a quadratic equation. Factoring is not used because it seldom works! Even with quadratics that have small integer coefficients, only a small percentage (less than 5%) can be factored over the rational numbers. The Quadratic Formula is powerful because it always works and because it is not difficult to learn.

Both calculator and graphing technology lend themselves to the subject matter of this chapter. Unlike lines, the curves of parabolas are very difficult to graph accurately by hand; instead we strongly encourage the use of automatic graphers. Calculators are also a necessity for this chapter. The numbers which arise from actual data are not always amenable to paper-and-pencil calculation. A decimal estimate may have more meaning in the context of the original problem than an exact value written with radicals. If the solutions to a quadratic equation in a motion problem are exactly $1 \pm 2\sqrt{7}$ meters, it is almost always necessary to estimate the solutions as −4.3 meters and 6.3 meters in order to have some idea of what is happening.

Lesson 9-1

Objectives

A Solve quadratic equations of the form $ax^2 = k$.
E Use quadratic equations to solve problems about paths of projectiles.
F Graph equations of the form $y = ax^2$ and interpret these graphs.

Resources

From the ***Teacher's Resource File***
- Lesson Master 9-1A or 9-1B
- Answer Master 9-1
- Teaching Aids
  28 Four-Quadrant Graph Paper
  95 Warm-up
  98 Graph of $y = x^2$
- Activity Kit, Activity 20
- Technology Sourcebook
  Computer Master 22

Additional Resources
- Visuals for Teaching Aids 28, 95, 98
- Flashlight with parabolic reflector
- Spreadsheet software

Teaching Lesson 9-1

Warm-up

1. Graph $y = -2x^2$.

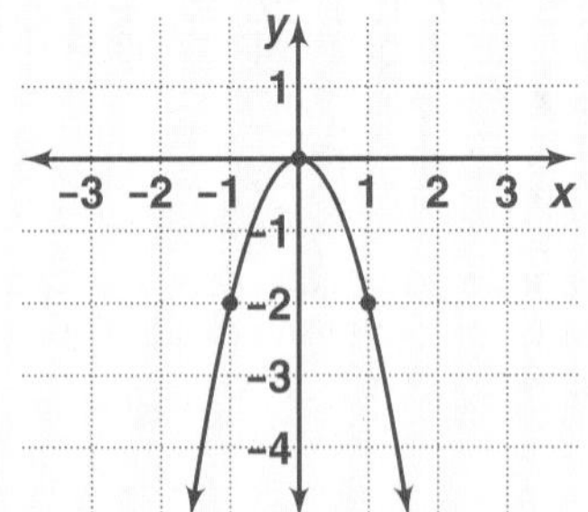

LESSON

9-1

Graphing $y = ax^2$

Shot in slow motion. *This time-lapse photo shows the parabolic path of a basketball as it travels toward the basket.*

Graphing $y = x^2$

The simplest quadratic equation is $y = x^2$. A table of values for $y = x^2$ is given below on the left. Below on the right is a graph of the equation.

x	y
-4	16
-3	9
-2	4
-1	1
0	0
1	1
2	4
3	9
4	16

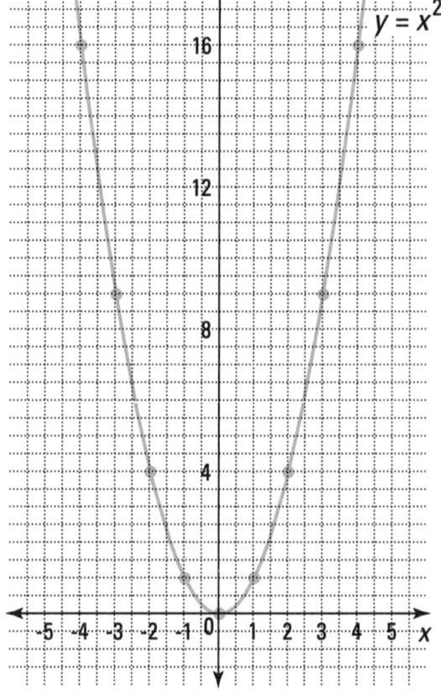

The graph of $y = x^2$ is a *parabola.* Notice that the y-axis separates this parabola into two halves. If you fold the parabola along the y-axis, the two halves coincide. For every point to the left of the y-axis, there is a matching point to the right. This is because the right half is the *reflection* or *mirror image* of the left half. For this reason we say this parabola is **symmetric** to the y-axis. The y-axis is called the **axis of symmetry** of the parabola. Because of this symmetry, every positive number is the y-coordinate for two points on the graph. For instance, 9 is the y-coordinate for both points (3, 9) and (-3, 9).

The intersection of a parabola with its axis of symmetry is called the **vertex** of the parabola. The vertex of the graph of $y = x^2$ is (0, 0).

Lesson 9-1 Overview

Broad Goals In earlier lessons, students worked with graphs of linear and exponential equations. This lesson introduces graphing of the simplest equations whose graphs are parabolas.

Perspective The equations in this lesson are all of the form $y = ax^2$. In all of the examples, the effect of the coefficient a is studied only in relation to its sign. It is also common to say that the value of a determines how "wide" the parabola looks. However, this is only true if the graphs are on the same axes. In UCSMP *Functions, Statistics, and Trigonometry*, we prove that all parabolas are similar by using a size-change transformation. Thus, they all have the same shape and differ only in their size. As automatic graphers so nicely show, the more you zoom in on a parabola, the wider it looks. You can think of it this way: the reason the graph in **Example 2** looks wider than the graph in **Example 1** is that you are seeing less of the entire parabola in **Example 2**. If you saw as much of that parabola as you see of the parabola in **Example 1**, the graph in **Example 2** would look bigger but not wider.

This lesson flows nicely into the next lesson; even if you do not finish this lesson in a single day, you may be able to catch up on the next day.

Graphing $y = ax^2$

The equation $y = x^2$ is of the form $y = ax^2$, with $a = 1$. You should be able to draw a graph of any equation of this form.

Example 1

a. Graph $y = 2x^2$ by plotting points where x goes from -2.5 to 2.5 in intervals of 0.5.

b. Does the graph have an axis of symmetry? If so, what is an equation for the axis of symmetry?

Solution

a. Make a table of values. When evaluating $2x^2$, remember the order of operations. Square x before multiplying by 2. So when $x = -2.5$, $y = 2(-2.5)^2 = 2(6.25) = 12.5$. A table of values and graph are shown below.

x	y
-2.5	12.5
-2	8
-1.5	4.5
-1	2
-0.5	0.5
0	0
0.5	0.5
1	2
1.5	4.5
2	8
2.5	12.5

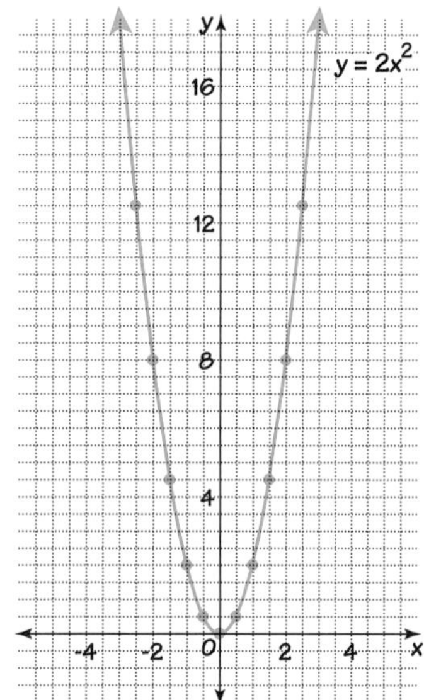

b. Examine the graph. Notice that for each point on the graph in the first quadrant, there is a corresponding point the same distance from the y-axis in the second quadrant and having the same y-coordinate. **Yes, the graph has an axis of symmetry. It is the y-axis, the line with equation x = 0.**

Equations of the form $y = ax^2$ all yield parabolas whose axis of symmetry is the y-axis. When $a > 0$ the parabola *opens up,* that is, like a cup or bowl standing upright. When $a < 0$, the graph *opens down,* like an upside-down bowl. This can be seen in Example 2.

2. Compare the graph you drew with the graph of $y = 2x^2$ in **Example 1** on page 549. What do you notice about the graphs? **They are reflection images of each other.**

3. Compare the graph you drew with the graph shown in **Example 2** on page 550. What do you notice? **Both graphs open down.**

Notes on Reading

Students who have studied *Transition Mathematics* will have seen symmetry and related it to reflections, so you can expect this language to be review.

Students will need graph paper or **Teaching Aid 28** for this lesson. The graph of $y = x^2$ is on **Teaching Aid 98**, so you can easily display it when discussing the lesson. Focus on some essential characteristics of the parabola. You might plot points close to (0, 0) to show that the parabola is really quite flat near the vertex. Point out the symmetry of a parabola to its axis, and use the idea of symmetry to show how to find points on the parabola. Make sure students understand that the sign of the coefficient a determines whether the parabola opens up or down. Emphasize that students should make smooth curves when they draw these graphs.

Optional Activities

Activity 1

You can use *Activity Kit, Activity 20,* to introduce the lesson. In this activity students generated data which when graphed approximate a parabola of the form $y = ax^2$.

Activity 2

The chapter opener mentions some real-life situations in which parabolas may occur. Ask students to suggest other situations. [Samples: Various antennas for receiving radio or TV signals have a parabolic cross section; archways and arched supports in architecture are often parabolic; the cross sections of some contact lenses are also parabolic.]

Activity 3 Technology Connection

You may wish to assign *Technology Sourcebook, Computer Master 22.* Students use a spreadsheet or a BASIC program to construct tables that give points on parabolas.

Additional Examples

1. Graph $y = 1.5x^2$.

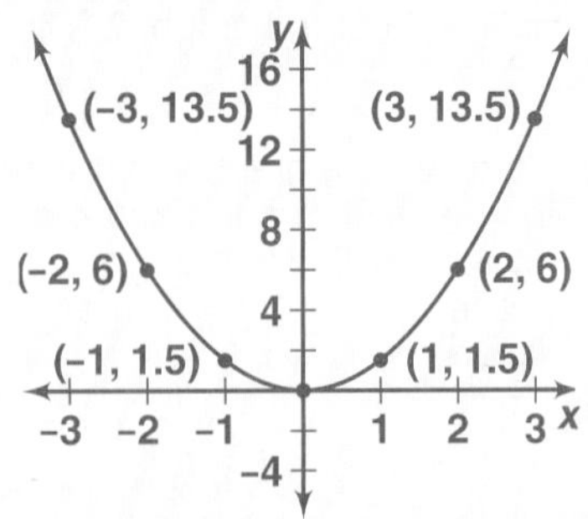

2. a. Graph $y = -1.5x^2$. **The graph is the reflection image of the graph for $y = 1.5x^2$ over the x-axis.**
   b. In what quadrant does this graph have points? **Quadrants 3 and 4**
   c. What are the coordinates of the vertex of this parabola? **(0, 0)**
3. Use your graph in Question 1 above. At what points does the graph of $y = 7$ intersect the graph of $y = 1.5x^2$? **Solve $1.5x^2 = 7$ to get $x = \sqrt{\frac{7}{1.5}} \approx 2.16$ or $x = -\sqrt{\frac{7}{1.5}} \approx -2.16$.**

You might ask students to check their answer to Question 3 by graphing $y = 7$ and looking at the points of intersection of the line and the parabola.

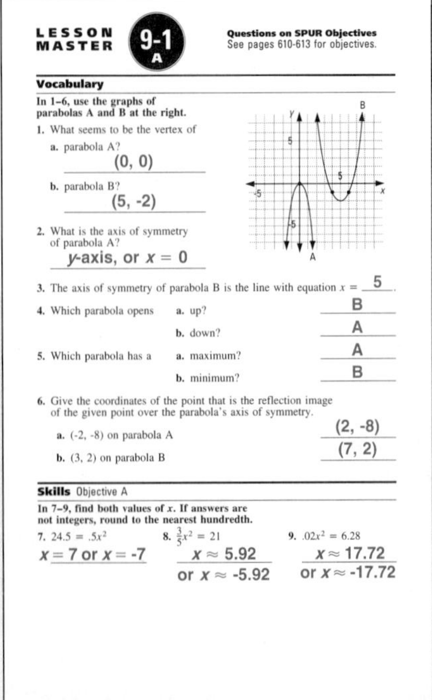

LESSON MASTER 9-1 A

Questions on SPUR Objectives
See pages 610-613 for objectives.

Vocabulary

In 1–6, use the graphs of parabolas A and B at the right.

1. What seems to be the vertex of
   a. parabola A? **(0, 0)**
   b. parabola B? **(5, -2)**
2. What is the axis of symmetry of parabola A? **y-axis, or $x = 0$**
3. The axis of symmetry of parabola B is the line with equation $x =$ **5**.
4. Which parabola opens a. up? **B** b. down? **A**
5. Which parabola has a a. maximum? **A** b. minimum? **B**
6. Give the coordinates of the point that is the reflection image of the given point over the parabola's axis of symmetry.
   a. (-2, -8) on parabola A **(2, -8)**
   b. (3, 2) on parabola B **(7, 2)**

Skills Objective A

In 7–9, find both values of x. If answers are not integers, round to the nearest hundredth.

7. $24.5 = .5x^2$ **$x = 7$ or $x = -7$**
8. $\frac{3}{5}x^2 = 21$ **$x \approx 5.92$ or $x \approx -5.92$**
9. $.02x^2 = 6.28$ **$x \approx 17.72$ or $x \approx -17.72$**

Example 2

a. Graph $y = -\frac{1}{5}x^2$.
b. In which quadrants does the graph have points?
c. What are the coordinates of the vertex of the parabola?

Solution

a. Make a table of values. Below we use $-5 \le x \le 5$. Plot the points and then connect them with a smooth curve.

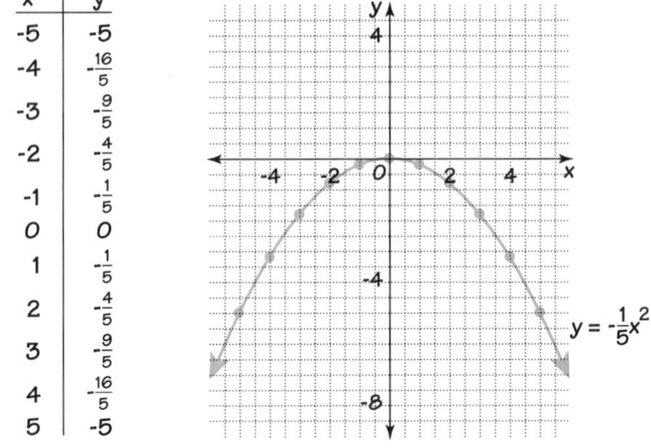

x	y
-5	-5
-4	$-\frac{16}{5}$
-3	$-\frac{9}{5}$
-2	$-\frac{4}{5}$
-1	$-\frac{1}{5}$
0	0
1	$-\frac{1}{5}$
2	$-\frac{4}{5}$
3	$-\frac{9}{5}$
4	$-\frac{16}{5}$
5	-5

b. Except when $x = 0$, $-\frac{1}{5}x^2$ is negative. So except for the origin, the parabola lies below the x-axis. It has points in the 3rd and 4th quadrants.
c. The vertex is (0, 0).

Notice that if the parabola opens up, the vertex is the lowest point on the graph or the **minimum.** If the parabola opens down, the vertex is the highest point or **maximum.** Below is a summary of properties of the graph of $y = ax^2$.

> The graph of $y = ax^2$, where $a \ne 0$, has the following properties:
> 1. It is a parabola symmetric to the y-axis.
> 2. Its vertex is (0, 0).
> 3. If $a > 0$, the parabola opens up.
>   If $a < 0$, the parabola opens down.

Finding Points on the Graph of $y = ax^2$

In Examples 1 and 2, points were found on the graph $y = ax^2$ by first choosing values of x and then finding values of y. You should also be able to work in the reverse order. That is, if you know the y-coordinate of a point on the graph of a parabola, you should be able to find the x-coordinate or coordinates.

Adapting to Individual Needs

Extra Help

Be sure students understand that a parabola has either a maximum or a minimum, not both. In **Example 1,** point out that the vertex is the lowest point, or minimum. The sides of the parabola have arrows designating that the graph continues on and on; there is no highest point, or maximum. In **Example 2** the vertex is the highest point, or maximum; this graph continues on and on downward; there is no minimum.

Example 3

Find the x-coordinates of the points on the graph of $y = -\frac{1}{5}x^2$ when $y = -3$.

Solution

Substitute $y = -3$ in the equation $y = -\frac{1}{5}x^2$ and solve for x.

$$-3 = -\frac{1}{5}x^2$$

$$-5(-3) = -5\left(-\frac{1}{5}x^2\right) \quad \text{Multiply each side by -5, the reciprocal of } -\frac{1}{5}.$$

So,

$$15 = x^2$$

$$x = \sqrt{15} \text{ or } x = -\sqrt{15}.$$

$$x \approx 3.87 \text{ or } x \approx -3.87.$$

Check

Draw a graph of $y = -3$ on the same axes as used in Example 2.

Notice that $y = -\frac{1}{5}x^2$ and $y = -3$ intersect at two points, labeled P and Q at the right. The x-coordinate of P is about -3.9; the x-coordinate of Q is about 3.9. It checks.

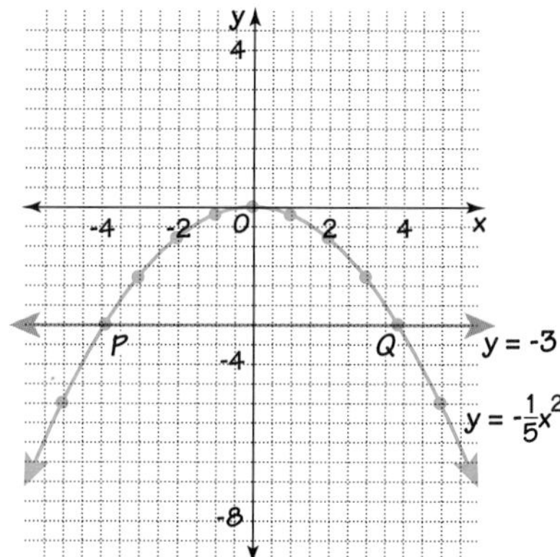

The curved path of the water from the mouth of the triceratops is part of a parabola.

QUESTIONS

Covering the Reading

1. a. Give an example of a quadratic equation found in this lesson.
   b. Give an example of a quadratic equation not found in this lesson.
   a) Sample: $y = x^2$; b) Sample: $y = 7x^2$
2. What is the origin of the word *quadratic?*
   The word *quadratic* comes from the Latin word for square.
3. What is the shape of the graph of every quadratic equation in this lesson? parabola
4. Name two instances in nature or in manufactured objects where parabolas occur. Samples: path of a basketball shot at a hoop, path of a stream of water from a fountain

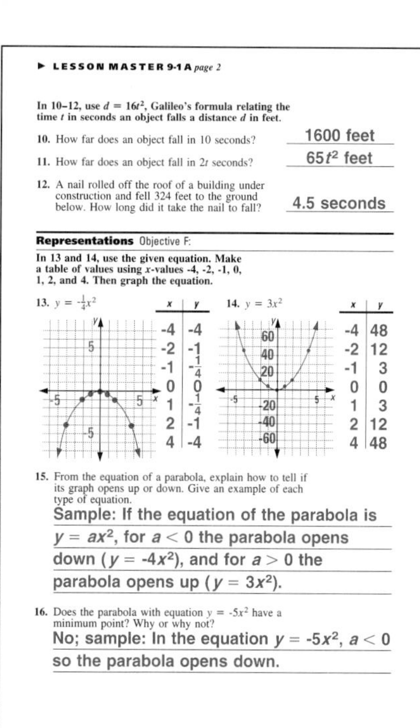

▶ **LESSON MASTER 9-1 A** *page 2*

In 10–12, use $d = 16t^2$, Galileo's formula relating the time t in seconds an object falls a distance d in feet.

10. How far does an object fall in 10 seconds? 1600 feet
11. How far does an object fall in $2t$ seconds? $65t^2$ feet
12. A nail rolled off the roof of a building under construction and fell 324 feet to the ground below. How long did it take the nail to fall? 4.5 seconds

Representations Objective F:

In 13 and 14, use the given equation. Make a table of values using x-values -4, -2, -1, 0, 1, 2, and 4. Then graph the equation.

13. $y = -\frac{1}{4}x^2$

x	y
-4	-4
-2	-1
-1	$-\frac{1}{4}$
0	0
1	$-\frac{1}{4}$
2	-1
4	-4

14. $y = 3x^2$

x	y
-4	48
-2	12
-1	3
0	0
1	3
2	12
4	48

15. From the equation of a parabola, explain how to tell if its graph opens up or down. Give an example of each type of equation.
    Sample: If the equation of the parabola is $y = ax^2$, for $a < 0$ the parabola opens down ($y = -4x^2$), and for $a > 0$ the parabola opens up ($y = 3x^2$).
16. Does the parabola with equation $y = -5x^2$ have a minimum point? Why or why not?
    No; sample: In the equation $y = -5x^2$, $a < 0$ so the parabola opens down.

Adapting to Individual Needs

English Language Development

You might want to use physical examples to define *maximum, minimum, vertex, parabola,* and *symmetric.* For example, a purse strap hanging on a person's shoulder suggests a parabola opening down, and the vertex is the part of the strap on the top of the shoulder. Since the vertex is the highest point, it is the maximum. Holding the purse so that the strap hangs down suggests a parabola opening up. The vertex is the lowest part of the strap and it is a minimum. You might demonstrate symmetry by folding a piece of paper and cutting out a symmetric shape.

Notes on Questions

Question 16 Error Alert Some students might draw a graph that is identical to the graph of $y = x^2$. Ask these students if the length of the side of a square could be negative.

Question 26 This kind of substitution is done in the next lesson. You could use it to introduce Lesson 9-2.

Question 27 Cooperative Learning This exploration question is suitable for **small-group work**. Good-quality flashlights have parabolic reflectors, but some inexpensive models may not.

Additional Answers

5a.

x	y
-4	8
-3	4.5
-2	2
-1	0.5
0	0
1	0.5
2	2
3	4.5
4	8

5b.

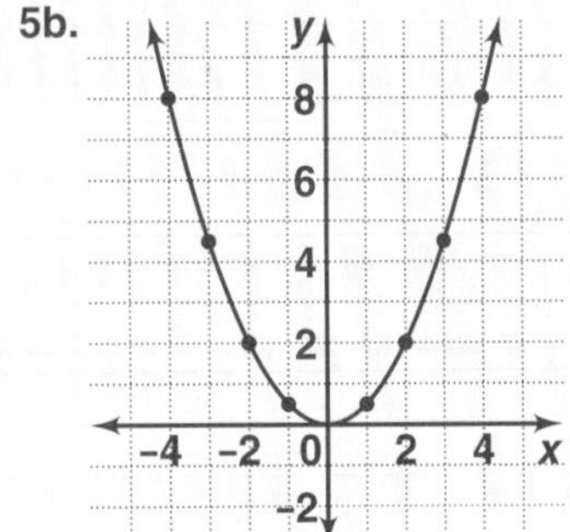

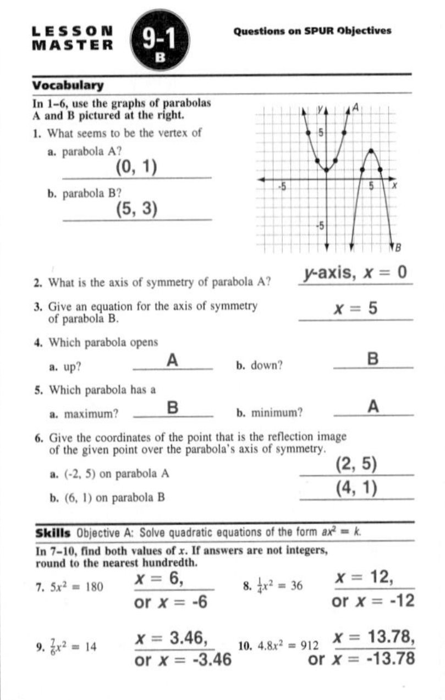

LESSON MASTER 9-1 B Questions on SPUR Objectives

Vocabulary

In 1–6, use the graphs of parabolas A and B pictured at the right.

1. What seems to be the vertex of
   a. parabola A? (0, 1)
   b. parabola B? (5, 3)
2. What is the axis of symmetry of parabola A? y-axis, $x = 0$
3. Give an equation for the axis of symmetry of parabola B. $x = 5$
4. Which parabola opens
   a. up? A b. down? B
5. Which parabola has a
   a. maximum? B b. minimum? A
6. Give the coordinates of the point that is the reflection image of the given point over the parabola's axis of symmetry.
   a. (-2, 5) on parabola A (2, 5)
   b. (6, 1) on parabola B (4, 1)

Skills Objective A: Solve quadratic equations of the form $ax^2 = k$.

In 7–10, find both values of x. If answers are not integers, round to the nearest hundredth.

7. $5x^2 = 180$ $x = 6$, or $x = -6$
8. $\frac{1}{4}x^2 = 36$ $x = 12$, or $x = -12$
9. $\frac{7}{6}x^2 = 14$ $x = 3.46$, or $x = -3.46$
10. $4.8x^2 = 912$ $x = 13.78$, or $x = -13.78$

5c) a parabola which opens up, and whose axis of symmetry is x = 0 and vertex is (0, 0)

6c) a parabola which opens down, and whose axis of symmetry is x = 0 and vertex is (0, 0)

10b)

x	y
-2	12
-1.5	6.75
-1	3
-0.5	0.75
0	0
0.5	0.75
1	3
1.5	6.75
2	12

c)

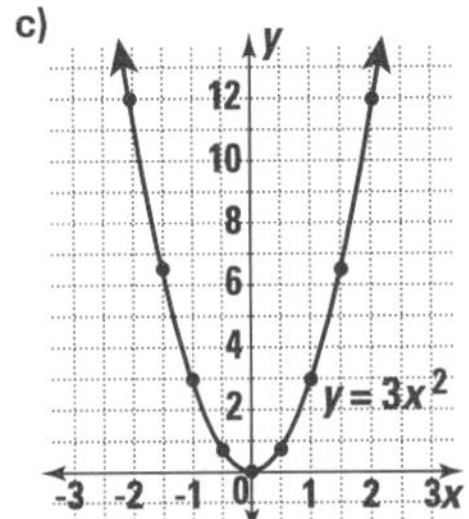

In 5 and 6, an equation is given.

a. Make a table of x- and y-values for the x-values -4, -3, -2, -1, 0, 1, 2, 3, and 4. **See margin.**

b. Graph the equation. **See margin.**

c. Describe the graph. **See left.**

5. $y = \frac{1}{2}x^2$ **6.** $y = -\frac{1}{2}x^2$

In 7–9, consider the graph of the equation $y = ax^2$, when $a \neq 0$.

7. What is the vertex of this parabola? **(0, 0)**

8. What is an equation for its axis of symmetry? **x = 0**

9. a. If a is positive, the parabola opens __?__. **up**
b. If a is negative, the parabola opens __?__. **down**

10. Think about the equation $y = 3x^2$.
a. Without plotting any points, sketch what you think the graph will look like. **Graphs will vary.**
b. Make a table of values satisfying this equation. Use x = -2, -1.5, -1, -0.5, 0, 0.5, 1, 1.5, 2. **See left.**
c. Draw a graph of this equation. **See left.**
d. Tell if the vertex is a minimum or maximum. **minimum**
e. For what values of x does $y = 12$? **2 and -2**
f. Find the x-coordinates of the points where $y = 15$. **$\sqrt{5}$ and $-\sqrt{5}$**

11. Refer to the parabola at the right.
a. Points A and B are symmetric to the y-axis. What are the coordinates of B? **(6, 10)**
b. Does this parabola open up or open down? **up**
c. Does this parabola have a maximum or does it have a minimum? **minimum**

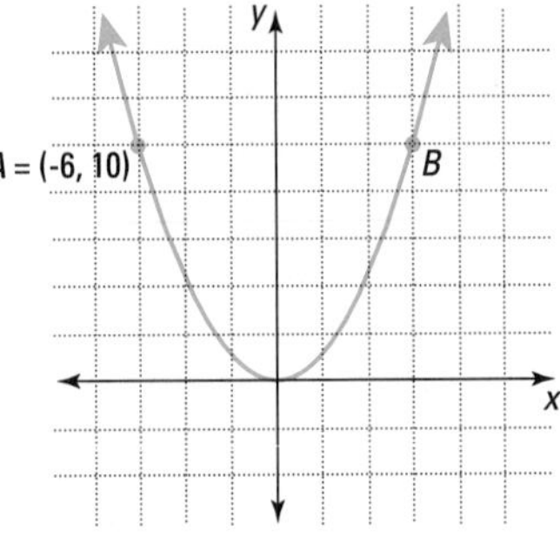

In 12 and 13, refer to the table, graph, and equation on page 547.

12. How far does an object fall in the first 3 seconds? **144 ft**

13. How far does an object fall in the first 2.5 seconds? **100 ft**

Additional Answers, continued

6a.

x	y
-4	-8
-3	-4.5
-2	-2
-1	-0.5
0	0
1	-0.5
2	-2
3	-4.5
4	-8

6b.

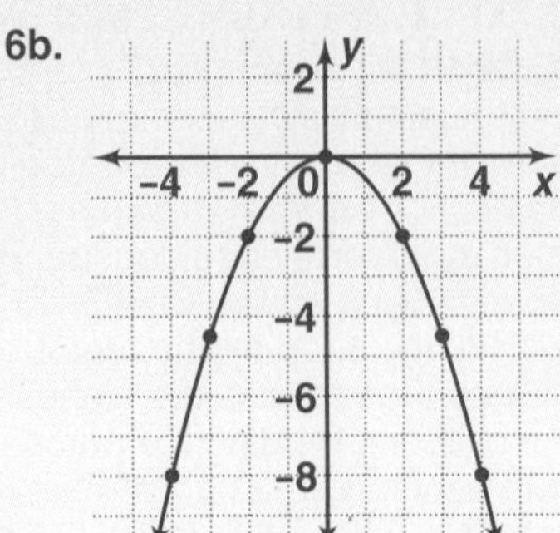

16a)

b) The right half of the graph is identical. There cannot be any negative values for the length of a side of a square, so the graph of $A = s^2$ lies entirely in the first quadrant.

22a) Sample: Adam bought 21 comic books for a total of $16. If each comic book has the same price, how much more will he have to pay to add 50 more comic books to his collection?

Enlightened science. *Scientists use the math principles of this lesson to produce the reflectors in flashlights, headlights, and searchlights.*

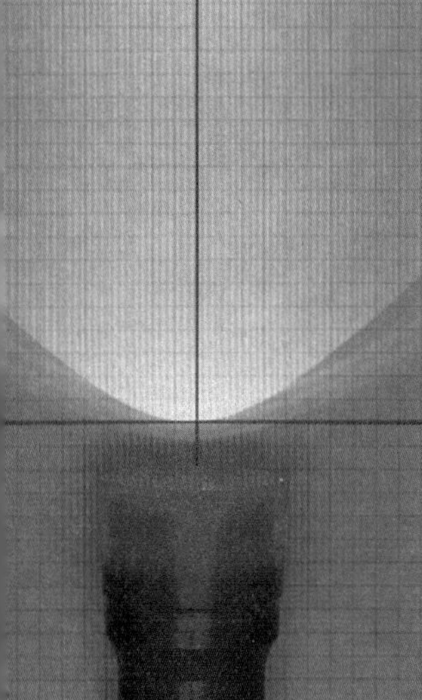

Applying the Mathematics

In 14 and 15, suppose a stone falls from a cliff.

14. How long does it take the stone to fall the first 200 feet? about 3.54 seconds

15. How long does it take the stone to fall the next 200 feet? about 1.46 seconds

16. Consider the formula $A = s^2$ for the area A of a square with a side of length s. See left.
  a. Graph all possible values of s and A on a coordinate plane.
  b. Explain how the graph in part **a** is like and unlike the graph of $y = x^2$ at the beginning of this lesson.

In 17 and 18, solve.

17. $16 = \frac{1}{9}(x + 1)^2$ 11; -13 **18.** $\pi r^2 = 40$ 3.568; -3.568

Review

In 19–21, simplify. *(Lessons 8-5, 8-7, 8-8)*

19. $x^4 \cdot y \cdot x^3 \cdot y^3$ x^7y^4 **20.** $\frac{9a^9}{6a^6}$ $\frac{3}{2}a^3$ **21.** $\left(\frac{2a}{5}\right)^3$ $\frac{8a^3}{125}$

22. **a.** Make up a real question that can be answered by solving $\frac{16}{21} = \frac{n}{50}$.
  b. Answer your question. *(Lesson 6-8)* $38.10
  a) See left.

23. Julio lives 3 km from Martina and 8 km from Natasha. From this information, what can you conclude about the distance between Martina's home and Natasha's home? Justify your answer. *(Lesson 4-8)* It is at least 5 km and at most 11 km since 8 – 3 = 5 and 8 + 3 = 11.

24. *Multiple choice.* The graph below pictures the solutions to which inequality? *(Lessons 2-7, 4-3)* c

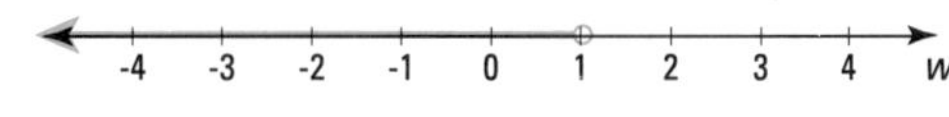

(a) $w + 6 < 5$ (b) $w - 6 < 5$
(c) $w + 5 < 6$ (d) $w - 5 < 6$

25. A house that covers 1800 square feet of ground is being built on a 7500-square-foot lot. The driveway will cover another k square feet. Write an expression for the area that is left for the lawn. *(Lesson 4-2)* 7500 – 1800 – k or 5700 – k

26. If $d = x + 0.05x^2$, find d when $x = 15$. *(Lesson 1-4)* 26.25

Exploration

27. Draw a set of axes on graph paper. Hold a lit flashlight at the origin so the light is centered on an axis as shown. What is the shape of the lighted area? Keep the lit end of the flashlight in the same position and tilt the flashlight to raise the other end. How does the shape of the lighted area change? Parabola; the parabola becomes more narrow.

Follow-up for Lesson 9-1

Practice

For more questions on SPUR Objectives, use **Lesson Master 9-1A** (shown on pages 550–551) or **Lesson Master 9-1B** (shown on pages 552–553).

Assessment

Oral/Written Communication Ask students to explain how they can tell if a parabola opens up or opens down. Then have them write equations for two different parabolas that open up and two that open down. [Students understand that the value of a in $y = ax^2$ determines if a parabola opens up or down.]

Extension

As an extension of **Question 16**, ask students to find other common formulas that have equations of the form $y = ax^2$. [Sample: Area of a circle: $A = \pi r^2$; Surface area of a cube: $A = 6s^2$; Area of an equilateral triangle: $A = \frac{\sqrt{3}}{4}s^2$]

Project Update Project 1, *Interview with Galileo*, Project 3, *Parabolas from Folding Paper*, and Project 6, *Parabolas in Manufactured Objects*, on pages 605–606, relate to the content of this lesson.

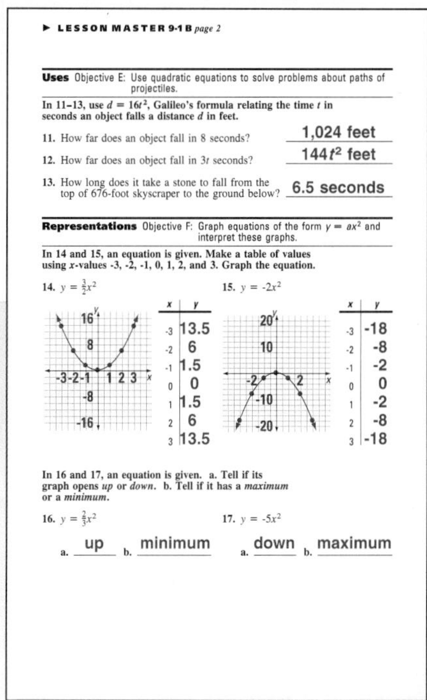

▸ LESSON MASTER 9-1 B *page 2*

Uses Objective E: Use quadratic equations to solve problems about paths of projectiles.

In 11–13, use $d = 16t^2$, Galileo's formula relating the time t in seconds an object falls a distance d in feet.

11. How far does an object fall in 8 seconds? 1,024 feet

12. How far does an object fall in 3t seconds? $144t^2$ feet

13. How long does it take a stone to fall from the top of 676-foot skyscraper to the ground below? 6.5 seconds

Representations Objective F: Graph equations of the form $y = ax^2$ and interpret these graphs.

In 14 and 15, an equation is given. Make a table of values using x-values -3, -2, -1, 0, 1, 2, and 3. Graph the equation.

14. $y = \frac{3}{2}x^2$

x	y
-3	13.5
-2	6
-1	1.5
0	0
1	1.5
2	6
3	13.5

15. $y = -2x^2$

x	y
-3	-18
-2	-8
-1	-2
0	0
1	-2
2	-8
3	-18

In 16 and 17, an equation is given. a. Tell if its graph opens *up* or *down*. b. Tell if it has a *maximum* or a *minimum*.

16. $y = \frac{2}{3}x^2$ a. up b. minimum

17. $y = -5x^2$ a. down b. maximum

Adapting to Individual Needs

Challenge

Materials: Graph paper or **Teaching Aid 28**

Have students graph $y = (x - 4)^2$ using integer x-values from -1 to 8. Then, on the same coordinate plane, have them graph $y = (x + 4)^2$ using integer x-values from -7 to 1. Ask them to make a general statement about the location of the vertex for any equation of the form $y = (x - a)^2$. [The vertex is at the point $(a, 0)$.]

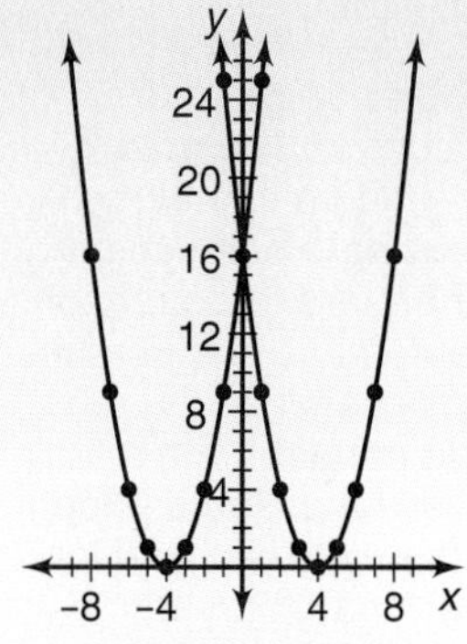

Lesson 9-2

Objectives

F Graph equations of the form $y = ax^2 + bx + c$ and interpret these graphs.

Resources

From the ***Teacher's Resource File***
- Lesson Master 9-2A or 9-2B
- Answer Master 9-2
- Teaching Aids
  - 28 Four-Quadrant Graph Paper
  - 37 Spreadsheet
  - 95 Warm-up
  - 99 Examples 1 and 2
  - 100 Questions 12, 15–18
  - 101 Optional Activity
  - 102 Challenge

Additional Resources
- Visuals for Teaching Aids 28, 37, 95, 99–102
- Automatic graphers
- Spreadsheet software

Teaching Lesson 9-2

Warm-up

Writing Without graphing, write a short paragraph describing the graph of the equation $y = -3x^2$. **Sample: The graph is a parabola that opens down. It is symmetric to the y-axis. Its vertex is (0, 0).**

Notes on Reading

Consumer Connection Students might be interested in the 4-second rule that enables a driver to approximate the stopping distance under ideal conditions at any speed. Use the following steps: (1) Choose a

LESSON

9-2 Graphing $y = ax^2 + bx + c$

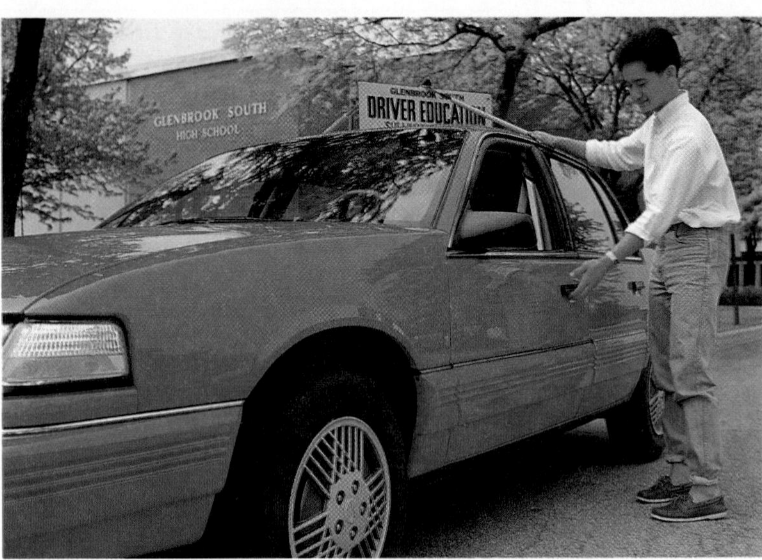

Keep a safe distance. *Driver education classes teach defensive driving techniques, such as estimating stopping distances and safe following distances at various speeds.*

How Much Distance Is Needed to Stop a Car?

When a driver decides to stop a car, it takes time to react and put the foot on the brake. Then it takes time for the car to slow down. The total distance traveled during this time is called the **stopping distance** of the car. Of course, the faster a car is traveling, the longer its stopping distance will be. A formula relating the speed x (in mph) of some cars and the stopping distance d (in feet) is

$$d = .05x^2 + x.$$

This formula is used by those who study automobile performance and safety. It is also important for determining the distance that should be maintained between a car and the car in front of it.

To find the distance needed to stop a car traveling 50 mph, you can substitute $x = 50$ in the above equation.

$$\begin{aligned} d &= .05(50)^2 + 50 \\ &= .05(2500) + 50 \\ &= 125 + 50 \\ &= 175 \end{aligned}$$

Thus, a car traveling 50 mph takes about 175 feet to come to a complete stop after the driver decides to stop.

At the top of page 555 is a table of values and a graph for the stopping distance formula. The situation makes no sense for negative values of x or d, so the graph has points in the first quadrant only. Notice that the graph is parabolic in shape.

Lesson 9-2 Overview

Broad Goals Continuing the ideas in Lesson 9-1, this lesson discusses the graphs of equations of the form $y = ax^2 + bx + c$. Students work from tables, from general properties of the equation, and from computer programs.

Perspective There are two parts to this lesson. The first part involves obtaining the graph of an equation of the form $y = ax^2 + bx + c$. This is done by hand in **Example 1.** It is done with a computer program to generate a table of values in **Example 2.** Another way to graph—and one that more and more people are using—is to use an automatic grapher. There are three reasons for not emphasizing automatic graphers in this lesson: they are discussed in the next lesson, they usually do not generate a table of values, and we believe that some paper-and-pencil work to plot points is useful for appreciating the power of a grapher.

The second part of the lesson involves interpreting the graph of a quadratic equation. We want students to be able to identify the axis of symmetry, the vertex, and the x- and y-intercepts by looking at the graph. As shown in **Example 2,** students should also recognize the symmetry line and the other key points by looking at a table of values. **Teaching Aid 99** can help in the discussion.

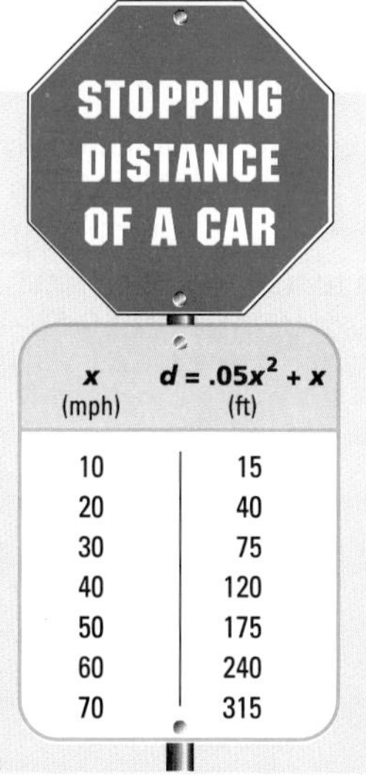

x (mph)	$d = .05x^2 + x$ (ft)
10	15
20	40
30	75
40	120
50	175
60	240
70	315

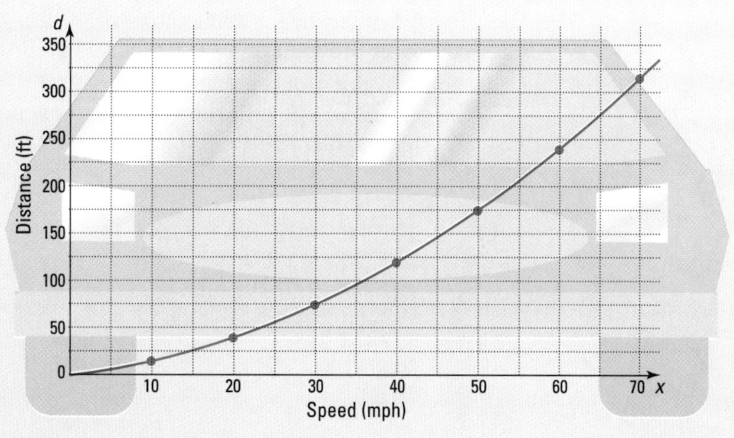

Properties of the Graph of $y = ax^2 + bx + c$

The equation $d = .05x^2 + x$ is of the form $y = ax^2 + bx + c$, with $a = .05$, $b = 1$, and $c = 0$. Many people are surprised to learn that the graph of every equation of the form $y = ax^2 + bx + c$ (with $a \neq 0$) is a parabola. The values of *a, b,* and *c* determine where the parabola is positioned in the plane, and whether it opens up or down.

The equations you studied in Lesson 9-1 are also of the form $y = ax^2 + bx + c$. As with those equations, if $a > 0$ the parabola opens up. If $a < 0$, the parabola opens down.

Example 1

a. Graph $y = -x^2 - 4$.
b. Tell whether the parabola opens up or down.
c. Identify its axis of symmetry and vertex.

Solution

a. Form a table of values and plot. Recall that $-x^2$ means $-1(x^2)$.

x	y
-2	-8
-1	-5
0	-4
1	-5
2	-8

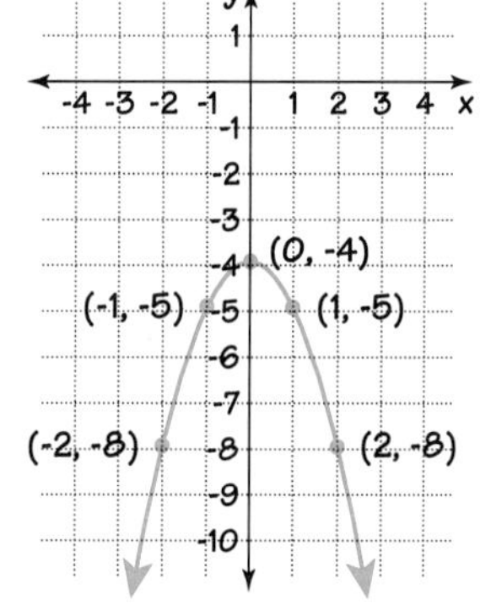

b. The parabola opens down. You can see this from the graph. You could also have predicted this from the equation, because the coefficient of x^2 is negative.
c. Examine the graph. The y-axis is the axis of symmetry. The vertex is (0, -4).

fixed point ahead where you think you could stop your vehicle. (2) Count off 4 seconds by counting from one thousand one to one thousand four. (3) Check your vehicle's position. If you have just reached the fixed point, assume the distance you estimated was the approximate distance it would have taken you to stop the vehicle. You might have students practice estimating stopping distances at various speeds while they are passengers in vehicles and report to the class about it.

It is critical to connect each equation with its graph. To do this, it is helpful to have access to an automatic grapher with a projection device. If you do not have such a device, graph the parabolas in **Examples 1 and 2** in advance, or use **Teaching Aid 99.** Then show students some of the indicated points on the graph, and also show a few additional points. For instance, you can extend the graph of **Example 1** to show that (3, –13) and (-3, –13) are on the graph. Going the other way, you can begin with the graph in **Example 2**, and show that (5, 16) satisfies the equation.

These graphs have two aspects that are not obvious. First, for a particular pair of axes, the size and direction of the opening of the graphs are determined only by *a*; the graph of $y = 2x^2 - 8x + 6$ is congruent to the graph of $y = 2x^2$. Second, all parabolas get wider and wider as one goes away from the vertex. (Parabolas do not have asymptotes.) Consider, for instance, the parabola of **Example 2**; ask students where it intersects the line $x = 100$. [At (100, 19,206)]

Optional Activities

Materials: **Teaching Aid 101**

After students complete the lesson, they might enjoy drawing the parabola described in this activity. Give each student **Teaching Aid 101** and point out that the lines are parallel and that the concentric circles are drawn so that the radius of each succeeding circle is one unit greater than that of the preceding one. Have students **work in pairs.** Give them these directions and questions:

1. Mark the point that is on both line 1 and circle 1.
2. Mark the two points that are on both line 2 and circle 2.
3. Continue the process by marking the two points on line 3 and circle 3, and so on with all the lines and circles.
4. Draw a smooth curve through the points.
5. What is the curve called? [A parabola]
6. Describe the line of symmetry. [The line of symmetry goes through the center of the circles and is perpendicular to the parallel lines.]

Students might extend the activity by using different colors to draw other parabolas. For example, they might begin with the same point on line 1 and circle 1, but then mark points on both line 2 and circle 3, line 3 and circle 4, and so on.

❶ For each run of the program, line 40 must be changed to reflect the different coefficients, and line 30 must be changed to accommodate different choices of x-values. Here is a program that emphasizes the coefficients a, b, and c in the Quadratic Formula.

```
10 PRINT "GIVE THE
   COEFFICIENTS A, B, C OF
   A QUADRATIC EQUATION"
12 INPUT A, B, C
15 PRINT "GIVE THE STARTING
   AND ENDING VALUES FOR X"
18 INPUT ST, ED
20 PRINT "GIVE THE INCREMENT
   FOR X"
22 INPUT IX
25 PRINT "X", "Y"
30 FOR X = ST TO ED STEP IX
40 LET Y = A*X*X + B*X + C
```

The remainder of the program is the same one that is given in the text.

Additional Examples

Students will need graph paper or **Teaching Aid 28.**

1. a. Graph $y = x^2 - 1$ for $-3 \le x \le 3$.

Sample:

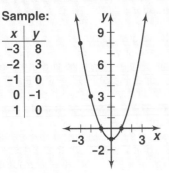

x	y
-3	8
-2	3
-1	0
0	-1
1	0

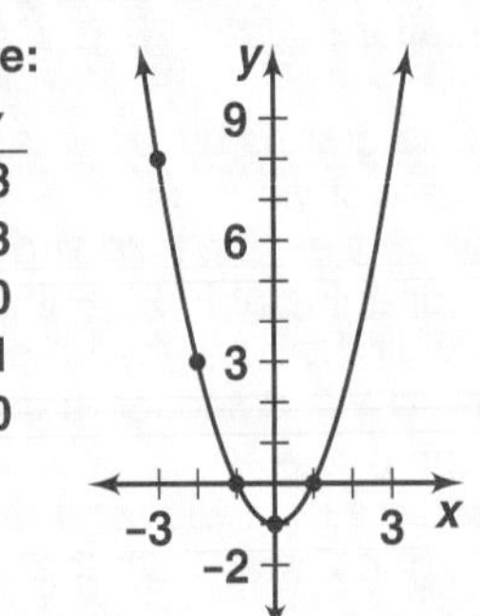

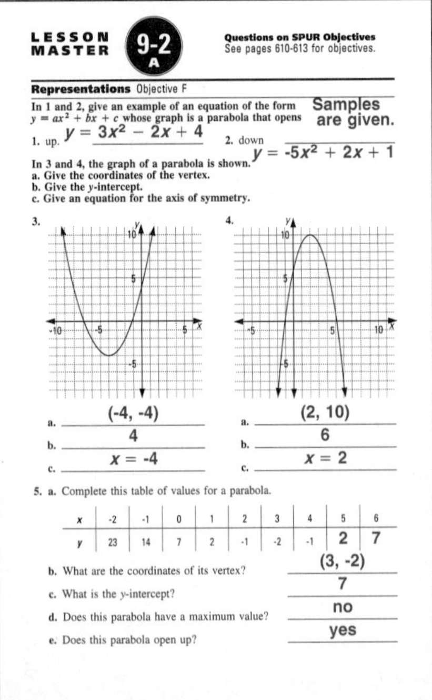

LESSON MASTER 9-2 A — **Questions on SPUR Objectives** See pages 610-613 for objectives.

Representations Objective F

In 1 and 2, give an example of an equation of the form $y = ax^2 + bx + c$ whose graph is a parabola that opens Samples are given.

1. up. $y = 3x^2 - 2x + 4$ **2.** down $y = -5x^2 + 2x + 1$

In 3 and 4, the graph of a parabola is shown.
a. Give the coordinates of the vertex.
b. Give the y-intercept.
c. Give an equation for the axis of symmetry.

3. a. (-4, -4) b. 4 c. $x = -4$

4. a. (2, 10) b. 6 c. $x = 2$

5. a. Complete this table of values for a parabola.

x	-2	-1	0	1	2	3	4	5	6
y	23	14	7	2	-1	-2	-1	2	7

b. What are the coordinates of its vertex? (3, -2)

c. What is the y-intercept? 7

d. Does this parabola have a maximum value? no

e. Does this parabola open up? yes

Using Tables to Graph $y = ax^2 + bx + c$

To save time when making tables of values, you can use a calculator or computer program or a spreadsheet. The BASIC program below prints a table of values for the quadratic equation $y = 2x^2 - 8x + 6$. In line 30 the command STEP tells the computer how much to add to x each time through the FOR/NEXT loop. The value of x increases in "steps" of 0.5, from -1 to 4.

❶

```
10 PRINT "TABLE FOR Y = 2X ^ 2 - 8X + 6"
20 PRINT "X", "Y"
30 FOR X = -1 TO 4 STEP.5
40 LET Y = 2 * X ^ 2 - 8 * X + 6
50 PRINT X, Y
60 NEXT X
70 END
```

Here is what the computer prints when the program is run.

TABLE FOR Y = 2X ^ 2 − 8X + 6

X	Y
-1	16
-.5	10.5
0	6
.5	2.5
1	0
1.5	-1.5
2	-2
2.5	-1.5
3	0
3.5	2.5
4	6

You can use the symmetry of the graph of a quadratic equation to answer questions about it.

Example 2

Consider the equation $y = 2x^2 - 8x + 6$ and use the table of values produced by the BASIC program above.

a. What is the vertex of the parabola?
b. What is the axis of symmetry of the parabola?
c. Find the y-intercept of the parabola.
d. Graph the parabola.
e. What are the x-coordinates of the two points where $y = 16$?

Solution

a. Look in the table for points with the same y-values. Notice how the pairs occur on either side of (2, -2). The vertex is (2, -2).

b. The axis of symmetry is the vertical line through the vertex. The axis of symmetry is $x = 2$.

c. The y-intercept is the y-coordinate of the point where $x = 0$. From the table you can see that the y-intercept is 6.

▶

Adapting to Individual Needs

Extra Help

Some students might be bothered by graphs for which the axis of symmetry is not the y-axis. These students might be equating symmetry with centering on the coordinate system. Point out that the axis of symmetry is always the vertical line passing through the vertex of the parabola. Use the overhead projector to show students three different parabolas with different vertices. Have volunteers name the coordinates of each vertex and use a colored marker to draw the vertical line through each vertex. Then have another volunteer give the equation for the line of symmetry. Point out that this equation names the x-coordinate of the vertex.

d. Plot the points in the table and connect them with a smooth curve.

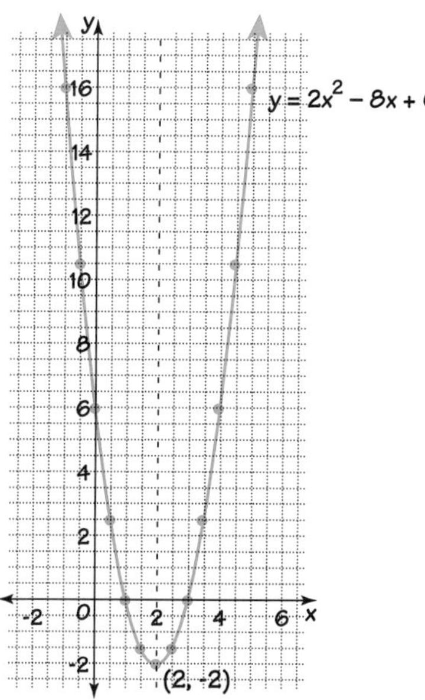

e. (-1, 16) is in the table. Because of symmetry there is another x-value for $y = 16$. Continue the pattern in the table. This results in (5, 16).

X	Y
-1	16
-.5	10.5
0	6
.5	2.5
1	0
1.5	-1.5
2	-2
2.5	-1.5
3	0
3.5	2.5
4	6
4.5	10.5
5	16

When y = 16, x = 5 or x = -1.

b. Tell whether the parabola opens up or down. **Up**

c. Identify its axis of symmetry and its vertex. ***y*-axis; (0, -1)**

2. Write a computer program to print values for $y = x^2 + 8x + 13$ for $x = -7, -6, -5, \dots 1$, and make a table of values. **See below.**

```
10 PRINT "TABLE OF VALUES
   FOR A PARABOLA"
20 PRINT "X", "Y"
30 for X = -7 to 2 step 1
40 Let y = X * X + 8 * X + 13
50 NEXT X
60 END
```

a. What is the vertex of the parabola? **(-4, -3)**

b. What is its axis of symmetry? $x = -4$

c. Find the y-intercept. **13**

d. Draw the graph. **See below.**

e. What are the x-coordinates of the two points where $y = 13$? $x = 0$ **and** $x = -8$

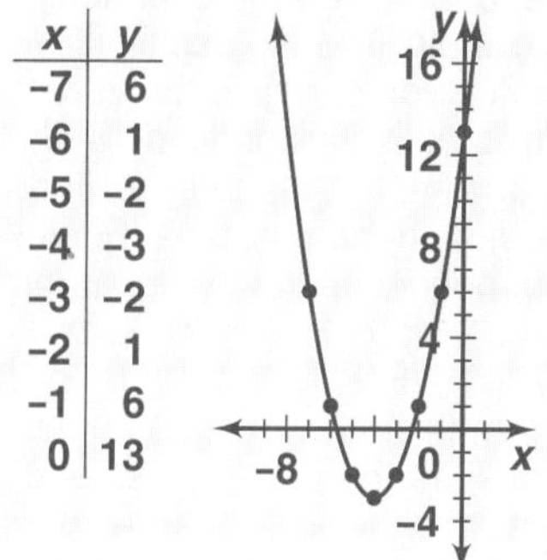

x	y
-7	6
-6	1
-5	-2
-4	-3
-3	-2
-2	1
-1	6
0	13

(Additional Examples continue on page 558.)

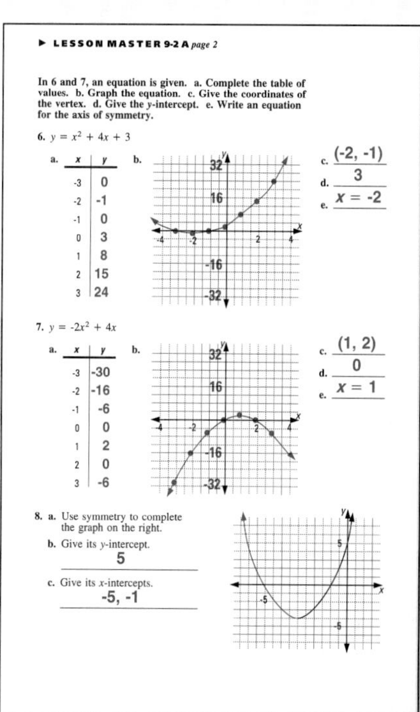

▸ **LESSON MASTER 9-2 A** *page 2*

In 6 and 7, an equation is given. a. Complete the table of values. b. Graph the equation. c. Give the coordinates of the vertex. d. Give the *y*-intercept. e. Write an equation for the axis of symmetry.

6. $y = x^2 + 4x + 3$

a.

x	y
-3	0
-2	-1
-1	0
0	3
1	8
2	15
3	24

b. [graph]

c. (-2, -1)

d. 3

e. $x = -2$

7. $y = -2x^2 + 4x$

a.

x	y
-3	-30
-2	-16
-1	-6
0	0
1	2
2	0
3	-6

b. [graph]

c. (1, 2)

d. 0

e. $x = 1$

8. a. Use symmetry to complete the graph on the right.

b. Give its y-intercept. 5

c. Give its x-intercepts. -5, -1

Adapting to Individual Needs

English Language Development

You might suggest that non-English-speaking students add *axis of symmetry* to their index-card files. Suggest that they include a graphic example with the definition. Encourage students to refer to their index cards while doing homework.

3. a. Graph $y = .5x^2 - 2x$.

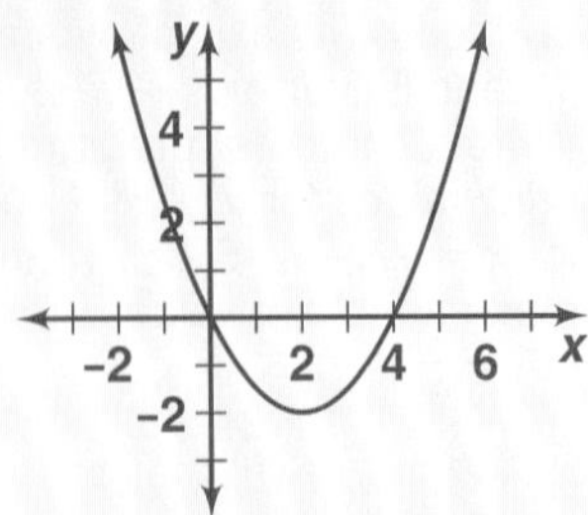

b. Does the graph have an axis of symmetry? If so, give an equation for it. **Yes; $x = 2$**

c. Name the vertex. **(2, -2)**

d. Identify the y-intercept. **0**

Notes on Questions

As you go over the students' graphs, you will probably find that most students have made tables of values. As they show their graphs, ask what information about the graph could be found using the symmetry of the parabola. To help explain the symmetry, you can give the vertex and three points on one side of the axis of symmetry; then ask for three more points on the parabola. For instance, suppose you find the following points on a graph of $y = ax^2 + bx + c$: (7, −6), (8, −1), (9, 2), (10, 3), (11, 2). From this, (10, 3) must be the vertex because the points (9, 2) and (11, 2) are on either side and have the same y-value. By the symmetry, (12, −1) and (13, −6) must also be on the parabola.

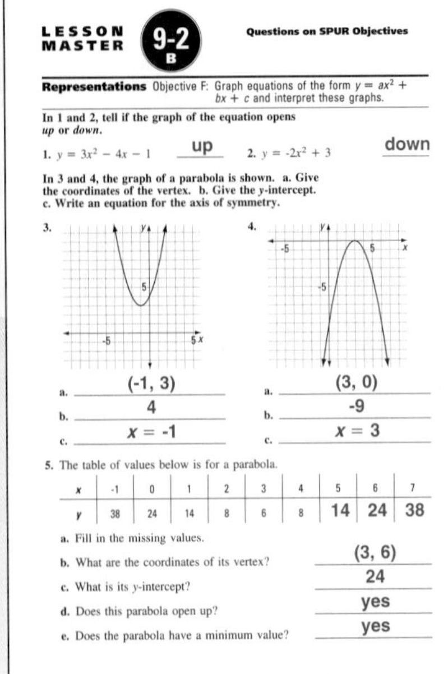

LESSON MASTER 9-2 B — Questions on SPUR Objectives

Representations Objective F: Graph equations of the form $y = ax^2 + bx + c$ and interpret these graphs.

In 1 and 2, tell if the graph of the equation opens *up* or *down*.

1. $y = 3x^2 - 4x - 1$ **up** 2. $y = -2x^2 + 3$ **down**

In 3 and 4, the graph of a parabola is shown. a. Give the coordinates of the vertex. b. Give the y-intercept. c. Write an equation for the axis of symmetry.

3. a. **(-1, 3)** b. **4** c. **$x = -1$**

4. a. **(3, 0)** b. **-9** c. **$x = 3$**

5. The table of values below is for a parabola.

x	-1	0	1	2	3	4	5	6	7
y	38	24	14	8	6	8	14	24	38

a. Fill in the missing values.

b. What are the coordinates of its vertex? **(3, 6)**

c. What is its y-intercept? **24**

d. Does this parabola open up? **yes**

e. Does the parabola have a minimum value? **yes**

A driving rain. *The equation for stopping distance given in this lesson is for dry pavement. Motorists should allow more distance for stopping on wet pavement.*

QUESTIONS

Covering the Reading

In 1–4, use the formula for automobile stopping distances in this lesson.

1. Define *stopping distance.* **Sample: Stopping distance is the distance a car travels after the driver decides to stop.**
2. Find the stopping distance for a car traveling 40 mph. **120 ft**
3. Find the stopping distance for a car traveling 55 mph. **206.25 ft**
4. *True or false.* The stopping distance for a car traveling 60 mph is double the stopping distance of a car traveling 30 mph. **False**
5. *True or false.* The equation $d = 0.05x^2 + x$ is of the form $y = ax^2 + bx + c$. **True**

In 6–8, an equation is given.

a. Make a table of x- and y-values for the equation when x equals -3, -2, -1, 0, 1, 2, and 3. **6a, b) See left. 7a, b and 8a, b) See page 559.**

b. Graph the equation.

c. Identify the vertex and tell whether it is a minimum or maximum.

d. Identify the y-intercept.

6. $y = x^2 + 5$ **d) 5** **c) (0, 5); minimum**
7. $y = x^2 - 2x - 3$ **d) -3** **c) (1, -4); minimum**
8. $y = -3x^2 + 5$ **d) 5** **c) (0, 5); maximum**

9. a. What are the coordinates of the vertex of the parabola at the left? **a) (-3, 1)**
   b. Write an equation of its axis of symmetry. **$x = -3$**

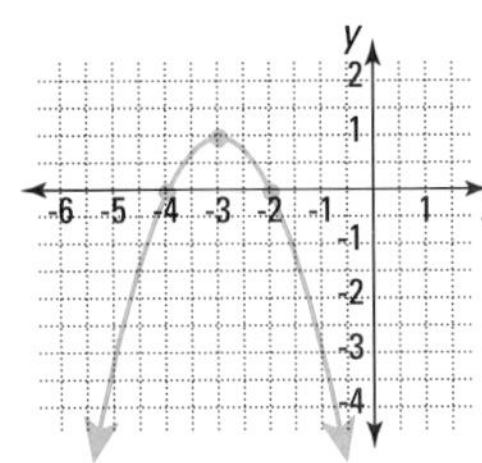

6a)

x	y
-3	14
-2	9
-1	6
0	5
1	6
2	9
3	14

b)

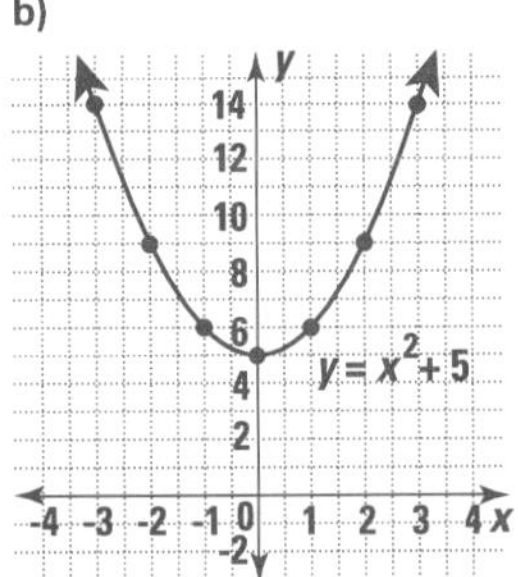

In 10 and 11, consider the BASIC program below.

```
10 PRINT "TABLE FOR Y = X ^ 2 − 2X"
20 PRINT "X", "Y"
30 FOR X = −1 TO 3 STEP.5
40 LET Y = X ^2 − 2 * X
50 PRINT X, Y
60 NEXT X
70 END
```

10) TABLE FOR Y = X^2 − 2X

X	Y
-1	3
-0.5	1.25
0	0
0.5	-0.75
1	-1
1.5	-0.75
2	0
2.5	1.25
3	3

10. What will this program print when run?
11. What is the axis of symmetry for this parabola? **$x = 1$**
12. Consider this table of values for a parabola.

x	-8	-7	-6	-5	-4	-3	-2	-1	0
y	-13	5	19	29	35	29	?	?	?

a. What is the vertex of the parabola? **(-4, 35)**

b. What is the axis of symmetry of the parabola? **$x = -4$**

c. Write the y-values that are missing from the table. **19; 5; -13**

13. Explain how you can tell by looking at an equation of the form $y = ax^2 + bx + c$, whether its graph will open up or down. **If $a > 0$, the parabola opens up. If $a < 0$, the parabola opens down.**

Adapting to Individual Needs

Challenge

Materials: **Teaching Aid 102**

Give each student a copy of **Teaching Aid 102**. Tell them to write an equation for each parabola. If they need a hint, tell them to study the equations and parabolas in **Exercises 15–18**. Then have them explain how they made their decisions.

[Equations: A: $y = -x^2 + 4$; B: $y = -x^2 + 1$; C: $y = -x^2 - 3$; D: $y = x^2 + 3$; E: $y = x^2 - 2$; F: $y = x^2 - 5$. Sample explanation: Each equation is of the form $y = ax^2 + bx + c$, where $b = 0$; $a = 1$ if the parabola opens up and $a = -1$ if the parabola opens down; c is determined by the y-coordinate of the vertex.]

7a)

x	y
-3	12
-2	5
-1	0
0	-3
1	-4
2	-3
3	0

b)

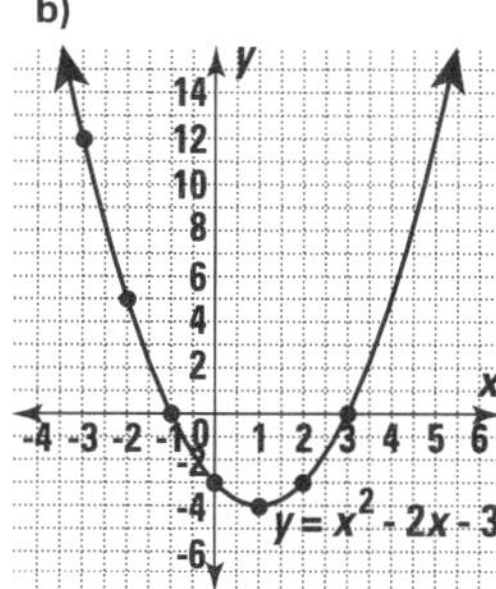

8a)

x	y
-3	-22
-2	-7
-1	2
0	5
1	2
2	-7
3	-22

b)

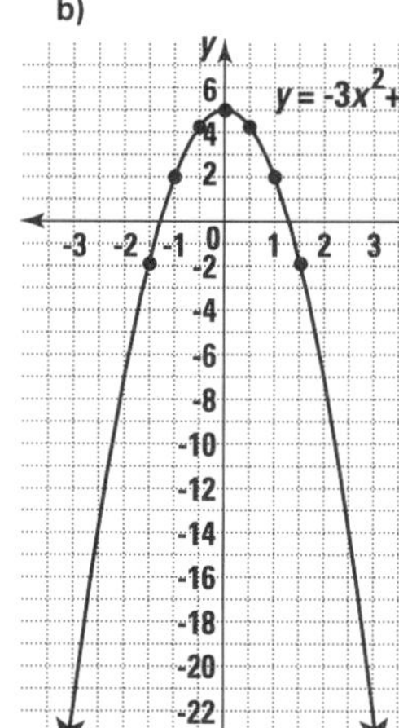

Applying the Mathematics

14. **a.** Copy the graph below on graph paper. Then use symmetry to graph more of the parabola.
b. Identify the vertex. (6, 9)
c. Give an equation for the axis of symmetry. $x = 6$
d. At what points does the parabola cross the x-axis? (These are the x-intercepts of the parabola.) (3, 0) (9, 0)

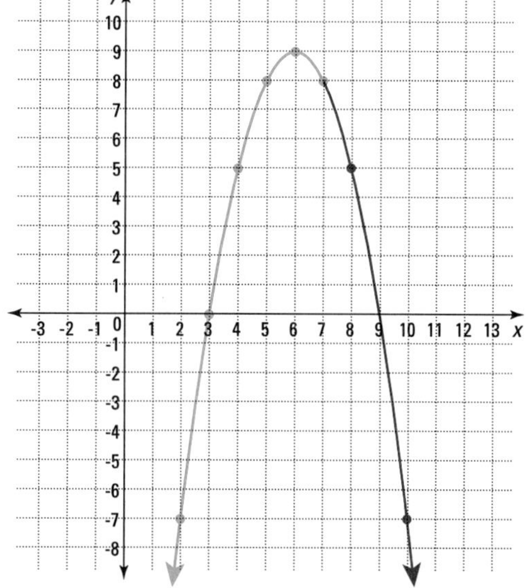

In 15–18, *multiple choice.* Match the graph with its equation.
(a) $y = x^2$ (b) $y = x^2 + 1$ (c) $y = x^2 - 1$ (d) $y = -x^2 - 1$

15. c

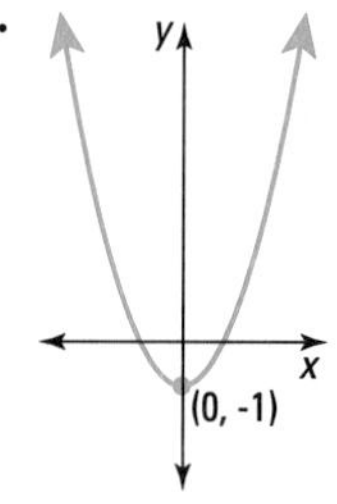

16. d

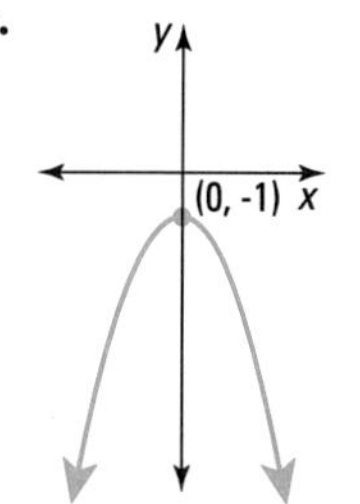

17. a

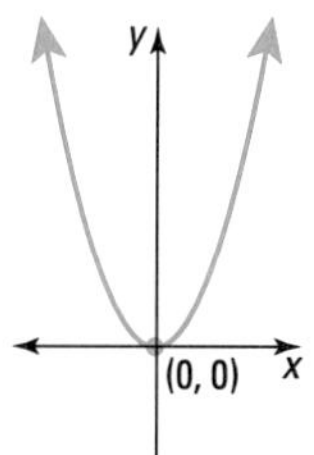

18. b

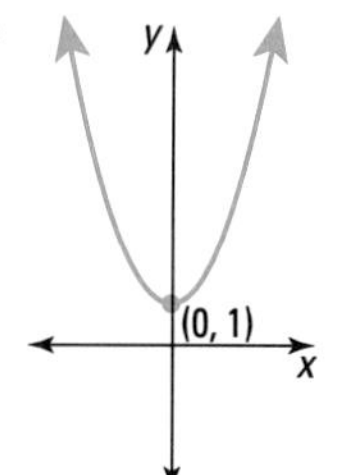

Questions 6–8 You might use the program in **Example 2,** or the one in these notes, to generate one or two tables for these questions. Having the tables will speed up the homework in this lesson. However, leave some of the problems to be done value-by-value by your students.

Question 12 Error Alert Some students have difficulty identifying the symmetry of a parabola in a table. Use **Teaching Aid 100**, or have a transparency available with these points plotted for those students who need visual cues.

Question 14 This item reverses the process of **Question 12.**

Questions 15–18 These graphs are shown on **Teaching Aid 100.** We expect students to compute a few values, and then to deduce which parabola matches which equation. It is often convenient to examine the vertex and the y-intercept.

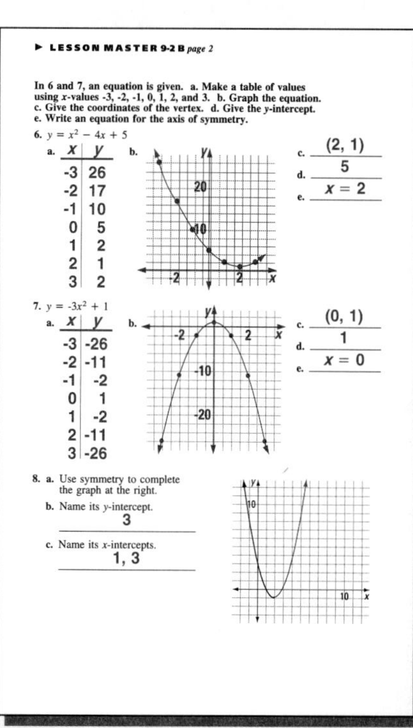

► **LESSON MASTER 9-2 B** *page 2*

In 6 and 7, an equation is given. a. Make a table of values using x-values -3, -2, -1, 0, 1, 2, and 3. b. Graph the equation. c. Give the coordinates of the vertex. d. Give the y-intercept. e. Write an equation for the axis of symmetry.

6. $y = x^2 - 4x + 5$

a.

x	y
-3	26
-2	17
-1	10
0	5
1	2
2	1
3	2

b. [graph]

c. (2, 1)
d. 5
e. $x = 2$

7. $y = -3x^2 + 1$

a.

x	y
-3	-26
-2	-11
-1	-2
0	1
1	-2
2	-11
3	-26

b. [graph]

c. (0, 1)
d. 1
e. $x = 0$

8. a. Use symmetry to complete the graph at the right.
b. Name its y-intercept. 3
c. Name its x-intercepts. 1, 3

Notes on Questions

Question 26 If students do not have access to a computer, they could use the spreadsheet on **Teaching Aid 37** and manually compute the information.

Follow-up for Lesson 9-2

Practice

For more questions on SPUR Objectives, use **Lesson Master 9-2A** (shown on pages 556–557) or **Lesson Master 9-2B** (shown on pages 558–559).

Assessment

Group Assessment Have students **work in groups.** Ask each member of the group to write a quadratic equation using any of the following numbers for the coefficients *a, b,* and *c:* -3, -2, -1, 1, 2, 3. Have each student exchange the equation with another member of the group and graph the other student's equation on graph paper, separate from the equation. Then have groups exchange equations and graphs and match each equation with its graph. [Students demonstrate the ability to write and graph quadratic equations and to match an equation to its graph.]

Extension

Project Update Project 2, *Quadratic Designs,* Project 3, *Parabolas from Folding Paper,* and Project 6, *Parabolas in Manufactured Objects,* on pages 605–606, relate to the content of this lesson.

19. *Multiple choice.* Which of these could be the graph of $y = x^2 - 6x + 8$? Justify your answer. b; Sample: if x = 1, y = 1 − 6 + 8 = 3. Graph b contains this point.

(a)

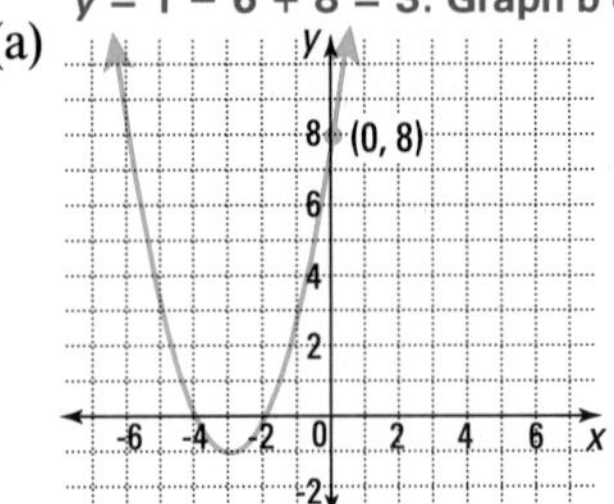

(b)

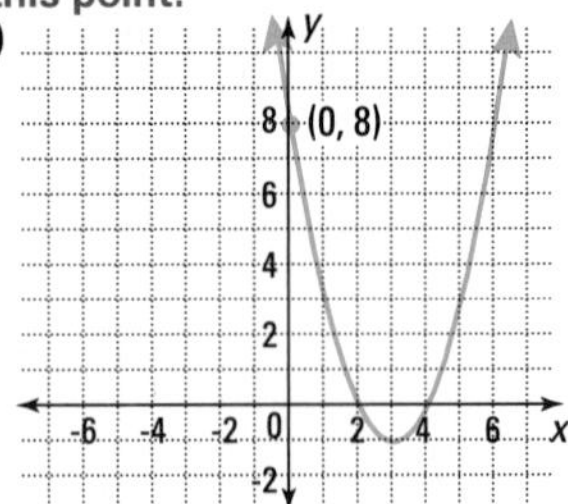

26)

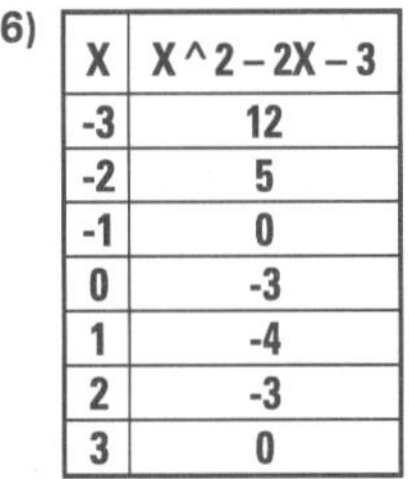
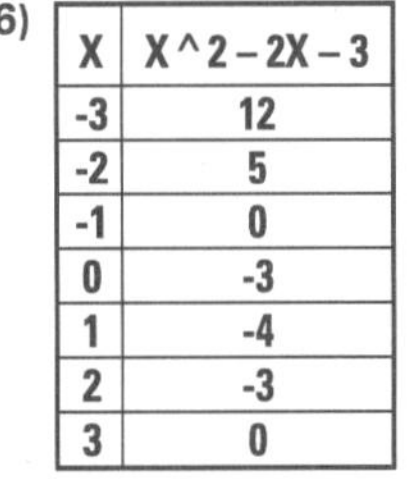

X	X^2-2X-3
-3	12
-2	5
-1	0
0	-3
1	-4
2	-3
3	0

Review

20. Consider the equation $d = 16t^2$ from page 547. *(Lesson 9-1)*

a. Find t when $d = 500$. $t \approx 5.59$

b. Write a question involving distance and time that can be answered using part **a.** Sample: How long does it take a stone to fall 500 ft from the top of a cliff?

21.

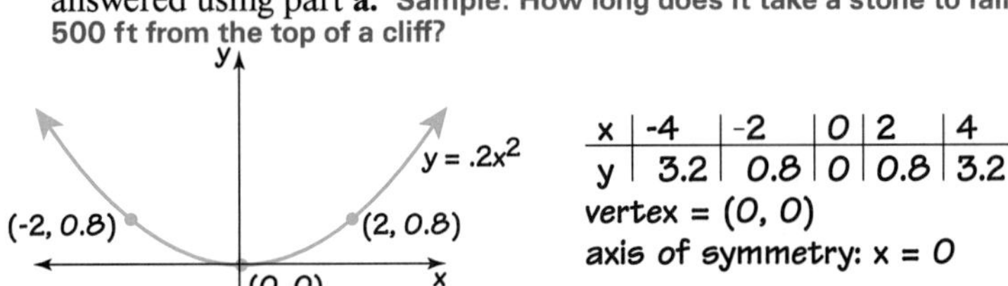

x	-4	-2	0	2	4
y	3.2	0.8	0	0.8	3.2

vertex = (0, 0)
axis of symmetry: x = 0

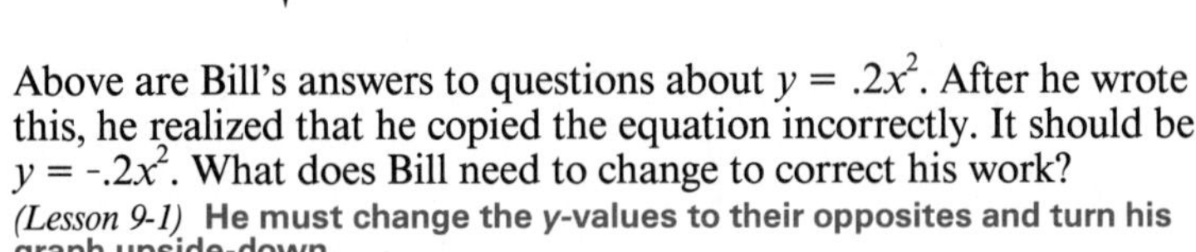

Above are Bill's answers to questions about $y = .2x^2$. After he wrote this, he realized that he copied the equation incorrectly. It should be $y = -.2x^2$. What does Bill need to change to correct his work? *(Lesson 9-1)* He must change the *y*-values to their opposites and turn his graph upside-down.

In 22 and 23, solve. *(Lessons 6-8, 9-1)*

22. $-3n^2 = -1200$ $n = 20$ or $n = -20$

23. $\frac{16}{x} = \frac{x}{40}$ $x = \sqrt{640} \approx 25.3$ or $x = -\sqrt{640} \approx -25.3$

In 24 and 25, suppose an animal shelter has only cats, dogs, and rabbits. Three of the cats are Manx cats, which have no tails. All the dogs and rabbits have tails. Let c = the number of cats, d = number of dogs, and r = number of rabbits. *(Lessons 3-1, 4-2, 6-3)*

24. What is the ratio of the number of cats to the total number of animals?

25. What is the ratio of the number of animal heads to the number of animal tails? $\frac{c + d + r}{(c - 3) + d + r}$

24) $\frac{c}{c + d + r}$

Tale of the Manx. *The Manx cat is named after the Isle of Man in the Irish Sea where the breed originated. There are four varieties of Manx cats—rumpy, rumpy-riser, stumpy, and longie. The rumpy is the only cat without a tail.*

Exploration

26. Use a spreadsheet to generate a table of values for one of the quadratic equations used in this lesson. See left for sample.

Setting Up Lesson 9-3

Materials If automatic graphers are available, we recommend that students use them for Lesson 9-3 and for the rest of the lessons in this chapter.

Introducing Lesson 9-3

Graphing Parabolas with an Automatic Grapher

IN-CLASS ACTIVITY

Materials: Automatic Grapher

When you want to work with several graphs at a time, an automatic grapher is very helpful.

1 **a.** On the default window for your grapher, graph $y = ax^2$ for each of these values of a: 0.5, 1, 2, and 3. You should see something similar to what is shown here. Which graph is which?

From outer to inner, the parabolas are as follows: $y = 0.5x^2$, $y = x^2$, $y = 2x^2$, $y = 3x^2$.

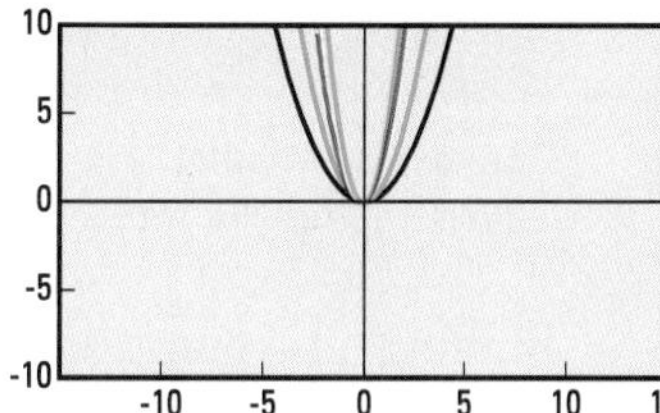

b. What happens to the graph of $y = ax^2$ as a gets larger? **It narrows.**

c. Predict what the graph of $y = 7x^2$ will look like in relation to the graphs in part **a.** Then check your prediction by graphing. **$y = 7x^2$ will be a graph of a parabola that is narrower than any of those in part a.**

2 Zoom in on the vertex (0, 0) of these parabolas. What happens to the way the parabolas look? **Close to the vertex, the parabolas tend to look more alike.**

3 Clear the screen. Graph $y = x^2 - 36x + 20$. Adjust the window of your grapher so that the graph looks something like this. Describe the window. **A window of $10 \le x \le 25$ and $-400 \le y \le -200$ seems to show a similar graph.**

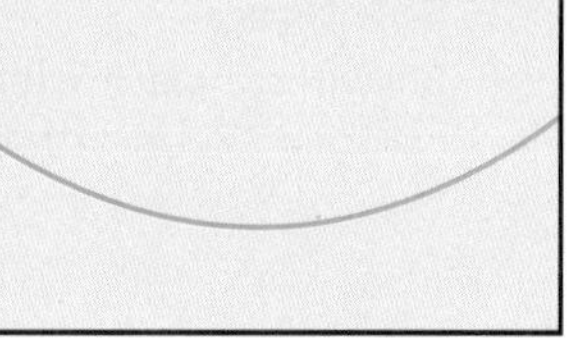

561

In-class Activity

Resources

From the ***Teacher's Resource File***

- Answer Master 9-3

Additional Resources

- Automatic graphers

This activity should be done with automatic graphers. If you do not have a class set, then you can demonstrate with a single automatic grapher attached to a projection device.

Lesson 9-3

Objectives

F Graph equations of the form $y = ax^2 + bx + c$ and interpret these graphs.

Resources

From the ***Teacher's Resource File***

- Lesson Master 9-3A or 9-3B
- Answer Master 9-3
- Assessment Sourcebook: Quiz for Lessons 9-1 through 9-3
- Teaching Aids
  55 Automatic Grapher Grids
  95 Warm-up
- Technology Sourcebook
  Computer Master 23

Additional Resources

- Visuals for Teaching Aids 55, 95
- GraphExplorer or other automatic graphers

Teaching Lesson 9-3

Warm-up

Tell what you know about *a*, *b*, or *c* in the equation $y = ax^2 + bx + c$ if

1. its vertex is its minimum point. $a > 0$
2. the *y*-axis is the axis of symmetry of the graph. $b = 0$
3. The point (0, 6) is on the graph. $c = 6$

Notes on Reading

One way to cover the reading is to treat each of the examples as an activity. Have students go through the activity with their own automatic graphers; they should try to obtain graphs similar to those shown in the

LESSON

9-3 Graphing Parabolas with an Automatic Grapher

Curves in nature. *A waterfall, such as this one at Iguazu Falls in Argentina, follows a parabolic path. A rainbow, however, is a circular arc.*

Finding an Appropriate Window

Parabolas are infinite in extent and they contain no circular arcs. This makes it quite difficult to draw an accurate parabola by hand. So we recommend that you use an automatic grapher to graph parabolas whenever you can. However, if you are not told what window to use, you may have to experiment to find values that give a good graph. You may find it convenient to try your default window first.

Example 1

Find a window that shows a graph of $y = 0.5x^2 - 20x + 100$, including its vertex, its *y*-intercept, and its *x*-intercepts.

Solution

Shown is a graph with the window $-15 \le x \le 15$, $-10 \le y \le 10$.

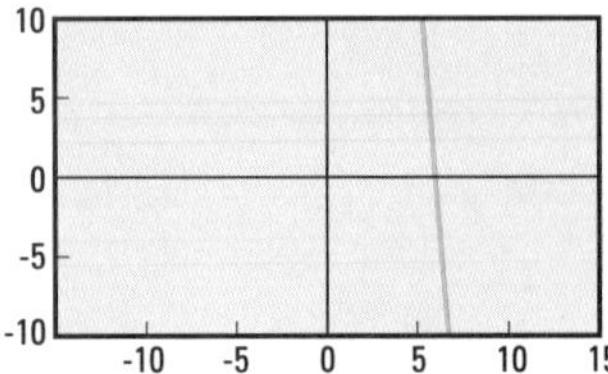

Since the equation is of the form $y = ax^2 + bx + c$, the graph will be a parabola. This window shows part of the parabola, including one *x*-intercept when *x* is a little larger than 5. However, this window does not show the other *x*-intercept, the *y*-intercept, or the vertex. The *y*-intercept can be calculated directly. Just substitute 0 in the equation.

$$y = 0.5(0)^2 - 20(0) + 100 = 100.$$

Lesson 9-3 Overview

Broad Goals In this lesson automatic graphers are employed to compare parabolas and to estimate the coordinates of the vertex and intercepts of a parabola.

Perspective The equipment you have available will determine how you approach this lesson and the lessons that follow. An automatic grapher that displays multiple graphs on an overhead is most helpful for demonstration.

If your school has a computer lab, demonstrate (or model) one or two examples to show students how to enter exponents on the grapher and to remind them how to change the viewing window. Then you might have students use the computers to answer the questions for the lesson.

If you do not have automatic graphers available, most of the work in this lesson can be done by hand. The instructions about window settings also apply to graph paper. It is still important that students know about automatic graphers since they will probably encounter them at some point.

Even if students draw their graphs on a calculator or computer, you may want them to also sketch the results on paper. In sketching the graph, students should note the *x*- and *y*-intercepts and the vertex.

The vertex will be to the right of $x = 15$ and below $y = -10$. So next we graph the equation on the window $0 \le x \le 35$, $-50 \le y \le 100$. Our graph is shown at the right.

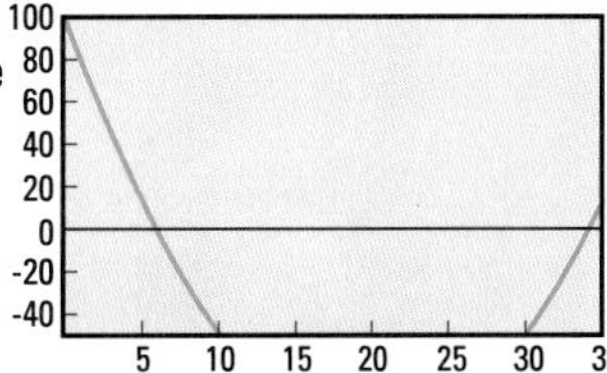

This window shows the y-intercept and the two x-intercepts, but not the vertex. It looks as if the axis of symmetry might be near $x = 20$. To check this we draw our next graph using $0 \le x \le 40$. To see the vertex we must use even smaller values of y. **Below is a graph on the window $0 \le x \le 40$, $-150 \le y \le 100$.**

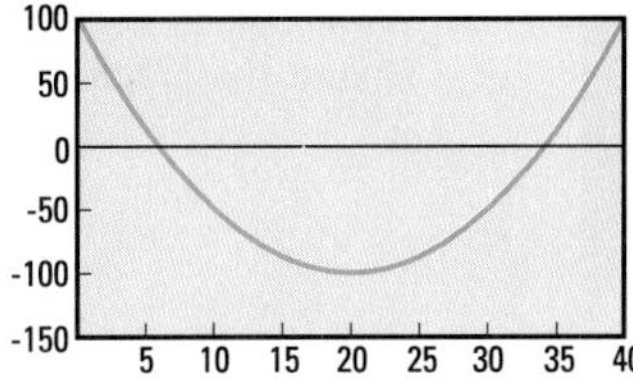

Locating Key Points on a Graph

You can use the trace feature to locate points on a parabola if it has been graphed with an automatic grapher. Key points are the parabola's vertex and the intercepts.

Example 2

Estimate the coordinates of the vertex and the x-intercepts of the graph of $y = 0.5x^2 - 20x + 100$.

Solution

You can see from the graph above that the vertex is near (20, -100), the smaller x-intercept is between 5 and 10, and the larger is between 30 and 35. Use the trace feature of your automatic grapher. **A grapher shows the vertex to be about (20, -100), and the x-intercepts to be about 5.9 and 34.1.**

In the parabola used for Examples 1 and 2, the average of the x-intercepts gives the x-coordinate of the vertex.

$$\frac{5.9 + 34.1}{2} = \frac{40}{2} = 20$$

This always happens. If a parabola with equation $y = ax^2 + bx + c$ has x-intercepts x_1 and x_2, then the x-coordinate of the vertex is

$$\frac{x_1 + x_2}{2}.$$

When asked to make a graph with an automatic grapher, you should either sketch the graph your grapher draws or make a printout.

lesson. If you ask students to sketch the results on paper, you may wish to give them **Teaching Aid 55**.

Since the term *x-intercept* has not been used earlier in the chapter, you may want to remind students of its meaning. Point out that every graph of the form $y = ax^2 + bx + c$ has exactly one y-intercept, which is c, and 0, 1, or 2 x-intercepts. Looking for the x-intercept(s) will set up the solving of the equation $ax^2 + bx + c = 0$, which is discussed in the next lesson.

Additional Examples

1. Find a window that shows a graph of $y = 2x^2 - 30x + 50$, including its y-intercept and its x-intercept. **Sample window:**

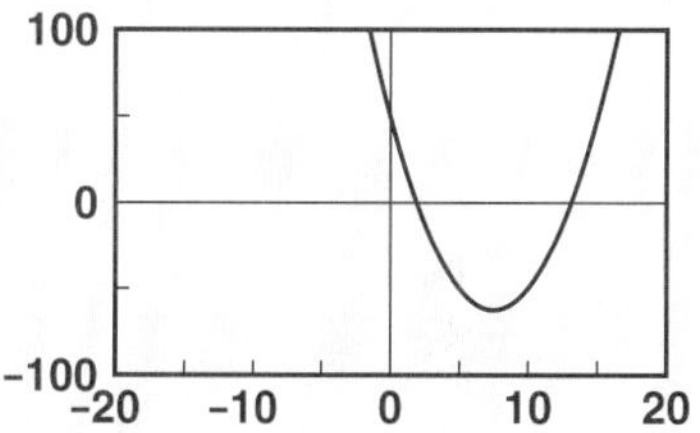

2. Estimate the coordinates of the vertex and the x-intercepts of the graph of $y = 2x^2 - 30x + 50$. **Sample: vertex about (7, -62); x-intercepts about 2 and 13**
3. The sum of the first n whole numbers can be found using the equation $s = .5n^2 + .5n$.
   a. Plot this equation using an automatic grapher. **A sample window is shown on page 564.**

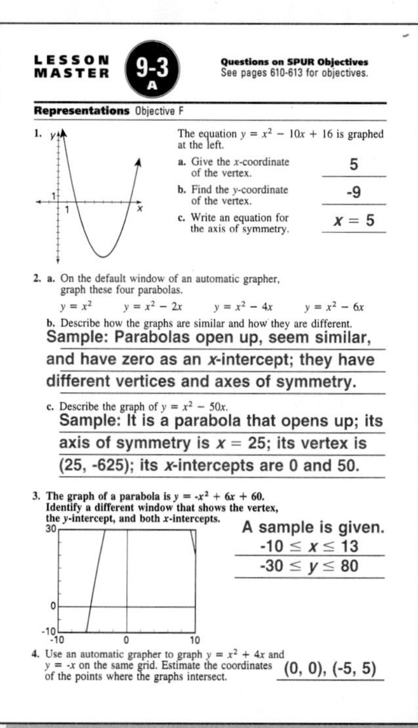

LESSON MASTER 9-3 A

Questions on SPUR Objectives
See pages 610-613 for objectives.

Representations Objective F

1. The equation $y = x^2 - 10x + 16$ is graphed at the left.
   a. Give the x-coordinate of the vertex. **5**
   b. Find the y-coordinate of the vertex. **-9**
   c. Write an equation for the axis of symmetry. **$x = 5$**

2. a. On the default window of an automatic grapher, graph these four parabolas.
   $y = x^2$ $y = x^2 - 2x$ $y = x^2 - 4x$ $y = x^2 - 6x$
   b. Describe how the graphs are similar and how they are different.
   Sample: Parabolas open up, seem similar, and have zero as an x-intercept; they have different vertices and axes of symmetry.
   c. Describe the graph of $y = x^2 - 50x$.
   Sample: It is a parabola that opens up; its axis of symmetry is $x = 25$; its vertex is (25, -625); its x-intercepts are 0 and 50.

3. **The graph of a parabola is $y = -x^2 + 6x + 60$. Identify a different window that shows the vertex, the y-intercept, and both x-intercepts.**
   A sample is given.
   $-10 \le x \le 13$
   $-30 \le y \le 80$

30
0
-10
-10 0 10

4. Use an automatic grapher to graph $y = x^2 + 4x$ and $y = -x$ on the same grid. Estimate the coordinates of the points where the graphs intersect. **(0, 0), (-5, 5)**

Optional Activities

Activity 1 Technology Connection
In *Technology Sourcebook, Computer Master 23,* students use *GraphExplorer* or similar software to graph quadratic equations. Students then use the trace feature to find the vertex, x-intercepts, and line of symmetry of a parabola.

Activity 2 Writing After completing this lesson, you might have students summarize Lessons 9-1 to 9-3 and include the following information: how to graph an equation of the form $y = ax^2 + bx + c$; how to determine the vertex, y-intercept, and x-intercepts (if any); and how to determine whether the parabola opens up or down.

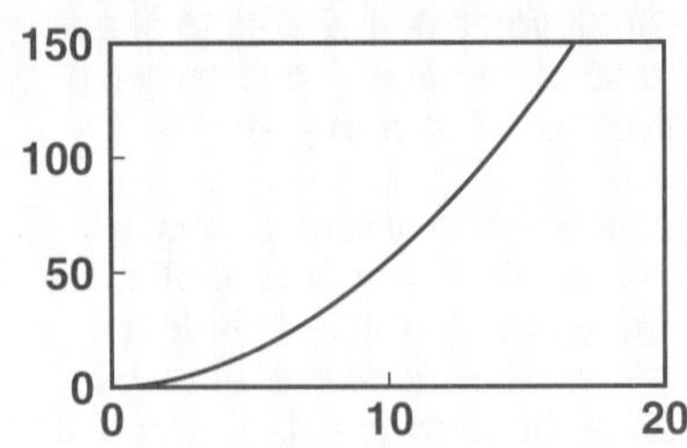

b. Use the trace to find the sum of the first 16 whole numbers. **136**

4. Graph $y = (x - d)^2$ three times: when $d = -8$, when $d = 0$, and when $d = 5$. Use the window $-15 \le x \le 15$ and $-5 \le y \le 25$. What happens as d increases? **The parabola moves to the right.**

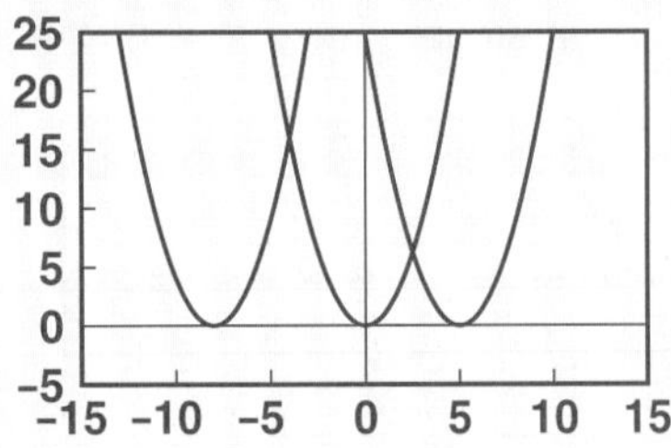

Notes on Questions

For classes with no access to graphing technology, you may want to omit **Questions 6–9 and 12** at this time.

Question 3 Cooperative Learning You may want the class to decide on a convenient window ahead of time. If students use different windows, the shapes of the graphs may differ, but the answer to **part b** will not be affected.

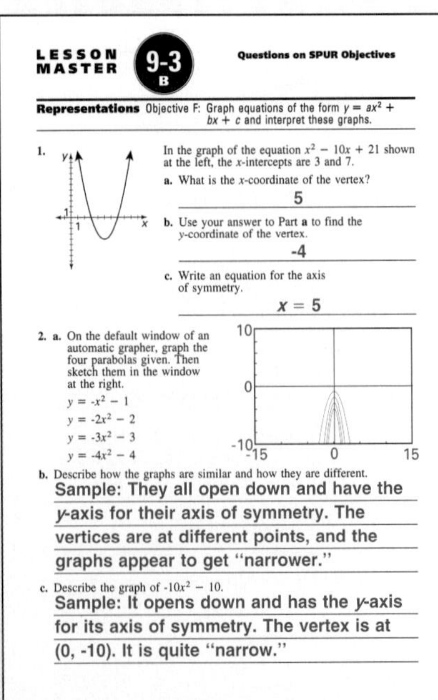

LESSON MASTER 9-3 B — Questions on SPUR Objectives

Representations Objective F: Graph equations of the form $y = ax^2 + bx + c$ and interpret these graphs.

1. In the graph of the equation $x^2 - 10x + 21$ shown at the left, the x-intercepts are 3 and 7.
   a. What is the x-coordinate of the vertex? **5**
   b. Use your answer to Part **a** to find the y-coordinate of the vertex. **-4**
   c. Write an equation for the axis of symmetry. **$x = 5$**

2. a. On the default window of an automatic grapher, graph the four parabolas given. Then sketch them in the window at the right.
   $y = -x^2 - 1$
   $y = -2x^2 - 2$
   $y = -3x^2 - 3$
   $y = -4x^2 - 4$
   b. Describe how the graphs are similar and how they are different. **Sample: They all open down and have the y-axis for their axis of symmetry. The vertices are at different points, and the graphs appear to get "narrower."**
   c. Describe the graph of $-10x^2 - 10$. **Sample: It opens down and has the y-axis for its axis of symmetry. The vertex is at (0, -10). It is quite "narrow."**

Bent out of shape. *Car insurance companies usually require written estimates from auto body shops before repair work is authorized.*

An Example of a Quadratic Model

Example 3

An insurance company reports that the equation

$$y = 0.4x^2 - 36x + 1000$$

relates the age of a driver x (in years) to the driver's accident rate y (number of accidents per 50 million miles driven) when $16 \le x \le 74$.

a. Plot this equation using an automatic grapher.

b. According to this model, at what age do drivers have the fewest accidents per mile driven? How many accidents do they have?

Solution

a. We chose to use the window $0 \le x \le 100$, $-100 \le y \le 2000$ to ensure the graph would fit. A graph is shown below.

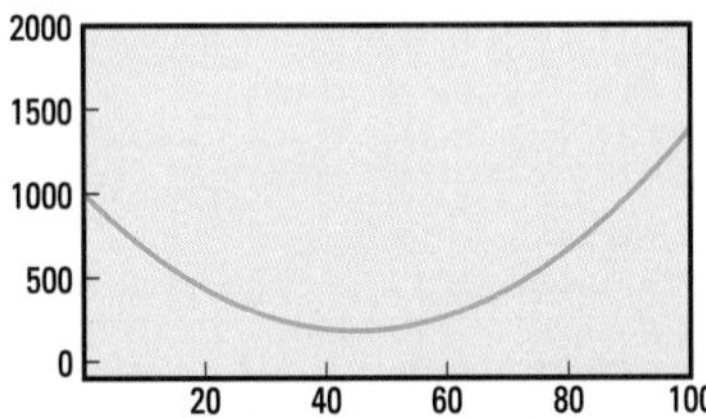

b. Use the trace feature and look for the smallest value of y. We find $y = 190.04$ when $x = 44.68$. **Drivers have the fewest accidents per mile at age 44. Then they have about 190 accidents per 50 million miles, which is about 1 accident per 260,000 miles driven.** In contrast, drivers at the age of 16 have about 1 accident per 95,000 miles driven.

2)

$y = 1x^2$ $y = 2x^2$ $y = 4x^2$ $y = 3x^2$ $y = .5x^2$

If you are graphing an equation to help answer a question about a real problem as in Example 3, the portion of the graph you will want to examine depends upon the question. However, if the goal is simply to draw a graph of a parabola, you will want to draw one that, like the third graph in the solution to Example 1, shows the vertex, y-intercept, and x-intercept or intercepts, if any.

QUESTIONS

Covering the Reading

In 1 and 2, refer to the activity on page 561.

1. Describe what happens to the graphs as a increases. **As the value of a increases, the graphs of $y = ax^2$ look narrower and narrower.**

2. Copy or print the graphs, and add a sketch of $y = 4x^2$ to them. **See left.**

Optional Activities

Activity 3

Materials: Automatic graphers

If, after completing the lesson, you want to give students more practice using automatic graphers, you might use this activity. Tell students to graph each pair of equations on the same coordinate plane. If the equations intersect, have students use the zoom and trace to approximate the coordinates of the point or points of intersection.

1. $y = x^2 + 4$, $y = -x^2 + 4$ [(0, 4)]
2. $y = x^2 + 1$, $y = -x^2 + 2x + 1$ [(0, 1), (1, 2)]
3. $y = x^2 + 1$, $y = -x^2$ [None]
4. $y = x^2 - 1$, $y = -x + 5$ [(-3, 8), (2, 3)]

Students might also **work with partners.** Have them find the points of intersection of the graphs of equations that they have written.

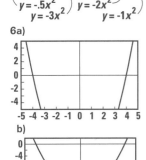

3a)

$y = -.5x^2$ $y = -3x^2$ $y = -2x^2$ $y = -1x^2$

6a) $-5 \le x \le 5$, $-5 \le y \le 5$

b)

3. a. Graph $y = ax^2$ when $a = -\frac{1}{2}, -1, -2,$ and -3.

b. What happens to the graph as a gets smaller? **As a gets smaller, the graph of $y = ax^2$ looks thinner and thinner.**

In 4 and 5, refer to Example 3.

4. At about what age do drivers have the fewest accidents per mile driven? **≈ 45 years of age**

5. At about what ages are there 500 accidents per 50 million miles driven? **≈ 17 and 73 years of age**

6. Graph $y = x^2 - 16$ on the following windows. **See left.**

a. $-5 \le x \le 5, -5 \le y \le 5$

b. a window that shows the vertex, the x-intercepts, and the y-intercept of the parabola

In 7 and 8, an equation and a graph of a parabola are given. Find a different window that shows the vertex and both x-intercepts.

7. $y = x^2 - 10x + 20$ **Sample: $-10 \le x \le 10$; $-10 \le y \le 30$**

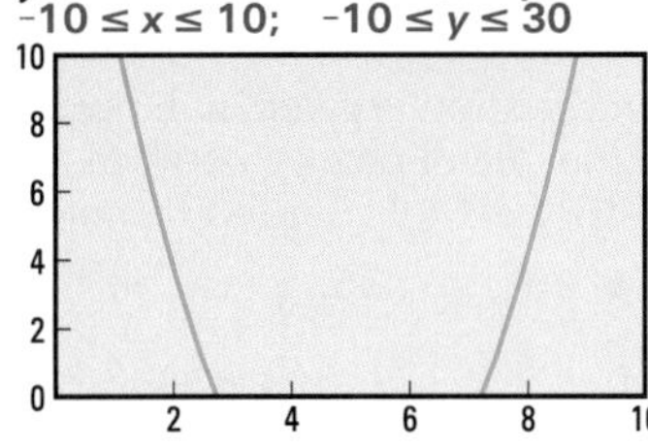

8. $y = -x^2 + 10x - 20$ **Sample: $-5 \le x \le 10$; $-25 \le y \le 10$**

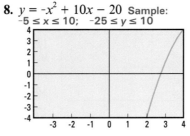

9. a. Graph $y = 2x^2 + 38x + 168$ on a window that shows the key points discussed in the lesson. **See left.**

b. Estimate the coordinates of the vertex of the graph. **(-9, -12)**

c. Give an equation of the axis of symmetry. **$x = -9$**

d. Estimate the x-intercepts. **-12; -7**

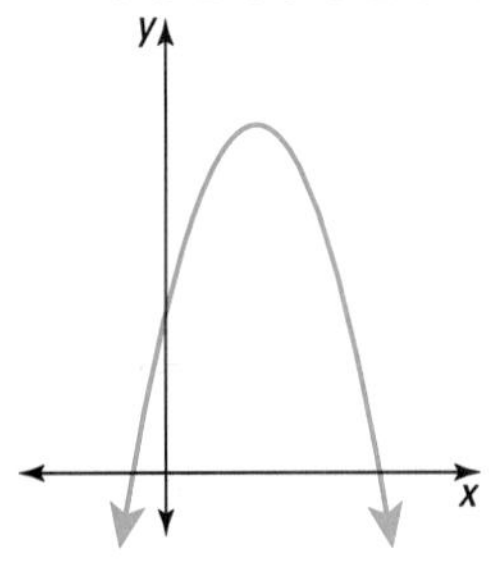

10. At the left is a sketch of the graph of $y = -x^2 + 6x + 7$. The x-intercepts are -1 and 7.

a. What is the x-coordinate of the vertex? **3**

b. What is the y-coordinate of the vertex? **16**

9a)

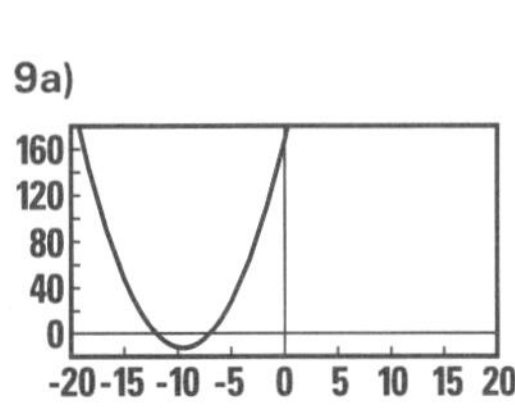

Applying the Mathematics

11. a. Graph on your default window. **See margin for a, d.**

$$y = x^2 - 3$$
$$y = x^2$$
$$y = x^2 + 3$$

b. Describe the graphs. **See below.**

c. Predict what the graph of $y = x^2 + 6$ will look like on the same window. **See below.**

d. Test your prediction by graphing $y = x^2 + 6$.

b) Sample: All the parabolas look alike in the window. They have different vertices and are translation images of each other.

c) It will be congruent to those in part a. It will be a translation image of them and have vertex (0, 6).

Questions 4–5 Multicultural Connection In most states in the United States, a teenager can obtain a driver's license at age 16, after taking an approved driver-education course. In France, teens may begin practice driving at 16 if they have had a minimum of 20 hours of lessons and 3000 kilometers of practice accompanied by an adult. The Czech Republic requires 45 hours of instruction and 16 hours of on-the-road practice. New Zealand has a licensing program in which drivers go through a series of theory classes and driving practice before they are eligible for a lifetime driver's license. If you have students from countries other than the United States in your class, you might ask them what the licensing requirements are in that country.

Questions 7–8 Students are expected to use trial and error for these questions. Ask how these two parabolas are related to each other. [They are reflection images of each other over the x-axis.] Ask how this idea can be used to determine a window for **Question 8** when you know the window for **Question 7**. [Use the same horizontal specifications; change the vertical specifications to their opposites.]

Question 9a Emphasize that there is not a single correct answer.

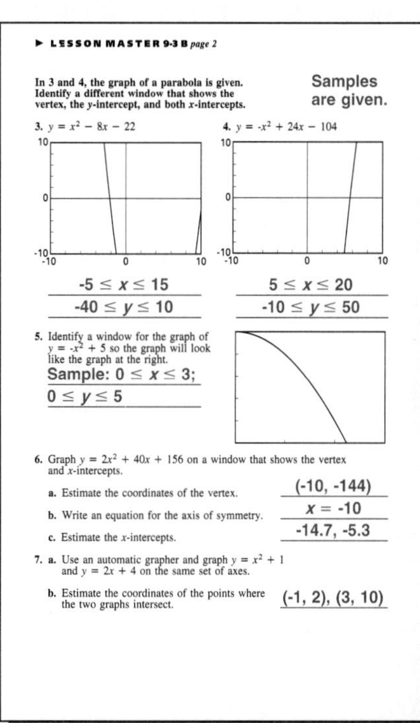

► LESSON MASTER 9-3 B *page 2*

In 3 and 4, the graph of a parabola is given. Identify a different window that shows the vertex, the y-intercept, and both x-intercepts. **Samples are given.**

3. $y = x^2 - 8x - 22$ — **$-5 \le x \le 15$; $-40 \le y \le 10$**

4. $y = -x^2 + 24x - 104$ — **$5 \le x \le 20$; $-10 \le y \le 50$**

5. Identify a window for the graph of $y = -x^2 + 5$ so the graph will look like the graph at the right. **Sample: $0 \le x \le 3$; $0 \le y \le 5$**

6. Graph $y = 2x^2 + 40x + 156$ on a window that shows the vertex and x-intercepts.

a. Estimate the coordinates of the vertex. **(-10, -144)**

b. Write an equation for the axis of symmetry. **$x = -10$**

c. Estimate the x-intercepts. **-14.7, -5.3**

7. a. Use an automatic grapher and graph $y = x^2 + 1$ and $y = 2x + 4$ on the same set of axes.

b. Estimate the coordinates of the points where the two graphs intersect. **(-1, 2), (3, 10)**

Adapting to Individual Needs

Extra Help

To help students understand why the x-coordinate of the vertex is the average of the x-intercepts, review the ideas that the line of symmetry of a parabola passes through the vertex and that when a parabola is folded along the line of symmetry, the two sides "match." The line of symmetry is halfway between the corresponding points on the parabola just as the average of two numbers is halfway between the numbers.

Additional Answers

11a.

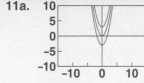

11d.

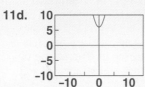

Notes on Questions

Question 13 This is a preview of solving systems of equations. Have students verify that the coordinates of the points of intersection satisfy each equation that is given.

Question 20 Multicultural Connection The changeless, unhurried and wise old turtle is a symbol used in many cultures. To some it is the symbol of longevity and eternal life. To others the turtle represents Mother Earth. Interested students might research other meanings.

Follow-up for Lesson 9-3

Practice

For more questions on SPUR Objectives, use **Lesson Master 9-3A** (shown on page 563) or **Lesson Master 9-3B** (shown on pages 564–565).

Assessment

Quiz A quiz covering Lessons 9-1 through 9-3 is provided in the *Assessment Sourcebook.*

Oral Communication Ask students to explain how the coordinates of the vertex of a parabola can be found when the *x*-intercepts are known. [Students understand how to find the *x*-coordinate of the vertex and substitute the value of *x* in the equation to find the *y*-coordinate.]

Extension

Ask students to explain how they would graph $y \leq x^2$. [Graph the parabola $y = x^2$. Test a point on one side of the parabola to see whether it makes $y < x^2$ true or false. Based on the test, shade the appropriate side.] Then ask them to explain how the graph of $y \leq x^2$ would differ from that of $y < x^2$. [The first graph includes the points on the parabola, so it would be drawn as a solid curve. The second graph does not include the points on the parabola, so it would be a dashed curve.] Finally, have students generalize the idea to the graphing of $y < ax^2 + bx + c$. [The region below the parabola]

Project Update Project 2, *Quadratic Designs*, on page 605, relates to the content of this lesson.

13a)

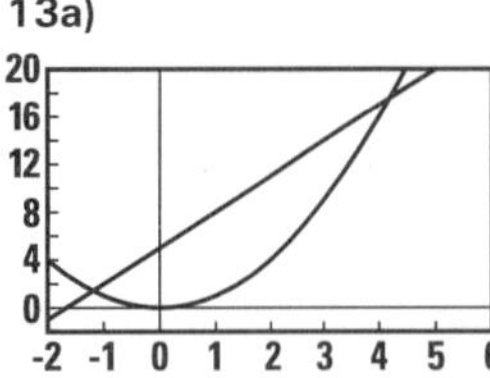

14a)

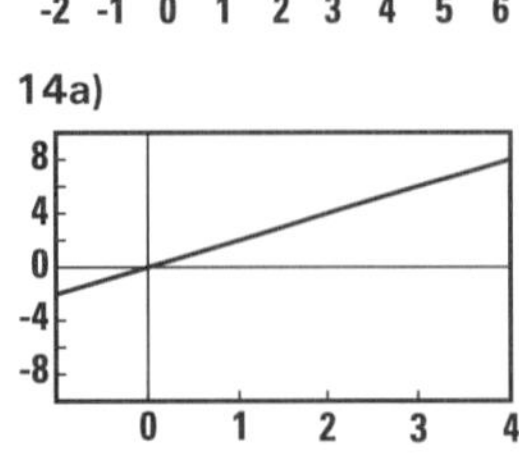

15a)

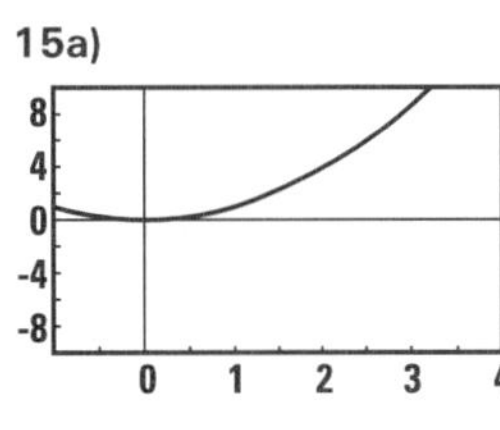

16a)

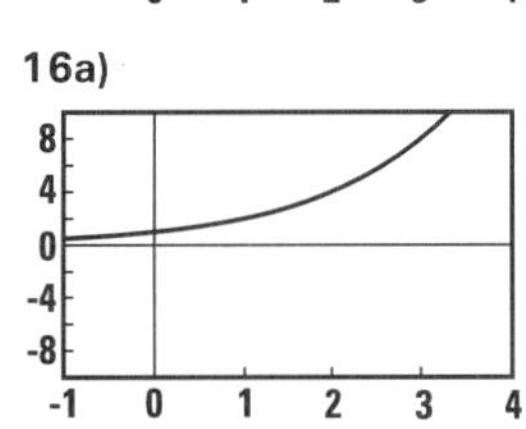

12. The parabola $y = x^2 - 20x + 104$ has vertex (10, 4). Find a window that will produce each of the following views of this parabola.

a.

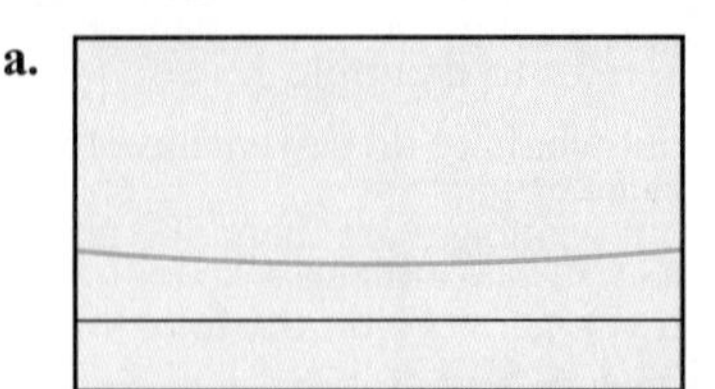

Sample: $9 \leq x \leq 11$; $-10 \leq y \leq 15$

b.

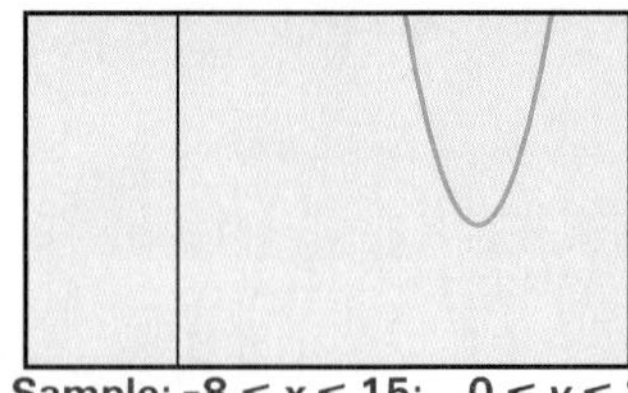

Sample: $-8 \leq x \leq 15$; $0 \leq y \leq 8$

13. a. Graph $y = 3x + 5$ and $y = x^2$ on the same axes. **See left.**

b. Estimate the two points where these graphs intersect. **(-1.25, 1.25); (4, 17)**

Review

In 14–16, an equation is given. **a.** Draw a graph when $-1 \leq x \leq 4$. **b.** Calculate the rate of change between $x = 0$ and $x = 1$. *(Lessons 7-1, 7-5, 8-2, 8-3, 9-1)* **See left for graphs.**

14. $y = 2x$ **b) 2** **15.** $y = x^2$ **b) 1** **16.** $y = 2^x$ **b) 1**

17. Solve $3x^2 + 9 = 156$. *(Lessons 5-9, 9-1)* **$x = 7$ or $x = -7$**

18. Rewrite $7x^2 + 2x + 5 = x^2 - 10x$ in the form $ax^2 + bx + c = 0$. *(Lesson 5-7)* **$6x^2 + 12x + 5 = 0$**

19. In Belleville a taxi costs \$.45 plus \$1.20 for each mile. In Carrolton a taxi ride costs \$1.25 plus \$1.00 for each mile. **a) 4 miles**

a. For what distance(s) traveled do rides in these cities cost the same?

b. When is it cheaper to travel in Belleville? Explain how you found your answer and why you know it is right. *(Lessons 5-2, 5-3, 5-4)* **for distances less than 4 miles; $.45 + 1.20x < 1.25 + 1.00x$; $x < 4$**

20. A tortoise is walking at a rate of 3 $\frac{\text{ft}}{\text{minute}}$. Assume this rate continues.

a. How long will it take the tortoise to travel 20 feet? **$6\frac{2}{3}$ minutes**

b. How long will it take the tortoise to travel *f* feet? *(Lesson 2-4)* **f/3 minutes**

21. *Skill sequence.* If $a = -7$, $b = 6$, and $c = 15$, find the value of each expression. *(Lessons 1-4, 1-6)*

a. $-4ac$ **420** **b.** $b^2 - 4ac$ **456** **c.** $\sqrt{b^2 - 4ac}$ **≈ 21.4**

Exploration

22b) As the value of *b* increases, the graph shifts to the left and down.

22. a. Draw graphs of $y = x^2 + bx + 2$ when $b = 0, 1, 2,$ and 3. **See margin.**

b. How does the value of *b* affect the graph? **See above.**

c. From your answer to part **b,** predict what the graph of $y = x^2 + 100x + 2$ will look like. **See left.**

d. When *b* is negative, where would you expect the graph of $y = x^2 + bx + 2$ to lie? Choose a negative value for *b* and check your prediction. **The vertex will lie in the first quadrant. See margin for check.**

22c) It will have the same shape as the graphs in part a, but will be far off to the left with a vertex in the third quadrant.

566

Adapting to Individual Needs

Challenge

Materials: Automatic graphers

Have students use a grapher to graph $y = (x - 3)^2 + 4$ and $y = (x + 2)^2 - 1$ on the same coordinate plane. Then have them describe the graph of any equation of the form $y = (x - a)^2 + b$. [When the equation is in the form $y = (x - a)^2 + b$, the graph is a parabola with vertex (a, b). The shape of the graph is the same as that for $y = x^2$.]

Additional Answers

22a.

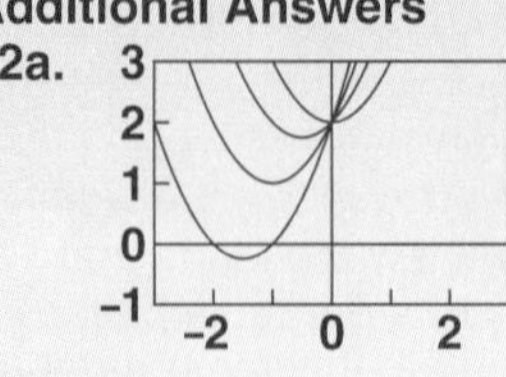

22d.

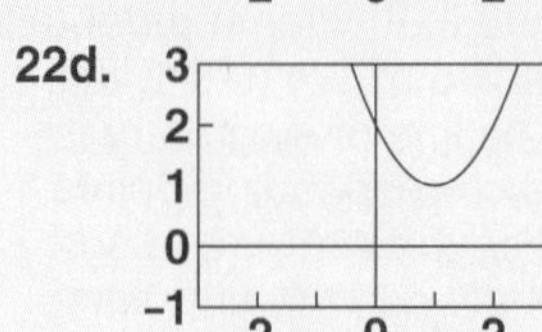

LESSON

9-4

Quadratic Equations and Projectiles

Projectile palette. *Firework flames follow parabolic paths.*

A projectile is an object that is thrown or dropped or shot, and then proceeds with no additional force of its own. Balls, bullets, and some rockets are projectiles.

Equations for Paths of Projectiles

Consider a quarterback who tosses a football to a receiver 40 yards downfield. If the ball is thrown and caught six feet above the ground and is 16 ft above the ground at the peak of the throw (the vertex), then the path of the ball can be graphed as it is below. The path is part of a parabola.

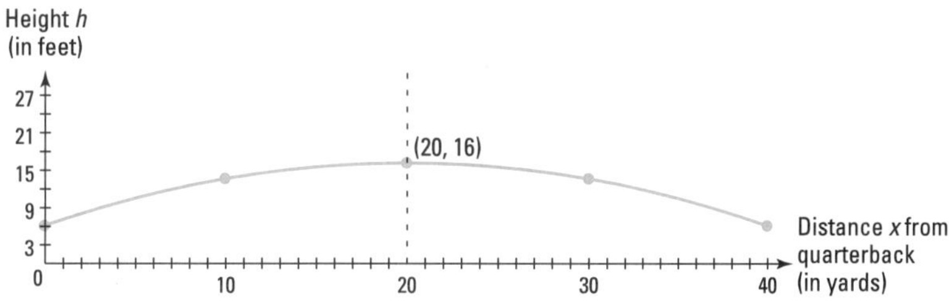

The ball reaches its maximum height halfway between the passer and the receiver, in this case after traveling 20 yards. So this parabola is symmetric to the vertical line $x = 20$. If the ball is at a height of h feet when it has traveled x yards, its path is a piece of the parabola described by the equation

$$h = -0.025x^2 + x + 6.$$

Lesson 9-4

Objectives

E Use quadratic equations to solve problems about paths of projectiles.

Resources

From the ***Teacher's Resource File***
- Lesson Master 9-4A or 9-4B
- Answer Master 9-4
- Teaching Aids
  96 Warm-up
  103 Graphs for Examples 1 and 2
- Technology Sourcebook
  Computer Master 24

Additional Resources
- Visuals for Teaching Aids 96, 103
- GraphExplorer or other automatic graphers

Teaching Lesson 9-4

Warm-up

Find the coordinates of the vertex of the graph of each equation. Tell if the vertex is a minimum or a maximum.

1. $y = -x^2 + 4x$ **(2, 4); maximum**
2. $y = 6x^2$ **(0, 0); minimum**
3. $y = x^2 + 9$ **(0, 9); minimum**
4. $y = -5x^2 + 20x - 13$ **(2, 7); maximum**

Lesson 9-4 Overview

Broad Goals This lesson discusses the use of quadratic equations in describing the height of a projectile.

Perspective Lessons 9-4, 9-5, and 9-6 are very closely related. In Lesson 9-4, students make and interpret graphs to find what x-values will produce a given value in a quadratic equation. In Lesson 9-5, they use the Quadratic Formula to answer the same question algebraically. Lesson 9-6 relates the two approaches and examines when a quadratic equation has one or no solutions.

Two applications of quadratic equations are presented in this lesson. The first is the relation between the *height of a projectile* and its *horizontal distance along the ground.* This quadratic equation describes the *path* of the projectile. The opening situation and **Example 1** illustrate this use of quadratics.

Many people think that the path of a projectile is not symmetric—that once a projectile reaches its peak, it drops down rather quickly (see Figure 1 on page 568). Leonardo da Vinci was unique among his contemporaries in showing that cannon balls follow a curved path (Figure 2), but his sketches were not widely circulated in Europe, and his pictures showing the paths seem to be more circular than parabolic.

(Overview continues on page 568.)

Notes on Reading

You can use **Teaching Aid 103** when discussing **Examples 1–3**.

Whether you read this lesson with students in class or the students read it on their own, emphasize the relation between the solutions that are obtained by substituting in an equation, by "eyeballing" the coordinates of a point on a graph, and by tracing along a graph on an automatic grapher.

You might want to use Activity 1 in *Optional Activities* on page 569 to introduce this lesson.

Reading Mathematics The paragraph above **Example 3** is very important. In **Example 1**, the parabola is the path of the ball, and both the *x*-coordinate and *y*-coordinate represent distances. In **Examples 2 and 3**, the parabola represents the height of the ball over time; thus the *y*-coordinate represents height, but the *x*-coordinate represents time. By finding the coordinates of the vertex of the parabola, you can use either graph to tell how high the ball will go. However, the first graph tells you *where* this will happen; the second tells you *when* this will happen.

Sports Connection Walter Chauncey Camp is known as the father of American football. As a college student, he suggested modifications which revolutionized the game—the scrimmage (today's beginning of a play) was introduced, the number of players was reduced from 15 to 11, and the position of quarterback was created. As coach, Camp was credited for suggestions that led to four downs, the gridiron pattern of the field, and permission to tackle below the waist.

Example 1

Refer to the situation on page 567. Suppose a defender is 3 yards in front of the receiver. This means the defender is 37 yards from the quarterback. Will he be able to deflect or catch the ball?

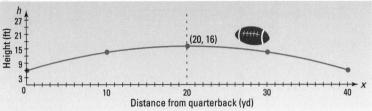

Solution

The answer depends on how high the ball is when it reaches him. The height is represented by h in the above equation. Substitute 37 for x in the equation.

$$\begin{aligned} h &= -0.025x^2 + x + 6 \\ &= -0.025 \cdot 37^2 + 37 + 6 \\ &= -0.025 \cdot 1369 + 37 + 6 \\ &= -34.225 + 37 + 6 \\ &= 8.775 \end{aligned}$$

The ball will be 8.775 feet above the ground when it reaches the defender. Since $0.775 \cdot 12$ inches ≈ 9 inches, this is approximately 8 feet 9 inches. To deflect or intercept the ball, the defender would have to reach a height of 8 feet 9 inches. With a well-timed jump, this is possible for most defenders.

Equations for the Heights of Projectiles over Time

A parabola can also describe the relation between the length of time since a projectile has been thrown or shot into the air and its height above the ground.

Example 2

The equation $h = -16t^2 + 32t + 6$ gives the height h in feet of another ball t seconds after being thrown from a height of 6 feet with an initial upward velocity of 32 feet per second.

a. How high will the ball be a half second after it is thrown?

b. What is the maximum height this ball reaches?

Solution

a. Substitute 0.5 for t in $h = -16t^2 + 32t + 6$.

$$\begin{aligned} h &= -16(.5)^2 + 32(.5) + 6 \\ &= -16(.25) + 16 + 6 \\ &= -4 + 22 \\ &= 18 \end{aligned}$$

In half a second, the ball will be 18 feet high. ▶

Lesson 9-4 Overview, continued

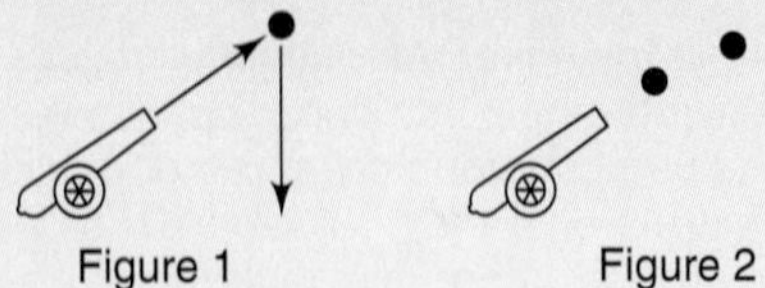

Figure 1 Figure 2

It took the work of the late 16th century scientists Tartaglia (1551) and Galileo (1600) to connect the physical motion of objects with the parabola and with quadratics. That these paths *must be* quadratics (in a situation where gravity is constant) was deduced by Isaac Newton from assumptions about the force of gravity. Since the paths of projectiles befuddled some of the great minds of the middle ages and early Renaissance, we should expect that the concept of parabolic motion may be quite difficult for algebra students. However, present-day students have some advantages that were not available even to the best minds in the past. Today virtually all students have seen slow motion pictures of thrown baseballs, basketballs, and footballs.

The second application of quadratics which is studied in this lesson, is the relationship between the *height of a projectile* and the *time* it has been in the air. **Example 2** illustrates this idea. The graph, while a parabola, is *not a graph of the path* of the projectile. In many books, the first equations for parabolas are like those in this example.

b. Use an automatic grapher. Plot $y = -16x^2 + 32x + 6$. A graph of this equation using the window $0 \le x \le 3$, $0 \le y \le 25$ is drawn below.

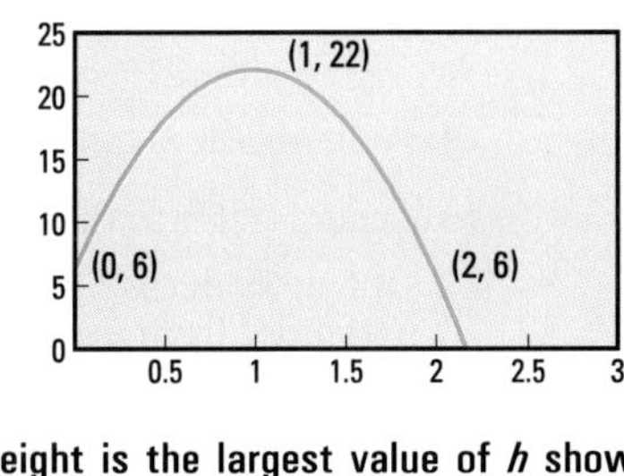

The maximum height is the largest value of h shown on the graph, the y-coordinate of the vertex. Trace along the graph; read the y-coordinate as you go. When we did this using a window $0 \le x \le 3$, $0 \le y \le 25$, our trace showed that (.989, 21.993) and (1.01, 21.998) are on the graph. So try $t = 1$. This gives $h = 22$. **The maximum height reached is 22 ft.**

Check

We verify the maximum height by noting that (0, 6) and (2, 6) are on the graph. This means the axis of symmetry is $t = 1$ and so the vertex is (1, 22).

Caution: The graph in Example 1 describes the actual path of an object, the height at a given *distance* from the start. The graph in Example 2 describes an object's height at a specific *time*. It is essential to read the labels of the axes to know what a graph represents.

Example 3

Refer to the situation in Example 2. Estimate when the ball is 12 feet high.

Solution 1

The values of t corresponding to $h = 12$ must be found. Draw a horizontal line at $h = 12$ feet. Read the x-coordinates at the points of intersection.

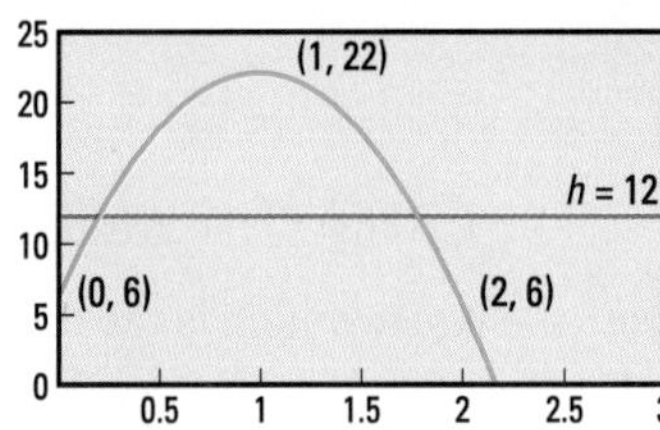

Our grapher showed that at $x \approx 0.21$ and $x \approx 1.79$, $y = 12$. The ball is 12 feet off the ground at about 0.2 second and 1.8 seconds after being thrown. Notice that both these times are 0.8 second away from 1 second, where the vertex of the parabola is located.

It is natural to want to find an exact answer to the question of Example 3. The method for finding it is discussed in Lesson 9-5.

Additional Examples

Because of the coefficients in the equations, we recommend that students use automatic graphers. If students do not have graphers, we recommend that you either provide a partial table of values or allow more time for this work.

A model rocket is shot at an angle into the air from the launch pad.

1. The height of the rocket when it has traveled horizontally x feet from the launch pad is given by $h = -.163x^2 + 11.43x$.

a. Graph this equation.

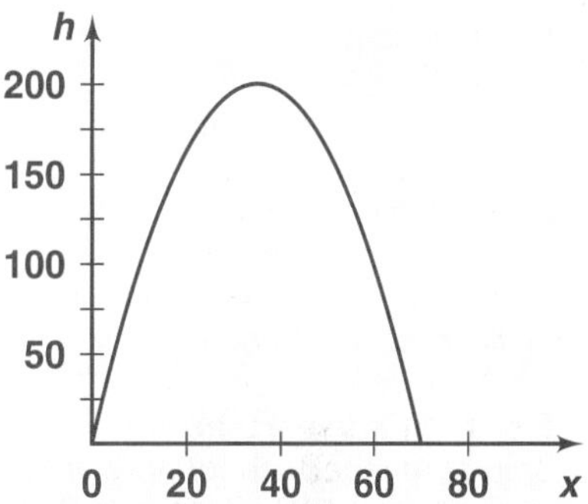

b. A 75-foot tree, 10 feet from the launch pad, is in the path of the rocket. Will the rocket clear the top of the tree? **Yes; at 10 feet from the launch pad, the rocket is 98 feet high.**

c. Estimate the maximum height that the rocket will reach. **About 200 feet**

(Additional Examples continue on page 570.)

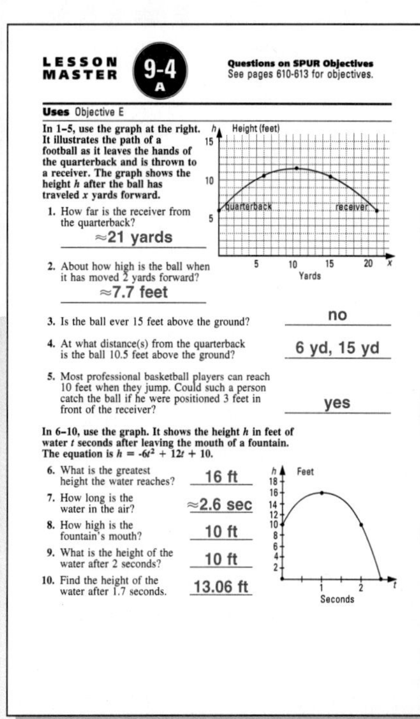

LESSON MASTER 9-4 A — **Questions on SPUR Objectives** See pages 610-613 for objectives.

Uses Objective E

In 1–5, use the graph at the right. It illustrates the path of a football as it leaves the hands of the quarterback and is thrown to a receiver. The graph shows the height h after the ball has traveled x yards forward.

1. How far is the receiver from the quarterback? $\approx$21 yards
2. About how high is the ball when it has moved 2 yards forward? $\approx$7.7 feet
3. Is the ball ever 15 feet above the ground? no
4. At what distance(s) from the quarterback is the ball 10.5 feet above the ground? 6 yd, 15 yd
5. Most professional basketball players can reach 10 feet when they jump. Could such a person catch the ball if he were positioned 3 feet in front of the receiver? yes

In 6–10, use the graph. It shows the height h in feet of water t seconds after leaving the mouth of a fountain. The equation is $h = -6t^2 + 12t + 10$.

6. What is the greatest height the water reaches? 16 ft
7. How long is the water in the air? $\approx$2.6 sec
8. How high is the fountain's mouth? 10 ft
9. What is the height of the water after 2 seconds? 10 ft
10. Find the height of the water after 1.7 seconds. 13.06 ft

Optional Activities

Activity 1

Materials: Rubber ball or tennis ball

You can use this activity to introduce the lesson. Partition a long chalkboard into rectangles using a few reference lines that are approximately equally spaced. Have one student toss a ball from one end of the board to a student at the other end of the board. The ball should be tossed so that it goes just to the top of the board. With a little experimentation, your students should see a very broad parabolic path.

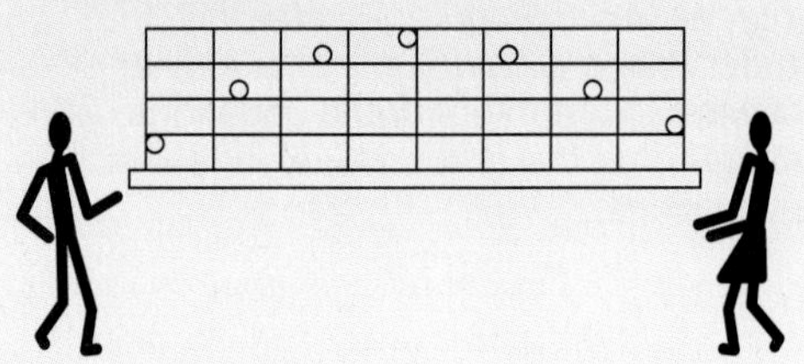

2. The rocket's height h at t seconds after launch is given by $h = -22.2t^2 + 133t$.
   a. Graph this equation.

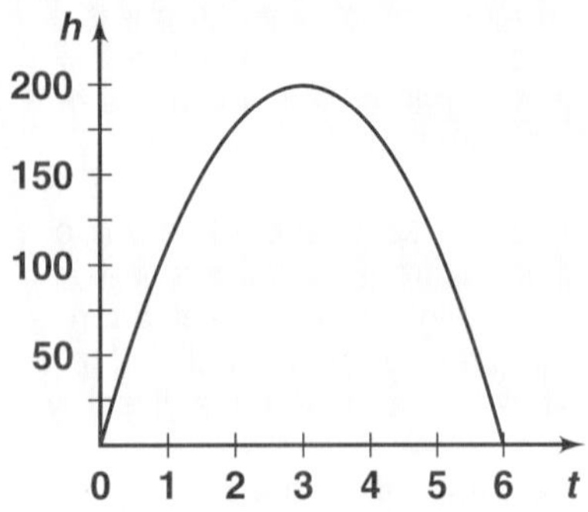

   b. How high is the rocket at 2 seconds? **From the formula, 177.2 ft; from the graph, about 175 ft**
   c. Use the graph to estimate how many seconds it will take for the rocket to reach its maximum height. **About 3 seconds**

3. Refer to Example 2 above. After being launched, about how many seconds will it take for the rocket to return to the ground? **After about 6 seconds**

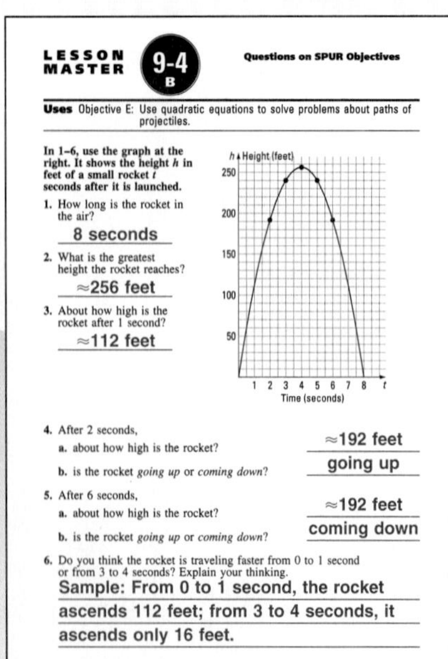

LESSON MASTER 9-4 B — Questions on SPUR Objectives

Uses Objective E: Use quadratic equations to solve problems about paths of projectiles.

In 1–6, use the graph at the right. It shows the height h in feet of a small rocket t seconds after it is launched.

1. How long is the rocket in the air? **8 seconds**
2. What is the greatest height the rocket reaches? **≈256 feet**
3. About how high is the rocket after 1 second? **≈112 feet**
4. After 2 seconds,
   a. about how high is the rocket? **≈192 feet**
   b. is the rocket *going up* or *coming down*? **going up**
5. After 6 seconds,
   a. about how high is the rocket? **≈192 feet**
   b. is the rocket *going up* or *coming down*? **coming down**
6. Do you think the rocket is traveling faster from 0 to 1 second or from 3 to 4 seconds? Explain your thinking. **Sample: From 0 to 1 second, the rocket ascends 112 feet; from 3 to 4 seconds, it ascends only 16 feet.**

QUESTIONS

Covering the Reading

1. What is the shape of the path of a tossed ball? **a parabola**

In 2 and 3, refer to the tossed football at the beginning of the lesson.

2. If a defender is 5 yards in front of the receiver, how far is the defender from the quarterback? **35 yards**
3. What is the height of the ball 39 yards from the quarterback? Write your answer to the nearest inch. **about 7 ft 0 in.**

In 4 and 5, a quarterback throws the ball to a receiver 60 yards away. Suppose the ball's height h in feet x yards away is

$$h = -\frac{x^2}{45} + \frac{4x}{3} + 5.$$

4. At what height does the ball reach the receiver? **5 feet**
5. A defender is 5 yards in front of the receiver.
   a. How far is he from the quarterback? **55 yards**
   b. How high would he have to reach to deflect the ball? **≈ 11.1 feet**
6. When the horizontal axis shows time and the vertical axis shows the height of a tossed ball, what shape is the graph? **a parabola**

In 7–11, refer to the graph below. It shows h, the height in yards of a ball t seconds after it is kicked into the air.

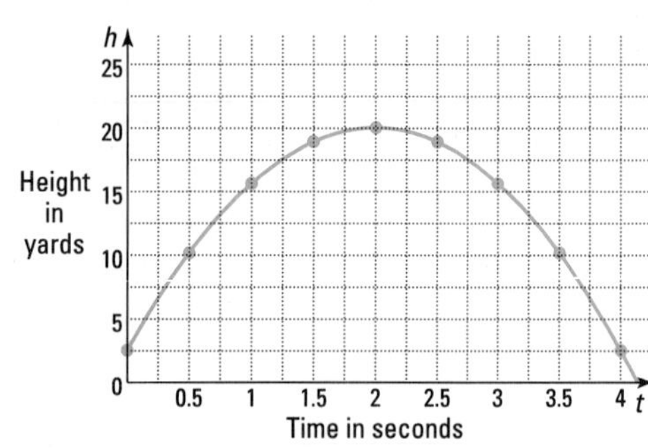

World Cup winner. *Shown is Zinho, a member of the Brazilian soccer team that won the XV World Cup in 1994. The championship was Brazil's fourth since the World Cup was started in 1930.*

7. What is the greatest height the ball reaches? **20 yards**
8. How high is the ball 1 second after it is kicked? **15 yards**
9. At what times is the ball 18 yards high? **≈ 1.5 and 2.5 seconds**
10. How long is the ball in the air? **a little more than 4 seconds**
11. For how many seconds is the ball more than 15 yards above the ground? **2 seconds**

Optional Activities

Activity 2 Technology Connection
You may wish to assign *Technology Sourcebook, Computer Master 24.* Students use *GraphExplorer* or similar software to graph quadratic equations and interpret the graphs for a variety of applications.

Adapting to Individual Needs

English Language Development
Define *projectile* as anything thrown forcibly or shot forward, and give a physical example by throwing a crumpled ball of paper into the trash can. Write the word *projectile* on the board. Then ask students to give other examples of projectiles. [Samples: a soccer ball being kicked, a basketball being thrown, a volley ball being served.]

Applying the Mathematics

In 12–15, a small rocket is shot from the edge of a cliff. Suppose that after t seconds, the rocket is y meters above the cliff where $y = 30t - 5t^2$. This equation is graphed below.

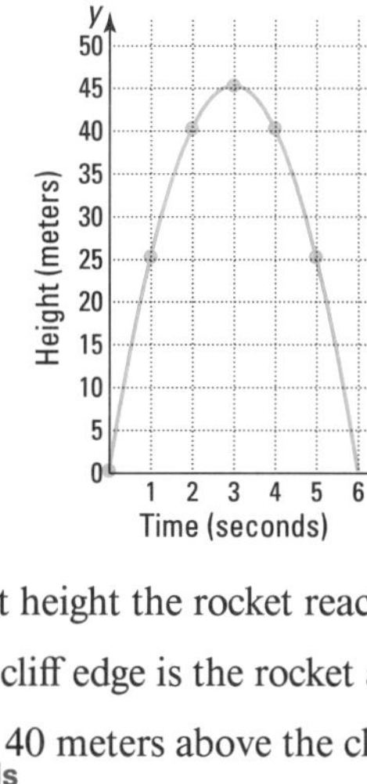

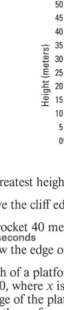

12. What is the greatest height the rocket reaches? **45 meters**

13. How far above the cliff edge is the rocket after 5 seconds? **25 meters**

14. When is the rocket 40 meters above the cliff edge? **after 2 and 4 seconds**

15. How far below the edge of the cliff is the rocket after 7 seconds? **35 meters**

In 16–18, the path of a platform diver is described by the equation $y = -2x^2 + x + 10$, where x is the horizontal distance (in meters) of the diver from the edge of the platform, and y is the height of the diver (in meters) above the surface of the water.

16. a. Evaluate y when $x = 1$. **$y = 9$**

b. Write a sentence about the diver that describes what you found out in part **a.** **Sample: When the diver is one meter from the edge of the diving board, he is 9 meters above the surface of the water.**

17. What is the maximum height the diver reaches? **10.125 m above the surface of the water.**

18. In horizontal distance, how far in front of the platform will the diver hit the water? **2.5 m**

Review

In 19 and 20 an equation is given. **a.** Make a table of values for integer values of x from -3 to 3. **b.** Graph the equation. *(Lessons 9-1, 9-2)*

19. $y = \frac{1}{2}x^2$

20. $y = 6x - x^2$ **See margin.**

19a)

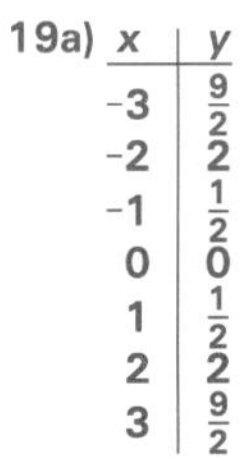

x	y
-3	$\frac{9}{2}$
-2	2
-1	$\frac{1}{2}$
0	0
1	$\frac{1}{2}$
2	2
3	$\frac{9}{2}$

b)

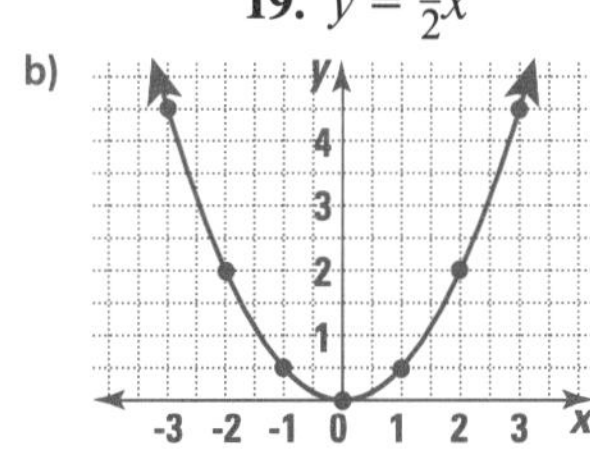

Notes on Questions

Emphasize throughout the discussion of these questions that several solution strategies are possible: numerical approaches involving substitution and graphical approaches involving reading coordinates of points. Encourage students to check the results of one approach against the other.

Questions 12–15 In order to simplify calculations in these problems, we have replaced $g \approx 9.8$ m/sec^2 with $g \approx 10$ m/sec^2. This accounts for the coefficient of t^2 being 5, rather than 4.9.

Question 15 This situation has meaning for a negative height; it is a height below the edge of the cliff. Accordingly, the graph continues down past the point where $x = 6$. However, in this problem it does not make sense to include negative heights before time zero.

Additional Answers

20a, b

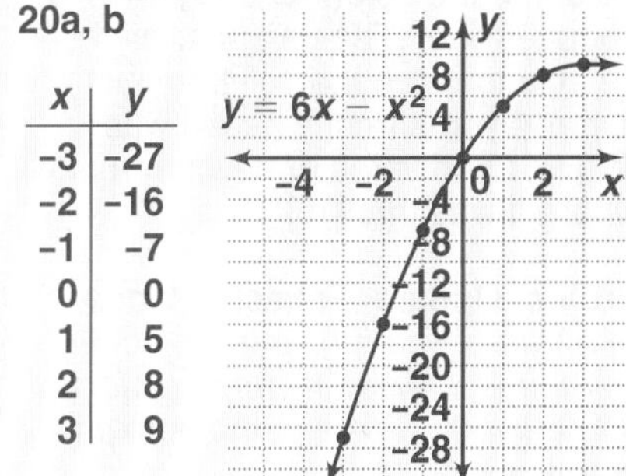

x	y
-3	-27
-2	-16
-1	-7
0	0
1	5
2	8
3	9

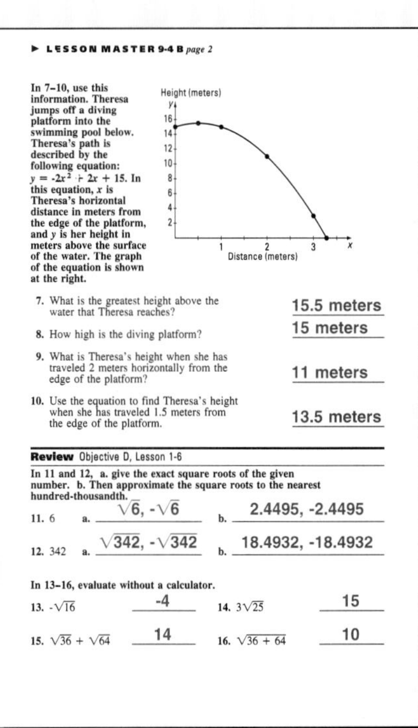

▶ LESSON MASTER 9-4 B *page 2*

In 7–10, use this information. Theresa jumps off a diving platform into the swimming pool below. Theresa's path is described by the following equation: $y = -2x^2 + 2x + 15$. In this equation, x is Theresa's horizontal distance in meters from the edge of the platform, and y is her height in meters above the surface of the water. The graph of the equation is shown at the right.

7. What is the greatest height above the water that Theresa reaches? **15.5 meters**
8. How high is the diving platform? **15 meters**
9. What is Theresa's height when she has traveled 2 meters horizontally from the edge of the platform? **11 meters**
10. Use the equation to find Theresa's height when she has traveled 1.5 meters from the edge of the platform. **13.5 meters**

Review Objective D, Lesson 1-6

In 11 and 12, **a.** give the exact square roots of the given number. **b.** Then approximate the square roots to the nearest hundred-thousandth.

11. 6 a. $\sqrt{6}, -\sqrt{6}$ b. 2.4495, -2.4495
12. 342 a. $\sqrt{342}, -\sqrt{342}$ b. 18.4932, -18.4932

In 13–16, evaluate without a calculator.

13. $-\sqrt{16}$ **-4**
14. $3\sqrt{25}$ **15**
15. $\sqrt{36} + \sqrt{64}$ **14**
16. $\sqrt{36 + 64}$ **10**

Adapting to Individual Needs

Extra Help

Some students might be confused by the use of variables other than x and y in the equations and graphs of this lesson. Remind them of earlier work that they have done using equations to solve problems, and that they have frequently chosen a variable that was the first letter of the unknown in the problem. Point out that the first graph in this lesson uses h to represent the y-axis because the y-coordinate of each point on the graph represents height. The graph used in Examples 2 and 3 uses h and t to represent height and time. Be sure students understand that the choice of variables is theirs but that using the first letter of the word for the quantity involved often helps relate the equation and the graph to the problem.

Practice

For more questions on SPUR Objectives, use **Lesson Master 9-4A** (shown on page 569) or **Lesson Master 9-4B** (shown on pages 570–571).

Assessment

Oral/Written Communication Have students draw parabolas with two x-intercepts, one x-intercept, and no x-intercept. [Drawings of parabolas illustrate knowledge of the meaning of x-intercept.]

Extension

As an extension of **Example 1**, have students identify all locations where a defender who is 6 feet 6 inches tall might be able to catch the ball. (Assume the defender can jump about 3 feet). [The defender might be able to catch the ball from just under 4 yards from where it was thrown to 36 yards or beyond from where it was thrown.]

Project Update Project 4, *Video of a Parabolic Path*, and Project 6, *Parabolas in Manufactured Objects* on page 606, relate to the content of this lesson.

In 21 and 22, use the graph of the parabola at the right. *(Lesson 9-2)*

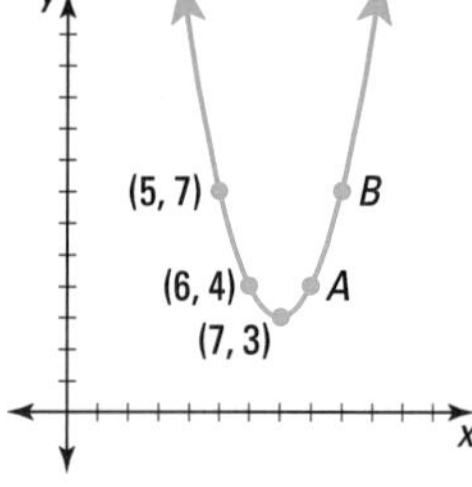

21. Use the symmetry of the parabola to find the coordinates of points A and B.
A = (8, 4); B = (9, 7)

22. Write an equation of the axis of symmetry.
x = 7

23. *Multiple choice.* Which of the following is *not* equal to ab^2? *(Lessons 8-5, 8-8)* **c**

(a) $a \cdot b \cdot b$ (b) $a(b^2)$ (c) $(ab)^2$ (d) $a(b)^2$

24. *Skill sequence.* Write as a single fraction. *(Lesson 3-9)*

a. $2 + \frac{7}{11}$ $\frac{29}{11}$ **b.** $2 + \frac{x}{y}$ $\frac{2y + x}{y}$ **c.** $a + \frac{b}{c}$ $\frac{ac + b}{c}$

25. What is the image of (-6, 0) after a slide 3 units to the right and 5 units down? *(Lesson 3-4)* **(-3, -5)**

26. *Multiple choice.* What situation can the graph at right represent? *(Lesson 3-3)* **b**

(a) the cost c of h pencils at 10¢ each
(b) the cost c of h pencils at 2 for 25¢
(c) the cost c of h pencils at 25¢ each
(d) the cost c of h pencils at 50¢ each

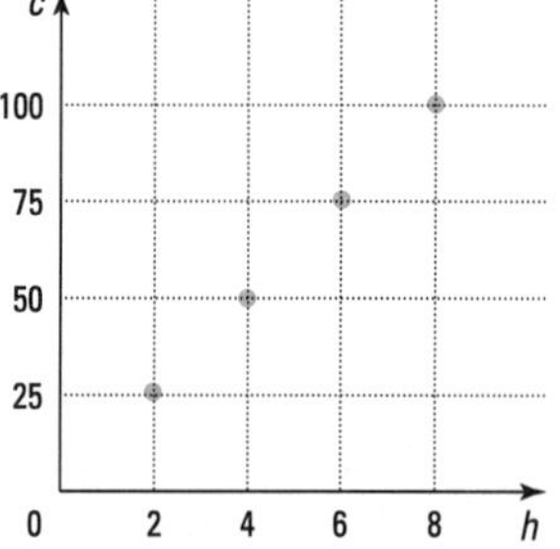

27. Find $\frac{8 + 4\sqrt{2}}{4}$ to the nearest tenth. *(Lesson 1-6)* **≈ 3.4**

28. Let $a = 6$, $b = 4$, and $c = -2$. Evaluate each expression. *(Lessons 1-4, 1-6)*

a. $\sqrt{b^2 - 4ac}$ **8**

b. $\frac{-b + \sqrt{b^2 - 4ac}}{2a}$ **$\frac{1}{3} \approx 0.33$**

c. $\frac{-b - \sqrt{b^2 - 4ac}}{2a}$ **-1**

Exploration

29. A pitched ball in baseball does not always follow the path of a parabola. Why not? **Samples: spinning of the ball; air-resistance; wind**

Adapting to Individual Needs

Challenge

Have students use a reference book, such as an advanced algebra book, to look up the meaning of the term *conic section*. Then have them draw a diagram to show how a parabola can be obtained by slicing a cone by a plane parallel to an edge of the cone. [A sample drawing is shown at the right.]

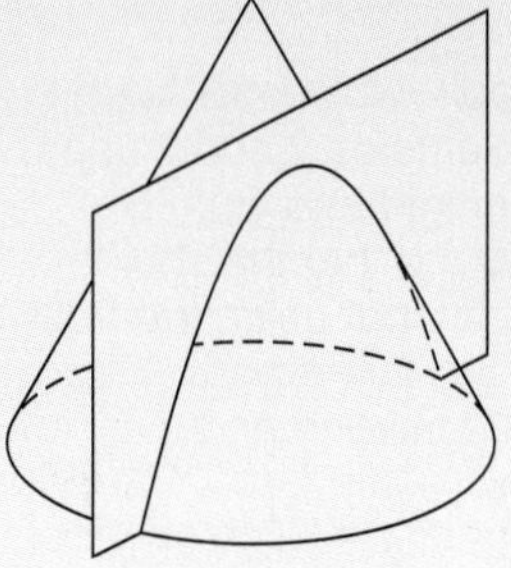

Setting Up Lesson 9-5

Discuss **Questions 16–18** so that students are familiar with the use of quadratic equations to describe the path of a diver. Point out that a diver, too, can be thought of as a projectile. The opening situation for Lesson 9-5 concerns divers in Acapulco, Mexico.

LESSON

9-5

The Quadratic Formula

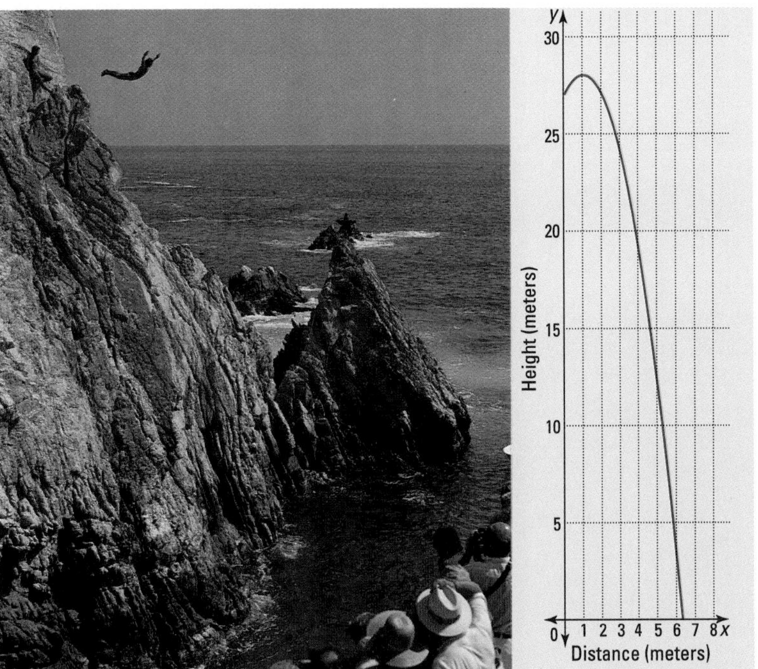

Quadratics at La Quebrada. *Divers at La Quebrada must time their dives to hit a large wave surge to ensure that the water is deep enough. The dive's path can be described with a quadratic equation.*

A Situation Leading to a Quadratic Equation

When a person dives, he or she becomes a projectile, so the person's path is part of a parabola. In Acapulco, Mexico, there is a famous place called "La Quebrada" ("the break in the rocks"). There, a diver dives from a height of approximately 27 meters to the waters below. If the diver is x meters away from the cliff and y meters above the water, then under certain conditions $y = -x^2 + 2x + 27$ describes the path of the dive.

When the diver enters the water, the height $y = 0$. So solving

$$-x^2 + 2x + 27 = 0$$

will give the number of meters from the cliff the diver enters the water. From the graph you can see that x is between 6 and 7. The equation $-x^2 + 2x + 27 = 0$ is an example of a quadratic equation. A **quadratic equation** is an equation that can be written in the form

$$ax^2 + bx + c = 0.$$

The Quadratic Formula

You can find solutions to *any* quadratic equation by using the *Quadratic Formula.* The formula tells how to calculate x using a, b, and c, the coefficients of the terms. It states that there are possibly two solutions.

Lesson 9-5

Objectives

A Solve quadratic equations using the Quadratic Formula.

E Use quadratic equations to solve problems about paths of projectiles.

Resources

From the ***Teacher's Resource File***

- Lesson Master 9-5A or 9-5B
- Answer Master 9-5
- Teaching Aids
  - 96 Warm-up
  - 104 Path of a Diver

Additional Resources

- Visuals for Teaching Aids 96, 104

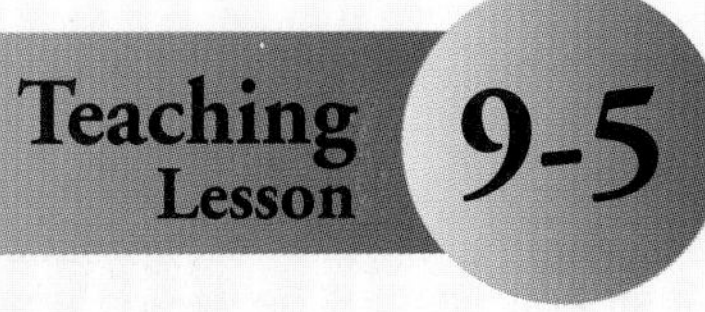

Warm-up

Evaluate each expression when $a = 4$, $b = -5$, and $c = 1$.

1. $-b$ 5
2. $b^2 - 4ac$ 9
3. $\sqrt{b^2 - 4ac}$ 3
4. $-b + \sqrt{b^2 - 4ac}$ 8
5. $-b - \sqrt{b^2 - 4ac}$ 2
6. $\frac{-b + \sqrt{b^2 - 4ac}}{2a}$ 1
7. $\frac{-b - \sqrt{b^2 - 4ac}}{2a}$ $\frac{1}{4}$

Lesson 9-5 Overview

Broad Goals The goal is for students to use the Quadratic Formula $x = \frac{-b \pm \sqrt{b^2 - 4ac}}{2a}$ to solve equations of the form $ax^2 + bx + c = 0$.

Perspective Since the calculator has made the Quadratic Formula even more important than it was 20 years ago, we suggest that students memorize the formula as quickly as possible. Older algebra texts had to stay away from applications of the quadratic formula because of the difficulty of evaluating the radical in the formula. However, today we do not have to omit expressions that would be difficult to compute by hand.

Since students have calculators, we expect them to carry out their computations to get approximations of noninteger solutions when solving problems about real situations. A solution of $\frac{-2 + \sqrt{112}}{-2}$ meters, such as the one found in **Example 1**, conveys little meaning; it is difficult to relate to the graph of $y = -x^2 + 2x + 27$ without a decimal approximation.

Notes on Reading

Reading Mathematics You may want to read this lesson together with your students in class.

The motivation to solve a quadratic equation is piqued by asking the question, "If we know how high the diver is, what is his or her location?" For some heights, there are two values of x distance from the cliff. Most divers actually start their dives by leaping up. Students will understand this more readily if you point out that a swimming-pool diving board has a spring to it. In the case of the Acapulco divers, the vertex is reached very quickly because the divers are leaping from rocks, rather than from a springboard.

Social Studies Connection The harbor of the present-day city of Acapulco was discovered in 1531 by Hernando Cortes. By the late 1500s, it had become a main depot for Spanish colonial fleets traveling between Mexico and the Orient. It was also a major export point for coffee, sugar, and other local products for steamship lines between Panama and San Francisco. Today, the harbor is considered one of the best on the Pacific coast of Mexico, and it is one of the finest natural places in the world for ships to anchor.

❶ To help students understand this problem, you might refer to the graph of $y = -x^2 + 2x + 27$ that is shown at the beginning of the lesson and also on **Teaching Aid 104**. The negative solution to **Example 1** can be related to the graph; however, the path of the diver is only part of the parabola. The diver's path does not include the portion of the parabola for which $x < 0$.

If $ax^2 + bx + c = 0$, then

$$x = \frac{-b + \sqrt{b^2 - 4ac}}{2a} \quad \text{or} \quad x = \frac{-b - \sqrt{b^2 - 4ac}}{2a}$$

The two expressions differ in only one way: $\sqrt{b^2 - 4ac}$ is added to $-b$ in the numerator of the first, while it is subtracted from $-b$ in the numerator of the second.

To apply the Quadratic Formula, notice that a is the coefficient of x^2, b is the coefficient of x, and c is the constant term.

The work in calculating the two solutions is almost the same. To save writing, both solutions can be written in one expression using the symbol $\pm$, which means *plus or minus.* It shows that you should do the calculation twice, once by adding and once by subtracting.

Quadratic Formula
If $ax^2 + bx + c = 0$ and $a \neq 0$, then

$$x = \frac{-b \pm \sqrt{b^2 - 4ac}}{2a}$$

The Quadratic Formula is one of the most famous formulas in all of mathematics. *You should memorize it today.*

Caution: Many calculators have a +/– key. That key takes the opposite of a number. It does *not* perform the two operations + and – required in the Quadratic Formula.

Applying the Quadratic Formula

❶ **Example 1**

Solve $-x^2 + 2x + 27 = 0$ to find the distance of the Acapulco diver from the cliff when he enters the water.

Solution

Recall that $-x^2 = -1x^2$. So rewrite the equation as

$$-1x^2 + 2x + 27 = 0.$$

Apply the Quadratic Formula with $a = -1$, $b = 2$, and $c = 27$.

$$x = \frac{-b \pm \sqrt{b^2 - 4ac}}{2a}$$

$$x = \frac{-2 \pm \sqrt{2^2 - 4 \cdot -1 \cdot 27}}{2 \cdot -1}$$

Using order of operations, work under the radical sign first.

$$= \frac{-2 \pm \sqrt{4 - (-108)}}{-2}$$

$$= \frac{-2 \pm \sqrt{112}}{-2}$$

Optional Activities

The form of the Quadratic Formula that is found in the United States is not universal. In some places, the standard form for a quadratic equation is $x^2 + 2bx + c = 0$, which yields the formula $x = -b \pm \sqrt{b^2 - c}$. The work it takes to get the quadratic into the form $x^2 + 2bx + c = 0$ seems well worth the effort, given the comparative simplicity of this alternate formula.

After students complete the lesson, you might have them **work in groups** and put the equation in **Example 3** in the form $x^2 + 2bx + c = 0$, and then solve it using $x = -b \pm \sqrt{b^2 - c}$. [Write $m^2 - 3m = 14$ as $m^2 + (-3)m - 14 = 0$. Then write the coefficient of m as 2 times some number: $-3 = 2(-\frac{3}{2})$. Now the equation can be written as $m^2 + 2(-\frac{3}{2})m + (-14) = 0$ and the formula can be used to find m.

$$m = \frac{3}{2} \pm \sqrt{\left(-\frac{3}{2}\right)^2 - (-14)}$$

$$= \frac{3}{2} \pm \sqrt{\frac{9}{4} + 14}$$

$$= \frac{3}{2} \pm \sqrt{\frac{65}{4}}$$

$$\text{So } x = \frac{-2 + \sqrt{112}}{-2} \quad \text{or} \quad x = \frac{-2 - \sqrt{112}}{-2}.$$

These are exact solutions. Using a calculator gives approximations. Since $\sqrt{112} \approx 10.6$,

$$x \approx \frac{-2 + 10.6}{-2} \quad \text{or} \quad x \approx \frac{-2 - 10.6}{-2}.$$

$$\text{So } x \approx -4.3 \quad \text{or} \quad x \approx 6.3.$$

The diver cannot land a negative number of meters from the cliff, so the solution $x \approx -4.3$ does not make sense in this situation. The diver will enter the water about 6.3 meters away from the cliff.

Check

Does 6.3 work in the equation $-x^2 + 2x + 27 = 0$? Substitute 6.3 for x.

$$-(6.3)^2 + 2(6.3) + 27 = -39.69 + 12.6 + 27 = -0.09.$$

This is close enough to zero, given that 6.3 is an approximation.

When the coefficients of a quadratic equation are integers and the number under the radical sign in the Quadratic Formula equals a perfect square, its solution can be written without a radical sign.

Example 2

Solve $3x^2 - 6x - 45 = 0$.

❷ **Solution**

First relate the equation to $ax^2 + bx + c = 0$ and identify a, b, and c.

$$3x^2 - 6x - 45 = 0$$

$a = 3$, $b = -6$, and $c = -45$

Now substitute these values in the Quadratic Formula.

$$x = \frac{-b \pm \sqrt{b^2 - 4ac}}{2a}$$

$$x = \frac{-(-6) \pm \sqrt{(-6)^2 - 4 \cdot 3 \cdot -45}}{2 \cdot 3}$$

$$x = \frac{6 \pm \sqrt{36 - (-540)}}{6}$$

Now simplify under the radical sign.

$$x = \frac{6 \pm \sqrt{576}}{6}$$

$$x = \frac{6 \pm 24}{6}$$

$$x = \frac{6 + 24}{6} = \frac{30}{6} = 5 \quad \text{or} \quad x = \frac{6 - 24}{6} = \frac{-18}{6} = -3$$

Check

To check, substitute each of the solutions in the original equation.

Substitute 5 for x. $3(5)^2 - 6(5) - 45 = 75 - 30 - 45 = 0$.

Substitute -3 for x. $3(-3)^2 - 6(-3) - 45 = 27 + 18 - 45 = 0$.

Both solutions check.

❷ Learning to use the Quadratic Formula requires practice, but practice without knowing whether answers are correct or not will often hide errors. Students have to learn ways of checking their answers. Thus, while reading, students should ask themselves if they can find ways to check that are different from those in the lesson.

As a second check for **Example 2,** notice that both sides of the original equation can be easily divided by 3. This gives $x^2 - 2x - 15 = 0$. Using the Quadratic Formula on that equation gives:

$$x = \frac{-(-2) \pm \sqrt{(-2)^2 - 4 \cdot 1 \cdot (-15)}}{2 \cdot 1}$$

$$= \frac{2 \pm \sqrt{4 + 60}}{2}$$

$$= 5 \text{ or } -3.$$

The answers check.

Adapting to Individual Needs

Extra Help

Some students might still make mistakes identifying the values of a, b, and c in a quadratic equation that involves subtraction. In the equation $2x^2 - 9x - 5 = 0$, they might say $b = 9$ and $c = 5$ instead of -9 and -5, respectively. Suggest that these students rewrite the equation changing all subtractions to additions. Remind them that subtracting any number is the same as adding its opposite. Use **Example 2** to illustrate this idea. Rewrite $3x^2 - 6x - 45 = 0$ as $3x^2 + -6x + -45 = 0$. Then write the equation for **Example 3** on the board, and have a volunteer rewrite it using addition. [$m^2 + -3m + -14 = 0$]

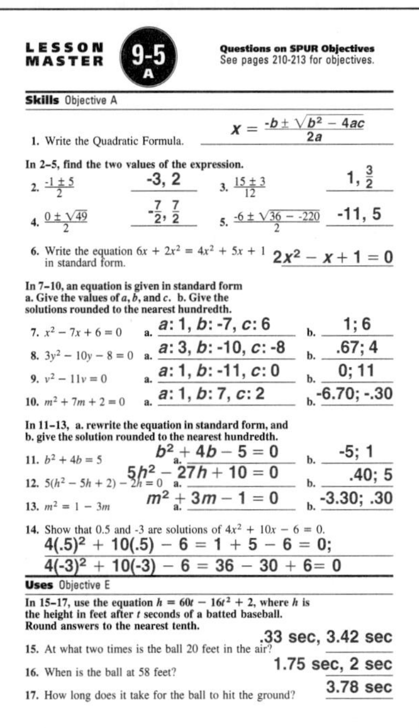

LESSON MASTER 9-5 A

Questions on SPUR Objectives
See pages 210-213 for objectives.

Skills Objective A

1. Write the Quadratic Formula. $x = \frac{-b \pm \sqrt{b^2 - 4ac}}{2a}$

In 2–5, find the two values of the expression.

2. $\frac{-1 \pm 5}{2}$ -3, 2
3. $\frac{15 \pm 3}{12}$ $1, \frac{3}{2}$
4. $\frac{0 \pm \sqrt{49}}{2}$ $-\frac{7}{2}, \frac{7}{2}$
5. $\frac{-6 \pm \sqrt{36 - -220}}{2}$ -11, 5
6. Write the equation $6x + 2x^2 = 4x^2 + 5x + 1$ in standard form. $2x^2 - x + 1 = 0$

In 7–10, an equation is given in standard form a. Give the values of a, b, and c. b. Give the solutions rounded to the nearest hundredth.

7. $x^2 - 7x + 6 = 0$ a. a: 1, b: -7, c: 6 b. 1; 6
8. $3y^2 - 10y - 8 = 0$ a. a: 3, b: -10, c: -8 b. .67; 4
9. $v^2 - 11v = 0$ a. a: 1, b: -11, c: 0 b. 0; 11
10. $m^2 + 7m + 2 = 0$ a. a: 1, b: 7, c: 2 b. -6.70; -.30

In 11–13, a. rewrite the equation in standard form, and b. give the solution rounded to the nearest hundredth.

11. $b^2 + 4b = 5$ a. $b^2 + 4b - 5 = 0$ b. -5; 1
12. $5(h^2 - 5h + 2) - 2h = 0$ a. $5h^2 - 27h + 10 = 0$ b. .40; 5
13. $m^2 = 1 - 3m$ a. $m^2 + 3m - 1 = 0$ b. -3.30; .30
14. Show that 0.5 and -3 are solutions of $4x^2 + 10x - 6 = 0$.
$4(.5)^2 + 10(.5) - 6 = 1 + 5 - 6 = 0;$
$4(-3)^2 + 10(-3) - 6 = 36 - 30 + 6 = 0$

Uses Objective E

In 15–17, use the equation $h = 60t - 16t^2 + 2$, where h is the height in feet after t seconds of a batted baseball. Round answers to the nearest tenth.

15. At what two times is the ball 20 feet in the air? .33 sec, 3.42 sec
16. When is the ball at 58 feet? 1.75 sec, 2 sec
17. How long does it take for the ball to hit the ground? 3.78 sec

For **Example 3**, the check for decimal approximations will not come out exactly, and it may require some discussion. For instance, checking 5.53 in $m^2 - 3m = 14$ yields 13.9909, not 14. (This idea is considered in **Question 9.)**

Additional Examples

1. Refer to the situation about the diver. Solve $10 = -x^2 + 2x + 27$ to find out how far the diver is away from the cliff when he or she is 10 meters above the water. **About 5.2 meters**
2. Solve $2r^2 - 11r + 12 = 0$. **$r = 1.5$ or $r = 4$**
3. Solve $5k^2 + 5k - 2 = 2k^2 - 2k$. **$k = \frac{-7 \pm \sqrt{73}}{6} \approx 0.26$ or -2.59**

LESSON MASTER 9-5 B — **Questions on SPUR Objectives**

Skills Objective A: Solve quadratic equations using the Quadratic Formula.

1. Write the *quadratic formula* and explain what *a*, *b*, and *c* represent.
   $x = \frac{-b \pm \sqrt{b^2 - 4ac}}{2a}$; *a*, *b*, and *c* are the coefficients *a*, *b*, and *c* in the equation $ax^2 + bx + c^2 = 0$.

In 2–7, find the two values of the expression. Round answers that are *not* integers to the nearest tenth.

2. $\frac{-2 \pm 8}{2}$ **3, -5**
3. $\frac{9 \pm 5}{4}$ **3.5, 1**
4. $\frac{-6 \pm 15}{-6}$ **3.5, -1.5**
5. $\frac{0 \pm \sqrt{81}}{3}$ **3, -3**
6. $\frac{-4 \pm \sqrt{24 - -120}}{2}$ **4, -8**
7. $\frac{-2 \pm \sqrt{4 - -10}}{2 \cdot 15}$ **0.1, -0.2**

In 8–13, write each equation in standard form.

8. $x^2 + 40 + 4x = 0$ **$x^2 + 4x + 40 = 0$**
9. $3t^2 + 9t = 2$ **$3t^2 + 9t - 2 = 0$**
10. $4v^2 = 80 + 6v^2 - 14v$ **$2v^2 - 14x + 80 = 0$**
11. $3w^2 - 4w - 18 = -w^2 + w + 1$ **$4w^2 - 5w - 19 = 0$**
12. $12y - y^2 = 0$ **$y^2 - 12y = 0$**
13. $16x - x^2 + 3 = 16x$ **$x^2 - 3 = 0$**

In Example 3, the equation has to be put into $ax^2 + bx + c = 0$ form before the formula can be applied. This form is called the **standard form for a quadratic equation.**

Example 3

Solve $m^2 - 3m = 14$. Give *m* to the nearest hundredth.

Solution

To apply the Quadratic Formula, you must write the equation in the form $ax^2 + bx + c = 0$. To put $m^2 - 3m = 14$ in this form add -14 to both sides.

$$m^2 - 3m - 14 = 0$$

Now the formula can be applied with $a = 1$, $b = -3$, $c = -14$.

$$m = \frac{-(-3) \pm \sqrt{(-3)^2 - 4 \cdot 1 \cdot -14}}{2 \cdot 1}$$

$$= \frac{3 \pm \sqrt{9 - (-56)}}{2}$$

$$= \frac{3 \pm \sqrt{65}}{2}$$

Thus $m = \frac{3 + \sqrt{65}}{2}$ or $m = \frac{3 - \sqrt{65}}{2}$.

$m \approx \frac{3 + 8.06}{2}$ or $m \approx \frac{3 - 8.06}{2}$.

So $m \approx 5.53$ or $m \approx -2.53$.

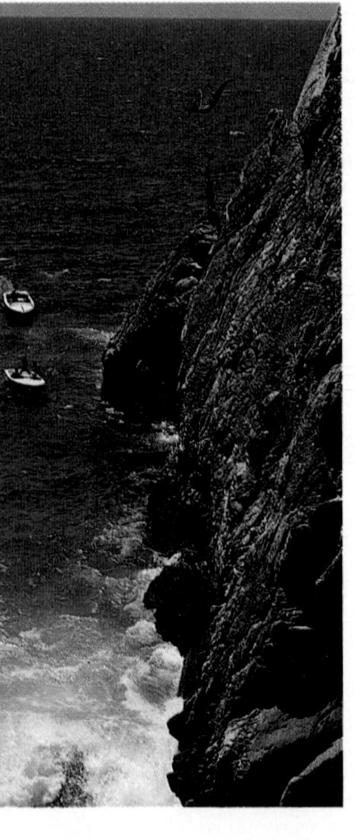

QUESTIONS

Covering the Reading

1. State the Quadratic Formula. **See below.**
2. *True or false.* The Quadratic Formula can be used to solve any quadratic equation. **True**
3. Find the two values of $\frac{-10 \pm 4}{2}$. **-3; -7**
4. Suppose the cliff diver at the beginning of the lesson dove from a cliff 22 meters high. Then under certain conditions, solving the equation $0 = -x^2 + 2x + 22$ gives the number of meters the diver is from the cliff when entering the water. **a) $a = -1$; $b = 2$; $c = 22$**
   a. Give the values of *a*, *b*, and *c* for use in the quadratic formula.
   b. Apply the Quadratic Formula to give the exact solutions to the equation. **$x = \frac{-2 \pm \sqrt{92}}{-2}$**
   c. Approximate the solutions to the nearest tenth. **$x \approx -3.8$ or $x \approx 5.8$**
   d. How far away from the cliff does the diver enter the water? **≈ 5.8 m**

1) If $ax^2 + bx + c = 0$ and $a \neq 0$, then $x = \frac{-b \pm \sqrt{b^2 - 4ac}}{2a}$.

Adapting to Individual Needs

English Language Development

Your might want to pay close attention to the way students read the Quadratic Formula. Make sure that they read the ± symbols as *plus or minus.* Also make sure they understand that this means there are two possible solutions. For example, 6 ± 2 means 6 + 2 or 6 – 2.

In 5–8, each sentence is equivalent to a quadratic equation of the form $ax^2 + bx + c = 0$. **a.** Give the values of a, b, and c. **b.** Give the exact solutions to the equation. **c.** Check the solutions. **b, c) See left.**

5. $12x^2 + 7x + 1 = 0$
a) $a = 12$ $b = 7$ $c = 1$

6. $3n^2 + n - 2 = 0$
a) $a = 3$ $b = 1$ $c = -2$

7. $x^2 + 6x + 9 = 0$
a) $a = 1$ $b = 6$ $c = 9$

8. $-x^2 + 4 = 0$
a) $a = -1$ $b = 0$ $c = 4$

9. Check both decimal answers to Example 3. **See left.**

In 10–13, a quadratic equation is given. **a.** Rewrite each equation in the form $ax^2 + bx + c = 0$. **b.** Solve the equation using the Quadratic Formula. Round solutions to the nearest hundredth. **See below for b.**

10. $20m^2 - 6m = 2$
$20m^2 - 6m - 2 = 0$

11. $3w^2 - w = 5$
$3w^2 - w - 5 = 0$

12. $3 - x = 2x^2$ $2x^2 + x - 3 = 0$

13. $3p^2 - 10p = 14 + 9p$
$3p^2 - 19p - 14 = 0$

10b) 0.50 or -0.20
11b) 1.47 or -1.14 12b) -1.50 or 1.00 13b) 7.00 or -0.67

Applying the Mathematics

14. The solutions to $ax^2 + bx + c = 0$ are the x-intercepts of the graph of $y = ax^2 + bx + c$.
a. Use the Quadratic Formula to find the x-intercepts of the graph of $y = 2x^2 + 3x - 2$. $\frac{1}{2}$; -2
b. Check by using an automatic grapher. **See left.**

15. When a ball on the moon is thrown upward with an initial velocity of 6 meters per second, its approximate height y after t seconds is given by $y = -0.8t^2 + 6t$.
a. At what *two* times will it reach a height of 10 m? Give your answer to the nearest tenth of a second. 2.5; 5.0
b. Graph $y = -0.8t^2 + 6t$. Use a domain of $0 \le t \le 7.5$. **See left.**
c. Use the graph to check your answer to part **a.**

16. Some students noticed that the equation of Example 2 could be simplified by dividing both sides by 3.
a. What is this simpler equation? $x^2 - 2x - 15 = 0$
b. Solve the simpler equation using the Quadratic Formula. 5; -3

Review

In 17 and 18, a rock is tossed up from a cliff with upward velocity of 10 meters per second. The formula $h = 10t - 4.9t^2$ gives the height h of the rock *above the cliff* after t seconds.

17. How high above the cliff is the rock 1 second after it is tossed?
5.1 meters above the cliff

18. How high above the cliff is the rock 3 seconds after it is tossed? *(Lesson 9-4)* 14.1 meters below the cliff

5b) $x = -\frac{1}{4}$ or $-\frac{1}{3}$
c) $12\left(-\frac{1}{4}\right)^2 + 7\left(-\frac{1}{4}\right) + 1 =$
$\frac{3}{4} + \frac{-7}{4} + 1 = 0$
$12\left(-\frac{1}{3}\right)^2 + 7\left(-\frac{1}{3}\right) + 1 =$
$\frac{4}{3} + \frac{-7}{3} + 1 = 0$

6b) $x = \frac{2}{3}$ or -1
c) $3\left(\frac{2}{3}\right)^2 + \frac{2}{3} - 2 =$
$\frac{4}{3} + \frac{2}{3} - 2 = 0$
$3(-1)^2 + (-1) - 2 =$
$3 + (-1) - 2 = 0$

7b) $x = -3$
c) $(-3)^2 + 6(-3) + 9 =$
$9 - 18 + 9 = 0$

8b) $x = -2$ or 2
c) $-(-2)^2 + 4 = -4 + 4 = 0$
$-(2)^2 + 4 = -4 + 4 = 0$

9) $(5.53)^2 - 3(5.53) = 14?$
$30.5809 - 16.59 = 14?$
$13.9909 \approx 14$
Yes, it checks.
$(-2.53)^2 - 3(-2.53) = 14?$
$6.4009 + 7.59 = 14?$
$13.9909 \approx 14$
Yes, it checks.

14b)

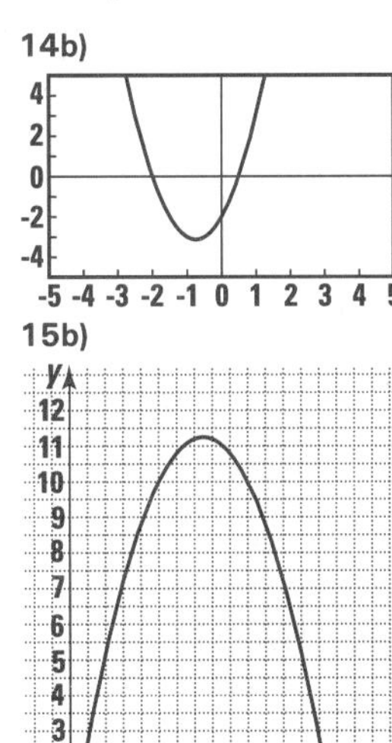

15b)

Notes on Questions

Question 5 Error Alert Some students will change the fractions $-\frac{1}{4}$ and $-\frac{1}{3}$ to decimals for checking; discourage this practice. Not only is an exact check easy with the fractions, but it also gives students practice with fractions and provides the satisfaction that the check is exact.

Questions 10–13 Asking students for both exact solutions and decimal approximations helps them to distinguish between the two.

Question 16b This method could also be used in **Example 2**.

Question 18 Emphasize that there is a physical interpretation of negative values of h. When h is negative, the rock is below the level of the cliff-top.

► **LESSON MASTER 9-5 B** *page 2*

In 14–17, an equation in standard form is given. a. Identify a, b, and c. b. Give the solutions rounded to the nearest hundredth.

14. $x^2 - 10x + 16 = 0$
a. $a =$ 1 $b =$ -10 $c =$ 16
b. $x = 8$ or $x = 2$

15. $-v^2 + 14v + 33 = 0$
a. $a =$ -1 $b =$ 14 $c =$ 33
b. $v = 16.06$ or $v = -2.06$

16. $2s^2 + 5s - 3 = 0$
a. $a =$ 2 $b =$ 5 $c =$ -3
b. $s = 0.5$ or $s = -3$

17. $y^2 + 12y = 0$
a. $a =$ 1 $b =$ 12 $c =$ 0
b. $y = 0$ or $y = -12$

In 18–21, an equation is given. a. Rewrite the equation in standard form. b. Give the solutions rounded to the nearest hundredth.

18. $a^2 - 9a = 18$
a. $a^2 - 9a - 18 = 0$
b. $a = 10.68$ or $a = -1.68$

19. $-2d^2 = 3d - 15$
a. $-2d^2 - 3d + 15 = 0$
b. $d = 2.09$ or $d = -3.59$

20. $x^2 + 36 = 12x$
a. $x^2 - 12x + 36 = 0$
b. $x = 6$

21. $4(n^2 + 2n) + 3 = 10$
a. $4n^2 + 8n - 7 = 0$
b. $x = .66$ or $x = -2.66$

22. Write a check for your solutions to Question 16.
$2(0.5)^2 + 5(0.5) - 3 = 0.5 + 2.5 - 3 = 0$
$2(-3)^2 + 5(-3) - 3 = 18 - 15 - 3 = 0$

Uses Objective E: Use quadratic equations to solve problems about paths of projectiles.

In 23–25, use this information: A rocket is launched from a cliff 30 feet above the ground. Its height h in feet above the ground t seconds after it is launched is given by the equation $h = -16t^2 + 192t + 30$.

23. At what two times is the rocket 100 feet above the ground? ≈ 11.6 sec, $\approx .4$ sec

24. When is the rocket 606 feet above the ground? after 6 sec

25. When does the rocket hit the ground? in ≈ 12.2 sec

Adapting to Individual Needs

Challenge

For any quadratic equation that is written in the form $ax^2 + bx + c = 0$, the sum of the roots is $-\frac{b}{a}$ and the product of roots is $\frac{c}{a}$. Have students verify this idea for each of the following equations.

1. $x^2 - 3x - 4 = 0$ [$x = 4$ or $x = -1$;
Sum: $-\frac{-3}{1} = 3$; $4 + (-1) = 3$
Product: $\frac{-4}{1} = -4$; $4 \cdot -1 = -4$]

2. $4x^2 - 11x - 3 = 0$ [$x = -\frac{1}{4}$ or $x = 3$;
Sum: $-\frac{-11}{4} = 2\frac{3}{4}$; $-\frac{1}{4} + 3 = 2\frac{3}{4}$
Product: $\frac{-3}{4} = -\frac{3}{4}$; $-\frac{1}{4} \cdot 3 = -\frac{3}{4}$]

3. $6x^2 - 28x + 30 = 0$ [$x = 1\frac{2}{3}$ or $x = 3$;
Sum: $-\frac{-28}{6} = 4\frac{2}{3}$; $1\frac{2}{3} + 3 = 4\frac{2}{3}$
Product: $\frac{30}{6} = 5$; $1\frac{2}{3} \cdot 3 = 5$]

Follow-up for Lesson 9-5

Practice

For more questions on SPUR Objectives, use **Lesson Master 9-5A** (shown on page 575) or **Lesson Master 9-5B** (shown on pages 576–577).

Assessment

Written Communication Have each student write two questions—one question involving the formula given in **Question 4** and the other question involving the formula given in **Question 15**. One question should require using the Quadratic Formula. Have students provide answers to their questions. [Students' questions can be answered using the equations. Students understand how to use the quadratic formula.]

Extension

The idea in **Question 22** can be used to determine a general method for finding the vertex of a parabola $y = ax^2 + bx + c$. From the Quadratic Formula, the two x-intercepts are $\frac{-b + \sqrt{b^2 - 4ac}}{2a}$ and $\frac{-b - \sqrt{b^2 - 4ac}}{2a}$. The x-coordinate of the vertex is their mean, $-\frac{b}{2a}$. Once the x-coordinate of the vertex is known, the y-coordinate can be found by evaluating $ax^2 + bx + c$ for that value of x.

Your students may enjoy using this method to find maximum or minimum points of parabolas. The formula for the height y in feet above the ground of a ball thrown vertically with an initial velocity of v ft/sec is $y = vx - 16x^2$. You might ask students to find the maximum heights for different initial velocities using this method and to verify their answers using an automatic grapher.

Project Update Project 5, *Computer Programs for Solutions of Quadratics*, on page 606, relates to the content of this lesson.

In 19–21, use the graph below. It shows $h = -.12x^2 + 2x + 5$, the path of a basketball free throw, where h is the height of the ball when it has moved x feet forward. *(Lesson 9-3)*

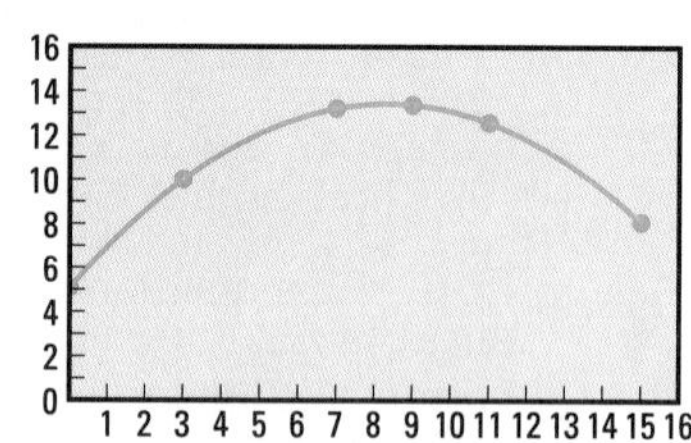

19. What is the height of the ball when it has moved 3 feet forward? **≈ 10 ft**

20. About how far has the ball moved forward when it was at a height of 12 feet? **5 ft or ≈ 11.7 ft**

21. Estimate the greatest height the ball reaches. **≈ 13.5 ft**

22. If the x-intercepts of a parabola are 3 and -1, what is the x-coordinate of the vertex? *(Lesson 9-2)* **1**

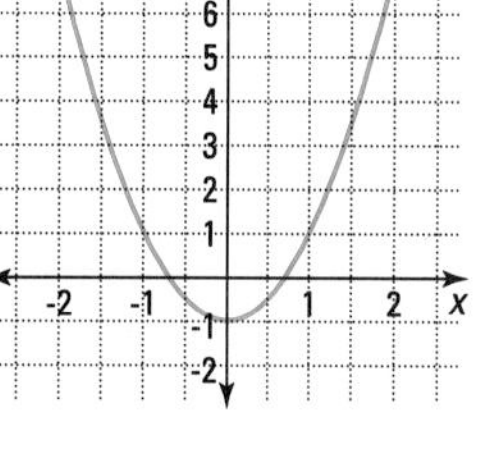

23. *Multiple choice.* Which equation is graphed at the left? *(Lesson 9-2)* **d**

(a) $y = 2x$ (b) $y = 2x^2$
(c) $y = 2x^2 + 1$ (d) $y = 2x^2 - 1$

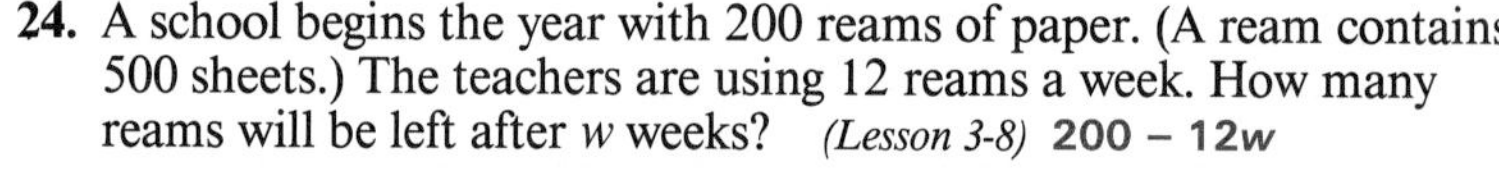

24. A school begins the year with 200 reams of paper. (A ream contains 500 sheets.) The teachers are using 12 reams a week. How many reams will be left after w weeks? *(Lesson 3-8)* **200 − 12w**

25. Mr. Robinson is ordering a new car. He can choose from 5 models, 12 exterior colors, and 9 interior colors. How many combinations of models, interiors, and exteriors are possible? *(Lesson 2-9)* **540**

26. Find the volume of a box that is $15a$ inches in length, $8b$ inches in width, and $6c$ inches in height. *(Lesson 2-1)* **720abc in^3**

Exploration

27. *A team's opening batter named Nero*
Squared his number of hits, the hero!
After subtracting his score,
He took off ten and two more,
And the number resulting was zero.

How many hits did Nero have? **4**

Video

Wide World of Mathematics The segment, *La Quebrada Divers,* provides dramatic footage of these divers. The segment may be used to introduce or extend a lesson on the quadratic formula. Related questions and an investigation are provided in videodisc stills and in the Video Guide. A related CD-ROM activity is also available.

Videodisc Bar Codes

Search Chapter 44

Play

LESSON

9-6

Analyzing Solutions to Quadratic Equations

Diving dolphins. *Trained dolphins perform stunts such as leaping 20 ft to ring a bell. By graphing an equation of the path of a dive, you can find a maximum dive height.*

Using Graphs to Determine the Number of Real Solutions

In the last lesson, the path of a diver from a cliff was described by the equation

$$y = -x^2 + 2x + 27,$$

where y is the height of the diver when he or she is x meters out from the cliff. As the graph on page 573 shows, when the diver first pushes off he or she arches upward and then descends. In this lesson you will learn several ways to determine whether or not the diver reaches a particular height. For instance, will the diver ever reach a height of 27.5 meters above the water? Will the diver ever reach a height of 28 meters? Will he or she ever reach a height of 29 meters?

First let's look at how you can use graphs to answer these questions. Below on the left is a graph of $y = -x^2 + 2x + 27$ on the window $0 \le x \le 8$, $0 \le y \le 40$. This is the parabola graphed on page 573.

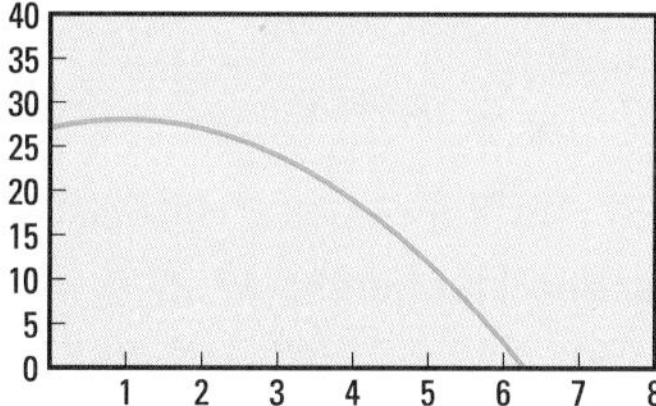

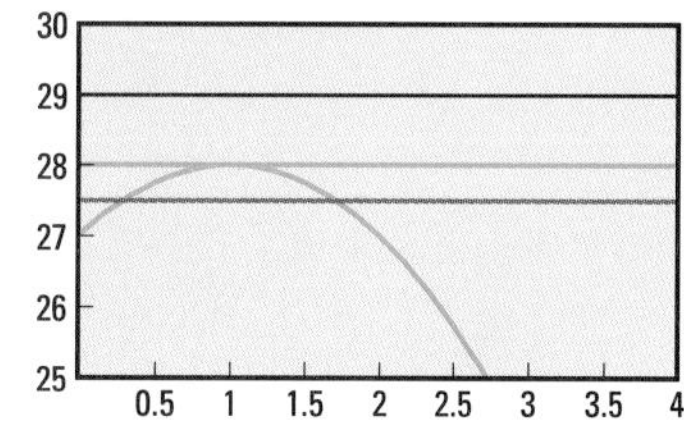

Can you tell from the graph if the height y ever reaches the values 27.5, 28, or 29?

Lesson 9-6

Objectives

D Identify and use the discriminant of a quadratic equation.

Resources

From the ***Teacher's Resource File***

- Lesson Master 9-6A or 9-6B
- Answer Master 9-6
- Assessment Sourcebook: Quiz for Lessons 9-4 through 9-6
- Teaching Aids
  - 96 Warm-up
  - 104 Path of a Diver

Additional Resources

- Visuals for Teaching Aids 96, 104
- Automatic graphers

Teaching Lesson 9-6

Warm-up

1. Write the Quadratic Formula. **If $ax^2 + bx + c = 0$ and $a \neq 0$, then $x = \frac{-b \pm \sqrt{b^2 - 4ac}}{2a}$.**
2. Compare what you wrote with the Quadratic Formula given on page 574. If necessary, make changes to what you wrote.
3. Turn your paper over and rewrite the formula from memory.

Lesson 9-6 Overview

Broad Goals This lesson provides another day to work on the Quadratic Formula and on the relationship between the graph of a quadratic equation and solutions to a quadratic equation. The only new content is the analysis of the discriminant $b^2 - 4ac$ to determine the number of solutions to the quadratic equation.

Perspective In Lesson 9-5, students learned that a quadratic equation may have zero, one, or two solutions. Here they learn that the key to the number of solutions is found in the discriminant. This idea is related not only to the Quadratic Formula, but also to the graph of the equation and to the application of the cliff diver. The ties between these various representations are important.

To prepare students and to pique their curiosity for the nonreal solutions that they will study in their next algebra course, you should inform them that we are talking about solutions that are real numbers. If the coefficients a, b, and c are real, and the discriminant D is negative, then there are two nonreal solutions. These solutions can be written using the square roots of negative numbers that appear in the Quadratic Formula.

Notes on Reading

When discussing the reading, you might want to use **Teaching Aid 104** which contains the graph on page 573. You can connect it to the graphs on page 579 and to the situations in the examples. A knowledge of the parabola of the diver's path helps students to see why **Example 2** has only one solution and why **Example 3** has no solution. Note that the horizontal lines for $y = 29$ and $y = 28$ have been shown. Students are, in effect, finding the intersection of a line and a parabola.

From the graph on the left on page 579, you should be able to see that if the diver ever does get to these heights, it will be when x is between 0 and 4. So on the right we have zoomed in on the top of the parabola. This gives another graph of $y = -x^2 + 2x + 27$, this time on the window $0 \le x \le 4$, $25 \le y \le 30$. To help us see whether there are heights of 27.5, 28, and 29, we have also graphed the three horizontal lines $y = 27.5$, $y = 28$, and $y = 29$.

From the graph on the right on page 579, we can make the following conclusions:

1. The line $y = 27.5$ crosses the parabola twice. So the diver reaches a height of 27.5 meters at *two* times.
2. The line $y = 28$ appears to intersect the parabola once at the vertex of the parabola. So the diver reaches a height of 28 meters *once.*
3. The line $y = 29$ never intersects the parabola. It is completely above the parabola. So the diver *never* reaches a height of 29 meters.

Using the Quadratic Formula to Determine the Number of Real Solutions

We can answer the same questions about the diver by using the Quadratic Formula.

Example 1

Will the diver ever reach a height of 27.5 meters above the water? If so, where does this event occur?

Solution

Let $y = 27.5$ in the equation $y = -x^2 + 2x + 27$.

$$27.5 = -x^2 + 2x + 27$$

Add -27.5 to both sides to put the equation in standard form.

$$0 = -x^2 + 2x - 0.5$$

Substitute $a = -1$, $b = 2$, and $c = -0.5$ in the Quadratic Formula.

$$x = \frac{-2 \pm \sqrt{2^2 - 4 \cdot -1 \cdot -0.5}}{2 \cdot -1}$$

$$x = \frac{-2 \pm \sqrt{2}}{-2}$$

$$x \approx \frac{-2 + 1.414}{-2} \quad \text{or} \quad x \approx \frac{-2 - 1.414}{-2}$$

$$x \approx 0.3 \quad \text{or} \quad x \approx 1.7$$

The diver reaches the height of 27.5 meters twice, at about 0.3 meters from the cliff (on the way up) and about 1.7 meters from the cliff (on the way down).

Optional Activities

You might use this activity after discussing the table in the middle of page 582. Have each student **work with a partner**. For each of the questions, tell them to compare b^2 and $4ac$ by replacing "?" with $>$, $<$, or $=$. Then have them write and solve an equation that satisfies the relationship between b^2 and $4ac$.

1. An equation has exactly two real solutions when b^2? $4ac$. [$>$; Sample equation: $2x^2 + 5x + 3 = 0$; $x = -1$ or $-\frac{3}{2}$]
2. An equation has exactly one solution when b^2 ? $4ac$. [$=$; Sample equation: $4x^2 - 4x + 1 = 0$; $x = \frac{1}{2}$]
3. An equation has no real solutions when b^2? $4ac$. [$<$; Sample equation: $3x^2 + 2x + 6 = 0$]

Example 2

Will the diver ever reach a height of 28 meters above the water? If so, where does this occur?

Solution

Let $y = 28$ in the equation $y = -x^2 + 2x + 27$.

$$28 = -x^2 + 2x + 27$$

Write the equation in standard form by adding -28 to both sides.

$$0 = -x^2 + 2x - 1$$

Substitute $a = -1$, $b = 2$, and $c = -1$ in the Quadratic Formula.

$$x = \frac{-2 \pm \sqrt{2^2 - 4 \cdot -1 \cdot -1}}{2 \cdot -1}$$

$$x = \frac{-2 \pm \sqrt{0}}{-2}$$

$$x = \frac{-2 \pm 0}{-2}$$

Because 0 is added and subtracted, the two values for x are identical.

$$x = \frac{-2 + 0}{-2} \quad \text{or} \quad x = \frac{-2 - 0}{-2}$$

$$x = \frac{-2}{-2} \quad \text{or} \quad x = \frac{-2}{-2}$$

$$x = 1 \quad \text{or} \quad x = 1$$

Thus when $y = 28$, there is just one value of x. **The diver reaches a height of 28 meters once, when he or she is 1 meter from the cliff.** This agrees with the graph, where the height of the vertex is 28.

Example 3

Will the diver ever reach a height of 29 meters? If so, where does this occur?

Solution

Solve

$$29 = -x^2 + 2x + 27.$$

$$0 = -x^2 + 2x - 2$$

Substitute $a = -1$, $b = 2$, and $c = -2$ in the Quadratic Formula.

$$x = \frac{-2 \pm \sqrt{2^2 - 4 \cdot -1 \cdot -2}}{2 \cdot -1}$$

$$x = \frac{-2 \pm \sqrt{-4}}{-2}$$

Since no real number multiplied by itself equals -4, there is no square root of -4 in the real number system. So $29 = -x^2 + 2x + 27$ does not have a solution. **The diver never reaches a height of 29 meters.** This should not be a surprise, since we have already seen that there was no point on the parabola with a height of 29.

Adapting to Individual Needs

Extra Help

Some students might need further help understanding why there is no real solution to a quadratic equation when $b^2 - 4ac$ is a negative number. Ask students what number multiplied by itself equals 16. [4, -4] Then remind them that the radical, or square root sign, implies the positive square root, so $\sqrt{16} = 4$. Then ask if they can find the square root of -16—that is, can they find a number which, when multiplied by itself, equals -16. Some students might pick 4 and -4. Remind them that 4 and -4 are two different numbers, an idea that can be illustrated using a number line.

Notes on Reading

Emphasize a limitation of the discriminant. Though the discriminant allows you to determine how many solutions a quadratic equation has, it doesn't tell if the solutions make sense in the situation which gave rise to the quadratic equation.

Additional Examples

Suppose that a ball is thrown into the air, and its height h (in feet) after t seconds is given by the formula $h = -16t^2 + 32t + 4$.

1. After how many seconds is the ball 12 feet in the air? **0.29 sec and 1.71 sec**
2. When will the ball reach a height of 20 feet? **After 1 second**
3. Will the ball ever reach a height of 25 feet? Why or why not? **No; the discriminant is negative so there is no real solution to $-16t^2 + 32t + 4 = 25$.**
4. Without solving, determine how many real solutions the equation $25x^2 - 10x + 1 = 0$ has. **1**

In general, the number of solutions to a quadratic equation is related to the number $b^2 - 4ac$ in the Quadratic Formula. This is the number under the radical sign in the expression $\frac{-b \pm \sqrt{b^2 - 4ac}}{2a}$. In Example 1, $b^2 - 4ac$ equals 2, which is positive. Adding $\sqrt{2}$ and subtracting $\sqrt{2}$ gives two different results when computing x. So there are two solutions. In Example 2, $b^2 - 4ac$ is 0. So $x = \frac{-2 + 0}{-2}$ and $x = \frac{-2 - 0}{-2}$ give the same result, 1. That quadratic equation has just one solution. There is no solution to the equation in Example 3 because $b^2 - 4ac = -4$, and $\sqrt{-4}$ is not a real number.

The Discriminant of a Quadratic Equation

Here is a summary of the examples.

height	equation	$b^2 - 4ac$	number of real solutions	number of times parabola intersects line
27.5 meters	$-x^2 + 2x + 27 = 27.5$	positive	2	$y = 27.5$, twice
28 meters	$-x^2 + 2x + 27 = 28$	zero	1	$y = 28$, once
29 meters	$-x^2 + 2x + 27 = 29$	negative	0	$y = 29$, never

Because the value of $b^2 - 4ac$ *discriminates* among the various possible numbers of real-number solutions to a quadratic equation, it is called the **discriminant** of the equation $ax^2 + bx + c = 0$. The examples above are instances of the following general property.

Discriminant Property
If $ax^2 + bx + c = 0$ and a, b, and c are real numbers with $a \neq 0$, then:
When $b^2 - 4ac > 0$, the equation has exactly two real solutions.
When $b^2 - 4ac = 0$, the equation has exactly one solution.
When $b^2 - 4ac < 0$, the equation has no real solutions.

Example 4

Without solving, determine how many real solutions the equation $8x^2 - 5x + 2 = 0$ has.

Solution

Find the value of the discriminant, $b^2 - 4ac$.
Here $a = 8$, $b = -5$, and $c = 2$.

$$\begin{aligned} \text{So, } b^2 - 4ac &= (-5)^2 - 4 \cdot 8 \cdot 2 \\ &= 25 - 64 \\ &= -39. \end{aligned}$$

By the Discriminant Property, Since $b^2 - 4ac$ is negative, there are no real solutions.

Adapting to Individual Needs

English Language Development
Because this lesson contains so many applications throughout the reading and questions, encourage non-English-speaking students to use their bilingual dictionaries.

QUESTIONS

Covering the Reading

In 1–4, refer to the path of the Acapulco cliff diver described in this lesson.

1. What equation can be solved to determine how far away from the cliff the diver will be when he or she is 28 meters above the water? **$28 = -x^2 + 2x + 27$ or $-x^2 + 2x - 1 = 0$**
2. Will the diver reach a height of 28.5 meters above the water? Justify your answer by referring to a quadratic equation. **No; $28.5 = -x^2 + 2x + 27$ has no solutions because its discriminant (-2) is negative.**
3. Will the diver ever be 27 meters above the water? If so, where does this occur? **Yes; when $x = 0$ and when $x = 2$**
4. When will the diver be 10 meters above the water? **when $x \approx 5.24$**
5. Tell what the discriminant is, and what use it has in solving quadratic equations. **$b^2 - 4ac$; the discriminant can be evaluated to find the number of real solutions to a quadratic equation.**
6. Give the number of real solutions to a quadratic equation when its discriminant is **a.** positive. **2** **b.** negative. **0** **c.** zero. **1**

In 7–10, a quadratic equation is given.

a. Calculate the value of the discriminant. **7) 0; 8) -119; 9) 0; 10) -11**

b. Give the number of real solutions. **7) 1; 8) 0; 9) 1; 10) 0**

c. Find all the real solutions. (If there are none, write "no real solutions.")

7. $w^2 - 16w + 64 = 0$ **c) 8**
8. $4x^2 - 3x + 8 = 0$ **c) no real solutions**
9. $25y^2 = 10y - 1$ **c) $\frac{1}{5}$**
10. $13z = 5z^2 + 9$ **c) no real solutions**

Applying the Mathematics

11. Suppose an equation which describes a diver's path when diving from a platform is $d = -5t^2 + 10t + 5$, where d is the distance above the water (in meters) and t is the time from the beginning of the dive (in seconds). Round answers to the nearest tenth of a second.
    a. How high is the diving platform? (Hint: let $t = 0$ seconds). **5 meters**
    b. After how many seconds is the diver 8 meters above the water?
    c. After how many seconds does the diver enter the water? **$\approx$ 2.4 seconds**
    b) 0.4 second and 1.6 seconds
12. **a.** Solve $4x^2 + 8x = 5$. **$\frac{1}{2}$; $-\frac{5}{2}$** **b) ≈ -0.13 or -1.86**
    b. Solve $4x^2 + 8x = -1$.
    c. Solve $4x^2 + 8x = -10$. **no real solutions**
13. For what value(s) of h does $x^2 + 6x + h = 0$ have exactly one solution? **9**

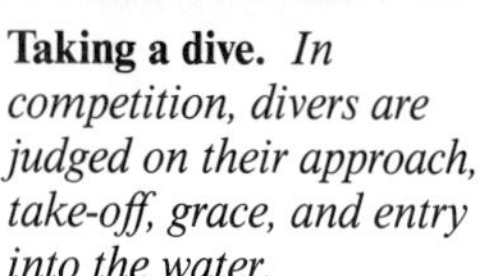

Taking a dive. *In competition, divers are judged on their approach, take-off, grace, and entry into the water.*

Notes on Questions

Question 11 Point out the difference between the quadratic in this question and the quadratic that has been used to describe the path of the Acapulco diver. The quadratic in this question gives height in relation to *time*. The Acapulco-diver quadratic gives height in relation to *distance from the cliff.* The latter parabola is a picture of the actual path of the diver. If they want to, students may graph the parabola in this question, but they should realize that it is not a picture of the dive itself.

Question 13 Error Alert This kind of question is usually quite difficult for students the first time they see it. If they have difficulty, suggest that students look at the solution graphically, as well as use the discriminant.

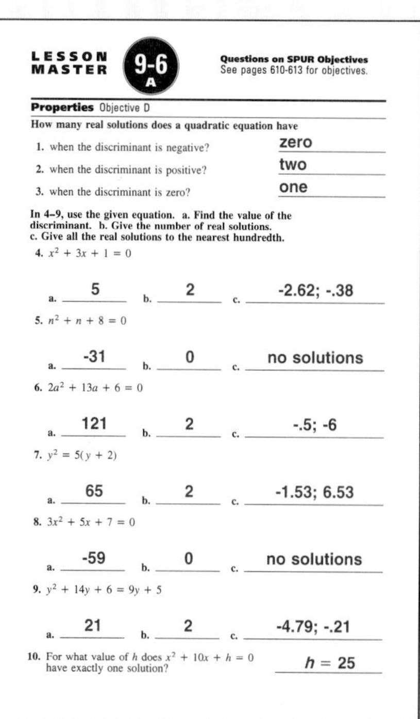

LESSON MASTER 9-6 A

Questions on SPUR Objectives
See pages 610-613 for objectives.

Properties Objective D

How many real solutions does a quadratic equation have

1. when the discriminant is negative? **zero**
2. when the discriminant is positive? **two**
3. when the discriminant is zero? **one**

In 4–9, use the given equation. a. Find the value of the discriminant. b. Give the number of real solutions. c. Give all the real solutions to the nearest hundredth.

4. $x^2 + 3x + 1 = 0$
   a. **5** b. **2** c. **-2.62; -.38**
5. $n^2 + n + 8 = 0$
   a. **-31** b. **0** c. **no solutions**
6. $2a^2 + 13a + 6 = 0$
   a. **121** b. **2** c. **-.5; -6**
7. $y^2 = 5(y + 2)$
   a. **65** b. **2** c. **-1.53; 6.53**
8. $3x^2 + 5x + 7 = 0$
   a. **-59** b. **0** c. **no solutions**
9. $y^2 + 14y + 6 = 9y + 5$
   a. **21** b. **2** c. **-4.79; -.21**
10. For what value of h does $x^2 + 10x + h = 0$ have exactly one solution? **$h = 25$**

Adapting to Individual Needs

Challenge

Have students look up the definition of a parabola, draw a picture to go with the definition, and make up an example equation and graph which shows the focus and directrix. [Sample: A parabola is a set of points equidistant from a fixed point (called the focus) and a fixed line (called the directrix). Example problems will vary.]

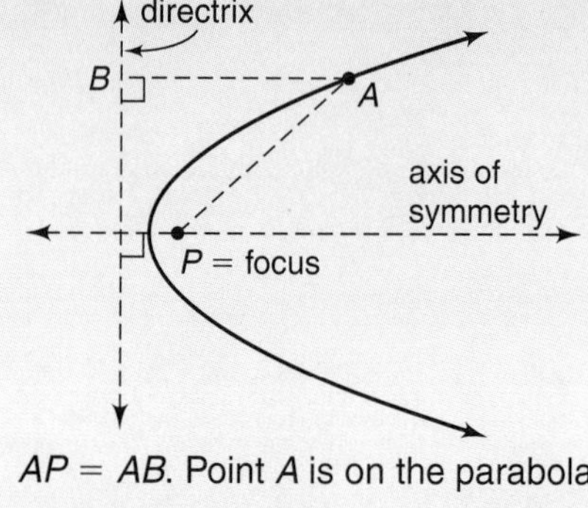

$AP = AB$. Point A is on the parabola.

Notes on Questions

Questions 17–18 Sports Connection In 1887 George W. Hancock of Chicago developed the game of softball as an indoor sport. In 1895, Lewis Robert of Minneapolis adapted the game for outdoor play. In 1933, the Amateur Softball Association of America was founded, and it established a set of rules that are now used by softball teams in all parts of the world. You might have interested students research the differences between the games of softball and baseball and give a short report in class.

Question 24 This *Skill Sequence* culminates with the sum of the roots of the general quadratic. The fact that the sum of the roots is $-\frac{b}{a}$ can be used to check solutions, and it can be related to the parabola. The average of the roots gives the x-value of the vertex of the parabola.

Follow-up for Lesson 9-6

Practice

For more questions on SPUR Objectives, use **Lesson Master 9-6A** (shown on page 583) or **Lesson Master 9-6B** (shown on pages 584–585).

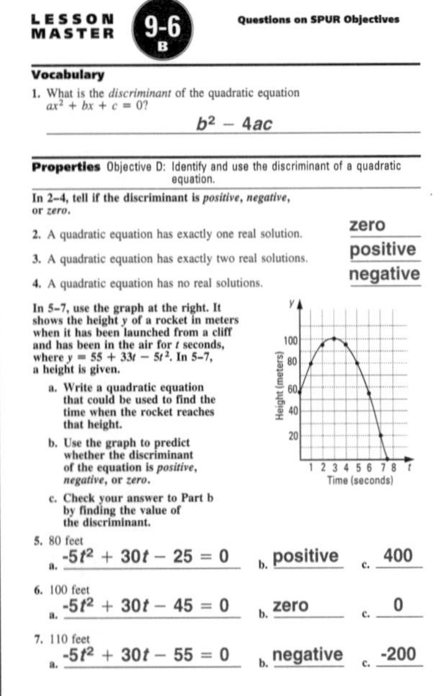
LESSON MASTER 9-6 B Questions on SPUR Objectives

Vocabulary

1. What is the *discriminant* of the quadratic equation $ax^2 + bx + c = 0$? $b^2 - 4ac$

Properties Objective D: Identify and use the discriminant of a quadratic equation.

In 2–4, tell if the discriminant is *positive*, *negative*, or *zero*.

2. A quadratic equation has exactly one real solution. zero
3. A quadratic equation has exactly two real solutions. positive
4. A quadratic equation has no real solutions. negative

In 5–7, use the graph at the right. It shows the height y of a rocket in meters when it has been launched from a cliff and has been in the air for t seconds, where $y = 55 + 33t - 5t^2$. In 5–7, a height is given.

a. **Write a quadratic equation that could be used to find the time when the rocket reaches that height.**

b. **Use the graph to predict whether the discriminant of the equation is *positive*, *negative*, or *zero*.**

c. **Check your answer to Part b by finding the value of the discriminant.**

5. 80 feet
a. $-5t^2 + 30t - 25 = 0$ b. positive c. 400

6. 100 feet
a. $-5t^2 + 30t - 45 = 0$ b. zero c. 0

7. 110 feet
a. $-5t^2 + 30t - 55 = 0$ b. negative c. -200

A test of strength. *The shot-put is a field event in which an athlete heaves a metal ball called a shot. The weight of the shot used by women is 8 pounds 13 ounces. In 1994, the world's record for the women's shot-put event was 74 ft 3 in.*

In 14 and 15, use the graph below. It shows the height h (in feet) of a shot t seconds after a shot putter released it. An equation for this path is $y = -16t^2 + 28t + 6$.

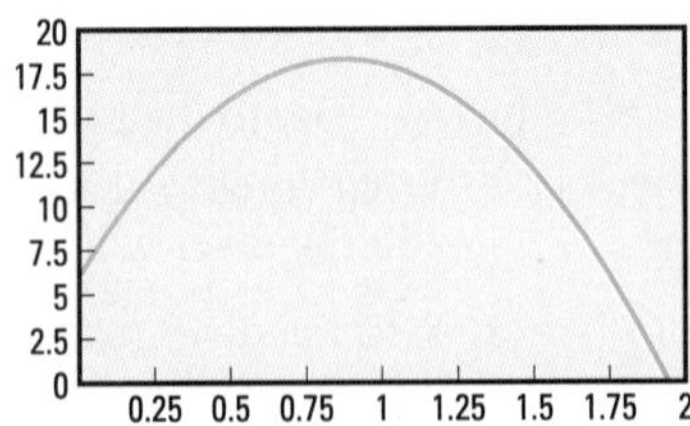

14. a. Write an equation that can be used to find if the shot is ever 10 feet above the ground. $0 = -16t^2 + 28t - 4$
b. Is the value of the discriminant positive, negative, or zero? positive
c. How many solutions does the equation have? 2

15. a. Write an equation that should be used to find if the shot is ever 18 feet above the ground. $0 = -16t^2 + 28t - 12$
b. Is the value of the discriminant positive, negative, or zero? positive
c. How many solutions does the equation have? 2

Review

16. Solve $60x^2 - 120 = 0$. *(Lessons 9-1, 9-5)* $\pm\sqrt{2} \approx \pm 1.414$

In 17 and 18, use this information: A softball pitcher tosses a ball to her catcher 50 ft away. The height h (in feet) when it is x feet from the pitcher is given by $h = -.016x^2 + .8x + 2$. *(Lesson 9-4)*

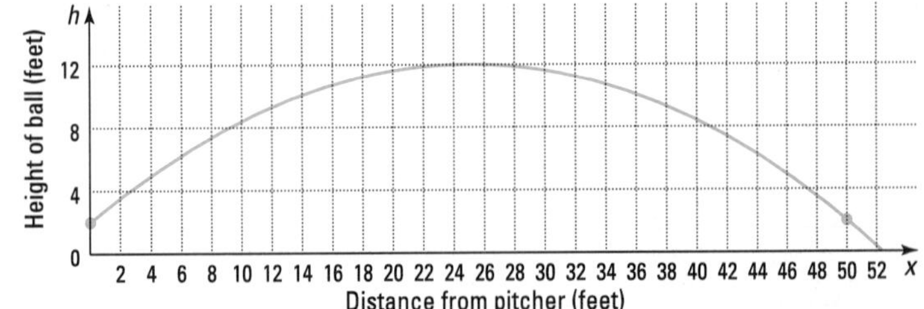

17. How high is the ball at its peak? 12 feet

18. a. If the batter is 2 ft in front of the catcher, how far is she from the pitcher? 48 feet
b. How high is the ball when it reaches the batter? $\approx$ 3.5 feet

Parabolic pitches. *Pitchers in slow-pitch softball games must throw the ball slowly enough to make it arch on its way to the batter. Only underhand pitching is allowed.*

In 19 and 20, state whether the parabola opens up or down. *(Lessons 9-1, 9-2)*

19. $y = -\frac{1}{5}x^2 - 3x + 2$ **down** **20.** $x^2 + x = y$ **up**

21. Give an equation for a parabola which has vertex at (0, 0) and opens down. *(Lesson 9-1)* **Sample: $y = -x^2$**

22. *Skill sequence.* Solve. *(Lesson 5-9)*
a. $\sqrt{x} = 5$ **$x = 25$** **b.** $\sqrt{y + 4} = 5$ **$y = 21$** **c.** $\sqrt{2z - 3} = 5$ **$z = 14$**

23. Solve. *(Lesson 3-10)*
a. $a - \frac{1}{3} < 0$ **$a < \frac{1}{3}$** **b.** $\frac{1}{3} - b < 0$ **$b > \frac{1}{3}$** **c.** $-\frac{1}{3}c < 0$ **$c > 0$**

24. *Skill sequence.* Add and simplify. *(Lesson 3-9)*
a. $\frac{-5 + x}{2a} + \frac{-5 - x}{2a}$ **$\frac{-5}{a}$**
b. $\frac{-b + y}{2a} + \frac{-b - y}{2a}$ **$\frac{-b}{a}$**
c. $\frac{-b + \sqrt{z}}{2a} + \frac{-b - \sqrt{z}}{2a}$ **$\frac{-b}{a}$**
d. $\frac{-b + \sqrt{b^2 - 4ac}}{2a} + \frac{-b - \sqrt{b^2 - 4ac}}{2a}$ **$\frac{-b}{a}$**

25. Calculate $\frac{15!}{8!7!}$. *(Lesson 2-10)* **6435**

26. Which numbers between 10 and 100 are perfect squares? *(Lesson 1-6)* **16, 25, 36, 49, 64, 81**

27)

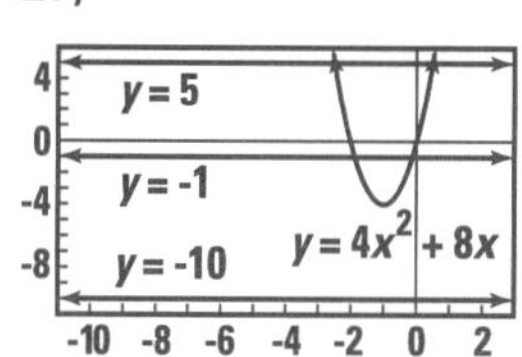

For this parabola, when $y = 5$, $x = \frac{1}{2}$ or $\frac{-5}{2}$; when $y = -1$, $x \approx -0.13$ or -1.86; There is no x value associated with $y = -10$.

Exploration

27. Graph the parabola $y = 4x^2 + 8x$ with an automatic grapher or as accurately as you can by hand. Explain the answers you get to Question 12 by referring to this parabola. **See left.**

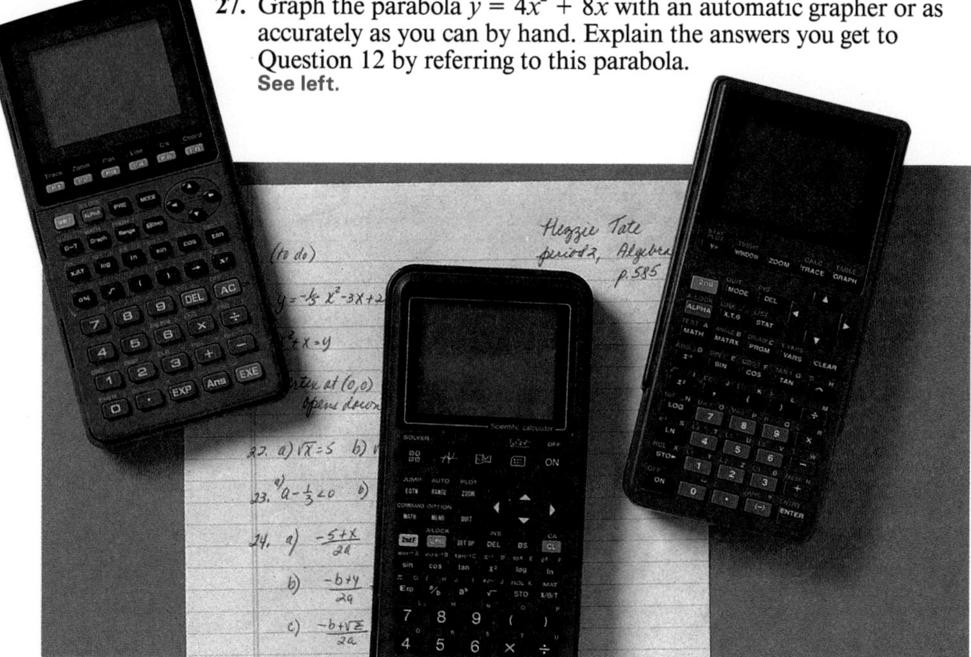

Assessment

Quiz A quiz covering Lessons 9-4 through 9-6 is provided in the *Assessment Sourcebook.*

Oral Communication Ask students if the discriminant will always tell the number of sensible solutions for a situation that can be described using a quadratic equation. Have them explain their answers. [Students recognize the difference between real-number solutions that are determined by the discriminant and sensible solutions that are determined by the situation which is described by a quadratic equation.]

Extension

Question 12 can be extended to ask, "For what values of k will the equation $4x^2 + 8x = k$ have two solutions?" [$k > -4$; the question can be answered by examining the graph of $y = 4x^2 + 8x$ (see **Question 27**) or by analyzing the discriminant of $4x^2 + 8x - k = 0$.]

Project Update Project 4, *Video of a Parabolic Path*, and Project 5, *Computer Programs for Solutions of Quadratics*, on page 606, relate to the content of this lesson.

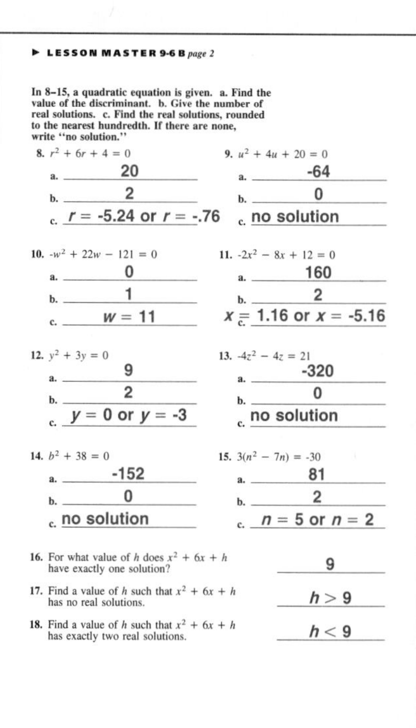

▶ **LESSON MASTER 9-6 B** *page 2*

In 8–15, a quadratic equation is given. a. Find the value of the discriminant. b. Give the number of real solutions. c. Find the real solutions, rounded to the nearest hundredth. If there are none, write "no solution."

8. $r^2 + 6r + 4 = 0$
a. 20
b. 2
c. $r = -5.24$ or $r = -.76$

9. $u^2 + 4u + 20 = 0$
a. -64
b. 0
c. no solution

10. $-w^2 + 22w - 121 = 0$
a. 0
b. 1
c. $w = 11$

11. $-2x^2 - 8x + 12 = 0$
a. 160
b. 2
c. $x = 1.16$ or $x = -5.16$

12. $y^2 + 3y = 0$
a. 9
b. 2
c. $y = 0$ or $y = -3$

13. $-4z^2 - 4z = 21$
a. -320
b. 0
c. no solution

14. $b^2 + 38 = 0$
a. -152
b. 0
c. no solution

15. $3(n^2 - 7n) = -30$
a. 81
b. 2
c. $n = 5$ or $n = 2$

16. For what value of h does $x^2 + 6x + h$ have exactly one solution? 9

17. Find a value of h such that $x^2 + 6x + h$ has no real solutions. $h > 9$

18. Find a value of h such that $x^2 + 6x + h$ has exactly two real solutions. $h < 9$

Setting Up Lesson 9-7

Discuss **Question 26** so that students review the meaning of *perfect square* which is used in Lesson 9-7.

Lesson 9-7

Objectives

B Simplify square roots.

Resources

From the ***Teacher's Resource File***

- Lesson Master 9-7A or 9-7B
- Answer Master 9-7
- Teaching Aids
  97 Warm-up
  105 Additional Examples

Additional Resources

- Visuals for Teaching Aids 97, 105

Teaching Lesson 9-7

Warm-up

1. Use a calculator to write the square roots of the integers from 1 to 20, truncating each answer at the thousandths place.
   $\sqrt{1} = 1$ $\sqrt{2} \approx 1.414$
   $\sqrt{3} \approx 1.732$ $\sqrt{4} = 2$
   $\sqrt{5} \approx 2.236$ $\sqrt{6} \approx 2.449$
   $\sqrt{7} \approx 2.645$ $\sqrt{8} \approx 2.828$
   $\sqrt{9} = 3$ $\sqrt{10} \approx 3.162$
   $\sqrt{11} \approx 3.316$ $\sqrt{12} \approx 3.464$
   $\sqrt{13} \approx 3.605$ $\sqrt{14} \approx 3.741$
   $\sqrt{15} \approx 3.872$ $\sqrt{16} = 4$
   $\sqrt{17} \approx 4.123$ $\sqrt{18} \approx 4.242$
   $\sqrt{19} \approx 4.358$ $\sqrt{20} \approx 4.472$
2. **Diagnostic** Which of the square roots in Question 1 are multiples of $\sqrt{2}$? $\sqrt{2} = 1\sqrt{2}$, $\sqrt{8} = 2\sqrt{2}$, $\sqrt{18} = 3\sqrt{2}$

LESSON

9-7 *Square Roots and Products*

Garden roots. *Shown are three sets of stairs in the Botanical Gardens at the Huntington Library in San Marino, CA. You can find the length of the hypotenuse of each right triangle formed by these steps using square roots.*

A Problem with Two Answers?

The figure below shows a side view of some stairs planned for a garden. The contractor needs to figure out the length of $\overline{LN}$ so he can build a form to make a concrete footing for the stairs. Can you determine LN?

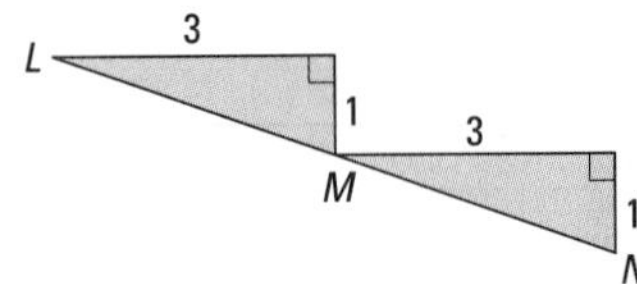

One way to solve this problem is to reason that the length of $\overline{LN}$ is twice the length of either $\overline{LM}$ or $\overline{MN}$. You can find either of these by using the Pythagorean Theorem.

$$3^2 + 1^2 = (LM)^2$$
$$10 = (LM)^2$$
$$\sqrt{10} = LM$$

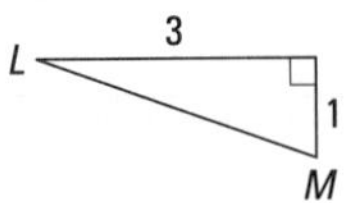

Therefore, $LN = 2(LM) = 2\sqrt{10}$.

Another technique is to imagine one large right triangle with hypotenuse $\overline{LN}$. Its legs are 6 and 2. Now use the Pythagorean Theorem on this right triangle.

$$6^2 + 2^2 = (LN)^2$$
$$36 + 4 = (LN)^2$$
$$40 = (LN)^2$$
$$\sqrt{40} = LN$$

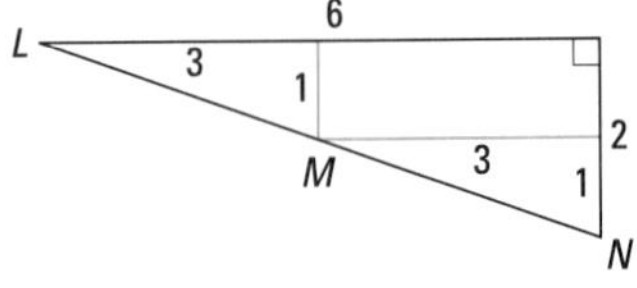

Lesson 9-7 Overview

Broad Goals This lesson discusses the uses of the Product of Square Roots Property: for all nonnegative real numbers a and b, $\sqrt{a} \cdot \sqrt{b} = \sqrt{ab}$.

Perspective With the advent of hand-held calculators, simplifying radicals as an aid for calculating decimal approximations is outdated. On a calculator, it is easier to evaluate $\sqrt{500}$ than to evaluate $10\sqrt{5}$. However, there are still cases where simplifying is useful. For instance, the solutions to **Example 1** of Lesson 9-5 are $\frac{-2 \pm \sqrt{112}}{-2}$ from the Quadratic Formula, but, since $\sqrt{112} = 4\sqrt{7}$, the solutions become $1 \pm 2\sqrt{7}$. In addition, there are situations where simplifying radicals is necessary. In some situations, such as the one that is given at the beginning of this lesson, patterns do not become apparent unless the numbers are simplified. Yet another reason to simplify radicals is reconciling answers that are computed using different methods.

Each of the six examples has a different role in this lesson, but all are related to the Product of Square Roots Property. **Examples 1 and 2** are direct instances. **Example 1** goes from the product to the single square root; **Example 2** goes from the single square root to the product.

It looks as though there are two different values for LN: $2\sqrt{10}$ and $\sqrt{40}$. Are these two names for the same number? Example 1 shows that they are.

Example 1

Verify that $\sqrt{40} = 2\sqrt{10}$.

Solution 1

Evaluate $2\sqrt{10}$ and $\sqrt{40}$ on a calculator. On our calculator we get

$$2\sqrt{10} \approx 6.325$$
$$\sqrt{40} \approx 6.325$$

They appear to be equal.

Solution 2

Square each number. Show that the results are equal. Recall that squaring means to multiply a number by itself,

so $(\sqrt{40})^2 = \sqrt{40} \cdot \sqrt{40} = 40.$

Also,
$$\begin{aligned}(2\sqrt{10})^2 &= 2\sqrt{10} \cdot 2\sqrt{10} \\ &= 2 \cdot 2 \cdot \sqrt{10} \cdot \sqrt{10} \\ &= 4 \cdot 10 \\ &= 40.\end{aligned}$$

So $2\sqrt{10} = \sqrt{40}$.

Multiplying Square Roots

In general, you may multiply square roots or rewrite square roots as products using the following property.

Product of Square Roots Property
For all nonnegative real numbers a and b,
$$\sqrt{a} \cdot \sqrt{b} = \sqrt{ab}.$$

In words, the square root of the product of two positive numbers is the product of their square roots.

You can multiply square roots by working from the left side of the property to the right side. For instance,

$$\sqrt{4} \cdot \sqrt{10} = \sqrt{4 \cdot 10} = \sqrt{40}.$$

Or working from the right side of the property to the left side, you can rewrite a square root as a product. For instance,

$$\begin{aligned}\sqrt{40} &= \sqrt{4 \cdot 10} \\ &= \sqrt{4} \cdot \sqrt{10} \\ &= 2\sqrt{10}.\end{aligned}$$

Notes on Reading

Advise students that the directions *simplify, give the exact answer,* and *rewrite the expression with a smaller integer under the radical symbol* mean that they should leave their answers as a radical. However, calculators can still be used to check answers. While reading the lesson, students should use a calculator to get decimal approximations for $\sqrt{40}$ and $2\sqrt{10}$.

Examples 3 and 4 exhibit the two most common places where this property is used in school mathematics: with the Pythagorean Theorem and with the Quadratic Formula. **Examples 5 and 6** are also direct instances of the Product of Square Roots Property, but these examples involve variables.

Notes on Reading

Ask students to explain why the answer to **Example 6** would be incorrect if x were negative and y were positive. [$3xy$ would be negative, yet the radical sign stands for the positive square root.] Give an instance with specific values, for example, −3 for x and 5 for y.

Example 6 asks students for a simplification of an expression of the form $\sqrt{kx^ay^b}$, but here k is a perfect square, $a = 2$, and $b = 2$. If your students are not having trouble, you may want to take them through a sequence of questions that considers other positive integer values of k, a, and b. See the *Extension* on page 591 for suggestions.

"Simplifying" Radicals

Some people consider $2\sqrt{10}$ to be simpler than $\sqrt{40}$ because it has a smaller integer under the radical sign. The process of rewriting a square root as the product of a whole number and a smaller square root is called **simplifying the radical.** The key to simplifying a radical is to find a perfect-square factor of the original number under the radical sign.

Example 2

Simplify $\sqrt{50}$.

Solution

A perfect-square factor of 50 is 25.

$$\begin{aligned}\sqrt{50} &= \sqrt{25 \cdot 2}\\ &= \sqrt{25} \cdot \sqrt{2} \quad \text{Product of Square Roots Property}\\ &= 5\sqrt{2}\end{aligned}$$

Check

Use a calculator. On ours, 50 [√] gives 7.0710678.
5 [×] 2 [√] [=] also gives 7.0710678.
It checks.

Is $5\sqrt{2}$ really simpler than $\sqrt{50}$? It all depends. For estimating and calculating answers to real problems, $\sqrt{50}$ is simpler. But for seeing patterns, $5\sqrt{2}$ may be easier. In the next example, the answer $8\sqrt{2}$ is related to the given information in a simple way. You would not see that without "simplifying" $\sqrt{128}$.

Example 3

Each leg of a right triangle is 8 inches long. Find the exact length of the hypotenuse.

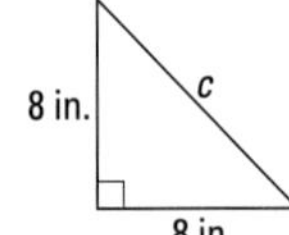

Solution

Use the Pythagorean Theorem.

$$\begin{aligned}c^2 &= 8^2 + 8^2\\ c^2 &= 128\\ c &= \sqrt{128}\end{aligned}$$

Now use the Product of Square Roots Property. Note that 64 is a perfect square factor of 128.

$$\begin{aligned}c &= \sqrt{64} \cdot \sqrt{2}\\ &= 8\sqrt{2}\end{aligned}$$

The exact length of the hypotenuse is $\sqrt{128}$, or $8\sqrt{2}$ inches.

Can you see how the lengths of the legs and the hypotenuse are related? You are asked to generalize the pattern in the Questions.

Optional Activities

You might use this activity after students have completed the lesson. In this lesson, students learn that $\sqrt{a} \cdot \sqrt{b} = \sqrt{ab}$. Many students will automatically assume that $\sqrt{a} + \sqrt{b} = \sqrt{a+b}$. Have students use a calculator to test several cases to determine if this second statement is always true. [No. As a counterexample, use $\sqrt{9} + \sqrt{16}$ and $\sqrt{9+16}$. The first expression is $3 + 4$ or 7. The second expression is $\sqrt{25}$ or 5.]

Applying the Product of Square Roots Property

You can combine the Product of Square Roots Property with other properties of real numbers to simplify many expressions.

Example 4

One solution to a quadratic equation is $\frac{6 + \sqrt{28}}{2}$. Simplify this expression.

Solution

A perfect-square factor of 28 is 4, so rewrite $\sqrt{28}$ as $\sqrt{4} \cdot \sqrt{7}$.

$$\frac{6 + \sqrt{28}}{2} = \frac{6 + \sqrt{4} \cdot \sqrt{7}}{2}$$
$$= \frac{6 + 2 \cdot \sqrt{7}}{2}$$

Now use the Adding Fractions form of the Distributive Property.

$$= \frac{6}{2} + \frac{2\sqrt{7}}{2}$$
$$= 3 + \sqrt{7}$$

The Product of Square Roots Property also applies to expressions containing variables.

Example 5

Assume m and n are positive. Multiply $\sqrt{7m} \cdot \sqrt{7n}$ and simplify the result.

Solution

$$\sqrt{7m} \cdot \sqrt{7n} = \sqrt{7m \cdot 7n}$$
$$= \sqrt{7^2} \cdot \sqrt{mn}$$
$$= 7\sqrt{mn}$$

Check

Does $\sqrt{7m} \cdot \sqrt{7n} = 7\sqrt{mn}$? Substitute values for m and n.
For instance, let $m = 3$ and $n = 2$.

Does $\sqrt{7 \cdot 3} \cdot \sqrt{7 \cdot 2} = 7\sqrt{3 \cdot 2}$?

Does $\sqrt{21} \cdot \sqrt{14} = 7\sqrt{6}$?

$\sqrt{21} \cdot \sqrt{14} = \sqrt{294} = \sqrt{49 \cdot 6} = 7\sqrt{6}$

It checks.

Example 6

Simplify $\sqrt{9x^2y^2}$. Assume x and y are both positive.

Solution

$$\sqrt{9x^2y^2} = \sqrt{9} \cdot \sqrt{x^2} \cdot \sqrt{y^2}$$
$$= 3 \cdot x \cdot y$$
$$= 3xy$$

▶

Additional Examples

These examples are also given on **Teaching Aid 105.**

1. Verify that $\sqrt{500} = 10\sqrt{5}$
   a. by finding decimal approximations. **Use a calculator: $\sqrt{500} \approx 22.36$ and, $10\sqrt{5} \approx 10 \cdot 2.236 = 22.36$.**
   b. by squaring each number. **$(\sqrt{500})^2 = 500$ and $(10\sqrt{5})^2 = 10 \cdot 10 \cdot \sqrt{5} \cdot \sqrt{5} = 500$.**
2. Simplify $\sqrt{150}$. **$5\sqrt{6}$**
3. Find the exact length of the hypotenuse of this right triangle. Simplify the answer. **$2\sqrt{13}$**

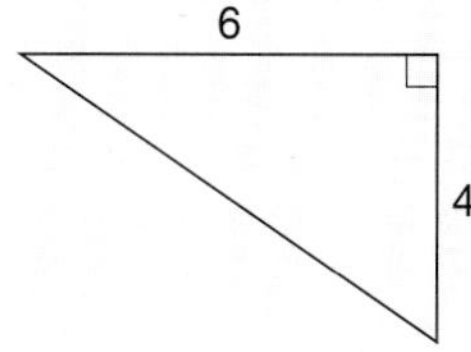

4. Simplify $\frac{-4 \pm \sqrt{48}}{8}$. **$\frac{-1 \pm \sqrt{3}}{2}$**
5. Assume a and b are positive. Find $\sqrt{12a} \cdot \sqrt{3b}$ and simplify the result. **$6\sqrt{ab}$**
6. Assume a and b are positive. Simplify $\sqrt{32a^2b^6}$. **$4ab^3\sqrt{2}$**
7. Given that $n \geq 0$, simplify $\sqrt{11n^2}$. **$n\sqrt{11}$**
8. Solve $y^2 + 9 = 33$ and rewrite the answer with the smallest integer possible under the radical sign. **$2\sqrt{6}$**
9. Simplify $\sqrt{2} \cdot \sqrt{3} \cdot \sqrt{4} \cdot \sqrt{5} \cdot \sqrt{6}$. **$12\sqrt{5}$**

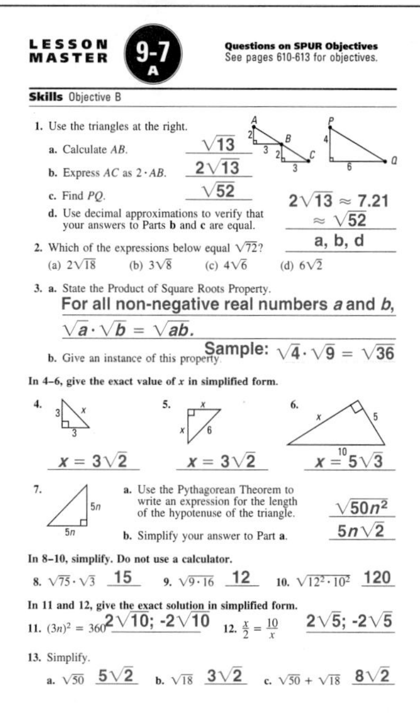

LESSON MASTER 9-7 A

Questions on SPUR Objectives
See pages 610-613 for objectives.

Skills Objective B

1. Use the triangles at the right.
   a. Calculate AB. **$\sqrt{13}$**
   b. Express AC as $2 \cdot AB$. **$2\sqrt{13}$**
   c. Find PQ. **$\sqrt{52}$**
   d. Use decimal approximations to verify that your answers to Parts **b** and **c** are equal. **$2\sqrt{13} \approx 7.21 \approx \sqrt{52}$**

2. Which of the expressions below equal $\sqrt{72}$? **a, b, d**
   (a) $2\sqrt{18}$ (b) $3\sqrt{8}$ (c) $4\sqrt{6}$ (d) $6\sqrt{2}$

3. a. State the Product of Square Roots Property. **For all non-negative real numbers a and b, $\sqrt{a} \cdot \sqrt{b} = \sqrt{ab}$.**
   b. Give an instance of this property. **Sample: $\sqrt{4} \cdot \sqrt{9} = \sqrt{36}$**

In 4–6, give the exact value of x in simplified form.

4. **$x = 3\sqrt{2}$**
5. **$x = 3\sqrt{2}$**
6. **$x = 5\sqrt{3}$**

7. a. Use the Pythagorean Theorem to write an expression for the length of the hypotenuse of the triangle. **$\sqrt{50n^2}$**
   b. Simplify your answer to Part **a**. **$5n\sqrt{2}$**

In 8–10, simplify. Do not use a calculator.

8. $\sqrt{75} \cdot \sqrt{3}$ **15**
9. $\sqrt{9 \cdot 16}$ **12**
10. $\sqrt{12^2 \cdot 10^2}$ **120**

In 11 and 12, give the exact solution in simplified form.

11. $(3n)^2 = 360$ **$2\sqrt{10}$; $-2\sqrt{10}$**
12. $\frac{x}{2} = \frac{10}{x}$ **$2\sqrt{5}$; $-2\sqrt{5}$**

13. Simplify.
    a. $\sqrt{50}$ **$5\sqrt{2}$**
    b. $\sqrt{18}$ **$3\sqrt{2}$**
    c. $\sqrt{50} + \sqrt{18}$ **$8\sqrt{2}$**

Adapting to Individual Needs

English Language Development

Have students add *The Product of Square Roots Property* to their index cards.

Extra Help

Be sure students remember and understand the meaning of *perfect square* as it is used in **Examples 2–4**. Help them to recall that a perfect square is a number whose square root is an integer. Ask a volunteer to name all the perfect squares from 1 through 100. [1, 4, 9, 16, 25, 36, 49, 64, 81, 100]

Notes on Questions

Questions 11–12 Error Alert These kinds of questions are frequently hard for students. Some of them forget the Distributive Property, and divide only the first term of the numerator by 2. Others miscalculate when dividing the radical by 2. Encourage students to use their calculators to check their work.

Question 21 Two solution methods are possible. One method is to transform the equation into $4a^2 = 48$ and solve. The other method is to use $2a = \pm\sqrt{48}$, simplify $\sqrt{48}$, and then solve for a. You may wish to show both solutions.

Questions 21–22 Students should look for both positive and negative solutions.

Question 23 The three parts of this question show that $\sqrt{a} + \sqrt{b}$ is not, in general, equal to $\sqrt{a + b}$. This idea is considered in *Optional Activities* on page 588.

(Notes on Questions continue on page 592.)

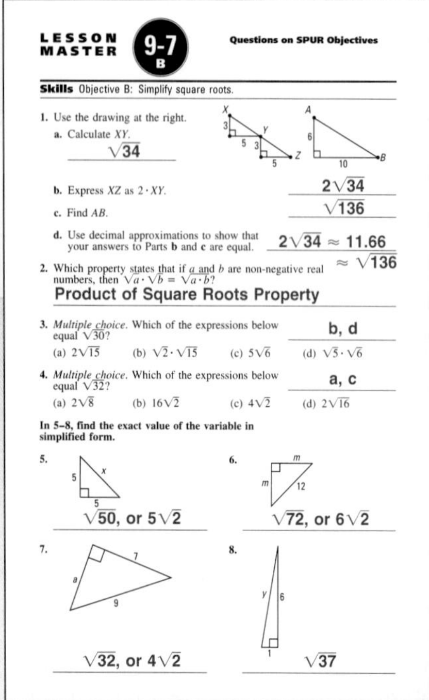

LESSON MASTER 9-7 B — **Questions on SPUR Objectives**

Skills Objective B: Simplify square roots.

1. Use the drawing at the right.
   a. Calculate XY. $\sqrt{34}$
   b. Express XZ as $2 \cdot XY$. $2\sqrt{34}$
   c. Find AB. $\sqrt{136}$
   d. Use decimal approximations to show that your answers to Parts **b** and **c** are equal. $2\sqrt{34} \approx 11.66 \approx \sqrt{136}$

2. Which property states that if a and b are non-negative real numbers, then $\sqrt{a} \cdot \sqrt{b} = \sqrt{a \cdot b}$? **Product of Square Roots Property**

3. *Multiple choice.* Which of the expressions below equal $\sqrt{30}$? **b, d**
   (a) $2\sqrt{15}$ (b) $\sqrt{2} \cdot \sqrt{15}$ (c) $5\sqrt{6}$ (d) $\sqrt{5} \cdot \sqrt{6}$

4. *Multiple choice.* Which of the expressions below equal $\sqrt{32}$? **a, c**
   (a) $2\sqrt{8}$ (b) $16\sqrt{2}$ (c) $4\sqrt{2}$ (d) $2\sqrt{16}$

In 5–8, find the exact value of the variable in simplified form.

5. $\sqrt{50}$, or $5\sqrt{2}$
6. $\sqrt{72}$, or $6\sqrt{2}$
7. $\sqrt{32}$, or $4\sqrt{2}$
8. $\sqrt{37}$

Check

Substitute for the variables in $\sqrt{9x^2y^2}$ and $3xy$. We let $x = 2$, $y = 4$. With these values, $\sqrt{9x^2y^2} = \sqrt{9(2)^2(4)^2} = \sqrt{9 \cdot 4 \cdot 16} = \sqrt{576} = 24$, and $3xy = 3 \cdot 2 \cdot 4 = 24$. It checks.

QUESTIONS

Covering the Reading

1. Refer to the drawing at the right.
   a. Find GI by first finding the length of $\overline{GH}$. $2\sqrt{5}$
   b. Find GI by using a right triangle with $\overline{GI}$ as hypotenuse. $\sqrt{20}$
   c. Verify that your answers to parts **a** and **b** are equal. $2\sqrt{5} \approx 4.472$; $\sqrt{20} \approx 4.472$

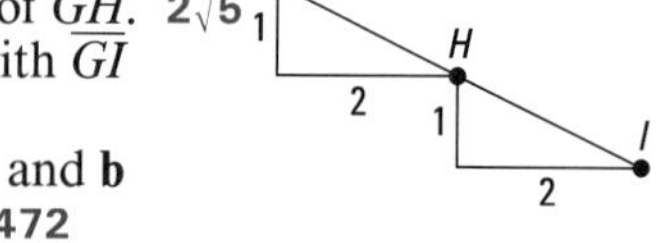

2. Show that $5\sqrt{3} = \sqrt{75}$ by each method.
   a. using a calculator $5\sqrt{3} \approx 8.66$; $\sqrt{75} \approx 8.66$
   b. squaring each of $5\sqrt{3}$ and $\sqrt{75}$ $(5\sqrt{3})^2 = 25 \cdot 3 = 75$; $(\sqrt{75})^2 = 75$
   c. simplifying $\sqrt{75}$ using the Product of Square Roots Property. $\sqrt{75} = \sqrt{25} \cdot \sqrt{3} = 5\sqrt{3}$
3. By the Product of Square Roots Property, what must $\sqrt{p} \cdot \sqrt{q}$ equal? $\sqrt{pq}$
4. Use a calculator to estimate the following to two decimal places.
   a. $\sqrt{7}$ 2.65 b. $\sqrt{5}$ 2.24 c. $\sqrt{7} \cdot \sqrt{5}$ 5.92 d. $\sqrt{35}$ 5.92
5. To use the Product of Square Roots Property to rewrite $\sqrt{ab}$, both a and b must be __?__. **greater than or equal to zero**

In 6–8, a radical sign is given. **a.** Find a perfect-square factor of the number under the radical sign. **b.** Simplify.

6. $\sqrt{20}$ 4; $2\sqrt{5}$
7. $\sqrt{50}$ 25; $5\sqrt{2}$
8. $\sqrt{700}$ 100; $10\sqrt{7}$
9. Check Example 3 by using a calculator to verify that $\sqrt{128} = 8\sqrt{2}$. $\sqrt{128} \approx 11.31$ $8\sqrt{2} \approx 8 \cdot 1.41 \approx 11.31$
10. Each leg of a right triangle is 10 cm. Find the exact length of the hypotenuse. $10\sqrt{2}$ cm

In 11 and 12, simplify each expression.

11. $\frac{6 + \sqrt{18}}{2}$ $3 + \frac{3}{2}\sqrt{2}$
12. $\frac{12 \pm \sqrt{12}}{2}$ $6 \pm \sqrt{3}$

In 13 and 14, assume a and b are both positive and simplify the result.

13. $\sqrt{6a} \cdot \sqrt{24b}$ $12\sqrt{ab}$
14. $\sqrt{36a^2b^2}$ $6ab$

Adapting to Individual Needs

Challenge

Give students the following question which is an extension of the *Warm-up:* which square roots of integers between 1 and 100 are multiples of $\sqrt{2}$? $\sqrt{3}$? $\sqrt{5}$? $\sqrt{7}$?
[Multiples of $\sqrt{2}$: $\sqrt{2}$, $\sqrt{8}$, $\sqrt{18}$, $\sqrt{32}$, $\sqrt{50}$, $\sqrt{72}$, $\sqrt{98}$;
multiples of $\sqrt{3}$: $\sqrt{3}$, $\sqrt{12}$, $\sqrt{27}$, $\sqrt{48}$, $\sqrt{75}$;
multiplies of $\sqrt{5}$: $\sqrt{5}$, $\sqrt{20}$, $\sqrt{45}$, $\sqrt{80}$;
multiples of $\sqrt{7}$: $\sqrt{7}$, $\sqrt{28}$, $\sqrt{63}$]

590

21) $(2(2\sqrt{3}))^2 = (4\sqrt{3})^2 = 16 \cdot 3 = 48$
$(2(-2\sqrt{3}))^2 = (-4\sqrt{3})^2 = 16 \cdot 3 = 48$

22) $\frac{9}{3\sqrt{6}} = \frac{3}{\sqrt{6}} \approx 1.225$
$\frac{3\sqrt{6}}{6} = \frac{\sqrt{6}}{2} \approx 1.225$
$\frac{9}{-3\sqrt{6}} = \frac{-3}{\sqrt{6}} \approx -1.225$
$\frac{-3\sqrt{6}}{6} = -\frac{\sqrt{6}}{2} \approx -1.225$

Applying the Mathematics

In 15 and 16, find the exact value of the variable. Simplify any radicals in your answers.

15.

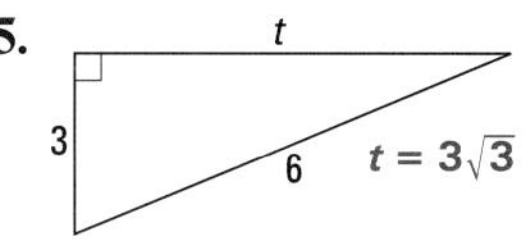

$t = 3\sqrt{3}$

16.

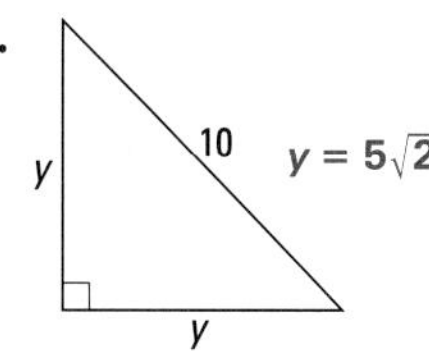

$y = 5\sqrt{2}$

In 17–20, simplify in your head. Be able to explain what you did.

17. $\sqrt{2} \cdot \sqrt{18}$ 6

18. $\sqrt{20} \cdot \sqrt{5}$ 10

19. $\sqrt{5^2 \cdot 11^2}$ 55

20. $\sqrt{100 \cdot 81 \cdot 36}$ 540

In 21 and 22, solve and check. Give your answer in simplified radical form.

21. $(2a)^2 = 48$ $a = \pm 2\sqrt{3}$
See left for check.

22. $\frac{9}{x} = \frac{x}{6}$ $x = \pm 3\sqrt{6}$

23. Simplify each expression.
a. $\sqrt{75}$ $5\sqrt{3}$
b. $\sqrt{12}$ $2\sqrt{3}$
c. $\sqrt{75} + \sqrt{12}$ $7\sqrt{3}$

24. a. Find the area of a rectangle with base $\sqrt{18}$ and height $5\sqrt{2}$. 30
b. Which is longer, the base or the height? Justify your answer.
See below.

25. The area of a square is $20w^2$. Write the exact length of one side in two ways. Sample: $w\sqrt{20}$, $2w\sqrt{5}$

26. Solve $x^2 + 4x - 9 = 0$, and write the solutions in simplified radical form. $-2 \pm \sqrt{13}$

24b) the height; $\sqrt{18} = 3\sqrt{2} < 5\sqrt{2}$

NASA vacuum chamber. *Shown is a test model satellite in vacuum Chamber "A" of the Space Environment Simulation Laboratory at the Johnson Space Center. Engineers conducted tests to ensure that the satellite's 30-foot parabolic antenna would unfold properly in a space vacuum.*

Review

27. A quadratic equation has only one solution. What can you conclude about its discriminant? *(Lesson 9-6)* It equals zero.

In 28–30, use this information. In a vacuum chamber on Earth, an object will drop d meters in approximately t seconds, where $d = 4.9t^2$. *(Lessons 1-6, 9-1)*

28. How far will the object drop in 3 seconds? 44.1 meters

29. How far will an object drop in $\sqrt{10}$ seconds? 49 meters

30. a. Write an equation that could be used to find the number of seconds it takes an object to drop 10 meters. $10 = 4.9t^2$
b. Solve the equation from part **a.**
≈ 1.43 seconds

Follow-up for Lesson 9-7

Practice

For more questions on SPUR Objectives, use **Lesson Master 9-7A** (shown on page 589) or **Lesson Master 9-7B** (shown on pages 590–591).

Assessment

Group Assessment Have each student **work with a partner**. Tell each student to write five expressions that can be simplified using the Product of Square Roots Property. Have students exchange their papers with their partners and simplify the expressions. [Students demonstrate the ability to simplify square roots.]

Extension

Give students the following property: For all nonnegative real numbers a and all positive real numbers b, $\frac{\sqrt{a}}{\sqrt{b}} = \sqrt{\frac{a}{b}}$. Then have students simplify the following expressions. Assume all variables are positive.

1. $\frac{\sqrt{450}}{\sqrt{75}}$ [$\sqrt{6}$]
2. $\frac{\sqrt{27m^3n^6}}{\sqrt{3mn^5}}$ [$3m\sqrt{n}$]
3. $\frac{\sqrt{48x^2y^3}}{\sqrt{2xy^2}}$ [$2\sqrt{6xy}$]
4. $\frac{\sqrt{210a^{15}b^{18}}}{\sqrt{14a^3b^8}}$ [$a^6b^5\sqrt{15}$]

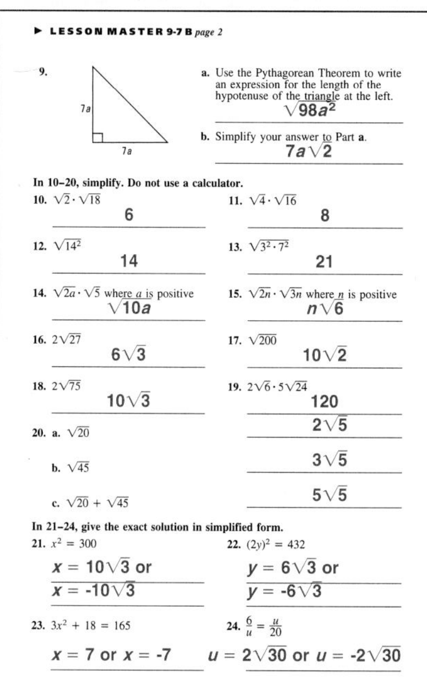

► LESSON MASTER 9-7 B *page 2*

9. a. Use the Pythagorean Theorem to write an expression for the length of the hypotenuse of the triangle at the left. $\sqrt{98a^2}$
b. Simplify your answer to Part a. $7a\sqrt{2}$

In 10–20, simplify. Do not use a calculator.

10. $\sqrt{2} \cdot \sqrt{18}$ 6
11. $\sqrt{4} \cdot \sqrt{16}$ 8
12. $\sqrt{14^2}$ 14
13. $\sqrt{3^2 \cdot 7^2}$ 21
14. $\sqrt{2a} \cdot \sqrt{5}$ where a is positive $\sqrt{10a}$
15. $\sqrt{2n} \cdot \sqrt{3n}$ where n is positive $n\sqrt{6}$
16. $2\sqrt{27}$ $6\sqrt{3}$
17. $\sqrt{200}$ $10\sqrt{2}$
18. $2\sqrt{75}$ $10\sqrt{3}$
19. $2\sqrt{6} \cdot 5\sqrt{24}$ 120
20. a. $\sqrt{20}$ $2\sqrt{5}$
b. $\sqrt{45}$ $3\sqrt{5}$
c. $\sqrt{20} + \sqrt{45}$ $5\sqrt{5}$

In 21–24, give the exact solution in simplified form.

21. $x^2 = 300$ $x = 10\sqrt{3}$ or $x = -10\sqrt{3}$
22. $(2y)^2 = 432$ $y = 6\sqrt{3}$ or $y = -6\sqrt{3}$
23. $3x^2 + 18 = 165$ $x = 7$ or $x = -7$
24. $\frac{6}{u} = \frac{u}{20}$ $u = 2\sqrt{30}$ or $u = -2\sqrt{30}$

Notes on Questions

Question 36 Multicultural Connection The oldest type of scale is the balance scale. One type, *the equal-arm balance,* was first used by the ancient Egyptians around 2500 B.C. This kind of balance consists of a horizontal bar with a pan suspended from each end. Another type of balance scale, the *steelyard balance*, was developed by the ancient Romans. The horizontal bar on this scale has arms of unequal length. A hook on the shorter arm holds the load. A weight moves along the longer arm until it balances the load; markings on the arm indicate the weight.

Question 37 This question is a forerunner of the theorem that students will likely study in geometry: in an isosceles right triangle, the hypotenuse is $\sqrt{2}$ times either leg.

Using Physical Models You might have students use paper strips, straws, or pasta to model right triangles with equal legs. Have students look for examples of right triangles with equal legs in their homes and tell the class about what they found. If the samples are toys or other small objects, encourage students to bring them to class.

In 31–34, determine whether the equation has 0, 1, or 2 real solutions. (You do not need to solve the equation.) *(Lessons 3-7, 9-6)*

31. $x^2 + x - 1 = 0$ **2**

32. $x^2 + x + 1 = 0$ **0**

33. $2x^2 - 12x - 18 = 0$ **2**

34. $-2x - 18 = 0$ **1**

35. Betsy has q quarters and d dimes. She has at least \$5.20. Write an inequality that describes this situation. *(Lessons 3-8, 3-10, 5-6)* **$0.25q + 0.10d \geq 5.20$**

36. On the scale below, the boxes are equal in weight. The other objects are one-kilogram weights. What can you conclude about the weight of one box? Justify your answer. *(Lesson 5-6)* **Each box weighs 1.5 kilograms; the picture shows $B + 4 = 3B + 1$; $B = 1.5$.**

Exploration

37. Generalize Example 3 and Question 10 from this lesson. That is, if the lengths of two legs of a right triangle are equal, what must be true about the hypotenuse? Justify your answer.

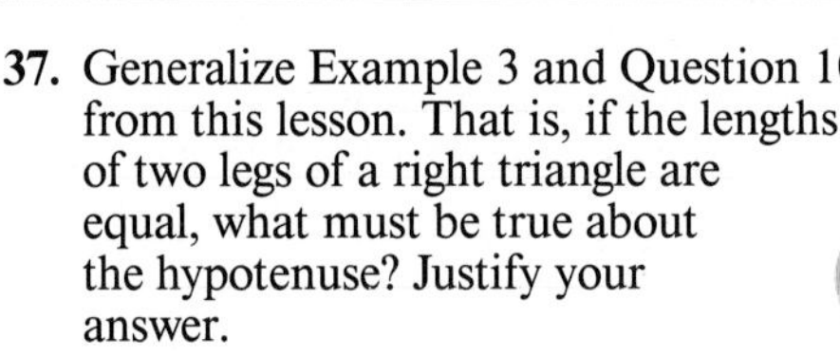

37) If the lengths of the two legs of a right triangle are equal, the hypotenuse has length equal to the product of the length of a leg and $\sqrt{2}$. Suppose x is the length of each of the legs. Then $(\text{hypotenuse})^2 = x^2 + x^2 = 2x^2$. Thus, the length of the hypotenuse is $x\sqrt{2}$.

LESSON

9-8

Absolute Value, Distance, and Square Roots

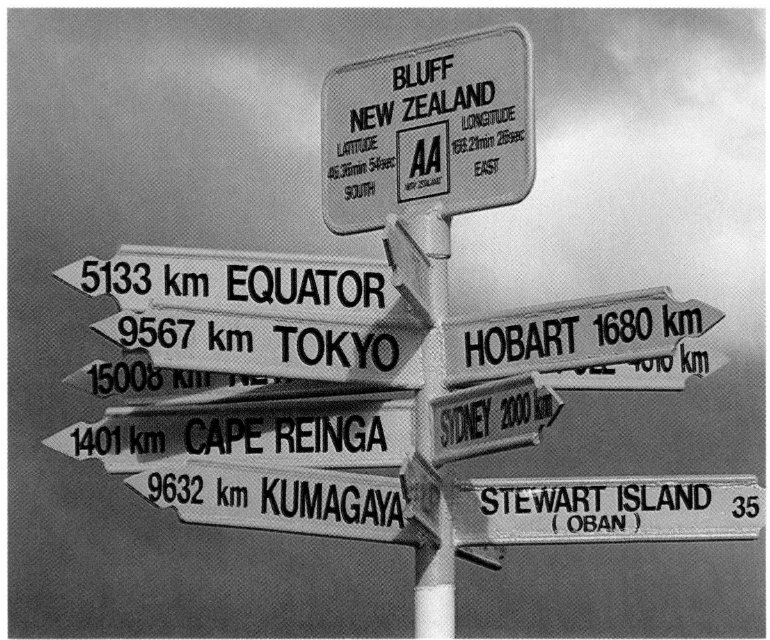

Absolutely positive. *New Zealand is in the South Pacific, far from all continents but Australia. No matter in which direction you go, the distance you would travel is a positive amount.*

What Is the Absolute Value of a Number?

Recall that every real number has an opposite, and that two numbers which are opposites of each other are both the same distance from the origin. For instance, 5 and -5 are opposites; each is 5 units from the origin. The numbers $\frac{3}{2}$ and $-\frac{3}{2}$ are opposites; the distance between each and the origin is $\frac{3}{2}$.

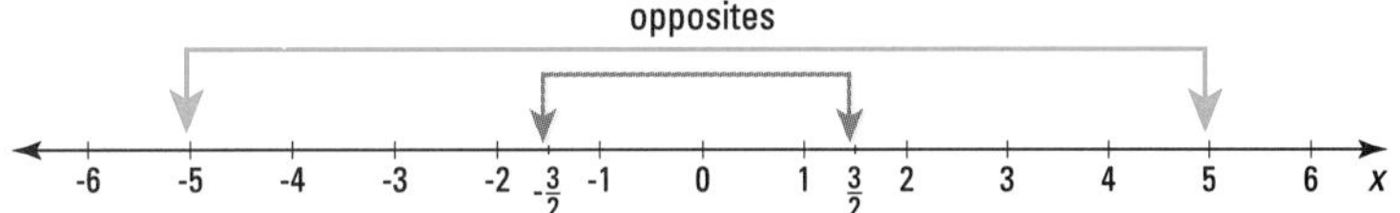

The **absolute value** of a number is the distance between its corresponding point on a number line and the origin. Because it is a distance, an absolute value is never negative. The concept of the absolute value of a number is so important in mathematics that it has its own symbol. The absolute value of x is written $|x|$. In computer languages or spreadsheets the absolute value of x is often written ABS(X).

Thus, $|5| = 5$ and $|-5| = 5$.

Also, $\left|\frac{3}{2}\right| = \frac{3}{2}$ and $\left|-\frac{3}{2}\right| = \frac{3}{2}$.

In BASIC, ABS(3.9) = ABS(-3.9) = 3.9, and ABS(0) = 0.

Lesson 9-8

Objectives

C Evaluate expressions and solve equations using absolute value.

G Calculate and represent distances on the number line.

Resources

From the ***Teacher's Resource File***

- Lesson Master 9-8A or 9-8B
- Answer Master 9-8
- Teaching Aid 97: Warm-up

Additional Resources

- Visual for Teaching Aid 97
- Automatic graphers

Teaching Lesson 9-8

Warm-up

In 1–6, how far is it between 0 and the given number on a number line?

1. 8 **8 units**
2. −8 **8 units**
3. −23 **23 units**
4. 23 **23 units**
5. $\sqrt{5}$ **$\sqrt{5}$ units**
6. $-\sqrt{5}$ **$\sqrt{5}$ units**
7. Graph 3, −3, $\sqrt{3}$, and $-\sqrt{3}$ on a number line.

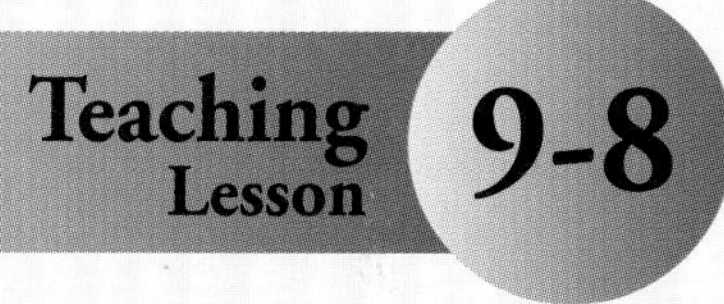

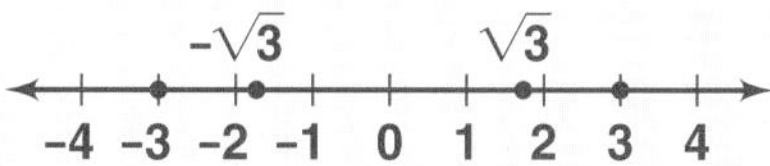

Lesson 9-8 Overview

Broad Goals This lesson introduces absolute value and two of its most important consequences: that the distance between two points with coordinates a and b on a number line is $|a - b|$, and that for all real numbers x, $|x| = \sqrt{x^2}$.

Perspective We place absolute value here because of its connections with square roots and because the simplest absolute-value equations, like quadratics, may have zero, one, or two solutions. Because finding the absolute value of a number is so easy, students may wonder why the idea is needed. Here are four places that the concept may be used: (1) Arithmetic: Absolute value helps in stating rules for addition of positive and negative integers. Students who have studied from *Transition Mathematics* will have seen these rules: To add two numbers with the same sign, add their absolute values, and use that sign in the answer. To add two numbers with different signs, subtract their absolute values and use the sign of the number with the larger absolute value in the answer. (2) Geometry: Distance is a model for absolute value on the number line in the same way that rate is a model for division. This idea is discussed before **Example 3** and it is used there. (3) Computer work: To avoid rounding or truncation errors, computer programmers use absolute value. For *(Overview continues on page 594.)*

Notes on Reading

Reading Mathematics Again, it is a good idea to ask students to identify what each example illustrates. **Example 1** is a direct calculation of absolute value. **Example 2** is a simple equation that involves absolute value. **Example 3** uses absolute value to find the distance between two points on a number line. The ideas in **Examples 2 and 3** provide the two solutions to the more complicated absolute-value equation of **Example 4.**

Example 1

Evaluate the following.

a. $|-4.5|$ **b.** $|1| - |-7|$ **c.** $|1 - 7|$

Solution

a. Think about the number line. The distance between the point -4.5 and the origin is 4.5, so

$$|-4.5| = 4.5.$$

b. Evaluate each absolute value from left to right.

$$|1| - |-7| = 1 - 7 = -6$$

c. The absolute value symbol is also a grouping symbol. So you must evaluate the expression inside it first.

$$|1 - 7| = |-6| = 6$$

Notice that when x is *negative,* $|x| = -x$. The absolute value changes the negative number to its opposite, which is a positive number.

The graph of the equation $y = |x|$ has an interesting shape. Here are a table of values and a graph.

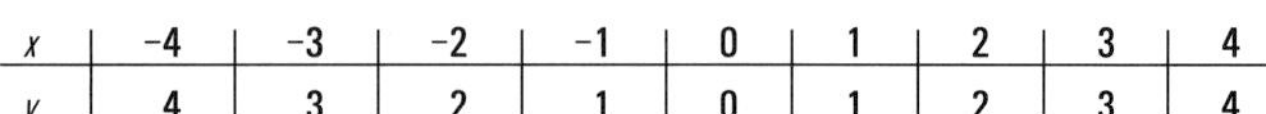

x	-4	-3	-2	-1	0	1	2	3	4
y	4	3	2	1	0	1	2	3	4

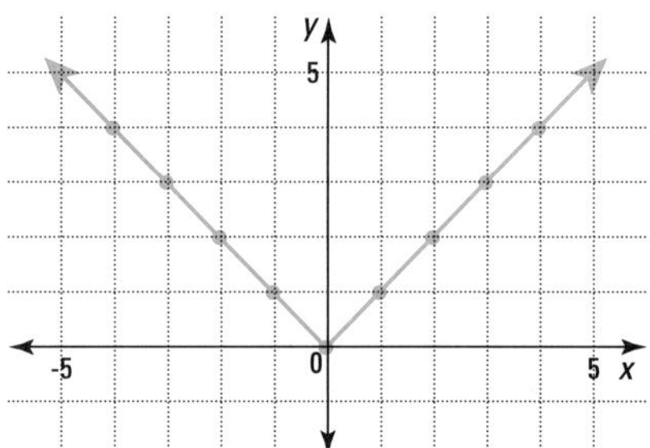

The graph has the shape of the letter V. The graph is an angle with vertex at (0, 0).

Solving Equations Involving Absolute Value

You can solve equations involving absolute value either geometrically or algebraically.

Example 2

Solve $|x| = 10$.

Solution 1

Think geometrically. What numbers on a number line are at a distance of 10 from zero? Answer: 10 and -10. So,

if $|x| = 10$, then $x = 10$ or -10.

Lesson 9-8 Overview, continued

instance, the instruction IF ABS(K – 3) <.001 THEN . . . means the number K is within one-thousandth of 3. Then a further instruction will be given. This is similar to the idea found in **Question 29**. (4) Statistics: If a set of data has a mean of 51, the deviation of any score x is $|x - 51|$. This idea is used in Lesson 12-7. Absolute values can also be used to compute a range without distinguishing between the maximum and minimum values: "If a and b are extreme values of a set of data, then the range is $|a - b|$." **Question 28c** is similar.

We avoid the traditional two-part definition of absolute value (if $x \geq 0$, then $|x| = x$; if $x < 0$, then $|x| = -x$) here because it can be misunderstood by so many students. However, you may wish to introduce it.

The graph of the solution set consists of two points.

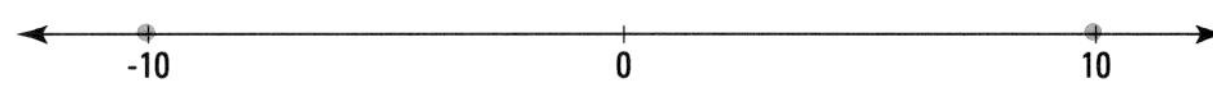

Solution 2

Think algebraically: $|-10| = 10$ and $|10| = 10$. So, if $|x| = 10$, then $x = 10$ or $x = -10$.

Absolute Value and Distance

These ideas can be extended to determine the distance between any two points on a number line. You know you can subtract to find out by how much two numbers differ. By then taking the absolute value, you can guarantee that the result is positive.

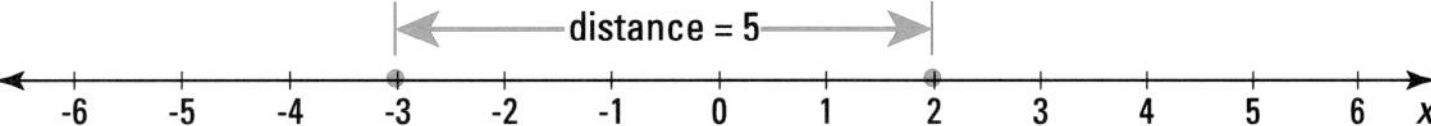

For instance, consider the distance between the points with coordinates -3 and 2. Subtracting the coordinates gives either $-3 - 2 = -5$ or $2 - (-3) = 5$. But taking the absolute value after subtracting, $|-3 - 2| = |-5| = 5$, and $|2 - -3| = |5| = 5$.

Distance Formula on a Number Line
If x_1 and x_2 are the coordinates of two points on a number line, then the **distance** between them is $|x_2 - x_1|$.

Example 3

Find the distance d between the points with coordinates -17 and 8.

Solution

Use the distance formula. $d = |8 - -17| = |8 + 17| = |25| = 25$

Check 1

Draw a number line and count.

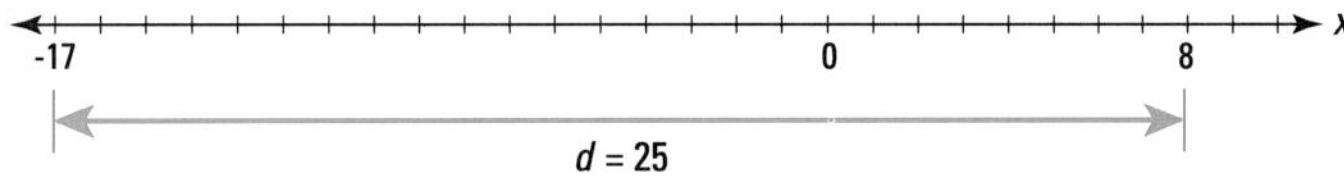

Check 2

Do the calculation in the other order. $d = |-17 - 8| = |-25| = 25.$

Optional Activities

Activity 1

Career Connection Technicians in many industries need to determine if objects, such as washers, bolts, or gaskets, fall within prescribed tolerances. As an extension of **Question 30,** you might have students find the allowable range of measurement for each of the machine parts shown in the table at the right.

Designed Length	Tolerance	Range x units to y units
0.23 in.	± 0.01 in.	[0.22 in. to 0.24 in.]
$3\frac{1}{2}$ in.	$\pm\frac{1}{4}$ in.	[$3\frac{1}{4}$ in. to $3\frac{3}{4}$ in.]
10 cm	± 1 mm	[9.9 cm to 10.1 cm]
6.5 cm	± 0.001 cm	[6.499 cm to 6.501 cm]
4 cm	± .1 mm	[3.99 cm to 4.00 cm]

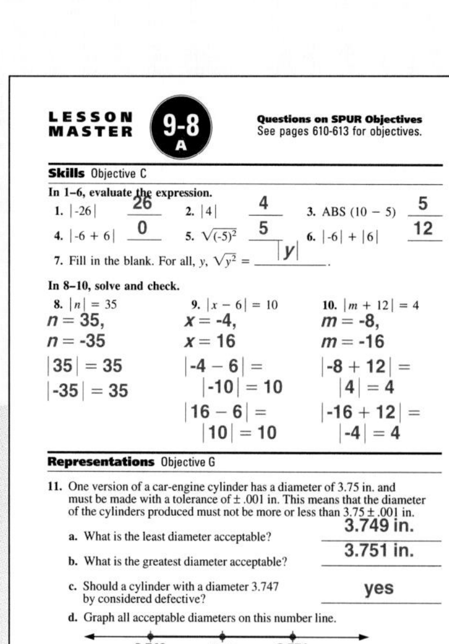

In **Example 4,** point out how the algebraic and geometric solutions are related. In Solution 2, emphasize the importance of carefully arranging one's work.

The last part of the lesson relates absolute value to the graphing of lines and to the square roots that students have seen in previous lessons. You may want to use the graph on page 594 to introduce the following characterization of absolute value, which is often used as a definition of absolute value:

If $x \geq 0$, then $|x| = x$. (The $|x|$ is the part of the graph on and to the right of the y-axis.)

If $x < 0$, then $|x| = -x$. (The $|x|$ is the part of the graph to the left of the y-axis.)

Additional Examples

1. Evaluate the following.
   a. $|-82|$ **82**
   b. $|-82 + 6|$ **76**
   c. $|-82| + |6|$ **88**
2. Solve.
   a. $|m| = \frac{1}{2}$ $m = \frac{1}{2}$ or $-\frac{1}{2}$
   b. $|x| = -8$ **No solution**
3. Find the distance between the points with coordinates -4 and -193. **189**
4. Solve $|x - 8| = 25$. Graph the solutions on a number line.
   $x = 33$ or $x = -17$

-17 33

-20 -10 0 10 20 30 40 x

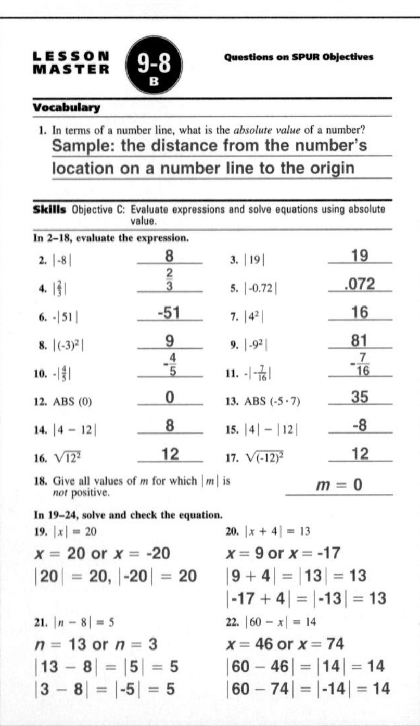
LESSON MASTER 9-8 B Questions on SPUR Objectives

Vocabulary

1. In terms of a number line, what is the *absolute value* of a number?
Sample: the distance from the number's location on a number line to the origin

Skills Objective C: Evaluate expressions and solve equations using absolute value.

In 2–18, evaluate the expression.

2. $\lvert -8 \rvert$	8	3. $\lvert 19 \rvert$	19
4. $\lvert \frac{2}{3} \rvert$	$\frac{2}{3}$	5. $\lvert -0.72 \rvert$	.072
6. $-\lvert 51 \rvert$	-51	7. $\lvert 4^2 \rvert$	16
8. $\lvert (-3)^2 \rvert$	9	9. $\lvert -9^2 \rvert$	81
10. $-\lvert \frac{4}{5} \rvert$	$-\frac{4}{5}$	11. $-\lvert -\frac{7}{16} \rvert$	$-\frac{7}{16}$
12. ABS (0)	0	13. ABS (-5 · 7)	35
14. $\lvert 4 - 12 \rvert$	8	15. $\lvert 4 \rvert - \lvert 12 \rvert$	-8
16. $\sqrt{12^2}$	12	17. $\sqrt{(-12)^2}$	12

18. Give all values of m for which $|m|$ is *not* positive. $m = 0$

In 19–24, solve and check the equation.

19. $|x| = 20$
$x = 20$ or $x = -20$
$|20| = 20$, $|-20| = 20$

20. $|x + 4| = 13$
$x = 9$ or $x = -17$
$|9 + 4| = |13| = 13$
$|-17 + 4| = |-13| = 13$

21. $|n - 8| = 5$
$n = 13$ or $n = 3$
$|13 - 8| = |5| = 5$
$|3 - 8| = |-5| = 5$

22. $|60 - x| = 14$
$x = 46$ or $x = 74$
$|60 - 46| = |14| = 14$
$|60 - 74| = |-14| = 14$

Example 4

Solve $|x - 4| = 6$.

Solution 1

Think geometrically: $|x - 4|$ represents the distance between the points with coordinates x and 4. This distance must be 6. Measure 6 units to either side of 4.

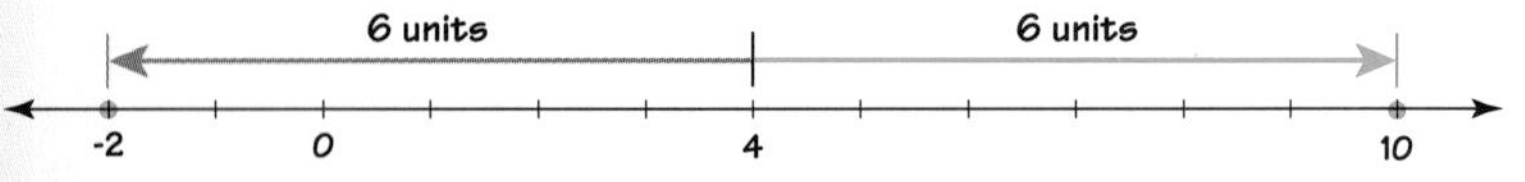

$x = -2$ or $x = 10$.

Solution 2

$|x - 4| = 6$

Think algebraically: use chunking. The only two numbers whose absolute value equals 6 are 6 itself and -6.

Either $x - 4 = 6$ or $x - 4 = -6$.

Solve each equation.

$x = 10$ or $x = -2$

Check

Substitute each solution in the equation $|x - 4| = 6$.
If $x = 10$, $|10 - 4| = |6| = 6$.
If $x = -2$, $|-2 - 4| = |-6| = 6$.
Each solution checks.

Absolute Value and Square Roots

Surprisingly, absolute values are related to square roots. Examine this table.

x	x^2	$\sqrt{x^2}$
-3	9	3
-2	4	2
-1	1	1
0	0	0
1	1	1
2	4	2
3	9	3

The general pattern is that the absolute value of a number equals the positive square root of its square.

Absolute Value–Square Root Property
For all x, $\sqrt{x^2} = |x|$.

Optional Activities

Activity 2

Questions 22–25 show an analogy between absolute value equations and quadratic equations, namely that there may be zero, one, or two real solutions. You could have students substitute $\sqrt{n^2}$ for $|n|$ in each equation in **Questions 22–24,** and substitute $|q|$ for $\sqrt{q^2}$ in **Question 25** to show that this analogy exists.

Adapting to Individual Needs

Extra Help

Some students might be confused by the idea that distance is an absolute value, so distance is never negative. They have learned to represent moves on the number line with positive and negative numbers and might insist that since -2 represents a move of two units to the left, then -2 represents a distance. Point out that a move on the number line consists of distance and direction. The positive or negative sign represents

QUESTIONS

Covering the Reading

1. On a number line, what does the absolute value of a number represent? **The distance between its corresponding point and the origin**

In 2–7, evaluate the expression.

2. $|-50|$ **50** **3.** $|1.3|$ **1.3** **4.** $|0|$ **0**

5. ABS(-16) **16** **6.** $|8 + -4|$ **4** **7.** $|8| + |-4|$ **12**

In 8 and 9, refer to the graph of $y = |x|$ in this lesson.

8. *True or false.* There are two points on the graph with y-coordinate 3. **True**

9. Name a point which is on the graph of $y = |x|$ but is not pictured in this lesson. **Samples: (6, 6) (-6, 6)**

10. a. Write a sentence that expresses the following: on a number line, the distance from the point with coordinate x and the origin is 11.
b. Solve the sentence from part **a.** $x = 11$ or $x = -11$
a) $|x| = 11$

In 11–14, solve and check.

11. $|t| = 25$ $t = -25$ or $t = 25$ **12.** $|x - 3| = 2$ $x = 1$ or $x = 5$

13. $|x - 3| = 42$ $x = -39$ or $x = 45$ **14.** $|300 - x| = 10$ $x = 290$ or $x = 310$

In 15 and 16, find the distance between the two points whose coordinates are given.

15. (number line with points -28 and 11) **39** **16.** (number line with points -81 and -57) **24**

17. Write an expression for the distance between the points whose coordinates are 17.5 and n. $|17.5 - n|$ or $|n - 17.5|$

In 18–21, evaluate.

18. $\sqrt{11^2}$ **11** **19.** $\sqrt{n^2}$ $|n|$ **20.** $\sqrt{(-11)^2}$ **11** **21.** $\sqrt{8^2}$ **8**

Applying the Mathematics

In 22–25, give the number of solutions to the equation. Then solve.

22. $|n| = 0$ 1; $n = 0$ **23.** $|n| = -6$ 0; no real solutions **24.** $|p| = \frac{1}{2}$ 2; $p = \frac{1}{2}$ or $p = -\frac{1}{2}$ **25.** $\sqrt{q^2} = 31$ 2; $q = 31$ or $q = -31$

In 26–28, find two points on the number line that are 7 units from the point with the given coordinate.

26. 3 -4; 10 **27.** -3 -10; 4 **28.** a $a - 7$; $a + 7$

Notes on Questions

Questions 6–7 These questions deal with the order of operations and absolute value. Inform students that the symbol | | acts as a grouping symbol; that is, it carries implicit parentheses | ( ) |. Thus, we first operate inside the absolute value sign.

Questions 11–14 Encourage those students who use an algebraic approach to check geometrically and vice versa.

Questions 18–21 Here is an additional question that you might give to students: Simplify $-\sqrt{(8)^2}$. [-8]

Questions 22–25 Activity 2 in *Optional Activities* on page 596 relates to these questions.

direction (negative left and positive right) and the absolute value of the number represents distance. In -2, the direction, left, is indicated by the negative sign, and the distance is represented by the absolute value of -2, which is 2.

You might ask a student to walk from his or her desk to the door, counting the steps taken to get there. Then have the student walk back to the desk. If it took 5 steps to walk from the desk to the door, ask if it took -5 steps to walk back from the door to the desk. Emphasize that it took 5 steps to walk in each direction; distance is given as a positive number.

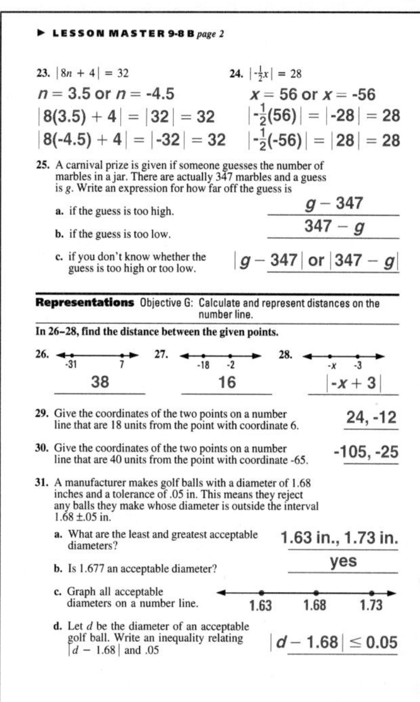

▶ LESSON MASTER 9-8 B *page 2*

23. $|8n + 4| = 32$
$n = 3.5$ or $n = -4.5$
$|8(3.5) + 4| = |32| = 32$
$|8(-4.5) + 4| = |-32| = 32$

24. $|-\frac{1}{2}x| = 28$
$x = 56$ or $x = -56$
$|-\frac{1}{2}(56)| = |-28| = 28$
$|-\frac{1}{2}(-56)| = |28| = 28$

25. A carnival prize is given if someone guesses the number of marbles in a jar. There are actually 347 marbles and a guess is g. Write an expression for how far off the guess is
a. if the guess is too high. $g - 347$
b. if the guess is too low. $347 - g$
c. if you don't know whether the guess is too high or too low. $|g - 347|$ or $|347 - g|$

Representations Objective G: Calculate and represent distances on the number line.

In 26–28, find the distance between the given points.

26. (-31, 7) 38 **27.** (-18, -2) 16 **28.** (-x, -3) $|-x + 3|$

29. Give the coordinates of the two points on a number line that are 18 units from the point with coordinate 6. 24, -12

30. Give the coordinates of the two points on a number line that are 40 units from the point with coordinate -65. -105, -25

31. A manufacturer makes golf balls with a diameter of 1.68 inches and a tolerance of .05 in. This means they reject any balls they make whose diameter is outside the interval 1.68 ±.05 in.
a. What are the least and greatest acceptable diameters? 1.63 in., 1.73 in.
b. Is 1.677 an acceptable diameter? yes
c. Graph all acceptable diameters on a number line. 1.63 1.68 1.73
d. Let d be the diameter of an acceptable golf ball. Write an inequality relating $|d - 1.68|$ and .05 $|d - 1.68| \le 0.05$

Follow-up for Lesson 9-8

Practice

For more questions on SPUR Objectives, use **Lesson Master 9-8A** (shown on page 595) or **Lesson Master 9-8B** (shown on pages 596–597).

Assessment

Written Communication Have each student write a short paragraph explaining in his or her own words how to find the distance between two points on a number line and how to check that the distance found is correct. [Students display understanding of absolute value and use the distance formula correctly.]

Extension

Students can consider whether there are distributive properties that involve absolute value. Have students **work in small groups** to investigate these questions. Encourage them to provide data or mathematical arguments to support their conjectures.

1. When does $|a + b| = |a| + |b|$? [When a and b have the same sign or one of them is zero; that is when $ab \geq 0$]
2. When does $|a - b| = |a| - |b|$? [When $ab > 0$ and $|a| \geq |b|$]
3. When does $|ab| = |a| \cdot |b|$? [This is always true.]
4. When does $\left|\frac{a}{b}\right| = \frac{|a|}{|b|}$? [This is always true.]

Projects Update Project 7, *Graphing Diamonds*, on page 606, relates to the content of this lesson.

These sensitive calipers measure the width of the metal plates made in an electric motor parts company. The width of each plate must be within specific tolerance levels.

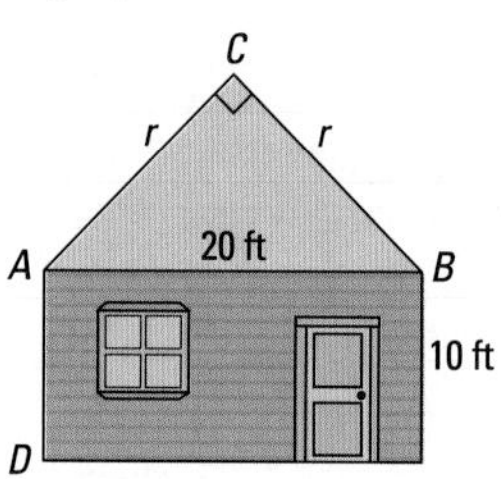

40a)

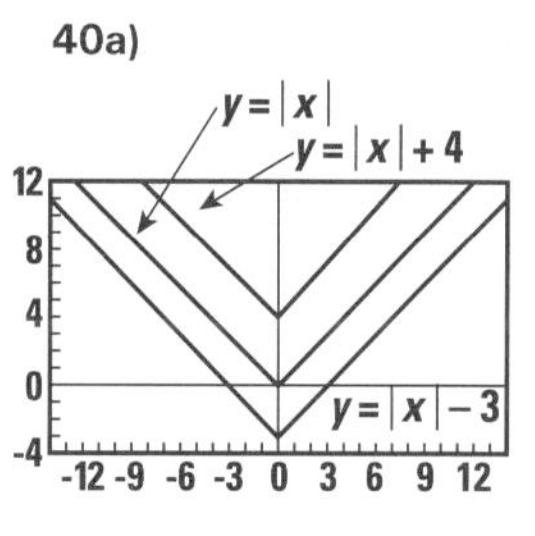

40c) Sample: When $k = 5$, the equation is $y = |x| + 5$. The graph should be a right angle with vertex at (0, 5), which it is.

29. Marika's age is x years and Wolfgang's age is y years. What is the difference in their ages, under the given conditions?
a. $x > y$ $x - y$ **b.** $y > x$ $y - x$
c. if you are not sure which of x or y is greater $|y - x|$ or $|x - y|$

30. Manufactured parts are allowed to vary from an accepted standard size by a specific amount called the **tolerance.** Consider a washer designed to have a diameter of 2 cm, with a tolerance of ± 0.001 cm. This means that the actual diameter may not fall outside the interval 2 ± 0.001 cm.
a. What is the smallest diameter allowed? 1.999 cm
b. What is the largest diameter allowed? 2.001 cm
c. Graph all allowable diameters on a number line. See below.
d. If d is the diameter of a washer that falls within the tolerance level above, what must be true about $|d - 2|$? $|d - 2| \leq 0.001$

30c)
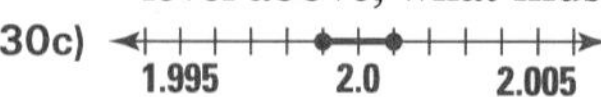

Review

In 31–33, simplify. *(Lesson 9-7)*

31. $\frac{\sqrt{175}}{5}$ $\sqrt{7}$ **32.** $\frac{\sqrt{300}}{6}$ $\frac{5\sqrt{3}}{3}$ **33.** $\frac{16 \pm \sqrt{288}}{4}$ $4 \pm 3\sqrt{2}$

34. *True or false.* $8\sqrt{10} = \sqrt{80}$. Explain how you got your answer. *(Lesson 9-7)* False; $\sqrt{80} = \sqrt{16 \cdot 5} = 4\sqrt{5} \neq 8\sqrt{10}$

35. The figure at the left represents a side view of a building plan for a garage. The garage is to be 20 ft wide. The rafters AC and BC of the roof are to be equal in length and to meet at right angles.
a. Find the length r of each rafter as a simplified radical. $10\sqrt{2}$ ft
b. Round the length of a rafter to the nearest tenth of a foot. 14.1 ft
c. Find BD. *(Lessons 1-8, 9-7)* $10\sqrt{5} \approx 22.4$ ft

36. How many real solutions does the equation $17p^2 + p = 20$ have? Justify your answer. *(Lesson 9-6)* 2; $b^2 - 4ac = 1 - 4(17)(-20) = 1361 > 0$

37. Solve $2x^2 - 16x - 4 = 0$. *(Lesson 9-5)* $x = 4 + 3\sqrt{2}$ or $x = 4 - 3\sqrt{2}$

38. Consider these points: (0, -3), (0, 0), (0, 3).
a. Write an equation for the line containing these three points. $x = 0$
b. On which axis do they lie? *(Lesson 5-1)* the y-axis

Exploration

See margin for a and b.

39. a. Make a table of values for $y = |x - 3|$, when $-6 \leq x \leq 6$.
b. Graph the points you found in part **a.**
c. Describe the set of *all* points for which $y = |x - 3|$. The graph is an angle with vertex (3, 0).

40. a. Use an automatic grapher to graph these three equations on the same set of axes. See left.
$y = |x|$ $y = |x| + 4$ $y = |x| - 3$
b. Make a conjecture about the shape of the graph $y = |x| + k$, for any real number k. The graph is a right angle with vertex at $(0, k)$.
c. Test your conjecture using a value of k other than 0, 4, or -3. See left.

Adapting to Individual Needs

Challenge

Materials: Automatic Graphers

Have students refer to **Exercises 39–40**. Have them use an automatic grapher to graph $y = |x - 3| + 4$. [A sample graph is shown at the right.] Ask them to name the vertex of the graph. [(3, 4)] Then have them graph other equations of the form $y = |x - a| + k$ and make a generalization for graphing any equation of this form.

[The graph will have the same shape as that for $y = |x|$ and the vertex will be (a, k).]

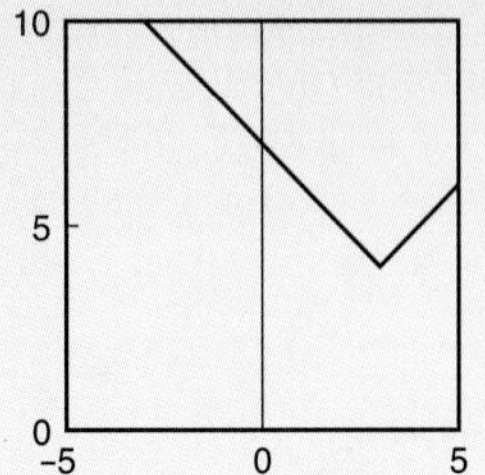

Additional Answers

39a, b

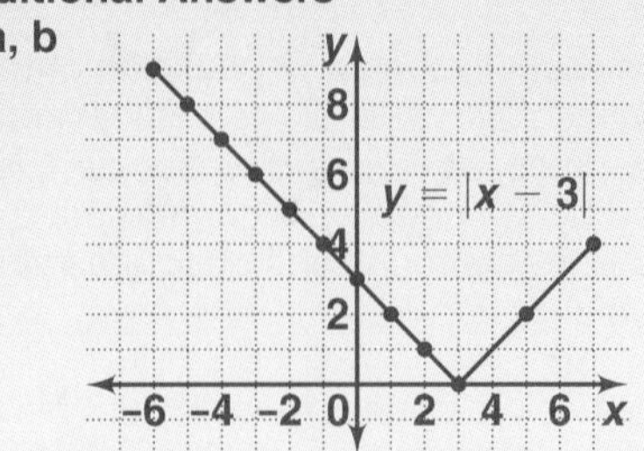

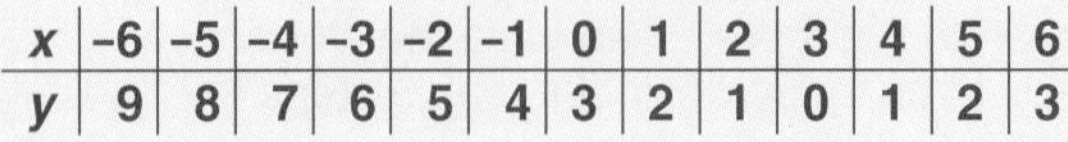

x	-6	-5	-4	-3	-2	-1	0	1	2	3	4	5	6
y	9	8	7	6	5	4	3	2	1	0	1	2	3

LESSON

9-9 Distances in the Plane

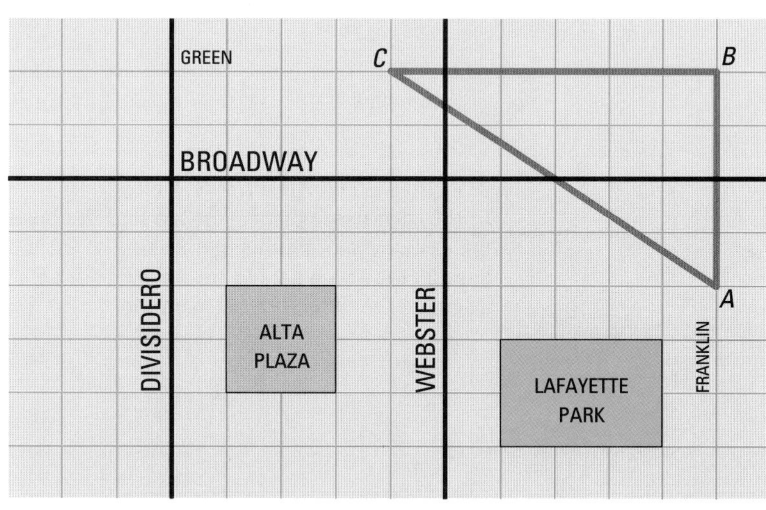

Distances Along Grids

Streets of many cities are laid out in a grid pattern like the one above. To get from one location to another by car you have to travel along streets. So the shortest distance from A to C is not the hypotenuse of triangle ABC, but the sum of the legs' lengths, $AB + BC$. But a bird or helicopter could go directly from A to C along the segment $\overline{AC}$ *"as the crow flies."*

Example 1

a. How far is it from A to C traveling by car along $\overline{AB}$ and $\overline{BC}$? (Each unit is a city block.)

b. How far is it from A to C as the crow flies?

Solution

a. Count the blocks between points A and B. There are 4 blocks. Count the blocks between B and C. There are 6 of them.

$$AB + BC = 4 + 6 = 10$$

Traveling by car, the distance between A and B is 10 blocks.

b. Use the Pythagorean Theorem on $\triangle ABC$ above.

$$\begin{aligned} AC^2 &= AB^2 + BC^2 \\ &= 4^2 + 6^2 \\ &= 16 + 36 \\ &= 52 \\ AC &= \sqrt{52} \\ &\approx 7.2 \text{ blocks} \end{aligned}$$

The distance as the crow flies from A to C is $\sqrt{52} \approx$ 7.2 blocks.

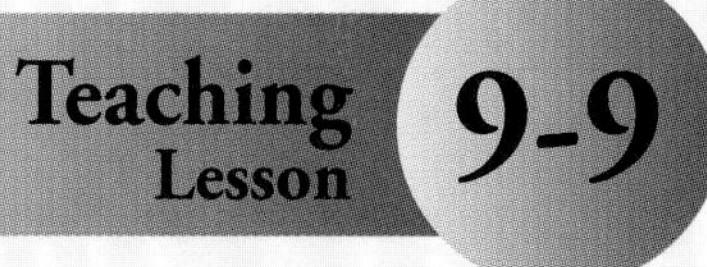

Objectives

G Calculate and represent distances in the plane.

Resources

From the ***Teacher's Resource File***
- Lesson Master 9-9A or 9-9B
- Answer Master 9-9
- Teaching Aids
  - 26 Graph Paper
  - 28 Four-Quadrant Graph Paper
  - 97 Warm-up
  - 106 Additional Examples
- Activity Kit, Activity 21

Additional Resources
- Visuals for Teaching Aids 26, 28, 97, 106

Teaching Lesson 9-9

Warm-up

Diagnostic Students need graph paper or **Teaching Aid 28.**

Draw a segment on a coordinate plane that is not parallel to either axis. Find the length of the segment. **Students can find the length of the segment using the Pythagorean Theorem or by using the distance formula. Sample: $AB = \sqrt{117}$.**

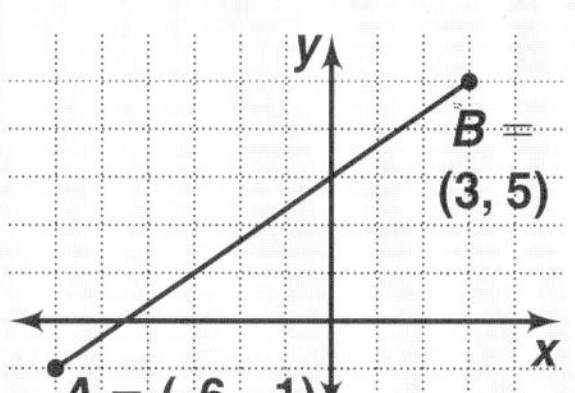

Lesson 9-9 Overview

Broad Goals This lesson introduces and applies the formula for the distance between two points in a plane.

Perspective The distance between two points in a plane is one of the most important ideas for later work in coordinate geometry, and it provides wonderful practice in algebra.

Rather than give students the distance formula at the beginning of the lesson, we have students find the distance between two points by drawing a right triangle whose hypotenuse is the segment connecting the two points. Having worked problems by drawing this triangle, students will better understand the traditional distance formula.

This lesson distinguishes between two types of distance—the street distance and the diagonal distance. This reinforces the Triangle Inequality Theorem. Unless the two points are on the same horizontal or vertical line, the distance "as the crow flies" is always shorter than the street distance.

Notes on Reading

To build understanding, we start with specific examples that use numbers, such as those in **Examples 1–3**; then we go to the general case. Some students may continue to draw right triangles even after they know the formula. This is fine; however, emphasize that the formula makes it unnecessary to draw a picture.

Reading Mathematics If you read this lesson aloud in class, remind students of the following: $\overline{AB}$ with the bar above it is read "the segment AB" or "the segment from A to B"; x_1 is read "x sub one" or just "x one"; $|x - y|$ is read "the absolute value of the quantity x minus y."

Discuss **Question 1** after **Example 1, Questions 4–6** before **Example 2, Questions 2–3** after **Example 2**, and **Questions 11–13** after **Example 3**.

Error Alert Some students may be confused by the fact that two triangles can be drawn in each example, one above the hypotenuse and one below. Note that each triangle is half of the same rectangle.

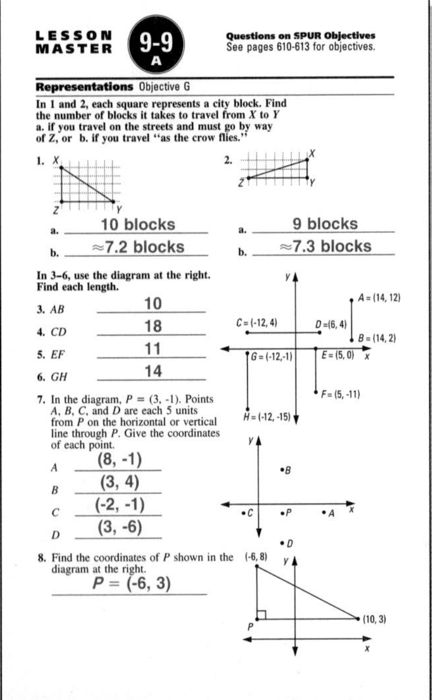

LESSON MASTER 9-9 A — **Questions on SPUR Objectives** See pages 610-613 for objectives.

Representations Objective G

In 1 and 2, each square represents a city block. Find the number of blocks it takes to travel from X to Y a. if you travel on the streets and must go by way of Z, or b. if you travel "as the crow flies."

1. a. 10 blocks
   b. ≈7.2 blocks

2. a. 9 blocks
   b. ≈7.3 blocks

In 3–6, use the diagram at the right. Find each length.

3. AB 10
4. CD 18
5. EF 11
6. GH 14

7. In the diagram, $P = (3, -1)$. Points A, B, C, and D are each 5 units from P on the horizontal or vertical line through P. Give the coordinates of each point.
   A (8, -1)
   B (3, 4)
   C (-2, -1)
   D (3, -6)

8. Find the coordinates of P shown in the diagram at the right.
   $P = (-6, 3)$

The map on page 599 looks like part of the coordinate plane. The idea used in Example 1 can be applied to find the distance between any two points in the coordinate plane. But first we examine the situations where the points are on the same horizontal or vertical line.

Distances Along Horizontal or Vertical Lines

In the coordinate plane, you can find the distance between two points on the same horizontal line by thinking of them as being on a number line parallel to the x-axis. For instance, to find RS in the figure below, calculate the difference between the x-coordinates of R and S.

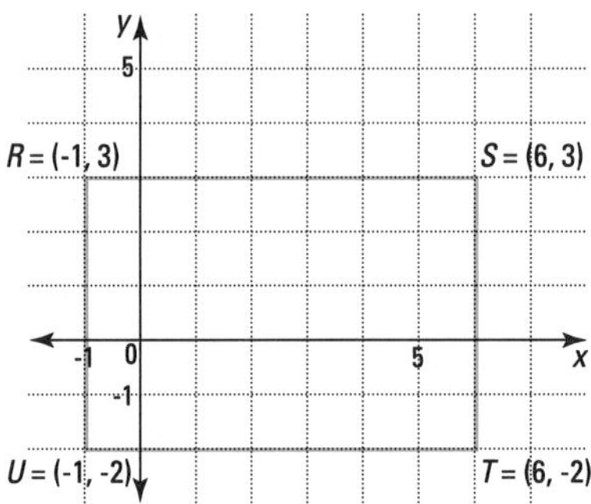

$$RS = |-1 - 6| = 7$$

Similarly, to find the distance ST, calculate the difference between the y-coordinates of S and T.

$$ST = |3 - -2| = 5$$

Distances Between Any Two Points

To find the distance between two points A and B on an oblique line, you can use the Pythagorean Theorem.

Example 2

Find AB in $\triangle ABC$ at the left.

Solution

AB is the length of the hypotenuse of a right triangle whose legs are parallel to the x- and y-axes.

$$AC = |2 - -1| = |3| = 3$$
$$BC = |1 - 7| = |-6| = 6$$

So,
$$AB^2 = AC^2 + BC^2 = 3^2 + 6^2 = 9 + 36 = 45.$$

Thus,
$$AB = \sqrt{45} = 3\sqrt{5}.$$

Optional Activities

Activity 1

Materials: Graph paper or **Teaching Aid 26**, city or regional maps

After completing the lesson, have each student use graph paper or **Teaching Aid 26** and make a map similar to that at the top of page 599. Students can make up an imaginary map or draw one that represents a real area, such as the neighborhood around the school or their homes. Tell them to draw key streets or roads and to identify several buildings, parks, or similar things. Then have them approximate distances between locations using both "street distances" and distances "as the crow flies."

Remind students that they must select a scale for their map. You might also suggest that they select streets or roads that more or less follow a grid pattern.

This method can be generalized to find the distance between any two points that lie on an oblique line. If the points are $A = (x_1, y_1)$ and $B = (x_2, y_2)$, then a right triangle can be formed with the third vertex at $C = (x_1, y_2)$.

Once the third vertex has been found, the lengths of the legs are $BC = |x_2 - x_1|$ and $AC = |y_2 - y_1|$. Now use the Pythagorean Theorem.

$$AB^2 = BC^2 + AC^2$$

Substitute. $$AB^2 = |x_2 - x_1|^2 + |y_2 - y_1|^2$$

Now take the square root of each side. The result is a formula for the distance between any two points in the plane.

Pythagorean Distance Formula in the Coordinate Plane
The distance AB between the points $A = (x_1, y_1)$ and $B = (x_2, y_2)$ in the coordinate plane is $AB = \sqrt{|x_2 - x_1|^2 + |y_2 - y_1|^2}$.

Example 3

Find the distance between the points (4, 3) and (6, -2).

Solution 1

Use the Pythagorean Distance Formula. Either point may be (x_1, y_1). Let $(x_1, y_1) = (4, 3)$ and $(x_2, y_2) = (6, -2)$.

$$\begin{aligned} \text{distance} &= \sqrt{|6 - 4|^2 + |-2 - 3|^2} \\ &= \sqrt{|2|^2 + |-5|^2} \\ &= \sqrt{2^2 + 5^2} \\ &= \sqrt{4 + 25} \\ &= \sqrt{29} \approx 5.4 \end{aligned}$$

Solution 2

Plot the points. Draw a right triangle whose hypotenuse has endpoints (4, 3) and (6, -2). At the right, the third vertex is at $C = (4, -2)$. First find the lengths of the legs.

$AC = |3 - -2| = |5| = 5$
$BC = |4 - 6| = |-2| = 2$

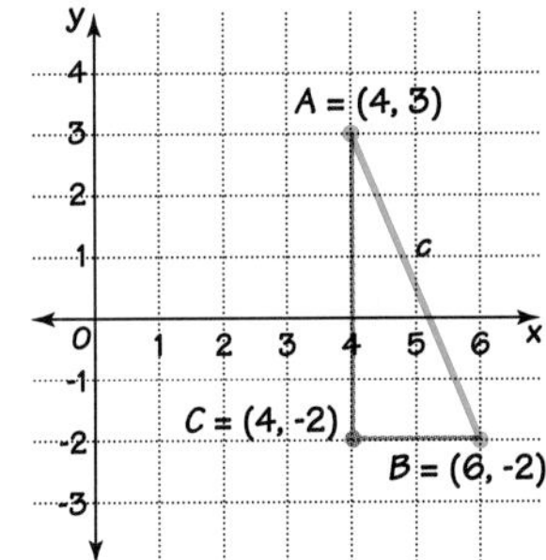

Now find AB by using the Pythagorean Theorem.

$$\begin{aligned} BC^2 + AC^2 &= AB^2 \\ 2^2 + 5^2 &= c^2 \\ 4 + 25 &= c^2 \\ 29 &= c^2 \\ \sqrt{29} &= c \end{aligned}$$

Thus, $AB = \sqrt{29} \approx 5.4$.

Additional Examples

These examples are also given on **Teaching Aid 106.**

1. Refer to the map on page 599. How far is it from the intersection of Divisidero and Broadway to the intersection of Franklin and Green
   a. traveling along Broadway and Franklin? **12 blocks**
   b. as the crow flies? **$\sqrt{104} \approx$ 10.2 blocks**
2. Triangle LMN has coordinates $L = (-6, -2)$, $M = (-6, 4)$, and $N = (6, -2)$. Find MN as a simplified radical. Draw the triangle on a grid if necessary. **$MN = 6\sqrt{5}$**
3. Find the distance between (-6, 2) and (-6, 7). **5**
4. Give the distance as a simplified radical and as a decimal.
   a. Let $C = (4, 2)$ and $K = (7, 11)$. Find CK. **$3\sqrt{10} \approx 9.49$**
   b. Let $N = (-5, 4)$ and $Q = (2, -2)$. Find NQ. **$\sqrt{85} \approx 9.22$**

(Additional Examples continue on page 602.)

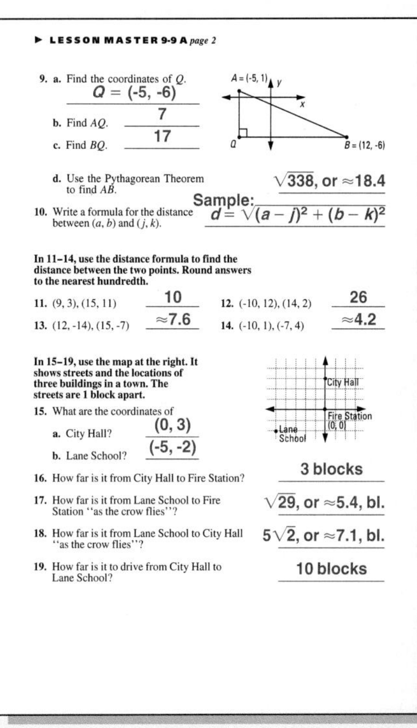

► LESSON MASTER 9-9 A *page 2*

9. a. Find the coordinates of Q. $Q = (-5, -6)$
   b. Find AQ. 7
   c. Find BQ. 17
   d. Use the Pythagorean Theorem to find AB. $\sqrt{338}$, or ≈ 18.4

10. Write a formula for the distance between (a, b) and (j, k). Sample: $d = \sqrt{(a - j)^2 + (b - k)^2}$

In 11–14, use the distance formula to find the distance between the two points. Round answers to the nearest hundredth.

11. (9, 3), (15, 11) 10
12. (-10, 12), (14, 2) 26
13. (12, -14), (15, -7) ≈ 7.6
14. (-10, 1), (-7, 4) ≈ 4.2

In 15–19, use the map at the right. It shows streets and the locations of three buildings in a town. The streets are 1 block apart.

15. What are the coordinates of
    a. City Hall? (0, 3)
    b. Lane School? (-5, -2)
16. How far is it from City Hall to Fire Station? 3 blocks
17. How far is it from Lane School to Fire Station "as the crow flies"? $\sqrt{29}$, or ≈ 5.4, bl.
18. How far is it from Lane School to City Hall "as the crow flies"? $5\sqrt{2}$, or ≈ 7.1, bl.
19. How far is it to drive from City Hall to Lane School? 10 blocks

Optional Activities

Activity 2 Social Studies Connection
The maximum distance D (in km) that a person can see from the top of a tall building with height h (in km) is approximately $111.7\sqrt{h}$. After completing the lesson, students might determine the distance they could see from the top of the following buildings.

1. CN Tower (Toronto): 555 m [About 83 km]
2. Sears Tower (Chicago): 443 m [About 74 km]
3. Central Plaza (Hong Kong): 313 m [About 62 km]
4. Eiffel Tower (Paris): 289 m [About 60 km]

5. Tony and Alicia each left camp on snowmobiles. Tony drove one mile north, then 5 miles west. Alicia drove 6 miles east, then 2 miles south. Make a diagram, and find the distance between Tony and Alicia.

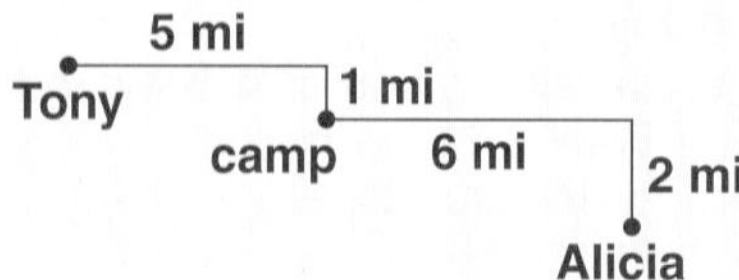

If Tony is at (0, 0), then Alicia is at (11, -3). The distance between those points is $\sqrt{11^2 + 3^2} = \sqrt{130} \approx 11.4$ miles

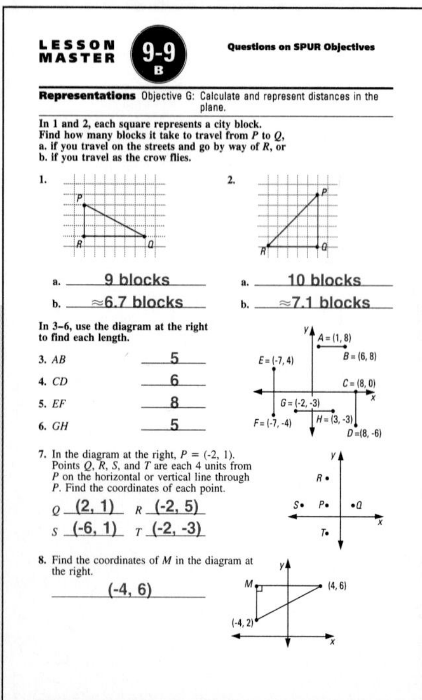

LESSON MASTER 9-9 B

Questions on SPUR Objectives

Representations Objective G: Calculate and represent distances in the plane.

In 1 and 2, each square represents a city block. Find how many blocks it take to travel from P to Q, a. if you travel on the streets and go by way of R, or b. if you travel as the crow flies.

1. a. 9 blocks
   b. ≈6.7 blocks

2. a. 10 blocks
   b. ≈7.1 blocks

In 3–6, use the diagram at the right to find each length.

3. AB 5
4. CD 6
5. EF 8
6. GH 5

A = (1, 8), B = (6, 8), C = (8, 0), D = (8, -6), E = (-7, 4), F = (-7, -4), G = (-2, -3), H = (3, -3)

7. In the diagram at the right, P = (-2, 1). Points Q, R, S, and T are each 4 units from P on the horizontal or vertical line through P. Find the coordinates of each point.

   Q (2, 1) R (-2, 5)
   S (-6, 1) T (-2, -3)

8. Find the coordinates of M in the diagram at the right.

   (-4, 6)

(4, 6), (-4, 2)

QUESTIONS

Covering the Reading

1. In the diagram below, each square in the grid represents a city block.
   a. How many blocks does it take to go from D to E by way of F? 11
   b. Use the Pythagorean Theorem to find the distance from D to E as the crow flies. $\sqrt{73} \approx 8.5$ blocks

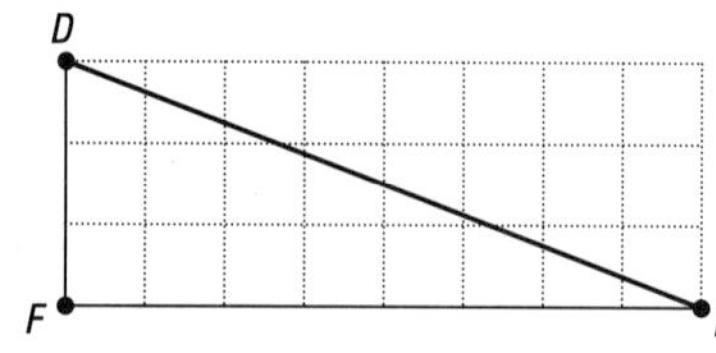

In 2 and 3, find the distance MN.

2. M = (40, 60) and N = (40, 18) 42

3. M = (-3, 10) and N = (11, 10) 14

In 4–6, refer to rectangle $RSTU$ on page 600. Find each length.

4. UT 7 **5.** RU 5 **6.** RT $\sqrt{74} \approx 8.6$

7. Use the graph at the right.
   a. Find the coordinates of P. (-4, -5)
   b. Find the length of $\overline{DP}$. 8
   c. Find the length of $\overline{GP}$. 5
   d. Use the Pythagorean Theorem to find DG. $\sqrt{89} \approx 9.4$

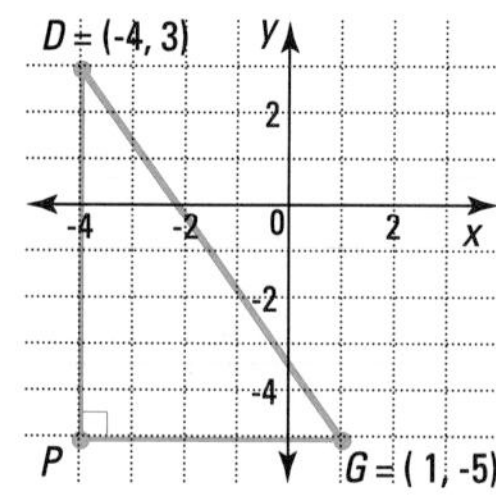

In 8 and 9, find the coordinates of point H. Then find FG.

8.

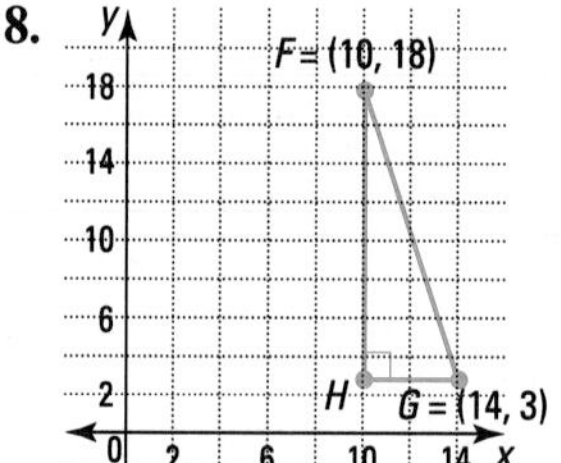

H = (10, 3); $GF = \sqrt{241} \approx 15.5$

9.

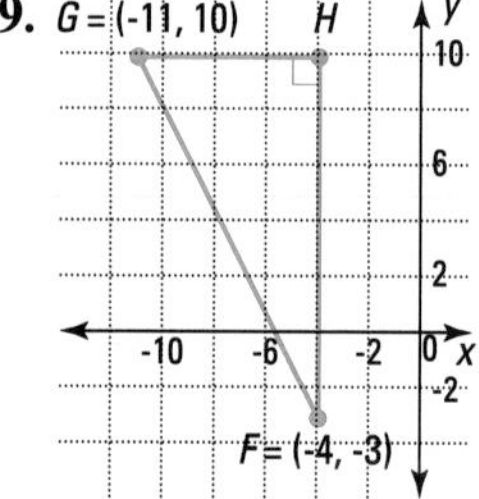

H = (-4, 10); $GF = \sqrt{218} \approx 14.8$

In 10–13, find the distance between the two points.

10. A = (0, 0); B = (-4, -3) 5

11. A = (4, 11); B = (0, 7) $4\sqrt{2} \approx 5.7$

12. D = (5, 1); C = (11, -7) 10

13. E = (-3, 5); F = (-1, -8) $\sqrt{173} \approx 13.2$

Optional Activities

Activity 3 You can use *Activity Kit, Activity 21,* before covering the lesson, or as an alternate approach to the beginning of the lesson through **Example 2.** In this activity, students explore techniques for finding distances in the coordinate plane.

Adapting to Individual Needs

English Language Development
You might suggest that students update their index cards or notes by checking the list of vocabulary words at the end of the chapter and adding words that are missing.

Applying the Mathematics

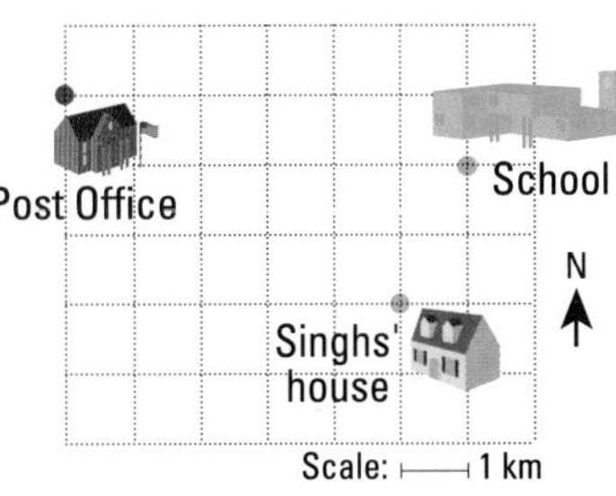

In 14–17, the map at the right shows the location of the Singhs' house, the post office, and the school. Each small square is 1 km on a side.

14. If the Singhs' house is at (0, 0), what are the coordinates of the school? **(1, 2)**

15. Find the distance one would travel by car to get from the post office to the school. **7 km**

16. How far is it from the post office to school as the crow flies? $\sqrt{37} \approx 6.1$ **km**

17. How far is it from the Singhs' house to the school as the crow flies? $\sqrt{5} \approx 2.2$ **km**

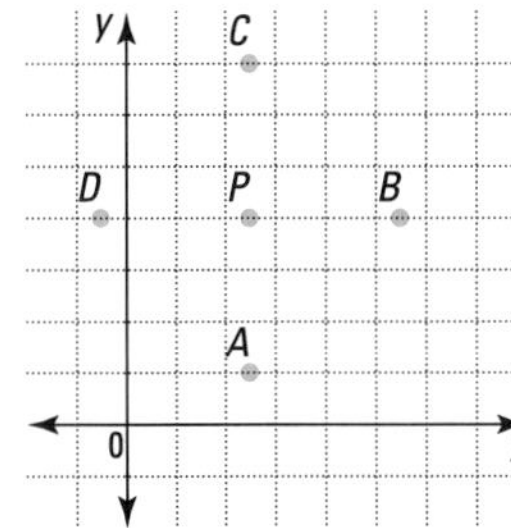

18. Refer to the diagram at the left. Point $P = (5, 8)$. Points A, B, C, and D are each 6 units from P on the horizontal or vertical line through P. Find the coordinates of these points. **A = (5, 2); B = (11, 8); C = (5, 14); D = (-1, 8)**

In 19 and 20, write an expression for the distance between the two points.

19. (0, 0) and (a, b) $\sqrt{|a|^2 + |b|^2}$

20. (a, b) and (c, d) $\sqrt{|a-c|^2 + |b-d|^2}$

21. Let $J = (-5, 0)$, $K = (1, 8)$, and $L = (16, 0)$. Find the length of each side of $\triangle JKL$. **JK = 10; KL = 17; JL = 21**

22. The distance between the points (1, 1) and (4, y) is 5. Find all possible values of y. **y = 5 or y = -3**

In 23 and 24, use this computer program that asks for coordinates of points (x_1, y_1) and (x_2, y_2) in the coordinate plane, and then calculates the distance between them.

```
10 PRINT "DISTANCE IN THE PLANE"
20 PRINT "BETWEEN (X1, Y1) AND (X2, Y2)"
30 PRINT "ENTER X1, Y1"
40 INPUT X1, Y1
50 PRINT "ENTER X2, Y2"
60 INPUT X2, Y2
70 LET H = ABS(X1 - X2)
80 LET V = ABS(Y1 - Y2)
90 LET DIST = SQR(H^2 + V^2)
100 PRINT "THE DISTANCE IS"; DIST
110 END
```

23. What value for DIST will the computer print when $(x_1, y_1) = (19, -5)$ and $(x_2, y_2) = (-3, 15)$? **29.732137**

24. What value for DIST will the computer print when $(x_1, y_1) = (0, 0)$ and $(x_2, y_2) = (1, 1)$? **1.4142136**

Notes on Questions

Questions 4–6 Ask students to compare their answers to these questions to the distances calculated on page 600. Then ask what properties of rectangles these numbers illustrate. [The opposite sides of a rectangle are equal in length. The diagonals are equal in length.]

Questions 10–13 These are key questions. Many students will plot the points, draw a triangle, and find the lengths of its sides. Make sure that they understand how to do the required steps, but then point out that by using the Pythagorean Distance Formula, the work can be shortened considerably.

Question 22 Some students may see that a 3-4-5 triangle is involved.

Questions 23–24 Here, the BASIC program computes the Pythagorean ("as the crow flies") distance. You can add a line to give street (often called the "taxicab") distance:

```
105 PRINT "THE STREET DISTANCE IS"; ABS(A) + ABS(B)
```

Adapting to Individual Needs

Extra Help

Be sure students understand that it does not matter which point is labeled (x_1, y_1) and which is labeled (x_2, y_2). Remind them that $|x_1 - x_2| = |x_2 - x_1|$. Use an example to reinforce the idea. If the coordinates of the two points are (3, 5) and (7, 4), then 3 and 7 are the two x coordinates. Show that $|3 - 7| = |-4| = 4$ and $|7 - 3| = |4| = 4$. Likewise, 5 and 4 are the y coordinates. Show that $|5 - 4| = |1| = 1$ and $|4 - 5| = |-1| = 1$. For students who get confused by the notation in the distance formula, the following description might help:

$$\text{Distance} = \sqrt{\left(\begin{matrix}\text{difference in}\\ \text{x-coordinates}\end{matrix}\right)^2 + \left(\begin{matrix}\text{difference in}\\ \text{y-coordinates}\end{matrix}\right)^2}$$

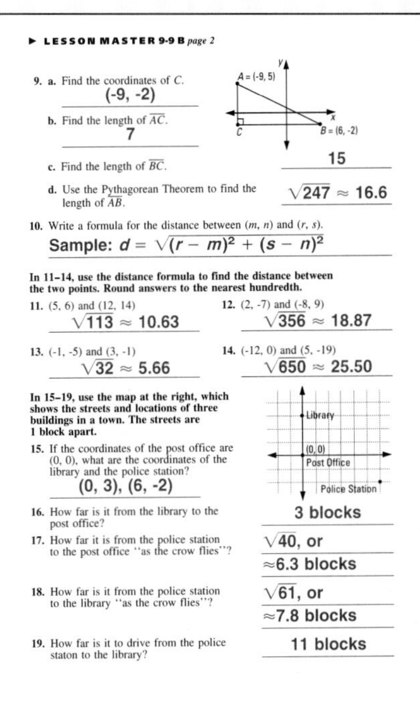

▶ **LESSON MASTER 9-9 B** *page 2*

9. a. Find the coordinates of C. (-9, -2)

b. Find the length of $\overline{AC}$. 7

c. Find the length of $\overline{BC}$. 15

d. Use the Pythagorean Theorem to find the length of $\overline{AB}$. $\sqrt{247} \approx 16.6$

A = (-9, 5)
B = (6, -2)

10. Write a formula for the distance between (m, n) and (r, s). Sample: $d = \sqrt{(r-m)^2 + (s-n)^2}$

In 11–14, use the distance formula to find the distance between the two points. Round answers to the nearest hundredth.

11. (5, 6) and (12, 14) $\sqrt{113} \approx 10.63$

12. (2, -7) and (-8, 9) $\sqrt{356} \approx 18.87$

13. (-1, -5) and (3, -1) $\sqrt{32} \approx 5.66$

14. (-12, 0) and (5, -19) $\sqrt{650} \approx 25.50$

In 15–19, use the map at the right, which shows the streets and locations of three buildings in a town. The streets are 1 block apart.

Library
(0, 0) Post Office
Police Station

15. If the coordinates of the post office are (0, 0), what are the coordinates of the library and the police station? (0, 3), (6, -2)

16. How far is it from the library to the post office? 3 blocks

17. How far is it from the police station to the post office "as the crow flies"? $\sqrt{40}$, or ≈ 6.3 blocks

18. How far is it from the police station to the library "as the crow flies"? $\sqrt{61}$, or ≈ 7.8 blocks

19. How far is it to drive from the police staton to the library? 11 blocks

Notes on Questions

Question 34 Multicultural Connection France is well known for its many delicious varieties of bread. Many French bakeries (*boulangerie*) bake bread by mixing the dough by hand and baking it in wood-fire ovens. Varieties include the crusty, long and thin *baguette* and the *boule,* which is short and fatter. A typical French breakfast might include soft rolls, such as *croissants* and *brioche* (round rolls made from a very light yeast mixture sometimes mixed with currants), served with butter and jam.

Question 35 Cooperative Learning This is an example of taxicab (street-distance) geometry. Students need graph paper or **Teaching Aid 28.** You may want to have students **work in small groups** to graph the points. If you do this, ask them to draw all points 8 units by car from this intersection. Then have them show all points 10 units by car from this intersection. Finally, ask for a generalization.

Follow-up for Lesson 9-9

Practice

For more questions on SPUR Objectives, use **Lesson Master 9-9A** (shown on pages 600–601) or **Lesson Master 9-9B** (shown on pages 602–603).

Assessment

Group Assessment Ask each group to prepare several questions that could be used on a quiz covering Lessons 9-7 through 9-9. You may want to use the questions written by each group as a review. [Students' questions include the key concepts covered in the lessons including the Products of Square Roots Property, absolute value, and distance formulas.]

Review

In 25 and 26, solve. *(Lesson 9-8)*

25. $|x| = 3.5$ **x = 3.5 or x = -3.5**

26. $2|x + 7| = 10$ **x = -12 or x = -2**

27. *Skill sequence.* Solve. *(Lessons 1-6, 9-8)*
a. $t^2 = 9$ **t = ±3**
b. $\sqrt{t} = 9$ **t = 81**
c. $\sqrt{t^2} = 9$ **t = 9 or t = -9**

28. Simplify the expression $\frac{-40 \pm \sqrt{20}}{12}$. *(Lesson 9-7)* $-\frac{10}{3} \pm \frac{\sqrt{5}}{6}$

29. The graph at the left is of the equation $y = ax^2 + bx + c$. How many solutions does the equation $ax^2 + bx + c = 10$ have? Justify your answer. *(Lessons 9-4, 9-6)* **2; the line y = 10 intersects the graph twice.**

30. *Skill sequence.* Solve. *(Lessons 3-5, 5-3, 9-5)*
a. $3x + 6 = 2$ $x = \frac{-4}{3}$
b. $3x + 6 = x + 2$ **x = -2**
c. $3x + 6 = x^2 + 2$ **x = -1 or x = 4**
d. $3x + 6 = x(x + 2)$ **x = -2 or x = 3**

31. Find the x-intercepts of the graph of $y = x^2 + 9x - 5$ using each method.
a. by letting $y = 0$ and using the Quadratic Formula
b. by using an automatic grapher and zooming in on the intercepts
(Lessons 9-2, 9-4) **a)** $\frac{-9 \pm \sqrt{101}}{2} \approx$ **0.52 or -9.52; b) See left.**

31b)

32. Copy quadrilateral $WXYZ$. Then graph the image of $WXYZ$ under a size change of magnitude $-\frac{1}{2}$. *(Lesson 6-7)*

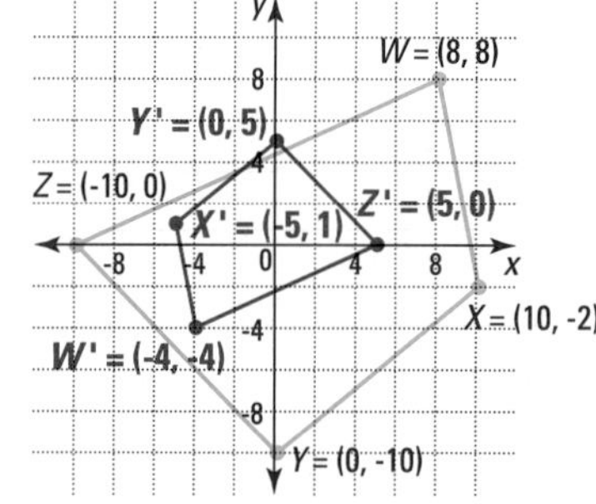

33. In the triangle at the right, find the value of x. *(Lessons 3-5, 4-7)* $x = 12\frac{2}{3}$

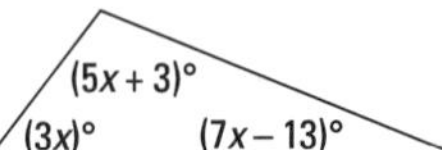

34. A bakery charges 70¢ for each croissant and 25¢ for a carry-out box. You have $5.00. At most, how many croissants can you buy if you also want them in a carry-out box? *(Lessons 3-8, 3-10)* **6**

Making dough. *Croissants originated in France as a breakfast food.* Croissant *is the French word for "crescent," the shape of these rolls.*

Exploration

35. The grid at the right pictures streets 1 block apart in a city. Graph all points 5 blocks by car from the intersection of Main and Central.

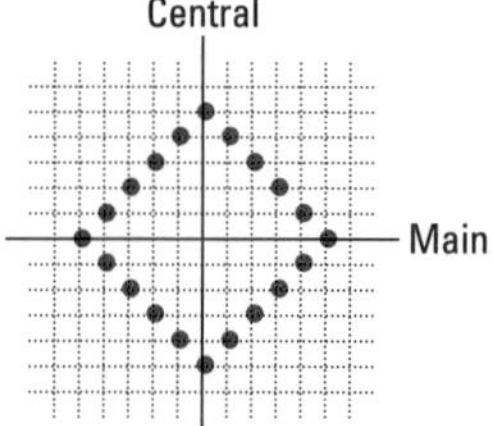

Adapting to Individual Needs

Challenge

Give students the following questions.

1. If you are given the coordinates of three points *P, Q,* and *R,* how can you tell if they are collinear? [Sample: you could find the distance between all the pairs of points to see if the sum of two of the distances equals the other. That is, if *Q* is between *P* and *R* and *PQ* + *QR* = *PR,* then *P, Q,* and *R* are collinear.]
2. Determine if the following points are collinear.
   a. *A*(–2, –1), *B*(1, 3), *C*(5, 6) [No]
   b. *D*(–5, 6), *E*(1, –2), *F*(4, –6) [Yes]

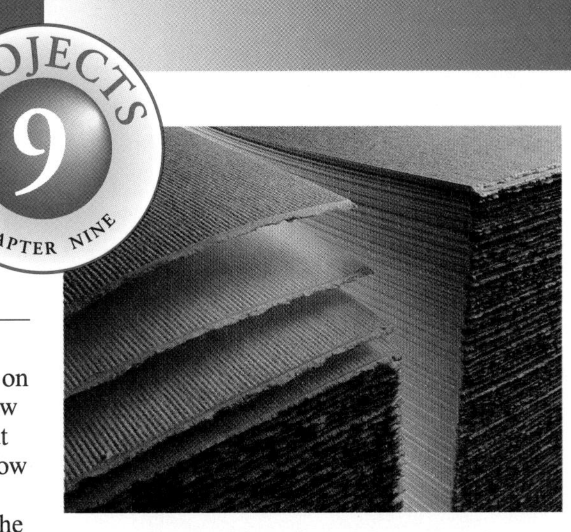

A project presents an opportunity for you to extend your knowledge of a topic related to the material of this chapter. You should allow more time for a project than you do for typical homework questions.

PROJECTS 9 CHAPTER NINE

1 Interview with Galileo

Use the library to do some research on Galileo. Write an imaginary interview with him in which you inquire about his work with falling objects. Also find out how square roots figured in his formula for *the period of a pendulum.* How did he discover the laws of the pendulum?

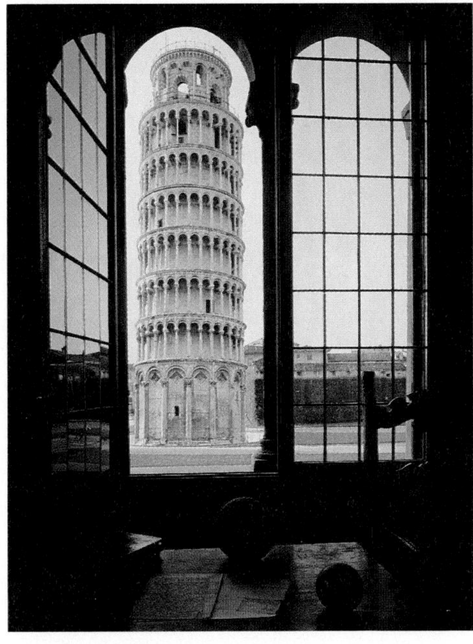

2 Quadratic Design

Make pleasing designs using parabolas (or parabolas and lines or exponential curves). For instance, McDonald's golden arch logo is a design that appears to be a parabola. Give the equations and window that you use for each design.

3 Parabolas from Folding Paper

Parabolas can be made without using equations. Follow these steps to see how to make a parabola by folding paper. You will need several sheets of unlined paper about $8\frac{1}{2}''$ by $11''$ or $9''$ by $12''$. Wax paper works very well with this activity.

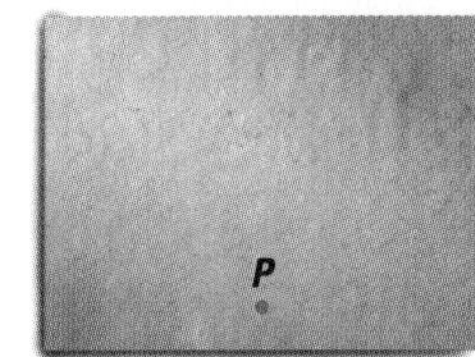

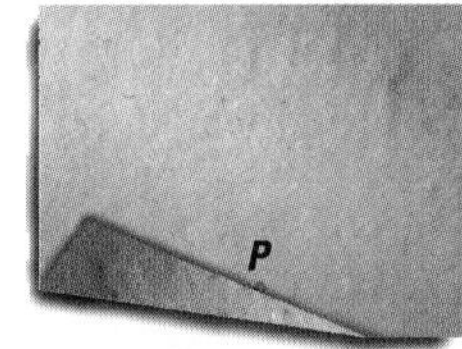

a. Mark a point P about one inch above the center of the lower edge. Fold the paper so that the lower edge touches P, make a careful crease as shown above, and then unfold the paper. Make 15 or 20 more folds each time folding in a different direction. The creases will outline a parabola. Trace the parabola.

b. Use another piece of paper the same size as you used in part **a.** This time mark P in about the center of the paper, and repeat the steps in part **a.**

c. Describe some ways in which the two parabolas are different.

Chapter 9 Projects

The projects relate to the content of the lessons of this chapter as follows:

Project	Lesson(s)
1	9-1
2	9-2, 9-3
3	9-1, 9-2
4	9-4, 9-6
5	9-5, 9-6
6	9-1, 9-2, 9-4
7	9-8

1 Interview with Galileo This is a rather straightforward project, since information about Galileo is easy to find. However, students may have to check several sources before finding all the information that is required for this project. Suggest that students write at least five questions for their interview, and stress that they should rewrite the information they find in their own words.

2 Quadratic Designs Students should be expected to show all the equations that they used in the design. You might want to establish a minimal expectation; for example, you could require that students involve at least five different parabolas in all the designs.

3 Parabolas from Folding Paper This activity can be done with regular typing paper, but the folds are easier to see when wax paper is used. Make sure students mark point *P* correctly and that they fold the paper from both sides of the point. The result of each fold is a tangent to a parabola with *focus P* and *directrix* at the edge of the paper. (See *Challenge* on page 583.) The result of all the folds is the set of all tangents to the parabola, which is called the *envelope of the parabola.*

Possible responses

1. **The following information is a sample of what students might include in their projects. Galileo Galilei (1564–1642) was a mathematician and scientist. He showed that heavier-than-air falling objects fell at the same rate regardless of their size and demonstrated this idea by dropping objects of different weights. Galileo discovered that the period of a pendulum varies with the square root of the length of the pendulum and not the amount of swing. For example, $\sqrt{16} = 2\sqrt{4}$, so the period of a 16-inch pendulum is twice as long as that of a 4-inch pendulum. Galileo discovered a law about the motion of a pendulum by observing the swing of a hanging lamp in a cathedral in Pisa. He observed that it took the same amount of time for the lamp to swing through a wide arc as it took it to swing through a narrow arc.**
2. **Responses will vary.**

(Responses continue on page 606.)

4 Video of a Parabolic Path This project may require more electronic equipment than some students have. If your school has a video camera, you may want to have students work in groups on this project. Stress the importance of covering the TV screen with a clear material before tracing the path of a ball or diver. This project can also be done by having students view football or diving videos. They can press the pause/still button on their VCR, plot the point on the plastic wrap covering the screen, and continue pausing and plotting until enough points have been marked so that they can trace the path taken by the ball or the diver.

5 Computer Programs for Solutions of Quadratics This project provides an opportunity for a student to write a program that requires if-then statements. If students have difficulty getting started, suggest that they write a program as if all values of a, b, and c would result in two real-number solutions. Then have them revise the program to show when $a = 0$ and when the discriminant is less than 0. Remind students that if $a = 0$, the equation is not a quadratic equation.

6 Parabolas in Manufactured Objects Students who are interested in finding out how things work should enjoy working on this project. Physics books often contain diagrams showing the reflective property of the parabola. Students may decide to use these diagrams as part of their posters. It is likely that students will come across the terms *focus* and *directrix* while doing research. (See *Challenge* on page 583.)

7 Graphing Diamonds The diamonds shown are indeed squares. The graph of the equation $y = a|x - h| + k$ is a right angle with vertex (h, k) when $a = 1$. To obtain angles that are not right angles, choose a value of a with $a \neq 1$ and graph $y = a|x - h| + k$. If a is negative, as with the parabola, the angle opens down. To get students started, you might refer them to **Example 4** in Lesson 9-8 on page 596. Ask them to graph $y = -|x|$ and explain how the graph changes. Then have them identify the vertices of the graphs shown in this project, and compare the graphs of $y = |x|$ and $y = -|x|$ to the graphs in the project.

(continued)

4 Video of a Parabolic Path

If you have access to a camcorder, take video shots of a football player throwing a football or of a diver doing a simple dive. Hold the camera steady so that you don't move the camera during the shot. When you play back the tape, cover the TV screen with plastic wrap or other clear material. Play back the tape in slow speed, using a marker to trace on the plastic wrap the path of the ball or diver (you should follow the waist of the diver if she/he does any twists or turns). Describe the shape of the path.

5 Computer Programs for Solutions of Quadratics

Write a program for a computer or calculator to find solutions for any quadratic equation of the form $ax^2 + bx + c = 0$. Your program should allow the user to type in the three values a, b, and c. The program should first test that $a \neq 0$. Then it should evaluate the discriminant. If the discriminant is less than zero, your program should print, "There are no real number solutions." Otherwise it should print the solutions.

6 Parabolas in Manufactured Objects

Parabolas appear in mirrors, flashlights, satellite dishes, and many other manufactured objects. Consult some reference books and write a short report or make a poster explaining where and how parabolas are used.

7 Graphing Diamonds

Graph equations involving absolute values to obtain the figures shown below. Tell what equations you used. Then make up some other designs using graphs of absolute value equations.

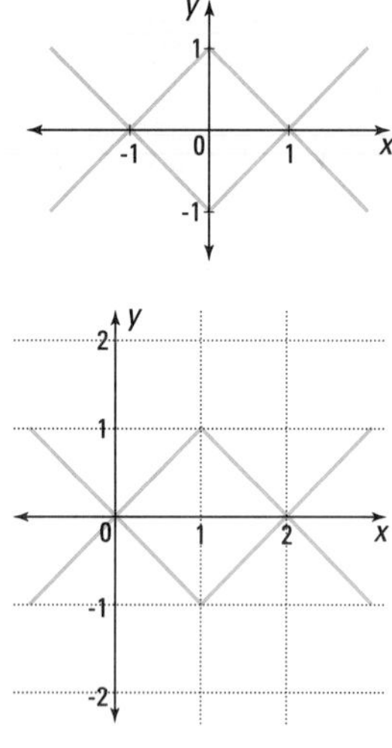

Additional responses, pages 605–606

3a. Sample:

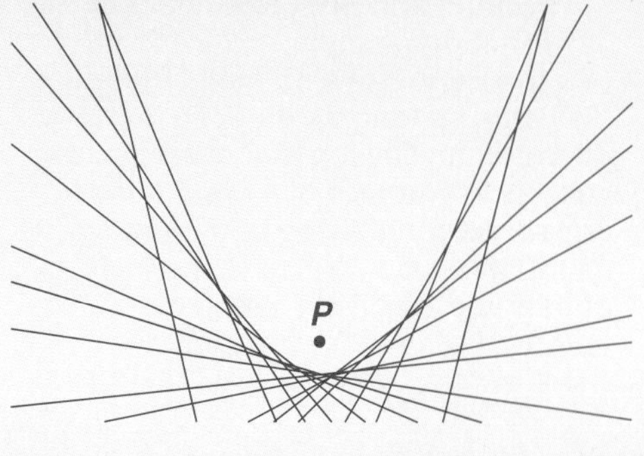

3b. Sample:

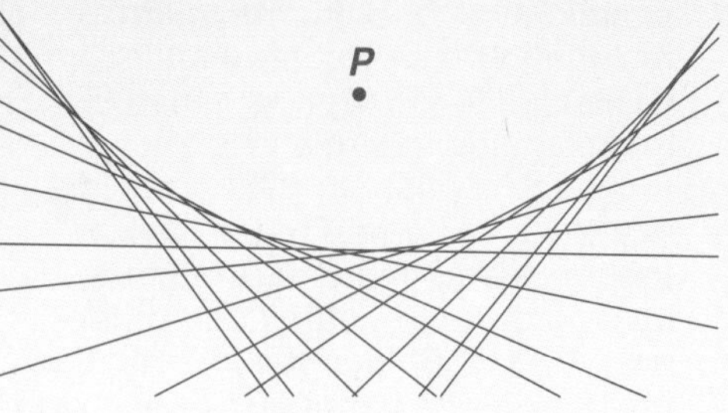

SUMMARY

The graph of $y = ax^2 + bx + c$ is a parabola. This parabola is symmetric to the vertical line through the vertex. If $a > 0$, the parabola opens up. If $a < 0$, the parabola opens down. If $b = 0$, the graph is symmetric to the y-axis.

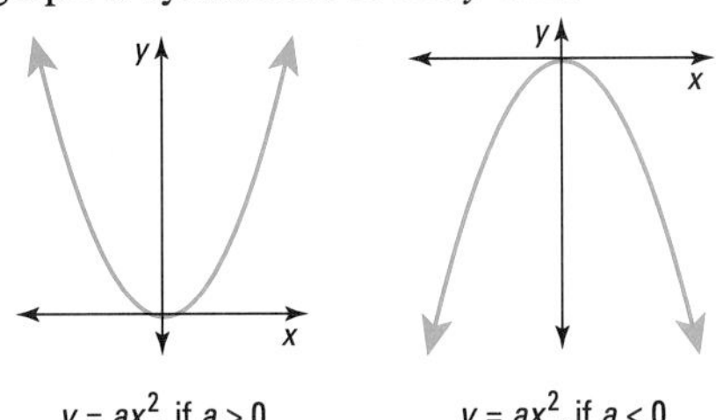

$y = ax^2$, if $a > 0$ $y = ax^2$, if $a < 0$

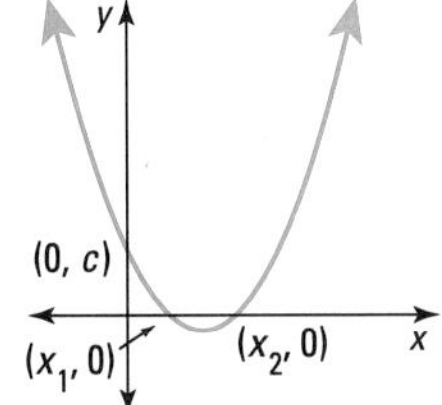

graph of $y = ax^2 + bx + c$ $(a > 0)$ with two x-intercepts, x_1 and x_2

To determine where this parabola crosses the horizontal line $y = k$, solve $ax^2 + bx + c = k$.

A quadratic equation (in one variable) is an equation that can be written in the form $ax^2 + bx + c = 0$, where $a \neq 0$. The solutions to the equation can be found by the Quadratic Formula,

$$x = \frac{-b \pm \sqrt{b^2 - 4ac}}{2a}.$$

The discriminant of the quadratic equation $ax^2 + bx + c = 0$ is $b^2 - 4ac$. If the discriminant is positive, there are two real solutions; if it is zero, there is one solution; if it is negative, there are no real solutions.

Solutions to quadratic equations involve square roots. Rewriting a square root so it is either a whole number or a product of a whole number and a smaller radical is called simplifying the radical. You can multiply or simplify square roots by using the Product of Square Roots Property.

If $a \geq 0$ and $b \geq 0$, then $\sqrt{a} \cdot \sqrt{b} = \sqrt{ab}$.

You can use square roots and absolute values to find distances on the number line or in the coordinate plane. There are formulas for both.

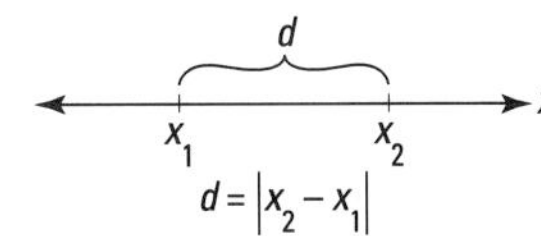

$d = |x_2 - x_1|$

Distance Formula for a Number Line

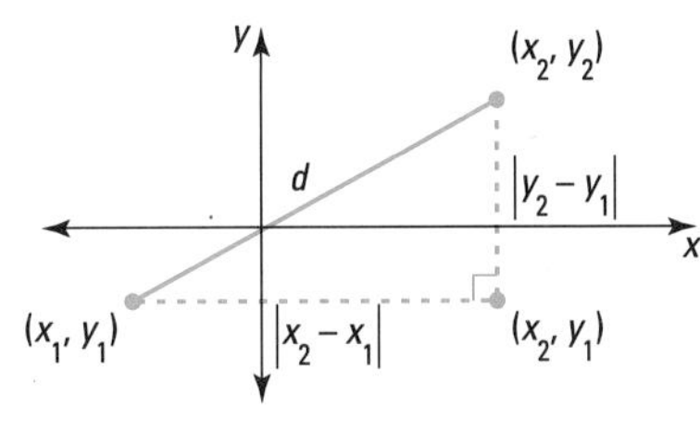

$$d = \sqrt{|x_2 - x_1|^2 + |y_2 - y_1|^2}$$

Pythagorean Distance Formula for the Coordinate Plane

Square roots and absolute value are related by the property, for all x, $\sqrt{x^2} = |x|$.

Summary

The Summary gives an overview of the entire chapter and provides an opportunity for students to consider the material as a whole. Thus, the Summary can be used to help students relate and unify the concepts presented in the chapter.

3c. **Sample response: The parabola in part a looks narrower than the parabola in part b. The distance between the vertex and point *P* in part a is less than the distance between the vertex and point *P* in part b. The crease lines in part a are closer together than they are in part b.**

4. **The shape of each path will be approximately parabolic.**

5. **Sample program:**

```
10  PRINT "SOLVING A*B^2 + B*X + C = 0"
20  PRINT "ENTER A, B, C"
30  INPUT A, B, C
40  IF A = 0 THEN 120
50  DISCRMT = B^2 – 4*A*C
60  IF DISCRMT < 0 THEN 140
70  X1 = (-B + SQR(DISCRMT))/(2*A)
80  X2 = (-B – SQR(DISCRMT))/(2*A)
90  IF X1 = X2 THEN 160
100 PRINT "X =";X1; "OR";X2;
110 GOTO 170
120 PRINT "NOT A QUADRATIC EQUATION"
130 GOTO 170
140 PRINT "THERE ARE NO REAL NUMBER SOLUTIONS."
150 GOTO 170
160 PRINT "X =";X1;
170 END
```

(Responses continue on page 608.)

Vocabulary

Terms, symbols, and properties are listed by lesson to provide a checklist of concepts that students must know. Emphasize that students should read the vocabulary list carefully before starting the Progress Self-Test. If students do not understand the meaning of a term, they should refer back to the indicated lesson.

Additional Answers, Progress Self-Test, page 609

7a, b

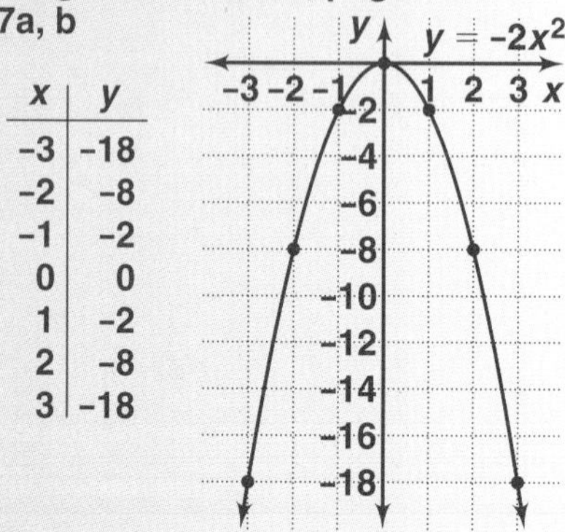

x	y
-3	-18
-2	-8
-1	-2
0	0
1	-2
2	-8
3	-18

8a, b

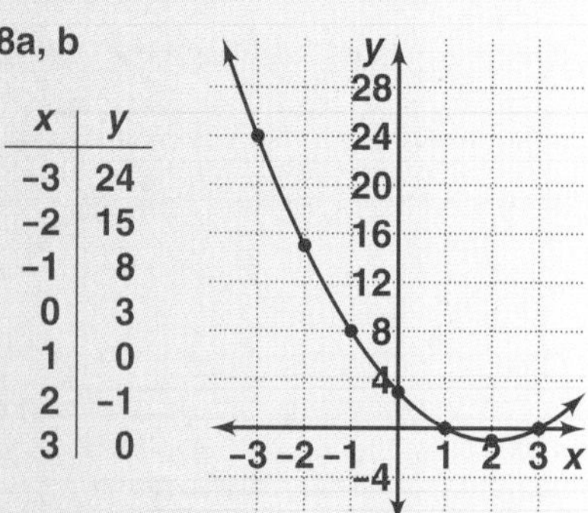

x	y
-3	24
-2	15
-1	8
0	3
1	0
2	-1
3	0

16.

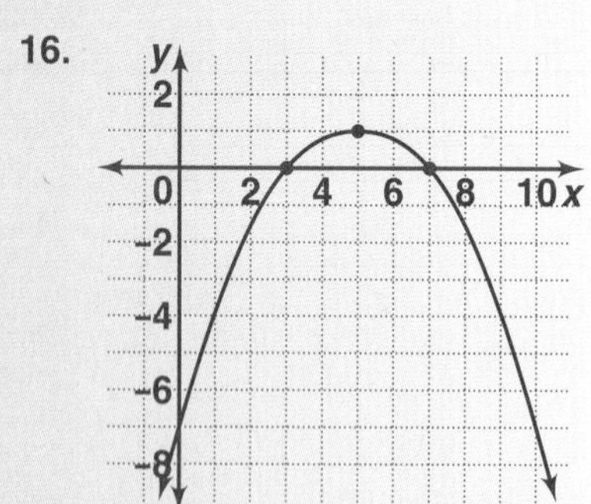

VOCABULARY

Below are the most important terms and phrases for this chapter. You should be able to give a general description and a specific example of each.

Lesson 9-1
parabola
axis of symmetry
symmetric
vertex, maximum, minimum
opens up
opens down

Lesson 9-2
STEP command
stopping distance

Lesson 9-4
projectile

Lesson 9-5
quadratic equation
Quadratic Formula
plus or minus, ±
standard form of a quadratic equation

Lesson 9-6
discriminant
Discriminant Property

Lesson 9-7
Product of Square Roots Property
simplifying the radical

Lesson 9-8
absolute value, $|n|$, ABS(N)
Distance Formula on a Number Line
Absolute Value–Square Root Property
tolerance

Lesson 9-9
"as the crow flies"
Pythagorean Distance Formula in the Coordinate Plane

Additional responses, page 606

5. **Sample program for TI-81:**
Bold print indicates a function key.

```
Disp "SOLVING QUADRATIC EQUATIONS"
Disp "ENTER THE COEFFICIENTS FOR AX² + BX + C."
Disp "ENTER A VALUE FOR A."
Input A
If A = 0
Disp "TRY ANOTHER A VALUE."
If A = 0
Input A
Disp "ENTER A VALUE FOR B."
Input B
Disp "ENTER A VALUE FOR C."
Input C
B² -4AC → D
If D < 0
Disp "NO REAL SOLUTIONS."
If D < 0
Stop
(-B – √D)/2A → X
(-B + √D)/2A → Y
If D = 0
Disp "THE SOLUTION IS"
If D = 0
Disp X
If D = 0
Stop
Disp "THE SOLUTIONS ARE"
Disp X
Disp Y
Stop
```

CHAPTER NINE

PROGRESS SELF-TEST

Take this test as you would take a test in class. Then check your work with the solutions in the Selected Answers section in the back of the book.

In 1–4, find all real solutions. Round answers to the nearest hundredth. If there is no real solution, state so. 1) 5; 4 2) ≈ 1.81; ≈ -1.21 3) no real solutions

1. $x^2 - 9x + 20 = 0$
2. $5y^2 - 3y = 11$
3. $8x^2 - 7x = -11$
4. $z^2 = 16z - 64$ 8
5. If the discriminant of a quadratic equation is 18, how many solutions does the equation have? 2
6. *Multiple choice.* Which of these is the graph of $y = 2.5x^2$? a

(a)

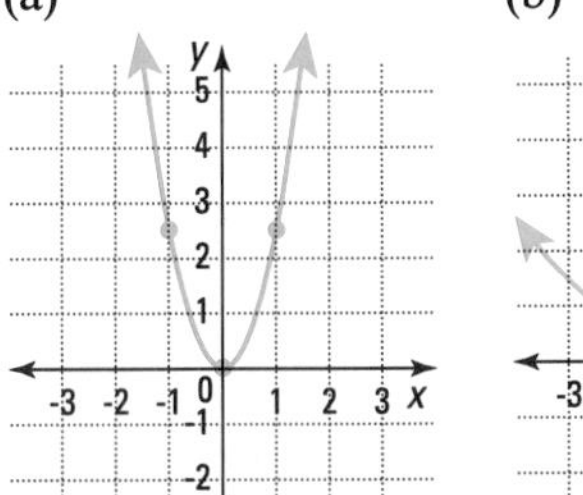

(b)

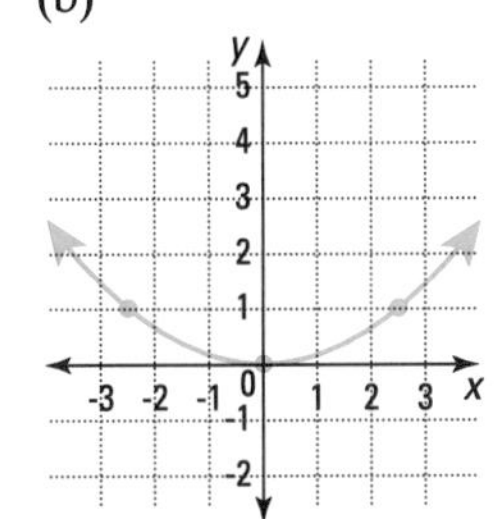

In 7 and 8, an equation is given. See margin.

a. Make a table of values of x and y for integer values of x from $x = -3$ to $x = 3$.

b. Graph the equation.

7. $y = -2x^2$
8. $y = x^2 - 4x + 3$
9. *True or false.* The parabola $y = 3x^2 - 7x - 35$ opens down. False

In 10–12, use the graph of the parabola at the right.

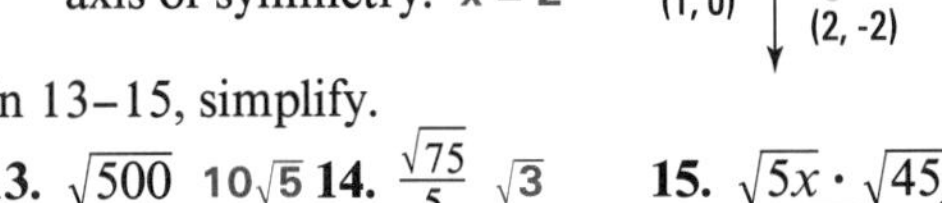

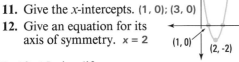

10. Give the coordinates of the vertex. (2, -2)
11. Give the x-intercepts. (1, 0); (3, 0)
12. Give an equation for its axis of symmetry. $x = 2$

In 13–15, simplify.

13. $\sqrt{500}$ $10\sqrt{5}$
14. $\frac{\sqrt{75}}{5}$ $\sqrt{3}$
15. $\sqrt{5x} \cdot \sqrt{45y}$ $15\sqrt{xy}$

16. Given the following two facts about a parabola, sketch a possible graph. See margin for sample.
(1) The vertex is (5, 1).
(2) When its equation is in the form $y = ax^2 + bx + c$, a is negative.

In 17–19, use the diagram at right.

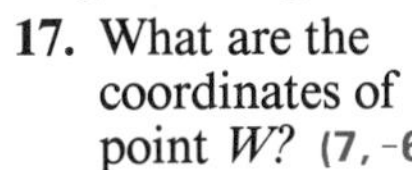

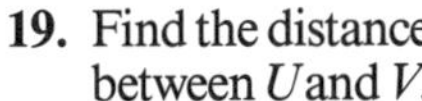

17. What are the coordinates of point W? (7, -6)
18. *True or false.* $VW = |4 - -6|$.
19. Find the distance between U and V.
20. Suppose $L = (3, -2)$ and $M = (x, y)$. Find LM, the distance between L and M.

18) True 19) $5\sqrt{5}$ 20) $\sqrt{|3 - x|^2 + |-2 - y|^2}$

In 21 and 22, a tennis ball is thrown from the top of a building. The graph shows $h = -16t^2 + 21t + 40$, giving the height h of the ball after t seconds.

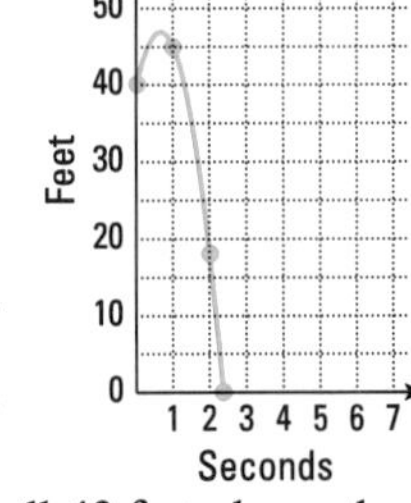

21. To the nearest tenth of a second, how long did it take the ball to reach the ground? ≈ 2.4 seconds
22. At what times was the ball 43 feet above the ground? Give your answer to the nearest tenth. 0.2 or 1.1 seconds

In 23 and 24, Harry tosses a ball to Fred who is 20 yards away. Melody is 2 yards in front of Fred. At x yards away from Harry, the ball is at a height h feet, where $h = -0.07x^2 + 1.4x + 5$.

23. Sketch the graph of this equation. Mark the positions of Harry, Fred, and Melody on your sketch. See margin.
24. How high is the ball when it passes above Melody? Explain how you get your answer. 7.52 ft; $-0.07(18)^2 + 1.4(18) + 5 = 7.52$

In 25–27, find all solutions.

25. $\frac{1}{3}x^2 = 10$ ≈ ±5.48
26. $|x| = 57$ ± 57
27. $|3 - n| = 0.5$ 2.5 or 3.5

Progress Self-Test

For the development of mathematical competence, feedback and correction, along with the opportunity to practice, are necessary. The Progress Self-Test provides the opportunity for feedback and correction; the Chapter Review provides additional opportunities and practice. We cannot overemphasize the importance of these end-of-chapter materials. It is at this point that the material "gels" for many students, allowing them to solidify skills and understanding. In general, student performance should be markedly improved after these pages.

Assign the Progress Self-Test as a one-night assignment. Worked-out *solutions* for all questions are in the Selected Answers section of the student book. Encourage students to take the Progress Self-Test honestly, grade themselves, and then be prepared to discuss the test in class.

Advise students to pay special attention to those Chapter Review questions (pages 610–613) which correspond to the questions that they missed on the Progress Self-Test.

Additional Answers, continued

23.

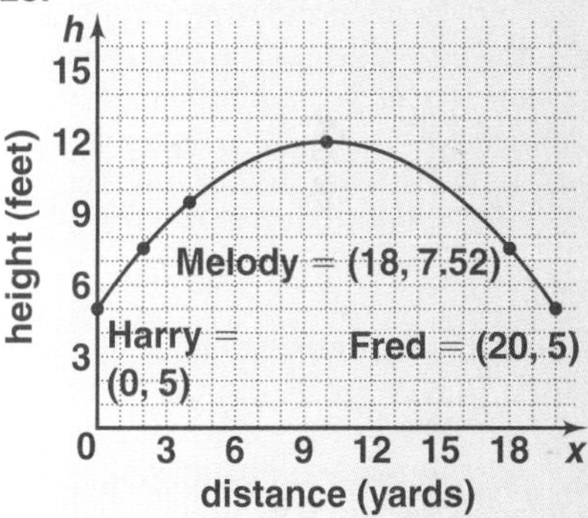

Students using the TI-82 will need to break some of *Disp* lines in order to view then on screen.

6. Sample response: Parabolas are used in reflecting telescopes, satellite dishes, vehicle headlights, searchlights, and radar antennae. The shape of the reflector in headlights and searchlights is formed by rotating a parabola about its axis of symmetry. When a light is placed at the focus of the parabola, the light rays that are reflected off the parabola are parallel to the axis of symmetry, resulting in light that shines straight out. (See Figure 1 on page 610.) Reflecting telescopes, satellite dishes, and radar antennae work in the opposite direction. Parallel light rays or signals come toward the parabola, bounce off, and accumulate at the focus of the parabola where they create a strong signal or light ray. (See Figure 2 on page 610.)

(Responses continue on page 610.)

Chapter 9 Review

Resources

From the ***Teacher's Resource File***

- Answer Master for Chapter 9 Review
- *Assessment Sourcebook:* Chapter 9 Test, Forms A–D Chapter 9 Test, Cumulative Form Comprehensive Test, Chapters 1–9

Additional Resources

- Quiz and Test Writer

The main objectives for the chapter are organized in the Chapter Review under the four types of understanding this book promotes—Skills, Properties, Uses, and Representations.

Whereas end-of-chapter material may be considered optional in some texts, in *UCSMP Algebra* we have selected these objectives and questions with the expectation that they will be covered. Students should be able to answer these questions with about 85% accuracy.

You may assign these questions over a single night to help students prepare for a test the next day, or you may assign the questions over a two-day period. If you do the latter, then we recommend assigning the *evens* for homework the first night so that students get feedback in class the next day, then assigning the *odds* the night before the test because answers are provided to the odd-numbered questions.

Additional Answers

44. If D > 0, the equation has exactly 2 real solutions. If D = 0, the equation has exactly one solution. If D < 0, the equation has no real solutions.

CHAPTER REVIEW

Questions on SPUR Objectives

SPUR stands for **S**kills, **P**roperties, **U**ses, and **R**epresentations. The Chapter Review questions are grouped according to the SPUR Objectives for this chapter.

SKILLS DEAL WITH THE PROCEDURES USED TO GET ANSWERS.

Objective A: *Solve quadratic equations.* *(Lessons 9-1, 9-5)*

In 1–4, solve without using the Quadratic Formula.

1. $16t^2 = 400$ **-5; 5** **2.** $18 = \frac{1}{2}x^2$ **-6; 6**

3. $g^2 + 21 = 70$ **-7; 7** **4.** $p^2 - 6 = 6$ **$2\sqrt{3}$; $-2\sqrt{3}$**

In 5–8, give the exact solutions to the equation.

5. $6y^2 + 7y - 20 = 0$ **6.** $x^2 + 7x + 12 = 0$ **-3; -4**

7. $14v - 49 = v^2$ **7** **8.** $14h - 3 = 2h^2$

In 9–14, use the Quadratic Formula to solve. Round answers to the nearest hundredth.

9. $k^2 - 7k = 2$ **10.** $2m^2 + m - 3 = 0$

11. $22a^2 + 2a + 3 = 0$ **12.** $0 = x^2 + 10x + 25$ **-5**

13. $30 + 10(m^2 - 5m) = 0$ **≈ 4.30 or ≈ 0.70**

14. $16p^2 + 8p = -5$ **no real solutions**

5) $-\frac{5}{2}$; $\frac{4}{3}$; **8)** $\frac{7 \pm \sqrt{43}}{2}$

Objective B: *Simplify square roots.* *(Lesson 9-7)*

In 15–23, simplify.

15. $\sqrt{7} \cdot \sqrt{28}$ **14** **16.** $\sqrt{4} \cdot \sqrt{9} - \sqrt{3} \cdot \sqrt{48}$ **-6**

17. $\sqrt{20^2 + 20^2}$ **$20\sqrt{2}$** **18.** $\sqrt{99}$ **$3\sqrt{11}$**

9) -0.27; 7.27 **10)** 1; -1.5; **11)** no real solutions

19. $\sqrt{500}$ **$10\sqrt{5}$** **20.** $\frac{\sqrt{150}}{5}$ **$\sqrt{6}$**

21. $3\sqrt{72}$ **$18\sqrt{2}$** **22.** $\frac{6 + \sqrt{128}}{2}$ **$3 + 4\sqrt{2}$**

23. $\frac{12 \pm 6\sqrt{24}}{4}$ **$3 \pm 3\sqrt{6}$**

In 24–28, simplify. Assume x and y are positive.

24. $\sqrt{24x} \cdot \sqrt{6x}$ **12x** **25.** $\sqrt{5x^2}$ **$x\sqrt{5}$**

26. $\sqrt{2x^2y^2}$ **$xy\sqrt{2}$** **27.** $\sqrt{(-17)^2}$ **17**

28. $-\sqrt{y^2}$ **-y**

Objective C: *Evaluate expressions and solve equations using absolute value.* *(Lesson 9-8)*

In 29–34, evaluate the expression.

29. $|-43|$ **43** **30.** $-|-1|$ **-1**

31. $|13 - 19|$ **6** **32.** ABS(1.6 − 1.8) **0.2**

33. $|3| - |8|$ **-5** **34.** $|-20| - |15| + |-2|$ **7**

In 35–40, find all solutions.

35. $|d| = 16$ **-16; 16** **36.** $|k| = -3$ **no solution**

37. $\sqrt{x^2} = 7$ **-7; 7** **38.** $12 = \sqrt{n^2}$ **-12; 12**

39. $|r - 10| = 5$ **5; 15** **40.** $|300 - s| = 23$ **277; 323**

PROPERTIES DEAL WITH THE PRINCIPLES BEHIND THE MATHEMATICS.

Objective D: *Identify and use properties of quadratic equations.* *(Lesson 9-6)*

41. Give the values of x that satisfy the equation $ax^2 + bx + c = 0$. **$\frac{-b \pm \sqrt{b^2 - 4ac}}{2a}$**

In 42 and 43, *true or false.*

42. Some quadratic equations have no real solutions. **True**

43. Any quadratic equation can be solved by using the Quadratic Formula. **True**

44. How can you use the discriminant to determine the number of real solutions a quadratic equation may have? **See margin.**

In 45–48, use the discriminant to determine the number of real solutions to the equation.

45. $2x^2 - 3x + 4 = 0$ **D = -23; no real solutions**

46. $a^2 = 3a + 8$ **D = 41; two real solutions**

47. $9d = 40 + 8d^2$ **D = -1199; no real solutions**

48. $n(n + 1) = -5$ **D = -19; no real solutions**

Additional responses, page 606

6.

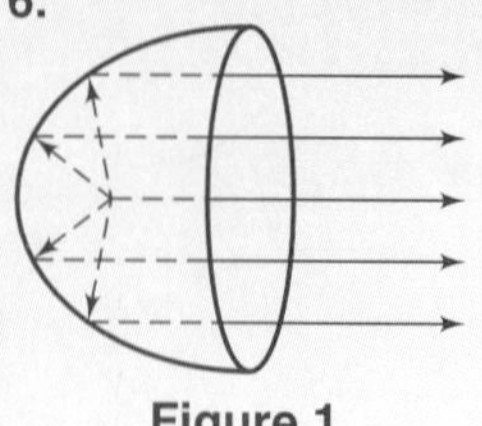

Figure 1

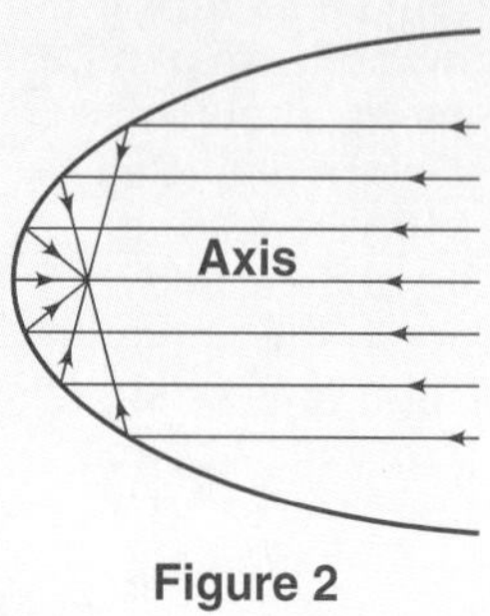

Figure 2

**7. Directions for the first figure:
Press the Y = button and enter the following equations:
Y = ABS X − 1
Y =(−) ABS X + 1
Set Range −2 < x < 2, −2 < y < 2, x and y scale = 0.5
Press Graph
Directions for the second figure:
Press the Y = button and enter the following equations:**

USES DEAL WITH APPLICATIONS OF MATHEMATICS IN REAL SITUATIONS.

Objective E: *Use quadratic equations to solve problems about paths of projectiles.* *(Lessons 9-1, 9-4, 9-5)*

In 49 and 50, when an object is dropped near the surface of a planet or moon, the distance d (in feet) it falls in t seconds is given by the formula $d = \frac{1}{2}gt^2$, where g is the acceleration due to gravity. Near Earth, $g \approx 32$ ft/sec^2; near Earth's moon, $g \approx 5.3$ ft/sec^2. **49a) 576 ft b) ≈ 11.2 seconds**

49. A sky diver jumps from a plane at an altitude of 5000 ft. He begins his descent in "free fall," that is, without opening the parachute.
- **a.** How far will the diver fall in 6 seconds?
- **b.** The diver plans to open the parachute after he has fallen 2000 feet. How many seconds after jumping will this take place?

50. An astronaut on the moon drops a hammer.
- **a.** How far will it fall in 4 seconds? **42.4 ft**
- **b.** How long will it take the hammer to fall 100 feet? **≈ 6.14 seconds**

51. One of the first astronauts who traveled to the moon hit a golf ball on the moon. Suppose that the height h in meters of a ball t seconds after it is hit is described by $h = -0.8t^2 + 10t$.
- **a.** Sketch the graph of this equation. **See margin.**
- **b.** Find the times when the ball is at a height of 20 meters. **2.5 and 10 seconds**
- **c.** Check your answer to part **b** by using the Quadratic Formula. **See margin.**

52. Consider again the quarterback in Lesson 9-4 who tosses a football to a receiver 40 yards down field. The ball is at height h feet, x yards down field, where $h = -0.025x^2 + x + 6$. Suppose a defender is 6 yards in front of the receiver. **a) 34 yards; b) See below**
- **a.** How far is he from the quarterback?
- **b.** Would the defender have a good chance to deflect the ball? Justify your answer.

53. Refer to the graph below of $h = -16t^2 + 64t$ which shows the height h in feet of a ball, t seconds after it is thrown from ground level at an initial upward velocity of 64 feet per second.

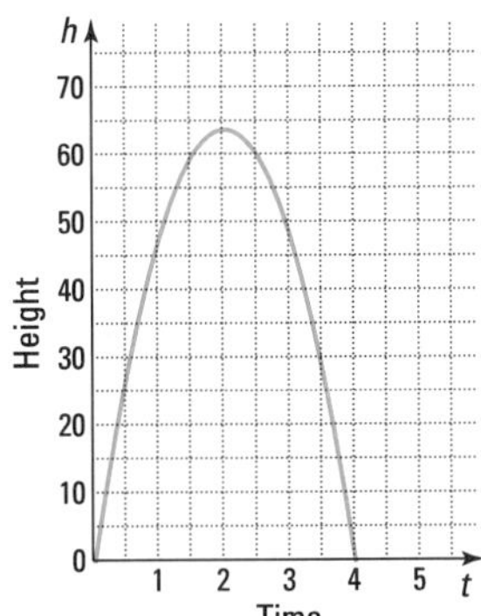

a) ≈ 48 ft **d) See margin.**
- **a.** Give the height of the ball after 1 second.
- **b.** Find when the ball will reach a height of 35 feet. **≈ 0.65 or 3.35 seconds**
- **c.** How many seconds is the ball in the air? **4**
- **d.** Make up a question whose answer is one of the coordinates of the vertex.

54. When a ball is thrown into the air with initial upward velocity of 20 meters per second, its approximate height y above the ground (in meters) after x seconds is given by $y = 20x - 5x^2$.
- **a.** When will the ball hit the ground?
- **b.** How high will the ball be at its highest point? **20 meters a) after 4 seconds**

52b) No, because 34 yards from the quarterback, the ball is at a height of 11.1 ft.


```
Y = ABS (x – 1) – 1
Y = (-) ABS (x – 1) + 1
Set Range -1 < x < 3, -2 < y < 2,
x and y scale = 0.5
Press Graph
```

Assessment

Evaluation The *Assessment Sourcebook* provides five forms of the Chapter 9 Test. Forms A and B present parallel versions in a short-answer format. Forms C and D offer performance assessment. The fifth test is Chapter 9 Test, Cumulative Form. About 50% of this test covers Chapter 9; 25% covers Chapter 8, and 25% covers earlier chapters. Comprehensive Test Chapters 1–9 gives roughly equal attention to all chapters covered thus far.

For information on grading see *General Teaching Suggestions; Grading* in the *Professional Sourcebook* which begins on page T20 in Part 1 of the Teacher's Edition.

Feedback After students have taken the test for Chapter 9 and you have scored the results, return the tests to students for discussion. Class discussion on the questions that caused trouble for most students can be very effective in identifying and clarifying misunderstandings. You might want to have them write down the items they missed and work either in groups or at home to correct them. It is important for students to receive feedback on every chapter test, and we recommend that students see and correct their mistakes before proceeding too far into the next chapter.

Additional Answers

51a.

h; 30, 24, 18, 12, 6, 0; height (feet); 3, 6, 9, 12, t; time (seconds)

51c. $\frac{-10 \pm \sqrt{10^2 - 4(-0.8)(-20)}}{2(-0.8)}$

$= \frac{-10 \pm \sqrt{100 - 64}}{-1.6}$

$= \frac{-10 \pm 6}{-1.6}$

$= 2.5$ or 10

53d. Sample: What is the maximum height the ball will reach?

CHAPTER NINE

REPRESENTATIONS DEAL WITH PICTURES, GRAPHS, OR OBJECTS THAT ILLUSTRATE CONCEPTS.

Objective F: *Graph equations of the form $y = ax^2 + bx + c$ and interpret these graphs. (Lessons 9-1, 9-2, 9-3)*

55. Use the parabola below.

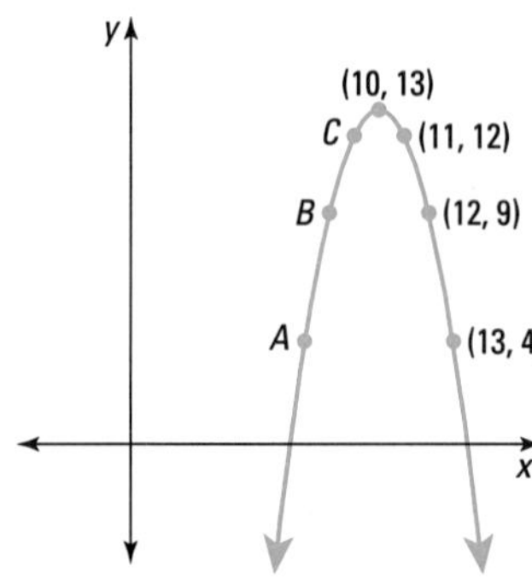

a. What are the coordinates of its vertex? (10, 13)

b. What is an equation for its axis of symmetry? $x = 10$ c) See below right.

c. Find the coordinates of points *A*, *B*, and *C*, the reflection images of the named points over the parabola's axis of symmetry.

56. A table of values for a parabola is given below.

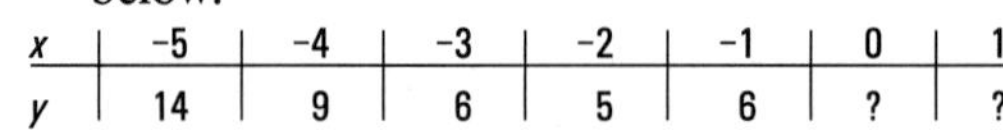

x	-5	-4	-3	-2	-1	0	1
y	14	9	6	5	6	?	?

a. Write an equation for its axis of symmetry.

b. What are the coordinates of its vertex?

c. Complete the table. a) $x = -2$ b) (-2, 5) c) 9; 14 58) True

In 57 and 58, *true or false.*

57. Every parabola has a minimum value. False

58. The parabola $y = -2x^2 + 3x + 1$ opens down.

59. What equation must you solve to find the *x*-intercepts of the parabola $ax^2 + bx + c = 0$

$$y = ax^2 + bx + c?$$

60. The parabola $y = 2x^2 - 16x + 24$ has *x*-intercepts 6 and 2. Find the coordinates of its vertex without graphing. (4, -8)

In 61–64, **a.** make a table of values. **b.** Graph the equation. See margin.

61. $y = 3x^2$ **62.** $y = -\frac{1}{2}x^2$

63. $y = x^2 + 5x + 6$ **64.** $y = x^2 - 4$

65. *Multiple choice.* Which equation has the given graph? b

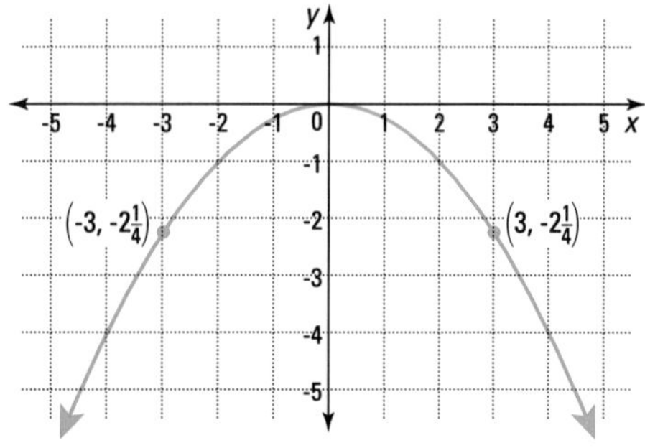

(a) $y = -4x^2$ (b) $y = -\frac{1}{4}x^2$

(c) $y = 4x^2$ (d) $y = \frac{1}{4}x^2$

66. *Multiple choice.* Which of these is the graph of $y = x^2 + 4x + 3$? a

(a) (b)

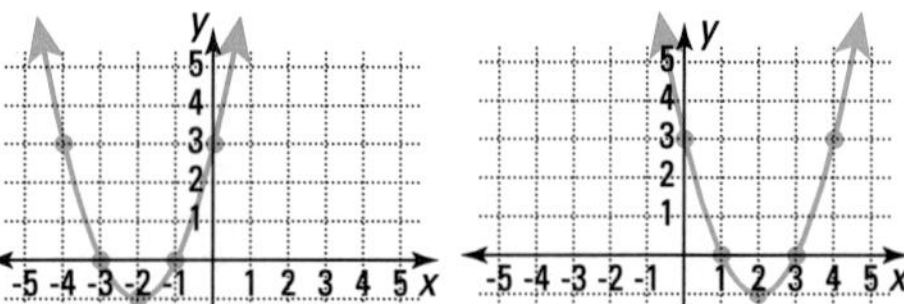

In 67 and 68, use an automatic grapher. **a.** Graph the equation. **b.** Estimate the coordinates of its vertex. **c.** Tell whether the vertex is a maximum or minimum point. See margin for graphs.

67. $y = -0.025x^2 + x + 6$, the equation of Question 52. b) (20, 16) c) maximum

68. $y = 20x - 5x^2$, the equation of Question 54. b) (2, 20) c) maximum

55c) $A = (7, 4)$; $B = (8, 9)$; $C = (9, 12)$

Additional Answers

61a, b

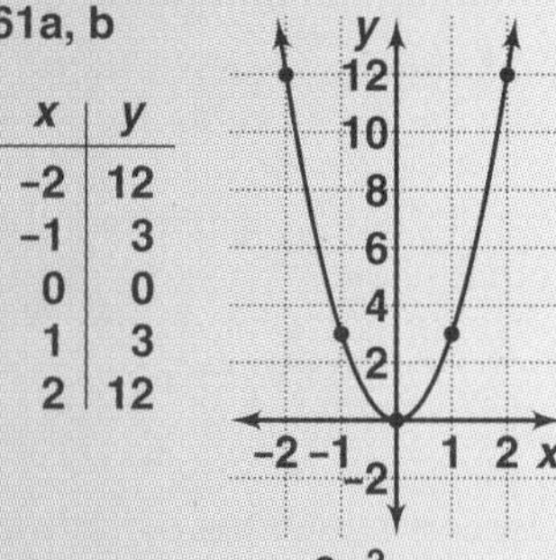

x	y
-2	12
-1	3
0	0
1	3
2	12

62a, b

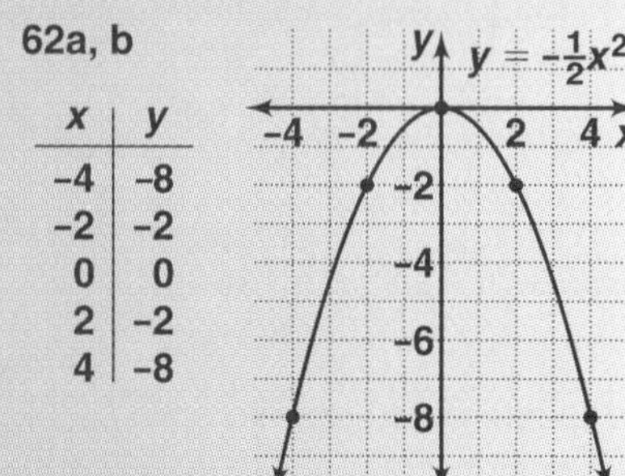

x	y
-4	-8
-2	-2
0	0
2	-2
4	-8

63a, b.

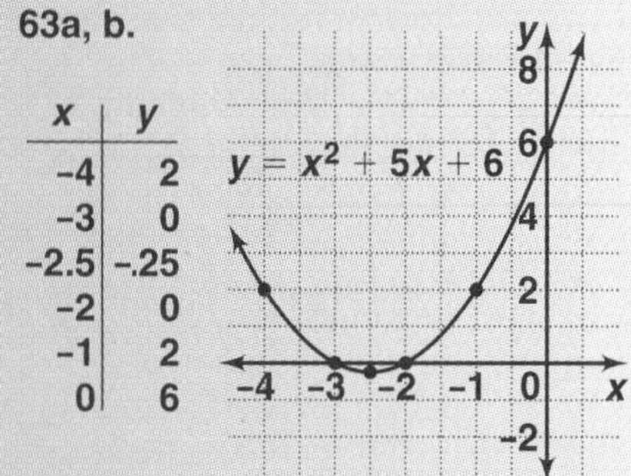

x	y
-4	2
-3	0
-2.5	-.25
-2	0
-1	2
0	6

64a, b

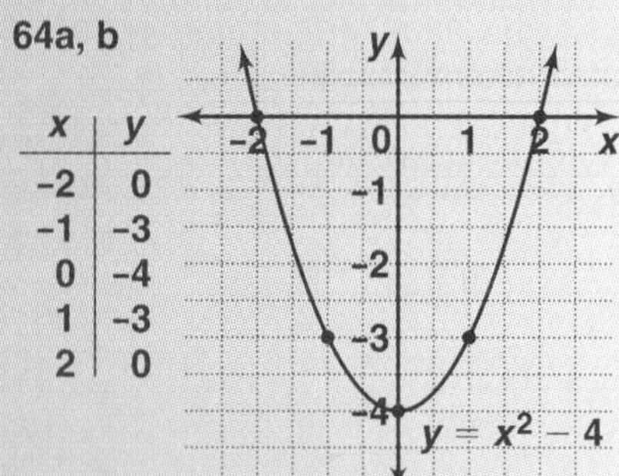

x	y
-2	0
-1	-3
0	-4
1	-3
2	0

67a.

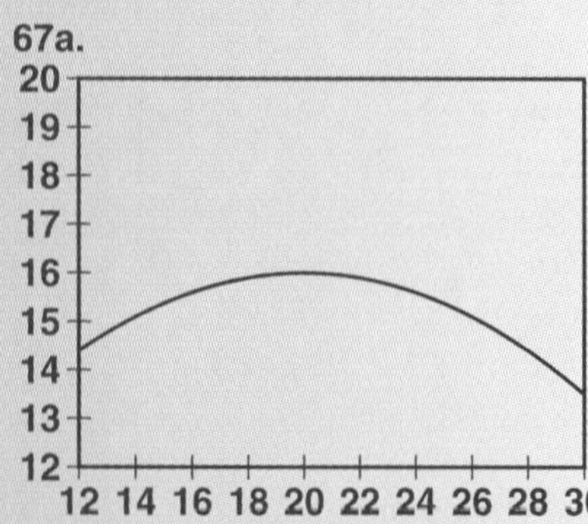

68a.

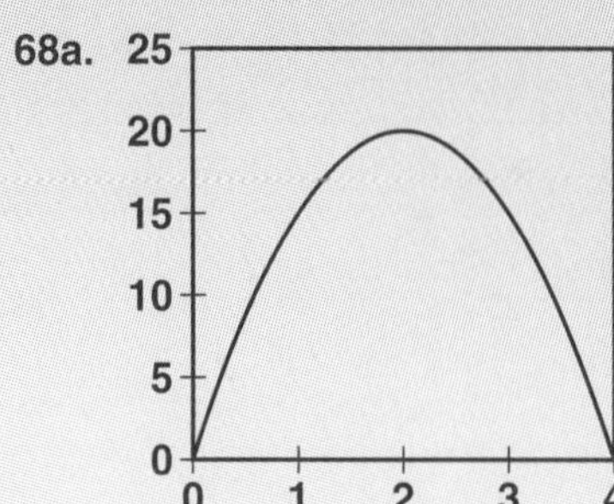

In 69 and 70, a graph of an equation is shown. Describe another window that will show the vertex and all intercepts.

69. Sample: $-5 \le x \le 15$; $-10 \le y \le 30$

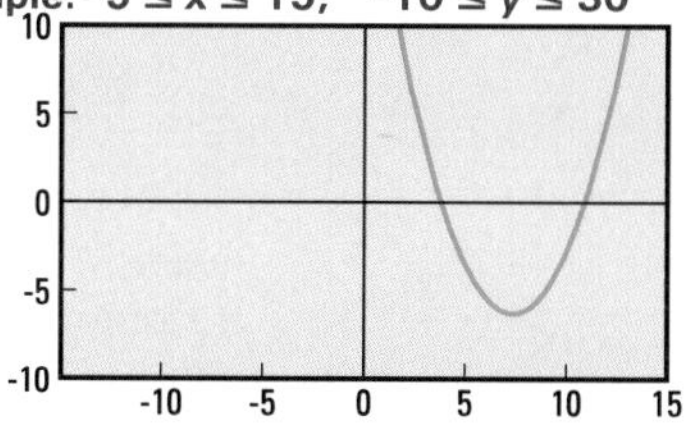

70. Sample: $-50 \le x \le 15$; $-50 \le y \le 1$

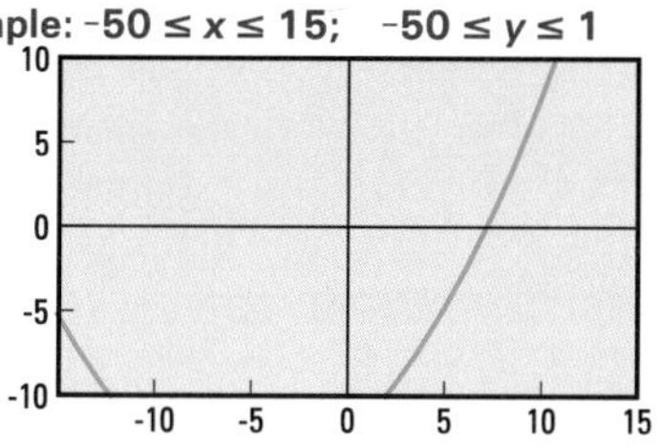

Objective G: *Calculate and represent distances on the number line or in the plane.* *(Lessons 9-8, 9-9)*

In 71 and 72, the coordinates of two points on the number line are given. Find the distance between them.

71. 

31

72. p and q $|p - q|$ or $|q - p|$

73. Find the coordinates of all points on a number line at a distance of 5 from the point -3. -8; 2

In 74–76, use the graph below.

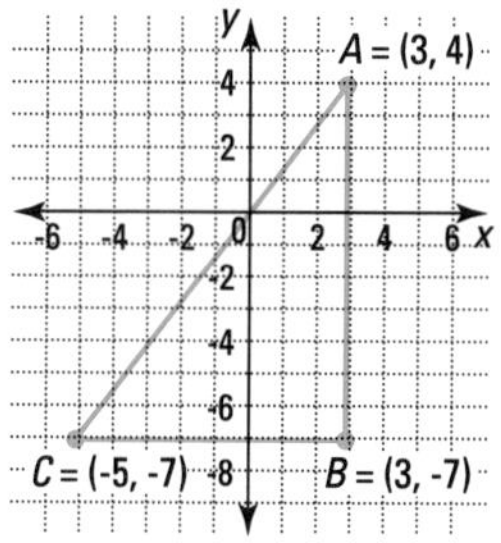

74. a. Find AB. 11 **b.** Find AC. $\sqrt{185} \approx 13.6$

75. Find the distance from A to the origin O. 5

76. Which is greater: BC or BO? Justify your answer. *BC*; $BC = 8$ while $BO = \sqrt{58} \approx 7.61$

In 77–82, find AB.

77. $A = (4, -10)$, $B = (-5, -10)$ 9

78. $A = (3, 20)$, $B = (3, 2)$ 18

79. $A = (50, 10)$, $B = (0, 0)$ $10\sqrt{26} \approx 50.99$

80. $A = (14, -20)$, $B = (-2, -8)$ 20

81. $A = (-4, -3)$, $B = (4, 12)$ 17

82. $A = (-2, 9)$, $B = (-5, -1)$ $\sqrt{109} \approx 10.44$

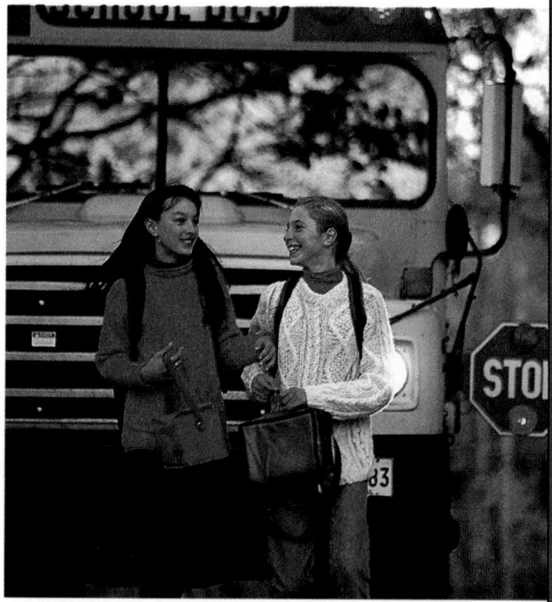

83. The school these girls attend is 3 miles east and 5 miles south of their homes. The library is 4 miles west and 2 miles north of their houses. The streets in their neighborhood are laid out in a square grid.

a. Draw a picture to represent this situation. See margin.

b. How far is it from the school to the library, as the crow flies? $7\sqrt{2} \approx 9.9$ miles

84. A right triangle has vertices at (0, 4), (2, 0), and (6, 2). Find the length of its hypotenuse. $2\sqrt{10} \approx 6.3$

Additional Answers

83a.

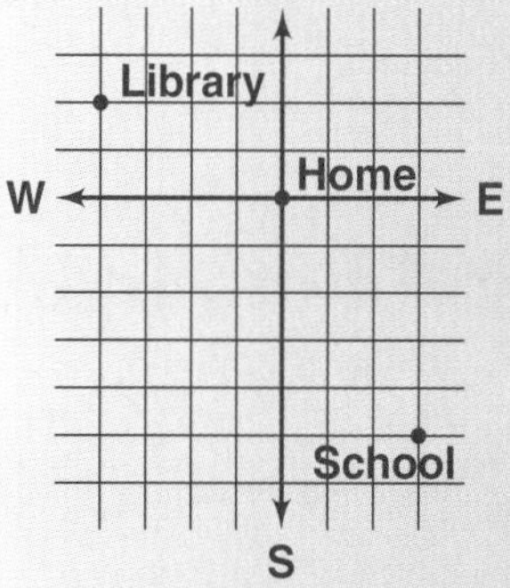

Setting Up Lesson 10-1

We recommend that you assign Lesson 10-1, both reading and some questions, for homework the evening of the test. It gives students work to do after they have completed the test and keeps the class moving.

Chapter 10 Planner

Adapting to Individual Needs

The student text is written for the vast majority of students. The chart at the right suggests two pacing plans to accommodate the needs of your students. Students in the Full Course should complete the entire text by the end of the year. Students in the Minimal Course will spend more time when there are quizzes and more time on the Chapter Review. Therefore, these students may not complete all of the chapters in the text.

Options are also presented to meet the needs of a variety of teaching and learning styles. For each lesson, the Teacher's Edition provides sections entitled: *Video* which describes video segments and related questions that can be used for motivation or extension; *Optional Activities* which suggests activities that employ materials, physical models, technology, and cooperative learning; and, *Adapting to Individual Needs* which regularly includes **Challenge** problems, **English Language Development** suggestions, and suggestions for providing **Extra Help.** The Teacher's Edition also frequently includes an **Error Alert,** an **Extension,** and an **Assessment** alternative. The options available in Chapter 10 are summarized in the chart below.

Chapter 10 Pacing Chart

Day	Full Course	Minimal Course
1	10-1	10-1
2	10-2	10-2
3	10-3	10-3
4	10-4	10-4
5	Quiz*; 10-5	Quiz*; begin 10-5.
6	10-6	Finish 10-5.
7	10-7	10-6
8	Self-Test	10-7
9	Review	Self-Test
10	Test*	Review
11		Review
12		Test*

*in the Teacher's Resource File

In the Teacher's Edition...

Lesson	Optional Activities	Extra Help	Challenge	English Language Development	Error Alert	Extension	Cooperative Learning	Ongoing Assessment
10-1	●	●	●	●		●	●	Group
10-2	●	●	●	●		●	●	Written
10-3	●	●	●	●	●	●		Oral/Written
10-4	●	●	●	●	●	●	●	Written
10-5	●	●	●		●	●	●	Oral
10-6	●	●	●		●	●		Written/Oral
10-7	●	●	●	●		●	●	Oral

In the Additional Resources...

	In the Teacher's Resource File								
Lesson	Lesson Masters, A and B	Teaching Aids*	Activity Kit*	Answer Masters	Technology Sourcebook	Assessment Sourcebook	Visual Aids**	Technology	Video Segments
10-1	10-1	26, 31, 107, 110, 111		10-1			26, 31, 107, 110, 111, AM		
10-2	10-2	107, 112, 113		10-2	Comp 25		107, 112, 113, AM	Spreadsheet	10-2
10-3	10-3	26, 31, 107, 114	22	10-3			26, 31, 107, 114, AM		
10-4	10-4	108, 115, 116		10-4		Quiz	108, 115, 116, AM		
In-class Activity		26, 31		10-5			26, 31, AM		
10-5	10-5	108, 117		10-5			108, 117, AM		
10-6	10-6	108, 118		10-6			108, 118, AM		
10-7	10-7	109, 119, 120	23	10-7	Comp 26		109, 119, 120, AM	Spreadsheet	
End of chapter				Review		Tests			

*Teaching Aids are pictured on pages 614C and 614D. The activities in the Activity Kit are pictured on page 614C.

**Visual Aids provide transparencies for all Teaching Aids and all Answer Masters.

Also available is the Study Skills Handbook which includes study-skill tips related to reading, note-taking, and comprehension.

Integrating Strands and Applications

	10-1	10-2	10-3	10-4	10-5	10-6	10-7
Mathematical Connections							
Algebra	●	●	●	●	●	●	●
Geometry	●	●	●	●	●	●	●
Measurement		●	●	●	●	●	●
Probability					●		●
Statistics/Data Analysis							●
Patterns and Functions		●		●	●	●	●
Interdisciplinary and Other Connections							
Art				●	●		
Science		●			●		●
Social Studies	●	●	●		●	●	●
Multicultural		●	●			●	
Technology		●					●
Career		●					
Consumer	●	●	●	●			
Sports			●				●

Teaching and Assessing the Chapter Objectives

Chapter 10 Objectives (Organized into the SPUR categories—Skills, Properties, Uses, and Representations)	**Lessons**	**Progress Self-Test Questions**	**Chapter Review Questions**	**In the Teacher's Resource File** **Chapter Test, Forms A and B**	**Chapter Test, Forms C**	**Chapter Test, Forms D**
Skills						
A: Add and subtract polynomials.	10-2	10, 11	1–6	8, 9	1	✓
B: Multiply polynomials.	10-4	9, 12	7–12	5	5	✓
C: Multiply a polynomial by a monomial or multiply two binomials.	10-3, 10-5, 10-6	4–7, 15	13–22	4, 6, 7	1	✓
D: Expand squares of binomials.	10-6	8	23–28	10, 12	3	
Properties						
E: Classify polynomials by their degree or number of terms.	10-1	1, 2, 3	29–34	2	1	
F: Write whole numbers as polynomials in base 10.	10-1	16	35–38	16		
Uses						
G: Translate investment situations into polynomials.	10-2	17, 18	39, 40	17, 18	2	
H: Use the chi-square statistic to determine whether or not an event is likely.	10-7	19–21	41–43	19	4	
Representations						
I: Represent areas and volumes of figures with polynomials.	10-1, 10-3, 10-4, 10-5, 10-6	13, 14, 22	44–53	13–15	5	✓

Multidimensional Assessment
Quiz for Lessons 10-1 through 10-4
Chapter 10 Test, Forms A–D
Chapter 10 Test, Cumulative Form

Quiz and Test Writer
Multiple forms of chapter tests and quizzes; Challenges

Activity Kit

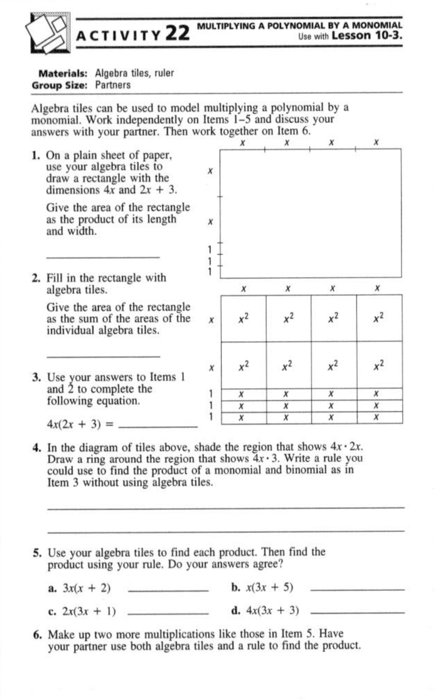
ACTIVITY 22 MULTIPLYING A POLYNOMIAL BY A MONOMIAL Use with **Lesson 10-3.**

Materials: Algebra tiles, ruler
Group Size: Partners

Algebra tiles can be used to model multiplying a polynomial by a monomial. Work independently on Items 1–5 and discuss your answers with your partner. Then work together on Item 6.

1. On a plain sheet of paper, use your algebra tiles to draw a rectangle with the dimensions $4x$ and $2x + 3$.
Give the area of the rectangle as the product of its length and width.

2. Fill in the rectangle with algebra tiles.
Give the area of the rectangle as the sum of the areas of the individual algebra tiles.

3. Use your answers to Items 1 and 2 to complete the following equation.
$4x(2x + 3) =$ ______

4. In the diagram of tiles above, shade the region that shows $4x \cdot 2x$. Draw a ring around the region that shows $4x \cdot 3$. Write a rule you could use to find the product of a monomial and binomial as in Item 3 without using algebra tiles.

5. Use your algebra tiles to find each product. Then find the product using your rule. Do your answers agree?
a. $3x(x + 2)$ b. $x(3x + 5)$
c. $2x(3x + 1)$ d. $4x(3x + 3)$

6. Make up two more multiplications like those in Item 5. Have your partner use both algebra tiles and a rule to find the product.

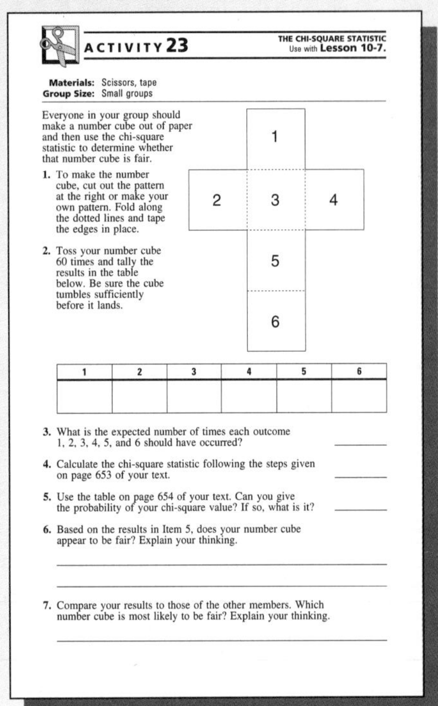
ACTIVITY 23 THE CHI-SQUARE STATISTIC Use with **Lesson 10-7.**

Materials: Scissors, tape
Group Size: Small groups

Everyone in your group should make a number cube out of paper and then use the chi-square statistic to determine whether that number cube is fair.

1. To make the number cube, cut out the pattern at the right or make your own pattern. Fold along the dotted lines and tape the edges in place.

2. Toss your number cube 60 times and tally the results in the table below. Be sure the cube tumbles sufficiently before it lands.

1	2	3	4	5	6

3. What is the expected number of times each outcome 1, 2, 3, 4, 5, and 6 should have occurred?

4. Calculate the chi-square statistic following the steps given on page 653 of your text.

5. Use the table on page 654 of your text. Can you give the probability of your chi-square value? If so, what is it?

6. Based on the results in Item 5, does your number cube appear to be fair? Explain your thinking.

7. Compare your results to those of the other members. Which number cube is most likely to be fair? Explain your thinking.

Teaching Aids

Teaching Aid 26, Graph Paper, (shown on page 140D) and **Teaching Aid 31, Algebra Tiles,** (shown on page 176) can be used with **Lessons 10-1, 10-3,** and the **In-class Activity.**

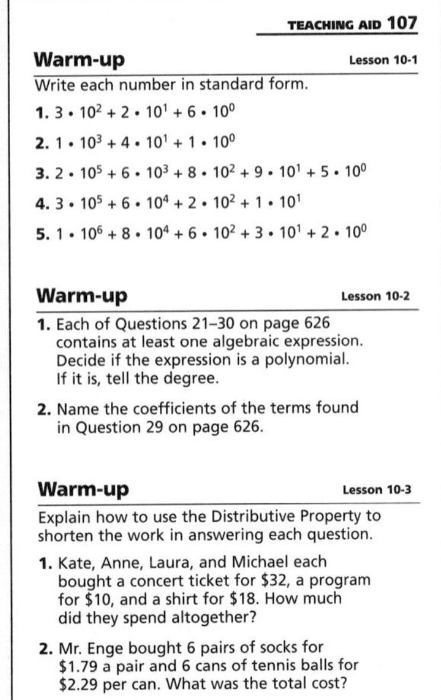
TEACHING AID 107

Warm-up Lesson 10-1
Write each number in standard form.
1. $3 \cdot 10^2 + 2 \cdot 10^1 + 6 \cdot 10^0$
2. $1 \cdot 10^3 + 4 \cdot 10^1 + 1 \cdot 10^0$
3. $2 \cdot 10^5 + 6 \cdot 10^3 + 8 \cdot 10^2 + 9 \cdot 10^1 + 5 \cdot 10^0$
4. $3 \cdot 10^5 + 6 \cdot 10^4 + 2 \cdot 10^2 + 1 \cdot 10^1$
5. $1 \cdot 10^6 + 8 \cdot 10^4 + 6 \cdot 10^2 + 3 \cdot 10^1 + 2 \cdot 10^0$

Warm-up Lesson 10-2
1. Each of Questions 21–30 on page 626 contains at least one algebraic expression. Decide if the expression is a polynomial. If it is, tell the degree.
2. Name the coefficients of the terms found in Question 29 on page 626.

Warm-up Lesson 10-3
Explain how to use the Distributive Property to shorten the work in answering each question.
1. Kate, Anne, Laura, and Michael each bought a concert ticket for \$32, a program for \$10, and a shirt for \$18. How much did they spend altogether?
2. Mr. Enge bought 6 pairs of socks for \$1.79 a pair and 6 cans of tennis balls for \$2.29 per can. What was the total cost?

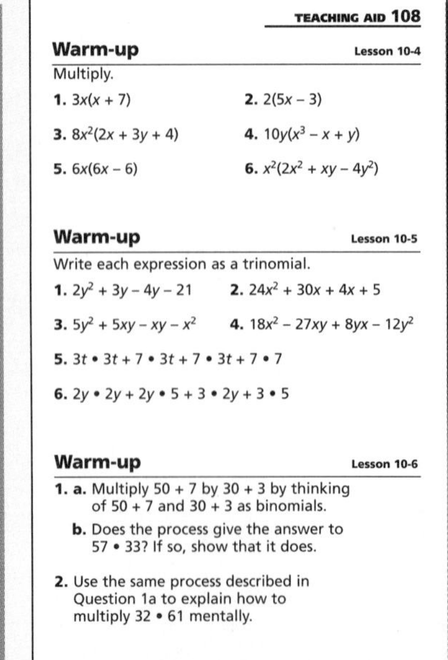
TEACHING AID 108

Warm-up Lesson 10-4
Multiply.
1. $3x(x + 7)$
2. $2(5x - 3)$
3. $8x^2(2x + 3y + 4)$
4. $10y(x^3 - x + y)$
5. $6x(6x - 6)$
6. $x^2(2x^2 + xy - 4y^2)$

Warm-up Lesson 10-5
Write each expression as a trinomial.
1. $2y^2 + 3y - 4y - 21$
2. $24x^2 + 30x + 4x + 5$
3. $5y^2 + 5xy - xy - x^2$
4. $18x^2 - 27xy + 8yx - 12y^2$
5. $3t \cdot 3t + 7 \cdot 3t + 7 \cdot 3t + 7 \cdot 7$
6. $2y \cdot 2y + 2y \cdot 5 + 3 \cdot 2y + 3 \cdot 5$

Warm-up Lesson 10-6
1. a. Multiply 50 + 7 by 30 + 3 by thinking of 50 + 7 and 30 + 3 as binomials.
b. Does the process give the answer to 57 • 33? If so, show that it does.
2. Use the same process described in Question 1a to explain how to multiply 32 • 61 mentally.

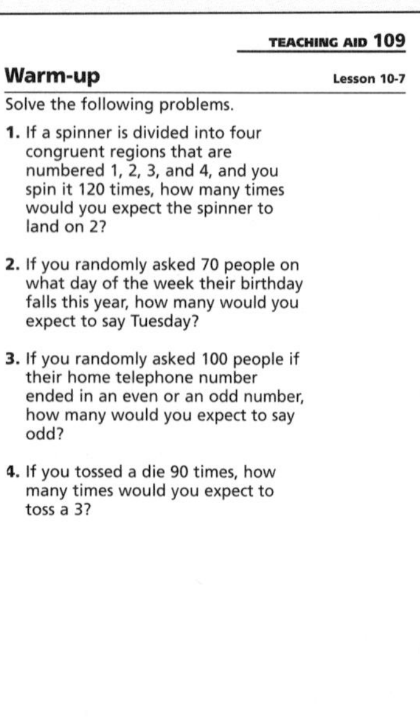
TEACHING AID 109

Warm-up Lesson 10-7
Solve the following problems.
1. If a spinner is divided into four congruent regions that are numbered 1, 2, 3, and 4, and you spin it 120 times, how many times would you expect the spinner to land on 2?
2. If you randomly asked 70 people on what day of the week their birthday falls this year, how many would you expect to say Tuesday?
3. If you randomly asked 100 people if their home telephone number ended in an even or an odd number, how many would you expect to say odd?
4. If you tossed a die 90 times, how many times would you expect to toss a 3?

TEACHING AID 110
Lesson 10-1

Classifying Polynomials

Degree	Name of Polynomial
1	linear
2	quadratic
3	cubic
4	quartic
5	quintic
6	no special name

Number of Terms	Name of Polynomial
1	monomial
2	binomial
3	trinomial
4	no special name
5	no special name
6	no special name

TEACHING AID 111
Lesson 10-1

Challenge

A	
1	9
3	11
5	13
7	15

B	
2	10
3	11
6	14
7	15

C	
4	12
5	13
6	14
7	15

D	
8	12
9	13
10	14
11	15

TEACHING AID 112
Lesson 10-2

	A	B	C	D	E
1	Year	Nellie's Deposit	Nellie's Balance	Joe's Deposit	Joe's Balance
2					
3	1	1000	1060.00	0	0
4	2	1000	2183.60	0	0
5	3	1000	3374.62	0	0
6	4	1000	4637.09	0	0
7	5	1000	5975.32	0	0
8	6	0	6333.84	1200	1272.00
9	7	0	6713.87	1200	2620.32
10	8	0	7116.70	1200	4049.54
11	9	0	7543.70	1200	5564.51
12	10	0	7996.32	1200	7170.38

TEACHING AID 113
Lesson 10-2

Polynomials from Mortgage Payments

Starting amount of mortgage: $50,000 Annual interest rate: 10%

Monthly payment for principal and interest: $ 439 Monthly interest rate: $\frac{10\%}{12} = .0083$

Month	Amount owed ($)	
0	50,000	= 50,000.00
1	$50{,}000(1.0083) - 439$	= 49,976.00
2	$[50{,}000(1.0083) - 439](1.0083) - 439$ or $50{,}000(1.0083)^2 - 439(1.0083) - 439$	= 49,951.80
3	$[50{,}000(1.0083)^2 - 439(1.0083) - 439](1.0083) - 439$ or $50{,}000(1.0083)^3 - 439(1.0083)^2 - 439(1.0083) - 439$	= 49,927.40
4	$50{,}000(1.0083)^4 - 439(1.0083)^3 - 439(1.0083)^2 - 439(1.0083) - 439$	= 49,902.80
5		
6		

TEACHING AID 114
Lesson 10-3

Example 1

CARDS Florist SHOES Boutique

h

L_1 L_2 L_3 L_4

Additional Examples

1. Give two equivalent expressions for the area pictured below.

	x	x	x	x
x	x^2	x^2	x^2	x^2
x	x^2	x^2	x^2	x^2
1	x	x	x	x
1	x	x	x	x
1	x	x	x	x
1	x	x	x	x

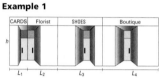

2. Multiply $-5y(y^3 - 6y^2 + 2y + 6)$.

3. Multiply $k^4(k^2 - 16km)$.

TEACHING AID 115
Lesson 10-4

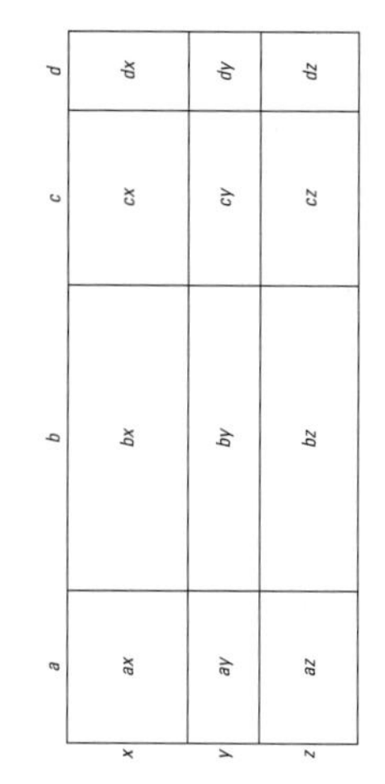

TEACHING AID 116
Lesson 10-4

Questions 1 and 9

1.

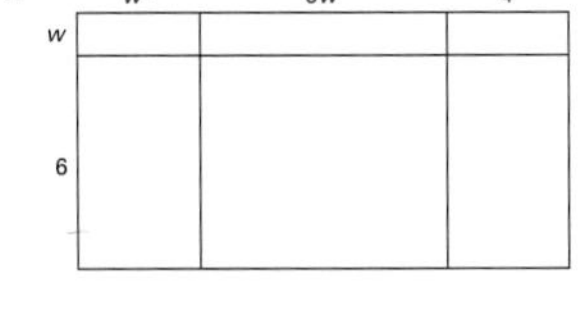

9.

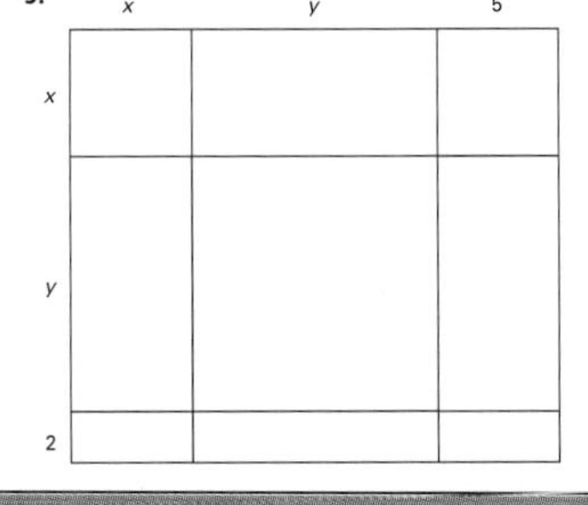

TEACHING AID 117
Lesson 10-5

Questions 1, 2, 13, and 14

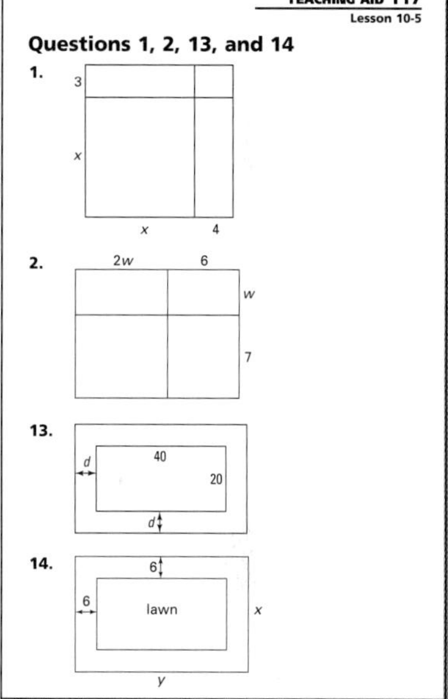

TEACHING AID 118
Lesson 10-6

Special Binomial Products

The Square of a Sum

The Square of a Difference

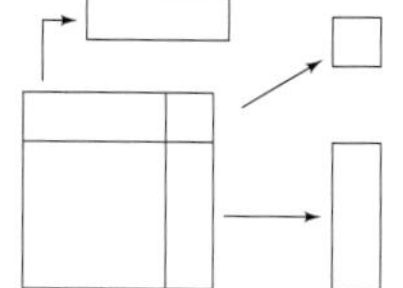

The Difference of Two Squares

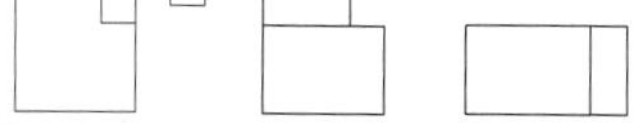

TEACHING AID 119
Lesson 10-7

Critical Chi-Square Values

$n - 1$	.10	.05	.01	.001
1	2.71	3.84	6.63	10.8
2	4.61	5.99	9.21	13.8
3	6.25	7.81	11.34	16.3
4	7.78	9.49	13.28	18.5
5	9.24	11.07	15.09	20.5
6	10.6	12.6	16.8	22.5
7	12.0	14.1	18.5	24.3
8	13.4	15.5	20.1	26.1
9	14.7	16.9	21.7	27.9
10	16.0	18.3	23.2	29.6
11	17.3	19.7	24.7	31.3
12	18.6	21.0	26.2	32.9
13	19.8	22.4	27.7	34.5
14	21.1	23.7	29.1	36.1
15	22.3	25.0	30.6	37.7
16	23.5	26.3	32.0	39.3
17	24.8	27.6	33.4	40.8
18	26.0	28.9	34.8	42.3
19	27.2	30.1	36.2	43.8
20	28.4	31.4	37.6	45.3
25	34.4	37.7	44.3	52.6
30	40.3	43.8	50.9	59.7
50	63.2	67.5	76.2	86.7

TEACHING AID 120
Lesson 10-7

Additional Examples

One die was tossed 60 times to see if it was fair. Here are the results of the tosses: 26225 23635 12224 12154 55351 42545 66465 46522 13224 23645 35552 44423.

1. What are the observed values for the events of getting a 1, 2, 3, 4, 5, and 6 in this situation?

2. What are the expected values of each event?

3. Use the chi-square statistic to give some evidence whether or not the die is fair.

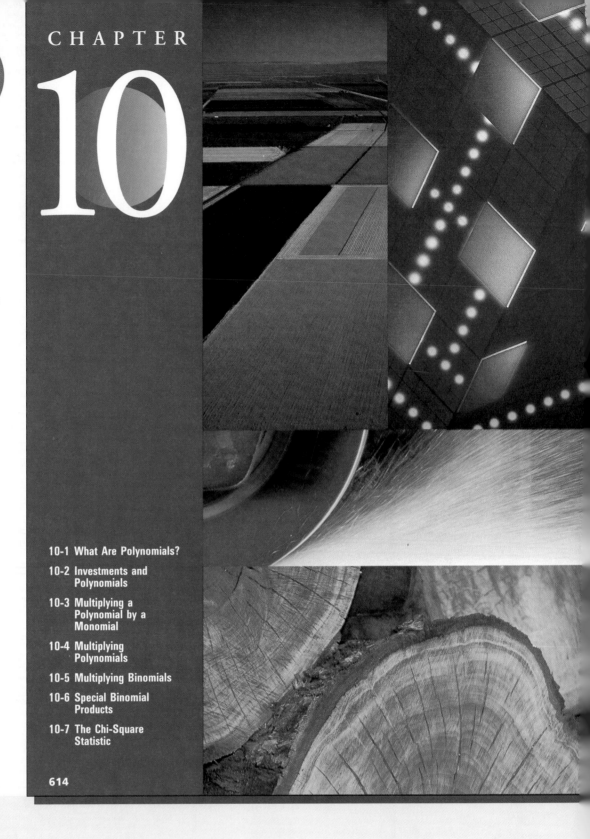

Chapter Opener

Pacing

All lessons in this chapter are designed to be covered in 1 day. At the end of the chapter, you should plan to spend 1 day to review the Progress Self-Test, 1 to 2 days for the Chapter Review, and 1 day for a test. You may want to spend a day on projects, and an extra day may be needed for quizzes. Therefore, this chapter should take 10 to 12 days. We strongly advise that you not spend more than 13 days on this chapter; it is not a difficult chapter, and most of its ideas are reviewed in Chapter 12.

Using Pages 614–615

The reading on these pages provides an advance organizer for the chapter. It introduces three kinds of applications that students will see in the early lessons—the use of polynomials in base 10 arithmetic; the use of polynomials to represent length, area, and volume; the use of polynomials with problems involving annuities.

Each of the applications on page 615 involves mathematics that students have seen before. You might want to mention that the material in this chapter is not entirely different from what they have seen and that they should not expect it to be difficult.

CHAPTER 10

Chapter 10 Overview

Two applications form the foundation of this chapter. For multiplying two polynomials, area models are used—both in the form of tiles to display multiplication of linear factors and in the more general form of areas of rectangles to display multiplications of sums. In particular, these models are found in Lesson 10-1 and in Lessons 10-3 through 10-6. For representing, adding, and subtracting polynomials of higher degrees, Lesson 10-2 introduces accounts into which periodic investments are made (annuities).

Two other applications are also mentioned. In Lesson 10-1, we review the representation of integers in base 10 as a polynomial in which 10 is substituted for the variable. The last lesson of this chapter, Lesson 10-7, shows a statistic that involves the squares of binomials. Although students learned how to add and subtract polynomials in Chapter 3, that skill is reviewed in enough detail in this chapter to be considered an objective.

The Distributive Property plays a major role in all the new algorithms of this chapter. In Lesson 10-3, it is used in multiplying a monomial by a polynomial:

$a(x + y + z) = ax + ay + az.$

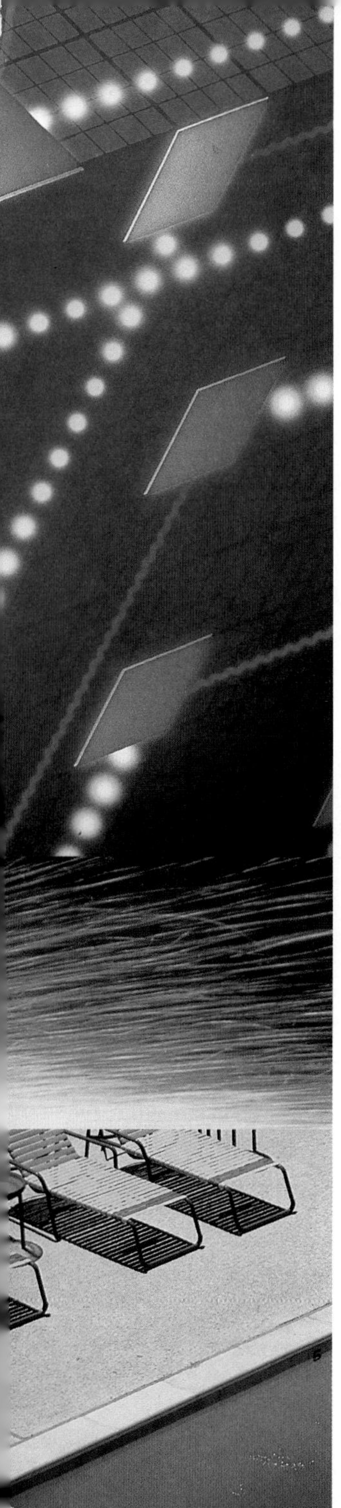

POLYNOMIALS

Expressions such as

$$6s^2,$$
$$2\ell + 2w + 4h,$$
$$\text{and } 1000x^4 + 500x^3 + 100x^2 + 200x$$

are *polynomials.*

Polynomials are found in geometry. For example, the monomial $6s^2$ represents the surface area of a cube with side s. The trinomial $2\ell + 2w + 4h$ represents the minimum length of ribbon needed to wrap a box of dimensions ℓ by w by h, as shown here.

Other polynomials arise from algebra. In Chapter 8, you calculated compound interest for a single deposit. When several deposits are made, the total amount of money accumulated can be expressed as a polynomial. For instance, the polynomial $1000x^4 + 500x^3 + 100x^2 + 250x$ represents the amount of money you would have if you invested \$1000 4 years ago, added \$500 to it 3 years ago, \$100 to it 2 years ago, and \$250 to it 1 year ago, all at the same rate of $x - 1$. The quadratic expressions you studied in Chapter 9 are all polynomials.

Polynomials also form the basic structure of our base 10 arithmetic. The expanded form of a number like 1796, $1 \cdot 10^3 + 7 \cdot 10^2 + 9 \cdot 10^1 + 6$, is like the polynomial $x^3 + 7x^2 + 9x + 6$ with the base 10 substituted for x.

In this chapter you will study these and other situations that give rise to polynomials, and how to add, subtract, and multiply polynomials.

A few examples of polynomials are given here. Ask students for other formulas or expressions that are polynomials and that describe certain quantities. [Examples: $\frac{1}{2}h(b_1 + b_2)$ for the area of a trapezoid; $a + b + c$ for the perimeter of a triangle; any of the quadratic expressions describing projectile motion; s^3 for the volume of a cube]

Photo Connections

The photo collage makes real-world connections to the content of the chapter: polynomials.

Farmland: The rectangular-shaped fields suggest the Area Model for Multiplication which, in this chapter, is extended to the multiplication of polynomials.

Cube: Another monomial found in geometry is s^3 which represents the volume of a cube with edge of length s.

Grinding Wheel: The Occupational Safety and Health Administration (OSHA) issues regulations and conducts investigations for violations of health standards that might contribute to unhealthy or unsafe conditions at the workplace. In Lesson 10-7, students use the chi-square statistic with respect to accidents at a factory.

Lumber: A cord is a measure of cut wood equal to 128 cubic feet. A pile of wood 4 feet wide, 4 feet high, and 8 feet long is a cord. In Lesson 10-2, a wood harvester is interested in the yearly growth factor of trees and what part of a cord a tree can provide after each year of growth.

Pool and Deck: If a contractor is given the dimensions of a pool and the maximum area of the pool and deck together, he or she could write an equation that uses the FOIL alogorithm, which is presented in Lesson 10-5, to find the maximum width of the deck.

Chapter 10 Projects

At this time, you might want to have students look over the projects on pages 657–658.

In Lesson 10-4 the Distributive Property is extended to multiply two polynomials:

$(a + b + c)(x + y + z) = ax + ay + az + bx + by + bz + cx + cy + cz.$

A special case of the Distributive Property is the multiplication of two binomials, which is discussed in Lesson 10-5:

$(a + b)(c + d) = ac + ad + bc + bd.$

Special cases of the multiplication of binomials are discussed in Lesson 10-6.

To some extent, this chapter and Chapter 12 can be considered companion chapters. Most of Chapter 12 is concerned with factoring those polynomials that are most easily factored, which reverses the process of multiplication that is studied here.

Lesson 10-1

Objectives

E Classify polynomials by their degree or number of terms.
F Write whole numbers as polynomials in base 10.
I Represent areas of figures with polynomials.

Resources

From the ***Teacher's Resource File***
- Lesson Master 10-1A or 10-1B
- Answer Master 10-1
- Teaching Aids
  - 26 Graph Paper
  - 31 Algebra Tiles
  - 107 Warm-up
  - 110 Classifying Polynomials
  - 111 Challenge

Additional Resources
- Visuals for Teaching Aids 26, 31, 107, 110, 111

Teaching Lesson 10-1

Warm-up

Write each number in standard form.
1. $3 \cdot 10^2 + 2 \cdot 10^1 + 6 \cdot 10^0$ **326**
2. $1 \cdot 10^3 + 4 \cdot 10^1 + 1 \cdot 10^0$ **1041**
3. $2 \cdot 10^5 + 6 \cdot 10^3 + 8 \cdot 10^2 + 9 \cdot 10^1 + 5 \cdot 10^0$ **206,895**
4. $3 \cdot 10^5 + 6 \cdot 10^4 + 2 \cdot 10^2 + 1 \cdot 10^1$ **360,210**
5. $1 \cdot 10^6 + 8 \cdot 10^4 + 6 \cdot 10^2 + 3 \cdot 10^1 + 2 \cdot 10^0$ **1,080,632**

LESSON

10-1

What Are Polynomials?

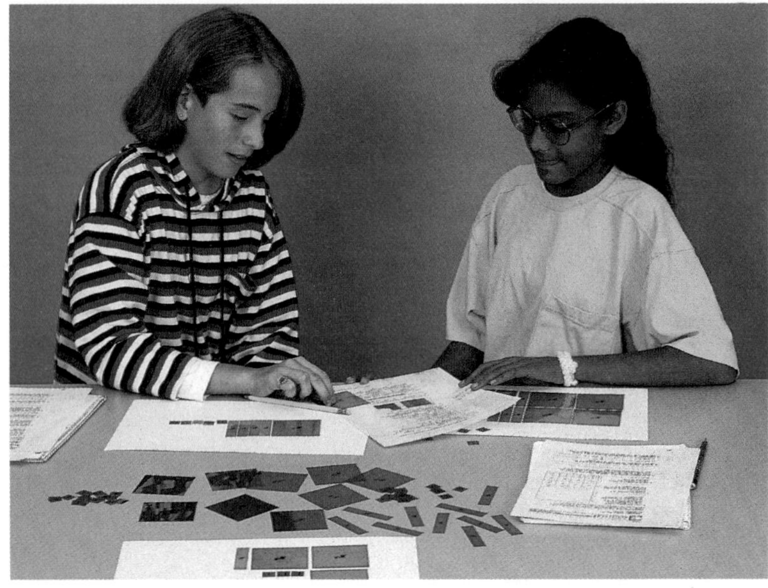

Picturing polynomials. *Polynomial expressions, such as $2x^2 + 7x + 3$, can be represented by algebra tiles. These tiles can also be used to illustrate operations with polynomials.*

You have worked with polynomials throughout your study of algebra. Now, because we are going to study polynomials in some detail, we introduce some terminology.

Monomials

Each of the expressions

$$3x, \quad 16t^2, \quad x^2y^4$$

is called a *monomial.* A **monomial** is an expression that can be written as a real number, a variable, or a product of a real number and one or more variables with nonnegative powers. The **degree of a monomial** is the sum of the exponents of the variables in the expression. For instance,

$3x$ has degree 1, (Think: $3x^1$.)
$16t^2$ has degree 2,
and x^2y^4 has degree 6.

Example 1

Tell whether the expression is a monomial or can be rewritten as a monomial. If so, identify its degree. If not, tell why not.
a. dw
b. -66
c. x^{-7}
d. $-8x^4 + 2x^4$

Lesson 10-1 Overview

Broad Goals This lesson covers the basic terminology of polynomials and the classification of polynomials by the number of terms or by a degree.

Perspective The definition of the degree of a polynomial comes from the fact that all polynomials are polynomials *in a variable or variables.* The polynomial $3x$ is a polynomial *in x.* Unless we are told otherwise, the polynomial x^2y^4 is a polynomial *in x and y*, and we add the degrees of x and y to determine that it is of degree 6. In certain circumstances, we might think of the polynomial x^2y^4 as a polynomial in y; then its degree is 4. If x^2y^4 is considered as a polynomial in x, then its degree is 2. Because we think of a, b, and c as constants (not variables) in $ax^2 + bx + c$, this is a polynomial in x and its degree is 2.

Polynomials are *in* a variable or variables, but they are also *over* the set from which their coefficients are to be chosen. This idea will come to play particularly in Chapter 12, which discusses factoring and analyzing solutions to quadratic equations. In this lesson, one might note that the polynomial forms which represent whole numbers in base 10 are over the set {0, 1, 2, 3, 4, 5, 6, 7, 8, 9}, whose elements we call *digits.*

Solution

a. $dw = d^1w^1$. So it is a product of variables with nonnegative exponents. Yes, dw is a monomial. Its degree is 1 + 1 = 2.

b. $-66 = -66x^0$. Zero is nonnegative. -66 is a monomial of degree 0.

c. $x^{-7} = \frac{1}{x^7}$. x^{-7} is not a monomial because it cannot be written as a product of variables with nonnegative exponents.

d. $-8x^4 + 2x^4 = (-8 + 2)x^4 = -6x^4$. Yes, $-8x^4 + 2x^4$ can be rewritten as a monomial of degree 4.

Classifying Polynomials by the Number of Terms

A **polynomial** is an expression that is either a monomial or a sum of monomials. Polynomials with two or three terms are used so often they have special names. A **binomial** is a polynomial which has two terms. A **trinomial** is a polynomial which has three terms. Here are some examples.

binomials: $x + 3$, $x^2 - y^2$, $7 - ab$

trinomials: $7x^2 - 3x + 5$, $a^2 + 2ab + b^2$, $ab + bc + cd$

The **degree of a polynomial** is the highest degree of any of its terms after the polynomial has been simplified.

Example 2

What is the degree of each polynomial?

a. $7x^2 - 3x + 5$

b. $p + q^2 + pq^2 + p^2q^3$

Solution

a. The terms have the following degrees:

$7x^2$ has degree 2;
$-3x$ has degree 1;
5 has degree 0.

$7x^2$ has degree 2, which is the highest in the polynomial. So the polynomial has degree 2.

b. Find the degree of each term by adding the exponents in that term.

$$\begin{array}{ccccccc} p & + & q^2 & + & pq^2 & + & p^2q^3 \\ \downarrow & & \downarrow & & \swarrow\downarrow & & \swarrow\downarrow \\ 1 & & 2 & & 1 + 2 = 3 & & 2 + 3 = 5 \end{array}$$

p^2q^3 has degree 5, which is the highest in the polynomial. So the polynomial has degree 5.

The polynomial

$$6x + 4x^5 + 8 + x^2$$

has degree 5. It is called a **polynomial in *x*,** because it is the sum of multiples of whole number powers of x. Its four terms could be rearranged in any order and the value of the polynomial would be the

Notes on Reading

Reading Mathematics As students read, you might want to make sure they understand these ideas.

1. In a polynomial, every exponent applied to a variable must be a nonnegative integer. For instance, $\frac{3}{x}$ is not a polynomial because it is equivalent to $3x^{-1}$. The term $4\sqrt{y}$ is not a polynomial because it involves taking a square root of a variable, and it is equal to $4y^{0.5}$.
2. When a polynomial has only one variable, it is easy to determine the degree of the polynomial by looking for the greatest exponent of the variable. However, if any term of the polynomial involves more than one variable, then the degrees of the individual variables in each term must be added.
3. Polynomials may be classified in two ways: by the number of terms and by the degree. The *Visual Organizers* on page 618 show these classifications.

It can be argued that algebra tiles are appropriate only for the representation of quadratic expressions because they use area representations. In this sense, only polynomials of degree 1 can be represented by lengths. Thus, although $4x + 3$ can be represented by algebra tiles, with four x-tiles and 3 unit tiles, the length of the sides is the key idea, and the area is not part of the representation.

LESSON MASTER 10-1 A

Questions on SPUR Objectives
See pages 661-663 for objectives.

Properties Objective E

1. After being simplified, which two of these expressions are binomials? **b and c**
(a) $x + 3x$ (b) $2(y^2 + 9)$ (c) $n + 3$ (d) $5a^2$
2. Give the degree of a. $6x^4$. **4** b. x^2y^5. **7**
3. Give an example of a monomial of degree 6 in which
a. the only variable is x. **Sample: $-4x^2$**
b. there are two variables, x and y. **Sample: $3xy^5$**
4. a. What is the degree of the monomial 12? **0**
b. Give an example of another monomial with the same degree as 12. **Sample: 9**

In 5–7, give the degree of the polynomial.

5. $x^5 - 2x^7$ **7**
6. $5n^2 + 7mn - 3$ **2**
7. $2x^4 + 5x^4 - 3x^6$ **6**
8. Give an example of a trinomial of degree 5. **Sample: $x^5 + xy - 4$**

Properties Objective F

In 9 and 10, write as a polynomial in base 10.

9. 7,007,007 **$7 \cdot 10^6 + 7 \cdot 10^3 + 7$**
10. 2,358 **$2 \cdot 10^3 + 3 \cdot 10^2 + 5 \cdot 10 + 8$**
11. Simplify $3 \cdot 10^5 + 4 \cdot 10^3 + 2 \cdot 10^2 + 3 \cdot 10^1$. **304,230**

Representations Objective I

12. Express the area of the figure as a polynomial. **$3x^2 + 3x + 5$**
13. Make a drawing to represent $2x + 6$. Use a rectangular arrangement of algebra tiles.

Optional Activities

Using Physical Models

Materials: Algebra tiles or **Teaching Aid 31;** graph paper or **Teaching Aid 26**

After completing the lesson, you might have each student **work with a partner**. Have students write a polynomial of the form $ax^2 + bx + c$ on a sheet of paper and then exchange papers with their partners. Then tell students to represent their partner's polynomial with algebra tiles and draw the representation on the paper.

same. It is customary to write the terms in *descending order* of the exponents of its terms, with the term with the largest exponent first:

$$4x^5 + x^2 + 6x + 8.$$

This is called the **standard form of a polynomial.** The numbers 4, 1, 6, and 8 are the *coefficients* of the polynomial. Notice that 1 is the coefficient of x^2. Since this polynomial has no x^3 or x^4 term, we say that 0 is the coefficient of x^3 and x^4. Written with these coefficients, here is a polynomial equal to the preceding one:

$$4x^5 + 0x^4 + 0x^3 + 1x^2 + 6x + 8.$$

Notice what happens when 10 is substituted for x in this polynomial. The value of the polynomial is

$$400{,}168.$$

The coefficients of the polynomial have become the digits of the decimal it equals. The expanded form of 400,168 is

$$4 \cdot 10^5 + 0 \cdot 10^4 + 0 \cdot 10^3 + 1 \cdot 10^2 + 6 \cdot 10 + 8.$$

This is a polynomial in a single variable with 10 substituted for the variable. Thus the normal way of writing whole numbers in base 10 is shorthand for a polynomial.

Classifying Polynomials by Their Degree

A polynomial of degree 1, such as $3x + 5$, is called **linear.** A polynomial of degree 2, such as $20 + 24t - 16t^2$, or $5xy$, is called **quadratic.** Linear and quadratic polynomials whose coefficients are small positive whole numbers can be represented by tiles, as you have seen. The tiles below represent the polynomial $x^2 + 4x + 5$ because this polynomial is the area of the figure.

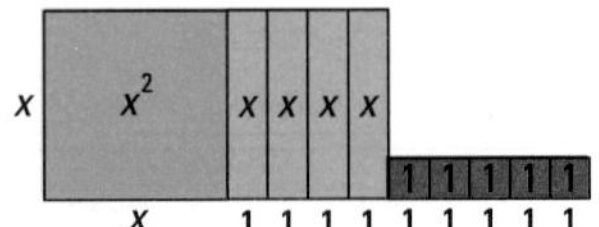

QUESTIONS

Covering the Reading

1. *Multiple choice.* Which of the following is *not* a monomial? **(d)**
   (a) x^4 (b) $4x$ (c) 4 (d) $x + 4$

2. Explain why $\frac{1}{y^2}$ is *not* a monomial.
   A monomial has variables with nonnegative powers. $\frac{1}{y^2} = y^{-2}$

Additional Examples

1. Tell if the expression is a monomial or if it can be rewritten as a monomial. If so, identify its degree. If not, tell why not.
   a. $15x^2$ **Yes, 2**
   b. 156 **Yes, 0**
   c. $\frac{5}{x^3}$ **No, the variable has a negative power.**
   d. $2x + 15x^2$ **No; it is a binomial.**
2. Give the degree of each polynomial.
   a. $1 + 3y + 4y^2 + 5y^3$ **3**
   b. $p - q + 3p - 4q^2 - p^2q^2$ **4**

Notes on Questions

Questions 9–13 These questions are designed to help students distinguish the difference between classifying polynomials by the number of terms and classifying them by degree. You might ask for an example of a binomial of degree 1, a binomial of degree 4, a trinomial of degree 4, and so on.

Question 17 This question anticipates using algebra tiles to represent the multiplication of a polynomial by a monomial, which will be presented in Lesson 10-3. Each row represents $2x^2 + 3x$; altogether, the tiles represent $(2x^2 + 3x) + (2x^2 + 3x) = 2(2x^2 + 3x) = 4x^2 + 6x$.
(Notes on Questions continue on page 620.)

LESSON MASTER 10-1 B **Questions on SPUR Objectives**

Properties Objective E: Classify polynomials by their degree or number of terms.

In 1–8, tell whether the expression is a *monomial* (M), *binomial* (B), *trinomial* (T), *polynomial* (P), or none of these (N). List all the terms that apply.

1. fgh **M, P**
2. -7.2 **M, P**
3. $3y^{-3}$ **N**
4. $\frac{3}{2} + 22ab$ **B, P**
5. $11x^2 - x - 12$ **T, P**
6. $\sqrt{5}$ **M, P**
7. $a^2 + ab - ab^2 + b^3$ **P**
8. 100π **M, P**

In 9–12, tell which of these expressions, after being simplified, are binomials. Write *yes* or *no*.

9. $4a + 5a$ **no**
10. $2(b^2 + 6)$ **yes**
11. $5h + 12u - 8h$ **yes**
12. $77x^2$ **no**

13. a. What is the degree of $3m^5$? **5**
    b. What is the degree of m^3n^7? **10**

14. Give an example of a monomial of degree 8 if
    a. the only variable is a. **Sample: $3a^8$**
    b. the monomial has two variables, a and b. **Sample: a^2b^6**

In 15–20, give the degree of the polynomial.

15. $3xy + 2x$ **2**
16. $3r^5 + 6.9$ **5**
17. $5m^8 - m^3$ **8**
18. $3e^5 - 2e^5 + 4e^7$ **7**
19. $a^4 + a^3 - ab^2 + ab^4$ **5**
20. $w^3 + 2w^2v^2 - 6$ **4**

Visual Organizers

Give students **Teaching Aid 110** which shows these charts of the two ways in which polynomials may be classified. You might ask students to give an example for each case and write it in a third column.

Degree	Name of Polynomial
1	linear
2	quadratic
3	cubic
4	quartic
5	quintic
6	no special name

Number of Terms	Name of Polynomial
1	monomial
2	binomial
3	trinomial
4	no special name
5	no special name
6	no special name

In 3–7, an expression is given. **a.** Tell whether the expression is a monomial. **b.** If it is a monomial, give its degree.

3. $5t^4$ Yes; 4 **4.** πr^2 Yes; 2 **5.** $5t^{-4}$ No **6.** $\frac{1}{2}bh$ Yes; 2 **7.** x^3y^6 Yes; 9

8. a. Is xyz a trinomial? **No**

b. Explain your reasoning. **A trinomial is the sum of three monomials; xyz is the product of three monomials.**

In 9–13, consider the following polynomials:

(a) $x^2 + 5$ (b) $x^2 + 5x + 6$

(c) $x^2 + 5xy + y^2$ (d) $x^3 + 5x^2 + 6x$

9. Which are binomials? **(a)**

10. Which are trinomials? **(b), (c), (d)**

11. Which have degree 2? **(a), (b), (c)**

12. Which have degree 3? **(d)**

13. Which are polynomials in x? **(a), (b), (d)**

In 14 and 15, rearrange the polynomial into standard form.

14. $16d^2 - 8d^4 + d^3$ **$-8d^4 + d^3 + 16d^2$** **15.** $x^9 + 2x + 4 - x^7$ **$x^9 - x^7 + 2x + 4$**

16. Jacob Lawrence, noted American painter, had his first one-man exhibition at the Harlem YMCA in 1938. Write 1938 as a polynomial with 10 substituted for the variable. **$1 \cdot 10^3 + 9 \cdot 10^2 + 3 \cdot 10 + 8$**

The Studio, *a self-portrait painted by Jacob Lawrence in 1977.*

In 17 and 18, what polynomial is represented by the tiles?

17.

x^2 x^2 x x x

x^2 x^2 x x x

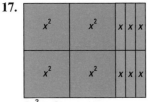

$4x^2 + 6x$

18.

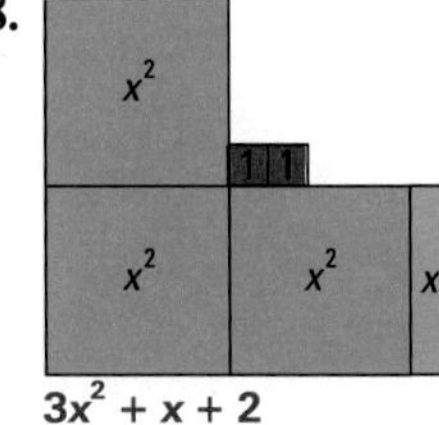

$3x^2 + x + 2$

In 19 and 20, draw a representation of the polynomial using tiles.

19. $2x + 3$ **See left.** **20.** $2x^2 + 3x + 1$ **See left.**

19)

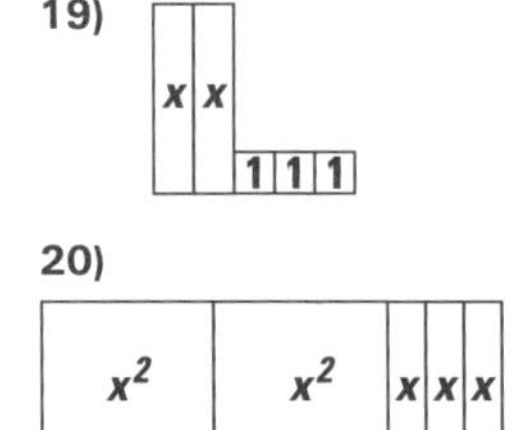

20)

x^2 x^2 x x x 1

Applying the Mathematics

In 21–25, an expression is given. **a.** Show that the expression can be simplified into a monomial. **b.** Give the degree of the monomial.

21. $3x + x$ **a) $4x$; b) 1** **22.** $3x \cdot x$ **a) $3x^2$; b) 2** **23.** $(5n^2)(-6n^2)$ **a) $-30n^4$; b) 4**

24. $2ab + 6ab$ **a) $8ab$; b) 2** **25.** $(2ab)^6$ **a) $64a^6b^6$; b) 12**

26. a. Write a monomial with one variable whose degree is 5.

b. Write a monomial with two variables whose degree is 5.

a) Sample: $6x^5$ b) Sample: x^2y^3

Follow-up for Lesson 10-1

Practice

For more questions on SPUR Objectives, use **Lesson Master 10-1A** (shown on page 617) or **Lesson Master 10-1B** (shown on pages 618–619).

Assessment

Group Assessment Have students **work in small groups** and take turns asking other group members to write a polynomial with a certain number of terms and of a certain degree. Have the group decide if the polynomials that are written are correct. [Students demonstrate the ability to classify polynomials.]

Extension

Write $50.91 = 5 \cdot 10^1 + 9 \cdot 10^{-1} + 1 \cdot 10^{-2}$ on the board. Explain that all finite decimals can be written as sums of multiples of powers of 10. Note that the representation on the board is not a polynomial in 10 because of the negative powers. Then have students represent 3.14159 (approximately π), 1609.344 m (exactly one mile), and another number that they chose in this way. [$3 \cdot 10^0 + 1 \cdot 10^{-1} + 4 \cdot 10^{-2} + 1 \cdot 10^{-3} + 5 \cdot 10^{-4} + 9 \cdot 10^{-5}$; $1 \cdot 10^3 + 6 \cdot 10^2 + 9 \cdot 10^0 + 3 \cdot 10^{-1} + 4 \cdot 10^{-2} + 4 \cdot 10^{-3}$]

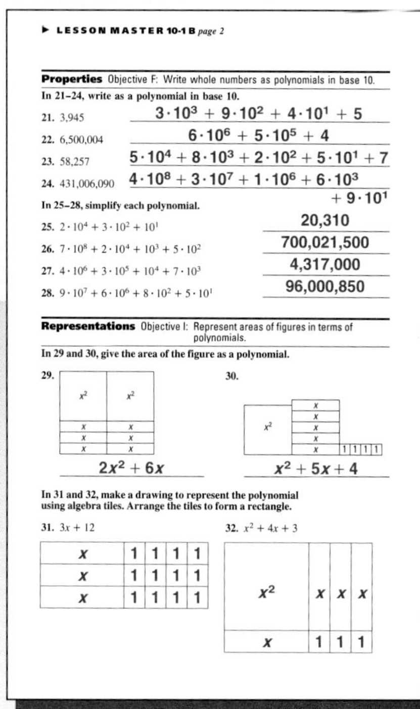

▶ **LESSON MASTER 10-1 B** *page 2*

Properties Objective F: Write whole numbers as polynomials in base 10.

In 21–24, write as a polynomial in base 10.

21. 3,945 **$3 \cdot 10^3 + 9 \cdot 10^2 + 4 \cdot 10^1 + 5$**

22. 6,500,004 **$6 \cdot 10^6 + 5 \cdot 10^5 + 4$**

23. 58,257 **$5 \cdot 10^4 + 8 \cdot 10^3 + 2 \cdot 10^2 + 5 \cdot 10^1 + 7$**

24. 431,006,090 **$4 \cdot 10^8 + 3 \cdot 10^7 + 1 \cdot 10^6 + 6 \cdot 10^3 + 9 \cdot 10^1$**

In 25–28, simplify each polynomial.

25. $2 \cdot 10^4 + 3 \cdot 10^2 + 10^1$ **20,310**

26. $7 \cdot 10^8 + 2 \cdot 10^4 + 10^3 + 5 \cdot 10^2$ **700,021,500**

27. $4 \cdot 10^6 + 3 \cdot 10^5 + 10^4 + 7 \cdot 10^3$ **4,317,000**

28. $9 \cdot 10^7 + 6 \cdot 10^6 + 8 \cdot 10^2 + 5 \cdot 10^1$ **96,000,850**

Representations Objective I: Represent areas of figures in terms of polynomials.

In 29 and 30, give the area of the figure as a polynomial.

29. x^2 x^2 x x x x x x **$2x^2 + 6x$**

30. x^2 x x x x x 1 1 1 1 **$x^2 + 5x + 4$**

In 31 and 32, make a drawing to represent the polynomial using algebra tiles. Arrange the tiles to form a rectangle.

31. $3x + 12$

x	1	1	1	1
x	1	1	1	1
x	1	1	1	1

32. $x^2 + 4x + 3$

x^2	x	x	x
x	1	1	1

Adapting to Individual Needs

Extra Help

When discussing the definition of *polynomial,* be sure students understand that an expression which is the difference of two monomials is included because a difference can be written as a sum. Show how these expressions can be rewritten as sums.

$x^2 - y^2 = x^2 + (-y^2)$

$7 - ab = 7 + (-ab)$

$7x^2 - 3x + 5 = 7x^2 + (-3x) + 5$

English Language Development

Because this lesson includes many new vocabulary words, have students write each word on an index card along with its definition and several examples of its usage. Discuss ways that will help students remember the meanings of the words. For example, relate *binomial* and *bicycle* (2 terms and 2 wheels), and *trinomial* and *tricycle* (3 terms and 3 wheels). Also explain that *mono-* means *one* and *poly-* means *many.*

Project Update Project 5, *Different Bases*, and Project 7, *Representing Positive Integers Using Powers*, on page 658, relate to the content of this lesson.

Notes on Questions

Questions 27–28 These questions anticipate the use of area models to express the square of a binomial, which students will encounter in Lesson 10-6.

Question 29 The digits in these numbers have been chosen so that the numbers can be added term by term, just as polynomials are.

Questions 31–33 Note that the inequality in **Question 33** has two variables, so it can be solved and graphed only in the coordinate plane. Contrast this with the inequalities in **Questions 31–32.**

Additional Answers

27a.

x^2	xy	xy	y^2

b.

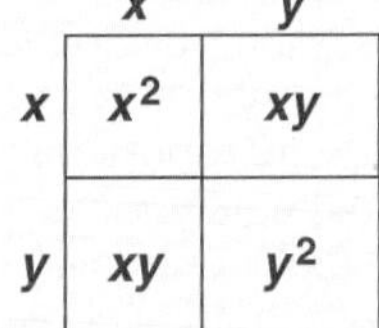

33.

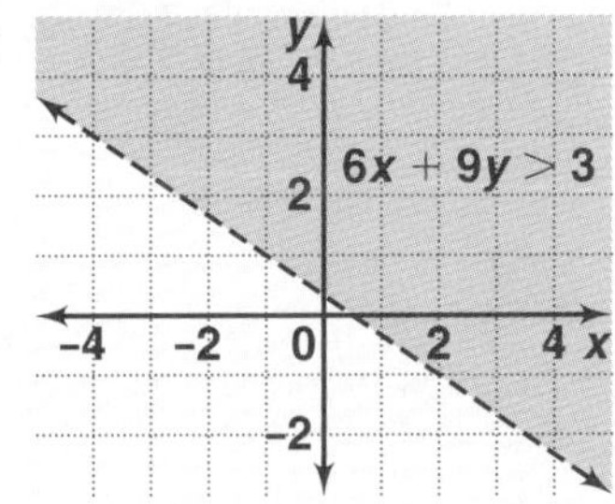

28)

x	1	1	1
x	1	1	1
x^2	x	x	x

31a)

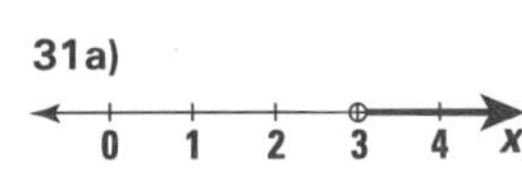

b)

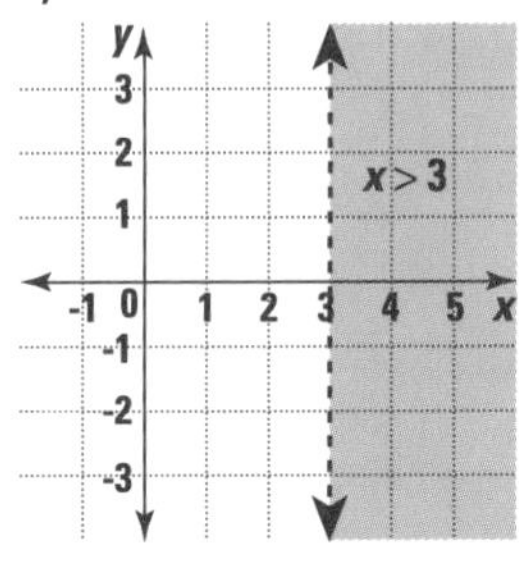

32a)

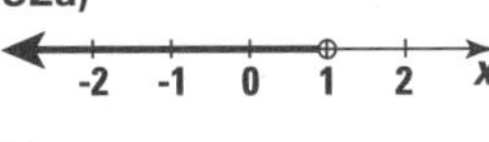

b)

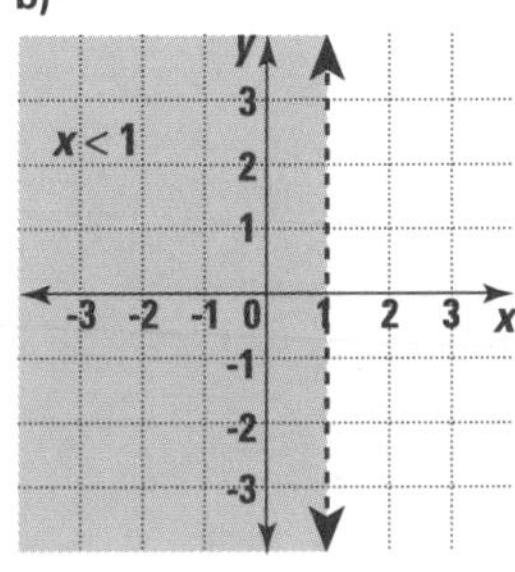

A three-winger. *This is the* Fokker Dr-1 *triplane, flown in World War I by the Red Baron.*

27. Suppose you have many tiles of each of the following sizes. **See margin.**

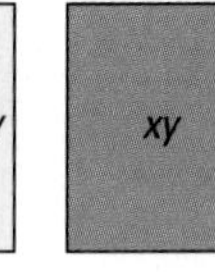

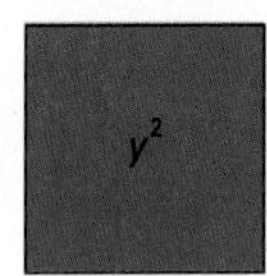

a. Draw a model for the polynomial $x^2 + 2xy + y^2$.
b. Show how the tiles in part **a** can be arranged to form a square.

28. Show how tiles representing the quantity $x^2 + 5x + 6$ can be arranged to form one large rectangle. **See left.**

29. a. Write 246 and 1032 as polynomials with 10 substituted for the variable. $2 \cdot 10^2 + 4 \cdot 10 + 6$; $1 \cdot 10^3 + 0 \cdot 10^2 + 3 \cdot 10 + 2$
b. Add the polynomials from part **a.** Is your sum equal to the sum of 246 and 1032? $1 \cdot 10^3 + 2 \cdot 10^2 + 7 \cdot 10 + 8 = 1278$; $246 + 1032 = 1278$; **Yes, the sums are equal.**

Review

30. If \$2000 is invested at an annual rate of 6%, compounded yearly, how much money will there be after each time period? *(Lesson 8-1)*
a. 1 year **\$2120** **b.** 2 years **\$2247.20** **c.** *n* years **\$2000(1.06)^n**

In 31–32, graph the solution set **a.** on a number line; **b.** in the coordinate plane. *(Lessons 1-1, 3-10, 7-9)* **See left.**

31. $x > 3$ **32.** $-6x + 9 > 3$

33. Graph in the coordinate plane: $6x + 9y > 3$. *(Lesson 7-9)* **See margin.**

In 34 and 35, consider a garage with a roof with a pitch of $\frac{4}{12}$. The garage is to be 30 feet wide. *(Lessons 1-7, 6-8, 7-3)*

34. What is the slope of $\overline{AC}$? $-\frac{1}{3}$

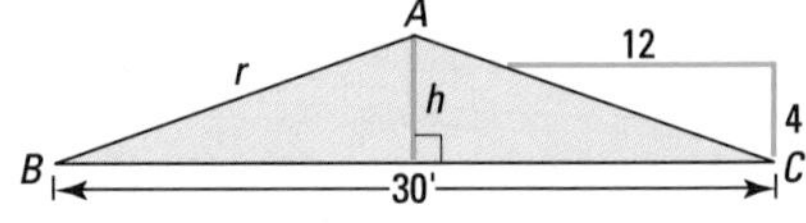

35. a. What is the height *h* of the roof? $h = 5$
b. Find the length *r* of one rafter. $r = 5\sqrt{10}$

Exploration

36. The words monomial, binomial, trinomial, and polynomial contain the prefixes mono-, bi-, tri-, and poly-, meaning one, two, three, and many, respectively. Find some other words, both mathematical and nonmathematical, that begin with these prefixes and give their meanings.
Samples: *monologue*—a long speech by one person in a group; *biped*—an animal with two feet; *triangle*—a three-sided figure; *polygraph*—an instrument such as a lie detector that records several pulsations (artery, vein, heart, etc.) at the same time

Adapting to Individual Needs

Challenge
Have students make four cards as shown below or give them **Teaching Aid 111.**

A		B		C		D	
1	9	2	10	4	12	8	12
3	11	3	11	5	13	9	13
5	13	6	14	6	14	10	14
7	15	7	15	7	15	11	15

Have a student think of a number from 1 through 15 and tell you on which cards this number appears. Add the numbers in the upper left corner of each card named and give the sum which is the number chosen. For example, if a person chooses 11, he or she will say that the number appears on cards A, B, and D. You add 1, 2, and 8 to get 11. Repeat the exercise several times without giving an explanation. Then ask students to figure out how you guessed the number.

Setting Up Lesson 10-2

Materials Students using the *Challenge* on page 625 will need encyclopedias or other reference books that discuss Babylonian and Mayan numeration systems.

Question 30 helps to set up Lesson 10-2.

LESSON

10-2

Investments and Polynomials

The largest money matters most adults commonly deal with are

salary or wages,
savings,
payments on loans for cars or trips or other items, and
home mortgages or rent.

Each of these items involves paying or receiving money each month, every few months, or every year. But what is the total amount paid or received? The answer is not easy to calculate because interest may be involved. Here is an example of this kind of situation.

Example 1

Each birthday from age 12 on, Maria has been receiving $500 from her grandparents to save for college. She is putting the money in an account that pays an annual yield of 7%. How much money will she have by the time she is 16?

Solution

It helps to write down how much Maria has on each birthday. On her 12th birthday she has $500. She then receives interest on that $500 and an additional $500 on her 13th birthday. At that time she will have

$$500(1.07) + 500 = \$1035.00.$$

Each year interest is paid on all the money previously saved and each year another $500 gift is added. The totals for her 12th through 16th birthdays are given in the chart below.

Birthday						Total
12th	500					= $500
13th	$500(1.07)$	$+ 500$				= $1035.00
14th	$500(1.07)^2$	$+ 500(1.07)$	$+ 500$			= $1607.45
15th	$500(1.07)^3$	$+ 500(1.07)^2$	$+ 500(1.07)$	$+ 500$		= $2219.97
16th	$500(1.07)^4$	$+ 500(1.07)^3$	$+ 500(1.07)^2$	$+ 500(1.07)$	$+ 500$	= $2875.37
	↑ from 12th birthday	↑ from 13th birthday	↑ from 14th birthday	↑ from 15th birthday	↑ from 16th birthday	

The total of $2875.37 she has by her 16th birthday is $375.37 more than the total $2500 she received as gifts because of the interest earned.

Lesson 10-2

Objectives

A Add and subtract polynomials.
G Translate investment situations into polynomials.

Resources

From the ***Teacher's Resource File***

- Lesson Master 10-2A or 10-2B
- Answer Master 10-2
- Teaching Aids
  107 Warm-up
  112 Example 3
  113 Extension
- Technology Sourcebook
  Computer Master 25

Additional Resources

- Visuals for Teaching Aids 107, 112, 113
- Spreadsheet software

Teaching Lesson 10-2

Warm-up

1. Each of **Questions 21–30** on page 626 contains at least one algebraic expression. Decide if the expression is a polynomial. If it is, tell the degree.
   21. Yes, degree 2
   22. Yes, degree 3
   23. Yes, degree 2
   24. Yes, degree 2
   25. Yes, degree 2
   26. No, but it simplifies to a first-degree polynomial.
   27. Not a polynomial
   28. Each part is a polynomial of degree 1.

(Warm-up continues on page 622.)

Lesson 10-2 Overview

Broad Goals Students should see how situations in which money is periodically placed in a savings account can lead to polynomials.

Perspective Polynomials arise in mathematics from many considerations: any continuous function can be approximated by a polynomial function; there is a polynomial function for any finite sequence; the set of polynomials with real coefficients under the usual operations forms a *ring*. These considerations are beyond the level of this book.

The work on compound interest and exponential growth in Chapter 8 is applied in this lesson. This application involves *annuities* (investments involving periodic deposits or withdrawals). In an annuity, the principal amounts can be different (have different coefficients in the compound-interest formula) and grow for different lengths of time (have different exponents in the compound-interest formula). That is, if x is $1 + r$, where r represents the annual yield, the polynomial $ax^2 + bx + c$ can stand for the total when an amount a has been invested (or received) for 2 time periods, an amount b has been in the annuity for 1 time period, and the amount c has just been invested. An advantage of this application is that each component of the polynomial has a clear meaning.

How Investments Lead to Polynomials

When $x = 1.07$, the following polynomial gives the amount of money Maria will have (in dollars)

$$500x^4 + 500x^3 + 500x^2 + 500x + 500.$$

This polynomial in x is useful because if the interest rate changes, you only have to substitute the new value for x. We call x in this expression a *scale factor*. For instance, if Maria's account had an annual yield of 4%, the scale factor would be 104% = 1.04. After 4 years Maria would have:

$$500(1.04)^4 + 500(1.04)^3 + 500(1.04)^2 + 500(1.04) + 500 \approx \$2708.16.$$

Example 2

❶ Cole's parents plan to give him \$50 on his 12th birthday, \$60 on his 13th, \$70 on his 14th, and \$80 on his 15th. If he invests the money in an account with a yearly scale factor x, how much money will he have on his 15th birthday?

Solution

The money from Cole's 12th birthday will earn three years worth of interest; from his 13th, two years of interest; and from his 14th, one year. The total amount of Coleman's birthday gifts will equal

$$50x^3 + 60x^2 + 70x + 80.$$

❷ Adding Polynomials

Suppose Cole's aunt gives him \$20 on each of these 4 birthdays. If he puts this money into the same account, the amount available from the aunt's gifts would be

$$20x^3 + 20x^2 + 20x + 20.$$

To find the total amount he would have, add these two polynomials.

$$(50x^3 + 60x^2 + 70x + 80) + (20x^3 + 20x^2 + 20x + 20)$$

Simplify the sum. First, use the Associative and Commutative Properties of Addition to rearrange the polynomials so that like terms are together.

$$= (50x^3 + 20x^3) + (60x^2 + 20x^2) + (70x + 20x) + (80 + 20)$$

Then use the Distributive Property to add like terms.

$$= (50 + 20)x^3 + (60 + 20)x^2 + (70 + 20)x + (80 + 20)$$
$$= 70x^3 + 80x^2 + 90x + 100$$

Notice what the answer means in relation to Cole's birthday presents. The first year he got \$70 (\$50 from his parents and \$20 from his aunt). The \$70 has 3 years to earn interest. The \$80 from his next birthday earns interest for 2 years, and so on. Also notice that in these examples we have written the polynomials in standard form. This is common practice.

29. Each expression is a polynomial; a. degrees 1 and 3; b. degree 3; c. degree 2; d. degrees 1 and 0

30. Yes, degree 1

2. Name the coefficients of the terms found in **Question 29** on page 626.

a. 4 and 4 b. 1 and 1
c. 3 and 6 d. 50 and 50

Notes on Reading

Reading Mathematics Spend time reading **Example 1** with students to ensure that they understand it. This example illustrates how polynomials arise, and it also demonstrates the value of writing *uncalculated* arithmetic expressions to describe situations. If students were to compute every step of the argument with their calculators, they would not see the polynomial pattern. However, students should check the values of the "Total" column with their calculators.

❶ Point out that this example develops a polynomial with different coefficients by varying information in the situation.

❷ This situation reviews the addition of polynomials that students first encountered in Chapter 3.

LESSON MASTER 10-2 A

Questions on SPUR Objectives
See pages 661-663 for objectives.

Skills Objective A

In 1–4, simplify the expression.

1. $(8x^2 + 5x - 4) + (6x^2 + 2x - 3)$ $14x^2 + 7x - 7$
2. $(5a^2 - 6a + 1) - (2a^2 - 3a - 2)$ $3a^2 - 3a + 3$
3. $(3n^2 - 5n) - (n^2 + 2n - 8)$ $2n^2 - 7n + 8$
4. $(m^2 + 3m - 12) + (m^2 - 8m + 3)$ $2m^2 - 5m - 9$

Uses Objective G

5. James Garcia won the lottery and will receive \$25,000 on June 1st each year for four years. He is planning to invest the money.
   a. Suppose James saves all the prize money, investing it in an account that earns interest at a scale factor x. Complete the table to show how much James will have on June 1 of each year. (Remember that the payments continue for only 4 years, although the money continues to earn interest.)

YEAR	AMOUNT
1	25,000
2	$25{,}000x + 25{,}000$
3	$25{,}000x^2 + 25{,}000x + 25{,}000$
4	$25{,}000x^3 + 25{,}000x^2 + 25{,}000x + 25{,}000$
5	$25{,}000x^4 + 25{,}000x^3 + 25{,}000x^2 + 25{,}000x + 25{,}000$
6	$25{,}000x^5 + 25{,}000x^4 + 25{,}000x^3 + 25{,}000x^2 + 25{,}000x + 25{,}000$

   b. Calculate how much James would have the sixth year if his interest rate were 5%. \$118,797.82
   c. How much he would have the sixth year if he could find an investment earning 12%? \$149,879.73
   d. Suppose the yearly prize had been \$250,000 instead of \$25,000. Explain how to use the Distributive Property and your answer to Part c to easily find out how much he would have the sixth year earning 12% if he saved all his prize money.
   Sample: The coefficient of each term for 6 years is 10 times as great, so multiply the 6-year sum by 10.

Optional Activities

Activity 1
You might want to use this activity after discussing **Example 3.** Have each student make up a plan for saving toward a goal such as college tuition, a new car, or a vacation. First, the student will have to decide on the amount of money that will be needed to achieve the goal. Then a plan will have to be developed that will enable the student to reach his or her goal.

Activity 2 Technology Connection
In *Technology Sourcebook, Computer Master 25,* students use a spreadsheet program to create tables that use polynomials in investment situations.

Comparing Investments

When comparing investments it is often useful to make a table.

Example 3

Nellie and Joe plan to save money for a round-the-world trip when they retire 10 years from now. Nellie plans to save $1000 per year for the first 5 years and then to stop making deposits. Joe plans to wait 5 years to begin saving but then to save $1200 per year for 5 years. They will deposit their savings at the beginning of the year into accounts earning 6% interest compounded annually. How much will each person have after 10 years?

Solution

Make a table showing the amount of money each person will have at the end of each year. At the end of the first year, Nellie will have $1.06(1000) =$ $1060. At the end of the second year she will have 106% of the sum of the previous balance and the new deposit of $1000. In all she will have $1.06(1060 + 1000) =$ $2183.60. This pattern continues. But after 5 years, she deposits no money. So her money accumulates only interest. Nellie's end-of-year balance in the spreadsheet below was computed by entering the formula = 1.06 * B3 into cell C3 and the formula = 1.06 * (C3 + B4) into cell C4. The formula in cell C4 was then replicated down column C to C12. A similar set of formulas generated Joe's end-of-year balance.

Lion's Gate Bridge in Vancouver, Canada.

	A	B	C	D	E
1	Year	Nellie's Deposit	Nellie's Balance	Joe's Deposit	Joe's Balance
2			(end of year)		(end of year)
3	1	1000	1060.00	0	0
4	2	1000	2183.60	0	0
5	3	1000	3374.62	0	0
6	4	1000	4637.09	0	0
7	5	1000	5975.32	0	0
8	6	0	6333.84	1200	1272.00
9	7	0	6713.87	1200	2620.32
10	8	0	7116.70	1200	4049.54
11	9	0	7543.70	1200	5564.51
12	10	0	7996.32	1200	7170.38

Ten years from now Nellie will have about $7996, and Joe will have about $7170.

Notice that even though Nellie deposits only $5000 and Joe deposits $6000, compounding interest over a longer period of time will give Nellie over $800 more than Joe. Here is what happens. After 10 years:

Nellie will have $1000x^{10} + 1000x^9 + 1000x^8 + 1000x^7 + 1000x^6$,
and Joe will have $1200x^5 + 1200x^4 + 1200x^3 + 1200x^2 + 1200x$.

When $x = 1.06$, as in Example 3, Nellie will have more than Joe.

Example 3 points out the advantages of people investing when they are young. You can also use this example to explain why people who already have some money find it easier to make more money in the long run. **Teaching Aid 112** shows this spreadsheet.

Additional Examples

1. As a New Year's resolution, Bert has decided to deposit $100 in a savings account every January 2nd. The account yields 3% interest annually. How much will his savings be worth when he makes his fourth deposit?
   $100(1.03)^3 + 100(1.03)^2 + 100(1.03) + 100 \approx$ $418.36
2. Janice has a savings account that has a scale factor of x. She makes deposits at regular yearly intervals. The first year she deposits $800, the second year $300, the third year $450, and the fourth year $775. What is her balance immediately after the fourth deposit?
   $800x^3 + 300x^2 + 450x + 775$
3. Which is more advantageous, to invest $50 per year for four years or to invest $100 in the first year and $100 in the fourth year? In both instances, the money earns 3% interest a year.
   The second investment yields $209.27 which is slightly better than the first which yields $209.18.

▸ LESSON MASTER 10-2 A *page 2*

In 8–10, use this information: Amy and Anton have just graduated and are each saving money for a down payment on a house. They will each deposit their savings into a special account earning 5.5% interest compounded annually. Their plans are described below.

Amy: **Wait and deposit $5,000 at the beginning of the 6th, 7th, and 8th years.**

Anton: **Deposit $4,000 at the beginning of the 1st, 2nd, and 3rd years.**

8. Complete the spreadsheet below.

	A	B	C	D	E
1	Year	Amy's Deposit	Amy's End of Year Balance	Anton's Deposit	Anton's End of Year Balance
2	1	0	**0**	4,000	**4,220**
3	2	0	**0**	4,000	8,672.10
4	3	0	**0**	4,000	**13,369.07**
5	4	0	**0**	0	**14,104.36**
6	5	0	**0**	0	**14,880.10**
7	6	5,000	**5,275**	0	**15,698.51**
8	7	5,000	**10,840.13**	0	**16,561.93**
9	8	5,000	**16,711.34**	0	**17,472.83**

9. What formula could be used to calculate the value in
   a. cell C9? **Samples: =1.055*(C8+B9)**
   B7*(1.055^3+1.055^2+1.055)
   b. cell E9? **Samples: =1.055*E8**
   D2*(1.055^8+1.055^7+1.055^6)

10. Who has more money at the end of 8 years? How much more?
   Anton; $761.49

Video

Wide World of Mathematics The segment, *Investing for College,* provides suggestions for saving for college and for obtaining funds from other sources. The use of investment rates presented in the segment provides motivation for introducing or extending a lesson on polynomials. Related questions and an investigation are provided in videodisc stills and in the Video Guide. A related CD-ROM activity is also available.

Videodisc Bar Codes

Search Chapter 49

Play

Notes on Questions

Questions 1–7 Cooperative Learning You will probably want to go through these questions in detail with your students.

Question 20 Science Connection Once a tree has been cut down, a person can "read" the history of the tree in the cross-section of the trunk. Most trees grow a layer of wood each year, forming an *annual ring*; the number of rings shows the age of the tree. The width of a ring indicates the amount of sun and moisture the tree received that year—the wider the ring, the more both of these elements were present. Black spots in the cross-section indicate fire damage, and "V" marks indicate points at which branches grew at one time.

(Notes on Questions continue on page 626.)

LESSON MASTER 10-2 B Questions on SPUR Objectives

Skills Objective A: Add and subtract polynomials.

In 1–6, simplify the expression.

1. $(3x^2 - 7x + 1) + (2x^2 + x - 8)$ $5x^2 - 6x + 7$
2. $(9b^2 - 3b - 2) - (b^2 + 4b - 6)$ $8b^2 - 7b + 4$
3. $(16 - 4m^2 + 2m) + (5m^2 - 3m - 10)$ $m^2 - m + 6$
4. $(a^2 - 12) - (a^2 - 4a + 2)$ $4a - 14$
5. $(2x^2 + 3x - 18) - (4x - 5)$ $2x^2 - x - 13$
6. $(p^3 + 2p^2 + 4) + (p^3 + 8p^2 + 7p)$ $2p^3 + 10p^2 + 7p + 4$

In 7 and 8, fill in the missing polynomial.

7. $(4x^2 - 7x + 11) + (\underline{x^2 + 2x + 4}) = 5x^2 - 5x + 15$
8. $(5e^2 + 12e - 3) - (\underline{-3e^2 + 12e - 4}) = 8e^2 + 1$

Uses Objective G: Translate investment situations into polynomials.

9. Jimmy has enrolled in a retirement plan in which he invests $2,000 at the beginning of each year.
   a. Suppose the money is invested in an account with a yearly scale factor x. Complete the chart to show how much Jimmy would have in his retirement account at the end of each year.

Year	Amount at End of Year
1	$2{,}000x$
2	$2{,}000x^2 + 2{,}000x$
3	$2{,}000x^3 + 2{,}000x^2 + 2{,}000x$
4	$2{,}000x^4 + 2{,}000x^3 + 2{,}000x^2 + 2{,}000x$
5	$2{,}000x^5 + 2{,}000x^4 + 2{,}000x^3 + 2{,}000x^2 + 2{,}000x$

   b. Calculate how much would be in the account after 5 years if the interest rate were 4%. $11,265.95
   c. Calculate how much would be in the account after 5 years if the interest rate were 10%. $13,431.22

QUESTIONS

Covering the Reading

1. Refer to Example 1. Suppose Maria is able to get an annual yield of only 5% on her investment.
   a. How much money will she have in her account by her 16th birthday? **$2,762.82**
   b. How much less is this than what she would have earned at 7% annual yield? **$112.55**

2. Ellery's children will give him a combined gift of $1200 on each of his birthdays from age 61 on.
   a. If he saves the money in an account paying an annual yield of 7%, how much will he have by the time he retires at age 65? **$6,900.89**
   b. How much will he have accumulated if his investment grows by a scale factor of x each year? $\mathbf{1200x^4 + 1200x^3 + 1200x^2 + 1200x + 1200}$

In 3 and 4, refer to Example 2.

3. Consider the polynomial $70x^3 + 80x^2 + 90x + 100$. What coefficient indicates the amount Cole received on his 14th birthday from his parents and his aunt? **90**

4. Suppose Cole also gets $15, $25, $35, and $45 from cousin Lilly on his four birthdays. He puts this money into his account also.
   a. By his 15th birthday, how much money will Cole have from just his cousin? $\mathbf{15x^3 + 25x^2 + 35x + 45}$
   b. What is the total Cole will have saved by his 15th birthday from all of his birthday presents? $\mathbf{85x^3 + 105x^2 + 125x + 145}$

In 5–7, refer to Example 3.

5. Which person—Nellie or Joe—will deposit more money in the 10 years? How much more? **Joe; $1,000**

6. Which person—Nellie or Joe—will have more money at the end of 10 years? How much more? **Nellie; $825.94**

7. Suppose Nellie and Joe could earn 7% on their investments. Recalculate the balances in the table and describe the end result. **Nellie, $8,630.31; Joe, $7,383.95**

Applying the Mathematics

8. Suppose in 1989 Carrie received $100 on her birthday. From 1990 to 1993 she received $150 on each birthday. She put the money in a shoebox. The money is still there.
   a. How much money did Carrie have after her 1993 birthday? **$700**
   b. How much more would she have had if she had invested her money at an annual yield of 6%? **$82.44**

Adapting to Individual Needs

English Language Development
Non-English-speaking students might benefit by being paired with native-English-speaking students when reading and discussing the lesson. Students should be encouraged to use their bilingual dictionaries.

Extra Help
Help students review the concept of interest by discussing experiences they may have had with savings accounts. Review the compound interest formula $T = P(1 + i)^n$, which was presented in Lesson 8-1. Remind students that T represents *total amount*, P represents the principal, i represents *annual yield* in decimal form, and n represents the number of years the money is allowed to accumulate.

In 9–12, Huey, Dewey, and Louie are triplets. They received cash presents on their birthdays as shown in the table.

Year	Huey	Dewey	Louie
In 1992	\$100	\$150	\$50
In 1993	\$200	\$150	\$400
In 1994	\$150	\$150	nothing

Each put all his money into a bank account which paid a 6% annual yield.

9. How much money did Huey have on his 1992 birthday? **\$100**

10. How much did Dewey have on his 1993 birthday? **\$309**

11. How much did Louie have on his 1994 birthday? **\$480.18**

12a) \$100
b) $100x + 200$
c) $100x^2 + 200x + 150$
d) $100x^3 + 200x^2 + 150x + 300$

12. In 1995, Huey received \$300 on his birthday. If he had invested all the money received from years 1992–1995 in an account with scale factor x, how much would he have had by his birthday in each of these years?
a. 1992 **b.** 1993 **c.** 1994 **d.** 1995
See left.

In 13–16, simplify the expressions.

13. $(12y^2 + 3y - 7) + (4y^2 - 2y - 10)$ $16y^2 + y - 17$

14. $(3 + 5k^2 - 2k) + (2k^2 - 3k - 10)$ $7k^2 - 5k - 7$

15. $(6w^2 - w + 14) - (4w^2 + 3)$ $2w^2 - w + 11$

16. $(x^3 - 4x + 1) - (5x^3 + 4x - 8)$ $-4x^3 - 8x + 9$

17. Solve the equation $(3x^2 + 2x + 4) + (3x^2 + 11x + 2) = 0$. $-\frac{3}{2}, -\frac{2}{3}$

In 18 and 19, find the missing polynomial.

18. $(9x^2 + 12x - 5) + (\underline{\ ?\ }) = 13x^2 + 6$ $4x^2 - 12x + 11$

19. $(2y^2 - y - 16) - (\underline{\ ?\ }) = -5y^2 - y + 31$ $7y^2 - 47$

A renewable resource. *This woman is reforesting an area in the Boise National Forest in Idaho.*

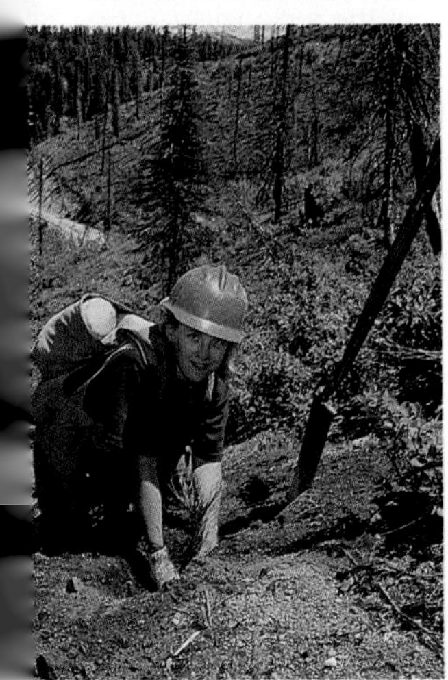

20. A wood harvester has planted trees in a forest each spring for four years, as shown in the following table.

Year	Number of trees planted
1	16,000
2	22,000
3	18,000
4	25,000

Suppose each tree when planted could provide .01 cord of wood and grows with a scale factor x each year. How many cords of wood could these trees provide in the fourth spring? $160x^3 + 220x^2 + 180x + 250$

Follow-up for Lesson 10-2

Practice

For more questions on SPUR Objectives, use **Lesson Master 10-2A** (shown on pages 622–623) or **Lesson Master 10-2B** (shown on pages 624–625).

Assessment

Written Communication Ask students to decide on an amount of money that they could put into a savings account on their birthday each year. Tell them to assume the account pays an annual yield of 4%, and have each student write a polynomial describing the amount of money in the account on the day after his or her 18th birthday. [Each student writes a polynomial with a scale factor of 1.04 and the correct number of terms.]

(Follow-up continues on page 626.)

► LESSON MASTER 10-2 B *page 2*

Suppose Jimmy's employer adds \$1,000 to Jimmy's retirement account each year.

d. Write a polynomial showing the portion of the account balance after 5 years due to the employer's contributions. Let the yearly scale factor be x.
$1{,}000x^5 + 1{,}000x^4 + 1{,}000x^3 + 1{,}000x^2 + 1{,}000x$

e. Write a polynomial showing the total amount in the account after 5 years with a yearly scale factor x. Give your answer in simplified form.
$3{,}000x^5 + 3{,}000x^4 + 3{,}000x^3 + 3{,}000x^2 + 3{,}000x$

In 10–12, Yuko will be ready for college in 6 years. Her mother is examining two plans for college savings. Each plan earns 6.5% interest compounded annually. Here are the plans.

Plan I: **Deposit \$4,000 at the beginning of the 1st, 2nd, and 3rd years.**

Plan II: **Wait and deposit \$5,000 at the beginning of the 4th, 5th, and 6th years.**

10. Fill in the spreadsheet below.

	A	B	C	D	E
1	Year	Plan I Deposit	Plan I End of Year Balance	Plan II Deposit	Plan II End of Year Balance
2	1	4,000	4,260	0	0
3	2	4,000	8,796.90	0	0
4	3	4,000	13,628.70	0	0
5	4	0	14,514.56	5,000	5,325
6	5	0	15,458.01	5,000	10,996.13
7	6	0	16,462.78	5,000	17,035.87

11. What formula can be used to calculate the value in

a. cell C7? Samples: $=4000*(1.065^6+1.065^5+1.065^4)$; $1.065*C6$

b. cell E7? Samples: $=5000*(1.065^3+1.065^2+1.065)$; $1.065*(E6+5000)$

12. Which plan yields more money after 6 years? How much more?
Plan II; \$573.09

Adapting to Individual Needs

Challenge Multicultural Connection
Materials: Encyclopedias

Ask students to explain in writing why our base-ten numeration system is a positional system. [Each digit in a base-ten numeral has a value which is determined by its position in the numeral, and the value is a multiple of some power of the base. For example, the 9 in 905 means $9 \cdot 10^2$, and the 9 in 9008 means $9 \cdot 10^3$.] Two ancient number systems that employed positional notation were the Babylonian system, which was *sexagesimal*, and the Mayan system, which was *vigesimal*. Have students look up the meanings of these words and then write a brief description of either the Babylonian or the Mayan numeration system. [Sexagesimal means based on the number 60, and vigesimal means based on the number 20.]

Extension

Give students **Teaching Aid 113.** Explain that polynomials arise from calculations in mortgages. The example on the Teaching Aid supposes that a family has a 30-year mortgage for $50,000 at 10% interest, and that payments each month are set at $439. Each month the interest rate is $\frac{10\%}{12} \approx$ .0083, and before the monthly payment is recorded, interest is assessed. Have students verify the calculations shown for the first four months. Note that most of the first amounts paid go for interest—for the first four months, only $24.00, $24.20, $24.40, and $24.60, respectively, are applied to the principal. The following polynomial can be used to calculate the principal p remaining on the mortgage after monthly payment n is recorded: $50{,}000(1.0083)^n - 439(1.0083)^{n-1} - 439(1.0083)^{n-2} - \ldots - 439(1.0083)^2 - 439(1.0083) - 439$.

Next, have students write the polynomials for months 5 and 6 and compute the principal remaining in each case. You might want to tell students that there is a formula which gives the sum of such polynomials when the coefficients are the same—it is the formula for the sum of a finite geometric series. This formula is found in three UCSMP texts: *Advanced Algebra; Functions, Statistics, and Trigonometry;* and *Precalculus and Discrete Mathematics.*

Project Update Project 1, *Lifelong Savings,* on page 657, relates to the content of this lesson.

Notes on Questions

Questions 21–23 The authors had particular situations in mind for answers to these questions, but students may have different ideas.

Question 32 Without a calculator or computer, this question is very difficult to answer. Point out that spreadsheets are often used to explore situations such as these.

21) Sample: area of a circle with a radius *r*; degree 2

22) Sample: volume of a cube with side *x*; degree 3

23) Sample: amount of money saved if $16 was invested at some rate *t* two years ago, and if $48 was added to the account one year ago; degree 2

Review

In 21–23, describe a situation that might yield the polynomial. Then give the degree of the polynomial. *(Lesson 10-1)* **See left.**

21. πr^2 **22.** x^3 **23.** $16t^2 + 48t$

24. a. Simplify $x(x + 5) - 10x$. $x^2 - 5x$
b. Solve $x(x + 5) = 10x$. *(Lessons 3-3, 3-9, 9-5)* $x = 0$ **or** $x = 5$

25. The formula $d = 0.042s^2 + 1.1s$ gives the approximate distance d in feet needed to stop a particular car traveling on dry pavement at a speed of s miles per hour. How much farther will this car travel before stopping if it is going at 65 mph instead of 55 mph? *(Lesson 9-2)* **61.4 ft**

26. Simplify $x^{-1} + -x - \frac{1}{x}$. *(Lesson 8-6)* $-x$

27. Find the slope of the line pictured at the right. *(Lesson 7-3)* $-\frac{b}{a}$

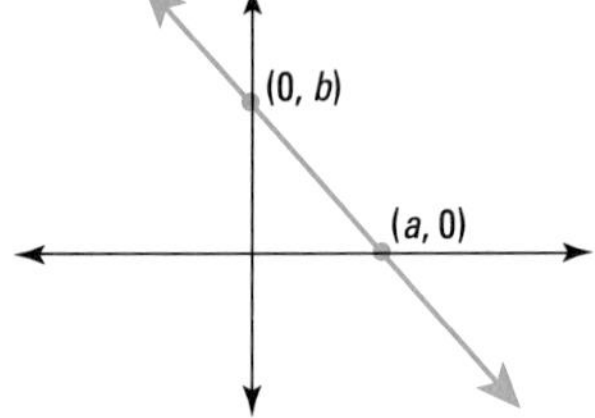

28. *Skill sequence.* Simplify. *(Lesson 3-7)*
a. $8(3y)$ $24y$ **b.** $8(3y + 7)$ $24y + 56$ **c.** $8(3y + 7 - 4x)$ $24y + 56 - 32x$

29. *Multiple choice.* Which is a pair of like terms? *(Lesson 3-2)* **(c)**
(a) $4b$ and $4b^3$ (b) x^3 and y^3
(c) $3m^2$ and $6m^2$ (d) $50x$ and 50

30. Simplify $-5n + 2n + k + k + 10n$. *(Lesson 3-2)* $7n + 2k$

31. The figure below shows two rectangles. If the area of the shaded region is 20, what are the missing dimensions of the rectangles? *(Lessons 2-1, 2-6, 3-2)* **.5 and 1.5**

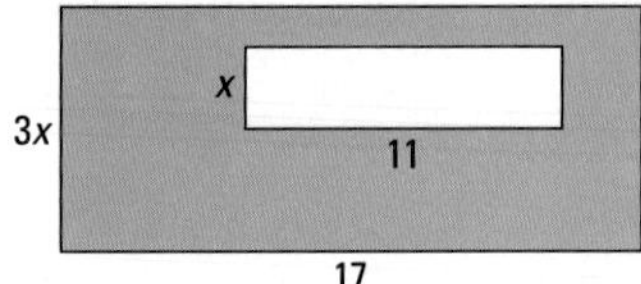

Exploration

32. Refer to Nellie and Joe in Example 3 and Questions 5–7.
a. For what interest rates will Nellie end up with more money than Joe? **for rates not less than 3.72%**
b. For what interest rates will Joe end up with more than Nellie? **for rates less than 3.72%**

Setting Up Lesson 10-3

The skills reviewed in **Question 28** are important for Lesson 10-3.

LESSON

10-3

Multiplying a Polynomial by a Monomial

Mall math. *To find the total area of these storefronts, you may find the area of each storefront and add the products, or multiply the height by the sum of the widths of the storefronts.*

You have already done several kinds of problems involving multiplication by a monomial. To multiply a monomial by a monomial, you can use properties of powers. For instance,

$$(2a^4b)(8a^3b^2) = 2 \cdot 8a^{(4+3)}b^{(1+2)} = 16a^7b^3.$$

To multiply a monomial by a binomial, you can use the Distributive Property, $a(b + c) = ab + ac$. For instance,

$$3x(2x + 1) = 3x \cdot 2x + 3x \cdot 1 = 6x^2 + 3x.$$

Using Area to Picture Multiplication by a Monomial

Some products of monomials and binomials can be pictured using the Area Model for Multiplication. For instance, the product $3x(2x + 1)$ is the area of a rectangle with dimensions $3x$ and $2x + 1$ as shown on the left.

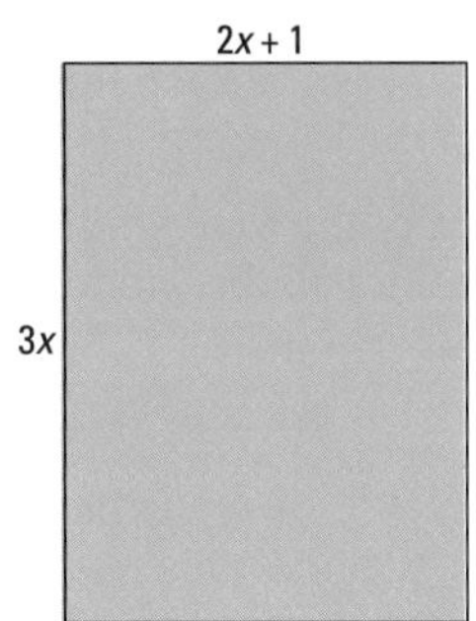

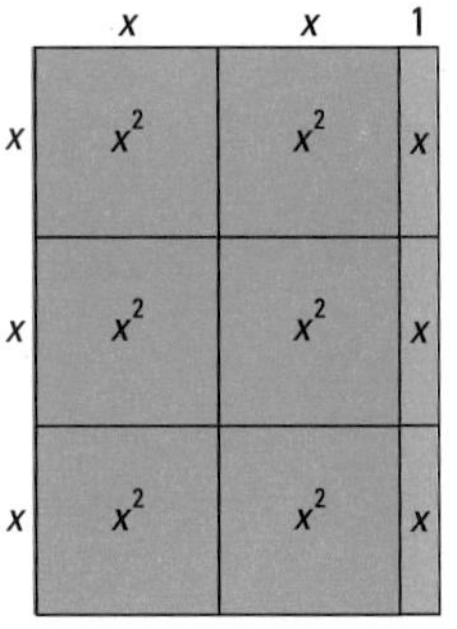

This rectangle can be split up into tiles as shown above at the right. The total area of the rectangle is $6x^2 + 3x$, which agrees with the result obtained using the Distributive Property.

Lesson 10-3

Objectives

C Multiply a polynomial by a monomial.

I Represent areas of figures with polynomials.

Resources

From the ***Teacher's Resource File***

- Lesson Master 10-3A or 10-3B
- Answer Master 10-3
- Teaching Aids
  - 26 Graph Paper
  - 31 Algebra Tiles
  - 107 Warm-up
  - 114 Example 1 and Additional Examples
- Activity Kit, Activity 22

Additional Resources

- Visuals for Teaching Aids 26, 31, 107, 114

Teaching Lesson 10-3

Warm-up

Explain how to use the Distributive Property to shorten the work in answering each question.

1. Kate, Anne, Laura, and Michael each bought a concert ticket for \$32, a program for \$10, and a shirt for \$18. How much did they spend altogether? **\$240; 4(32 + 18 + 10) = 4 × 60 = 240**
2. Mr. Enge bought 6 pairs of socks for \$1.79 a pair and 6 cans of tennis balls for \$2.29 per can. What was the total cost? **\$24.48; 6(1.79 + 2.29) = 6(4.08) = 24.48**

Lesson 10-3 Overview

Broad Goals Students studied the multiplication of a binomial by a monomial in Chapter 3. This lesson applies the Distributive Property to multiply a polynomial by a monomial. Area models are utilized to picture these products.

Perspective This lesson extends properties from earlier chapters. If a monomial is multiplied by a monomial, then only the Product of Powers Property and the Commutative and Associative Properties of Multiplication are needed. However, if a monomial is multiplied by a polynomial other than a monomial, then, because either addition or subtraction is involved, the Distributive Property must also be used.

Of the situations presented in this lesson, algebra tiles can represent only the product of polynomials of degree 0 or 1. These polynomials may seem so simple that nothing is accomplished, but it is enough so that some quadratics can be pictured.

Notes on Reading

You may want to use algebra tiles or **Teaching Aid 31** when discussing the lesson. Emphasize that there are no new properties in this lesson.

It is important that students understand the area representation for the storefront display. You can clarify it by using **Teaching Aid 114**, or by drawing the diagram on the board in two ways. First, show it as a single large rectangle.

h | $h(L_1 + L_2 + L_3 + L_4)$

$L_1 + L_2 + L_3 + L_4$

Then show it as the union of smaller rectangles.

h | hL_1 | hL_2 | hL_3 | hL_4

L_1 L_2 L_3 L_4

Example 1

Give two equivalent expressions for the total area pictured below.

Solution

Thinking of the total area as the sum of the areas of the individual tiles, the area is $4x^2 + 8x$. Thinking of the total area as length times width, the area is $4x(x + 2)$. So $4x(x + 2) = 4x^2 + 8x$.

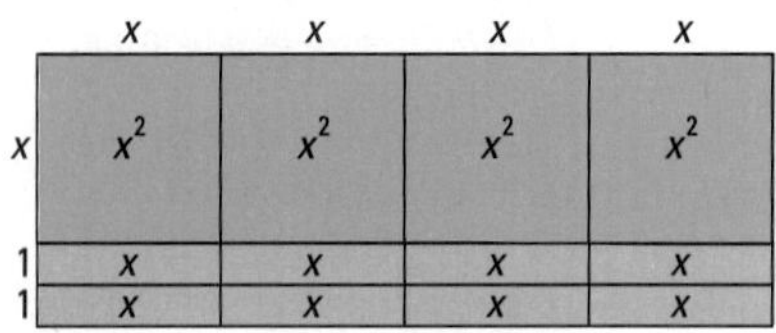

The area representation of a polynomial shows how to multiply a monomial by any other polynomial.

The pictures show a view of some storefronts at a shopping mall.

The displays in the windows are used to attract shoppers, so store owners and mall managers are interested in the areas of storefronts. Note that the height of each storefront h is a **monomial,** and the sum of the lengths of the storefronts $(L_1 + L_2 + L_3 + L_4)$ is a **polynomial.**

The total area of the four windows can be computed in two ways. One way is to consider all the windows together. They form one big rectangle with length $(L_1 + L_2 + L_3 + L_4)$ and height h. Thus, the total area equals

$$h \cdot (L_1 + L_2 + L_3 + L_4).$$

A second way is to compute the area of each storefront and add the results. Thus, the total area also equals

$$hL_1 + hL_2 + hL_3 + hL_4.$$

These areas are equal, so

$$h \cdot (L_1 + L_2 + L_3 + L_4) = hL_1 + hL_2 + hL_3 + hL_4.$$

Optional Activities

Activity 1
You can use *Activity Kit, Activity 22,* before covering the lesson or as an alternate approach to the material in the lesson through **Example 1.** In this activity, students use algebra tiles to model multiplying a polynomial by a monomial.

Activity 2
Materials: Graph paper or **Teaching Aid 26**

You may want to use this activity after discussing **Example 1**. Give students **Teaching Aid 26**, and ask them to draw several rectangles similar to those in **Example 1**. For each rectangle they make, they should write two equivalent expressions for the area that is pictured.

In general, to multiply a monomial by a polynomial, simply extend the Distributive Property: multiply the monomial by each term in the polynomial and add the results. As always, you must be careful with the signs in polynomials.

Example 2

Multiply $7x(x^2 + 4.5x + 1)$.

Solution

Multiply each term in the trinomial by the monomial $7x$.

$$7x(x^2 + 4.5x + 1) = 7x \cdot x^2 + 7x \cdot 4.5x + 7x \cdot 1$$
$$= 7x^3 + 31.5x^2 + 7x$$

Check

Test a special case. We let $x = 3$.

Does $7 \cdot 3(3^2 + 4.5 \cdot 3 + 1) = 7 \cdot 3^3 + 31.5 \cdot 3^2 + 7 \cdot 3$?

$21 \cdot 23.5 = 189 + 283.5 + 21$?

Yes, $493.5 = 493.5$.

Example 3

Multiply $-4x^2(2x^3 + y - 5)$.

Solution

Distribute $-4x^2$ over each term of the polynomial.

$$-4x^2(2x^3 + y - 5) = -4x^2 \cdot 2x^3 + -4x^2 \cdot y - -4x^2 \cdot 5$$
$$= -8x^5 - 4x^2y + 20x^2$$

Multiplying a Decimal by a Power of Ten

Recall the rule for multiplying a decimal by a power of 10: To multiply by 10^n, move the decimal point n places to the right. Multiplication of a monomial by a polynomial can show why this rule works. For instance, suppose 59,078 is multiplied by 1000. We write 59,078 in expanded form which looks like a polynomial. We think of 1000 as the monomial 10^3.

$$1000 \cdot 59{,}078 = 10^3 \cdot (5 \cdot 10^4 + 9 \cdot 10^3 + 7 \cdot 10 + 8)$$

Now we use the Distributive Property.

$$= 10^3 \cdot 5 \cdot 10^4 + 10^3 \cdot 9 \cdot 10^3 + 10^3 \cdot 7 \cdot 10 + 10^3 \cdot 8$$

The products can be simplified using the Product of Powers Property and the Commutativc and Associative Properties of Multiplication.

$$= 5 \cdot 10^7 + 9 \cdot 10^6 + 7 \cdot 10^4 + 8 \cdot 10^3$$

Now change the expanded form back to decimal form.

$$= 59{,}078{,}000$$

This same procedure could be repeated to explain the product of any decimal and any integer power of 10.

Additional Examples

These examples are also given on **Teaching Aid 114.**

1. Give two equivalent expressions for the area pictured below.

	x	x	x
x	x^2	x^2	x^2
x	x^2	x^2	x^2
1	x	x	x
1	x	x	x
1	x	x	x
1	x	x	x

$3x(2x + 4)$ and $6x^2 + 12x$

2. Multiply $-5y(y^3 - 6y^2 + 2y + 6)$.
$-5y^4 + 30y^3 - 10y^2 - 30y$
3. Multiply $k^4(k^2 - 16km)$
$k^6 - 16k^5m$

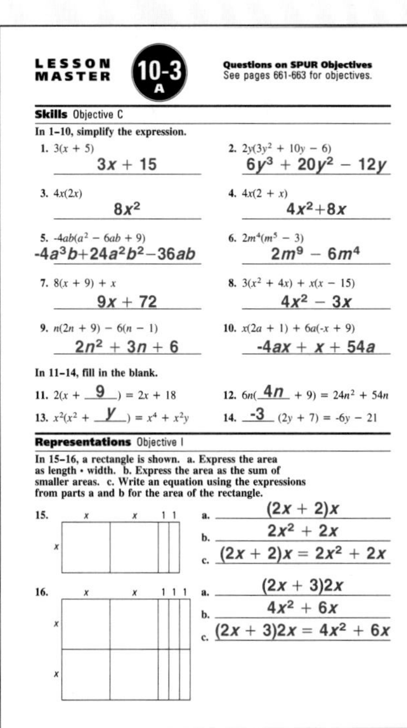

LESSON MASTER 10-3 A

Questions on SPUR Objectives
See pages 661-663 for objectives.

Skills Objective C

In 1–10, simplify the expression.

1. $3(x + 5)$ — $3x + 15$
2. $2y(3y^2 + 10y - 6)$ — $6y^3 + 20y^2 - 12y$
3. $4x(2x)$ — $8x^2$
4. $4x(2 + x)$ — $4x^2 + 8x$
5. $-4ab(a^2 - 6ab + 9)$ — $-4a^3b + 24a^2b^2 - 36ab$
6. $2m^4(m^5 - 3)$ — $2m^9 - 6m^4$
7. $8(x + 9) + x$ — $9x + 72$
8. $3(x^2 + 4x) + x(x - 15)$ — $4x^2 - 3x$
9. $n(2n + 9) - 6(n - 1)$ — $2n^2 + 3n + 6$
10. $x(2a + 1) + 6a(-x + 9)$ — $-4ax + x + 54a$

In 11–14, fill in the blank.

11. $2(x + \underline{9}) = 2x + 18$
12. $6n(\underline{4n} + 9) = 24n^2 + 54n$
13. $x^2(x^2 + \underline{y}) = x^4 + x^2y$
14. $\underline{-3}\,(2y + 7) = -6y - 21$

Representations Objective I

In 15–16, a rectangle is shown. a. Express the area as length • width. b. Express the area as the sum of smaller areas. c. Write an equation using the expressions from parts a and b for the area of the rectangle.

15. a. $(2x + 2)x$
b. $2x^2 + 2x$
c. $(2x + 2)x = 2x^2 + 2x$

16. a. $(2x + 3)2x$
b. $4x^2 + 6x$
c. $(2x + 3)2x = 4x^2 + 6x$

Activity 3

To give students more practice of the type in **Questions 9–12,** you could make a slight change in each question as shown below. Ask students to find the answer to each new question and to compare it with the original answer.

Question 9: $x(x^2 - x + 1)$
[$x^3 - x^2 + x$; the sign of the second term is changed.]

Question 10: $-5x^2(x^2 - 9x + 2)$
[$-5x^4 + 45x^3 - 10x^2$; the sign of each term is changed.]

Question 11: $3p^2(2 + p^2 + 5p^4)$
[$6p^2 + 3p^4 + 15p^6$; each term has one greater power of p.]

Question 12: $\frac{-wy}{3}(4y - 2w - 1)$
[$\frac{-4wy^2}{3} + \frac{2w^2y}{3} + \frac{wy}{3}$; each term is divided by 3 rather than multiplied by 3.]

Notes on Questions

Questions 9–12 For extra practice relating to these questions, you might use Activity 3 in *Optional Activities* on page 629.

Questions 14–15 Note that the first of these questions can be modeled easily with algebra tiles. Because of the subtraction, the second question cannot be modeled with tiles.

Question 16 This question can be done in several ways. Ask your students to share their strategies.

Multicultural Connection Hopscotch is an English name given to a game that is played in many countries. The object of the game is to hop over the "scotch," which is a line drawn on the ground. In one variation, the player kicks a stone while hopping from square to square. Or, the player can toss the stone into the first space and then hop on one foot into the space and kick the stone back across the base line and out of the diagram. In the German variation, *hinkspiel*, a player who completes the sequence turns away from the diagram and tosses the stone over his or her shoulder. The space where it lands is the player's "house," and other players must avoid the space unless its "owner" gives them permission to use it.

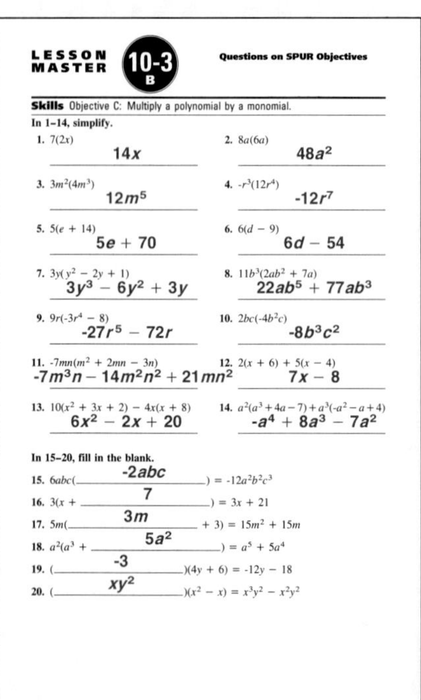

LESSON MASTER 10-3 B — Questions on SPUR Objectives

Skills Objective C: Multiply a polynomial by a monomial.

In 1–14, simplify.

1. $7(2x)$ **$14x$**
2. $8a(6a)$ **$48a^2$**
3. $3m^2(4m^3)$ **$12m^5$**
4. $-r^3(12r^4)$ **$-12r^7$**
5. $5(e + 14)$ **$5e + 70$**
6. $6(d - 9)$ **$6d - 54$**
7. $3y(y^2 - 2y + 1)$ **$3y^3 - 6y^2 + 3y$**
8. $11b^3(2ab^2 + 7a)$ **$22ab^5 + 77ab^3$**
9. $9r(-3r^4 - 8)$ **$-27r^5 - 72r$**
10. $2bc(-4b^2c)$ **$-8b^3c^2$**
11. $-7mn(m^2 + 2mn - 3n)$ **$-7m^3n - 14m^2n^2 + 21mn^2$**
12. $2(x + 6) + 5(x - 4)$ **$7x - 8$**
13. $10(x^2 + 3x + 2) - 4x(x + 8)$ **$6x^2 - 2x + 20$**
14. $a^2(a^3 + 4a - 7) + a^3(-a^2 - a + 4)$ **$-a^4 + 8a^3 - 7a^2$**

In 15–20, fill in the blank.

15. $6abc($ **$-2abc$** $) = -12a^2b^2c^3$
16. $3(x +$ **7** $) = 3x + 21$
17. $5m($ **$3m$** $+ 3) = 15m^2 + 15m$
18. $a^2(a^3 +$ **$5a^2$** $) = a^5 + 5a^4$
19. $($ **-3** $)(4y + 6) = -12y - 18$
20. $($ **xy^2** $)(x^2 - x) = x^3y^2 - x^2y^2$

3b)

4b)

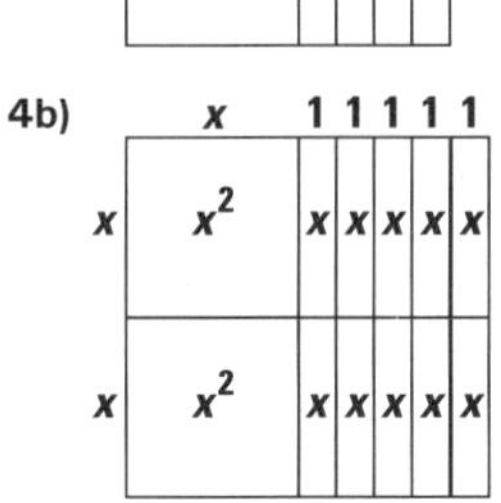

QUESTIONS

Covering the Reading

In 1 and 2, find the product.

1. $(3x)(4x)$ **$12x^2$**

2. $(3xy^3)(4x^2y)$ **$12x^3y^4$**

In 3 and 4, **a.** find the product; **b.** draw a rectangle to represent the product. **See left for diagrams.**

3. $3x(x + 4)$ **$3x^2 + 12x$**

4. $2x(x + 5)$ **$2x^2 + 10x$**

In 5 and 6, a large rectangle is made up of tiles as shown.
a. Express its area as the sum of smaller areas.
b. Express its area as length · width.
c. What equality is shown?

5.

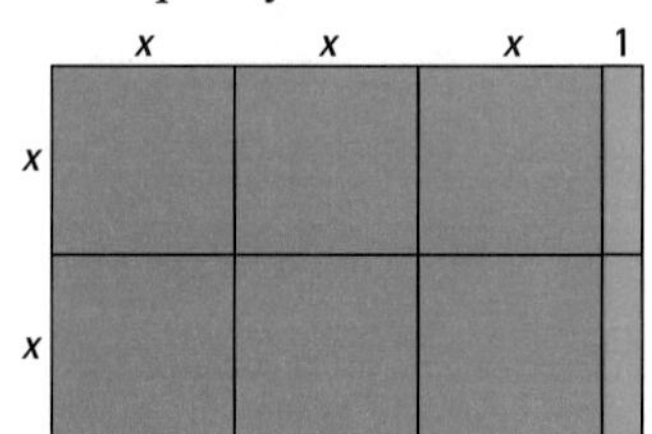

a) $6x^2 + 2x$ b) $(3x + 1) \cdot 2x$ c) $(3x + 1) \cdot 2x = 6x^2 + 2x$

6.

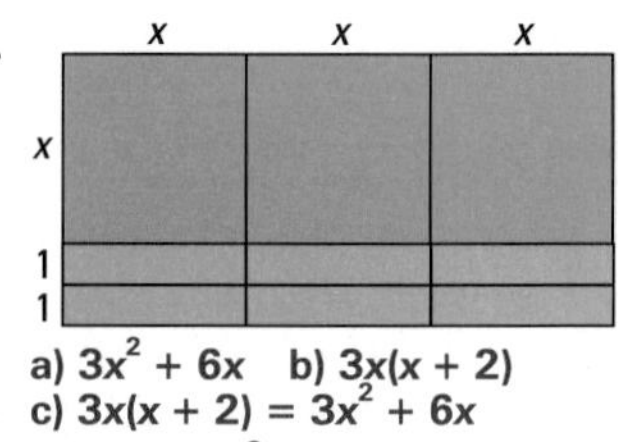

a) $3x^2 + 6x$ b) $3x(x + 2)$ c) $3x(x + 2) = 3x^2 + 6x$

7. Suppose the height of the stores mentioned in this lesson is $2h$. Write two different expressions for the total area of the group of storefronts in the diagram. **$2h(L_1 + L_2 + L_3 + L_4)$; $2hL_1 + 2hL_2 + 2hL_3 + 2hL_4$**

8. Use the Distributive Property to multiply $a(b + c + d)$. **$ab + ac + ad$**

In 9–12, multiply.

9. $x(x^2 + x + 1)$ **$x^3 + x^2 + x$**

10. $5x^2(x^2 - 9x + 2)$ **$5x^4 - 45x^3 + 10x^2$**

11. $3p(2 + p^2 + 5p^4)$ **$6p + 3p^3 + 15p^5$**

12. $-3wy(4y - 2w - 1)$ **$-12wy^2 + 6w^2y + 3wy$**

13. Use the multiplication of a monomial by a polynomial to explain why the product of 46329 and 10,000 is 463,290,000. **See below.**

Applying the Mathematics

In 14 and 15, suppose the width of a rectangle is w cm. Write two equivalent expressions for the area if the rectangle's length is as described below.

14. Its length is 7 cm more than twice the width. **$w(7 + 2w)$; $7w + 2w^2$**

15. Its length is 1 cm less than five times the width. **$w(5w - 1)$; $5w^2 - w$**

13)
$$10{,}000 \cdot 46{,}329 = 10^4 \cdot (4 \cdot 10^4 + 6 \cdot 10^3 + 3 \cdot 10^2 + 2 \cdot 10 + 9)$$
$$= 10^4 \cdot 4 \cdot 10^4 + 10^4 \cdot 6 \cdot 10^3 + 10^4 \cdot 3 \cdot 10^2 + 10^4 \cdot 2 \cdot 10 + 10^4 \cdot 9$$
$$= 4 \cdot 10^8 + 6 \cdot 10^7 + 3 \cdot 10^6 + 2 \cdot 10^5 + 9 \cdot 10^4$$
$$= 463{,}290{,}000$$

Adapting to Individual Needs

Extra Help

In order to understand this lesson, some students might need a review of the Area Model for Multiplication. Refer to the examples in Lesson 2-1. Then have students look at **Example 1** in this lesson. Be sure they understand that the two expressions $4x$ and $x + 2$ represent the length and width of a rectangle and their product represents the area of the rectangle.

Point out that **Examples 2 and 3** convey the same idea but in each case, the second expression has three terms. Be sure students understand that the Distributive Property works regardless of the number of terms.

Hopscotch. *Different versions of hopscotch are played the world over. Shown are children playing hopscotch in Soweto Township, South Africa.*

16. The formation of rectangles below is used by children in many countries for playing hopscotch. What is the total area? **$4x^2 + 4xy$**

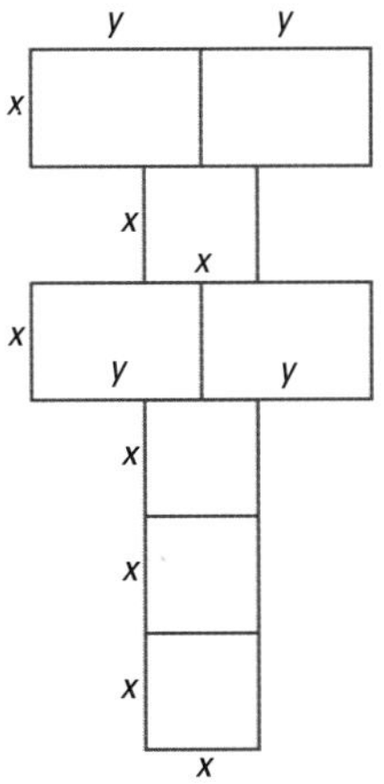

17. All angles in this figure are right angles. Find its area. **$8x^2 + 2x$**

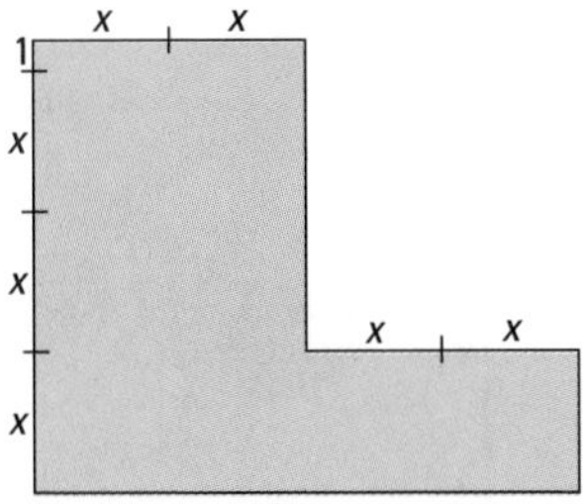

In 18–22, simplify.

18. $(5x)(2x) - (3x)(4x)$ **$-2x^2$**

19. $(a^2b)(2b^2c)(3ca)$ **$6a^3b^3c^2$**

20. $2(x^2 + 3x) - 3x^2$ **$6x - x^2$**

21. $a(y^2 - 2y) + y(a^2 + 2a)$ **$ay^2 + ya^2$**

22. $m^3(m^2 - 3m + 2) - m^2(m^3 - 5m^2 - 6)$ **$2m^4 + 2m^3 + 6m^2$**

23. Find the missing monomial:
$(\underline{\ ?\ }) \cdot (2n^2 + 5n - 6) = 12n^4 + 30n^3 - 36n^2$. **$6n^2$**

Review

24. After five years of birthdays, Wanda has received and saved

$$80x^4 + 60x^3 + 70x^2 + 45x + 50$$

dollars, having put the money in a savings account at a scale factor x.

a. How much did Wanda get on her most recent birthday? **$50**

b. How much did Wanda get on the first of these birthdays? **$80**

c. Give an example of a reasonable value for x in this problem. **Sample: $x = 1.065$**

d. If $x = 1$, how much has Wanda saved? **$305**

e. What does a value of 1 for x mean? *(Lesson 10-2)*
Wanda has put her money in a noninterest-bearing account.

Question 22 Error Alert Some students might simplify incorrectly because they distribute m^2 instead of $-m^2$. Remind these students to use the definition of subtraction first and then use the Distributive Property.

Adapting to Individual Needs

English Language Development

You might want to remind students what the term *distributive* means. Ask a student to *distribute* some papers, books, or pencils to class members. Write: $2x(3y + 2x - 5)$ on the board. Tell students that they *distribute* the $2x$ over $3y + 2x - 5$ when they multiply each term of $3y + 2x - 5$ by $2x$. Then have a student perform the multiplication. $[6xy + 4x^2 - 10x]$

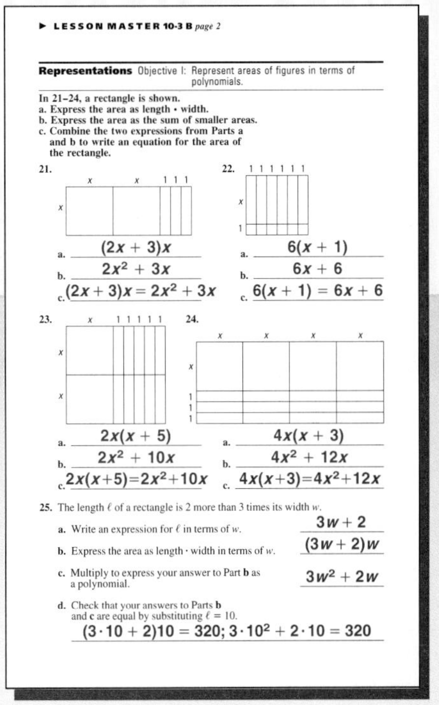

► LESSON MASTER 10-3 B *page 2*

Representations Objective I: Represent areas of figures in terms of polynomials.

In 21–24, a rectangle is shown.
a. Express the area as length · width.
b. Express the area as the sum of smaller areas.
c. Combine the two expressions from Parts a and b to write an equation for the area of the rectangle.

21. a. $(2x + 3)x$ b. $2x^2 + 3x$ c. $(2x + 3)x = 2x^2 + 3x$

22. a. $6(x + 1)$ b. $6x + 6$ c. $6(x + 1) = 6x + 6$

23. a. $2x(x + 5)$ b. $2x^2 + 10x$ c. $2x(x+5) = 2x^2 + 10x$

24. a. $4x(x + 3)$ b. $4x^2 + 12x$ c. $4x(x+3) = 4x^2 + 12x$

25. The length ℓ of a rectangle is 2 more than 3 times its width w.

a. Write an expression for ℓ in terms of w. $3w + 2$

b. Express the area as length · width in terms of w. $(3w + 2)w$

c. Multiply to express your answer to Part **b** as a polynomial. $3w^2 + 2w$

d. Check that your answers to Parts **b** and **c** are equal by substituting $\ell = 10$.
$(3 \cdot 10 + 2)10 = 320;\ 3 \cdot 10^2 + 2 \cdot 10 = 320$

Follow-up for Lesson 10-3

Practice

For more questions on SPUR Objectives, use **Lesson Master 10-3A** (shown on page 629) or **Lesson Master 10-3B** (shown on pages 630–631).

Assessment

Oral/Written Communication Draw the rectangle pictured in **Example 1** on page 628 on the board. Increase either the width or the length of the rectangle by units of 1 or *x*. Ask students to name two different expressions for the total area of the new rectangle. Repeat the activity several times. [Students display understanding of an area representation of a polynomial.]

Extension

Suppose that the area of a rectangle is $6x^2 + 4x$. Have students give possible dimensions of the rectangle. [Samples: x and $6x + 4$; $2x$ and $3x + 2$] Then tell them that the area of a square is $x^2 + 2x + 1$, and have them find the length of a side. [$x + 1$]

In 25 and 26, an expression is given. **a.** Tell whether the expression is a polynomial. **b.** If it is a polynomial, give its degree. If it is not a polynomial, tell why not. *(Lesson 10-1)* **See left for 25b.**

25b) $2x^{-3}$ cannot be written as a product of variables with nonnegative exponents.

25. $3x^2 + 2x^{-3}$ **No**

26. $a^2b + a^3b^2 + a^4b^3$ **Yes; degree 7**

27. Write a trinomial with degree 7 and constant term 1. *(Lesson 10-1)* **Sample: $-4x^7 + 5x + 1$**

In 28 and 29, simplify. *(Lessons 3-9, 8-7)*

28. $\frac{20x^2y}{2xy}$ **$10x$**

29. $\frac{3}{2v} + \frac{4v}{2v}$ **$\frac{3 + 4v}{2v}$**

In 30 and 31, write an equation for the line with the given characteristics.

30. contains the points (-5, 8) and (-1, 16) *(Lesson 7-6)* **$y = 2x + 18$**

31. slope 3, *y*-intercept 5 *(Lesson 7-4)* **$y = 3x + 5$**

In 32–34, an object is described. **a.** Calculate its speed. **b.** Make a guess about what the object is. *(Lesson 6-2)*

32. It flew 200 miles in $\frac{1}{2}$ hour. **a) 400 mph b) Sample: jet aircraft**

33. It slithered 14 meters in 7 seconds. **a) 2m/sec b) Sample: snake**

34. It took 3 days to creep 15 inches. **a) 5 in./day b) Sample: snail**

Hints. *Shown are some visual clues that may help you with Questions 32–34.*

Exploration

35. a. What is the rule for dividing a decimal by 1,000,000?
b. Make up an explanation like the one in this lesson to show why the rule works.

35a) Move the decimal point six places to the left.
b) Sample: $43{,}918.6 \div 10^6$
$= 43{,}918.6 \cdot 10^{-6}$
$= 10^{-6} \cdot (4 \cdot 10^4 + 3 \cdot 10^3 + 9 \cdot 10^2 + 1 \cdot 10 + 8 + 6 \cdot 10^{-1})$
$= 10^{-6} \cdot 4 \cdot 10^4 + 10^{-6} \cdot 3 \cdot 10^3 + 10^{-6} \cdot 9 \cdot 10^2 + 10^{-6} \cdot 1 \cdot 10 + 10^{-6} \cdot 8 + 10^{-6} \cdot 6 \cdot 10^{-1}$
$= 4 \cdot 10^{-2} + 3 \cdot 10^{-3} + 9 \cdot 10^{-4} + 1 \cdot 10^{-5} + 8 \cdot 10^{-6} + 6 \cdot 10^{-7}$
$= 0.0439186$

Adapting to Individual Needs

Challenge

Have students answer the following questions:

1. In how many ways can you separate a 5×2 rectangle into 1×2 rectangles? [8 ways, as shown at the right.]
2. How many ways are there if congruent diagrams are considered identical? [5 ways]

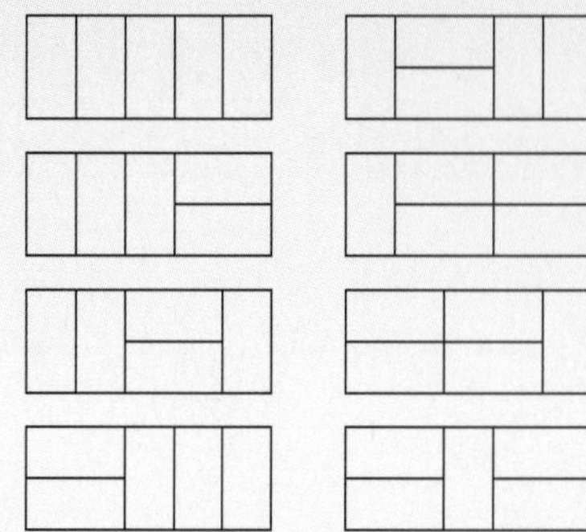

LESSON

10-4

Multiplying Polynomials

A Mondrian masterpiece. *Shown is a detail from the painting,* Composition with Red, Blue, Yellow, & Black *by the Dutch artist Piet Mondrian. The total area of the painting can be found in many ways.*

Using Area to Picture the Multiplication of Polynomials

The Area Model for Multiplication pictures how to multiply two polynomials with many terms. For instance, to multiply $a + b + c + d$ by $x + y + z$, draw a rectangle with length $a + b + c + d$ and width $x + y + z$.

	a	b	c	d
x	ax	bx	cx	dx
y	ay	by	cy	dy
z	az	bz	cz	dz

The area of the largest rectangle equals the sum of the areas of the twelve separate smaller rectangles.

Total area $= ax + ay + az + bx + by + bz + cx + cy + cz + dx + dy + dz$

But the area of the largest rectangle also equals the product of its length and width.

Total area $= (a + b + c + d) \cdot (x + y + z)$

Lesson 10-4

Objectives

B Multiply polynomials.
I Represent areas and volumes of figures with polynomials.

Resources

From the ***Teacher's Resource File***
- Lesson Master 10-4A or 10-4B
- Answer Master 10-4
- Assessment Sourcebook: Quiz for Lessons 10-1 through 10-4
- Teaching Aids
  108 Warm-up
  115 Multiplying Polynomials
  116 Questions 1 and 9

Additional Resources
- Visuals for Teaching Aids 108, 115, 116

Teaching Lesson 10-4

Warm-up

Multiply.
1. $3x(x + 7)$ $\mathbf{3x^2 + 21x}$
2. $2(5x - 3)$ $\mathbf{10x - 6}$
3. $8x^2(2x + 3y + 4)$ $\mathbf{16x^3 + 24x^2y + 32x^2}$
4. $10y(x^3 - x + y)$ $\mathbf{10x^3y - 10xy + 10y^2}$
5. $6x(6x - 6)$ $\mathbf{36x^2 - 36x}$
6. $x^2(2x^2 + xy - 4y^2)$ $\mathbf{2x^4 + x^3y - 4x^2y^2}$

Notes on Reading

Art Connection Piet Mondrian (1872–1944) was influenced by the cubism movement of modern art. He painted in a style called *neo-plasticism* in which forms are restricted to geometric shapes. Mondrian

Lesson 10-4 Overview

Broad Goals This lesson extends the work that students have done with multiplication of monomials to work with multiplication of polynomials having more terms.

Perspective The Area Model for Multiplication is applied here when multiplying a polynomial by a polynomial. In each example, before the multiplication is done, a note is made of the number of terms that will be in the result. This application of the Multiplication Counting Principle, which was discussed in Chapter 2, helps students to organize and check their work.

Notice that we do the more general pattern in this lesson before the special cases that occur in Lessons 10-5 and 10-6. There is always a decision to be made regarding whether it is better to give special cases and then generalize, or to give a general case and then specialize. We put the general case first because the algorithm for the multiplication of polynomials, as stated in the Extended Distributive Property, is in some sense easier than its special case. For the multiplication of binomials, every term is enclosed by parentheses. When polynomials with many terms are multiplied, the parentheses cause less distraction.

reduced painting to areas of white and flat primary colors separated by irregular black grids. One can see his theory of art in the clean lines of some of today's architecture and interior decoration.

The multiplication problem that opens this lesson is shown on **Teaching Aid 115**. You will probably want to show your students how each term in the first polynomial is multiplied by each term in the second polynomial by using both the diagram and the algebraic manipulation.

Error Alert As the number of multiplications increases, it is easy to lose track of the products. In our examples, we always distribute a term from the first polynomial over all terms of the second polynomial. Students are not as well organized. They may, for instance, start a problem like **Example 1** by distributing the $5x^2$ over the $(x + 7)$ and then distribute the x over $(5x^2 + 4x + 3)$. Using colored chalk may help to emphasize a correct pattern. See *Extra Help* on page 635.

❶ You may want to have students show their work for **Example 1** in a rectangular pattern as shown below.

	$5x^2$	$4x$	3
x	$5x^3$	$4x^2$	$3x$
7	$35x^2$	$28x$	21

The rectangle demonstrates how each term of the first polynomial is distributed over the terms of the second polynomial. Note that each small rectangle should be filled in as the products of terms are computed.

The Extended Distributive Property

The Distributive Property can be used to justify why the two expressions must be equal. Distribute the chunk $(x + y + z)$ over $(a + b + c + d)$ to get:

$$(a + b + c + d) \cdot (x + y + z)$$
$$= a(x + y + z) + b(x + y + z) + c(x + y + z) + d(x + y + z).$$

Four more applications of the Distributive Property lead to the result found above.

$$= ax + ay + az + bx + by + bz + cx + cy + cz + dx + dy + dz$$

Because of the multiple use of the Distributive Property, we call this an instance of the Extended Distributive Property.

> **The Extended Distributive Property**
> To multiply two sums, multiply each term in the first sum by each term in the second sum.

If one polynomial has m terms and the second n terms, there will be mn terms in their product. This is due to the Multiplication Counting Principle. If some of these are like terms, you should simplify the product by combining like terms.

Example 1

❶ Multiply $(5x^2 + 4x + 3)(x + 7)$.

Solution

Multiply each term in the first polynomial by each in the second. There will be six terms in the product before you simplify.

$$(5x^2 + 4x + 3)(x + 7)$$
$$= 5x^2 \cdot x + 5x^2 \cdot 7 + 4x \cdot x + 4x \cdot 7 + 3 \cdot x + 3 \cdot 7$$
$$= 5x^3 + 35x^2 + 4x^2 + 28x + 3x + 21$$

Now simplify by adding or subtracting like terms.

$$= 5x^3 + 39x^2 + 31x + 21$$

Check

Let $x = 2$. The check is left to you in the Questions.

After being simplified, the product of two polynomials can be a polynomial with fewer terms than one or both factors.

Optional Activities

Activity 1
Materials: **Teaching Aid 116**

When discussing **Questions 1 and 9,** you might also have students find the perimeters of the rectangles.
[**Question 1:** $2(w + 6) + 2(w^2 + 5w + 4) = 2w^2 + 12w + 20$
Question 9: $2(x + y + 2) + 2(x + y + 5) = 4x + 4y + 14$]

Activity 2
You might want to use the following activity after students have completed the lesson. Have students multiply several different 3-digit numbers by 1001 and describe a pattern. [To multiply a 3-digit number by 1001, first write the 3-digit number and then repeat the same 3 digits. For example, $384 \cdot 1001 = 384{,}384$ and $701 \cdot 1001 = 701{,}701$.]

Then have students use multiplication of polynomials to illustrate the pattern. That is, let $a \cdot 10^2 + b \cdot 10^1 + c$ represent any 3-digit number. Then multiply that polynomial by $10^3 + 1$.
[$(a \cdot 10^2 + b \cdot 10^1 + c)(10^3 + 1) = a \cdot 10^5 + b \cdot 10^4 + c \cdot 10^3 + a \cdot 10^2 + b \cdot 10^1 + c$]

Example 2

Multiply $x^2 - 2x + 2$ by $x^2 + 2x + 2$.

Solution

Each term of $x^2 + 2x + 2$ must be multiplied by x^2, $-2x$, and 2. There will be nine terms before you simplify.

$$\begin{aligned} & (x^2 - 2x + 2)(x^2 + 2x + 2) \\ &= x^2(x^2 + 2x + 2) - 2x(x^2 + 2x + 2) + 2(x^2 + 2x + 2) \\ &= x^4 + 2x^3 + 2x^2 - 2x^3 - 4x^2 - 4x + 2x^2 + 4x + 4 \\ &= x^4 + 4 \quad \text{Combine like terms.} \end{aligned}$$

Check

Let $x = 10$. (Ten is a nice value to use in checks, because powers of 10 are so easily calculated.) Then $x^2 - 2x + 2 = 82$ and $x^2 + 2x + 2 = 122$. Now $82 \cdot 122 = 10{,}004$, which is the value of $x^4 + 4$ when $x = 10$.

In these long problems it is particularly important to be neat and precise. Be extra careful with problems involving negatives or several variables.

Example 3

Multiply $(3x + y - 1)(x - 5y + 8)$.

Solution

❷ Each term of $x - 5y + 8$ must be multiplied by $3x$, y, and -1. At first there will be nine terms.

$$\begin{aligned} & (3x + y - 1)(x - 5y + 8) \\ &= 3x \cdot x + 3x \cdot -5y + 3x \cdot 8 + y \cdot x + y \cdot -5y + \\ & \quad y \cdot 8 + -1 \cdot x + -1 \cdot -5y + -1 \cdot 8 \end{aligned}$$

Watch the signs!

$$\begin{aligned} &= 3x^2 - 15xy + 24x + xy - 5y^2 + 8y - x + 5y - 8 \\ &= 3x^2 - 14xy + 23x - 5y^2 + 13y - 8 \end{aligned}$$

Check

A quick check of the coefficients can be found by letting all variables equal 1 in the last expression above.
Does $(3 + 1 - 1)(1 - 5 + 8) = 3 - 14 + 23 - 5 + 13 - 8$?
Does $3 \cdot 4 = 12$? Yes.
A better check requires using different values for both x and y.

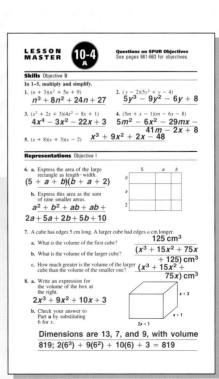

Adapting to Individual Needs

Extra Help

You might use colored chalk and underscoring to help students understand the Extended Distributive Property. For **Example 1,** you might double underline $5x^2$ in yellow or write the term in yellow to show how it has been distributed over $x + 7$. Use different colors for $4x$ and 3. Point out how both x and 7 are multiplied by each term in the trinomial.

$(\underline{\underline{5x^2}} + \underline{4x} + \mathbf{3})(x + 7) = \underline{\underline{5x^2}} \cdot x + \underline{\underline{5x^2}} \cdot 7 + \underline{4x} \cdot x + \underline{4x} \cdot 7 + \mathbf{3} \cdot x + \mathbf{3} \cdot 7$

❷ Point out that the subtractions are handled differently in **Examples 2 and 3**. In **Example 3,** each term multiplied by -1 is *added* to the sum. In **Example 2**, the subtractions are distributed.

Additional Examples

Multiply.

1. $(n - 5)(2n^2 - 3n + 7)$
   $\mathbf{2n^3 - 13n^2 + 22n - 35}$
2. $(w^2 + 4w + 6)(w^2 + w + 1)$
   $\mathbf{w^4 + 5w^3 + 11w^2 + 10w + 6}$
3. $(2a + 10b + 1)(4a + b + 11)$
   $\mathbf{8a^2 + 42ab + 26a + 10b^2 + 111b + 11}$
4. $3j - 5k + 6$ by $3j + 5k - 8$.
   $\mathbf{9j^2 - 6j - 25k^2 + 70k - 48}$

LESSON MASTER 10-4 A

Questions on SPUR Objectives
See pages 661-663 for objectives.

Skills Objective B

In 1–5, multiply and simplify.

1. $(n + 3)(n^2 + 5n + 9)$ $\quad n^3 + 8n^2 + 24n + 27$
2. $(y - 2)(5y^2 + y - 4)$ $\quad 5y^3 - 9y^2 - 6y + 8$
3. $(x^2 + 2x + 3)(4x^2 - 8x + 1)$ $\quad 4x^4 - 3x^2 - 22x + 3$
4. $(5m + x - 1)(m - 6x - 8)$ $\quad 5m^2 - 6x^2 - 29mx - 41m - 2x + 8$
5. $(x + 8)(x + 3)(x - 2)$ $\quad x^3 + 9x^2 + 2x - 48$

Representations Objective I

6. a. Express the area of the large rectangle as length · width.
   $(5 + a + b)(b + a + 2)$
   b. Express this area as the sum of nine smaller areas.
   $a^2 + b^2 + ab + ab + 2a + 5a + 2b + 5b + 10$

	5	a	b
b			
a			
2			

7. A cube has edges 5 cm long. A larger cube had edges a cm longer.
   a. What is the volume of the first cube? $\quad 125 \text{ cm}^3$
   b. What is the volume of the larger cube? $\quad (x^3 + 15x^2 + 75x + 125) \text{ cm}^3$
   c. How much greater is the volume of the larger cube than the volume of the smaller one? $\quad (x^3 + 15x^2 + 75x) \text{ cm}^3$
8. a. Write an expression for the volume of the box at the right.
   $2x^3 + 9x^2 + 10x + 3$
   b. Check your answer to Part **a** by substituting 6 for x.
   Dimensions are 13, 7, and 9, with volume 819; $2(6^3) + 9(6^2) + 10(6) + 3 = 819$

$x + 3$
$x + 1$
$2x + 1$

Notes on Questions

Teaching Aid 116 shows the rectangles for **Questions 1 and 9.**

Questions 4–7 These answers can be checked by substituting a number for the variable. The product in **Question 7** can be written as the sum and difference of two numbers: $[(x^2 + 8) + 4x][(x^2 + 8) - 4x]$. In Lesson 10-6, students will learn the Difference of Two Squares Pattern and the Perfect Square Pattern.

Question 13 Error Alert Some students may get answers of 0 or −32. They have probably made errors both in the multiplication and in the squaring. Suggest that they write out the polynomials as follows: $(x + 4)(x + 4) - (x - 16)(x - 16)$.

Question 15 This question can be tricky—there should be 16 terms in the product before it is simplified.

Follow-up for Lesson 10-4

Practice

For more questions on SPUR Objectives, use **Lesson Master 10-4A** (shown on page 635) or **Lesson Master 10-4B** (shown on pages 636–637).

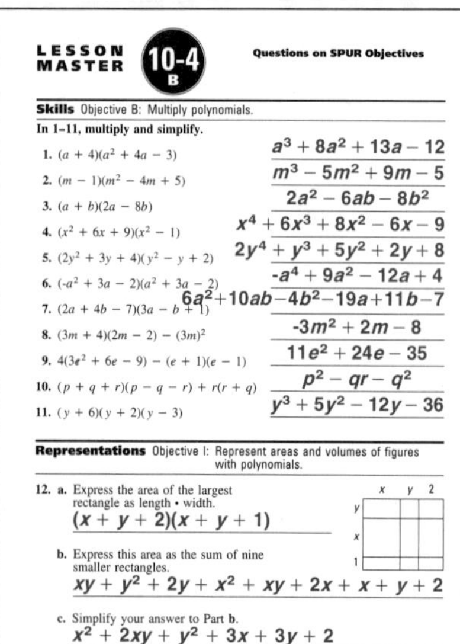

LESSON MASTER 10-4 B Questions on SPUR Objectives

Skills Objective B: Multiply polynomials.

In 1–11, multiply and simplify.

1. $(a + 4)(a^2 + 4a - 3)$ $a^3 + 8a^2 + 13a - 12$
2. $(m - 1)(m^2 - 4m + 5)$ $m^3 - 5m^2 + 9m - 5$
3. $(a + b)(2a - 8b)$ $2a^2 - 6ab - 8b^2$
4. $(x^2 + 6x + 9)(x^2 - 1)$ $x^4 + 6x^3 + 8x^2 - 6x - 9$
5. $(2y^2 + 3y + 4)(y^2 - y + 2)$ $2y^4 + y^3 + 5y^2 + 2y + 8$
6. $(-a^2 + 3a - 2)(a^2 + 3a - 2)$ $-a^4 + 9a^2 - 12a + 4$
7. $(2a + 4b - 7)(3a - b + 1)$ $6a^2 + 10ab - 4b^2 - 19a + 11b - 7$
8. $(3m + 4)(2m - 2) - (3m)^2$ $-3m^2 + 2m - 8$
9. $4(3e^2 + 6e - 9) - (e + 1)(e - 1)$ $11e^2 + 24e - 35$
10. $(p + q + r)(p - q - r) + r(r + q)$ $p^2 - qr - q^2$
11. $(y + 6)(y + 2)(y - 3)$ $y^3 + 5y^2 - 12y - 36$

Representations Objective I: Represent areas and volumes of figures with polynomials.

12. a. Express the area of the largest rectangle as length • width.
    $(x + y + 2)(x + y + 1)$
    b. Express this area as the sum of nine smaller rectangles.
    $xy + y^2 + 2y + x^2 + xy + 2x + x + y + 2$
    c. Simplify your answer to Part **b**.
    $x^2 + 2xy + y^2 + 3x + 3y + 2$

2) To multiply two sums, multiply each term in the first sum by each term in the second sum.

3) $(5 \cdot 4 + 4 \cdot 2 + 3) \cdot (2 + 7) = 31 \cdot 9 = 279$; $5 \cdot 2^3 + 39 \cdot 2^2 + 31 \cdot 2 + 21 = 40 + 156 + 62 + 21 = 279$

8) Does $[3(10) + 2 - 1] \cdot [10 - 5(2) + 8] = 3(10)^2 - 14(10)(2) + 23(10) - 5(2)^2 + 13(2) - 8$? $(30 + 1) \cdot (18 - 10) = 3(100) - 280 + 230 - 5(4) + 26 - 8$? $248 = 248$. Yes

16a) In a first degree polynomial, the highest exponent on a variable is 1. In a second degree polynomial, the highest exponent is 2. To find the degree of the product of these two polynomials, you add the exponents of unlike variables, or take the exponent of the product of like variables. In either case, the degree will be three.

636

QUESTIONS

Covering the Reading

1. a. What multiplication is pictured at the right? $(w^2 + 5w + 4)(w + 6)$
   b. Do the multiplication. $w^3 + 11w^2 + 34w + 24$

w^2 5w 4
w
6

2. State the Extended Distributive Property. See left.
3. Finish the check of Example 1. See left.

In 4–7, multiply and simplify.

4. $(y^2 + 7y + 2)(y + 6)$ $y^3 + 13y^2 + 44y + 12$
5. $(x + 1)(2x^2 + 3x - 1)$ $2x^3 + 5x^2 + 2x - 1$
6. $(m^2 + 10m + 3)(3m^2 - 4m - 2)$ $3m^4 + 26m^3 - 33m^2 - 32m - 6$
7. $(x^2 + 4x + 8)(x^2 - 4x + 8)$ $x^4 + 64$
8. Check Example 3 by letting $x = 10$ and $y = 2$. See left.
9. Find the area of the rectangle at the right, and simplify the result. $x^2 + 7x + 2xy + 7y + y^2 + 10$

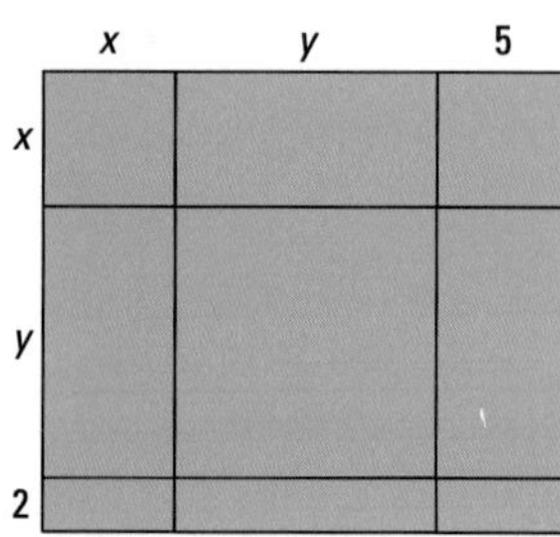

10. Multiply $(5c - 4d + 1)(c - 7d)$. $5c^2 - 39cd + 28d^2 + c - 7d$

Applying the Mathematics

11. Multiply $(n - 3)(n + 4)(2n + 5)$ by first multiplying $n - 3$ by $n + 4$. Then multiply their product by $2n + 5$. $2n^3 + 7n^2 - 19n - 60$
12. How much greater is the volume of a cube with sides of length $n + 1$ than the volume of a cube with sides of length n? $3n^2 + 3n + 1$

In 13–15, multiply and simplify.

13. $(x + 4)(x + 4) - (x - 16)^2$ $40x - 240$
14. $(a + b + c)(a + b - c) - (a + c)(a - c)$ $2ab + b^2$
15. $(m + 2n + 3p + 4q)(m - 2n - 3p - 4q)$ $m^2 - 4n^2 - 9p^2 - 16q^2 - 12np - 16nq - 24pq$
16. a. In Example 1, a 2nd degree polynomial is multiplied by a 1st degree polynomial. The product is a 3rd degree polynomial. Explain why this will always happen. See left.
    b. If a polynomial of degree m is multiplied by a polynomial of degree n, what must be true about the degree of the product? It must be of degree $m + n$.

Adapting to Individual Needs

English Language Development

You might define *to extend* as *to stretch out.* Then shake hands with students and, as you begin, emphasize how you extend your hand. Show students an extension cord or a picture of an extension ladder. Relate these ideas to the Extended Distributive Property.

Review

17. Which investment is worth more at the end of 10 years if the annual yield is 6%? Justify your answer. **Plan B**
Plan A: Deposit \$100 each year on January 2, beginning in 1990.
Plan B: Deposit \$200 every other year on January 2, beginning in 1990. *(Lesson 10-2)* **At the end of 10 years, Plan B is worth \$1437.86 and Plan A is worth \$1397.16.**

18. Suppose x is either 7, 6, 5, 3, -2, or -8. Find x, given the clues.
Clue 1: $x > -3$.
Clue 2: x is not the degree of $a^3 + 4a^2 + 7$.
Clue 3: x is not the coefficient of b^2 in the polynomial $4b^4 + 6b^2 - 3b + 9$.
Clue 4: x is not the constant term in $2a^2 + a - 3b + c + 5$.
(Lesson 10-1) **-2 or 7**

19. a. Simplify $\sqrt{200}$. $10\sqrt{2}$
b. Make up a question whose answer is $\sqrt{200}$. *(Lesson 9-7)*
Sample: What is the length of the diagonal of a square with side 10?

20. *Skill sequence.* Graph on the same coordinate grid. *(Lessons 9-1, 9-2)*
a. $y = 3x^2$ **b.** $y = 3x^2 - 2$ **c.** $y = 3x^2 + 4$ **See left.**

20)

In 21 and 22, simplify. *(Lesson 8-7)*

21. $\dfrac{14a^3b}{6ab^2}$ $\dfrac{7a^2}{3b}$ **22.** $\dfrac{-150m^5n^8}{100m^6n^3}$ $\dfrac{-3n^5}{2m}$

23. Solve $\frac{y}{9} = 9$. *(Lesson 6-1)* $y = 81$

24. Subtract $3x^2 + 5$ from $2x^2 - 3x + 40$. *(Lesson 3-6)* $-x^2 - 3x + 35$

Exploration

25. Multiply each of the polynomials in parts a–d by $x + 1$.
a. $x - 1$ $x^2 - 1$
b. $x^2 - x + 1$ $x^3 + 1$
c. $x^3 - x^2 + x - 1$ $x^4 - 1$
d. $x^4 - x^3 + x^2 - x + 1$ $x^5 + 1$
e. Look for a pattern and use it to multiply $(x + 1)(x^8 - x^7 + x^6 - x^5 + x^4 - x^3 + x^2 - x + 1)$. $x^9 + 1$
f. Predict what you think will be the product of $(x + 1)$ and $(x^{100} - x^{99} + x^{98} - x^{97} + \ldots + x^2 - x + 1)$ when simplified. Can you explain why your answer must be correct?
$x^{101} + 1$; **The degree of the product must be 100 + 1 = 101 and from parts a–e, we know that the only terms that will not cancel will be x^{101} and a 1 that will be positive since 101 is odd.**

Assessment

Quiz A quiz covering Lessons 10-1 through 10-4 is provided in the *Assessment Sourcebook.*

Written Communication Have each student write a paragraph that includes examples to explain what he or she should be able to do after completing this lesson. [Paragraphs explain the Extended Distributive Property, and examples reveal an understanding of multiplying polynomials.]

Extension

Explain that it is unusual for the product of two trinomials to be a binomial, as in **Example 2**. Your students may wonder if there are other examples such as this. Ask them to multiply: $(x^2 - ax + b)(x^2 + ax + b)$. $[x^4 + ax^3 + bx^2 - ax^3 - a^2x^2 - abx + bx^2 + bax + b^2 = x^4 + (2b - a^2)x^2 + b^2]$ Note that there will never be a term in x^3 or x. Then ask when the product will be a polynomial with only two terms. [When $a^2 = 2b$. For instance, let $a = 6$ and $b = 18$. Then $(x^2 - 6x + 18)(x^2 + 6x + 18) = x^4 + 324$.] Have students **work in groups** to decide if the product of two binomials can be a binomial. [Yes, for example, $(x + y)(x - y) = x^2 - y^2$.]

Project Update Project 6, *The Size of Products,* on page 658, relates to the content of this lesson.

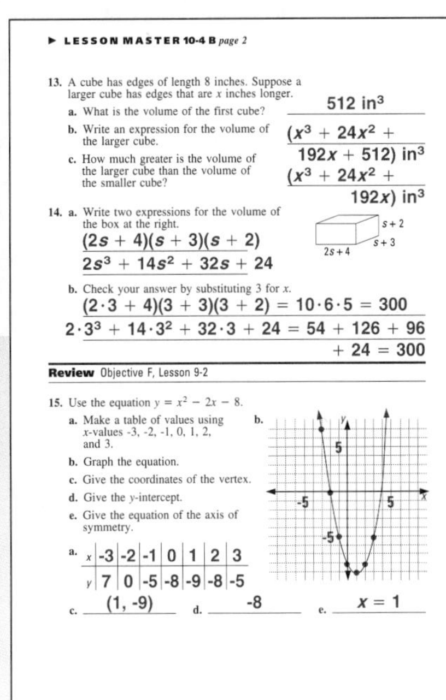
► LESSON MASTER 10-4 B *page 2*

13. A cube has edges of length 8 inches. Suppose a larger cube has edges that are x inches longer.
a. What is the volume of the first cube? 512 in^3
b. Write an expression for the volume of the larger cube. $(x^3 + 24x^2 + 192x + 512) \text{ in}^3$
c. How much greater is the volume of the larger cube than the volume of the smaller cube? $(x^3 + 24x^2 + 192x) \text{ in}^3$

14. a. Write two expressions for the volume of the box at the right.
$(2s + 4)(s + 3)(s + 2)$
$2s^3 + 14s^2 + 32s + 24$
b. Check your answer by substituting 3 for x.
$(2 \cdot 3 + 4)(3 + 3)(3 + 2) = 10 \cdot 6 \cdot 5 = 300$
$2 \cdot 3^3 + 14 \cdot 3^2 + 32 \cdot 3 + 24 = 54 + 126 + 96 + 24 = 300$

Review Objective F, Lesson 9-2

15. Use the equation $y = x^2 - 2x - 8$.
a. Make a table of values using x-values -3, -2, -1, 0, 1, 2, and 3.
b. Graph the equation.
c. Give the coordinates of the vertex.
d. Give the y-intercept.
e. Give the equation of the axis of symmetry.

a.

x	-3	-2	-1	0	1	2	3
y	7	0	-5	-8	-9	-8	-5

c. (1, -9) d. -8 e. $x = 1$

Adapting to Individual Needs

Challenge
Have students try to solve the following problem: Can the product of two polynomials, each with four terms, be a binomial? If so, give an example. [Yes; sample answer: $(x^3 - x^2 + x - 1)(x^5 + x^4 + x + 1) = x^8 - 1$]

Setting Up Lesson 10-5

Materials If you plan to use Activity 1 in *Optional Activities* on page 640, you will need heavy paper, cardboard, or plastic foam, rulers, and scissors that students can use to build a three-dimensional model.

In-class Activity

Resources

From the ***Teacher's Resource File***
- Answer Master 10-5
- Teaching Aids
  26 Graph Paper
  31 Algebra Tiles

Additional Resources
- Visuals for Teaching Aids 26, 31
- Rulers or **Geometry Templates**

Cooperative Learning We encourage the use of this activity before students read Lesson 10-5. The activity lends itself to **working in groups**. If you have time, ask the groups to share their patterns with the entire class. You will be able to relate these patterns to the FOIL algorithm in Lesson 10-5 and to the Square of a Binomial Theorem in Lesson 10-6.

Additional Answers

2.

	x	1	1	1	1
x	x^2	x	x	x	x
1	x	1	1	1	1
1	x	1	1	1	1

$x^2 + 6x + 8$

3.

	x	1	1	1
x	x^2	x	x	x
1	x	1	1	1
1	x	1	1	1

$x^2 + 5x + 6$

4.

	x	x	1
x	x^2	x^2	x
1	x	x	1
1	x	x	1
1	x	x	1
1	x	x	1
1	x	x	1

$2x^2 + 11x + 5$

5.

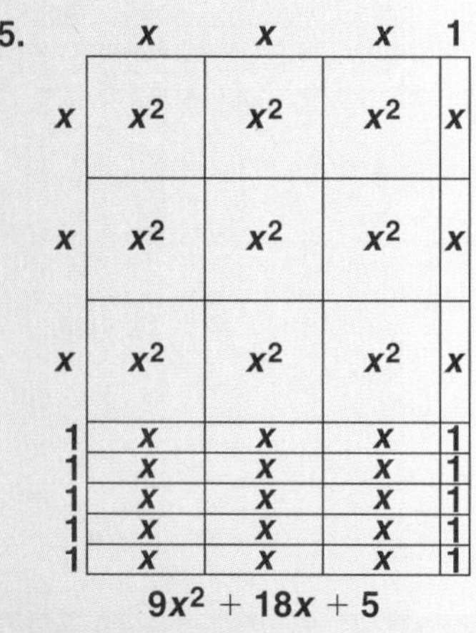

$9x^2 + 18x + 5$

IN-CLASS ACTIVITY

Introducing Lesson 10-5

Modeling Multiplication of Binomials with Algebra Tiles

Materials: Algebra tiles or ruler and graph paper

1 On a sheet of paper use tiles to form a rectangle with dimensions $2x + 1$ by $3x + 4$.

	x	x	x	1	1	1	1
x	x^2	x^2	x^2	x	x	x	x
x	x^2	x^2	x^2	x	x	x	x
1	x	x	x	1	1	1	1

a. Give the area of the rectangle as a product of its length and width. $A = (2x + 1)(3x + 4)$

b. Give the area of the rectangle by writing the polynomial represented by the tiles. $6x^2 + 11x + 4$

c. Use your answers to parts **a** and **b** to complete the following equation:
$(2x + 1)(3x + 4) = \underline{\ ?\ }$ $6x^2 + 11x + 4$

In 2–5, use algebra tiles to find each product. **See margin.**

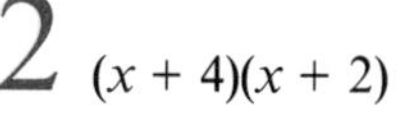

2 $(x + 4)(x + 2)$

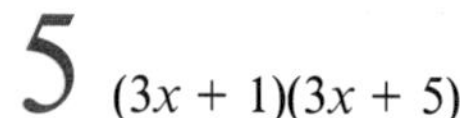

3 $(x + 3)(x + 2)$

4 $(2x + 1)(x + 5)$

5 $(3x + 1)(3x + 5)$

6 ***Draw conclusions.*** What patterns do you notice in the answers to Questions 1–5? **Answers will vary.**

7 a. On a sheet of paper use tiles to form a square with a side of length $2x + 3$. Give the area of the square using the formula $A = s^2$. $A = (2x + 3)^2$

	2x + 3				
x	x^2	x^2	x	x	x
2x + 3 x	x^2	x^2	x	x	x
1	x	x	1	1	1
1	x	x	1	1	1
1	x	x	1	1	1
	x	x	1	1	1

b. Give the area of the square by writing the polynomial represented by the tiles.

c. Use your answers to parts **a** and **b** to complete the following:
$(2x + 3)^2 = \underline{\ ?\ }$. **b and c)** $4x^2 + 12x + 9$

In 8 and 9, use tiles to find each product. **See margin.**

8 $(x + 4)^2$

9 $(3x + 1)^2$

8.

	x	1	1	1	1
x	x^2	x	x	x	x
1	x	1	1	1	1
1	x	1	1	1	1
1	x	1	1	1	1
1	x	1	1	1	1

$x^2 + 8x + 16$

9.

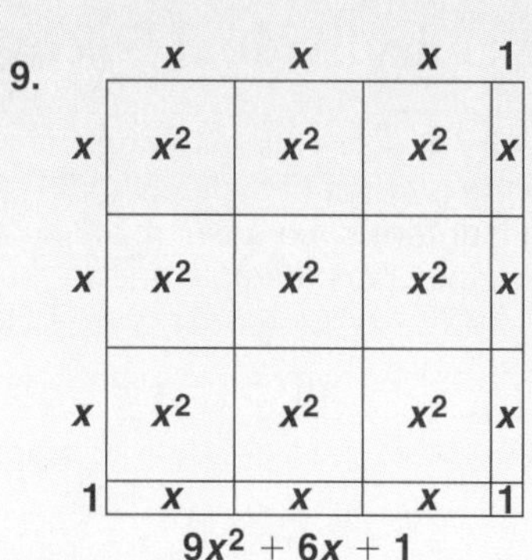

$9x^2 + 6x + 1$

638

LESSON

10-5

Multiplying Binomials

Painstaking calculation. *To find the total area, the amount of glass required for this stained-glass window, you could use a product of binomials. This kind of calculation is done in Example 3.*

Multiplying Two-Digit Numbers

Every multiplication with two two-digit numbers can be considered as the multiplication of two binomials. Consider 94 times 65. At the left is the way many people do it by hand. At the right the numbers have been rewritten as (90 + 4) and (60 + 5), and the multiplication has been done using the Extended Distributive Property.

Familiar Algorithm:

$$\begin{array}{r} 65 \\ \times\ 94 \\ \hline 260 \\ 585\ \ \\ \hline 6110 \end{array}$$

Extended Distributive Property:

$$\begin{aligned} &(90 + 4) \cdot (60 + 5) \\ &= 90 \cdot 60 + 90 \cdot 5 + 4 \cdot 60 + 4 \cdot 5 \\ &= 5400 + 450 + 240 + 20 \\ &= 6110 \end{aligned}$$

The Extended Distributive Property explains why the familiar algorithm works. The 260 at left is the sum of 4 · 60 and 4 · 5. The 585 is really 5850 with the 0 understood; it is the sum of 90 · 5 and 90 · 60.

Using Area to Picture the Multiplication of Two Binomials

The dimensions of the rectangle at the right are the binomials $(a + b)$ and $(c + d)$.

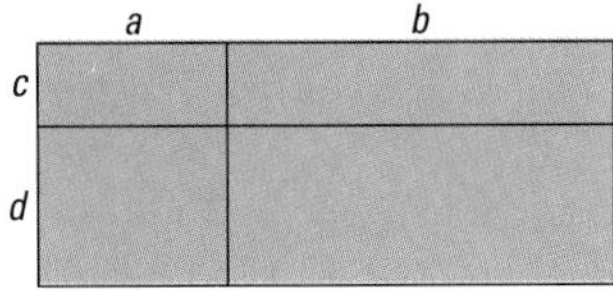

Its area is $(a + b) \cdot (c + d)$. But the area of the rectangle also must equal the sum of the areas of the four small rectangles inside it.

	a	b
c	ac	bc
d	ad	bd

The sum of the areas of the four small rectangles = $ac + ad + bc + bd$. So,

$$(a + b) \cdot (c + d) = ac + ad + bc + bd.$$

Lesson 10-5

Objectives

C Multiply two binomials.

I Represent the product of two binomials as an area.

Resources

From the ***Teacher's Resource File***

- Lesson Master 10-5A or 10-5B
- Answer Master 10-5
- Teaching Aids
  108 Warm-up
  117 Questions 1, 2, 13, and 14

Additional Resources

- Visuals for Teaching Aids 108, 117
- GraphExplorer or other automatic graphers

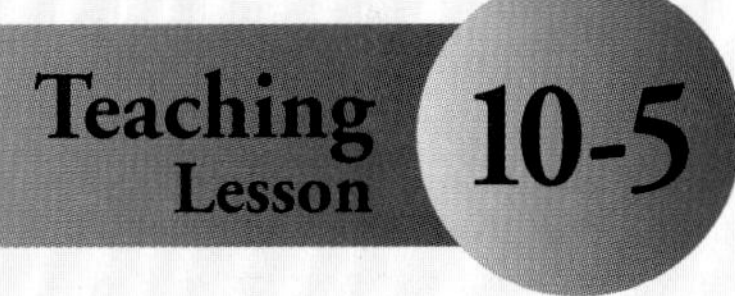

Warm-up

Write each expression as a trinomial.

1. $2y^2 + 3y - 4y - 21$
   $\mathbf{2y^2 - y - 21}$
2. $24x^2 + 30x + 4x + 5$
   $\mathbf{24x^2 + 34x + 5}$
3. $5y^2 + 5xy - xy - x^2$
   $\mathbf{5y^2 + 4xy - x^2}$
4. $18x^2 - 27xy + 8yx - 12y^2$
   $\mathbf{18x^2 - 19xy - 12y^2}$
5. $3t \cdot 3t + 7 \cdot 3t + 7 \cdot 3t + 7 \cdot 7$
   $\mathbf{9t^2 + 42t + 49}$
6. $2y \cdot 2y + 2y \cdot 5 + 3 \cdot 2y + 3 \cdot 5$
   $\mathbf{4y^2 + 16y + 15}$

Lesson 10-5 Overview

Broad Goals This lesson discusses the important special case of multiplying two polynomials, namely the case of multiplying a binomial by a binomial.

Perspective The most common mistake students make in multiplying $(a + b)$ and $(c + d)$ is to miss the two middle terms ad and bc. For instance, some students say $(x + 3)(x + 7) = x^2 + 21$. This lesson provides two tools to correct this mistake.

One is the FOIL mnemonic. The four letters of FOIL help students expect four terms in the product. The other tool is the visual image of the area model which shows the rectangle with dimensions $(a + b)$ and $(c + d)$ split into four parts.

Some teachers love the acronym FOIL. Other teachers detest it. They argue that because we have done the general case first, an efficient method is already in hand.

Our view is that FOIL is quite a useful mnemonic for remembering this particular skill. However, it is not a substitute for knowing the general properties that underlie the skill.

Notes on Reading

Multicultural Connection The process of staining glass can be traced back to ancient times in China and in Egypt. In the 1100s, Islamic countries used pieces of colored glass set in wooden or stucco frames. Stained-glass windows in Gothic cathedrals picture events in history and religious teachings. Some depict the coat of arms or the figure of the window's donor. The oldest surviving stained-glass windows in Europe can be found in the Augsburg Cathedral in Germany.

Some students may be a little confused because the product of two binomials in this lesson is seen to have four terms, but the final answer to each of the examples has less than four terms. Point out that many of the questions in Lesson 10-4 had a similar characteristic. Emphasize that each of these answers was derived by simplifying a previous line in which there were four terms. If this does not clear up the confusion, try the following sequence of multiplications. Note that the first three multiplications have four terms in the product, the next two have three terms, and the last has two terms.

$(a + b)(c + d) = ac + ad + bc + bd$
$(a + b)(a + d) = a^2 + ab + ad + bd$
$(a + b)(a + 3) = a^2 + 3a + ab + 3b$
$(a + 5)(a + 3) = a^2 + 8a + 15$
$(a + 5)(a - 7) = a^2 - 2a - 35$
$(a + 5)(a - 5) = a^2 - 25$

When discussing the examples, tell students that they will be expected to determine the factors from the product in Chapter 12. In this regard, **Example 5** is particularly important. It is an instance of a general pattern that is found in the next lesson.

The FOIL Algorithm

Another way to show the pattern above is true is to think of $(c + d)$ as a chunk and to distribute it over $(a + b)$ as follows:

$$(a + b) \cdot (\mathbf{c} + \mathbf{d}) = a(\mathbf{c} + \mathbf{d}) + b(\mathbf{c} + \mathbf{d}).$$

Now apply the Distributive Property twice more.

$$= ac + ad + bc + bd$$

Notice that the product of two binomials has four terms. To find them, multiply each term in the first binomial by each term in the second binomial. The face below may help you remember.

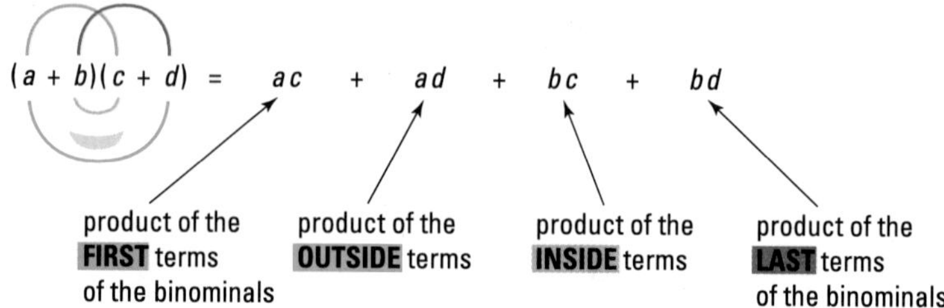

Because the product is composed of the First, Outside, Inside, and Last terms, the above pattern is sometimes called the **FOIL algorithm.** (Recall that an **algorithm** is a step-by-step recipe or procedure.)

Example 1

Multiply $(m + 4)(m + 3)$.

Solution 1

F O I L

$$(m + 4)(m + 3) = m \cdot m + m \cdot 3 + 4 \cdot m + 4 \cdot 3$$
$$= m^2 + 3m + 4m + 12$$
$$= m^2 + 7m + 12$$

Solution 2

Draw a rectangle with sides of $m + 4$ and $m + 3$. Find its area by dividing it into smaller rectangles. The areas of the smaller rectangles are m^2, $4m$, $3m$, and 12. So

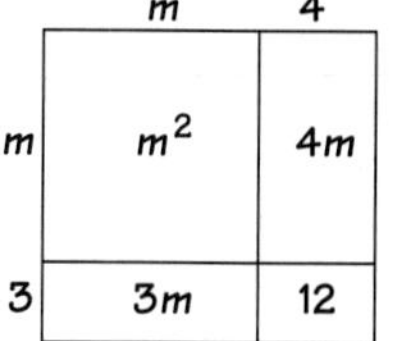

$$(m + 4)(m + 3) = m^2 + 4m + 3m + 12$$
$$= m^2 + 7m + 12.$$

Check

Test a special case. Let $m = 10$. Then $m + 4 = 14$, $m + 3 = 13$, and $m^2 + 7m + 12 = 100 + 70 + 12 = 182$. It checks, because $14 \cdot 13 = 182$.

Optional Activities

Activity 1 Using Physical Models
Materials: Heavy paper, cardboard, or plastic foam; rulers or **Geometry Templates;** scissors

You might want to use this activity before students answer **Question 33.** Have small **groups of students** use heavy paper, cardboard, or plastic foam to build a model for the figure in **Question 33**. The model should be made of eight parts; each part should be labeled to show its volume. The model can be used to illustrate $(a + b)(c + d)(e + f)$.

The FOIL algorithm applies to all binomials, including those with subtraction. You must, however, be careful with signs.

Example 2

Multiply $(x - 3)(x - 7)$.

Solution

Think of $x - 3$ as $x + -3$, and $x - 7$ as $x + -7$. Now use the FOIL algorithm.

$$(x + -3)(x + -7) = x^2 + -7x + -3x + 21$$
$$= x^2 + -10x + 21$$
$$\text{So } (x - 3)(x - 7) = x^2 - 10x + 21.$$

Check 1

Substitute a value for x in the original and final expression. We use $x = 5$.
$(x - 3)(x - 7) = (5 - 3)(5 - 7) = 2(-2) = -4$
$x^2 - 10x + 21 = 5^2 - 10 \cdot 5 + 21 = 25 - 50 + 21 = -4$.

It checks.

Check 2

If $(x - 3)(x - 7) = x^2 - 10x + 21$, then the graphs of $y = (x - 3)(x - 7)$ and $y = x^2 - 10x + 21$ should be identical. Below we show the output from our automatic grapher.

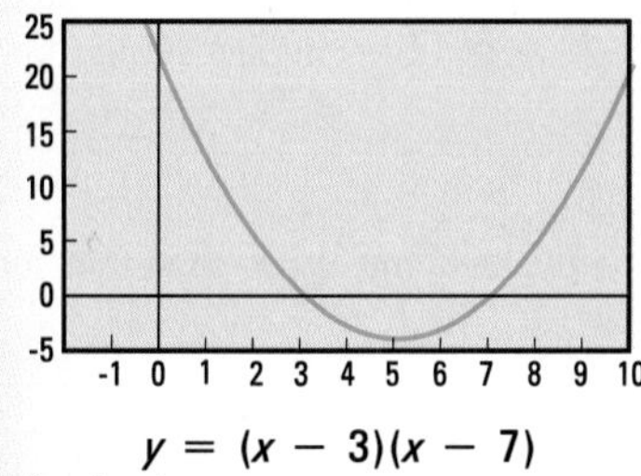
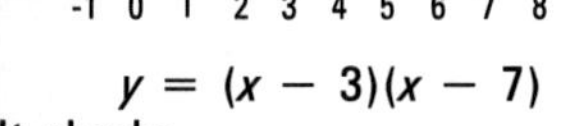

$y = (x - 3)(x - 7)$

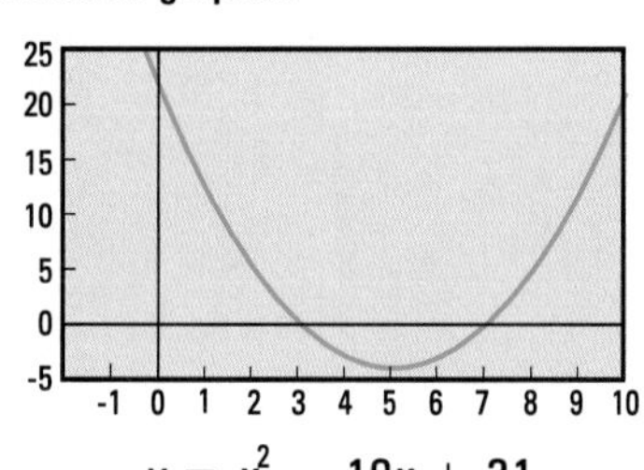

$y = x^2 - 10x + 21$

It checks.

The FOIL algorithm applies to all real numbers *a, b, c,* and *d*. You may substitute any algebraic expression for the numbers.

Additional Examples

1. Multiply $(3n + 10)(7n + 2)$.
   $\mathbf{21n^2 + 76n + 20}$
2. Multiply $(y - 5)(y - 8)$.
   $\mathbf{y^2 - 13y + 40}$
3. A 8″-by-10″ photograph is to be surrounded by a mat with width *w*. Sketch the photo and mat. What is the total area of the photo and mat?
   $\mathbf{4w^2 + 36w + 80}$ **square inches**

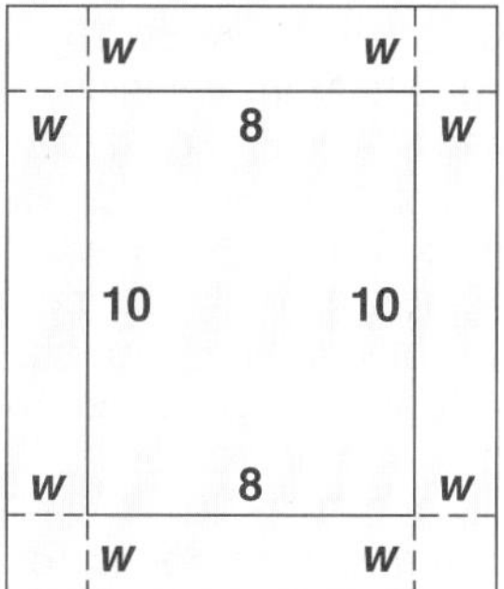

4. Multiply $(5a - 6x)(3a - 2x)$.
   $\mathbf{15a^2 - 28ax + 12x^2}$
5. Multiply $(7m + 2y)(7m - 2y)$.
   $\mathbf{49m^2 - 4y^2}$

Optional Activities

Activity 2
Materials: **Teaching Aid 31**

After finishing the lesson, you might have students **work in groups** to find quotients of polynomials using algebra tiles.

1. $\frac{x^2 + 5x + 4}{x + 1}$ $[x + 4]$
2. $\frac{2x^2 + 7x + 6}{2x + 3}$ $[x + 2]$
3. $\frac{6x^2 + 7x + 2}{2x + 1}$ $[3x + 2]$

If necessary, help students get started with Question 1. Explain that $x^2 + 5x + 4$ tells them that the rectangle is to contain one "x^2" tile, 5 "x" tiles, and 4 "unit" tiles, and $x + 1$ tells them the dimension of one side of the rectangle.

Some students might be interested in the algorithm for dividing polynominals shown at the right. Note how similar the procedure is to long division.

$$\begin{array}{r|l} x + 4 & \text{Divide } x^2 \text{ by } x. \text{ Write} \\ x + 1\,\overline{)x^2 + 5x + 4} & x \text{ in the quotient.} \\ \underline{x^2 + x} & \text{Multiply } x + 1 \text{ by } x. \\ 4x + 4 & \text{Subtract to get } 4x. \\ \underline{4x + 4} & \text{Bring down the 4.} \\ 0 & \text{Divide } 4x \text{ by } x. \text{ Write 4} \\ & \text{in the quotient. Multiply} \\ & x + 1 \text{ by 4. Subtract.} \end{array}$$

Notes on Questions

Teaching Aid 117 contains the diagrams for **Questions 1, 2, 13, and 14.**

Questions 7–12 Error Alert Watch for students who multiply only the first terms and the last terms. Remind them that the FOIL method involves finding four products, and that some of them might be like terms which can be combined. Except for **Question 10,** these situations are not easily represented by an area model.

Question 13 There are several possible strategies that students might use. They can find the total area of the deck and pool and subtract the area of the pool: $(2d + 20)(2d + 40) - 20 \cdot 40$. Or, they can divide the deck into non-overlapping rectangular pieces and find the sum of those areas—for example, $2d(20 + 2d) + 2d \cdot 40$.

Example 3

A 5″ by 7″ photo is to be surrounded on all sides by a mat of width w inches. Find an expression for the combined area of the photo and mat.

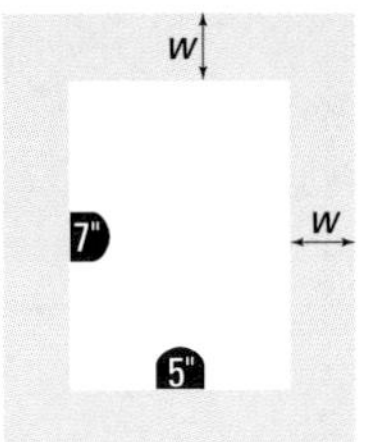

Solution

The region to be framed is a rectangle with base $5 + 2w$ and height $7 + 2w$. The area is $(5 + 2w)(7 + 2w)$. Think of $2w$ as a chunk and apply the FOIL algorithm.

$$\begin{aligned}(5 + 2w)(7 + 2w) &= 5 \cdot 7 + 5 \cdot 2w + 2w \cdot 7 + 2w \cdot 2w \\ &= 35 + 10w + 14w + 4w^2 \\ &= 35 + 24w + 4w^2\end{aligned}$$

Check

You are asked to check this result in the Questions.

Example 4 shows the FOIL algorithm in a situation where there is more than one variable in each binomial.

Example 4

Multiply $(7x + 5y)(x - 4y)$.

Solution

$$\begin{aligned}(7x + 5y)(x - 4y) &= 7x \cdot x + 7x \cdot -4y + 5y \cdot x + 5y \cdot -4y \\ &= 7x^2 - 28xy + 5yx - 20y^2\end{aligned}$$

Notice that $-28xy$ and $5yx$ are like terms. Combine them.

$$= 7x^2 - 23xy - 20y^2$$

Check

Substitute two values, one for x, the other for y. We let $x = 2$ and $y = 3$.
Does $(7 \cdot 2 + 5 \cdot 3)(2 - 4 \cdot 3) = 7 \cdot 2^2 - 23 \cdot 2 \cdot 3 - 20 \cdot 3^2$?
Does $29 \cdot -10 = 28 - 138 - 180$?

Yes, each side equals -290.

The product of two binomials generally has four terms, as in FOIL. After simplification, the products in Examples 1–4 have three terms each. Sometimes, after simplification, there are only two terms.

Example 5

Multiply $4x - 3$ by $4x + 3$.

Solution

$$\begin{aligned}(4x - 3)(4x + 3) &= 16x^2 + 12x - 12x - 9 \\ &= 16x^2 - 9\end{aligned}$$

Adapting to Individual Needs

Extra Help

In Lesson 10-1, we stated that the normal way of writing whole numbers in base 10 is shorthand for a polynomial. That is, $58 = 5x + 8$ when $x = 10$. This idea can be reinforced by comparing multiplication of two-digit numbers with multiplication of binomials in vertical form.

58	=	50 + 8	$5x + 8$
× 43	=	40 + 3	$4x + 3$
174	=	150 + 24	$15x + 24$
2320	=	2000 + 320	$20x^2 + 32x$
2494	=	2000 + 470 + 24	$20x^2 + 47x + 24$

QUESTIONS

Covering the Reading

1. Refer to the largest rectangle below.
a. What are its dimensions? $x + 4, x + 3$ **b.** What is its area? $x^2 + 7x + 12$

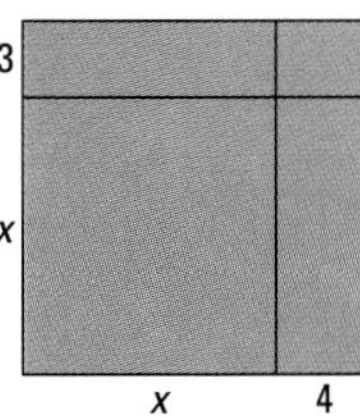

2. a. What multiplication is pictured below? $(2w + 6)(w + 7)$
b. Do the multiplication. $2w^2 + 20w + 42$

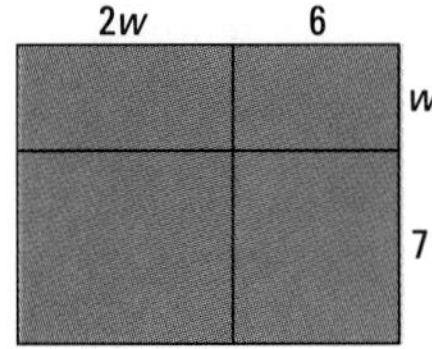

3. In the FOIL algorithm, what do the letters F, O, I, and L stand for?
First, Outside, Inside, Last

In 4 and 5, **a.** multiply; **b.** draw a picture to justify your answer.

4. $(a + b)(c + d)$ $ac + ad + bc + bd$ **5.** $(n + 4)(n + 1)$ $n^2 + 5n + 4$
See left for 4b and 5b.

4b)

	a	b
c	ac	bc
d	ad	bd

5b)

	n	4
n	n^2	4n
1	n	4

6. Refer to Example 3. Check the answer by letting $w = 2$.
Does $(5 + 2 \cdot 2)(7 + 2 \cdot 2) = 35 + 24 \cdot 2 + 4 \cdot 2^2$? $9 \cdot 11 = 35 + 48 + 16$? Yes, it checks.

In 7–12, multiply and simplify.

7. $(x - 3)(x - 4)$ $x^2 - 7x + 12$
8. $(a - 10)(a + 7)$ $a^2 - 3a - 70$
9. $(3m + 2n)(7m - 6n)$ $21m^2 - 4mn - 12n^2$
10. $(k + 3)(9k + 8)$ $9k^2 + 35k + 24$
11. $(a + 6)(a - 6)$ $a^2 - 36$
12. $(2x - 3y)(2x + 3y)$ $4x^2 - 9y^2$

13. A 20-ft by 40-ft pool is surrounded by a deck d ft wide.
a. What are the outer dimensions of the deck? $40 + 2d$ ft, $20 + 2d$ ft
b. What is the area of the deck? $4d^2 + 120d$ ft^2

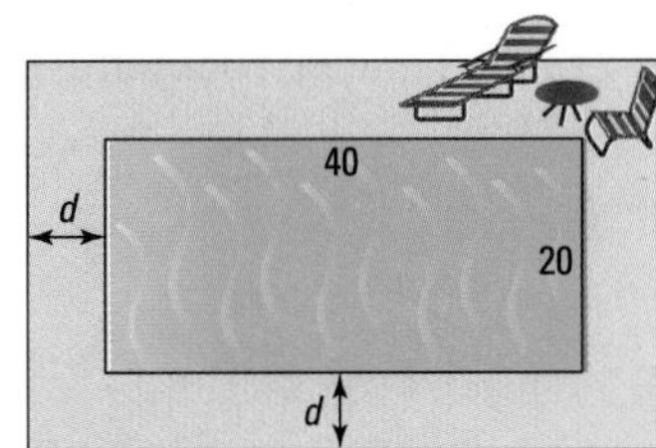

Adapting to Individual Needs

Challenge

Have students solve the following problem: A rectangular swimming pool is 3 yards longer than it is wide. The outer edge is lined with square paving stones, each 1 yard on a side. If 30 paving stones are needed to surround the pool, what are the dimensions of the pool? [8 yards by 5 yards]

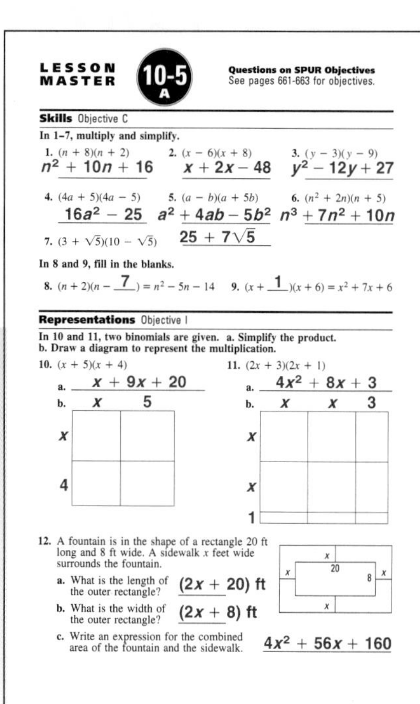

LESSON MASTER 10-5 A

Questions on SPUR Objectives
See pages 661-663 for objectives.

Skills Objective C

In 1–7, multiply and simplify.

1. $(n + 8)(n + 2)$ $n^2 + 10n + 16$
2. $(x - 6)(x + 8)$ $x + 2x - 48$
3. $(y - 3)(y - 9)$ $y^2 - 12y + 27$
4. $(4a + 5)(4a - 5)$ $16a^2 - 25$
5. $(a - b)(a + 5b)$ $a^2 + 4ab - 5b^2$
6. $(n^2 + 2n)(n + 5)$ $n^3 + 7n^2 + 10n$
7. $(3 + \sqrt{5})(10 - \sqrt{5})$ $25 + 7\sqrt{5}$

In 8 and 9, fill in the blanks.

8. $(n + 2)(n - \underline{7}) = n^2 - 5n - 14$
9. $(x + \underline{1})(x + 6) = x^2 + 7x + 6$

Representations Objective I

In 10 and 11, two binomials are given. a. Simplify the product. b. Draw a diagram to represent the multiplication.

10. $(x + 5)(x + 4)$
a. $x + 9x + 20$
b.

	x	5
x		
4		

11. $(2x + 3)(2x + 1)$
a. $4x^2 + 8x + 3$
b.

	x	x	3
x			
x			
1			

12. A fountain is in the shape of a rectangle 20 ft long and 8 ft wide. A sidewalk x feet wide surrounds the fountain.
a. What is the length of the outer rectangle? $(2x + 20)$ ft
b. What is the width of the outer rectangle? $(2x + 8)$ ft
c. Write an expression for the combined area of the fountain and the sidewalk. $4x^2 + 56x + 160$

Notes on Questions

Question 15 This question helps set up the next lesson.

Question 16 This question applies the Multiplication Counting Principle: Number of twosomes = (number of freshmen and sophomores) · (number of juniors and seniors). **Part a** does not appear to have anything to do with binomials, but the answer to **part b** is very instructive. Either the answer is $(f + p)(j + s)$, using the Multiplication Counting Principle once, or the answer is $fj + fs + pj + ps$ (that is, number of freshmen-junior pairs + number of freshmen-senior pairs, and so on).

Question 25 Error Alert If students think they need a numerical answer, point out that they are solving for x in terms of a, b, c, and d. Also note the advantage of chunking.

Question 27b Ask students how they would respond if the rocket were launched at the edge of a cliff. Then the height at 7 seconds (-106 feet) would represent the rocket's position 106 feet below the point of firing.

Question 31 This question previews the Zero Product Property that is covered in Chapter 12.

Question 33 You might want to use Activity 1 in *Optional Activities* on page 640 and relate it to this question.

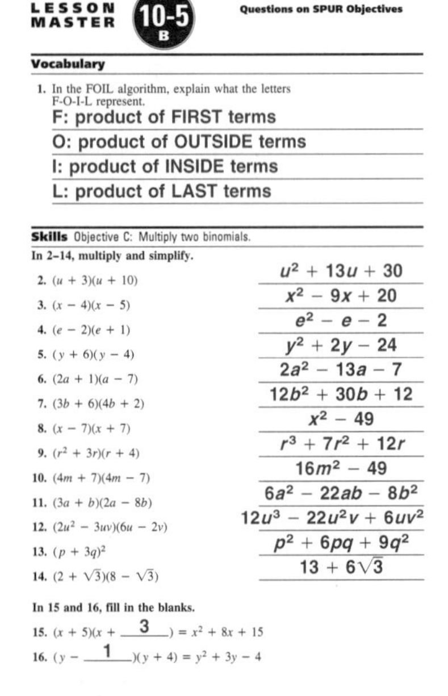

LESSON MASTER 10-5 B Questions on SPUR Objectives

Vocabulary

1. In the FOIL algorithm, explain what the letters F-O-I-L represent.
F: product of FIRST terms
O: product of OUTSIDE terms
I: product of INSIDE terms
L: product of LAST terms

Skills Objective C: Multiply two binomials.

In 2–14, multiply and simplify.

2. $(u + 3)(u + 10)$ — $u^2 + 13u + 30$
3. $(x - 4)(x - 5)$ — $x^2 - 9x + 20$
4. $(e - 2)(e + 1)$ — $e^2 - e - 2$
5. $(y + 6)(y - 4)$ — $y^2 + 2y - 24$
6. $(2a + 1)(a - 7)$ — $2a^2 - 13a - 7$
7. $(3b + 6)(4b + 2)$ — $12b^2 + 30b + 12$
8. $(x - 7)(x + 7)$ — $x^2 - 49$
9. $(r^2 + 3r)(r + 4)$ — $r^3 + 7r^2 + 12r$
10. $(4m + 7)(4m - 7)$ — $16m^2 - 49$
11. $(3a + b)(2a - 8b)$ — $6a^2 - 22ab - 8b^2$
12. $(2u^2 - 3uv)(6u - 2v)$ — $12u^3 - 22u^2v + 6uv^2$
13. $(p + 3q)^2$ — $p^2 + 6pq + 9q^2$
14. $(2 + \sqrt{3})(8 - \sqrt{3})$ — $13 + 6\sqrt{3}$

In 15 and 16, fill in the blanks.

15. $(x + 5)(x + \underline{\ 3\ }) = x^2 + 8x + 15$
16. $(y - \underline{\ 1\ })(y + 4) = y^2 + 3y - 4$

Shown are members of an organization called S.A.F.E. (St. Agatha Family Empowerment), which offers teens community-service jobs. These teens are beautifying a neighborhood garden in Chicago.

19a, b)

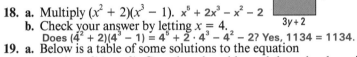

Applying the Mathematics

14. A rectangular public garden measuring x-ft by y-ft has a sidewalk surrounding the lawn. The sidewalk is 6 ft wide.
 a. What are the dimensions of the lawn? $(y - 12)$ ft, $(x - 12)$ ft
 b. What is the area of the lawn? $xy - 12y - 12x + 144$ ft^2

15. Write $(3x + 4)^2$ as a polynomial. (Hint: First convert the 2nd power to multiplication.) $9x^2 + 24x + 16$

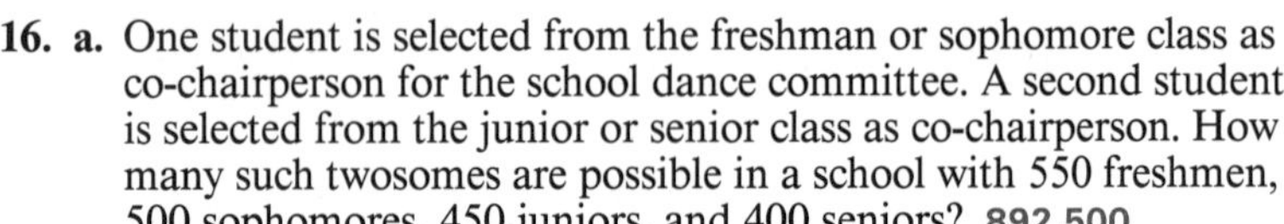

16. a. One student is selected from the freshman or sophomore class as co-chairperson for the school dance committee. A second student is selected from the junior or senior class as co-chairperson. How many such twosomes are possible in a school with 550 freshmen, 500 sophomores, 450 juniors, and 400 seniors? 892,500
 b. Repeat part **a** if there are f freshmen, p sophomores, j juniors, and s seniors. $(f + p)(j + s)$ or $fj + fs + pj + ps$

17. Write the area of the square at the right as a trinomial. $9y^2 + 12y + 4$

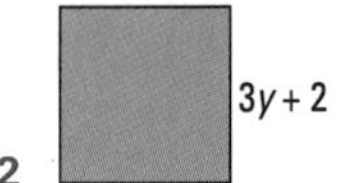

18. a. Multiply $(x^2 + 2)(x^3 - 1)$. $x^5 + 2x^3 - x^2 - 2$
 b. Check your answer by letting $x = 4$.
 Does $(4^2 + 2)(4^3 - 1) = 4^5 + 2 \cdot 4^3 - 4^2 - 2$? Yes, 1134 = 1134.

19. a. Below is a table of some solutions to the equation $y = (x + 2)(x - 3)$. Complete the table, and then plot the points. See left for graph.

x	-4	-3	-2	-1	0	1	2	3	4
y	14	6	0	-4	-6	-6	-4	0	6

 b. Make a table of solutions to the equation $y = x^2 - x - 6$ for integer values of x from -4 to 4. Plot the points. (same as part a)
 c. What is the relation between the tables and graphs in parts **a** and **b**? Use the content of this lesson to explain that relation. They are the same; the equation in part b is the expansion of the equation in part a.

20. a. Multiply $(\sqrt{5} - 2)(\sqrt{5} + 3)$ using the FOIL algorithm.
 b. Use your calculator to evaluate the expression in part **a**.
 c. Are your answers in parts **a** and **b** equal? How can you tell?
 a) $\sqrt{5} - 1$ b) ≈ 1.24 c) Yes; $0.236 \cdot 5.236 \approx \sqrt{5} - 1 \approx 1.2$.

21. A rectangular box has dimensions x, $x + 1$, and $x + 2$. What is its volume? $x^3 + 3x^2 + 2x$

In 22 and 23, work backwards. Fill in the blanks with the missing number.

22. $(x + 5)(x + \underline{\ ?\ }) = x^2 + \underline{\quad} x + 35$ 7, 12

23. $(y - \underline{\ ?\ })(y + 5) = y^2 + \underline{\quad} y - 10$ 2, 3

Review

24. Bill has 3 blue shirts, 2 red shirts, and 5 green shirts. He can wear them with 4 pairs of blue jeans and 2 pairs of shorts. **a.** How many outfits are possible? **b.** How is this related to the multiplication of polynomials? *(Lesson 10-4)* **a) 60 b) This problem can be viewed as the multiplication of the binomial (4 + 2) by the trinomial (3 + 2 + 5).**

25. If $3(a + b - c + d) + 2(a + b - c + d) = 5x$, solve for x. *(Lesson 10-3)* $a + b - c + d$

26. Multiply $3 \cdot 10^5 + 6 \cdot 10^4 + 2 \cdot 10^3 + 10^1 + 9$ by $8 \cdot 10^3 + 9 \cdot 10^2 + 8 \cdot 10 + 1$. *(Lessons 10-1, 10-3)* **3,251,292,639**

3 . . . 2 . . . 1 . . . liftoff! *The flight of a model rocket consists of several phases. Initially, the flight path is straight up, powered by a solid fuel. Next, it begins to arch as it coasts without fuel. Lastly, a parachute is released as the rocket descends to the ground.*

In 27 and 28, use the formula for the height h (in feet) of a model rocket t seconds after being fired straight up from a pad 6 feet off the ground, $h = 6 + 96t - 16t^2$.

27. a. Find the height of the rocket after 7 seconds. **-106 ft**

b. What does this answer mean? *(Lesson 9-4)* **The rocket hit the ground before seven seconds elapsed.**

28. The rocket reaches its maximum height at 3 seconds. What is its maximum height? *(Lesson 9-2)* **150 ft**

29. Write an equation for the line which passes through the points (-3, 7) and (-9, 4). *(Lesson 7-6)* $y = \frac{1}{2}x + \frac{17}{2}$

30. Find the probability that a point selected at random from the big square at the right will be in the shaded area. *(Lesson 6-6)* **0.91**

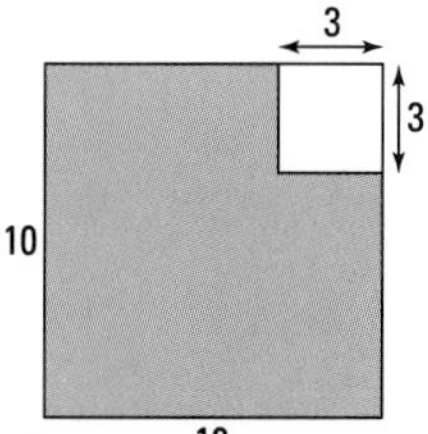

31. Solve $42(512 - x) = 0$ in your head. *(Lesson 5-8)* $x = 512$

32. Evaluate $(x + 1)(x - 3)(x + 4)$ in your head when $x = 3$. *(Lesson 2-2)* **0**

Exploration

33. The largest rectangular solid below has length $(a + b)$, width $(c + d)$, and height $(e + f)$. Give at least two ways of computing the volume of the box.

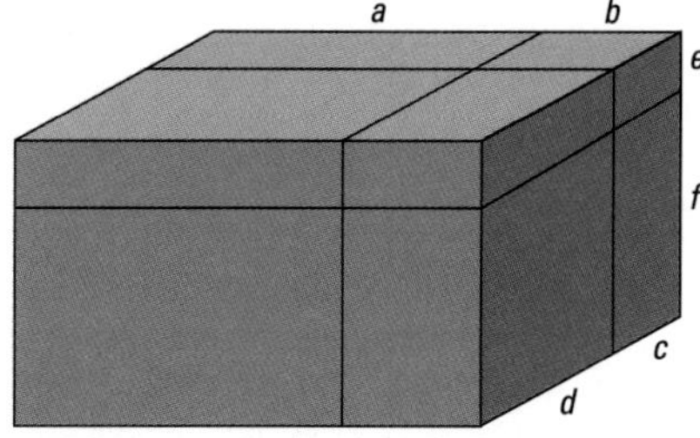

Sample: $(a + b)(c + d)(e + f) = ace + acf + ade + adf + bce + bcf + bde + bdf$. **Multiply the product of** $(a + b)(c + d)$ **by** $(e + f)$ **or multiply the product of** $(c + d)(e + f)$ **by** $(a + b)$.

Follow-up for Lesson 10-5

Practice

For more questions on SPUR Objectives, use **Lesson Master 10-5A** (shown on page 643) or **Lesson Master 10-5B** (shown on pages 644–645).

Assessment

Oral Communication Ask students to name two different ways to check multiplication of binomials. [Students choose between testing a special case and using an automatic grapher.]

Extension

Have students combine their knowledge of multiplying polynomials and solving equations to solve these equations.

1. $\frac{12}{x - 3} = 8$ [4.5]
2. $\frac{7}{x + 2} = x - 4$ [-3, 5]
3. $\frac{5x}{x - 2} = x - 6$ [1, 12]
4. $\frac{x - 7}{2x + 1} = x + 5$ [-2, -3]

Project Update Project 6, *The Size of Products,* on page 658, relates to the content of this lesson.

Setting Up Lesson 10-6

In **Example 5** of this lesson, the product is an instance of the Difference of Two Squares Pattern. In **Question 15**, the product is an instance of the Perfect Square Pattern. Both patterns are studied in Lesson 10-6.

If you plan to use the visual demonstrations shown in *Optional Activities* on pages 647–649, you might want to look over the visuals for **Teaching Aid 118** before using them.

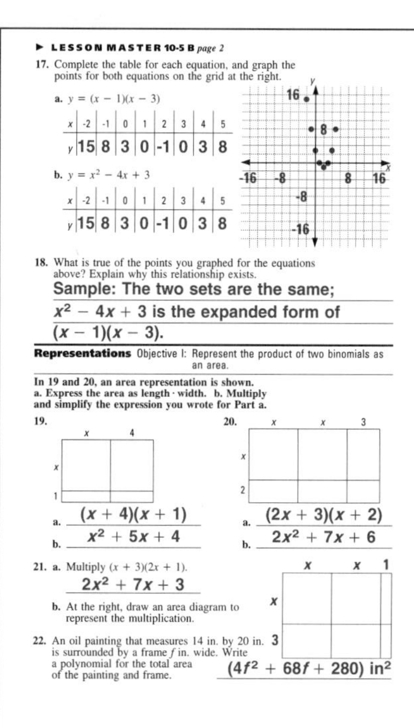

► **LESSON MASTER 10-5 B** *page 2*

17. Complete the table for each equation, and graph the points for both equations on the grid at the right.

a. $y = (x - 1)(x - 3)$

x	-2	-1	0	1	2	3	4	5
y	15	8	3	0	-1	0	3	8

b. $y = x^2 - 4x + 3$

x	-2	-1	0	1	2	3	4	5
y	15	8	3	0	-1	0	3	8

18. What is true of the points you graphed for the equations above? Explain why this relationship exists.
Sample: The two sets are the same; $x^2 - 4x + 3$ is the expanded form of $(x - 1)(x - 3)$.

Representations Objective I: Represent the product of two binomials as an area.

In 19 and 20, an area representation is shown. a. Express the area as length · width. b. Multiply and simplify the expression you wrote for Part a.

19.

a. $(x + 4)(x + 1)$
b. $x^2 + 5x + 4$

20.

a. $(2x + 3)(x + 2)$
b. $2x^2 + 7x + 6$

21. a. Multiply $(x + 3)(2x + 1)$.
$2x^2 + 7x + 3$

b. At the right, draw an area diagram to represent the multiplication.

22. An oil painting that measures 14 in. by 20 in. is surrounded by a frame f in. wide. Write a polynomial for the total area of the painting and frame. $(4f^2 + 68f + 280)$ in^2

Lesson 10-6

Objectives

C Multiply two binomials.
D Expand squares of binomials.
I Represent the square of a binomial as an area.

Resources

From the ***Teacher's Resource File***
- Lesson Master 10-6A or 10-6B
- Answer Master 10-6
- Teaching Aid
  108 Warm-up
  118 Special Binomial Products

Additional Resources
- Visuals for Teaching Aids 108, 118

Teaching Lesson 10-6

Warm-up

1. a. Multiply 50 + 7 by 30 + 3 by thinking of 50 + 7 and 30 + 3 as binomials. **(50 + 7)(30 + 3) = 50 · 30 + 50 · 3 + 7 · 30 + 7 · 3 = 1500 + 150 + 210 + 21 = 1881**
   b. Does the process give the answer to 57 · 33? If so, show that it does. **Yes; 57 · 33 = 1881, which is the same as the answer in part a.**
2. Use the same process described in Question 1a to explain how to multiply 32 · 61 mentally. **(30 + 2)(60 + 1) = 1800 + 30 + 120 + 2 = 1952**

LESSON

10-6 *Special Binomial Products*

Some binomial products are used so frequently that they are given special names. In this lesson you will study two of them: Perfect Squares and the Difference of Two Squares.

The Square of a Sum

The expression $(a + b)^2$ (read "*a* plus *b*, the quantity squared") is the square of the binomial $(a + b)$. This product can be *expanded* by writing it as the product $(a + b)(a + b)$ and then using the Distributive Property or its special case, the FOIL algorithm.

$$(a + b)(a + b) = a^2 + ab + ba + b^2$$
$$= a^2 + 2ab + b^2 \text{ (}ab\text{ and }ba\text{ are like terms.)}$$

Geometrically, $(a + b)^2$ can be interpreted as the area of a square with side $a + b$. As the figure at the right shows, its area is $a^2 + 2ab + b^2$.

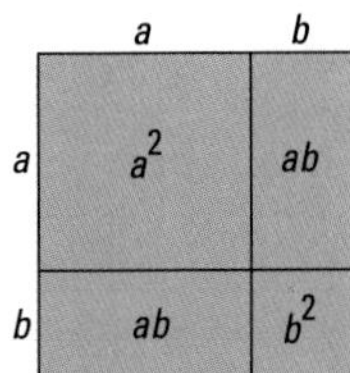

Example 1

The area of a square with side $5n + 4$ is $(5n + 4)^2$. Expand this binomial.

Solution 1

Rewrite the square as a multiplication and apply the FOIL algorithm.

$$(5n + 4)^2 = (5n + 4)(5n + 4)$$
$$= 25n^2 + 20n + 20n + 16$$
$$= 25n^2 + 40n + 16$$

Solution 2

Use the pattern for $(a + b)^2$ with $a = 5n$ and $b = 4$.

$$(a + b)^2 = a^2 + 2ab + b^2$$
$$(5n + 4)^2 = (5n)^2 + 2 \cdot 5n \cdot 4 + 4^2$$
$$= 25n^2 + 40n + 16$$

Solution 3

Draw a square with side $5n + 4$. Subdivide it into smaller rectangles and find the sum of their areas.

$$\text{Area} = (5n + 4)^2$$
$$= (5n)^2 + 2(4 \cdot 5n) + 4^2$$
$$= 25n^2 + 40n + 16$$

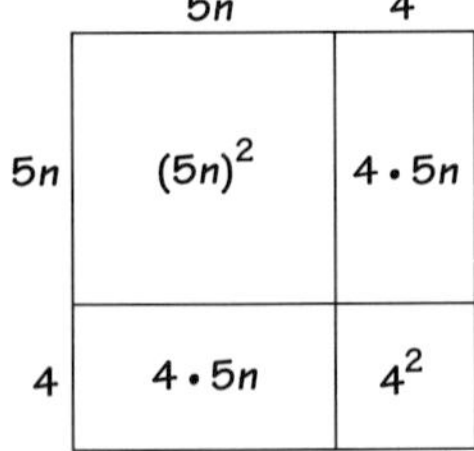

Lesson 10-6 Overview

Broad Goals Two patterns of binomial multiplication occur more often than any others: the square of a binomial and the difference of two squares. Each of these patterns can be applied to do arithmetic multiplication mentally. They illustrate one way in which algebra can contribute to increased arithmetic proficiency.

Perspective The perfect square pattern $(a + b)^2 = a^2 + 2ab + b^2$ has a variety of applications, including arithmetic shortcuts, proofs of geometry theorems, completing the square, and generalizations to powers of the binomials leading to Pascal's Triangle. The first two applications are found in this lesson. The difference of squares pattern is also very important, having all of the applications mentioned above except Pascal's Triangle.

Some teachers prefer to give only the perfect square pattern $(a + b)^2 = a^2 + 2ab + b^2$ because the second pattern is merely a special case. We think it helps if students are given both patterns, so they can see one more time that the square of a binomial is a trinomial.

The Square of a Difference

To square the difference $(a - b)^2$, think of $a - b$ as $a + -b$.

$$\begin{aligned}(a - b)^2 &= (a - b)(a - b) \\ &= (a + -b)(a + -b) \\ &= a^2 - ab - ba + b^2 \\ &= a^2 - 2ab + b^2\end{aligned}$$

Notice that after simplifying, the square of a binomial has three terms. It is a trinomial. Trinomials of the form $a^2 + 2ab + b^2$ or $a^2 - 2ab + b^2$ are called **perfect square trinomials** because each is the result of squaring a binomial.

> **Perfect Square Patterns**
> For all numbers a and b,
> $$(a + b)^2 = a^2 + 2ab + b^2,$$
> $$(a - b)^2 = a^2 - 2ab + b^2.$$

You need to remember these patterns, and you should realize that they can be derived from the multiplication of binomials. The algebraic description is short, but many people remember these patterns in words. The square of a binomial is:

the square of its first term,
plus twice the product of its terms,
plus the square of its last term.

Example 2

Expand $(w - 3)^2$.

Solution 1

Follow the Perfect Square Pattern for $(a - b)^2$. Here $a = w$ and $b = 3$.

$$\begin{aligned}(a - b)^2 &= a^2 - 2 \cdot a \cdot b + b^2 \\ (w - 3)^2 &= w^2 - 2 \cdot w \cdot 3 + 3^2 \\ &= w^2 - 6w + 9\end{aligned}$$

Solution 2

Change the square to the multiplication of $w - 3$ by itself.

$$\begin{aligned}(w - 3)^2 &= (w - 3)(w - 3) \\ &= w^2 - 3w - 3w + 9 \\ &= w^2 - 6w + 9\end{aligned}$$

Check

Test a special case. Let $w = 5$. Then $(w - 3)^2 = (5 - 3)^2 = 2^2 = 4$, and $w^2 - 6w + 9 = 5^2 - 6 \cdot 5 + 9 = 25 - 30 + 9 = 4$. It checks.

Notes on Reading

This lesson defines the term *expand* only for the square of a binomial. More generally, expanding the power of a polynomial means to rewrite that power in the standard form for a polynomial. The following everyday use of the word *expand* may help your students remember its mathematical meaning. *Expand* means "express in full or write out in greater detail."

When discussing the square of a sum in the opening paragraphs, you may want to use **Teaching Aid 118** which contains the diagram on page 646.

A visual demonstration of $(a - b)^2$ is shown in Activity 1 in *Optional Activities* below. You can also use this pattern to demonstrate the specific case given in **Example 2.**

The proof of the Pythagorean Theorem that is given on page 648 is only one of the many proofs of that theorem which utilize the idea of area. Another proof is found in Project 3 on page 657.

For a visual interpretation of the Difference of Two Squares Pattern, see Activity 2 in *Optional Activities* on pages 648–649.

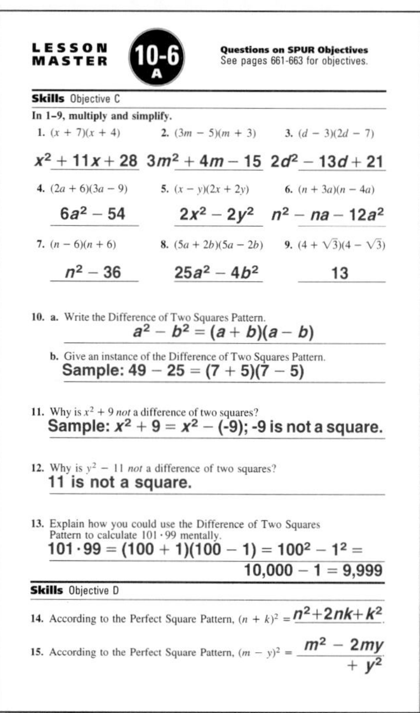

LESSON MASTER 10-6 A — **Questions on SPUR Objectives** See pages 661-663 for objectives.

Skills Objective C

In 1–9, multiply and simplify.

1. $(x + 7)(x + 4)$ — $x^2 + 11x + 28$
2. $(3m - 5)(m + 3)$ — $3m^2 + 4m - 15$
3. $(d - 3)(2d - 7)$ — $2d^2 - 13d + 21$
4. $(2a + 6)(3a - 9)$ — $6a^2 - 54$
5. $(x - y)(2x + 2y)$ — $2x^2 - 2y^2$
6. $(n + 3a)(n - 4a)$ — $n^2 - na - 12a^2$
7. $(n - 6)(n + 6)$ — $n^2 - 36$
8. $(5a + 2b)(5a - 2b)$ — $25a^2 - 4b^2$
9. $(4 + \sqrt{3})(4 - \sqrt{3})$ — 13

10. a. Write the Difference of Two Squares Pattern. $a^2 - b^2 = (a + b)(a - b)$
    b. Give an instance of the Difference of Two Squares Pattern. Sample: $49 - 25 = (7 + 5)(7 - 5)$
11. Why is $x^2 + 9$ *not* a difference of two squares? Sample: $x^2 + 9 = x^2 - (-9)$; -9 is not a square.
12. Why is $y^2 - 11$ *not* a difference of two squares? 11 is not a square.
13. Explain how you could use the Difference of Two Squares Pattern to calculate $101 \cdot 99$ mentally. $101 \cdot 99 = (100 + 1)(100 - 1) = 100^2 - 1^2 = 10{,}000 - 1 = 9{,}999$

Skills Objective D

14. According to the Perfect Square Pattern, $(n + k)^2 =$ $n^2 + 2nk + k^2$.
15. According to the Perfect Square Pattern, $(m - y)^2 =$ $m^2 - 2my + y^2$

Optional Activities

Activity 1
Materials: **Teaching Aid 118**

After discussing **Example 2**, you may want to derive the Perfect Square Pattern $(a - b)^2$ by using the following demonstration. Start with a square of side a. Remove a square of side b and two rectangles as shown at the right. We assume $a > b$. This demonstration shows that $(a - b)^2 = a^2 - 2b(a - b) - b^2 = a^2 - 2ab + 2b^2 - b^2 = a^2 - 2ab + b^2$.

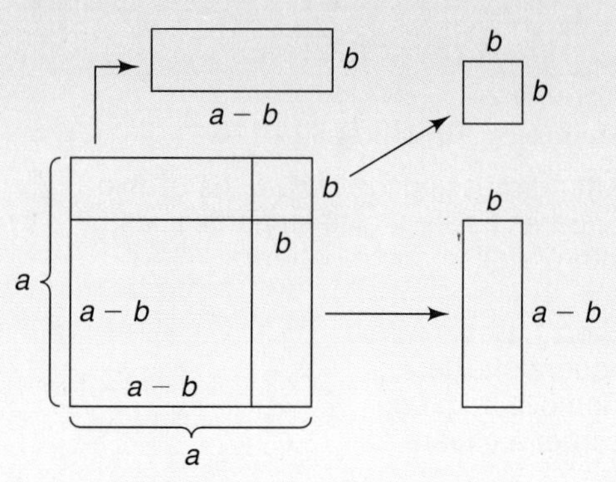

Additional Examples

1. The length of the side of a square is $y + 7$.
   a. Write the area of the square in expanded form.
   $y^2 + 14y + 49$
   b. Draw the square and show how the expanded form relates to the figure.

	y	7
y	y^2	$7y$
7	$7y$	49

$y^2 + 7y + 7y + 49 =$
$y^2 + 14y + 49$

2. Expand $(2n - 5)^2$.
   $4n^2 - 20n + 25$
3. Multiply $(10n - 7)(10n + 7)$.
   $100n^2 - 49$
4. Compute 51^2 in your head.
   Think: $(50 + 1)^2 =$
   $50^2 + 2 \cdot 50 \cdot 1 + 1^2 =$
   $2500 + 100 + 1 = 2601$

Notes on Questions

Question 13 Reading Mathematics
Recognizing forms is an important part of mathematics. Here, you may want to emphasize the subtle similarities and differences between the words and symbols:

difference of squares	$a^2 - b^2$
square of a difference	$(a - b)^2$

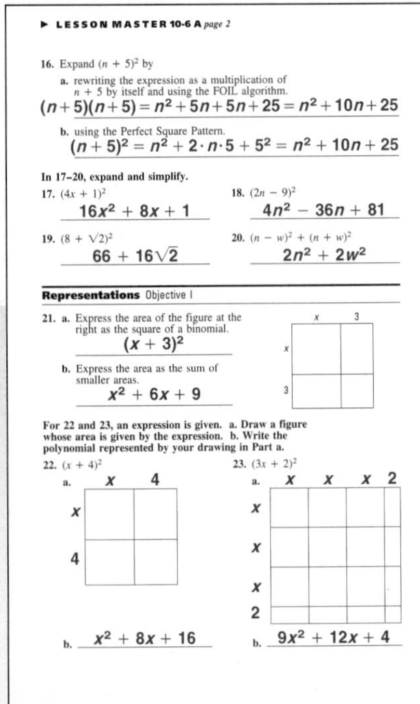
▶ LESSON MASTER 10-6 A *page 2*

16. Expand $(n + 5)^2$ by
   a. rewriting the expression as a multiplication of $n + 5$ by itself and using the FOIL algorithm.
   $(n + 5)(n + 5) = n^2 + 5n + 5n + 25 = n^2 + 10n + 25$
   b. using the Perfect Square Pattern.
   $(n + 5)^2 = n^2 + 2 \cdot n \cdot 5 + 5^2 = n^2 + 10n + 25$

In 17–20, expand and simplify.

17. $(4x + 1)^2$ $16x^2 + 8x + 1$
18. $(2n - 9)^2$ $4n^2 - 36n + 81$
19. $(8 + \sqrt{2})^2$ $66 + 16\sqrt{2}$
20. $(n - w)^2 + (n + w)^2$ $2n^2 + 2w^2$

Representations Objective I

21. a. Express the area of the figure at the right as the square of a binomial.
   $(x + 3)^2$
   b. Express the area as the sum of smaller areas.
   $x^2 + 6x + 9$

For 22 and 23, an expression is given. a. Draw a figure whose area is given by the expression. b. Write the polynomial represented by your drawing in Part a.

22. $(x + 4)^2$
   a.
   b. $x^2 + 8x + 16$
23. $(3x + 2)^2$
   a.
   b. $9x^2 + 12x + 4$

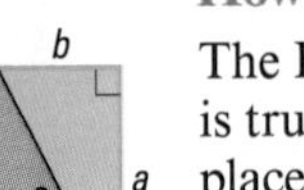

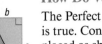

How Do We Know the Pythagorean Theorem Is True?

The Perfect Square Patterns help explain why the Pythagorean Theorem is true. Consider four right triangles with legs a and b and hypotenuse c placed as shown at the left. The triangles form an outer square of side $a + b$ and an inner tilted square of side c. Each triangle has area $\frac{1}{2}ab$. How is the area of the tilted square related to a and b?

The tilted square has side c. Notice that its area equals the area of the outer square with side $a + b$ minus the areas of the four right triangles.

$$c^2 = (a + b)^2 - 4\left(\frac{1}{2}ab\right)$$

Now use the Perfect Square Pattern to expand the binomial.

$$c^2 = a^2 + 2ab + b^2 - 2ab$$
$$c^2 = a^2 + b^2$$

This final equation is the Pythagorean Theorem.

The Product of the Sum and the Difference of the Same Numbers

Another interesting binomial pattern involves sums and differences of numbers. Consider two numbers a and b. Multiply their sum $a + b$ by their difference $a - b$.

$$(a + b)(a - b) = a^2 + ab - ab - b^2$$

The two middle terms are opposites, so their sum is 0.

$$= a^2 - b^2$$

In words, the product of the sum and difference of two numbers is the **difference of squares** of the two numbers.

Difference of Two Squares Pattern
For all numbers a and b,
$$(a + b)(a - b) = a^2 - b^2.$$

Example 3

Multiply $(3y + 7)(3y - 7)$.

Solution

The binomial factors are the sum and difference of the same terms. Use the Difference of Two Squares Pattern and chunking with $a = 3y$ and $b = 7$.

$$(3y + 7)(3y - 7) = (3y)^2 - 7^2$$
$$= 9y^2 - 49$$

Check

Test a special case. Let $y = 4$.

$(3 \cdot 4 + 7)(3 \cdot 4 - 7) = 19 \cdot 5 = 95$

Does $9 \cdot 4^2 - 49 = 95$? Yes.

Optional Activities

Activity 2
Materials: **Teaching Aid 118**

After discussing the Difference of Two Squares Pattern, give students a visual interpretation of the pattern.

Step 1: Begin with a square of side a. Remove from one corner a square of side b. Ask students to give the area of the remaining part. $[a^2 - b^2]$

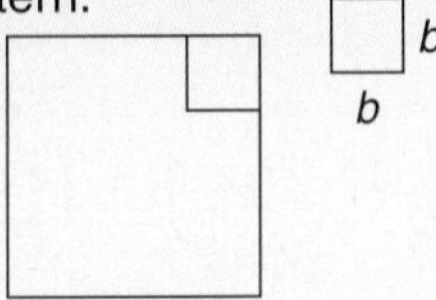

Step 2: Split the remaining part in two parts and ask students for the dimensions of the two rectangles. [b and $a - b$; a and $a - b$]

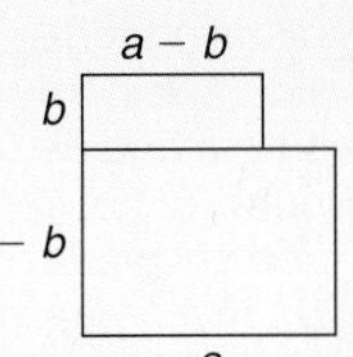

The Perfect Square and Difference of Two Squares Patterns can be applied to do mental arithmetic.

Example 4

Compute in your head: **a.** 79^2; **b.** $51 \cdot 49$.

Solution

a. Use the Perfect Square Pattern. Think of 79 as $80 - 1$.

$$79^2 = (80 - 1)^2$$
$$= 80^2 - 2 \cdot 80 \cdot 1 + 1^2$$
$$= 6400 - 160 + 1 = 6241$$

b. Think of 51 as $50 + 1$, and 49 as $50 - 1$. Use the Difference of Squares Pattern.

$$51 \cdot 49 = (50 + 1)(50 - 1)$$
$$= 50^2 - 1^2$$
$$= 2500 - 1 = 2499$$

QUESTIONS

Covering the Reading

In 1 and 2, expand the binomial.

1. $(x + y)^2$ $x^2 + 2xy + y^2$ **2.** $(x - y)^2$ $x^2 - 2xy + y^2$

3. What is a *perfect square trinomial?* trinomials of the form $a^2 + 2ab + b^2$ or $a^2 - 2ab + b^2$

4. a. What is the area of a square with side of length $n + 3$? $n^2 + 6n + 9$

b. Justify your answer in part **a** by drawing a picture. See left.

c. Check your answer by letting $n = 2$. Does $(2 + 3)^2 = 2^2 + 6 \cdot 2 + 9$? Yes; $25 = 25$; it checks.

4b)

	n	3
n	n^2	$3n$
3	$3n$	9

5. Write an expression for the area of a square with side $3x + 7$,

a. as the square of a binomial; $(3x + 7)^2$

b. as a perfect square trinomial. $9x^2 + 42x + 49$

In 6–9, expand the binomial.

6. $(m - 6)^2$ $m^2 - 12m + 36$ **7.** $(m + 12)^2$ $m^2 + 24m + 144$ **8.** $(2x + 5)^2$ $4x^2 + 20x + 25$ **9.** $(3y - 4)^2$ $9y^2 - 24y + 16$

10. Fill in the blank: $a^2 - b^2 = (a - b) \cdot$ __?__. $(a + b)$

In 11 and 12, multiply and simplify.

11. $(x + 13)(x - 13)$ $x^2 - 169$ **12.** $(3 + 8p)(3 - 8p)$ $9 - 64p^2$

13. *Multiple choice.* Which is *not* the difference of two squares? (b)

(a) $9 - w^2$ (b) $(x - y)^2$

(c) $x^2y^2 - 1$ (d) $121m^2 - n^2$

In 14–17, compute in your head using the methods in Example 4.

14. $(41)^2$ **15.** $(37)^2$ **16.** $69 \cdot 71$ **17.** $85 \cdot 95$

14) $(40 + 1)^2 = 1600 + 80 + 1 = 1681$

15) $(30 + 7)^2 = 900 + 420 + 49 = 1369$

16) $(70 - 1)(70 + 1) = 4900 - 1 = 4899$

17) $(90 - 5)(90 + 5) = 8100 - 25 = 8075$

Follow-up for Lesson 10-6

Practice

For more questions on SPUR Objectives, use **Lesson Master 10-6A** (shown on pages 647–648) or **Lesson Master 10-6B** (shown on pages 649–650).

Assessment

Written/Oral Communication Have students choose negative integer values for a and b, and check that the Difference of Two Squares Pattern is true for the numbers they chose. Then, using the same values for a and b, have students check if $(ax + b)(ax - b) = (ax)^2 - b^2$. Have students compare and discuss their results with other students in the class. [Students display understanding of the Difference of Two Squares Pattern.]

Extension

Project Update Project 2, *Powers of Binomials,* and Project 3, *Another Proof of the Pythagorean Theorem,* on page 657, relate to the content of this lesson.

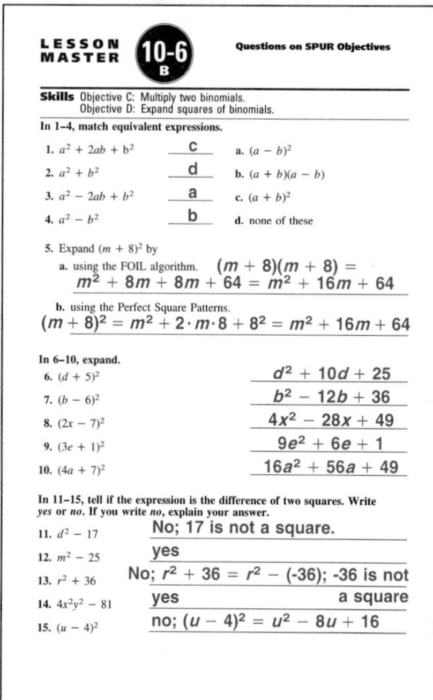
LESSON MASTER 10-6 B — Questions on SPUR Objectives

Skills Objective C: Multiply two binomials. Objective D: Expand squares of binomials.

In 1–4, match equivalent expressions.

1. $a^2 + 2ab + b^2$ — c
2. $a^2 + b^2$ — d
3. $a^2 - 2ab + b^2$ — a
4. $a^2 - b^2$ — b

a. $(a - b)^2$
b. $(a + b)(a - b)$
c. $(a + b)^2$
d. none of these

5. Expand $(m + 8)^2$ by
a. using the FOIL algorithm. $(m + 8)(m + 8) = m^2 + 8m + 8m + 64 = m^2 + 16m + 64$
b. using the Perfect Square Patterns. $(m + 8)^2 = m^2 + 2 \cdot m \cdot 8 + 8^2 = m^2 + 16m + 64$

In 6–10, expand.

6. $(d + 5)^2$ — $d^2 + 10d + 25$
7. $(b - 6)^2$ — $b^2 - 12b + 36$
8. $(2x - 7)^2$ — $4x^2 - 28x + 49$
9. $(3e + 1)^2$ — $9e^2 + 6e + 1$
10. $(4a + 7)^2$ — $16a^2 + 56a + 49$

In 11–15, tell if the expression is the difference of two squares. Write *yes* or *no*. If you write *no*, explain your answer.

11. $d^2 - 17$ — No; 17 is not a square.
12. $m^2 - 25$ — yes
13. $r^2 + 36$ — No; $r^2 + 36 = r^2 - (-36)$; -36 is not a square
14. $4x^2y^2 - 81$ — yes
15. $(u - 4)^2$ — no; $(u - 4)^2 = u^2 - 8u + 16$

Adapting to Individual Needs

Step 3: Rearrange the rectangles to make a large rectangle. Ask students to give the dimensions and the area of this rectangle [$a + b$ and $a - b$; $(a + b)(a - b)$] Then ask how the area of the figures in Steps 1 and 3 are related. [They are equal.]

$a - b$

$a + b$

Extra Help

Some students may confuse the patterns for the Square of a Difference and the Difference of Two Squares. Students who want to equate $(a - b)^2$ to $a^2 - b^2$ should write $(a - b)^2$ as $(a - b)(a - b)$ and then use the FOIL algorithm until they understand the two patterns.

Notes on Questions

Questions 18–19 Error Alert Some students persist in writing $(a + b)^2 = a^2 + b^2$ even after seeing many examples. They are over-generalizing the Distributive Property to apply to exponents. These questions are designed to counteract that tendency.

Question 23 Take some time to show the power of using a pattern. $(x + \sqrt{11})(x - \sqrt{11}) = x^2 - (\sqrt{11})^2$ is easier than $(x + \sqrt{11})(x - \sqrt{11}) = x^2 - x\sqrt{11} + x\sqrt{11} - (\sqrt{11})^2$, even though both methods are correct.

Question 24 Have students check the answer by approximating the square roots and multiplying with a calculator. Again, students may be surprised to find that this technique works.

Question 25 Multicultural Connection The wedding cake has its origins in ancient times. In Greece, eating a cake made from sesame-seed meal and honey served as the final act of the marriage ceremony. In Rome, the marriage ceremony was called *confarreatio*, which is the name of a wheat cake that the couple ate together. In Western Europe, early marriage cakes were small unleavened biscuits. However, as baking techniques developed, the biscuits evolved into large cakes that were elaborately decorated.

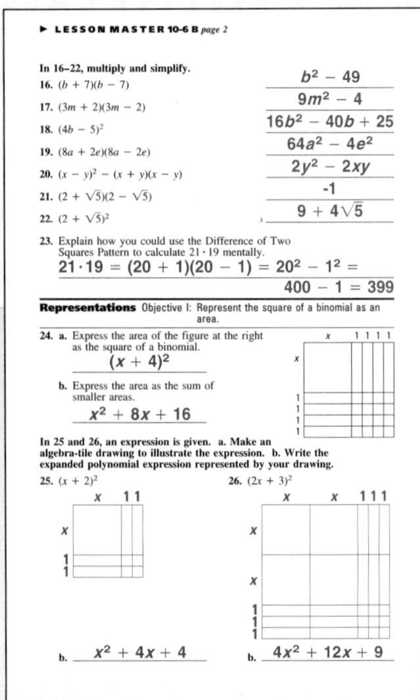

▶ **LESSON MASTER 10-6 B** *page 2*

In 16–22, multiply and simplify.

16. $(b + 7)(b - 7)$ $b^2 - 49$
17. $(3m + 2)(3m - 2)$ $9m^2 - 4$
18. $(4b - 5)^2$ $16b^2 - 40b + 25$
19. $(8a + 2e)(8a - 2e)$ $64a^2 - 4e^2$
20. $(x - y)^2 - (x + y)(x - y)$ $2y^2 - 2xy$
21. $(2 + \sqrt{5})(2 - \sqrt{5})$ -1
22. $(2 + \sqrt{5})^2$ $9 + 4\sqrt{5}$
23. Explain how you could use the Difference of Two Squares Pattern to calculate $21 \cdot 19$ mentally. $21 \cdot 19 = (20 + 1)(20 - 1) = 20^2 - 1^2 = 400 - 1 = 399$

Representations Objective I: Represent the square of a binomial as an area.

24. a. Express the area of the figure at the right as the square of a binomial. $(x + 4)^2$

 b. Express the area as the sum of smaller areas. $x^2 + 8x + 16$

In 25 and 26, an expression is given. a. Make an algebra-tile drawing to illustrate the expression. b. Write the expanded polynomial expression represented by your drawing.

25. $(x + 2)^2$ b. $x^2 + 4x + 4$

26. $(2x + 3)^2$ b. $4x^2 + 12x + 9$

18) Sample: $(3 + 5)^2 = 8^2 = 64$; $3^2 + 2 \cdot 3 \cdot 5 + 5^2 = 9 + 30 + 25 = 64$. So, $(3 + 5)^2 = 3^2 + 2 \cdot 3 \cdot 5 + 5^2$.

19) Sample: $(1 + 2)^2 = 3^2 = 9$; $1^2 + 2^2 = 1 + 4 = 5$. So, $(1 + 2)^2 \neq 1^2 + 2^2$

25d) Does $2 \cdot 4^3 + 8 \cdot 4^2 + 25 \cdot 4 = 9 \cdot 9 \cdot 4 + 4 \cdot 4 \cdot 2$? Yes, 356 = 356.

Applying the Mathematics

18–19) See left.

18. By substituting numbers for a and b, verify that $(a + b)^2 = a^2 + 2ab + b^2$.

19. By substituting numbers for a and b, verify that $(a + b)^2 \neq a^2 + b^2$.

20. Does $(x - 3)^2$ equal $(3 - x)^2$ for all values of x? Explain your reasoning. **Yes, they have the same expansion, $x^2 - 6x + 9$.**

In 21–24, write the expression as a simplified polynomial.

21. $4(9 + y)^2$ **$324 + 72y + 4y^2$**

22. $(x + y)^2 - (x - y)^2$ **$4xy$**

23. $(x + \sqrt{11})(x - \sqrt{11})$ **$x^2 - 11$**

24. $(\sqrt{7} - \sqrt{2})(\sqrt{7} + \sqrt{2})$ **5**

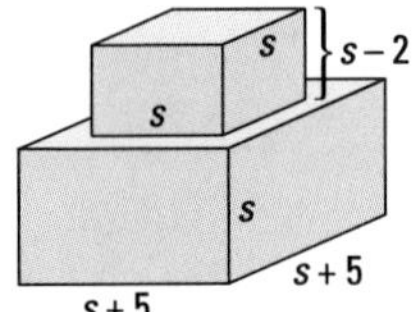

25. A wedding cake is made from two layers, each in the shape of a rectangular solid. The bottom layer has dimensions $s + 5$ by $s + 5$ by s. The top layer has dimensions s by s by $s - 2$.
 a. What is the volume of the bottom layer?
 b. What is the volume of the top layer? **$s^3 - 2s^2$**
 c. Find the volume of the cake. **$2s^3 + 8s^2 + 25s$**
 d. Check your answer by letting $s = 4$. **See left.**
 a) **$s^3 + 10s^2 + 25s$**

Review

In 26 and 27, multiply and simplify. *(Lesson 10-5)*

26. $(z - 11)(z + 8)$ **$z^2 - 3z - 88$**

27. $(2c - 7d)(c + 5d)$ **$2c^2 + 3cd - 35d^2$**

28. *Skill sequence.* Multiply. *(Lessons 10-3, 10-4, 10-5)*
 a. $x(x + 3)$ **$x^2 + 3x$**
 b. $(x + 4)(x + 3)$ **$x^2 + 7x + 12$**
 c. $(x + 4 - y)(x + 3)$ **$x^2 - xy + 7x - 3y + 12$**

29. Solve. a. $x = 2x$ **$x = 0$** b. $x > 2x$ **$x < 0$** *(Lessons 6-6, 6-7)*

30. In 1992, the total sales of General Electric increased 3.3% to \$62.2 billion. What were the company's sales in 1991? *(Lessons 5-4, 6-3)* **\$60.2 billion**

31. Four consecutive integers (such as 7, 8, 9, and 10) can be represented by the expressions n, $n + 1$, $n + 2$, $n + 3$. The sum of four consecutive integers is 250. What are the integers? *(Lessons 3-7, 3-8)* **61, 62, 63, 64**

32. Simplify $(x^3 - 4x + 1) - (5x^3 + 4x - 8)$. *(Lesson 3-6)* **$-4x^3 - 8x + 9$**

Exploration

33b) Sample: 9, 10, 11 yields 100 and 99; 24, 25, 26 yields 625 and 624; -2, -1, 0 yields 1 and 0.

33. a. Write three consecutive integers. **Sample: 3, 4, 5**
 Square the second number. **16**
 Find the product of the first and the third. **15**
 b. Repeat part **a** with 3 more sets of consecutive integers. **See left.**
 c. What do you notice? **The square of the second number is one more than the product of surrounding integers.**
 d. Explain why this will always happen. (Hint: Let n equal the second number.) **Three consecutive integers can be written as $n - 1, n, n + 1$. $n^2 = (n - 1)(n + 1) + 1$, since $(n - 1)(n + 1) = n^2 - 1$.**

Adapting to Individual Needs

Challenge

Demonstrate this number game and then have students explain why it works. Have a student choose a whole number greater than 1, multiply one more than the number by one less than the number, and tell you the result. You tell the student the original number by taking the square root of 1 plus the number given to you. [If the original number is n, then $(n + 1)(n - 1) = n^2 - 1$. Adding 1 gives n^2; the square root of n^2 is n.]

Setting Up Lesson 10-7

Materials If students use Activity 1 or 2 in the *Optional Activities* in Lesson 10-7, they will need dice. Students using the *Challenge* will need almanacs.

LESSON

10-7

The Chi-Square Statistic

Safety first! *Steelworkers follow safety guidelines to prevent accidents.*

In this lesson, we consider a statistic different from any you have yet seen. This statistic compares actual frequencies with the frequencies that would be expected by calculating probabilities. It is one of the most often used of all statistics. Here is a typical use, though the data are made up.

In a large factory, over a two-year interval, there were 180 accidents. Someone noticed that more accidents seemed to occur on Mondays. So the accidents were tabulated by the day of the week. Below is the table. Does Monday seem to be special?

Day of the week	M	Tu	W	Th	F
Actual numbers of accidents	55	38	23	22	42

The expected number of accidents is the mean number of accidents for a given day that is predicted by a probability. If accidents occurred randomly, then the expected number for each day would be the same. So the **expected number** of accidents for each day is $\frac{180}{5}$, or 36.

Day of the week	M	Tu	W	Th	F
Expected numbers of accidents	36	36	36	36	36

Recall, however, that even if events occur randomly, it is not common for all events to occur with the same frequency. When you toss a coin 10 times, you would not usually get exactly 5 heads even if the coin were fair. Similarly, if there were 35 accidents on each of 4 days and 39 on the 5th day, that would not seem to be much of a variance from the expected numbers. So we ask: Do the actual numbers deviate so much from the expected numbers that we should think the accidents are not happening randomly? Or, are the numbers so close that the differences are probably due to chance?

Lesson 10-7

Objectives

H Use the chi-square statistic to determine whether or not an event is likely.

Resources

From the ***Teacher's Resource File***

- Lesson Master 10-7A or 10-7B
- Answer Master 10-7
- Teaching Aids
  109 Warm-up
  119 Critical Chi-Square Values
  120 Additional Examples
- Activity Kit, Activity 23
- Technology Sourcebook
  Computer Master 26

Additional Resources

- Visuals for Teaching Aids 109, 119, 120
- Spreadsheet software

Teaching Lesson 10-7

Warm-up

Solve the following problems.

1. If a spinner is divided into four congruent regions that are numbered 1, 2, 3, and 4, and you spin it 120 times, how many times would you expect the spinner to land on 2? **30 times**
2. If you randomly asked 70 people on what day of the week their birthday falls this year, how many would you expect to say Tuesday? **10**

(*Warm-up continues on page 652.*)

Lesson 10-7 Overview

Broad Goals This lesson shows students how to use the chi-square statistic to determine if a set of discrete data differs significantly from values that would be expected when certain probabilities are given.

Perspective The statistics that students have seen in earlier lessons (and earlier courses) are *descriptive statistics*—they describe data. In this lesson, we introduce an *inferential statistic*, one that is used to help in decision-making. The chi-square statistic is the easiest inferential statistic to calculate (it can be done by hand, but is easier with a calculator). It can be used on data that are quite accessible to students, and its calculation involves only simple probability and elementary algebra.

The idea behind the chi-square statistic is a set of distributions known as chi-square distributions. Each distribution in the set displays the probability of obtaining a chi-square value that is greater than a particular number for a specified number of degrees of freedom (the value $n - 1$ in the table of Critical Chi-Square Values on page 653 or on **Teaching Aid 119**). In all of the examples and questions in this book, the chi-square statistic is being used to compare a set of n observed values with a corresponding set of n expected values. The number of degrees of freedom is $n - 1$.

3. If you randomly asked 100 people if their home telephone number ended in an even or an odd number, how many would you expect to say odd? **50 people**
4. If you tossed a die 90 times, how many times would you expect to toss a 3? **15 times**

Notes on Reading

Reading Mathematics You might want to read this lesson with your students.

❶ The table of Critical Chi-Square Values on page 653 is also on **Teaching Aid 119.** Emphasize that the table should be read one row at a time. Note that we examine row 4 because $n = 5$, and so $n - 1 = 4$. Suppose that accidents were randomly assigned to days. If so, then one would expect a chi-square value greater than 7.78 only 10% of the time, a chi-square value greater than 9.49 about 5% of the time, a value greater than 13.28 about 1% of the time, and a value greater than 18.5 only $\frac{1}{10}$% of the time. Thus, although the actual number of accidents given in the lesson *could* happen, the chi-square value of 21.3 found for the data indicates that such a value would happen very rarely if the events were occurring randomly.

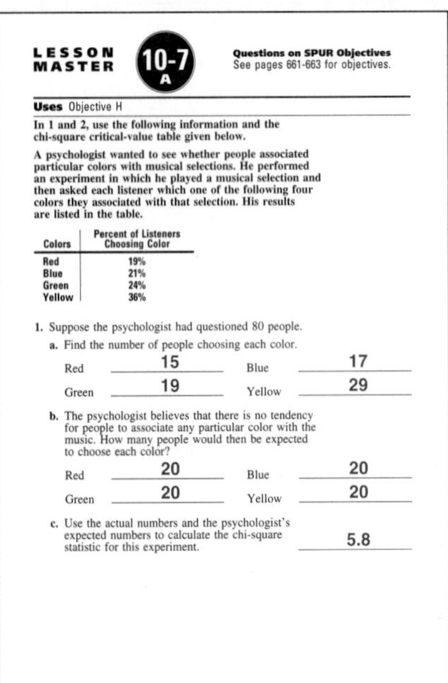

LESSON MASTER 10-7 A

Questions on SPUR Objectives
See pages 661-663 for objectives.

Uses Objective H

In 1 and 2, use the following information and the chi-square critical-value table given below.

A psychologist wanted to see whether people associated particular colors with musical selections. He performed an experiment in which he played a musical selection and then asked each listener which one of the following four colors they associated with that selection. His results are listed in the table.

Colors	Percent of Listeners Choosing Color
Red	19%
Blue	21%
Green	24%
Yellow	36%

1. Suppose the psychologist had questioned 80 people.
   a. Find the number of people choosing each color.
      Red 15 Blue 17
      Green 19 Yellow 29
   b. The psychologist believes that there is no tendency for people to associate any particular color with the music. How many people would then be expected to choose each color?
      Red 20 Blue 20
      Green 20 Yellow 20
   c. Use the actual numbers and the psychologist's expected numbers to calculate the chi-square statistic for this experiment. 5.8

If we let an expected number be e and an actual observed number be a, then $|e - a|$, the absolute value of the difference between these numbers, is called the deviation of a from e. For instance, the deviation of the actual from the expected number of accidents on Monday is $|36 - 55|$, or 19.

In 1900, the English statistician Karl Pearson developed a method of determining whether the differences in two frequency distributions is greater than that expected by chance. This method uses a number called the chi-square statistic. ("Chi" is pronounced "ky" as in "sky.") The algorithm for calculating this statistic uses the squares of deviations, which is why we study it here.

Step 1: Count the number of events. Call this number n.
In the above situation, there are 5 events, one each for M, Tu, W, Th, and F.

Step 2: Let a_1, a_2, a_3, a_4, and a_5 be the actual frequencies.
In our case, $a_1 = 55$, $a_2 = 38$, $a_3 = 23$, $a_4 = 22$, and $a_5 = 42$.

Step 3: Let e_1, e_2, e_3, e_4, and e_5 be the expected frequencies.
Here $e_1 = 36$, $e_2 = 36$, $e_3 = 36$, $e_4 = 36$, and $e_5 = 36$.

Step 4: Calculate $\frac{(a_1 - e_1)^2}{e_1}$, $\frac{(a_2 - e_2)^2}{e_2}$, $\frac{(a_3 - e_3)^2}{e_3}$, and so on. Each number is the square of the deviation, divided by the expected frequency.

$$\frac{(a_1 - e_1)^2}{e_1} = \frac{(55 - 36)^2}{36} = \frac{361}{36} \qquad \frac{(a_2 - e_2)^2}{e_2} = \frac{(38 - 36)^2}{36} = \frac{4}{36}$$

$$\frac{(a_3 - e_3)^2}{e_3} = \frac{(23 - 36)^2}{36} = \frac{169}{36} \qquad \frac{(a_4 - e_4)^2}{e_4} = \frac{(22 - 36)^2}{36} = \frac{196}{36}$$

$$\frac{(a_5 - e_5)^2}{e_5} = \frac{(42 - 36)^2}{36} = \frac{36}{36}$$

Step 5: Add the n numbers found in Step 4.
This sum is the chi-square statistic.

$$\frac{361}{36} + \frac{4}{36} + \frac{169}{36} + \frac{196}{36} + \frac{36}{36} = \frac{766}{36} \approx 21.3$$

The chi-square statistic measures how different a set of actual observed scores is from a set of expected scores. The larger the differences are, the greater the chi-square statistic. But is 21.3 unusually large? You can find that out by looking in a chi-square table such as the one shown on page 653. That table gives the values for certain values of n and certain probabilities. In the table, n is the number of events. The other columns of the table correspond to probabilities of .10 (an event expected to happen $\frac{1}{10}$ of the time), .05 (or $\frac{1}{20}$ of the time), .01 (or $\frac{1}{100}$ of the time), and .001 (or $\frac{1}{1000}$ of the time). You are not expected to know how the values in the table were calculated. The mathematics needed to calculate them is beyond that normally studied before college.

Optional Activities

Activity 1
Materials: Dice

You might want to use this activity after you discuss *Additional Examples* on page 654. Distribute one die to every two students. Have students **work in pairs** and repeat the experiment discussed in the example. Ask if any of the students have dice that seem to be unfair. If so, have someone else do the tossing with that die and perform a second experiment.

Activity 2
You can use *Activity Kit, Activity 23,* as a follow-up to the lesson. In this activity, students make a paper number cube and then use the chi-square statistic to test whether their cube is fair.

How to Read This Table

The left column, titled $n - 1$, is one less than the number of events. This is because once $n - 1$ observed frequencies are known, the last frequency can be calculated by subtracting from the total of expected frequencies. For instance, once the numbers of accidents are known for Monday through Thursday, you could subtract from 180 to find the frequency for Friday. The number $n - 1$ is known as the number of **degrees of freedom** for this statistic.

The other columns give the probabilities that chi-square values as large as these will occur. For instance, examine the number 14.1, which appears in column .05, row 7. This means that, with 8 events, a chi-square value greater than 14.1 occurs with probability .05 or less.

Critical Chi-Square Values

❶

$n - 1$	.10	.05	.01	.001
1	2.71	3.84	6.63	10.8
2	4.61	5.99	9.21	13.8
3	6.25	7.81	11.34	16.3
4	7.78	9.49	13.28	18.5
5	9.24	11.07	15.09	20.5
6	10.6	12.6	16.8	22.5
7	12.0	14.1	18.5	24.3
8	13.4	15.5	20.1	26.1
9	14.7	16.9	21.7	27.9
10	16.0	18.3	23.2	29.6
11	17.3	19.7	24.7	31.3
12	18.6	21.0	26.2	32.9
13	19.8	22.4	27.7	34.5
14	21.1	23.7	29.1	36.1
15	22.3	25.0	30.6	37.7
16	23.5	26.3	32.0	39.3
17	24.8	27.6	33.4	40.8
18	26.0	28.9	34.8	42.3
19	27.2	30.1	36.2	43.8
20	28.4	31.4	37.6	45.3
25	34.4	37.7	44.3	52.6
30	40.3	43.8	50.9	59.7
50	63.2	67.5	76.2	86.7

With the data on factory accidents, we obtained a chi-square value of 21.3 with $n = 5$ events. So we look in row $n - 1$, which is row 4. A value as large as 18.5 (the largest value in row 4) would occur with probability less than .001. Thus, a value of 21.3 is even less likely. Since .001 is a very small probability, we have evidence that the accidents are not evenly distributed among the days of the week. The factory should try to determine why there are more accidents on Monday. Perhaps it is because people come back tired from a weekend.

Suppose the frequencies of the accidents had led to a chi-square value of 8.62. Then, looking across row 4, we would see that this value is between the values 7.78 and 9.49. So 8.62 has a probability between .10 and .05.

Point out that if data such as the accident data were collected each week for a period of years (so that one might have data from hundreds of weeks), it would not be so unlikely to have *one* week, or maybe several weeks, like the one in the text. Even rare events can become expected if there are enough repetitions of an experiment.

Students may wonder how the chi-square values are calculated. That process is well beyond the scope of this course; the theory requires calculus and college-level statistics, and it is often not studied until graduate school.

Emphasize that the chi-square statistic can never indicate that something is *certainly* unlikely, but it does help to give evidence one way or the other.

Optional Activities

Activity 3 Technology Connection

You may wish to assign *Technology Sourcebook, Computer Master 26.* Students use the chi-square statistic to create a spreadsheet program that compares observed and expected values.

Adapting to Individual Needs

English Language Development

You might suggest that students add the following vocabulary words to their index cards: *expected number*, *actual frequency*, *expected frequency*, *deviation*, and *chi-square statistic.*

► **LESSON MASTER 10-7 A** *page 2*

Critical Chi-Square Values

$n - 1$	.10	.05	.01	.001
1	2.71	3.84	6.63	10.8
2	4.61	5.99	9.21	13.8
3	6.25	7.81	11.34	16.3
4	7.78	9.49	13.28	18.5
5	9.24	11.07	15.09	20.5
6	10.6	12.6	16.8	22.5
7	12.0	14.1	18.5	24.3
8	13.4	15.5	20.1	26.1
9	14.7	16.9	21.7	27.9
10	16.0	18.3	23.2	29.6

d. Refer to the Critical Chi-Square Values table above. Using your answer to Part **c**, what can you conclude about the psychologist's belief?

Sample: The psychologist's belief is invalid, as there is a greater than 10% chance the results would occur in the equally-likely cases.

2. Suppose the psychologist had asked 200 people. Answer the same questions that you did for Question 1 for 200 people.

a. Red 38 Blue 42

Green 48 Yellow 72

b. Red 50 Blue 50

Green 50 Yellow 50

c. 13.92

d. Sample: The psychologist's belief is valid, as there is less than 1% chance the results would occur in the equally-likely cases.

3. From Questions 1 and 2, what can you say about the relationship between sample size and the conclusions reached in the experiment?

Sample: Because the chi-square value is greater for a greater sample size, it is more likely that the conclusions are correct.

❷ **History Connection** Millard Fillmore inherited the presidency in 1850 after the death of Zachary Taylor. During Fillmore's time in office, he helped pass the Compromise of 1850—a series of laws designed to end the quarreling between anti-slavery advocates in the North and the slave owners in the South. However, because of Fillmore's enforcement of the Compromise provision for returning runaway slaves, many Northerners did not support him, and he was not nominated to run for President in 1852.

Additional Examples

These examples are also given on **Teaching Aid 120.**

One die was tossed 60 times to see if it was fair. Here are the results of the tosses: 26225 23635 12224 12154 55351 42545 66465 46522 13224 23645 35552 44423.

1. What are the observed values for the events of getting a 1, 2, 3, 4, 5, and 6 in this situation? **5 ones, 16 twos, 7 threes, 11 fours, 14 fives, 7 sixes**
2. What are the expected values of each event? **10 for each event**
3. Use the chi-square statistic to give some evidence whether or not the die is fair. **Chi-square value: 9.6; a value this large would happen more than 5% of the time; statisticians would not generally consider this to be enough evidence to conclude that the die is not fair.**

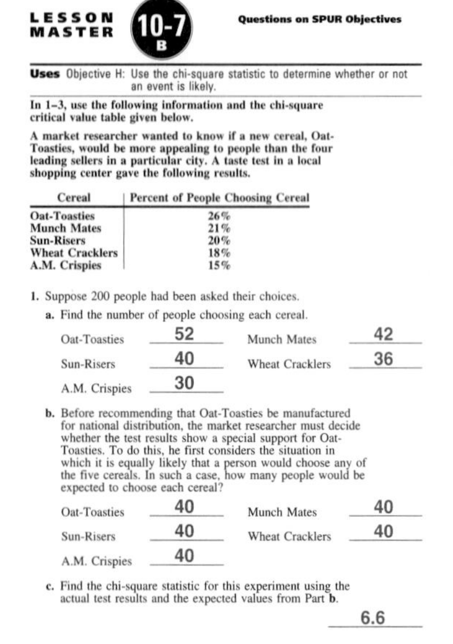

LESSON MASTER 10-7 B Questions on SPUR Objectives

Uses Objective H: Use the chi-square statistic to determine whether or not an event is likely.

In 1–3, use the following information and the chi-square critical value table given below.

A market researcher wanted to know if a new cereal, Oat-Toasties, would be more appealing to people than the four leading sellers in a particular city. A taste test in a local shopping center gave the following results.

Cereal	Percent of People Choosing Cereal
Oat-Toasties	26%
Munch Mates	21%
Sun-Risers	20%
Wheat Cracklers	18%
A.M. Crispies	15%

1. Suppose 200 people had been asked their choices.
   a. Find the number of people choosing each cereal.
   Oat-Toasties 52 Munch Mates 42
   Sun-Risers 40 Wheat Cracklers 36
   A.M. Crispies 30
   b. Before recommending that Oat-Toasties be manufactured for national distribution, the market researcher must decide whether the test results show a special support for Oat-Toasties. To do this, he first considers the situation in which it is equally likely that a person would choose any of the five cereals. In such a case, how many people would be expected to choose each cereal?
   Oat-Toasties 40 Munch Mates 40
   Sun-Risers 40 Wheat Cracklers 40
   A.M. Crispies 40
   c. Find the chi-square statistic for this experiment using the actual test results and the expected values from Part **b**.
   6.6

That means that a chi-square value as high as 8.62 would occur between $\frac{1}{10}$ and $\frac{1}{20}$ of the time. Statisticians normally do not consider this probability to be low enough to think there is reason to question the expected values.

The cutoff value is usually taken as the value with probability .05 or .01. In this book, we use the probability .05. When a chi-square value as large as the one found would occur with probability less than .05, we then question the assumptions leading to the expected values. This is the case with the chi-square value for the accidents. Because a value as large as 21.3 would occur with probability less than .05, we suspect that the accidents are not occurring randomly among the days.

Millard Fillmore, Abraham Lincoln, and Martin Van Buren

❷

Example

Sixty people were asked to name the U.S. President in 1850 from the names below.

14 picked Millard Fillmore.
25 picked Abraham Lincoln.
21 picked Martin Van Buren.

Is there evidence to believe the people were just guessing?

Solution

Calculate the chi-square statistic following the steps given on page 652.

1. The number of events $n = 3$.
2. Identify the actual observed values. $a_1 = 14$; $a_2 = 25$; $a_3 = 21$.
3. Calculate the expected values. If people were just guessing, we would expect each of the three names to be picked by the same number of people. Since there were 60 people in all, each name would be picked by 20. So $e_1 = 20$; $e_2 = 20$; $e_3 = 20$.
4. Calculate $\frac{(a_1 - e_1)^2}{e_1}$, $\frac{(a_2 - e_2)^2}{e_2}$, and $\frac{(a_3 - e_3)^2}{e_3}$.

$$\frac{(a_1 - e_1)^2}{e_1} = \frac{(14 - 20)^2}{20} = \frac{36}{20} \qquad \frac{(a_2 - e_2)^2}{e_2} = \frac{(25 - 20)^2}{20} = \frac{25}{20}$$

$$\frac{(a_3 - e_3)^2}{e_3} = \frac{(21 - 20)^2}{20} = \frac{1}{20}$$

5. The sum of the numbers in step 4 is $\frac{36 + 25 + 1}{20} = \frac{62}{20} = 3.1$.

Now examine the table. When $n = 3$, $n - 1 = 2$, so look at the 2nd row. The number 3.1 is smaller than the value 5.99 that would occur with probability .05. This is not a high enough chi-square value to question the way the expected values were calculated. It is quite possible that the people were guessing.

The chi-square statistic can be used whenever there are actual frequencies and you have some way of calculating expected frequencies. However, the chi-square statistic should not be used when there is an expected frequency that is less than 5.

Adapting to Individual Needs

Extra Help

Some students might need a review of relative frequency and probability. Remind them that a relative frequency is a statistic derived from an experiment. It will vary with subsequent experiments. Contrast this with the idea that probability is a number derived from the assumption that all outcomes are equally likely or from other assumptions.

Also, help students recall that *randomly* describes equal likelihood of outcomes. That is, if outcomes occur randomly, nothing exists that would cause one outcome to be more likely than the other.

QUESTIONS

Covering the Reading

7) No; a chi-square value of 8.4 would occur less than 5% of the time. This is enough evidence to question whether students were guessing randomly.

1. What does the chi-square statistic measure? how different a set of actual observed scores is from a set of expected scores
2. When and by whom was the chi-square statistic developed? 1900; Karl Pearson

In 3–6, suppose frequencies of accidents for a stretch of highway are as given for each weekday. **a.** Calculate the chi-square statistic assuming that accidents occur on random days. **b.** Is there evidence to believe that the accidents are not occurring randomly?

3.

M	T	W	T	F
22	18	20	17	23

a) 1.3 b) No

4.

M	T	W	T	F
23	22	20	18	17

a) 1.3 b) No

5.

M	T	W	T	F
25	15	15	30	15

a) 10 b) Yes

6.

M	T	W	T	F
20	20	20	20	20

a) 0 b) No

Forty seconds that shook L.A. *Shown is one result of the devastating earthquake that struck in the Los Angeles area on January 17, 1994.*

7. Suppose in the Example of this lesson that 30 students had picked Abraham Lincoln, 18 had picked Millard Fillmore, and 12 had picked Martin Van Buren. Would there still be evidence that students were guessing randomly? See left.
8. For what expected frequencies should the chi-square statistic not be used? less than 5

Applying the Mathematics

9. *The World Almanac and Book of Facts 1994* lists 59 major earthquakes since 1940. Here are their frequencies by season of the year.
   Autumn, 13 Winter, 13 Spring, 12 Summer, 21
   Use the chi-square statistic to determine whether these figures support a view that more earthquakes occur at certain times of the year than at others. See below.

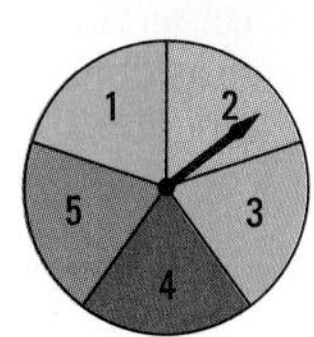

10. You build a spinner as shown at the left and spin it 40 times with the following outcomes.

Outcome	1	2	3	4	5
Frequency	10	6	4	6	14

Use the chi-square statistic to determine whether or not the spinner seems to be fair. Chi-square value = 8. Such a value would occur almost 10% of the time. There is not enough evidence to call the spinner unfair.

9) Chi-square value = 3.58. Such a value would occur over 10% of the time. There is not enough evidence to say earthquakes occur more in certain seasons.

Notes on Questions

Questions 3–4 These questions point out that switching the order of the data does not affect the chi-square value, since all expected frequencies are the same.

Question 6 Since there is no deviation of the observed values from the expected values, the chi-square value is 0. This rarely occurs in actual data.

Question 7 You might try a similar question with your students. Ask who was the President of the United States in 1900—Grover Cleveland, William McKinley, or Theodore Roosevelt. [William McKinley] Tabulate the values for the class.

Question 9 Science Connection We found these data by looking through an almanac. We did not expect the data to have one season in which earthquakes were so much more prevalent than others. Many students may think that this means earthquakes occur in the summertime. This may be the case, but the chi-square statistic points out that the data are not so far off from what might be expected. More data would be needed before one could conclude that earthquakes usually occur in the summer.

▶ LESSON MASTER 10-7 B *page 2*

Critical Chi-Square Values

$n-1$	.10	.05	.01	.001
1	2.71	3.84	6.63	10.8
2	4.61	5.99	9.21	13.8
3	6.25	7.81	11.34	16.3
4	7.78	9.49	13.28	18.5
5	9.24	11.07	15.09	20.5
6	10.6	12.6	16.8	22.5
7	12.0	14.1	18.5	24.3
8	13.4	15.5	20.1	26.1
9	14.7	16.9	21.7	27.9
10	16.0	18.3	23.2	29.6

d. Refer to the Critical Chi-Square Values Table above. Using your answer to Part **c**, do you think the researcher should recommend that the cereal be manufactured for national distribution? Explain your thinking.
Sample: No; there is a greater than 10% chance the results would occur in the equally-likely cases.

2. Suppose 1,000 people had been asked their opinion, and their percents were those given in the table on page 179.
   a. Calculate the chi-square statistic for these test results. (Follow the steps suggested by Parts **a** and **b** in Question 1.) 33
   b. Refer to the Critical Chi-Square Values Table above. Using your answer to Part **a**, do you think the researcher should recommend that the cereal be manufactured for national distribution? Explain your thinking.
   Sample: Yes; there is less than .1% chance the results would occur in the equally-likely cases.

3. From your answers to Questions 1 and 2, what can you say about the relationship between sample size and the conclusions reached by the market researcher?
   Sample: When the sample size is greater, there is more support for Oat-Toasties and the researcher's conclusion is correct.

Adapting to Individual Needs

Challenge Social Studies Connection
Materials: Almanac

Have students use an almanac to make up a problem similar to **Question 9.** For example, the *World Almanac and Book of Facts 1994* contains a wealth of information that could be used to make up problems.

Notes on Questions

Question 11 Overall, one might expect fewer runs in the last three innings, because in about half of all games, the home team does not bat in the ninth inning (when it is ahead).

Follow-up for Lesson 10-7

Practice

For more questions on SPUR Objectives, use **Lesson Master 10-7A** (shown on pages 652–653) or **Lesson Master 10-7B** (shown on pages 654–655).

Assessment

Oral Communication Ask students if the chi-square statistic can be used to prove that an event is occurring more often than expected. [Students understand that the chi-square statistic can give evidence that an event is occurring more often than expected, but it can not be used to prove that that is actually the case.]

Extension

Project Update Project 4, *Testing Astrology,* on page 658, relates to the content of this lesson.

11. Here are the total runs scored in each of the first nine innings for the 15 Major League Baseball games played on May 1, 1994.

Inning	1	2	3	4	5	6	7	8	9	Total
Runs	24	18	17	11	8	21	17	14	8	138

Group innings 1–3, 4–6, and 7–9 together to represent the beginning, middle, and end of a game. Use the chi-square statistic to help in answering this question: Do baseball teams tend to score more runs in the beginning, middle, or end of a game, or do the runs appear to be scored equally in these sections of the game? **Chi-square value ≈ 5.52. Such a value would occur over 5% of the time. There is not enough evidence to say that more runs are scored in one part of the game.**

Review

12. a. Calculate 71^2 in your head by thinking of it as $(70 + 1)^2$. **5041**

b. Calculate 69^2 in your head by thinking of it as $(70 - 1)^2$. **4761** *(Lesson 10-6)*

In 13 and 14, expand. *(Lessons 10-5, 10-6)* **See left.**

13. a. $(3a - b)^2$ **b.** $(3a + b)^2$ **c.** $(3a - b)(3a + b)$

14. a. $(5 - y)(y - 5)$ **b.** $(4x^2 + 8xy + 4y^2)^2$

13a) $9a^2 - 6ab + b^2$
b) $9a^2 + 6ab + b^2$
c) $9a^2 - b^2$

14a) $-25 + 10y - y^2$
b) $16x^4 + 64x^3y + 96x^2y^2 + 64xy^3 + 16y^4$

8p

p + 1

4p + 2

3p

15. A rectangle with dimensions $3p$ and $p + 1$ is contained in a rectangle with dimensions $8p$ and $4p + 2$, as in the figure at the left. *(Lesson 10-3)*

a. Write an expression for the area of the big rectangle. **$8p(4p + 2)$**

b. Write an expression for the area of the little rectangle. **$3p(p + 1)$**

c. Write a simplified expression for the area of the shaded region. **$8p(4p + 2) - 3p(p + 1) = 29p^2 + 13p$**

16. After five years of putting money into a retirement account at a scale factor x, a worker has $1000x^4 + 1100x^3 + 1200x^2 + 1400x + 1500$ dollars.

a. How much did the worker put in during the most recent year? **a) $1500**

b. How much did the worker put in during the first year? **$1000**

c. Give an example of a reasonable value for x in this problem, and evaluate the polynomial for that value of x. *(Lesson 10-2)* **Sample: 1.06; $6904.91**

In 17 and 18, simplify. *(Lessons 3-6, 3-7)*

17. $(12y^4 - 3y^3 + y) + (5y^3 - 7y^2 - 2y + 1) + (2y - 3y^2 + 6)$ **$12y^4 + 2y^3 - 10y^2 + y + 7$**

18. $(x^2 - 4x + 1) - (3x^2 - 2x) - 2(7x + 4)$ **$-2x^2 - 16x - 7$**

19. There are 6 girls, 8 boys, 4 women, and 3 men on a community youth board. How many different leadership teams consisting of one adult and one child could be formed from these people? *(Lesson 2-9)* **98 teams**

Exploration

20. Repeat Question 11 with more recent data. Compare your results with those you found in Question 11. **Answers will vary.**

656

A project presents an opportunity for you to extend your knowledge of a topic related to the material of this chapter. You should allow more time for a project than you do for typical homework questions.

1 Lifelong Savings

Sara and Sheila are twins who began to work at age 20 with identical jobs and identical salaries. At the end of each year they received identical bonuses of \$2000. In other ways the twins were not identical. Early in life Sara was conservative. Each year she invested the \$2000 bonus in a savings program earning 9% interest compounded annually. At age 30, Sara decided to have some fun and began spending her \$2000 bonus, but she let her earlier investment continue to earn interest. This continued until she was 65.

In contrast, for the first 10 years she worked, Sheila spent her \$2000 bonuses. At age 30, she began to invest her bonus every year in an account paying 9% annual interest. This continued until Sheila was 65 years old.

a. Which sister deposited more of her own money into her account? How much more?
b. Create a spreadsheet to determine how much each sister had in her account in each year.
c. How much had each sister accumulated at age 65?
d. What are the advantages of Sara's savings plan? What are the advantages of Sheila's savings plan?
e. There is a moral to the story of Sara and Sheila. What is the moral? What do you think of their story?

2 Powers of Binomials

You studied perfect square trinomials:

$$(a + b)^2 = a^2 + 2ab + b^2.$$

a. Expand $(a + b)^3$ and $(a + b)^4$.
b. What patterns do you observe in the coefficients?
c. Continue with higher powers if you can.

3 Another Proof of the Pythagorean Theorem

In the diagram below, a right triangle with legs of lengths a and b and hypotenuse of length c is replicated four times. The four triangles are placed so that the hypotenuses form a large square. Explain how to use the diagram to prove the Pythagorean Theorem.

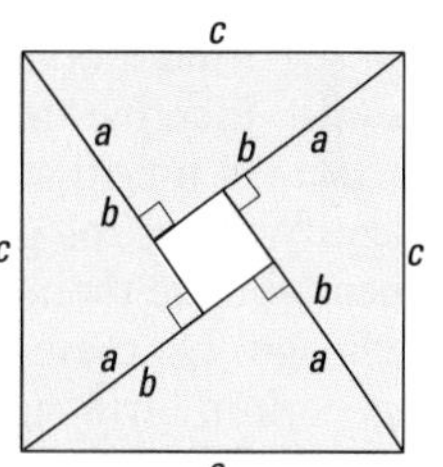

Chapter 10 Projects

The projects relate to the content of the lessons of this chapter as follows:

Project	Lesson(s)
1	10-2
2	10-6
3	10-6
4	10-7
5	10-1
6	10-4, 10-5
7	10-1

1 Lifelong Savings The financial records of both Sheila and Sara are divided into two parts. For the first ten years, Sara's end-of-year balance can be found by adding \$2000 to the previous end-of-year balance and then multiplying that sum by 1.09. For the remaining 35 years, her end-of-year balance can be found by multiplying the previous year's end-of-year balance by 1.09. Sheila's end-of-year balance remains at zero until the tenth year. From then on, the end-of-year balance can be found in the same way Sara's end-of-year balances for the first ten years were found. You might ask students to find an interest rate that would cause Sheila to have more money at age 65. [Interest at 6.21% or under compounded annually gives the late saver more money at age 65. Rates at 6.22% or higher gives the early saver more money.]

2 Powers of Binomials If students do not know how to begin expanding $(a + b)^3$, remind them that $(a + b)^3 = (a + b)(a + b)^2$. If they have difficulty recognizing that terms such as $2a^2b$ and ba^2 are like terms and can be combined to equal $3a^2b$, point out that ba^2 is the same as $1a^2b$. In **part b**, if students write the coefficients of the expansions of $(a + b)^0$, $(a + b)^1$, and $(a + b)^2$ and examine them along with those generated in **part a**, they may see Pascal's Triangle forming. The patterns give rise to the Binomial Theorem, a topic that is covered in second-year algebra texts.

3 Another Proof of the Pythagorean Theorem This is a relatively easy project. The proof is similar to one that is given in Lesson 10-6. Explanations of similar proofs can be found in many geometry texts including UCSMP *Geometry*. Hundreds of proofs may be found in *The Pythagorean Proposition,* by Elisha

Possible responses

1a. **Sheila; \$50,000**
b. **See the spreadsheet beginning on page 658.**
c. **Sara: \$676,122.60**
Sheila: \$470,249.45
d. **Sample: Sara invested her bonuses for ten years and then was able to spend them for the remaining 25 years. Sheila was able to enjoy her bonuses immediately—probably when her salary was lower; she started investing when her income was larger.**
e. **Sample: It is important to start saving money early in your career; one should look to the future rather than the present. Students' opinion of the story will vary.**

(Responses continue on page 658.)

Loomis. This is one of the classic books in the mathematics education series available from the National Council of Teachers of Mathematics.

4 Testing Astrology The celebrities shown at the top of the page from left to right are Oprah Winfrey, Arnold Schwarzenegger, Paula Abdul, Bill Cosby, and Barbra Streisand. Encourage students to find data on famous people in an area in which they are interested. The reference section of most libraries contains books listing birthdays of athletes, actresses and actors, artists, politicians, and musicians. If students collect baseball cards or any other type of cards, they can use them to develop a list of birthdays.

There is not a shred of scientific evidence to support any belief in astrology. Because the planets and stars are so far away from us, the gravitational influence on humans is less than our gravitational effects on each other.

5 Different Bases Students will be able to find information about the duodecimal system in some encyclopedias. Descriptions of the hexagesimal system can be found in some computer-science books.

6 The Size of Products Some students may find the largest product of two integers for several values of m and n, write the product as the sum or difference of powers of 10, and then try to find a general rule. Other students may recognize that the largest m-digit number is $10^m - 1$, and the largest n-digit number is $10^n - 1$; they might multiply the binomials together to find a general rule. Students will have to review the properties of exponents to find the formula for the smallest product of an m-digit and an n-digit number.

7 Representing Positive Integers Using Powers The first part is fairly straightforward and should not be difficult for most students. Suggest that they write each number as the sum of the values of powers of 2 as well as the sum of powers of 2. For example, 6 would be written as $6 = 2^2 + 2^1 = 4 + 2$, and 19 would be written as $19 = 2^4 + 2^1 + 2^0 = 16 + 2 + 1$. Writing numbers in this manner will provide a check to insure that each number is represented correctly.

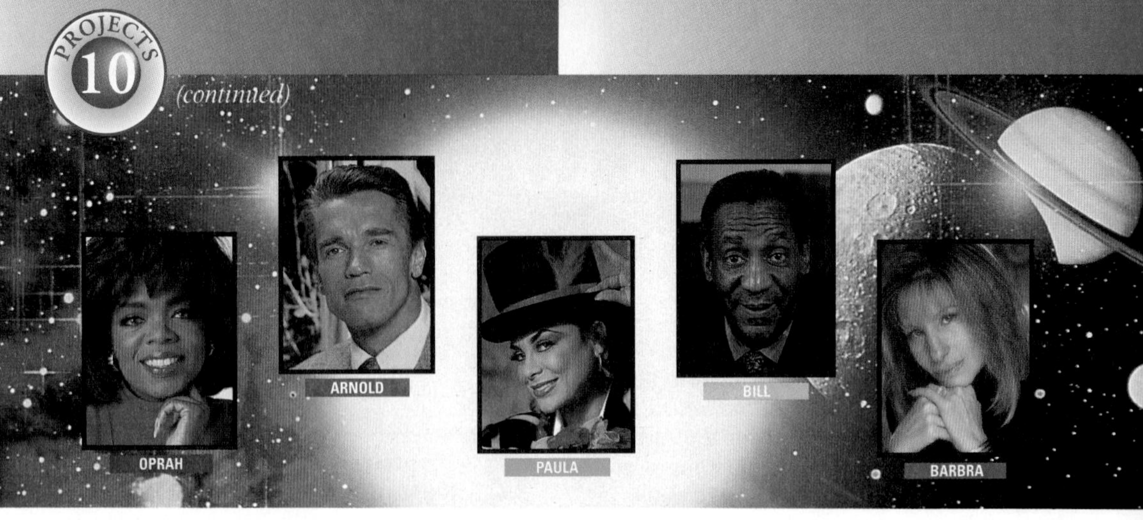

4 Testing Astrology

Find a book (like a *Who's Who*) that identifies at least 100 famous people in a field and gives their birthdates. Identify the astrological sign of each person. Then tabulate the number of people with each sign. Do these data lead you to believe that certain birth signs are more likely to produce famous people? Use a chi-square statistic assuming that a random distribution of the birthdays among the 12 astrological signs is expected. Since $n - 1 = 11$, refer to row 11 from the chi-square table on page 653.

5 Different Bases

Report on how positive integers are represented in the duodecimal system (base 12) or the hexadecimal system (base 16). Include examples on how to convert from base 10 into each system, and vice versa.

6 The Size of Products

The largest product of a 2-digit integer and a 3-digit integer is 99×999, or 98901. This equals $10^5 - 10^3 - 10^2 + 1$. The smallest product is 10×100, which equals 10^3.

a. Find a formula for the largest product of an m-digit and an n-digit integer.

b. Find a formula for the smallest product of these integers.

c. Explain why your formulas are correct.

7 Representing Positive Integers Using Powers

Do **a.** or **b.**

a. Every positive integer can be represented as a sum of different powers of 2. For instance, $100 = 2^6 + 2^5 + 2^2$. What may surprise you is that there is only one such representation. Find this power-of-2 representation for all the integers from 1 through 99. Try to explain why there is only one such representation.

b. Every positive integer can be represented as a sum or difference of different powers of 3. For instance, $100 = 3^4 + 3^3 - 3^2 + 3^0$. What may surprise you is that there is again only one such representation. Find this power-of-3 representation for all the integers from 1 through 99. Try to explain why there is only one such representation.

Additional responses, page 657

1. In the following spreadsheet
   A = Year, B = Sara's deposit,
   C = Sara's end-of-year balance,
   D = Sheila's deposit, and
   E = Sheila's end-of-year balance.

A	B	C	D	E
1	2000	2180.00	0	0.00
2	2000	4556.20	0	0.00
3	2000	7146.26	0	0.00
4	2000	9969.42	0	0.00
5	2000	13046.67	0	0.00
6	2000	16400.87	0	0.00
7	2000	20056.95	0	0.00
8	2000	24042.07	0	0.00
9	2000	28385.86	0	0.00
10	2000	33120.59	0	0.00
11	0	36101.44	2000	2180.00
12	0	39350.57	2000	4556.20

SUMMARY

A monomial is a product of terms. The degree of a monomial is the sum of the exponents of its variables. A polynomial is an expression that is either a monomial or a sum of monomials. The degree of a polynomial is taken to be the largest degree of its monomial terms. Linear expressions are polynomials of degree 1. Quadratic expressions are polynomials of degree 2. This chapter extends these ideas to consideration of polynomials of higher degree.

Polynomials emerge from a variety of situations. Our customary way of writing whole numbers in base 10 can be considered as a polynomial with 10 substituted for the variable. If different amounts of money are invested each year at a scale factor x, the total amount after several years is a polynomial in x. When the dimensions of a geometric figure are given as linear expressions, then areas or volumes related to the figure may be polynomials.

Addition and subtraction of polynomials are based on the Like Terms form of the Distributive Property, which you studied earlier in this book. Multiplication of polynomials is also justified by the Distributive Property. To multiply one polynomial by a second, multiply each term in the first polynomial by each term in the second, then add the products. For instance:

monomial by a polynomial:
$a(x + y + z) = ax + ay + az$

two polynomials: $(a + b + c)(x + y + z) = ax + ay + az + bx + by + bz + cx + cy + cz$

two binomials:
$(a + b)(c + d) = ac + ad + bc + bd$

perfect square patterns:
$(a + b)^2 = (a + b)(a + b) = a^2 + 2ab + b^2$
$(a - b)^2 = (a - b)(a - b) = a^2 - 2ab + b^2$

difference of two squares:
$(a + b)(a - b) = a^2 - b^2$

The square of the difference of actual and expected values in an experiment, $(a - e)^2$, appears in the calculation of the chi-square statistic. This statistic can help you to decide whether the assumptions that led to the expected values are correct.

VOCABULARY

Below are the most important terms and phrases for this chapter. You should be able to give a general description and a specific example of each.

Lesson 10-1
monomial, binomial, trinomial
degree of a monomial
polynomial
degree of a polynomial
polynomial in x
standard form of a polynomial
linear polynomial, quadratic polynomial

Lesson 10-4
Extended Distributive Property

Lesson 10-5
algorithm
FOIL algorithm

Lesson 10-6
expanding a binomial
perfect square trinomial
Perfect Square Patterns
difference of squares
Difference of Two Squares Pattern

Lesson 10-7
expected number
deviation
Chi-square statistic
degrees of freedom

Summary

The Summary gives an overview of the entire chapter and provides an opportunity for students to consider the material as a whole. Thus, the Summary can be used to help students relate and unify the concepts that are presented in the chapter.

Vocabulary

Terms, symbols, and properties are listed by lesson to provide a checklist of concepts that students must know. Emphasize that students should read the vocabulary list carefully before starting the Progress Self-Test. If students do not understand the meaning of a term, they should refer back to the indicated lesson.

A	B	C	D	E
13	0	42892.12	2000	7146.26
14	0	46752.41	2000	9969.42
15	0	50960.13	2000	13046.67
16	0	55546.54	2000	16400.87
17	0	60545.73	2000	20056.95
18	0	65994.84	2000	24042.07
19	0	71934.38	2000	28385.86
20	0	78408.47	2000	33120.59

A	B	C	D	E
21	0	85465.24	2000	38281.44
22	0	93157.11	2000	43906.77
23	0	101541.25	2000	50038.38
24	0	110679.96	2000	56721.83
25	0	120641.16	2000	64006.80
26	0	131498.86	2000	71947.41
27	0	143333.76	2000	80602.68
28	0	156233.80	2000	90036.92

A	B	C	D	E
29	0	170294.84	2000	100320.24
30	0	185621.37	2000	111529.06
31	0	202327.30	2000	123746.68
32	0	220536.75	2000	137063.88
33	0	240385.06	2000	151579.63
34	0	262019.72	2000	167401.79
35	0	285601.49	2000	184647.95
36	0	311305.63	2000	203446.27

(Responses continue on page 660.)

Progress Self-Test

The Progress Self-Test provides the opportunity for feedback and correction; the Chapter Review provides additional opportunities and practice.

Assign the Progress Self-Test as a one-night assignment. Worked-out *solutions* for all questions are in the Selected Answers section of the student book. Encourage students to take the Progress Self-Test honestly, grade themselves, and then be prepared to discuss the test in class.

Advise students to pay special attention to those Chapter Review questions which correspond to the questions that they missed on the Progress Self-Test.

Additional Answers, page 660

10. $15x^3 - 4x^2 - 9x - 1$

11. $3t^3 + 8t^2 - 7t + 1$

12. $ax + bx + 2x + ay + by + 2y + 5a + 5b + 10$

14.

	a	b
c	ac	bc
d	ad	bd
d	ad	bd

$ac + bc + 2ad + 2bd$

21. Yes; the number of events n is 4; $n - 1$ is 3. Look at the third row of the table. The Chi-Square statistic 23.2 is greater than the critical value 16.3 that would occur with probability .001. So it is very unlikely that the sophomores were given fewer lines by chance.

CHAPTER TEN

PROGRESS SELF-TEST

Take this test as you would take a test in class. Then check your work with the solutions in the Selected Answers section in the back of the book.

In 1–3, consider the polynomial $4x^2 - 7x + 9x^2 - 12 - 11$.

1. Write this polynomial in standard form. $13x^2 - 7x - 23$

2. What is the degree of this polynomial? 2

3. Is the simplified polynomial a monomial, binomial, trinomial, or none of these? trinomial

In 4–9, perform the indicated operations and simplify.

4. Multiply $3v^2 - 9 + 2v$ by 4. $12v^2 + 8v - 36$

5. $-5z(z^2 - 7z + 8)$ $-5z^3 + 35z^2 - 40z$

6. $(3x - 8)(3x + 8)$ $9x^2 - 64$

7. $(4y - 2)(3y - 16)$ $12y^2 - 70y + 32$

8. Expand $(d - 12)^2$. $d^2 - 24d + 144$

9. $(x - 3)(x^2 - 6x + 9)$ $x^3 - 9x^2 + 27x - 27$

In 10–12, write as a single polynomial. See margin.

10. $(3x^2 - 10x) + (15x^3 - 7x^2 + x - 1)$

11. $8t^3 + t^2 - 7t + 1 - (5t^3 - 7t^2)$

12. $(x + y + 5)(a + b + 2)$

13. Write the area of the shaded region as a polynomial in standard form. Each polygon is a rectangle. $11x^2 + x + 2$

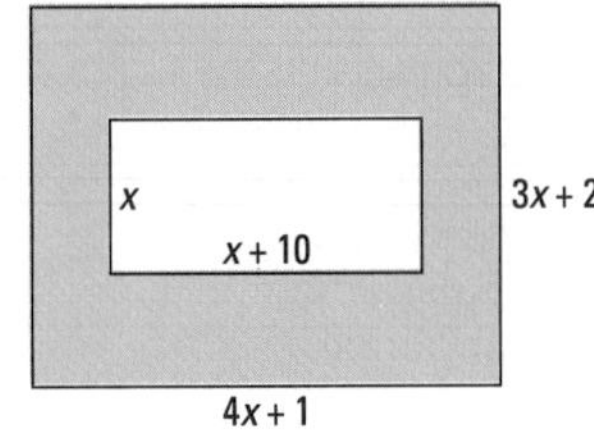

14. Represent the product $(a + b)(c + 2d)$ using areas of rectangles. See margin.

15. Show how you can compute $29 \cdot 31$ mentally. $(30 - 1)(30 + 1) = 900 - 1 = 899$

16. Write 26,384 as a polynomial in standard form with 10 substituted for the variable x. $2 \cdot 10^4 + 6 \cdot 10^3 + 3 \cdot 10^2 + 8 \cdot 10 + 4$

In 17 and 18, use this information. On his 18th birthday, Hank received \$80. He received \$60 on his 19th birthday and \$90 on his 20th birthday.

17. If he had invested all this money at a scale factor x, how much total money would he have on his 20th birthday? $80x^2 + 60x + 90$

18. Evaluate your answer to Question 15 when $x = 1.04$. \$238.93

In 19–21, consider this situation. The Vulcan High School newspaper is supposed to give equal coverage to each of its four classes, freshmen, sophomores, juniors, and seniors. The sophomores thought that they were being shortchanged. One student totaled the numbers of lines devoted to students in each of the classes in the newspaper this year. Here is what the student found.

Class	Number of lines
Freshmen	861
Sophomores	748
Juniors	812
Seniors	939

19. Suppose the expected number of lines for each class had been the same. What would that expected number have been? 840

20. Calculate the chi-square statistic for this situation, assuming the expected number of lines for each class had been the same. 20) ≈ 23.2

21. Use the Critical Chi-Square Values table on page 653. Does this support the view of the sophomores? Why or why not? See margin.

22. Write a simplified polynomial expression for the volume of the figure below. $10y^3 + 84y^2 - 54y$

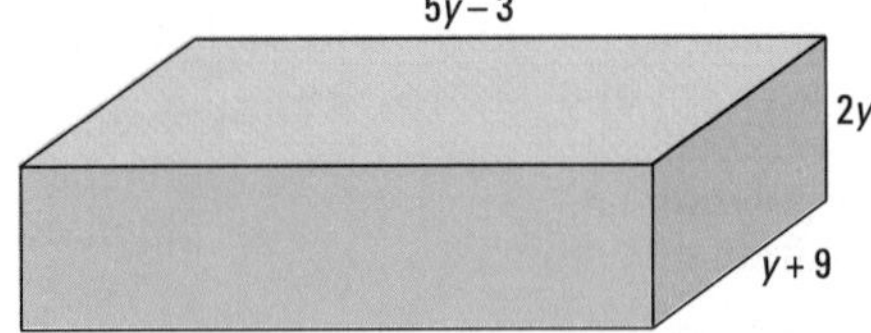

Additional responses, pages 657–658

A	B	C	D	E
37	0	339323.13	2000	223936.43
38	0	369862.21	2000	246270.71
39	0	403149.81	2000	270615.08
40	0	439433.30	2000	297150.43
41	0	478982.29	2000	326073.97
42	0	522090.70	2000	357600.63
43	0	569078.86	2000	391964.69

A	B	C	D	E
44	0	620295.96	2000	429421.51
45	0	676122.60	2000	470249.45
Total Profit		656122.60		400249.45

2a. $(a + b)^3 = a^3 + 3a^2b + 3ab^2 + b^3$
$(a + b)^4 = a^4 + 4a^3b + 6a^2b^2 + 4ab^3 + b^4$

b. Students might observe that in each term of the expansion of $(a + b)^n$, the sum of the exponents of a and b is n; all powers of a from a^n to a^0 occur in descending order while the powers of b from b^0 to b^n occur in ascending order; the pattern of coefficients is found in Pascal's Triangle. The coefficients of the expansion of $(a + b)^n$ are the nth row of Pascal's Triangle.

CHAPTER REVIEW

Questions on SPUR Objectives

SPUR stands for **S**kills, **P**roperties, **U**ses, and **R**epresentations. The Chapter Review questions are grouped according to the SPUR Objectives for this chapter.

SKILLS DEAL WITH THE PROCEDURES USED TO GET ANSWERS.

Objective A: *Add and subtract polynomials.* *(Lesson 10-2)* **2) $8m^4 + 10m^3 - 6m^2 - 3m$; 4**

In 1 and 2, **a.** simplify the expression; **b.** give the degree of the simplified expression.

1. $(5x^2 - 3x) + (2x^2 + 7x + 1)$ **$7x^2 + 4x + 1$; 2**
2. $(8m^4 - 2m^3) + (12m^3 - 6m^2 - 3m)$
3. Add $1.3x^2 + 14$, $4.7x - 1$, and $2.6x^2 - 3x + 6$.
4. Subtract $4y^5 - 2y^3 + 8y$ from $4y^5 - 6y^3 + 4y + 2$. **$-4y^3 - 4y + 2$; 3**

In 5 and 6, simplify. **3) $3.9x^2 + 1.7x + 19$; 2**

5. $(k - 4) - (k^2 + 1)$ **$-k^2 + k - 5$**
6. $(5p^2 - 1) - (6p^2 - p)$ **$-p^2 + p - 1$**

Objective B: *Multiply polynomials.* *(Lesson 10-4)*

In 7–12, write as a single polynomial.

7. $(a + b + 1)(c + d + 1)$ **See below.**
8. $x^2(x^2 + 4) - x(x^3 - 3)$ **$4x^2 + 3x$**
9. $(y - 1)(y - 1)(y + 1)$ **$y^3 - y^2 - y + 1$**
10. $(x^2 + x - 1)(x + 3)$ **$x^3 + 4x^2 + 2x - 3$**
11. $(x + 1)(a + b + 3)$ **$ax + bx + 3x + a + b + 3$**
12. $(y - 2)(y^2 + 2y + 3)$ **$y^3 - y - 6$**

7) $ac + ad + a + bc + bd + b + c + d + 1$

Objective C: *Multiply a polynomial by a monomial or multiply two binomials.* *(Lessons 10-3, 10-5, 10-6)* **13–19) See margin.**

In 13–22, multiply and simplify, if possible.

13. $3k(k^2 + 4k - 1)$
14. $5xy(x + 3y^2)$
15. $2(4x^2 - x - 4) + 4(3x - 7)$
16. $(x - 3)(x + 7)$
17. $(y + 1)(y - 13)$
18. $(a - b)(c - d)$
19. $(a + 15)(a - 15)$
20. $(12b + m)(12b - m)$ **$144b^2 - m^2$**
21. $(4z + 1)(-z - 1)$ **$-4z^2 - 5z - 1$**
22. $(a + 3)(a^2 - 1)$ **$a^3 + 3a^2 - a - 3$**

Objective D: *Expand squares of binomials.* *(Lesson 10-6)*

In 23–28, expand. **23–26) See below.**

23. $(d - 1)^2$
24. $(2t + 3)^2$
25. $3(4x + 5)^2$
26. $(a_1 - e_1)^2$
27. $(x)(x + 1)^2$ **$x^3 + 2x^2 + x$**
28. $(m + 3n)^2 - (m - 3n)^2$ **$12mn$**

23) $d^2 - 2d + 1$ **24) $4t^2 + 12t + 9$**
25) $48x^2 + 120x + 75$ **26) $a_1^2 - 2a_1e_1 + e_1^2$**

PROPERTIES DEAL WITH THE PRINCIPLES BEHIND THE MATHEMATICS.

Objective E: *Classify polynomials by their degree or number of terms.* *(Lesson 10-1)*

In 29–32, consider the following polynomials:

(a) $x^2 - 7$ (b) $x^3 - 5x^2 + 6$
(c) $x^2 + 8xy + 15y^2$ (d) $x^2 - 5x + 6$

29. Which are binomials? **(a)**
30. Which are trinomials? **(b), (c), (d)**
31. Which have degree 2? **(a), (c), (d)**
32. Which have degree 3? **(b)**
33. Give an example of a monomial of degree 4.
34. Give an example of a trinomial of degree 4.

33) Sample: x^4 **34) Sample: $a^4 - 2a^2b + b^2$**

Chapter 10 Review

Resources

From the ***Teacher's Resource File***

- Answer Master for Chapter 10 Review
- Assessment Sourcebook Chapter 10 Test, Forms A–D Chapter 10 Test, Cumulative Form

Additional Resources

- Quiz and Test Writer

The main objectives for the chapter are organized in the Chapter Review under the four types of understanding this book promotes—Skills, Properties, Uses, and Representations.

Whereas end-of-chapter material may be considered optional in some texts, in *UCSMP Algebra* we have selected these objectives and questions with the expectation that they will be covered. Students should be able to answer these questions with about 85% accuracy.

You may assign these questions over a single night to help students prepare for a test the next day, or you may assign the questions over a two-day period. If you work the questions over two days, then we recommend assigning the *evens* for homework the first night so that students get feedback in class the next day, then assigning the *odds* the night before the test because answers are provided to the odd-numbered questions.

(Review continues on page 662.)

Additional Answers, page 661

13. $3k^3 + 12k^2 - 3k$
14. $5x^2y + 15xy^3$
15. $8x^2 + 10x - 36$
16. $x^2 + 4x - 21$
17. $y^2 - 12y - 13$
18. $ac - ad - bc + bd$
19. $a^2 - 225$

c. $(a + b)^5 = a^5 + 5a^4b + 10a^3b^2 + 10a^2b^3 + 5ab^4 + b^5$
$(a + b)^6 = a^6 + 6a^5b + 15a^4b^2 + 20a^3b^3 + 15a^2b^4 + 6ab^5 + b^6$
$(a + b)^7 = a^7 + 7a^6b + 21a^5b^2 + 35a^4b^3 + 35a^3b^4 + 21a^2b^5 + 7ab^6 + b^7$

3. Sample: The area of the large square is c^2. The area also equals the area of the small square with side $b - a$ plus the areas of the four right triangles.

$$c^2 = (b - a)^2 + 4(\tfrac{1}{2}ab) = b^2 - 2ab + a^2 + 2ab = a^2 + b^2$$

4. Responses will vary. The chi-square values will differ, but students should find that no one birth sign is more likely to produce famous people.

(Responses continue on page 662.)

CHAPTER TEN

It is effective to ask students which questions they still do not understand and use the day or days as a total class discussion of the material that the class finds most difficult.

Assessment

Evaluation The *Assessment Sourcebook* provides five forms of the Chapter 10 Test. Forms A and B present parallel versions in a short-answer format. Forms C and D offer performance assessment. The fifth test is Chapter 10 Test, Cumulative Form. About 50% of this test covers Chapter 10; 25% covers Chapter 9, and 25% covers earlier chapters.

For information on grading, see *General Teaching Suggestions; Grading* in the *Professional Sourcebook*, which begins on page T20 in Part 1 of the Teacher's Edition.

Feedback After students have taken the test for Chapter 10 and you have scored the results, return the tests to students for discussion. Class discussion on the questions that caused trouble for most students can be very effective in identifying and clarifying misunderstandings. It is important for students to receive feedback on every chapter test, and we recommend that students see and correct their mistakes before proceeding too far into the next chapter.

Additional Answers, page 662

42c. Yes. The chi-square value 4.67 is less than the critical value 9.49 that would occur with probability .05.

43c. Yes. The chi-square statistic 7.11 is less than the critical value 19.7 that would occur with probability .05. This is not a high enough chi-square value to support a claim that the temperatures are different throughout the year.

Objective F: *Write whole numbers as polynomials in base 10.* (*Lesson 10-1*)

In 35 and 36, simplify.

35. $3 \cdot 10^7 + 2 \cdot 10^5 + 9 \cdot 10^2 + 1$ 30,200,901

36. $10^4 + 2 \cdot 10^3 + 2 \cdot 10^2 + 10$ 12,210

In 37 and 38, write as a polynomial in base 10.

37. 98,103 $9 \cdot 10^4 + 8 \cdot 10^3 + 1 \cdot 10^2 + 3$

38. 4,005,600 $4 \cdot 10^6 + 5 \cdot 10^3 + 6 \cdot 10^2$

USES DEAL WITH APPLICATIONS OF MATHEMATICS IN REAL SITUATIONS.

Objective G: *Translate investment situations into polynomials.* (*Lesson 10-2*)

39. Each birthday from age 11 on Katherine has received \$250. She puts the money in a savings account with a scale factor of x.

a. Write an expression which shows how much Katherine will have after her 15th birthday. $250x^4 + 250x^3 + 250x^2 + 250x + 250$

b. If the bank pays 8% interest a year, calculate how much Katherine will have after her 13th birthday. \$811.60

40. Jose received \$25 on his 12th birthday, \$50 on his 13th birthday and \$75 on his 14th birthday, which he invested at a scale factor y. He kept his money in the same account at the same scale factor for 4 more years.

a. How much money did he have in this account at the end of that time?

b. If $y = 1.05$, how much money did he have in the account on his 15th birthday? \$162.82

a) $25y^6 + 50y^5 + 75y^4$ dollars

Objective H: *Use the chi-square statistic to determine whether or not an event is likely.* (*Lesson 10-7*)

41. In a taste test of two colas, 100 people were asked which cola they preferred. 56 preferred cola A and 44 preferred cola B.

a. If the colas are of equal taste, what are the expected numbers of preference for colas A and B? 50

b. Calculate the chi-square statistic for this situation, using the actual numbers and the expected numbers from part **a.** 1.44

c. Use the Critical Chi-Square Values table on page 653. Does the evidence support the fact that cola A is preferred by more people than cola B? Explain why or why not. No; the chi-square value 1.44 is less than the critical value 3.84 that would occur with probability .05.

42. A company was open only Monday through Friday. Because it was not open Saturday or Sunday, it expected that it would get about the same amount of mail Tuesday through Friday, but three times this amount on Monday. However, some people thought there was too much mail coming on Monday. When the numbers of pieces of mail for each day for a few weeks were totaled, here were the numbers on each day.

Day of week	Mon	Tue	Wed	Thu	Fri
Pieces of mail	143	38	51	40	36

a. How many pieces of mail did the company expect each day? Mon: 132; Tue–Fri: 44

b. Calculate the chi-square statistic for this situation, using the actual numbers and the expected numbers from part **a.** 4.67

c. Use the Critical Chi-Square Values table on page 653. Does the evidence support the company's expectations on how much mail to expect? Explain why or why not. 42c, 43c) See margin.

43. Some people say that the temperature of Los Angeles is "the same the year round."

Month	J	F	M	A	M	J	J	A	S	O	N	D
Actual Mean	57	58	60	61	65	69	74	75	72	68	63	58

a. What mean temperature would be expected each month if the temperature stayed the same all year? 65°; b) ≈ 7.11

b. Calculate the chi-square statistic for this situation, using the means given above and the expected number from part **a.**

c. Use the Critical Chi-Square Values table on page 653. Use $n = 12$. Does the chi-square statistic support the claim that the temperature is the same all year? See margin.

5. Sample response: The duodecimal system is a base 12 system. It requires two more symbols. Often the letters T (ten) and E (eleven) are used. The values for the first five places in the duodecimal system are shown in the following table:

12^4	12^3	12^2	12^1	12^0
20,736	1728	144	12	1

To convert 1000 (base ten) into the duodecimal system, note that it is between 1728 and 144, so it has three digits. Dividing 1000 by 144 shows that there are 6 groups of 144 with a remainder of 136, dividing 136 by 12 shows that there are 11 groups of 12 with 4 left over. Therefore 1000 (base ten) = 6E4 (base twelve). To convert a duodecimal number to a base ten number, use duodecimal place value: 20TE = $2 \cdot 1728 + 0 \cdot 144 + 10 \cdot 12 + 11 \cdot 1 = 3587$.

Converting to and from the hexagesimal system is similar, but five new symbols are needed. The first five places in the hexagesimal system are shown below:

16^4	16^3	16^2	16^1	16^0
65,536	4096	256	16	1

CHAPTER TEN

REPRESENTATIONS DEAL WITH PICTURES, GRAPHS, OR OBJECTS THAT ILLUSTRATE CONCEPTS.

Objective I: *Represent areas and volumes of figures with polynomials.*
(Lessons 10-1, 10-3, 10-4, 10-5, 10-6)

In 44 and 45, a rectangle is given.

a. Write the area of the figure as a polynomial.

b. Write the area as a product of polynomials.

44.

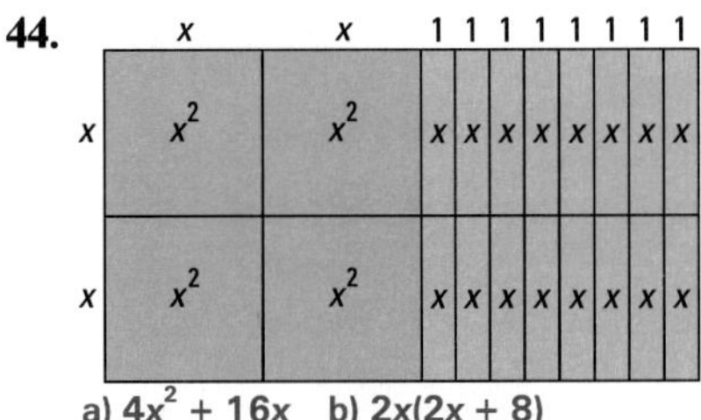

a) $4x^2 + 16x$ b) $2x(2x + 8)$

45.

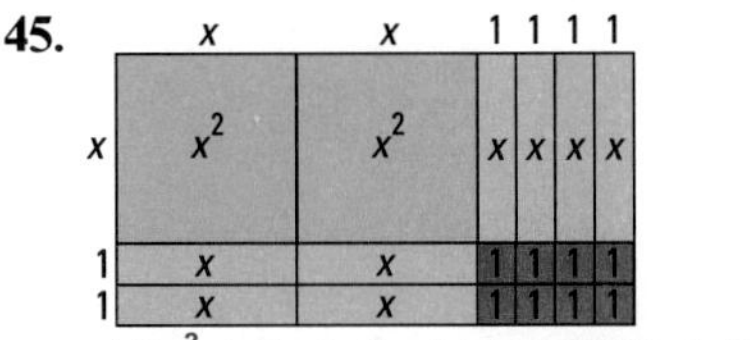

a) $2x^2 + 8x + 8$ b) $(x + 2)(2x + 4)$

46. Represent $(a + b)(c + d)$ using areas of rectangles. See margin.

47. a. Write the area of rectangle *ABCD* below as the sum of 4 terms. $xy + 3y + 2x + 6$

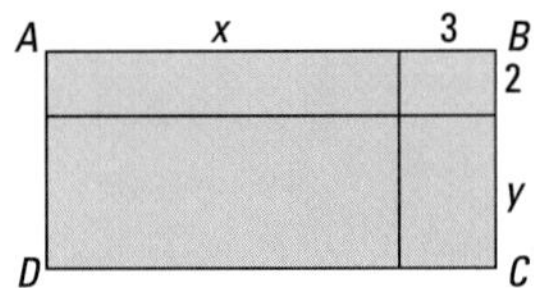

b. Write the area of *ABCD* as the product of 2 binomials. $(x + 3)(y + 2)$

c. Are the answers to parts **a** and **b** equal? Yes

In 48–51, express the area of the shaded region as a polynomial.

48.

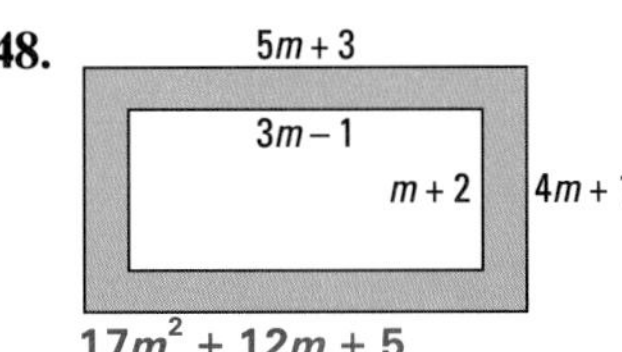

$17m^2 + 12m + 5$

49.

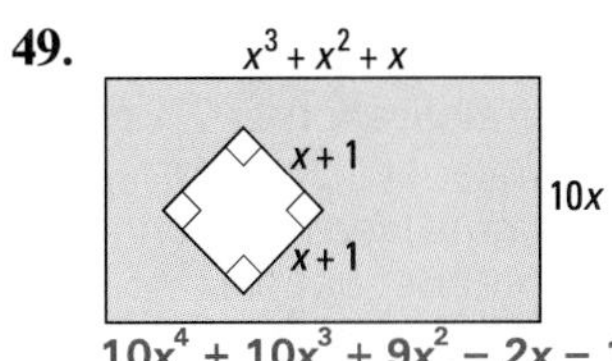

$10x^4 + 10x^3 + 9x^2 - 2x - 1$

50.

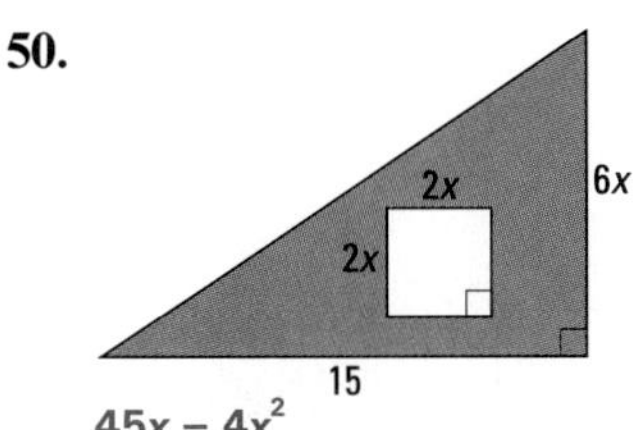

$45x - 4x^2$

51.

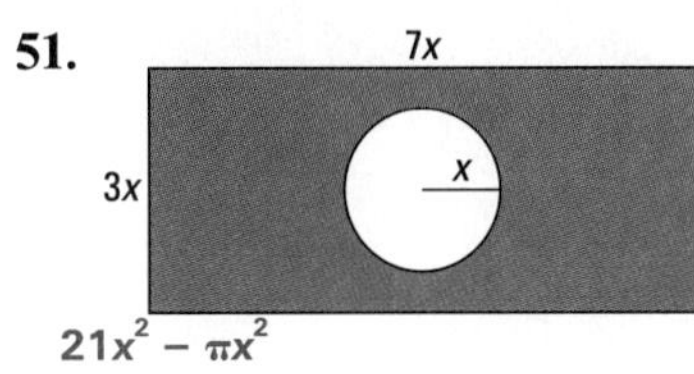

$21x^2 - \pi x^2$

52. Write a polynomial for the volume of the figure below. $24x^3 + 291x^2 + 36x$

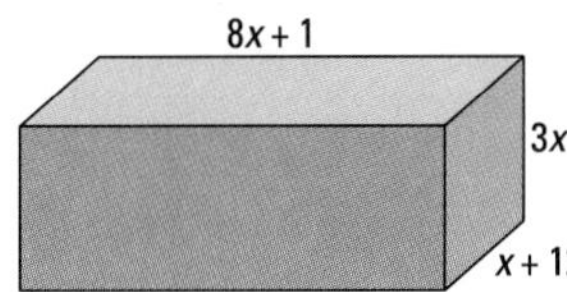

53. A box has dimensions x, $x + 1$, and $x - 1$. Write its volume as a polynomial. $x^3 - x$

6. In the following responses, m = the number of digits of one number, and n = the number of digits of the other number.
   a. The largest product of an m-digit and an n-digit number is $10^{m+n} - 10^m - 10^n + 1$.
   b. The smallest product of these integers is 10^{m+n-2}.
   c. The quantity $(10^m - 1)$ is the largest m-digit number, the quantity $(10^n - 1)$ is the largest n-digit number: $(10^m - 1)(10^n - 1) = 10^{m+n} - 10^m - 10^n + 1$. The smallest m-digit number is 10^{m-1}, the smallest n-digit number is 10^{n-1}: $(10^{m-1})(10^{n-1}) = 10^{m+n-2}$

(Responses continue in the side column.)

Addtional responses, continued, page 658

7a. Representations for integers 1–16 as the sum of powers of 2 are shown below.

$1 = 2^0$
$2 = 2^1$
$3 = 2^1 + 2^0$
$4 = 2^2$
$5 = 2^2 + 2^0$
$6 = 2^2 + 2^1$
$7 = 2^2 + 2^1 + 2^0$
$8 = 2^3$
$9 = 2^3 + 2^0$
$10 = 2^3 + 2^1$
$11 = 2^3 + 2^1 + 2^0$
$12 = 2^3 + 2^2$
$13 = 2^3 + 2^2 + 2^0$
$14 = 2^3 + 2^2 + 2^1$
$15 = 2^3 + 2^2 + 2^1 + 2^0$
$16 = 2^4$

7b. Representations for integers 1–16 as sums/differences of powers of 3 are shown below.

$1 = 3^0$
$2 = 3^1 - 3^0$
$3 = 3^1$
$4 = 3^1 + 3^0$
$5 = 3^2 - 3^1 - 3^0$
$6 = 3^2 - 3^1$
$7 = 3^2 - 3^1 + 3^0$
$8 = 3^2 - 3^0$
$9 = 3^2$
$10 = 3^2 + 3^0$
$11 = 3^2 + 3^1 - 3^0$
$12 = 3^2 + 3^1$
$13 = 3^2 + 3^1 + 3^0$
$14 = 3^3 - 3^2 - 3^1 - 3^0$
$15 = 3^3 - 3^2 - 3^1$
$16 = 3^3 - 3^2 - 3^1 + 3^0$

Sample: In part a, each integer can be uniquely represented by the sum of different powers of 2. This explains why any integer can be uniquely represented in base 2. In part b, the sum of the first n powers of 3 equals half of one less than the $(n + 1)$st power of 3. If the numbers from 1 to 3^n can be represented uniquely, then the numbers from the first half of the interval from 3^n to 3^{n+1} can be represented by adding these numbers to 3^n. The second half can be represented by subtracting these representations from 3^{n+1}.

Additional Answers, page 663

46.

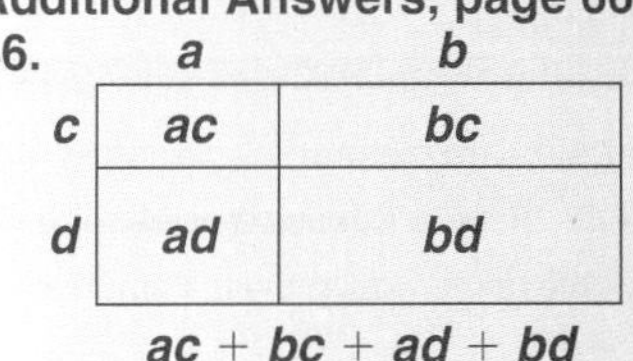

	a	b
c	ac	bc
d	ad	bd

$ac + bc + ad + bd$

Chapter 11 Planner

Adapting to Individual Needs

The student text is written for the vast majority of students. The chart at the right suggests two pacing plans to accommodate the needs of your students. Students in the Full Course should complete the entire text by the end of the year. Students in the Minimal Course will spend more time when there are quizzes and more time on the Chapter Review. Therefore, these students may not complete all of the chapters in the text.

Options are also presented to meet the needs of a variety of teaching and learning styles. For each lesson, the Teacher's Edition provides sections entitled: *Video* which describes video segments and related questions that can be used for motivation or extension; *Optional Activities* which suggests activities that employ materials, physical models, technology, and cooperative learning; and, *Adapting to Individual Needs* which regularly includes **Challenge** problems, **English Language Development** suggestions, and suggestions for providing **Extra Help.** The Teacher's Edition also frequently includes an **Error Alert,** an **Extension,** and an **Assessment** alternative. The options available in Chapter 11 are summarized in the chart below.

Chapter 11 Pacing Chart

Day	Full Course	Minimal Course
1	11-1	11-1
2	11-2	11-2
3	11-3	11-3
4	Quiz*; 11-4	Quiz*; begin 11-4.
5	11-5	Finish 11-4.
6	11-6	11-5
7	Quiz*; 11-7	11-6
8	11-8	Quiz*; begin 11-7.
9	Self-Test	Finish 11-7.
10	Review	11-8
11	Test*	Self-Test
12		Review
13		Review
14		Test*

*in the Teacher's Resource File

In the Teacher's Edition...

Lesson	Optional Activities	Extra Help	Challenge	English Language Development	Error Alert	Extension	Cooperative Learning	Ongoing Assessment
11-1	●	●	●	●	●	●	●	Group
11-2	●	●	●			●	●	Oral
11-3	●	●	●	●		●	●	Group
11-4	●	●	●	●	●	●		Written
11-5	●	●	●	●		●	●	Written
11-6	●	●	●	●		●		Oral
11-7	●	●	●	●		●	●	Group
11-8	●	●	●	●		●		Oral

In the Additional Resources...

	In the Teacher's Resource File								
Lesson	Lesson Masters, A and B	Teaching Aids*	Activity Kit*	Answer Masters	Technology Sourcebook	Assessment Sourcebook	Visual Aids**	Technology	Video Segments
Opener		124					124		
11-1	11-1	28, 121, 124, 125	24	11-1			28, 121, 124, 125, AM		
11-2	11-2	28, 121		11-2			28, 121, AM		
11-3	11-3	28, 122		11-3		Quiz	28, 122, AM		
11-4	11-4	28, 122, 126		11-4			28, 122, 126, AM		
11-5	11-5	122		11-5			122, AM		11-5
11-6	11-6	28, 123	25	11-6	Comp 27	Quiz	28, 123, AM	GraphExplorer	
11-7	11-7	28, 123		11-7			28, 123, AM		
11-8	11-8	28, 123		11-8	Comp 28		28, 123, AM	GraphExplorer	
End of chapter				Review		Tests			

*Teaching Aids are pictured on pages 664C and 664D. The activities in the Activity Kit are pictured on page 664C.

**Visual Aids provide transparencies for all Teaching Aids and all Answer Masters.

Also available is the Study Skills Handbook which includes study-skill tips related to reading, note-taking, and comprehension.

Integrating Strands and Applications

	11-1	11-2	11-3	11-4	11-5	11-6	11-7	11-8
Mathematical Connections								
Number Sense	●							
Algebra	●	●	●	●	●	●	●	●
Geometry	●	●		●	●	●	●	●
Measurement					●			
Logic and Reasoning	●				●			●
Probability					●			
Patterns and Functions						●	●	
Discrete Mathematics								●
Interdisciplinary and Other Connections								
Art			●					
Music					●		●	●
Science			●	●	●	●		
Social Studies	●	●	●	●	●		●	
Multicultural	●		●	●			●	●
Technology		●			●	●		●
Career				●			●	●
Consumer		●	●	●	●	●	●	●
Sports	●					●		●

Teaching and Assessing the Chapter Objectives

Chapter 11 Objectives (Organized into the SPUR categories—Skills, Properties, Uses, and Representations)	**Lessons**	**Progress Self-Test Questions**	**Chapter Review Questions**	**In the Teacher's Resource File** Chapter Test, Forms A and B	Chapter Test, Forms C	Chapter Test, Forms D
Skills						
A: Solve systems using substitution.	11-2, 11-3	1, 2, 5	1–10	2, 3	5	
B: Solve systems by addition.	11-4	3	11–16	1		
C: Solve systems by multiplying.	11-5	4	17–22	4	4	
Properties						
D: Recognize sentences with no solution, one solution, or all real numbers as solutions.	11-7	15, 16	23–28	12–14		
E: Determine whether a system has no solution, one solution, or infinitely many solutions.	11-6	6, 7	29–36	7, 8	3	
Uses						
F: Use systems of linear equations to solve real-world problems.	11-2, 11-3, 11-4, 11-5, 11-6, 11-7	10–13	37–44	9, 11	2	✓
G: Use systems of linear inequalities to solve real-world problems.	11-8	14	45–48	10		✓
Representations						
H: Find solutions to systems of equations by graphing.	11-1, 11-6	8	49–54	5	1	
I: Graphically represent solutions to systems of linear inequalities.	11-8	9, 14	55–60	6	5	✓

Multidimensional Assessment
Quiz for Lessons 11-1 through 11-3
Quiz for Lessons 11-4 through 11-7
Chapter 11 Test, Forms A–D
Chapter 11 Test, Cumulative Form

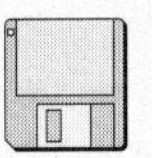

Quiz and Test Writer
Multiple forms of chapter tests and quizzes; Challenges

Activity Kit

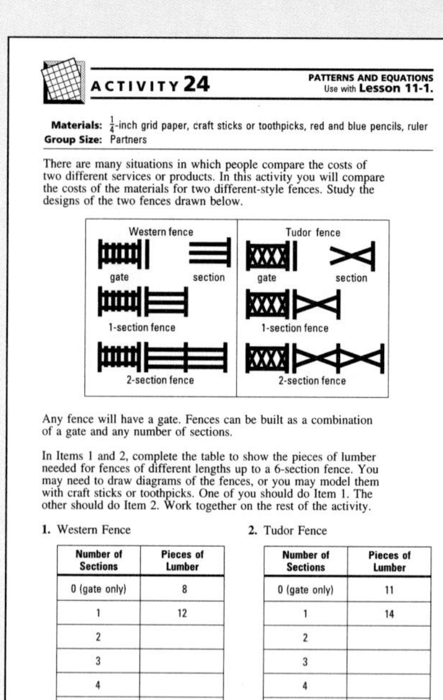

ACTIVITY 24 — PATTERNS AND EQUATIONS
Use with **Lesson 11-1.**

Materials: $\frac{1}{4}$-inch grid paper, craft sticks or toothpicks, red and blue pencils, ruler
Group Size: Partners

There are many situations in which people compare the costs of two different services or products. In this activity you will compare the costs of the materials for two different-style fences. Study the designs of the two fences drawn below.

Any fence will have a gate. Fences can be built as a combination of a gate and any number of sections.

In Items 1 and 2, complete the table to show the pieces of lumber needed for fences of different lengths up to a 6-section fence. You may need to draw diagrams of the fences, or you may model them with craft sticks or toothpicks. One of you should do Item 1. The other should do Item 2. Work together on the rest of the activity.

1. Western Fence

Number of Sections	Pieces of Lumber
0 (gate only)	8
1	12
2	
3	
4	
5	
6	

2. Tudor Fence

Number of Sections	Pieces of Lumber
0 (gate only)	11
1	14
2	
3	
4	
5	
6	

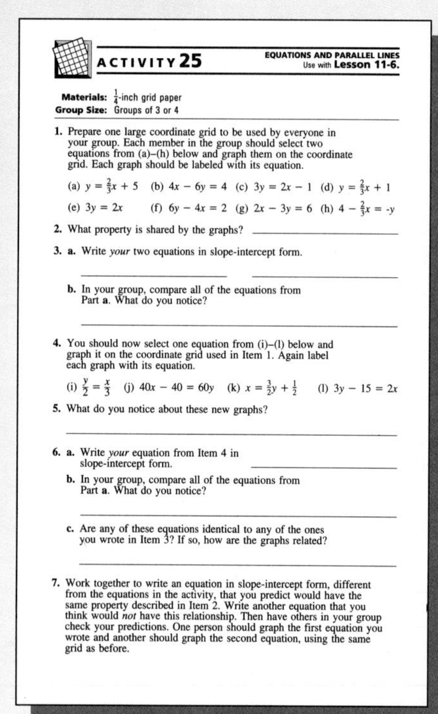

ACTIVITY 25 — EQUATIONS AND PARALLEL LINES
Use with **Lesson 11-6.**

Materials: $\frac{1}{4}$-inch grid paper
Group Size: Groups of 3 or 4

1. Prepare one large coordinate grid to be used by everyone in your group. Each member in the group should select two equations from (a)–(h) below and graph them on the coordinate grid. Each graph should be labeled with its equation.

(a) $y = \frac{2}{3}x + 5$ (b) $4x - 6y = 4$ (c) $3y = 2x - 1$ (d) $y = \frac{2}{3}x + 1$

(e) $3y = 2x$ (f) $6y - 4x = 2$ (g) $2x - 3y = 6$ (h) $4 - \frac{2}{3}x = -y$

2. What property is shared by the graphs? ____________

3. a. Write *your* two equations in slope-intercept form.

____________ ____________

b. In your group, compare all of the equations from Part **a**. What do you notice?

____

4. You should now select one equation from (i)–(l) below and graph it on the coordinate grid used in Item 1. Again label each graph with its equation.

(i) $\frac{y}{2} = \frac{x}{3}$ (j) $40x - 40 = 60y$ (k) $x = \frac{3}{2}y + \frac{1}{2}$ (l) $3y - 15 = 2x$

5. What do you notice about these new graphs?

____

6. a. Write *your* equation from Item 4 in slope-intercept form. ____________

b. In your group, compare all of the equations from Part **a**. What do you notice?

____

c. Are any of these equations identical to any of the ones you wrote in Item 3? If so, how are the graphs related?

____

7. Work together to write an equation in slope-intercept form, different from the equations in the activity, that you predict would have the same property described in Item 2. Write another equation that you think would *not* have this relationship. Then have others in your group check your predictions. One person should graph the first equation you wrote and another should graph the second equation, using the same grid as before.

Teaching Aids

Teaching Aid 28, Four-Quadrant Graph Paper, (shown on page 140D) can be used with **Lessons 11-1** through **11-8.**

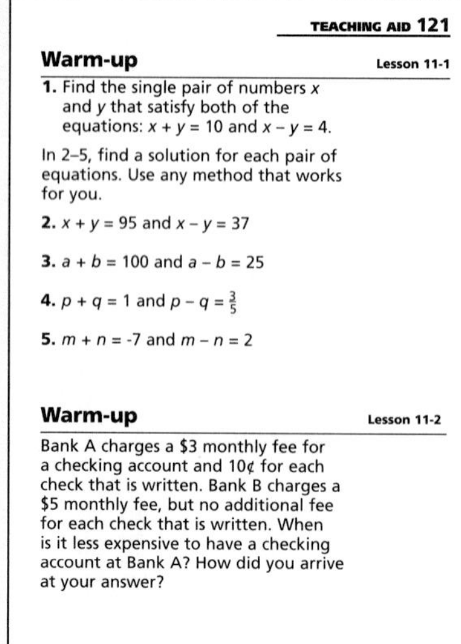

TEACHING AID 121

Warm-up — Lesson 11-1

1. Find the single pair of numbers x and y that satisfy both of the equations: $x + y = 10$ and $x - y = 4$.

In 2–5, find a solution for each pair of equations. Use any method that works for you.

2. $x + y = 95$ and $x - y = 37$

3. $a + b = 100$ and $a - b = 25$

4. $p + q = 1$ and $p - q = \frac{3}{5}$

5. $m + n = -7$ and $m - n = 2$

Warm-up — Lesson 11-2

Bank A charges a $3 monthly fee for a checking account and 10¢ for each check that is written. Bank B charges a $5 monthly fee, but no additional fee for each check that is written. When is it less expensive to have a checking account at Bank A? How did you arrive at your answer?

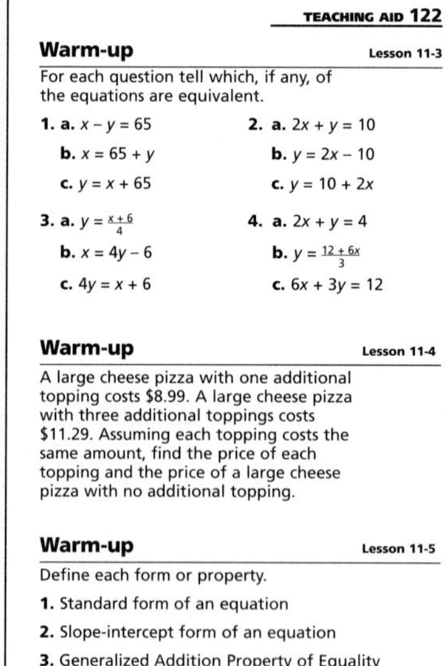

TEACHING AID 122

Warm-up — Lesson 11-3

For each question tell which, if any, of the equations are equivalent.

1. a. $x - y = 65$
b. $x = 65 + y$
c. $y = x + 65$

2. a. $2x + y = 10$
b. $y = 2x - 10$
c. $y = 10 + 2x$

3. a. $y = \frac{x + 6}{4}$
b. $x = 4y - 6$
c. $4y = x + 6$

4. a. $2x + y = 4$
b. $y = \frac{12 + 6x}{3}$
c. $6x + 3y = 12$

Warm-up — Lesson 11-4

A large cheese pizza with one additional topping costs $8.99. A large cheese pizza with three additional toppings costs $11.29. Assuming each topping costs the same amount, find the price of each topping and the price of a large cheese pizza with no additional topping.

Warm-up — Lesson 11-5

Define each form or property.

1. Standard form of an equation

2. Slope-intercept form of an equation

3. Generalized Addition Property of Equality

4. Multiplication Property of Equality

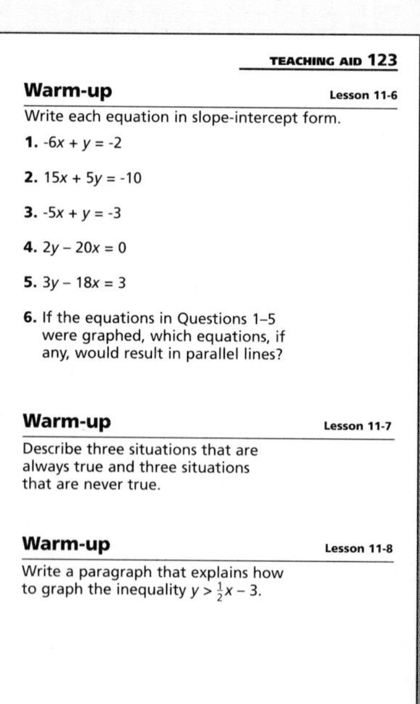

TEACHING AID 123

Warm-up — Lesson 11-6

Write each equation in slope-intercept form.

1. $-6x + y = -2$

2. $15x + 5y = -10$

3. $-5x + y = -3$

4. $2y - 20x = 0$

5. $3y - 18x = 3$

6. If the equations in Questions 1–5 were graphed, which equations, if any, would result in parallel lines?

Warm-up — Lesson 11-7

Describe three situations that are always true and three situations that are never true.

Warm-up — Lesson 11-8

Write a paragraph that explains how to graph the inequality $y > \frac{1}{2}x - 3$.

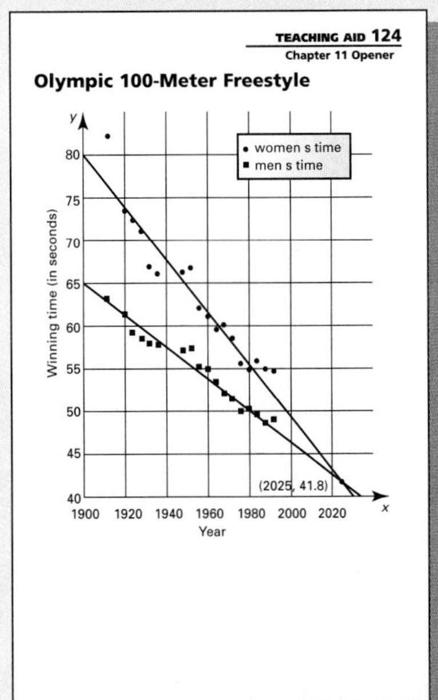

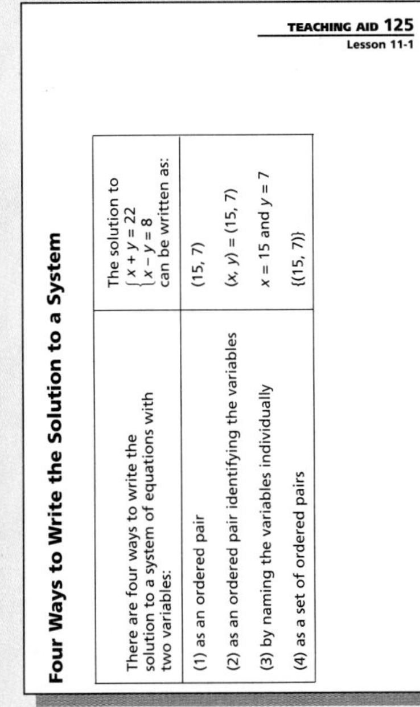

TEACHING AID 125
Lesson 11-1

Four Ways to Write the Solution to a System

There are four ways to write the solution to a system of equations with two variables:	The solution to $\begin{cases} x + y = 22 \\ x - y = 8 \end{cases}$ can be written as:
(1) as an ordered pair	(15, 7)
(2) as an ordered pair identifying the variables	$(x, y) = (15, 7)$
(3) by naming the variables individually	$x = 15$ and $y = 7$
(4) as a set of ordered pairs	{(15, 7)}

TEACHING AID 126
Lesson 11-4

Additional Examples

1. Going with the wind, a blimp flies 360 miles to an air show. The trip takes 4 hours. The return trip, flying against the wind, takes 9 hours. How fast is the blimp flying in still air? What is the speed of the wind?

2. Solve this system.
$$\begin{cases} 2x - 5y = 18 \\ 4x - 5y = -4 \end{cases}$$

3. At the 1987 annual meeting of the National Council of Teachers of Mathematics, the system below was found on a menu card at a restaurant in California. (Yes, the system really did appear on the bottom of the menu card.) How much did each item in column A cost, and how much did each item in column B cost?

Multiple Choice
2 or more entrées
Choose 1 from column A and 1 from column B — $5.49
Choose 1 from column A and 2 from column B — $6.99

A	B
• Chicken Dijon	• Fried Chicken (2 pieces)
• Top Sirloin Steak	• Battered Cod
• Steak Dijon	• Shrimp
• Fried Chicken	• Chicken Strips

A + B = $5.49
A + 2B = $6.99

Chapter Opener

Pacing

All lessons in this chapter are designed to be covered in one day. At the end of the chapter, you should plan to spend 1 day to review the Progress Self-Test, 1 to 2 days for the Chapter Review, and 1 day for a test. You may want to spend a day on projects, and a day may be needed for quizzes. Therefore, this chapter should take 11 to 14 days. We recommend that you not spend more than 15 days on the chapter.

Using Pages 664–665

You can use the discussion about Olympic swimming times as an advance organizer to introduce important ideas in the chapter. **Teaching Aid 124** contains the graph on page 665. Point out that the two lines in the graph represent the winning times for men's and women's Olympic 100-meter freestyle swimming champions. An equation can be found for each of the lines by using methods that students studied in Chapter 7. Together the two equations form a *system of equations.* The solution to this system is given by the point on the graph where the two lines intersect. This point is very close to (2025, 41.8). The coordinates of the point are the year 2025, when the men's time and women's time will be the same if the trends continue, and the time in that year, namely 41.8 seconds.

Students *should* be skeptical about the use of lines to represent these data. The points identifying the winning times seem to be bottoming out. There has been only a one-second decrease in the winning times for both sexes in the past 16 years. You might inform students that before the 1992 Olympics, the lines of best fit met at the point (2013, 44.3). Since the first edition of this book, the point of intersection has been moved farther into the future.

However, there was only a one-second decrease in the men's times from 1932 to 1952. So there can be times when it seems as if a bottom is being approached; then a new technique or training style may make it possible for swimmers to better their times. Note that today, women's

CHAPTER 11

664

Chapter 11 Overview

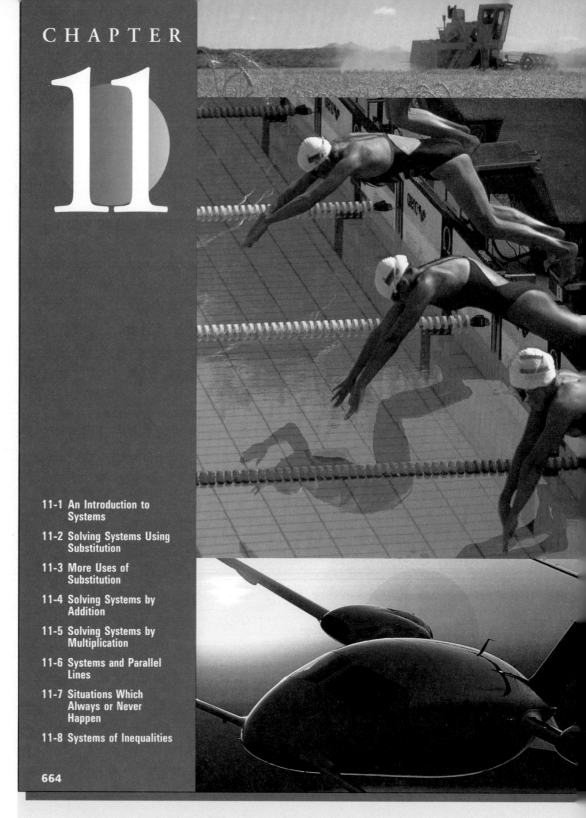

The four dimensions of the SPUR approach are all quite evident in this chapter. The chapter begins with the well-known representation of the solutions to systems as the intersection of the graphs of the equations in the system (Lesson 11-1). Then algebraic methods—the traditional skills—are covered. The first method discussed is substitution (Lessons 11-2 and 11-3), and then addition and multiplication are presented (Lessons 11-4 and 11-5). The chapter then moves on to the properties which help to determine when a system has a solution. Parallel lines are related to systems of linear equations that have no solutions (Lessons 11-6 and 11-7). The chapter closes with a discussion of systems of linear inequalities (Lesson 11-8). Throughout, the student sees realistic applications of this important idea.

LINEAR SYSTEMS

The table and graph below give the men's and women's winning times in the Olympic 100-meter freestyle swimming race.

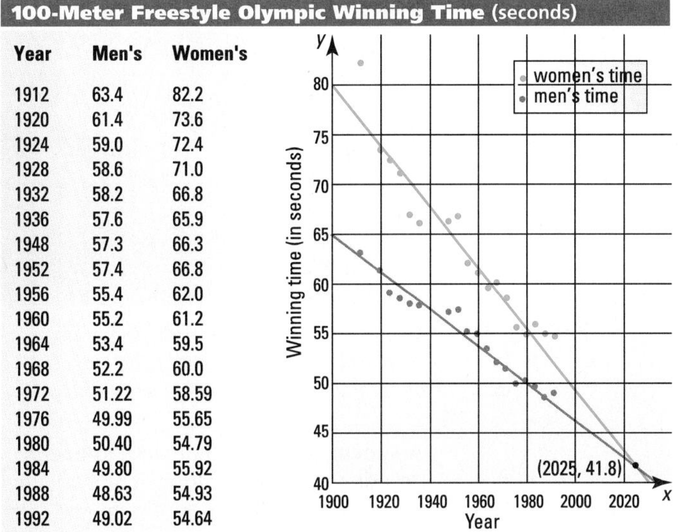

100-Meter Freestyle Olympic Winning Time (seconds)

Year	Men's	Women's
1912	63.4	82.2
1920	61.4	73.6
1924	59.0	72.4
1928	58.6	71.0
1932	58.2	66.8
1936	57.6	65.9
1948	57.3	66.3
1952	57.4	66.8
1956	55.4	62.0
1960	55.2	61.2
1964	53.4	59.5
1968	52.2	60.0
1972	51.22	58.59
1976	49.99	55.65
1980	50.40	54.79
1984	49.80	55.92
1988	48.63	54.93
1992	49.02	54.64

Notice that both men's and women's Olympic winning times have been generally decreasing since 1912. Also, the women's winning time has been decreasing faster than the men's. Lines have been fitted to the data. These lines have the following equations:

$$y = -0.176x + 398.204 \text{ (for men)}$$
$$\text{and } y = -0.299x + 647.308 \text{ (for women)}$$

where x is the year and y is the winning time in seconds. The lines intersect near (2025, 41.8). This means that if the winning times continue to decrease at these rates, the women's winning time will be faster than the men's in the Olympic year 2028. The winning times then will each be about 42 seconds.

Finding points of intersection of lines or other curves by working with their equations is called **solving a system.** In this chapter you will learn various ways of solving **linear systems.**

Why does one solve systems? One reason is found on these pages—to determine when two curves (in this case lines) meet. When the curves represent data, as they do here, their point of intersection has some meaning, and the solution to the system gives us some information about the situation.

Students may remember that, essentially, they solved systems earlier in this course, such as when they determined when two light bulbs would cost the same and in other similar examples beginning in Lesson 5-4. The difference is that students will now learn algebraic methods for solving these systems.

winning times are faster than the men's winning times were in 1960. Point out that even if the winning times are never the same, it is still interesting to look at the two sets of records.

Photo Connections

The photo collage makes real-world connections to the content of the chapter: linear systems.

Farm: According to the U.S. Department of Agriculture, the average value of an acre of farmland in 1992 was $685. A system of linear equations can be used to solve problems involving the number of acres, the price per acre, and the total land value.

Swimmers: The two equations on page 665 descibing the Olympic winning times of men and women in the 100-meter freestyle events are an example of a system of equations. On a graph, each solution of a system is a point of intersection of the graphs of the sentences.

Taxi: Local city laws determine the maximum rates that taxicabs are allowed to charge. Some cities have zonal fares; additional fares are charged when going from one zone into another. A system of equations can be used to find when two taxicabs will cost the same.

Airplane: In 1991, there were about 692,000 certified active pilots in the United States; approximately 293,000 were pilots of private aircraft. A system of equations involving headwinds and tailwinds is solved in Lesson 11-4.

Band: Marching bands, some with as many as 300 members, have to arrange themselves in a variety of formations while playing an instrument. The director can use a system of equations to decide if he or she has enough people to form repetitive groupings, such as the ones given in Example 3 in Lesson 11-5.

Chapter 11 Projects

At this time you might want to have students look over the projects on pages 711–712.

Lesson 11-1

Objectives

H Find solutions to systems of equations by graphing.

Resources

From the ***Teacher's Resource File***
- Lesson Master 11-1A or 11-1B
- Answer Master 11-1
- Teaching Aids
  - 28 Four-Quadrant Graph Paper
  - 121 Warm-up
  - 124 Olympic 100-Meter Freestyle
  - 125 Four Ways to Write the Solution to a System
- Activity Kit, Activity 24

Additional Resources
- Visuals for Teaching Aids 28, 121, 124, 125
- GraphExplorer or other automatic graphers

Teaching Lesson 11-1

Warm-up

1. Find the single pair of numbers x and y that satisfy both of the equations: $x + y = 10$ and $x - y = 4$. $x = 7$ and $y = 3$

In 2–5, find a solution for each pair of equations. Use any method that works for you.

2. $x + y = 95$ and $x - y = 37$ $x = 66, y = 29$
3. $a + b = 100$ and $a - b = 25$ $a = 62.5, y = 37.5$

(Warm-up continues on page 666.)

LESSON

11-1 An Introduction to Systems

On your marks. *The 1992 U.S. women's 400-meter freestyle relay team won a gold medal with a record time of 3:39.46 at the Olympic games in Barcelona. Often members of the relay team also participate in the 100-meter freestyle race.*

A **system** is a set of sentences joined by the word "and," which together describe a single situation. The two equations on page 665 describing the Olympic winning times of men and women in the 100-meter freestyle events are an example of a system of equations.

Systems are often signified by a single left-hand brace $\{$ in place of the word "and." So we can write this system as

$$\begin{cases} y = -0.176x + 398.204 \\ y = -0.299x + 647.308. \end{cases}$$

What Is a Solution to a System?

Each sentence in a system is sometimes called a **condition of the system.** Thus, the system above has two conditions.

A **solution to a system** of sentences with two variables is a pair of numbers which satisfies all the conditions of the system. On a graph, each solution of a system is a point of intersection of the graphs of the sentences.

Lesson 11-1 Overview

Broad Goals This lesson indicates what it means to solve a system numerically and graphically.

Perspective Although this chapter emphasizes using symbolic algorithms to solve systems, students should be continually reminded that they can use the graphical interpretation for indicating numbers of solutions and approximating the solutions. Advances in technology make solving systems from graphs easier than it was in the past.

It is important to note that in this chapter we concentrate on linear systems because exact solutions can be found algebraically. However, there are systems (not linear) that cannot be easily solved algebraically but that can be easily solved graphically. For instance, in comparing linear increase with exponential growth as was discussed in Chapter 8, situations can lead to solving a system such as $y = 1000(1.06)^x$ *and* $y = 2000 + .06x$. (In how many years x will $1000 saved at 6% compounded yearly grow to the same amount as $2000 saved at 6% simple interest? What will that amount y be?)

Example 1

Verify that the point (2025, 41.8) is a good estimate of the solution to the system of equations on page 666.

Solution

Substitute $x = 2025$ and $y = 41.8$ into each of
$y = -0.176x + 398.204$
and $y = -0.299x + 647.308$.

men: $y = -0.176x + 398.204$
Does $41.8 = -0.176 \cdot 2025 + 398.204$?
$41.8 \approx 41.804$? Yes.

women: $y = -0.299x + 647.308$
Does $41.8 = -0.299 \cdot 2025 + 647.308$?
$41.8 \approx 41.833$? Yes.

Solving Systems by Graphing

You can find the solutions to any system of equations with two variables by graphing each equation and finding the coordinates of the point(s) of intersection of the graphs.

Example 2

Find two numbers whose sum is 22 and whose difference is 8.

Solution

Translate the conditions into a system of two equations.
Let x and y be the two numbers. Then
$x + y = 22$ and $x - y = 8$.
Graph each equation and identify the point of intersection.
As shown at the right, the solution is (15, 7).

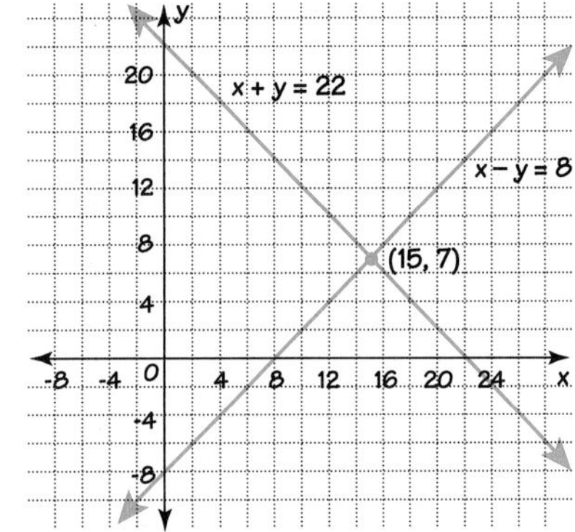

❶ **Check**

To check that (15, 7) is a solution, $x = 15$ and $y = 7$ must be checked in both conditions.
Does $x + y = 22$? $15 + 7 = 22$? Yes. Does $x - y = 8$? $15 - 7 = 8$? Yes.

❷ In general, there are four ways to write the solution to a system. They are shown below using the solution to the system in Example 2.

as an ordered pair	(15, 7)
as an ordered pair identifying the variables	$(x, y) = (15, 7)$
by naming the variables individually	$x = 15$ and $y = 7$
as a set of ordered pairs	{(15, 7)}

4. $p + q = 1$ and $p - q = \frac{3}{5}$
$p = \frac{4}{5}, q = \frac{1}{5}$

5. $m + n = -7$ and $m - n = 2$
$m = -2.5, n = -4.5$

Students may complain that they do not know how to solve the pairs of equations in the *Warm-up*. That's exactly the point! They will learn how to solve them in this chapter.

Notes on Reading

❶ Here and throughout the chapter, stress the importance of the checking process. Since solving a system involves several steps, it is very easy to make a mistake. Also, each time a solution is checked, students use the idea that the solution works in both equations. One way to emphasize the importance of the check is to ask not only for the solution to the system but also for the verification that the answer is correct. You might want to give partial credit for a wrong solution if the student recognizes that it is wrong because it doesn't check.

❷ **Teaching Aid 125** uses a table to show four ways to write the solution to a system.

You might note that $x = 15$ and $y = 7$ is a system because these are equations for a vertical line and a horizontal line whose intersection is (15, 7). When students get $x = 15$ and $y = 7$, they are getting an *equivalent system*, just as they get equivalent equations when they solve $3x + 5 = 8$ to find that $x = 1$.

Optional Activities

Activity 1 Using Physical Models

If students' desks can be put in rows and columns, you might introduce the lesson as follows. Assign each student in the class a point (x, y) with $-3 \le x \le 3$ and $-3 \le y \le 3$. Designate a seat near the center of the room as (0, 0), and have each student identify his or her coordinates in relation to seat (0, 0). Then give the following equations.

A. $x + y = 3$ B. $x - y = 1$

Ask students to stand if their seat is a solution to Equation A. Ask what type of geometric figure is formed by the students who are standing. [A line] Then ask students to stand if their seat is a solution to Equation B. How many people stood both times? [1] This situation illustrates two lines intersecting at one point. It also helps to reinforce the meaning of *and*. Repeat the activity with other pairs of equations.

Activity 2

You can use *Activity Kit, Activity 24* before covering the lesson. This activity offers students a real-world situation that leads to an understanding of systems and finding solutions by graphing.

❸ Relate this paragraph to the situation in the chapter opener. When the intersection point does not have integer coordinates, you can estimate the coordinates from a graph; however, it may be impossible to read the exact solution from the graph.

Emphasize that the equations in a system do not have to be linear equations and that there can be more than two sentences. For example, here is a graph that compares a linear function with an exponential function. The line and curve intersect at two points.

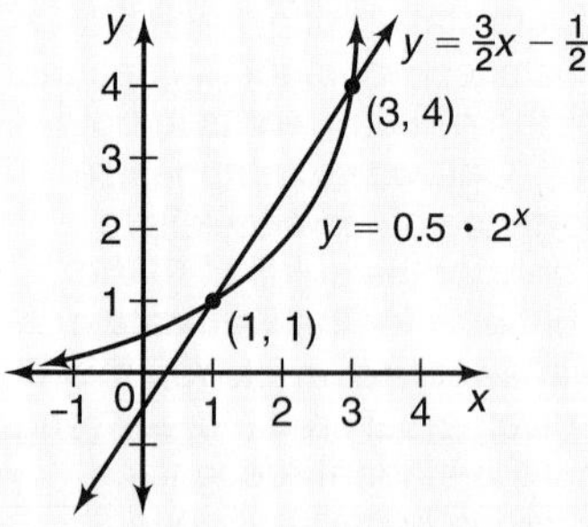

Additional Examples

Students will need graph paper or **Teaching Aid 28.**

1. Verify that $(\frac{44}{23}, -\frac{3}{23})$ is the solution to the system.
   $\begin{cases} 3x - 2y = 6 \\ 4x + 5y = 7 \end{cases}$
   Does $3(\frac{44}{23}) - 2(-\frac{3}{23}) = 6$?
   $\frac{132}{23} + \frac{6}{23} = \frac{138}{23} = 6$? Yes
   Does $4(\frac{44}{23}) + 5(-\frac{3}{23}) = 7$?
   $\frac{176}{23} - \frac{15}{23} = \frac{161}{23} = 7$? Yes
2. The sum of two numbers is 14 and their product is 40.
   a. What system can be solved to find the numbers? **$m + n = 14$ and $mn = 40$**
   b. Graph each equation. What are the numbers? **10 and 4** **See the graph on page 669.**

Systems with No Solutions

When the sentences in a system have no solutions in common, we say that there is *no solution* to the system. We cannot write the solution as an ordered pair or by listing the elements. The solution set is the set with no elements { }, or ø.

Example 3

Find all solutions to the system $\begin{cases} y = 2x + 1 \\ y = 2x - 3 \end{cases}$.

Solution

Draw the graph of each equation and look for all intersection points.

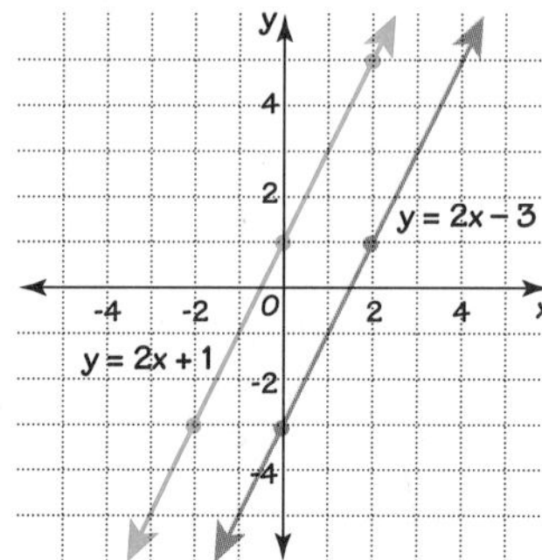

Notice that each line has slope 2, and a different y-intercept. So they are parallel and do not intersect. Thus, there is no pair of numbers that makes both equations true. There is no solution to the system. *The solution set is ø.*

Check

Twice a number plus 1 could never give the same value as twice the same number minus 3. That is, for all real numbers x, $2x + 1 \neq 2x - 3$.

❸ Graphing can help you find exact solutions to a system, as was shown in Example 2. If solutions do not have integer coordinates, however, it is likely that reading a graph will only give you an estimate. In the following lessons of this chapter, you will learn other techniques to find exact solutions to systems.

QUESTIONS

Covering the Reading

1. **a.** Define *system.*
   b. Give an example of a system not in this lesson. **See left.**
2. What does the brace { represent in a system? **the word "and"**

1a) a set of sentences joined by the word "and" which together describe a single situation

b) Sample: $\begin{cases} y = -2x + 3 \\ y = -3x + 6 \end{cases}$

Adapting to Individual Needs

Extra Help

Some students are not comfortable with problems that have no solution. Explain that the Venn diagram at the right shows the intersection of E and M. A number cannot be both even and odd. Then show how the following system of equations has no solution: $y = x + 3$ *and* $y = x - 3$. There is no (x, y) that makes both sentences true.

E = the Positive Even Numbers Less than 10

M = the Positive Odd Numbers Less than 10

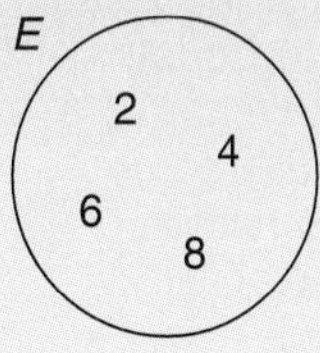

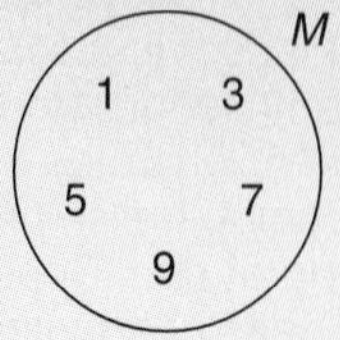

5c) 2(-1) + 4 = -2 + 4 = 2; -1 + 2(2) = -1 + 4 = 3

8b)

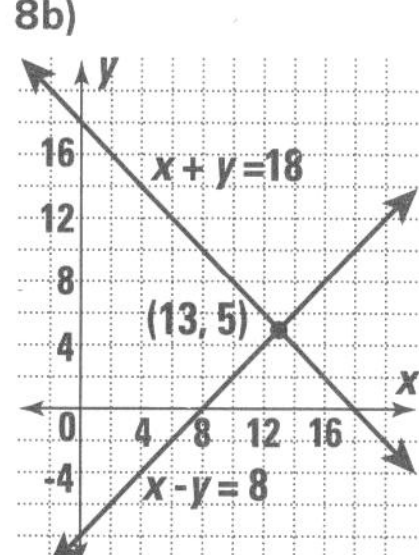

10a)

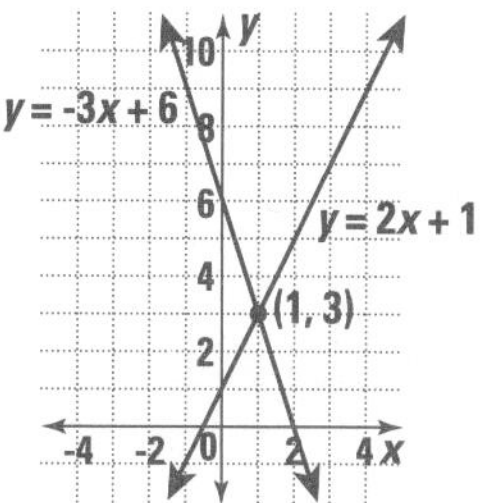

11a)

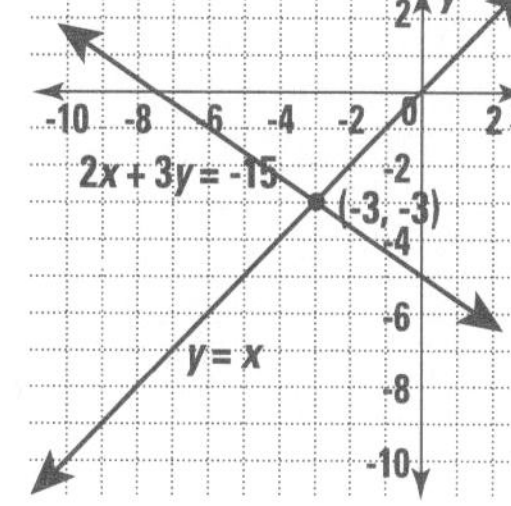

3. *True or false.* When a system has two variables, each solution is an ordered pair. **True**

4. What does the solution set to a system represent in a coordinate plane? **the intersection of the solution sets for each condition in the system**

5. Refer to the graph at the right.
   a. What system is represented? $\begin{cases} y = 2x + 4 \\ x + 2y = 3 \end{cases}$
   b. What is the solution to the system? **(-1, 2)**
   c. Verify your answer to part **b**. **See left.**

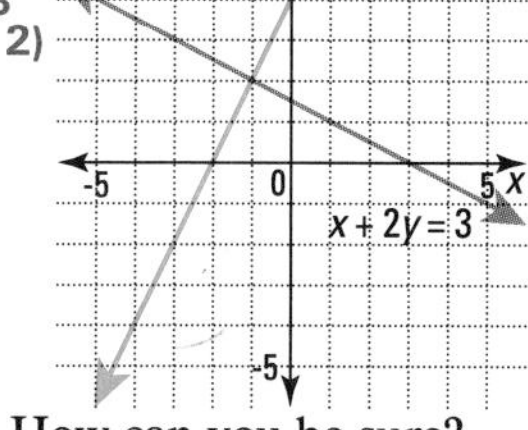

6. a. Verify that the solution to the system $\begin{cases} y = 9x \\ y = 2x - 7 \end{cases}$ is (-1, -9). **9(-1) = -9; 2(-1) - 7 = -9**
   b. Write this solution in two other ways. **Samples: x = -1 and y = -9; {(-1, -9)}**

7. Is (4, 8) a solution to the following system? How can you be sure?
   $\begin{cases} 10x - y = 32 \\ y - x = 4 \end{cases}$ **Yes. 10 · 4 − 8 = 32; 8 − 4 = 4**

8. The sum of two numbers is 18 and their difference is 8.
   a. If the numbers are x and y, translate the two conditions of the sentence above into two equations. **x + y = 18; x − y = 8**
   b. Graph both of these equations on the same coordinate system. **See left.**
   c. What are the numbers? **13 and 5**

9. Find all solutions to the system $\begin{cases} y = 4x - 2 \\ y = 4x + 5 \end{cases}$. **no solution**

In 10 and 11, a system is given.
a. Solve each system by graphing. **See left.**
b. Check your work.

10. $\begin{cases} y = 2x + 1 \\ y = -3x + 6 \end{cases}$
    b) 2(1) + 1 = 3; -3(1) + 6 = 3

11. $\begin{cases} y = x \\ 2x + 3y = -15 \end{cases}$
    b) -3 = -3; 2(-3) + 3(-3) = -6 − 9 = -15

Applying the Mathematics

In 12 and 13, a system is given.
a. Graph the system.
b. Find the coordinates of the point(s) of intersection. (Hint: In Question 13, one graph is not a line.)

12. $\begin{cases} x = -5 \\ x - y = 3 \end{cases}$ **b) (-5, -8)**

a)

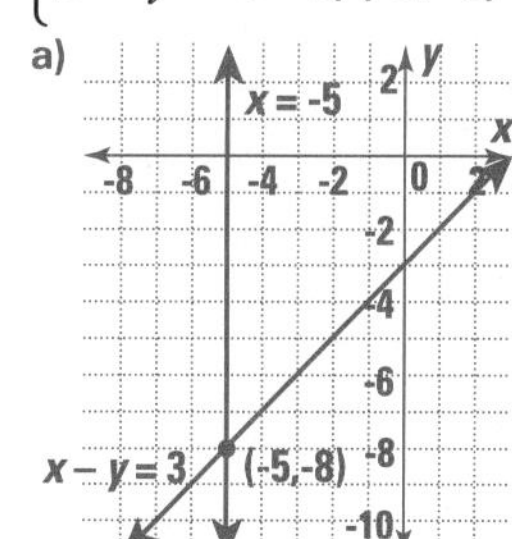

13. $\begin{cases} y = -\frac{1}{2}x^2 \\ y = \frac{1}{2}x - 1 \end{cases}$ **b) (-2, -2); (1, -.5)**

a)

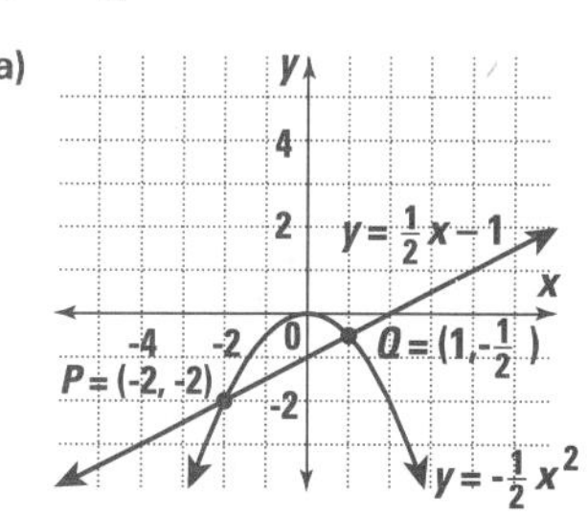

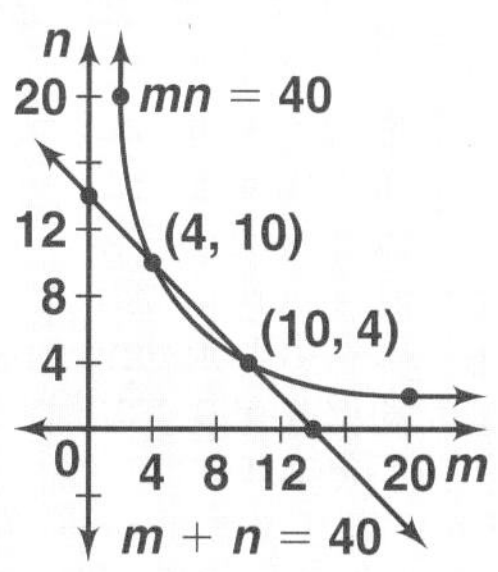

3. By graphing, find all solutions to the system.
   $\begin{cases} y = -3x + 1 \\ -2x + 2y = -14 \end{cases}$

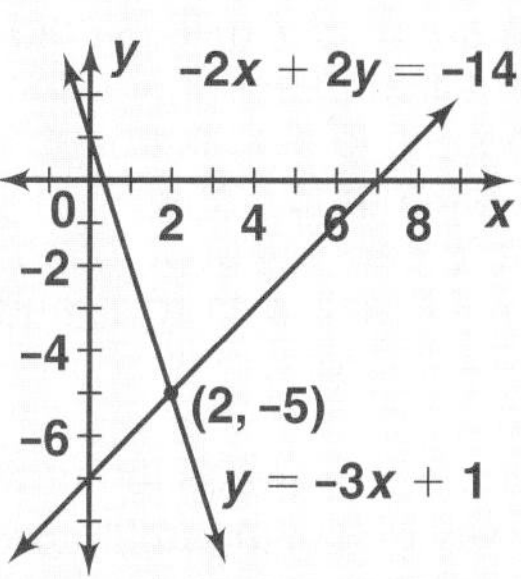

Notes on Questions

Question 13 Note that one of the equations is not linear. Solutions might be obtained by trial and error or by zooming and tracing on an automatic grapher. An advantage of graphing is that you can easily see the number of solutions.

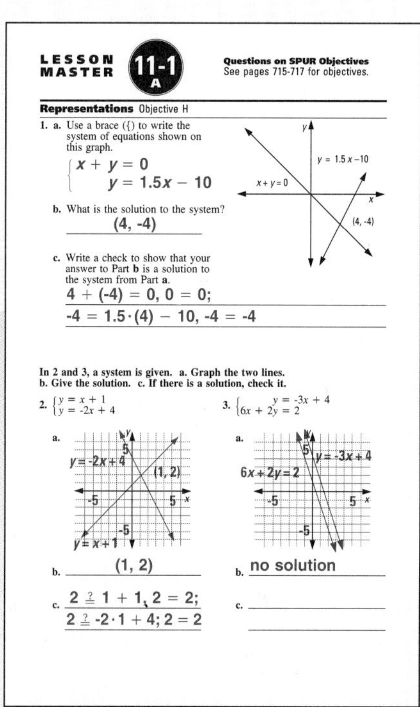

LESSON MASTER 11-1 A

Questions on SPUR Objectives
See pages 715-717 for objectives.

Representations Objective H

1. a. Use a brace ({) to write the system of equations shown on this graph.
   $\begin{cases} x + y = 0 \\ y = 1.5x - 10 \end{cases}$
   b. What is the solution to the system? (4, -4)
   c. Write a check to show that your answer to Part **b** is a solution to the system from Part **a**.
   4 + (-4) = 0, 0 = 0;
   -4 = 1.5·(4) − 10, -4 = -4

In 2 and 3, a system is given. a. Graph the two lines. b. Give the solution. c. If there is a solution, check it.

2. $\begin{cases} y = x + 1 \\ y = -2x + 4 \end{cases}$
   a. (graph: y = -2x + 4, y = x + 1, (1, 2))
   b. (1, 2)
   c. 2 ≟ 1 + 1, 2 = 2; 2 ≟ -2·1 + 4; 2 = 2

3. $\begin{cases} y = -3x + 4 \\ 6x + 2y = 2 \end{cases}$
   a. (graph: y = -3x + 4, 6x + 2y = 2)
   b. no solution
   c. ______

Adapting to Individual Needs

English Language Development
You might use equations like $x + y = 25$ and $x - y = 9$ to define the terms *system, condition of a system,* and *solution to a system.* Explain that the word *system* can mean "parts forming a whole." In this case, we consider the equations together, so we call them a system. Each equation states a condition or a "requirement" of the system: $x + y = 25$ requires that the sum of the two numbers is 25, and $x - y = 9$ requires that the difference of the same two numbers is 9. The solution of the system is a pair of numbers that solves *both* equations. In this case, the pair of numbers is 8 and 17.

Notes on Questions

Question 15 Error Alert Some students may not notice that Olympic times are missing for 1940 and 1944. Students must be careful when labeling the axes and plotting the points. If your automatic grapher or statistics package can fit lines to data, you may want to have students compare the solutions they got by hand to the line of best fit as determined by your software.

Question 19 Multicultural Connection In the United States, there are about 257 million people and 143 million passenger cars; the ratio of people to cars is about 1.8 to 1. In France and Canada, the ratio of people to cars is about 2.5 to 1. The ratio in Brazil is 11 to 1, and in China, it is 730 to 1.

Question 24 This question provides another way of looking at the swimming data. It also helps point out that data sets can often be interpreted in many ways. **Teaching Aid 124** may be used when discussing this question.

Question 25 Cooperative Learning This question is suitable for small group work. Have students work with computers or calculators to find pairs of equations that give the desired graphs.

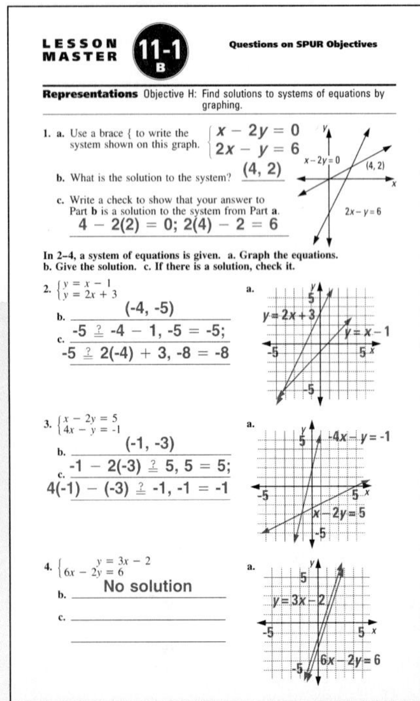

LESSON MASTER 11-1 B

Questions on SPUR Objectives

Representations Objective H: Find solutions to systems of equations by graphing.

1. a. Use a brace { to write the system shown on this graph. $\begin{cases} x - 2y = 0 \\ 2x - y = 6 \end{cases}$

b. What is the solution to the system? (4, 2)

c. Write a check to show that your answer to Part **b** is a solution to the system from Part **a**. $4 - 2(2) = 0$; $2(4) - 2 = 6$

In 2–4, a system of equations is given. a. Graph the equations. b. Give the solution. c. If there is a solution, check it.

2. $\begin{cases} y = x - 1 \\ y = 2x + 3 \end{cases}$

b. (-4, -5)

c. $-5 \stackrel{?}{=} -4 - 1$, $-5 = -5$; $-5 \stackrel{?}{=} 2(-4) + 3$, $-8 = -8$

3. $\begin{cases} x - 2y = 5 \\ 4x - y = -1 \end{cases}$

b. (-1, -3)

c. $-1 - 2(-3) \stackrel{?}{=} 5$, $5 = 5$; $4(-1) - (-3) \stackrel{?}{=} -1$, $-1 = -1$

4. $\begin{cases} y = 3x - 2 \\ 6x - 2y = 6 \end{cases}$

b. No solution

c.

14b) $2^2 = 4$, $2^2 = 4$; $4^2 = 16$, $2^4 = 16$; $(-.75)^2 \approx .6$, $2^{-.75} \approx .6$

14. Below is a graph of the system $\begin{cases} y = x^2 \\ y = 2^x \end{cases}$.

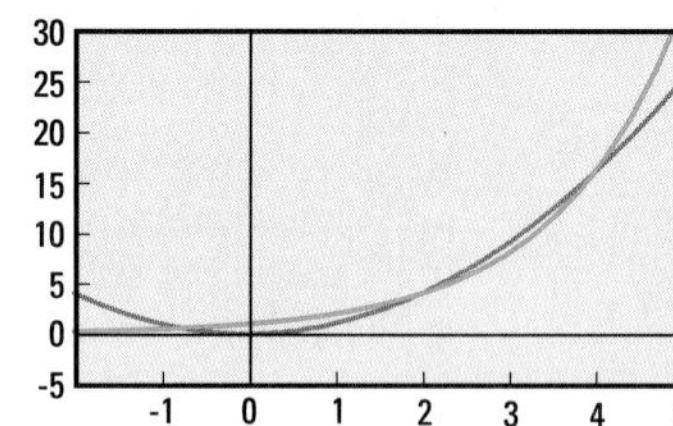

a. Use a calculator or automatic grapher to find three solutions to this system. Samples: (2, 4); (4, 16); (-.75, .6)

b. Check each solution. See left.

15. Below are a table and graph of the winning times in seconds for the Olympic men's and women's 100-meter backstroke events.

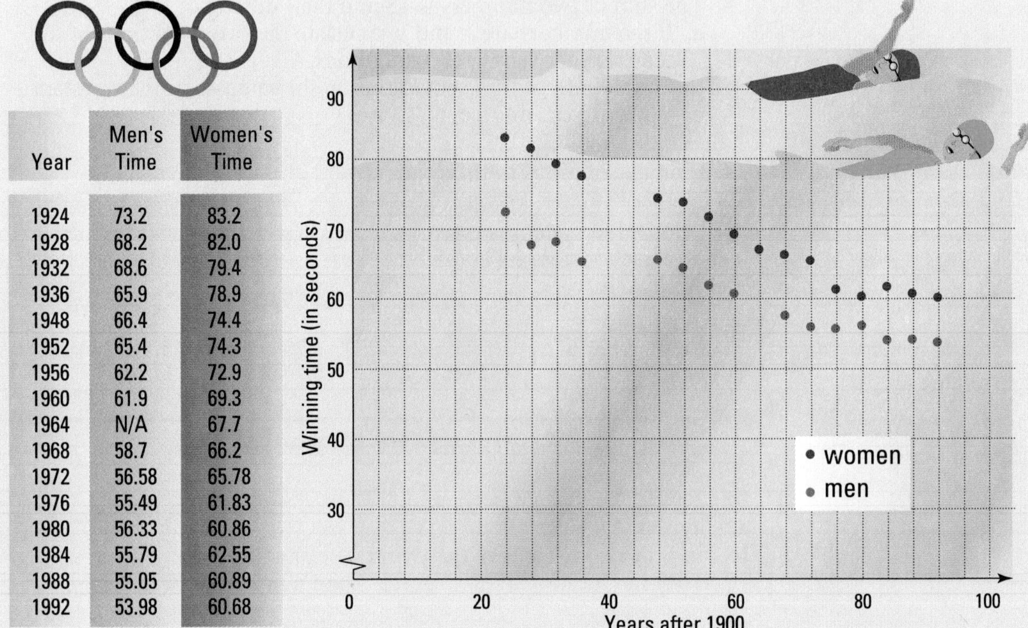

Year	Men's Time	Women's Time
1924	73.2	83.2
1928	68.2	82.0
1932	68.6	79.4
1936	65.9	78.9
1948	66.4	74.4
1952	65.4	74.3
1956	62.2	72.9
1960	61.9	69.3
1964	N/A	67.7
1968	58.7	66.2
1972	56.58	65.78
1976	55.49	61.83
1980	56.33	60.86
1984	55.79	62.55
1988	55.05	60.89
1992	53.98	60.68

Based on these data, do you think the women's winning time will ever equal the men's winning time in the 100-meter backstroke? If yes, estimate the year when this will happen. If no, explain why not. Sample: Yes. Since the two lines are not parallel and the women's times are decreasing faster than the men's times, the times will be equal in 2044.

670

Adapting to Individual Needs

Challenge

Materials: Graph paper or **Teaching Aid 28**, or automatic graphers

Have students find the points of intersection for the graphs of $y = 2x - 3$, $y = -x + 12$, and $y = 1$. Then have them find the area of the triangle having these points as vertices. [In the graph at the right; the vertices of the triangle are (2, 1), (11, 1), and (5, 7); the area is 27.]

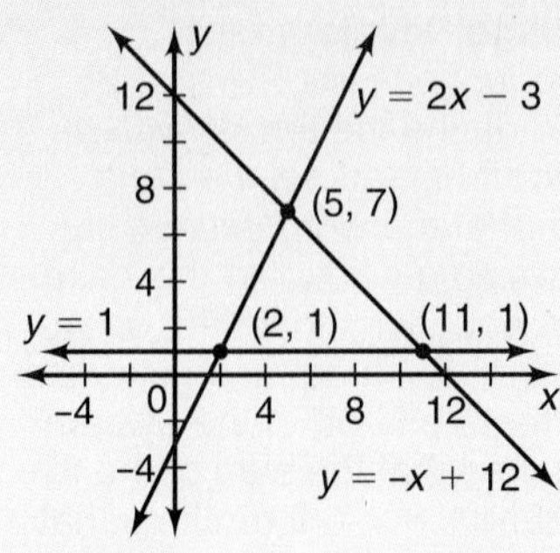

24a)

1912; 0.77	1960; 0.90
1920; 0.83	1964; 0.90
1924; 0.81	1968; 0.87
1928; 0.83	1972; 0.87
1932; 0.87	1976; 0.90
1936; 0.87	1980; 0.92
1948; 0.86	1984; 0.89
1952; 0.86	1988; 0.89
1956; 0.89	1992; 0.90

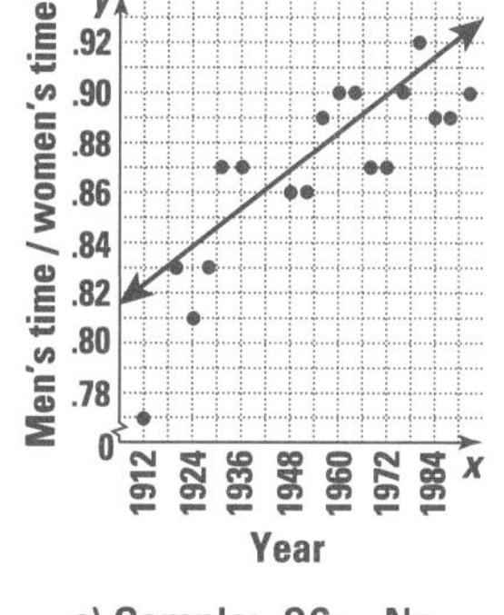

Houston, Texas

Review

16. *Skill sequence.* Solve. *(Lessons 3-7, 5-3, 9-5, 10-6)*

a. $3x + 8 = x - 12$ $x = -10$

b. $3(x + 8) = -4(x - 12)$ $x = \frac{24}{7} = 3\frac{3}{7}$

c. $3(x + 8)^2 = -3(x - 12)$ $x = -4$ or $x = -13$

17. a. Evaluate $\frac{y_2 - y_1}{x_2 - x_1}$ where $y_2 = 7$, $y_1 = -1$, $x_2 = 8$, and $x_1 = 10$. -4

b. What have you calculated in part **a?** *(Lessons 1-4, 7-2)*
the slope of the line that passes through the points (10, -1) and (8, 7)

18. Suppose that on a map 1 inch represents 325 miles. The map distance from Los Angeles to Houston is $4\frac{3}{4}$ inches. Suppose you want the actual distance in miles from Los Angeles to Houston.

a. Write a proportion that will help solve the problem. Sample: $\frac{325}{1} = \frac{x}{4\frac{3}{4}}$

b. Find the distance. *(Lesson 6-8)* 1543.75 miles

19. In 1993, American families owned an average of 1.8 cars per family, and drove an average of 18,600 miles. How many miles was the typical American family car driven that year? *(Lesson 6-2)*
≈ 10,333 miles

20. Simplify $7\pi \div \left(\frac{2\pi}{3}\right)$. *(Lesson 6-1)* $\frac{21}{2}$

21. Solve $2x + y = 7$ for y. *(Lesson 5-7)* $y = 7 - 2x$

22. Eight more than three times a number is two more than six times the number. What is the number? *(Lesson 5-3)* 2

23. If a country has population P now and the population is increasing by X people per year, what will its population be in Y years? *(Lesson 3-8)*
$P + XY$

Exploration

24b)

24. Some experts believe that even though the women's swim times are decreasing faster than the men's, it is the ratio of the times that is the key to predictions.

a. Compute the ratio of the men's time to the women's time for the 100-meter freestyle for each Olympic year. See above left.

b. Graph your results. (Plot Olympic year on the horizontal axis and the ratio of times on the vertical axis.)

c. What do you think the ratio will be in 2025? Does this agree with the prediction on page 665?

c) Sample: .96; No, for the women's time to be faster than the men's time, the ratio must be greater than 1.

25. Question 13 describes a system in which a line and a parabola intersect in two points. Sketch examples of the following systems.

a. a line and a parabola with no points of intersection See margin.

b. a line and a parabola with exactly one point of intersection

c. two parabolas intersecting in two points

d. two parabolas intersecting in exactly one point

Additional Answers

25a.

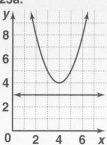

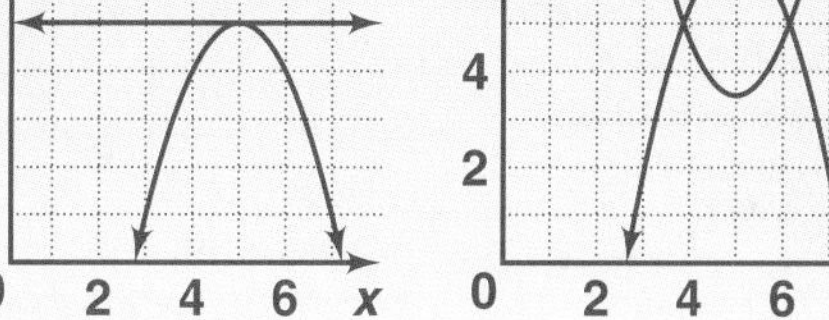

b.

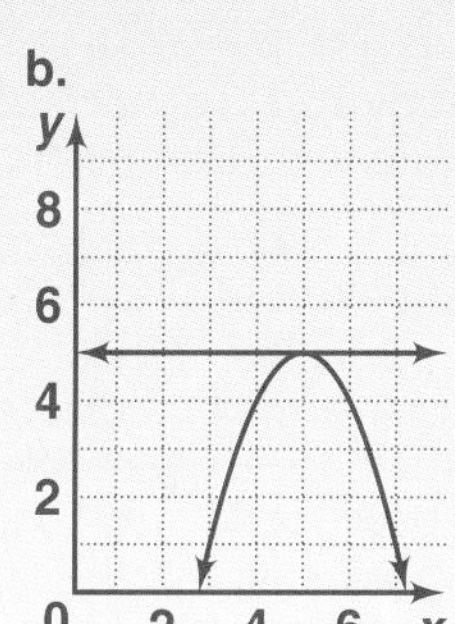

c.

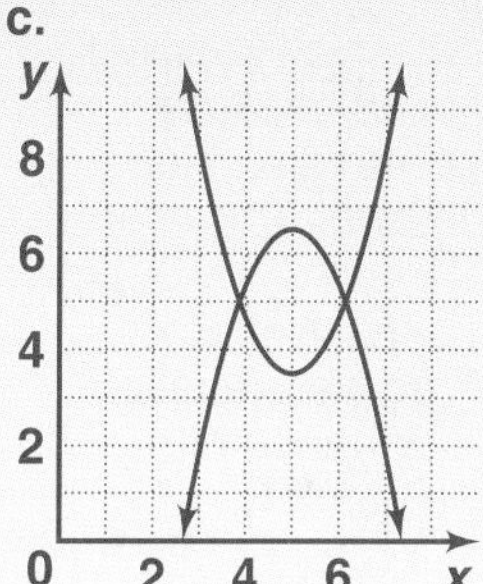

d.

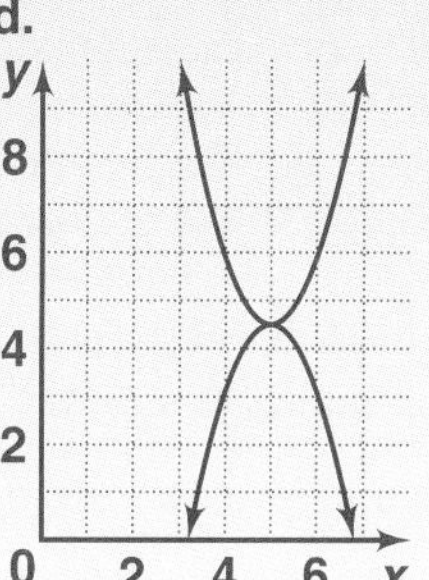

Follow-up for Lesson 11-1

Practice

For more questions on SPUR Objectives, use **Lesson Master 11-1A** (shown on page 669) or **Lesson Master 11-1B** (shown on pages 670–671).

Assessment

Group Assessment Have each student pick two numbers and write a system of two equations that describes the two numbers. Have students exchange papers and graph the system to find the numbers selected by their partners. [Students identify the point of intersection of the graphs of two equations as the solution to a system of equations.]

Extension

Have students write answers to **Exercises 9–11** in each of the four ways described on page 667. [**9:** no solution; no solution; no solution; { } or ø. **10:** (1, 3); $(x, y) = (1, 3)$; $x = 1$, $y = 3$; {(1, 3)}. **11:** (-3, -3); $(x, y) = (-3, -3)$; $x = -3$, $y = -3$; {(-3, -3)}]

Project Update Project 1, *Olympic Records*, and Project 5, *Life Expectancy*, on pages 711–712, relate to the content of this lesson.

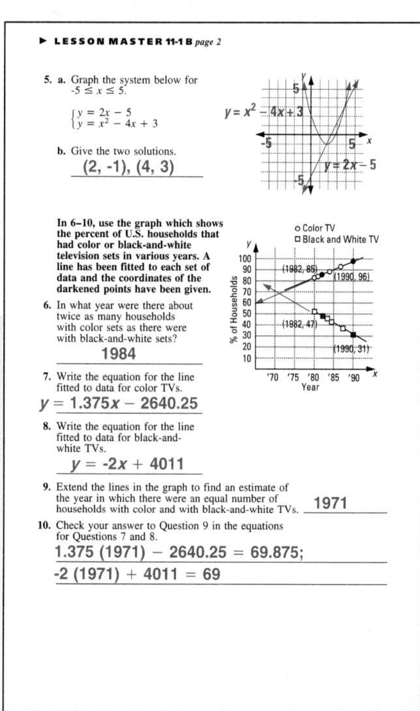

▶ **LESSON MASTER 11-1 B** *page 2*

5. a. Graph the system below for $-5 \le x \le 5$.

$$\begin{cases} y = 2x - 5 \\ y = x^2 - 4x + 3 \end{cases}$$

b. Give the two solutions. (2, -1), (4, 3)

In 6–10, use the graph which shows the percent of U.S. households that had color or black-and-white television sets in various years. A line has been fitted to each set of data and the coordinates of the darkened points have been given.

6. In what year were there about twice as many households with color sets as there were with black-and-white sets? 1984

7. Write the equation for the line fitted to data for color TVs. $y = 1.375x - 2640.25$

8. Write the equation for the line fitted to data for black-and-white TVs. $y = -2x + 4011$

9. Extend the lines in the graph to find an estimate of the year in which there were an equal number of households with color and with black-and-white TVs. 1971

10. Check your answer to Question 9 in the equations for Questions 7 and 8.
1.375 (1971) − 2640.25 = 69.875;
-2 (1971) + 4011 = 69

Lesson 11-2

Objectives

A Solve systems using substitution.
F Use systems of linear equations to solve real-world problems.

Resources

From the ***Teacher's Resource File***

- Lesson Master 11-2A or 11-2B
- Answer Master 11-2
- Teaching Aid
  28 Four-Quadrant Graph Paper
  121 Warm-up

Additional Resources

- Visuals for Teaching Aids 28, 121

Teaching Lesson 11-2

Warm-up

Bank A charges a $3 monthly fee for a checking account and 10¢ for each check that is written. Bank B charges a $5 monthly fee, but no additional fee for each check that is written. When is it less expensive to have a checking account at Bank A? How did you arrive at your answer? **When less than 20 checks are written per month; explanations will vary.**

LESSON

11-2 *Solving Systems Using Substitution*

Taxi! *A system of equations can be used to compare the fees charged by two different taxicab companies. See Example 2.*

Consider the system

$$\begin{cases} y = 5x - 25 \\ y = -8x + 27 \end{cases}$$

graphed below. It appears that the x-coordinate of the point of intersection is 4 and that the y-coordinate is -5.

1

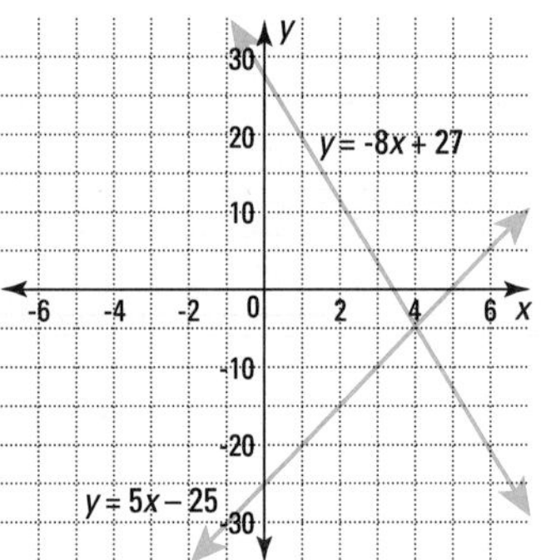

When the coordinates of the intersection point are not integers, it is difficult to find their exact values from a graph. However, you can find the exact coordinates of the point of intersection by using *substitution.* To solve a system using substitution, take the expression equal to y in one equation, and substitute it for y in the other equation. Example 1 illustrates this technique.

Lesson 11-2 Overview

Broad Goals This lesson covers systems of two linear equations where both equations are of the form $y = mx + b$.

Perspective In this lesson, the systems are of the form $y = ax + b$ and $y = cx + d$, and the justification for $ax + b = cx + d$ could be either substitution (substitute $cx + d$ for y in the first equation) or the Symmetric and Transitive Properties of Equality ($ax + b = y$ by the Symmetric Property and then $ax + b = cx + d$ by the Transitive Property). We opt for calling it substitution because the justification is more direct and because it helps to set up the next lesson.

Point out the general form here: $A = B$ and $A = C$ is the given system, so we are able to say $B = C$. Notice the importance of chunking $ax + b$ as a single quantity.

Optional Activities

Activity 1 Technology Connection
Materials: Computers

The following computer program asks for equations in slope-intercept form and gives the solution. It requires that the slopes of the two lines be different so that there is a unique solution. After reading and discussing the lesson, students can use this program for the examples and questions. You might want to ask the more proficient

Example 1

Solve the system $\begin{cases} y = 5x - 25 \\ y = -8x + 27 \end{cases}$ using substitution.

Solution

Substitute $-8x + 27$ for y in the first equation.

$$-8x + 27 = 5x - 25$$

Now solve for x.

$$27 = 13x - 25$$
$$52 = 13x$$
$$4 = x$$

To find y, substitute 4 for x in either of the original equations. We use the first equation.

$$y = 5x - 25$$
$$y = 5 \cdot 4 - 25 = -5$$

The solution is (4, -5).

Check

The point is on both lines, as substitution shows.
$-5 = 5 \cdot 4 - 25$ and $-5 = -8 \cdot 4 + 27$

Suppose two quantities are increasing or decreasing at different constant rates. Then each quantity can be described by an equation of the form $y = ax + b$. To find out when the quantities are equal, you can solve a system by using substitution. Example 2 illustrates this idea.

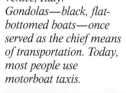

Water taxis. *Canals take the place of streets in Venice, Italy. Gondolas—black, flat-bottomed boats—once served as the chief means of transportation. Today, most people use motorboat taxis.*

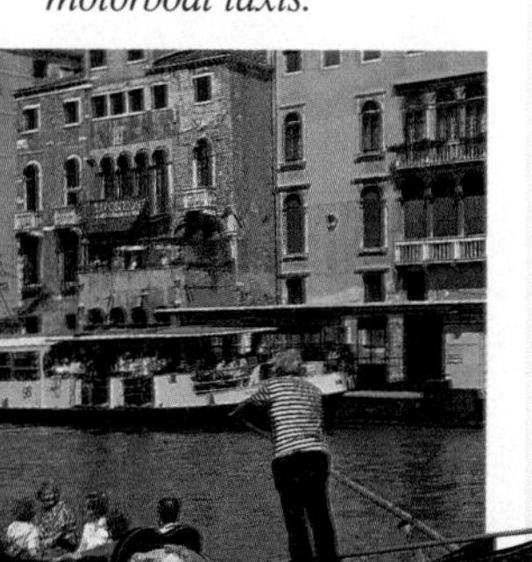

❷ ### Example 2

A taxi ride in Burford costs \$1.50 plus 20¢ for each $\frac{1}{10}$ mile traveled. In Spotswood, taxi rides cost 90¢ plus 25¢ for each $\frac{1}{10}$ mile traveled. For what distance do rides cost the same?

Solution

Let d = the distance of a cab ride in tenths of a mile.
Let C = the cost of a cab ride of distance d.

In Burford: $C = 1.50 + .20d$
In Spotswood: $C = .90 + .25d$

The rides cost the same when the values of C and d in Burford equal the values in Spotswood. So there is a system to solve. We substitute $1.50 + .20d$ for C in the second equation.

$$1.50 + .20d = .90 + .25d$$

Now solve as usual. Add -.90 and -.20d to both sides.

$$.60 = .05d$$
$$d = 12$$

A ride of a distance 12 tenths of a mile, or 1.2 miles, will cost the same in both cities.

▶

Notes on Reading

❶ This example is straightforward. Notice that because the scales on the axes in the graph are quite different, the value of y is not easily found.

❷ Taxi rates provide a nice setting for systems. Notice that we put the distance in tenths of a mile. This is to make it easier to obtain the equations in the system, but it means that the solution is a little more difficult to interpret.

Social Studies Connection Today there are about 260,000 taxicabs in the United States. The first electrically driven taxis appeared in 1898; these vehicles were replaced by gasoline-powered automobiles starting in 1907.

Additional Examples

1. Solve by substitution.
   $\begin{cases} y = 5x + 9 \\ y = -3x + 37 \end{cases}$ **(3.5, 26.5)**
2. Mrs. Janeski wants to have a clown deliver helium balloons to her daughter's birthday party. Two companies offer this service. Company A charges 80¢ a balloon, plus a \$6 delivery fee. Company B charges 95¢ per balloon but has no delivery fee. For how many balloons is the cost the same? (Let x = the number of balloons. Let y = the cost of balloons.) **40 balloons**

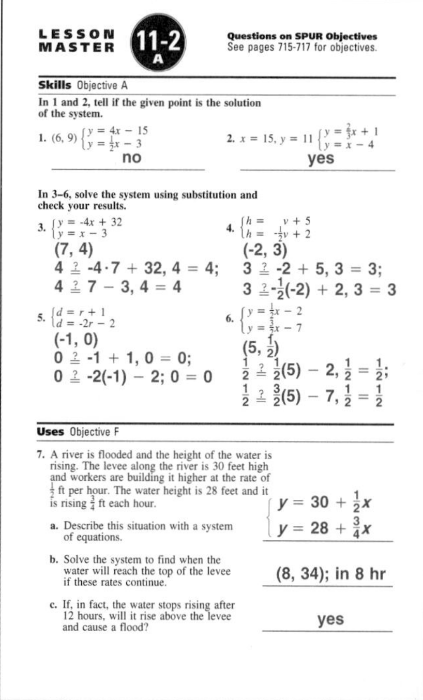

LESSON MASTER 11-2 A **Questions on SPUR Objectives** See pages 715-717 for objectives.

Skills Objective A

In 1 and 2, tell if the given point is the solution of the system.

1. (6, 9) $\begin{cases} y = 4x - 15 \\ y = \frac{1}{2}x - 3 \end{cases}$ **no**
2. $x = 15, y = 11$ $\begin{cases} y = \frac{2}{3}x + 1 \\ y = x - 4 \end{cases}$ **yes**

In 3–6, solve the system using substitution and check your results.

3. $\begin{cases} y = -4x + 32 \\ y = x - 3 \end{cases}$
   (7, 4)
   $4 \stackrel{?}{=} -4 \cdot 7 + 32, 4 = 4;$
   $4 \stackrel{?}{=} 7 - 3, 4 = 4$
4. $\begin{cases} h = v + 5 \\ h = -\frac{1}{2}v + 2 \end{cases}$
   (-2, 3)
   $3 \stackrel{?}{=} -2 + 5, 3 = 3;$
   $3 \stackrel{?}{=} -\frac{1}{2}(-2) + 2, 3 = 3$
5. $\begin{cases} d = r + 1 \\ d = -2r - 2 \end{cases}$
   (-1, 0)
   $0 \stackrel{?}{=} -1 + 1, 0 = 0;$
   $0 \stackrel{?}{=} -2(-1) - 2; 0 = 0$
6. $\begin{cases} y = \frac{1}{2}x - 2 \\ y = \frac{3}{2}x - 7 \end{cases}$
   $(5, \frac{1}{2})$
   $\frac{1}{2} \stackrel{?}{=} \frac{1}{2}(5) - 2, \frac{1}{2} = \frac{1}{2};$
   $\frac{1}{2} \stackrel{?}{=} \frac{3}{2}(5) - 7, \frac{1}{2} = \frac{1}{2}$

Uses Objective F

7. A river is flooded and the height of the water is rising. The levee along the river is 30 feet high and workers are building it higher at the rate of $\frac{1}{2}$ ft per hour. The water height is 28 feet and it is rising $\frac{3}{4}$ ft each hour.
   a. Describe this situation with a system of equations. $\begin{cases} y = 30 + \frac{1}{2}x \\ y = 28 + \frac{3}{4}x \end{cases}$
   b. Solve the system to find when the water will reach the top of the levee if these rates continue. **(8, 34); in 8 hr**
   c. If, in fact, the water stops rising after 12 hours, will it rise above the levee and cause a flood? **yes**

programmers in your class to modify the program so that it tests for equal slope.

```
10 PRINT "FIND INTERSECTION OF
   LINES"
20 PRINT "ENTER SLOPE, INTERCEPT
   OF FIRST LINE"
30 INPUT M1,B1
40 PRINT "ENTER SLOPE, INTERCEPT
   OF SECOND LINE"
50 INPUT M2,B2
60 PRINT "SYSTEM IS"
70 PRINT "Y = "; M1; "X+ "; B1
80 PRINT "Y = "; M2; "X+ "; B2
90 LET X = (B2 – B1)/(M1 – M2)
100 LET Y = M1*X+B1
110 PRINT "SOLUTION IS X = ";X;
    "AND Y = ";Y
120 END
```

Students might add these steps to test for equal slopes.

```
85 IF M1 = M2 THEN 115
115 PRINT "THERE IS NO SOLUTION."
```

Notes on Questions

Questions 6–7 A graph of the equations in **Example 2** provides an alternate way of answering these questions. Still another way is to test a distance other than 1.2 miles (where the costs are equal). This test is similar to testing a point to decide which side of the boundary to shade when graphing an inequality.

Questions 14–16 These are all constant-increase situations that are similar to **Example 2**.

Question 20 Solving **part a** might alert students to the fact that in **parts b and c**, it is possible to first divide both sides by 2.

Question 27 Cooperative Learning You might want to have students exchange their questions with a partner. Have students comment constructively on the clarity of the questions. Then have them exchange their revised questions with another classmate to solve.

Additional Answers

6.

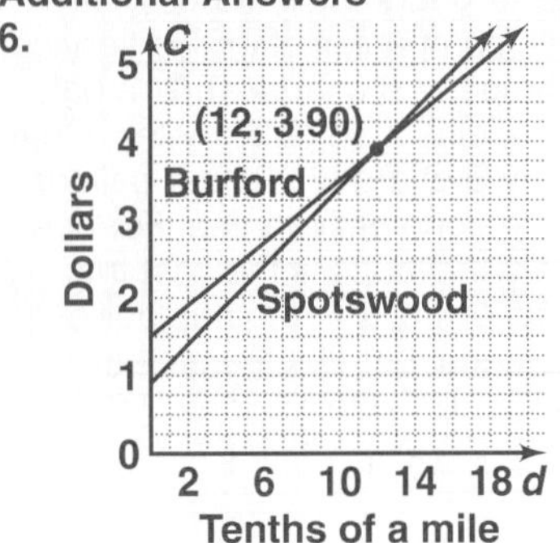

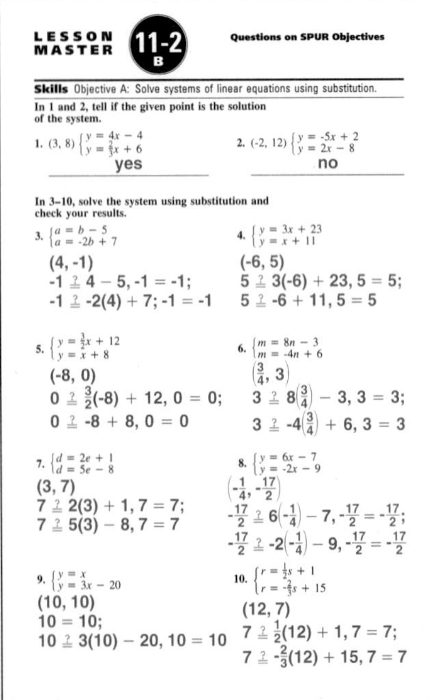

LESSON MASTER 11-2 B **Questions on SPUR Objectives**

Skills Objective A: Solve systems of linear equations using substitution.

In 1 and 2, tell if the given point is the solution of the system.

1. (3, 8) $\begin{cases} y = 4x - 4 \\ y = \frac{2}{3}x + 6 \end{cases}$ yes

2. (-2, 12) $\begin{cases} y = -5x + 2 \\ y = 2x - 8 \end{cases}$ no

In 3–10, solve the system using substitution and check your results.

3. $\begin{cases} a = b - 5 \\ a = -2b + 7 \end{cases}$
(4, -1)
$-1 \stackrel{?}{=} 4 - 5, -1 = -1;$
$-1 \stackrel{?}{=} -2(4) + 7; -1 = -1$

4. $\begin{cases} y = 3x + 23 \\ y = x + 11 \end{cases}$
(-6, 5)
$5 \stackrel{?}{=} 3(-6) + 23, 5 = 5;$
$5 \stackrel{?}{=} -6 + 11, 5 = 5$

5. $\begin{cases} y = \frac{3}{2}x + 12 \\ y = x + 8 \end{cases}$
(-8, 0)
$0 \stackrel{?}{=} \frac{3}{2}(-8) + 12, 0 = 0;$
$0 \stackrel{?}{=} -8 + 8, 0 = 0$

6. $\begin{cases} m = 8n - 3 \\ m = -4n + 6 \end{cases}$
$\left(\frac{3}{4}, 3\right)$
$3 \stackrel{?}{=} 8\left(\frac{3}{4}\right) - 3, 3 = 3;$
$3 \stackrel{?}{=} -4\left(\frac{3}{4}\right) + 6, 3 = 3$

7. $\begin{cases} d = 2e + 1 \\ d = 5e - 8 \end{cases}$
(3, 7)
$7 \stackrel{?}{=} 2(3) + 1, 7 = 7;$
$7 \stackrel{?}{=} 5(3) - 8, 7 = 7$

8. $\begin{cases} y = 6x - 7 \\ y = -2x - 9 \end{cases}$
$\left(-\frac{1}{4}, -\frac{17}{2}\right)$
$-\frac{17}{2} \stackrel{?}{=} 6\left(-\frac{1}{4}\right) - 7, -\frac{17}{2} = -\frac{17}{2};$
$-\frac{17}{2} \stackrel{?}{=} -2\left(-\frac{1}{4}\right) - 9, -\frac{17}{2} = -\frac{17}{2}$

9. $\begin{cases} y = x \\ y = 3x - 20 \end{cases}$
(10, 10)
$10 = 10;$
$10 \stackrel{?}{=} 3(10) - 20, 10 = 10$

10. $\begin{cases} r = \frac{1}{2}s + 1 \\ r = -\frac{2}{3}s + 15 \end{cases}$
(12, 7)
$7 \stackrel{?}{=} \frac{1}{2}(12) + 1, 7 = 7;$
$7 \stackrel{?}{=} -\frac{2}{3}(12) + 15, 7 = 7$

2a) (10, 11)
b) $3 \cdot 10 - 19 = 30 - 19 = 11;$
$10 + 1 = 11$

3a) (-55, -610)
b) $12 \cdot -55 + 50 = -660 + 50 = -610;$
$10 \cdot -55 - 60 = -550 - 60 = -610$

London's black cabs are rated best in the world. Drivers must pass an exam that includes going to the most obscure addresses on a bicycle.

12a) $\left(-\frac{21}{10}, \frac{13}{10}\right)$ b) Does $\frac{1}{3} \cdot -\frac{21}{10} + 2 = \frac{13}{10}$? $-\frac{21}{30} + \frac{60}{30} = \frac{39}{30} = \frac{13}{10}$? Yes. Does $-3\left(-\frac{21}{10}\right) - 5 = \frac{13}{10}$? Yes.

13a) (0, 0); (4, 4)
b) Does 0 = 0? Yes. Does $0 = \frac{1}{4}(0)^2$? Yes. Does 4 = 4? Yes. Does $4 = \frac{1}{4}(4)^2$? Yes.

Check

Check to see if the cost will be the same for a ride of 12 tenths of a mile.
The cost in Burford is $1.50 + .20 \cdot 12 = 1.50 + 2.40 = 3.90$.
The cost in Spotswood is $.90 + .25 \cdot 12 = .90 + 3.00 = 3.90$.
The cost is \$3.90 in each city, so it checks.

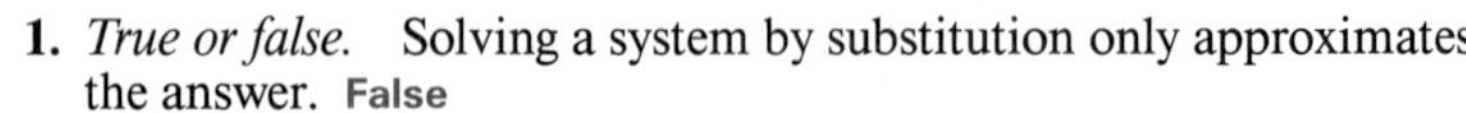

QUESTIONS

Covering the Reading

1. *True or false.* Solving a system by substitution only approximates the answer. False

In 2 and 3, a system is given. **a.** Use substitution to find the point of intersection of the two lines. **b.** Check your answer. See left.

2. $\begin{cases} y = 3x - 19 \\ y = x + 1 \end{cases}$

3. $\begin{cases} y = 12x + 50 \\ y = 10x - 60 \end{cases}$

In 4–9, refer to Example 2.

4. What would it cost for a 3-mile taxi ride in Burford? (Hint: convert to tenths of a mile.) \$7.50

5. What would it cost for a 2.5-mile taxi ride in Spotswood? \$7.15

6. Check the answer to Example 2 by graphing the lines $C = 1.50 + .20d$ and $C = .90 + .25d$ on the same axes. (Let d be the first coordinate and C be the second coordinate.) See margin.

7. For what distances are taxi rides more expensive in Burford than in Spotswood? Taxi rides are more expensive in Burford for distances less than 1.2 miles.

8. For what distances are taxi rides more expensive in Spotswood than in Burford? Taxi rides in Spotswood are more expensive for distances greater than 1.2 miles.

9. Suppose that in Manassas, a taxi ride costs \$1.70 plus 15¢ each $\frac{1}{10}$ mile.
   a. What does it cost to ride d tenths of a mile? $1.70 + .15d$ dollars
   b. At what distance does a ride in Manassas cost the same as a ride in Spotswood? $\frac{8}{10}$ mile

Applying the Mathematics

In 10–13, **a.** solve each system; **b.** check your answers.

10. $\begin{cases} y = 21 - x \\ y = 3 + x \end{cases}$ a) (9, 12) b) Does 21 – 9 = 12? Yes. Does 3 + 9 = 12? Yes.

11. $\begin{cases} y = \frac{1}{2}x - 5 \\ y = -\frac{3}{4}x + 10 \end{cases}$ a) (12, 1) b) Does $\frac{1}{2}(12) - 5 = 1$? Yes. Does $-\frac{3}{4}(12) + 10 = 1$? Yes.

12. $\begin{cases} y = \frac{1}{3}x + 2 \\ y = -3x - 5 \end{cases}$ See left.

13. $\begin{cases} y = x \\ y = \frac{1}{4}x^2 \end{cases}$ See left.

Optional Activities

Activity 2 Social Studies Connection
Materials: Almanacs

After discussing **Question 16**, you might ask students to find data for large cities near you, and answer **parts a and b** using those data. The situation is most dramatic when one city is losing population, and a second is gaining population, but the same process could be used if the populations of both cities were increasing (or decreasing).

Adapting to Individual Needs

Extra Help
In **Example 1,** point out that y, $5x - 25$, and $-8x + 27$ all represent the same number. Thus, any one of the three names can be substituted for any one of the other names. Illustrate by using 75%, $\frac{3}{4}$, .75, $\sqrt{\frac{9}{16}}$, and so on. No matter which name is used, the number is the same.

Phoenix mystery. *This building, called* Mystery Castle, *is located in Phoenix, Arizona.*

14. Jana has \$290 and saves \$5 a week. Dana has \$200 and saves \$8 a week.
- **a.** After how many weeks will they each have the same amount of money? **30 weeks**
- **b.** How much money will each one have? **\$440**

15. One plumbing company charges \$45 for the first half-hour of work and \$23 for each additional half-hour. Another company charges \$35 for the first half-hour and then \$28 for each additional half-hour. For how many hours of work will the cost of each company be the same? **1.5 hours**

16. In 1990 the population of New Orleans, Louisiana, was about 497,000 and was decreasing by about 6,000 people a year. The 1990 population of Phoenix, Arizona, was about 983,000 and was increasing by 20,000 people a year.
- **a.** If these trends had been going on for some time, how many years before 1990 did each city have the same population?
- **b.** What was this population? **about 609,000**

a) 18.69 years ≈ 18 years, 8 months ago

Review

In 17 and 18, solve by graphing. *(Lesson 11-1)* **See margin.**

17. the system of Question 11 **18.** the system of Question 12

19. a. How many solutions does this system have? **2**

$$\begin{cases} y = |x| \\ y = 5 \end{cases}$$

b. Find the solutions. *(Lessons 9-8, 11-1)* **(5, 5); (-5, 5)**

20. *Skill sequence.* Solve for x. *(Lessons 3-5, 9-1, 9-5)*
- **a.** $2x - 18 = 0$ **$x = 9$**
- **b.** $2x^2 - 18 = 0$ **$x = 3$ or $x = -3$**
- **c.** $2x^2 - 18x = 0$ **$x = 0$ or $x = 9$**

21. Multiply $2p(p^2 + 3p + 1)$. *(Lesson 10-3)* **$2p^3 + 6p^2 + 2p$**

22. Simplify $\frac{7m^4n^5}{343m^3n^6}$. *(Lesson 8-7)* **$\frac{m}{49n}$**

23. Graph all points (x, y) for which $y > -x + 5$. *(Lesson 7-9)* **See left.**

24. *Skill sequence.* Determine the y-intercept in your head. *(Lesson 7-4)*
- **a.** $y = 7x - 2$ **-2**
- **b.** $2x + y = 7$ **7**
- **c.** $7 + y = 2x$ **-7**

25. What is the cost of x videotapes at \$4 each and y computer disks at \$2 each? *(Lessons 2-4, 3-1)* **$4x + 2y$ dollars**

23) [graph of $y > -x + 5$]

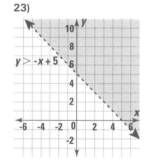

27) Sample: Ace Moving Company charges \$90 per hour and a one-time fee of \$45. Midway Movers charge \$80 per hour and a one-time fee of \$60. For how many hours of work will the cost of each company be the same? Answer: For 1.5 hours, both companies charge \$180.

Exploration

26. Find the taxi rates where you live or in a nearby community. How do these rates compare to those in Example 2? **Answers may vary.**

27. Make up a question comparing two quantities which are increasing or decreasing, each at its own constant rate (as in Example 2 and Questions 14–16 above). Use substitution to answer your question.

Follow-up for Lesson 11-2

Practice

For more questions on SPUR Objectives, use **Lesson Master 11-2A** (shown on page 673) or **Lesson Master 11-2B** (shown on pages 674–675).

Assessment

Oral Communication Ask students to explain why graphing is not always used to solve a system of linear equations. [Students understand that graphing does not always show exact solutions to a system. Sometimes it is faster and easier to use another method.]

Extension

You might want to show students how to locate an ordered triple (x, y, z) in a three-dimensional coordinate system. Use the top of your desk as the xy-plane and hold a yardstick, which represents the z-axis, perpendicular to the top of the desk.

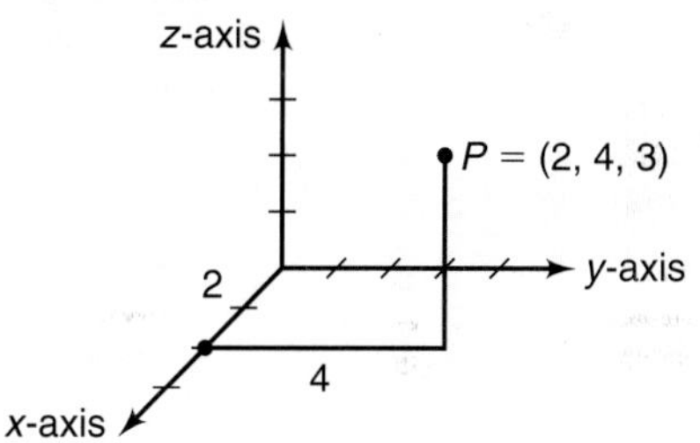

▸ LESSON MASTER 11-2 B *page 2*

Uses Objective F: Use systems of linear equations to solve real-world problems.

11. A cellular telephone company offers two plans to customers who use their mobile phone service. The monthly charges are given below.

Basic Plan: \$20 service fee plus \$.30 per minute of use
Frequent-Caller Plan: \$45 service fee plus \$.20 per minute of use

Let x = minutes of phone use and y = cost.

- **a.** Write an equation describing the basic plan. $y = .3x + 20$
- **b.** Write an equation describing the frequent-caller plan. $y = .2x + 45$
- **c.** Solve a system of equations to find the number of minutes of phone use that would cost the same under the two plans. 250 minutes
- **d.** If you estimate that you will use a cellular telephone for 100 minutes per month, which plan is better? Explain your reasoning. Basic Plan; The basic plan would cost \$50, while the frequent-caller plan would cost \$65.

12. CD Showcase Club charges a membership fee of \$15 and then \$10.50 for each CD purchased. CD Budget Club charges a membership fee of \$5 and then \$11.75 for each CD.

- **a.** Describe these charges with a system of equations. $\begin{cases} y = 10.5x + 15 \\ y = 11.75x + 5 \end{cases}$
- **b.** Solve the system to find the number of CDs purchased for which the total charges at each club are the same. 8 CDs
- **c.** If you think you will be buying many, many CDs, which club is less expensive? Explain your reasoning. CD Showcase Club; After 8 CDs, the charges at Showcase Club will be less than charges at Budget Club.

Adapting to Individual Needs

Challenge

Give students the following systems of three equations and have them use substitution to find the ordered triple (x, y, z) that satisfies all three equations in each system.

1. $\begin{cases} x + y + z = 9 \\ x + y - z = 3 \\ y = 2 \end{cases}$ **2.** $\begin{cases} x + 2y + z = 6 \\ x - 4y - z = 2 \\ z + 1 = 0 \end{cases}$

$[(x, y, z) = (4, 2, 3)]$ $[(x, y, z) = (5, 1, -1)]$

Additional Answers

17.

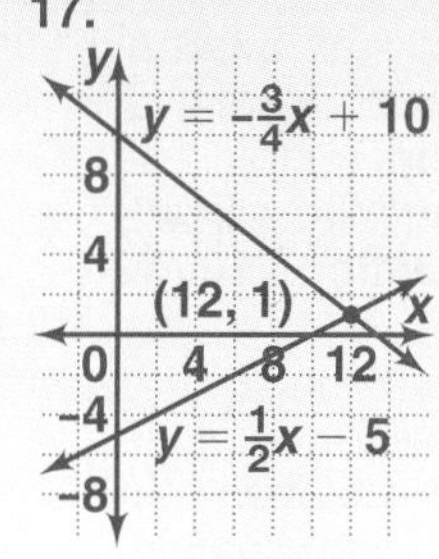

18.

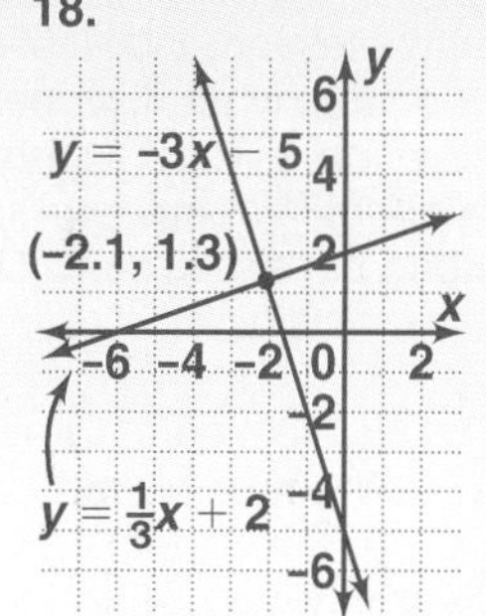

Lesson 11-3

Objectives

A Solve systems using substitution.
F Use systems of linear equations to solve real-world problems.

Resources

From the ***Teacher's Resource File***
- Lesson Master 11-3A or 11-3B
- Answer Master 11-3
- Assessment Sourcebook: Quiz for Lessons 11-1 through 11-3
- Teaching Aids
  28 Four-Quadrant Graph Paper
  122 Warm-up

Additional Resources
- Visuals for Teaching Aids 28, 122
- Colored chalk (Optional)

Teaching Lesson 11-3

Warm-up

For each question tell which, if any, of the equations are equivalent.

1. a. $x - y = 65$
   b. $x = 65 + y$
   c. $y = x + 65$ — a and b
2. a. $2x + y = 10$
   b. $y = 2x - 10$
   c. $y = 10 + 2x$ — none
3. a. $y = \frac{x+6}{4}$
   b. $x = 4y - 6$
   c. $4y = x + 6$ — a, b, and c
4. a. $2x + y = 4$
   b. $y = \frac{12+6x}{3}$
   c. $6x + 3y = 12$ — a and c

LESSON

11-3 More Uses of Substitution

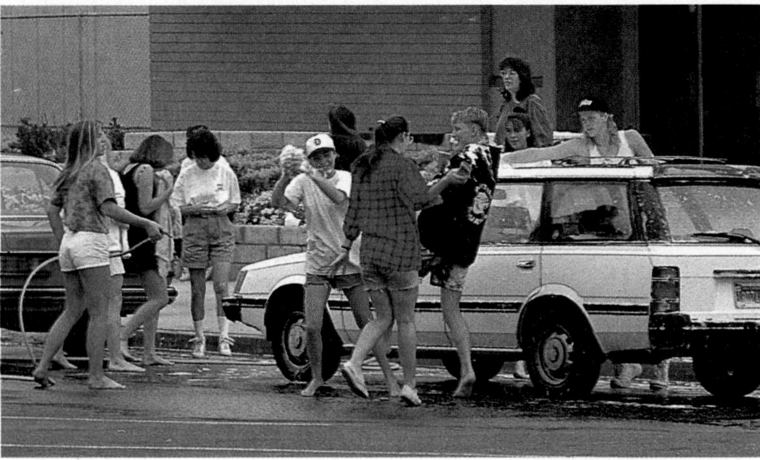

Pass the sponges. *Car washes are a common way for youth organizations to raise money for good causes.*

In the previous lesson, you saw how to use substitution to solve systems of equations when each was solved for *y*. Substitution may also be used when one or more of the equations is not solved for *y*.

Example 1

Some Boy Scouts and Girl Scouts sponsored a car wash that made \$109. There were twice as many girls as boys working, so a decision was made to give the girls twice as much money. How much did each group receive?

Solution

Translate each condition into an equation. Let B equal the proceeds received by the Boy Scouts and G equal the proceeds received by the Girl Scouts.

The total proceeds equal \$109. $B + G = 109$
The girls' proceeds are twice that of the boys'. $G = 2B$

Since $G = 2B$, you can substitute $2B$ for G in the first equation.

$$B + G = 109$$
$$B + 2B = 109$$
$$3B = 109$$
$$B = 36\tfrac{1}{3}$$

To find G, substitute $36\frac{1}{3}$ for B in either equation. We use the second equation because it is solved for G.

$$G = 2B$$
$$= 2 \cdot 36\tfrac{1}{3}$$
$$= 72\tfrac{2}{3}$$

Lesson 11-3 Overview

Broad Goals This lesson considers systems in which one of the equations in the system is of the form $y = mx + b$, but not all equations are of this form.

Perspective The various kinds of systems come about as a result of different sorts of problems. In Lesson 11-2, the system equated two constant-increase situations; those situations lend themselves to the slope-intercept form $y = mx + b$.

In this lesson, the application in **Example 1** illustrates what is sometimes called a *ratio situation.* In a ratio situation, a total is separated into various parts, and the ratio of those parts is known. **Questions 10–12** are of this type.

The application in **Example 3** combines a putting-together addition situation ($x + y = 100$) with a linear combination ($20x + 30y = 2450$) for the value of the parts. This kind of system can be solved efficiently either by substitution or by multiplication and addition. The latter process will be shown in Lesson 11-5.

So the solution is $(B, G) = \left(36\frac{1}{3}, 72\frac{2}{3}\right)$. The Boy Scouts will receive $\$36\frac{1}{3}$, or about \$36.33, and the Girl Scouts will get $\$72\frac{2}{3}$, or about \$72.67.

Check

Are both conditions satisfied? Will the groups receive a total of \$109? Yes, \$36.33 + \$72.67 = \$109. Will the girls get twice as much as the boys? Yes, \$72.67 is about twice as much as \$36.33.

In the equation $G = 2B$ from Example 1, G is given in terms of B. Substitution is usually a good method to use when one variable is given in terms of others. In Example 2, one equation gives y in terms of x.

Example 2

Two lines have equations $3x + 2y = 10$ and $y = 4x + 1$. Where do they intersect?

Solution

Notice that the equation $y = 4x + 1$ is solved for y. Substitute $4x + 1$ for y in the other equation and solve the linear equation that results.

$$\begin{aligned} 3x + 2y &= 10 \\ 3x + 2(4x + 1) &= 10 \\ 3x + 8x + 2 &= 10 \\ 11x + 2 &= 10 \\ 11x &= 8 \\ x &= \frac{8}{11} \end{aligned}$$

Thus $\frac{8}{11}$ is the x-coordinate of the point of intersection. To find the y-coordinate, substitute $\frac{8}{11}$ for x in one of the equations. We use the second equation because it is solved for y.

$$\begin{aligned} y &= 4 \cdot \frac{8}{11} + 1 \\ &= \frac{32}{11} + 1 \\ &= \frac{43}{11} \end{aligned}$$

The lines intersect at $\left(\frac{8}{11}, \frac{43}{11}\right)$.

Check

You could graph the lines to see if your solution is reasonable. But to produce an exact check, substitute $\frac{8}{11}$ for x and $\frac{43}{11}$ for y in both equations.

Does $3\left(\frac{8}{11}\right) + 2\left(\frac{43}{11}\right) = 10$?

Does $4\left(\frac{8}{11}\right) + 1 = \left(\frac{43}{11}\right)$?

You should verify that they do.

Notes on Reading

Reading Mathematics You may want to spend extra time reading and explaining the examples in this lesson. Some students find the concepts and algebraic manipulations difficult to understand. It may help these students to follow the process as it unfolds in class. You might want to use colored chalk to help students understand the substitution step.

For systems of equations that use variables other than x and y, it is customary to write and graph ordered pairs with the variables in alphabetical order, or in the order of cause (independent variable) and effect (dependent variable).

In **Example 1,** you can illustrate substituting $2B$ for G by writing $B + G = 109$ on the board and then erasing G and writing $2B$ in its place (or under it). You may want to ask students if they can solve this example by using only one variable. Have them discuss how the single-variable solution is related to the two-variable solution.

Students may ask, "I substituted for x and not for y. Does that matter?" Point out that it doesn't matter, but sometimes one choice will ease the manipulations that are required in solving.

Point out the two types of situations that are represented by **Examples 1 and 3.** The *Overview* on page 676 discusses them in more detail.

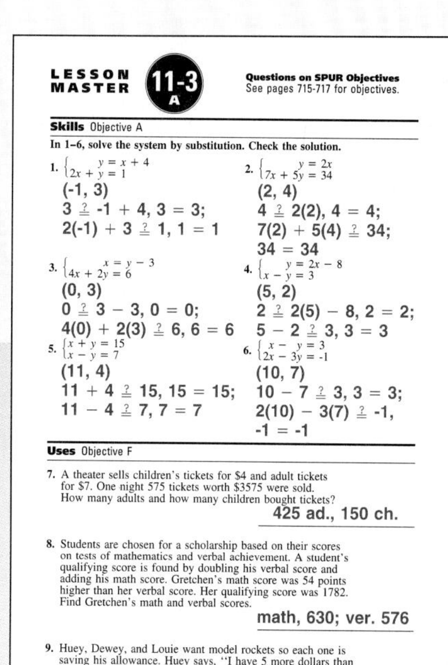

LESSON MASTER 11-3 A

Questions on SPUR Objectives
See pages 715-717 for objectives.

Skills Objective A

In 1–6, solve the system by substitution. Check the solution.

1. $\begin{cases} y = x + 4 \\ 2x + y = 1 \end{cases}$
   (-1, 3)
   $3 \stackrel{?}{=} -1 + 4, 3 = 3;$
   $2(-1) + 3 \stackrel{?}{=} 1, 1 = 1$

2. $\begin{cases} y = 2x \\ 7x + 5y = 34 \end{cases}$
   (2, 4)
   $4 \stackrel{?}{=} 2(2), 4 = 4;$
   $7(2) + 5(4) \stackrel{?}{=} 34;$
   $34 = 34$

3. $\begin{cases} x = y - 3 \\ 4x + 2y = 6 \end{cases}$
   (0, 3)
   $0 \stackrel{?}{=} 3 - 3, 0 = 0;$
   $4(0) + 2(3) \stackrel{?}{=} 6, 6 = 6$

4. $\begin{cases} y = 2x - 8 \\ x - y = 3 \end{cases}$
   (5, 2)
   $2 \stackrel{?}{=} 2(5) - 8, 2 = 2;$
   $5 - 2 \stackrel{?}{=} 3, 3 = 3$

5. $\begin{cases} x + y = 15 \\ x - y = 7 \end{cases}$
   (11, 4)
   $11 + 4 \stackrel{?}{=} 15, 15 = 15;$
   $11 - 4 \stackrel{?}{=} 7, 7 = 7$

6. $\begin{cases} x - y = 3 \\ 2x - 3y = -1 \end{cases}$
   (10, 7)
   $10 - 7 \stackrel{?}{=} 3, 3 = 3;$
   $2(10) - 3(7) \stackrel{?}{=} -1,$
   $-1 = -1$

Uses Objective F

7. A theater sells children's tickets for \$4 and adult tickets for \$7. One night 575 tickets worth \$3575 were sold. How many adults and how many children bought tickets?
   425 ad., 150 ch.

8. Students are chosen for a scholarship based on their scores on tests of mathematics and verbal achievement. A student's qualifying score is found by doubling his verbal score and adding his math score. Gretchen's math score was 54 points higher than her verbal score. Her qualifying score was 1782. Find Gretchen's math and verbal scores.
   math, 630; ver. 576

9. Huey, Dewey, and Louie want model rockets so each one is saving his allowance. Huey says, "I have 5 more dollars than you do, Louie." Dewey says, "I have twice as much as you do, Louie." Louie says, "Together we have \$31.00. That's enough to buy a rocket." How much does each have?
   H, \$11.50; D, \$13; L, \$6.50

Optional Activities

Writing Consumer Connection
As an extension of **Example 3** and **Question 8,** you might have students **work in groups** to find the cost of vacant land or commercial space in your area. If you live in a rural area, students can find the value of farmland. If you live in an urban area, they might find the value of commercial space—this is often quoted in number of dollars per square foot. To find this information, ask three or four volunteers to interview local real-estate agents and report their findings to the class. Then all students could use their information to write problems that are similar to those in **Example 3** or **Question 8**.

Additional Examples

1. Without graphing, find the point where the graphs for $7x - 2y = -3$ and $y = 6x - 1$ intersect. **(1, 5)**
2. The Wolff family bought two chairs. One cost $15 less than the other. Together the chairs cost $374. Find the price of each chair. **$179.50, $194.50**
3. The recipe for a vegetable sauce calls for 5 parts mayonnaise to one part mustard. How much of each ingredient is needed to make 16 ounces of sauce? **$13\frac{1}{3}$ oz mayonnaise, $2\frac{2}{3}$ oz mustard**

Follow-up for Lesson 11-3

Practice

For more questions on SPUR Objectives, use **Lesson Master 11-3A** (shown on page 677) or **Lesson Master 11-3B** (shown on pages 678–679).

Assessment

Quiz A quiz covering Lessons 11-1 through 11-3 is provided in the *Assessment Sourcebook.*

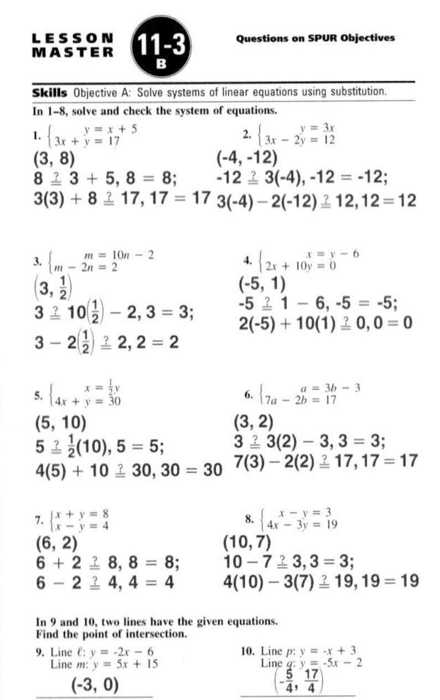
LESSON MASTER 11-3 B Questions on SPUR Objectives

Skills Objective A: Solve systems of linear equations using substitution.

In 1–8, solve and check the system of equations.

1. $\begin{cases} y = x + 5 \\ 3x + y = 17 \end{cases}$
(3, 8)
$8 \stackrel{?}{=} 3 + 5$, $8 = 8$;
$3(3) + 8 \stackrel{?}{=} 17$, $17 = 17$

2. $\begin{cases} y = 3x \\ 3x - 2y = 12 \end{cases}$
(-4, -12)
$-12 \stackrel{?}{=} 3(-4)$, $-12 = -12$;
$3(-4) - 2(-12) \stackrel{?}{=} 12$, $12 = 12$

3. $\begin{cases} m = 10n - 2 \\ m - 2n = 2 \end{cases}$
$\left(3, \frac{1}{2}\right)$
$3 \stackrel{?}{=} 10\left(\frac{1}{2}\right) - 2$, $3 = 3$;
$3 - 2\left(\frac{1}{2}\right) \stackrel{?}{=} 2$, $2 = 2$

4. $\begin{cases} x = y - 6 \\ 2x + 10y = 0 \end{cases}$
(-5, 1)
$-5 \stackrel{?}{=} 1 - 6$, $-5 = -5$;
$2(-5) + 10(1) \stackrel{?}{=} 0$, $0 = 0$

5. $\begin{cases} x = \frac{1}{2}y \\ 4x + y = 30 \end{cases}$
(5, 10)
$5 \stackrel{?}{=} \frac{1}{2}(10)$, $5 = 5$;
$4(5) + 10 \stackrel{?}{=} 30$, $30 = 30$

6. $\begin{cases} a = 3b - 3 \\ 7a - 2b = 17 \end{cases}$
(3, 2)
$3 \stackrel{?}{=} 3(2) - 3$, $3 = 3$;
$7(3) - 2(2) \stackrel{?}{=} 17$, $17 = 17$

7. $\begin{cases} x + y = 8 \\ x - y = 4 \end{cases}$
(6, 2)
$6 + 2 \stackrel{?}{=} 8$, $8 = 8$;
$6 - 2 \stackrel{?}{=} 4$, $4 = 4$

8. $\begin{cases} x - y = 3 \\ 4x - 3y = 19 \end{cases}$
(10, 7)
$10 - 7 \stackrel{?}{=} 3$, $3 = 3$;
$4(10) - 3(7) \stackrel{?}{=} 19$, $19 = 19$

In 9 and 10, two lines have the given equations. Find the point of intersection.

9. Line ℓ: $y = -2x - 6$
Line m: $y = 5x + 15$
(-3, 0)

10. Line p: $y = -x + 3$
Line q: $y = -5x - 2$
$\left(-\frac{5}{4}, \frac{17}{4}\right)$

In Example 3, one equation can easily be solved for a variable.

These algebra books were printed between the years 1877 and 1923.

Example 3

(This problem is taken from an 1881 algebra text; the prices are out of date, but the situation is not.) A farmer purchased 100 acres of land for $2450. For a part of it he paid $20 an acre, and $30 an acre for the rest of it. How many acres were there in each part?

Solution

You want to find two amounts, so use two variables.
Let x = the number of acres at $20/acre,
and y = the number of acres at $30/acre.

A total of 100 acres was purchased, so

$$x + y = 100.$$

The total cost was $2450. So

$$20x + 30y = 2450.$$

Solve the system $\begin{cases} x + y = 100 \\ 20x + 30y = 2450 \end{cases}$.

Note that neither equation is solved for a variable. But notice that the first equation can be rewritten as

$$y = 100 - x.$$

Now substitute $100 - x$ for y in the second equation.

$$\begin{aligned} 20x + 30y &= 2450 \\ 20x + 30(100 - x) &= 2450 \\ 20x + 3000 - 30x &= 2450 \\ 3000 - 10x &= 2450 \\ -10x &= -550 \\ x &= 55 \end{aligned}$$

To find y, substitute $x = 55$ in either of the original equations. We use the first equation because it is simpler.

$$\begin{aligned} x + y &= 100 \\ 55 + y &= 100 \\ y &= 45 \end{aligned}$$

The farmer bought 55 acres at $20/acre, and 45 acres at $30/acre.

Check

Substitute $x = 55$, $y = 45$ in both equations.
Does $x + y = 55 + 45 = 100$? Yes, it checks.
Does $20x + 30y = 20(55) + 30(45) = 1100 + 1350 = 2450$? Yes, it checks.

Adapting to Individual Needs

Extra Help

Some students might still have trouble translating data from a problem into an equation. In problems like **Example 1,** they might incorrectly write $2G = B$ instead of $G = 2B$. Refer these students back to the problem and ask, "Who received more money, the Girl Scouts or the Boy Scouts?" [Girl Scouts] Then point out that the amount received by the girls was 2 times the amount received by the boys, so $G = 2B$ is the correct equation. Contrast this with the situation in **Question 6** where the boys received three times as much as the girls; here the equation would be $B = 3G$. Also point out that an answer should always be checked by going back to the problem to see if it fits the data that are given.

1) $3\left(\frac{8}{11}\right) + 2\left(\frac{43}{11}\right) = \frac{24}{11} + \frac{86}{11} = \frac{110}{11} = 10$; $4\left(\frac{8}{11}\right) + 1 = \frac{32}{11} + \frac{11}{11} = \frac{43}{11}$; Yes, it checks.

2b) Does 12 + 2 = 14? Yes. Does 12 = 6 · 2? Yes.

3b) Does 12(0) − 5(-6) = 30? Yes. Does -6 = 2(0) -6? Yes.

This 1855 painting, Guarding the Corn Fields *by Seth Eastman, depicts Dakota Indians frightening birds away from crops by banging sticks together.*

4b) Does 6 = 8 − 2? Yes. Does -4(8) + 7(6) = 10? -32 + 42 = 10? Yes.

5b) Does $3\left(\frac{-3}{11}\right) + 4\left(\frac{-39}{11}\right) = -15$? $\frac{-9}{11} + \frac{-156}{11} = \frac{-165}{11} = -15$? Yes. Does $\frac{-39}{11} = 2\left(\frac{-3}{11}\right) - 3$? $\frac{-6}{11} - \frac{33}{11} = \frac{-39}{11}$? Yes.

QUESTIONS

Covering the Reading

1. Complete the check of Example 2. **See left.**

In 2–5, **a.** find the point of intersection of the lines without graphing; **b.** check your answer. **See left for checks.**

2. $\begin{cases} x + y = 14 \\ x = 6y \end{cases}$ **(12, 2)**

3. $\begin{cases} 12x - 5y = 30 \\ y = 2x - 6 \end{cases}$ **(0, -6)**

4. $\begin{cases} y = x - 2 \\ -4x + 7y = 10 \end{cases}$ **(8, 6)**

5. $\begin{cases} 3x + 4y = -15 \\ y = 2x - 3 \end{cases}$ $\left(\frac{-3}{11}, \frac{-39}{11}\right)$

6. Suppose the service club made \$180 on a car wash. If the boys were to get three times as much money as the girls, how much would each group receive? **boys, \$135; girls, \$45**

7. a. Solve the system $\begin{cases} B = 7t \\ B - t = 30 \end{cases}$. **$t = 5$, $B = 35$**
b. Check your answer. **Does 35 = 7 · 5? Yes. Does 35 − 5 = 30? Yes.**

8. Here is another problem from the 1881 algebra book. A farmer bought 100 acres of land, part at \$37 and part at \$45 an acre, paying for the whole \$4220. How much land was there in each part? **35 acres at \$37 per acre and 65 acres at \$45 per acre.**

Applying the Mathematics

9. Profits of a company were up \$200,000 this year over last year. This was a 25% increase. If T and L are the profits (in dollars) for this year and last year, then:

$$\begin{cases} T = L + 200{,}000 \\ T = 1.25L \end{cases}$$ **T = \$1,000,000; L = \$800,000**

Find the profits for this year and last year.

10. A will states that John is to get 3 times as much money as Mary. The total amount they will receive is \$11,000. **$J = 3M$; $J + M = 11{,}000$**
a. Write a system of equations describing this situation.
b. Solve to find the amounts of money John and Mary will get. **Mary gets \$2750; John gets \$8250**

11. Solve for A, B, and K. $\begin{cases} A = 40K \\ B = 30K \\ A + B = 1400 \end{cases}$ **$A = 800$, $B = 600$, $K = 20$**

12. A homemade sealer to use on furniture after it is stained can be made by mixing one part shellac with five parts denatured alcohol. To make a pint (16 fluid ounces) of sealer, how many fluid ounces s of shellac and how many fluid ounces A of denatured alcohol are needed? (Hint: $\frac{A}{s} = \frac{5}{1}$.) **$s = 2\frac{2}{3}$ fl oz.; $A = 13\frac{1}{3}$ fl oz**

In 13 and 14, solve the system using substitution.

13. $\begin{cases} 4x - 3y = 8 \\ 2x + y = -1 \end{cases}$ $\left(\frac{1}{2}, -2\right)$

14. $\begin{cases} 8a - 1 = 4b \\ 3a = b + 1 \end{cases}$ $\left(\frac{3}{4}, \frac{5}{4}\right)$

Group Assessment Have students **work in small groups** to write a real-world problem that can be answered by setting up and solving a system of two linear equations. Have groups exchange problems and solve them with a system. [Students demonstrate they understand how to use a system in a real-world situation. Students show that they can solve a system.]

Extension

Solutions of systems involving quadratics are usually left to an advanced algebra course. The problems that students should master in this course are those that involve linear relationships. However, students will solve systems of the form $xy = a$ *and* $x + y = b$ in Lesson 12-6. You may want to show your class the following system to point out the power of the substitution method and to reinforce their ability to chunk terms.

$$\begin{cases} y = x^2 \\ 2x + y + 1 = 4 \end{cases}$$

Use x^2 in place of y in the second equation.

$$2x + y + 1 = 4$$
$$2x + x^2 + 1 = 4$$
$$x^2 + 2x + 1 = 4$$
$$(x + 1)^2 = 4$$

So $x + 1 = 2$ or $x + 1 = -2$.

The two solutions are $x = 1$ and $x = -3$. The corresponding y values are $y = 1$ and $y = 9$. The solution set is {(1, 1), (−3, 9)}. Students can graph the two equations as a check.

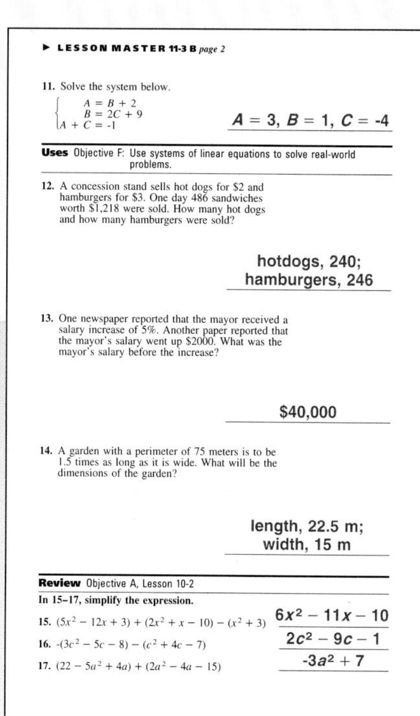

▸ **LESSON MASTER 11-3 B** *page 2*

11. Solve the system below.
$\begin{cases} A = B + 2 \\ B = 2C + 9 \\ A + C = -1 \end{cases}$ **$A = 3$, $B = 1$, $C = -4$**

Uses Objective F: Use systems of linear equations to solve real-world problems.

12. A concession stand sells hot dogs for \$2 and hamburgers for \$3. One day 486 sandwiches worth \$1,218 were sold. How many hot dogs and how many hamburgers were sold? **hotdogs, 240; hamburgers, 246**

13. One newspaper reported that the mayor received a salary increase of 5%. Another paper reported that the mayor's salary went up \$2000. What was the mayor's salary before the increase? **\$40,000**

14. A garden with a perimeter of 75 meters is to be 1.5 times as long as it is wide. What will be the dimensions of the garden? **length, 22.5 m; width, 15 m**

Review Objective A, Lesson 10-2

In 15–17, simplify the expression.

15. $(5x^2 - 12x + 3) + (2x^2 + x - 10) - (x^2 + 3)$ **$6x^2 - 11x - 10$**

16. $-(3c^2 - 5c - 8) - (c^2 + 4c - 7)$ **$2c^2 - 9c - 1$**

17. $(22 - 5a^2 + 4a) + (2a^2 - 4a - 15)$ **$-3a^2 + 7$**

Adapting to Individual Needs

English Language Development

Several of the examples and questions in this lesson involve the translation of conditions from words into equations. You might want to pair non-English-speaking students with English-speaking students. Once the system of equations is written, many non-English-speaking students will have no trouble solving them.

Challenge

The equations of the following system are not linear. However, if c is substituted for $\frac{1}{x}$ and d is substituted for $\frac{1}{y}$, the system becomes linear in c and d. Have students solve the system. [$x = 8$, $y = 6$]

$$\begin{cases} \frac{1}{x} + \frac{1}{y} = \frac{7}{24} \\ \frac{3}{x} - \frac{2}{y} = \frac{1}{24} \end{cases}$$

Notes on Questions

Question 16 Science Connection A hot-air balloon consists of a bag made of nylon or polyester and a *payload*—a basket, passengers, equipment, and supplies. The size of the bag that is needed depends on the weight of the payload. For example, to carry two people, the bag must have an inflated volume of about 60,000 cubic feet. (A room 60 ft wide, 100 ft long, and 10 ft high has this volume.)

Question 24 You can use this question to set up Lesson 11-5. Ask students to suppose that they have 20 nickels and dimes with a total value of \$1.40. Then the system becomes $n + d = 20$ and $5n + 10d = 140$. Students can solve this system now by substitution to get $n = 12$ and $d = 8$. In Lesson 11-5 they will learn how to solve it using multiplication and addition.

Question 25 This question relates to **Example 1** with *cities* replacing *scouts.* However, there are ways to split the costs other than by population. Where will the conference center be located? What roads or other improvements must be made for the center? The scout situation is less complicated than this one.

Question 26 Try to discuss this question before students read Lesson 11-4. It has the same context as **Example 1** in that lesson.

Science Connection The homing pigeon is a domestic pigeon that can return to its home loft after traveling great distances. Just how homing pigeons do this is not known; they seem to have the ability to orient themselves with the earth's magnetic field, to guide themselves by the sun or stars, and to recall landmarks.

Full of hot air. *Pictured are balloons at a fiesta in Albuquerque. Hot-air balloons rise because the air inside the bag is warmer, and therefore lighter, than the surrounding air.*

21)

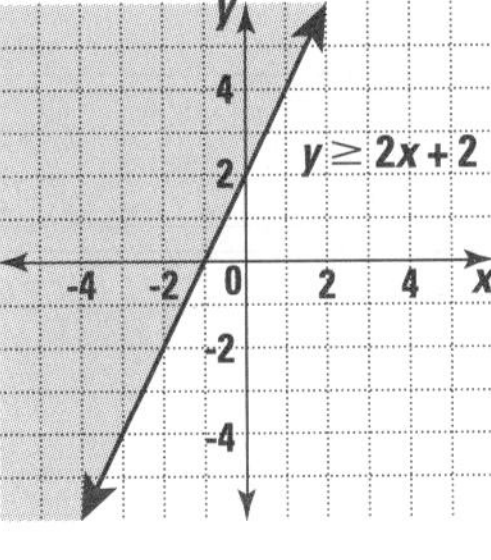

25) Elylov, \$6,750,166; Lovely, \$3,249,834; $L + E = 10$ million and $\frac{L}{35,729} = \frac{E}{74,212}$; $74212L = 35729E$; $L = \frac{35729}{74212}E$; Substitute this value in $L + E = 10$ million and solve for E.

15. Noah got the results of tests on mathematics and verbal achievement. His verbal score is 70 points less than his mathematics score. His total score for the two parts is 1250.

a. Let v = Noah's verbal score, and m = his mathematics score. Write a system of equations for this situation. $m - 70 = v$; $m + v = 1250$

b. Find Noah's two scores. $v = 590$; $m = 660$

Review

16. One hot-air balloon takes off from Albuquerque, New Mexico, and rises at a rate of 120 ft/min. At the same time, another balloon takes off from Santa Fe, NM, and rises at a rate of 75 ft/min. The altitude of Albuquerque is about 4950 ft; the altitude of Santa Fe is about 6950 ft.

a. When are the two balloons at the same altitude? $\approx$ 44.44 minutes

b. What is their altitude at that time? *(Lessons 3-8, 5-2, 5-3, 11-2)* $\approx$ 10,283 ft

In 17 and 18, a system is given. See margin.

a. Graph or use substitution to find the solution to the system.

b. Check your answer. *(Lessons 11-1, 11-2)*

17. $\begin{cases} y = x + 1 \\ y = -2x + 13 \end{cases}$

18. $\begin{cases} y = 3x - 2 \\ y = 3x + 3 \end{cases}$

In 19 and 20, simplify. *(Lessons 4-5, 10-6)*

19. $(3a - 2b) - (a + b)$ $2a - 3b$

20. $x^2 + y^2$ when $y = x - 1$ $2x^2 - 2x + 1$

21. Graph $y \geq 2x + 2$ on a coordinate plane. *(Lesson 7-9)* See left.

22. Solve $5x + 9y = 7$ for y. *(Lesson 7-4)* $y = \frac{-5}{9}x + \frac{7}{9}$

23. If you travel 120 miles in 40 minutes, what is your average speed in miles per hour? *(Lesson 6-2)* 180 mph

24. Suppose you have n nickels and d dimes and these are the only coins you have. **a.** How many coins do you have? **b.** What is the total value of the coins in cents? *(Lessons 2-4, 3-1)*
a) $n + d$ b) $5n + 10d$ cents

Exploration

25. Neighboring cities Lovely (population 35,729) and Elylov (population 74,212) are building a joint conference center that will cost \$10 million. How would you suggest they split the cost? Why? See left.

26. A homing pigeon has been clocked doing 94.3 mph in still air.

a. How fast can it fly *with* the wind if the wind speed is s mph?

b. How fast can it fly *against* the wind if the wind speed is s mph?

c. How far can it fly in m minutes in still air? $\frac{94.3m}{60}$ miles

d. If it is flying down the highway (where the speed limit is 65 mph) against a 25 mph headwind, would you give it a ticket?
Yes, its rate of 69.3 mph is faster than the speed limit.

a) $94.3 + s$ b) $94.3 - s$

Additional Answers

17a.

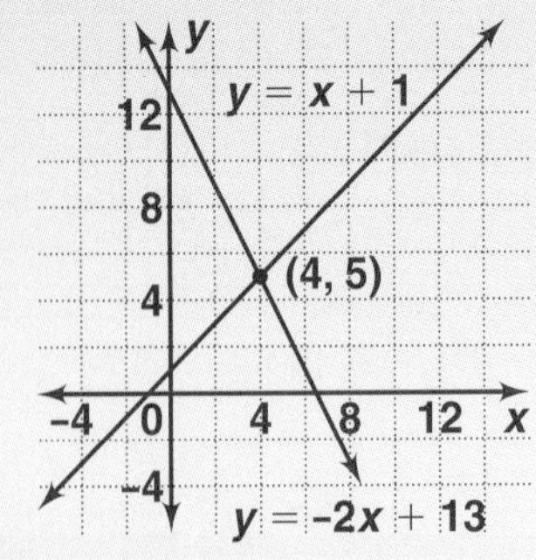

b. Does $5 = 4 + 1$? Yes.
Does $5 = -2(4) + 13 = -8 + 13 = 5$?
Yes, it checks.

18.

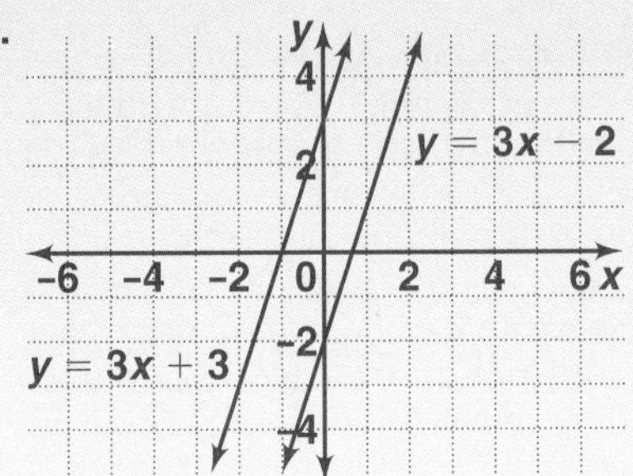

No solution; no check.

LESSON

11-4

Solving Systems by Addition

Young pilot. *In 1994, 12-year-old Vicki Van Meter, accompanied by her flight instructor Curt Arnspiger, became the youngest female pilot to cross the Atlantic Ocean. Systems of equations can help find plane and wind speeds. See Example 1.*

The numbers $\frac{3}{4}$ and 75% are equal even though they may not look equal. So are $\frac{1}{5}$ and 20%. If you add pairs of these numbers (as seen below), the sums are equal.

$$\frac{3}{4} = 75\%$$

$$\frac{1}{5} = 20\%$$

So $\qquad \frac{3}{4} + \frac{1}{5} = 75\% + 20\%.$

Simplifying each side, we get $\qquad \frac{19}{20} = 95\%.$

This is one instance of the following generalization of the Addition Property of Equality.

Generalized Addition Property of Equality
For all numbers or expressions a, b, c, and d:
If $a = b$
and $c = d$,
then $a + c = b + d$.

The Generalized Addition Property of Equality can be used to solve some systems. Consider this situation:

The sum of two numbers is 63.
Their difference is 12. What are the numbers?

If x and y are the two numbers, with x the larger, we can write the following system.

$$\begin{cases} x + y = 63 \\ x - y = 12 \end{cases}$$

Lesson 11-4

Objectives

B Solve systems by addition.
F Use systems of linear equations to solve real-world problems.

Resources

From the ***Teacher's Resource File***

- Lesson Master 11-4A or 11-4B
- Answer Master 11-4
- Teaching Aids
  - 28 Four-Quadrant Graph Paper
  - 122 Warm-up
  - 126 Additional Examples

Additional Resources

- Visuals for Teaching Aids 28, 122, 126

Teaching Lesson 11-4

Warm-up

A large cheese pizza with one additional topping costs \$8.99. A large cheese pizza with three additional toppings costs \$11.29. Assuming each topping costs the same amount, find the price of each topping and the price of a large cheese pizza with no additional topping.
\$1.15; \$7.84

Lesson 11-4 Overview

Broad Goals In this lesson, we consider systems for which either addition or subtraction leads to an equation in one variable.

Perspective The systems in this lesson have one variable in which the coefficients in one of the equations are either opposite or equal to the coefficients in the other equation. In the former case, "adding the equations," which is short for "adding the sides of one equation to the sides of the other equation," yields an equation in one variable and thus solves the system quite easily. In the latter case, we multiply one of the equations by -1 and then add. Some teachers prefer to avoid the multiplication step, and subtract one equation from the other; others feel that this leads to too many sign errors. Still other teachers like to teach both methods.

As the examples indicate, there are some situations leading to each of these forms. Also, students have to be able to solve these forms in order to solve more general systems, such as those found in Lesson 11-5.

Notes on Reading

Compare the Generalized Addition Property of Equality to the original Addition Property of Equality that is presented in Lesson 3-2 (pages 149–154). There, the *same* expression was added to both sides of an equation. Here, *equal* expressions are added.

❶ Note that it is easy to follow the steps in this algorithm because the equal signs are aligned and the variables are written in neat columns. Advise students that they are less likely to make errors if they follow the same format. To emphasize the benefit of properly organizing the variables and symbols, you might want to solve a system slightly more complicated, such as the one below. It looks as if addition is a good strategy, but the $2x$ on the right gets in the way.

$$\begin{cases} 4x - 3y = 20 \\ -4x + 36 + 4y = 2x \end{cases}$$

❷ If students do not see how 150 miles per hour for the speed with the wind is obtained, have them solve this proportion: $\frac{120 \text{ miles}}{\frac{4}{5}\text{ hour}} = \frac{x \text{ miles}}{1 \text{ hour}}$.

Additional Examples

These examples are also given on **Teaching Aid 126**.

1. Going with the wind, a blimp flies 360 miles to an air show. The trip takes 4 hours. The return trip, flying against the wind, takes 9 hours. How fast is the blimp flying in still air? What is the speed of the wind? **Let b represent the speed of the blimp in still air, and let w represent the speed of the wind.**
$$\begin{cases} b + w = 90 \\ b - w = 40 \end{cases}$$
65 mph, 25 mph

Notice what happens when the left sides are added and the right sides are added.

$$\begin{array}{r} x + y = 63 \\ + \; x - y = 12 \\ \hline 2x + 0 = 75 \end{array}$$

Because y and $-y$ add to 0, the sum is an equation with only one variable. Solve $2x = 75$ as usual.

$$x = 37.5$$

To find y, substitute 37.5 for x in one of the original equations.

❶
$$\begin{aligned} x + y &= 63 \\ 37.5 + y &= 63 \\ y &= 25.5 \end{aligned}$$

The ordered pair (37.5, 25.5) checks in both equations, so the solution to the system $\begin{cases} x + y = 63 \\ x - y = 12 \end{cases}$ is (37.5, 25.5).

Using the Generalized Addition Property of Equality to eliminate one variable from a system is sometimes called the **addition method** for solving a system. The addition method is an efficient way to solve systems when the coefficients of the same variable are opposites.

They can wing it. *These small planes are from the Lion's Club Fly-In Fish Boil on Washington Island, Wisconsin. Planes like these were often used for training pilots during World War II.*

Example 1

A pilot flew a small plane 120 miles from Washington Island, Wisconsin, to Appleton, Wisconsin, in one hour against the wind. The pilot returned to Washington Island in 48 minutes $\left(\frac{48}{60} = \frac{4}{5}\text{ hour}\right)$ with the wind at the plane's back. How fast was the plane flying (without wind)? What was the speed of the wind?

Solution

Let A be the average speed of the airplane without wind and W be the speed of the wind, both in miles per hour. The total speed against the wind is then $A - W$, and the speed with the wind is $A + W$. There are two conditions given on these total speeds.

From Washington Island to Appleton the rate was $\frac{120 \text{ miles}}{1 \text{ hour}}$, or $120 \frac{\text{miles}}{\text{hour}}$.

This was against the wind, so A – W = 120.

❷ From Appleton to Washington Island the rate was $\frac{120 \text{ miles}}{\frac{4}{5}\text{hour}}$, or $150 \frac{\text{miles}}{\text{hour}}$.

This was with the wind, so A + W = 150.

Now solve the system. Since the coefficients of W are opposites (1 and -1), add the equations.

$$\begin{array}{r} A - W = 120 \\ + \quad A + W = 150 \\ \hline 2A = 270 \\ A = 135 \end{array}$$

Optional Activities

Activity 1 Using Physical Models
As you discuss **Question 18**, you might want to picture the situation as shown at the right; w = Wadlow's height and t = Thumb's height. Diagrams 1 and 2 show the equations, $w - t = 67$ and $w + t = 147$. Diagram 3 shows the result of adding them; $(w - t) + w + t = 67 + 147$ is the same as $2w = 214$. So, $w = 107$ and $t = 40$. Wadlow was 107 in. tall (8 ft 11 in.) and Thumb was 40 in. tall.

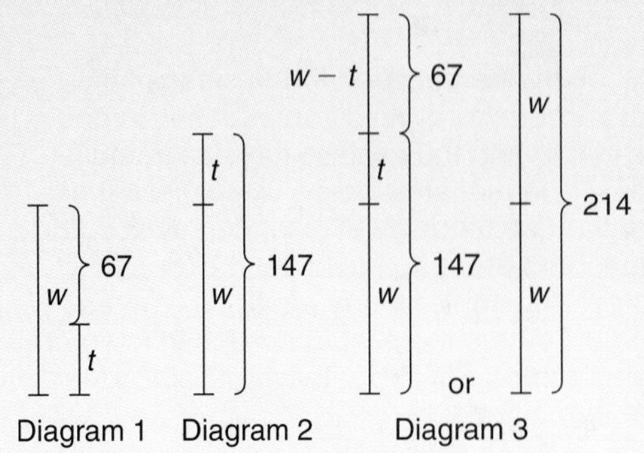

Activity 2
After you discuss **Question 29,** you might have students write other repeating decimals as fractions. (Sample: $d = .\overline{9}$ and $10d = 9.\overline{9}$, so $10d - d = 9.\overline{9} - .\overline{9}$. Then $9d = 9$ and $d = 1$.) Many students will be surprised at the result. As further verification, they can compute $.\overline{3} + .\overline{6} = .\overline{9}$ by converting to fractions. Caution them that $.\overline{4} + .\overline{7} \neq 1.\overline{1}$.

Substitute 135 for A in either of the original equations. We choose the second equation.

$$135 + W = 150$$
$$W = 15$$

The average speed of the airplane was 135 mph and the speed of the wind was 15 mph.

Check

Refer to the original question. Against the wind, the plane flew at 135 − 15, or 120 mph. With the wind the plane flew at 135 + 15, or 150 mph. At that rate, in 48 minutes the pilot flew $\frac{48}{60}$ hr · 150 $\frac{\text{mi}}{\text{hr}}$ = 120 miles, which checks with the given conditions.

Sometimes the coefficients of the same variable are equal. In this case, use the Multiplication Property of Equality to multiply both sides of one of the equations by -1. This changes all the numbers in that equation to their opposites. Then you can use the addition method to find solutions to the system.

Example 2

Solve this system.

$$\begin{cases} 4x + 13y = 40 \\ 4x + 3y = -40 \end{cases}$$

Solution

We rewrite the equations and number them.

$$4x + 13y = 40 \quad \#1$$
$$4x + 3y = -40 \quad \#2$$

Notice that the coefficients of x in the two equations are equal. Multiply the second equation by -1. Call the resulting equation #3.

$$-4x - 3y = 40 \quad \#3$$

Now use the addition method with the first and third equations.

$$\begin{array}{rl} 4x + 13y = 40 & \#1 \\ + \ -4x - 3y = 40 & \#3 \\ \hline 10y = 80 & \#1 + \#3 \\ y = 8 & \end{array}$$

To find x, substitute 8 for y in one of the original equations. We use the first equation.

$$4x + 13(8) = 40$$
$$4x + 104 = 40$$
$$4x = -64$$
$$x = -16$$

So (x, y) = (-16, 8).

Check

Substitute -16 for x and 8 for y in both equations.
Does $4x + 13y = 4 \cdot -16 + 13 \cdot 8 = 40$? Yes.
Does $4x + 3y = 4 \cdot -16 + 3 \cdot 8 = -40$? Yes.

2. Solve this system.
$$\begin{cases} 2x - 5y = 18 \\ 4x - 5y = -4 \end{cases}$$ **(-11, -8)**

3. At the 1987 annual meeting of the National Council of Teachers of Mathematics, the system below was found on a menu card at a restaurant in California. (Yes, the system really did appear on the bottom of the menu card.) How much did each item in column *A* cost, and how much did each item in column *B* cost?
***A* = $3.99; *B* = $1.50**

Multiple Choice
2 or more entrées

Choose 1 from column A and
1 from column B — $5.49
Choose 1 from column A and
2 from column B — $6.99

A	B
• Chicken Dijon	• Fried Chicken (2 pieces)
• Top Sirloin Steak	• Battered Cod
• Steak Dijon	• Shrimp
• Fried Chicken	• Chicken Strips

A + B = $5.49
A + 2B = $6.99

Adapting to Individual Needs

Extra Help

Some students might wonder why systems can be solved by subtracting one equation from another when the principle being used is the Generalized *Addition* Property of Equality. Remind them that subtraction is defined in terms of addition.

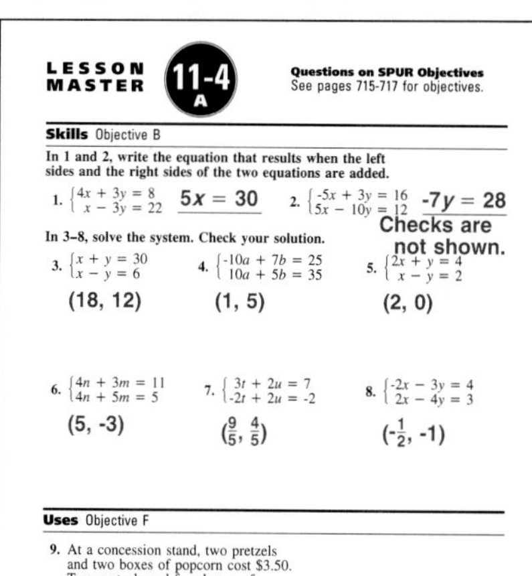
LESSON MASTER 11-4 A
Questions on SPUR Objectives
See pages 715-717 for objectives.

Skills Objective B

In 1 and 2, write the equation that results when the left sides and the right sides of the two equations are added.

1. $\begin{cases} 4x + 3y = 8 \\ x - 3y = 22 \end{cases}$ **5x = 30**
2. $\begin{cases} -5x + 3y = 16 \\ 5x - 10y = 12 \end{cases}$ **-7y = 28**

In 3–8, solve the system. Check your solution. **Checks are not shown.**

3. $\begin{cases} x + y = 30 \\ x - y = 6 \end{cases}$ **(18, 12)**
4. $\begin{cases} -10a + 7b = 25 \\ 10a + 5b = 35 \end{cases}$ **(1, 5)**
5. $\begin{cases} 2x + y = 4 \\ x - y = 2 \end{cases}$ **(2, 0)**
6. $\begin{cases} 4n + 3m = 11 \\ 4n + 5m = 5 \end{cases}$ **(5, -3)**
7. $\begin{cases} 3t + 2u = 7 \\ -2t + 2u = -2 \end{cases}$ **$(\frac{9}{5}, \frac{4}{5})$**
8. $\begin{cases} -2x - 3y = 4 \\ 2x - 4y = 3 \end{cases}$ **$(-\frac{1}{2}, -1)$**

Uses Objective F

9. At a concession stand, two pretzels and two boxes of popcorn cost $3.50. Two pretzels and four boxes of popcorn cost $6.00. Find the cost of each item. **pr., $0.50; p.c.,$1.25**

10. Beth and Carol had dinner at a cafe. The total bill was $14.40. Beth's meal cost $2.00 more than Carol's. Find the cost of each person's meal. **B, $8.20; C, $6.20**

11. When Rosa flew from Boston to Cleveland, the plane was going against the wind and traveled only 120 mph. However, on the return trip, the plane traveled with the wind at 230 mph. What was the plane's speed without wind? What was the average speed of the wind? **plane, 175 mph; wind, 55 mph**

Notes on Questions

Questions 14–15 Error Alert Both systems have the variables in different orders in the equations. Students who commute the addition in one of the equations will probably solve the systems correctly. You will have to make sure that students who have not commuted in the first equations have not equated $2x - 3y$ with $3y - 2x$, or equated $4z - 5w$ with $5w - 4z$.

Questions 16–17 Students may be able to solve **Question 16** without algebra by reasoning that the extra egg costs $2.70 – $1.80 = $0.90, so the bacon costs $1.80 – $0.90 = $0.90. **Question 17** can be answered with similar reasoning. It is worthwhile discussing these nonalgebraic methods and relating them to the algebraic methods which employ essentially the same steps using variables.

Consumer Connection Between 1982 and 1991, total energy consumption worldwide increased 24%. The U.S., the former U.S.S.R., and China are the leading producers as well as the leading consumers of the world's primary energy sources—petroleum, natural gas, coal, hydroelectricity, and nuclear electricity. In 1991, these three countries produced 47% and consumed 48% of the total energy used worldwide. Since the 1980s, petroleum has been the world's most heavily used source of energy. World production

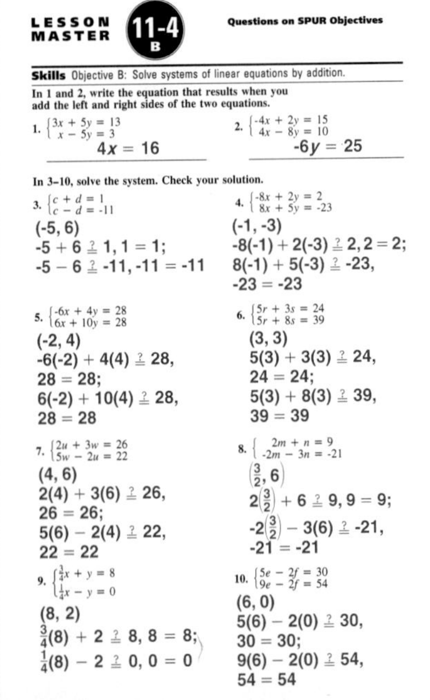

LESSON MASTER 11-4 B Questions on SPUR Objectives

Skills Objective B: Solve systems of linear equations by addition.

In 1 and 2, write the equation that results when you add the left and right sides of the two equations.

1. $\begin{cases} 3x + 5y = 13 \\ x - 5y = 3 \end{cases}$ $4x = 16$

2. $\begin{cases} -4x + 2y = 15 \\ 4x - 8y = 10 \end{cases}$ $-6y = 25$

In 3–10, solve the system. Check your solution.

3. $\begin{cases} c + d = 1 \\ c - d = -11 \end{cases}$
(-5, 6)
$-5 + 6 \stackrel{?}{=} 1, 1 = 1;$
$-5 - 6 \stackrel{?}{=} -11, -11 = -11$

4. $\begin{cases} -8x + 2y = 2 \\ 8x + 5y = -23 \end{cases}$
(-1, -3)
$-8(-1) + 2(-3) \stackrel{?}{=} 2, 2 = 2;$
$8(-1) + 5(-3) \stackrel{?}{=} -23,$
$-23 = -23$

5. $\begin{cases} -6x + 4y = 28 \\ 6x + 10y = 28 \end{cases}$
(-2, 4)
$-6(-2) + 4(4) \stackrel{?}{=} 28,$
$28 = 28;$
$6(-2) + 10(4) \stackrel{?}{=} 28,$
$28 = 28$

6. $\begin{cases} 5r + 3s = 24 \\ 5r + 8s = 39 \end{cases}$
(3, 3)
$5(3) + 3(3) \stackrel{?}{=} 24,$
$24 = 24;$
$5(3) + 8(3) \stackrel{?}{=} 39,$
$39 = 39$

7. $\begin{cases} 2u + 3w = 26 \\ 5w - 2u = 22 \end{cases}$
(4, 6)
$2(4) + 3(6) \stackrel{?}{=} 26,$
$26 = 26;$
$5(6) - 2(4) \stackrel{?}{=} 22,$
$22 = 22$

8. $\begin{cases} 2m + n = 9 \\ -2m - 3n = -21 \end{cases}$
$\left(\frac{3}{2}, 6\right)$
$2\left(\frac{3}{2}\right) + 6 \stackrel{?}{=} 9, 9 = 9;$
$-2\left(\frac{3}{2}\right) - 3(6) \stackrel{?}{=} -21,$
$-21 = -21$

9. $\begin{cases} \frac{3}{4}x + y = 8 \\ \frac{1}{4}x - y = 0 \end{cases}$
(8, 2)
$\frac{3}{4}(8) + 2 \stackrel{?}{=} 8, 8 = 8;$
$\frac{1}{4}(8) - 2 \stackrel{?}{=} 0, 0 = 0$

10. $\begin{cases} 5e - 2f = 30 \\ 9e - 2f = 54 \end{cases}$
(6, 0)
$5(6) - 2(0) \stackrel{?}{=} 30,$
$30 = 30;$
$9(6) - 2(0) \stackrel{?}{=} 54,$
$54 = 54$

Alternatively, in Example 2, you could subtract the second equation from the first. This again gives $10y = 80$ and from there you can continue as in the solution. The solution to Example 3 also uses subtraction.

Example 3

A resort hotel offers two weekend specials.

Plan *A:* 3 nights with 6 meals for $264
Plan *B:* 3 nights with 2 meals for $218

At these rates, what is the cost of one night's lodging and what is the average cost per meal? (Assume there is no discount for 6 meals.)

Solution

Let N = price of one night's lodging.
Let M = average price of one meal.
From Plan A: 3N + 6M = 264
From Plan B: 3N + 2M = 218

The coefficients of *N* are the same; so subtract the second equation from the first equation.

4M = 46
M = 11.5

Substitute 11.5 for *M* in either equation. We select the first equation.

3N + 6(11.5) = 264
3N + 69 = 264
3N = 195
N = 65

Thus, (N, M) = (65, 11.5). One night's lodging costs $65.00; and an average meal costs $11.50.

Check

In the Questions, you are asked to check that at these rates the totals for Plans *A* and *B* are correct.

QUESTIONS

Covering the Reading

1. **a.** When is adding equations an appropriate method to solve systems? **when the coefficients of the same variable are opposites**
   b. What is the goal in adding equations to solve systems? **to obtain an equation that contains only one variable**
2. Which property allows you to add corresponding sides of two equations to get a new equation? **the Generalized Addition Property of Equality**

In 3 and 4, **a.** solve the system; **b.** check your solution. **See left for checks.**

3. $\begin{cases} 2x + 8y = 2 \\ -2x - 4y = 6 \end{cases}$ **(-7, 2)**

4. $\begin{cases} a + b = 11 \\ a - b = 4 \end{cases}$ $\left(\frac{15}{2}, \frac{7}{2}\right)$

5. The sum of two numbers is 90; their difference is 75. Find the numbers. **7.5 and 82.5**
6. Find two numbers whose sum is -1 and whose difference is 5. **2 and -3**

3b) Does $2(-7) + 8(2) = 2$? Does $-14 + 16 = 2$? Yes, it checks. Does $-2(-7) - 4(2) = 6$? Does $14 - 8 = 6$? Yes, it checks.

4b) Does $\frac{15}{2} + \frac{7}{2} = 11$? Does $\frac{22}{2} = 11$? Yes, it checks. Does $\frac{15}{2} - \frac{7}{2} = 4$? Does $\frac{8}{2} = 4$? Yes, it checks.

Adapting to Individual Needs

English Language Development
You might want to explain that *flying with the wind* means that the wind is blowing in the same direction as the plane. Conversely, *flying against the wind* means that the plane and the wind are going in opposite directions.

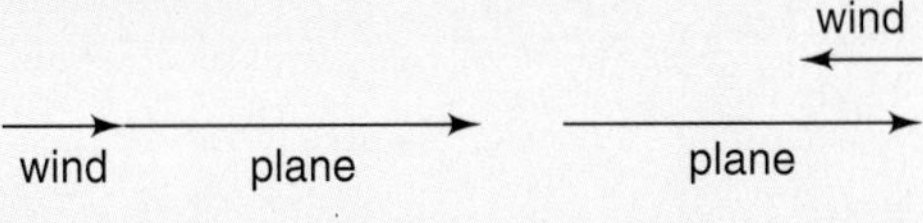

7. In Example 1, suppose it took 50 minutes to fly from Washington Island to Appleton against the wind and 40 minutes for the return flight with the wind. Find the speed of the plane and the speed of the wind in miles per hour under these conditions. **The average speed of the plane was 162 mph and the average wind speed was 18 mph.**

8. When is it useful to multiply an equation by -1 or subtract as a first step in solving a system? **when coefficients of the same variable are equal**

In 9 and 10, solve the system.

9. $\begin{cases} 2x - 3y = 5 \\ 5x - 3y = 11 \end{cases}$ $\left(2, -\frac{1}{3}\right)$

10. $\begin{cases} 2m + n = -5 \\ 2m + 3n = 7 \end{cases}$ $m = -\frac{11}{2}$, $n = 6$

11. Check Example 3. **Does 3(65) + 6(11.5) = 264? Does 195 + 69 = 264? Yes. Does 3(65) + 2(11.5) = 218? Does 195 + 23 = 218? Yes.**

12. A hotel offers the following specials: Plan *A* is two nights and one meal for \$153. Plan *B* is 2 nights and 4 meals for \$195. What price is the hotel charging per night and per meal? **\$69.50 per night and \$14 per meal**

Applying the Mathematics

13. As you know, $\frac{3}{4} = 75\%$ and $\frac{1}{5} = 20\%$. Is it true that $\frac{3}{4} - \frac{1}{5} = 55\%$? Justify your answer. **Yes; by the Generalized Addition Property of Equality, $\frac{3}{4} - \frac{1}{5} = \frac{3}{4} + -\frac{1}{5} = 75\% + -20\% = 75\% - 20\% = 55\%$.**

In 14 and 15, solve.

14. $\begin{cases} 2x - 3y = 17 \\ 3y + x = 1 \end{cases}$ $\left(6, -\frac{5}{3}\right)$

15. $\begin{cases} 4z - 5w = 15 \\ 2w + 4z = -6 \end{cases}$ $z = 0$, $w = -3$

16. Suppose two eggs with bacon cost \$2.70. One egg with bacon costs \$1.80. At these rates, what should bacon alone cost? **\$0.90**

17. Five gallons of regular unleaded gas and eight gallons of premium gas cost \$17.15. Five gallons of regular unleaded and two gallons of premium gas cost \$8.75. Find the cost per gallon of each kind of gasoline. **regular, \$1.19 per gallon; premium, \$1.40 per gallon**

18. The tallest man known to have lived was Robert Wadlow (1918–1940). One of the most famous dwarfs was Charles Sherwood Stratton, alias "General Tom Thumb" (1838–1883). If they stood next to each other, Wadlow would have been 67″ taller. If one had stood on the other's head, they would have stood 12′3″ tall. How tall was each man? **Wadlow, 8′11″; Stratton, 3′4″**

General Tom Thumb. *Charles Sherwood Stratton is shown here with his wife Lavinia Warren. Stratton stopped growing when he was 6 months old. He remained 25 in. (0.6 m) tall until his teens when he began growing again.*

Review

In 19 and 20, solve by using any method. *(Lessons 11-1, 11-2, 11-3)*

19. $\begin{cases} y = 2x - 1 \\ y = 9x + 6 \end{cases}$ $(-1, -3)$

20. $\begin{cases} Q = 4z \\ R = -5z \\ 4R + Q = 40 \end{cases}$ $z = -\frac{5}{2}$, $Q = -10$, $R = \frac{25}{2}$

21. a. Solve $x^2 + 5x - 14 = 0$. **$x = -7$ or $x = 2$**
    b. Find the x-intercepts of the graph of $y = x^2 + 5x - 14$. *(Lesson 9-5)* **-7; 2**

of crude oil and natural gas is greater than 65 million barrels per day.

Question 18 You might want to picture this situation as shown in Activity 1 in *Optional Activities*, on page 682.

Adapting to Individual Needs

Challenge

Ask students to solve the following nonlinear system.

$$\begin{cases} 2x - y^2 = -2 \\ x + y^2 = 14 \end{cases}$$

[Adding gives $3x = 12$; $x = 4$. Substituting in the first equation gives $8 - y^2 = -2$ or $10 = y^2$. So $y = \pm\sqrt{10}$. The two parabolas intersect at points $(4, \sqrt{10})$ and $(4, -\sqrt{10})$.]

▶ **LESSON MASTER 11-4 B** *page 2*

Uses Objective F: Use systems of linear equations to solve real-world problems.

11. In the school bookstore, four pencils and an eraser cost 65¢. Two pencils and an eraser cost 45¢. Find the cost of each item. **pencil, 10¢; eraser, 25¢**

12. Joanie weighs 8 pounds more than Jennie does. Together they weigh 212 pounds. Find the weight of each girl. **Joanie, 110 lb; Jennie, 102 lb**

13. When Brad flew from Indianapolis to St. Louis, he had the wind with him and was traveling at 260 mph. However, on the return trip, he was going against the wind and traveled only 170 mph. What was the plane's speed (without wind)? What was the average speed of the wind? **215 mph; 45 mph**

14. Mark has one less than twice the number of tapes as Felipe has. Together they have 65 tapes. How many tapes does each boy have? **Mark, 43 tapes; Felipe, 22 tapes**

15. On Saturday, Katie earned \$51 for mowing 3 lawns and weeding 3 gardens. On Sunday, she earned \$25 for mowing 1 lawn and weeding 3 gardens. How much does she earn for each lawn she mows and for each garden she weeds? **lawn, \$13; weed, \$4**

16. At the university dormitory, two plans are offered.
    Plan 1: Room and board and 13 meals per week for \$5,110
    Plan 2: Room and board and 19 meals per week for \$5,146
    At these rates what is the cost for room and board alone? What is the cost per meal? **\$5032; \$6**

17. A sandwich with 2 slices of bread and 4 slices of ham has 350 calories. A sandwich with 2 slices of bread and 2 slices of ham has 240 calories. How many calories are in each slice of bread and in each slice of ham? **ham, 55 cal; bread, 65 cal**

Notes on Questions

Question 24 The concept of using a linear combination in this question is also used in the next lesson in **Example 3**, where the problem is about squares and hexagons.

Question 25 This question reviews rewriting an equation in standard form. You should mention that this is sometimes a useful step. Problems involving this idea appear in the next lesson.

Question 29 Students of *Transition Mathematics* may have studied this idea in some detail. It is again mentioned in Lesson 12-6 and in Activity 2 in *Optional Activities* on page 682.

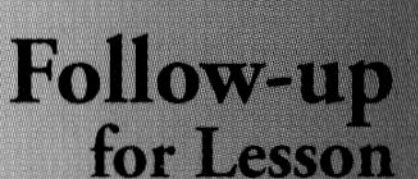

Follow-up for Lesson 11-4

Practice

For more questions on SPUR Objectives, use **Lesson Master 11-4A** (shown on page 683) or **Lesson Master 11-4B** (shown on pages 684–685).

Assessment

Written Communication Have each student write a paragraph explaining which method he or she would use to solve a system of equations in which the coefficients of the same variable are equal. Ask students to include reasons for their choices. [Explanations are clear and students choose either subtracting one equation from another or multiplying one equation by -1 and adding the equations.]

Extension

Project Update Project 2, *Restaurant Prices*, and Project 4, *Adding and Subtracting Equations*, on pages 711–712, relate to the content of this lesson.

In 22 and 23, let $P = (-3, -7)$ and $Q = (5, -1)$. *(Lessons 7-2, 9-8)*

22. Find the distance between P and Q. **10**

23. Find the slope of $\overline{PQ}$. $\frac{3}{4}$

24. Ida is playing with toothpicks. It takes 5 toothpicks to make a pentagon and 6 toothpicks to make a hexagon. She has 100 toothpicks. She wants to make P pentagons and H hexagons.

a. Give three different possible pairs of values of P and H that use all the toothpicks. **Sample: $P = 20, H = 0$; $P = 14, H = 5$; $P = 8, H = 10$**

b. Write a sentence that expresses the relation between the number of pentagons and hexagons she can make with 100 toothpicks. *(Lesson 7-8)* **$5P + 6H = 100$**

25. Rewrite $-15 - 8x = y$ in the form $Ax + By = C$. *(Lesson 7-8)* **$8x + y = -15$ or $-8x - y = 15$**

26. When jewelry is made of 14K gold, $\frac{14}{24}$ of the jewelry is gold and the rest is made of other materials. How many grams of gold are in a 14K gold necklace weighing 30 grams? *(Lesson 6-3)* **17.5 grams**

27. *Skill sequence.* Simplify. *(Lessons 3-2, 4-5, 5-9)*

a. $100 - (80 - 4p) + 2(p + 5)$ **$30 + 6p$**

b. $\frac{100}{p+2} - \frac{80 - 4p}{p+2} + \frac{2(p+5)}{p+2}$ **$\frac{30 + 6p}{p + 2}$**

28. a. How many quarts are there in 5 gallons? **20 qt**

b. How many quarts are there in n gallons? *(Previous course)* **$4n$ qt**

All that glitters. *Most women in Sri Lanka own at least one piece of gold jewelry. The jewelry serves as a form of insurance as well as an investment and symbol of wealth. Because of the constant demand for gold, there are goldsmiths throughout the country.*

Exploration

29. Subtracting equations is part of a process that can be used to find simple fractions for repeating decimals. For instance, to find a fraction for $.\overline{39}$, first let

$$d = .\overline{39}.$$

Then multiply both sides of the equation by 10^2 because $.\overline{39}$ has a two-digit block that repeats.

$$100d = 39.\overline{39} \quad \#1$$
$$d = .\overline{39} \quad \#2$$

Subtract the second equation from the first.

$$99d = 39 \quad \#1 - \#2$$

Solve for d. $d = \frac{39}{99}$

Simplify the fraction. $d = \frac{13}{33}$

A calculator shows that $\frac{13}{33}$ has a decimal equivalent 0.393939 . . .

a. Use the above process to find a simple fraction equal to $.\overline{81}$ **$\frac{9}{11}$**

b. Modify the process to find a simple fraction equal to $.\overline{003}$. **$\frac{1}{333}$**

c. Find a simple fraction equal to $3.89\overline{5}$. **$\frac{1753}{450}$**

Setting Up Lesson 11-5

Questions 24 and 25 should be discussed to help students with the material that will be presented in Lesson 11-5.

LESSON

11-5

Solving Systems by Multiplication

. . . and the band played on. *Musicians in a marching band often arrange themselves into various formations. Many marching bands include drum majors, baton twirlers, and pompon squads. See Example 3.*

Recall that there are two common forms for equations of lines.

Name of form	General form	Sample
standard form	$Ax + By = C$	$3x + 8y = 20$
slope-intercept form	$y = mx + b$	$y = -2x + 1$

The substitution method described in Lessons 11-2 and 11-3 is convenient for solving systems in which one or both equations are in slope-intercept form. The addition method studied in Lesson 11-4 is convenient for solving systems in which both equations are in standard form, and a pair of coefficients are either equal or opposites. However, not all systems fall into one of these two categories.

Consider the system $\begin{cases} 5x + 8y = 21 \\ 10x - 3y = -15 \end{cases}$.

Substitution could be used, but you would first have to solve one of the equations for one of the variables. This is messy! Adding or subtracting the two equations will not result in an equation with just one variable, because neither the x nor the y terms are equal or opposites.

To solve this system, you can use the Multiplication Property of Equality to create an equivalent system of equations. **Equivalent systems** are systems with exactly the same solutions. Notice that if you multiply each side of the first equation by -2, the x-terms of the resulting system have opposite coefficients.

$$\begin{cases} 5x + 8y = 21 \\ 10x - 3y = -15 \end{cases} \rightarrow \begin{cases} -10x - 16y = -42 \\ 10x - 3y = -15 \end{cases}$$

Lesson 11-5

Objectives

C Solve systems by multiplying.
F Use systems of linear equations to solve real-world problems.

Resources

From the ***Teacher's Resource File***
- Lesson Master 11-5A or 11-5B,
- Answer Master 11-5
- Teaching Aid 122: Warm-up

Additional Resources
- Visual for Teaching Aid 122

Teaching Lesson 11-5

Warm-up

Define each form or property.

1. Standard form of an equation **Any equation in the form $Ax + By = C$**
2. Slope-intercept form of an equation **Any equation in the form $y = mx + b$**
3. Generalized Addition Property of Equality **If $a = b$ and $c = d$, then $a + c = b + d$**
4. Multiplication Property of Equality **For all real numbers a, b, and c, if $a = b$, then $ca = cb$**

Lesson 11-5 Overview

Broad Goals Students learn the multiplication method for solving a system.

Perspective This lesson illustrates the following problem-solving strategy: If you cannot solve a problem, transform it into a simpler one that you can solve. In one step,any linear system of the form $ax + by = c$ and $dx + ey = f$ can be transformed into a system in which the addition process of the previous lesson can be utilized.

The multiplication method is a very powerful way to solve systems that have integer coefficients. But even systems with non-integer coefficients can be solved with the multiplication approach. Consider system 1 shown in the next column. If the second equation is multiplied by 3, the fractions are cleared and the result, shown in system 2, is a system which is ready for the addition method.

(1) $\begin{cases} 2x + y = 3 \\ x - \frac{1}{3}y = 4 \end{cases}$ (2) $\begin{cases} 2x + y = 3 \\ 3x - y = 12 \end{cases}$

This technique is valuable when doing problems involving interest rates or money. For instance, $0.08x + 0.05y = 25.32$ may look easier to students when both sides are multiplied by 100: $8x + 5y = 2532$.

Notes on Reading

Cooperative Learning This is a good lesson to ask students to compare and contrast the techniques used to solve each example.

❶ **Example 1** requires that only one equation be multiplied by a number. The two solutions show students that either equation can be multiplied to eliminate the variable *x*. The answer is the same regardless of which way the problem is solved.

❷ **Example 2** illustrates the strategy in which *both* equations are multiplied by some number. You might want to mention that they don't both have to be multiplied by some number. If you multiplied the first equation by $-\frac{2}{3}$, then the resulting equation would be $-2a - \frac{10}{3}b = -\frac{16}{3}$, and the two equations could be added. In the text, we multiply both equations by numbers to avoid fractions. Since computers are not bothered by complicated numbers, only one equation is multiplied when computers are programmed to use the multiplication method.

You might want to go back to the system on page 687 and show students that solving this system by eliminating *y* involves two multiplications, as shown below. Multiply $5x + 8y = 21$ by 3 and $10x - 3y = -15$ by 8.

$$\begin{aligned} 15x + 24y &= 63 \\ 80x - 24y &= -120 \\ \hline 95x &= -57 \\ x &= -\tfrac{57}{95} = -\tfrac{3}{5} \end{aligned}$$

These systems may look different, but they have the same solutions. As Example 1 shows, the system on the right at the bottom of page 687 can be solved using the addition method.

Example 1

Solve the following system.

$$\begin{cases} 5x + 8y = 21 \\ 10x - 3y = -15 \end{cases}$$

Solution 1

❶ Multiply the first equation by -2.

$$\begin{cases} -10x - 16y = -42 \\ 10x - 3y = -15 \end{cases}$$

Add the equations. $-19y = -57$

Solve for *y*. $y = 3$

Substitute $y = 3$ in one of the original equations to find *x*.

$$\begin{aligned} 5x + 8y &= 21 \\ 5x + 8 \cdot 3 &= 21 \\ 5x + 24 &= 21 \\ 5x &= -3 \\ x &= -\tfrac{3}{5} \end{aligned}$$

So the solution is $(x, y) = \left(-\frac{3}{5}, 3\right)$.

Solution 2

Multiply the second equation by $-\frac{1}{2}$. This also makes the coefficients of *x* opposites.

$$\begin{cases} 5x + 8y = 21 \\ 10x - 3y = -15 \end{cases} \rightarrow \begin{cases} 5x + 8y = 21 \\ -5x + 1.5y = 7.5 \end{cases}$$

Add. $9.5y = 28.5$

$y = 3$

Proceed as in Solution 1 to find *x*.

Again $(x, y) = \left(-\frac{3}{5}, 3\right)$.

Check

Does $5 \cdot -\frac{3}{5} + 8 \cdot 3 = 21$? Yes, $-3 + 24 = 21$.

Does $10 \cdot -\frac{3}{5} - 3 \cdot 3 = -15$? Yes, $-6 - 9 = -15$.

Example 1 shows that the solution is the same no matter which equation is multiplied by a number. The goal is to obtain opposite coefficients for one of the variables in the two equations. Then the resulting equations can be added to eliminate that variable. This technique is sometimes called the **multiplication method** for solving a system.

Optional Activities

Technology Connection
Materials: Computers

After reading the lesson, students might use this program (which is like the one in the Lesson 11-2 notes) to find solutions to linear systems. The program works for equations that are written in standard form. However, it does not test situations in which there are no solutions or in which there are an infinite number of solutions.

```
10 PRINT "SOLVE A SYSTEM OF
   EQUATIONS"
20 PRINT "FIRST EQUATION AX +
   BY = C"
30 PRINT "ENTER A, B, AND C."
40 INPUT A, B, C
50 PRINT "SECOND EQUATION DX +
   EY = F"
60 PRINT "ENTER D, E, AND F"
70 INPUT D, E, F
80 DT = A * E - D * B
90 PRINT "X = ";(C * E - B * F)/DT
100 PRINT "Y = ";(A * F - D * C)/DT
110 END
```

You might have students check the examples or some of the problems using this program. You may want to ask the better programmers to modify the program and test for situations in which there are no solutions or in which there are an infinite number of solutions.

Sometimes it is easier to multiply **each** equation by a number before adding.

Example 2

Solve the system $\begin{cases} 3a + 5b = 8 \\ 2a + 3b = 4.6. \end{cases}$

❷ **Solution**

You have your choice of eliminating a or b. In either case, you can multiply by a number so that the resulting system has a pair of opposite coefficients. To make the coefficients of a opposites, multiply the first equation by 2 and the second equation by -3.

$$\begin{cases} 6a + 10b = 16 \\ -6a - 9b = -13.8 \end{cases}$$

Now add. $b = 2.2$

Substitute 2.2 for b in the first equation to find the value of a.

$$3a + 5(2.2) = 8$$
$$3a + 11 = 8$$
$$3a = -3$$
$$a = -1$$

Therefore, $(a, b) = (-1, 2.2)$.

Check

Does $3(-1) + 5(2.2) = 8$? Yes, $-3 + 11 = 8$.
Does $2(-1) + 3(2.2) = 4.6$? Yes, $-2 + 6.6 = 4.6$.

Many situations naturally lead to linear equations in standard form. This results in a linear system that can be solved using multiplication.

❸ **Example 3**

A marching band has 52 musicians (M) and 24 people in the pompon squad (P). They wish to form hexagons and squares like those diagrammed below. Can it be done with no people left over? If so, how many hexagons and how many squares can be made?

It's a great day for a parade. *This pompon squad is marching in Chicago's annual St. Patrick's Day Parade.*

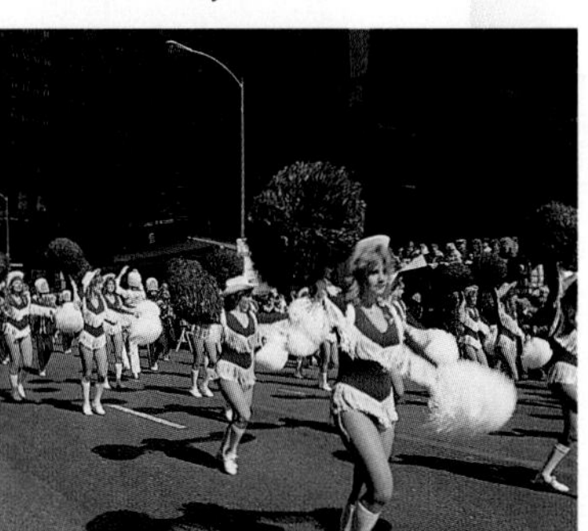

Hexagon
pompon person in center

Square
musician in center

▶

Frequently there is a choice of multipliers which will keep the products small and minimize the arithmetic that is needed for the rest of the solution. Do not expect your students to make the ideal choice every time. Students need experience with the algorithm before they become skilled at selecting the most efficient multipliers. Stress that a good strategy is to look for coefficients of the same variable that are multiples of each other. If such a pair of coefficients exists, then only one multiplication is needed.

❸ In **Example 3,** emphasize that the multiplication method will always find a solution to the *system* if one exists but that the *situation* here requires that the solution be an integer. You can modify the situation slightly by asking if the band and pompon people can form hexagons and squares if one of the band members is sick and can't march. The system can still be solved, but the number of hexagons is $\frac{180}{23}$. You might ask students how many band members should be sent to the sidelines if one band member is ill. This example relates to **Questions 7, 8, 26, and 27.**

Adapting to Individual Needs

Extra Help

Be sure students understand that using any of the properties of equality produces an equivalent equation—one that may look different but has the same solution. Remind them that in a system, the first goal is to eliminate a term and get an equation in one variable that can be solved. They have learned to do this by substitution, by addition (or subtraction), and now by multiplication followed by addition. The solution for the second variable can always be found by substitution. Point out to students that they have to closely compare the x terms with each other and the y terms with each other in order to find the easiest way to eliminate a term.

Additional Examples

1. Solve the following system.
$$\begin{cases} 5x + 2y = 11 \\ x + 6y = 19 \end{cases} \quad (1, 3)$$
2. Solve the following system.
$$\begin{cases} 5a + 3b = -15 \\ a + .5b = -3 \end{cases} \quad (-3, 0)$$
3. The accounting department of a company bought 4 staplers and 5 boxes of paper for $51.50. The planning department bought 10 staplers and one box of paper for $65.50. What is the cost of one stapler and one box of paper?
$$\begin{cases} 4x + 5y = 51.50 \\ 10x + y = 65.50 \end{cases}$$
Stapler: $6, paper: $5.50
The cost of one stapler and one box of paper is $11.50.

Notes on Questions

Questions 5–6 These questions emphasize that a system can be solved by eliminating either variable. It is useful to compare and contrast alternate strategies.

Solution

Consider the formation to include h hexagons and s squares. There are two conditions in the system: one for musicians and one for pompon people.

There are 6 $\frac{\text{musicians}}{\text{hexagon}}$ and 1 $\frac{\text{musician}}{\text{square}}$, so $6h + s = 52$.

There is 1 $\frac{\text{pompon person}}{\text{hexagon}}$ and 4 $\frac{\text{pompon people}}{\text{square}}$, so $h + 4s = 24$.

Multiply the first equation by -4 and add the result to the second equation.

$$\begin{aligned} -24h - 4s &= -208 \\ h + 4s &= 24 \\ \hline -23h &= -184 \\ h &= 8 \end{aligned}$$

Substitute 8 for h in one of the original equations. We use the first equation.

$$\begin{aligned} 6 \cdot 8 + s &= 52 \\ 48 + s &= 52 \\ s &= 4 \end{aligned}$$

Since h and s are both positive integers, the formations can be done with no people left over. The band can make 8 hexagons and 4 squares.

Check

Making 8 hexagons would use 48 musicians and 8 pompon people. Making 4 squares would use 4 musicians and 16 pompon people. This setup uses exactly $48 + 4 = 52$ musicians and $8 + 16 = 24$ pompon people.

When a system involves equations that are not in either standard or slope-intercept form, it is wisest to rewrite the equations in one of these forms before proceeding. For example, to solve the system

$$\begin{cases} 5b = 8 - 3a \\ 2a + 3b = 4.6 \end{cases}$$

you could add $3a$ to both sides of the first equation. The result is the system which was solved in Example 2.

QUESTIONS

Covering the Reading

1. Which property allows you to multiply both sides of an equation by a number without affecting the solutions? **Multiplication Property of Equality**
2. a. What are equivalent systems? **systems that have the same solutions**
   b. Give an example of two equivalent systems in this lesson.
   Sample: $\begin{cases} 5x + 8y = 21 \\ 10x - 3y = -15 \end{cases}$ **and** $\begin{cases} -10x - 16y = -42 \\ 10x - 3y = -15 \end{cases}$

Adapting to Individual Needs

English Language Development

You might want to remind students that in mathematics, *equivalent* equations are equations that have the same solution. Note that equivalent equations may not look the same; $x = 2$, $x + 327 = 329$ and $\frac{1}{2}x - 32 = -31$ are equivalent equations. Similarly, *equivalent systems* are systems that have the same solution.

3. Consider this system. $\begin{cases} 6u - 5v = 2 \\ 12u - 8v = 5 \end{cases}$

   a. If the first equation is multiplied by __?__, then adding the equations will eliminate u. **-2**

   b. Solve the system. **$u = \frac{3}{4}$, $v = \frac{1}{2}$**

4. In Example 2, suppose the first equation is multiplied by 3 and the second equation is multiplied by -5.

   a. What is the resulting system? **$\begin{cases} 9a + 15b = 24 \\ -10a - 15b = -23 \end{cases}$**

   b. Use the system from part **a** to verify the answer to Example 2. **See left.**

4b) $a = -1$. To find b: $3(-1) + 5b = 8$; $5b = 11$; $b = \frac{11}{5} = 2.2$; Yes, it checks.

5. Consider this system. $\begin{cases} 3a - 2b = 20 \\ 9a + 4b = 40 \end{cases}$

   a. If the first equation is multiplied by -3, and the result is added to the second equation, what is the resulting equation? **$10b = -20$**

   b. If the first equation is multiplied by 2, and the result is added to the second equation, what is the resulting equation? **$15a = 80$**

   c. Use one of these methods to solve the system. **$a = \frac{16}{3}$, $b = -2$**

6. Solve the system $\begin{cases} x + 3y = 19 \\ 2x + y = 3. \end{cases}$

   a. by eliminating x first. **$-5y = -35$; $y = 7$; $x + 3 \cdot 7 = 19$; $x = -2$**

   b. by eliminating y first. **$5x = -10$; $x = -2$; $2 \cdot -2 + y = 3$; $y = 7$**

7. Use the hexagon and square formations of Example 3. Will there be an exact fit if the marching band consists of 100 musicians and 42 pompon people? Why or why not? **See left.**

7) No; solving the system using this many people shows h and s will not be integers. The formation cannot be done without having people left over.

8. A marching band has 67 musicians and 47 pompon people. They wish to form pentagons and squares like those diagramed below. Will every person have a spot? If so, how many of each formation will be needed? **Yes; make 7 pentagons and 8 squares.**

Canadian bands. *These marchers are in a parade at the Calgary Stampede, an annual exhibition and rodeo in Alberta, Canada. It lasts several days and includes many unusual rodeo events.*

In 9–12, solve the system.

9. $\begin{cases} 5x + y = 30 \\ 3x - 4y = 41 \end{cases}$ **$x = 7$, $y = -5$**

10. $\begin{cases} 4a + b = 38 \\ 2a + 3b = 24 \end{cases}$ **$a = 9$, $b = 2$**

11. $\begin{cases} 2m - 5n = 0 \\ 6m + n = 0 \end{cases}$ **$m = 0$, $n = 0$**

12. $\begin{cases} 6m - 7n = 6 \\ 7m - 8n = 15 \end{cases}$ **$m = 57$, $n = 48$**

Adapting to Individual Needs

Challenge

Have each student write a system to solve this problem: The sum of three numbers is 72. The sum of the first and second number is twice the third number. The sum of the first and third number is 6 more than twice the second number. Find the numbers.

[Let a = the first number, b = the second number, and c = the third number.

$$\begin{cases} a + b + c = 72 \\ a + b = 2c \\ a + c = 2b + 6 \end{cases}$$

$a = 26$, $b = 22$, $c = 24$]

LESSON MASTER 11-5 A

Questions on SPUR Objectives
See pages 715-717 for objectives.

Skills Objective C

1. Solve the system $\begin{cases} x - 2y = 1 \\ 4x + y = 22 \end{cases}$

   a. by multiplying and adding to eliminate y. **(5, 2)**

   b. by multiplying and adding to eliminate x. **(5, 2)**

2. Consider the system $\begin{cases} 5a - 4b = 14 \\ 2a + 3b = 1 \end{cases}$.

   a. Tell what to multiply each equation by to eliminate a. **1st: -2; 2nd: 5**

   b. Tell what to multiply each equation by to eliminate b. **1st: 3; 2nd: 4**

   c. Solve the system and check your solution. **(2, -1) Checks are not given.**

In 3–8, solve the system, and check your solution.

3. $\begin{cases} 2e + 3f = 18 \\ 5e - f = 11 \end{cases}$ **(3, 4)**

4. $\begin{cases} 2x + 6y = -2 \\ 5x - 3y = 31 \end{cases}$ **(5, -2)**

5. $\begin{cases} 10x + 3y = 23 \\ 5x + 12y = 22 \end{cases}$ **(2, 1)**

6. $\begin{cases} 4w + 5z = 7 \\ 6w - 2z = -18 \end{cases}$ **(-2, 3)**

7. $\begin{cases} 3m + 2n = 10 \\ 2m + 5n = 3 \end{cases}$ **(4, -1)**

8. $\begin{cases} -8r + 3s = 10 \\ 10r - 2s = -2 \end{cases}$ **(1, 6)**

Uses Objective F

9. A manager of an apartment building needs to buy 45 air conditioners, one for each apartment. She has planned to spend \$6,000 on the air conditioners. Two models are available, one at \$110 and the other at \$160. How many of each should she buy to spend \$6,000? **\$110, 24; \$160, 21**

10. For Mother's Day a florist sells two bouquets. A bouquet of 5 roses and 10 carnations costs \$8.75. A larger bouquet has a dozen roses and 15 carnations. It costs \$16.50. Find the cost of one rose and of one carnation. **rose, \$0.75; car., \$0.50**

Notes on Questions

Question 17 Despite the situation that dresses it up, this problem is not derived from a real situation. The wildlife managers would not be expected to know the numbers of heads or feet without knowing how many animals they have. It is a puzzle problem and a famous type of problem at that. You might ask students for a way to solve this problem without solving a system. [The 25 animals have 50 feet, counting 2 per animal. There are 24 feet not counted from the deer, so there are 12 deer, and the rest are birds.]

Question 25 Science Connection The *Voyager,* an experimental American aircraft, took off and landed at Edwards Air Force Base in southern California. When the plane started the journey, it contained 8,934 pounds of fuel—around four times the weight of the airplane itself. Almost all of the fuel was used up during the flight. Today *Voyager* is in the National Air and Space Museum in Washington, D.C.

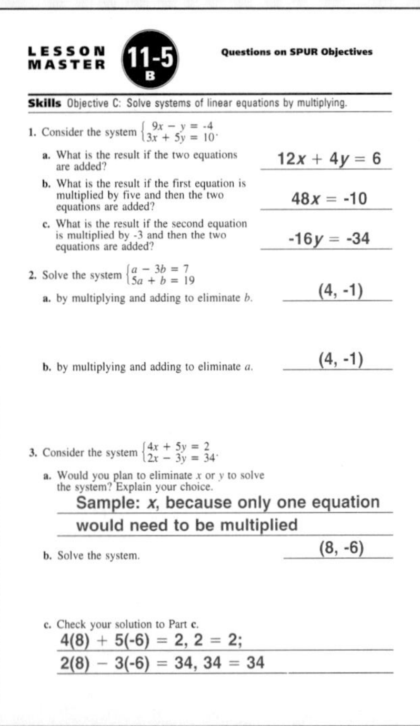

LESSON MASTER 11-5 B — Questions on SPUR Objectives

Skills Objective C: Solve systems of linear equations by multiplying.

1. Consider the system $\begin{cases} 9x - y = -4 \\ 3x + 5y = 10 \end{cases}$.
   a. What is the result if the two equations are added? **$12x + 4y = 6$**
   b. What is the result if the first equation is multiplied by five and then the two equations are added? **$48x = -10$**
   c. What is the result if the second equation is multiplied by -3 and then the two equations are added? **$-16y = -34$**

2. Solve the system $\begin{cases} a - 3b = 7 \\ 5a + b = 19 \end{cases}$
   a. by multiplying and adding to eliminate b. **(4, -1)**
   b. by multiplying and adding to eliminate a. **(4, -1)**

3. Consider the system $\begin{cases} 4x + 5y = 2 \\ 2x - 3y = 34 \end{cases}$.
   a. Would you plan to eliminate x or y to solve the system? Explain your choice. **Sample: x, because only one equation would need to be multiplied**
   b. Solve the system. **(8, -6)**
   c. Check your solution to Part c. **$4(8) + 5(-6) = 2, 2 = 2;$ $2(8) - 3(-6) = 34, 34 = 34$**

13) $\begin{cases} 3x - 4y = 2 \\ 9x - 5y = 7 \end{cases}$, $\left(\frac{6}{7}, \frac{1}{7}\right)$

14) $\begin{cases} 3a - 2b = 5 \\ a - 4b = 6 \end{cases}$ $a = 0.8, b = -1.3$

17) 13 birds and 12 deer; sample reasoning: Let b = number of birds and d = number of deer. Since all animals have one head, $b + d = 25$. Since birds have two feet and deer have four feet, $2b + 4d = 74$. Solving the system gives $b = 13$ and $d = 12$.

Caretakers. *Pictured is a park ranger from Cumberland Island, Georgia. National park rangers often find injured animals and arrange for their care.*

Applying the Mathematics

In 13 and 14, solve the system by first rewriting each equation in standard form. **See left.**

13. $\begin{cases} 3x = 4y + 2 \\ 9x - 5y = 7 \end{cases}$ **14.** $\begin{cases} 3a = 2b + 5 \\ a - 4b = 6 \end{cases}$

15. The sum of two numbers is 45. Three times the first number plus seven times the second is 115. Find the two numbers. **(50, -5)**

16. A test has m multiple-choice questions and t true-false questions. If multiple choice questions are worth 7 points each and the true-false questions 2 points each, the test will be worth a total of 185 points. If multiple-choice and true-false questions are all worth 4 points apiece, the test will be worth a total of 200 points. Find m and t. **$m = 17, t = 33$**

17. A wildlife management station is to care for sick animals. At present they have only birds and deer. The animals being cared for have 25 heads and 74 feet. (No animal is missing a foot!) How many of each type of animal are at the station? Explain your reasoning. **See left.**

18. Solve the system $\begin{cases} 4x - 3y = 2x + 5 \\ 8y = 5x - 13 \end{cases}$. **(1, -1)**

Review

In 19 and 20, solve the system using any method. *(Lessons 11-1, 11-2, 11-3, 11-4)*

19. $\begin{cases} 10x + 5y = 32 \\ 8x + 5y = 10 \end{cases}$ **$\left(11, \frac{-78}{5}\right)$** **20.** $\begin{cases} y = \frac{1}{2}x + 3 \\ y = \frac{1}{3}x - 2 \end{cases}$ **(-30, -12)**

21. The two diagrams below illustrate a system of equations.

x x x y | 29 x x x | y 19

a. Write an equation for the diagram on the left. **$3x + y = 29$**
b. Write an equation for the diagram on the right. **$3x = y + 19$**
c. Solve the system for x and y. **(8, 5)**
d. Check your work. *(Lessons 3-1, 11-2, 11-4)* **Does 3(8) + 5 = 29? Does 24 + 5 = 29? Yes, it checks. Does 3(8) = 5 + 19? Yes, it checks.**

22. Molly has \$400 and saves \$25 a week. Vince has \$1400 and spends \$25 a week.
a. How many weeks from now will they each have the same amount of money? **20 weeks**
b. What will this amount be? *(Lesson 11-2)* **\$900**

Video

Wide World of Mathematics The segment, *Ballooning,* suggests some problems about changes in altitude, wind speed, and direction that could be solved with systems of equations. The segment could be used to extend a lesson on solving systems of equations. Related questions and an investigation are provided in videodisc stills and in the Video Guide. A related CD-ROM activity is also available.

Videodisc Bar Codes

Search Chapter 54

Play

23a) PYTHAGOREAN TRIPLES
ENTER M
5
ENTER N
3
A = 16
B = 30
C = 34

23. **Pythagorean triples** are three whole numbers *A, B,* and *C,* such that $A^2 + B^2 = C^2$. Here is a program that will generate a Pythagorean triple from two positive integers *M* and *N,* where $M > N$. *(Lesson 7-5)*

```
10 PRINT "PYTHAGOREAN TRIPLES"
20 PRINT "ENTER M"
30 INPUT M
40 PRINT M
50 PRINT "ENTER N"
60 INPUT N
70 PRINT N
80 LET A = M * M - N * N
90 LET B = 2 * M * N
100 LET C = M * M + N * N
110 PRINT "A ="; A
120 PRINT "B ="; B
130 PRINT "C ="; C
140 END
```

23b) PYTHAGOREAN TRIPLES
ENTER M
7
ENTER N
1
A = 48
B = 14
C = 50

c) Does $16^2 + 30^2 = 34^2$? Does 256 + 900 = 1156? Yes.
Does $48^2 + 14^2 = 50^2$? Does 2304 + 196 = 2500? Yes.

a. What will the program print when $M = 5$ and $N = 3$?
b. What will the program print when $M = 7$ and $N = 1$?
c. Verify that the answers to parts **a** and **b** satisfy the Pythagorean Theorem.

24. *True or false.* *(Lesson 6-4)*
a. Probabilities are numbers from 0 to 1. **True**
b. A probability of 1 means that an event must occur. **True**
c. A relative frequency of -1 cannot occur. **True**

25. In December, 1986, Jeana Yeager and Dick Rutan flew the *Voyager* airplane nonstop around the earth without refueling. The average rate for the 24,987-mile trip was 116 mph. How many days long was this flight? *(Lesson 6-2)* **about 9 days**

Heavy load. *The fuel for the* Voyager *on its famous flight around the world was stored in the fuselage, wings, and other frame elements. The fuel weighed four times as much as the airplane!*

Exploration

26. If your school has a band with a pompon squad, determine whether the members could fit exactly into the formations of Example 3.
Answers will vary.

27. Create formations that would require all members of a band consisting of 80 musicians and a 30-member pompon squad.
Samples:

```
M   M          M       M
  P                P
M   M        M   P   P   M
                   P
               M       M
```


Follow-up for Lesson 11-5

Practice

For more questions on SPUR Objectives, use **Lesson Master 11-5A** (shown on page 691) or **Lesson Master 11-5B** (shown on pages 692–693).

Assessment

Written Communication Have each student write and solve two systems of equations, one that can be solved using the addition method and one that can be solved using the multiplication method. [Solutions are correct. Examples using addition have coefficients of the same variable that are opposites. Examples using multiplication are multiplied by an appropriate number to create an equivalent system that can be solved using addition.]

Extension

Have students solve this system.

$$\begin{cases} 2x - 6y + 3z = 2 \\ -3x + 2y - 4z = 5 \\ 2x - y + 4z = -1 \end{cases}$$

[(*x, y, z*) = (-5, -1, 2)]

Project Update Project 2, *Restaurant Prices*, on page 711, relates to the content of this lesson.

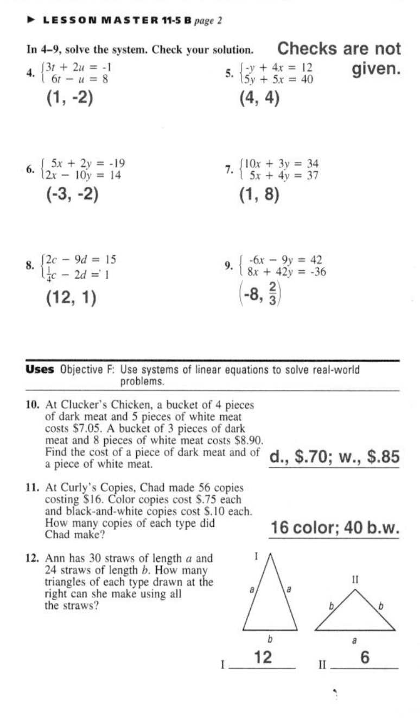

▶ **LESSON MASTER 11-5 B** *page 2*

In 4–9, solve the system. Check your solution. **Checks are not given.**

4. $\begin{cases} 3t + 2u = -1 \\ 6t - u = 8 \end{cases}$ **(1, -2)**

5. $\begin{cases} -y + 4x = 12 \\ 5y + 5x = 40 \end{cases}$ **(4, 4)**

6. $\begin{cases} 5x + 2y = -19 \\ 2x - 10y = 14 \end{cases}$ **(-3, -2)**

7. $\begin{cases} 10x + 3y = 34 \\ 5x + 4y = 37 \end{cases}$ **(1, 8)**

8. $\begin{cases} 2c - 9d = 15 \\ \frac{1}{4}c - 2d = 1 \end{cases}$ **(12, 1)**

9. $\begin{cases} -6x - 9y = 42 \\ 8x + 42y = -36 \end{cases}$ $\left(-8, \frac{2}{3}\right)$

Uses Objective F: Use systems of linear equations to solve real-world problems.

10. At Clucker's Chicken, a bucket of 4 pieces of dark meat and 5 pieces of white meat costs $7.05. A bucket of 3 pieces of dark meat and 8 pieces of white meat costs $8.90. Find the cost of a piece of dark meat and of a piece of white meat. **d., $.70; w., $.85**

11. At Curly's Copies, Chad made 56 copies costing $16. Color copies cost $.75 each and black-and-white copies cost $.10 each. How many copies of each type did Chad make? **16 color; 40 b.w.**

12. Ann has 30 straws of length *a* and 24 straws of length *b*. How many triangles of each type drawn at the right can she make using all the straws?

I: sides *a*, *a*, *b*; II: sides *b*, *b*, *a*

I **12** II **6**

Lesson 11-6

Objectives

E Determine whether a system has no solution, one solution, or infinitely many solutions.

F Use systems of linear equations to solve real-world problems.

H Find solutions to systems of equations by graphing.

Resources

From the ***Teacher's Resource File***

- Lesson Master 11-6A or 11-6B
- Answer Master 11-6
- Assessment Sourcebook: Quiz on Lessons 11-4 through 11-6
- Teaching Aids
  28 Four-Quadrant Graph Paper
  123 Warm-up
- Activity Kit, Activity 25
- Technology Sourcebook
  Computer Master 27

Additional Resources

- Visuals for Teaching Aids 28, 123
- GraphExplorer or other automatic graphers

Teaching Lesson 11-6

Warm-up

Write each equation in slope-intercept form.

1. $-6x + y = -2$ **$y = 6x - 2$**
2. $15x + 5y = -10$ **$y = -3x - 2$**
3. $-5x + y = -3$ **$y = 5x - 3$**
4. $2y - 20x = 0$ **$y = 10x$**
5. $3y - 18x = 3$ **$y = 6x + 1$**
6. If the equations in **Questions 1–5** were graphed, which equations, if any, would result in parallel lines? **Equations 1 and 5**

LESSON

11-6 Systems and Parallel Lines

Drawing a parallel. *The parallel bars used in men's gymnastics competitions were invented by Friedrich Jahn. The bars are 1.7 m (5 ft 5 in.) high and are from 42 cm (16.5 in.) to 48 cm (19 in.) apart.*

Recall that parallel lines "go in the same direction." All vertical lines are parallel to each other. So are all horizontal lines. But not all oblique lines are parallel. For oblique lines to be parallel, they must have the same slope.

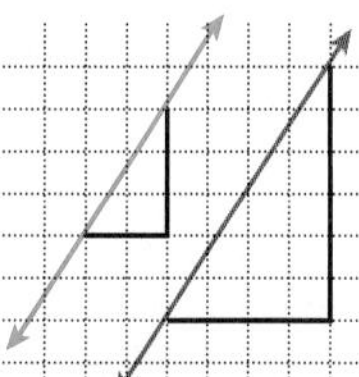

> **Slopes and Parallel Lines Property**
> If two lines have the same slope, then they are parallel.

Nonintersecting Parallel Lines

You have learned that when two lines intersect in exactly one point, the coordinates of the point of intersection can be found by solving a system. But what happens when the lines are parallel? Consider this linear system.

$$\begin{cases} 2x + 3y = -6 \\ 4x + 6y = 24 \end{cases}$$

You can solve the system by multiplying the first equation by -2 and adding the result to the second equation.

$$\begin{array}{r} -4x - 6y = 12 \\ + \ 4x + 6y = 24 \\ \hline 0 = 36. \end{array}$$

When you add you get $0 = 36$.

This is impossible! The false statement 0 = 36 signals that the system has no solution. The graphs have no point of intersection. To check, graph the lines. The graph on page 695 shows that the lines are parallel with no points in common.

Lesson 11-6 Overview

Broad Goals This lesson returns to the graphical representation of equations as lines. The slope-intercept form of the equation is used to show if the lines are parallel. This idea is related to the solving of systems.

Perspective Our use of the term *parallel* includes identical lines as being parallel. This is very common in higher mathematics; for example, equivalent vectors are those with the same length lying on parallel lines, and, under a translation, lines are parallel to their images. Our definition also allows us to simply say: If two lines have the same slope, then they are parallel. The familiar postulates and theorems of geometry still hold with this broader notion of parallel. There is exactly one line that is parallel to a given line through a point (whether or not the point is on the line). Two lines perpendicular to the same line are parallel. (The lines may or may not be the same line.) Lines in a plane either intersect in one point or they are parallel.

As another check, rewrite the equations for the lines in slope-intercept form.

$$\text{line } \ell: \quad 2x + 3y = -6$$
$$3y = -2x - 6$$
$$y = -\tfrac{2}{3}x - 2$$
$$\text{line } m: \quad 4x + 6y = 24$$
$$6y = -4x + 24$$
$$y = -\tfrac{2}{3}x + 4$$

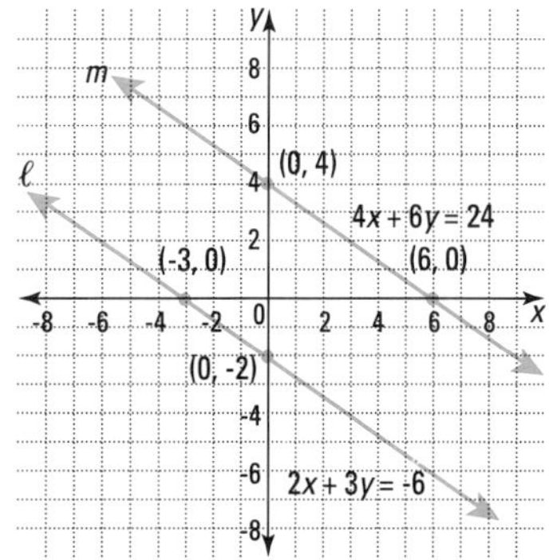

Lines ℓ and m have the same slope, $-\frac{2}{3}$. Thus they are parallel.

Planely parallel. *Uneven parallel bars are used in women's gymnastic competitions. While the bars are parallel to each other, the plane that contains the bars is not parallel to the floor.*

When an equation with no solution (such as $0 = 36$) results from correct applications of the addition and multiplication methods on a system of linear equations, the system has no solutions. That is, there are no pairs of numbers that work in *both* equations. The graph of the system is two parallel nonintersecting lines.

Coincident Lines

Lines that coincide are sometimes called **coincident.** They too are parallel because they go in the same direction and have the same slope.

Example

Solve the system $\begin{cases} y = 1.6x - 6 \\ 16x - 10y = 60 \end{cases}$.

Solution 1

Draw a graph of each equation. As shown below, the graphs are the same line. No matter how much you zoom with an automatic grapher, the two lines are identical. The solution set consists of all ordered pairs on these lines.

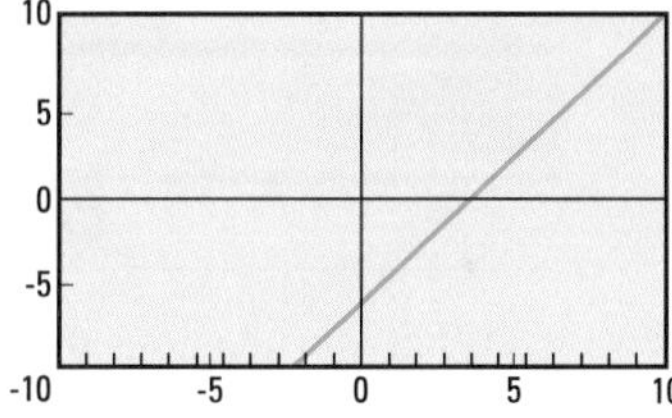

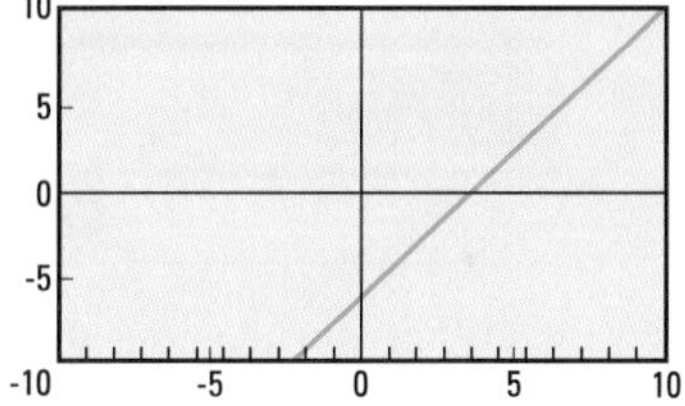

Notes on Reading

The lesson shows that solving a system can lead to a false statement such as $0 = 36$. When this occurs, the system has no solution. It may not be clear to your students that a tautology like $0 = 0$ does have solutions. Ask them to think of that sentence as $0x = 0x$ or $0x = 0$, and ask which values of x work.

Emphasize that if combining two equations in a system results in a statement that is false, then there are no solutions. If combining two equations results in a statement that is always true, then there are infinitely many solutions. However, caution students to be careful in the latter case—if one sentence in a system is subtracted from itself, then $0 = 0$ arises. That would normally signify that the lines coincide; however, the other sentence was not used. *All sentences in a system* must contribute in order for these specific situations to have meaning.

Reading Mathematics Some people like to use the words *consistent* and *inconsistent* to describe systems in which there is at least one solution (consistent) or no solution (inconsistent). We do not use these terms because we feel there is already enough vocabulary associated with these ideas (parallel, intersecting, non-intersecting, coincident). However, if you feel your students would be comfortable with these words, it is fine to use them.

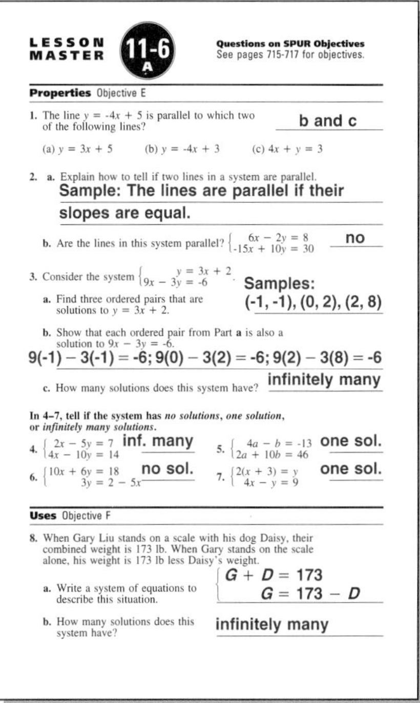

LESSON MASTER 11-6 A

Questions on SPUR Objectives
See pages 715-717 for objectives.

Properties Objective E

1. The line $y = -4x + 5$ is parallel to which two of the following lines? **b and c**

   (a) $y = 3x + 5$ (b) $y = -4x + 3$ (c) $4x + y = 3$

2. a. Explain how to tell if two lines in a system are parallel.
   Sample: The lines are parallel if their slopes are equal.

   b. Are the lines in this system parallel? $\begin{cases} 6x - 2y = 8 \\ -15x + 10y = 30 \end{cases}$ **no**

3. Consider the system $\begin{cases} y = 3x + 2 \\ 9x - 3y = -6 \end{cases}$.

   a. Find three ordered pairs that are solutions to $y = 3x + 2$. **Samples: (-1, -1), (0, 2), (2, 8)**

   b. Show that each ordered pair from Part **a** is also a solution to $9x - 3y = -6$.
   9(-1) – 3(-1) = -6; 9(0) – 3(2) = -6; 9(2) – 3(8) = -6

   c. How many solutions does this system have? **infinitely many**

In 4–7, tell if the system has *no solutions*, *one solution*, or *infinitely many solutions*.

4. $\begin{cases} 2x - 5y = 7 \\ 4x - 10y = 14 \end{cases}$ **inf. many**

5. $\begin{cases} 4a - b = -13 \\ 2a + 10b = 46 \end{cases}$ **one sol.**

6. $\begin{cases} 10x + 6y = 18 \\ 3y = 2 - 5x \end{cases}$ **no sol.**

7. $\begin{cases} 2(x + 3) = y \\ 4x - y = 9 \end{cases}$ **one sol.**

Uses Objective F

8. When Gary Liu stands on a scale with his dog Daisy, their combined weight is 173 lb. When Gary stands on the scale alone, his weight is 173 lb less Daisy's weight.

   a. Write a system of equations to describe this situation. $\begin{cases} G + D = 173 \\ G = 173 - D \end{cases}$

   b. How many solutions does this system have? **infinitely many**

Optional Activities

Activity 1

Some people like to examine the systems by taking ratios of the corresponding coefficients; this is the subject of **Question 25**. Students might use this approach for other systems in the lesson. For this approach: (1) If all the ratios are equal, then the system has infinitely many solutions (the graphs are identical). This occurs in the Example on page 695 where the ratios all equal $\frac{1}{10}$. (2) If the ratios of coefficients of the variables are equal, but the ratios of the constants are different, then the lines are parallel and nonintersecting. This is what happens in the system on page 694 where the ratios of the coefficients of the variables are $\frac{1}{2}$ and where the ratio of the coefficients of the constants is $-\frac{1}{4}$. (3) If the ratios of the coefficients of the variables are unequal, there is exactly one solution.

Additional Examples

Solve each system.

1. $\begin{cases} y = 6x - 7 \\ 24x - 4y = 28 \end{cases}$

   infinitely many solutions

2. $\begin{cases} y = 5x - 7 \\ 20x - 4y = 9 \end{cases}$ **no solution**

Notes on Questions

Question 16 Students should write a system of equations to help them answer this question. They do not have to solve the system; they can just look at the graphs or at the slopes and intercepts. You might want to go over the problem and show how, by dividing the given numbers in the first equation by 3 and in the second equation by 4, one could answer the question without solving the system.

Question 17 Science Connection An Englishman, Frederick Walton, invented linoleum around 1860. He discovered that when linseed oil was exposed to air it became a rubber-like material that could be used to cover floors. The oil came from the flax plant, and Walton called the product linoleum from the Latin words *linum* (flax) and *oleum* (oil).

(Notes on Questions continue on page 698.)

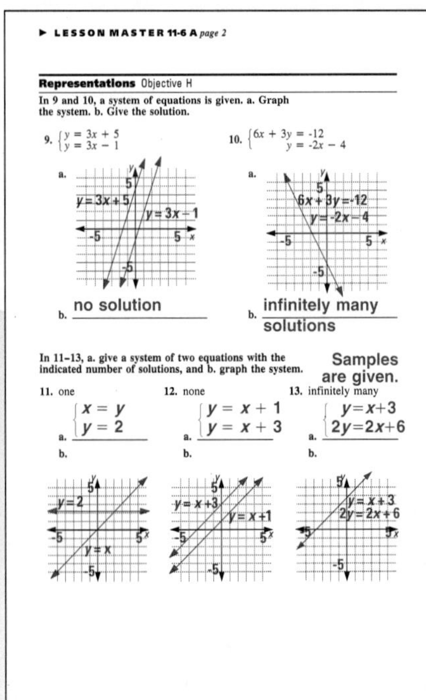

► LESSON MASTER 11-6 A *page 2*

Representations Objective H

In 9 and 10, a system of equations is given. a. Graph the system. b. Give the solution.

9. $\begin{cases} y = 3x + 5 \\ y = 3x - 1 \end{cases}$

   b. no solution

10. $\begin{cases} 6x + 3y = -12 \\ y = -2x - 4 \end{cases}$

    b. infinitely many solutions

In 11–13, a. give a system of two equations with the indicated number of solutions, and b. graph the system. Samples are given.

11. one a. $\begin{cases} x = y \\ y = 2 \end{cases}$

12. none a. $\begin{cases} y = x + 1 \\ y = x + 3 \end{cases}$

13. infinitely many a. $\begin{cases} y = x + 3 \\ 2y = 2x + 6 \end{cases}$

Solution 2

Rewrite the second equation in slope-intercept form.

$$16x - 10y = 60$$
$$-10y = -16x + 60 \quad \text{Add -16x to each side.}$$
$$y = 1.6x - 6 \quad \text{Divide each side by -10.}$$

Notice that this equation is identical to the first. So, any ordered pair that is a solution to one equation is also a solution to the other equation.

Solution 3

Substitute $1.6x - 6$ for y in the second equation.

$$16x - 10(1.6x - 6) = 60$$
$$16x - 16x + 60 = 60$$
$$60 = 60$$

The sentence $60 = 60$ is always true. So, any ordered pair that is a solution to one equation is a solution to the other equation.

Check

Find an ordered pair that satisfies $y = 1.6x - 6$, say (0, -6). Check that it also satisfies the second equation. Does $16 \cdot 0 - 10 \cdot -6 = 60$? Yes, $60 = 60$.

When a sentence that is always true ($60 = 60$) occurs from correct work with a system of linear equations, the system has infinitely many solutions.

The table summarizes the ways that two lines in the plane can be related and gives the corresponding solutions for a system of their equations.

Graph of system	Number of solutions to system	Slopes of lines
Two intersecting lines	One (the point of intersection)	Different
Two parallel and nonintersecting lines	Zero	Equal
One line (parallel and coincident lines)	Infinitely many	Equal

QUESTIONS

Covering the Reading

3a, b) Sample

1. What is true about the slopes of parallel lines? **They are equal.**
2. Which two lines are parallel? **(a) and (c)**
   (a) $y = 3x + 5$ (b) $y = 2x + 5$
   (c) $y = 3x + 6$ (d) $x = 2y + 5$
3. **a.** Graph the line $y = -2x + 5$. **See left for 3a, b.**
   b. Draw a line parallel to it through the origin.
   c. What is an equation of the line you drew in part **b**? $y = -2x$
4. Give an example of a system with two nonintersecting lines.
   Sample: $\begin{cases} 2x + 3y = -6 \\ 4x + 6y = 24 \end{cases}$

Optional Activities

Activity 2

You can use *Activity Kit, Activity 25,* before covering the lesson. In this activity, students are led to the generalization that if two lines have the same slope, then they are parallel.

Activity 3 Technology Connection

In *Technology Sourcebook, Computer Master 27,* students use *GraphExplorer* or similar software to identify systems that have no solution or are always true.

Activity 4 Writing

You might use this activity after discussing the lesson. For each type of solution listed below, tell students to give an example of a system of two equations, and explain how they know that the answer is correct. [Responses will vary.]

1. The system has no solution.
2. The system has infinitely many solutions.
3. The system has one solution.

5) Sample: $\begin{cases} 8x - 7y = 3 \\ 16x - 14y = 6 \end{cases}$

6b)

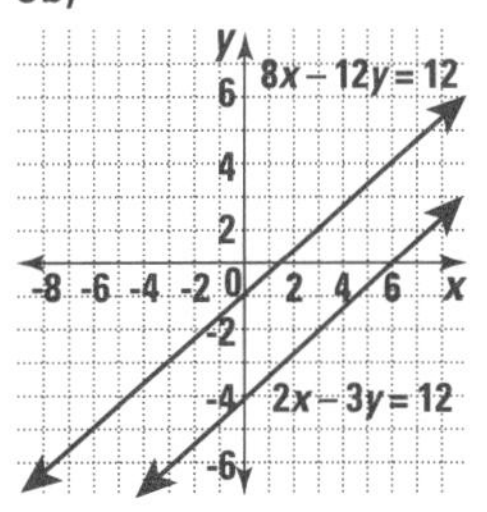

7b)

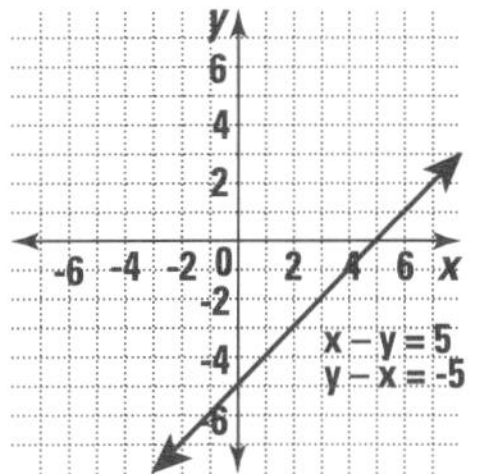

8b) Does $4 \cdot 0 - 2 \cdot (-1) = 2$? Yes. Does $4 \cdot 1 - 2 \cdot 1 = 2$? Yes. Does $4 \cdot 2 - 2 \cdot 3 = 2$? Yes.

c) By substituting $y = 2x - 1$ in the equation $4x - 2y = 2$, $4x - 2(2x - 1) = 2$; $2 = 2$. Therefore, the system has infinitely many solutions.

5. Give an example of a system with two coincident lines. **See left.**

In 6 and 7, a system is given. **a.** Determine whether the system includes nonintersecting or coincident lines. **b.** Check by graphing. **See left for graphs.**

6. $\begin{cases} 2x - 3y = 12 \\ 8x - 12y = 12 \end{cases}$ **a) nonintersecting**

7. $\begin{cases} x - y = 5 \\ y - x = -5 \end{cases}$ **a) coincident**

8. Consider the system $\begin{cases} y = 2x - 1 \\ 4x - 2y = 2 \end{cases}$.

a. Give the coordinates of 3 ordered pairs that are solutions of $y = 2x - 1$. **Sample: (0, -1); (1, 1) (2, 3)**

b. Show that each pair you gave also is a solution of $4x - 2y = 2$. **See left.**

c. Use the substitution method to solve this system. **See left.**

In 9–11, match the description of the graph with the number of solutions to the system.

9. lines intersect in one point **(c)**

10. lines nonintersecting **(a)**

11. lines coincident **(b)**

(a) no solution

(b) infinitely many solutions

(c) one solution

Applying the Mathematics

In 12–15, *multiple choice.* Describe the graph of the system.

(a) two intersecting lines

(b) two parallel non-intersecting lines

(c) one line

12. $\begin{cases} 3u + 2t = 7 \\ 14 - 2t = 6u \end{cases}$ **(a)**

13. $\begin{cases} 6a + 2b = 9 \\ 9a + 3b = 12 \end{cases}$ **(b)**

14. $\begin{cases} 2x - 5y = -3 \\ -4x + 10y = 6 \end{cases}$ **(c)**

15. $\begin{cases} \frac{1}{2}x - \frac{3}{2} = y \\ 2x + y = 3 \end{cases}$ **(a)**

16. Could the given situation have happened? Justify your answer by using a system of equations.

A pizza parlor sold 39 pizzas and 21 gallons of soda for $396. The next day, at the same prices, they sold 52 pizzas and 28 gallons of soda for $518. **No; the system $\begin{cases} 39p + 21s = 396 \\ 52p + 28s = 518 \end{cases}$ represents parallel, nonintersecting lines.**

Review

17) No, with a 10% discount, you should have paid only $9,000. The full price would have been $10,000. You paid only $400 less, getting a 4% discount.

17. Suppose you negotiated a deal to buy 500 square yards of carpet and 500 square yards of linoleum for a total of $9600, which is the price after receiving a 10% discount, or so you've been told. You later learn that another customer bought 60 square yards of the same carpet and 30 square yards of the same linoleum for $1080. Still another customer purchased 50 square yards of the carpet and 10 square yards of the linoleum for $840. If the others paid full price, did you receive the discount you were promised? Explain. *(Lesson 11-5)* **See left.**

Follow-up for Lesson 11-6

Practice

For more questions on SPUR Objectives, use **Lesson Master 11-6A** (shown on pages 695–696) or **Lesson Master 11-6B** (shown on pages 697–698).

Assessment

Quiz A quiz covering Lessons 11-4 to 11-6 is provided in the *Assessment Sourcebook.*

Oral Communication Ask students if they think a system of two linear equations can have exactly two solutions. Have them explain their reasoning. [Students understand that such a system can have zero, one, or infinitely many solutions; and they understand that if a system has two solutions, it must have infinitely many other solutions as well.]

Extension

Have students investigate oblique perpendicular lines to discover how their slopes are related. [For oblique perpendicular lines, the slope of one line is the opposite of the reciprocal of the slope of the other line. That is, if $y = m_1x + b_1$ and $y = m_2x + b_2$ are equations of oblique perpendicular lines, then $m_2 = \frac{-1}{m_1}$.]

Adapting to Individual Needs

English Language Development

Point out that *coincident* can mean "occupying the same space." Draw a line on the board with white chalk. Then, with a different color chalk, draw another line on top of the first line. Explain that the lines coincide or are coincident. Ask students to show points that the lines have in common. Ask if they can find a point on the white line that is not on the colored line.

Extra Help

Students who wonder why there is no solution for systems that represent nonintersecting parallel lines might need to focus on what it means to solve a system of equations. Ask how the solutions of a system are represented on the graph. [By points of intersection]

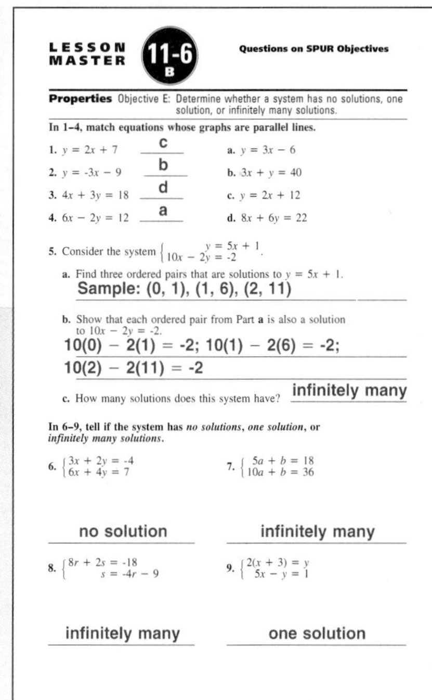
LESSON MASTER 11-6 B

Questions on SPUR Objectives

Properties Objective E: Determine whether a system has no solutions, one solution, or infinitely many solutions.

In 1–4, match equations whose graphs are parallel lines.

1. $y = 2x + 7$ c
2. $y = -3x - 9$ b
3. $4x + 3y = 18$ d
4. $6x - 2y = 12$ a

a. $y = 3x - 6$

b. $3x + y = 40$

c. $y = 2x + 12$

d. $8x + 6y = 22$

5. Consider the system $\begin{cases} y = 5x + 1 \\ 10x - 2y = -2 \end{cases}$.

a. Find three ordered pairs that are solutions to $y = 5x + 1$. Sample: (0, 1), (1, 6), (2, 11)

b. Show that each ordered pair from Part **a** is also a solution to $10x - 2y = -2$. $10(0) - 2(1) = -2$; $10(1) - 2(6) = -2$; $10(2) - 2(11) = -2$

c. How many solutions does this system have? infinitely many

In 6–9, tell if the system has *no solutions*, *one solution*, or *infinitely many solutions*.

6. $\begin{cases} 3x + 2y = -4 \\ 6x + 4y = 7 \end{cases}$ no solution

7. $\begin{cases} 5a + b = 18 \\ 10a + b = 36 \end{cases}$ infinitely many

8. $\begin{cases} 8r + 2s = -18 \\ s = -4r - 9 \end{cases}$ infinitely many

9. $\begin{cases} 2(x + 3) = y \\ 5x - y = 1 \end{cases}$ one solution

Project Update Project 1, *Olympic Records*, on page 711, relates to the content of this lesson.

Notes on Questions

Question 22 This question reviews a skill that will be used in Lesson 11-8.

Additional Answers

18. Let P = pentagon formation, S = square formation. Solving the system $\begin{cases} 5P + 3S = 94 \\ P + 2S = 28 \end{cases}$ gives $S = 6\frac{4}{7}$ and $P = 14\frac{6}{7}$. These are not integer values. This combination of people will not fit exactly in these formations.

19a. $\begin{cases} t + u = 20 \\ t + 3u = 32 \end{cases}$

b. Sample: Substitute $t = 20 - u$ in the equation $t + 3u = 32$ to get $20 - u + 3u = 32$. Solve this equation for u to get $u = 6$. Substitute $u = 6$ in the equation $t = 20 - u$ to get $t = 14$.

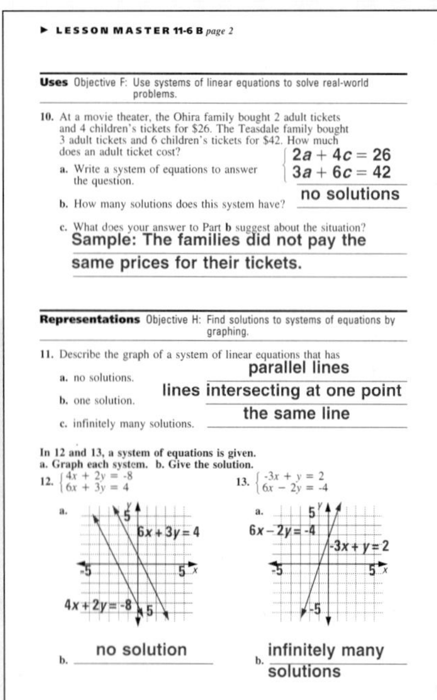

▶ LESSON MASTER 11-6 B *page 2*

Uses Objective F: Use systems of linear equations to solve real-world problems.

10. At a movie theater, the Ohira family bought 2 adult tickets and 4 children's tickets for \$26. The Teasdale family bought 3 adult tickets and 6 children's tickets for \$42. How much does an adult ticket cost?

a. Write a system of equations to answer the question. $\begin{cases} 2a + 4c = 26 \\ 3a + 6c = 42 \end{cases}$

b. How many solutions does this system have? no solutions

c. What does your answer to Part **b** suggest about the situation? Sample: The families did not pay the same prices for their tickets.

Representations Objective H: Find solutions to systems of equations by graphing.

11. Describe the graph of a system of linear equations that has

a. no solutions. parallel lines

b. one solution. lines intersecting at one point

c. infinitely many solutions. the same line

In 12 and 13, a system of equations is given.
a. Graph each system. b. Give the solution.

12. $\begin{cases} 4x + 2y = -8 \\ 6x + 3y = 4 \end{cases}$

13. $\begin{cases} -3x + y = 2 \\ 6x - 2y = -4 \end{cases}$

12b. no solution

13b. infinitely many solutions

18. A band has 94 musicians (M) and 28 flag bearers (F). They plan to form the following formations:

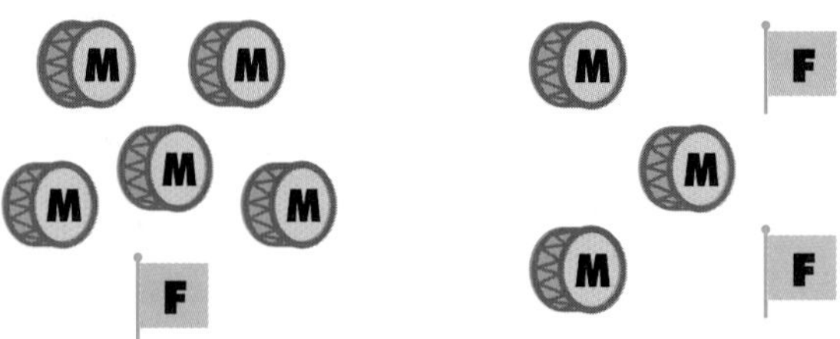

Can all members be fit into these formations? **No** Why or why not? *(Lesson 11-5)* **See margin.**

19. Each diagram at the left represents an equation involving lengths t and u.

a. Write a pair of equations describing these relations. **a, b) See margin.**

b. Use either your equations or the diagrams to find the lengths of t and u. Explain your reasoning.

c. Check your work. *(Lesson 11-4)* **Does 14 + 6 = 20? Yes. Does 14 + 3(6) = 32? Yes.**

20. Solve by any method. $\begin{cases} 4x + 3y = 7 \\ x - 1.5y = 7 \end{cases}$ *(Lessons 11-1, 11-3, 11-5)* $\left(\frac{7}{2}, -\frac{7}{3}\right)$

21. a. Write 1872 as a polynomial with 10 substituted for the variable.

b. Write 1872 in scientific notation. *(Lesson 10-1, Appendix)* **a)** $10^3 + 8 \times 10^2 + 7 \times 10 + 2$; **b)** 1.872×10^3

22. Draw the graph of $y \le 2x - 3$. *(Lesson 7-9)* **See left.**

23. A sweater costs a store owner \$15.50. If the profit is to be 40% of the cost, what is the selling price? *(Lessons 3-7, 6-5)* **\$21.70**

24. a. Give an equation of the horizontal line through (8, -12). $y = -12$

b. Give an equation of the vertical line through (8, -12). *(Lesson 5-1)* $x = 8$

22)

$y \le 2x - 3$

Exploration

25. Here are some examples of parallel lines, both non-intersecting and coincident. The first system in the first group is from the lesson.

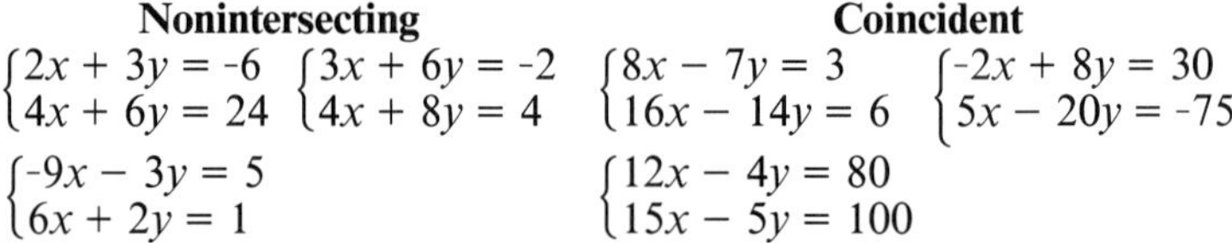

Nonintersecting		Coincident	
$\begin{cases} 2x + 3y = -6 \\ 4x + 6y = 24 \end{cases}$	$\begin{cases} 3x + 6y = -2 \\ 4x + 8y = 4 \end{cases}$	$\begin{cases} 8x - 7y = 3 \\ 16x - 14y = 6 \end{cases}$	$\begin{cases} -2x + 8y = 30 \\ 5x - 20y = -75 \end{cases}$
$\begin{cases} -9x - 3y = 5 \\ 6x + 2y = 1 \end{cases}$		$\begin{cases} 12x - 4y = 80 \\ 15x - 5y = 100 \end{cases}$	

Look for patterns in the two columns of equations. (Hint: Compare the ratios of the coefficients of x, the coefficients of y, and the constant terms.) Use the patterns you find to explain how to recognize parallel lines directly from their equations. **See left.**

25) Sample: Suppose $A_1x + B_1y = C_1$ and $A_2x + B_2y = C_2$. If $\frac{A_1}{A_2} = \frac{B_1}{B_2} = \frac{C_1}{C_2}$, the lines coincide. If $\frac{A_1}{A_2} = \frac{B_1}{B_2}$, but $\frac{A_1}{A_2} \ne \frac{C_1}{C_2}$, then the lines are parallel. If $\frac{A_1}{A_2} \ne \frac{B_1}{B_2}$, the lines intersect.

Adapting to Individual Needs

Challenge

Materials: Automatic graphers

For three distinct lines in the coordinate plane, one of these four cases exists: (A) All the lines are parallel. (B) Two of the lines are parallel, and the third intersects both of them. (C) None of the lines are parallel, and they intersect in three points. (D) The lines are *concurrent*. That is, none of the lines are parallel, and they intersect in one point.

For each system, have students decide which case is illustrated. You might suggest that they use automatic graphers.

1. $\begin{cases} y = 3x + 5 \\ y = 4x + 4 \\ y = 6x + 2 \end{cases}$ [D]

2. $\begin{cases} y = 3x + 3 \\ y = 3x + 2 \\ y = 3x + 5 \end{cases}$ [A]

3. $\begin{cases} y = 2x - 5 \\ y = 2x - 2 \\ y = -x + 4 \end{cases}$ [B]

4. $\begin{cases} y = 4x - 2 \\ y = -2x + 3 \\ y = \frac{1}{2}x - 2 \end{cases}$ [C]

LESSON

11-7

Situations Which Always or Never Happen

Believe it or not. *Lumiere, the talking candelabra from Walt Disney's* Beauty and the Beast, *sings and dances atop a cake. This scene is clearly a situation that never occurs in real life.*

Which job would you take?

Job 1	**Job 2**
Starting wage \$5.60/hour; every 3 months the wage increases \$.10/hour.	Starting wage \$5.50/hour; every 3 months the wage increases \$.10/hour.

Of course, the answer is obvious. Job 1 will always pay better than Job 2. But what happens when this is solved algebraically? Let n = number of 3-month periods worked.

Wage at Job 1	Wage at Job 2
$5.60 + 0.10n$	$5.50 + 0.10n$

When is the pay at Job 1 better than the pay at Job 2? You must solve

$$5.60 + 0.10n > 5.50 + 0.10n.$$

Add $-0.10n$ to each side.

$$5.60 + 0.10n - 0.10n > 5.50 + 0.10n - 0.10n$$
$$5.60 > 5.50$$

As was the case in solving systems representing parallel lines, the variable has disappeared. Since $5.60 > 5.50$ is always true, n can be any real number. Job 1 will always pay a better wage than Job 2, as expected. For any equation or inequality the following generalization is true.

> **If, in solving a sentence, you get a sentence which is *always* true, then the original sentence is always true.**

Lesson 11-7

Objectives

D Recognize sentences with no solution, one solution, or all real numbers as solutions.

F Use systems of linear equations to solve real-world problems.

Resources

From the ***Teacher's Resource File***

- Lesson Master 11-7A or 11-7B
- Answer Master 11-7
- Teaching Aids
  - 28 Four-Quadrant Graph Paper
  - 123 Warm-up

Additional Resources

- Visuals for Teaching Aids 28, 123

Teaching Lesson 11-7

Warm-up

Writing Describe three situations that are always true and three situations that are never true.

Sample answers are given. Always true: (1) Monday precedes Tuesday. (2) Six inches is shorter than 1 foot. (3) The sun sets in the west.

Never true: (1) You are older than your biological grandmother. (2) The Fourth of July is celebrated in May. (3) A dime is worth more than a quarter.

Lesson 11-7 Overview

Broad Goals The meaning of equivalent sentences that are either always true or never true is applied here to equations and inequalities with one variable.

Perspective This lesson begins by returning to equations and inequalities in one variable. This may be surprising, but there are three reasons why this is done. First, the single-variable situation is often harder for students than the two-variable systems in which these special cases appear. The representation of the ways that two lines can be related has a vividness that cannot be duplicated with single-variable equations. In Lesson 11-6, there were simple ways to identify the special cases of no solution or an infinite number of solutions; these ways are not used for one-dimensional equations and inequalities.

Second, we want to return to inequalities. By focusing on familiar inequalities in one variable, students are better prepared for the inequalities in two variables that they will see in Lesson 11-8.

Third, we want students to realize that not only can algebraic methods give solutions when there indeed are solutions but that algebraic methods also can tell them when there are *no solutions* in a situation. This is an important idea students often miss.

Notes on Reading

In the two real-life choices in this lesson, the answers are obvious. Thus, algebra is not very useful in these cases. However, the point of this lesson is to show that the real-life obviousness has a counterpart in the algebraic solution. Algebra can tell you whether there is no solution, only one solution, or infinitely many solutions; it covers all the possibilities.

Note that **Example 3** could be solved as a system in which the equation for Town 1 is $P = 23{,}000 + 1000t$, and the equation for Town 2 is $P = 23{,}000 + 1200t$. Solving the system, we see that the lines intersect when $t = 0$. You may want to make a transparency or use your overhead automatic grapher to illustrate this idea.

The distinction between zero solutions and the solution zero should be emphasized. Because students first saw the ø notation for the null set in Chapter 1, you might use this opportunity to review the difference between {0} and ø.

When does Job 1 pay less than Job 2? To answer this, you could solve:

$$5.60 + 0.10n < 5.50 + 0.10n.$$
$$5.60 + 0.10n - 0.10n < 5.50 + 0.10n - 0.10n$$
$$5.60 < 5.50$$

It is never true that 5.60 is less than 5.50. So Job 1 never pays less than Job 2, something which was obvious from the pay rates. The following generalization is also true.

> **If, in solving a sentence, you get a sentence which is *never* true, then the original sentence is never true.**

Suppose the sentence you are solving has only one variable. If it leads to a sentence (such as 0 = 0) which is always true, the solution set is the set of all real numbers. When a false statement (such as 0 = 2) arises, there is no real solution to the original equation or inequality.

Example 1

Solve $5 + 3x = 3(x - 2)$.

Solution

$$5 + 3x = 3x - 6$$
$$5 + 3x + -3x = -3x + 3x + -6$$
$$5 = -6$$

The statement 5 = -6 is never true, *so the original sentence has no solution.*

Example 2

Solve $8(2y + 5) < 16y + 60$.

Solution

$$16y + 40 < 16y + 60$$
$$-16y + 16y + 40 < -16y + 16y + 60$$
$$40 < 60$$

This is always true, so the original sentence is true for every possible value of y. Thus, *y may be any real number.*

Check

Substitute any real number for y, say -3. Is it true that $8(2 \cdot -3 + 5) < 16 \cdot -3 + 60$? Is $8(-1) < -48 + 60$? Yes. $-8 < 12$. It checks.

Optional Activities

Activity 1 Career Connection
You might use this activity after discussing the introductory situation. Have students **work in groups.** Tell each group to select two jobs and to investigate starting salaries and potential increases for each job. Then have them compare these numbers for both short-term and long-term employment. Students might interview siblings, parents or guardians, or appropriate business personnel to compile their information.

Activity 2
After discussing **Examples 1 and 2**, you might have students **work in small groups** and solve related equations.

For **Example 1**, ask what happens when "=" in $5 + 3x = 3(x - 2)$ is replaced with ">" and then with "<." [$5 + 3x > 3(x - 2)$ is always true; x can be any real number. $5 + 3x < 3(x - 2)$ is never true; it has no solution.]

For **Example 2**, ask what happens when "<" in $8(2y + 5) < 16$ is replaced with ">" and then with "=." [$8(2y + 5) > 16y + 60$ is never true; there is no solution. $8(2y + 5) = 16y + 60$ is never true; it has no solution.]

CAUTION: Here is a problem that looks like the one at the beginning of the lesson but is different.

Example 3

Suppose Town 1 and Town 2 each have populations of 23,000 at present. Town 1's population is growing by 1000 people per year; Town 2 is growing by 1200 per year. When will the population of the towns be the same?

Solution

Let t = number of years from now.

Population of Town 1	Population of Town 2
$23{,}000 + 1000t$	$23{,}000 + 1200t$

The populations will be the same when

$$23{,}000 + 1000t = 23{,}000 + 1200t.$$
$$23{,}000 = 23{,}000 + 200t$$
$$0 = 200t$$
$$0 = t$$

The solution $t = 0$ means The populations of the towns are the same now.

Notice that having zero for a solution, as in Example 3, is different from having no solution at all.

1b) Let y = number of years employed. $6.20 + 1.00y > 6.00 + 1.00y$. Since $6.20 > 6.00$ is always true, y may be any real number.

QUESTIONS

Covering the Reading

1. The Tri-City Amusement Park pays a starting salary of \$6.20 an hour, and each year increases it by \$1.00 an hour. Molly's Supermart starts at \$6.00 an hour, and also increases \$1.00 an hour per year.
   a. When does Tri-City pay more? always
   b. Show how algebra can be used to represent this situation. See left.

2. a. Add $-2x$ to both sides of the sentence $2x + 10 < 2x + 8$. What sentence results? $10 < 8$
   b. What should you write to describe the solutions to this sentence? No solution

3. a. Add $5y$ to both sides of the sentence $-5y + 9 = 3 - 5y + 6$. What sentence results? $9 = 3 + 6$
   b. What should you write to describe the solution(s) to this sentence? b) y may be any real number.
   c. Check your solution(s). Sample: Substitute 3 for y. $-5 \cdot 3 + 9 = 3 - 5 \cdot 3 + 6$; $-15 + 9 = 3 - 15 + 6$; $-6 = -6$.

In 4–9, solve.

4. $2(2y - 5) \leq 4y + 6$ any real number
5. $3x + 5 = 5 + 3x$ any real number
6. $-2m = 3 - 2m$ no solution
7. $2A - 10A > 4(1 - 2A)$ no solution
8. $\frac{1}{2}x + 6 = 3\left(\frac{1}{4}x + 2\right)$ $x = 0$
9. $7 + 4y > 7 - y$ any positive number

Ticket to ride. *These rides are at Knott's Berry Farm in California. Amusement parks were first developed in the United States in the late 1800s.*

Additional Examples

In 1–2, solve the sentence.

1. $9x - 5x - 2(2x + 1) = 15$
   No solution
2. $(20y + 17) - (7 + 20y) < 11$
   y may be any real number.
3. Town A has a population of 50,000 and is growing at a rate of 3000 people a year. Town B has a population of 40,000 and is growing at a rate of 3000 people a year. When will these towns have the same population?
   $50{,}000 + 3000t = 40{,}000 + 3000t$; the towns will never have the same population.

Notes on Questions

Questions 2–3 By asking students to write the resulting sentence in **part a,** we emphasize its importance.

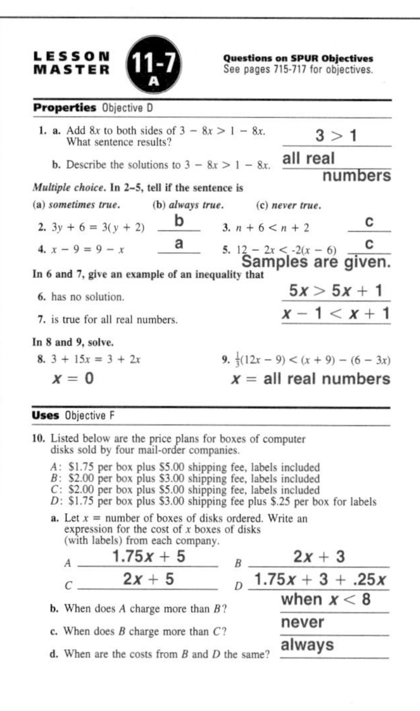

LESSON MASTER 11-7 A

Questions on SPUR Objectives
See pages 715-717 for objectives.

Properties Objective D

1. a. Add $8x$ to both sides of $3 - 8x > 1 - 8x$. What sentence results? $3 > 1$
   b. Describe the solutions to $3 - 8x > 1 - 8x$. all real numbers

Multiple choice. **In 2–5, tell if the sentence is**
(a) *sometimes true.* (b) *always true.* (c) *never true.*

2. $3y + 6 = 3(y + 2)$ b
3. $n + 6 < n + 2$ c
4. $x - 9 = 9 - x$ a
5. $12 - 2x < -2(x - 6)$ c

Samples are given.

In 6 and 7, give an example of an inequality that

6. has no solution. $5x > 5x + 1$
7. is true for all real numbers. $x - 1 < x + 1$

In 8 and 9, solve.

8. $3 + 15x = 3 + 2x$
   $x = 0$
9. $\frac{1}{3}(12x - 9) < (x + 9) - (6 - 3x)$
   x = all real numbers

Uses Objective F

10. Listed below are the price plans for boxes of computer disks sold by four mail-order companies.
    A: \$1.75 per box plus \$5.00 shipping fee, labels included
    B: \$2.00 per box plus \$3.00 shipping fee, labels included
    C: \$2.00 per box plus \$5.00 shipping fee, labels included
    D: \$1.75 per box plus \$3.00 shipping fee plus \$.25 per box for labels
    a. Let x = number of boxes of disks ordered. Write an expression for the cost of x boxes of disks (with labels) from each company.
       A $1.75x + 5$ *B* $2x + 3$
       C $2x + 5$ *D* $1.75x + 3 + .25x$
    b. When does *A* charge more than *B*? when $x < 8$
    c. When does *B* charge more than *C*? never
    d. When are the costs from *B* and *D* the same? always

Adapting to Individual Needs

Extra Help

Remind students that the basic purpose of solving sentences is to answer the question, "For what values of the variable is this sentence true?" Have them solve $6 + m = 6$ and $6 + m = m$. Explain that for $6 + m = 6$ "0" is the solution; in set notation this is shown as {0}. For $6 + m = m$, there is no solution; this can be shown as { } or ø. Emphasize that {0} shows a set with one element, namely 0; { } shows a set with no elements.

English Language Development

You might want to discuss situations that *always* happen, *never* happen, and *sometimes* happen. For example, it is always true that if you are x years old, then a year from now you will be $x + 1$ years old. The sentence $y^2 = y$ is sometimes true and $z > z + 1$ is never true.

Notes on Questions

Question 10 If students are alert, they will not need algebra to recognize that the cities now have the same population.

Questions 11–13 Apartment A is always cheaper. Again, the point of this situation is not to necessarily use algebra to answer the question but to show how algebra mirrors the real situation.

Questions 14–16 Cooperative Learning Have students share their responses with the class.

Question 21 Unlike many systems, this one can be solved easily by using any one of the three methods studied in this chapter. You may want to poll your class to see how many students prefer each of the methods.

Question 22 Notice that you can find each vertex by solving a system of two equations. You can use this question to introduce the next lesson by asking how you could describe the interior of the triangle with a system.

Question 28 Cooperative Learning This is a good question for **small group work**. You might tell students that the most desirable answer is an equation or equations that relate some or all of the variables *a*, *b*, *c*, and *d*. Even with the hint, this question is difficult.

LESSON MASTER 11-7 B Questions on SPUR Objectives

Properties Objective D: Recognize sentences with no solutions, one solution, or all real numbers as solutions.

1. a. Add $6x$ to both sides of $10 - 6(x + 1) = 4 - 6x$. What sentence results? $4 = 4$
   b. Describe the solutions to $10 - 6(x + 1) = 4 - 6x$. all real numbers
2. a. Add $-7x$ to both sides of $7x + 9 < 7 + 7x$. What sentence results? $9 < 7$
   b. Describe the solutions to $7x + 9 < 7 + 7x$. no solutions

In 3–12, *multiple choice*. Tell if the sentence is
(a) *sometimes true.* (b) *always true.* (c) *never true.*

3. $h + 4 > h + 1$ b
4. $2x + 8 = 2(x + 8)$ c
5. $-7a = 6 - 7a$ c
6. $2x + 9 + 7x \geq 9(1 + x)$ b
7. $m + 17 = -m - 1$ a
8. $-3u + 12 < 8u - 10$ a
9. $2(b + 7) = 6b + 14$ a
10. $-4(x + 8) + 20 = 3x - (12 + 7x)$ b
11. $-12a + 19 < 4a - 16a + 15$ c
12. $7n + 16 \leq 3(n + 5) + 1$ a

In 13–15, write an inequality that Samples are given.

13. has no solutions. $x + 5 < x + 4$
14. has $x > 0$ as its solution. $2x + 10 > x + 10$
15. is true for all real numbers. $x + 8 > x + 7$

10. The population of Homsburg is about 200,000 and growing at about 5,000 people a year. Prairieville has a population of about 200,000 and is growing at about 4,000 people a year.
   a. In y years, what will be the population of Homsburg?
   b. In y years what will be the population of Prairieville?
   c. When will their populations be the same? **They are the same now.**
   a) 200,000 + 5,000*y*; b) 200,000 + 4,000*y*

Applying the Mathematics

In 11–13, consider the following information. Apartment A rents for \$375 per month including utilities. Apartment B rents for \$315 per month but the renter must pay \$60 per month for utilities, and a one-time \$25 fee for a credit check.

11. a. What sentence could you solve to find out when apartment A is cheaper? **$375m < 315m + 60m + 25$**
   b. Solve this sentence. **$0 < 25$; *m* can be any number of months.**

12. a. What sentence could you solve to find out when apartment B is cheaper? **$375m > 315m + 60m + 25$**
   b. Solve this sentence. **$0 > 25$; no solution**

13. If you wanted to rent one of these apartments for two years, which one would be cheaper? **A**

In 14–16, make up an example of an equation different from those in this lesson with the given solution.

14. The only solution is $x = 0$. **Sample: $4x + 1 = 2x + 1$**

15. There is no real solution. **Sample: $y - 2 = y - 1$**

16. The equation is true for all real numbers.
Sample: $x^2 + 2x + 1 = (x + 1)^2$

Review

17. Without drawing any graphs, explain how you can tell whether the graphs of $11x - 10y = 102$ and $12x - 10y = 101$ are two intersecting lines, one line, or two nonintersecting parallel lines. *(Lesson 11-6)* **two intersecting lines because the slopes, $\frac{11}{10}$ and $\frac{12}{10}$, are not equal**

In 18 and 19, solve by using any method. *(Lessons 11-1, 11-2, 11-4, 11-6)*

18. $\begin{cases} 4x - 3y = 12 \\ 8x - 6y = 24 \end{cases}$ **infinitely many solutions**

19. $\begin{cases} y = 5x - 7 \\ y = 7 - 5x \end{cases}$ **$\left(\frac{7}{5}, 0\right)$**

20. Two hungry football players go through a cafeteria line. One orders 3 slices of lasagna and 3 salads and pays \$8.64. The other orders 4 slices of lasagna and 2 salads, and pays \$9.54. How much would two slices of lasagna and one salad cost? *(Lesson 11-5)* **\$4.77**

Adapting to Individual Needs

Challenge

Give students these problems.

1. Suppose that the area and perimeter of a rectangle are the same number. Write an equation showing this equality. [Let x = width and y = length: $2x + 2y = xy$.]
2. Solve this equation for y in terms of x and then for x in terms of y. [$y = \frac{2x}{x - 2}$ and $x = \frac{2y}{y - 2}$]
3. Determine what values are impossible for the length and width. [Neither x or y can be less than or equal to 2.]
4. Only two rectangles with whole-number dimensions have the same number for their area and perimeter. Find the dimensions of these rectangles. [4 by 4 and 6 by 3]

21. In parts **a-c**, solve the system 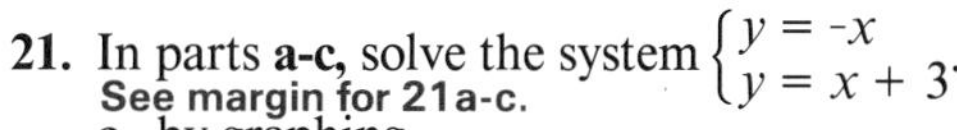 $\begin{cases} y = -x \\ y = x + 3 \end{cases}$. **See margin for 21a-c.**
   a. by graphing.
   b. by substitution.
   c. by the addition method.
   d. Which method in parts **a-c** do you prefer? Why? *(Lessons 11-1, 11-2, 11-3)* **Sample: The addition method requires the fewest steps.**

22. The graph of the equations $x + y = 5$, $x = -1$, and $y = 1$ form a triangle.
   a. Find the vertices of the triangle. **(-1, 6); (4, 1); (-1, 1)**
   b. Find the length of each side of the triangle. **5, 5, $5\sqrt{2}$**
   c. Find the area of the triangle. *(Lessons 1-8, 9-9, 11-1)* **12.5 sq units**

23. Simplify so that no parentheses are needed. *(Lesson 8-8)*
   a. $\left(\frac{2}{3}\right)^2$ **$\frac{4}{9}$** b. $(5d^2g)^2$ **$25d^4g^2$** c. $\frac{1}{4}\left(\frac{c}{a}\right)^2$ **$\frac{c^2}{4a^2}$**

24. Graph the half-plane $x + 2y > 0$. *(Lesson 7-9)* **See left.**

25. An Aztec calendar is being placed on a rectangular mat that is twice as wide and 1.5 times as high as the calendar. What percent of the mat is taken up by the calendar? *(Lessons 6-1, 6-6)* **about 26%**

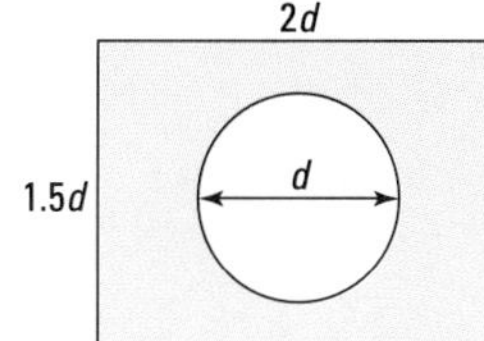

Ancient dating. *The Aztec "Calendar Stone" that appears on this stamp was uncovered in Mexico City in 1790. The Aztec calendar consists of 18 months of 20 days each, plus an additional five days called* nemontemi. *The* nemontemi *were considered very unlucky.*

26. Mrs. Chang wants to lease about 2,500 square meters of floor space for a business. As she walked by a set of vacant stores, she wondered if the 4 stores together would meet this requirement. Their widths are given in the floor plan below. How deep must these stores be to give the required area? *(Lessons 2-1, 2-5)* **about 35.7m**

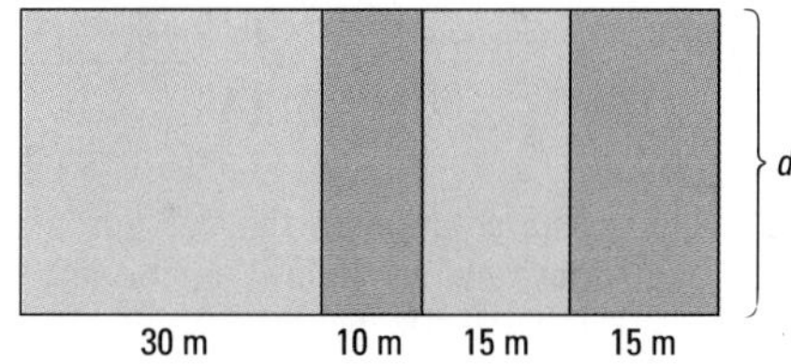

27. Suppose a book has an average of 25 lines per page, w words per line, and 21 pages per chapter. Estimate the number of words per chapter in this book. *(Lesson 2-4)* **525w**

24) $x + 2y > 0$

Exploration

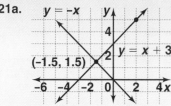

28. When you solve the equation $ax + b = cx + d$ for x, you may find no solution, exactly one solution, or infinitely many solutions. What must be true about a, b, c, and d to guarantee each of the following?
   a. There is exactly one solution. **$a \neq c$**
   b. There are no solutions. **$a = c$ and $b \neq d$**
   c. There are infinitely many solutions. **$a = c$ and $b = d$**

Follow-up for Lesson 11-7

Practice

For more questions on SPUR Objectives, use **Lesson Master 11-7A** (shown on page 701) or **Lesson Master 11-7B** (shown on pages 702–703).

Assessment

Group Assessment Have students **work in pairs**. Tell each student to write the answers for **Questions 14–16** in any order on a piece of paper. Have students exchange papers with their partners and identify each sentence as having no real solution, $x = 0$ as a solution, or all real numbers as a solution. [Students recognize when a sentence is always true, and demonstrate the ability to differentiate between no real solution and $x = 0$ as a solution.]

Extension

Have students choose a rent for Apartment B in **Questions 11–13** so that it is cheaper than Apartment A

1. from the beginning. [Sample: B's rent is \$280.]
2. after one year? [Sample: B's rent is \$313. If the rent for B is \$313 during the first year, the total rent paid for A is \$1 less than B. However, after the rent is paid for month 13, Apartment A will cost more than B.]

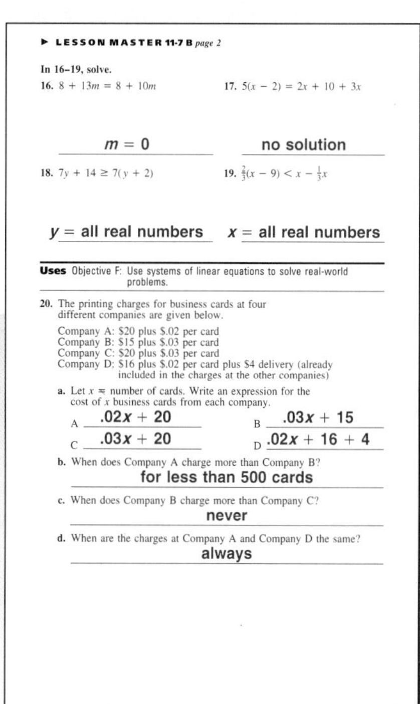

LESSON MASTER 11-7 B *page 2*

In 16–19, solve.

16. $8 + 13m = 8 + 10m$ — $m = 0$

17. $5(x - 2) = 2x + 10 + 3x$ — no solution

18. $7y + 14 \geq 7(y + 2)$ — $y =$ all real numbers

19. $\frac{2}{3}(x - 9) < x - \frac{1}{3}x$ — $x =$ all real numbers

Uses Objective F: Use systems of linear equations to solve real-world problems.

20. The printing charges for business cards at four different companies are given below.

Company A: \$20 plus \$.02 per card
Company B: \$15 plus \$.03 per card
Company C: \$20 plus \$.03 per card
Company D: \$16 plus \$.02 per card plus \$4 delivery (already included in the charges at the other companies)

a. Let x = number of cards. Write an expression for the cost of x business cards from each company.

A $.02x + 20$ B $.03x + 15$
C $.03x + 20$ D $.02x + 16 + 4$

b. When does Company A charge more than Company B? for less than 500 cards

c. When does Company B charge more than Company C? never

d. When are the charges at Company A and Company D the same? always

Additional Answers

21a. $y = -x$, $y = x + 3$, $(-1.5, 1.5)$

b. $-x = x + 3$; $-2x = 3$; $x = -1.5$, $y = 1.5$

c. $2y = 3$; $y = 1.5$, $x = -1.5$

Setting Up Lesson 11-8

Materials In Lesson 11-8, students graph solutions to systems of linear inequalities. You may want to make transparencies to use to show overlapping regions.

Lesson 11-8

Objectives

G Use systems of linear inequalities to solve real-world problems.

I Graphically represent solutions to systems of linear inequalities.

Resources

From the ***Teacher's Resource File***

- Lesson Master 11-8A or 11-8B
- Answer Master 11-8
- Teaching Aids
  28 Four-Quadrant Graph Paper
  123 Warm-up
- Technology Connection
  Computer Master 28

Additional Resources

- Visuals for Teaching Aids 28, 123
- GraphExplorer or other automatic graphers

Teaching Lesson 11-8

Warm-up

Writing Write a paragraph that explains how to graph the inequality $y > \frac{1}{2}x - 3$. **Sample response: First, draw a dotted or dashed line for the equation $y = \frac{1}{2}x - 3$. This shows that the points on the line do not satisfy the inequality. Then test a point on either side of the dotted line. Shade the half-plane that contains the point (0, 0) because it satisfies the condition. The points in the half-plane containing (0, −4) do not satisfy the inequality.**

LESSON

11-8 *Systems of Inequalities*

In Lesson 7–9, you graphed linear inequalities like $x > -5$ and $y \le 2x + 3$ on a plane. These sentences describe half-planes. Now you will graph regions described by two or more inequalities. Solving a system of inequalities involves finding the common solutions of two or more inequalities. As with systems of equations, in this course we will concentrate on linear sentences.

Example 1

Graph all solutions to the system $\begin{cases} x \ge 0 \\ y \ge 0 \end{cases}$.

Solution

First graph the solution to $x \ge 0$. It is shown on the left below. Then on it superimpose the graph of $y \ge 0$, shown by itself at the right below.

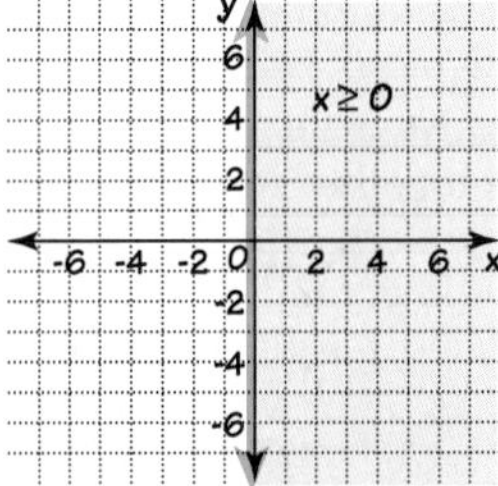

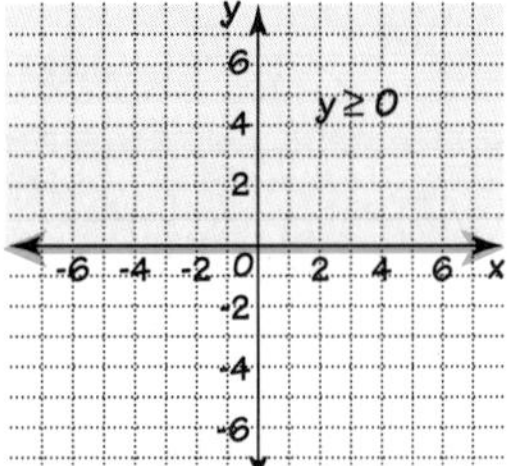

The solution to the system is the set of points common to both of the sets above. Below on the left we show the solutions to the two inequalities superimposed. On the right is the solution to the system. Notice that it is the intersection of the two solution sets above. It consists of the first quadrant and the nonnegative *x*- and *y*-axes.

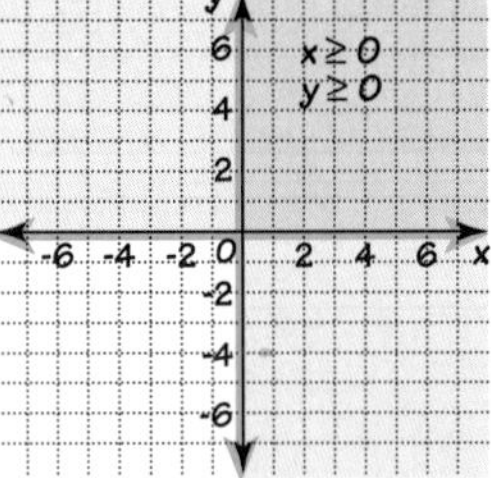

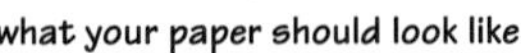

what your paper should look like

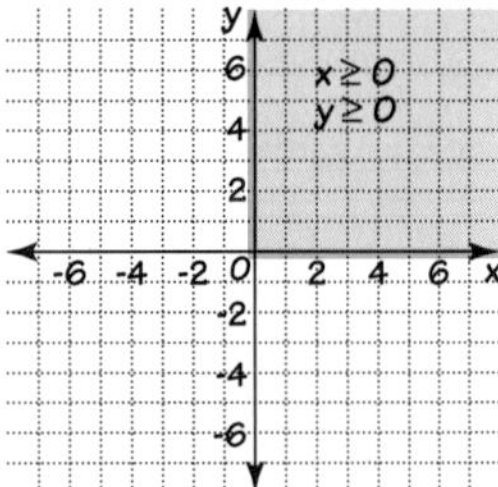

the solution to the system

Lesson 11-8 Overview

Broad Goals In this lesson, systems of linear inequalities, with each inequality in the system of the form $ax + by > c$, are graphed.

Perspective Graphing systems of inequalities gives another visual image for the idea that solutions of systems must satisfy all the conditions. Systems of inequalities have applications like those in **Example 4**, and they are also the basis of linear programming, a topic that is studied in some detail in UCSMP *Advanced Algebra.*

Example 1 shows that the first quadrant can be described as the solution to the system of linear inequalities $x > 0$ and $y > 0$. The system in **Question 2** gives the second quadrant.

Example 2 involves the same idea, but the boundary lines are oblique. **Examples 3 and 4** show how systems of linear inequalities can describe the interiors of polygons. The discrete situation in **Example 4** is shown because applications often involve discrete quantities. You might want to inform students that the points whose coordinates are integers are called *lattice points.*

Recall that in general, the graph of $Ax + By < C$ is a half-plane, and that it lies on one side of the boundary line $Ax + By = C$.

Example 2

Graph all solutions to the system $\begin{cases} y \geq -3x + 2 \\ y < x - 2 \end{cases}$.

Solution

First graph the boundary line $y = -3x + 2$ for the first inequality. The graph of $y \geq -3x + 2$ consists of points on or above the line with equation $y = -3x + 2$.

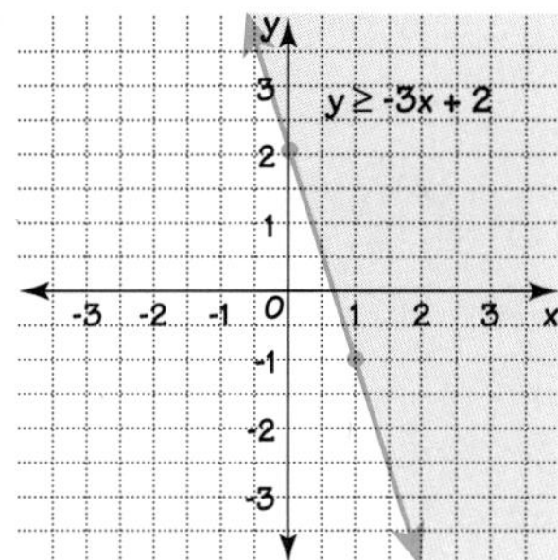

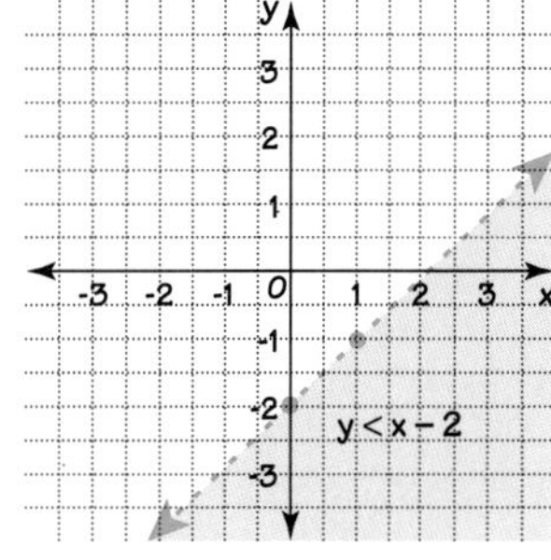

Next, graph the boundary line $y = x - 2$ for the second inequality. The graph of $y < x - 2$ consists of points below the line given by $y = x - 2$. Points on the line are excluded, as shown at the right above.

We have drawn the half-planes on different axes only to make it easier to see them. You should draw them on the same axes. The part of the plane marked with both types of shading is the solution set for the system.

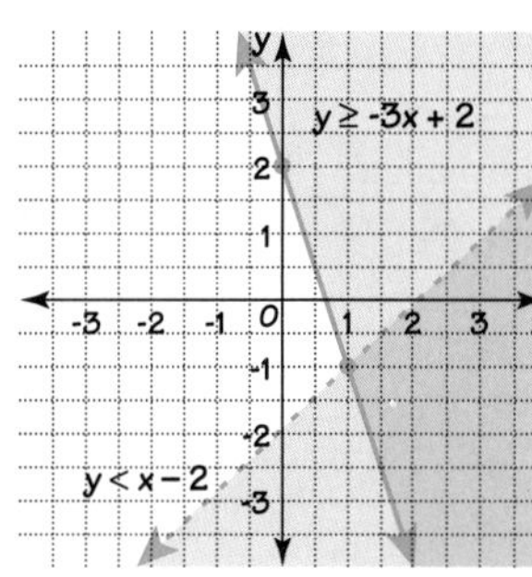

what your paper should look like

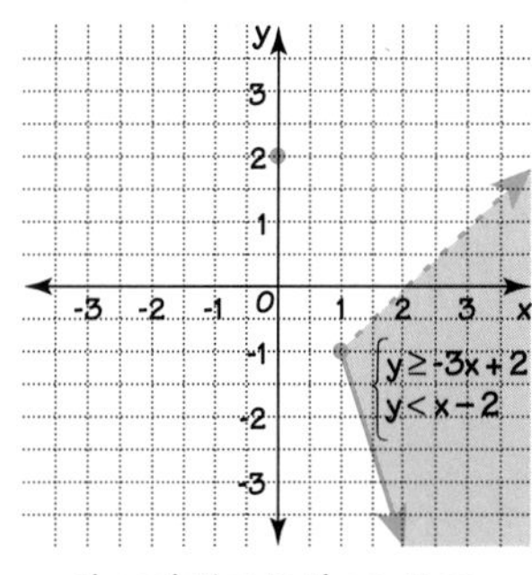

the solution to the system

In Example 2 notice that the point (1, -1) of intersection of the two boundary lines was seen easily from the graph. If an intersection point cannot be found easily from a graph, you can use any of the other techniques learned in this chapter to find its coordinates.

Example 3 shows a system of four linear inequalities.

Notes on Reading

For ideas on how you might extend the examples, see Activity 1 in *Optional Activities* below.

For many students, graphing systems of inequalities is not easy. Transparencies are very useful in showing the overlap on an overhead projector. Shade the graphs of two inequalities in different colors. Show them separately to make sure that students understand which half-plane should be shaded for each inequality. Then place both transparencies on the overhead so students can see that the region where the individual graphs overlap is the solution region.

Point out that the solution to **Example 3** on page 706 is the interior of a trapezoid (because the lines for $x + y = 25$ and $x + y = 50$ are parallel). Ask students for the coordinates of a point in this interior. Then show that those coordinates satisfy the four inequalities of the system. Ask for the coordinates of a point in the exterior of the trapezoid. Show that those coordinates fail to satisfy *at least one* of the sentences in the system. **Question 28** relates to **Example 3.**

You might point out that the figures drawn in computer games are often described by giving inequalities. This is how the computer is programmed to shade regions rather than to just use stick figures.

Optional Activities

Activity 1

As you discuss **Examples 1–3**, you might ask students how the graphs would change if certain conditions changed.

Example 1: How would this graph change if $x \geq 0$ and $y \geq 0$ are changed to

a. $x > 0$ and $y > 0$? [The nonnegative x- and y-axes would not be included.]

b. $x \leq 0$ and $y \leq 0$? [It would consist of the third quadrant and the negative x- and y-axes.]

Example 2: How would this graph change

a. if the second equation is changed to $y \leq x - 2$? [The points on line given by $y = x - 2$ would be included.]

b. if the conditions $x > 0$ and $y > 0$ were included? [The solution would be the portion of the graph in the first quadrant.]

Example 3: How would this graph change

a. if ">" is changed to "≥"? [The boundary lines would be included.]

b. if the equations $x > 0$ and $y > 0$ were not included? [The entire region between the lines given by $x + y = 50$ and $x + y = 25$ would be included.]

The graph in **Example 4** does not show all of the boundary lines. The graph of $3f + p = 20$ is not shown. Point out that when the solution set is a discrete set, none or some of the points may be on the boundary lines. We have shown the boundary lines in the answers to some of the more complicated systems. If your students draw the boundary lines, be sure they understand that only the indicated points on the lines are solutions. See the answers to **Questions 17a and 18c.**

Additional Examples

Students will need graph paper or **Teaching Aid 28** for the Additional Examples.

1. Describe the set of points in the fourth quadrant with a system of inequalities.

$$\begin{cases} x > 0 \\ y < 0 \end{cases}$$

2. Graph all solutions to this system.

$$\begin{cases} 3x + y \leq 4 \\ x - y > 1 \end{cases}$$

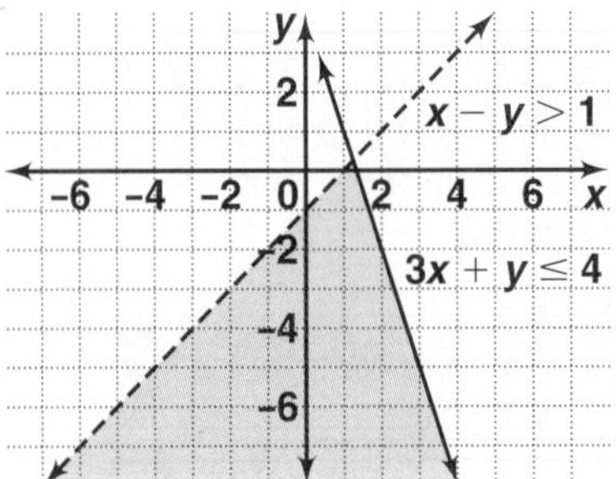

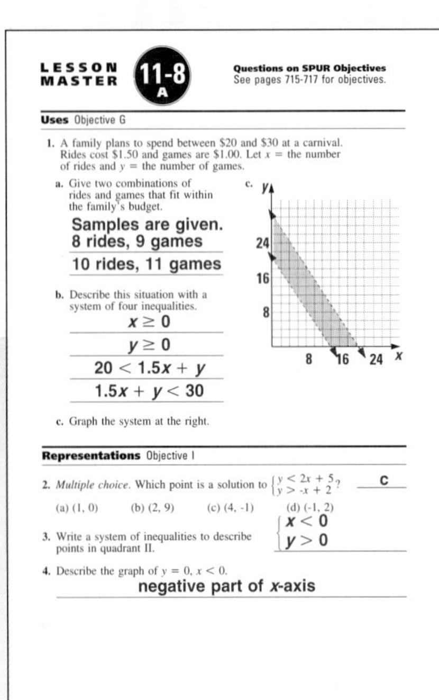

LESSON MASTER 11-8 A — Questions on SPUR Objectives. See pages 715-717 for objectives.

Uses Objective G

1. A family plans to spend between \$20 and \$30 at a carnival. Rides cost \$1.50 and games are \$1.00. Let x = the number of rides and y = the number of games.
   a. Give two combinations of rides and games that fit within the family's budget.
   Samples are given. 8 rides, 9 games; 10 rides, 11 games
   b. Describe this situation with a system of four inequalities.
   $x \geq 0$; $y \geq 0$; $20 < 1.5x + y$; $1.5x + y < 30$
   c. Graph the system at the right.

Representations Objective I

2. *Multiple choice.* Which point is a solution to $\begin{cases} y < 2x + 5 \\ y > -x + 2 \end{cases}$? **c**
   (a) (1, 0) (b) (2, 9) (c) (4, -1) (d) (-1, 2)
3. Write a system of inequalities to describe points in quadrant II. $\begin{cases} x < 0 \\ y > 0 \end{cases}$
4. Describe the graph of $y = 0, x < 0$. **negative part of x-axis**

Example 3

Suppose the sum of two positive numbers x and y is less than 50 and greater than 25. Show all possible values for x and y graphically.

Solution

Because x and y are positive, $x > 0$ and $y > 0$. The desired numbers are the solution to this system of inequalities.

$$\begin{cases} x > 0 \\ y > 0 \\ x + y < 50 \\ x + y > 25 \end{cases}$$

The graph of the solutions to the system is the intersection of the four sets graphed below.

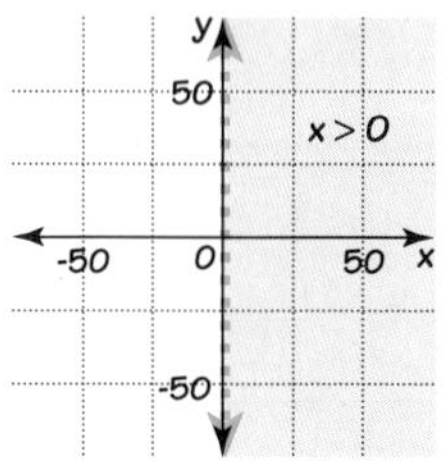

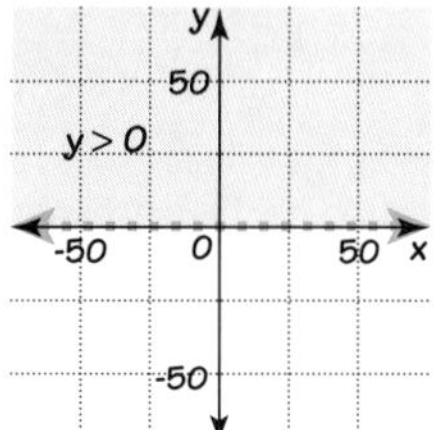

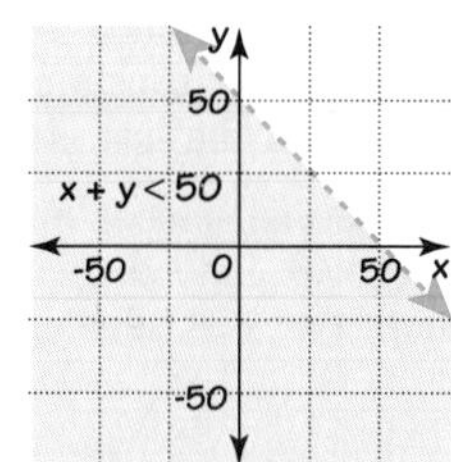

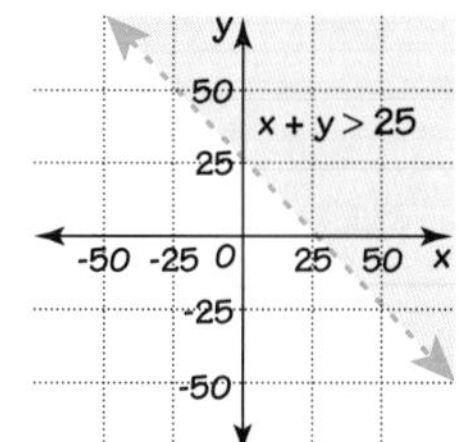

Here is the intersection.

$$\begin{cases} x > 0 \\ y > 0 \\ x + y < 50 \\ x + y > 25 \end{cases}$$

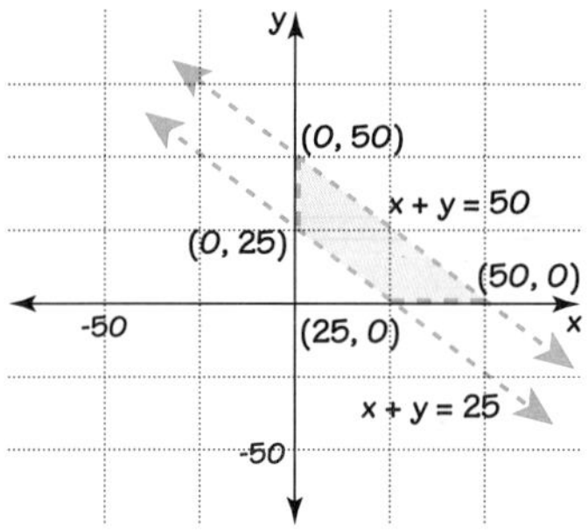

Because there are infinitely many solutions to the system, they cannot be listed. But the graph is easy to describe. It is the interior of the quadrilateral with vertices (0, 50), (50, 0), (25, 0), and (0, 25).

Optional Activities

Activity 2 Writing
After students have completed the lesson, you might have them make up situations like the ones in **Questions 16–18**. Ask volunteers to read their problems to the class. Have the class select one of the problems and solve it.

Activity 3 Technology Connection
You may wish to assign *Technology Sourcebook, Computer Master 28.* Students use *GraphExplorer* or similar software to graph systems of inequalities.

Get a kick out of this. *George Blanda, kicker and quarterback, holds the record for scoring the most points, 2002, in the NFL. Blanda played 26 seasons, then retired at age 48 in 1975.*

Systems of inequalities arise from many different kinds of situations. Here is one.

Example 4

In football, a kicker scores 3 points for a field goal and 1 point for a point after touchdown. Suppose a kicker has scored no more than 20 points in a season. How many field goals and points after touchdowns might the kicker have made?

Solution

You could answer this question using trial-and-error method, but there are a lot of possibilities. It is much easier to show the answers on a graph.

Let f = the number of field goals kicked, and
p = the number of points after touchdowns.

The numbers f and p must be nonnegative integers, so

$$f \geq 0 \text{ and } p \geq 0.$$

Since the total number of points is less than or equal to 20,

$$3f + p \leq 20.$$

Since this is a discrete situation, only points with integer coordinates are possible solutions. These points must be on or above the f-axis, on or to the right of the p-axis, and on or below the line with equation $3f + p = 20$.

The graph below is the solution to this system of inequalities

$$\begin{cases} f \geq 0 \\ p \geq 0 \\ 3f + p \leq 20. \end{cases}$$

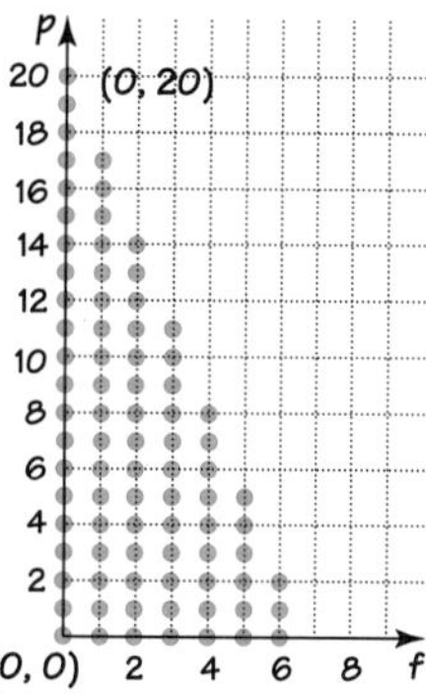

There are 84 points on the graph representing the 84 ways the kicker might have scored no more than 20 points.

Check

Select a point from the solution set, such as (5, 2), and see if it checks. The point (5, 2) represents the possibility of the kicker making 5 field goals and 2 points after touchdowns. This yields $5 \cdot 3 + 2 = 17$, which is fewer than 20. Also $5 \geq 0$ and $2 \geq 0$. Thus (5, 2) is on the graph of the solution.

If the above situation allowed fractions, the entire triangle would be shaded.

3. Suppose two positive numbers x and y have a sum that is less than 20 and a difference that is greater than 10. Show all possible values for x and y graphically.

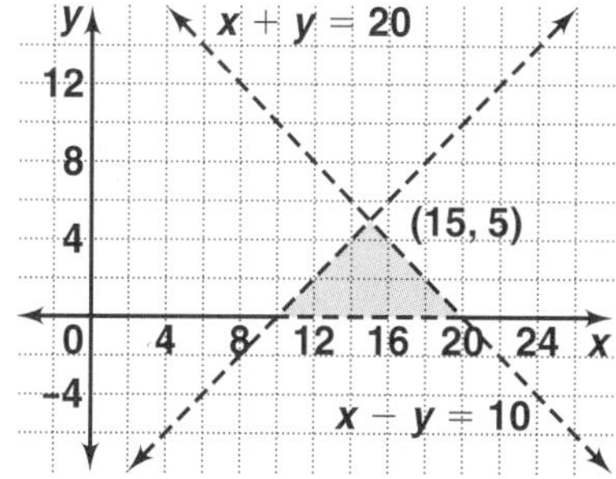

4. Miss Moffat is planning refreshments for a party. She decides to have at least 10 but no more than 15 pounds of salad—some potato salad and some fruit salad. She wants to have at least four times as much fruit salad as potato salad.

a. Write a system of inequalities to describe this situation. Let p and f be the number of pounds of potato and fruit salad, respectively.
The inequalities are: $p > 0$, $f > 0$, $p + f \geq 10$, $p + f \leq 15$, and $f \geq 4p$.

(Additional Examples continue on page 708.)

Adapting to Individual Needs

Extra Help

Be sure students recall the significance of the solid and dashed boundary lines as indicated in **Example 2**. Explain that $y \geq -3x + 2$ is a short way of writing $y > -3x + 2$ or $y = -3x + 2$; therefore, the line is included in the graph, and it is shown with a solid line. For the graph of $y < x - 2$ a dashed line is drawn because $y = x - 2$ is not included.

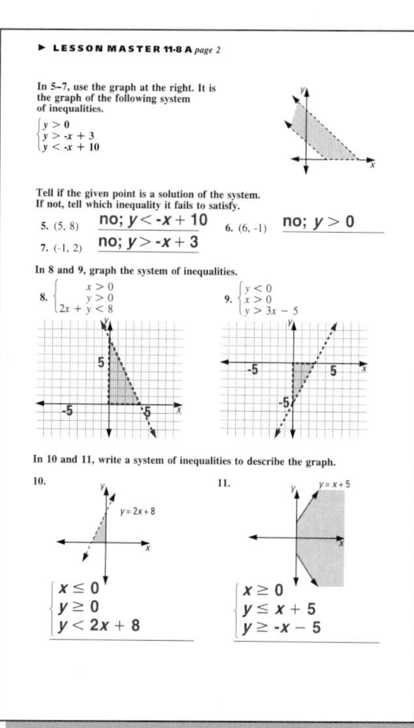

LESSON MASTER 11-8 A *page 2*

In 5–7, use the graph at the right. It is the graph of the following system of inequalities.

$$\begin{cases} y > 0 \\ y > -x + 3 \\ y < -x + 10 \end{cases}$$

Tell if the given point is a solution of the system. If not, tell which inequality it fails to satisfy.

5. (5, 8) **no; $y < -x + 10$**
6. (6, -1) **no; $y > 0$**
7. (-1, 2) **no; $y > -x + 3$**

In 8 and 9, graph the system of inequalities.

8. $\begin{cases} x > 0 \\ y > 0 \\ 2x + y < 8 \end{cases}$

9. $\begin{cases} y < 0 \\ x > 0 \\ y > 3x - 5 \end{cases}$

In 10 and 11, write a system of inequalities to describe the graph.

10. $y = 2x + 8$
$\begin{cases} x \leq 0 \\ y \geq 0 \\ y < 2x + 8 \end{cases}$

11. $y = x + 5$
$\begin{cases} x \geq 0 \\ y \leq x + 5 \\ y \geq -x - 5 \end{cases}$

b. Graph the number of pounds of each kind of salad that she can have.

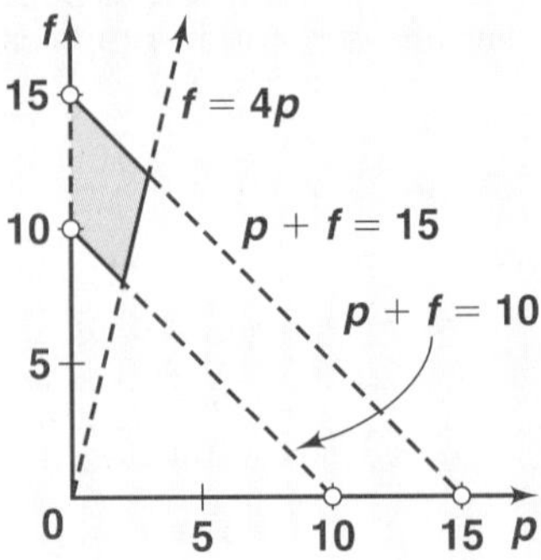

The vertices of the shaded region are (0, 10), (0, 15), (2, 8), and (3, 12). Any point in the interior of this region or on its boundary, except for the f-axis), is a solution. For instance, she can have 2 pounds of potato salad and 10 pounds of fruit salad.

5. Graph all solutions to this system.

$\begin{cases} x - y \le -3 \\ 2x - 2y > 4 \end{cases}$

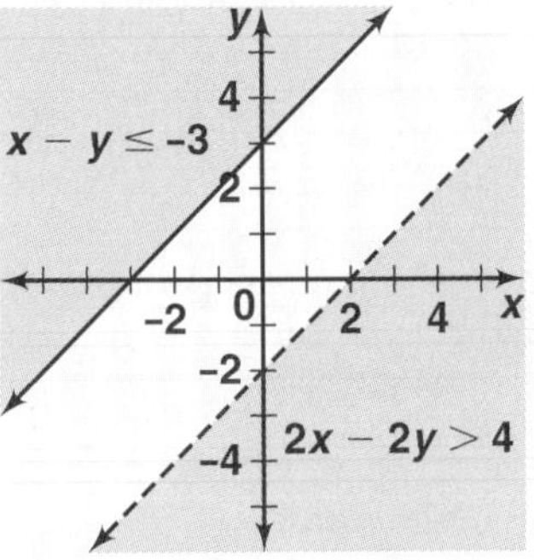

There are no solutions.

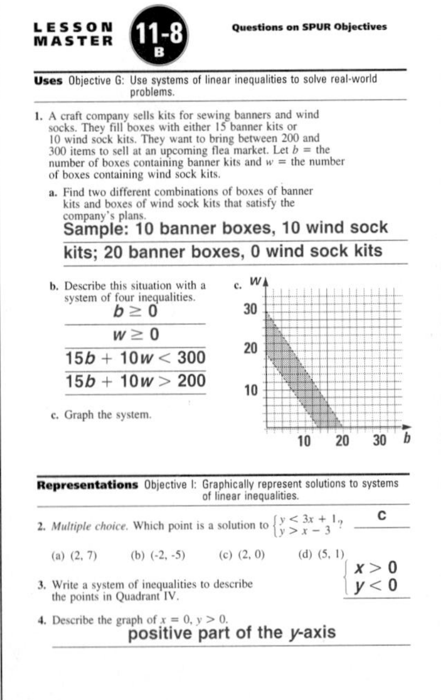

LESSON MASTER **11-8** B

Questions on SPUR Objectives

Uses Objective G: Use systems of linear inequalities to solve real-world problems.

1. A craft company sells kits for sewing banners and wind socks. They fill boxes with either 15 banner kits or 10 wind sock kits. They want to bring between 200 and 300 items to sell at an upcoming flea market. Let b = the number of boxes containing banner kits and w = the number of boxes containing wind sock kits.

a. Find two different combinations of boxes of banner kits and boxes of wind sock kits that satisfy the company's plans. Sample: 10 banner boxes, 10 wind sock kits; 20 banner boxes, 0 wind sock kits

b. Describe this situation with a system of four inequalities.
$b \ge 0$
$w \ge 0$
$15b + 10w < 300$
$15b + 10w > 200$

c. Graph the system.

Representations Objective I: Graphically represent solutions to systems of linear inequalities.

2. *Multiple choice.* Which point is a solution to $\begin{cases} y < 3x + 1 \\ y > x - 3 \end{cases}$? c

(a) (2, 7) (b) (-2, -5) (c) (2, 0) (d) (5, 1)

3. Write a system of inequalities to describe the points in Quadrant IV. $\begin{cases} x > 0 \\ y < 0 \end{cases}$

4. Describe the graph of $x = 0$, $y > 0$. positive part of the y-axis

2)

x < 0
y > 0

3) It is the intersection of the two half-planes described by $y \ge -3x + 2$ and $y < x - 2$.

5) The "equal to" sign means that points on the boundary satisfy the conditions of the sentence.

QUESTIONS

Covering the Reading

1. The graph of $\begin{cases} x > 0 \\ y > 0 \end{cases}$ consists of all points in which quadrant? first

2. Graph the system $\begin{cases} x < 0 \\ y > 0 \end{cases}$. See left.

In 3–6, refer to Example 2.

3. How is the graph of all solutions to the system

$$\begin{cases} y \ge -3x + 2 \\ y < x - 2 \end{cases}$$

related to the graphs of $y \ge -3x + 2$ and $y < x - 2$? See left.

4. The graph of $y < x - 2$ is a __?__ half-plane

5. Why does the graph of $y \ge -3x + 2$ include its boundary line? See left.

6. Is (2, 1) a solution to this system? How can you tell? No; $1 < 2 - 2$ is not true, so (2, 1) does not satisfy the second condition.

In 7 and 8, consider the system in the solution to Example 3.

$$\begin{cases} x > 0 \\ y > 0 \\ x + y < 50 \\ x + y > 25 \end{cases}$$

7. The graph of all solutions to this system is the interior of a quadrilateral with what vertices? (0, 50); (50, 0); (25, 0); (0, 25)

8. Name two points that are solutions to the system. Sample: (1, 26); (26, 1)

In 9 and 10, refer to Example 4.

9. Give at least three possible combinations of field goals and points after touchdowns that would total exactly 20 points. Samples: (0, 20); (1, 17); (5, 5); (6, 2)

10. Suppose the player made at least five field goals. How many possibilities are there then for $3f + p \le 20$? 9

In 11 and 12, graph the solution set. See below.

11. $\begin{cases} x > 0 \\ y > 0 \\ x + y < 6 \end{cases}$

12. $\begin{cases} x \ge -1 \\ y \le 2 \\ y \ge x - 1 \end{cases}$

11)

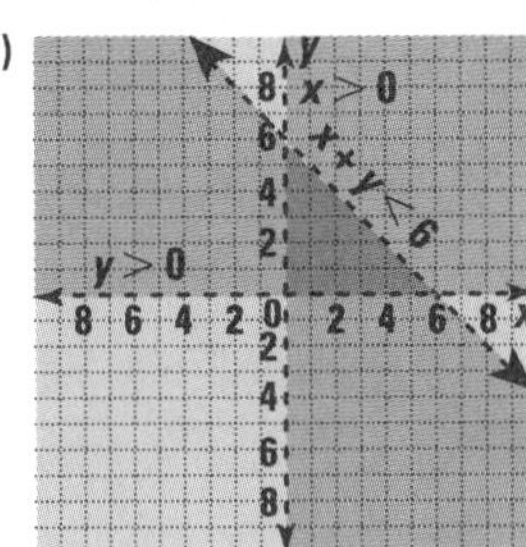

12)

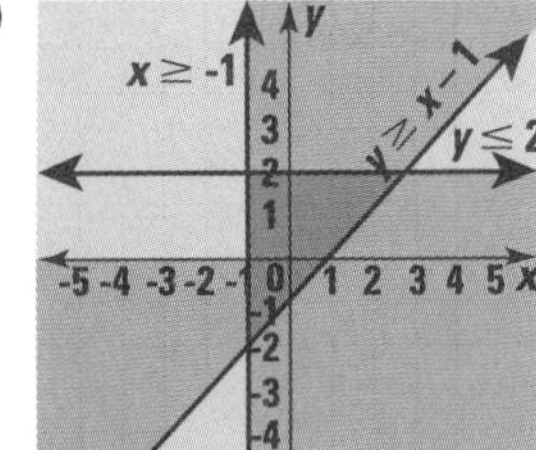

Adapting to Individual Needs

English Language Development

For the portions of this lesson in which students must write systems, non-English-speaking students might benefit by being paired with English-speaking students. Once the systems have been written, many non-English-speaking students will be able to work on their own.

15) Sample: 
$\begin{cases} x \geq -1 \\ x \leq 1 \\ y \geq -2 \\ y \leq 2 \end{cases}$

16a) $\begin{cases} 10L + 8P \leq 60 \\ L \geq 0 \\ P \geq 0 \end{cases}$

b)

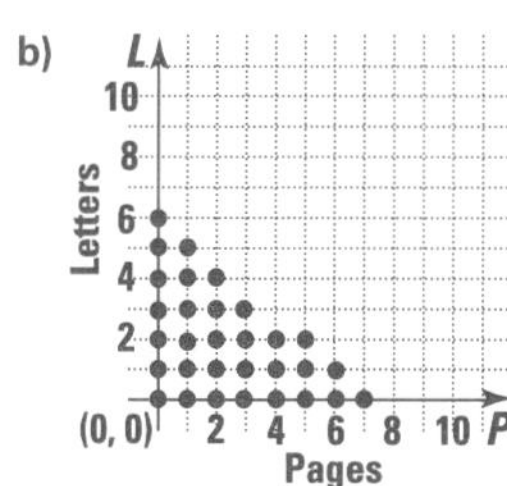

17a)

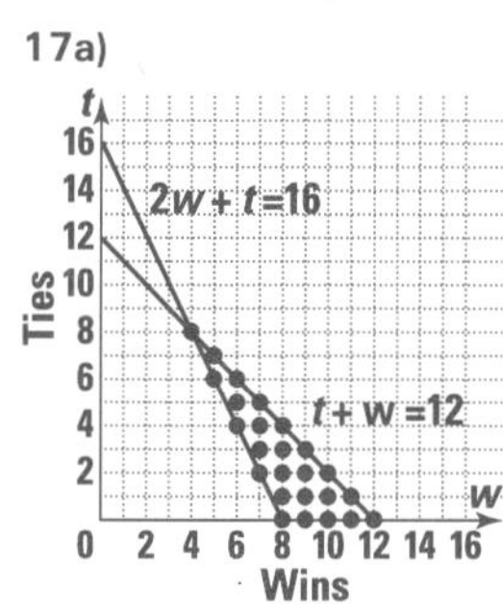

Action! Rita Moreno is shown here in the film version of West Side Story.

Applying the Mathematics

In 13 and 14, describe the shaded region with a system of inequalities.

13.

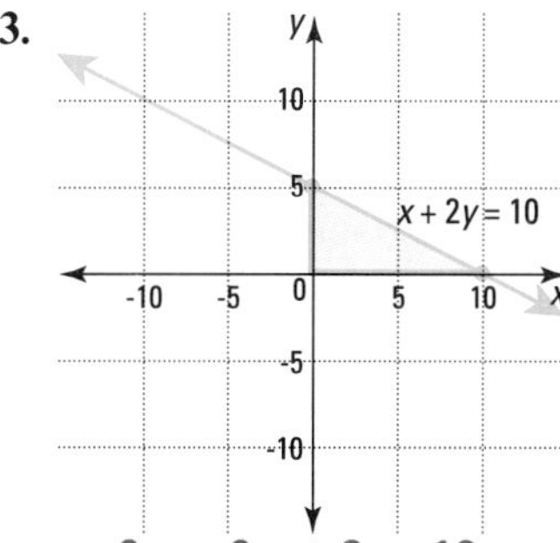

$x \geq 0, y \geq 0, x + 2y \leq 10$

14.

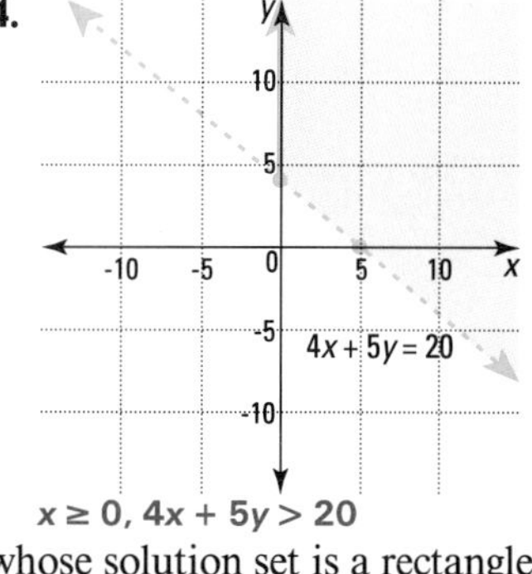

$x \geq 0, 4x + 5y > 20$

15. Make up a system of inequalities whose solution set is a rectangle and all the points inside it. See left.

16. It takes a good typist about 10 minutes to type a letter of moderate length and about 8 minutes to type a normal double-spaced page.
  a. Write a system of inequalities that describes the total number of letters L and pages P a typist can do in an hour or less. See left.
  b. Accurately graph the set of points that satisfies the system. See left.

17. A hockey team is scheduled to play 12 games during a season. The coach estimates that the team needs at least 16 points to make the playoffs. A win is worth 2 points and a tie is worth 1 point.
  a. Make a graph of all the combinations of wins w and ties t that will get the team into the playoffs. See left.
  b. How many ways are there for the team to make the playoffs? 25

18. An actress is paid \$250 per day to understudy a part and \$500 per day to perform the role before an audience. During one run, an actress earned between \$3000 and \$5000 as Maria in *West Side Story.*
  a. What is the maximum number of times she might have performed the role of Maria? 10
  b. What is the maximum number of times she might have been an understudy? 20
  c. Graph all possible ways she might have earned her salary.

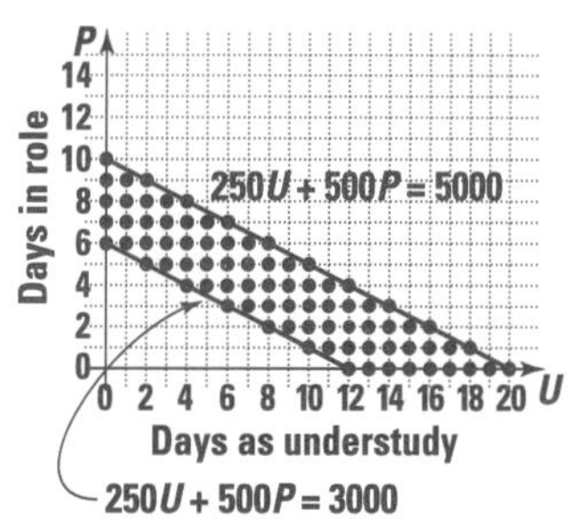

Notes on Questions

Question 10 Show how the graph is affected by adding the condition that $f > 5$. By limiting the number of possible field goals, it is reasonable to count the possibilities from the graph.

Questions 11–12 Notice that in **Question 11**, all of the boundaries are dashed lines; in **Question 12**, all of the boundaries are solid lines.

Questions 16–18 Remind students that when variables other than x and y are used, it is customary to plot the one that comes first in the alphabet on the horizontal axis and to plot the other on the vertical axis. Notice that all of the situations here are discrete.

Question 18 Music Connection *West Side Story*, with music composed by Leonard Bernstein, was a hit Broadway show in 1957. The story is a modern-day *Romeo and Juliet* involving a feud between lower-Manhattan street gangs.

Multicultural Connection The oldest form of traditional Japanese drama is the *no* play, which developed during the 1300s. Actors are masked and perform the stories with carefully controlled gestures and movements; every movement of the hands and feet, every vocal intonation, and every detail of costume and makeup follows a rule. Like the ancient Greek tragedy, a *no* drama is accompanied by music, dance, and choral speaking.

Adapting to Individual Needs

Challenge

Show students the graph at the right and have them describe the shaded region with a system of inequalities.

$\left[\begin{cases} y \leq -x + 4 \\ y \leq x + 4 \\ y \geq -x - 4 \\ y \geq x - 4 \end{cases}\right]$

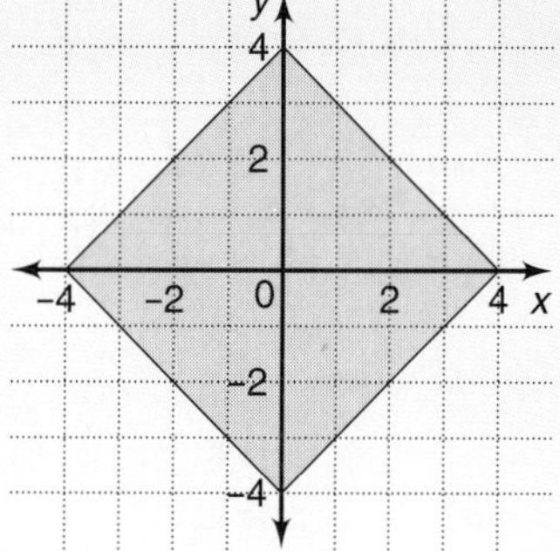

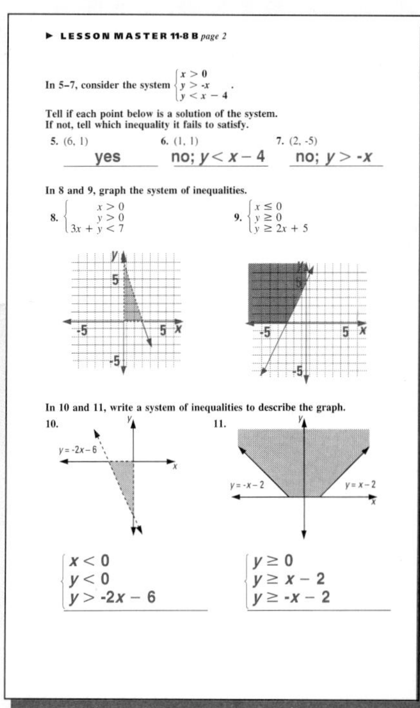

▸ LESSON MASTER 11-8 B *page 2*

In 5–7, consider the system $\begin{cases} x > 0 \\ y > -x \\ y < x - 4 \end{cases}$.

Tell if each point below is a solution of the system. If not, tell which inequality it fails to satisfy.

5. (6, 1) yes
6. (1, 1) no; $y < x - 4$
7. (2, -5) no; $y > -x$

In 8 and 9, graph the system of inequalities.

8. $\begin{cases} x > 0 \\ y > 0 \\ 3x + y < 7 \end{cases}$

9. $\begin{cases} x \leq 0 \\ y \geq 0 \\ y \geq 2x + 5 \end{cases}$

In 10 and 11, write a system of inequalities to describe the graph.

10. $y = -2x - 6$

$\begin{cases} x < 0 \\ y < 0 \\ y > -2x - 6 \end{cases}$

11. $y = -x - 2$, $y = x - 2$

$\begin{cases} y \geq 0 \\ y \geq x - 2 \\ y \geq -x - 2 \end{cases}$

Notes on Questions

Question 20 Students can answer this question by simplifying the left side of the equation to get $15 = 15$ or by using a proportion.

Follow-up for Lesson 11-8

Practice

For more questions on SPUR Objectives, use **Lesson Master 11-8A** (shown on pages 706–707) or **Lesson Master 11-8B** (shown on pages 708–709).

Assessment

Oral Communication For **Example 3**, ask students to describe the graph that shows all possible values for x and y if x and y are (1) any real numbers, (2) any integers, and (3) any whole numbers. [Descriptions are accurate and differences between real numbers, integers, and whole numbers are understood.]

Extension

Have students graph all pairs of positive numbers whose sum and product are less than 20.

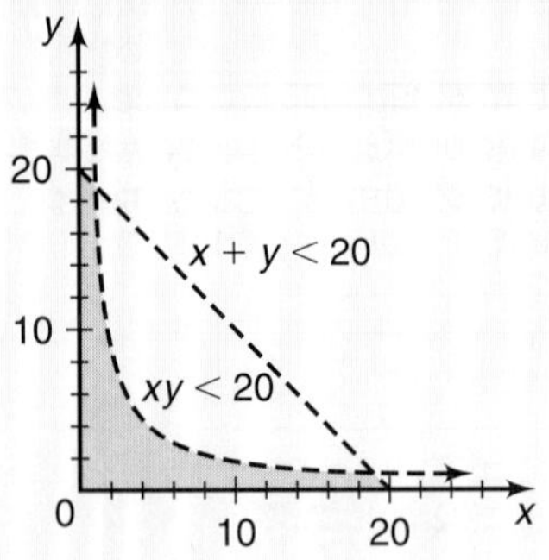

Project Update Project 3, *Systems with Equations and Inequalities*, on page 712, relates to the content of this lesson.

Review

23a) $\begin{cases} d + q = 27 \\ 0.1d + 0.25q = 5.10 \end{cases}$

19. Solve. *(Lesson 11-7)*
a. $6(3 - 2x) = -12(x - 3)$ no solution
b. $6(3 - 2y) = -12\left(y - \frac{3}{2}\right)$ y may be any real number.

20. Consider the equation $\frac{150(z - 3)}{10(z - 3)} = 15$. *(Lesson 11-7)*
a. What value can z not have? 3
b. Solve for z.
z can be any value except 3.

In 21 and 22, determine whether the lines are parallel and non-intersecting, coincident, or intersecting in only one point. *(Lesson 11-6)*

21. $\begin{cases} 4x - y = 8 \\ 8x - y = 8 \end{cases}$ intersecting

22. $\begin{cases} y = 3x + 9 \\ 6x - 2y = -27 \end{cases}$ parallel

23. A commuter has 27 coins consisting only of dimes and quarters. The total value of the coins is \$5.10. Let d = the number of dimes, and q = the number of quarters. *(Lessons 2-4, 3-1, 11-4)*
a. Write two equations determined by the information above.
b. How many of each coin does the commuter have?
a) See left. b) 11 dimes and 16 quarters

Loose change. *These automated ticket machines are in a subway station in Washington, D.C.*

24. Given the system $\begin{cases} w = -9z \\ z - 2w = 323 \end{cases}$, a student substituted $-9z$ for w in the second equation. The student wrote $z - 18z = 323$.
a. Is the student's work correct? No
b. If it is correct, finish solving the system. If not, describe what is wrong with it. *(Lessons 11-2, 11-3)* $z - 2(-9z) = z + 18z$

25. Write $(2x - 5)^2$ as a perfect square trinomial. *(Lesson 10-6)*
$4x^2 - 20x + 25$

26. *Skill sequence.* Solve. *(Lessons 1-6, 9-1, 9-5)*
a. $x^2 = 121$ 11, -11
b. $x^2 + 21 = 121$ 10, -10
c. $4x^2 + 21 = 121$ 5, -5
d. $4x^2 + 24x + 157 = 121$ -3

27. Suppose a bank offers an 8.5% annual yield. What would be the amount in an account after 5 years if \$1000 is invested? *(Lesson 8-1)*
\$1503.66

Exploration

28a) Sample: An isosceles trapezoid is a trapezoid with base angles of equal measure.

28. The graph of the solution to the system of inequalities in Example 3 on page 706 is a special type of quadrilateral called an *isosceles trapezoid.*
a. Look in a dictionary for a definition of isosceles trapezoid. See left.
b. Make up another system of inequalities whose solution set is the interior of an isosceles trapezoid. Sample: $\begin{cases} y < x \\ y < 5 - x \\ y > 0 \\ y < 2 \end{cases}$

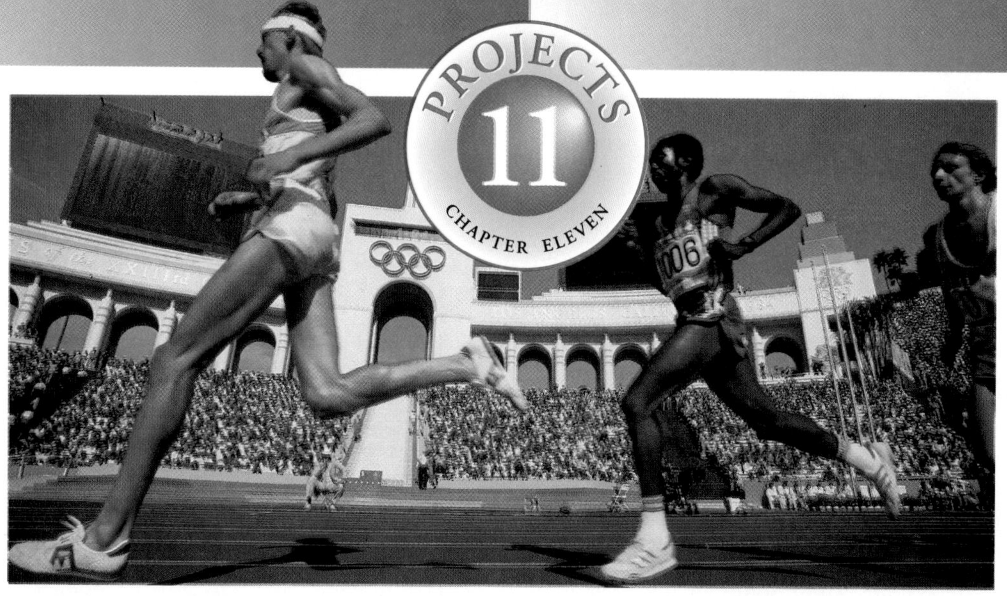

A project presents an opportunity for you to extend your knowledge of a topic related to the material in this chapter. You should allow more time for a project than you do for typical homework questions.

1 Olympic Records

Find the winning times in an Olympic sport in which both men and women participate.

a. Graph the data with the year on the horizontal axis and winning times on the vertical axis.

b. What trends do you observe in the data?

c. Based on the data, does it seem reasonable that in some year in your lifetime the women's winning time will be better than the men's? If so, when do you predict this will happen? Explain your reasoning.

2 Restaurant Prices

Check a restaurant's breakfast menu. Compare the costs of combination meals with the costs of purchasing items separately.

a. How much does the restaurant charge for two eggs and bacon? How much does it charge for one egg and bacon? At these rates, what should be the price of one egg alone? . . . one serving of bacon alone? How do your calculations compare to the restaurant's prices for a "side" order?

b. Investigate the costs of some other combination meals. Are the costs of individual items consistent with the costs of these items if purchased in a combination meal? Explain your reasoning.

Chapter 11 Projects

The projects relate chiefly to the content of the lessons of this chapter as follows:

Project	Lesson(s)
1	11-1, 11-6
2	11-4, 11-5
3	11-8
4	11-4
5	11-1

1 Olympic Records Students can find the data that are needed for this project in almanacs and encyclopedias. Winter events for both men and women include speedskating, downhill skiing, luge, and slalom skiing. Students can choose from a variety of swimming and running events from the summer Olympics. Suggest that students fit a line to each set of data they graph, find an equation for each line, and solve the system of equations to find when the men's time and the women's time will be the same. They should compare the answer with the graphs of their data.

2 Restaurant Prices Suggest that students check the prices of several restaurants in their area. They should be aware that side orders such as toast, potatoes, and/or grits are often served with a breakfast entree. Students should take this into account when doing this project.

Possible responses

1. a. Responses will vary.
   b. For most events, both women's and men's times are decreasing; women's times are decreasing faster than the men's times.
   c. Responses will vary.

2a–b Usually, students will find that a side order is more expensive if they have taken into consideration all items served with the meal.

3 Systems with Equations and Inequalities This project shows students how segments and other kinds of figures can result as solutions to systems. **Parts a and b** are straightforward. Have students name the endpoints of the line segment that is the solution to each part. Then have them check that points on the line segment satisfy all conditions of each system. For the first two questions in **part c**, you might remind students that they can use the given points to find the slope of the line and then use the slope-intercept to find the equation. Graphing the equations and highlighting the ray in (i), the segment in (ii), and the part of the parabola in (iii) should help students determine systems that satisfy the conditions given.

Suggest that students work backward in **part d**. That is, have them decide on the type of solution they want their system to have, and then have them write equations and inequalities to fit the solution. You may want to use several of these systems as a class assignment.

4 Adding and Subtracting Equations Suggest that students start this project with two equations whose graphs are non-parallel lines. Then have them repeat the project with two equations whose graphs are parallel lines. In **part d**, emphasize that each equation in **part a** can be multiplied by a different number; then **parts b and c** repeat. Automatic graphers should be used if they are available to students.

5 Life Expectancy Life expectancy is the average number of years that a group of people of a certain age might be expected to live. Tables of changes in life expectancy over time can be found in some almanacs and encyclopedias. It is estimated that in the year 1650, the average newborn baby could have been expected to live 25 years (40% of them died before they were a year old). In 1900, a newborn could have been expected to live more than 47 years, and by 1990, life expectancy was about 75 years. You might ask students for reasons why life expectancy has been increasing. [Sample reasons: development of medicines to combat disease in infants; public health awareness by governments; higher standards of living]

(continued)

3 Systems with Equations and Inequalities

Linear equations and inequalities can both be present in a single system. When they are, the solution to the system may be graphed as a line, a ray, a line segment, or a point. The system may also have no solution. Graph the solutions to the following systems, and describe the solution in words.

a. $\begin{cases} y \ge \frac{1}{3}x - 3 \\ y \le \frac{1}{3}x + 2 \\ y = -\frac{2}{3}x \end{cases}$ **b.** $\begin{cases} 2x + y \le 5 \\ 4x - 5y \le 10 \\ 2x + 5y = 5 \end{cases}$

c. Write systems whose solutions are each of the following.
   i. the ray that begins at (3, 5) and contains the point (5, 9)
   ii. the segment with endpoints (-6, 1) and (4, 0)
   iii. the part of the graph of the parabola $y = x^2$ between (-4, 16) and (4, 16)

d. Make up a system of equations and inequalities whose solution is different than those in parts **a-c.**

4 Adding and Subtracting Equations

Select two linear equations and do the following:

a. Graph the two linear equations.
b. Graph the equation found by adding the two selected equations.
c. Graph the equation found by subtracting them.
d. Do multiplications, additions, and subtractions and continue to graph the resulting equations.
e. Do the graphs have any common features? If so, describe them.
f. Verify your calculations starting with a different pair of equations.

5 Life Expectancy

In this century, the number of years of life expected at birth has gradually been increasing over time in most countries of the world, for both men and women of all ethnicities. Almanacs and other reference books contain tables of what has been the life expectancy for many years. Graph the life expectancy of males and females in the United States since 1940 or earlier. Find lines of fit for both males and females. According to your lines, in what year will or did males and females have the same life expectancy? You may wish to consider the life expectancies in other nations, or of people in various ethnic groups.

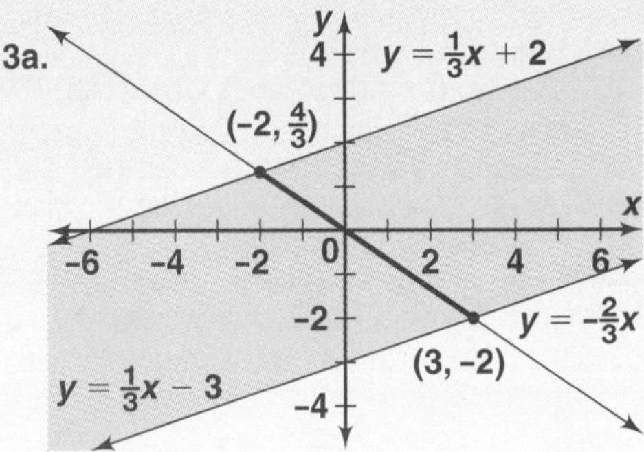

3. Sample graphs are shown at the right and on page 713.
   a. The solution is a line segment with endpoints $(-2, \frac{4}{3})$ and (3, -2).
   b. The solution is a ray with endpoint (2.5, 0) passing through (0, 1).

3a.
$y = \frac{1}{3}x + 2$
$(-2, \frac{4}{3})$
$y = -\frac{2}{3}x$
(3, -2)
$y = \frac{1}{3}x - 3$

SUMMARY

A system is a set of sentences which together describe a single situation. The solution set to a system is the set of all solutions common to all of the sentences in the system. A solution to a system of two linear equations is an ordered pair (x, y) that satisfies each equation. Systems of two linear equations may have no solution, one solution, or infinitely many solutions.

One way to solve a system is by graphing. By looking for the intersection point(s) on a graph you can quickly tell if there are any solutions to the system. There are as many solutions as intersection points. Graphing is also a way to describe solutions to systems that have infinitely many solutions. For instance, overlapping half-planes, which arise from systems of linear inequalities, and coincident lines have infinitely many solutions.

However, graphing does not always yield exact solutions. In this chapter, three strategies are presented for finding exact solutions to systems of linear equations. They are substitution, addition, and multiplication. Substitution is a good method to use if at least one equation is given in $y = mx + b$ form. Addition is appropriate if the same term has opposite signs in the two equations in the system. Multiplication is a good method when both equations are in $Ax + By = C$ form. Each method changes the system into an equivalent system whose solutions are the same as those of the original system.

Any kind of situation that leads to a linear equation can lead to a linear system. All that is needed is more than one condition which must be satisfied.

VOCABULARY

Below are the new terms and phrases for this chapter. You should be able to give a general description and a specific example for each.

Lesson 11-1
system
condition of a system
solution to a system
linear systems

Lesson 11-2
substitution

Lesson 11-4
Generalized Addition Property of Equality
addition method for solving a system

Lesson 11-5
equivalent systems
multiplication method for solving a system

Lesson 11-6
coincident lines

Lesson 11-8
system of inequalities

Summary

The Summary gives an overview of the entire chapter and provides an opportunity for students to consider the material as a whole. Thus, the Summary can be used to help students relate and unify the concepts that are presented in the chapter.

Vocabulary

Terms, symbols, and properties are listed by lesson to provide a checklist of concepts that students must know. Emphasize that students should read the vocabulary list carefully before starting the Progress Self-Test. If students do not understand the meaning of a term, they should refer back to the indicated lesson.

3b.

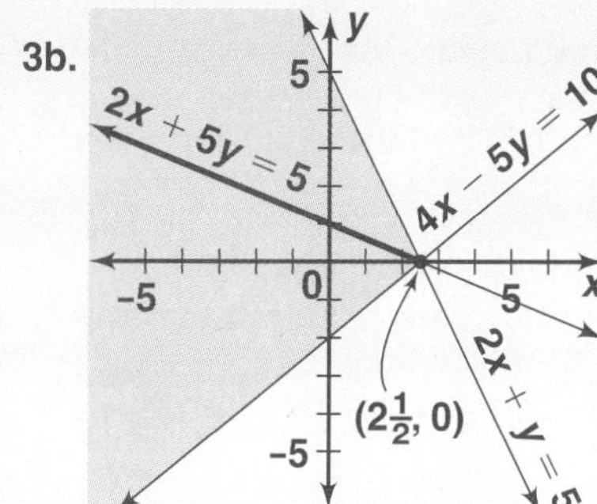

3c. Sample systems:

(Part i) $\begin{cases} x \geq 3 \\ y \geq 5 \\ y = 2x - 1 \end{cases}$

(Part ii) $\begin{cases} x \leq 4 \\ y \geq -6 \\ y = -0.1x + 0.4 \end{cases}$

(Part iii) $\begin{cases} -4 < x < 4 \\ y < 16 \\ y = x^2 \end{cases}$

d. Responses will vary.

4. Sample equations and graphs:
a. $2x + y = 6$ and $-3x + y = -1$
b. Sum: $-x + 2y = 5$
c. Difference: $5x = 7$
d. Multiply $2x + y = 6$ by 2:
$4x + 2y = 12$
Multiply $-3x + y = -1$ by -1:
$3x - y = 1$
Sum: $7x + y = 13$
Difference: $x + 3y = 11$

(Responses continue on page 714.)

Progress Self-Test

For the development of mathematical competence, feedback and correction, along with the opportunity to practice, are necessary. The Progress Self-Test provides the opportunity for feedback and correction; the Chapter Review provides additional opportunities and practice. We cannot overemphasize the importance of these end-of-chapter materials. It is at this point that the material "gels" for many students, allowing them to solidify skills and understanding. In general, student performance should be markedly improved after these pages.

Assign the Progress Self-Test as a one-night assignment. Worked-out *solutions* for all questions are in the Selected Answers section of the student book. Encourage students to take the Progress Self-Test honestly, grade themselves, and then be prepared to discuss the test in class.

Advise students to pay special attention to those Chapter Review questions (pages 715–717) which correspond to the questions that they missed on the Progress Self-Test.

Additional Answers, page 714

8.

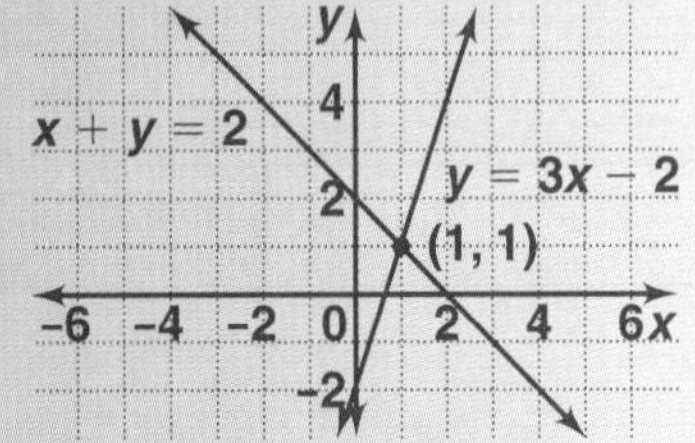

9.

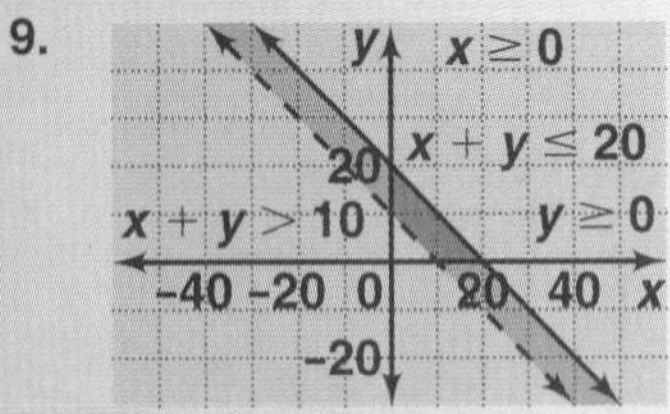

14.

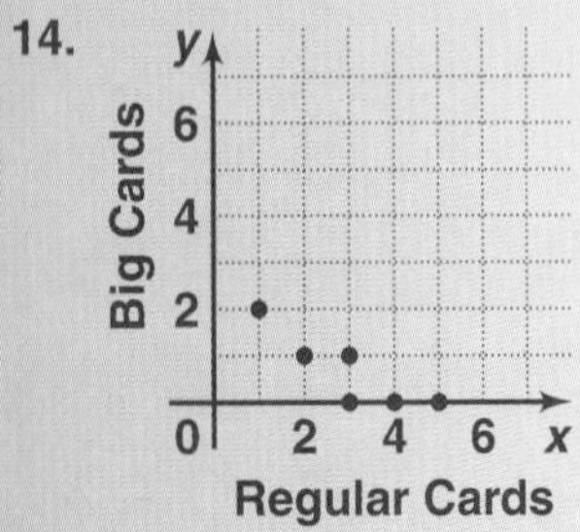

PROGRESS SELF-TEST

Take this test as you would take a test in class. Then check your work with the solutions in the Selected Answers section in the back of the book.

In 1–4, solve by using any method.

1. $\begin{cases} a - 3b = -8 \\ b = 3a \end{cases}$ $a = 1$, $b = 3$

2. $\begin{cases} p = 5r + 80 \\ p = -7r - 40 \end{cases}$ $p = 30$, $r = -10$

3. $\begin{cases} m - n = -1 \\ -m + 2n = 4 \end{cases}$ $m = 2$, $n = 3$

4. $\begin{cases} 7x + 3y = 1 \\ 4x - y = 6 \end{cases}$ $x = 1$, $y = -2$

5. Line ℓ has equation $y = -5x - 15$. Line m has equation $y = x - 3$. Find the point of intersection. (-2, -5)

In 6 and 7, determine whether the lines coincide, intersect in one point, or are parallel and nonintersecting.

6. $\begin{cases} 5 = 2A + 7B \\ 10 = 4A + 14B \end{cases}$ coincide

7. $\begin{cases} y = 2x + 7 \\ y - 2x = 3 \end{cases}$ parallel

In 8 and 9, solve the system by graphing. See margin.

8. $\begin{cases} y = 3x - 2 \\ x + y = 2 \end{cases}$

9. $\begin{cases} x \ge 0 \\ y \ge 0 \\ x + y \le 20 \\ x + y > 10 \end{cases}$

10. Lisa weighs 4 times as much as her baby sister. Together they weigh 95 pounds. How much does each person weigh? See below.

11. The Reid family went to a restaurant and ordered 3 hamburgers and 4 small salads. Without tax, the bill was \$21.30. At the same restaurant, the Millers ordered 5 hamburgers and 2 small salads. Their bill without tax was \$22.90. What was the cost of a small salad? \$2.70

12. If the cost per unit is constant, could the following situation have happened? Why or why not? 10 roses and 15 daffodils were sold for \$35. 2 roses and 3 daffodils were sold for \$8. No. Let r = price of a rose and d = price of a daffodil. $5(2r + 3d) = 5 \cdot 8$; $10r + 15d = 40$; $35 \neq 40$.

13. Massachusetts had a 1990 population of about 6,016,000 and was growing at a rate of about 30,000 people each year. Minnesota had a 1990 population of about 4,375,000 and was growing at a rate of about 30,000 people each year. If these rates continue, in how many years after 1990 will these two states have the same population? Explain how you arrived at your answer. See below.

14. Joseph needs to buy at least 3 birthday cards, and he has \$5 to spend. If the big ones cost \$2 and the regular ones cost \$1, show on a graph how many of each kind he might buy. See margin

In 15 and 16, solve and check.

15. $12z + 8 = 12z - 3$
$12z + 8 - 12z = -3$; $8 = -3$; z has no solution.

16. $-19p < 22 - 19p$
$0 < 22 - 19p + 19p$; $0 < 22$; p may be any real number.

Field of daffodils.

10) Lisa, 76 lb; baby, 19 lb
13) They will never have the same population because $6{,}016{,}000 + 30{,}000y = 4{,}375{,}000 + 30{,}000y$ has no solution. The lines are parallel.

Additional responses, page 712

4.

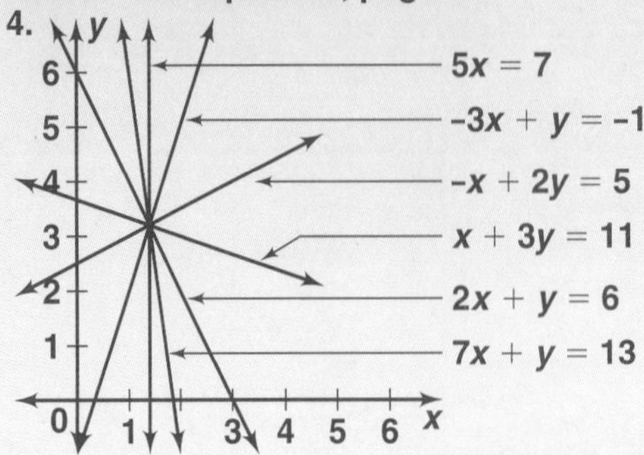

4e–f If the graphs of the equations in part a are not parallel lines, the graphs of all of the equations will be concurrent (intersect at one point). If the graphs of the equations in part a are parallel lines, then the graphs of all of the other equations will be lines that are parallel to those chosen in part a.

CHAPTER REVIEW

Questions on SPUR Objectives

SPUR stands for **S**kills, **P**roperties, **U**ses, and **R**epresentations. The Chapter Review questions are grouped according to the SPUR Objectives for this chapter.

SKILLS DEAL WITH THE PROCEDURES USED TO GET ANSWERS.

Objective A: *Solve systems using substitution.* *(Lessons 11-2, 11-3)* See below for 5-8.

1. Find (a, b). (15, 45) $\begin{cases} b = 3a \\ 60 = a + b \end{cases}$

2. Find (x, y). (17, 4) $\begin{cases} x - y = 13 \\ x = 6y - 7 \end{cases}$

3. Solve for p and q. $\begin{cases} p + q = 50 \\ p + 6q = 200 \end{cases}$ $p = 20$, $q = 30$

4. Solve the system. $\begin{cases} 10y = 20x + 20 \\ 2x + 4y = 29 \end{cases}$ $(x, y) = (2.1, 6.2)$

In 5–8, solve the system.

5. $\begin{cases} y = x + 5 \\ 300 + 10x = 35y \end{cases}$

6. $\begin{cases} q = z + 8 \\ 14z + 89 = 13q \end{cases}$

7. $\begin{cases} a = 2b + 3 \\ a = 3b + 20 \end{cases}$

8. $\begin{cases} 16 - 2x = y \\ x + 4 = y \end{cases}$

In 9 and 10, two lines have the given equations. Find the point of intersection, if any.

9. Line ℓ: $y = 7x + 20$
Line m: $y = 3x - 16$ (-9, -43)

10. Line ℓ: $y = \frac{2}{3}x - \frac{1}{6}$
Line m: $y = \frac{1}{3}x + \frac{1}{3}$ $\left(\frac{3}{2}, \frac{5}{6}\right)$

5) $(x, y) = (5, 10)$ 6) $(q, z) = (23, 15)$
7) $(a, b) = (-31, -17)$ 8) $(x, y) = (4, 8)$
12) $(a, c) = (15, 55)$ 14) $(x, y) = (12, -4)$

Objective B: *Solve systems by addition.* *(Lesson 11-4)* See left for 12 and 14.

In 11–16, solve. 11) $(m, b) = (-22, 77)$

11. $\begin{cases} 3m + b = 11 \\ -4m - b = 11 \end{cases}$

12. $\begin{cases} 6a + 2c = 200 \\ 9a - 2c = 25 \end{cases}$

13. $\begin{cases} 0.6x - 0.4y = 1.1 \\ 0.2x - 0.4y = 2.3 \end{cases}$ $(x, y) = (-3, -7.25)$

14. $\begin{cases} \frac{1}{2}x + 3y = -6 \\ \frac{1}{2}x + y = 2 \end{cases}$

15. $\begin{cases} 4f + g = 15 \\ 3g - 4f = -3 \end{cases}$ $(f, g) = (3, 3)$

16. $\begin{cases} s + \frac{2}{3}t = 3 \\ \frac{2}{3}t - 6s = 10 \end{cases}$ $(s, t) = (-1, 6)$

Objective C: *Solve systems by multiplying.* *(Lesson 11-5)* See margin for 17 and 18.

In 17 and 18, **a.** multiply one of the equations by a number which makes it possible to solve the system by adding. **b.** Solve the system.

17. $\begin{cases} 5x + y = 30 \\ 3x - 4y = 41 \end{cases}$

18. $\begin{cases} 5u + 6v = -295 \\ u - 9v = 400 \end{cases}$

In 19–22, solve the system. 19) $(y, z) = (3, 3)$ 20) $(m, n) = (4, 7)$

19. $\begin{cases} 3y - 2z = 3 \\ 2y + 5z = 21 \end{cases}$

20. $\begin{cases} 7m - 4n = 0 \\ 9m - 5n = 1 \end{cases}$

21. $\begin{cases} a + b = 3 \\ 5b - 3a = -17 \end{cases}$ $(a, b) = (4, -1)$

22. $\begin{cases} 46 = 2t + u \\ 20 = 8t - 4u \end{cases}$ $(t, u) = \left(\frac{51}{4}, \frac{41}{2}\right)$

PROPERTIES DEAL WITH THE PRINCIPLES BEHIND THE MATHEMATICS.

Objective D: *Recognize sentences with no solution, one solution, or all real numbers as solutions.* *(Lesson 11-7)*

In 23–26, solve.

23. $2a + 4 < 2a + 3$

24. $12c < 6(3 + 2c)$

25. $7x - x = 12x$

26. $-10x = 15 - 10x$

23) No solutions
24) All real numbers are solutions.
25) $x = 0$ (one solution)
26) No solution

27. Is $2k - 7 = 2k$ ever true? Explain why or why not.

28. Is $100d < 100(d - 1)$ ever true? Explain why or why not. No. $100d < 100d - 100$ gives $0 < -100$ which is never true; so the original sentence is never true.

27) No. Subtract $2k$. $-7 = 0$ is never true; so the original sentence is never true.

Chapter 11 Review

Resources

From the ***Teacher's Resource File***

- Answer Master for Chapter 11 Review
- Assessment Sourcebook: Chapter 11 Test, Forms A–D Chapter 11 Test, Cumulative Form

Additional Resources

- Quiz and Test Writer

The main objectives for the chapter are organized in the Chapter Review under the four types of understanding this book promotes—Skills, Properties, Uses, and Representations.

Whereas end-of-chapter material may be considered optional in some texts, in *UCSMP Algebra* we have selected these objectives and questions with the expectation that they will be covered. Students should be able to answer these questions with about 85% accuracy after studying the chapter.

You may assign these questions over a single night to help students prepare for a test the next day, or you may assign the questions over a two-day period. If you work the questions over two days, then we recommend assigning the *evens* for homework the first night so that students get feedback in class the next day, then assigning the *odds* the night before the test because answers are provided to the odd-numbered questions.

Additional Answers, page 715

17. a. Sample: Multiply equation (1) by 4.
b. $(x, y) = (7, -5)$

18. a. Sample: Multiply equation (2) by -5.
b. $(u, v) = (-5, -45)$

5. Sample graph is shown at the right. If decade points are used from 1920 to 1990, the lines of best fit are approximately:
Women of all races:
$a = 0.3326y - 580.7$
Men of all races:
$a = 0.2462y - 417.1$
Where y = year and a = age (life expectancy)

U.S. Life Expectancy (1920–Present)

a
80
70
60
50
Age
$a = 0.3326y - 580.7$
$a = 0.2462y - 417.1$
• Women
○ Men
1920 1940 1960 1980 2000 y
Year

Student's conclusions will vary. According to the lines of best fit, men and women had the same life expectancy of about 49 years in 1894.

If students have considered the life expectancy of another nation or of people in a specific ethnic group, their responses will vary. Responses will probably follow the same format as those presented for the United States.

Assessment

Evaluation The *Assessment Sourcebook* provides five forms of the Chapter 11 Test. Forms A and B present parallel versions in a short-answer format. Forms C and D offer performance assessment. The fifth test is the Chapter 11 Test, Cumulative Form. About 50% of this test covers Chapter 11; 25% covers Chapter 10, and 25% covers earlier chapters.

For information on grading, see *General Teaching Suggestions; Grading* in the *Professional Sourcebook*, which begins on page T20 in Part 1 of the Teacher's Edition.

Feedback After students have taken the test for Chapter 11 and you have scored the results, return the tests to students for discussion. Class discussion on the questions that caused trouble for most students can be very effective in identifying and clarifying misunderstandings. You might want to have them write down the items they missed and work either in groups or at home to correct them. It is important for students to receive feedback on every chapter test, and we recommend that students see and correct their mistakes before proceeding too far into the next chapter.

Additional Answers

29. No solution when $b \neq c$. If the slopes are equal, but the y-intercepts are different, the lines are parallel and the system has no solution.
30. When the first equation is multiplied by -3, the result is the second equation. The lines coincide and the system has infinitely many solutions.
31. Coincide
32. Parallel
33. Coincide
34. Intersect

CHAPTER ELEVEN

Objective E: *Determine whether a system has no solution, one solution, or infinitely many solutions.* (Lesson 11-6) See margin for 29-34.

29. When will the following system have no solution? $\begin{cases} y = mx + b \\ y = mx + c \end{cases}$ Explain how you know this.
30. Tell how you can determine the number of solutions to the system $\begin{cases} r - s = 1 \\ 3s - 3r = -3 \end{cases}$.

In 31–34, determine whether each system describes lines that coincide, intersect, or are parallel and nonintersecting.

31. $\begin{cases} 2x + 4y = 7 \\ 10x + 20y = 35 \end{cases}$
32. $\begin{cases} y - 2x = 5 \\ y = 2x + 4 \end{cases}$
33. $\begin{cases} 6 = m - n \\ -6 = n - m \end{cases}$
34. $\begin{cases} a - 3b = 2 \\ a - 4b = 2 \end{cases}$
35. *Multiple choice.* Two straight lines *cannot* intersect in (c)
    (a) exactly one point.
    (b) no points.
    (c) exactly two points.
    (d) infinitely many points.
36. Parallel, nonintersecting lines have the same __?__ but different y-intercepts. slope

USES DEAL WITH APPLICATIONS OF MATHEMATICS IN REAL SITUATIONS.

Objective F: *Use systems of linear equations to solve real-world problems.* (Lessons 11-2, 11-3, 11-4, 11-5, 11-6, 11-7)

37. Suppose Joe earned three times as much as Marty during the summer. Together they earned $210. How much did each earn? Marty, $52.50; Joe, $157.50
38. Renting a car for a day from company C costs $39 plus $.10 a mile. Renting a car for a day from company D costs $22.95 plus $.25 a mile. At what distance does renting the cars cost the same? 107 miles
39. The starting salary on Job (1) is $7.00 an hour and every 6 months increases $0.50 an hour. For Job (2) the starting salary is $7.20 an hour and every 6 months increases $0.50 an hour. When does Job (2) pay more than Job (1)? always
40. From 1980 to 1990, Tucson, Arizona, grew at about 7,500 people a year, to a population of about 405,000. Mesa, Arizona, grew about 13,500 people a year to a population of 290,000. If these rates of increase stay the same, in about how many years will Mesa and Tucson have the same population? ≈19 yrs
41. In her restaurant Charlene sells 2 eggs and a muffin for $1.80. She sells 1 egg with a muffin for $1.35. At these rates, how much is she charging for the egg and how much for the muffin? egg, $0.45; muffin, $0.90
42. A hotel offers two weekend packages. Plan A, which costs $315, gives one person 3 nights lodging and 2 meals. Plan B gives 2 nights lodging and 1 meal and costs $205. At these rates, what is the charge for a room for one night? $95
43. If the cost per unit is constant, could the given situation have happened? Why or why not? Lydia bought 16 pencils and 5 erasers for $8.00. Then she bought 32 pencils and 10 erasers for $16.00. See below.
44. Tickets to a school play cost $2.50 for students and $4.00 for adults. The school treasurer reported that 850 tickets were sold, and the total revenue was $2395. How many student tickets were sold? 670

43) Yes; $16p + 5e = 8$ is equivalent to $32p + 10e = 16$. There are infinitely many solutions.

This scene is from Godspell, *performed by students at Anaheim High School in California.*

45.

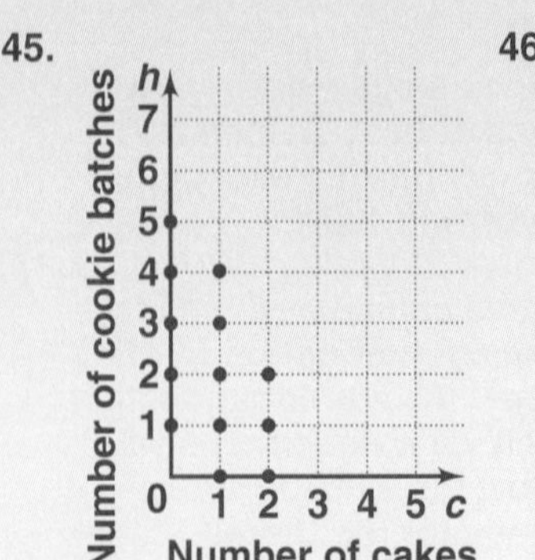

46.

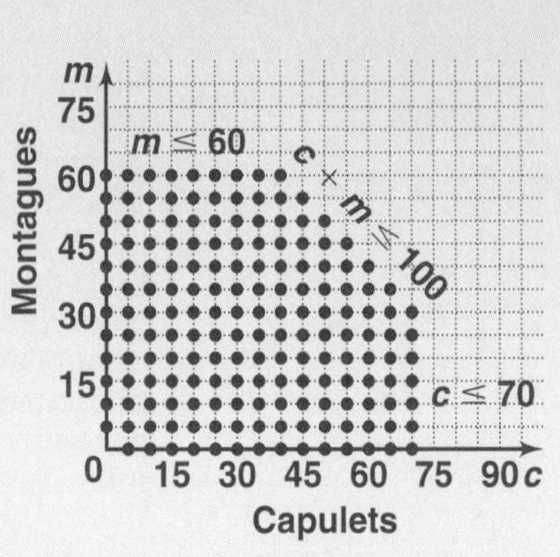

47a.

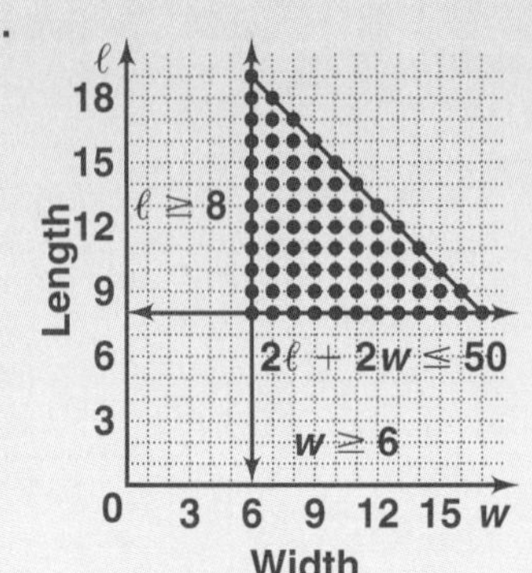

48a.

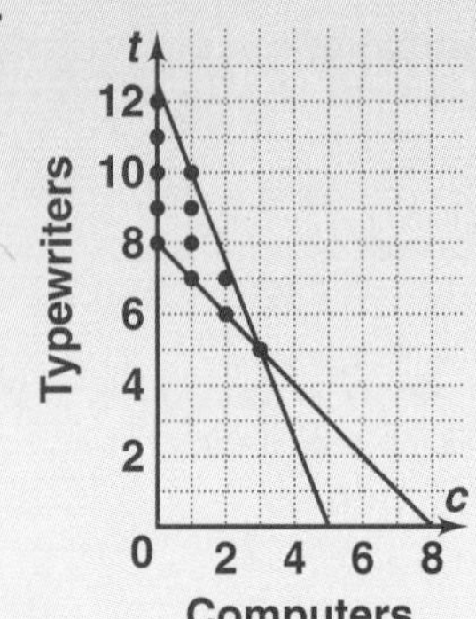

See margin for 45-48.

Objective G: *Use systems of linear inequalities to solve real-world problems.* (Lesson 11-8)

45. Jean has 11 cups of flour on hand, plenty of cookie sheets, but only two cake pans. If it takes 2 cups of flour to make a batch of cookies, and 3 cups to make a cake, what can Jean bake? Make a graph showing all the possibilities.

46. Romeo wants no more than 70 Capulets at the wedding, while Juliet insists that there be no more than 60 Montagues. The hall for the reception is big enough to hold only 100 people. Use a graph to show how many people from each family could attend.

60)

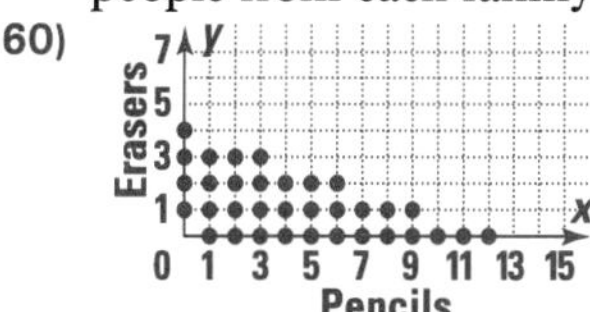

CHAPTER ELEVEN

47. Rochelle bought 50 ft of fence for a rectangular pen. She wants to make the pen at least 8 ft long and 6 ft wide.

a. Draw a graph to show all possible dimensions (to the nearest foot) of the pen.

b. At most, how long could the pen be? 19 ft

48. Rhiann budgeted \$5,000 to buy office machines for 8 new typists. A computer costs \$1,000 and a typewriter costs \$400.

a. Use a graph to show all possible combinations of computers and typewriters Rhiann could buy if she must buy at least 8 machines.

b. At most, how many computers can Rhiann buy if she must provide all 8 new typists with at least one computer or one typewriter? 3 computers

REPRESENTATIONS DEAL WITH PICTURES, GRAPHS, OR OBJECTS THAT ILLUSTRATE CONCEPTS.

See margin for 49-56.

Objective H: *Find solutions to systems of equations by graphing.* (Lessons 11-1, 11-6)

In 49–54, solve each system by graphing.

49. $\begin{cases} y = 4x + 6 \\ y = \frac{1}{2}x - 1 \end{cases}$

50. $\begin{cases} y = x - 4 \\ y = -3x \end{cases}$

51. $\begin{cases} y = x + 3 \\ -2x + 3y = 4 \end{cases}$

52. $\begin{cases} 2y - 4x = 1 \\ y = 2x + \frac{1}{2} \end{cases}$

53. $\begin{cases} 3x - 3y = 3 \\ \frac{1}{2}x - \frac{1}{2}y = -1 \end{cases}$

54. $\begin{cases} x + 3y = 5 \\ y = 3x + 5 \end{cases}$

Objective I: *Graphically represent solutions to systems of linear inequalities.* (Lesson 11-8)

In 55–58, graph all solutions to the system.

55. $\begin{cases} y \le \frac{1}{2}x + 4 \\ y \ge -x + 1 \end{cases}$

56. $\begin{cases} -x + 2y > 4 \\ x + \frac{1}{2}y < 2 \end{cases}$

See below for 57-59.

57. $\begin{cases} x > 0 \\ y > 0 \\ x + y < 2 \end{cases}$

58. $\begin{cases} x \ge 0 \\ y \ge 0 \\ x + y \le 6 \\ x + y \ge 4 \end{cases}$

In 59 and 60, accurately graph the set of points that satisfies each situation.

59. A small elevator in a building has a capacity of 280 kg. If a child averages 40 kg and an adult 70 kg, how many children C and adults A can the elevator hold without being overloaded?

60. A person wants to buy x pencils at 5¢ each and y erasers at 15¢ each and cannot spend more than 60¢. What are the possible values of x and y? See above left.

57)

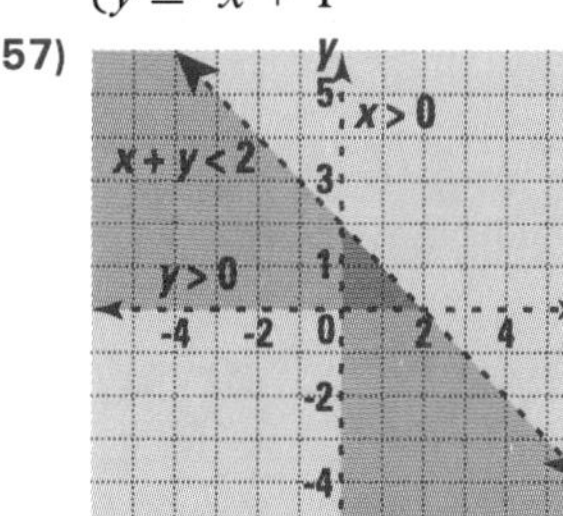

58)

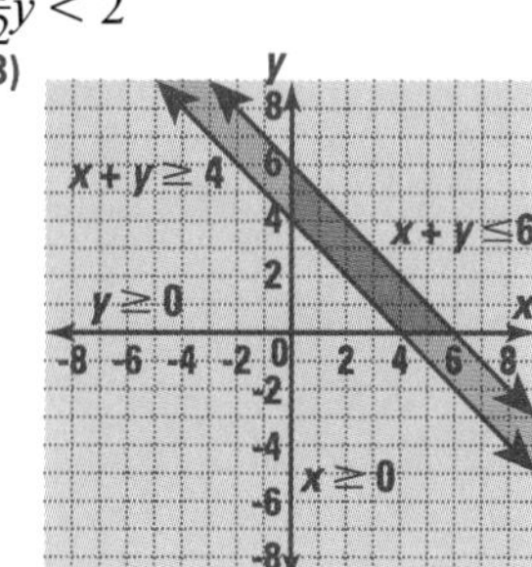

59)

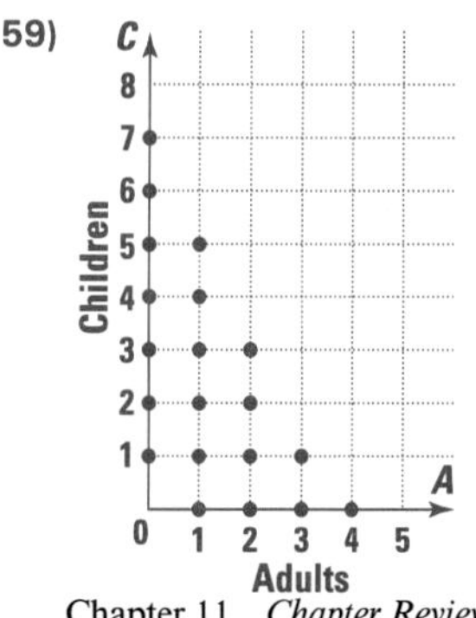

52. coincident, infinitely many solutions

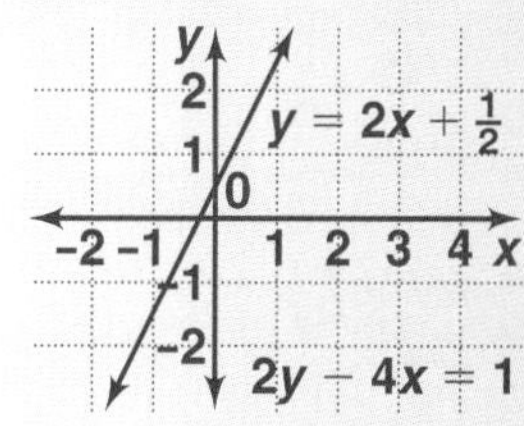

53. parallel, no solutions

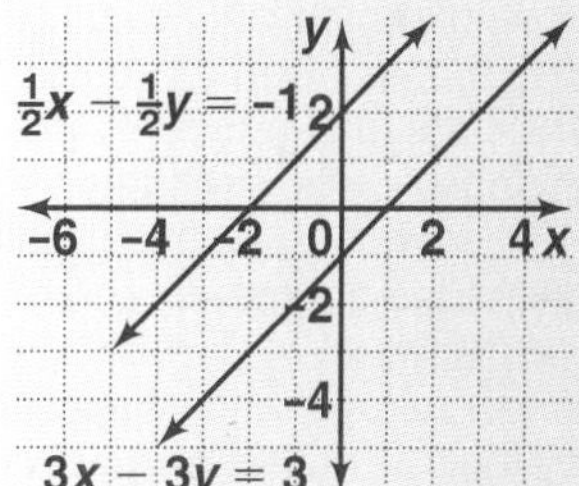

54.

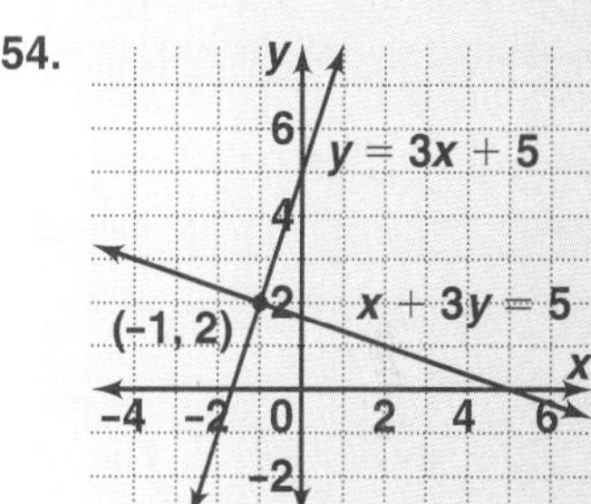

55.

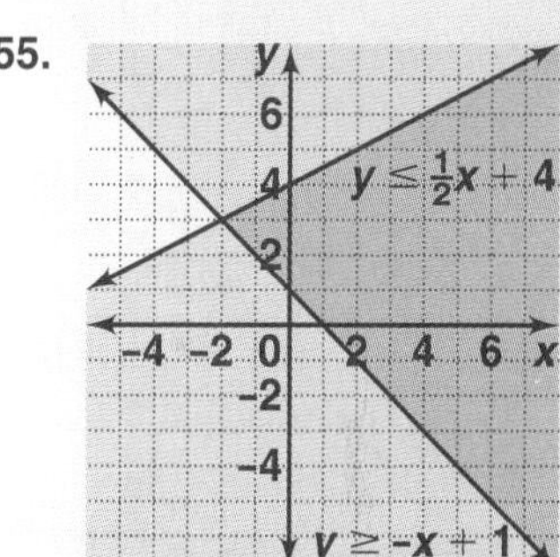

56.

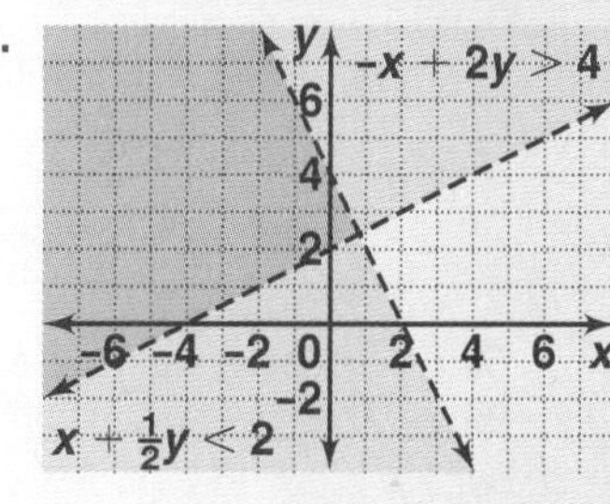

49.

y = 4x + 6

y = ½x − 1

(−2, −2)

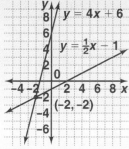

50.

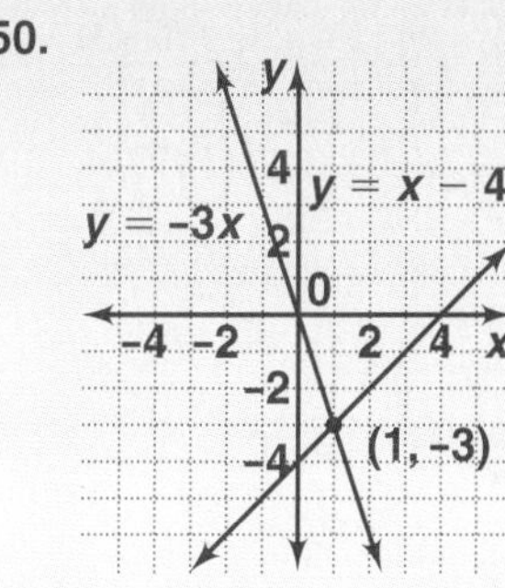

51.

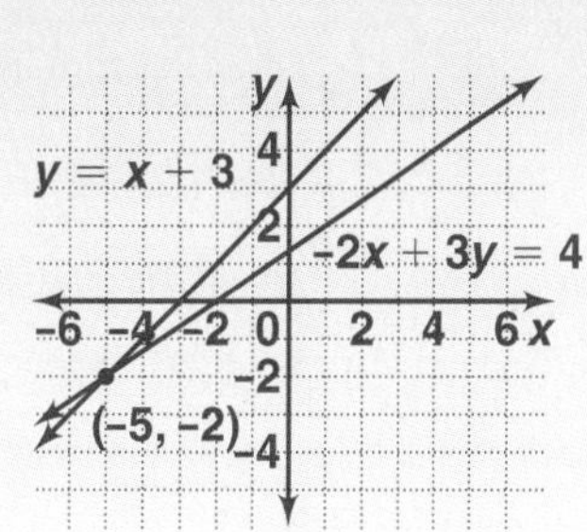

Setting Up Lesson 12-1

We recommend that you assign Lesson 12-1, both reading and some questions, for homework the evening of the test.

Chapter 12 Planner

Adapting to Individual Needs

The student text is written for the vast majority of students. The chart at the right suggests two pacing plans to accommodate the needs of your students. Students in the Full Course should complete the entire text by the end of the year. Students in the Minimal Course will spend more time when there are quizzes and more time on the Chapter Review. Therefore, these students may not complete all of the chapters in the text.

Options are also presented to meet the needs of a variety of teaching and learning styles. For each lesson, the Teacher's Edition provides sections entitled: *Video* which describes video segments and related questions that can be used for motivation or extension; *Optional Activities* which suggests activities that employ materials, physical models, technology, and cooperative learning; and, *Adapting to Individual Needs* which regularly includes **Challenge** problems, **English Language Development** suggestions, and suggestions for providing **Extra Help.** The Teacher's Edition also frequently includes an **Error Alert,** an **Extension,** and an **Assessment** alternative. The options available in Chapter 12 are summarized in the chart below.

Chapter 12 Pacing Chart

Day	Full Course	Minimal Course
1	12-1	12-1
2	12-2	12-2
3	12-3	12-3
4	Quiz*; 12-4	Quiz*; begin 12-4.
5	12-5	Finish 12-4.
6	12-6	12-5
7	Quiz*; 12-7	12-6
8	12-8	Quiz*; begin 12-7.
9	Self-Test	Finish 12-7.
10	Review	12-8
11	Test*	Self-Test
12		Review
13		Review
14		Test*

*in the Teacher's Resource File

In the Teacher's Edition...

Lesson	Optional Activities	Extra Help	Challenge	English Language Development	Error Alert	Extension	Cooperative Learning	Ongoing Assessment
12-1	●	●	●	●		●	●	Oral
12-2	●	●	●	●	●	●	●	Group
12-3	●	●	●	●	●	●	●	Written
12-4	●	●	●	●		●	●	Written
12-5	●	●	●			●	●	Written
12-6	●	●	●			●	●	Group
12-7	●		●	●	●	●		Written
12-8	●	●	●	●	●	●	●	Group

In the Additional Resources...

	In the Teacher's Resource File								
Lesson	Lesson Masters, A and B	Teaching Aids*	Activity Kit*	Answer Masters	Technology Sourcebook	Assessment Sourcebook	Visual Aids**	Technology	Video Segments
12-1	12-1	127		12-1	Comp 29		127, AM	BASIC	12-1
12-2	12-2	26, 31, 127		12-2			26, 31, 127, AM		
In-class Activity		31					31, AM		
12-3	12-3	31, 128	26	12-3		Quiz	31, 128, AM		
12-4	12-4	128, 130		12-4			128, 130, AM		
12-5	12-5	128		12-5	Comp 30		128, AM	GraphExplorer	
12-6	12-6	129		12-6		Quiz	129, AM		
12-7	12-7	4, 129	27	12-7			4, 129, AM		
12-8	12-8	129		12-8	Comp 31		129, AM	Spreadsheet	
End of chapter				Review		Tests			

*Teaching Aids are pictured on pages 718C and 718D. The activities in the Activity Kit are pictured on page 718C.

**Visual Aids provide transparencies for all Teaching Aids and all Answer Masters.

Also available is the Study Skills Handbook which includes study-skill tips related to reading, note-taking, and comprehension.

Integrating Strands and Applications

	12-1	12-2	12-3	12-4	12-5	12-6	12-7	12-8
Mathematical Connections								
Number Sense	●			●	●			●
Algebra	●	●	●	●	●	●	●	●
Geometry	●	●	●	●	●	●	●	●
Measurement		●		●	●	●	●	●
Logic and Reasoning	●				●	●	●	
Statistics/Data Analysis	●		●	●				
Patterns and Functions	●	●	●		●	●		
Discrete Mathematics	●	●	●	●				●
Interdisciplinary and Other Connections								
Science	●	●	●	●	●	●		●
Social Studies	●		●			●	●	
Multicultural				●		●	●	
Technology	●	●			●			●
Consumer		●	●					
Sports	●			●	●			●

Teaching and Assessing the Chapter Objectives

				In the Teacher's Resource File		
Chapter 12 Objectives (Organized into the SPUR categories—Skills, Properties, Uses, and Representations)	**Lessons**	**Progress Self-Test Questions**	**Chapter Review Questions**	**Chapter Test, Forms A and B**	**Chapter Test, Forms C**	**Chapter Test, Forms D**
Skills						
A: Factor positive integers into primes.	12-1	1	1–4	1	1	✓
B: Find common monomial factors of polynomials.	12-2	3–6	5–12	2, 6	3	✓
C: Factor quadratic expressions.	12-3, 12-5	7–10	13–30	7–10	4	✓
D: Solve quadratic equations by factoring.	12-4, 12-5	15, 17	31–40	14, 15	6	✓
Properties						
E: Apply the definitions and properties of primes and factors.	12-1	2, 19	41–44	18	1	✓
F: Recognize and use the Zero Product Property.	12-4	14, 16	45–52	11, 13	6	✓
G: Determine whether a quadratic polynomial can be factored over the integers.	12-3, 12-8	11, 12, 21	53–59	12, 16		✓
H: Apply the definitions and properties of rational and irrational numbers.	12-7	18, 20	60–69	4, 5	2	✓
Uses						
I: Solve quadratic equations in real situations.	12-4, 12-6	22–24	70–75	17, 19, 20		✓
Representations						
J: Represent quadratic expressions and their factorizations with areas.	12-2, 12-3	13	76–79	3	5	✓

Multidimensional Assessment
Quiz for Lessons 12-1 through 12-3
Quiz for Lessons 12-4 through 12-6
Chapter 12 Test, Forms A–D
Chapter 12 Test, Cumulative Form

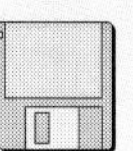

Quiz and Test Writer
Multiple forms of chapter tests and quizzes; Challenges

Activity Kit

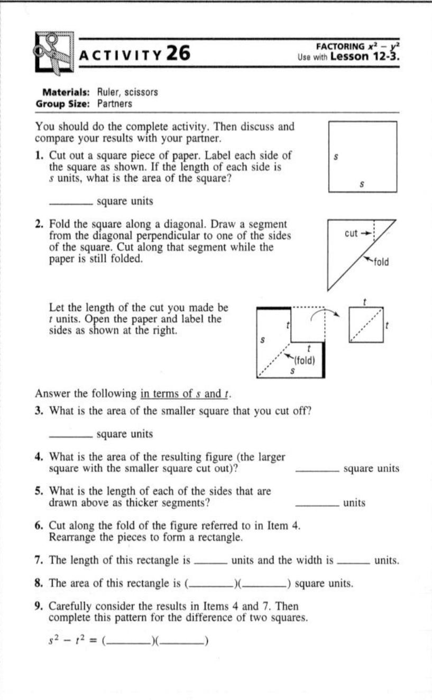

ACTIVITY 26 FACTORING $x^2 - y^2$ Use with **Lesson 12-3.**

Materials: Ruler, scissors
Group Size: Partners

You should do the complete activity. Then discuss and compare your results with your partner.

1. Cut out a square piece of paper. Label each side of the square as shown. If the length of each side is s units, what is the area of the square?

   ______ square units

2. Fold the square along a diagonal. Draw a segment from the diagonal perpendicular to one of the sides of the square. Cut along that segment while the paper is still folded.

   Let the length of the cut you made be t units. Open the paper and label the sides as shown at the right.

Answer the following in terms of s and t.

3. What is the area of the smaller square that you cut off?

   ______ square units

4. What is the area of the resulting figure (the larger square with the smaller square cut out)? ______ square units

5. What is the length of each of the sides that are drawn above as thicker segments? ______ units

6. Cut along the fold of the figure referred to in Item 4. Rearrange the pieces to form a rectangle.

7. The length of this rectangle is ____ units and the width is ____ units.

8. The area of this rectangle is (______)(______) square units.

9. Carefully consider the results in Items 4 and 7. Then complete this pattern for the difference of two squares.

   $s^2 - t^2 =$ (______)(______)

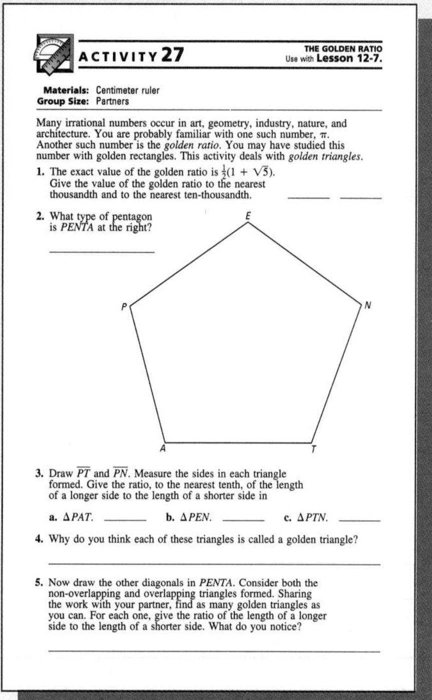

ACTIVITY 27 THE GOLDEN RATIO Use with **Lesson 12-7.**

Materials: Centimeter ruler
Group Size: Partners

Many irrational numbers occur in art, geometry, industry, nature, and architecture. You are probably familiar with one such number, π. Another such number is the *golden ratio*. You may have studied this number with golden rectangles. This activity deals with *golden triangles*.

1. The exact value of the golden ratio is $\frac{1}{2}(1 + \sqrt{5})$. Give the value of the golden ratio to the nearest thousandth and to the nearest ten-thousandth. ______ ______

2. What type of pentagon is *PENTA* at the right?

   ______

3. Draw $\overline{PT}$ and $\overline{PN}$. Measure the sides in each triangle formed. Give the ratio, to the nearest tenth, of the length of a longer side to the length of a shorter side in

   a. ΔPAT. ______ **b.** ΔPEN. ______ **c.** ΔPTN. ______

4. Why do you think each of these triangles is called a golden triangle?

   ______

5. Now draw the other diagonals in *PENTA*. Consider both the non-overlapping and overlapping triangles formed. Sharing the work with your partner, find as many golden triangles as you can. For each one, give the ratio of the length of a longer side to the length of a shorter side. What do you notice?

   ______

Teaching Aids

Teaching Aid 4, Real Numbers, (shown on page 4D) can be used with **Lesson 12-7. Teaching Aid 26, Graph Paper,** (shown on page 140D) can be used with **Lesson 12-2. Teaching Aid 31, Algebra Tiles,** (shown on page 176) can be used with **Lessons 12-2** and **12-3** and the **In-class Activity.**

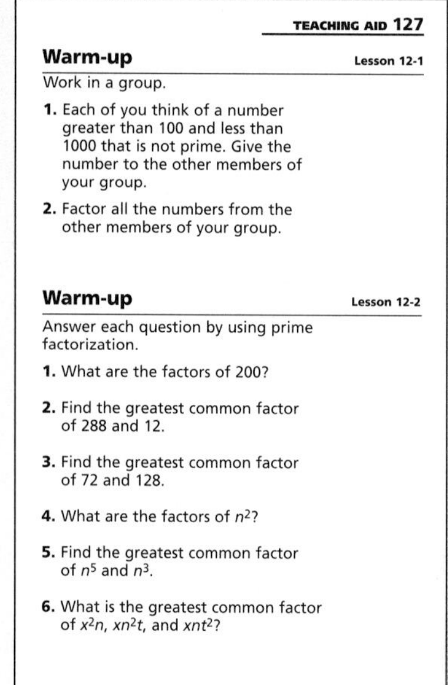

TEACHING AID 127

Warm-up Lesson 12-1

Work in a group.

1. Each of you think of a number greater than 100 and less than 1000 that is not prime. Give the number to the other members of your group.
2. Factor all the numbers from the other members of your group.

Warm-up Lesson 12-2

Answer each question by using prime factorization.

1. What are the factors of 200?
2. Find the greatest common factor of 288 and 12.
3. Find the greatest common factor of 72 and 128.
4. What are the factors of n^2?
5. Find the greatest common factor of n^5 and n^3.
6. What is the greatest common factor of x^2n, xn^2t, and xnt^2?

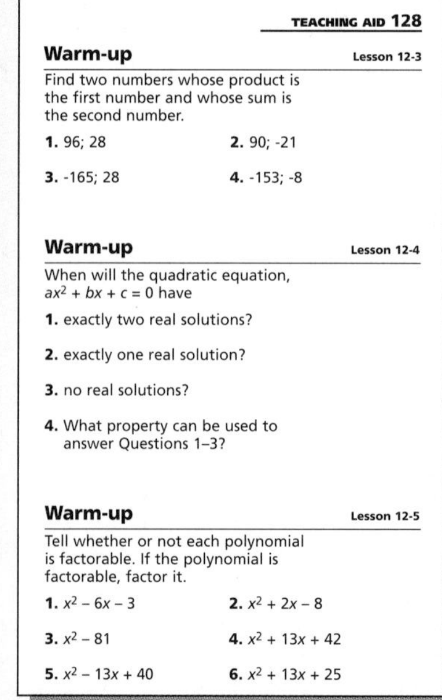

TEACHING AID 128

Warm-up Lesson 12-3

Find two numbers whose product is the first number and whose sum is the second number.

1. 96; 28 **2.** 90; -21
3. -165; 28 **4.** -153; -8

Warm-up Lesson 12-4

When will the quadratic equation, $ax^2 + bx + c = 0$ have

1. exactly two real solutions?
2. exactly one real solution?
3. no real solutions?
4. What property can be used to answer Questions 1–3?

Warm-up Lesson 12-5

Tell whether or not each polynomial is factorable. If the polynomial is factorable, factor it.

1. $x^2 - 6x - 3$ **2.** $x^2 + 2x - 8$
3. $x^2 - 81$ **4.** $x^2 + 13x + 42$
5. $x^2 - 13x + 40$ **6.** $x^2 + 13x + 25$

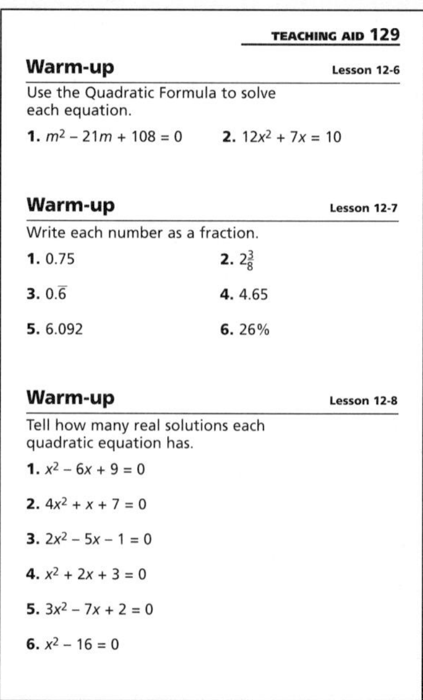

TEACHING AID 129

Warm-up Lesson 12-6

Use the Quadratic Formula to solve each equation.

1. $m^2 - 21m + 108 = 0$ **2.** $12x^2 + 7x = 10$

Warm-up Lesson 12-7

Write each number as a fraction.

1. 0.75 **2.** $2\frac{3}{8}$
3. $0.\overline{6}$ **4.** 4.65
5. 6.092 **6.** 26%

Warm-up Lesson 12-8

Tell how many real solutions each quadratic equation has.

1. $x^2 - 6x + 9 = 0$
2. $4x^2 + x + 7 = 0$
3. $2x^2 - 5x - 1 = 0$
4. $x^2 + 2x + 3 = 0$
5. $3x^2 - 7x + 2 = 0$
6. $x^2 - 16 = 0$

Question 32

Diameter (in.)	Volume (ft^3)	Diameter (in.)	Volume (ft^3)
8.3	10.3	12.9	33.8
8.6	10.3	13.3	27.4
8.8	10.2	13.7	25.7
10.5	16.4	13.8	24.7
10.7	18.8	14.0	34.5
11.0	15.6	14.2	31.7
11.0	18.2	14.5	36.3
11.1	22.6	16.3	42.6
11.2	19.9	17.3	55.4
11.3	24.2	17.5	55.7
11.4	21.0	17.9	58.3
11.4	21.4	18.0	51.0
11.7	21.3	18.0	51.5
12.0	19.1	20.6	77.0
12.9	22.2		

y
80
60
40
20
0
Volume (cubic feet)
8
12
16
20
24
x
Diameter (inches)
4.5 ft above ground level

Chapter 12 Opener

Pacing

All of the lessons in this chapter are designed to be covered in one day. At the end of the chapter, you should plan to spend 1 day to review the Progress Self-Test, 1 to 2 days for the Chapter Review, and 1 day for a test. You may want to spend a day on projects, and a day may be needed for quizzes. Therefore, this chapter should take 11 to 14 days. We recommend that you not spend more than 15 days on the chapter.

Using Pages 718–719

We try to open each chapter with an application or applications of the concepts that are presented in the chapter. However, that is difficult with factoring because factoring is more an *algorithm* by which equations are solved and expressions are simplified than it is a *concept*. For instance, if there is a situation that leads to a quadratic equation, the equation can always be solved by using the Quadratic Formula; only rarely can it be solved by factoring. The application is the quadratic, not the process which solves it. Nonetheless, in recent years, an application of the factoring of very large integers has arisen; that application is shown on page 719.

History Connection The number RSA-129 is named for the initials of its three inventors—Ronald L. Rivest, Adi Shamir, and Leonard Adleman—all professors at MIT. They presented a challenge to factor RSA-129 in the August, 1977, issue of *Scientific American*. At that time, it was predicted that the number might not be factored for 40 quadrillion years! A reward of $100 was offered to the

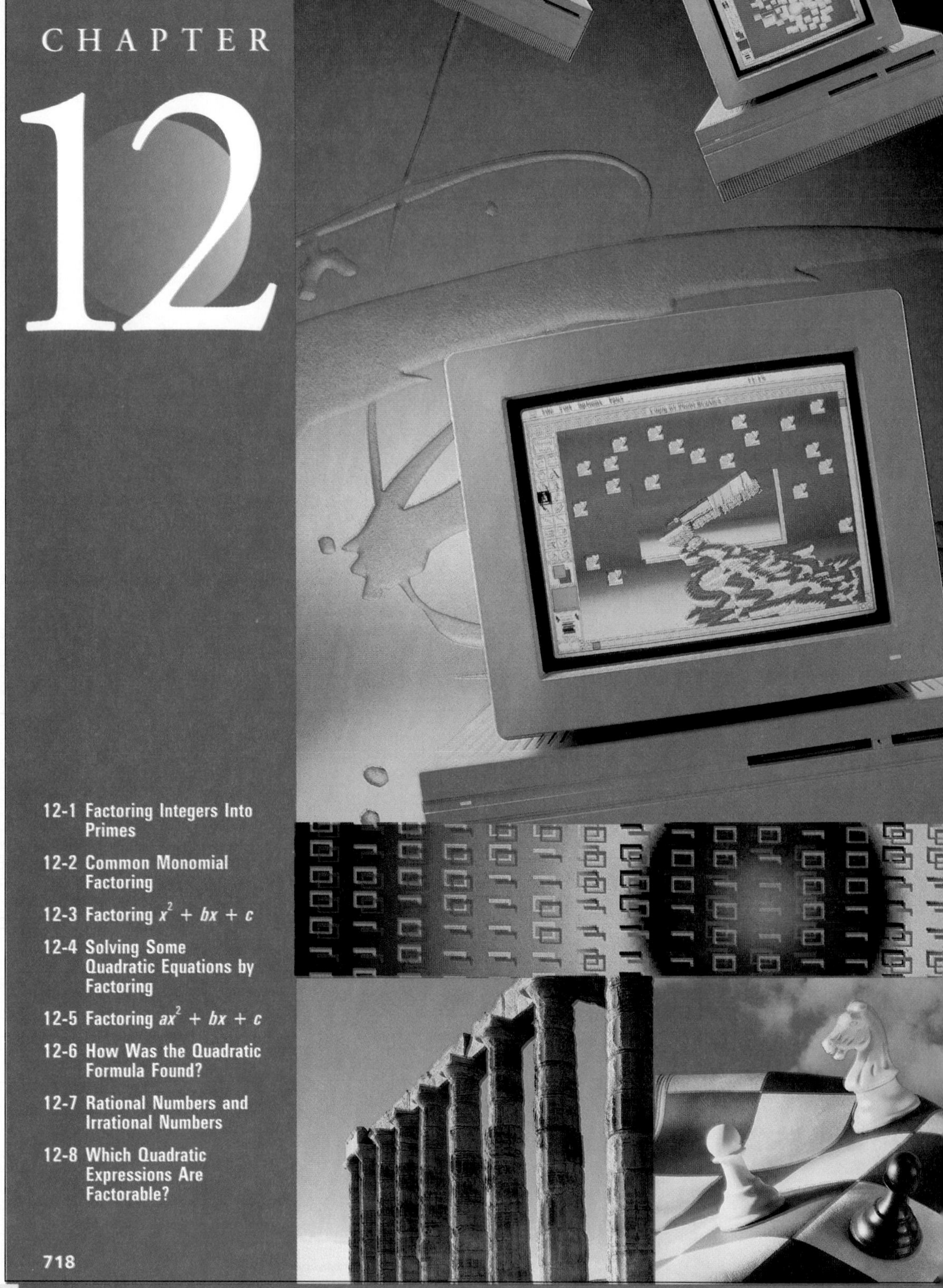

CHAPTER 12

718

Chapter 12 Overview

Two types of factoring provide the backbone for this chapter: the factoring of integers into primes and the factoring of quadratic trinomials. These topics are usually studied separately and years apart; here they are presented together in the analysis of roots to quadratic equations.

The first two lessons present a discussion of the basic types of factoring. Lesson 12-1 covers the factoring of integers into primes and should be a review for students. Lesson 12-2 covers the simplest form of algebraic factoring, the application of the Distributive Property known as common monomial factoring.

Lessons 12-3 through 12-6 deal with the factoring of quadratic trinomials. Lesson 12-3 deals with trinomials in which the coefficient of the quadratic term is 1, while Lesson 12-4, gives the important application of this factoring to the solving of equations. Lesson 12-5 covers trinomials in which the coefficient of the quadratic term is an arbitrary integer. An extra day of practice for this more general type of factoring is given in Lesson 12-6. In this lesson, quadratics are related to systems, and a proof of the Quadratic Formula is presented.

The last part of the chapter discusses rational and irrational numbers. Lesson 12-7

FACTORING

In March, 1994, six hundred people on five continents using 1,600 computers, having worked over a period of 8 months, completed a task that was thought to be too difficult to do in our day. They found the prime factorization of this 129-digit number:

114,381,625,757,888,867,669,235,779,976,146,612,010,218,296, 721,242,362,562,561,842,935,706,935,245,733,897,830,597,123, 563,958,705,058,989,075,147,599,290,026,879,543,541.

The fact that so many people collaborated and used so many computers for so long on a single task indicates that the task was quite important. Numbers of this size are used as codes to protect information that is held within computers. This information might include a company's recipe for making a food, a military secret, private information about individuals, or code numbers for money accounts. For instance, when a person goes to an automatic teller to obtain money, the person needs to enter a code number. Prime numbers are involved in the process by which this code number is checked to ensure that it goes with the correct account. Because prime numbers are involved in the encoding and decoding, ways of factoring large numbers into primes are studied by some mathematicians.

By breaking the code for this number, the 600 people demonstrated that a 129-digit number was not large enough to be used for coding, and that information thought to be protected might no longer be so safe. Larger numbers are needed.

In this chapter, you will review the prime factorization of positive integers and then extend these ideas to factoring polynomials and other algebraic expressions. These factorizations help in factoring large numbers, simplifying fractions, and solving equations.

introduces these numbers and uses prime factorizations to show that certain square roots of integers are irrational. Lesson 12-8 applies this knowledge to show that certain quadratics cannot be factored over the integers.

Historically, factoring was often related to area. Thus, area is a strong representation throughout much of the chapter. Because factoring is the reverse process of multiplication, the Distributive Property plays a major role in all of the algorithms of this chapter. Additionally, the Zero Product Property is introduced as a way to solve equations. The application contexts that were introduced in Chapter 10 are continued here, and a couple of new contexts are introduced. There is also information on the graphing of quadratic equations. Consequently, this chapter provides a rather in-depth review of Chapter 10.

individual or team that could find the factors. (One should not expect to get rich doing this sort of thing!)

The effort to factor the number was led by Arjen K. Lenstra, a scientist at Bell Communications Research. He was assisted by three computer hobbyists and over 600 volunteers working on Internet. The search involved approximately 100 quadrillion computer instructions, and it is considered one of the largest and most difficult single computations ever performed. Internet participants donated time on 1600 computers to create 8.2 million pieces of data. The information was stored at Bellcore, where a supercomputer took two days to churn out the 64-digit and 65-digit factors. A summary of the work was presented in the July, 1994, issue of *Scientific American*.

Photo Connections

The photo collage makes real-world connections to the contents of the chapter: factoring.

Computers: Computer use is not limited to research facilities. In 1981, about 18% of U.S. public elementary and secondary schools had computers available for student use. By 1990, the percentage had risen to about 97%.

Cubes: A composite number is an integer that has more than two factors. If the large cube represents an integer, then the small cubes might represent the idea of factors of the number.

Digital Code: A digital computer represents all data in the form of numbers, usually 1 and 0. The digital code pictured here might represent a portion of the 129-digit number given on this page.

Temple of Poseidon: The geometry that ancient Greeks applied to construction also formed the basis for their exploration of quadratic equations and irrational numbers.

Chess: In Lesson 12-4, a quadratic equation is used to determine the number of players in a chess tournament.

Chapter 12 Projects

At this time you might want to have students look over the projects on pages 765–766.

Lesson 12-1

Objectives

A Factor positive integers into primes.
E Apply the definitions and properties of primes and factors.

Resources

From the ***Teacher's Resource File***
- Lesson Master 12-1A or 12-1B
- Answer Master 12-1
- Teaching Aid 127: Warm-up
- Technology Connection Computer Master 29

Additional Resources
- Visual for Teaching Aid 127

Teaching Lesson 12-1

Warm-up

Diagnostic Work in a group.

1. Each of you think of a number greater than 100 and less than 1000 that is not prime. Give the number to the other members of your group. **Answers will vary. Such composite numbers can be found by multiplying a two-digit number by a one-digit number, or by multiplying two relatively small two-digit numbers.**
2. Factor all the numbers from the other members of your group.

Any number under 1000 can be factored by trying the prime divisors up to 31. Nonetheless, the process should convince students that trying to find such factors by trial and error is a difficult process.

LESSON

12-1 *Factoring Integers into Primes*

Olympic array. *Multiplication helps in calculating the number of band members at the opening ceremonies of the 1984 Olympics in Los Angeles.*

You have probably studied prime numbers in earlier mathematics classes. Here we review some information and perhaps show you some things you have not seen before.

Factors and Multiples

Throughout this lesson, the domain of each variable is the set of positive integers {1, 2, 3, 4, . . . }. When $ab = c$, we say that a and b are **factors** of c. We also say that c is a **multiple** of a, and c is a multiple of b. For instance, since $3 \cdot 15 = 45$,

3 and 15 are factors of 45,
45 is a multiple of 3, and
45 is a multiple of 15.

We also say that 45 **is divisible by** 15. Because for every number n, $n \cdot 1 = n$, every number is a factor of itself and every number is also a multiple of itself.

Suppose two numbers are divisible by the same number. For instance, both 45 and 522 are divisible by 3. We say that 3 is a **common factor** of 45 and 522. The sum 45 + 522, or 567, is also divisible by 3, and so is the difference 522 − 45 = 477. The reason this works for any integers can be deduced by using the Distributive Property.

Suppose c is a multiple of a. Then there is a number m with $c = ma$.
If d is also a multiple of a, then there is a number n with $d = na$.
Add the equations, as you did with systems: $c + d = ma + na$.
But, by the Distributive Property, $ma + na = (m + n)a$.
Thus $c + d = (m + n)a$, and so $c + d$ is a multiple of a.

This argument can be repeated for $c - d$.

Lesson 12-1 Overview

Broad Goals This lesson reviews the prime factorization of numbers and uses prime factorizations to simplify fractions.

Perspective We assume that students are familiar with the definitions of primes and multiples. (They are studied in UCSMP *Transition Mathematics*; students should have seen them in that course or another previous course.) However, two aspects of this lesson may be new to students. The first concept that may be new is the Common Factor Sum Property on page 721. When a number is divisible by a number, this property enables us to generate other numbers that are divisible by the same number. The second concept is the representation of prime numbers by arrays.

Optional Activities

Activity 1 Cooperative Learning Most likely, students will have used divisibility tests like those that are given in the teacher's note for **Question 15.** Have students discuss divisibility tests they know. Give them those rules from the following list that they do not mention.

Common Factor Sum Property
If a is a common factor of b and c, then it is a factor of $b + c$.

As a special case of this property, if a number is divisible by x, then you can repeatedly add x to the number to generate many other numbers that are divisible by x. For instance, 2600 is divisible by 13. Therefore, so is

2613,	2626,	2639,	and so on.
(2600 + 13)	(2600 + 13 + 13)	(2600 + 13 + 13 + 13)	

The same idea works with subtraction. By repeatedly subtracting 13, you can conclude that 2587, 2574, 2561, and so on, are also divisible by 13.

Primes and Composites

Every integer greater than 1 has at least two factors, the number itself and 1. But some integers have only two factors. One such integer is 47. Numbers with this property are the *prime numbers.* A **prime number** is an integer greater than 1 whose only integer factors are itself and 1. Here is a list of the prime numbers less than 50:

2, 3, 5, 7, 11, 13, 17, 19, 23, 29, 31, 37, 41, 43, 47.

Most integers are not prime. They are *composite.* A **composite number** is an integer that has more than two factors. For instance, the number 60 has 12 different factors: 1, 2, 3, 4, 5, 6, 10, 12, 15, 20, 30, and 60. You can use the Area Model for Multiplication to show each factor as the number of dots on a side of a rectangular array containing 60 dots. Here are four of these arrays.

1-by-60 array:

●●

2-by-30 array:

●●●●●●●●●●●●●●●●●●●●●●●●●●●●●●
●●●●●●●●●●●●●●●●●●●●●●●●●●●●●●

3-by-20 array:

●●●●●●●●●●●●●●●●●●●●
●●●●●●●●●●●●●●●●●●●●
●●●●●●●●●●●●●●●●●●●●

4-by-15 array:

●●●●●●●●●●●●●●●
●●●●●●●●●●●●●●●
●●●●●●●●●●●●●●●
●●●●●●●●●●●●●●●

So you can think of a prime number as a number p of dots that cannot be arranged in any array other than a 1-by-p array.

To determine whether or not an integer is prime, you can divide it by the primes less than it. For small numbers, there are not as many divisions needed as you might think.

Notes on Reading

Two main points are emphasized in the reading. The first is that any determination of primeness must consider the domain over which the factoring is done. When we think of factoring integers, we almost always consider factoring *over* the integers. In that domain, 7 is a prime number. However, when factoring over the set of *rational numbers*, 7 is not prime, since $7 = 14 \cdot \frac{1}{2}$. This idea is applied in later lessons when we consider factoring of polynomials because it is common to consider the factoring of polynomials over at least three different domains.

The second point to emphasize is that any factorization is unique. To illustrate this idea, factor a number such as 60 in more than one way.

$60 = 5 \cdot 12 = 5 \cdot 3 \cdot 4 = 5 \cdot 3 \cdot 2 \cdot 2$
$60 = 2 \cdot 30 = 2 \cdot 5 \cdot 6 = 2 \cdot 5 \cdot 2 \cdot 3$
$60 = 6 \cdot 10 = 2 \cdot 3 \cdot 2 \cdot 5$

Except for the order of the factors, the factorization is the same. Thus, there is essentially only one factorization, so we call it *unique*. In order to make all of the answers look the same, we write the factorization in standard form. Then the order is increasing by the size of the factors, and repetitions of factors are indicated with exponents: $60 = 2^2 \cdot 3 \cdot 5$.

You may want to note the *number* of prime factors of a number as preparation for Lesson 12-7. In that lesson, the number of prime factors of a perfect square is employed to prove that certain numbers are irrational. For instance, 60 has 4 prime factors—2, 2, 3, and 5.

A number is divisible by
- 2 if it ends in an even digit.
- 3 if the sum of its digits is divisible by 3.
- 4 if the number formed by its last two digits is divisible by 4.
- 5 if it ends in 5 or 0.
- 8 if the number formed by its last three digits is divisible by 8.
- 9 if the sum of its digits is divisible by 9.
- 11 if, when – and + operations are alternated between its digits, the result is divisible by 11. For example, 8943 is divisible by 11, since $8 - 9 + 4 - 3 = 0$, and 0 is divisible by 11.

A number is divisible by a composite number if and only if it is divisible by each of the powers of the primes in the factorization of the composite.

Have students use these properties to determine all prime numbers between 100 and 200. [101, 103, 107, 109, 113, 127, 131, 137, 139, 149, 151, 157, 163, 167, 173, 179, 181, 191, 193, 197, 199]

Ask if there are more prime numbers between 100 and 200 than there are between 1 and 100. [There are 25 prime numbers between 1 and 100 and 21 between 101 and 200. Generally, primes become less frequent as one considers larger numbers.]

The application of factoring in **Example 3b** to simplify fractions is important. Later in the chapter, this application will also be extended to polynomials (though we do not emphasize it in beginning algebra).

Additional Examples

1. Is 177 a prime number? **No, it is divisible by 3.**
2. Write the prime factorization of 432,432 in standard form. **$2^4 \cdot 3^3 \cdot 7 \cdot 11 \cdot 13$**
3. Use the prime factorizations of 60 and 28 to
   a. write the prime factorization of $60 \cdot 28$. **$(2^2 \cdot 3 \cdot 5) \cdot (2^2 \cdot 7) = 2^4 \cdot 3 \cdot 5 \cdot 7$**
   b. write $\frac{28}{60}$ in lowest terms. **$\frac{28}{60} = \frac{2^2 \cdot 7}{2^2 \cdot 3 \cdot 5} = \frac{7}{3 \cdot 5} = \frac{7}{15}$**

Follow-up for Lesson 12-1

Practice

For more questions on SPUR Objectives, use **Lesson Master 12-1A** (shown on page 723) or **Lesson Master 12-1B** (shown on pages 724–725).

Assessment

Oral Communication Refer to **Example 1.** Ask students to explain why 113 was not divided by 4, 6, 8, 9, or 10 to decide if it was prime. [Students understand that if a number is not divisible by a prime number, it will not be divisible by a multiple of that prime number.]

Example 1

Is 113 a prime number?

Solution

Divide 113 by the primes less than 113. If any quotient is an integer, then 113 has a prime factor and so is not prime.

$$113 \div 2 = 56.5$$
$$113 \div 3 = 37.6\ldots$$
$$113 \div 5 = 22.6$$
$$113 \div 7 = 16.1\ldots$$
$$113 \div 11 = 10.2\ldots$$

We need go no further. For any prime divisor greater than 11, the quotient will be less than the divisor. So it would have been found as a factor earlier. 113 is prime.

Secret codes. *Prime numbers are used to code information. Pictured are PFC Preston Toledo and his cousin, PFC Frank Toledo. During World War II, they transmitted orders in their own Navajo language. This was like a code because the Japanese did not know the language.*

Prime Factorizations

When a number is composite, it can be factored into primes. Here is how this is done. Consider factoring 60. Because 60 is composite, it can be factored into two or more factors, with each factor less than the original number. For instance,

$$60 = 2 \cdot 30.$$

Since 2 is prime, we leave it alone. But 30 is not prime. Pick two factors that multiply to 30 (but don't pick 1 and 30). We pick 5 and 6.

$$60 = 2 \cdot 5 \cdot 6$$

Now 2 and 5 are prime, but 6 is composite. Factor 6.

$$60 = 2 \cdot 5 \cdot 2 \cdot 3$$

This is the *prime factorization* of 60. A **prime factorization** of n is the writing of n as a product of primes. To put the prime factorization in **standard form,** order the primes and use exponents if a prime is repeated.

$$60 = 2^2 \cdot 3 \cdot 5$$

If you started with different factors of 60, would you get a different final result? No. This was proved by the ancient Greeks, who seem to have been the first people to study prime numbers in a systematic way. Every integer has a unique factorization into primes. This statement is so important that it is sometimes called the Fundamental Theorem of Arithmetic.

> **Unique Factorization Theorem**
> Every integer can be represented as a product of primes in exactly one way, disregarding order of the factors.

Activity 2 Technology Connection In *Technology Sourcebook, Computer Master 29,* students factor integers into primes.

Video

Wide World of Mathematics The segment, *Breaking the German Code,* provides footage on how the D-Day invasion was enhanced by the breaking of the German *Enigma* by a team of cryptographers, mathematicians, and others. The segment provides motivation to introduce a lesson on primes. Related questions and an investigation are provided in videodisc stills and in the Video Guide. A CD-ROM activity is also available.

Videodisc Bar Codes

Search Chapter 59

Play

Example 2

Write the prime factorization of 40,768 in standard form.

Solution

Don't be psyched out by the size of the number! Since 2 is obviously a factor, divide by it.

$$40{,}768 = 2 \cdot 20384$$

Keep dividing by 2 until an odd number results.

$$\begin{aligned} 40{,}768 &= 2 \cdot 2 \cdot 10{,}192 \\ &= 2 \cdot 2 \cdot 2 \cdot 5096 \\ &= 2 \cdot 2 \cdot 2 \cdot 2 \cdot 2548 \\ &= 2 \cdot 2 \cdot 2 \cdot 2 \cdot 2 \cdot 1274 \\ &= 2 \cdot 2 \cdot 2 \cdot 2 \cdot 2 \cdot 2 \cdot 637 \end{aligned}$$

To determine whether 637 is prime, divide it by primes larger than 2.
$637 \div 3 = 212.3\ldots$
637 is obviously not divisible by 5. Do you know why?
$637 \div 7 = 91$. So 637 can be factored.

$$40{,}768 = 2 \cdot 2 \cdot 2 \cdot 2 \cdot 2 \cdot 2 \cdot 7 \cdot 91$$

You may know that $91 = 7 \cdot 13$, or you would try 7 again to find this out.

$$40{,}768 = 2 \cdot 2 \cdot 2 \cdot 2 \cdot 2 \cdot 2 \cdot 7 \cdot 7 \cdot 13$$

This is the prime factorization. Now write it in standard form.

$$40{,}768 = 2^6 \cdot 7^2 \cdot 13$$

By factoring two numbers into primes, you may be able to multiply and divide them more easily.

Example 3

Use the prime factorizations of 40,768 and 294 to do the following:

a. Write the prime factorization of $40{,}768 \cdot 294$.

b. Write $\frac{40{,}768}{294}$ in lowest terms.

Solution

a. From Example 2,

$$40{,}768 = 2^6 \cdot 7^2 \cdot 13.$$

We can find, by the process used in Example 2, that

$$294 = 2 \cdot 3 \cdot 7^2.$$

Consequently, by substitution,

$$40{,}768 \cdot 294 = (2^6 \cdot 7^2 \cdot 13) \cdot (2 \cdot 3 \cdot 7^2).$$

Now use the Product of Powers Property to multiply the powers with the same base.

$$40{,}768 \cdot 294 = 2^7 \cdot 3 \cdot 7^4 \cdot 13$$

b. Use the prime factorizations from part **a**.

$$\frac{40{,}768}{294} = \frac{2^6 \cdot 7^2 \cdot 13}{2 \cdot 3 \cdot 7^2}$$

Now use the Quotient of Powers Property.

$$= \frac{2^5 \cdot 13}{3}$$

$$= \frac{416}{3}$$

Is it prime? *As of September, 1994, the largest known prime number was $2^{859433} - 1$. This number is 258,716 digits long! If set in type this size, the digits would fill 52 pages of standard notebook paper.*

Extension

History Connection In 1742, a German mathematician, Christian Goldbach, observed that it seemed that every even integer, with the exception of 2, could be represented as the sum of two prime numbers. For example, $4 = 2 + 2$, $6 = 3 + 3$, $8 = 3 + 5$, and so on. This conjecture, called Goldbach's conjecture, has never been proven. Have students give at least ten examples to support the conjecture. [Possible answers: $10 = 3 + 7$, $12 = 5 + 7$, $14 = 3 + 11$, $16 = 3 + 13$, $18 = 5 + 13$, $20 = 7 + 13$, $22 = 3 + 19$, $24 = 5 + 19$, $26 = 3 + 23$, $28 = 5 + 23$]

Project Update Project 2, *Perfect, Abundant, and Deficient Numbers*, Project 3, *Public Key Cryptography*, and Project 5, *Packing Boxes* on pages 765–766, relate to the content of this lesson.

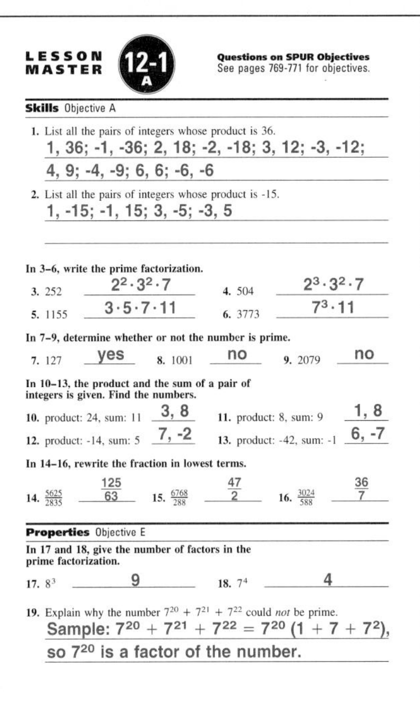

LESSON MASTER 12-1 A — Questions on SPUR Objectives
See pages 769-771 for objectives.

Skills Objective A

1. List all the pairs of integers whose product is 36.
1, 36; -1, -36; 2, 18; -2, -18; 3, 12; -3, -12; 4, 9; -4, -9; 6, 6; -6, -6

2. List all the pairs of integers whose product is -15.
1, -15; -1, 15; 3, -5; -3, 5

In 3–6, write the prime factorization.

3. 252 $2^2 \cdot 3^2 \cdot 7$ 4. 504 $2^3 \cdot 3^2 \cdot 7$
5. 1155 $3 \cdot 5 \cdot 7 \cdot 11$ 6. 3773 $7^3 \cdot 11$

In 7–9, determine whether or not the number is prime.

7. 127 yes 8. 1001 no 9. 2079 no

In 10–13, the product and the sum of a pair of integers is given. Find the numbers.

10. product: 24, sum: 11 3, 8 11. product: 8, sum: 9 1, 8
12. product: -14, sum: 5 7, -2 13. product: -42, sum: -1 6, -7

In 14–16, rewrite the fraction in lowest terms.

14. $\frac{5625}{2835}$ $\frac{125}{63}$ 15. $\frac{6768}{288}$ $\frac{47}{2}$ 16. $\frac{3024}{588}$ $\frac{36}{7}$

Properties Objective E

In 17 and 18, give the number of factors in the prime factorization.

17. 8^3 9 18. 7^4 4

19. Explain why the number $7^{20} + 7^{21} + 7^{22}$ could *not* be prime.
Sample: $7^{20} + 7^{21} + 7^{22} = 7^{20}(1 + 7 + 7^2)$, so 7^{20} is a factor of the number.

Adapting to Individual Needs

Extra Help

When factoring an integer into its prime factors, some students find it helpful to construct a *factor tree* as follows: (1) Write the given number as a product of two factors. (2) Write any factor that is not prime as a product. (3) Continue this process until all the factors are prime. Explain to students that as long as they factor completely into primes, the result will be the same set of prime factors.

Here are two different factor trees for 120.

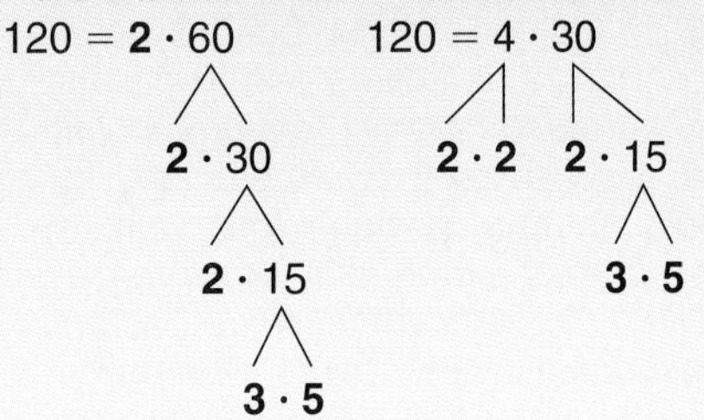

The common result is $120 = 2 \cdot 2 \cdot 2 \cdot 3 \cdot 5$.

Notes on Questions

Question 9 You might discuss the following related property: If *a* is divisible by *c*, and *b* is *not* divisible by *c*, then $a + b$ is *not* divisible by *c*. Applying this property, we know without calculating that $7^3 + 4$ is not divisible by 2, 4, or 7.

Question 15 Students may ask if they are expected to know which numbers between 1 and 100 are prime. You might tell them that while it is not necessary to memorize a list of primes, it is useful to know which numbers are composites. This skill helps in writing fractions in lowest terms. If students know the multiplication tables up to 9 times 9, it is not as hard to remember such a list as they might think. Only odd numbers that end in 1, 3, 7, and 9 can be prime. For numbers that are divisible by 3, the sum of the digits must be a multiple of 3. For numbers divisible by 9, the sum of the digits must be a multiple of 9. Numbers divisible by 11 that are less than 100 have repeated digits. Aside from these numbers, the only other number that needs to be tested is 7. If one knows the multiplication tables, the only composite that remains to be found is $7 \cdot 13$, or 91. Activity 1 in *Optional Activities* on pages 720–721 deals with divisibility rules, including those mentioned above.

LESSON MASTER 12-1 B — Questions on SPUR Objectives

Skills Objective A: Factor positive integers into primes.

In 1–4, list all the pairs of integers whose product is the given integer.

1. 42 1, 42; -1, -42; 2, 21; -2, -21; 3, 14; -3, -14; 6, 7; -6, -7
2. 67 1, 67; -1, -67
3. -32 1, -32; -1, 32; 2, -16; -2, 16; 4, -8; -4, 8
4. -55 1, -55; -1, 55; 5, -11; -5, 11

In 5–7, list all the common factors of the given pair of integers.

5. 12 and 20 1, 2, 4
6. 6 and 27 1, 3
7. 8 and 25 1

In 8–12, write the prime factorization of the given integer in standard form.

8. 72 $2^3 \cdot 3^2$
9. 177 $3 \cdot 59$
10. 540 $2^2 \cdot 3^3 \cdot 5$
11. 4653 $3^2 \cdot 11 \cdot 47$
12. 5775 $3 \cdot 5^2 \cdot 7 \cdot 11$

In 13–18, determine if the given number is prime.

13. 71 yes 14. 91 no
15. 131 yes 16. 1149 no
17. 2001 no 18. 5537 no

19. Show with an array that 30 is a composite number.
Sample:

```
• • • • • •
• • • • • •
• • • • • •
• • • • • •
• • • • • •
```

QUESTIONS

Covering the Reading

1. Because $13 \cdot 17 = 221$, 13 is a __?__ of 221, and 221 is a __?__ of 13. **Factor; multiple**

In 2–5, give four values of the variable that satisfy the sentence.

2. 60 is divisible by *b*. **Sample: 2; 3; 4; 5**
3. *c* is a factor of 35. **1; 5; 7; 35**
4. *d* is a multiple of 17. **Sample: 17; 34; 51; 68**
5. *e* is a common factor of 48 and 60. **1; 2; 4; 12**
6. Explain why 9 is not a prime number. **9 has more than two different factors: 1, 3, and 9.**
7. Determine whether or not 133 is a prime number. **No; because $133 = 7 \cdot 19$**
8. Draw an array to show that 27 is a composite number. **See left.**
9. Without calculating, how do you know that $7^2 + 3 \cdot 7^3$ is divisible by 7? **See left.**
10. Give the prime factorization of 3216 and put it in standard form. **$2^4 \cdot 3 \cdot 67$**

In 11–13, use this information. The prime factorization of 2025 is $3^4 \cdot 5^2$. The prime factorization of 735 is $3 \cdot 5 \cdot 7^2$.

11. What is the prime factorization of $2025 \cdot 735$? **$3^5 \cdot 5^3 \cdot 7^2$**
12. Write $\frac{735}{2025}$ in lowest terms. **$\frac{49}{135}$**
13. Write $\frac{2025}{735}$ in lowest terms. **$\frac{135}{49}$**
14. To what important practical use have prime numbers been put? **encoding and decoding**

Applying the Mathematics

15. List the prime numbers between 50 and 100. **53, 59, 61, 67, 71, 73, 79, 83, 89, 97**

In 16 and 17, find all solutions.

16. *e* is a multiple of 3 and a factor of 300. **3, 6, 12, 15, 30, 60, 75, 150, 300**
17. *f* is a factor of 20 and divisible by 20. **20**
18. Give an argument to show that the following is true for all positive integers *a*, *b*, and *c*: If *a* is a common factor of *b* and *c*, then it is a factor of $b - c$. **See left.**
19. a. Explain why the 13-digit number $2^{40} + 332$ must be divisible by 4.
 b. Is this number divisible by 8? Why or why not? **No; 332 is not divisible by 8.**

8)

```
• • • • • • • • •
• • • • • • • • •
• • • • • • • • •
```

9) 7 is a factor of both 7^2 and $3 \cdot 7^3$, so by the Common Factor Sum Property, it is a factor of $7^2 + 3 \cdot 7^3$.

18) Sample: If *a* is a factor of *b*, then there is a number *m* with $am = b$. If *a* is a factor of *c*, then there is a number *n* with $an = c$. Subtracting equations: $am - an = b - c$. By the Distributive Property, $am - an = a(m - n)$. Thus, $a(m - n) = b - c$, and so *a* is a factor of $b - c$.

19a) Because $2^{40} = 4 \cdot 2^{38}$ and $332 = 4 \cdot 83$, then 4 is a common factor of 2^{40} and 332. Using the Common Factor Sum Property, we know 4 is also a factor of $2^{40} + 332$.

Adapting to Individual Needs

English Language Development
This lesson includes many vocabulary words that might be new to non-English-speaking students: *factor, common factor, multiple, is divisible by, prime number, composite number, prime factorization, standard form of a factorization, Common Factor Sum Property, and Unique Factorization Theorem.* Have students write the words or phrases on one side of an index card and the definition and examples on the other side. Encourage them to use the index cards to study vocabulary and to bring the cards to the next class session to use for reference.

Review

20. Draw a picture of the following multiplication using tiles.
$2x(3x + 5) = 6x^2 + 10x$. *(Lesson 10-3)* **See left.**

21. Find all values of d that satisfy $(d^2)^2 - 25d^2 + 144 = 0$.
(Lessons 5-9, 9-5) **$d = 3$; -3; 4; -4**

22. The orbits of Venus and Earth around the Sun are not circles but are reasonably close to circles. Venus averages a distance of about 108 million kilometers from the Sun. The Earth averages about 150 million kilometers from the Sun. Venus takes 225 days to go around the Sun. Which planet travels farther in an Earth year? *(Previous course, Lesson 6-8)* **Venus**

23. The Cleveland Indians won 12 of its first 19 games in 1994. At this rate, how many games would you expect them to win in a 162-game season? *(Lesson 6-8)* **about 102 games**

24. A student has scored 75%, 80%, and 90% on three exams and has to take one more exam.
  a. What are the highest and lowest mean percents the student can have on the four exams? **highest: 86.25%; lowest: 61.25%**
  b. What are the highest and lowest median percents the student can have? **highest: 85%; lowest: 77.5%**
  c. What are the highest and lowest mode percents if there is to be a single mode? *(Previous course)* **highest: 90%; lowest: 75%**

25. On page 719, mention was made of a 129-digit number. Which of these numbers has 129 digits? *(Appendix B)* **a**
(a) $2 \cdot 10^{128}$ (b) $2 \cdot 10^{129}$ (c) $2 \cdot 10^{130}$ (d) $(2 \cdot 10)^{64}$

Play ball! *The Cleveland Indians began the 1994 baseball season in a new ballpark,* Jacobs Field. *The Indians won that game for their first opening day victory in five years.*

20)

	x	x
x	x^2	x^2
x	x^2	x^2
x	x^2	x^2
1	x	x
1	x	x
1	x	x
1	x	x
1	x	x

Exploration

26. A statement proved by Pierre Fermat in 1675, known as Fermat's Little Theorem, is that, if p is any prime number and a is any positive integer, then $a^p - a$ is divisible by p. Verify this theorem for five different pairs of values of a and p larger than 2.
Sample:
$a = 3$; $p = 3$; $a^p - a = 3^3 - 3 = 27 - 3 = 24 = 3 \cdot 8$
$a = 3$; $p = 5$; $a^p - a = 3^5 - 3 = 243 - 3 = 240 = 5 \cdot 48$
$a = 4$; $p = 5$; $a^p - a = 4^5 - 4 = 1020 = 5 \cdot 204$
$a = 4$; $p = 7$; $a^p - a = 4^7 - 4 = 16380 = 7 \cdot 2340$
$a = 5$; $p = 13$; $a^p - a = 5^{13} - 5 = 1220703120 = 13 \cdot 93900240$

Question 18 To answer this question, students are expected to imitate the argument given in this lesson.

Question 22 There is no simple formula for the perimeter of an ellipse; even if more data about the orbits were given, some sort of approximation would be needed.

Question 23 History Connection The game of baseball most likely developed from an old English sport called *rounders*. Rounders involved hitting a ball with a bat and advancing around bases. In rounders, fielders threw the ball at the runners; if the ball hit a runner who was not on a base, the runner was out. In the 1700s, American colonists played a game similar to rounders, but the rules varied according to local customs. Gradually, this game evolved into baseball. One of the key changes was replacing the practice of hitting runners with the ball with the present practice of tagging them.

You might mention that the 1994 season was cut short by a players' strike. Cleveland ended the season with a 66-47 record, 1 game behind the first-place Chicago White Sox. At that rate one could expect Cleveland to win about 95 games in the 162-game season.

► **LESSON MASTER 12-1 B** *page 2*

In 20–23, the product and sum of a pair of integers is given. Find the numbers.

20. product = 12, sum = 8 **2 and 6**
21. product = 36, sum = 37 **1 and 36**
22. product = -16, sum = 6 **8 and -2**
23. product = -24, sum = -2 **-6 and 4**

In 24–27, rewrite the fraction in lowest terms.

24. $\frac{168}{196}$ $\frac{6}{7}$
25. $\frac{484}{1331}$ $\frac{4}{11}$
26. $\frac{3528}{140}$ $\frac{126}{5}$
27. $\frac{4455}{189}$ $\frac{165}{7}$

Properties Objective E: Apply the definitions and properties of primes and factors.

In 28–30, give the number of factors in the prime factorization.

28. 5^3 **3** **29.** 27^4 **12** **30.** $2^3 \cdot 25^2$ **7**

31. Explain why the number $11^4 + 11^{15} + 11^{23}$ could not be prime.
Sample: $11^4 + 11^{15} + 11^{23} = 11^4(1 + 11^{11} + 11^{19})$, so 11^4 is a factor of the number.

Review Objective B, Lesson 8-5

In 32–39, simplify.

32. $y^2 \cdot y^5$ y^7
33. $x^7 \cdot x^7$ x^{14}
34. $(m^5)^3$ m^{15}
35. $(e^5)^5$ e^{25}
36. $a^2 \cdot a^4 \cdot a^4$ a^{10}
37. $c \cdot c^5$ c^6
38. $(m^2 \cdot m^6)^3$ m^{24}
39. $x^4 \cdot y \cdot x^3 \cdot y^3$ x^7y^4

Adapting to Individual Needs

Challenge

Have students solve the following problem: A number such as 2992, which reads the same both forward and backward, is called a *palindrome.* A certain prime number is a factor of every 4-digit palindrome. What is this prime number? [11]

Setting Up Lesson 12-2

Materials Student will need algebra tiles for Lesson 12-2. You may have a supply from earlier lessons, or you can have students use **Teaching Aid 31.**

Lesson 12-2

Objectives

B Find common monomial factors of polynomials.

J Represent quadratic expressions and their factorizations with area.

Resources

From the ***Teacher's Resource File***

- Lesson Master 12-2A or 12-2B
- Answer Master 12-2
- Teaching Aids
  - 26 Graph Paper
  - 31 Algebra Tiles
  - 127 Warm-up

Additional Resources

- Visuals for Teaching Aids 26, 31, 127

Teaching Lesson 12-2

Warm-up

Answer each question by using prime factorization.

1. What are the factors of 200? 1, 2, 4, 5, 8, 10, 20, 25, 40, 50, 100, 200
2. Find the greatest common factor of 288 and 12. 12
3. Find the greatest common factor of 72 and 128. 8
4. What are the factors of n^2? 1, n, n^2
5. Find the greatest common factor of n^5 and n^3. n^3
6. What is the greatest common factor of x^2n, xn^2t, and xnt^2? xn

LESSON

12-2 Common Monomial Factoring

A tight squeeze. *There are three ways to pack books into a box where all are facing the same way: flat, on end, and on their sides. The books will fit snugly in all three ways only if all the dimensions of the book are common factors of the dimensions of the box.*

When two or more numbers are multiplied, the result is a single number. Factoring is the reverse process. In factoring, one begins with a single number and expresses it as the product of two or more numbers. For instance, the product of 7 and 4 is 28. So, when we factor 28, we get $28 = 7 \cdot 4$. In Lesson 12–1, you used the process of factoring to obtain the prime factorizations of integers. Now we turn our attention to polynomials.

Example 1 shows how to find factors of a monomial.

1 **Example 1**

What are the factors of $25x^3$?

Solution

The factors of 25 are 1, 5, and 25. The factors of x^3 are 1, x, x^2, and x^3. The factors of $25x^3$ are the 12 possible products formed from those factors:

$$1,\ 5,\ 25,\ x,\ 5x,\ 25x,\ x^2,\ 5x^2,\ 25x^2,\ x^3,\ 5x^3,\ 25x^3.$$

The **greatest common factor** (GCF) of two or more monomials is found by multiplying the greatest common factor of their coefficients by the greatest common factor of their variables.

Example 2

Find the greatest common factor of $24x^2y$ and $6x$.

Solution

The GCF of 24 and 6 is 6. The GCF of x^2 and x is x. Since the factor y does not appear in all terms, y does not appear in the GCF. So, The GCF of $24x^2y$ and $6x$ is $6 \cdot x$, which is $6x$.

Lesson 12-2 Overview

Broad Goals This lesson introduces the idea of factoring a polynomial and concentrates on common monomial factors and their application to the division of a polynomial by a monomial.

Perspective Factoring is the reverse process of multiplication. Specifically, in the Distributive Property $a(b + c) = ab + ac$, the process from left to right is multiplication, and the process from the right to left is factoring. In other words, common monomial factoring is the reverse of multiplying a number by a sum or difference.

Since the factoring of integers into primes over the set of integers is unique, we would like the factorizations of polynomials into prime polynomials over the integers also to be unique. However, real numbers provide a problem. For instance, consider the expression $12\pi r^2 + 12\pi rh$, which gives the total surface area of 6 cylinders, each with radius r and height h. It can be rewritten as $12\pi r(r + h)$ or as $2\pi r(6r + 6h)$. Here, since π is not an integer, we cannot think of its prime factorization over the integers. This is why the Unique Factorization Theorem for Polynomials allows the disregarding of real-number integer multiples of its factors.

In this book, all the polynomials we factor will have integer coefficients. These are called **polynomials over the integers.** Unless you are told otherwise, you should factor over the integers.

As with integers, the result of factoring a polynomial is called a **factorization.** Here is a factorization of $8x^2 + 12x$.

$$8x^2 + 12x = 2x(4x + 6)$$

Again, as with integers, a factorization with two factors means that a rectangular figure can be formed with the factors as its dimensions. Here is a picture of the above factorization.

❷

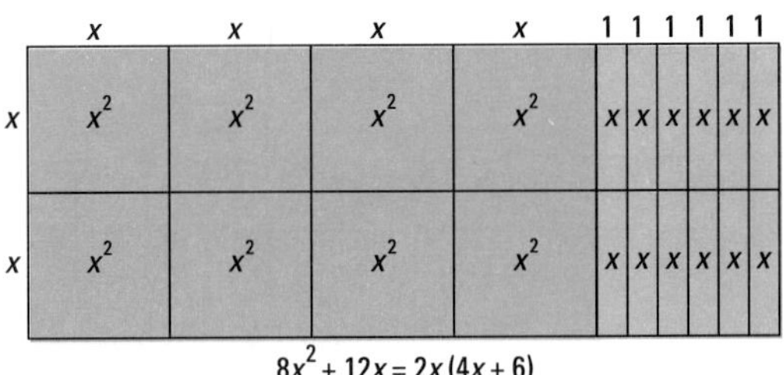

$8x^2 + 12x = 2x(4x + 6)$

Because the terms in $4x + 6$ have the common factor 2, another factorization of $8x^2 + 12x$ is possible.

$$\begin{aligned} 8x^2 + 12x &= 2x(4x + 6) \\ &= 2x \cdot 2(2x + 3) \\ &= 4x(2x + 3) \end{aligned}$$

Monomials such as $4x$, and binomials such as $2x + 3$ which cannot be factored into polynomials of a lower degree are called **prime polynomials.** To factor a polynomial completely means to factor it into prime polynomials. When there are no common numerical factors in the terms of any of the prime polynomials, the result is called a **complete factorization.** The complete factorization of $8x^2 + 12x$ is $4x(2x + 3)$.

The complete factorization of a polynomial has a uniqueness property much like that of the prime factorization of an integer.

Unique Factorization Theorem for Polynomials
The complete factorization of a polynomial is unique, disregarding the order of the factors and multiplication by -1.

Notes on Reading

❶ Note the similarity between **Examples 1 and 2** and the answers to Questions 1 and 2 in the *Warm-up.* (Substitute 2 for x in **Example 1** to get Question 1; substitute 2 for x and 3 for y in **Example 2** to get Question 2.)

❷ Point out that there are three factorizations of the number 12 into two integer factors: $1 \cdot 12$, $2 \cdot 6$, and $3 \cdot 4$. Thus, a rectangle with area 12 can be drawn with three different pairs of integer dimensions. Similarly, when there is more than one way to factor a polynomial, it can be pictured in more than one way. Ask students for another way to draw an area model for $8x^2 + 12x$. [Samples: $x(8x + 12)$—a rectangle x units wide and $8x + 12$ units long; $4x(2x + 3)$—a rectangle $4x$ units wide and $2x + 3$ units long.]

Optional Activities

You might use this activity after discussing the meaning of greatest common factor. For an expression like $145x + 696$, it is hard to quickly determine if 145 and 696 have a common factor greater than 1. A process called the *Euclidean algorithm* can be used to determine the greatest common factor of two numbers.

1. Divide the larger number (696) by the smaller number (145). $\quad 696 \div 145 = 4\text{R}116$
2. Divide the previous divisor (145) by the remainder (116). $\quad 145 \div 116 = 1\text{R}29$
3. Repeat step 2 until the remainder is 0. $\quad 116 \div 29 = 4\text{R}0$
4. When the remainder is 0, the greatest common factor is the divisor in that step. The greatest common factor of 145 and 696 is 29; $145x + 696 = 29(5x + 24)$.

Have students use the Euclidean algorithm to factor each expression.

$51y^2 + 272y$	$[17(3y + 16)]$
$65x^2 + 559$	$[13(5x^2 + 43)]$
$203x - 435$	$[29(7x - 15)]$
$483y^3 - 736y^2$	$[23y^2(21y - 32)]$

❸ In the examples and exercises, the complete factoring is accomplished by searching for the *greatest* common factor. However, any problem of this type can be done by simply finding *some* common factor and factoring until all the factors have been found. For instance, with $-30x^3y + 20x^4 + 50x^2y^5$ in **Example 5,** a student could first factor out x, then factor out 10, and finally factor out x again.

❹ If $xy = z$, then x and y are factors of z. When z is divided by either of the factors, the result is the other factor. This shows that factoring and division are closely related. For instance, in **Example 6,** since $5n^2 + 3n = n(5n + 3)$, we have either $\frac{5n^2 + 3n}{5n + 3} = n$ or $\frac{5n^2 + 3n}{n} = 5n + 3$.

Example 3

Factor $24x^2y + 6x$ completely.

Solution

From Example 2, we know that *6x is the greatest common factor of $24x^2y$ and of 6x.* So it is a factor of $24x^2y + 6x$. Now find the binomial which when multiplied by $6x$ gives $24x^2y + 6x$.

$$24x^2y + 6x = 6x(\underline{\ ?\ } + \underline{\ ?\ })$$

Divide each term by $6x$ to fill in the factors.

$$\frac{24x^2y}{6x} = 4xy \qquad \frac{6x}{6x} = 1$$

$$24x^2y + 6x = 6x(4xy + 1)$$

Check

Test a special case. Let $x = 3$, $y = 4$. Follow the order of operations.
$24x^2y + 6x = 24 \cdot 3^2 \cdot 4 + 6 \cdot 3 = 864 + 18 = 882.$
$6x(4xy + 1) = 6 \cdot 3(4 \cdot 3 \cdot 4 + 1) = 18 \cdot 49 = 882.$ It checks.

Three or more polynomials can also have common factors.

Example 4

Find the greatest common factor of $-30x^3y$, $20x^4$, and $50x^2y^5$.

Solution

The greatest common factor of -30, 20, and 50 is 10. The GCF of x^3, x^4, and x^2 is x^2. Since the variable y does not appear in all terms, y does not appear in the GCF. *The GCF of $-30x^3y$, $20x^4$, and $50x^2y^5$ is $10x^2$.*

Notice the pattern from Examples 3 and 4. The GCF of two or more terms includes the GCF of the coefficients of the terms. It also includes any common variable raised to the *lowest* exponent of that variable found in the terms.

❸ **Example 5**

Factor $-30x^3y + 20x^4 + 50x^2y^5$ completely.

Solution

In Example 4, we found that the GCF of the three terms of this polynomial is $10x^2$. Thus

$$-30x^3y + 20x^4 + 50x^2y^5 = 10x^2(\underline{\ ?\ } + \underline{\ ?\ } + \underline{\ ?\ }).$$

Now divide to find the terms in parentheses. Experts do these steps in their heads.

$$\frac{-30x^3y}{10x^2} = -3xy \qquad \frac{20x^4}{10x^2} = 2x^2 \qquad \frac{50x^2y^5}{10x^2} = 5y^5$$

So $\quad -30x^3y + 20x^4 + 50x^2y^5 = 10x^2(-3xy + 2x^2 + 5y^5).$

Adapting to Individual Needs

Extra Help

Sometimes students are confused by the fact that a monomial such as $20x^3$ can be factored in more than one way depending on the situation. For example, suppose students are asked to factor $20x^3 + 8x^2$, as well as $20x^3 + 10x$. In the first polynomial, $20x^3$ is factored as $4x^2 \cdot 5x$ because $4x^2$ is the greatest common factor of $20x^3$ and $8x^2$. In the second polynomial, $20x^3$ is factored as $10x \cdot x^2$ because $10x$ is the greatest common factor of $20x^3$ and $10x$.

$$20x^3 + 8x^2 = 4x^2(5x + 2)$$
$$20x^3 + 10x = 10x(2x^2 + 1)$$

Point out that 20 can be factored into prime factors in only one way—as $2 \cdot 2 \cdot 5$. But when $20x^3$ is added to another monomial (as in the two examples above), we look for a common factor in each of the monomials.

Factoring provides a way of simplifying some fractions.

❹ **Example 6**

Simplify $\frac{5n^2 + 3n}{n}$. (Assume $n \neq 0$.)

Solution 1

Factor the numerator and simplify the fraction.

$$\frac{5n^2 + 3n}{n} = \frac{n(5n + 3)}{n}$$
$$= 5n + 3$$

Solution 2

Separate the given expression into the sum of two fractions and divide.

$$\frac{5n^2 + 3n}{n} = \frac{5n^2}{n} + \frac{3n}{n}$$
$$= 5n + 3$$

QUESTIONS

Covering the Reading

1. List all the factors of $14x^4$. $1, 2, 7, 14, x, x^2, x^3, x^4, 2x, 7x, 14x, 2x^2, 7x^2, 14x^2, 2x^3, 7x^3, 14x^3, 2x^4, 7x^4, 14x^4$
2. What does the abbreviation GCF represent? **greatest common factor**

In 3 and 4, find the GCF.

3. $15x^3$ and $10x^2$ $5x^2$
4. $30ab^2$ and $24a^2$ $6a$
5. Picture the factorization $8x^2 + 12x = 4x(2x + 3)$ with algebra tiles. See left.
6. a. Factor $3x^2 + 6x$ by first finding the greatest common factor of the terms. GCF is $3x$; $3x(x + 2)$
   b. Illustrate the factorization by drawing a rectangle whose sides are the factors. See left.
7. Tell whether or not the polynomial is prime.
   a. $9x^3 + 12x^2$ No b. $9x^2 + 12x$ No c. $9x + 12$ Yes d. $9 + 12x$ Yes
8. In parts a–c, complete the products.
   a. $20x^3 + 10x^2 = 5(\underline{?} + \underline{?})$ $4x^3$; $2x^2$
   b. $20x^3 + 10x^2 = 10(\underline{?} + \underline{?})$ $2x^3$; x^2
   c. $20x^3 + 10x^2 = 10x(\underline{?} + \underline{?})$ $2x^2$; x
   d. Are any of the products in parts a–c a complete factorization of $20x^3 + 10x^2$? No

In 9–11, copy and fill in the blanks.

9. $8x^3 + 40x = 8x(\underline{?} + \underline{?})$ x^2; 5
10. $4p^2 - 3p = p(\underline{?} - \underline{?})$ $4p$; 3
11. $8a^2b + 4ab^2 = 4ab(\underline{?} + \underline{?})$ $2a$; b

5)

	x	x	1	1	1
x	x^2	x^2	x	x	x
x	x^2	x^2	x	x	x
x	x^2	x^2	x	x	x
x	x^2	x^2	x	x	x

6b)

	x	1	1
x	x^2	x	x
x	x^2	x	x
x	x^2	x	x

Additional Examples

1. What are the factors of $18m^2$?
   $1, 2, 3, 6, 9, 18, m, 2m, 3m, 6m, 9m, 18m, m^2, 2m^2, 3m^2, 6m^2, 9m^2, 18m^2$
2. Find the greatest common factor of $16a^5m$, $-12a^3m^3$, and $4a^2m^5$.
   $4a^2m$
3. Factor each expression.
   a. $18y^3 - 6y^2 - 15y$
   $3y(6y^2 - 2y - 5)$
   b. $28c^2d - 14d^5$ **$14d(2c^2 - d^4)$**
4. Simplify $\frac{-10z^2 + 5z}{5z}$. **$-2z + 1$**
5. Find the greatest common factor of $12a^2b^3$, $6ab^4$, and $24ab^5$.
   $6ab^3$
6. Factor $12a^2b^3 + 6ab^4 - 24ab^5$.
   $6ab^3(2a + b - 4b^2)$

Notes on Questions

Questions 5, 6, and 19 In these questions, students can use graph paper (**Teaching Aid 26**) or algebra tiles (**Teaching Aid 31**) to picture factorization.

Adapting to Individual Needs

English Language Development
There are several terms in this lesson that you might want to emphasize for non-English-speaking students: *monomial, polynomial, greatest common factor, prime polynomials, complete factorization*, and *unique factorization*. Since students have encountered several of the individual words in these phrases before, you might begin by reviewing the meaning of each word. Then define the phrase using examples. Encourage students to add these words and phrases to their index cards.

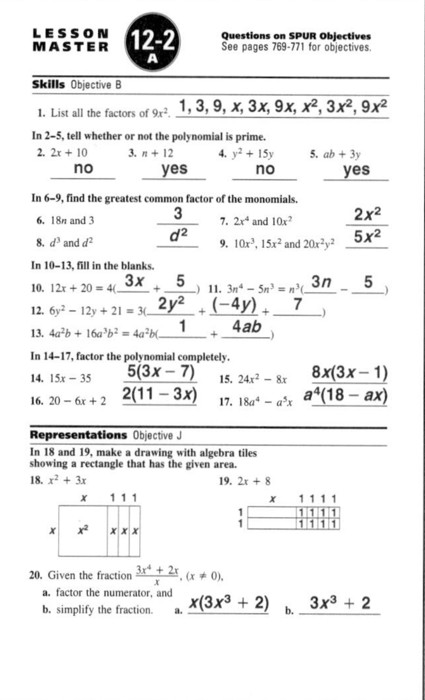

LESSON MASTER 12-2 A

Questions on SPUR Objectives
See pages 769-771 for objectives.

Skills Objective B

1. List all the factors of $9x^2$. $1, 3, 9, x, 3x, 9x, x^2, 3x^2, 9x^2$

In 2–5, tell whether or not the polynomial is prime.

2. $2x + 10$ no
3. $n + 12$ yes
4. $y^2 + 15y$ no
5. $ab + 3y$ yes

In 6–9, find the greatest common factor of the monomials.

6. $18n$ and 3 3
7. $2x^4$ and $10x^2$ $2x^2$
8. d^3 and d^2 d^2
9. $10x^3$, $15x^2$ and $20x^2y^2$ $5x^2$

In 10–13, fill in the blanks.

10. $12x + 20 = 4(\underline{3x} + \underline{5})$
11. $3n^4 - 5n^3 = n^3(\underline{3n} - \underline{5})$
12. $6y^2 - 12y + 21 = 3(\underline{2y^2} + \underline{(-4y)} + \underline{7})$
13. $4a^2b + 16a^3b^2 = 4a^2b(\underline{1} + \underline{4ab})$

In 14–17, factor the polynomial completely.

14. $15x - 35$ $5(3x - 7)$
15. $24x^2 - 8x$ $8x(3x - 1)$
16. $20 - 6x + 2$ $2(11 - 3x)$
17. $18a^4 - a^5x$ $a^4(18 - ax)$

Representations Objective J

In 18 and 19, make a drawing with algebra tiles showing a rectangle that has the given area.

18. $x^2 + 3x$

	x	1	1	1
x	x^2	x	x	x

19. $2x + 8$

	x	1	1	1	1
1		1	1	1	1
1		1	1	1	1

20. Given the fraction $\frac{3x^4 + 2x}{x}$, $(x \neq 0)$,
   a. factor the numerator, and
   b. simplify the fraction.
   a. $x(3x^3 + 2)$ b. $3x^3 + 2$

Notes on Questions

Question 15 Error Alert Some students may not know how to factor $23a^3 + a^2$ because they fail to realize that $a^2 = 1a^2$. Remind students that after they find the GCF, they should divide each term of the polynomial by it.

Question 23 Consumer Connection Cans made from steel covered with a thin coat of tin are used to package a variety of food products while a large share of aluminum alloy goes into the packaging industry for use as beverage cans. Beverage cans are made from wide sheets of aluminum alloy from which circular discs are die stamped into cylindrical molds. After the cans are sterilized and the outsides labeled, the cans are filled with a beverage, carbon dioxide is added, and circular lids with ring tops are stamped on and sealed.

Question 31 Depending on how fast and powerful your computer or calculator is, the program might run for 5 to 20 minutes. Here is a way of rewriting the expression in line 20 that uses fewer operations and shortens the computing time.

((50 * X + 50) * X + 50) * X + 50

LESSON MASTER 12-2 B — Questions on SPUR Objectives

Skills Objective B: Find common monomial factors of polynomials.

In 1 and 2, list all the factors of the given monomial.

1. $5x^2$ — $1, 5, x, x^2, 5x, 5x^2$
2. $49ax^2$ — $1, 7, 49, a, x, ax, x^2, ax^2, 7a, 7x, 7ax, 7x^2, 7ax^2, 49a, 49x, 49ax, 49x^2, 49ax^2$

In 3–8, tell if the polynomial is prime.

3. $4x + 8$	no	4. $3n + 10$	yes
5. $x^2 + 7x$	no	6. $pq + 2y$	yes
7. $2m + 3mn$	no	8. $2x^2y + bx^2$	no

In 9–14, find the greatest common factor of the given monomials.

9. $20x$ and 5	5	10. $12x^2$ and $18x^3$	$6x^2$
11. y^5 and y^3	y^3	12. $5t$ and 32	1
13. $5x^3$, $12x^2$, and $20x^2y$	x^2	14. $15ab^2$, $21a^2b^2$, and $6a^2b$	$3ab$

In 15–19, fill in the blanks.

15. $16x + 24 = 8(\underline{2x} + \underline{3})$
16. $7b^5 - 12b^2 = b^2(\underline{7b^3} - \underline{12})$
17. $20uv^2 + 28u^2v = 4uv(\underline{5v} + \underline{7u})$
18. $12x^4 + 18x^3 - 30x^2 = 6x^2(\underline{2x^2} + \underline{3x} - \underline{5})$
19. $3m^2n + 6m^2n^2 = \underline{3m^2n}(1 + 2n)$

12. Simplify $\frac{12n^2 + 15n}{3n}$. $4n + 5$

13. Find the greatest common factor of $6x^2y^2$, $-9xy^3$, and $12x^2y^4$. $3xy^2$

In 14–17, factor the polynomial completely.

14. $27b^3 - 27c^3 + 27bc$
15. $23a^3 + a^2$ $a^2(23a + 1)$
16. $12a + 16a^2$ $4a(3 + 4a)$
17. $14p^3 - 21p^2q$ $7p^2(2p - 3q)$

14) $27(b^3 - c^3 + bc)$

Applying the Mathematics

Drums of steel. *Steel drums, originating in the 20th century in Trinidad, West Indies, are made from the unstoppered end and part of the wall of a metal shipping drum.*

18. The area of a rectangle is $100y^3 - 55y^2 + 30y$. One dimension is $5y$. What is the other dimension? $20y^2 - 11y + 6$

19. Use algebra tiles to draw two different rectangles each with area equal to $12x^2 + 8x$. See margin for samples.

20. a. Graph $y = 2x^2 - 6x$. See left.
   b. Find an equation of the form $y = (x - 3) \cdot \underline{\ ?\ }$ which has the same graph as the one in part **a.** $y = (x - 3) \cdot 2x$

In 21 and 22, simplify.

21. $\frac{-3x^2y + 6xy - 9xy^2}{3xy}$ $-x + 2 - 3y$

22. $\frac{n^{14} - 3n^{10} + 5n^6 - n^2}{n^2}$ $n^{12} - 3n^8 + 5n^4 - 1$

23. The circular top of a drum is cut out from a square piece of metal, as pictured at the right.

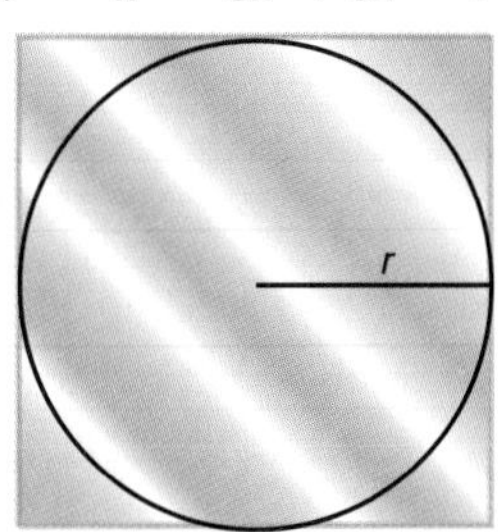

   a. What is the area of the square? $4r^2$
   b. What is the area of the circle? πr^2
   c. How much metal is not used? $(4 - \pi)r^2$
   d. If a dozen drum tops are needed, how much metal will not be used?
   e. If the unused metal in part **d** could be recycled, how many complete drum tops could be made? 3

d) $(48 - 12\pi)r^2$

Review

20a)

24. a. Give the prime factorization of 1001. $7 \cdot 11 \cdot 13$
   b. Give the prime factorization of 91. $7 \cdot 13$
   c. Give the prime factorization of $1001 \cdot 91$. $7^2 \cdot 11 \cdot 13^2$
   d. Write $\frac{91}{1001}$ in lowest terms. *(Lesson 12-1)* $\frac{1}{11}$

25. a. Multiply $(x + 14)(x - 11)$. $x^2 + 3x - 154$
   b. Solve: $(x + 14)(x - 11) = 0$. *(Lessons 9-5, 10-5)* $x = 11$ or $x = -14$

26. A person is on a supervised diet. On the third day of the diet, the person weighed 93.4 kg. On the 10th day, the person weighed 91.2 kg.
   a. What was the average rate of change in weight per day?
   b. Express the rate in part **a** in lb/week. *(Lessons 2-4, 7-1)*

a) ≈ −.31 kg per day b) ≈ −4.84 lb/wk

Adapting to Individual Needs

Challenge

Writing Explain that one way to find the greatest common factor of two or more numbers is to use their prime factorizations. Have students write a paragraph explaining how this can be done. Tell them to use an example in their explanations. [Sample: Write the prime factorization of each number. The greatest common factor of two or more numbers is the product of all their *common* prime factors with each factor using the least exponent that is found in the factorizations. For example, to find the GCF of 252, 360, and 408, show the following steps.

$252 = 2^2 \cdot 3^2 \cdot 7$
$360 = 2^3 \cdot 3^2 \cdot 5$
$408 = 2^3 \cdot 3 \cdot 17$
$\text{GCF} = 2^2 \cdot 3 = 12$]

27. Simplify $(3a + 5) \cdot a + (3a + 5) \cdot 2$. *(Lesson 3-7)* $3a^2 + 11a + 10$

28. Solve $3(2x^2 + 2x) - 6(x^2 - 2) = 18$. *(Lessons 3-5, 3-6, 3-7)* $x = 1$

In 29 and 30, find x if $AB = 12$. *(Lesson 3-2)*

29. A —3— C —x— B $x = 9$

30. A —5— D —x + 1— E —4— B $x = 2$

Computer room. *One of the early computers, AVIDAC's Arithmetic Unit, contained over 1500 electronic tubes. Its memory unit appears at the left, and its power distribution cabinets are in the background.*

Exploration

31. Computers often have to do millions of calculations, so reducing the number of calculations can save money. The following program asks the computer to evaluate

$$50x^4 + 50x^3 + 50x^2 + 50x + 50$$

for the 20,000 values of x from 1.00001 to 1.20000, increasing by steps of 0.00001. You saw a use for this polynomial in Chapter 10.

```
10 FOR X = 1.00001 TO 1.2 STEP 0.00001
20 LET P = 50 * X ^ 4 + 50 * X ^ 3 + 50 * X ^ 2 +
   50 * X + 50
30 NEXT X
40 PRINT P
50 END
```

a. If, on the average, each operation takes the computer a millionth of a second, how long will it take this program to run? 0.22 second

b. If the expression in line 20 is rewritten in factored form, how long will it take the program to run? 0.16 second

c. If each second of running time costs $.25, how much will factoring save? $0.015 per second

d. Write a problem that leads to the need to evaluate this polynomial.
Sample: Each birthday from age 10 on, David has received $50 from his uncle. He saves the money in an account that pays an annual yield of $(x - 1)$%. How much money will he have by the time he is 14?

Computer palm.
Compared to the AVIDAC, any of today's hand-held computers has far more capability and speed.

Follow-up for Lesson 12-2

Practice

For more questions on SPUR Objectives, use **Lesson Master 12-2A** (shown on page 729) or **Lesson Master 12-2B** (shown on pages 730–731).

Assessment

Group Assessment Ask each student to (1) write a fraction that can be simplified by factoring $2n$ from both the numerator and denominator and (2) write another fraction that can be simplified by factoring $5xy$ from both the numerator and the denominator. Have students exchange papers and check that the fractions are correct. [Students demonstrate understanding of common monomial factoring by writing and simplifying fractions correctly.]

Extension

Give students the following situation: In a physical education class, the teacher divides the students into teams of differing sizes according to the activity. For groups of 3, 4, or 6, there is always one person left over. For groups of 7, no one is left over. What is the least possible number of students in the class? [49]

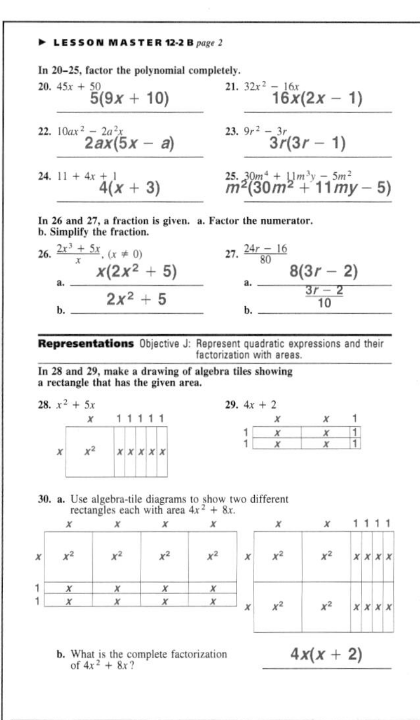

► **LESSON MASTER 12-2 B** *page 2*

In 20–25, factor the polynomial completely.

20. $45x + 50$ — $5(9x + 10)$

21. $32x^2 - 16x$ — $16x(2x - 1)$

22. $10ax^2 - 2a^2x$ — $2ax(5x - a)$

23. $9r^2 - 3r$ — $3r(3r - 1)$

24. $11 + 4x + 1$ — $4(x + 3)$

25. $30m^4 + 11m^3y - 5m^2$ — $m^2(30m^2 + 11my - 5)$

In 26 and 27, a fraction is given. a. Factor the numerator. b. Simplify the fraction.

26. $\frac{2x^3 + 5x}{x}$, $(x \neq 0)$
a. $x(2x^2 + 5)$
b. $2x^2 + 5$

27. $\frac{24r - 16}{80}$
a. $8(3r - 2)$
b. $\frac{3r - 2}{10}$

Representations Objective J: Represent quadratic expressions and their factorization with areas.

In 28 and 29, make a drawing of algebra tiles showing a rectangle that has the given area.

28. $x^2 + 5x$

	x	1	1	1	1	1
x	x^2	x	x	x	x	x

29. $4x + 2$

	x	x	1
1	x	x	1
1	x	x	1

30. a. Use algebra-tile diagrams to show two different rectangles each with area $4x^2 + 8x$.

	x	x	x	x
x	x^2	x^2	x^2	x^2
1	x	x	x	x
1	x	x	x	x

	x	x	1	1	1	1
x	x^2	x^2	x	x	x	x
x	x^2	x^2	x	x	x	x

b. What is the complete factorization of $4x^2 + 8x$? $4x(x + 2)$

Additional Answers

19a.

	x	x	x	1	1
x	x^2	x^2	x^2	x	x
x	x^2	x^2	x^2	x	x
x	x^2	x^2	x^2	x	x
x	x^2	x^2	x^2	x	x

b.

	x	x	x	x	x	x	1	1	1	1
x	x^2	x^2	x^2	x^2	x^2	x^2	x	x	x	x
x	x^2	x^2	x^2	x^2	x^2	x^2	x	x	x	x

In-class Activity

Resources

From the ***Teacher's Resource File***

- Answer Master 12-3
- Teaching Aid 31: Algebra Tiles

Additional Resources

- Visual for Teaching Aid 31

Cooperative Learning You might have students **work in small groups** so that they can discuss their work. After students complete the activity, ask each group to share the patterns they found with the class.

Additional Answers

4a.

	x	x	1
x	x^2	x^2	x
1	x	x	1
1	x	x	1
1	x	x	1

5.

	x	x	1	1	1
x	x^2	x^2	x	x	x
1	x	x	1	1	1
1	x	x	1	1	1

6.

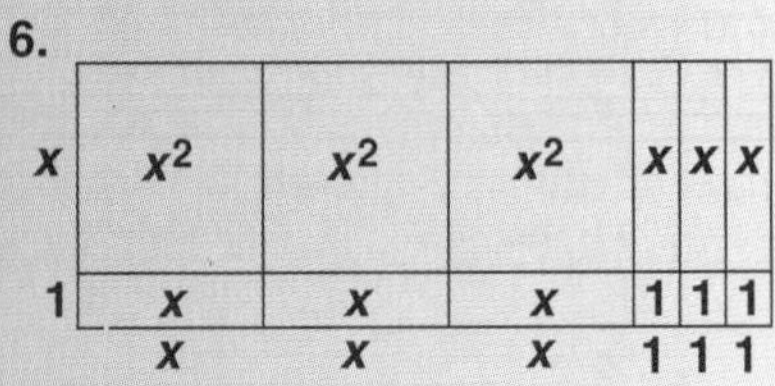

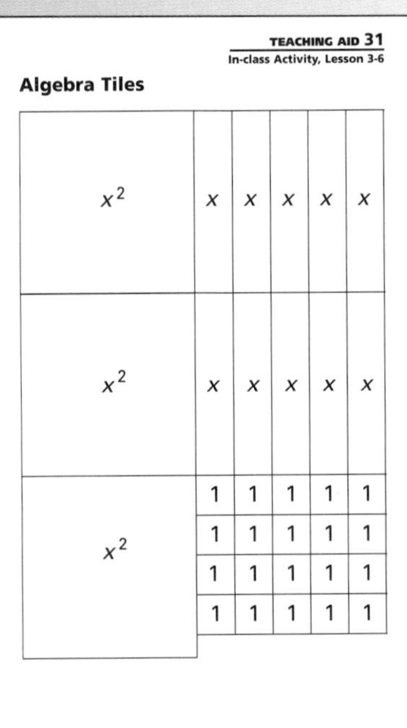
TEACHING AID 31
In-class Activity, Lesson 3-6

Algebra Tiles

x^2	x	x	x	x	x
x^2	x	x	x	x	x
x^2	1	1	1	1	1
	1	1	1	1	1
	1	1	1	1	1
	1	1	1	1	1

Introducing Lesson 12-3

Factoring Trinomials

IN-CLASS ACTIVITY

1-3) See below for tile arrangements.
4-7) See margin for tile arrangements.

Materials: algebra tiles
Work in small groups.

Factoring of some polynomials can be shown with algebra tiles.

1 Displayed at the right are the algebra tiles for $x^2 + 3x + 2$. **a) (x + 2)(x + 1)**

x^2	x	x	x	1	1

a. Rearrange the tiles to form a rectangle. What are its dimensions?

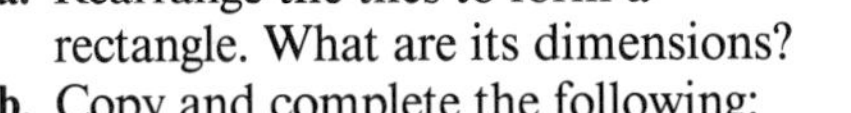

b. Copy and complete the following: $x^2 + 3x + 2 = (x + \underline{?})(x + \underline{?})$. **2; 1**

In 2 and 3, follow the idea of Question 1 to factor the expression.

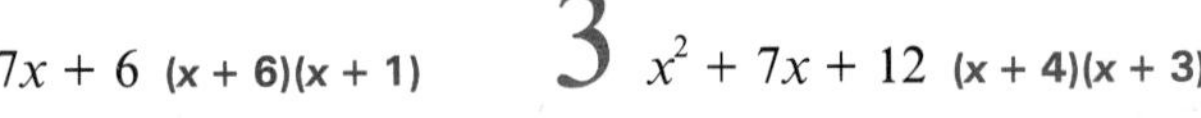

2 $x^2 + 7x + 6$ **(x + 6)(x + 1)**

3 $x^2 + 7x + 12$ **(x + 4)(x + 3)**

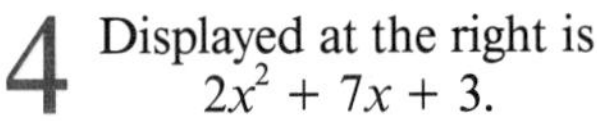

4 Displayed at the right is $2x^2 + 7x + 3$.

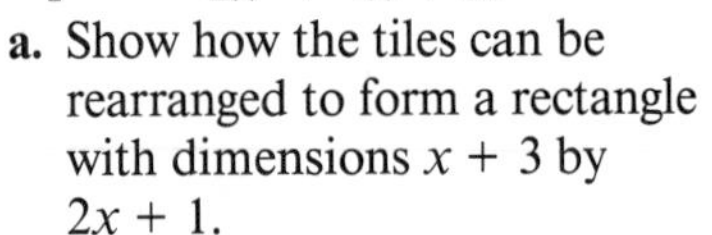

a. Show how the tiles can be rearranged to form a rectangle with dimensions $x + 3$ by $2x + 1$.

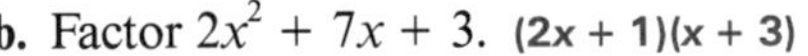

b. Factor $2x^2 + 7x + 3$. **(2x + 1)(x + 3)**

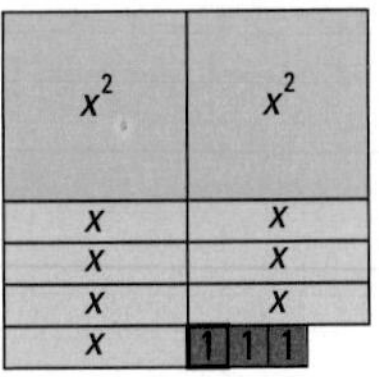

In 5 and 6, follow the idea of Question 4 to factor the expression.

5 $2x^2 + 7x + 6$ **(2x + 3)(x + 2)**

6 $3x^2 + 6x + 3$ **(3x + 3)(x + 1)**

7 **a.** Display algebra tiles for $4x^2 + 16x + 16$.

b. Arrange the tiles to form a square. What is area of the square? What is the length of a side? **$4x^2 + 16x + 16$; $2x + 4$**

c. Copy and complete the following: $4x^2 + 16x + 16 = (\underset{\mathbf{2x}}{\underline{?}} + \underset{\mathbf{4}}{\underline{?}})^2$.

8 Follow the idea of Question 7 to show that $9x^2 + 12x + 4$ is a perfect square trinomial. **$(3x + 2)^2$**

1a)

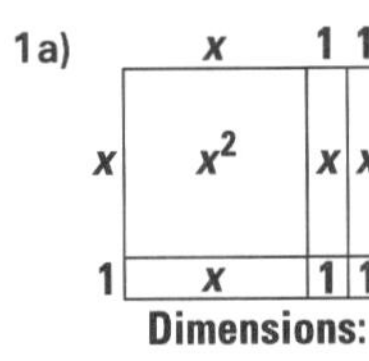

2)

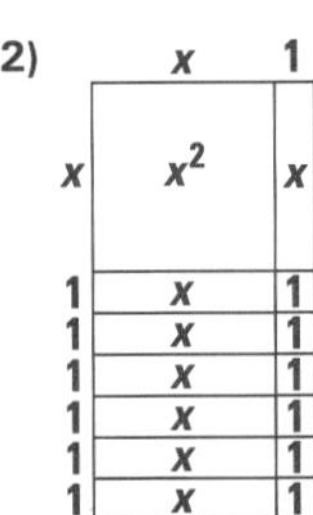

3)

	x	1	1	1	1
x	x^2	x	x	x	x
1	x	1	1	1	1
1	x	1	1	1	1
1	x	1	1	1	1

732

Additional Answers, continued

7a.

x^2	x^2	x	x	x	x	x	x	x	x				
x^2	x^2	x	x	x	x	x	x	x	x	1 1 1 1	1 1 1 1	1 1 1 1	1 1 1 1

7b.

	x	x	1	1	1	1
x	x^2	x^2	x	x	x	x
x	x^2	x^2	x	x	x	x
1	x	x	1	1	1	1
1	x	x	1	1	1	1
1	x	x	1	1	1	1
1	x	x	1	1	1	1

LESSON

12-3

Factoring $x^2 + bx + c$

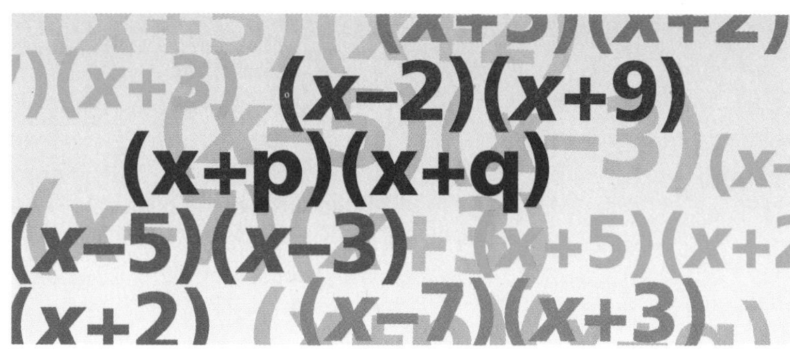

The terms in trinomials of the form $x^2 + bx + c$ have no common factors. Yet some of these trinomials can be factored. Notice the pattern which results from the multiplication of the binomials $(x + p)$ and $(x + q)$.

	Square Term		Linear Term		Constant Term
$(x + 5)(x + 2) = x^2 + 2x + 5x + 10 =$	x^2	+	$7x$	+	10
$(x - 5)(x - 3) = x^2 - 3x - 5x + 15 =$	x^2	−	$8x$	+	15
$(x - 2)(x + 9) = x^2 + 9x - 2x - 18 =$	x^2	+	$7x$	−	18
$(x - 7)(x + 3) = x^2 + 3x - 7x - 21 =$	x^2	−	$4x$	−	21
$(x + p)(x + q) = x^2 + qx + px + pq =$	x^2	+	$(p + q)x$	+	pq

The constant term of each trinomial above is the product of the last terms of the binomial factors. The coefficient in the linear term is the sum of the last terms. This pattern suggests a way to factor trinomials in which the leading coefficient is 1.

Example 1

Factor $x^2 + 11x + 18$.

Solution

To factor, we need to identify two binomials, $(x + p)(x + q)$, whose product equals $x^2 + 11x + 18$. We must find p and q, two numbers whose product is 18 and whose sum is 11. Since the product is positive and the sum is positive, both p and q are positive. List the positive factors of 18; then calculate their sums.

Factors of 18	Sum of Factors
1, 18	19
2, 9	11
3, 6	9

The sum of the numbers 2 and 9 is 11. So
$x^2 + 11x + 18$ can be factored into $(x + 9)(x + 2)$.

Check

Factoring can always be checked by multiplication.
$(x + 9)(x + 2) = x^2 + 2x + 9x + 18 = x^2 + 11x + 18$. It checks.

Lesson 12-3

Objectives

C Factor quadratic expressions.
G Determine whether a quadratic polynomial can be factored over the integers.
J Represent quadratic expressions and their factorizations with areas.

Resources

From the ***Teacher's Resource File***
- Lesson Master 12-3A or 12-3B
- Answer Master 12-3
- Assessment Sourcebook: Quiz for Lessons 12-1 through 12-3
- Teaching Aids
  31 Algebra Tiles
  128 Warm-up
- Activity Kit, Activity 26

Additional Resources
- Visuals for Teaching Aids 31, 128
- Automatic graphers

Teaching Lesson 12-3

Warm-up

Find two numbers whose product is the first number and whose sum is the second number.

1. 96; 28 **4 and 24**
2. 90; −21 **−15 and −6**
3. −165; 28 **33 and −5**
4. −153; −8 **−17 and 9**

Lesson 12-3 Overview

Broad Goals This lesson covers the factoring of quadratic trinomials where the coefficient of x^2 is 1. Lesson 12-5 considers the more general case of factoring $ax^2 + bx + c$.

Perspective There are two questions to ask about the factorization of $x^2 + bx + c$. First, is it factorable? Then, if it is, what are the factors? The first of these questions is critical because most of the expressions of this type (even when b and c are small integers) are not factorable over the integers. In this lesson, students must realize this fact.

In the set of polynomials with real coefficients, any quadratic of the form $x^2 + bx + c$ is factorable. Specifically, for all b and c, $x^2 + bx + c =$
$\left(x + \frac{b + \sqrt{b^2 - 4c}}{2}\right)\left(x + \frac{b - \sqrt{b^2 - 4c}}{2}\right)$.

More generally, for all a, b, and c, $ax^2 + bx + c =$
$\left(x + \frac{b + \sqrt{b^2 - 4ac}}{2a}\right)\left(x + \frac{b - \sqrt{b^2 - 4ac}}{2a}\right)$.

In a later lesson, students will learn that the value of the discriminant $b^2 - 4ac$ is critical in determining if the quadratic is factorable over the set of polynomials with integer coefficients.
(Overview continues on page 734.)

Notes on Reading

Even though the idea is the same, students need practice on the four different combinations of additions and subtractions in $x^2 \pm bx \pm c$. Every time one of the factorizations is done, students should check it by multiplication. The practice of factoring and checking by multiplication should lead to the following generalizations:

1. When the second operation is addition (c is positive in $ax^2 + bx + c$), the numbers p and q in the factorization $(x + p)(x + q)$ must have the same sign, and that sign is determined by the sign of b.
2. When the second operation is subtraction (c is negative in $ax^2 + bx + c$), the numbers p and q must have different signs, and the sign of b will correspond to the value of p or q that has the greater absolute value.

❶ You might use an overhead automatic grapher to illustrate Check 2 of **Example 2.** Plot the graph that corresponds to an incorrect factorization such as $y = (x - 3)(x + 4)$. Show that it does not coincide with the graph of $y = x^2 - x - 12$.

❷ Point out Solution 2 in **Example 4** so that students realize the process works even when $b = 0$.

At this time you might want to use Activity 1 in *Optional Activities* below. In the activity, students use algebra tiles to demonstrate the difference between a perfect square trinomial, a factorable trinomial that is not a perfect square, and a trinomial that is not factorable over the integers.

Pacing Lesson 12-4 applies the kind of factoring that is presented here, so there is no need to spend more than one day on this lesson.

Example 2

Factor $x^2 - x - 12$.

Solution

Think of this trinomial as $x^2 + -1x + -12$.

We need two numbers whose product is -12 and whose sum is -1. Since the product is negative, one of the factors is negative. List the possibilities.

Factors of -12	Sum of Factors
-1, 12	11
-2, 6	4
-3, 4	1
-4, 3	-1
-6, 2	-4
-12, 1	-11

The only two factors of -12 whose sum is -1 are -4 and 3. So

$$x^2 - x - 12 = (x - 4)(x + 3).$$

Check 1

$(x - 4)(x + 3) = x^2 + 3x - 4x - 12 = x^2 - x - 12$. It checks.

Check 2

Graph $y = x^2 - x - 12$ and $y = (x - 4)(x + 3)$. The graphs should be identical. Below we show the output from our automatic grapher.

❶

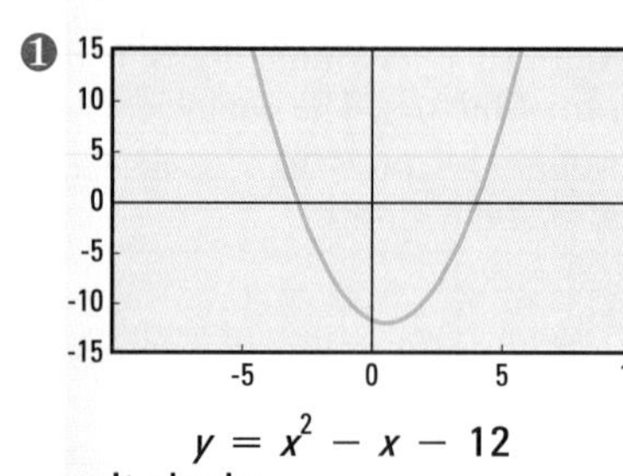

$y = x^2 - x - 12$

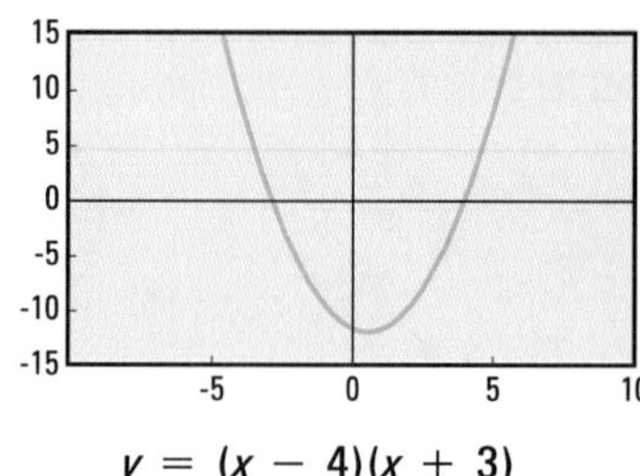

$y = (x - 4)(x + 3)$

It checks.

Example 3

Factor $t^2 - 6t + 9$.

Solution

Think: Which factors of 9 have a sum of -6? Because the product is positive and the sum negative, both numbers are negative. You need only consider negative factors of 9.

Factors of 9	Sum of Factors
-9, -1	-10
-3, -3	-6

So $t^2 - 6t + 9 = (t - 3)(t - 3)$.

Another way to write the answer is as $(t - 3)^2$.

Lesson 12-3 overview, continued

The trinomial $x^2 + bx + c$ is factorable into $(x + p)(x + q)$ if and only if $pq = c$ and $p + q = b$. If there are no integers p and q with this property, then the quadratic is not factorable. Suppose you want to factor $x^2 + bx + c$. Using algebra tiles, begin with one big square, b rectangles, and c units. Try to arrange these tiles into a rectangle. If a rectangle can be formed, its dimensions are $x + p$ and $x + q$. Therefore, $p + q$ must equal b, and pq must equal c.

Optional Activities

Activity 1 Using Physical Models
Materials: Algebra tiles or **Teaching Aid 31**

After discussing the examples, you might have students **work in small groups**. Have them use algebra tiles to demonstrate the difference between a perfect square trinomial, a factorable trinomial that is not a perfect square, and a trinomial that is not factorable over the integers. Examples that you might use are given at the right.

1. $x^2 + 6x + 9$ can be represented with a square having sides $x + 3$. It is a perfect square trinomial.
2. $x^2 + 6x + 8$ cannot be represented by putting the tiles in a square. Yet it can be shown as a rectangle with sides $x + 2$ and $x + 4$. So even though the trinomial is not a perfect square, it is factorable.
3. $x^2 + 6x + 1$ cannot be represented with a rectangle; it illustrates a prime polynomial.

Example 3 shows that $t^2 - 6t + 9$ is a *perfect square trinomial.*

Not all trinomials of the form $x^2 + bx + c$ can be factored into polynomials with integer coefficients. For instance, to factor $x^2 + 8x + 14$ as two binomials, $(x + p)(x + q)$, you would have to find two numbers p and q whose product is 14 and whose sum is 8.

Factors of 14	Sum of Factors
1, 14	15
2, 7	9

None of the pairs of factors have a sum of 8, so $x^2 + 8x + 14$ is not factorable over the integers. It is *prime.*

Binomials that are differences of squares can be factored.

❷ **Example 4**

Factor $d^2 - 49$.

Solution 1

Both d^2 and 49 are perfect squares. Use the Difference of Squares pattern from Lesson 10-6.

$$d^2 - 49 = d^2 - 7^2 = (d + 7)(d - 7)$$

Solution 2

Think of $d^2 - 49$ as $d^2 + 0d + -49$. Two numbers are needed whose product is -49 and whose sum is 0. They must be opposites, so use 7 and -7.

The factors of $d^2 - 49$ are $(d + 7)(d - 7)$.

Check

Multiply: $(d + 7)(d - 7) = d^2 - 7d + 7d - 49 = d^2 - 49$.

Some trinomials have both monomial and binomial factors. If the terms of a trinomial have a greatest common factor other than 1, you should first factor out the GCF.

Example 5

Factor $3r^3 + 30r^2 + 75r$.

Solution

$3r$ is the GCF of the three terms.

$$3r^3 + 30r^2 + 75r = 3r(r^2 + 10r + 25)$$

To factor $r^2 + 10r + 25$, we need binomials whose constant terms have a product of 25 and a sum of 10.

Factors of 25	Sum of Factors
1, 25	26
5, 5	10

$$r^2 + 10r + 25 = (r + 5)(r + 5).$$

So $\quad 3r^3 + 30r^2 + 75r = 3r(r + 5)(r + 5)$ or $3r(r + 5)^2$. ▸

Additional Examples

In 1–5, a quadratic trinomial is given. Factor it completely over the integers and then check by either multiplying or using an automatic grapher.

1. $x^2 + 50x + 96$ **$(x + 48)(x + 2)$**
   Check: $(x + 48)(x + 2) = x^2 + 2x + 48x + 96 = x^2 + 50x + 96$. It checks.
2. $y^2 - 3y - 54$ **$(y - 9)(y + 6)$**
   Check: $(y - 9)(y + 6) = y^2 + 6y - 9y - 54 = y^2 - 3y - 54$. It checks.
3. $z^2 - 9z + 20$ **$(z - 4)(z - 5)$**
   Check: $(z - 4)(z - 5) = z^2 - 5z - 4z + 20 = z^2 - 9z + 20$. It checks.
4. $c^2 - 16$ **$(c - 4)(c + 4)$**
   Check: $(c - 4)(c + 4) = c^2 + 4c - 4c - 16 = c^2 - 16$. It checks.
5. $d^3 - 16d$ **$d(d - 4)(d + 4)$**
   Check: $d(d - 4)(d + 4) = d(d^2 + 4d - 4d - 16) = d(d^2 - 16) = d^3 - 16d$. It checks.
6. Explain why $c^2 + 16$ is not factorable over the integers. **There are no two integers whose product is 16 and whose sum is 0.**

Activity 2

You might want to use *Activity Kit, Activity 26,* before covering **Example 4** or as a follow-up to the lesson. This activity offers a geometric interpretation of the difference of two squares.

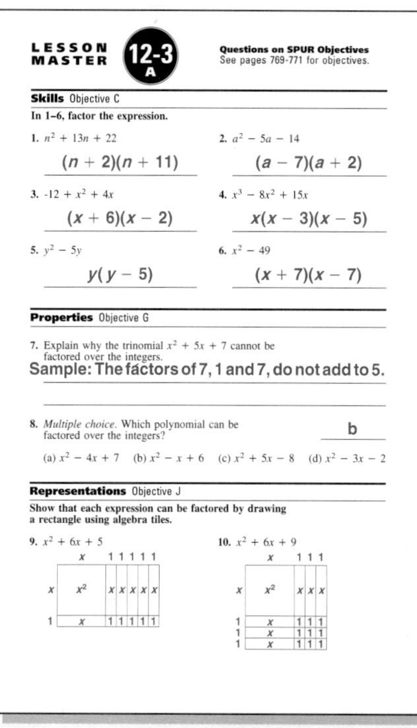
LESSON MASTER 12-3 A — **Questions on SPUR Objectives** See pages 769-771 for objectives.

Skills Objective C

In 1–6, factor the expression.

1. $n^2 + 13n + 22$ — $(n + 2)(n + 11)$
2. $a^2 - 5a - 14$ — $(a - 7)(a + 2)$
3. $-12 + x^2 + 4x$ — $(x + 6)(x - 2)$
4. $x^3 - 8x^2 + 15x$ — $x(x - 3)(x - 5)$
5. $y^2 - 5y$ — $y(y - 5)$
6. $x^2 - 49$ — $(x + 7)(x - 7)$

Properties Objective G

7. Explain why the trinomial $x^2 + 5x + 7$ cannot be factored over the integers.
Sample: The factors of 7, 1 and 7, do not add to 5.

8. *Multiple choice.* Which polynomial can be factored over the integers? b
(a) $x^2 - 4x + 7$ (b) $x^2 - x + 6$ (c) $x^2 + 5x - 8$ (d) $x^2 - 3x - 2$

Representations Objective J

Show that each expression can be factored by drawing a rectangle using algebra tiles.

9. $x^2 + 6x + 5$

	x	1	1	1	1	1
x	x^2	x	x	x	x	x
1	x	1	1	1	1	1

10. $x^2 + 6x + 9$

	x	1	1	1
x	x^2	x	x	x
1	x	1	1	1
1	x	1	1	1
1	x	1	1	1

Notes on Questions

Questions 5–10 Encourage students to check their work by multiplying the two binomials or by using an automatic grapher.

Questions 17 and 20 Error Alert Commutativity does not affect the factorization, but it can confuse students. Rewriting the polynomials in standard form will help students.

Question 18 You might use the trinomial in **Question 5** and consider including a second variable in order to show the pattern. For instance, write $x^2 + 7x + 10$ as $x^2 + 7xy + 10y^2$. Then factor it: $(x + 5y)(x + 2y)$.

Question 32 Simplifying rational polynomial fractions by factoring is not an objective of this course. Still, it is nice to show students that the methods which are used to write fractions in lowest terms have analogies with polynomials.

Follow-up for Lesson 12-3

Practice

For more questions on SPUR Objectives, use **Lesson Master 12-3A** (shown on page 735) or **Lesson Master 12-3B** (shown on pages 736–737).

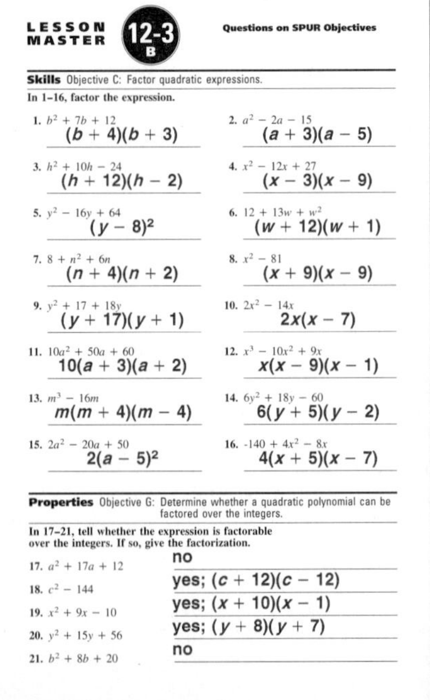

LESSON MASTER 12-3 B — Questions on SPUR Objectives

Skills Objective C: Factor quadratic expressions.

In 1–16, factor the expression.

1. $b^2 + 7b + 12$ — $(b + 4)(b + 3)$	2. $a^2 - 2a - 15$ — $(a + 3)(a - 5)$
3. $h^2 + 10h - 24$ — $(h + 12)(h - 2)$	4. $x^2 - 12x + 27$ — $(x - 3)(x - 9)$
5. $y^2 - 16y + 64$ — $(y - 8)^2$	6. $12 + 13w + w^2$ — $(w + 12)(w + 1)$
7. $8 + n^2 + 6n$ — $(n + 4)(n + 2)$	8. $x^2 - 81$ — $(x + 9)(x - 9)$
9. $y^2 + 17 + 18y$ — $(y + 17)(y + 1)$	10. $2x^2 - 14x$ — $2x(x - 7)$
11. $10a^2 + 50a + 60$ — $10(a + 3)(a + 2)$	12. $x^3 - 10x^2 + 9x$ — $x(x - 9)(x - 1)$
13. $m^3 - 16m$ — $m(m + 4)(m - 4)$	14. $6y^2 + 18y - 60$ — $6(y + 5)(y - 2)$
15. $2a^2 - 20a + 50$ — $2(a - 5)^2$	16. $-140 + 4x^2 - 8x$ — $4(x + 5)(x - 7)$

Properties Objective G: Determine whether a quadratic polynomial can be factored over the integers.

In 17–21, tell whether the expression is factorable over the integers. If so, give the factorization.

17. $a^2 + 17a + 12$ — no
18. $c^2 - 144$ — yes; $(c + 12)(c - 12)$
19. $x^2 + 9x - 10$ — yes; $(x + 10)(x - 1)$
20. $y^2 + 15y + 56$ — yes; $(y + 8)(y + 7)$
21. $b^2 + 8b + 20$ — no

▸ **Check**

$$3r(r + 5)(r + 5) = (3r^2 + 15r)(r + 5)$$
$$= 3r^3 + 15r^2 + 15r^2 + 75r$$
$$= 3r^3 + 30r^2 + 75r \quad \text{It checks.}$$

QUESTIONS

Covering the Reading

1. What was the coefficient of the square term in all the trinomials in Examples 1 to 3? **1**

2. a. How can you check that you have found the correct factors of a quadratic trinomial? **Multiply the factors and check that you obtain the original trinomial.**
   b. Perform the check for Example 3. **$(t - 3)(t - 3) = t^2 - 3t - 3t + 9 = t^2 - 6t + 9$**

3. a. In order to factor $x^2 + 8x + 12$, list the possible integer factors of the last term, and their sums. **1, 12, 13; 2, 6, 8; 3, 4, 7**
   b. Factor $x^2 + 8x + 12$.
   c. Check your work.
   b) $(x + 2)(x + 6)$ c) $(x + 2)(x + 6) = x^2 + 2x + 6x + 12 = x^2 + 8x + 12$

4. Suppose $(x + p)(x + q) = x^2 + bx + c$.
   a. What must pq equal? ***c***
   b. What must $p + q$ equal? ***b***

In 5–10, write the trinomial as the product of two binomials.

5. $x^2 + 7x + 10$ **$(x + 5)(x + 2)$**
6. $p^2 + 14p + 13$ **$(p + 1)(p + 13)$**
7. $t^2 - t - 30$ **$(t + 5)(t - 6)$**
8. $a^2 - 81$ **$(a + 9)(a - 9)$**
9. $q^2 - 10q + 16$ **$(q - 2)(q - 8)$**
10. $n^2 + 4n - 12$ **$(n - 2)(n + 6)$**

11. Explain why the trinomial $x^2 + 2x + 3$ cannot be factored over the integers. **None of the pairs of integer factors of 3 has a sum of 2.**

12. a. Factor $x^3 - 5x^2 + 6x$ into the product of a monomial and a trinomial. **$x(x^2 - 5x + 6)$**
    b. Complete the factoring by finding factors of the trinomial in part **a.** **$x(x - 2)(x - 3)$**

In 13 and 14, factor completely.

13. $2m^2 + 18m + 36$ **$2(m + 3)(m + 6)$**
14. $4y^3 - 20y^2 - 24y$ **$4y(y + 1)(y - 6)$**

15. The diagram at the left uses algebra tiles to show the factorization of $x^2 + 4x + 3$. Make a drawing of algebra tiles to show the factorization of $x^2 + 5x + 6$. **$(x + 3)(x + 2)$ See left for diagram.**

15)

	x	1	1	1
x	x^2	x	x	x
1	x	1	1	1
1	x	1	1	1

16b)

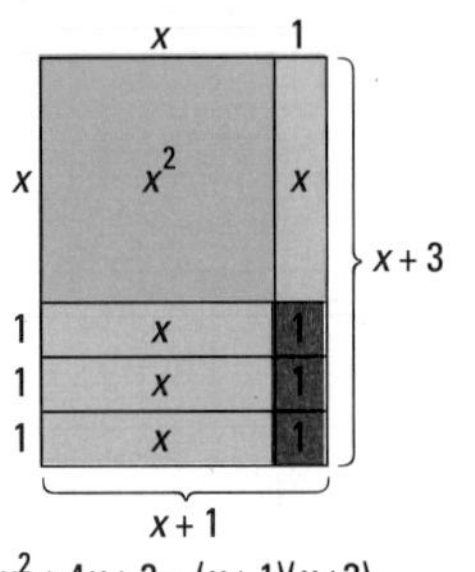

x 1
x x^2 x
1 x 1
1 x 1
1 x 1
x+3
x+1

$x^2 + 4x + 3 = (x + 1)(x + 3)$

Applying the Mathematics

16. a. Find an equation of the form $y = (x + p)(x + q)$ whose graph is identical to the graph of $y = x^2 - 10x + 21$. **$y = (x - 3)(x - 7)$**
    b. Check your work by graphing both equations on the same set of axes. **See left.**

In 17–20, factor.

17. $35 - 12x + x^2$ **$(7 - x)(5 - x)$**
18. $a^2 - 8ab + 7b^2$ **$(a - b)(a - 7b)$**

Adapting to Individual Needs

Extra Help

A common mistake that students make when factoring is forgetting to look for a common factor. **Questions 12–14** and **Questions 19–20** all involve polynomials in which each term has a common factor. Stress the importance of first checking for a common factor. Since some students think they are finished after they factor out a common factor, emphasize that the remaining polynomial can often be factored again.

English Language Development

Encourage non-English-speaking students to review their index cards. You might pair each of these students with an English-speaking student, and have them review the definitions together.

19. $40 - 10x^2$ **$10(2 + x)(2 - x)$**

20. $10r^2 + 5r^3 + 5r$ **$5r(r + 1)(r + 1)$**

21. Notice that $1{,}032{,}060 = 10^6 + 32 \cdot 10^3 + 60$. Use this information to help find the prime factorization of 1,032,060. (Hint: Begin by letting $x = 10^3$.) **See left.**

22. Explain how $28^2 - 22^2$ can be calculated in your head. **$(28 + 22)(28 - 22) = 50 \cdot 6 = 300$**

Review

23. Phone numbers of businesses often end in 0. But does this apply to state tourism offices? We wondered, so we examined the phone numbers of the in-state tourism offices in the 50 states. At the left are the frequencies for the digits in which they ended. Use a chi-square test to test whether these ending digits depart unusually from what would be expected if last digits occurred at random. Explain your reasoning. *(Lesson 10-7)* **See left.**

Ending Digit	Frequency
0	10
1	7
2	2
3	3
4	4
5	3
6	5
7	7
8	2
9	7

In 24 and 25, multiply. *(Lessons 10-5, 10-6)*

24. $(3x + 1)(2x - 4)$ **$6x^2 - 10x - 4$**

25. $(8a + 1)(8a - 1)$ **$64a^2 - 1$**

26. Solve $x^2 + 5x + 4 = 0$. *(Lesson 9-5)* **$x = -4$ or $x = -1$**

In 27 and 28, a thermometer is taken from a room temperature of 70°F to a point outside where it is 45°F. After t minutes, the thermometer reading is $24(1 - 0.04t)^2 + 45$. *(Lesson 9-3)*

27. About what temperature will the thermometer be after 10 minutes? **53.64°**

28. About how long will it take the thermometer to get down to 45°F? **25 minutes**

In 29 and 30, solve. *(Lesson 6-8)*

29. $\frac{3}{5} = \frac{x}{15}$ **$x = 9$**

30. $\frac{d}{9} = \frac{5}{d}$ **$d = \pm 3\sqrt{5}$**

31. The World Wide Wrench Company increased prices by 12%. Their basic wrench now sells for \$9.25. What did it sell for before? *(Lesson 6-5)* **\$8.26**

21) $1{,}032{,}060$
$= 10^6 + 32 \cdot 10^3 + 60$
$= (10^3 + 2)(10^3 + 30)$
$= 1002 \cdot 1030$
$= 2 \cdot 501 \cdot 2 \cdot 515$
$= 2^2 \cdot 3 \cdot 167 \cdot 5 \cdot 103$
$= 2^2 \cdot 3 \cdot 5 \cdot 103 \cdot 167$

23) The chi-square value is 12.8. A chi-square value this size would be expected more than 10% of the time, so this is not an unusual distribution of digits.

Exploration

32. The fraction $\frac{x^2 - 4x + 3}{x^2 - 9}$ can be simplified when $x \neq 3$ and $x \neq -3$ by the following process: **See margin.**

$$\frac{x^2 - 4x + 3}{x^2 - 9} = \frac{(x - 3)(x - 1)}{(x - 3)(x + 3)} \quad \text{Factoring}$$
$$= \frac{x - 1}{x + 3} \quad \text{Equal Fractions Property}$$

a. Check this answer by letting $x = 2$ in the original and final fractions.
b. Write another fraction with quadratic expressions in its numerator and denominator that can be simplified.
c. Simplify your expression in part **b** and check your work.

Assessment

Quiz A quiz covering Lessons 12-1 through 12-3 is provided in the *Assessment Sourcebook.*

Written Communication Have students draw an area model that shows the factorization of $x^2 + 8x + 15$. Then ask them to write a different polynomial and draw a diagram showing its factorization. [Students demonstrate the ability to represent factoring using area models and writing a polynomial that is factorable over the integers.]

Extension

Point out to students that polynomials of the form $x^2 + bx + c$ can be factored even if b and c are not integers. Have students factor each of the following expressions.

1. $x^2 + \frac{2}{9}x + \frac{1}{81}$ $[(x + \frac{1}{9})(x + \frac{1}{9})]$
2. $x^2 + x + \frac{1}{4}$ $[(x + \frac{1}{2})(x + \frac{1}{2})]$
3. $x^2 + \frac{7}{3}x + \frac{2}{3}$ $[(x + 2)(x + \frac{1}{3})]$

Project Update Project 1, *What Percentage of Some Simple Quadratics are Factorable?,* on page 765, relates to the content of this lesson.

Additional Answers
32. See page 771.

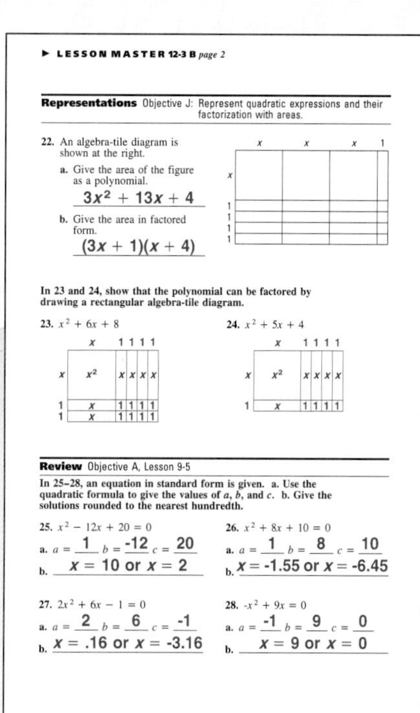
► LESSON MASTER 12-3 B *page 2*

Representations Objective J: Represent quadratic expressions and their factorization with areas.

22. An algebra-tile diagram is shown at the right.
a. Give the area of the figure as a polynomial. **$3x^2 + 13x + 4$**
b. Give the area in factored form. **$(3x + 1)(x + 4)$**

In 23 and 24, show that the polynomial can be factored by drawing a rectangular algebra-tile diagram.

23. $x^2 + 6x + 8$

24. $x^2 + 5x + 4$

Review Objective A, Lesson 9-5

In 25–28, an equation in standard form is given. a. Use the quadratic formula to give the values of a, b, and c. b. Give the solutions rounded to the nearest hundredth.

25. $x^2 - 12x + 20 = 0$
a. $a =$ **1** $b =$ **-12** $c =$ **20**
b. **$x = 10$ or $x = 2$**

26. $x^2 + 8x + 10 = 0$
a. $a =$ **1** $b =$ **8** $c =$ **10**
b. **$x = -1.55$ or $x = -6.45$**

27. $2x^2 + 6x - 1 = 0$
a. $a =$ **2** $b =$ **6** $c =$ **-1**
b. **$x = .16$ or $x = -3.16$**

28. $-x^2 + 9x = 0$
a. $a =$ **-1** $b =$ **9** $c =$ **0**
b. **$x = 9$ or $x = 0$**

Adapting to Individual Needs

Challenge

Ask students the following questions.

1. Suppose a quadratic trinomial has the form $x^2 + bx + 12$. For what values of b is this trinomial factorable over the integers? [For the sum of any two integers whose product is 12. The only possible sums are: $1 + 12 = 13$, $2 + 6 = 8$, $3 + 4 = 7$, $-1 + -12 = -13$, $-2 + -6 = -8$, and $-3 + -4 = -7$. So the possible values of b are 13, 8, 7, −13, −8, and −7.]

2. Now suppose a quadratic trinomial has the form $x^2 + 12x + c$. For what values of c is this trinomial factorable over the integers? [The product of any two integers whose sum is 12 is a possible value of c; there are an infinite number of these products. Three samples are $5 \cdot 7 = 35$, $-9 \cdot 21 = -189$, and $52 \cdot -40 = -2080$.]

Lesson 12-4

Objectives

D Solve quadratic equations by factoring.

F Recognize and use the Zero Product Property.

I Solve quadratic equations in real situations.

Resources

From the ***Teacher's Resource File***

- Lesson Master 12-4A or 12-4B
- Answer Master 12-4
- Teaching Aids
  128 Warm-up
  130 Question 32

Additional Resources

- Visuals for Teaching Aids 128, 130

Teaching Lesson 12-4

Warm-up

When will the quadratic equation $ax^2 + bx + c = 0$ have

1. exactly two real solutions? **When $b^2 - 4ac > 0$**
2. exactly one real solution? **When $b^2 - 4ac = 0$**
3. no real solutions? **When $b^2 - 4ac < 0$**
4. What property can be used to answer Questions 1–3? **Discriminant Property**

LESSON

12-4 *Solving Some Quadratic Equations By Factoring*

Side out. *Hits (volleys) in volleyball, except when straight down or straight up, follow the path of a parabola.*

Suppose a ball is thrown upward from waist level at a speed of 48 feet per second. If there were no pull of gravity, the distance above the waist after t seconds would be $48t$, and the ball would never come down. Gravity pulls the ball down $16t^2$ feet in t seconds. So the distance d above the waist after t seconds is given by

$$d = 48t - 16t^2.$$

How long after being thrown does the ball return to waist level, where it can be caught? At waist level, the distance d is zero, so to answer this question you need to solve the equation $48t - 16t^2 = 0$.

In Chapter 9 you learned two ways to solve this equation. You can make a graph of $d = 48t - 16t^2$, and find the values of t when $d = 0$. Below is a graph of $y = 48x - 16x^2$ on the window $0 \le x \le 5$, $0 \le y \le 40$.

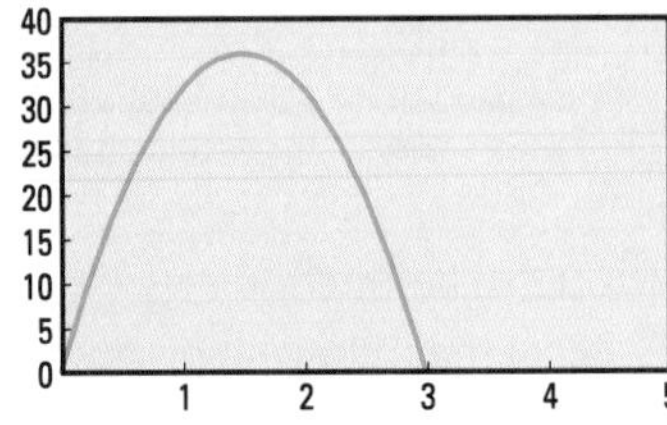

The graph shows that $y = 0$ when $x = 0$ or $x = 3$. So the ball returns to waist level 3 seconds after being thrown upward.

Lesson 12-4 Overview

Broad Goals The factoring of the previous lesson is applied to the solution of those quadratics that can be factored over the integers.

Perspective In Chapter 9, we emphasized the use of the Quadratic Formula as the primary method for solving quadratic equations because it applies in all cases. We also showed students how to use a graph to solve a quadratic equation. In this lesson, we explain how to use factoring to solve quadratics; students should be able to solve equations by factoring when the factoring is easy.

This lesson also provides a second day of practice on trinomial factoring itself, and it reviews some of the projectile motion ideas from Chapter 9.

You can also solve the equation $48t - 16t^2 = 0$ by using the Quadratic Formula. It helps to rewrite the equation in the form $-16t^2 + 48t = 0$, so you can see that $a = -16$, $b = 48$, and $c = 0$.

$$t = \frac{-48 \pm \sqrt{48^2 - 4(-16) \cdot 0}}{2(-16)}$$
$$= \frac{-48 \pm \sqrt{48^2}}{-32}$$
$$= \frac{-48 \pm 48}{-32}$$

So $t = \frac{-48 + 48}{-32} = 0$ or $t = \frac{-48 - 48}{-32} = \frac{-96}{-32} = 3$

A third way to solve $48t - 16t^2 = 0$ is to use factoring and the *Zero Product Property.*

❶ **The Zero Product Property**
If the product of two real numbers a and b is 0, then $a = 0$ or $b = 0$.

Numbers may be represented by expressions. So in words, the Zero Product Property is: If the product of two *expressions* is zero, one or the other (or both) must be zero.

Example 1

Refer to the situation described on the preceding page. Use the Zero Product Property to find how long the ball is in the air.

Solution

We need to solve $48t - 16t^2 = 0$.

$t(48 - 16t) = 0$ Factor.

$t = 0$ or $48 - 16t = 0$ Zero Product Property

$t = 0$ or $-16t = -48$

$t = 0$ or $t = \frac{-48}{-16} = 3$

The solution $t = 0$ means the ball was at waist level at the start.
The solution $t = 3$ means that it returned to waist level again after 3 seconds.
It was in the air for 3 seconds.

In Chapter 9 you learned that the Quadratic Formula can be used to solve all quadratic equations. Factoring and the Zero Product Property can be useful tools to solve those quadratic equations which factor over the integers.

Notes on Reading

❶ Emphasize these two points. First, one side of the equation must be zero before the Zero Product Property can be used, and second, students should look for a common factor first.

Optional Activities

Materials: Automatic graphers

You might use this activity with **Questions 4–7** to give students more practice using an automatic grapher. Remind students that they can multiply each side of an equation by a number or add a number to each side of an equation to generate an equivalent equation (an equation with the same solution). Then, for one or more of the equations in **Questions 4–7**, have students write two equivalent equations and graph the three equations on an automatic grapher to verify that the equations have the same solutions.

❷ To show that all three methods yield the same solution, you may also want to solve **Example 2** by using the Quadratic Formula and an automatic grapher.

❸ In this example, students may be surprised at the answer—they may assume that the number of games is proportional to the number of players. Examining the formula should show that the rate is more complex than a simple proportion.

Multicultural Connection The game of chess is believed to have originated in India in the seventh century A.D., and by the year 1000, it had spread throughout much of Europe. The modern game of chess dates from the 1500s when the moves of the game were developed. Philidor, a Frenchman who played chess in the 1700s, is regarded as the first world champion chess player. In 1972, Bobby Fischer became the first American to win the official world chess championship.

Additional Examples

1. Suppose a ball is thrown upward at a speed of 20 feet per second. How long will the ball be in the air? **1.25 seconds**
2. Solve $x^2 - 14x + 33 = 0$ by factoring. **$x = 3$ or $x = 11$**
3. In a sports league that has n teams, if each team is to play all of the other teams twice (once on its home field, once on the other team's field), $n^2 - n$ games must be played. If a baseball league finds that it has time slots available for 72 games, how many teams can be formed? **$n^2 - n = 72$; 9 teams**
4. Solve $10t^3 + 80t^2 - 200t = 0$. **$t = 0, 2,$ or -10**

❷ **Example 2**

Solve $r^2 - 11r - 12 = 0$ by factoring.

Solution

$r^2 - 11r - 12 = 0$

$(r - 12)(r + 1) = 0$ Factor the trinomial.

$r - 12 = 0$ or $r + 1 = 0$ Use the Zero Product Property.

$r = 12$ or $r = -1$

Check

Let $r = 12$: $12^2 - 11 \cdot 12 - 12 = 144 - 132 - 12 = 0.$

Let $r = -1$: $(-1)^2 - 11 \cdot -1 - 12 = 1 + 11 - 12 = 0.$

It checks.

Recall that in order to solve a quadratic equation with the Quadratic Formula, one side of the equation must be 0. This is also true if you wish to solve an equation by factoring.

Checkmate. *Shown is "Josh Waitzkin" from the movie,* Searching for Bobby Fischer, *who aspires to play chess like Bobby Fischer. At age 14, Fischer won his first U.S. National Championship. At 15, he was the youngest player in the world ever to attain the rank of grand master.*

❸ **Example 3**

In a round-robin chess tournament with n players, $\frac{n^2 - n}{2}$ games are needed. If 55 games were played, how many players were entered in the tournament?

Solution

Here $\frac{n^2 - n}{2} = 55.$

This is a quadratic equation, but it is not in standard form. Multiply both sides by 2 and then subtract 110 from both sides.

$n^2 - n = 110$

$n^2 - n - 110 = 0$

$(n - 11)(n + 10) = 0$ Factor.

$n - 11 = 0$ or $n + 10 = 0$ Zero Product Property

$n = 11$ or $n = -10$

A negative number of players does not make sense in this situation. So 11 players were entered.

Check

Substitute. Does $\frac{11^2 - 11}{2} = 55$? Yes, it does.

The Zero Product Property applies to products of more than two expressions or numbers.

Adapting to Individual Needs

Extra Help

Some students will try to solve an equation such as $(x + 3)(x - 1) = 1$ by letting $x + 3 = 1$ and $x - 1 = 1$. Of course, this would yield $x = -2$ and $x = 2$, neither of which checks when substituted in the equation. This illustrates why checking is so important. Even though students resist checking, it is important that they develop the habit.

English Language Development

Non-English-speaking students may be able to solve equations using the Zero Product Property but they may have difficulty verbalizing their work. You might model or list key phrases that explain the examples and then give students several equations to solve. Observe students as they solve equations and have them explain each step to you.

Example 4

For what values of y does $y(2y + 3)(y - 2) = 0$?

Solution

$$y(2y + 3)(y - 2) = 0$$

$$y = 0 \quad \text{or} \quad 2y + 3 = 0 \quad \text{or} \quad y - 2 = 0 \qquad \text{Zero Product Property}$$

$$y = -\frac{3}{2} \quad \text{or} \quad y = 2$$

So y is 0, 2, or $-\frac{3}{2}$.

Check

Let $y = 0$: $0(2 \cdot 0 + 3)(0 - 2) = 0 \cdot 3 \cdot -2 = 0$.
Let $y = 2$: $2(2 \cdot 2 + 3)(2 - 2) = 2 \cdot 7 \cdot 0 = 0$.
Let $y = -\frac{3}{2}$: $-\frac{3}{2}(2 \cdot -\frac{3}{2} + 3)(-\frac{3}{2} - 2) = -\frac{3}{2} \cdot 0 \cdot -\frac{7}{2} = 0$.

Note: to use the Zero Product Property, you must have a product of *zero.* This property does *not* work for equations such as $(x + 3)(x - 2) = 1$ or $y(2y + 3)(y - 2) = -3$. Also, this property does not work for equations like $x^2 + 3x + 50 = 0$, because $x^2 + 3x + 50$ is prime over the integers.

Service charge. *Shown is tennis star Steffi Graf. Many tennis players bounce the ball in preparation for the service toss.*

QUESTIONS

Covering the Reading

1. State the Zero Product Property. **If the product of two real numbers *a* and *b* is 0, then *a* = 0 or *b* = 0.**
2. a. For what values of k does $(k + 4)(k - 1) = 0$? **$k = -4$ or $k = 1$**
   b. Check your answers. **$(-4 + 4)(-4 - 1) = 0 \cdot -5 = 0$; $(1 + 4)(1 - 1) = 5 \cdot 0 = 0$**
3. A ball bounces upward from ground level at 64 feet per second. The distance d above the ground after t seconds is $d = 64t - 16t^2$.
   a. Use factoring and the Zero Product Property to determine after how many seconds the ball will hit the ground. **after 4 seconds**
   b. Explain how part **a** could be answered by graphing. **The x-intercepts of the graph, if any, are times when the ball is at ground level.**

In 4–9, solve by factoring.

4. $x^2 + 2x - 3 = 0$ **$x = -3$ or $x = 1$**
5. $y^2 - 4y - 5 = 0$ **$y = -1$ or $y = 5$**
6. $x^2 - 12x + 27 = 0$ **$x = 3$ or $x = 9$**
7. $a^2 + 13a + 36 = 0$ **$a = -4$ or $a = -9$**
8. $t^2 - 64 = 0$ **$t = -8$ or $t = 8$**
9. $v^2 + 14v + 49 = 0$ **$v = -7$**

In 10 and 11, use Example 3. **10) See left.**

10. Why was the answer -10 not accepted as a solution to the question?
11. If 21 chess games are played, how many players are entered? **7**
12. To solve a quadratic equation by factoring, first make sure the equation is in standard form. Then **a.** _?_ the quadratic expression and apply the **b.** _?_ Property to solve. **factor; Zero Product**

10) There cannot be a negative number of players entered in a chess tournament.

Notes on Questions

Question 3 Relate this question to the first page of the lesson. Notice that the initial vertical velocity of the ball determines how long it will be in the air. (This does not apply to lighter-than-air objects, such as feathers, or objects that can generate lift, such as Frisbees.) However, the initial vertical velocity does not determine how far the ball will travel. In football, a punter has the dual task of keeping the ball in the air as long as possible and also kicking it as far down field as possible. Kicking it down the field cuts down on the *vertical* component of the velocity that is needed to keep the football in the air.

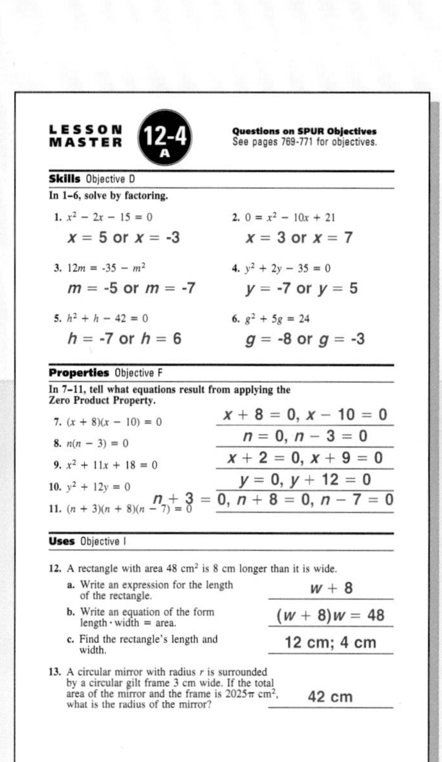

LESSON MASTER 12-4 A — **Questions on SPUR Objectives** See pages 769-771 for objectives.

Skills Objective D

In 1–6, solve by factoring.

1. $x^2 - 2x - 15 = 0$ — $x = 5$ or $x = -3$
2. $0 = x^2 - 10x + 21$ — $x = 3$ or $x = 7$
3. $12m = -35 - m^2$ — $m = -5$ or $m = -7$
4. $y^2 + 2y - 35 = 0$ — $y = -7$ or $y = 5$
5. $h^2 + h - 42 = 0$ — $h = -7$ or $h = 6$
6. $g^2 + 5g = 24$ — $g = -8$ or $g = -3$

Properties Objective F

In 7–11, tell what equations result from applying the Zero Product Property.

7. $(x + 8)(x - 10) = 0$ — $x + 8 = 0, x - 10 = 0$
8. $n(n - 3) = 0$ — $n = 0, n - 3 = 0$
9. $x^2 + 11x + 18 = 0$ — $x + 2 = 0, x + 9 = 0$
10. $y^2 + 12y = 0$ — $y = 0, y + 12 = 0$
11. $(n + 3)(n + 8)(n - 7) = 0$ — $n + 3 = 0, n + 8 = 0, n - 7 = 0$

Uses Objective I

12. A rectangle with area 48 cm² is 8 cm longer than it is wide.
    a. Write an expression for the length of the rectangle. — $w + 8$
    b. Write an equation of the form length · width = area. — $(w + 8)w = 48$
    c. Find the rectangle's length and width. — 12 cm; 4 cm
13. A circular mirror with radius r is surrounded by a circular gilt frame 3 cm wide. If the total area of the mirror and the frame is 2025π cm², what is the radius of the mirror? — 42 cm

Adapting to Individual Needs

Challenge

Ask students to write a quadratic equation that has the given solution set. The equation should be written in the form $ax^2 + bx + c = 0$ where a, b, and c are integers.

1. $\{3, -4\}$ $[x^2 + x - 12 = 0]$
2. $\{-5, -6\}$ $[x^2 + 11x + 30 = 0]$
3. $\{6, -2\}$ $[x^2 - 4x - 12 = 0]$
4. $\{\frac{1}{2}, \frac{2}{3}\}$ $[6x^2 - 7x + 2 = 0]$
5. $\{\frac{3}{5}, -\frac{1}{4}\}$ $[20x^2 - 7x - 3 = 0]$
6. $\{-\frac{1}{2}, -\frac{1}{2}\}$ $[4x^2 + 4x + 1 = 0]$

If students need a hint, note if -3 and 7 are solutions, then the line previous to "$x = -3$ or $x = 7$" could be "$x + 3 = 0$ or $x - 7 = 0$." This tells the factors of the quadratic equation. So, $(x + 3)(x - 7) = 0$ and therefore, $x^2 - 4x - 21 = 0$ is an equation with the solution set $\{-3, 7\}$.

Notes on Questions

Questions 17–22 Cooperative Learning Students might **work in small groups** to find at least two ways to solve each equation. Then poll the students about the method that they prefer.

Questions 26, 29, and 31 Advise students that it is advantageous to check by testing a special case.

Question 32 Teaching Aid 130 can be used for this question.

Question 34 Because of the Difference of Squares Pattern, students know a second way to represent the area, namely $(a - b)(a + b)$.

Question 35 The questions in the *Extension* on page 743 extend this idea.

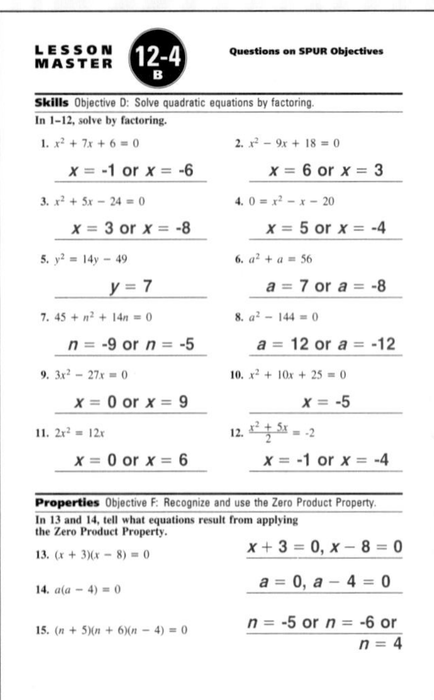

LESSON MASTER 12-4 B Questions on SPUR Objectives

Skills Objective D: Solve quadratic equations by factoring.

In 1–12, solve by factoring.

1. $x^2 + 7x + 6 = 0$ $x = -1$ or $x = -6$
2. $x^2 - 9x + 18 = 0$ $x = 6$ or $x = 3$
3. $x^2 + 5x - 24 = 0$ $x = 3$ or $x = -8$
4. $0 = x^2 - x - 20$ $x = 5$ or $x = -4$
5. $y^2 = 14y - 49$ $y = 7$
6. $a^2 + a = 56$ $a = 7$ or $a = -8$
7. $45 + n^2 + 14n = 0$ $n = -9$ or $n = -5$
8. $a^2 - 144 = 0$ $a = 12$ or $a = -12$
9. $3x^2 - 27x = 0$ $x = 0$ or $x = 9$
10. $x^2 + 10x + 25 = 0$ $x = -5$
11. $2x^2 = 12x$ $x = 0$ or $x = 6$
12. $\frac{x^2 + 5x}{2} = -2$ $x = -1$ or $x = -4$

Properties Objective F: Recognize and use the Zero Product Property.

In 13 and 14, tell what equations result from applying the Zero Product Property.

13. $(x + 3)(x - 8) = 0$ $x + 3 = 0, x - 8 = 0$
14. $a(a - 4) = 0$ $a = 0, a - 4 = 0$
15. $(n + 5)(n + 6)(n - 4) = 0$ $n = -5$ or $n = -6$ or $n = 4$

16a) $a = \frac{5 \pm \sqrt{25 + 200}}{2} = \frac{5 \pm 15}{2}$
$a = 10$ or $a = -5$

b) $(a + 5)(a - 10) = 0$
$a = -5$ or $a = 10$

c) Sample: Factoring; it is quicker and requires fewer calculations.

30) Sample: $(3 + 4)(3 - 4) = 7 \cdot -1 = -7$; $(\sqrt{9})^2 - (\sqrt{16})^2 = 9 - 16 = -7$; $(3 + 4)(3 - 4) = 7(3 - 4) = 21 - 28 = -7$

31a)

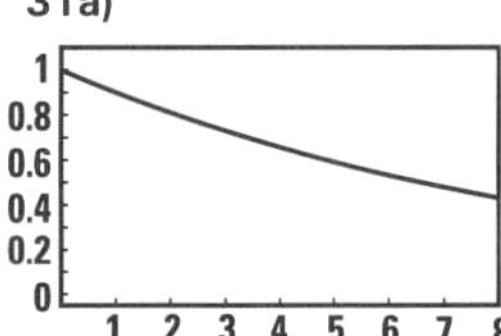

White light. *When a beam of white light, which contains all the wavelengths of visible light, passes through a prism, the beam is split up to form the band of colors called a spectrum.*

13. Why can't the Zero Product Property be used on the equation $(w + 1)(w + 2) = 3$? **Because the product of factors is not zero.**

In 14 and 15, solve.

14. $(y - 15)(9y - 8) = 0$ **$y = 15$ or $y = \frac{8}{9}$**

15. $r(2r + 5)(r - 6) = 0$ **$r = 0$ or $r = \frac{-5}{2}$ or $r = 6$**

Applying the Mathematics

16. Consider the equation $a^2 - 5a - 50 = 0$ **See left.**
    a. Solve this equation by using the Quadratic Formula.
    b. Solve this equation by using factoring.
    c. Which solution strategy, the one in part **a** or part **b**, do you prefer? Why?

In 17–22, solve by using any method.

17. $y^2 - 4y = 5$ **$y = 5$ or $y = -1$**

18. $26 + b^2 = -15b$ **$b = -2$ or $b = -13$**

19. $7p^2 + 7p - 84 = 0$ **$p = 3$ or $p = -4$**

20. $g^2 - 100 = 0$ **$g = 10$ or $g = -10$**

21. $2v^2 = 3v$ **$v = 0$ or $v = \frac{3}{2}$**

22. $0 = x^3 - 4x^2 + 4x$ **$x = 0$ or $x = 2$**

23. The sum of the consecutive even integers from 2 to $2n$ is $n^2 + n$.
    a. Show that the formula is true for $n = 5$.
    b. How many consecutive even integers need to be added in order to reach a sum of 132? **11**

    a) $5^2 + 5 = 30$; $2 + 4 + 6 + 8 + 10 = 30$

24. If $a \neq 0$, solve $ax^2 - ax = 0$ for x. **$x = 0$ or $x = 1$**

25. *Multiple choice.* Which number must be a solution to the equation $ax^2 + bx = 0$, if $a \neq 0$? **d**
    (a) 1 (b) a (c) b (d) 0

Review

26. Factor $2x^3 - 18x^2 + 16x$. *(Lesson 12-3)* **$2x(x - 1)(x - 8)$**

In 27 and 28, simplify. *(Lessons 10-5, 12-2)*

27. $\frac{13x^2 - 14x}{x}$ (Assume $x \neq 0$.) **$13x - 14$**

28. $(u - v)^2 - (u^2 - v^2)$ **$2v^2 - 2uv$**

29. Find the area of a square which is $4e + 1$ units on a side. *(Lesson 10-5)* **$16e^2 + 8e + 1$ square units**

30. Describe at least three different ways to evaluate the expression $(\sqrt{9} + \sqrt{16})(\sqrt{9} - \sqrt{16})$. *(Lessons 1-6, 9-7, 10-6)* **See left.**

31. A certain glass allows 90% of the light hitting it to pass through. The fraction y of light passing through x thicknesses of glass is $y = (0.9)^x$.
    a. Draw a graph of this equation for $0 \leq x \leq 8$. **See left.**
    b. Use the graph to estimate the number of thicknesses you would need to allow only half the light hitting it to pass through. *(Lesson 8-4)* **about 6.6 thicknesses**

742

Diam. (in.)	Vol. (ft³)
8.3	10.3
8.6	10.3
8.8	10.2
10.5	16.4
10.7	18.8
11.0	15.6
11.0	18.2
11.1	22.6
11.2	19.9
11.3	24.2
11.4	21.0
11.4	21.4
11.7	21.3
12.0	19.1
12.9	22.2
12.9	33.8
13.3	27.4
13.7	25.7
13.8	24.7
14.0	34.5
14.2	31.7
14.5	36.3
16.3	42.6
17.3	55.4
17.5	55.7
17.9	58.3
18.0	51.0
18.0	51.5
20.6	77.0

32. Foresters in the Allegheny National Forest were asked to estimate the volume of timber in the black cherry trees in the forest. They had to do it without cutting all the black cherry trees down. They did cut down 29 trees of varying sizes. They measured the diameter of each tree 4.5 ft above ground level and the volume (in cubic feet) of wood which the tree produced. At the left are their data. A scatterplot is shown below.

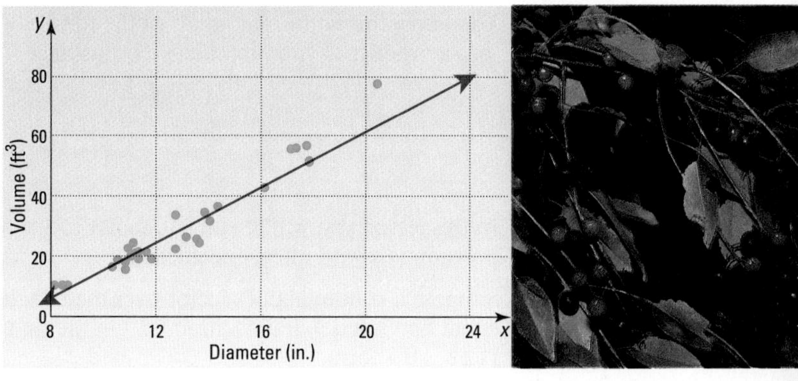

Fruit from a black cherry tree.

a. Fit a line to the data by eye.

b. Find an equation for your line. **Sample: $y = 4.45x - 29.94$**

c. Estimate the volume of a black cherry tree with diameter 15 inches. *(Lesson 7-7)* **about 36.8 ft³**

In 33 and 34, calculate the area of the shaded region. *(Lessons 2-1, 4-2)*

33.

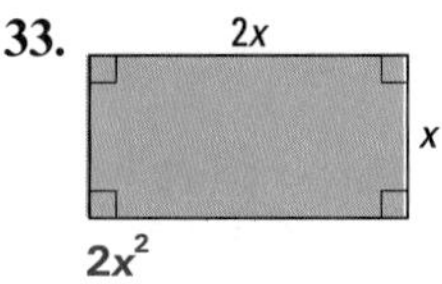

$2x^2$

34.

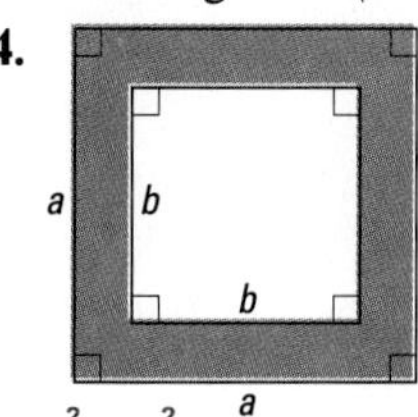

$a^2 - b^2$

Exploration

35. Some polynomials of degree higher than 2 can be solved by using chunking and the method of this lesson. For example,

$$x^4 - 3x^2 - 4 = (x^2)^2 - 3(x^2) - 4 = (x^2 - 4)(x^2 + 1).$$

Use this idea to solve the following equations.

a. $x^4 - 3x^2 - 4 = 0$ **$x = 2$ or $x = -2$**

b. $m^4 - 13m^2 + 36 = 0$ **$m = 2$ or $m = -2$ or $m = 3$ or $m = -3$**

Follow-up for Lesson 12-4

Practice

For more questions on SPUR Objectives, use **Lesson Master 12-4A** (shown on page 741) or **Lesson Master 12-4B** (shown on pages 742–743).

Assessment

Written Communication Have students write a paragraph that describes three methods which can be used to solve a quadratic equation. [Students understand that quadratic equations can be solved by graphing, using the Quadratic Formula, or using factoring and the Zero Product Property.]

Extension

You might extend **Question 35** by substituting other expressions for x and using chunking. For instance, have students solve $x^2 - 5x + 4 = 0$. [4, 1] Then have them try to solve these equations.

1. $(2y + 1)^2 - 5(2y + 1) + 4 = 0$ [0, 1.5]
2. $|x|^2 - 5|x| + 4 = 0$ [1, -1, 4, -4]
3. $(\sqrt{z})^2 - 5\sqrt{z} + 4 = 0$ [1, 16]

▶ **LESSON MASTER 12-4 B** *page 2*

16. a. Give four pairs of numbers whose product is 12. **Sample: 1, 12; 2, 6; 3, 4; -3, -4**

b. Give four pairs of numbers whose product is 0. **Sample: 1, 0; 8, 0; -6, 0; 100, 0**

c. Explain why there is a Zero Product Property but there is no Twelve Product Property. **Sample: To get a product of zero, zero must be a factor; to get a product of 12, 12 does not have to be a factor.**

In 17–20, if the polynomial is factorable over the integers, factor and solve the equation using the Zero Product Property. If it is not factorable, use the Quadratic Formula.

17. $x^2 + 9x + 12 = 0$ **$x = -1.63$ or $x = -7.37$**

18. $x^2 - 8x - 20 = 0$ **$x = 10$ or $x = -2$**

19. $t^2 + 9t + 14 = 0$ **$t = -2$ or $t = -7$**

20. $x^2 - 12x = 0$ **$x = 0$ or $x = 12$**

Uses Objective I: Solve quadratic equations in real situations.

21. A rectangular park is 3 blocks longer than it is wide. Its area is 40 square blocks. Let w = width.

a. Write an expression for the length of the rectangle. **$w + 3$**

b. Write an equation of the form length · width = area. **$(w + 3)w = 40$**

c. Find the length and width of the park. **8 blocks, 5 blocks**

22. At a party with n guests, every guest shook hands with every other guest once, for $\frac{n^2 - n}{2}$ handshakes. If there were 66 handshakes, how many guests were there? **12 guests**

Lesson 12-5

Objectives

C Factor quadratic expressions.
D Solve quadratic equations by factoring.

Resources

From the ***Teacher's Resource File***
- Lesson Master 12-5A or 12-5B
- Answer Master 12-5
- Teaching Aid 128: Warm-up
- Technology Connection Computer Master 30

Additional Resources
- Visual for Teaching Aid 128
- GraphExplorer or other automatic graphers

Teaching Lesson 12-5

Warm-up

Tell whether or not each polynomial is factorable. If the polynomial is factorable, factor it.

1. $x^2 - 6x - 3$ **not factorable**
2. $x^2 + 2x - 8$ $(x - 2)(x + 4)$
3. $x^2 - 81$ $(x + 9)(x - 9)$
4. $x^2 + 13x + 42$ $(x + 6)(x + 7)$
5. $x^2 - 13x + 40$ $(x - 5)(x - 8)$
6. $x^2 + 13x + 25$ **not factorable**

LESSON

12-5

Factoring $ax^2 + bx + c$

You know that some trinomials of the form $x^2 + bx + c$ can be factored into a product of two binomials. For instance,

$$x^2 - y^2 = (x + y)(x - y)$$
$$x^2 + 6x + 9 = (x + 3)(x + 3)$$
$$x^2 - 2x + 1 = (x - 1)(x - 1)$$
$$x^2 - 6x - 16 = (x + 2)(x - 8).$$

In this lesson you will see how to factor $ax^2 + bx + c$ in the set of polynomials with integer coefficients.

Suppose a trinomial does factor over the integers. On the left side of the equal sign is a quadratic trinomial. On the right side are two binomials. Here is the form.

$$ax^2 + bx + c = (dx + e)(fx + g)$$

The product of d and f, from the first terms of the binomials, is a. The product of e and g, the last terms of the binomials, is c. The task is to find these numbers so that the rest of the multiplication gives b.

Example 1

Factor $2x^2 + 11x + 14$.

Solution

The idea is to rewrite the expression as a product of two binomials.

$$2x^2 + 11x + 14 = (dx + e)(fx + g)$$

So we need to find integers d, e, f, and g.

The coefficient of $2x^2$ is 2, so $df = 2$. Thus one of d or f is 2, and the other is 1. Now you know

$$2x^2 + 11x + 14 = (2x \_\_ e)(x \_\_ g).$$

The product of e and g is 14, so $eg = 14$. Thus, e and g might equal 1 and 14, or 2 and 7, in either order. Try all four possibilities.

Can e and g be 1 and 14?

$$(2x + 1)(x + 14) = 2x^2 + 29x + 14$$
$$(2x + 14)(x + 1) = 2x^2 + 16x + 14$$

No, we want $b = 11$, not 29 or 16.

Can e and g be 2 and 7?

$$(2x + 2)(x + 7) = 2x^2 + 16x + 14$$
$$(2x + 7)(x + 2) = 2x^2 + 11x + 14$$

The last product works.

$$2x^2 + 11x + 14 = (2x + 7)(x + 2)$$

►

Lesson 12-5 Overview

Broad Goals The ideas of the preceding two lessons are extended here to cover situations in which the coefficient of the square term is not 1.

Perspective A study of advanced algebra, precalculus, and calculus books indicates that factoring is seldom used except in simplifying and operating with rational expressions and in finding limits of rational functions. In almost all cases, it appears that the examples which are given are contrived for the sole purpose of providing practice with factoring. Thus, we have the following circular situation: factoring quadratic trinomials is included in books to provide work with rational polynomial expressions; rational polynomial expressions are included in books to provide practice with factoring. These remarks do not apply to either the difference of squares or to the factoring of perfect squares, both of which are seen in many contexts.

The trial-and-error method of **Example 2** does not satisfy the many students who want an automatic algorithm. One algorithm for factoring a trinomial uses the Quadratic Formula. Consider the equation of **Example 3:** $6y^2 - 7y - 5 = 0$. Use the Quadratic

Check 1

Substitute a value for x, say 4. Then does

$$2x^2 + 11x + 14 = (2x + 7)(x + 2)?$$
$$2 \cdot 4^2 + 11 \cdot 4 + 14 = (2 \cdot 4 + 7)(4 + 2)?$$
$$32 + 44 + 14 = 15 \cdot 6?$$

Yes. Each side equals 90.

Check 2

Graph $y = 2x^2 + 11x + 14$ and $y = (2x + 7)(x + 2)$ on the same set of axes. The graphs should be identical. At the right is the output of our automatic grapher. It checks.

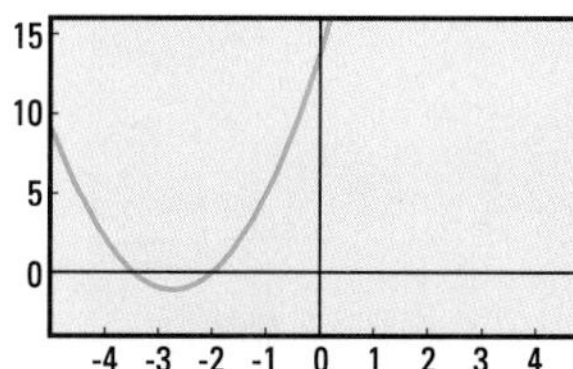

In Example 1, because the coefficient of x^2 is 2 and all terms are positive, there are only a few possible factors. Example 2 has more possibilities, but the idea is still the same.

Example 2

Factor $6y^2 - 7y - 5$.

Solution

First put down the form. $6y^2 - 7y - 5 = (ay + b)(cy + d)$

Now $ac = 6$. Thus either a and c are 3 and 2 or they are 1 and 6. The product $bd = -5$. So b and d are either 1 and -5, or -1 and 5. Here are all the possible factors with $a = 3$ and $c = 2$.

$$(3y + 1)(2y - 5)$$
$$(3y - 1)(2y + 5)$$
$$(3y - 5)(2y + 1)$$
$$(3y + 5)(2y - 1)$$

Here are all the possible factors with $a = 1$ and $c = 6$.

$$(y + 1)(6y - 5)$$
$$(y - 1)(6y + 5)$$
$$(y - 5)(6y + 1)$$
$$(y + 5)(6y - 1)$$

At most, you need to do these eight multiplications. If one of them gives $6y^2 - 7y - 5$, then that is the correct factorization.

We show all eight multiplications. Notice that the third one is correct.

$$(3y + 1)(2y - 5) = 6y^2 - 13y - 5$$
$$(3y - 1)(2y + 5) = 6y^2 + 13y - 5$$
$$(3y - 5)(2y + 1) = 6y^2 - 7y - 5$$
$$(3y + 5)(2y - 1) = 6y^2 + 7y - 5$$
$$(y + 1)(6y - 5) = 6y^2 + y - 5$$
$$(y - 1)(6y + 5) = 6y^2 - y - 5$$
$$(y - 5)(6y + 1) = 6y^2 - 29y - 5$$
$$(y + 5)(6y - 1) = 6y^2 + 29y - 5$$

So $6y^2 - 7y - 5 = (3y - 5)(2y + 1)$.

Notes on Reading

Reading Mathematics The questions students should ask themselves as they read this material are the same as those in previous lessons.

1. Is there a common monomial factor?
2. Can the quadratic be factored over the integers? (See the first paragraph and **Example 1.**)
3. How does a person obtain the factors? (The method used here is sophisticated trial and error; the same method is found in both **Examples 2 and 3.** The strategy is to first consider all the possibilities based on the factors of a and c and then to check each possibility by multiplying until the desired factors are found.)
4. How is this procedure applied in solving equations? (See **Example 3.**)

For **Example 1**, we have written both what a student might write and what a student might think. You may want to take the material in the writing font and display it without the intervening notes. It is important to repeat the process with other examples.

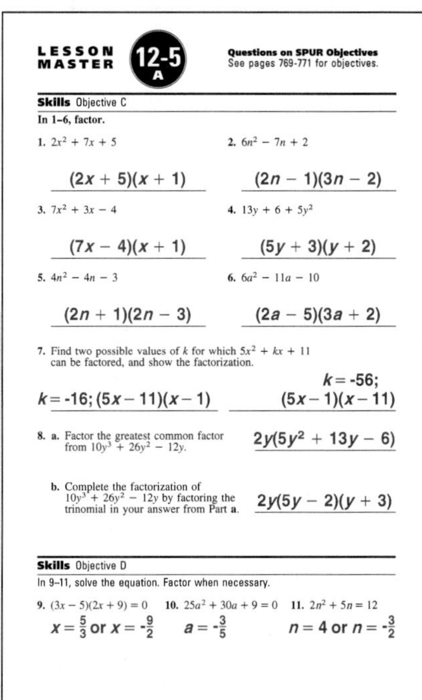
LESSON MASTER 12-5 A

Questions on SPUR Objectives
See pages 769-771 for objectives.

Skills Objective C

In 1–6, factor.

1. $2x^2 + 7x + 5$ — $(2x + 5)(x + 1)$
2. $6n^2 - 7n + 2$ — $(2n - 1)(3n - 2)$
3. $7x^2 + 3x - 4$ — $(7x - 4)(x + 1)$
4. $13y + 6 + 5y^2$ — $(5y + 3)(y + 2)$
5. $4n^2 - 4n - 3$ — $(2n + 1)(2n - 3)$
6. $6a^2 - 11a - 10$ — $(2a - 5)(3a + 2)$

7. Find two possible values of k for which $5x^2 + kx + 11$ can be factored, and show the factorization.
$k = -16$; $(5x - 11)(x - 1)$ $k = -56$; $(5x - 1)(x - 11)$

8. a. Factor the greatest common factor from $10y^3 + 26y^2 - 12y$. $2y(5y^2 + 13y - 6)$

b. Complete the factorization of $10y^3 + 26y^2 - 12y$ by factoring the trinomial in your answer from Part a. $2y(5y - 2)(y + 3)$

Skills Objective D

In 9–11, solve the equation. Factor when necessary.

9. $(3x - 5)(2x + 9) = 0$ — $x = \frac{5}{3}$ or $x = -\frac{9}{2}$
10. $25a^2 + 30a + 9 = 0$ — $a = -\frac{3}{5}$
11. $2n^2 + 5n = 12$ — $n = 4$ or $n = -\frac{3}{2}$

Formula to find the solutions $\frac{5}{3}$ and $-\frac{1}{2}$. Now reverse the steps to obtain the factorization of $6y^2 - 7y - 5$. An advantage of this algorithm is that it works for factorizations of polynomials with irrational coefficients.

A second algorithm that students will see in later mathematics courses involves the Rational Root Theorem, which enables one to find all the rational roots. A third algorithm for factoring uses graphing. To factor $ax^2 + bx + c$, graph $y = ax^2 + bx + c$, and estimate the x-intercepts r and s by tracing along the graph.

Emphasize to students that before they try to factor a trinomial, a question they should ask themselves is whether the trinomial is factorable. Why waste time trying to factor if it is not factorable? Also point out that multiplying is not necessarily a good way to check a problem solved by factoring because the factors were probably found by multiplying in the first place.

Pacing Traditional algebra courses spend a great deal of time on this topic. We suggest that you spend two days on these ideas, one day on this lesson and one day on Lesson 12-6. You may find that you will spend another day on the ideas during the Chapter Review.

Additional Examples

1. Is $2x^2 + 5x + 3$ factorable? If it is, factor it. If not, tell why not. **Yes, $(2x + 3)(x + 1)$**
2. Factor each trinomial.
   a. $2n^2 - 3n - 20$. **$(2n + 5)(n - 4)$**
   b. $6y^2 - 29y - 5$ **$(y - 5)(6y + 1)$**
3. Solve each equation.
   a. $2n^2 - 3n - 20 = 0$ **4, -2.5**
   b. $6y^2 - 29y - 5 = 0$ **5, $-\frac{1}{6}$**

LESSON MASTER 12-5 B Questions on SPUR Objectives

Skills Objective C: Factor quadratic expressions.

1. Consider factoring the polynomial $4x^2 + 4x - 3$ into two binomials $(ax + b)(cx + d)$.
   a. List all the possible pairs of numbers for a and c. **1, 4; 2, 2**
   b. List all the possible pairs of numbers for b and d. **3, -1; -3, 1**
   c. List all the possible binomials $(ax + b)(cx + d)$. **$(4x + 3)(x - 1)$; $(x - 3)(4x + 1)$; $(4x - 3)(x + 1)$; $(x + 3)(4x - 1)$; $(2x + 1)(2x - 3)$; $(2x - 1)(2x + 3)$**
   d. Factor $4x^2 + 4x - 3$. **$(2x + 3)(2x - 1)$**

In 2–15, factor.

2. $2x^2 + 9x + 9$ **$(2x + 3)(x + 3)$**
3. $3x^2 - 14x - 5$ **$(3x + 1)(x - 5)$**
4. $5x^2 + 6 - 17x$ **$(5x - 2)(x - 2)$**
5. $6x^2 + 5x + 1$ **$(3x + 1)(2x + 1)$**
6. $6x^2 + 13x + 2$ **$(6x + 1)(x + 2)$**
7. $4n^2 - 3n - 7$ **$(n + 1)(4n - 7)$**
8. $10a^2 + 11a - 6$ **$(2a + 3)(5a - 2)$**
9. $20x + 4 + 9x^2$ **$(9x + 2)(x + 2)$**
10. $3x^2 + 14x - 49$ **$(x + 7)(3x - 7)$**
11. $4y^2 + 4y + 1$ **$(2y + 1)^2$**

If at first you don't succeed . . . *To find your way out of the* Maze, *displayed at the Museum of Science and Industry in Chicago, you would probably use trial and error, just as you might do in factoring.*

In Example 2, notice that each choice of factors gives a product that differs only in the coefficient of y (the middle term). If the original problem were to factor $6y^2 - 100y - 5$, the products listed show that no factors with integer coefficients will work. The quadratic $6y^2 - 100y - 5$ is prime in the set of polynomials with integer coefficients.

Factoring quadratic trinomials using trial and error is a skill many people learn to do in their heads. But it takes practice. This skill can save you time in solving some quadratic equations or in recognizing equivalent forms of expressions.

Example 3

Solve $6y^2 - 7y - 5 = 0$.

Solution

Use the factorization of $6y^2 - 7y - 5$ found in Example 2.

$$6y^2 - 7y - 5 = 0$$
$$(3y - 5)(2y + 1) = 0$$
So $$3y - 5 = 0 \quad \text{or} \quad 2y + 1 = 0$$
$$y = \frac{5}{3} \quad \text{or} \quad y = -\frac{1}{2}$$

QUESTIONS

Covering the Reading

1. Perform the multiplications in **a–d.** **a) $5x^2 + 38x + 21$ c) $5x^2 + 106x + 21$**
   a. $(5x + 3)(x + 7)$
   b. $(5x + 7)(x + 3)$ **$5x^2 + 22x + 21$**
   c. $(5x + 1)(x + 21)$
   d. $(5x - 1)(x - 21)$ **$5x^2 - 106x + 21$**
   e. Explain how these multiplications are related to factoring $5x^2 + 26x + 21$. **See left.**

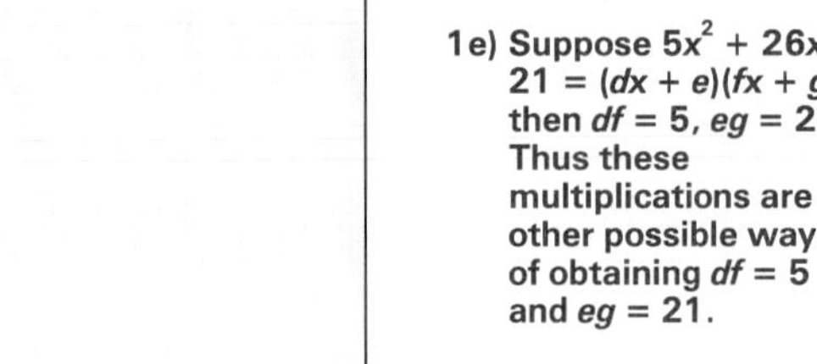

1e) Suppose $5x^2 + 26x + 21 = (dx + e)(fx + g)$; then $df = 5$, $eg = 21$. Thus these multiplications are other possible ways of obtaining $df = 5$ and $eg = 21$.

2. Suppose $ax^2 + bx + c = (dx + e)(fx + g)$ for all values of x.
   a. The product of d and f is __?__. ***a***
   b. The product of __?__ and __?__ is c. ***e, g***
3. Give two ways to check the factorization of a quadratic trinomial. **Substitute a number; graph**

In 4–9, factor the trinomial, if possible.

4. $2x^2 + 7x + 5$ **$(2x + 5)(x + 1)$**
5. $7x^2 - 36x + 5$ **$(7x - 1)(x - 5)$**
6. $y^2 + 10y + 9$ **$(y + 9)(y + 1)$**
7. $4x^2 - 12x - 7$ **$(2x - 7)(2x + 1)$**
8. $3x^2 + 11x - 4$ **$(3x - 1)(x + 4)$**
9. $10k^2 - 23k + 12$ **$(5k - 4)(2k - 3)$**
10. Check the solutions of Example 3 by substitution. **See below.**
11. Solve $2x^2 + 11x + 14 = 0$ by factoring. **$(2x + 7)(x + 2) = 0$; $x = \frac{-7}{2}$ or $x = -2$**

10) $6(\frac{5}{3})^2 - 7(\frac{5}{3}) - 5 = 0$? $6 \cdot \frac{25}{9} - 7 \cdot \frac{5}{3} - 5 = \frac{50}{3} - \frac{35}{3} - 5 = \frac{15}{3} - 5 = 0$.
$6(-\frac{1}{2})^2 - 7(-\frac{1}{2}) - 5 = 0$? $6 \cdot \frac{1}{4} - 7 \cdot -\frac{1}{2} - 5 = \frac{6}{4} + \frac{7}{2} - 5 = \frac{10}{2} - 5 = 0$.

Optional Activities

Activity 1 Cooperative Learning
You might use this activity after students have read the lesson. Have three students each name an integer, one for a, one for b, and a third for c. Then write a trinomial of the form $ax^2 + bx + c$ on the board, using the values that are given. Have students try to factor the trinomial. Repeat the activity for other sets of three numbers.

Activity 2 Technology Connection
You may wish to assign *Technology Sourcebook, Computer Master 30.* Students use *GraphExplorer* or similar software to factor quadratic expressions.

Applying the Mathematics

14a, i) $t = \frac{5 \pm \sqrt{25 + 504}}{12}$
$= \frac{5 \pm 23}{12}$
$t = \frac{7}{3}$ or $t = -\frac{3}{2}$
ii) $(2t + 3)(3t - 7) = 0$
$t = -\frac{3}{2}$ or $t = \frac{7}{3}$

b) Sample: the Quadratic Formula, because there are many possible factors to consider.

15a, i)

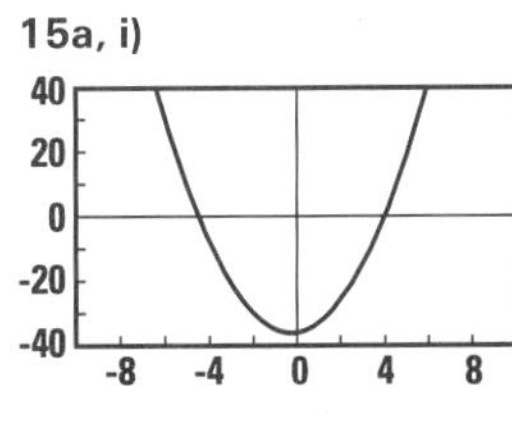

12. Without graphing, what can you say about the graphs of $y = (2x - 5)(x + 1)$ and $y = 2x^2 - 3x - 5$? Justify your answer. **The graphs are the same since $(2x - 5)(x + 1) = 2x^2 - 3x - 5$.**
13. Find k if $8x^2 + 10x - 25 = (2x + k)(4x - k)$. **$k = 5$**
14. a. Solve the equation $6t^2 - 5t - 21 = 0$ **See left.**
    i. by using the Quadratic Formula.
    ii. by factoring.
    b. Which method in part **a** do you prefer? Why?
15. a. Find the x-intercepts of the graph of $y = 2x^2 + x - 36$
    i. by making a graph and estimating the coordinates. **See left.**
    ii. by factoring. **$(x - 4)(2x + 9) = 0$; $x = 4$ or $x = -\frac{9}{2}$**
    b. Which method in part **a** do you prefer? Why? **Sample: factoring, because it is more precise**
16. a. Factor $20x^3 - 70x^2 + 60x$ into the product of a monomial and a trinomial. **$10x(2x^2 - 7x + 6)$**
    b. Give the complete factorization of $20x^3 - 70x^2 + 60x$. **$10x(x - 2)(2x - 3)$**

In 17 and 18, find the complete factorization.

17. $9p^4 + 12p^3 + 4p^2$ **$p^2(3p + 2)^2$**
18. $2x^2 - 5xy + 3y^2$ **$(2x - 3y)(x - y)$**

Review

19. State the Zero Product Property. *(Lesson 12-4)* **If the product of two real numbers a and b is zero, then $a = 0$ or $b = 0$.**

In 20–23, solve by using any method. *(Lessons 9-5, 12-4)*

20. $x^2 - 8x + 7 = 0$ **$x = 1$ or $x = 7$**
21. $y^2 + 9y + 7 = 0$ **$y \approx -8.14$ or $y \approx -0.86$**
22. $d^2 + 16 = 8d$ **$d = 4$**
23. $100v^2 - 400 = 0$ **$v = 2$ or $v = -2$**
24. a. Factor $x^2 + 8x + 7$. **$(x + 1)(x + 7)$**
    b. Use the factoring of part **a** to find the prime factorization of 187. *(Lessons 12-1, 12-3)* **$187 = 10^2 + 8 \cdot 10 + 7 = (10 + 1)(10 + 7) = 11 \cdot 17$**
25. Rewrite $x^4 - 81$ as the product of
    a. two binomials. **$(x^2 + 9)(x^2 - 9)$**
    b. three binomials. *(Lessons 10-6, 12-3)* **$(x^2 + 9)(x + 3)(x - 3)$**
26. Which holds more?
    a. a cube with edges of length 4, or a rectangular box with dimensions 3 by 4 by 5 **the cube**
    b. a cube with edges of length e, or a box with dimensions $e - 1$ by e by $e + 1$. *(Lessons 2-1, 10-6)* **the cube**

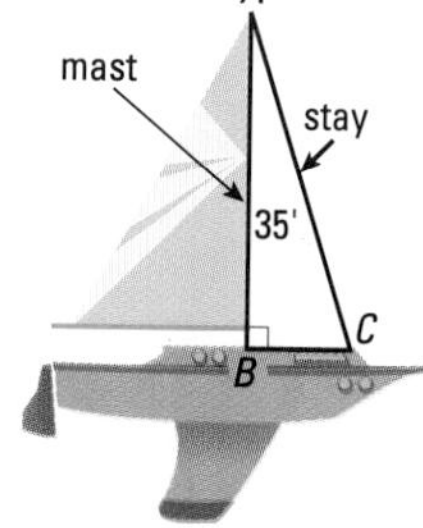

In 27 and 28, a 35-foot mast on a sailboat is strengthened by a wire called a *stay*, as shown at the left.

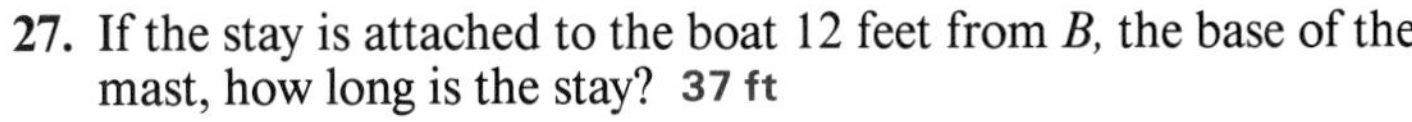

27. If the stay is attached to the boat 12 feet from B, the base of the mast, how long is the stay? **37 ft**

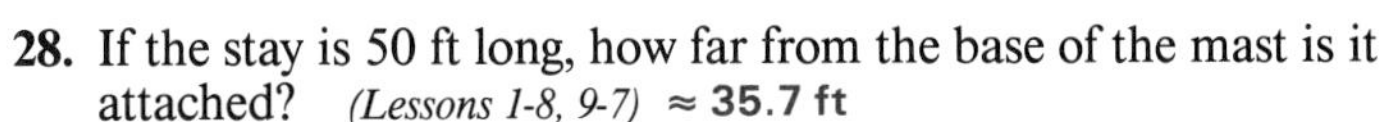

28. If the stay is 50 ft long, how far from the base of the mast is it attached? *(Lessons 1-8, 9-7)* **≈ 35.7 ft**

Notes on Questions

Questions 4–9 In Lesson 12-7, students will learn to use the discriminant to test for "factorability." This prevents the frustration and uncertainty caused by unsuccessfully searching for factors. Thus you might want to use these questions as motivation for Lesson 12-7.

Questions 12–13 You might have students check their answers by using an automatic grapher.

Questions 14–15 Expect that students will not necessarily agree. Encourage discussion of the relative merits of each method to help students realize that it is useful to learn all of the methods.

Question 25 Students have to use chunking here, considering x^2 as the chunk. Many students are surprised that **part b** is possible.

Question 26 You might also ask students for the difference in the volumes. [a. 4; b. e]

Questions 27–28 Sports Connection Sailing requires considerable skill and an intuitive knowledge of geometry. For instance, to take advantage of the wind blowing across the sails, sails should be 45° from the wind. To determine the relationship of the wind to the setting of the sails, a sailor observes the direction that the masthead pennant is blowing.

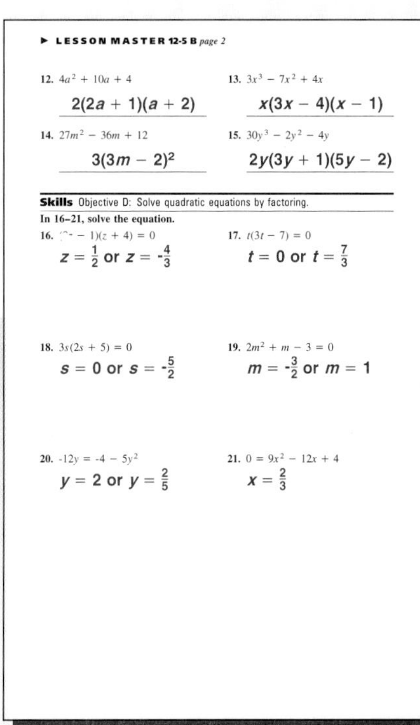

► LESSON MASTER 12-5 B *page 2*

12. $4a^2 + 10a + 4$ — $2(2a + 1)(a + 2)$
13. $3x^3 - 7x^2 + 4x$ — $x(3x - 4)(x - 1)$
14. $27m^2 - 36m + 12$ — $3(3m - 2)^2$
15. $30y^3 - 2y^2 - 4y$ — $2y(3y + 1)(5y - 2)$

Skills Objective D: Solve quadratic equations by factoring.

In 16–21, solve the equation.

16. $(2z - 1)(z + 4) = 0$ — $z = \frac{1}{2}$ or $z = -\frac{4}{3}$
17. $t(3t - 7) = 0$ — $t = 0$ or $t = \frac{7}{3}$
18. $3s(2s + 5) = 0$ — $s = 0$ or $s = -\frac{5}{2}$
19. $2m^2 + m - 3 = 0$ — $m = -\frac{3}{2}$ or $m = 1$
20. $-12y = -4 - 5y^2$ — $y = 2$ or $y = \frac{2}{5}$
21. $0 = 9x^2 - 12x + 4$ — $x = \frac{2}{3}$

Adapting to Individual Needs

Extra Help

A very important thing for students to remember in factoring trinomials is that they can always check the results of factoring by multiplying. The FOIL algorithm was introduced in Lesson 10-5. Some students may have to review that lesson. If they cannot successfully apply the FOIL algorithm, they will probably have trouble factoring trinomials.

Notes on Questions

Question 30 If scales were not given on both axes, the answer could be either choice c or choice d.

Follow-up for Lesson 12-5

Practice

For more questions on SPUR Objectives, use **Lesson Master 12-5A** (shown on page 745) or **Lesson Master 12-5B** (shown on pages 746–747).

Assessment

Written Communication Ask each student to write a paragraph showing what he or she is able to do after completing this lesson. [Students demonstrate the ability to solve quadratic equations by factoring and are able to determine if a polynomial can be factored over the integers.]

Extension

Have students use the procedure of **Question 24** to determine which of the following numbers can be written as $x^2 + bx + c$ and then factored to find the prime factorization.

1. 143 [$143 = 10^2 + 4 \cdot 10 + 3 = (10 + 3)(10 + 1)$; $13 \cdot 11$]
2. 121 $10^2 + 2 \cdot 10 + 1 = (10 + 1)(10 + 1)$; $11 \cdot 11$]
3. 129 [129 cannot be factored using this method. However, students can find the prime factors of 129 which are 3 and 43.]
4. 165 [$10^2 + 6 \cdot 10 + 5 = (10 + 5)(10 + 1)$; $15 \cdot 11 = 3 \cdot 5 \cdot 11$]

Project Update Project 6, *Factors and Graphs*, on page 766, relates to the content of this lesson.

29b) Yes; $(2x + 2y)(x - y) = 2x^2 - 2xy + 2xy - 2y^2 = 2x^2 - 2y^2 = 2(x^2 - y^2)$

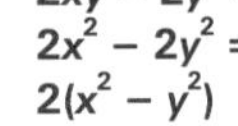

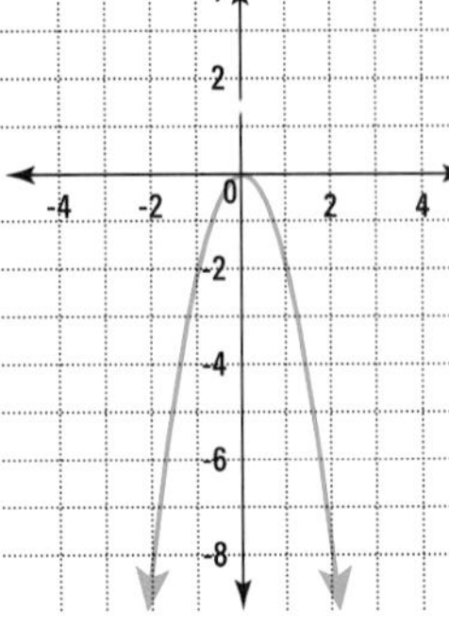

29. **a.** Here are 3 instances of a pattern. Describe the general pattern using two variables, x and y. *(Lessons 1-7, 10-6)* $(2x + 2y)(x - y) = 2(x^2 - y^2)$

$$(72 + 100)(36 - 50) = 2(36^2 - 50^2)$$
$$(30 + 20)(15 - 10) = 2(15^2 - 10^2)$$
$$(2 + 1)(1 - 0.5) = 2(1^2 - 0.5^2)$$

b. Does your general pattern hold for all real values of x and y? Justify your answer. **See left.**

30. *Multiple choice.* Which equation has the graph at the left? *(Lesson 9-1)* c

(a) $y = \frac{1}{2}x^2$ (b) $y = 2x^2$

(c) $y = -2x^2$ (d) $y = -\frac{1}{2}x^2$

In 31 and 32, give the slope and y-intercept for each line. *(Lesson 7-4)*

31. $y = \frac{1}{2}x$ slope $= \frac{1}{2}$, y-intercept = 0

32. $8x - 5y = 1$ slope $= \frac{8}{5}$, y-intercept $= -\frac{1}{5}$

33. If "One picture is worth a thousand words," what is the worth of 325 words? *(Lesson 6-8)* $\frac{13}{40}$, or .325 of a picture

One picture is worth a thousand words.

Exploration

34. The polynomial $x^3 + 6x^2 + 11x + 6$ has the factorization over the integers

$$(x + a)(x + b)(x + c).$$

Find a, b, and c. (Hint: What is abc?)
$abc = 6$; $a = 1$, $b = 2$, $c = 3$

Adapting to Individual Needs

Challenge

Point out to students that the sum of two cubes and the difference of two cubes can be factored as follows:

$$a^3 + b^3 = (a + b)(a^2 - ab + b^2)$$
$$a^3 - b^3 = (a - b)(a^2 + ab + b^2)$$

Then have students factor each of the following expressions.

1. $x^3 - 8$
   [$(x - 2)(x^2 + 2x + 4)$]
2. $y^3 + 27$
   [$(y + 3)(y^2 - 3y + 9)$]
3. $64x^3 - 125y^3$
   [$(4x - 5y)(16x^2 + 20xy + 25y^2)$]
4. $216x^3 + 1$
   [$(6x + 1)(36x^2 - 6x + 1)$]

LESSON

12-6

How Was the Quadratic Formula Found?

Heavy notebooks. *Pictured is a 14th century B.C. Babylonian clay tablet showing a seal dedicated to King Kidin-Marduk. Much of the ancient Babylonian work in mathematics was recorded on clay tablets.*

What Problem First Led to Quadratics?

Our knowledge of ancient civilizations is based only on what survives today. The earliest known problems that led to quadratic equations are on Babylonian tablets dating from 1700 B.C. These problems are to find two numbers x and y that satisfy the system

$$x + y = b$$
$$xy = c.$$

This suggests that the Babylonian scribes were interested in finding the length x and width y of a rectangle with a given area c and a given ❶ perimeter $2b$. The historian Victor Katz suggests that maybe there were some people who believed that if you knew the area of a rectangle, then you knew its perimeter. In solving these problems, the Babylonian scribes may have been trying to show that there are many rectangles with different dimensions and the same area.

Example

Find the dimensions of a rectangular field whose perimeter is 200 meters and whose area is 2475 square meters.

Solution

Let L and W be the length and width of this rectangle.

Then
$$2L + 2W = 200$$
$$LW = 2475.$$

This system can be solved by substitution. First solve the first equation for W.

$$2W = 200 - 2L$$
$$W = 100 - L$$

Now, substitute $100 - L$ for W in the second equation.

$$L(100 - L) = 2475$$

This is a quadratic equation and so it can be solved.

▶

Lesson 12-6

Objectives

I Solve quadratic equations in real situations.

Resources

From the ***Teacher's Resource File***

- Lesson Master 12-6A or 12-6B
- Answer Master 12-6
- Assessment Sourcebook: Quiz for Lessons 12-4 through 12-6
- Teaching Aid 129: Warm-up

Additional Resources

- Visual for Teaching Aid 129

Teaching Lesson 12-6

Warm-up

Use the Quadratic Formula to solve each equation.

1. $m^2 - 21m + 108 = 0$
   $m = 9$ or $m = 12$
2. $12x^2 + 7x = 10$
   $x = \frac{2}{3}$ or $x = -\frac{5}{4}$

Notes on Reading

❶ It is interesting to note that teachers, just like the Babylonian scribes, want to point out that there exist rectangular fields of different shapes that have the same area.

Lesson 12-6 Overview

Broad Goals The type of problem leading to the first known solutions of quadratics, a general description of the process used to solve these problems, and a proof of the Quadratic Formula are presented in this lesson.

Perspective A good source for the history of this lesson is *A History of Mathematics*, by Victor Katz (HarperCollins, 1993). This book is probably the best recent source for the history of mathematics; it gives more information on the contributions of non-Western cultures than most other histories.

The lesson begins with a problem that is very much like that of finding the factors of $x^2 + bx + c$, namely to find two numbers whose sum is b and whose product is c. The difference is that the Babylonians considered situations in which the numbers were not integers. The geometric approach that they used is very much like the area approach that we have been using. Students may think that the Babylonian method is a trick, but in fact, it is a kind of procedure that is often used in mathematics. Geometrically, it is equivalent to moving the graph of the quadratic so that its axis is the y-axis, which makes its roots symmetric to the y-axis. This is the algebraic equivalent to the analytic geometry of moving a *(Overview continues on page 750.)*

❷ **Multicultural Connection** You might want to use a globe to show students where the ancient places that are mentioned in this lesson were located: Babylonia (now Iraq), the Greek empire (now several countries), and China. Although the locations seem quite distant from one another, particularly considering the methods of travel available to people in ancient times, it is quite possible that the ideas of one culture were picked up by other cultures through trade.

History Connection Mathematicians of the 16th century considered the following problem: Find two numbers whose sum is 10 and whose product is 40. Solving the quadratic $(5 + x)(5 - x) = 40$ that arises (by calling the numbers $5 + x$ and $5 - x$), led to $x^2 = -15$. Thus, the numbers become $5 + \sqrt{-15}$ and $5 - \sqrt{-15}$. Complex numbers first arose from these solutions.

❸ Point out the three parts in the derivation of the Quadratic Formula. In the first part we write the equation so that x appears just once. This is accomplished in Steps 1–4 by working with the left side of the equation until it is a perfect square and then factoring it. In the second part (Step 5), knowing that the square of an expression involves x, we take the two square roots. Finally, in Steps 6 and 7, we solve each of the resulting equations for x.

Students should not be expected to memorize the derivation of the Quadratic Formula, but they should be able to see what was done to get from one step to the next. Point out that this is *proof* that these values of x will work. Students will study proof in more detail when they study geometry.

Activity 1

Solve the quadratic equation of the Example using either the Quadratic Formula or factoring. $L = 45$ or $L = 55$; Question 3 asks students to show their work for this Activity.

❷ How Were the First Quadratics Solved?

The Babylonians, like the Greeks who came after them, used a geometric approach to solve such problems. Using today's algebraic language and notation, here is what they did. It is a very sneaky way to solve this sort of problem. Refer back to the Example.

Since $L + W = 100$, the average of L and W is 50. This means that L is as much bigger than 50 as W is smaller than 50. So let $L = 50 + x$ and $W = 50 - x$. Substitute these values into the second equation.

$$(50 + x)(50 - x) = 2475$$
$$2500 - x^2 = 2475$$
$$x^2 = 25$$

So $x = 5$ or $x = -5$.
$L = 50 + x$, so $L = 50 + 5 = 55$ or $L = 50 + -5 = 45$.
$W = 50 - x$, so $W = 50 - 5 = 45$ or $W = 50 - -5 = 55$.

Either solution tells us that the rectangle has dimensions 45 meters by 55 meters.

This stamp from the former Soviet Union honors al-Khwarizmi.

Activity 2

Use the Babylonian method to find two numbers whose sum is 64 and whose product is 903. (Hint: Let one of the numbers be $32 + x$, the other $32 - x$.) 43 and 21; Question 4 asks students to show their work for this Activity.

Notice what the Babylonians did. They took a complicated quadratic equation and, with a clever substitution, reduced it to an equation of the form $x^2 = k$. That equation is easy to solve. Then they substituted back.

The work of the Babylonian scribes was lost for many years. In 825 A.D., about 2500 years after the Babylonian tablets were carved, the general method that is similar to today's Quadratic Formula was written down in words by the Arab mathematician al-Khwarizmi in a book entitled *Hisab al-jabr w'al muqabala.* Al-Khwarizmi's techniques were more general than those of the Babylonians. He gave a method to solve any equation of the form $ax^2 + bx = c$, where a, b, and c are positive numbers. His book was very influential. From the second word in that title comes our modern word "algebra." From his name comes our word "algorithm."

How Do We Know the Quadratic Formula Is True?

Neither the Babylonians nor al-Khwarizmi worked with an equation of the form $ax^2 + bx + c = 0$, because they considered only positive numbers, and if a, b, and c are positive, the equation has no positive solutions.

Lesson 12-6 Overview, continued

figure to a convenient location. Since $x^2 = k$ can be solved even when k is not a perfect square, the Babylonians were able to consider situations in which the numbers to be found were not integers. For instance, suppose we want to find two numbers whose sum is 8 and whose product is 13. Call the numbers $4 + x$ and $4 - x$. Now the product is $(4 + x)(4 - x)$, which equals $16 - x^2$. So $16 - x^2 = 13$ and $x = \sqrt{3}$. Therefore, the numbers are $4 + \sqrt{3}$ and $4 - \sqrt{3}$. The general strategy to solve any quadratic which derives the Quadratic Formula is naturally a little more difficult. The proof we offer on page 751 has the advantage of avoiding fractions as much as possible.

Not until the 1700s was the general solution to the quadratic given as you have learned it in this book. Now we show why the formula works.

Examine the argument below closely. See how each equation follows from the preceding equation. The idea is quite similar to the one used by the Babylonians, but a little more general. We work with the equation $ax^2 + bx + c = 0$ until the left side is a perfect square. Then the equation has the form $t^2 = k$, which you know how to solve for t.

The goal is to solve $ax^2 + bx + c = 0$.

❸ **Step 1:** Multiply both sides by $4a$. This makes the first term of the expression on the left equal to $4a^2x^2$, the square of $2ax$.	$4a^2x^2 + 4abx + 4ac = 0$
Step 2: When the quantity $2ax + b$ is squared, it equals $4a^2x^2 + 4abx + b^2$. The left and center terms of this trinomial match what is in the equation. We add b^2 to both sides of the equation to get all three terms into our equation.	$4a^2x^2 + 4abx + b^2 + 4ac = b^2$
Step 3: The first three terms are the square of $2ax + b$.	$(2ax + b)^2 + 4ac = b^2$
Step 4: Add $-4ac$ to both sides.	$(2ax + b)^2 = b^2 - 4ac$
Step 5: Now the equation has the form $t^2 = k$, with $t = 2ax + b$ and $k = b^2 - 4ac$. This is where the discriminant $b^2 - 4ac$ becomes so important. If $b^2 - 4ac \geq 0$, then there are real solutions. They are found by taking the square roots of both sides.	$2ax + b = \pm \sqrt{b^2 - 4ac}$
Step 6: It's beginning to look like the formula. Now add $-b$ to each side.	$2ax = -b \pm \sqrt{b^2 - 4ac}$
Step 7: Divide both sides by $2a$ to obtain the Quadratic Formula.	$x = \dfrac{-b \pm \sqrt{b^2 - 4ac}}{2a}$

What if $b^2 - 4ac < 0$? Then the quadratic equation has no real number solutions. The formula still works, but you have to take square roots of negative numbers to get solutions. You will study these non-real solutions in a later course.

QUESTIONS

Covering the Reading

1. *Multiple choice.* The earliest known problems that led to the solving of quadratic equations were considered about how many years ago? d
   (a) 1175 (b) 1700 (c) 2500 (d) 3700

Additional Examples

For 1–2, write a quadratic equation and solve it. Let L and W be the length and width of the paper.

1. Find the dimensions of a sheet of paper if its perimeter is 100 cm and its area is 600 cm². **$L(50 - L) = 600$ or $L^2 - 50L + 600 = 0$; 30 cm by 20 cm**
2. Find the dimensions of a sheet of paper if its perimeter is 100 cm and its area is 96 cm². **$L(50 - L) = 96$ or $L^2 - 50L + 96 = 0$; 48 cm by 2 cm**
3. Use the Babylonian method to find two numbers whose sum is 18 and whose product is 18. (Note: The numbers are not integers.) **$(9 + x)(9 - x) = 18$, $x^2 = 63$, $x = \pm 3\sqrt{7}$; the two numbers are $9 + 3\sqrt{7}$ and $9 - 3\sqrt{7}$.**

Optional Activities

You might use this activity after students have completed the lesson. Have students **work in pairs.** Tell students to pick two integers and give their partner the sum and the product of the numbers. Have the partner find the numbers and explain which method was used to find them.

LESSON MASTER 12-6 A **Questions on SPUR Objectives**
See pages 769-771 for objectives.

Uses Objective I

In 1 and 2, use this information: The area of a rectangular field is 9100 square meters and its perimeter is 400 meters.

1. Use the Babylonian method to find the dimensions of the field. Show your work.
   $\ell + w = 200$, so let $\ell = 100 + x$ and $w = 100 - x$.
   $(100 + x)(100 - x) = 9100$, $10{,}000 - x^2 = 9100$, $x^2 = 900$, $x = 30$
   $\ell = 100 + 30 = 130$, $w = 100 - 30 = 70$
   length, 130 m; width, 70 m
2. Use a modern method to find the dimensions of the field. Show your work.
   $2\ell + 2w = 400$, $2w = 400 - 2\ell$, $w = 200 - \ell$
   $\ell w = 9100$, $\ell(200 - \ell) = 9100$, $\ell^2 - 200\ell + 9100 = 0$, $(\ell - 130)(\ell - 70) = 0$
   $\ell = 130$ or $\ell = 70$
   length, 130 m; width, 70 m

In 3 and 4, use this information: A rectangular garden covering 1500 square feet is enclosed by 160 feet of fencing.

3. Use the Babylonian method to find the dimensions of the garden. Show your work.
   $\ell + w = 80$, so let $\ell = 40 + x$ and $w = 40 - x$.
   $(40 + x)(40 - x) = 1500$, $1600 - x^2 = 1500$, $x^2 = 100$, $x = 10$
   $\ell = 40 + 10 = 50$, $w = 40 - 10 = 30$
   length, 50 ft; width, 30 ft
4. Use a modern method to find the dimensions of the garden. Show your work.
   $2\ell + 2w = 160$, $\ell + w = 80$, $w = 80 - \ell$
   $\ell w = 1500$, $\ell(80 - \ell) = 1500$, $\ell^2 - 80\ell + 1500 = 0$, $(\ell - 50)(\ell - 30) = 1500$
   $\ell = 50$ or $\ell = 30$
   length, 50 ft; width, 30 ft

Notes on Questions

Question 14 We have purposely left the wording of this question as close to the original as possible. Because the problem asks for the dimensions of the door, the corners refer to opposite vertices of the rectangle.

History Connection This problem is from a book, *Jiuzhang suanshu* (*Nine Chapters on the Mathematical Art*), that was probably compiled around 200 B.C. However, it is believed that some of the material in the book was in existence before that date.

Question 19 This is not as fanciful a situation as it might seem. In most states with big lottery jackpots, a winner receives the money over a period of time that can be as long as 20 or 30 years.

Question 20 Health Connection In 1993, according to the U.S. Centers of Disease Control and Prevention, cigarette smoking cost the U.S. health system almost 50 billion dollars, or about \$2.06 for each of the 24 billion packs of cigarettes sold that year.

Question 21 The generalization is known in some countries of Europe as Viète's Theorem.

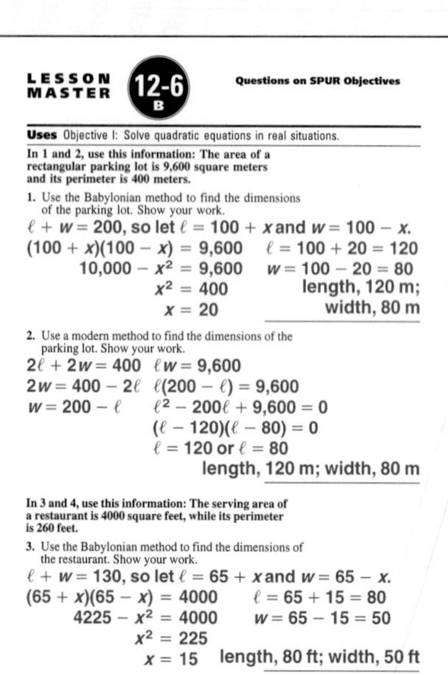

LESSON MASTER 12-6 B — Questions on SPUR Objectives

Uses Objective I: Solve quadratic equations in real situations.

In 1 and 2, use this information: The area of a rectangular parking lot is 9,600 square meters and its perimeter is 400 meters.

1. Use the Babylonian method to find the dimensions of the parking lot. Show your work.
$\ell + w = 200$, so let $\ell = 100 + x$ and $w = 100 - x$.
$(100 + x)(100 - x) = 9{,}600$ $\quad \ell = 100 + 20 = 120$
$10{,}000 - x^2 = 9{,}600$ $\quad w = 100 - 20 = 80$
$x^2 = 400$ $\quad$ length, 120 m;
$x = 20$ $\quad$ width, 80 m

2. Use a modern method to find the dimensions of the parking lot. Show your work.
$2\ell + 2w = 400$ $\quad \ell w = 9{,}600$
$2w = 400 - 2\ell$ $\quad \ell(200 - \ell) = 9{,}600$
$w = 200 - \ell$ $\quad \ell^2 - 200\ell + 9{,}600 = 0$
$(\ell - 120)(\ell - 80) = 0$
$\ell = 120$ or $\ell = 80$
length, 120 m; width, 80 m

In 3 and 4, use this information: The serving area of a restaurant is 4000 square feet, while its perimeter is 260 feet.

3. Use the Babylonian method to find the dimensions of the restaurant. Show your work.
$\ell + w = 130$, so let $\ell = 65 + x$ and $w = 65 - x$.
$(65 + x)(65 - x) = 4000$ $\quad \ell = 65 + 15 = 80$
$4225 - x^2 = 4000$ $\quad w = 65 - 15 = 50$
$x^2 = 225$
$x = 15$ $\quad$ length, 80 ft; width, 50 ft

3) Sample:
$L(100 - L) = 2475$
$L^2 - 100L + 2475 = 0$
$(L - 45)(L - 55) = 0$
$L = 45$ or $L = 55$

4) $(32 + x)(32 - x) = 903$
$32^2 - x^2 = 903$
$x^2 = 121$
$x = \pm 11$
So, the two numbers are 43 and 21.

7) The average of L and W is $\frac{17}{2}$, so let $L = \frac{17}{2} + x$ and $W = \frac{17}{2} - x$.
$(\frac{17}{2} + x)(\frac{17}{2} - x) = 60$;
$\frac{289}{4} - x^2 = 60$
$x^2 = \frac{49}{4}$
$x = \pm\frac{7}{2}$.
So $L = \frac{17}{2} + \frac{7}{2} =$ 12 yards and $W = \frac{17}{2} - \frac{7}{2} =$ 5 yards.

2. In what civilization do quadratic equations first seem to have been considered and solved? **Babylonian**

3. Show your work in Activity 1. **See left.**

4. Show your work in Activity 2. **See left.**

5. Suppose a rectangular room has a floor area of 60 square yards. Give two different lengths and widths that this floor might have. **Sample: 10 yards by 6 yards; 12 yards by 5 yards**

In 6 and 7, suppose a rectangular room has a floor area of 60 square yards and that the perimeter of its floor is 34 yards.

6. Find its length and width by solving a quadratic equation using the Quadratic Formula or factoring. **$LW = 60$ and $L + W = 17$ gives $L^2 - 17L + 60 = 0$. $(L - 12)(L - 5) = 0$, $L = 12$ or $L = 5$, and $W = 5$ or $W = 12$.**

7. Find its length and width using a more ancient method. **See left.**

8. Find two numbers whose sum is 12 and whose product is 9. **$6 + 3\sqrt{3}$, $6 - 3\sqrt{3}$**

9. What is the significance of the work of al-Khwarizmi in the history of the Quadratic Formula? **His techniques for solving a quadratic equation were more general than those of the Babylonians.**

10. Here are steps in the derivation of the Quadratic Formula. Tell what was done to get each step.
$$ax^2 + bx + c = 0.$$
a. $4a^2x^2 + 4abx + 4ac = 0$ **Multiply each side by $4a$.**
b. $4a^2x^2 + 4abx + 4ac + b^2 = b^2$ **Add b^2 to both sides.**
c. $4a^2x^2 + 4abx + b^2 = b^2 - 4ac$ **Subtract $4ac$ from both sides.**
d. $(2ax + b)^2 = b^2 - 4ac$ **Factor the left side.**
e. $2ax + b = \pm\sqrt{b^2 - 4ac}$ **Take the square root of both sides.**
f. $2ax = -b \pm \sqrt{b^2 - 4ac}$ **Add $-b$ to both sides.**
g. $x = \dfrac{-b \pm \sqrt{b^2 - 4ac}}{2a}$ **Divide both sides by $2a$.**

11. When the discriminant of a quadratic equation is negative, how many real solutions does the equation have? **none**

Applying the Mathematics

12) $6x^2 - 5x - 1 = 0$
$144x^2 - 120x - 24 = 0$
$144x^2 - 120x - 24 + 25 = 25$
$144x^2 - 120x + 25 = 49$
$(12x - 5)^2 = 49$
$12x - 5 = \pm\sqrt{49}$
$12x = 5 \pm 7$
$x = \frac{5 \pm 7}{12}$
$x = 1$ or $x = -\frac{1}{6}$

12. Solve the equation $6x^2 - 5x - 1 = 0$ by following the steps in the derivation of the Quadratic Formula. **See left.**
(Hint: The first step is to multiply both sides by $4 \cdot 6$, or 24.)

13. Explain why there are no real numbers x and y whose sum is 15 and whose product is 60. **The discriminant is negative.**

14. In a Chinese text that is thousands of years old, the following problem is given: The height of a door is 6.8 more than its width. The distance between its corners is 10. Find the height and width of the door.
height: 9.6; width: 2.8

Adapting to Individual Needs

Extra Help

For students having difficulty understanding the derivation of the Quadratic Formula you might write the proof on the board, making sure students understand why each step is valid. Remember, students are not expected to replicate the derivation of the formula; we just want them to have a feeling for its derivation and why it works.

Additional Answers

19. See page 771.

Review

In 15–17, factor. *(Lessons 12-2, 12-5)* **See left.**

15) $4ab(2c - b + 3d)$

16) $(2x - 3y)(2x + 3y)$

17) $(4m - 5)(5m - 4)$

15. $8abc - 4ab^2 + 12abd$ **16.** $4x^2 - 9y^2$ **17.** $20m^2 + 20 - 41m$

18. Solve: $12x^3 + 20x^2 + 3x = 0$. *(Lessons 12-2, 12-5)* $x = 0$ or $-\frac{3}{2}$ or $-\frac{1}{6}$

19. Refer to the cartoon below.

MISS PEACH

Use the prime factorization of one million to determine all possible whole numbers of years and amounts over which $1,000,000 could be spread and still give the winner of the lottery a whole number of dollars. *(Lesson 12-1)* **See margin.**

20. According to the National Center for Health Statistics, the following percentages of the adult population smoked in particular years. *(Lesson 7-7)*

Year	Men	Women
1965	52	34
1979	38	30
1983	35	30
1991	28	24

Assume the trends continue. **Sample method: fitting a line to the data by eye**

a. Use any method to predict what percentage of adult men will be smokers in the year 2000. **about 19%**

b. Use any method to predict when the same percentage of adult men and women will smoke. **about the year 2004**

c. Use any method to predict when no men or women will smoke. **about the year 2031**

Exploration

21) The answer may vary; Viète discovered that if x_1 and x_2 are two solutions to the quadratic equation $ax^2 + bx + c = 0$. Then $\begin{cases} x_1 + x_2 = -\frac{b}{a} \\ x_1x_2 = \frac{c}{a}. \end{cases}$

21. Solve a few quadratic equations of your own choosing. If the solutions are fractions, keep them that way. Find the sum of the two solutions to each equation. Then find the product of the two solutions. In the late 1500s, Viète discovered how the sum and product of the solutions are related to the coefficients *a, b,* and *c* of the equation. Try to rediscover what Viète discovered. **See left.**

Follow-up for Lesson 12-6

Practice

For more questions on SPUR Objectives, use **Lesson Master 12-6A** (shown on page 751) or **Lesson Master 12-6B** (shown on pages 752–753).

Assessment

Quiz A quiz covering Lessons 12-4 through 12-6 is provided in the *Assessment Sourcebook.*

Group Assessment Have students **work with a partner.** Tell students to write a problem that can be solved using the Quadratic Formula. Then have students solve their partner's problem. [Students understand when the Quadratic Formula can be used.]

Extension

Give students this problem from an old Babylonian tablet: The sum of the areas of two squares is 1525. The side of the second square is $\frac{2}{3}$ that of the first square plus 5. Find the sides of each square. [Let a = side of the first square and b = side of the second square. Then $a^2 + b^2 = 1525$ and $b = \frac{2}{3}a + 5$; the sides of the squares are 30 and 25.]

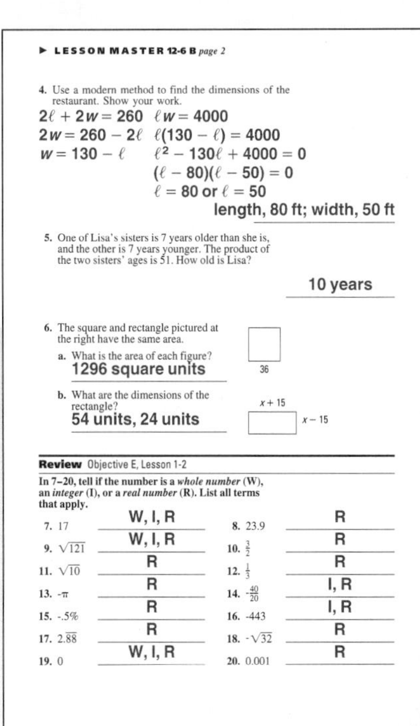
► **LESSON MASTER 12-6 B** *page 2*

4. Use a modern method to find the dimensions of the restaurant. Show your work.

$2\ell + 2w = 260$ $\ell w = 4000$

$2w = 260 - 2\ell$ $\ell(130 - \ell) = 4000$

$w = 130 - \ell$ $\ell^2 - 130\ell + 4000 = 0$

$(\ell - 80)(\ell - 50) = 0$

$\ell = 80$ or $\ell = 50$

length, 80 ft; width, 50 ft

5. One of Lisa's sisters is 7 years older than she is, and the other is 7 years younger. The product of the two sisters' ages is 51. How old is Lisa?

10 years

6. The square and rectangle pictured at the right have the same area.

a. What is the area of each figure? **1296 square units**

b. What are the dimensions of the rectangle? **54 units, 24 units**

36   $x + 15$ $x - 15$

Review Objective E, Lesson 1-2

In 7–20, tell if the number is a *whole number* (W), an *integer* (I), or a *real number* (R). List all terms that apply.

7. 17	W, I, R	8. 23.9	R
9. $\sqrt{121}$	W, I, R	10. $\frac{3}{2}$	R
11. $\sqrt{10}$	R	12. $\frac{1}{3}$	R
13. $-\pi$	R	14. $-\frac{40}{20}$	I, R
15. -.5%	R	16. -443	I, R
17. $2.\overline{88}$	R	18. $-\sqrt{32}$	R
19. 0	W, I, R	20. 0.001	R

Adapting to Individual Needs

Challenge

For each equation, have students describe the values of c such that the equation has two real-number solutions, one real-number solution, and no real-number solutions.

1. $x^2 - 8x + c = 0$
   [$c < 16$; $c = 16$; $c > 16$]
2. $x^2 + 10x + c = 0$
   [$c < 25$; $c = 25$; $c > 25$]
3. $5x^2 - 6x + c = 0$
   [$c < 1.8$; $c = 1.8$; $c > 1.8$]
4. $7x^2 - x + c = 0$
   [$c < \frac{1}{28}$; $c = \frac{1}{28}$; $c > \frac{1}{28}$]

Lesson 12-7

Objectives

H Apply the definitions and properties of rational and irrational numbers.

Resources

From the ***Teacher's Resource File***

- Lesson Master 12-7A or 12-7B
- Answer Master 12-7
- Teaching Aids
  4 Real Numbers
  129 Warm-up
- Activity Kit, Activity 27

Additional Resources

- Visuals for Teaching Aids 4, 129

Teaching Lesson 12-7

Warm-up

Write each number as a fraction.

1. 0.75 $\frac{3}{4}$ **2.** $2\frac{3}{8}$ $\frac{19}{8}$
3. $0.\overline{6}$ $\frac{2}{3}$ **4.** 4.65 $\frac{93}{20}$
5. 6.092 $\frac{1523}{250}$ **6.** 26% $\frac{13}{50}$

Notes on Reading

❶ **Reading Mathematics** Go over the example carefully. Note that it is important to write out a large number of decimal places for each repeating decimal so that students understand how to pick the number by which to multiply both sides. Some students may be skeptical of being able to multiply an infinite decimal by a number. Point out that they multiply $\frac{1}{3}$ by 2 to get $\frac{2}{3}$; likewise they can multiply 0.3333. . . by 2 to get 0.6666. . . .

LESSON

12-7 Rational Numbers and Irrational Numbers

THANK YOU FOR NOT USING VALUABLE TIME TO FIGURE OUT PI TO MORE THAN A MILLION DIGITS
S. Harris

Every number that can be represented as a decimal is a real number. All the real numbers are either rational or irrational. In this lesson you will learn to distinguish between these two types of real numbers.

What Are Rational Numbers?

Recall that a **simple fraction** is a fraction with integers in its numerator and denominator. For instance, $\frac{2}{3}$, $\frac{5488}{212}$, $\frac{-7}{-2}$, and $\frac{-43}{1}$ are simple fractions.

Some numbers are not written as simple fractions, but *equal* simple fractions. Any mixed number equals a simple fraction. For example, $3\frac{2}{7} = \frac{23}{7}$. Also, any integer equals a simple fraction. For example, $-10 = \frac{-10}{1}$. And any finite decimal equals a simple fraction. For instance, $3.078 = 3\frac{78}{1000} = \frac{3078}{1000}$. All these numbers are *rational numbers.* A **rational number** is a number that is a simple fraction or equals a simple fraction. All repeating decimals are also rational numbers.

❶ **Example**

Show that $5.7\overline{62}$ is a rational number.

Solution

Let $x = 5.7\overline{62} = 5.7626262626262\ldots$

Multiply both sides by 10^n, where n is the number of digits in the repetend. Here there are two digits in the repetend, so we multiply by 10^2, or 100.

$$100x = 576.2\overline{62} = 576.2626262626262\ldots$$

Subtract the top equation from the bottom equation. The key idea here is that the result is no longer an infinite repeating decimal; in this case, after the first decimal place the repeating parts subtract to zero.

▶

Lesson 12-7 Overview

Broad Goals The purpose of this lesson is to provide students with the criteria to distinguish rational numbers from irrational numbers, to show them how to draw some irrational lengths, and to demonstrate a proof that certain numbers are irrational.

Perspective The lesson is divided into two parts. The first part deals with the definition and the identification of rational numbers. The second part deals with the same ideas as they apply to irrational numbers. By treating these ideas together, we show students that both kinds of numbers are important and that both must be considered in algebra and geometry.

Because calculators more often give answers as decimals than in any other form, the skill of rewriting a repeating decimal as a rational number is more important today than it might have been a generation ago. Furthermore, it provides a different use of the Multiplication Property of Equality. It is a relatively easy skill, but it does take a little practice. Students who have studied UCSMP *Transition Mathematics* may be familiar with this skill.

$$\begin{aligned} 100x &= 576.2\overline{62} \\ -\quad x &= -\ \ 5.7\overline{62} \\ \hline 99x &= 570.5 \end{aligned}$$

Divide both sides by 99. $x = \frac{570.5}{99} = \frac{5705}{990} = \frac{1141}{198}$

Since x can be represented as a simple fraction, x is a rational number.

Rational numbers have nice properties. They can be added, subtracted, multiplied, and divided, and give answers that are rational numbers. They can always be written as terminating or repeating decimals.

What Are Irrational Numbers?

The ancient Greeks seem to have been the first to discover that there are numbers that are not rational numbers. They called them *irrational.* An **irrational number** is a real number that is not a rational number. Some of the most commonly found irrational numbers in mathematics are the square roots of integers that are not perfect squares. That is, numbers like $\sqrt{2}$, $\sqrt{3}$, $\sqrt{5}$, $\sqrt{6}$, $\sqrt{7}$, $\sqrt{8}$, $\sqrt{10}$, and so on, are irrational. As you know, these numbers arise from situations involving right triangles. At the left is a figure that you saw in Chapter 1.

❷ $\triangle ABC$ is a right triangle with legs of 1 and 1. By the Pythagorean Theorem,

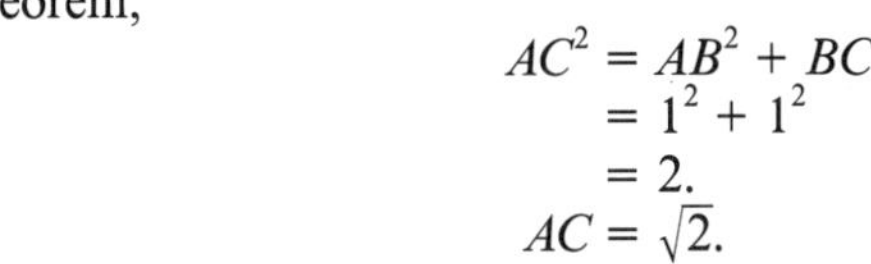

$$\begin{aligned} AC^2 &= AB^2 + BC^2 \\ &= 1^2 + 1^2 \\ &= 2. \end{aligned}$$

So $AC = \sqrt{2}.$

$\triangle ACD$ is drawn with leg $\overline{AC}$, and another leg $CD = 1$. So, by the Pythagorean Theorem,

$$\begin{aligned} AD^2 &= AC^2 + CD^2 \\ &= (\sqrt{2})^2 + 1^2 \\ &= 2 + 1 \\ &= 3. \end{aligned}$$

So $AD = \sqrt{3}.$

You are asked to find other lengths in this figure in the Questions.

Prime Factorizations of Perfect Squares

Consider the square of 28. The prime factorization of 28 is $2 \cdot 2 \cdot 7$. So the prime factorization of 28^2, or 784, is $(2 \cdot 2 \cdot 7) \cdot (2 \cdot 2 \cdot 7)$. Notice that 28 is a product of 3 primes, while its square is a product of 6 prime numbers.

Activity

Pick a perfect square other than 784. Find its prime factorization. Verify that the number is the product of an even number of primes. **Question 9 asks students to show their choice for this Activity.**

Explain that a number can *look* irrational but still be rational. For example, $\frac{\sqrt{45}}{\sqrt{5}}$ is rational because it is equal to 3.

❷ Most students will learn to divide a segment into *n* equal parts (shown in Activity 2 of *Optional Activities* on pages 756–757), and learn to draw segments with lengths $\sqrt{n}$ in geometry courses, so these ideas need not be given strong emphasis here. However, the drawing of segments with length $\sqrt{n}$ gives a concreteness to irrational numbers that is helpful to many students; thus, it should be demonstrated.

When *n* is not a perfect square, the proof that $\sqrt{n}$ is irrational (as given in this lesson) needs to be modified. For instance, consider $\sqrt{6}$. If there were numbers *a* and *b* with $\sqrt{6} = \frac{a}{b}$, then these numbers would satisfy $6b^2 = a^2$. Now consider the prime factorizations of a^2 and b^2. The factorization of a^2 would have to contain the number 2 an even number of times, which is twice the number of times that the number *a* contains this factor. Similarly, the factorization of b^2 has to contain the number 2 an even number of times. But the factorization of 6 is $3 \cdot 2$, and 2 appears an odd number of times. This means that the prime factorization of $6b^2$ has an odd number of 2s, while the prime factorization of a^2 has an even number of 2s. So we cannot have $6b^2 = a^2$, and we cannot have integers *a* and *b* with $\sqrt{6} = \frac{a}{b}$.

LESSON MASTER 12-7 A

Questions on SPUR Objectives
See pages 769–771 for objectives.

Properties Objective H

In 1–9, tell whether the number is *rational* or *irrational*.

1. $\frac{2}{3}$ rational	2. $\sqrt{2}$ irrational	3. $1.\overline{3}$ rational
4. 12 rational	5. π irrational	6. $\sqrt{25}$ rational
7. $\sqrt{5}$ irrational	8. 8.1 rational	9. $2.7\overline{3}$ rational

In 10–12, find a simple fraction equal to the number.

10. $21\frac{48}{61}$	11. $5.86\overline{7}$	12. $13.\overline{521}$
$\frac{1329}{61}$	$\frac{5281}{900}$	$\frac{13,508}{999}$

13. Determine whether the solutions to the equation $4x^2 - 1 = 0$ are *rational* or *irrational*. Explain your reasoning.
rational; Sample: The solutions to $4x^2 - 1 = 0$ are $x = \frac{1}{2}$ and $x = -\frac{1}{2}$, both of which are rational numbers.

14. Can rational numbers be found in real situations? If so, give an example.
yes; Sample: The number of students in a class

Optional Activities

Activity 1

You might use *Activity Kit, Activity 27,* after covering the lesson. In this activity, students explore the golden ratio in golden triangles.

❸ **History Connection** Johann Lambert, a Swiss-German mathematician, astronomer, physicist, and philosopher, was largely self-educated. Early in his career, he began geometric and astronomical investigations with instruments that he designed and built by himself.

Point out to students that over one billion digits for pi are known.

You might want to use the *Visual Organizer* on **Teaching Aid 4** to help students classify numbers.

Additional Examples

Tell whether the number is rational or irrational. If the number is rational, write it as the quotient of two integers.

1. 103.97 [rational; $\frac{10397}{100}$]
2. $-\sqrt{400}$ [rational; $-\frac{20}{1}$]
3. $\sqrt{18}$ [irrational]
4. $.\overline{731}$ [rational; $\frac{724}{990}$]

LESSON MASTER 12-7 B Questions on SPUR Objectives

Vocabulary

1. a. Explain what a *simple fraction* is.
Sample: a fraction with an integer in both the numerator and the denominator
b. List any three simple fractions. Sample: $\frac{4}{5}, \frac{100}{3}, -\frac{7}{2}$
c. List any three fractions that are *not* simple fractions. Sample: $2\frac{1}{4}, -3\frac{1}{3}, \frac{\sqrt{2}}{2}$

2. a. Explain what a *rational number* is.
Sample: a number that is a simple fraction or that equals a simple fraction
b. List any three rational numbers. Sample: $5, \frac{2}{3}, .\overline{6}$
c. List any three numbers that are *not* rational numbers. Sample: $\pi, \sqrt{3}, -\sqrt{10}$

Properties Objective H: Apply the definitions and properties of rational and irrational numbers.

In 3–14, tell whether the number is *rational* or *irrational*.

3. $\frac{4}{5}$ rational
4. $\sqrt{3}$ irrational
5. 8.7 rational
6. $6\frac{1}{2}$ rational
7. $2.\overline{6}$ rational
8. -0.4 rational
9. $\frac{8}{9}$ rational
10. 0 rational
11. $-\sqrt{81}$ rational
12. $\sqrt{50}$ irrational
13. $\sqrt{36}$ rational
14. $9.\overline{83}$ rational

In general, if a number is the product of n primes, then we can write the number as $x_1 \cdot x_2 \cdot x_3 \cdot \ldots \cdot x_n$. Its square is

$$(x_1 \cdot x_2 \cdot x_3 \cdot \ldots \cdot x_n) \cdot (x_1 \cdot x_2 \cdot x_3 \cdot \ldots \cdot x_n),$$

which is a product of $2n$ primes. Since n is an integer, $2n$ is an even number. The important conclusion here is: *If a number is a perfect square, then it is the product of an even number of primes.*

How Do We Know That Certain Numbers Are Irrational?

To show that $\sqrt{7}$ is an irrational number, we first show that there cannot be two perfect squares a^2 and b^2 with $a^2 = 7b^2$. (The same process works with the square root of every prime.) That is, no perfect square can be exactly 7 times another perfect square. (You can come very close; 64 is one more than 7 times 9.)

Here is the argument: If the numbers $7b^2$ and a^2 were the same number, the Unique Factorization Theorem says they must have the same prime factorization. The number a^2 is the product of an even number of primes, because a^2 is a perfect square. Since b^2 is a perfect square, it is also the product of an even number of primes. But $7b^2$ has one more prime factor (the number 7) than b^2 has. So $7b^2$ is the product of an odd number of primes. Therefore, a^2 and $7b^2$ cannot be the same number.

Now work backwards. Since there are no integers a and b with $7b^2 = a^2$, then there cannot be integers a and b with $7 = \frac{a^2}{b^2}$, and so there are no integers a and b with $\sqrt{7} = \frac{a}{b}$. This means that $\sqrt{7}$ cannot be written as a simple fraction, and so it is irrational.

Arguments like this one can be used to prove the following theorem.

Theorem
If a positive integer n is not a perfect square, then $\sqrt{n}$ is irrational.

Today we know that there are many irrational numbers. For instance, every number that has a decimal expansion that does not end or repeat is irrational. Among the irrational numbers is the famous number π. But the argument to show that π is irrational is far more difficult than the argument used above for some square roots of integers. It requires advanced mathematics, and was first given by the German mathematician Johann Lambert in 1767, over 2000 years after the Greeks had first discovered that some numbers were irrational.

There is a practical reason for knowing whether a number is rational or irrational. When a number is rational, arithmetic can be done with it rather easily. Just write the number as a simple fraction and work as you do with fractions. But if a number is irrational, then it is generally more difficult to do arithmetic with it.

Optional Activities

Activity 2 Before discussing how to show irrational numbers on the number line, you might discuss how to picture rational numbers. Explain that every rational number can be graphed on a number line by splitting the number line into equal intervals, as described at the right for $\frac{5}{7}$. Then have students use the procedure to graph other rational numbers, such as $\frac{2}{3}, \frac{4}{5}$, and $\frac{4}{11}$.

1. Draw $AB = 1$ unit. Then draw $\overrightarrow{AC}$, which is not on $\overline{AB}$. It is often convenient to make the angle about 60°.
2. Beginning at A, mark off seven segments of equal length on $\overrightarrow{AC}$. Let the last endpoint be D.
3. Draw $\overline{BD}$. Then draw segments parallel to $\overline{BD}$ that begin at the endpoints of the seven segments on $\overrightarrow{AC}$ and end on $\overline{AB}$.
4. Count 5 equal lengths from A and make the endpoint X. $AX = \frac{5}{7}$.

QUESTIONS

Covering the Reading

7) N', 1, $\sqrt{2}$, M, 1, N

In 1–3, give an example.

1. a simple fraction **Sample:** $\frac{2}{3}$
2. a fraction that is not a simple fraction **Sample:** $\frac{\sqrt{2}}{3}$
3. a rational number **Sample:** $1.\overline{3}$

In 4–6, write the number as a simple fraction.

4. 99.44 $\frac{9944}{100} = \frac{2486}{25}$ 5. $.\overline{15}$ $\frac{15}{99} = \frac{5}{33}$ 6. $14.8\overline{3}$ $\frac{1335}{90} = \frac{89}{6}$

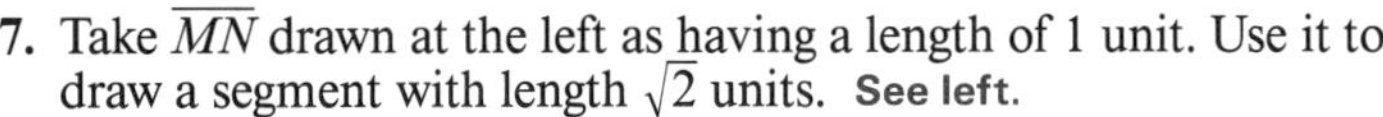

7. Take $\overline{MN}$ drawn at the left as having a length of 1 unit. Use it to draw a segment with length $\sqrt{2}$ units. **See left.**

8. Use the drawing on page 755.
   a. Find AE. $\sqrt{4} = 2$
   b. Determine whether AE is rational or irrational. **rational**
   c. Find AF. $\sqrt{5}$
   d. Determine whether AF is rational or irrational. **irrational**

9. Refer to the Activity on page 755. **Sample given.**
   a. What perfect square did you use for the Activity? **196**
   b. Give its prime factorization. $2 \cdot 2 \cdot 7 \cdot 7$
   c. How many prime factors are in the factorization? **4**

10. a. How many prime factors are in the factorization of 50^2? **6**
    b. How many prime factors are in the factorization of $13 \cdot 50^2$? **7**
    c. Why does the answer to part **b** guarantee that $\sqrt{13 \cdot 50^2}$ is irrational? **There is an odd number of prime factors.**

In 11–13, tell whether the number is a rational number, an irrational number, or neither.

11. π **irrational** 12. -45 **rational** 13. $\sqrt{64}$ **rational**

14. Who first discovered that there were numbers that were not rational numbers? **the Greeks**

Applying the Mathematics

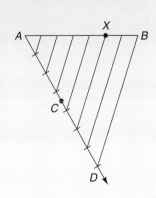

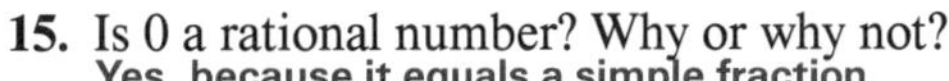

15. Is 0 a rational number? Why or why not? **Yes, because it equals a simple fraction.**
16. Is it possible for two irrational numbers to have a sum that is a rational number? Explain why or why not. **Yes, for example** $\sqrt{2} + -\sqrt{2} = 0$.
17. A square card table has a side of length 30″. Find the length of its diagonal and tell whether the length is rational or irrational. $30\sqrt{2}$ **in.; irrational**
18. If a circular table has a diameter of 120 cm, is its circumference rational or irrational? **irrational**

Notes on Questions

Questions 1–3 The purpose of these questions is to help students distinguish between a fraction, which is how a number might *look*, and a rational number, whose definition tells what the number *is*.

Question 10 This question provides another way to look at why square roots of numbers that are not perfect squares are irrational. If a number is not a perfect square, then it must have at least one prime factor that appears an odd number of times. Thus, when it is simplified, that prime factor will remain under the square root sign.

Question 16 Error Alert Many students look for complicated possibilities and miss the easy examples, such as an irrational number and its opposite.

Question 17 Some people believe that because all measurements are estimates, none of them are really irrational. However, measurements are estimates whether they are rational or irrational. In mathematics, when we speak of *the* length of an object, we are always idealizing that length. Here we are asking for that idealized length for the diagonal of the table; we could just as well be asking for that idealized length of half a side of the table.

▸ **LESSON MASTER 12-7 B** *page 2*

In 15–22, find a simple fraction for the number.

15. $8\frac{1}{3}$	$\frac{25}{3}$	16. 1.025	$\frac{41}{40}$
17. $16\frac{12}{31}$	$\frac{508}{31}$	18. $.\overline{7}$	$\frac{7}{9}$
19. 56%	$\frac{14}{25}$	20. $9.\overline{45}$	$\frac{104}{11}$
21. $.\overline{541}$	$\frac{541}{999}$	22. $6.84\overline{8}$	$\frac{1541}{225}$

In 23 and 24, determine whether the solutions are *rational* or *irrational*.

23. $4x^2 - 9 = 0$ **rational** 24. $x^2 - 4x - 1 = 0$ **irrational**

25. A tire has a diameter of 24 inches. Find its circumference and tell if the circumference is *rational* or *irrational*. 24π **in.; irrational**

26. A square window has a side of length of 18. Find the length of a diagonal and tell if the length is *rational* or *irrational*. $18\sqrt{2}$**; irrational**

27. A square picture has a side of length $12\sqrt{2}$. Find the length of a diagonal and tell if the length is *rational* or *irrational*. **24; rational**

A, X, B, C, D

Adapting to Individual Needs

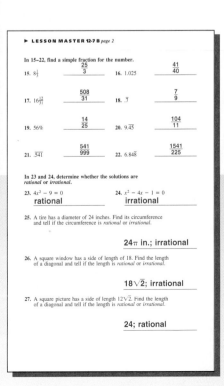

English Language Development

Write *rational* on the board. Ask students what the word *ratio* means to them. They should say division. Remind them that we can think of a division as a fraction. Then relate this idea to the definition of a rational number as a number that can be written as a quotient of two integers. Explain that *irrational* means not rational—an irrational number is a number that cannot be written as a quotient of two integers.

Notes on Questions

Question 25 Although the discussion on projectiles is in Lesson 9-4, the equation $d = 16t^2$ that students need to use is given on page 547.

Question 26 This construction of $\sqrt{n}$ uses the fact that the altitude to the hypotenuse of a right triangle is the mean proportional of the segments of the hypotenuse, a theorem students will encounter in UCSMP *Geometry* and most other geometry courses.

Follow-up for Lesson 12-7

Practice

For more questions on SPUR Objectives, use **Lesson Master 12-7A** (shown on page 755) or **Lesson Master 12-7B** (shown on pages 756–757).

Assessment

Written Communication Ask students to write an explanation of how to write a repeating decimal as a fraction. Have them include at least two examples. [Students' explanations are clear and display understanding of the use of algorithms to write a repeating decimal as a fraction.]

Extension

Some students might enjoy looking through an advanced algebra book to find out about numbers that are not real numbers.

Project Update Project 4, *Infinite Repeating Continued Fractions*, on page 766, relates to the content of this lesson.

Still waters run deep. *Pictured is a water well in Egypt.*

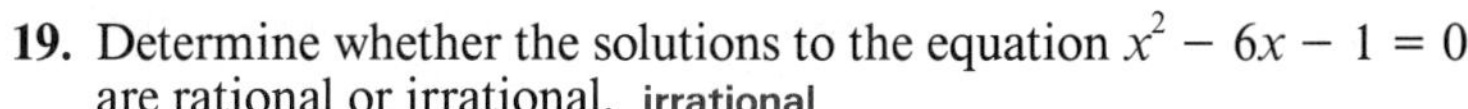

19. Determine whether the solutions to the equation $x^2 - 6x - 1 = 0$ are rational or irrational. **irrational**

Review

20. Find two numbers whose sum is 562 and whose product is 74,865. *(Lesson 12-6)* **217 and 345**

21. Suppose $12x^2 + 7xy - 12y^2 = (ax + b)(cx + d)$.
a. Find the value of $ad + bc$. **7y**
b. Find a, b, c, and d. *(Lesson 12-5)*
$a = 3$, $b = 4y$, $c = 4$, and $d = -3y$ or $a = 4$, $b = -3y$, $c = 3$, and $d = 4y$

22. Factor $9 + 6x + x^2$. *(Lesson 12-3)* **$(3 + x)^2$**

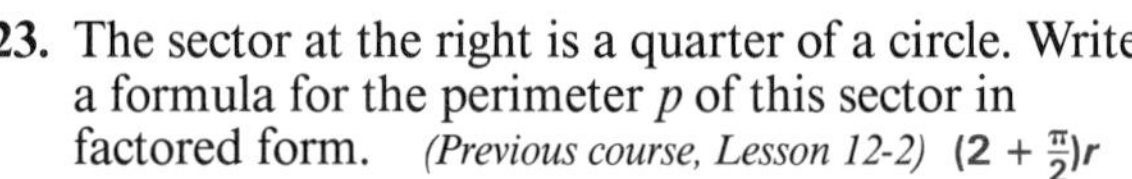

23. The sector at the right is a quarter of a circle. Write a formula for the perimeter p of this sector in factored form. *(Previous course, Lesson 12-2)* **$(2 + \frac{\pi}{2})r$**

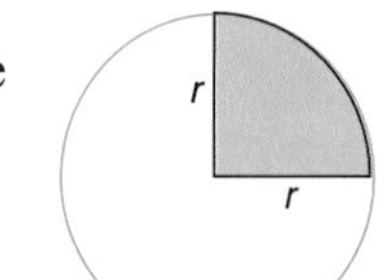

24. Find all values of x that satisfy the equation $bx^2 + cx + a = 0$, when $b \neq 0$. *(Lesson 9-5)* $\frac{-c \pm \sqrt{c^2 - 4ab}}{2b}$

25. With a stopwatch and a stone, you can estimate the depth of a well. If the stone takes 2.4 seconds to reach bottom, how deep is the well? *(Lesson 9-4)* **92.16 ft**

Exploration

26. Shown here is another way to draw a segment with length $\sqrt{n}$.

Step 1: Draw a segment $\overline{AB}$ with length 1, and then next to it a segment $\overline{BC}$ with length n. (In the drawing at the left, $n = 5$.)

Step 2: Find the midpoint M of segment $\overline{AC}$. Draw the circle with center M that contains A and C. ($\overline{AC}$ will be a diameter of this circle.)

Step 3: Draw a segment perpendicular to $\overline{AC}$ from B to the circle. This segment has length $\sqrt{n}$. (In our drawing it should have length $\sqrt{5}$.)

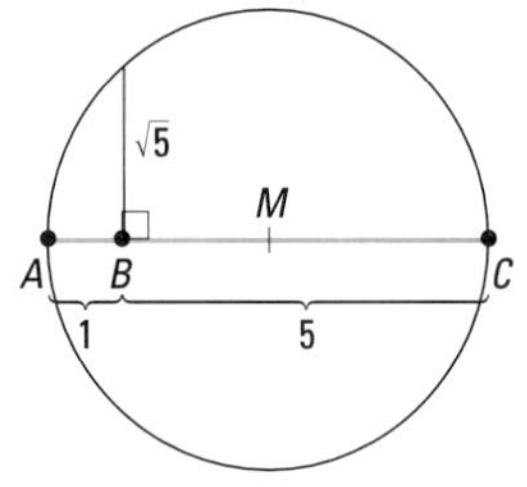

a. Try this algorithm to draw a segment with length $\sqrt{6}$. **See below.**
b. Measure the segment you get. **Sample: 2.5**
c. How close is its length to $\sqrt{6}$?
$\sqrt{6} \approx 2.4495$; the approximate difference is about .05.

a)

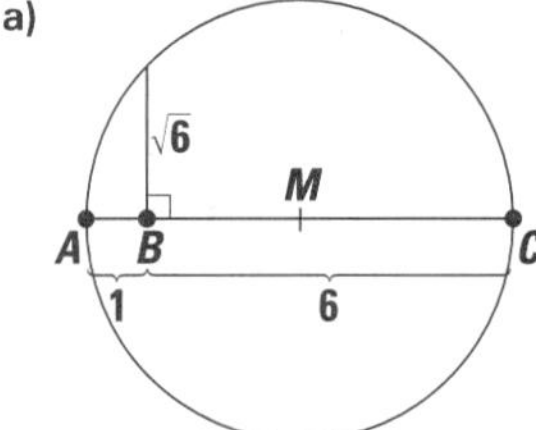

Adapting to Individual Needs

Challenge

In **Question 26** of Lesson 1-8, students extended a figure like the one on page 755 to show $\sqrt{7}$ and $\sqrt{8}$. This type of drawing is sometimes called the Wheel of Theodorus. Students can construct a Wheel of Theodorus as follows: Begin by drawing a right triangle with legs that are each 1 unit long and continue drawing right triangles with one leg equal to 1 and the other leg equal to the hypotenuse of the preceding triangle. Tell students to stop when the wheel starts to overlap and to determine the hypotenuse for the first triangle that overlaps. [$\sqrt{18}$]

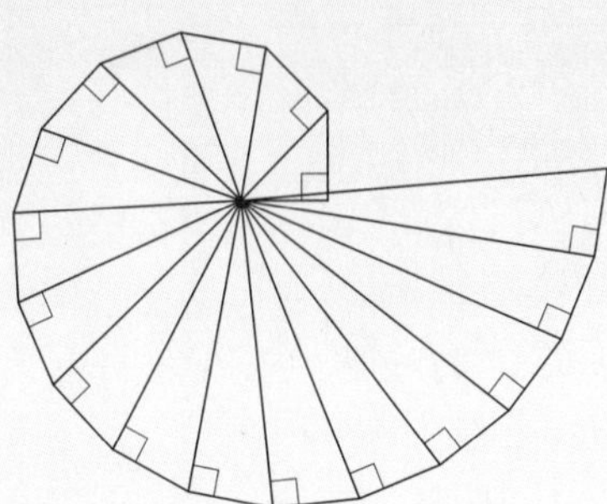

LESSON

12-8

Which Quadratic Expressions Are Factorable?

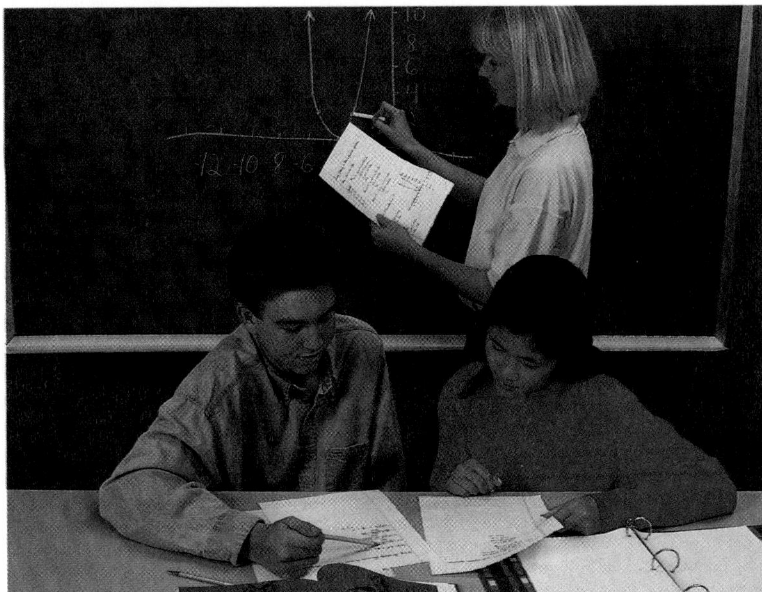

Factoring in what to do. *When solving a quadratic equation, you may try any of three methods: graphing, using the Quadratic Formula, or applying the Zero Product Property.*

This lesson connects three topics you have seen in this chapter: factoring, irrational numbers, and solutions to quadratic equations. These topics seem quite different. Their relationship to each other is an example of how what you learn in one part of mathematics is often useful in another part.

You have studied three ways to find the real-number values of x that satisfy $ax^2 + bx + c = 0$. (1) You can graph $y = ax^2 + bx + c$ and look for its x-intercepts. (2) You can use the Quadratic Formula. (3) You can factor $ax^2 + bx + c$ and use the Zero Product Property. The first two ways can always be done. But you know that it is not always possible to factor $ax^2 + bx + c$ over the integers. So it is useful to know when it is possible.

❶ A Quadratic Equation with Rational Solutions

Consider the equation $2x^2 + 5x - 12 = 0$.

By the Quadratic Formula,

Step 1: $x = \frac{-5 \pm \sqrt{25 - 4 \cdot 2 \cdot (-12)}}{4}$

Step 2: $= \frac{-5 \pm \sqrt{121}}{4}$

Step 3: $= \frac{-5 \pm 11}{4}$

Step 4: So, $x = \frac{-5 + 11}{4} = \frac{3}{2}$, or $x = \frac{-5 - 11}{4} = -4$.

Lesson 12-8

Objectives

G Determine whether a quadratic polynomial can be factored over the integers.

Resources

From the ***Teacher's Resource File***
- Lesson Master 12-8A or 12-8B
- Answer Master 12-8
- Teaching Aid 129: Warm-up
- Technology Connection Computer Master 31

Additional Resources
- Visual for Teaching Aid 129
- Spreadsheet software

Teaching Lesson 12-8

Warm-up

Tell how many real solutions each quadratic equation has.

1. $x^2 - 6x + 9 = 0$ **one**
2. $4x^2 + x + 7 = 0$ **none**
3. $2x^2 - 5x - 1 = 0$ **two**
4. $x^2 + 2x + 3 = 0$ **none**
5. $3x^2 - 7x + 2 = 0$ **two**
6. $x^2 - 16 = 0$ **two**

Notes on Reading

❶ You might want to put a specific case and the general form side by side so that students see the connection between them. Start with the equation $2x^2 + 5x - 12 = 0$. It has rational solutions $\frac{3}{2}$ and -4, which means that it can be factored as $(2x - 3)(x + 4) = 0$. Then use the

Lesson 12-8 Overview

Broad Goals This lesson develops the criterion that $b^2 - 4ac$ must be a perfect square in order for a quadratic equation to be factorable over the integers.

Perspective In Lesson 12-3, students had to determine if a particular quadratic expression was factorable over the integers by using trial and error. In this lesson, students combine their knowledge of the Quadratic Formula with what they have learned about rational and irrational numbers; they develop and use the discriminant to determine if a quadratic expression is factorable over the integers.

The logic of the development is given in the Discriminant Theorem. Here are the three *if and only if* statements which lead to that theorem.

1. A quadratic equation $ax^2 + bx + c = 0$ has rational solutions if and only if $ax^2 + bx + c = (dx + e)(fx + g)$ where d, e, f, and g are integers.
2. A quadratic equation has rational solutions if and only if $b^2 - 4ac$ is a perfect square.
3. A quadratic expression $ax^2 + bx + c$ is factorable if and only if it has factors of the form $(dx + e)(fx + g)$ where d, e, f, and g are integers.

equation $ax^2 + bx + c = 0$ where a, b, and c are integers. It has rational solutions $\frac{-b \pm \sqrt{b^2 - 4ac}}{2a}$ if and only if $b^2 - 4ac$ is a perfect square.

❷ Unless students know that a quadratic can be factored, we strongly urge them to calculate the discriminant before they engage in any sort of trial-and-error process to find the factors. If the discriminant is not a perfect square, students will have saved valuable time that would have been wasted otherwise. If the discriminant is a perfect square, then the quadratic can be factored, and the students are halfway toward the calculation of the solutions.

Notice that there are no radical signs in the solutions in Step 4. This is because the number 121 under the square root sign is a perfect square (Step 2). So, after the square root is calculated (Step 3), one integer is divided by another, and the solutions are rational numbers.

In general, for polynomials over the integers: If the discriminant $b^2 - 4ac$ of the quadratic equation $ax^2 + bx + c = 0$ is a perfect square, then the solutions are rational. Otherwise, the square root will not equal an integer, it will remain in the solutions, and the solutions will be irrational. These results can be summarized in one sentence. When a, b, and c are integers, the solutions to $ax^2 + bx + c = 0$ are rational numbers if and only if $b^2 - 4ac$ is a perfect square.

Now we connect this with factoring. Consider the same equation as before.

$$2x^2 + 5x - 12 = 0$$

Factor the left side.

$$(2x - 3)(x + 4) = 0.$$

Use the Zero Product Property.

$$2x - 3 = 0 \quad \text{or} \quad x + 4 = 0.$$

So, as before, $x = \frac{3}{2}$ or $x = -4$.

Relating the Solutions to $ax^2 + bx + c = 0$ to the Factorization of $ax^2 + bx + c$

Recall that polynomials can be factored in only one way, except for order and for real number multiples. Now you can see why this is true for quadratics. If $ax^2 + bx + c$ could be factored in more than one way, then you would wind up with different solutions to the equation $ax^2 + bx + c = 0$. And we already know there can be at most two solutions; they can be found by using the Quadratic Formula.

Notice how the solutions are related to the factors. When $(dx + e)$ is one of the factors of the left side, then we set $dx + e = 0$ and solve for x to obtain $x = \frac{-e}{d}$. Because d and e are integers, $\frac{-e}{d}$ is a simple fraction. This tells us: If a, b, c are integers and $ax^2 + bx + c$ is factorable over the integers, then each solution to $ax^2 + bx + c = 0$ is a rational number.

The argument can be reversed. Suppose each solution to $ax^2 + bx + c = 0$ is a rational number. Then the solutions can be written as $\frac{-e}{d}$ and $\frac{-g}{f}$, where d, e, f, and g are integers. Because $x = \frac{-e}{d}$ or $x = \frac{-g}{f}$, clearing the equations of fractions gives,

$$dx = -e \quad \text{or} \quad fx = -g.$$

Now, adding to obtain 0 on the right side of each equation, we get

$$dx + e = 0 \quad \text{or} \quad fx + g = 0.$$

Optional Activities

Activity 1
After completing this lesson, you might have students **work in groups** and look back at the quadratic equations in Chapter 9. Tell them to find three equations that are factorable over the set of polynomials with integer coefficients and three equations that are not factorable over the set of polynomials with integer coefficients. Ask them which types are easier to find.

[Sample equations that are factorable:
Page 556 Example 2: $2x^2 - 8x + 6 = 0$
Page 575 Example 2: $3x^2 - 6x - 45 = 0$
Page 581 Example 2: $28 = -x^2 + 2x + 27$

Sample equations that are not factorable:
Page 562 Example 1: $0.5x^2 - 20x + 100 = 0$
Page 568 Example 2: $-16t^2 + 32t + 6 = 0$
Page 574 Example 1: $-x^2 + 2x + 27 = 0$]

Activity 2 Technology Connection
You may wish to assign *Technology Sourcebook, Computer Master 31.* Students use spreadsheet software to determine which quadratic equations are factorable.

Consequently,

$$(dx + e)(fx + g) = 0.$$

This must be a factorization of $ax^2 + bx + c$, because all factorizations give the same solutions. Consequently, when a, b, and c are integers, if there are rational solutions to $ax^2 + bx + c = 0$, then $ax^2 + bx + c$ is factorable.

All of these arguments together show the importance of the discriminant $b^2 - 4ac$. You need only examine it to determine whether a quadratic expression is factorable.

❷ **Discriminant Theorem**
When a, b, and c are integers, with $a \neq 0$, either all three of the following conditions happen at exactly the same time, or none of these conditions happens.

(1) The solutions to $ax^2 + bx + c = 0$ are rational numbers.

(2) $b^2 - 4ac$ is a perfect square.

(3) $ax^2 + bx + c$ is factorable over the set of polynomials with integer coefficients.

Example 1

Is $5x + 1 + 4x^2$ factorable into polynomials with integer coefficients?

Solution

First rewrite this expression in the standard form of a polynomial: $4x^2 + 5x + 1$. Thus $a = 4$, $b = 5$, and $c = 1$, so $b^2 - 4ac = 9$. Since 9 is a perfect square, the expression is factorable.

Activity

Verify Example 1 by finding the factorization of $5x + 1 + 4x^2$.
$(4x + 1)(x + 1)$

Caution: The phrase "with integer coefficients" is necessary in Example 1, because every quadratic expression is factorable if noninteger coefficients are allowed. For instance, the quadratic expression $x^2 - 12 = x^2 - (\sqrt{12})^2 = (x - \sqrt{12})(x + \sqrt{12})$. The polynomial $x^2 - 12$ is prime over the set of polynomials with integer coefficients, but not over the set of all polynomials.

Adapting to Individual Needs

Extra Help

Emphasize to students that the Discriminant Theorem can save them time in factoring trinomials. Stress that they can always determine the discriminant first when they are confronted with a quadratic expression. If the discriminant is not a perfect square, there is no need to even try factoring.

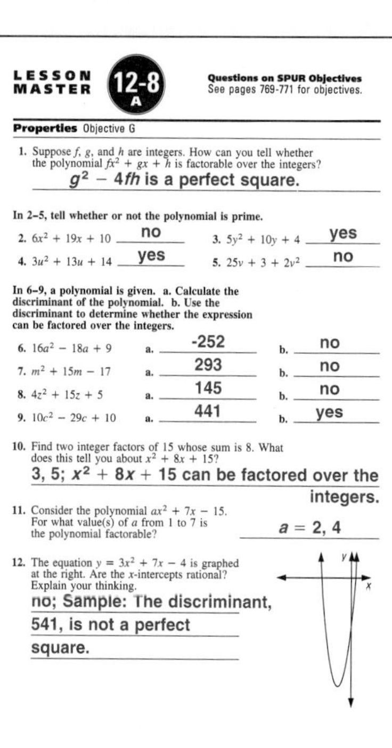

LESSON MASTER 12-8 A — **Questions on SPUR Objectives** See pages 769-771 for objectives.

Properties Objective G

1. Suppose f, g, and h are integers. How can you tell whether the polynomial $fx^2 + gx + h$ is factorable over the integers?
$g^2 - 4fh$ is a perfect square.

In 2–5, tell whether or not the polynomial is prime.

2. $6x^2 + 19x + 10$ no
3. $5y^2 + 10y + 4$ yes
4. $3u^2 + 13u + 14$ yes
5. $25v + 3 + 2v^2$ no

In 6–9, a polynomial is given. a. Calculate the discriminant of the polynomial. b. Use the discriminant to determine whether the expression can be factored over the integers.

	a.	b.
6. $16a^2 - 18a + 9$	-252	no
7. $m^2 + 15m - 17$	293	no
8. $4z^2 + 15z + 5$	145	no
9. $10c^2 - 29c + 10$	441	yes

10. Find two integer factors of 15 whose sum is 8. What does this tell you about $x^2 + 8x + 15$?
3, 5; $x^2 + 8x + 15$ can be factored over the integers.

11. Consider the polynomial $ax^2 + 7x - 15$. For what value(s) of a from 1 to 7 is the polynomial factorable?
$a = 2, 4$

12. The equation $y = 3x^2 + 7x - 4$ is graphed at the right. Are the x-intercepts rational? Explain your thinking.
no; Sample: The discriminant, 541, is not a perfect square.

❸ This example is related to Project 1, *What Percentage of Some Simple Quadratics Are Factorable?*, on page 765. From a practical point of view, few people would care if the height of the object is rational or irrational. You should point out, however, that in theory, the object reaches rational heights after irrational times. You might ask students why the object always reaches rational heights after irrational times. [The computation of the height involves only addition, multiplication, and squaring.]

Science Connection In most cases, the object in **Example 3** will rise to some maximum height, reverse direction, and fall back to earth. Space-age technology allows objects to overcome this limitation. At a critical speed called *escape speed,* an object can overcome gravity and escape the Earth. The *Apollo* moon missions successfully escaped Earth with a speed of about 11.2 km/sec.

Additional Examples

1. a. Is $-2 + 5t + 8t^2$ factorable into polynomials with integer coefficients? **No**

   b. Verify your answer by solving $-2 + 5t + 8t^2 = 0$.
   $t = \frac{-5 \pm \sqrt{89}}{16}$**, so both solutions are irrational; therefore,** $-2 + 5t + 8t^2$ **cannot be factored into polynomials with integer coefficients.**

LESSON MASTER 12-8 B **Questions on SPUR Objectives**

Properties Objective G: Determine whether a quadratic polynomial can be factored over the integers.

In 1 and 2, consider the polynomial $ax^2 + bx + c$.

1. Suppose $b^2 - 4ac$ is a perfect square.
   a. What must be true about the solutions to $ax^2 + bx + c = 0$? **They are rational.**
   b. Is $ax^2 + bx + c$ factorable over the set of polynomials with integer coefficients? **yes**

2. Suppose $b^2 - 4ac$ is not a perfect square.
   a. What must be true about the solutions to $ax^2 + bx + c = 0$? **They are irrational.**
   b. Is $ax^2 + bx + c$ factorable over the set of polynomials with integer coefficients? **no**

In 3–10, a polynomial is given. a. Calculate the discriminant of the polynomial. b. Use the discriminant to determine whether the expression can be factored over the integers. c. If possible, factor the polynomial.

3. $x^2 + 3x + 1$
   a. 5
   b. not factorable
   c. ______

4. $3x^2 + 8x + 4$
   a. 16
   b. factorable
   c. $(3x + 2)(x + 2)$

5. $2w^2 - 10w + 8$
   a. 36
   b. factorable
   c. $2(w - 4)(w - 1)$

6. $5y^2 + 4y - 3$
   a. 76
   b. not factorable
   c. ______

Applying the Discriminant Theorem

Knowing whether an expression is factorable can help determine what methods are available to solve an equation.

Example 2

Solve $m^2 - 31m - 24 = 0$ by any method.

Solution

Since the coefficient of m^2 is 1, it is reasonable to try to factor the left side. But first evaluate $b^2 - 4ac$ to see whether this is possible.
In this equation, $a = 1$, $b = -31$, and $c = -24$. So

$$b^2 - 4ac = (-31)^2 - 4 \cdot 1 \cdot (-24) = 1057.$$

This is not a perfect square, so the equation does not factor over the integers. So use the Quadratic Formula. This is easier to do because the discriminant has already been calculated.

$$x = \frac{31 \pm \sqrt{1057}}{2}$$

Check

$\frac{31 \pm \sqrt{1057}}{2} \approx 15.5 \pm 16.3 \approx 31.8$ or -0.8.

Substitute 31.8 for x in the original equation. Is $(31.8)^2 - 31 \cdot 31.8 - 24 = 0$? The left side equals 1.44, which is very close to 0 given the size of the numbers. You should check the second solution.

You may wonder how often quadratic expressions are factorable. Example 3 asks this for an application of quadratics. Recall that, neglecting such factors as air resistance, an object shot into the air at an upward velocity of v feet per second will, after t seconds, be at a height given by $h = vt - 16t^2$.

❸ Example 3

At what upward integer velocities from 80 to 100 feet per second can an object be shot and reach a height of 100 feet at a time that is a rational number of seconds?

Solution

Begin with equation $h = vt - 16t^2$. Here $h = 100$, so $100 = vt - 16t^2$. The question asks for the integer values of v so that t is rational. To find these values, put the equation into standard form.

$$16t^2 - vt + 100 = 0$$

Use the Discriminant Theorem. If the discriminant is a perfect square, then t will be rational. Calculate the discriminant.

$$b^2 - 4ac = v^2 - 4 \cdot 16 \cdot (100) = v^2 - 6400$$

▶

Adapting to Individual Needs

English Language Development
Have students add the Discriminant Theorem to their set of index cards. You might suggest that they update their index cards by checking the vocabulary words for Chapter 12 that are given on page 767.

	A	B	C
1	V	V*V−6400	SQRT(V*V−6400)
2	80	0	0
3	81	161	12.68857754
4	82	324	18
5	83	489	22.11334439
6	84	656	25.61249695
7	85	825	28.72281323
8	86	996	31.55946768
9	87	1169	34.19064199
10	88	1344	36.66060556
11	89	1521	39
12	90	1700	41.23105626
13	91	1881	43.37049688
14	92	2064	45.43126677
15	93	2249	47.42362281
16	94	2436	49.35585072
17	95	2625	51.23475383
18	96	2816	53.06599665
19	97	3009	54.85435261
20	98	3204	56.60388679
21	99	3401	58.31809325
22	100	3600	60

Now the problem requires substituting the numbers 80 to 100 for v and checking whether $v^2 - 6400$ is a perfect square for each value. This is tedious without some technology. However, with a spreadsheet, it is not difficult. At the left is a spreadsheet that shows the calculations. The top row of each column gives the formula for that column. For instance, the formulas for the fourth row, which starts with 82, are = A4*A4 − 6400 in cell B4, and = SQRT(B4) in cell C4.

At velocities of 80, 82, 89, and 100 feet per second the object will reach 100 feet at a rational number of seconds.

1) **Not possible; because $b^2 - 4ac$ = 233, which is not a perfect square.**

2) **Not possible; because $b^2 - 4ac$ = 72, which is not a perfect square.**

4) **Not possible; because $b^2 - 4ac$ = -8, which is not a perfect square.**

6) **$x = \frac{-b \pm \sqrt{b^2 - 4ac}}{2a}$; if $b^2 - 4ac$ is a perfect square, then $\sqrt{b^2 - 4ac}$ is an integer. Since a and b are also integers, x must be rational.**

QUESTIONS

Covering the Reading

In 1–4, a quadratic expression is given. If possible, factor the expression into polynomials with integer coefficients. If this is not possible, explain why not. **1, 2 and 4) See left.**

1. $7x^2 - 11x - 4$
2. $y^2 - 18$
3. $9n^2 - 12n + 4$ **$(3n - 2)^2$**
4. $3 + m^2 + 2m$
5. What factorization did you get as a result of the Activity of this lesson? **$(4x + 1)(x + 1)$**
6. Consider the equation $ax^2 + bx + c = 0$ when a, b, and c are integers. If $b^2 - 4ac$ is a perfect square, explain why x is rational. **See left.**
7. Give an example of a quadratic expression that can only be factored if noninteger coefficients are allowed. **Sample: $x^2 - 7$**

In 8 and 9, refer to Example 3. At what times will the ball reach a height of 100 feet if

8. the ball is shot into the air with an upward velocity of 89 feet per second? **1.5625 seconds and 4 seconds after the ball has been thrown**
9. the ball is thrown into the air with an upward velocity of 88 feet per second? **≈ 1.60 seconds and ≈ 3.90 seconds after the ball has been thrown**
10. What in this lesson tells you that the solutions to the quadratic equation in Question 9 are irrational? **The discriminant is not a perfect square.**

Applying the Mathematics

11. Find a value of k such that $4x^2 + kx - 3$ is factorable. **Sample: $k = 4$**
12. Suppose a, b, and c are integers. When will the x-intercepts of $y = ax^2 + bx + c$ be rational numbers? **when $b^2 - 4ac$ is a perfect square**

2. Solve $2x^2 - 3x = 2$ by any method. **The discriminant is 25, so either factoring or the Quadratic Formula is a reasonable method. $x = 2$ or $x = -\frac{1}{2}$**
3. Refer to **Example 3** in the lesson.
   a. At what integer velocities from 110 to 130 feet per second can an object be shot and reach a height of 100 feet at a time that is a rational number of seconds? **116 feet per second**
   b. At what times does the object reach this height? **After 1 second and 6.25 seconds**

Notes on Questions

Question 7 Any quadratic whose discriminant is not a perfect square is acceptable.

Questions 8–10 Error Alert Students may overlook the fact that the spreadsheet gives velocities but not times. To calculate the number of times that an object reaches this height for an initial upward velocity v, one must solve the quadratic equation $16t^2 - vt + 100 = 0$.

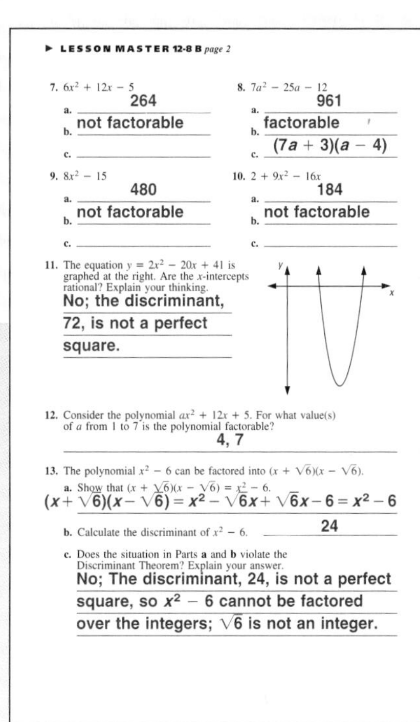
► LESSON MASTER 12-8 B *page 2*

7. $6x^2 + 12x - 5$
   a. 264
   b. not factorable
   c.
8. $7a^2 - 25a - 12$
   a. 961
   b. factorable
   c. $(7a + 3)(a - 4)$
9. $8x^2 - 15$
   a. 480
   b. not factorable
   c.
10. $2 + 9x^2 - 16x$
    a. 184
    b. not factorable
    c.
11. The equation $y = 2x^2 - 20x + 41$ is graphed at the right. Are the x-intercepts rational? Explain your thinking. No; the discriminant, 72, is not a perfect square.

12. Consider the polynomial $ax^2 + 12x + 5$. For what value(s) of a from 1 to 7 is the polynomial factorable? 4, 7
13. The polynomial $x^2 - 6$ can be factored into $(x + \sqrt{6})(x - \sqrt{6})$.
    a. Show that $(x + \sqrt{6})(x - \sqrt{6}) = x^2 - 6$. $(x + \sqrt{6})(x - \sqrt{6}) = x^2 - \sqrt{6}x + \sqrt{6}x - 6 = x^2 - 6$
    b. Calculate the discriminant of $x^2 - 6$. 24
    c. Does the situation in Parts **a** and **b** violate the Discriminant Theorem? Explain your answer. No; The discriminant, 24, is not a perfect square, so $x^2 - 6$ cannot be factored over the integers; $\sqrt{6}$ is not an integer.

Adapting to Individual Needs

Challenge

Remind students that every quadratic equation of the form $ax^2 + bx + c = 0$ has a related equation of the form $y = ax^2 + bx + c$ whose graph is a parabola. The vertex of this parabola has an x-coordinate that can be determined from the equation $x = -\frac{b}{2a}$. Have students use this formula to determine the x-coordinate of the vertex for each parabola and then use this x-value to determine the y-value of the vertex. They can check their answers by graphing the equation.

1. $x^2 - 6x - 16 = 0$ [(3, -25)]
2. $x^2 - 6x + 13 = 0$ [(3, 4)]
3. $x^2 - 10x + 25 = 0$ [(5, 0)]

Notes on Questions

Question 13 A natural response is to determine the times (which are rational). They are found by solving $4.9t^2 - 28t + 30 = 0$. The times are $\frac{30}{7}$ seconds and $\frac{10}{7}$ seconds.

Question 14 Solving the equation $x^2 + 6x + 7 = 0$ shows the relationship between the roots and the factors of this quadratic.

Questions 15–16 These questions reiterate the idea that whether a number is rational or irrational is not determined by how it is written.

Question 20 Students might find the answer using trial and error. Any fourth power of a prime integer that is greater than one will do, and 16 will be the first such power encountered.

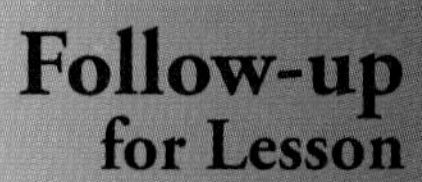

Follow-up for Lesson 12-8

Practice

For more questions on SPUR Objectives, use **Lesson Master 12–8A** (shown on page 761) or **Lesson Master 12–8B** (shown on pages 762–763).

Assessment

Group Assessment Have students **work in small groups**. Ask each group to prepare five questions for a chapter review. Then have the groups exchange questions and answer them. [Students' questions cover the objectives of this chapter, and students work together to answer each question.]

Extension

Have students **work in groups**. Tell them to imagine that for a quadratic equation the discriminant is 64; that is, $b^2 - 4ac = 64$. Then have them write and solve three equations for which this is true and that have integer solutions. [Sample equations: $x^2 + 6x - 7 = 0$; $x = 1$ or $x = -7$; $x^2 - 4x - 12 = 0$; $x = 6$ or $x = -2$; $3x^2 + 4x - 4 = 0$; $x = \frac{2}{3}$ or $x = -2$)]

Project Update Project 1, *What Percentage of Some Simple Quadratics are Factorable?*, on page 765, relates to the content of this lesson.

This skier is in a wind tunnel at the Calspan Corporation in Rochester, New York. The wind tunnel is used to find ways to streamline bodies in order to minimize air resistance.

14a) $(x + 3 + \sqrt{2})(x + 3 - \sqrt{2}) = (x + 3)^2 - (\sqrt{2})^2 = x^2 + 6x + 9 - 2 = x^2 + 6x + 7$

b) $b^2 - 4ac = 36 - 4 \cdot 1 \cdot 7 = 8$, not a perfect square

13. Try a version of Example 3 using metric units. Neglecting such factors as air resistance, an object shot into the air at an upward velocity of v meters per second will, after t seconds, be at a height $h = vt - 4.9t^2$. At what upward integer velocities from 25 to 40 meters per second can an object be shot and reach a height of 30 meters at a time that is a rational number of seconds? **only at 28 meters per second**

14. a. By multiplying, verify that $(x + 3 + \sqrt{2})(x + 3 - \sqrt{2}) = x^2 + 6x + 7$.
    b. Verify that the discriminant of the expression $x^2 + 6x + 7$ is not a perfect square. **(a, b) See left.**
    c. Part **a** indicates that $x^2 + 6x + 7$ is factorable, and yet its discriminant is not a perfect square. Does this situation contradict the Discriminant Theorem? **No, because $3 + \sqrt{2}$ and $3 - \sqrt{2}$ are not integers**

Review

15. Tell whether the number is rational or irrational. *(Lesson 12-7)*
    a. $\sqrt{25}$ **rational** b. $\sqrt{26}$ **irrational** c. $\sqrt{27}$ **irrational**

16. a. Is $\frac{\sqrt{3}}{\sqrt{12}}$ rational or irrational? **rational**
    b. Explain why or why not. *(Lesson 12-7)* **because $\frac{\sqrt{3}}{\sqrt{12}} = \frac{1}{2}$**

17. The winning percentage of a high school tennis team was $.58\overline{3}$. If the team had fewer than 20 matches, how many matches did they have and how many did they win? *(Lesson 12-7)* **12; 7**

18. Roseanne was surprised to learn that a farm with an area of 1 square mile had a perimeter of 4.5 miles. Explain to Roseanne how this is possible. *(Lesson 12-6)* **See margin.**

19. The surface area *S.A.* of a cylinder with radius r and height h is given by the formula $S.A. = 2\pi r^2 + 2\pi rh$.
    a. Factor the right side of this formula into prime factors. $2\pi r(r + h)$
    b. Calculate the surface area of a cylinder 10 cm in diameter and 8 cm high using either the given formula or its factored form. Which form do you think is easier? *(Lessons 1-5, 12-2)* **≈ 408.4 cm²; the factored form**

20. Find an integer that has exactly five integer factors. (The factors need not be prime.) *(Lesson 12-1)* **Sample: 16**

Exploration

21. Refer to Question 11. Find all values of k such that $4x^2 + kx - 3$ is factorable. Explain how you know that you have found all values. **$k = \pm 1, \pm 4$, or ± 11; The discriminant, $k^2 + 48$, must be a perfect square. Use a spreadsheet to determine the values.**

Additional Answers

18. There are two numbers whose sum is 2.25 and whose product is 1. They are $\frac{9}{8} - \frac{\sqrt{17}}{8}$ and $\frac{9}{8} + \frac{\sqrt{17}}{8}$, or approximately 0.61 and 1.64.

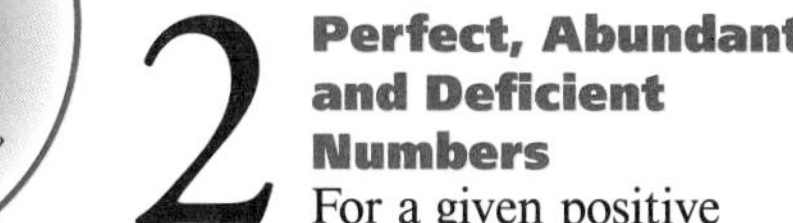

A project presents an opportunity for you to extend your knowledge of a topic related to the material in this chapter. You should allow more time for a project than you do for typical homework questions.

1 What Percentage of Some Simple Quadratics Are Factorable?

Use a spreadsheet or computer program to consider the discriminant of all quadratics of the form $x^2 + bx + c$, where b and c are integers from -10 to 10. Use this information to determine what percentage of these quadratics are factorable over the integers. If it is easy for you to do so, you might extend the ranges of b and c. When the absolute values of b and c are larger, is it more or less likely that the quadratic will be factorable?

2 Perfect, Abundant, and Deficient Numbers

For a given positive integer, consider the sum of all its factors that are less than the given integer. For instance, for the number 10, this sum is 8 because $1 + 2 + 5 = 8$. Positive integers are classified as perfect, abundant, and deficient according to whether this sum is equal to, is greater than, or is less than the integer itself. The number 10 is deficient because $8 < 10$. Classify each number from 1 to 100 as perfect, abundant, or deficient. Find patterns in the numbers that fall into these categories. What numbers are certain to be abundant? Which are certain to be deficient?

Chapter 12 Projects

The projects relate to the content of the lessons of this chapter as follows:

Project	Lesson(s)
1	12-3, 12-8
2	12-1
3	12-1
4	12-7
5	12-1
6	12-5

1 What Percentage of Some Simple Quadratics Are Factorable? There are 21^2, or 441, cases to be considered; so without technology, this can be a very time-consuming project. We suggest using a spreadsheet. Remind students that the value of a is equal to one for each of the equations they are considering. Therefore, the square root of $b^2 - 4c$ can be used to find which quadratics are factorable over the integers.

2 Perfect, Abundant, and Deficient Numbers Check that students understand the meaning of the terms *perfect, abundant,* and *deficient* before starting this project. Students who have studied UCSMP *Transition Mathematics* may be familiar with the content of this project.

A table listing each number and its factors will allow students to find patterns easily. They should discover that abundant numbers have at least four factors that are less than the number and that all powers of 2 are deficient by 1. You might tell students that the next three perfect numbers after 6 and 28 are 496, 8,128, and 33,550,336. The Greek mathematician Euclid knew of only the first four perfect numbers.

History Connection A nun named Hrotsvitha, who lived in Saxony in the 10th century, was a noted playwright, and in many of her plays she showed an interest in mathematics. For example, in one play a woman is asked the ages of her three daughters. The woman replies that their ages are a defective even number, a defective odd number, and an abundant number.

Possible responses

1. **About 17% of these quadratics, or 76 of the 441 equations, are factorable over the integers. In 21 equations, $c = 0$, in which case the equation is always factorable. Of the remaining 420 cases, 55 are factorable as shown in the chart at the right. With larger bounds, smaller percents of quadratics are factorable.**

b	c
0	-1, -4, -9
±1	-2, -6,
±2	1, -3, -8
±3	2, -4, -10
±4	3, 4, -5
±5	4, 6, -6

b	c
±6	5, 8, 9, -7
±7	6, 10, -8
±8	7, -9
±9	8, -10
±10	9

(Responses continue on page 766.)

3 Public-key Cryptography Inform students that *cryptography* is the process of writing or deciphering secret codes. The word cryptography is from the Greek *kryptos* which means secret or hidden and *graphia* which means writing. Although one-key systems have been in use for hundreds of years, two-key systems, often called public-key systems, are somewhat new. The first two-key system was developed at Stanford University in 1976. Students should find information about public-key cryptography in most recent books about cryptography.

4 Infinite Repeating Continued Fractions These complex fractions may initially look intimidating to some students. However, as they work at calculating the value of each of the first five terms of the sequence, they should recognize that each term is equal to a fraction with a numerator of 1 and a denominator that is equal to 2 plus the value of the previous term. Some students may need a hint to help them solve the equation in **part b**. Tell them to multiply both sides of the equation $x = \frac{1}{2 + x}$ by $(2 + x)$ and then solve the resulting equation.

5 Packing Boxes Each factorization of 144 with three factors gives a possible rectangular array of stapler boxes. Each stapler box is a rectangular prism with three different dimensions. Therefore, there are six different-sized crates that will fit the stapler boxes for each factorization in which none of the factors are the same and three different-sized crates for each factorization in which two of the factors are the same. The dimensions of each crate are found by multiplying the number of layers by 3, the number of rows by 8 (or 2), and the number of columns by 2 (or 8). All crates have the same volume. The crate with the least surface area is the one which most resembles a cube, so it will be the one whose dimensions are closest to being equal.

6 Factors and Graphs This is a relatively straightforward project. Suggest that students use automatic graphers. You should ask students to sketch the graphs they produce.

(continued)

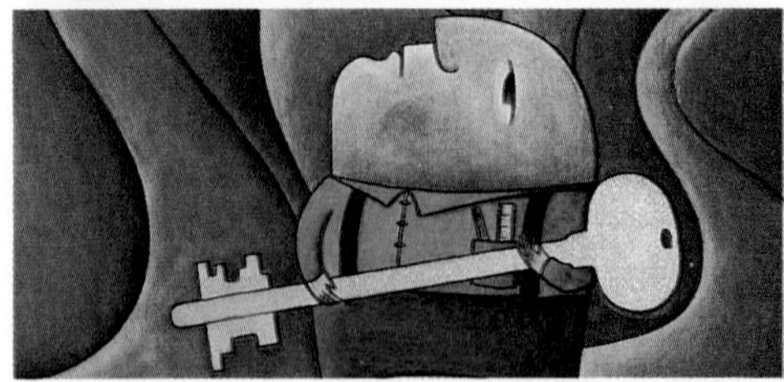

3 Public-key Cryptography

The use of codes based on prime numbers to protect information is called public-key cryptography. Do research in your library and write an essay describing how public-key cryptography works.

4 Infinite Repeating Continued Fractions

Consider this sequence of complex fractions.

$$\frac{1}{2}, \frac{1}{2 + \frac{1}{2}}, \frac{1}{2 + \frac{1}{2 + \frac{1}{2}}}, \frac{1}{2 + \frac{1}{2 + \frac{1}{2 + \frac{1}{2}}}}, \dots$$

a. Calculate the values of the first five terms of this sequence. (Four terms are shown.)

b. As you calculate more and more terms of this sequence, the sequence approaches the value of x, where

$$x = \frac{1}{2 + \frac{1}{2 + \frac{1}{2 + \frac{1}{2 + \frac{1}{2 + \dots}}}}}.$$

Then $x = \frac{1}{2 + x}$.

Solve this equation to find x.

c. Replace the 2s by 3s and repeat parts **a** and **b**.

d. If you can, generalize what you have found.

5 Packing Boxes

A rectangular box containing a stapler is 8″ long, 2″ wide, and 3″ high. The manufacturer wants to pack 144 (a *gross*) of these boxes in a crate for shipping to stores.

a. One crate that will hold 144 boxes without any space left over is to have 4 layers of 36 boxes, each layer having 6 boxes in each row and column. Find the dimensions of other crates that will hold 144 of these boxes without any space left over.

b. Which crate of those possible has the least surface area?

c. Which crate of those possible do you think the manufacturer should choose for packing, and why?

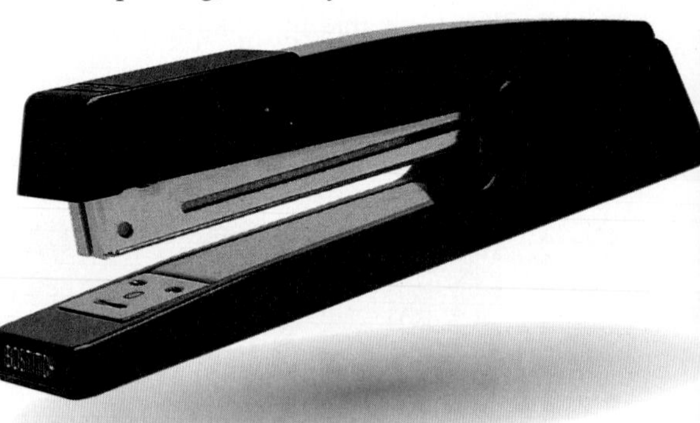

6 Factors and Graphs

a. Multiply $x - 1$ by a binomial of the form $ax + b$, where a and b are integers of your own choosing. Then graph the equation $y = (x - 1)(ax + b)$. Next, repeat this procedure three times. What do the graphs have in common? If you can, explain why they have this commonality.

b. Repeat part **a** but with some factor other than $x - 1$. What do the four new graphs have in common?

c. If you can, generalize the results you find in parts **a** and **b**.

Additional responses, page 765

2. Perfect numbers: 6, 28

Abundant numbers: 12, 18, 20, 24, 30, 36, 40, 42, 48, 54, 56, 60, 66, 70, 72, 78, 80, 84, 88, 90, 96

Deficient numbers: 2, 3, 4, 5, 7, 8, 9, 10, 11, 13, 14, 15, 16, 17, 19, 21, 22, 23, 25, 26, 27, 28, 29, 31, 32, 33, 34, 35, 37, 38, 39, 41, 43, 44, 45, 46, 47, 49, 50, 51, 52, 53, 55, 57, 58, 59, 61, 62, 63, 64, 65, 67, 68, 69, 71, 73, 74, 75, 76, 77, 79, 81, 82, 83, 85, 86, 87, 89, 91, 92, 93, 94, 95, 97, 98, 99

A multiple of an abundant or a perfect number is certain to be abundant. A prime number or an odd number is deficient.

SUMMARY

There are many similarities between the factoring of integers and the factoring of polynomials over the integers. A prime number is an integer greater than 1 that has exactly two integer factors, itself and 1. Every integer can be factored into primes in exactly one way, except for order. A prime polynomial is a polynomial that cannot be factored into polynomials of lower degree. The complete factorization of a polynomial is unique, except for the order of the factors. Prime factorizations are used nowadays in the construction of codes to protect information.

If an integer is a perfect square, then it is the product of an even number of primes. From this fact it can be deduced that certain square roots cannot be written as simple fractions. Consequently, such square roots are irrational numbers. More generally, any number that cannot be written as a finite or an infinitely repeating decimal is an irrational number.

If each of two integers has a common factor, then so does their sum, and the common factor can be factored out using the Distributive Property $ab + ac = a(b + c)$. Similarly, if each of the terms of a polynomial has a common monomial factor, then so does their sum, and it can be factored out using the Distributive Property.

Factoring the general quadratic trinomial $ax^2 + bx + c$ is more difficult, and may require trial and error procedures. However, when $a = 1$, then the quadratic is relatively easy to factor. Specifically, $x^2 + bx + c$ can be factored into $(x + p)(x + q)$ provided there are integers p and q such that $p + q = b$ and $pq = c$. In general, $ax^2 + bx + c$ can be factored over the integers if and only if its discriminant, $b^2 - 4ac$, is a perfect square.

The Zero Product Property states that if the product of two or more factors is zero, then at least one of the factors must be zero. Thus, if $ax^2 + bx + c = 0$ and $ax^2 + bx + c$ can be factored, then the solutions of the equation can be found quickly by setting each factor equal to zero and solving these simpler equations. This is one of the major uses for factoring, even though most quadratic expressions do not factor over the integers.

About 3700 years ago, Babylonian scribes showed how to solve certain quadratics. They used what we today call substitution to convert a quadratic equation into one of the form $x^2 = k$. In the 18th century, a similar idea was used to derive the Quadratic Formula we know today.

VOCABULARY

Below are the new terms and phrases for this chapter. You should be able to give a general description and a specific example for each.

Lesson 12-1
factor, multiple
is divisible by
common factor
Common Factor Sum Property
prime number
composite number
prime factorization
standard form of a factorization
Unique Factorization Theorem

Lesson 12-2
greatest common factor
polynomial over the integers
factorization of a polynomial
prime polynomial
complete factorization
Unique Factorization Theorem for Polynomials

Lesson 12-4
Zero Product Property

Lesson 12-7
simple fraction
rational number
irrational number

Lesson 12-8
Discriminant Theorem

Summary

The Summary gives an overview of the entire chapter and provides an opportunity for students to consider the material as a whole. Thus, the Summary can be used to help students relate and unify the concepts that are presented in the chapter.

Vocabulary

Terms, symbols, and properties are listed by lesson to provide a checklist of concepts that students must know. Emphasize that students should read the vocabulary list carefully before starting the Progress Self-Test. If students do not understand the meaning of a term, they should refer back to the indicated lesson.

3. The following information is a sample of what students might include in their projects. Public key cryptography is a two-key code system. A key is a set of instructions used to encode and decode messages. In a public key system, one key is used to encrypt information to be transmitted. This encrypting key is a public key and can be found in a directory. The second key is used to decrypt the original information. This key is kept secret and known only by the receiver of the information. The Rivest/Shamir/Adelman cryptoalgorithm, a public key system, uses very large numbers to make the private and public keys. A computer program is used to randomly select two large prime numbers about 100 digits long. The product of these two primes is the basis of the two keys. The security of this system is based on the difficulty of factoring such large integers. In 1976, Whitfield Diffie and Martin E. Hellman proposed a conceptual scheme for the system; Admiral Bobby Inman believes that the U. S. National Security Agency discovered it a decade earlier.

(Responses continue on page 768.)

CHAPTER TWELVE

Progress Self-Test

For the development of mathematical competence, feedback and correction, along with the opportunity to practice, are necessary. The Progress Self-Test provides the opportunity for feedback and correction; the Chapter Review provides additional opportunities and practice. We cannot overemphasize the importance of these end-of-chapter materials. It is at this point that the material "gels" for many students, allowing them to solidify skills and understanding. In general, student performance should be markedly improved after these pages.

Assign the Progress Self-Test as a one-night assignment. Worked-out *solutions* for all questions are in the Selected Answers section of the student book. Encourage students to take the Progress Self-Test honestly, grade themselves, and then be prepared to discuss the test in class.

Advise students to pay special attention to those Chapter Review questions (pages 769–771) which correspond to the questions that they missed on the Progress Self-Test.

PROGRESS SELF-TEST

2, 18, 22) See below right.

Take this test as you would take a test in class. You will need a calculator. Then check your work with the solutions in the Selected Answers section in the back of the book.

1. Write the prime factorization of 300 in standard form. $2^2 \cdot 3 \cdot 5^2$
2. Explain how you know that the number $6^{1000} + 36$ is divisible by 3.
3. Give the greatest common factor of $15a^2b^3$, $30a^2b$, and $25a^3b$. $5a^2b$
4. Simplify $\frac{8c^2 + 4c}{c}$. $8c + 4$

In 5–10, factor completely over the integers.

5. $12m - 2m^3$ $2m(6 - m^2)$
6. $500x^2y + 100xy + 50y$ $50y(10x^2 + 2x + 1)$
7. $z^2 - 81$ $(z + 9)(z - 9)$
8. $k^2 - 9k + 14$ $(k - 2)(k - 7)$
9. $3y^2 - 17y - 6$ $(3y + 1)(y - 6)$
10. $4x^2 - 20xy + 25y^2$ $(2x - 5y)^2$
11. *Multiple choice.* Which of the following can be factored over the integers? a
    (a) $x^2 + 7x + 12$
    (b) $x^2 + 7x - 12$
    (c) $x^2 + 12x - 7$
    (d) $x^2 - 12x - 7$
12. Explain how you can determine whether or not $ax^2 + bx + c$ can be factored over the integers. **If $b^2 - 4ac$ is a perfect square it can be factored.**
13. Show the factorization
    $$2x^2 + 11x + 15 = (x + 3)(2x + 5)$$
    using areas of rectangles.

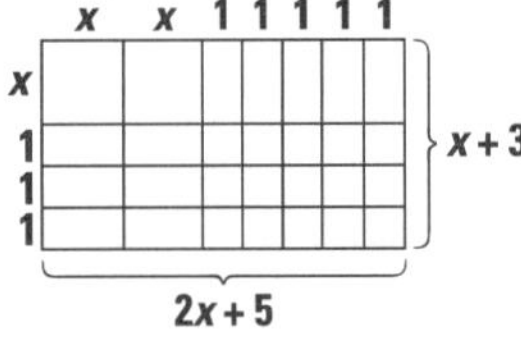

In 14–17, solve.

14. $(q - 7)^2 = 0$ $q = 7$
15. $d^2 - 20 = d$ $d = 5$ or $d = -4$
16. $(2a - 5)(3a + 1) = 0$ $a = \frac{5}{2}$ or $a = -\frac{1}{3}$
17. $x^3 + 6x^2 = 7x$ $x = 0$, or $x = -7$, or $x = 1$
18. Explain why 3.54 is a rational number.
19. How many prime factors does the number 26^2 have? 4
20. Give an example of an irrational number between 10 and 11. **Sample:** $\sqrt{105}$
21. Is the larger solution to
    $$5x^2 - 18x - 18 = 0$$
    rational or irrational? **irrational**
22. A square frame is d ft on a side. The artwork it holds is 1 ft shorter and 2 ft narrower than the frame. If the area of the artwork is 12 sq ft, how big is the frame?
23. A tennis ball bounces up from ground level at 8 meters per second. An equation that estimates the distance d above ground (in meters) after t seconds is $d = 8t - 5t^2$. After how many seconds will the ball return to the ground? **1.6 seconds**
24. Find two numbers whose product is 15.51 and whose sum is 8. **3.3; 4.7**

2) Both 6^{1000} and 36 are divisible by 3, so by The Common Factor Sum Property their sum is divisible by 3.

18) Because it can be written as the simple fraction $\frac{354}{100}$.

22) 5 ft by 5 ft

Additional responses, page 766

4a. $\frac{1}{2}, \frac{2}{5}, \frac{5}{12}, \frac{12}{29}, \frac{29}{70}$

b. The positive solution is $x = -1 + \sqrt{2}$

c. $\frac{1}{3}, \frac{3}{10}, \frac{10}{33}, \frac{33}{109}, \frac{109}{369}$; The positive solution of $x = \frac{1}{3 + x}$ is $x = \frac{\sqrt{13} - 3}{2}$.

d. Sample generalization: The numerator of each term is the same as the denominator of the preceding term. The denominator of each term is twice the denominator of the preceding term plus the numerator of the preceding term. The positive solution of $x = \frac{1}{a + x}$ is $x = \frac{-a + \sqrt{a^2 + 4}}{2}$

5a. Possible crate sizes and the resulting surface areas are shown in the chart at the right. Since the desired crate is one whose dimensions are closest to equal, no one-digit dimensions are shown. In the chart, LY, R, and C refer to layers, rows, and columns; H, L, and W refer to height (layers times 3), length (rows times 8) and width (columns times 2), and SA refers to surface area in square inches.

 CHAPTER TWELVE

CHAPTER REVIEW

Questions on SPUR Objectives

SPUR stands for **S**kills, **P**roperties, **U**ses, and **R**epresentations. The Chapter Review questions are grouped according to the SPUR Objectives for this chapter.

SKILLS DEAL WITH THE PROCEDURES USED TO GET ANSWERS.

Objective A: *Factor positive integers into primes.* (Lesson 12-1)

In 1–4, write the prime factorization of the given integer in standard form.

1. 175 $5^2 \cdot 7$ **2.** 8888 $2^3 \cdot 11 \cdot 101$

3. $441 \cdot 9$ $3^4 \cdot 7^2$ **4.** $1024 + 512$ $2^9 \cdot 3$

Objective B: *Find common monomial factors of polynomials.* (Lesson 12-2)

5. Copy and complete: $7x^4 + 49x = 7x(\underline{\ ?\ } + \underline{\ ?\ })$. x^3; 7

6. Find the greatest common factor of $8a^2$ and $12a$. $4a$

7. Find the greatest common factor of $27x^2y$, $12x^2y^2$, and $3x^3y$. $3x^2y$

8. Find the greatest common factor of $20ay^3$, $-15y^4$, and $35a^2y^6$. $5y^3$

In 9 and 10, factor.

9. $14m^4 + m^2$ $m^2(14m^2 + 1)$

10. $18b^3 - 21ab + 3b$ $3b(6b^2 - 7a + 1)$

In 11 and 12, simplify, assuming the denominator is not 0.

11. $\frac{6z^3 - z}{z}$ $6z^2 - 1$ **12.** $\frac{14x^2 + 12x}{2x}$ $7x + 6$

Objective C: *Factor quadratic expressions.* (Lessons 12-3, 12-5)

In 13–18, factor if possible.

13. $x^2 + 7x + 6$ **14.** $p^2 + 9p - 10$

15. $r^2 - 10r + 28$ **16.** $x^2 - 1$

17. $4L^3 + 28L^2 + 48L$ **18.** $d^2 - 8d - 20$

13) $(x + 6)(x + 1)$ 14) $(p + 10)(p - 1)$
15) not factorable 16) $(x + 1)(x - 1)$
17) $4L(L + 3)(L + 4)$ 18) $(d - 10)(d + 2)$

In 19 and 20, *multiple choice.*

19. $11a^2 + 26a - 21 =$ c
(a) $(11a - 7)(a - 3)$
(b) $(11a + 7)(a - 3)$
(c) $(11a - 7)(a + 3)$
(d) $(11a + 7)(a + 3)$

20. $24x^2 - 83x + 10 =$ c
(a) $(8x + 1)(3x + 10)$
(b) $(8x - 1)(3x + 10)$
(c) $(8x - 1)(3x - 10)$
(d) $(8x + 1)(3x - 10)$

In 21–24, factor.

21. $3y^2 + 2xy - 8x^2$ $(3y - 4x)(y + 2x)$

22. $10a^2 - 19a + 7$ $(5a - 7)(2a - 1)$

23. $12m^3 + 117m^2 + 81m$ $3m(4m + 3)(m + 9)$

24. $-3 - 2k + 8k^2$ $(4k - 3)(2k + 1)$

In 25 and 26, write each perfect square trinomial as the square of a binomial. 25) $(m + 8)^2$

25. $m^2 + 16m + 64$ **26.** $9a^2 - 24ab + 16b^2$

In 27–30, write each difference of squares as the product of two binomials. 28) $(b + 9m)(b - 9m)$

27. $a^2 - 4$ $(a + 2)(a - 2)$ **28.** $b^2 - 81m^2$

29. $4x^2 - 1$ $(2x + 1)(2x - 1)$ **30.** $25t^2 - 25$ $25(t + 1)(t - 1)$ or $(5t + 5)(5t - 5)$

26) $(3a - 4b)^2$

Objective D: *Solve quadratic equations by factoring.* (Lessons 12-4, 12-5)

34) 3 or 2
38) 0 or $\frac{-4}{3}$

In 31–40, solve by factoring.

31. $x^2 - 2x = 0$ 0 or 2 **32.** $z^2 + 7z = -12$ −3 or −4

33. $y^2 - 2y - 3 = 0$ **34.** $2r^2 - 10r + 12 = 0$

35. $b^2 - 48 = 2b$ −6 or 8 **36.** $k^2 = 9k - 14$ 2 or 7

37. $0 = m^2 - 16$ −4 or 4 **38.** $9w^2 + 12w = 0$

39. $6y^2 + y - 2 = 0$ $\frac{1}{2}$ or $\frac{-2}{3}$ **40.** $0 = 16m^2 - 8m + 1$ $\frac{1}{4}$

33) −1 or 3

Chapter 12 Review

Resources

From the ***Teacher's Resource File***

- Answer Master for Chapter 12 Review
- Assessment Sourcebook Chapter 12 Test, Forms A–D Chapter 12 Test, Cumulative Form

Additional Resources

- Quiz and Test Writer

The main objectives for the chapter are organized in the Chapter Review under the four types of understanding this book promotes—Skills, Properties, Uses, and Representations.

Whereas end-of-chapter material may be considered optional in some texts, in *UCSMP Algebra* we have selected these objectives and questions with the expectation that they will be covered. Students should be able to answer these questions with about 85% accuracy after studying the chapter.

You may assign these questions over a single night to help students prepare for a test the next day, or you may assign the questions over a two-day period. If you work the questions over two days, then we recommend assigning the *evens* for homework the first night so that students get feedback in class the next day, then assigning the *odds* the night before the test because answers are provided to the odd-numbered questions.

It is effective to ask students which questions they still do not understand and use the day or days as a total class discussion of the material that the class finds most difficult.

LY	R	C	H	L	W	SA
4	4	9	12	32	18	2352
4	3	12	12	24	24	2304
8	3	6	24	24	12	2304
6	3	8	18	24	16	2208
6	4	6	18	32	12	2352

LY	R	C	H	L	W	SA
4	6	6	12	48	12	2592
6	2	12	18	16	24	2208
12	2	6	36	16	12	2400
4	2	18	12	16	36	2400
8	2	9	24	16	18	2208
9	2	8	27	16	16	2240

5b. The crate with dimensions of 24 in. by 18 in. by 16 in. has the least surface area.

c. Responses will vary. Sample response: The manufacturer should choose a crate that uses the least amount of material and allows the user easy access to a stapler.

(Responses continue on page 770.)

 CHAPTER TWELVE

Assessment

Evaluation The *Assessment Sourcebook* provides five forms of the Chapter 12 Test. Forms A and B present parallel versions in a short-answer format. Forms C and D offer performance assessment. The fifth test is the Chapter 12 Test, Cumulative Form. About 50% of this test covers Chapter 12; 25% covers Chapter 11, and 25% covers earlier chapters.

For information on grading, see *General Teaching Suggestions; Grading* in the *Professional Sourcebook*, which begins on page T20 in Part 1 of the Teacher's Edition.

Feedback After students have taken the test for Chapter 12 and you have scored the results, return the tests to students for discussion. Class discussion on the questions that caused trouble for most students can be very effective in identifying and clarifying misunderstandings. You might want to have them write down the items they missed and work either in groups or at home to correct them. It is important for students to receive feedback on every chapter test, and we recommend that students see and correct their mistakes before proceeding too far into the next chapter.

PROPERTIES DEAL WITH THE PRINCIPLES BEHIND THE MATHEMATICS.

41-42) See below right.

Objective E: *Apply the definitions and properties of primes and factors.* (Lesson 12-1)

41. Is 203 prime? Explain why or why not.

42. Is 311 prime? Explain why or why not.

43. Give the number of factors in the prime factorization of 38^2. 4

44. Indicate why the number $3^{40} + 3^{39} + 3^{38}$ could not be prime. Each of the addends is divisible by 3, so the sum is divisible by 3.

Objective F: *Recognize and use the Zero Product Property.* (Lesson 12-4) See below right.

45. What is the Zero Product Property?

In 46–48, why can't the Zero Product Property be used on the given equation?

46. $(x + 3)(x + 4) = 5$ The product is not zero.

47. $(x + 3) + (x - 4) = 0$ There is no product.

48. $(x + 3)^2 = 25$ The product is not zero.

In 49–52, solve.

49. $5q(2q - 7) = 0$ 0 or $\frac{7}{2}$

50. $(m - 3)(m - 1) = 0$ 3 or 1

51. $(2w - 3)(3w + 5) = 0$ $\frac{3}{2}$ or $\frac{-5}{3}$

52. $(y - 3)(2y - 1)(2y + 1) = 0$ 3 or $\frac{1}{2}$ or $\frac{-1}{2}$

Objective G: *Determine whether a quadratic polynomial can be factored over the integers.* (Lessons 12-3, 12-8)

53. *Multiple choice.* Which polynomial can be factored over the integers? b

(a) $x^2 - 11$ (b) $x^2 - 121$

(c) $x^2 + 121$ (d) $x^2 + 112$

54. Suppose *m*, *n*, and *p* are integers. When will the quadratic expression $mx^2 + nx + p$ be factorable over the integers? See right.

In 55–57, use the discriminant to determine whether the expression can be factored over the integers.

55. $x^2 + 7x - 13$ No

56. $x^2 + 7x - 60$ Yes

57. $3r^2 + 2r - 21$ Yes

58. In attempting to factor $x^2 - 16x + 20$, Rachel made a list of pairs of factors of 20 and checked the sum of each pair.

Factors of 20	Sums of factors
-1, -20	-21
-2, -10	-12
-4, -5	-9

From this list she deduced that $x^2 - 16x + 20$ was not factorable over the integers. Determine whether she was right or wrong. Explain your answer. See below.

59. Find two integer factors of 24 whose sum is 10. What does this tell you about $x^2 + 10x + 24$?
6 and 4; It is factorable over the integers.

Objective H: *Apply the definitions and properties of rational and irrational numbers.* (Lesson 12-7)

In 60–67, tell whether the number is rational or irrational.

60. $\sqrt{50}$ irrational **61.** $\sqrt{100}$ rational

62. $2 + \sqrt{2}$ irrational **63.** $\frac{\sqrt{12}}{3}$ irrational

64. π irrational **65.** -100 rational

66. 3.14 rational **67.** $\frac{2}{87}$ rational

68. Show that $5.8\overline{7}$ is a rational number by finding a simple fraction equal to it.

69. Show that $0.\overline{428}$ is a rational number by finding a simple fraction equal to it.

68) $\frac{529}{90}$ 69) $\frac{428}{999}$

41) No; $203 = 7 \cdot 29$

42) Yes; because it only has two factors 1 and 311

45) For any two numbers *a* and *b*, if $ab = 0$, then $a = 0$ or $b = 0$.

54) when $n^2 - 4mp$ is a perfect square

58) She was right because none of the pairs of factors of 20 has a sum of -16.

Additional responses, page 766

6a. Graphs will vary. All graphs are parabolas. The graphs all contain the point (1, 0) because when $x = 1$, $(x - 1) = 0$ and by the multiplication property of zero, $y = 0$.

b. The graphs will contain the point $(r, 0)$ when the factor multiplied by $(ax + b)$ is $(x - r)$.

c. The graph of any equation of the form $y = (x - r)(ax + b)$, where *a*, *b*, and *r* are integers will contain the point $(r, 0)$.

CHAPTER TWELVE

USES DEAL WITH APPLICATIONS OF MATHEMATICS IN REAL SITUATIONS.

Objective I: *Solve quadratic equations in real situations.* *(Lessons 12-4, 12-6)*

70. A circular swimming pool with radius r feet will have a seating area 6 feet in width about the pool.

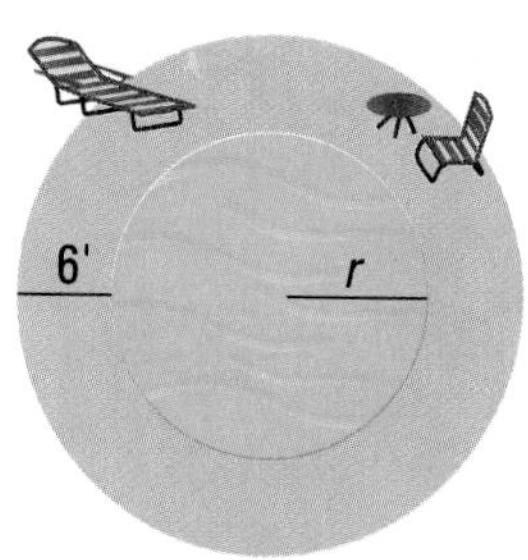

The total area of the pool and seating area is to be 256π square ft. What is the radius of the pool? **10 ft**

71. The area of a rectangular picture is 90 square inches. The length is 4 inches greater than the width. What are the dimensions of the picture? **≈ 7.7 in. by 11.7 in.**

In 72 and 73, use this information. When a golf ball is hit with an upward velocity of 80 feet per second, an equation that gives its height (in feet) above the ground after t seconds is $h = 80t - 16t^2$. **72) 5 seconds**

72. How long will the golf ball be in the air?

73. After how many seconds will the golf ball be 96 feet high? **after 2 seconds and 3 seconds**

74. The area of a rectangular field is 44,800 square feet. The perimeter of the field is 1640 feet. What are the dimensions of the field? **≈ 760 ft by 59 ft**

75. The area of a rectangular poster is 10,350 cm^2. The perimeter of the poster is 410 cm. What are thc dimensions of the poster? **115 cm by 90 cm**

REPRESENTATIONS DEAL WITH PICTURES, GRAPHS, OR OBJECTS THAT ILLUSTRATE CONCEPTS.

Objective J: *Represent quadratic expressions and their factorizations with areas.* *(Lessons 12-2, 12-3)*

In 76 and 77,

a. Write the area of the figure as a polynomial.

b. Write the area in factored form.

76.

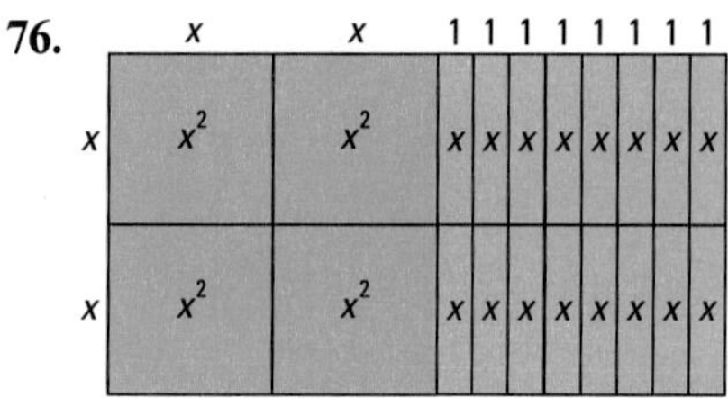

a) $4x^2 + 16x$
b) $2x(2x + 8)$ or $4x(x + 4)$

77. x x 1 1 1 1 | x: x^2 x^2 x x x x | 1: x x 1 1 1 1 | 1: x x 1 1 1 1

77–78. See below.

78. Show that $x^2 + 7x + 6$ can be factored by arranging tiles representing the polynomial in a rectangle. Sketch your arrangement.

79. A square has an area of $9a^2 + 30ab + 25b^2$. What is the length of a side of the square? **$3a + 5b$**

77a) $2x^2 + 8x + 8$

b) $(x + 2)(2x + 4)$ or $2(x + 2)^2$

78)

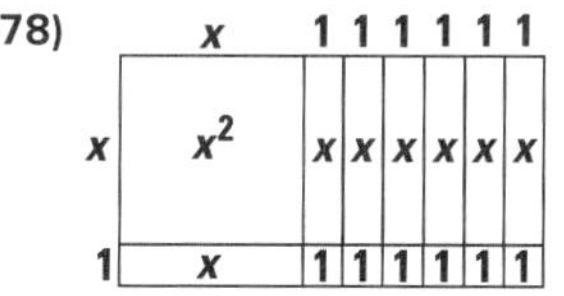

Additional Answers, page 753.

19. Each factor gives a possibility. The factors in the table are organized by the powers of 2, and then by multiplying each by a power of 5.

Years	Amount per Year
1	$1,000,000
2	500,000
4	250,000
8	125,000
16	62,500
32	31,250
64	15,625
5	200,000
25	40,000
125	8,000
625	1,600
3,125	320
15,625	64
10	100,000
50	20,000
250	4,000
1,250	800
6,250	160
31,250	32
20	50,000
100	10,000
500	2,000
2,500	400
12,500	80
62,500	16
40	25,000
200	5,000
1,000	1,000
5,000	200
25,000	40
125,000	8
80	12,500
400	2,500
2,000	500
10,000	100
50,000	20
250,000	4
160	6,250
800	1,250
4,000	250
20,000	50
100,000	10
500,000	2
320	3,125
1,600	625
8,000	125
40,000	25
200,000	5
1,000,000	1

Additional Answers, page 737

32a. Does $\frac{2^2 - 4 \cdot 2 + 3}{2^2 - 9} = \frac{2-1}{2+3}$?

$\frac{4 - 8 + 3}{4 - 9} = \frac{1}{5}$?

$\frac{-1}{-5} = \frac{1}{5}$?

Yes, it checks.

b. Sample: $\frac{x^2 - 1}{x^2 + 7x + 6}$

when $x \neq -1$ and $x \neq -6$

c. $\frac{(x-1)(x+1)}{(x+1)(x+6)} = \frac{x-1}{x+6}$

Check: Let $x = 2$.

Does

$\frac{2^2 - 1}{2^2 + 7 \cdot 2 + 6} = \frac{2-1}{2+6}$?

$\frac{3}{24} = \frac{1}{8}$?

Yes, it checks.

Chapter 13 Planner

Adapting to Individual Needs

The student text is written for the vast majority of students. The chart at the right suggests two pacing plans to accommodate the needs of your students. Students in the Full Course should complete the entire text by the end of the year. Students in the Minimal Course will spend more time when there are quizzes and more time on the Chapter Review. Therefore, these students may not complete all of the chapters in the text.

Options are also presented to meet the needs of a variety of teaching and learning styles. For each lesson, the Teacher's Edition provides sections entitled: *Video* which describes video segments and related questions that can be used for motivation or extension; *Optional Activities* which suggests activities that employ materials, physical models, technology, and cooperative learning; and, *Adapting to Individual Needs* which regularly includes **Challenge** problems, **English Language Development** suggestions, and suggestions for providing **Extra Help.** The Teacher's Edition also frequently includes an **Error Alert,** an **Extension,** and an **Assessment** alternative. The options available in Chapter 13 are summarized in the chart below.

Chapter 13 Pacing Chart

Day	Full Course	Minimal Course
1	13-1	13-1
2	13-2	13-2
3	13-3	13-3
4	Quiz*; 13-4	Quiz*; begin 13-4.
5	13-5	Finish 13-4.
6	13-6	13-5
7	Quiz*; 13-7	13-6
8	13-8	Quiz*; begin 13-7.
9	Self-Test	Finish 13-7.
10	Review	13-8
11	Test*	Self-Test
12	Comprehensive Test*	Review
13		Review
14		Test*
15		Comprehensive Test*

*in the Teacher's Resource File

In the Teacher's Edition...

Lesson	Optional Activities	Extra Help	Challenge	English Language Development	Error Alert	Extension	Cooperative Learning	Ongoing Assessment
13-1	●	●	●	●	●	●	●	Written
13-2	●	●	●	●	●	●	●	Oral
13-3	●	●	●		●	●	●	Group
13-4	●	●	●	●		●	●	Written
13-5	●	●	●		●	●	●	Written
13-6	●	●	●	●	●	●	●	Oral
13-7	●	●	●	●	●	●		Written
13-8	●	●	●	●		●		Written

In the Additional Resources...

	In the Teacher's Resource File								
Lesson	Lesson Masters, A and B	Teaching Aids*	Activity Kit*	Answer Masters	Technology Sourcebook	Assessment Sourcebook	Visual Aids**	Technology	Video Segments
Opener		134					134		
13-1	13-1	28, 131, 135		13-1			28, 131, 135, AM		
13-2	13-2	28, 131	28	13-2			28, 131, AM		
13-3	13-3	28, 131, 136		13-3		Quiz	28, 131, 136, AM		
13-4	13-4	132, 137		13-4			132, 137, AM		13-4
In-class Activity		28, 61		13-5			28, 61, AM		
13-5	13-5	28, 61, 132, 138		13-5	Comp 32		28, 61, 132, 138, AM	Spreadsheet	
13-6	13-6	28, 132, 139	29	13-6	Comp 33	Quiz	28, 132, 139, AM	GraphExplorer	
In-class Activity				13-7			AM		
13-7	13-7	133, 140, 141		13-7			133, 140, 141, AM		
13-8	13-8	133, 142	30	13-8			133, 142, AM		
End of chapter				Review		Tests			

*Teaching Aids are pictured on pages 772C and 772D. The activities in the Activity Kit are pictured on page 772C.

**Visual Aids provide transparencies for all Teaching Aids and all Answer Masters.

Also available is the Study Skills Handbook which includes study-skill tips related to reading, note-taking, and comprehension.

Integrating Strands and Applications

	13-1	13-2	13-3	13-4	13-5	13-6	13-7	13-8
Mathematical Connections								
Number Sense				●	●	●		●
Algebra	●	●	●	●	●	●	●	●
Geometry	●	●	●	●	●	●	●	●
Measurement	●				●		●	●
Probability				●	●	●	●	
Statistics/Data Analysis	●							
Patterns and Functions	●	●	●	●	●	●	●	●
Discrete Mathematics				●				
Interdisciplinary and Other Connections								
Art						●		
Science	●	●	●	●	●			●
Social Studies		●		●	●			
Multicultural					●			
Technology		●	●	●	●	●	●	●
Career	●	●						
Consumer	●	●	●	●		●		
Sports			●					

Teaching and Assessing the Chapter Objectives

				In the Teacher's Resource File		
Chapter 13 Objectives (Organized into the SPUR categories—Skills, Properties, Uses, and Representations)	**Lessons**	**Progress Self-Test Questions**	**Chapter Review Questions**	**Chapter Test, Forms A and B**	**Chapter Test, Forms**	
					C	**D**
Skills						
A: Evaluate functions and solve equations involving function notation.	13-2, 13-3	1	1–10	1, 20	3	
B: Use function keys on a calculator.	13-7, 13-8	4	11–20	3		
Properties						
C: Determine whether a set of ordered pairs is a function.	13-1	2, 6, 7	21–30	2, 6, 19	2	
D: Find the domain and the range of a function from its formula, graph, or rule.	13-4	8	31–40	7	1	✓
Uses						
E: Use function notation and language in real situations.	13-2, 13-3, 13-4	14, 17	41–46	11, 12, 15, 16	3	✓
F: Determine values of probability functions.	13-5	11, 12	47–50	8, 13	4	
G: Find lengths of sides or tangents of angles in right triangles using the tangent function.	13-7	3, 5	51–54	4, 5, 20	5	
Representations						
H: Determine whether or not a graph represents a function.	13-1	9, 10	55–60	9, 10	2	
I: Graph functions.	13-1, 13-2, 13-3, 13-5	13, 15, 16	61–67	14	1	✓
J: Graph polynomial functions.	13-6	18, 19	68–71	17		

Multidimensional Assessment
Quiz for Lessons 13-1 through 13-3
Quiz for Lessons 13-4 through 13-6
Chapter 13 Test, Forms A–D
Chapter 13 Test, Cumulative Form
Comprehensive Test, Chapters 1–13

Quiz and Test Writer
Multiple forms of chapter tests and quizzes; Challenges

Activity Kit

ACTIVITY 28 — FUNCTION NOTATION Use with **Lesson 13-2.**

Materials: Typing paper, tape, yardsticks, grid paper, ruler
Group Size: Partners

Each of you should roll a sheet of typing paper into a tube. Tape it in several places. Be careful when handling the tube to maintain its circular shape.

Position a yardstick with the smaller numbers at the top. Tape it to the wall at about the height of a student.

Each person should do the entire activity and have his or her own set of data and graphs. For the first three items, one of you should view the yardstick through the tube while the other helps with the measuring, computing, and recording. Switch roles and repeat the first three items.

1. If you have the tube, stand with your toes exactly 1 foot away from the wall with the yardstick. Then, look through the tube and tell how many inches on the yardstick are visible through the tube. Call this the *viewable height.* Take the time to accurately determine this measure to the nearest $\frac{1}{4}$ inch. Then record it in the table on the next page.
2. The area of the circular region that is seen through the tube is called the *viewable area.* Calculate the viewable area to the nearest square inch and record it in the table on the next page.
3. Now move back to 2 feet from the wall and record the viewable height and viewable area. Repeat for 3, 4, 5, and 6 feet.

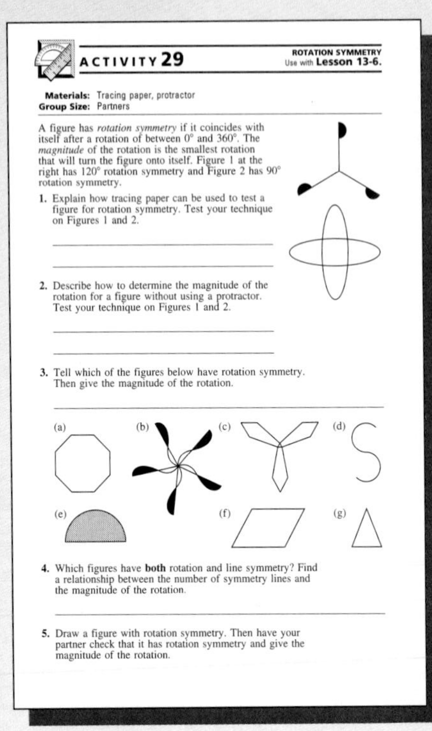
ACTIVITY 29 — ROTATION SYMMETRY Use with **Lesson 13-6.**

Materials: Tracing paper, protractor
Group Size: Partners

A figure has *rotation symmetry* if it coincides with itself after a rotation of between 0° and 360°. The *magnitude* of the rotation is the smallest rotation that will turn the figure onto itself. Figure 1 at the right has 120° rotation symmetry and Figure 2 has 90° rotation symmetry.

1. Explain how tracing paper can be used to test a figure for rotation symmetry. Test your technique on Figures 1 and 2.
2. Describe how to determine the magnitude of the rotation for a figure without using a protractor. Test your technique on Figures 1 and 2.
3. Tell which of the figures below have rotation symmetry. Then give the magnitude of the rotation.

(a) (b) (c) (d) (e) (f) (g)

4. Which figures have **both** rotation and line symmetry? Find a relationship between the number of symmetry lines and the magnitude of the rotation.
5. Draw a figure with rotation symmetry. Then have your partner check that it has rotation symmetry and give the magnitude of the rotation.

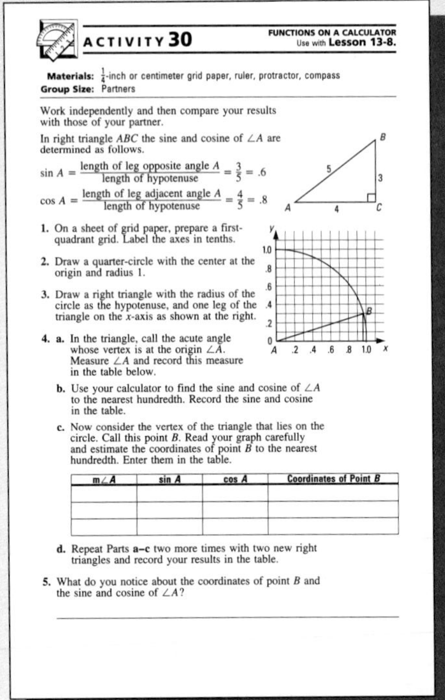
ACTIVITY 30 — FUNCTIONS ON A CALCULATOR Use with **Lesson 13-8.**

Materials: $\frac{1}{4}$-inch or centimeter grid paper, ruler, protractor, compass
Group Size: Partners

Work independently and then compare your results with those of your partner.

In right triangle *ABC* the sine and cosine of ∠*A* are determined as follows.

$$\sin A = \frac{\text{length of leg opposite angle } A}{\text{length of hypotenuse}} = \frac{3}{5} = .6$$

$$\cos A = \frac{\text{length of leg adjacent angle } A}{\text{length of hypotenuse}} = \frac{4}{5} = .8$$

1. On a sheet of grid paper, prepare a first-quadrant grid. Label the axes in tenths.
2. Draw a quarter-circle with the center at the origin and radius 1.
3. Draw a right triangle with the radius of the circle as the hypotenuse, and one leg of the triangle on the *x*-axis as shown at the right.
4. a. In the triangle, call the acute angle whose vertex is at the origin ∠*A*. Measure ∠*A* and record this measure in the table below.
   b. Use your calculator to find the sine and cosine of ∠*A* to the nearest hundredth. Record the sine and cosine in the table.
   c. Now consider the vertex of the triangle that lies on the circle. Call this point *B*. Read your graph carefully and estimate the coordinates of point *B* to the nearest hundredth. Enter them in the table.

m∠*A*	sin *A*	cos *A*	Coordinates of Point *B*

   d. Repeat Parts **a–c** two more times with two new right triangles and record your results in the table.
5. What do you notice about the coordinates of point *B* and the sine and cosine of ∠*A*?

Teaching Aids

Teaching Aid 28, Four-Quadrant Graph Paper, (shown on page 140D) can be used with **Lessons 13-1** through **13-3,** the **In-class Activity** preceding **Lesson 13-5,** and **Lessons 13-5** and **13-6. Teaching Aid 61, Two-Dice Outcomes,** (shown on page 368) can be used with the **In-class Activity** preceding **Lesson 13-5,** and with **Lesson 13-5.**

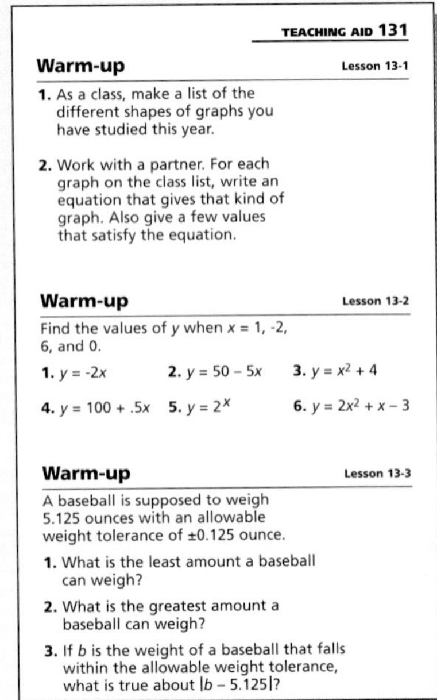
TEACHING AID 131

Warm-up Lesson 13-1

1. As a class, make a list of the different shapes of graphs you have studied this year.
2. Work with a partner. For each graph on the class list, write an equation that gives that kind of graph. Also give a few values that satisfy the equation.

Warm-up Lesson 13-2

Find the values of *y* when *x* = 1, -2, 6, and 0.

1. $y = -2x$
2. $y = 50 - 5x$
3. $y = x^2 + 4$
4. $y = 100 + .5x$
5. $y = 2^x$
6. $y = 2x^2 + x - 3$

Warm-up Lesson 13-3

A baseball is supposed to weigh 5.125 ounces with an allowable weight tolerance of ±0.125 ounce.

1. What is the least amount a baseball can weigh?
2. What is the greatest amount a baseball can weigh?
3. If *b* is the weight of a baseball that falls within the allowable weight tolerance, what is true about $|b - 5.125|$?

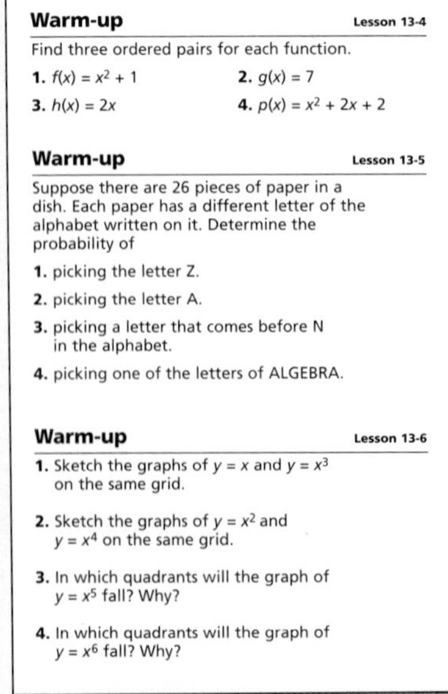
TEACHING AID 132

Warm-up Lesson 13-4

Find three ordered pairs for each function.

1. $f(x) = x^2 + 1$
2. $g(x) = 7$
3. $h(x) = 2x$
4. $p(x) = x^2 + 2x + 2$

Warm-up Lesson 13-5

Suppose there are 26 pieces of paper in a dish. Each paper has a different letter of the alphabet written on it. Determine the probability of

1. picking the letter Z.
2. picking the letter A.
3. picking a letter that comes before N in the alphabet.
4. picking one of the letters of ALGEBRA.

Warm-up Lesson 13-6

1. Sketch the graphs of $y = x$ and $y = x^3$ on the same grid.
2. Sketch the graphs of $y = x^2$ and $y = x^4$ on the same grid.
3. In which quadrants will the graph of $y = x^5$ fall? Why?
4. In which quadrants will the graph of $y = x^6$ fall? Why?

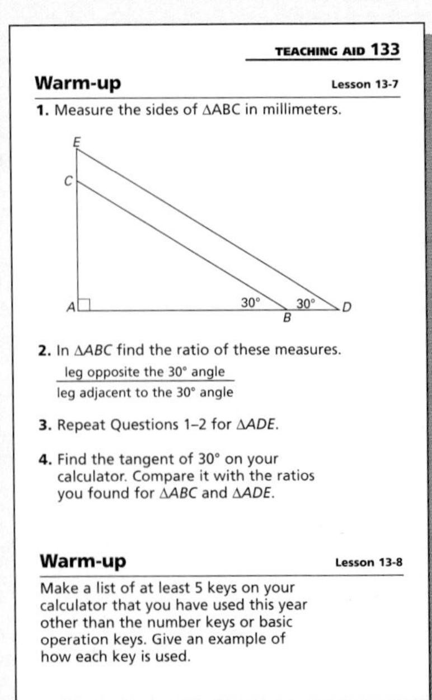
TEACHING AID 133

Warm-up Lesson 13-7

1. Measure the sides of △ABC in millimeters.

2. In △*ABC* find the ratio of these measures.
$$\frac{\text{leg opposite the 30° angle}}{\text{leg adjacent to the 30° angle}}$$
3. Repeat Questions 1–2 for △*ADE*.
4. Find the tangent of 30° on your calculator. Compare it with the ratios you found for △*ABC* and △*ADE*.

Warm-up Lesson 13-8

Make a list of at least 5 keys on your calculator that you have used this year other than the number keys or basic operation keys. Give an example of how each key is used.

Functions

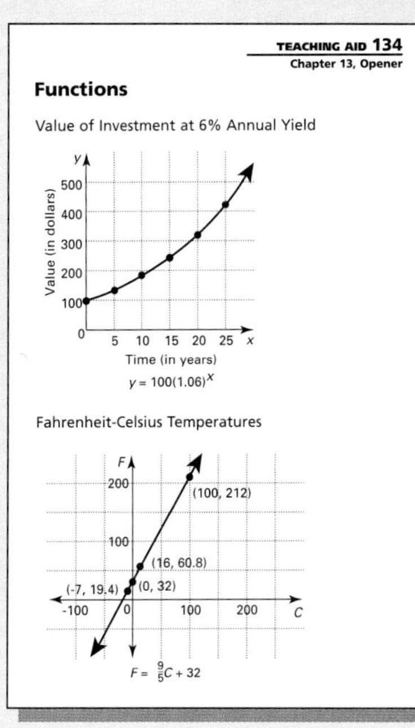

Additional Examples

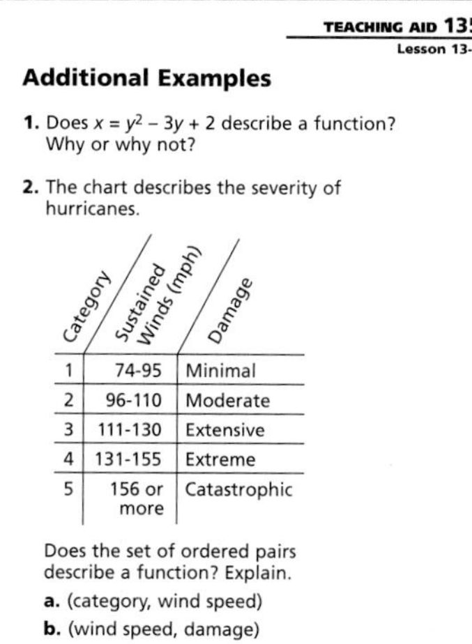

1. Does $x = y^2 - 3y + 2$ describe a function? Why or why not?
2. The chart describes the severity of hurricanes.

Category	Sustained Winds (mph)	Damage
1	74-95	Minimal
2	96-110	Moderate
3	111-130	Extensive
4	131-155	Extreme
5	156 or more	Catastrophic

Does the set of ordered pairs describe a function? Explain.

a. (category, wind speed)
b. (wind speed, damage)
c. (damage, category)

Test-Flight Graph

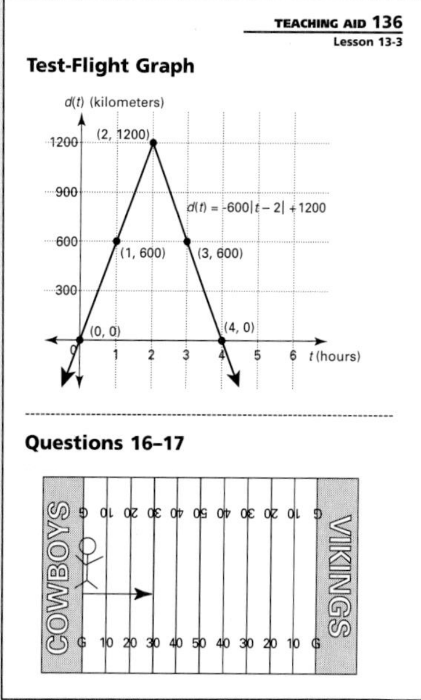

Questions 16–17

Domain and Range

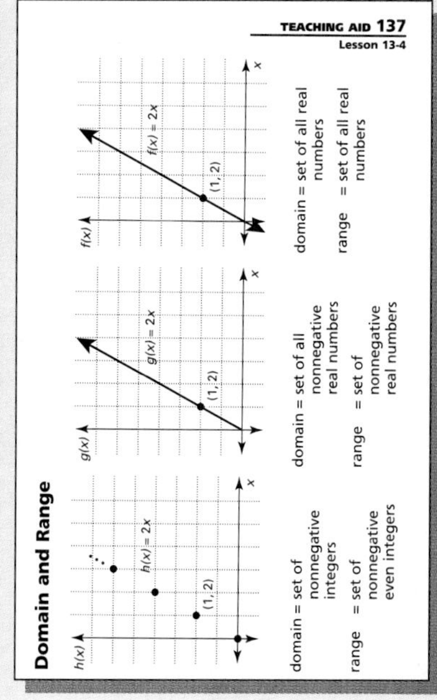

Additional Examples

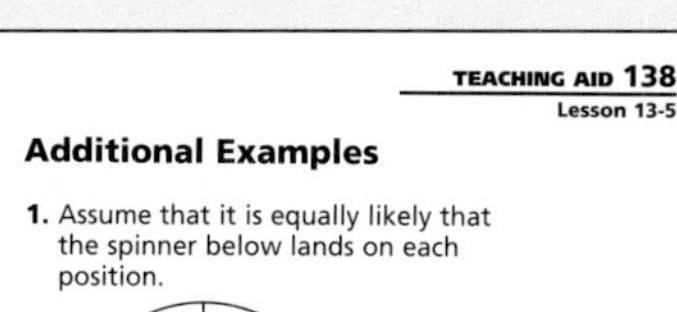

1. Assume that it is equally likely that the spinner below lands on each position.

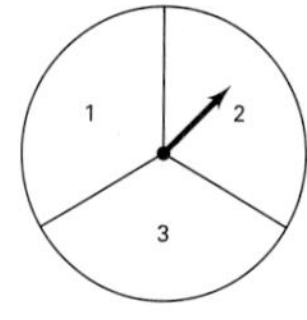

a. Let $P(n)$ = the probability of landing in region n. Graph the function P.
b. Give an equation for the function.

2. Assume that four fair coins are tossed. Let $P(t)$ = the probability of tossing t tails. Give the ordered pairs of this function. Hint: First make a list of all possible outcomes.

Examples 1–3

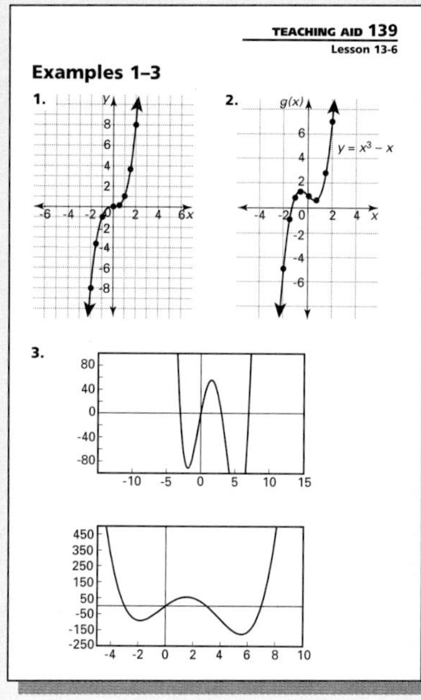

Activity

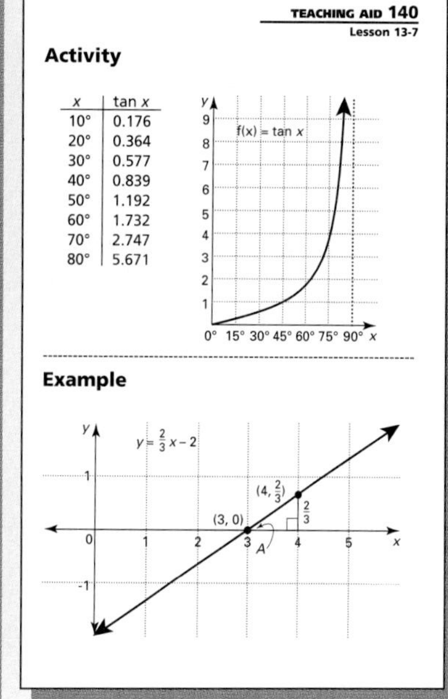

x	tan x
10°	0.176
20°	0.364
30°	0.577
40°	0.839
50°	1.192
60°	1.732
70°	2.747
80°	5.671

Example

Additional Examples

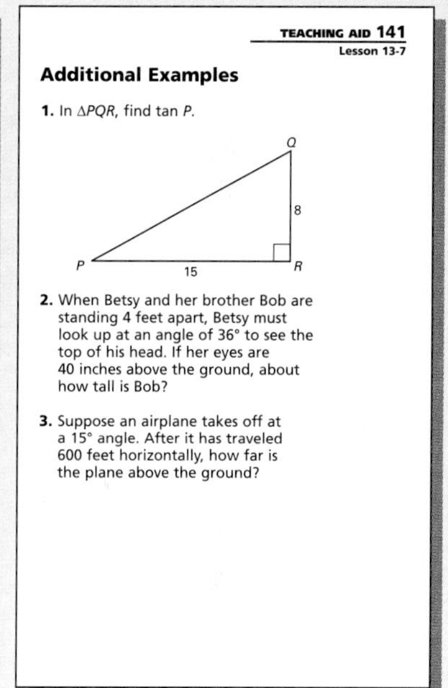

1. In ΔPQR, find tan P.
2. When Betsy and her brother Bob are standing 4 feet apart, Betsy must look up at an angle of 36° to see the top of his head. If her eyes are 40 inches above the ground, about how tall is Bob?
3. Suppose an airplane takes off at a 15° angle. After it has traveled 600 feet horizontally, how far is the plane above the ground?

Graphs of y = sin x and y = log x

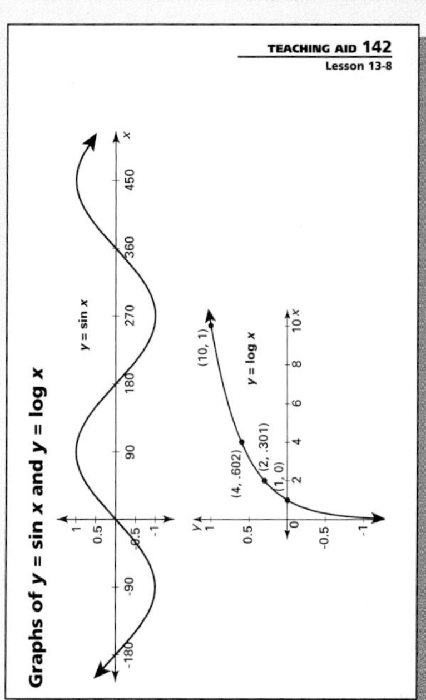

Chapter Opener

Pacing

All lessons in this chapter are designed to be covered in one day. At the end of the chapter, you should plan to spend 1 day to review the Progress Self-Test, 1 to 2 days for the Chapter Review, and 1 day for a test. You may want to spend a day on projects, and possibly a day is needed for quizzes. Therefore, this chapter should take 11 to 14 days.

Using Pages 772–773

When discussing this chapter opener, you might want to use **Teaching Aid 134** which contains the graphs on page 773.

A *function* is a correspondence in which the value of the first variable determines the value of the second variable. This does not mean that there must be a formula for the second variable in terms of the first variable. As these pages point out, students have been studying functions throughout this course. So, although the *word* function might be new, the *idea* is not.

CHAPTER 13

772

Chapter 13 Overview

The Chapter Opener introduces two ways of describing functions: by a graph and by an equation.

The idea of a function is one of the unifying themes in mathematics. Similarly, in this chapter, functions summarize many of the ideas that have been covered in the book. The language and symbolism of functions make it easier to communicate ideas about them. The placement of these ideas in the last chapter of this text gives students the opportunity to review Skills, Properties, Uses, and Representations of the following families of functions.

Lessons 13-1 to 13-3 discuss functions whose graphs have already been studied, including linear, exponential, quadratic, and absolute-value functions. These lessons provide students with sufficient examples to make the study of domain and range in Lesson 13-4 a reasonable aspect of the analysis of functions.

Earlier chapters covered probability and polynomials. In Lessons 13-5 and 13-6, functions that are related to those topics are treated; this causes us to look at them somewhat differently. The function idea in particular lends itself to a graphical representation.

FUNCTIONS

In earlier chapters, you studied the graphs of many equations. Here are two of them.

Value of Investment at 6% Annual Yield

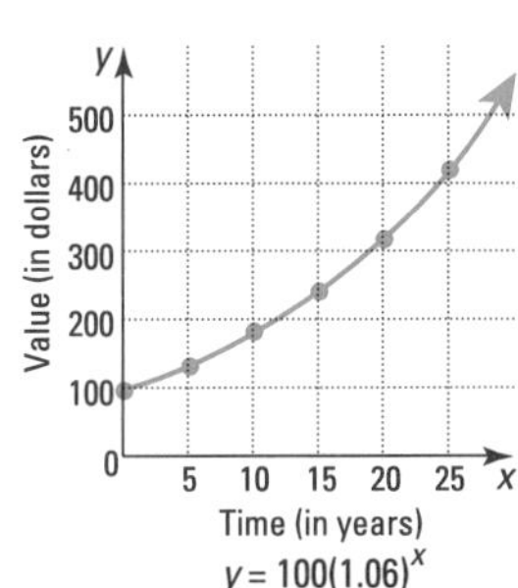

$y = 100(1.06)^x$

Fahrenheit-Celsius Temperatures

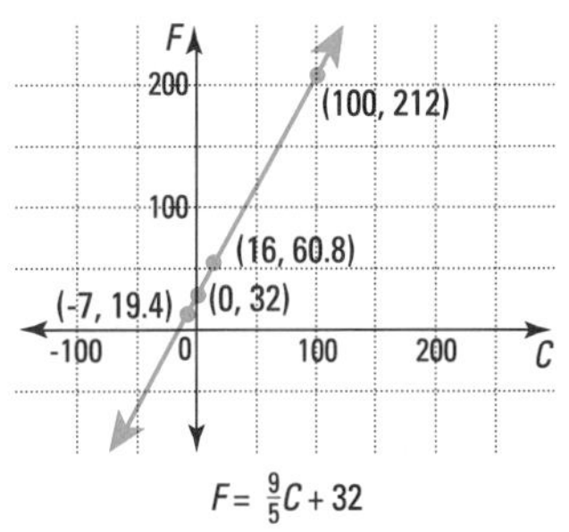

$F = \frac{9}{5}C + 32$

Despite their differences, these graphs describe situations that have several things in common. In each situation there are two variables, and every value of the first variable determines exactly one value of the second variable. In the investment situation, the length of time that the money has been invested determines the value of the investment. In the Fahrenheit-Celsius situation, the Celsius temperature determines the Fahrenheit temperature. When a first variable determines a second, we call the relationship between the two variables a *function.*

Functions are found in every branch of mathematics and its applications. As a result, there are many ways to describe functions: by graphs; by equations; by lists of ordered pairs; by rules written in words. The analysis of functions is extremely important in mathematics, and entire courses are often devoted to them. In this chapter, you will use the language of functions to review many ideas you have seen in earlier chapters. You will also encounter some important functions you have not seen before.

Photo Connections

The photo collage makes real-world connections to the content of the chapter: functions.

Flags: The Model United Nations (MUN) is a simulation of UN activities by high school and college students. In mathematics, a computer can simulate the tossing of two dice as in the activity on page 794. This activity can help students recognize the connection between relative frequencies and probability.

Wave Graph: Variations of the sine curve serve as a background for paper airplanes. The sine function and other trigonometric functions are introduced in Lesson 13-8.

Time: The graph on this page, *Value of Investment,* shows that the longer the investment period, the greater the value of the investment. Here the value of the investment is a function of a measurable period, time.

Babies: Every 2 seconds, 9 babies are born and 3 people die. The result is a net increase of 3 people each second, or a growth in world population of 93 million people each year. In Lesson 13-5, probability and relative frequency functions are used to examine the probability that the number of male babies born each year is the same as the number of female babies born.

Chapter 13 Projects

At this time you might want to have students look over the projects on pages 817–818.

Lessons 13-7 and 13-8 introduce students to some of the functions that they will encounter in later courses—the trigonometric functions and the common logarithm function.

Lesson 13-1

Objectives

C Determine whether a set of ordered pairs is a function.
H Determine whether or not a graph represents a function.
I Graph functions.

Resources

From the ***Teacher's Resource File***

- Lesson Master 13-1A or 13-1B
- Answer Master 13-1
- Teaching Aids
  - 28 Four-Quadrant Graph Paper
  - 131 Warm-up
  - 135 Additional Examples

Additional Resources

- Visuals for Teaching Aids 28, 131, 135
- Automatic graphers

Teaching Lesson 13-1

Warm-up

Cooperative Learning

1. As a class, make a list of the different shapes of graphs you have studied this year. **Sample: line, angle, parabola, exponential curve**
2. **Work with a partner**. For each graph on the class list, write an equation that gives that kind of graph. Also give a few values that satisfy the equation. **Responses will vary.**

LESSON

13-1 *What Is a Function?*

A function is a relationship between two variables in which the first variable determines the second variable. Here is a function you first saw in Chapter 9 while studying quadratic equations. In this function, the horizontal distance of the La Quebrada cliff diver from the cliff determines the height of the diver above the water. This function can be described nicely by a graph or by an equation. The table gives some, but not all, values of the function. The description in words is quite long.

La Quebrada Cliff-Diver Function

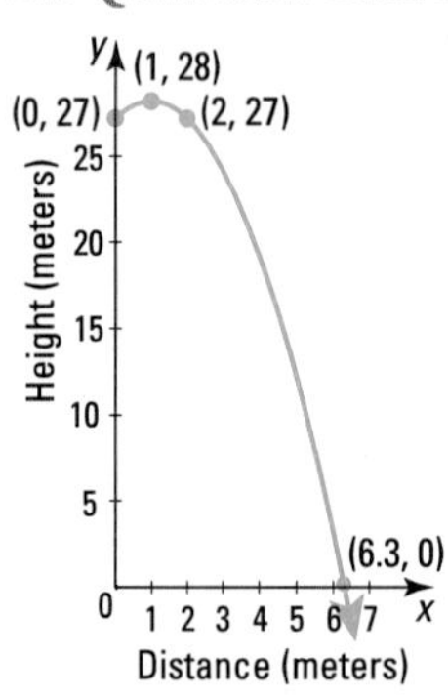

Graph: (at the left)

Equation: $y = -x^2 + 2x + 27$

Table:

x	0	1	2	3	...
y	27	28	27	24	...

Words:
The height of the diver is 27 meters plus twice the diver's horizontal distance from the cliff less the square of that distance.

Two Definitions of Function

In general, you can think of functions as special kinds of correspondences or as special sets of ordered pairs. A correspondence between x and y is shown in the table of the La Quebrada cliff-diver function. A **function** is a correspondence between two variables in which each value of the first variable corresponds to *exactly one* value of the second variable.

The graph shows the function as a set of ordered pairs. A **function** is a set of ordered pairs in which each first coordinate appears with *exactly one* second coordinate. That is, once you know the value of the first variable (often called x), then there is only one value for the second variable (often called y). The value of the second variable is called a **value of the function.**

Example 1

Does the equation $3x + 4y = 12$ describe a function? Why or why not?

Solution 1

Solve the equation for y.

$$4y = -3x + 12$$
$$y = -\frac{3}{4}x + 3$$

Because each value of x corresponds to just one value of y, the equation describes a function.

Lesson 13-1 Overview

Broad Goals This lesson discusses the distinction between a function and a relation. In the examples, graphs and equations are examined to determine if they describe functions.

Perspective The key idea in a function is the notion that the first coordinate of the ordered pair determines the second coordinate. But this does not have to be a causal determination. For instance, there is a function of the set of ordered pairs (name, age) for the set of people in your class, but a name does not cause a certain age.

Of the two conceptions of functions given on page 774, the one found most often in today's school mathematics books is *static*: a function is a special relation. The correspondence conception is *dynamic*: a function is a special kind of mapping.

For functions whose domain and range are sets of real numbers, the static conception has the advantage of lending itself more easily to the representation of a function by a graph. The dynamic conception has the advantage of more easily fitting with the idea of independent and dependent variables.

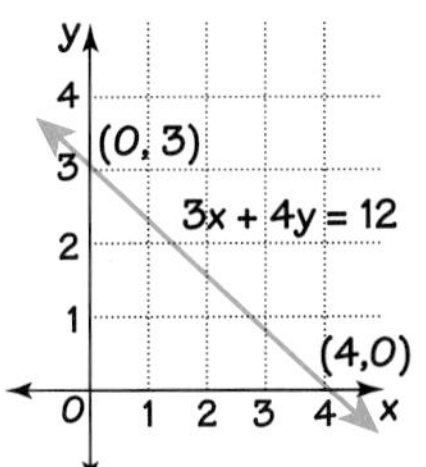

Solution 2

Graph $3x + 4y = 12$. Because no two points have the same first coordinate, the graph describes a function.

What Are Some Types of Functions?

The function of Example 1 is a **linear function.** Linear functions have equations of the form $y = mx + b$. Their graphs are lines. A **quadratic function** has an equation of the form $y = ax^2 + bx + c$. Its graph is a parabola. The cliff-diver graph on page 774 is an example of a quadratic function. An **exponential function** has an equation of the form $y = ab^x$. The investment graph on page 773 is an example of an exponential function.

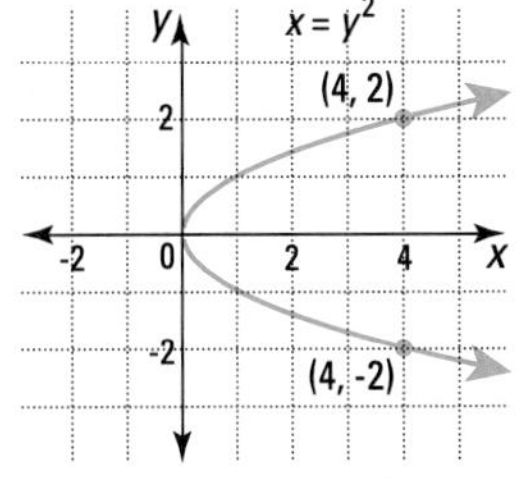

A Relation That Is Not a Function

At the left is a graph of $x = y^2$. This equation does *not* describe a function. When $x = y^2$, the value $x = 4$ corresponds to two different values for y, 2 and -2. Since the points (4, 2) and (4, -2) are both on the graph, the set of ordered pairs satisfying $x = y^2$ is not a function.

By solving $x = y^2$ for y, you can tell without graphing that this equation is not a function. $x = y^2$ implies $y = \pm\sqrt{x}$. Because in $y = \pm\sqrt{x}$ every positive value of x corresponds to two values of y, the equation $y^2 = x$ does not describe a function.

Located at the southern end of Moscow's Red Square is St. Basil's Cathedral. Each of its 10 domes differs in design and color.

Example 2

This scatterplot from page 458 gives latitudes and April mean high temperatures (°F) for 10 selected cities. Does it describe a function? Why or why not?

Solution

No. The latitudes for Copenhagen and Moscow are the same, but the temperatures are different. These ordered pairs have the same first coordinate but different second coordinates. This scatterplot does not describe a function.

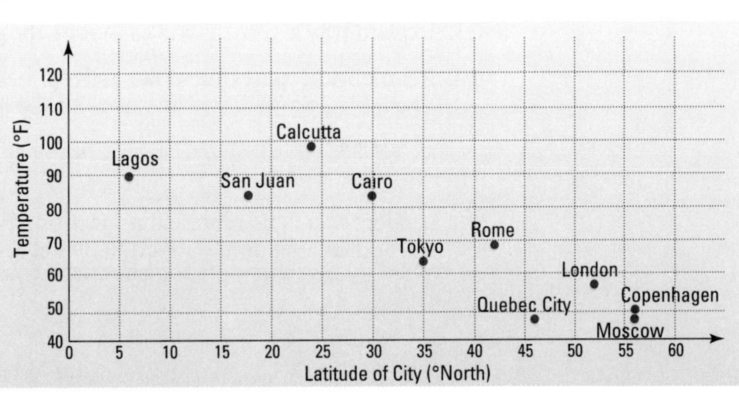

Notes on Reading

Students can use the Vertical-Line Test to decide if a graph describes a function. If it is possible to draw a vertical line that intersects more than one point of a graph, the graph is *not* the graph of a function.

Additional Examples

These examples are also given on **Teaching Aid 135.**

1. Does $x = y^2 - 3y + 2$ describe a function? Why or why not? **No, a value of x may correspond to more than one value of y.**
2. The chart describes the severity of hurricanes.

Category	Sustained Winds (mph)	Damage
1	74–95	Minimal
2	96–110	Moderate
3	111–130	Extensive
4	131–155	Extreme
5	156 or more	Catastrophic

Does the set of ordered pairs describe a function? Explain.

a. (category, wind speed) **No, for each category there is more than one wind speed.**
b. (wind speed, damage) **Yes, for each wind speed there is only one damage type.**
c. (damage, category) **Yes, for each type of damage there is only one category.**

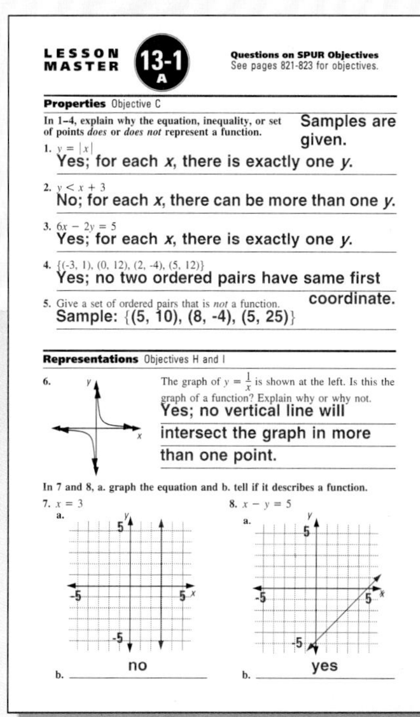

LESSON MASTER 13-1 A

Questions on SPUR Objectives
See pages 821-823 for objectives.

Properties Objective C

In 1–4, explain why the equation, inequality, or set of points *does* or *does not* represent a function. Samples are given.

1. $y = |x|$
Yes; for each x, there is exactly one y.

2. $y < x + 3$
No; for each x, there can be more than one y.

3. $6x - 2y = 5$
Yes; for each x, there is exactly one y.

4. {(-3, 1), (0, 12), (2, -4), (5, 12)}
Yes; no two ordered pairs have same first coordinate.

5. Give a set of ordered pairs that is *not* a function.
Sample: {(5, 10), (8, -4), (5, 25)}

Representations Objectives H and I

6. The graph of $y = \frac{1}{x}$ is shown at the left. Is this the graph of a function? Explain why or why not.
Yes; no vertical line will intersect the graph in more than one point.

In 7 and 8, a. graph the equation and b. tell if it describes a function.

7. $x = 3$
b. no

8. $x - y = 5$
b. yes

Optional Activities

You might use this activity to introduce the lesson. Ask students these questions:

1. Is there a function relating a person with his or her age? [Yes, at any given moment each person has a unique age.]
2. Is there a function relating an age with a person? [No, each age may relate to many people.]
3. Is there a function relating the length of the side of a rectangle with its width? [No, a given length can be associated with infinitely many widths.]
4. Is there a function relating the length of the side of a square with its area? [Yes, a given length determines a unique area.]

Now have students work in groups and make up other statements that illustrate relationships which are functions and relationships which are not functions.

Notes on Questions

Students will need **Teaching Aid 28** or graph paper.

Question 9 For **parts a, b, and c,** there is a simple equation that each point satisfies: (a) $y = x$, (b) $x + y = 8$, and (c) $x^2 + y^2 = x + y$.

Question 13 This is the one type of linear relation that is not a function.

Question 14 Error Alert Students who have difficulty recognizing an equation that relates *h* and *w* should be encouraged to either rewrite the table with all *h* values in increasing order or to graph the data.

Questions 15–16 Point out the connections between the ordered pairs, the graph, and the equation.

Questions 17–18 These questions point out the relationships among verbal descriptions, formulas, and sets of ordered pairs.

Question 24 Error Alert If students have difficulty understanding this question, you might tell them to imagine the fractions as part of a pie. Ask, "When $m > n$, do you get more pie with *m* people or with *n* people?" [With *n* people because $\frac{1}{n} > \frac{1}{m}$.]

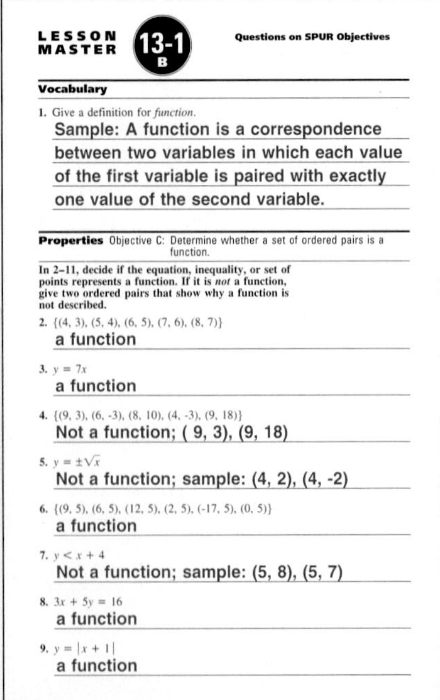

LESSON MASTER 13-1 B — Questions on SPUR Objectives

Vocabulary

1. Give a definition for *function*.
Sample: A function is a correspondence between two variables in which each value of the first variable is paired with exactly one value of the second variable.

Properties Objective C: Determine whether a set of ordered pairs is a function.

In 2–11, decide if the equation, inequality, or set of points represents a function. If it is *not* a function, give two ordered pairs that show why a function is not described.

2. {(4, 3), (5, 4), (6, 5), (7, 6), (8, 7)}
a function

3. $y = 7x$
a function

4. {(9, 3), (6, -3), (8, 10), (4, -3), (9, 18)}
Not a function; (9, 3), (9, 18)

5. $y = \pm\sqrt{x}$
Not a function; sample: (4, 2), (4, -2)

6. {(9, 5), (6, 5), (12, 5), (2, 5), (-17, 5), (0, 5)}
a function

7. $y < x + 4$
Not a function; sample: (5, 8), (5, 7)

8. $3x + 5y = 16$
a function

9. $y = |x + 1|$
a function

QUESTIONS

Covering the Reading

1. Name four ways in which a function can be described.
by a graph, by an equation, with a table, or as a set of ordered pairs
2. Define *function* using the definition you prefer. **See left.**

2) Sample: A function is a correspondence between two variables in which each value of the first variable corresponds to exactly one value of the second variable.

3. In your own words explain why $x = y^2$ is not an equation for a function. **Sample: Any particular positive value for x will give two different values for y.**

In 4 and 5, a relation and a first coordinate in the relation are given. **a.** Find all corresponding second coordinates. **b.** Is the relation also a function?

4. cliff-diver graph on page 774; $x = 4$ **a) y = 19; b) Yes**
5. latitude-high temperature relation on page 775; latitude = 56
a) Temperature = 47°F and 50°F; b) No
6. Does the graph of $x - 3y = 6$ describe a function? Why or why not?
Yes, $y = \frac{1}{3}x - 2$. Each value of x corresponds to one value of y.

In 7 and 8, give an equation for the type of function.

7. exponential
Sample: $y = 100 \cdot (1.06)^x$
8. quadratic
Sample: $y = x^2 + 2x + 1$

Applying the Mathematics

9. *Multiple choice.* Which set of ordered pairs is *not* a function? **c**
(a) {(0, 0), (1, 1), (2, 2)}
(b) {(3, 5), (5, 3), (4, 4)}
(c) {(0, 0), (1, 0), (0, 1)}
(d) $\left\{\left(\frac{1}{2}, 1\right), (\sqrt{7}, \sqrt{8}), \left(6, \frac{-9}{23}\right)\right\}$

In 10–12, does the graph represent a function?

10. **No**
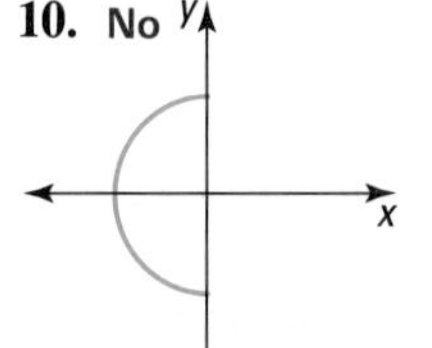

11. **Yes**
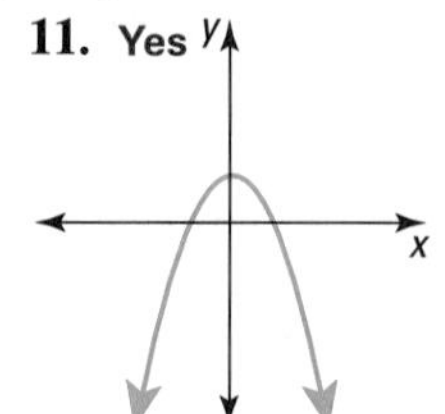

12. **Yes**
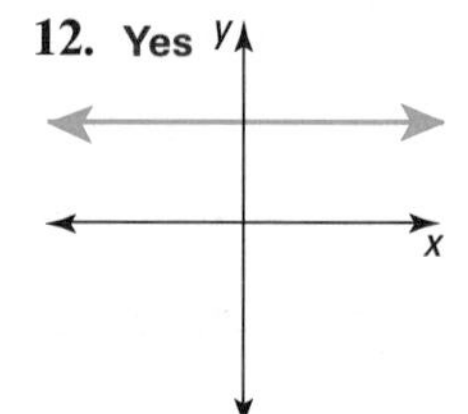

13. Explain why a vertical line cannot be the graph of a function.
One x value is paired with infinitely many y values.
14. Rennie works part-time doing jobs for people in her neighborhood. The table below shows her wages for the past six weeks.

h = hours worked	12	10	18	9	4	16
w = wages earned ($)	54	45	81	40.50	18	72

a. Does this table represent a function? Why or why not?
b. Find an equation to describe the relation between *w* and *h*.
a) Yes, because each *h* value corresponds to only one *w* value. b) *w* = 4.5 *h*

Adapting to Individual Needs

Extra Help

Discuss the Vertical-Line Test. Then have students use it to determine which of these graphs represents a function. [2 and 3]

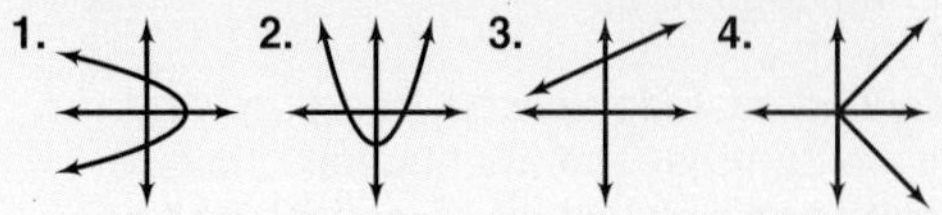

English Language Development

Draw several circles on the board and identify the radius of each circle. Explain that the circumference of a circle depends on, or is determined by, the radius. We say that the circumference is a *function* of the radius. Likewise, in an equation such as $y = 2x + 3$, the value of *y* depends on, or is a *function* of, the value of *x*.

15a) Sample: (-2, 5), (0, 3), (2, 5)

b)

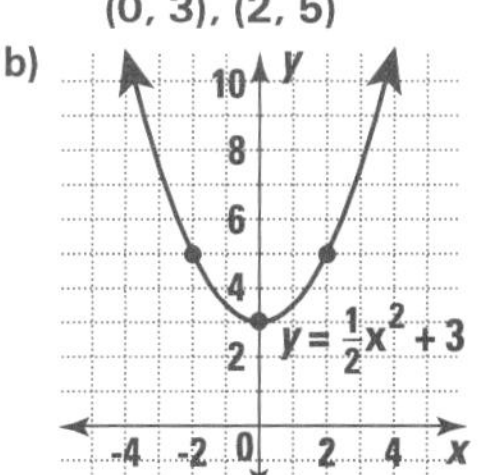

In 15 and 16, an equation for a function is given. **a.** Identify three points of the function. **b.** Graph the function. **See left.**

15. $y = \frac{1}{2}x^2 + 3$

16. $x = y + 400$

In 17 and 18, a written rule for a function is given. **a.** Translate the written description into an equation. **b.** Name three ordered pairs of the function. **17a, 18a) See below.**

17. The cost of wrapping and mailing a package is fifty cents a pound plus a three-dollar handling charge. **b) Sample: (0, 3.00), (1, 3.50), (2, 4.00)**

18. To find the volume of a sphere, multiply the cube of the radius by $\frac{4}{3}\pi$. **b) Sample: $(1, \frac{4}{3}\pi)$, $(2, \frac{32}{3}\pi)$, $(3, 36\pi)$**

Review

19. The Drama Club sold 520 tickets to the school musical. Adult tickets cost \$5 and student tickets cost \$3. If club members sold \$1840 in tickets, how many of each kind did they sell? *(Lesson 11-5)* **140 adult tickets and 380 student tickets**

20. Write $2^{-3} + 4^{-3}$ as a simple fraction. *(Lesson 8-6)* $\frac{9}{64}$

21. *True or false.* The slope of the line through (x_1, y_1) and (x_2, y_2) is the opposite of the slope of the line through (x_2, y_2) and (x_1, y_1). *(Lesson 7-2)* **False**

22. A class of 24 students contains 3% of all the students in the school. How many students are in the school? *(Lesson 6-5)* **800 students**

23. A case contains c cartons. Each carton contains b boxes. Each box has 100 paper clips. How many paper clips are in the case? *(Lesson 2-4)* **100*bc***

24. When $m > n > 0$, which is larger, $\frac{1}{m}$ or $\frac{1}{n}$? *(Previous course)* $\frac{1}{n}$

This scene is from Anaheim High School's production of The King and I.

16a) Sample: (400, 0), (0, -400), (200, -200)

b)

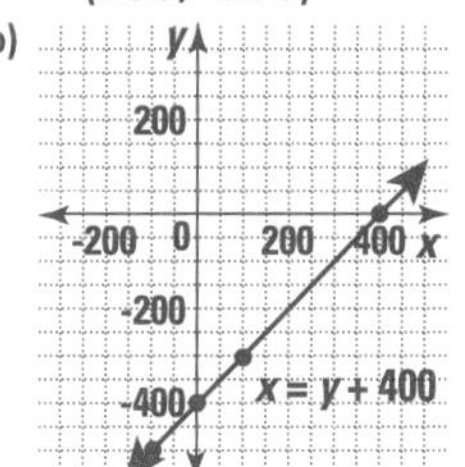

Exploration

25. A function contains the ordered pairs (1, 1) and (2, 4). **Samples given.**

a. Find a possible linear equation describing this function. $y = 3x - 2$

b. Find a possible quadratic equation describing this function. $y = x^2$

c. Find a third possible equation describing this function that is not equivalent to those in **a** and **b.** $y = -x^2 + 6x - 4$

17a) Let cost = y and weight in pounds = x. $y = 0.50x + 3.00$

18a) Let volume = y and radius = x. $y = \frac{4}{3}\pi x^3$

Follow-up for Lesson 13-1

Practice

For more questions on SPUR Objectives, use **Lesson Master 13-1A** (shown on page 775) or **Lesson Master 13-1B** (shown on pages 776–777).

Assessment

Written Communication Ask each student to draw two graphs that describe functions and two graphs that do not describe functions. Students should label their graphs as either "a function" or "not a function." [Students' graphs demonstrate understanding of the difference between graphs of relations that are functions and graphs of relations that are not functions.]

Extension

Have students determine if $x^2 + y^2 = 36$ and $x^2 + 4y^2 = 36$ are equations of functions and then explain why or why not. [Neither equation is a function because for each value of x there are two values of y. The graph of $x^2 + y^2 = 36$ is a circle which crosses the x-axis at −6 and 6 and the y-axis at −6 and 6; $x^2 + 4y^2 = 36$ is an ellipse which crosses the x-axis at −6 and 6 and the y-axis at −3 and 3.] See *Challenge* below.

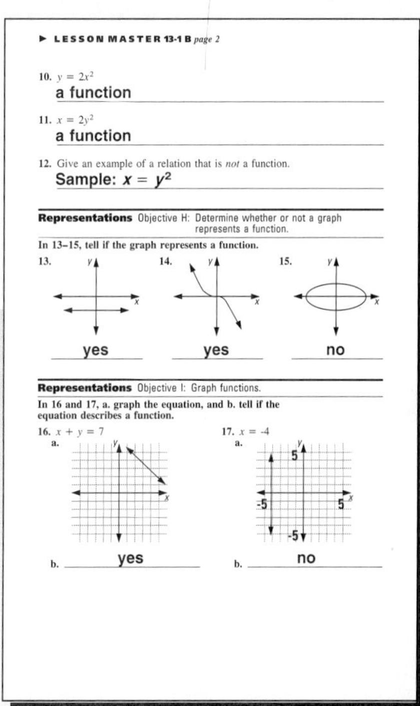

▶ **LESSON MASTER 13-1 B** *page 2*

10. $y = 2x^2$
a function

11. $x = 2y^2$
a function

12. Give an example of a relation that is *not* a function.
Sample: $x = y^2$

Representations Objective H: Determine whether or not a graph represents a function.

In 13–15, tell if the graph represents a function.

13. yes

14. yes

15. no

Representations Objective I: Graph functions.

In 16 and 17, a. graph the equation, and b. tell if the equation describes a function.

16. $x + y = 7$
a.
b. yes

17. $x = -4$
a.
b. no

Adapting to Individual Needs

Challenge

Give students the equation of a circle in a Cartesian plane and have them explain what r, h, and k represent.

$$(x - h)^2 + (y - k)^2 = r^2$$

[The radius is r and (h, k) is the center of the circle.]

Setting Up Lesson 13-2

Materials If automatic graphers are available, students can use them for Lesson 13-2 and for the remaining lessons in the chapter.

Lesson 13-2

Objectives

A Evaluate functions and solve equations involving function notation.
E Use function notation and language in real situations.
I Graph functions.

Resources

From the ***Teacher's Resource File***

- Lesson Master 13-2A or 13-2B
- Answer Master 13-2
- Teaching Aids
  28 Four-Quadrant Graph Paper
  131 Warm-up
- Activity Kit, Activity 28

Additional Resources

- Visuals for Teaching Aids 28, 131
- Automatic graphers

Teaching Lesson 13-2

Warm-up

Find the values of y when $x = 1, -2, 6$, and 0.

1. $y = -2x$ **-2, 4, -12, 0**
2. $y = 50 - 5x$ **45, 60, 20, 50**
3. $y = x^2 + 4$ **5, 8, 40, 4**
4. $y = 100 + .5x$ **100.5, 99, 103, 100**
5. $y = 2^x$ **$2, \frac{1}{4}, 64, 1$**
6. $y = 2x^2 + x - 3$ **0, 3, 75, -3**

LESSON

13-2 *Function Notation*

Press left foot (**A**) on pedal (**B**) which pulls down handle (**C**) on tire pump (**D**). Pressure of air blows whistle (**E**)—goldfish (**F**) believes this is dinner signal and starts feeding on worm (**G**). The pull on string (**H**) releases brace (**I**), dropping shelf (**J**), leaving weight (**K**) without support. Naturally, hatrack (**L**) is suddenly extended and boxing glove (**M**) hits punching bag (**N**) which, in turn, is punctured by spike (**O**). Escaping air blows against sail (**P**) which is attached to page of music (**Q**), which turns gently and makes way for the next outburst of sweet or sour melody.

A function machine. *This 1929 cartoon entitled* Automatic Sheet Music Turner, *is by Rube Goldberg who specialized in drawing absurdly connected machines. Here the musician presses a pedal and a page is turned. With a function, a number or other thing is input and out comes a value of the function.*

What Is Function Notation?

Ordered pairs in functions need not be numbers. You are familiar with the abbreviation $P(E)$, read "the probability of E," or even shorter, "P of E." In $P(E)$, the letter E names an event and $P(E)$ names the probability of that event. An event can have only one probability, so any set of events and their probabilities is a function.

This kind of abbreviation is used for all functions. For instance, we can use the shorthand $s(x)$, read "s of x," for the *square of* x. Then for each number x, the value of the function $s(x)$ is its square. The function is named s and called the *squaring function.* In the abbreviation $s(x)$, as in $P(E)$, the parentheses do *not* mean multiplication. The abbreviation $s(x)$ stands for the square of x.

The most common letter used to name a function is f. For example, $f(x) = -5x + 40$ represents the linear function f with slope -5 and y-intercept 40. It is read "f of x equals negative 5 times x plus 40." We call this description **function notation.** The $f(x)$ function notation was first used by the great mathematician Leonhard Euler (pronounced "oiler") in the 1700s.

Lesson 13-2 Overview

Broad Goals This lesson introduces the function notation $f(x)$ and gives examples of its use in computer programs.

Perspective The $f(x)$ function notation is not completely new to students; they have seen $P(E)$ used to describe the probability of an event and $N(S)$ used to represent the number of elements in a set. They have also used this symbolism in some BASIC commands. On some automatic graphers, they may have been required to use function notation to graph functions.

It is natural for students to wonder why $f(x)$ notation is needed. After all, they have graphed $y = x^2$, and it seems that $f(x) = x^2$ is just a more complicated way of saying the same thing. Here are three reasons for having this notation: (1) The notation clearly tells about the dependence of the second value on the first value; that is, $s(x)$ depends on x, as shown in **Example 1**. (2) The use of $s(x)$ rather than y allows the introduction of a letter that describes the situation—in this case squaring. This use is often found with computer functions where more than one letter is used, as in $ABS(X)$. **Example 2** shows this advantage of the notation. (3) When two or more quantities depend on x, they can be handled as $a(x)$, $b(x)$, and so on. This is illustrated in **Example 3**.

Example 1

Suppose $s(x) = x^2$.

a. Evaluate $s(3)$.

b. Give the value of $s(-3)$.

Solution

a. $s(3)$ means "the square of 3." Substitute 3 for x in the formula $s(x) = x^2$. $s(3) = 3^2 = 9$

b. Substitute -3 for x in the formula $s(x) = x^2$. $s(-3) = (-3)^2 = 9$

Here are some values of the squaring function and a graph of $s(x) = x^2$. You are familiar with this function. In previous lessons it was described by $y = x^2$. Its graph is a parabola.

x	$s(x)$
-4	16
-3	9
0	0
1	1
1.7	2.89
2	4
3	9
4	16

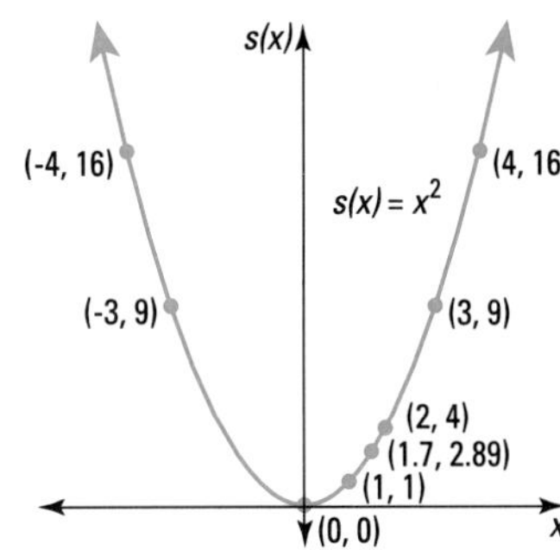

Function Notation in Calculators and Computers

Calculator and computer programs take advantage of function notation. Recall that in BASIC,

SQR(x)	means	the square *root* of *x*.
ABS(x)	means	the absolute value of *x*.

In this way, BASIC uses function notation. The names of the functions are SQR and ABS. Many automatic graphers use function notation.

Example 2

Evaluate the following.

a. ABS(-3.4)

b. SQR(4 + 9)

Solution

a. ABS(-3.4) = $|-3.4| = 3.4$

b. Work within parentheses before applying the function.
SQR(4 + 9) = $\sqrt{4 + 9} = \sqrt{13} \approx 3.6$

Function notation is particularly useful when one variable determines two or more values. The function notation helps you distinguish between the values.

Notes on Reading

Reading Mathematics Even if students have read the opening paragraphs as homework, you might want to read them aloud in class.

Students have seen $P(E)$ and $N(S)$ notation as precursors to $f(x)$ notation. It is much easier for students to understand $f(x)$ notation when they have already seen the notation in other contexts. Nevertheless, $f(x)$ can be more difficult for them than $P(E)$ or $N(S)$ because both f and x are variables; in addition, f stands for something the students have not yet encountered. You may want to emphasize that a function f signifies a correspondence and that $f(x)$ is the number that corresponds to x.

Error Alert Some students may confuse function notation with multiplication. This is a logical mistake since an expression like $n(2)$ was previously used to indicate a product. Refer again and again to $N(S)$ and $ABS(t)$ as well as to other instances where students have seen the notation, and remind them that these are function notations which do not indicate multiplication.

Before 1960, it was common to call $f(x)$ a function in all mathematics books. Now it is more common to distinguish the name of the function f from its values represented by $f(x)$.

Optional Activities

Activity 1

Materials: Graph paper or **Teaching Aid 28**

You can use this activity to give students more experience in reading and describing functions. Have each student **work with a partner**, graph each equation, tell if it is a function, and, if so, write it in function notation.

1. $y = x + 1$
2. $x = y + 1$
3. $y = x^2 + 1$
4. $x = y^2 + 1$

Equations 1 and 2 are functions; the graphs of $f(x) = x + 1$ and $f(x) = x - 1$ are shown at the right.

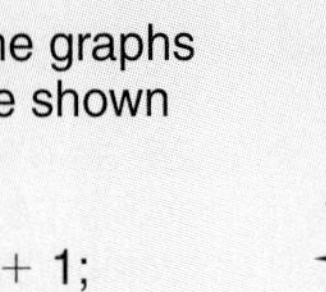

Equation 3 is a function: $f(x) = x^2 + 1$; equation 4 is not a function. The graphs for equations 3 and 4 are shown at the far right.

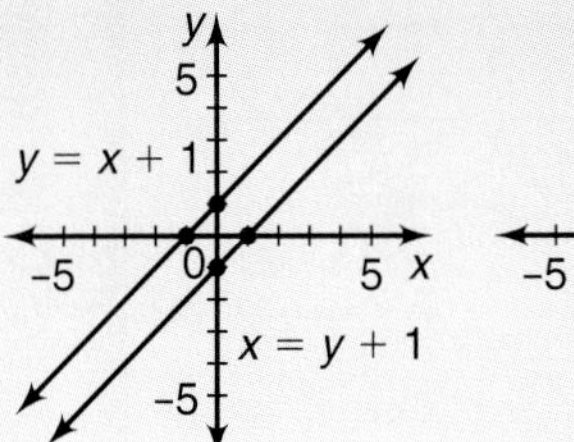

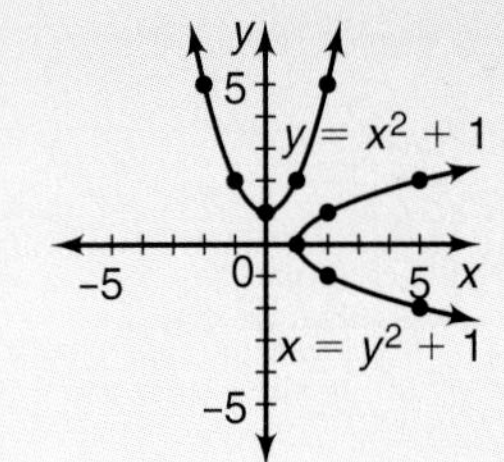

Additional Examples

1. Suppose $f(x) = \frac{1}{2}x$.
   a. Find the value of $f(40)$. **20**
   b. Evaluate $f(-7)$. **-3.5**
2. Evaluate the following.
   a. $ABS(47) + ABS(-5)$ **52**
   b. $SQR(9) - SQR(4)$ **1**
3. The average height in inches of boys at age a years can be approximated with the formula $h(a) = -0.1a^2 + 4.3a + 24.6$.
   a. Find $h(5)$. **43.6 in.**
   b. Find $h(10)$. Tell what it represents. **57.6 in.; the average height of a 10-year-old boy**
   c. Find the amount an average boy grows between the ages of 5 and 10. **14 in.**
4. On what interval does the function in Question 3 seem accurate? (Students might graph the function using an automatic grapher to help them determine the interval.) **Sample answer: between ages 5 and 20**

Check, please. *Waiters usually receive an hourly or weekly wage in addition to tips.*

Example 3

Khalil is offered two jobs as a waiter. His earnings at each restaurant depend on the total value x of the meals he serves. He thinks he can make 15% in tips at the Comfy Cafe but only 10% at the Dreamy Diner. He estimates his weekly wages at the Comfy Cafe and the Dreamy Diner would be given by these formulas, where 160 and 200 represent base pay.

Comfy Cafe: $c(x) = 160 + 0.15x$

Dreamy Diner: $d(x) = 200 + 0.10x$.

a. Evaluate $c(500)$ and $d(500)$, and tell what the values represent.

b. For what values of x is $c(x) > d(x)$? Explain what your answer means in words.

Solution

a. To find $c(500)$, substitute $x = 500$ into $160 + 0.15x$.

$$c(500) = 160 + 0.15(500) = 160 + 75 = 235$$

Similarly, $$d(500) = 200 + 0.1(500) = 200 + 50 = 250.$$

If Khalil serves \$500 worth of meals in one week, he will earn \$235 at the Comfy Cafe and \$250 at the Dreamy Diner.

b. To solve $c(x) > d(x)$, use their formulas and substitute.

$$c(x) > d(x)$$
$$160 + 0.15x > 200 + 0.10x$$
$$0.15x > 40 + 0.10x$$
$$0.05x > 40$$
$$x > 800$$

If he serves more than \$800 worth of meals each week, Khalil will earn more at the Comfy Cafe than at the Dreamy Diner.

Check

b. Draw the graphs of $c(x) = 160 + 0.15x$ and $d(x) = 200 + 0.10x$ on the same set of axes. Identify the y-coordinates of the points where the graph for Comfy Cafe's wages is higher than the graph of Dreamy Diner's wages.

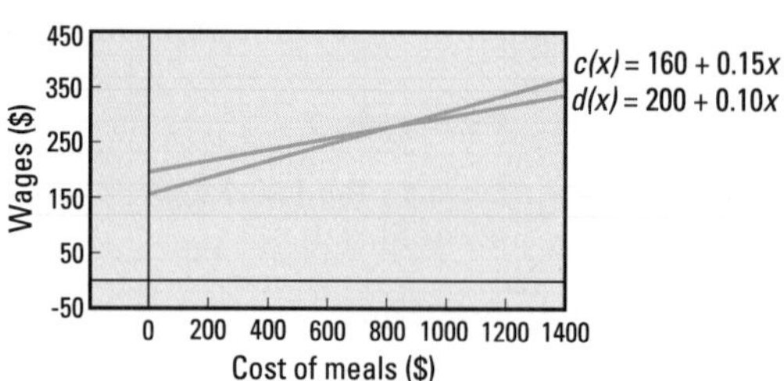

The graph shows that when $x > 800$, $c(x)$ is higher than $d(x)$.

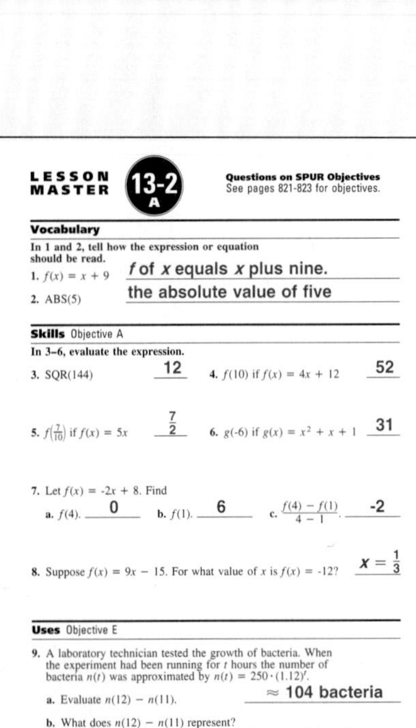

LESSON MASTER 13-2 A

Questions on SPUR Objectives
See pages 821-823 for objectives.

Vocabulary

In 1 and 2, tell how the expression or equation should be read.

1. $f(x) = x + 9$ — *f* of *x* equals *x* plus nine.
2. ABS(5) — the absolute value of five

Skills Objective A

In 3–6, evaluate the expression.

3. SQR(144) — 12
4. $f(10)$ if $f(x) = 4x + 12$ — 52
5. $f(\frac{7}{10})$ if $f(x) = 5x$ — $\frac{7}{2}$
6. $g(-6)$ if $g(x) = x^2 + x + 1$ — 31
7. Let $f(x) = -2x + 8$. Find
   a. $f(4)$. 0
   b. $f(1)$. 6
   c. $\frac{f(4) - f(1)}{4 - 1}$. -2
8. Suppose $f(x) = 9x - 15$. For what value of x is $f(x) = -12$? $x = \frac{1}{3}$

Uses Objective E

9. A laboratory technician tested the growth of bacteria. When the experiment had been running for t hours the number of bacteria $n(t)$ was approximated by $n(t) = 250 \cdot (1.12)^t$.
   a. Evaluate $n(12) - n(11)$. ≈ 104 bacteria
   b. What does $n(12) - n(11)$ represent? Sample: the approximate number of new cells between 11 hours and 12 hours

Optional Activities

Activity 2

You might want to use *Activity Kit, Activity 28,* as a follow-up to the lesson. In this activity, students' experimentation with looking through paper tubes leads to two functions, one linear and the other quadratic. Students analyze their graphs and practice using function notation.

Advantages of Function Notation

In Example 3, if we had used $y = 160 + .15x$ and $y = 200 + .10x$, it would have been more difficult to remember which equation stood for which restaurant. By using a different letter for each function, c and d, the first letters of *Comfy* and *Dreamy*, the functions are easier to distinguish. Another advantage of function notation is that it is shorter than a verbal description. For instance,

$d(500) > c(500)$ means: Khalil's earnings at Dreamy Diner will be greater than his earnings at Comfy Cafe if he serves \$500 worth of meals during the week.

QUESTIONS

Covering the Reading

In 1–3, write out how each symbol is read. **1) *P* of *E*, or probability of *E***

1. $P(E)$ **2.** SQR(x) **square root of x** **3.** $f(x) = 100 - x$ ***f* of *x* equals 100 minus *x***

In 4–6, let $s(x) = x^2$. Give the value of:

4. $s(8)$ **64** **5.** $s(-8)$ **64** **6.** $s\left(\frac{2}{5}\right)$ $\frac{4}{25}$

In 7–9, evaluate.

7. SQR(40) **≈ 6.32** **8.** ABS(-2.5) **2.5** **9.** SQR(9) − ABS(-9) **-6**

In 10–12, an equation for a function is given. Find $f(-2)$.

10. $f(x) = 4x$ **-8** **11.** $f(x) = x^4$ **16** **12.** $f(x) = 4^x$ $\frac{1}{16}$

Car shopping.
81.3 million American households owned at least one motor vehicle in 1990.

In 13 and 14, refer to Example 3.

13. a. Evaluate $c(650)$. **257.5**
b. What does $c(650)$ mean for Khalil? **If Khalil serves meals worth \$650, his weekly wage at the Comfy Cafe will be \$257.50.**

14. a. Which is larger, $c(1000)$ or $d(1000)$? **$c(1000)$**
b. What does the answer to part **a** mean for Khalil? **If Khalil serves meals worth \$1000, his weekly wage will be greater at Comfy Cafe.**

Applying the Mathematics

15. Peggy has to choose between buying two autos, one new and one used. Including the cost of gas, maintenance, and insurance, she figured the cost of owning each as

new: $n(t) = 11{,}300 + 160t$
used: $u(t) = 6{,}500 + 200t$,

where t is the number of months she owns the auto.
a. What is meant by the sentence $n(24) > u(24)$?
b. How long would Peggy have to keep the new auto for it to be a better deal than the used one? **more than 120 months**
a) After 24 months, the cost of owning the new car is greater than the cost of owning the used car.

Notes on Questions

We suggest going through all the questions in class. A variety of functions are presented, and it is important that the notation is clear to students.

Questions 1–3 Reading Mathematics Students must learn to read these symbols correctly in order to use them well and to communicate effectively. You might have students read them aloud.

Question 15 Note how the choice of appropriate letters makes it easier to keep track of these functions. If x and y were used in both equations instead of $n(t)$, $u(t)$, and t, a person could easily become confused.

Consumer Connection Worldwide, just over 47 million motor vehicles were produced in 1992. Japan produced about 12.5 million vehicles and continued to be the world production leader. The United States was second with 9.2 million vehicles, and Germany was third with 5.2 million vehicles.

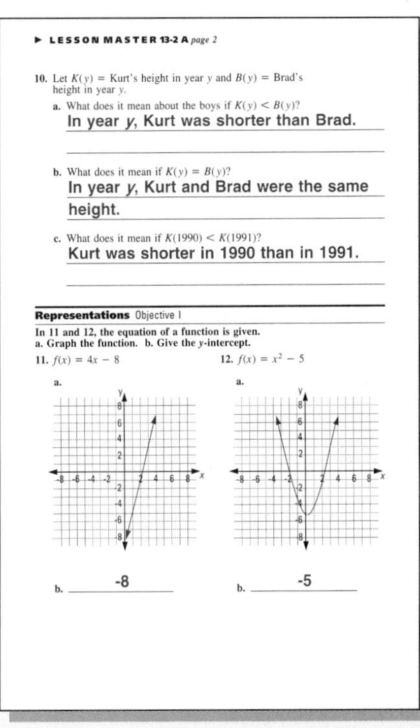

▸ LESSON MASTER 13-2 A *page 2*

10. Let $K(y)$ = Kurt's height in year y and $B(y)$ = Brad's height in year y.
a. What does it mean about the boys if $K(y) < B(y)$? **In year *y*, Kurt was shorter than Brad.**
b. What does it mean if $K(y) = B(y)$? **In year *y*, Kurt and Brad were the same height.**
c. What does it mean if $K(1990) < K(1991)$? **Kurt was shorter in 1990 than in 1991.**

Representations Objective I

In 11 and 12, the equation of a function is given.
a. Graph the function. b. Give the *y*-intercept.

11. $f(x) = 4x - 8$
a.
b. **-8**

12. $f(x) = x^2 - 5$
a.
b. **-5**

Adapting to Individual Needs

English Language Development
You may want to ask both English-speaking and non-English-speaking students to read **Examples 1 and 2** aloud. This will enable you to check for any possible problems they have with reading and/or interpretation.

Extra Help
Be sure students realize that a function is a set of ordered pairs $(x, f(x))$. If the function is denoted by a rule such as $f(x) = x^2$, then the ordered pair is (x, x^2). Refer students to **Questions 4–6**. Have them write the ordered pair associated with each function value. [**4.** (8, 64), **5.** (−8, 64), **6.** $\left(\frac{2}{5}, \frac{4}{25}\right)$]

Notes on Questions

Questions 18–19 The goal here is to give students a feel for the notation by using reasonable abbreviations and by looking at patterns in answers. If *f* is always used for the name of the function, or if students calculate values without thinking about operations or patterns, this goal is not met.

Question 29 By this time, we expect students to know that the population of the U.S. is between 250 million and 300 million and the population of the world is between 5 billion and 7 billion.

Social Studies Connection In some pyramid schemes, you receive a list of names, send a dollar to the name at the top of the list, erase that name, put your name on the bottom of the list, and send your letter to six people. In theory, it looks as if you could make $60,466,176. However, pyramid schemes are illegal because they dupe recipients into believing that everyone makes money. If everyone cooperated, people would soon be receiving large numbers of letters asking them to send money and they would have to reply before they would receive any money.

Question 30 Be sure students understand that there are four functions to be graphed. Similar questions regarding other functions are presented later in this chapter.

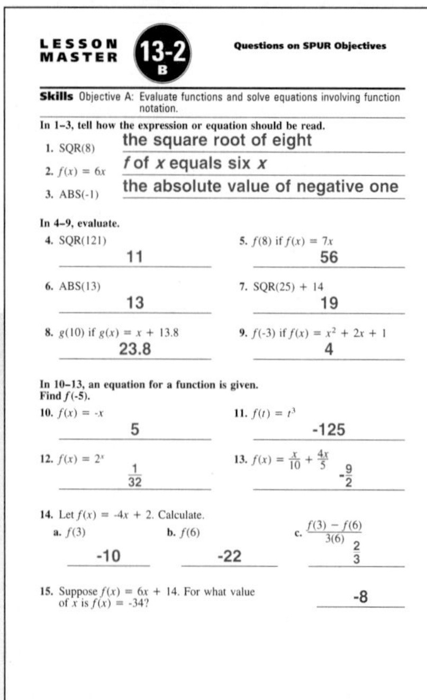

LESSON MASTER 13-2 B

Questions on SPUR Objectives

Skills Objective A: Evaluate functions and solve equations involving function notation.

In 1–3, tell how the expression or equation should be read.

1. SQR(8) the square root of eight
2. $f(x) = 6x$ f of x equals six x
3. ABS(-1) the absolute value of negative one

In 4–9, evaluate.

4. SQR(121) 11
5. $f(8)$ if $f(x) = 7x$ 56
6. ABS(13) 13
7. SQR(25) + 14 19
8. $g(10)$ if $g(x) = x + 13.8$ 23.8
9. $f(-3)$ if $f(x) = x^2 + 2x + 1$ 4

In 10–13, an equation for a function is given. Find $f(-5)$.

10. $f(x) = -x$ 5
11. $f(t) = t^3$ -125
12. $f(x) = 2^x$ $\frac{1}{32}$
13. $f(x) = \frac{x}{10} + \frac{4x}{5}$ $-\frac{9}{2}$

14. Let $f(x) = -4x + 2$. Calculate.
a. $f(3)$ -10
b. $f(6)$ -22
c. $\frac{f(3) - f(6)}{3(6)}$ $\frac{2}{3}$

15. Suppose $f(x) = 6x + 14$. For what value of x is $f(x) = -34$? -8

16a)

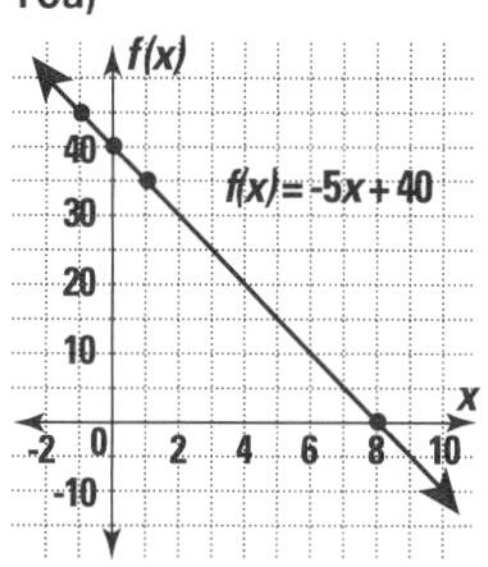

17a)

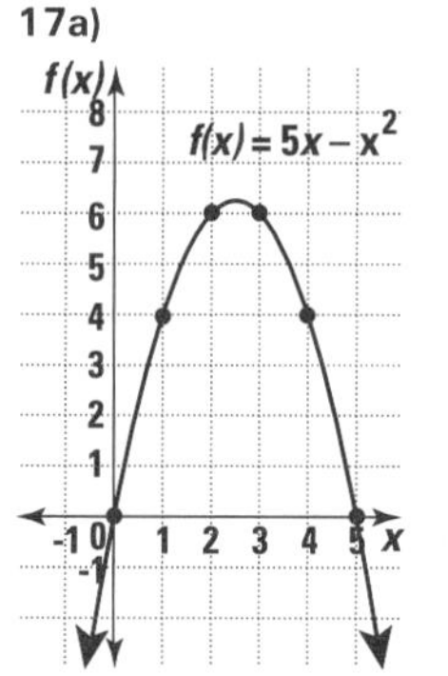

20b)

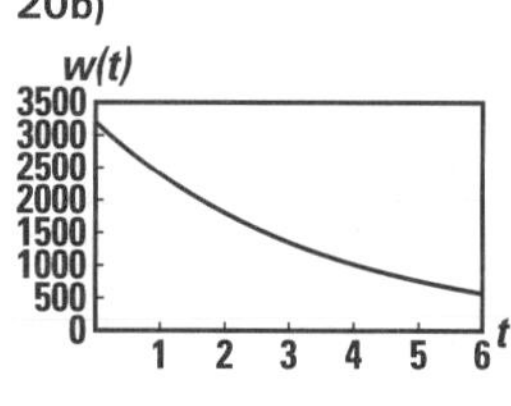

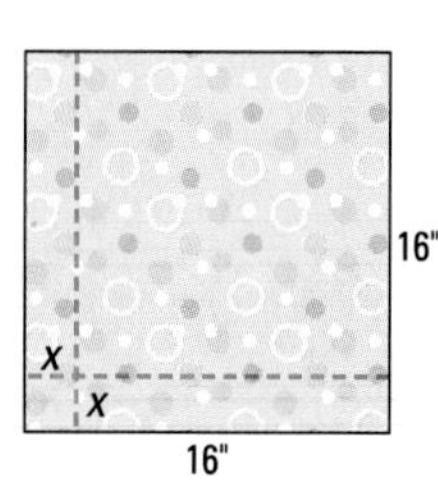

In 16 and 17, an equation for a function is given. **a.** Graph the function. **b.** Identify its *x*- and *y*-intercepts. a) See left.

16. $f(x) = -5x + 40$ b) 8; 40

17. $f(x) = 5x - x^2$ b) 0, 5; 0

18. Let $c(n) = n^3$.
a. Calculate $c(1)$, $c(2)$, $c(3)$, $c(4)$, and $c(5)$. 1; 8; 27; 64; 125
b. What might be an appropriate name for *c*? cubing function

19. Let $s(p)$ = the number of sisters of a person *p*. Let $b(p)$ = the number of brothers of a person *p*. a) Answers will vary.
a. If you are the person *p*, give the values of $s(p)$ and $b(p)$.
b. What does $s(p) + b(p) + 1$ stand for? total number of children in *p*'s family

20. A computer purchased for $3200 is estimated to depreciate at a rate of 25% per year. That is, its worth $W(t)$ after *t* years is given by $W(t) = 3200(0.75)^t$. a) 1350; after 3 years the computer is worth $1350.
a. Evaluate $W(3)$ and explain what the value means.
b. Graph the function for $0 \le t \le 6$. See left.
c. Use your graph to solve $W(t) < 1000$, and explain what your answer means. $t > 4$; the computer is worth less than $1000 after about 4 years.

21. Suppose $L(x) = 17x + 10$.
a. Calculate $L(5)$. 95
b. Calculate $L(2)$. 44
c. Calculate $\frac{L(5) - L(2)}{5 - 2}$. 17
d. What have you calculated in part **c**? the slope of the graph of *L*

Review

In 22 and 23, **a.** graph the equation; **b.** tell whether the graph represents a function. *(Lessons 13-1, 9-8)* a) See below.

22. $y = |x|$ b) Yes

23. $x = |y|$ b) No

24. *Multiple choice.* From a 16"-by-16" sheet of wrapping paper, rectangles of width x" are cut off two adjacent sides. What is the area in square inches of the large, square region that remains? *(Lesson 10-5)* d
(a) $16 - x$ (b) $256 + x^2$ (c) $256 - x^2$ (d) $256 - 32x + x^2$

In 25–27, simplify. *(Lessons 9-7, 8-6, 1-6)*

25. $6 \cdot 3^{-2}$ $\frac{2}{3}$

26. $(\sqrt{7})^2$ 7

27. $\frac{6 \pm \sqrt{24}}{2}$ $3 \pm \sqrt{6}$

28. *Skill sequence.* Write as a single fraction. *(Lesson 3-9)*
a. $3 + \frac{2}{5}$ $\frac{17}{5}$
b. $3 + \frac{7}{5}$ $\frac{22}{5}$
c. $3 + \frac{k}{5}$ $\frac{15 + k}{5}$

22a)

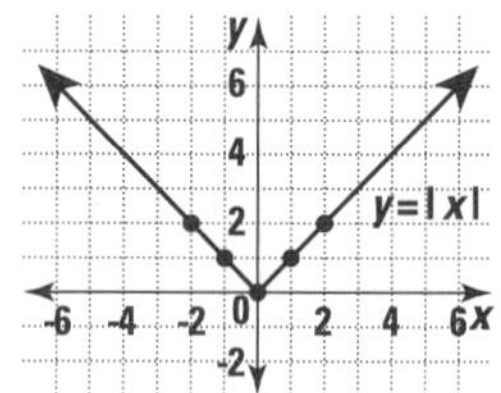

23a)

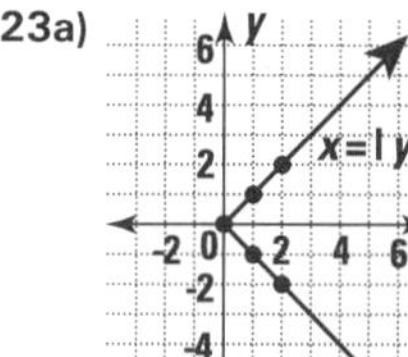

Adapting to Individual Needs

Challenge

Explain that an operation called *composition of functions* can be performed when you have two functions *f* and *g*. The *composite* of *f* and *g*, written $g \circ f$, is the result of first applying *f* and then applying *g* to the result. Thus, $(g \circ f)(x) = g(f(x))$. As an example, tell students to suppose that $f(x) = x^2$ and $g(x) = x - 2$. To find $(g \circ f)(4)$, first find $f(4)$: $f(4) = 16$. Then apply function *g* to this result: $g(16) = 16 - 2 = 14$. Thus, $(g \circ f)(4) = g(f(4)) = g(16) = 14$. Then have students do the following problems.

If $f(x) = 3x + 1$ and $g(x) = x^3$, find
1. $g(f(2))$ [343]
2. $f(g(2))$ [25]

If $f(x) = x^2 + 1$ and $g(x) = 4x$, find
3. $g(f(3))$ [40]
4. $f(g(3))$ [145]

5. In general, does it seem that $f(g(x)) = g(f(x))$? [No]

29. In June 1994, the U.S. Postal Service put out an advisory containing crime-prevention tips. Among these was a warning against participation in chain letters that guarantee money with one small investment. Chain letters can violate federal mail fraud laws. Here is what the advisory said: "A typical scheme may require you to mail the chain letter, along with a specified amount of money to six people, each of whom must then mail letters to six more people, and so on. But looking at the chart, you can see that more participants are required than there are people in the entire world!"

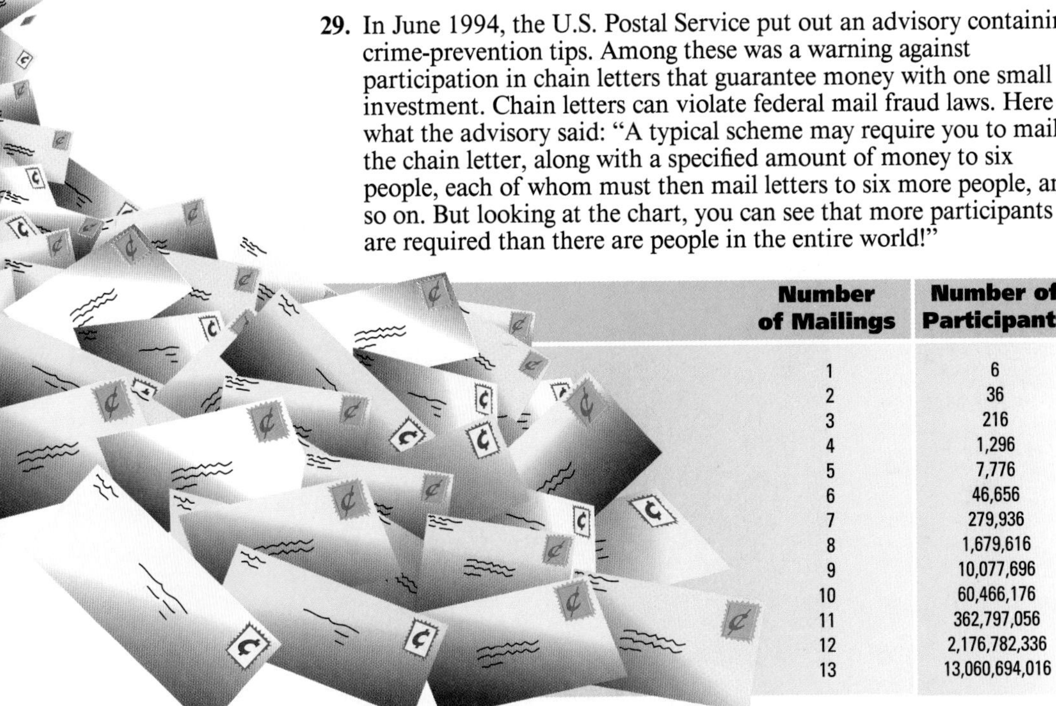

Number of Mailings	Number of Participants
1	6
2	36
3	216
4	1,296
5	7,776
6	46,656
7	279,936
8	1,679,616
9	10,077,696
10	60,466,176
11	362,797,056
12	2,176,782,336
13	13,060,694,016

The table suggests the name sometimes given to this chain letter idea: a *pyramid scheme.*

a. The table defines a function because the number of mailings m determines the number of participants p. Give an equation relating m and p that describes the function. $p = 6^m$

b. What kind of function is this? exponential

c. For what number of mailings does the number of participants first surpass the U.S. population? 11

d. For what number of mailings does the number of participants first surpass the world population? *(Lessons 13-1, 8-2)* 13

Exploration

30)

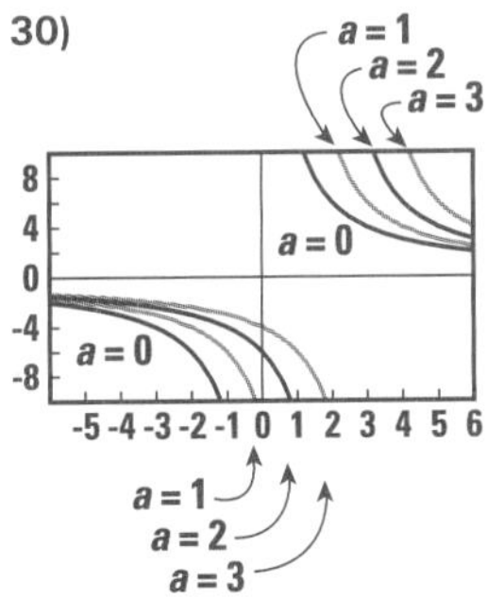

30. Let $f(x) = \frac{12}{x - a}$. Use an automatic grapher to graph the function f from $x = -5$ to $x = 5$ when $a = 0$, $a = 1$, $a = 2$, and $a = 3$. What do the graphs have in common? How are they different? **See left for graph. The graphs all have two branches and a value of x for which y is undefined. The undefined value occurs at $x = a$, which is different for each graph.**

Follow-up for Lesson 13-2

Practice

For more questions on SPUR Objectives, use **Lesson Master 13-2A** (shown on pages 780–781) or **Lesson Master 13-2B** (shown on pages 782–783).

Assessment

Oral Communication Have students explain why $s(3)$ and $s(-3)$ in **Example 1** can have the same value even though $s(x)$ is a function. [Students understand that each value of x is paired with exactly one value of y but that the same value of y can be paired with more than one value of x.]

Extension

Modify **Question 19** as follows. Suppose $b(p)$ = the eldest brother of a person p and $m(p)$ = the mother of a person p. Then ask students to find another name for each of the following functions:

1. $b(m(p))$ [Uncle of p]
2. $m(b(p))$ [Mother of p]
3. $m(m(p))$ [Maternal grandmother of p]

Project Update Project 5, *Birthday Cubics*, and Project 7, *Functions of Two Variables*, on page 818, relate to the content of this lesson.

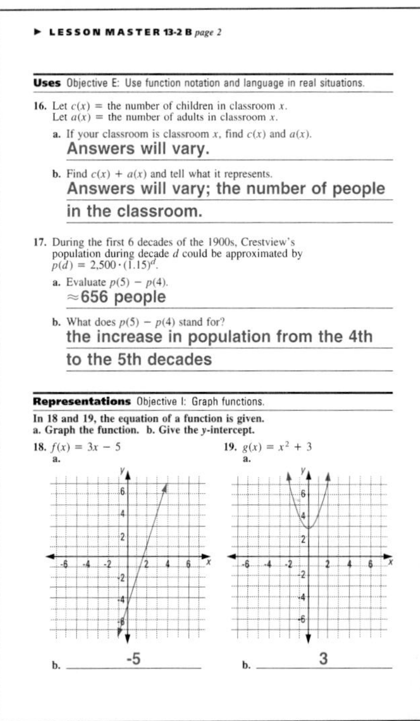

► **LESSON MASTER 13-2 B** *page 2*

Uses Objective E: Use function notation and language in real situations.

16. Let $c(x)$ = the number of children in classroom x.
Let $a(x)$ = the number of adults in classroom x.

a. If your classroom is classroom x, find $c(x)$ and $a(x)$.
Answers will vary.

b. Find $c(x) + a(x)$ and tell what it represents.
Answers will vary; the number of people in the classroom.

17. During the first 6 decades of the 1900s, Crestview's population during decade d could be approximated by $p(d) = 2{,}500 \cdot (1.15)^d$.

a. Evaluate $p(5) - p(4)$.
≈656 people

b. What does $p(5) - p(4)$ stand for?
the increase in population from the 4th to the 5th decades

Representations Objective I: Graph functions.

In 18 and 19, the equation of a function is given. a. Graph the function. b. Give the y-intercept.

18. $f(x) = 3x - 5$
a.
b. -5

19. $g(x) = x^2 + 3$
a.
b. 3

Setting Up Lesson 13-3

Materials Students will need almanacs or similar resource materials for the *Assessment* in Lesson 13-3.

If you want your class to do the experiment in the **Example** on page 785, you will need stopwatches. Your science or physical education department may have stopwatches that you can use.

You might use **Questions 22–23** to introduce Lesson 13-3.

Lesson 13-3

Objectives

A Evaluate functions and solve equations involving absolute value notation.
E Use function notation and language in real situations.
I Graph absolute value functions.

Resources

From the ***Teacher's Resource File***
- Lesson Master 13-3A or 13-3B
- Answer Master 13-3
- Assessment Sourcebook: Quiz for Lessons 13-1 through 13-3
- Teaching Aids
  - 28 Four-Quadrant Graph Paper
  - 131 Warm-up
  - 136 Test-Flight Graph and Questions 16–17

Additional Resources
- Visuals for Teaching Aids 28, 131, 136
- GraphExplorer or other automatic graphers
- Stopwatches (Example)
- Almanacs

Teaching Lesson 13-3

Warm-up

A baseball is supposed to weigh 5.125 ounces with an allowable weight tolerance of ± 0.125 ounce.

1. What is the least amount a baseball can weigh? **5.0 oz**
2. What is the greatest amount a baseball can weigh? **5.25 oz**

LESSON

13-3 *Absolute Value Functions*

It's about time. *The line drawn in the street is the Greenwich Meridian. This line has longitude 0° and is the starting point for the world's 24 time zones. See the Example on page 785 concerning people's perception of time.*

The Function $f(x) = |x|$

The function with equation

$$f(x) = |x|,$$

or $f(x) =$ ABS(x), is the simplest example of an *absolute value function.* By substituting, you can find ordered pairs for this function.

x	$f(x) = \|x\|$	ordered pair
2	$f(2) = \|2\| = 2$	(2, 2)
-8	$f(-8) = \|-8\| = 8$	(-8, 8)
0	$f(0) = \|0\| = 0$	(0, 0)

Here is a graph of $f(x) = |x|$. You saw this graph as $y = |x|$ in Lesson 9-8.

In general, when x is positive, $f(x) = x$, so the graph is part of the line $y = x$. When x is negative, then $f(x) = -x$, and the graph is part of the line $y = -x$. The result is that the graph of the function is an angle. The angle has vertex at the origin (0, 0) and has measure 90°. It is a right angle.

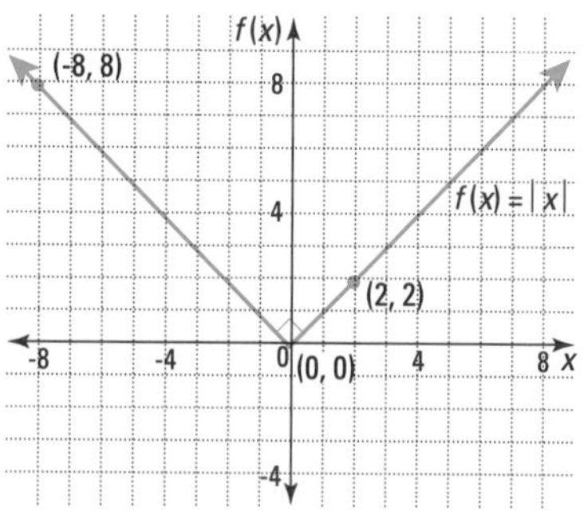

Lesson 13-3 Overview

Broad Goals This lesson discusses variants of the absolute value function.

Perspective In Lesson 9-8, the absolute value function was connected with distance and square root and applied to measuring error and finding distances on the number line. Here, instead of finding the error or distance for specific values, we graph all possible errors or distances.

Optional Activities

Activity 1
Materials: Graph paper or **Teaching Aid 28**

After discussing the lesson, you might have students graph each of these functions and tell how it is related to the graph of $f(x) = |x|$.

1. $f(x) = |x + 4|$ **2.** $f(x) = |x - 4|$

3. $f(x) = |x| + 4$ **4.** $f(x) = |x| - 4$

Some Uses of Absolute Value Functions

One place absolute value functions are used is in the study of error.

❶ **Example**

A psychologist was studying people's perceptions of time. Some people were asked to estimate the length of a minute. With each person, the psychologist rang a bell. The person waited until he or she thought a minute was up, and then rang the bell again. The estimate x(in seconds) has the error $|60 - x|$. Graph the function with equation

$$f(x) = |60 - x|.$$

Solution

Since x represents time in seconds, $x \geq 0$. Draw axes with units of 10. Make a table.

x	error f(x) = |60 − x|
0	60
40	20
50	10
60	0
70	10
100	40

These and other points are plotted below. You can see that the graph again is a right angle, but its vertex is at (60, 0).

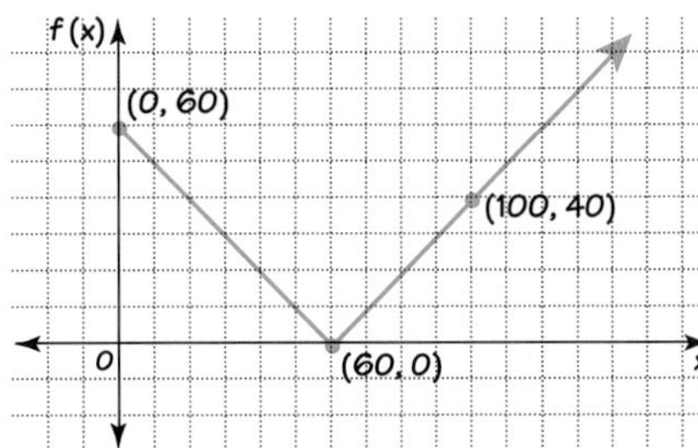

The symmetry of the graph to the vertical line $x = 60$ shows that the error has the same values on either side of 60. The slopes of -1 and 1 for the graph are due to the fact that error increases by 1 as the estimate differs by 1 more from 60.

Another reason for absolute value functions is that they help to explain some complicated situations. They are particularly useful in some situations involving distance.

3. If b is the weight of a baseball that falls within the allowable weight tolerance, what is true about $|b - 5.125|$?
$|b - 5.125| < 0.125$

Notes on Reading

Students often draw parabolas that look like graphs of absolute value functions. You might illustrate their difference with a tennis ball. When a ball is tossed to someone, the direction of the ball creates a parabolic path—the ball gradually slows its climb and starts to fall. A V-shape, like that in the absolute value graph, can be illustrated by rolling the ball along a floor so that it bounces off a wall at an angle—the ball follows a straight line and then suddenly changes direction.

❶ The two-part description of absolute value, $|a| = a$ if $a \geq 0$ and $|a| = -a$ if $a < 0$, can be related to the **Example.** If the guess for x is too low, then $60 - x$ is positive, and $y = |60 - x| = 60 - x$. This is an equation for the line that contains the left ray of the angle. If the guess for x is too high, then $60 - x$ is negative, and $y = |60 - x| = -(60 - x) = x - 60$. This is an equation for the line that contains the right ray of the angle.

Activity Your class might enjoy doing the experiment in the **Example.** In addition to calculating each student's error, the class might plot the function N, where $N(e) =$ the number of students with error e.

LESSON MASTER 13-3 A

Questions on SPUR Objectives
See pages 821-823 for objectives.

Skills Objective A

1. If $f(x) = 2|3x + 2|$, find $f(-2)$. **8**
2. If $g(x) = -|5x|$, find $g(-1)$. **-5**

In 3 and 4, solve.

3. $|x + 2| = 3$ **$x = 1$ or $x = -5$**
4. $|3m - 2| + 1 = 5$ **$m = 2$ or $m = -\frac{2}{3}$**

Uses Objective E

5. *Multiple choice.* The actual inside diameter of certain sewer tiles is slightly more or less than 0.8 meter. Let $f(d)$ = the error in a tile with inside diameter d. Which equation relates $f(d)$ and d? **b**
(a) $f(d) = |d|$ (b) $f(d) = |0.8 - d|$ (c) $f(d) = -|d + 0.8|$
6. *Multiple choice.* An incubator was programmed so that its temperature inside starts to rise at 84°F and then rise for 4 hours at a constant rate of 2°F per hour until reaching a maximum temperature of 92°F. The temperature then falls at the same rate. Which equation gives the temperature $f(t)$ in terms of time t? **c**
(a) $f(t) = |t|$ (b) $f(t) = |92 - t|$ (c) $f(t) = -2|t - 4| + 92$

Representations Objective I

In 7 and 8, graph the function with the given equation.

7. $f(x) = |2x|$
8. $y = -|x - 2| + 2$

[Graph 1 is 4 units left of the graph of $f(x) = |x|$ and graph 2 is 4 units right.]

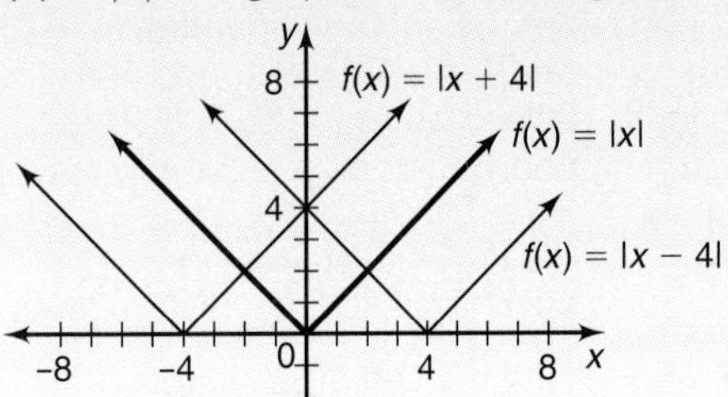

[Graph 3 is 4 units up from the graph of $f(x) = |x|$ and graph 4 is 4 units down.]

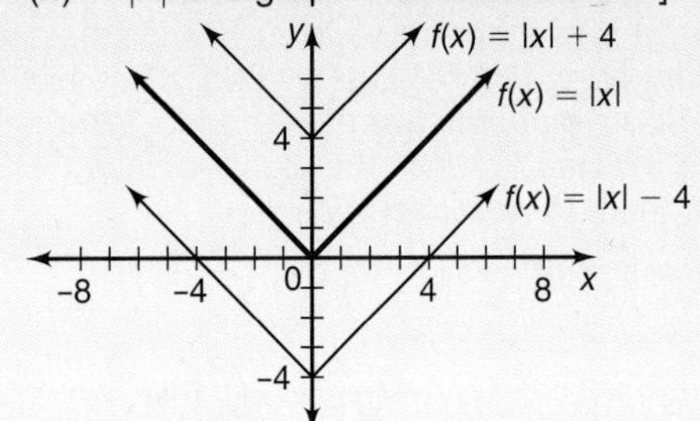

❷ **Teaching Aid 136** shows the test-flight graph that is discussed on page 786. We do not expect students to be able to write an equation for a situation like this one. We only expect them to see that the equation works and to verify the values of the function for any given value of t.

Additional Examples

Students will need graph paper or **Teaching Aid 28.**

1. A bus driver is expected to drive a route in 35 minutes. If x represents the actual number of minutes it takes for the route, then $|35 - x|$ is the number of minutes the driver is off schedule. Graph $f(x) = |35 - x|$.

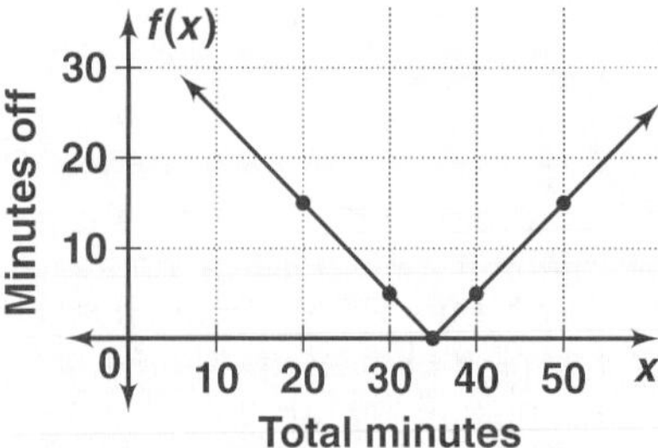

2. A train is traveling 80 miles per hour. Three hours after leaving the station, it crosses a river and continues. The distance between the train and the river t hours after leaving the station is given by $d(t) = 80|3 - t|$. Graph this function. **The graph is shown on page 787.**

LESSON MASTER 13-3 B **Questions on SPUR Objectives**

Skills Objective A: Evaluate functions and solve equations involving absolute-value notation.

In 1–4, let $A(x) = |x|$. Calculate.

1. $A(-12)$ 12
2. $A(8)$ 8
3. $A(-5.9)$ 5.9
4. $A\left(\frac{6}{5}\right)$ $\frac{6}{5}$

In 5–8, let $g(x) = |3x - 4|$. Calculate.

5. $g(6)$ 14
6. $g(1)$ 1
7. $g\left(\frac{4}{3}\right)$ 0
8. $g(-5)$ 19

In 9–12, let $f(x) = -|2x|$. Calculate.

9. $f(-4)$ -8
10. $f\left(\frac{1}{2}\right)$ -1
11. $f(-11)$ -22
12. $f(-0.05)$ -0.1

In 13–16, solve.

13. $|x + 3| = 5$ $x = 2$ or $x = -8$
14. $|y - 4| + 1 = 3$ $y = 6$ or $y = 2$
15. $3|2z| = 30$ $z = 5$ or $z = -5$
16. $-4|5a + 2| = -28$ $a = 1$ or $a = -\frac{9}{5}$

Uses Objective E: Use function notation and language in real situations.

17. *Multiple choice.* When 2-inch nails are manufactured, the actual length is usually slightly more or less than 2 inches. Let $f(x)$ = the error in a nail with length x. Which equation relates $f(x)$ and x? b

(a) $f(x) = \frac{|2 + x|}{2}$ (b) $f(x) = |2 - x|$

(c) $f(x) = |x|$ (d) $f(x) = -|x + 2|$

Suppose that in a test flight, a plane going due east crosses a checkpoint. It flies east at 600 km/h for 2 hours and then returns flying due west.

We let $d(t)$ = the plane's distance from the checkpoint t hours after crossing it.

When $t = 0$, the plane is at the checkpoint. So $d(0) = 0$.
When $t = 1$, the plane is 600 km east. So $d(1) = 600$.
When $t = 2$, the plane is 1200 km east. So $d(2) = 1200$.

All this time the plane has been going at a constant rate. Then it turns back.

When $t = 3$, the plane is again 600 km east. So $d(3) = 600$.
When $t = 4$, the plane crosses the checkpoint again. So $d(4) = 0$.

The graph below results from graphing these and other ordered pairs. Notice how the graph describes the situation.

❷

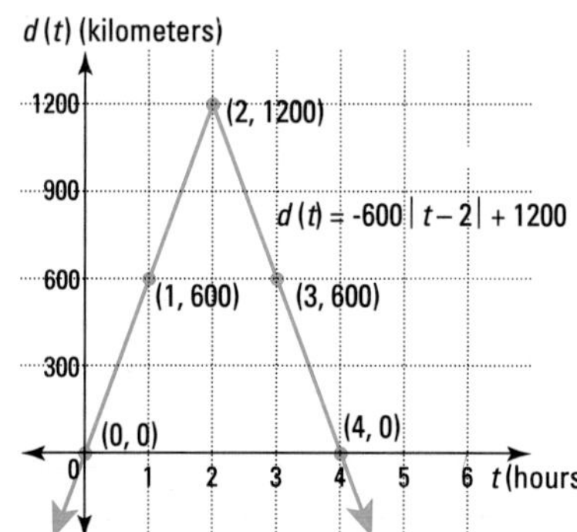

Because the graph is an angle, you should expect a formula for $d(t)$ to involve the absolute value function. And it does.

$$d(t) = -600|t - 2| + 1200.$$

Activity

Check that the formula $d(t) = -600|t - 2| + 1200$ does work for the 5 values we found and for 3 other values of t. **The formula works for the 5 values. Other values (samples) are: $d(2.5) = 900$, $d(6) = -1200$, $d(-1) = -600$.**

Without knowing the graph of the absolute value function $f(x) = |x|$, you probably would never think that the test-flight situation could involve absolute value. The formula

$$d(t) = -600|t - 2| + 1200$$

is a special case of the **general absolute value function**

$$f(x) = a|x - h| + k,$$

where $a = -600$, $h = 2$, and $k = 1200$. In the Questions, you are asked to explore what the graph of this function looks like for various values of a, h, and k. In your later study of mathematics, you will learn how to determine such formulas.

Adapting to Individual Needs

Extra Help

Students sometimes have to be reminded that $f(x)$ is another name for y. For example, students saw $f(x) = |x|$ as $y = |x|$ in Lesson 9-8. Some students are more comfortable using y instead of $f(x)$. If that is the case, allow them to rewrite $f(x)$ as y.

Challenge

Point out that the *inverse* of a function is obtained by interchanging the variables of the function. For example, suppose a function is defined as $f(x) = 3x - 6$. This function can be written as $y = 3x - 6$, and its inverse is $x = 3y - 6$ or $y = \frac{1}{3}x + 2$. Have students determine and graph the inverse of each of the following functions and then tell if the inverse is a function.

Earlier in this book, you would have seen the formula $d = -600|t - 2| + 1200$ without function notation. Using the $d(t)$ function notation makes it clear that the value of d depends on t. There may be other quantities that depend on the time. We could write

$a(t)$ = altitude of the plane t hours after crossing the checkpoint,

$f(t)$ = amount of fuel used t hours after crossing the checkpoint, and so on.

Using $a(t)$ and $f(t)$ makes it clear that the altitude and fuel used by the plane depends on time.

Flight of the Eagle. *On January 21, 1987, Lois McCallin pedaled this plane, the* Eagle, *10 miles in 37 min 38 sec. This was the longest such flight by a woman. Until 1988 the* Eagle *held the distance record—37.3 miles—for human-powered flight.*

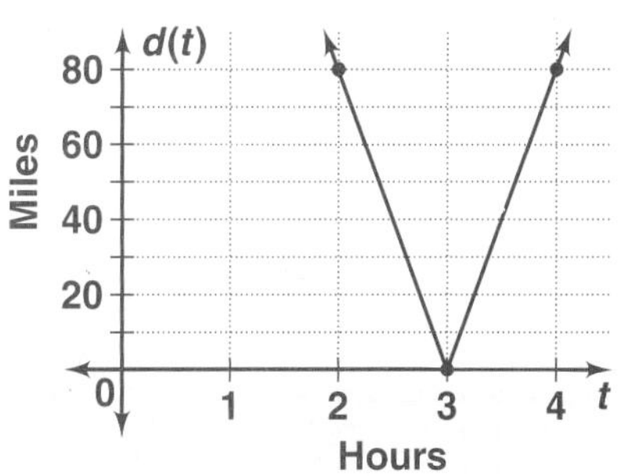

QUESTIONS

Covering the Reading

In 1–4, let $f(x) = |x|$. Calculate.

1. $f(-3)$ 3 **2.** $f(2)$ 2 **3.** $f\left(-\frac{3}{4}\right)$ $\frac{3}{4}$ **4.** $f(0)$ 0

5. The graph of an absolute value function is _?_. an angle

6. Name two reasons for studying absolute value functions. They are used in the study of error and in some situations involving distance.

In 7–9, let f be the function of the Example.

7. If $x = 90$, $f(x) =$ _?_. 30

8. The graph of f has vertex _?_. (60, 0)

9. $f(x)$ stands for the absolute difference between the actual and estimated values of _?_. one minute.

In 10–13, let $d(t) = -600|t - 2| + 1200$.

10. Calculate $d(0)$, $d(1)$, $d(2)$, and $d(3)$. 0; 600; 1200; 600

11. Describe a situation that can lead to the function d. See left.

12. The function d contains (1.5, 900). What does this point represent? 1.5 hours after crossing the checkpoint, the plane is 900 km east of the checkpoint.

13. The function d contains (5, -600). What could this point represent? 5 hours after crossing the checkpoint, the plane is 600 km west of the checkpoint.

Applying the Mathematics

In 14 and 15, graph the function with the given equation.

14. $f(x) =$ ABS(3x) **15.** $y = |x - 10| + 7$

11) Sample: In a test flight, a plane crosses a checkpoint going due east. It then flies 600 km/h for 2 hours and returns flying due west. $d(t)$ represents the plane's distance from the checkpoint t hours after crossing it.

14) 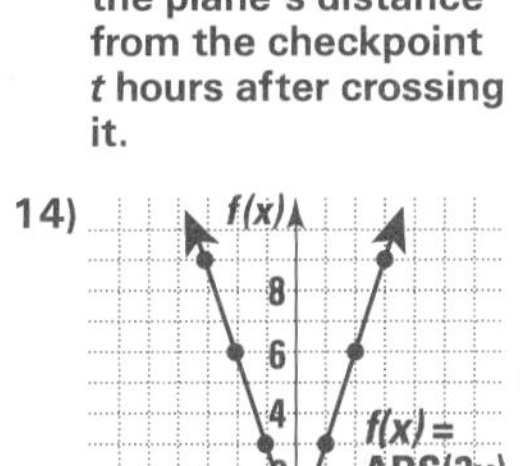

15) 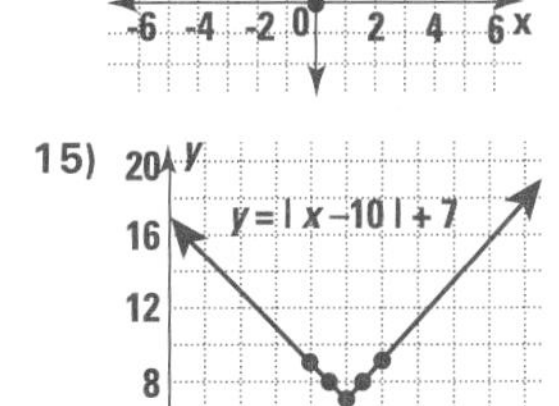

Notes on Questions

Question 15 You might ask what values of x are possible [Any real number] and what values of y are possible. [Any real number greater than or equal to 7] Ask similar questions for **Question 16.** [w can be any number from 0 to 100, while y can be any number from 0 to 50.]

Questions 16–17 You might want to use **Teaching Aid 136** when discussing these questions that appear on page 788.

Sports Connection A game similar to football as we know it today took place in 1874 when a team from Harvard met a team from McGill University in Montreal. Harvard played a soccer-style football game which involved kicking the ball. McGill played a rugby-style game which involved running and tackling. At their meeting, the teams played two games, one under McGill's rules and the other under Harvard's. Harvard liked the rugby-style game and soon running, tackling, and kicking became part of football in the U. S.

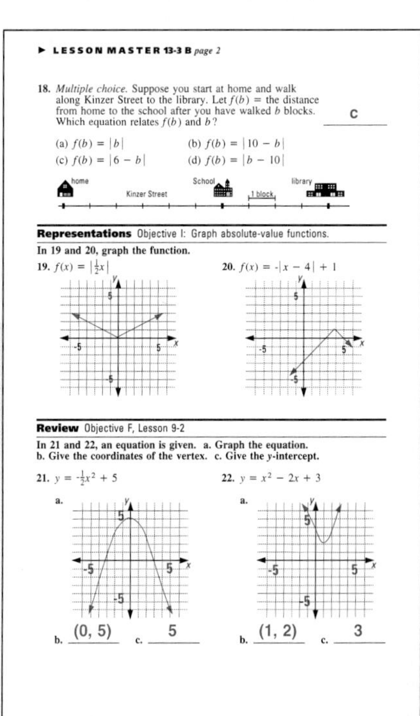

► LESSON MASTER 13-3 B *page 2*

18. *Multiple choice.* Suppose you start at home and walk along Kinzer Street to the library. Let $f(b)$ = the distance from home to the school after you have walked b blocks. Which equation relates $f(b)$ and b? c

(a) $f(b) = |b|$ (b) $f(b) = |10 - b|$
(c) $f(b) = |6 - b|$ (d) $f(b) = |b - 10|$

Representations Objective I: Graph absolute-value functions.

In 19 and 20, graph the function.

19. $f(x) = |\frac{1}{2}x|$ 20. $f(x) = -|x - 4| + 1$

Review Objective F, Lesson 9-2

In 21 and 22, an equation is given. a. Graph the equation. b. Give the coordinates of the vertex. c. Give the y-intercept.

21. $y = -\frac{1}{2}x^2 + 5$ b. (0, 5) c. 5

22. $y = x^2 - 2x + 3$ b. (1, 2) c. 3

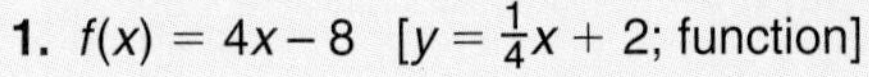

1. $f(x) = 4x - 8$ [$y = \frac{1}{4}x + 2$; function]
2. $f(x) = |x|$ [$x = |y|$; not a function]
3. $f(x) = x^2$ [$x = y^2$; not a function]

1. 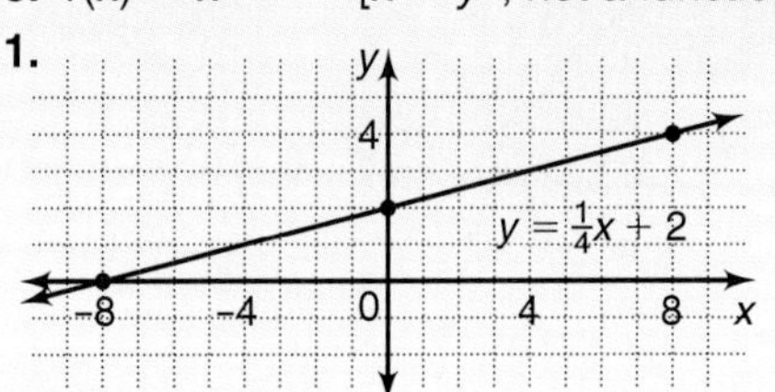

2. 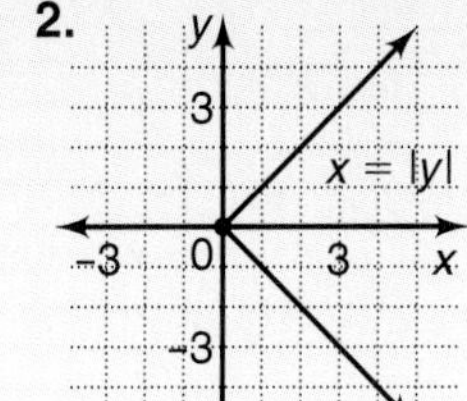

3. 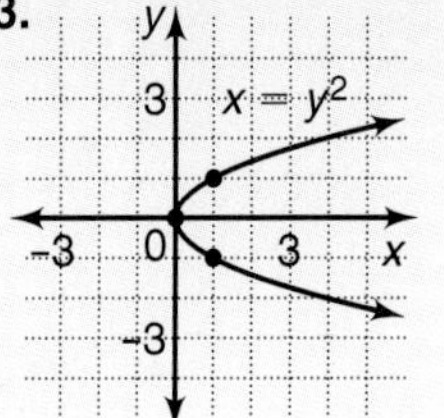

Notes on Questions

Questions 24–25 Error Alert Some students may not see how *a* and *h* affect the graph of $f(x) = a|x - h|$. Suggest that students show all four functions on the same graph.

Follow-up for Lesson 13-3

Practice

For more questions on SPUR Objectives, use **Lesson Master 13-3A** (shown on page 785) or **Lesson Master 13-3B** (shown on pages 786–787).

Assessment

Quiz A quiz covering Lessons 13-1 through 13-3 is provided in the *Assessment Sourcebook.*

Group Assessment
Materials: Almanacs
Have students **work in groups.** Tell each student to find the age of a famous living person. Have each of the other members of the group guess the person's age. Have them describe the possible error using absolute value and then graph the absolute value function, making sure each student's guess is a point on the graph. [Students write and graph an absolute function equation of the form $y(x) = |(\text{age of person}) - x|$ where x is the age guessed by others in the group.]

Extension

Have students graph inequalities involving absolute value.

1. $y > |x|$ **2.** $y \leq |x|$
3 $y \geq |x| + 2$ **4.** $y \leq |x - 2| + 3$

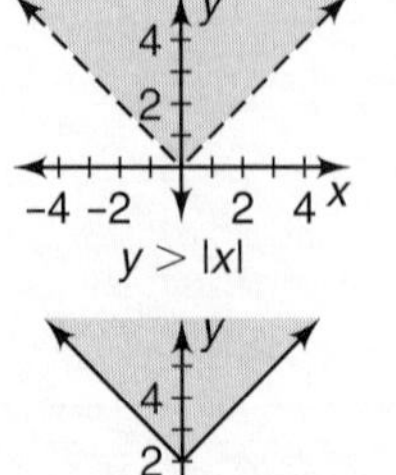

$y > |x|$

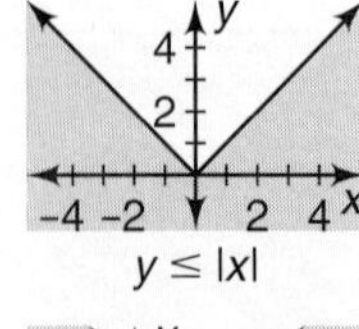

$y \leq |x|$

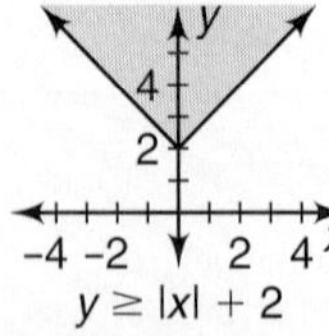

$y \geq |x| + 2$

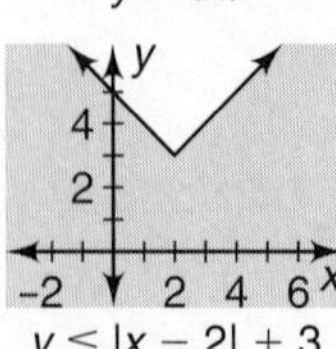

$y \leq |x - 2| + 3$

Project Update Project 3, *Absolute Value Functions*, on page 817, relates to the content of this lesson.

In 16 and 17, suppose you start at the goal line of a football field and walk to the other goal line. After you have walked *w* yards, you will be on the *y* yard line.

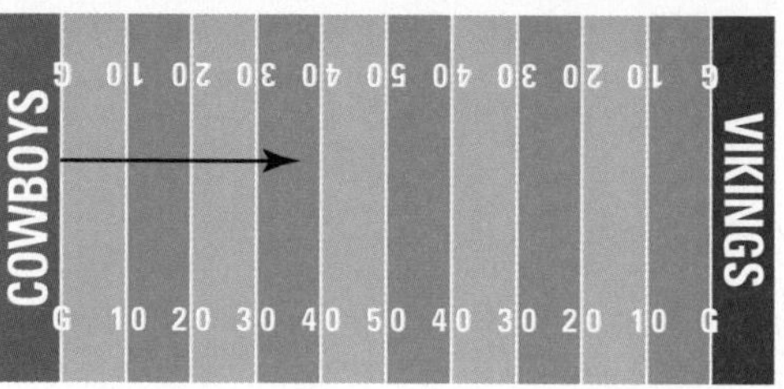

16. *Multiple choice.* Which equation relates *w* and *y*? **d**
(a) $y = w$
(b) $y = |w|$
(c) $y = |50 - w| + w$
(d) $y = -|w - 50| + 50$

17. Let f be the function relating w and y. Graph f. **See left.**

17)

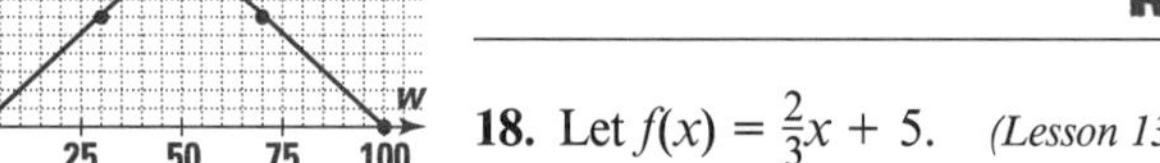

Review

18. Let $f(x) = \frac{2}{3}x + 5$. *(Lesson 13-2)*
a. Calculate $f(120)$. **85**
b. Calculate $f(-120)$. **-75**
c. Describe the graph of f. **a line with slope $\frac{2}{3}$ and y-intercept 5**

19. Let $A(x)$ = the April mean high temperature for city x, as shown in Lesson 13-1. What is A(Moscow)? *(Lesson 13-1)* **47°F**

20. *Skill sequence.* Factor. *(Lesson 12-3)*
a. $x^2 - 16$ **$(x + 4)(x - 4)$**
b. $y^2 - 6y - 16$ **$(y - 8)(y + 2)$**
c. $a^2 - 6ab - 16b^2$ **$(a - 8b)(a + 2b)$**

21. What value(s) can v not have in the expression $\frac{(v - 1)(v - 3)}{(v - 2)(v - 4)}$? *(Lessons 12-4, 6-2)* **2 and 4**

22. If 10 pencils and 7 erasers cost \$4.23 and 3 pencils and 1 eraser cost \$0.95, what is the cost of two erasers? *(Lesson 11-5)* **\$0.58**

23. *Skill sequence.* Simplify. *(Lesson 3-9)*
a. $x + \frac{x}{2}$ **$\frac{3x}{2}$**
b. $\frac{x}{3} + \frac{x}{2}$ **$\frac{5x}{6}$**
c. $\frac{x}{3} + \frac{y}{2}$ **$\frac{2x + 3y}{6}$**

Exploration

In 24 and 25, consider absolute value functions of the form $f(x) = a|x - h|$.

24. Fix $a = 1$. Then vary the value of h, choosing any numbers you wish. For instance, if you let $h = 3$, then $f(x) = |x - 3|$.
a. Graph the function f for four different values of h. **See margin.**
b. How does the value of h affect the graphs? **Sample: h is the x-intercept of the graph $f(x) = |x - h|$.**

25. Fix $h = 2$. Now vary the value of a, choosing any numbers you wish. For instance, if you let $a = 4$, then $f(x) = 4|x - 2|$.
a. Graph the function f for four different values of a. **See margin.**
b. How does the value of a affect the graphs? **Samples: If $a > 0$, then the angle opens upward. If $a < 0$, then the angle opens downward. The larger $|a|$ is, the smaller the measure of the angle.**

Additional answers

24a. Sample: $h = -1, 0, 1, 2$

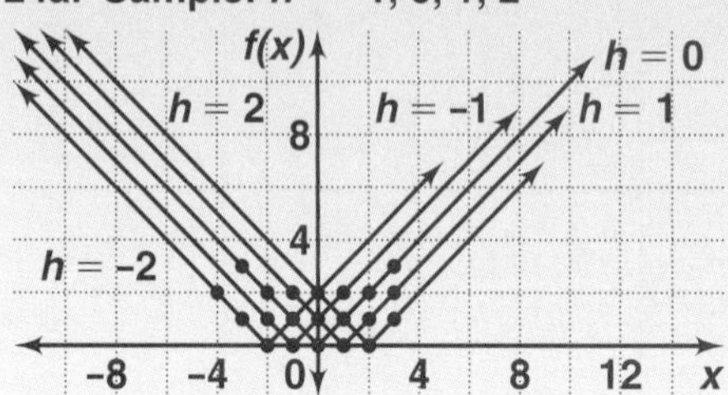

25a. Sample: $a = -3, -1, 1, 3$

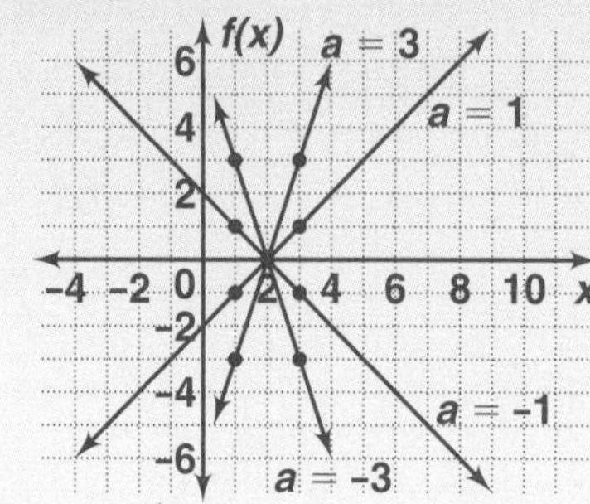

LESSON

13-4

Domain and Range

Peacemakers. *Pictured is part of the United Nations headquarters in New York City. The primary objective of the U.N. is the maintenance of international peace and security.*

What Are the Domain and the Range of a Function?

Every function can be thought of as a set of ordered pairs. The set of first coordinates of these pairs is called the **domain** of the function. The set of second coordinates of these pairs is called the **range** of the function. If the function is a finite set of ordered pairs, then you can list the elements of the domain and the range.

Example 1

The set of ordered pairs {(1945, 51), (1965, 117), (1985, 159), (1993, 184)} associates a year with the number of members of the United Nations that year. Give the domain and the range of the function.

Solution

Only four points are given for this function.
The domain is the set of first coordinates {1945, 1965, 1985, 1993}.
The range is the set of second coordinates {51, 117, 159, 184}.

When a function is expressed as an equation or other rule involving x and y, or x and $f(x)$, then the domain is the replacement set for x, and the range is the replacement set for y or $f(x)$. When no domain is given, the domain is assumed to be the largest set possible in the situation.

To find the range, think about what y-values (or $f(x)$-values) are possible, or make a graph. If the problem involves a real use, you will also have to think about the situation.

Lesson 13-4

Objectives

D Find the domain and the range of a function from its formula, graph, or rule.

E Use function notation and language in real situations.

Resources

From the ***Teacher's Resource File***

- Lesson Master 13-4A or 13-4B
- Answer Master 13-4
- Teaching Aids
  132 Warm-up
  137 Domain and Range

Additional Resources

- Visuals for Teaching Aids 132, 137
- GraphExplorer or other automatic graphers

Teaching Lesson 13-4

Warm-up

Find three ordered pairs for each function. **Answers will vary. Sample ordered pairs are given.**

1. $f(x) = x^2 + 1$ **(0, 1), (2, 5), (-2, 5)**
2. $g(x) = 7$ **(0, 7), (2, 7), (-2, 7)**
3. $h(x) = 2x$ **(0, 0), (2, 4), (-2, -4)**
4. $p(x) = x^2 + 2x + 2$ **(0, 2) (2, 10), (-2, 2)**

Lesson 13-4 Overview

Broad Goals This lesson introduces the language of domain and range of a function.

Perspective While the language of domain and range is being introduced, an additional day is available for students to practice function notation and graphing functions. As with the broad ideas of functions, students also have encountered the idea of domain and range earlier in the year. In fact, the domain of a function is the same as the replacement set for the variable, an idea that students encountered in Chapter 1.

Experience with automatic graphers helps in explaining domain and range. The domain is the set of values that needs to be covered by the horizontal dimensions of the window if you want to see all of the graph; the range is the set of values that has to be covered by the vertical dimensions of the window.

Optional Activities

You can use this activity with the examples. Explain that the *minimum value* of a function is the least value of the range (if there is one). The *maximum value* of a function is the greatest value of the range (if there is one). These points can often be determined from an equation or graph. Pick functions from the examples in the lesson. Have students **work in groups** to determine which functions have maximum values and which have minimum values.

Notes on Reading

Reading Mathematics You might want to read and discuss the lesson in detail with students.

The domain and range are most easily found for a small set in which you can see the ordered pairs. This is shown in **Example 1.**

Social Studies Connection In April 1945, representatives from various nations that opposed Germany, Italy, and Japan in World War II met in San Francisco to discuss establishing an organization to help keep peace in the world. The plan that emerged was described in a document called the *Charter of the United Nations*. The 50 nations that signed the charter became the first United Nations members.

❶ Sometimes questions about real-world situations can be answered by thinking about what is reasonable, as in the functions *f, g,* and *h,* which are shown here. **Teaching Aid 137** contains these graphs.

❷ Often the range is determined by finding minimum or maximum values, as in **Example 3**. Here the range of $y = -x^2 + 6x - 4$ is found by examining the graph. If students square a binomial, they can see that this equation is equivalent to $y = -(x - 3)^2 + 5$. Because $(x - 3)^2$ is always nonnegative, *y* must be less than or equal to 5.

❸ We do not expect students to be able to find the range in an example such as this one. However, with an automatic grapher, the graph of this function is easily seen to be a *hyperbola*.

As shown in **Examples 4 and 5,** finding the domain in an abstract problem involves looking for special numbers, such as zero and negative numbers, that restrict the domain.

Functions with the Same Rule but with Different Domains and Ranges

❶ Here are three linear functions that have the same equation, but have different domains and ranges because of the different situations.

Let $h(x)$ be the number of children in x sets of twins. Then $h(x) = 2x$.

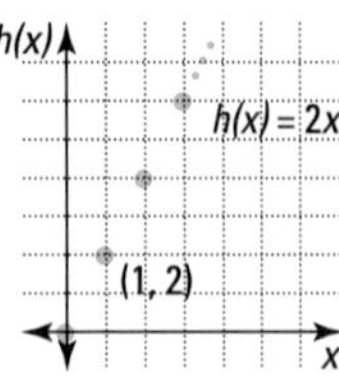

domain = set of all nonnegative integers
range = set of nonnegative even integers

Let $g(x)$ be the length of two ribbons if the length of one ribbon is x. Then $g(x) = 2x$.

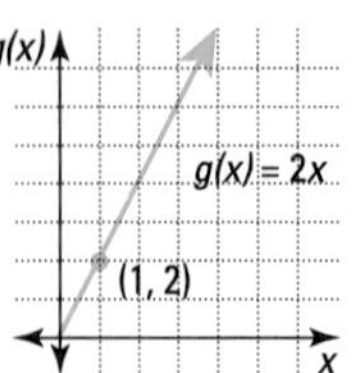

domain = set of nonnegative real numbers
range = set of nonnegative real numbers

Let $f(x)$ be twice x. Then $f(x) = 2x$.

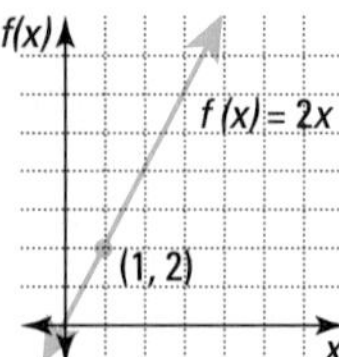

domain = set of all real numbers
range = set of all real numbers

Seeing double. *This Twinsburg school bus certainly lives up to its name!*

Functions with Restricted Range

Absolute value functions and quadratic functions allow any real number to be in their domains. However, not all real numbers are in their ranges.

Example 2

Let $f(x) = |x|$. What are the domain and the range of f?

Solution 1

The domain is the set of possible values of x. x can be any real number. So, **the domain is the set of all real numbers.**
The range is the set of possible values of $f(x)$. $f(x) = |x|$ may equal any nonnegative real number. Thus **the range is the set of nonnegative real numbers.**

Solution 2

Examine the graph of $y = |x|$. This is the same as $f(x) = |x|$.

The graph shows that any value of x is possible, but only nonnegative values of y are possible. This leads to the same answers as in Solution 1.

Optional Activities

[Example 1 minimum: 51; maximum: 184
$f(x) = 2x$ no minimum or maximum
$g(x) = 2x$ minimum: 0; no maximum
$h(x) = 2x$ minimum: 0; no maximum
Example 2 minimum: 0; no maximum
Example 3 no minimum; maximum: 5
Example 4 no minimum or maximum
Example 5 minimum: 4; no maximum]

Video

Wide World of Mathematics The segment *Rube Goldberg Machines,* provides a motivational way to introduce a lesson on domain and range. In the segment, a domain value is input into a machine. The resulting range value then becomes a domain value for the next "function," and so on. Related questions and an investigation are provided in videodisc stills and in the Video Guide. A CD-ROM activity is also available.

Videodisc Bar Codes

Search Chapter 64

Play

Example 3

Find the domain and the range of the function $f(x) = -x^2 + 6x - 4$.

Solution

Any value can be substituted for x, so **the domain = the set of all real numbers.**

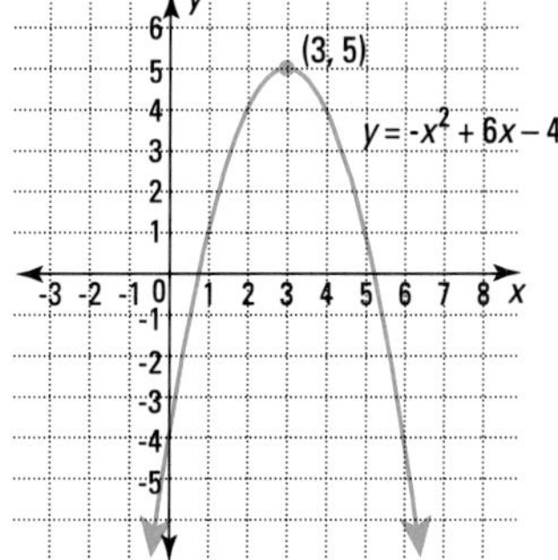

❷ Find the range by examining its graph. In the graph, y could be any number less than or equal to 5. So **the range is the set of real numbers less than or equal to 5, or $y \le 5$.**

Functions with Restricted Domain

In functions with rules that have variables in the denominator or square roots with variables, the domain sometimes cannot contain particular values.

❸ ### Example 4

What is the domain of the function f with rule $f(x) = \frac{x+1}{x-2}$?

Solution

The domain is the set of allowable values for x. The numerator can be any number, so it can be ignored. However, the denominator cannot be 0. Therefore, **the domain is the set of all real numbers except 2.**

Example 5

Determine the domain and the range of the function f with equation $f(x) = \sqrt{x} + 4$.

Solution

You can take the square root only of a nonnegative number. **The domain is the set of nonnegative real numbers.** The number $\sqrt{x}$ can be any positive number or zero. Thus $\sqrt{x} + 4$ can be any number greater than or equal to 4. **The range is the set of numbers greater than or equal to 4.**

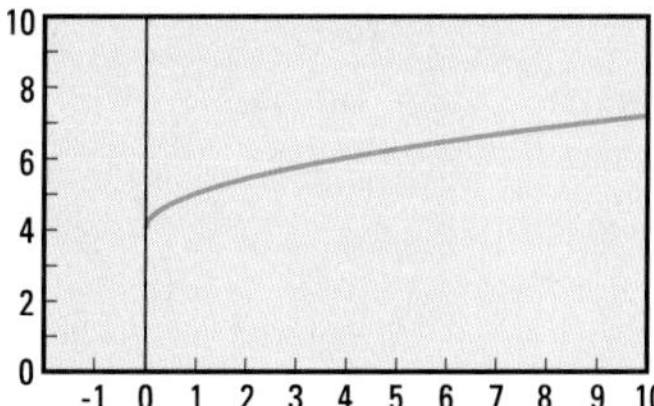

Check

A graph of $f(x) = \sqrt{x} + 4$ with an automatic grapher is shown at the right. Notice that it contains no points for negative values of x or values of $f(x)$ less than 4. This verifies the answer.

Additional Examples

1. The function {(Pacific, 64.0), (Atlantic, 31.8), (Indian, 25.3), (Arctic, 5.4)} relates oceans and their approximate areas in millions of square miles. Give the domain and range.
   Domain = {Pacific, Atlantic, Indian, Arctic}
   Range = {64.0, 31.8, 25.3, 5.4}
2. What are the domain and range of f with rule $f(x) = |x| - 5$?
   Domain = set of real numbers
   Range = set of real numbers greater than or equal to -5 ($f(x) \ge -5$)
3. Give the domain and range of the function with equation $g(x) = 3x^2 + 12x + 6$.
   Domain = set of real numbers
   Range = set of real numbers greater than or equal to -6 ($g(x) \ge -6$)
4. What is the domain of f if $f(x) = \frac{x}{2x+5}$? **The set of all real numbers except $-\frac{5}{2}$**
5. Give the domain and range of the function with equation $t(x) = \sqrt{x-5} + 8$.
   Domain = set of real numbers greater than or equal to 5
   Range = set of real numbers greater than or equal to 8

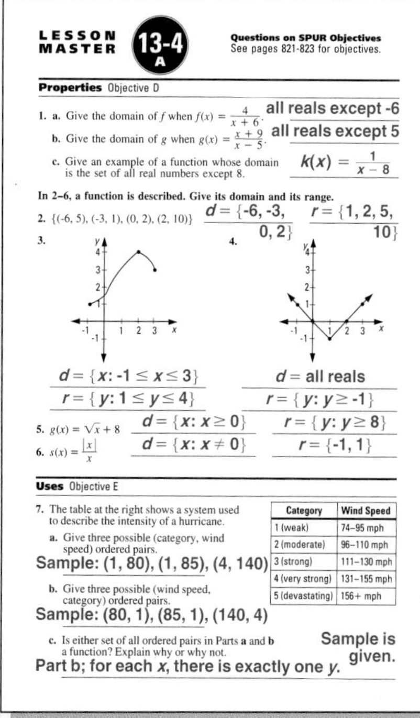

LESSON MASTER 13-4 A

Questions on SPUR Objectives
See pages 821-823 for objectives.

Properties Objective D

1. a. Give the domain of f when $f(x) = \frac{4}{x+6}$. **all reals except -6**
   b. Give the domain of g when $g(x) = \frac{x+9}{x-5}$. **all reals except 5**
   c. Give an example of a function whose domain is the set of all real numbers except 8. **$k(x) = \frac{1}{x-8}$**

In 2–6, a function is described. Give its domain and its range.

2. {(-6, 5), (-3, 1), (0, 2), (2, 10)} **$d = \{-6, -3, 0, 2\}$** **$r = \{1, 2, 5, 10\}$**
3. [graph] **$d = \{x: -1 \le x \le 3\}$** **$r = \{y: 1 \le y \le 4\}$**
4. [graph] **$d =$ all reals** **$r = \{y: y \ge -1\}$**
5. $g(x) = \sqrt{x} + 8$ **$d = \{x: x \ge 0\}$** **$r = \{y: y \ge 8\}$**
6. $s(x) = \frac{|x|}{x}$ **$d = \{x: x \ne 0\}$** **$r = \{-1, 1\}$**

Uses Objective E

7. The table at the right shows a system used to describe the intensity of a hurricane.

Category	Wind Speed
1 (weak)	74–95 mph
2 (moderate)	96–110 mph
3 (strong)	111–130 mph
4 (very strong)	131–155 mph
5 (devastating)	156+ mph

   a. Give three possible (category, wind speed) ordered pairs.
   Sample: (1, 80), (1, 85), (4, 140)
   b. Give three possible (wind speed, category) ordered pairs.
   Sample: (80, 1), (85, 1), (140, 4)
   c. Is either set of all ordered pairs in Parts **a** and **b** a function? Explain why or why not.
   Part b; for each x, there is exactly one y. **Sample is given.**

Adapting to Individual Needs

Extra Help

Sometimes students confuse the words *domain* and *range*. Remind these students that domain comes before range when the words are placed in alphabetical order, just as domain represents the set of first coordinates of a function and range represents the set of second coordinates.

English Language Development

When discussing the restricted domain of a function, you might want to explain that *restricted* means *limited*. For example, voting in the United States is *restricted* to citizens who are 18 years old or older. Write examples like $f(x) = 3 + \sqrt{x}$ and $f(x) = \frac{x+6}{x-4}$ on the board. Discuss what restrictions must be placed on the domain. [$x \not< 0$; $x \ne 4$]

Notes on Questions

Question 13 You might use an automatic grapher to show students the graph of the function.

Question 14 History Connection Students must know a little U.S. history to do **part e**. If you wish to review fitting a line to data, you might ask students to predict how many daily newspapers will be in circulation in the U.S. in the year 2000.

Question 15 This question can be answered by using the Multiplication Counting Principle. The number -1 can be matched with any of the three elements in the range, then 2 can be matched with any of the two remaining elements, and then 3 is matched with the 1 remaining element. Therefore, there are $3 \cdot 2 \cdot 1$, or 6, possible functions.

Follow-up for Lesson 13-4

Practice

For more questions on SPUR Objectives, use **Lesson Master 13-4A** (shown on page 791) or **Lesson Master 13-4B** (shown on pages 792–793).

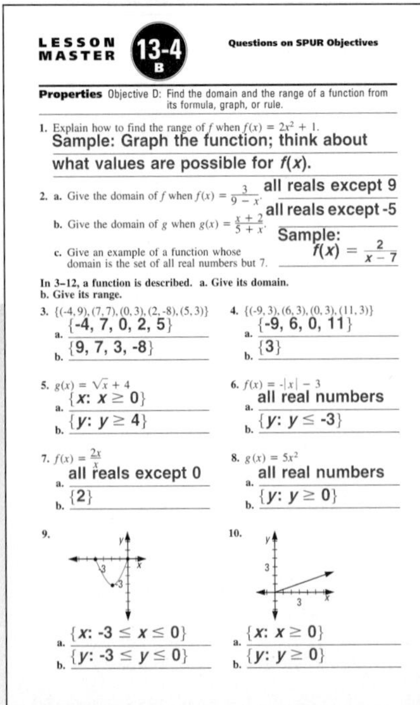

LESSON MASTER 13-4 B — Questions on SPUR Objectives

Properties Objective D: Find the domain and the range of a function from its formula, graph, or rule.

1. Explain how to find the range of f when $f(x) = 2x^2 + 1$. **Sample: Graph the function; think about what values are possible for $f(x)$.**

2. a. Give the domain of f when $f(x) = \frac{3}{9 - x}$. **all reals except 9**
   b. Give the domain of g when $g(x) = \frac{x + 2}{5 + x}$. **all reals except -5**
   c. Give an example of a function whose domain is the set of all real numbers but 7. **Sample: $f(x) = \frac{2}{x - 7}$**

In 3–12, a function is described. a. Give its domain. b. Give its range.

3. $\{(-4, 9), (7, 7), (0, 3), (2, -8), (5, 3)\}$
   a. $\{-4, 7, 0, 2, 5\}$
   b. $\{9, 7, 3, -8\}$

4. $\{(-9, 3), (6, 3), (0, 3), (11, 3)\}$
   a. $\{-9, 6, 0, 11\}$
   b. $\{3\}$

5. $g(x) = \sqrt{x} + 4$
   a. $\{x: x \ge 0\}$
   b. $\{y: y \ge 4\}$

6. $f(x) = -|x| - 3$
   a. all real numbers
   b. $\{y: y \le -3\}$

7. $f(x) = \frac{2x}{x}$
   a. all reals except 0
   b. $\{2\}$

8. $g(x) = 5x^2$
   a. all real numbers
   b. $\{y: y \ge 0\}$

9. [graph]
   a. $\{x: -3 \le x \le 0\}$
   b. $\{y: -3 \le y \le 0\}$

10. [graph]
    a. $\{x: x \ge 0\}$
    b. $\{y: y \ge 0\}$

1) the set of first coordinates of the ordered pairs in the function

2) the set of second coordinates of the ordered pairs in the function

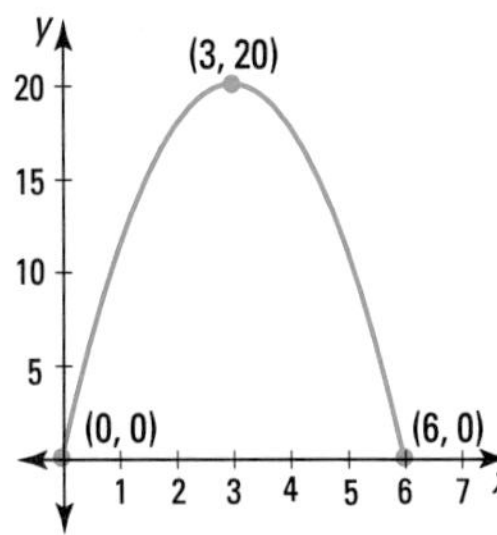

9a,b) set of all real numbers

10a) set of all real numbers
b) set of all real numbers ≥ 100

11a) set of all real numbers
b) set of all real numbers ≥ -3

12a) set of all real numbers ≥ 0
b) set of all real numbers ≥ -5

14a)

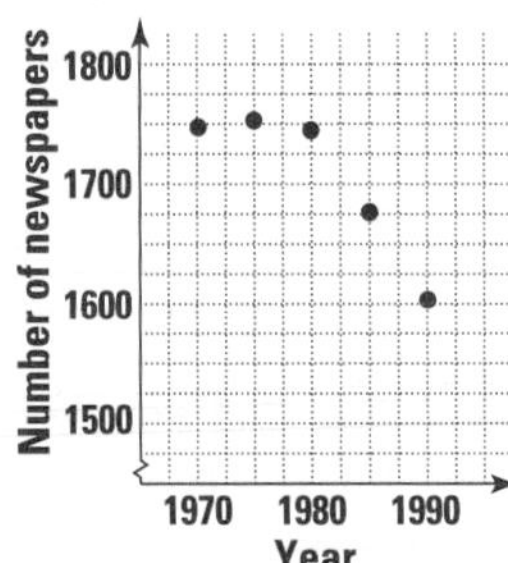

15a)

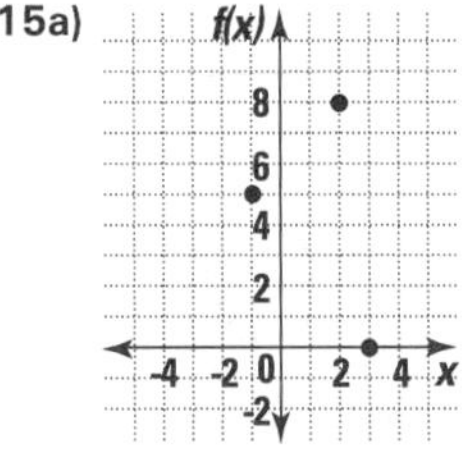

QUESTIONS

Covering the Reading

1. Define *domain* of a function. **See left.**
2. Define *range* of a function.
3. Refer to Example 1. How many more members did the United Nations have in 1993 than when it was founded in 1945? **133**

In 4–6, a function is given. State its domain and its range.

4. $\{(1, 2), (3, 4), (5, 7)\}$ **domain = {1, 3, 5}; range = {2, 4, 7}**
5. the function with equation $f(x) = 4x$ **domain: set of real numbers; range: set of real numbers**
6. the function whose entire graph is shown at the left **domain: set of numbers x with $0 \le x \le 6$; range: set of numbers y with $0 \le y \le 20$**
7. When no domain is given for a function, what can you assume about the domain? **The domain is assumed to be the largest set possible.**
8. Let $t(x)$ = the price of four tires if one costs x dollars. Let $b(x)$ = the number of people in x foursomes for bridge.
   a. Give a formula for t. **$t(x) = 4x$** b. Give a formula for b. **$b(x) = 4x$**
   c. How do t and b differ? **The range of t is the set of nonnegative rational numbers, while the range of b is the set of whole numbers.**

In 9–12, an equation for a function is given. **a.** Determine the domain of the function. **b.** Determine its range. **See left.**

9. $f(x) = 3x + 1$
10. $s(x) = |x| + 100$
11. $h(x) = 2x^2 - 3$
12. $f(x) = \sqrt{x} - 5$
13. What number is not in the domain of the function g with rule $g(x) = \frac{x - 2}{x - 3}$? **$x = 3$**

Applying the Mathematics

14. Let $n(y)$ equal the number of daily newspapers in circulation in the U.S. in year y. The table below gives some values of this function.

year	1970	1975	1980	1985	1990
number of newspapers	1748	1756	1745	1676	1611

   a. Plot these ordered pairs. **a) See left.**
   b. What trend do you notice in the plot? **b) Since 1975, the number of newspapers has decreased.**
   c. What is the largest value of the domain listed in the table? **1990**
   d. What is the largest value of the range listed in the table? **1756**
   e. Suppose this function is defined for the U.S. from the year of the Declaration of Independence until the present. State the domain of n. **Sample: the set of whole numbers with $1776 \le y \le$ current year**

15. a. Graph a function with domain {-1, 2, 3} and range {5, 8, 0}.
    b. How many such functions are possible?
    a) See left for sample. b) 6
16. Sketch a graph or give an equation for a function that has all real numbers for its domain and all real numbers less than or equal to -1 for its range. **See page 793 for sample.**

Adapting to Individual Needs

Challenge

Materials: Automatic graphers

Have students give the name of the kind of graph shown in **Exercise 18** and the meaning of *asymptote*. [The graph is a hyperbola. An asymptote is a line that a graph gets closer and closer to.] Now have students graph the hyperbolas in **Example 4** and **Question 13,** and give equations for the asymptotes.

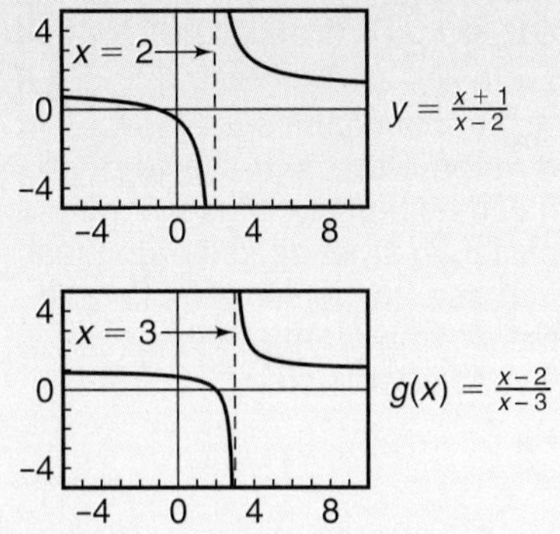

16)

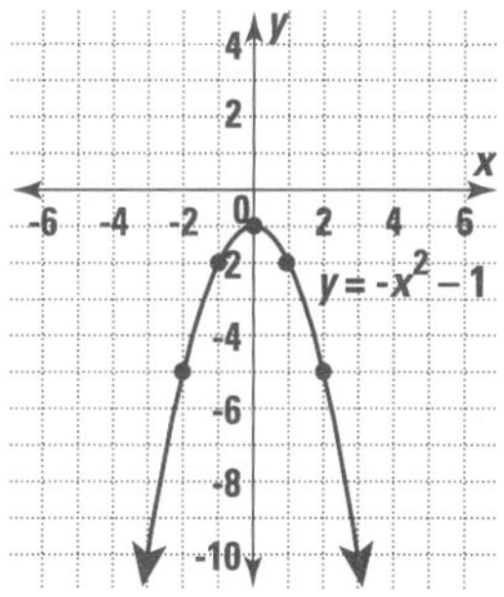

In 17–19, state the domain and the range of the function.

17. the function with equation $y = 2^x$ graphed below at the left
domain: set of all real numbers; range: set of all real numbers > 0

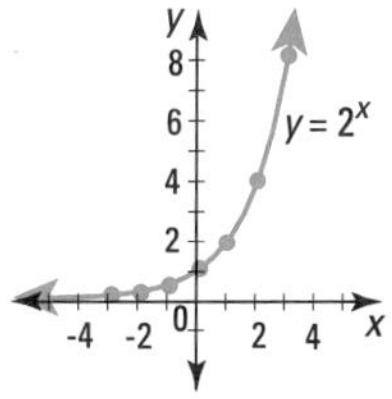

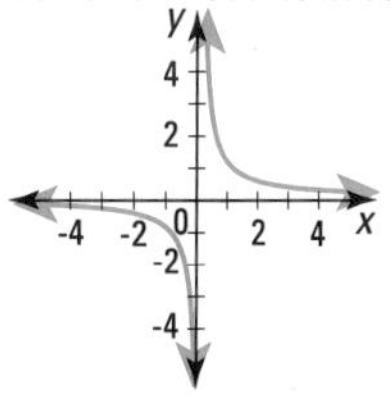

18. the function with equation $y = \frac{1}{x}$ graphed above at the right **domain: set of all nonzero real numbers; range: set of all nonzero real numbers**

19. the investment function described on page 773
domain: set of all real numbers ≥ 0; range: set of all real numbers ≥ 100

Review

20. Let $f(x) = (x - 1)(x + 2)$. Calculate each value. *(Lesson 13-2)*
a. $f(1)$ **0** **b.** $f(2)$ **4** **c.** $f\left(\frac{7}{3}\right)$ $\mathbf{\frac{52}{9}}$

In 21 and 22, tell whether or not the graph represents a function. *(Lesson 13-1)*

21.

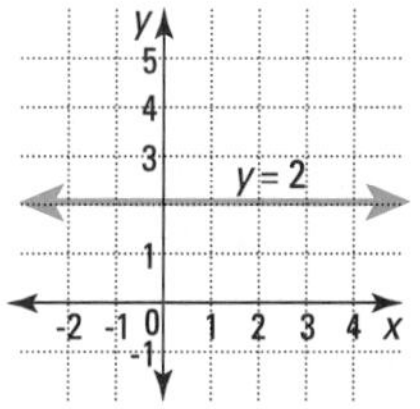

Yes

22.

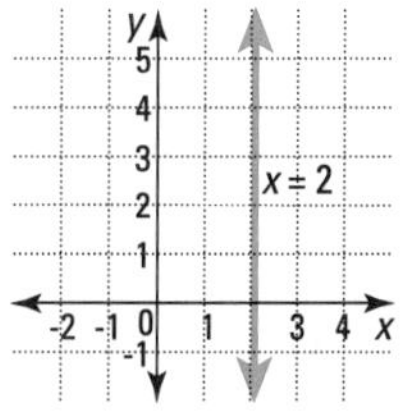

No

23. The price of a hat was increased by $\frac{1}{3}$ to the new price of $10.00.
a. What equation can be solved to find the former price?
b. What was that price? *(Lesson 6-7)* **$7.50 a) $(1 + \frac{1}{3})x = 10.00$**

24. What is the probability that one toss of a fair die does not show a 6? *(Lesson 6-4)* $\mathbf{\frac{5}{6}}$

Havana, Cuba. *Pictured is the Vedado District in Havana, the capital of Cuba.*

25. In 1991, Cuba had a population of about 10.7 million living on about 44,218 square miles of land. Mexico had a population of 90.0 million living on 761,604 square miles. **a.** Which country was more densely populated? **b.** How can you tell? *(Lesson 6-2)*
a) Cuba; b) See below.

26. Solve for x: $y = 2x + 6$. *(Lesson 3-5)* $\mathbf{x = \frac{1}{2}y - 3}$

25b) $\frac{10{,}700{,}000}{44{,}218} > \frac{90{,}000{,}000}{761{,}604}$, **or 242 > 118**

Exploration

27. Consider the absolute value functions of the form $f(x) = |x| + k$.
a. How does the value of k affect the graph of f? (You may need to graph f for different values of k.)
a) k moves the graph vertically along the $f(x)$ axis.
b. Describe the range of f in terms of k.
set of all real numbers ≥ k

Assessment

Written Communication Ask each student to write an equation for a function that has a domain which is the set of all real numbers except 3. [Students demonstrate the ability to find the domain of functions.]

Extension

Have students use automatic graphers to show $y = x^2$ and $y = \frac{1}{x^2}$. Then, for each graph, have them tell if it is a function, and if so, give the domain and range along with the maximum and minimum values of the range. [$y = x^2$: function; domain: real numbers; range: nonnegative real numbers; no maximum value; minimum value: 0. $y = \frac{1}{x^2}$: function; domain: real numbers; range: real numbers greater than 0; no maximum value; minimum value approaches but does not reach 0.]

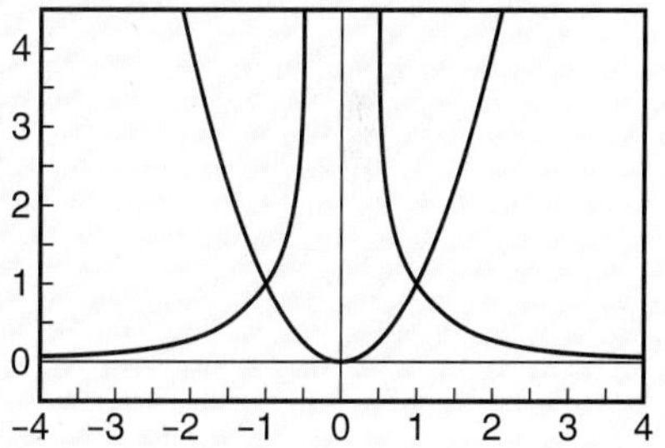

Project Update Project 6, *Functions in the Real World*, on page 818, relates to the content of this lesson.

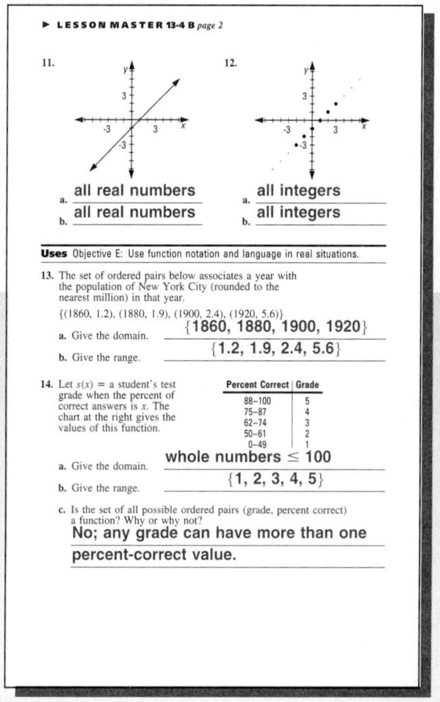

► LESSON MASTER 13-4 B *page 2*

11. a. all real numbers b. all real numbers

12. a. all integers b. all integers

Uses Objective E: Use function notation and language in real situations.

13. The set of ordered pairs below associates a year with the population of New York City (rounded to the nearest million) in that year.
{(1860, 1.2), (1880, 1.9), (1900, 2.4), (1920, 5.6)}
a. Give the domain. {1860, 1880, 1900, 1920}
b. Give the range. {1.2, 1.9, 2.4, 5.6}

14. Let $s(x)$ = a student's test grade when the percent of correct answers is x. The chart at the right gives the values of this function.

Percent Correct	Grade
88–100	5
75–87	4
62–74	3
50–61	2
0–49	1

a. Give the domain. whole numbers ≤ 100
b. Give the range. {1, 2, 3, 4, 5}
c. Is the set of all possible ordered pairs (grade, percent correct) a function? Why or why not?
No; any grade can have more than one percent-correct value.

Setting Up Lesson 13-5

Materials Students will need six-sided dice for the In-class Activity on page 794. They will need eight-sided (octahedral) dice for Activity 2 in *Optional Activities* on page 797. If you do Activity 1 on page 796, do not assign **Question 24** from Lesson 13-5.

In-class Activity

Resources

From the ***Teacher's Resource File***
- Answer Master 13-5
- Teaching Aids
  28 Four-Quadrant Graph Paper
  61 Two-Dice Outcomes

Additional Resources
- Visuals for Teaching Aids 28, 61
- Dice, calculator

You might want to use **Teaching Aid 61** to show students all the possible sums for two dice.

After students complete the activity, you might ask them to explain how each of the probabilities in the table or graph was obtained.

You might also ask students to consider the tossing of three dice which is the topic of Project 1 on page 817.

Introducing Lesson 13-5

Results of Tossing Two Dice

IN-CLASS

ACTIVITY

Materials: Dice, graph paper, and a calculator
Work in pairs.

1 **a.** Each pair should toss two dice at least 50 times and record the sum after each toss. Answers may vary.
b. Tally the number of times each sum from 2 to 12 comes up, and make a table of frequencies. After you have tallied the frequency of each sum, calculate its relative frequency. For instance, if you toss the dice 50 times and get a sum of 2 three times, your table should have the following entries. Answers may vary.

sum	frequency	relative frequency
2		$\frac{3}{50} = .06$
⋮	⋮	⋮
12		
total	50	

2 All the pairs in the class should combine their data in one table as in Step 1. Graph the function whose ordered pairs are (sum, relative frequency of that sum). Answers may vary.

3 To calculate the probabilities, answer the following.
a. When one fair die is tossed, how many outcomes are possible? What are they? 6; 1, 2, 3, 4, 5, 6
b. When two fair dice are tossed, how many outcomes are possible? List them all. See margin.
c. How many of these outcomes give a sum of 2? What is the probability of getting a sum of 2 when two dice are tossed? 1; $\frac{1}{36}$
d. How many of these outcomes give a sum of 5? What is the probability of getting a sum of 5 when two dice are tossed? 4; $\frac{4}{36}$
e. Copy and complete this table of probabilities. See margin.

sum	probability
2	.
.	.
.	.
12	.

4 ***Draw conclusions.*** Do your relative frequencies approximate the probabilities? Explain why or why not. Answers may vary.

794

Additional Answers

3b. (1,1) (1,2) (1,3) (1,4) (1,5) (1,6)
(2,1) (2,2) (2,3) (2,4) (2,5) (2,6)
(3,1) (3,2) (3,3) (3,4) (3,5) (3,6)
(4,1) (4,2) (4,3) (4,4) (4,5) (4,6)
(5,1) (5,2) (5,3) (5,4) (5,5) (5,6)
(6,1) (6,2) (6,3) (6,4) (6,5) (6,6)
There are 36 possible outcomes.

e.

Sum	Probability
2	$\frac{1}{36}$
3	$\frac{2}{36}$
4	$\frac{3}{36}$
5	$\frac{4}{36}$
6	$\frac{5}{36}$
7	$\frac{6}{36}$

Sum	Probability
8	$\frac{5}{36}$
9	$\frac{4}{36}$
10	$\frac{3}{36}$
11	$\frac{2}{36}$
12	$\frac{1}{36}$

LESSON

13-5 Probability Functions

A Probability Function for Two Dice

In many board games, two dice are tossed and the sum of the numbers that appear is used to make a move. Since the outcome of the game depends on landing or not landing on particular spaces, it is helpful to know the probability of obtaining each sum. The following diagram shows the 36 possibilities for two fair dice.

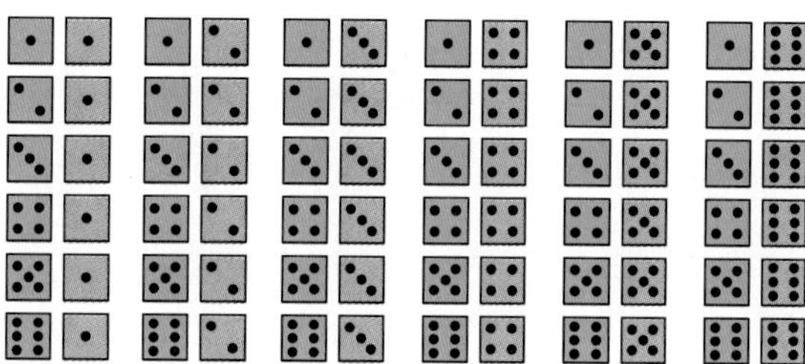

If the dice are fair, then each of the 36 outcomes has a probability of $\frac{1}{36}$. Let $P(n)$ = the probability of getting a sum of n. The domain of P is the set of possible values for n, namely {2, 3, 4, 5, 6, 7, 8, 9, 10, 11, 12}. By counting, you can find the values of $P(n)$ given in the table below. The range of the function P is thus $\left\{\frac{1}{36}, \frac{2}{36}, \frac{3}{36}, \frac{4}{36}, \frac{5}{36}, \frac{6}{36}\right\}$. This probability function is graphed below.

n	$P(n)$
2	$1/36 = .02\overline{7}$
3	$2/36 = .0\overline{5}$
4	$3/36 = .08\overline{3}$
5	$4/36 = .\overline{1}$
6	$5/36 = .13\overline{8}$
7	$6/36 = .1\overline{6}$
8	$5/36 = .13\overline{8}$
9	$4/36 = .\overline{1}$
10	$3/36 = .08\overline{3}$
11	$2/36 = .0\overline{5}$
12	$1/36 = .02\overline{7}$

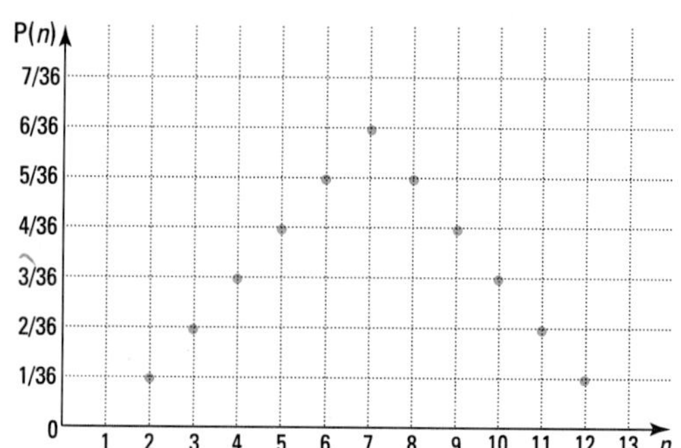

Your table and graph in Step 2 from the In-class Activity on page 794 should approximate these. Notice that the graph is part of an angle.

In general, a **probability function** is a function whose domain is a set of outcomes in a situation, and in which each ordered pair of the function contains an outcome and its probability.

Lesson 13-5

Objectives

F Determine values of probability functions.

I Graph probability functions.

Resources

From the ***Teacher's Resource File***

- Lesson Master 13-5A or 13-5B
- Answer Master 13-5
- Teaching Aids
  - 28 Four-Quadrant Graph Paper
  - 61 Two-Dice Outcomes
  - 132 Warm-up
  - 138 Additional Examples
- Technology Connection Computer Master 32

Additional Resources

- Visuals for Teaching Aids 28, 61, 132, 138
- Spreadsheet software

Teaching Lesson 13-5

Warm-up

Suppose there are 26 pieces of paper in a dish. Each paper has a different letter of the alphabet written on it. Determine the probability of

1. picking the letter Z. $\frac{1}{26}$
2. picking the letter A. $\frac{1}{26}$
3. picking a letter that comes before N in the alphabet. $\frac{1}{2}$
4. picking one of the letters of ALGEBRA. $\frac{6}{26}$ or $\frac{3}{13}$

Lesson 13-5 Overview

Broad Goals This lesson considers probability functions—those functions that map the set of all possible outcomes in a situation onto their corresponding probabilities.

Perspective Probability is another topic that students have studied earlier in this course and that can be reviewed using the language and symbolism of functions. The symbolism $P(E)$ has been used since Lesson 6-4. Here the probabilities of all the possible outcomes of a situation are considered. A function is a natural way with which to summarize the information.

Each example has a bonus. The probability function for the sum of the numbers that appear when two fair dice are tossed is an absolute value function as well (note **Question 24**). The graph of **Example 1** is part of a horizontal line. The values in **Example 2** are related to Pascal's Triangle, though this is beyond the scope of the course. The graphs on page 797 show how probabilities and relative frequencies are related. Taken together, these examples provide a wonderful illustration of the ways in which diverse topics in mathematics can come together.

Notes on Reading

Multicultural Connection
Monopoly, a real-estate game, is the best-selling privately patented board game in history. While the game originated in the United States, it is also popular in many other parts of the world. In the game sets that are sold in North America, the properties are named for streets in Atlantic City, New Jersey, while sets that are sold in other countries may be adapted to represent local cities. For example, in England, the names of London streets are used.

Explain that there are many types of probability functions. The functions in this lesson have been chosen for their relationship to content that was studied in earlier chapters.

To make sure students understand **Example 1**, discuss **Question 6.** For **Example 2**, you could also discuss tossing two fair coins. [The ordered pairs of the function are $(0, \frac{1}{4})$, $(1, \frac{2}{4})$, and $(2, \frac{1}{4})$.]

Note that the sum of the range values for every probability function is 1. This can be seen by adding the fractions on page 795 and in **Examples 1 and 2**. This is true because the probability function must cover every possible outcome exactly once.

LESSON MASTER 13-5 A — **Questions on SPUR Objectives** See pages 821-823 for objectives.

Uses Objective F

1. Gregor Mendel was the first scientist to connect probability to genetics. He crossed snapdragons that have red flowers with snapdragons that have white flowers. He obtained three kinds of offspring: snapdragons with red flowers, snapdragons with white flowers, and snapdragons with pink flowers. He found $P(\text{red}) = \frac{1}{4}$ and $P(\text{white}) = \frac{1}{4}$ where $P(\text{color})$ means the probability that the offspring of snapdragons will have flowers of color x.
   a. What is $P(\text{pink})$? $\frac{1}{2}$
   b. What is the range of this function? $\{\frac{1}{4}, \frac{1}{2}\}$

2. Two dice are tossed. Find $P(\text{sum is even})$ and $P(\text{sum is odd})$. Explain your method.
   $P(\text{sum is even}) = \frac{1}{2}$; $P(\text{sum is odd}) = \frac{1}{2}$; sample explanation: There are 36 outcomes, with 18 even sums and 18 odd sums; $\frac{18}{36} = \frac{1}{2}$.

3. Suppose for families with 2 children, if $P(n)$ = the probability that n children in the family can roll their tongue, then $P(1) = \frac{1}{8}$ and $P(2) = \frac{1}{64}$.
   a. What does $P(0)$ represent?
   Sample: the probability that neither child can roll his or her tongue
   b. Find $P(0)$. $\frac{55}{64}$

Other Probability Functions

Example 1

Consider the spinner at the right. Assume all regions have the same probability of the spinner landing in them. Let $P(n)$ = the probability of the spinner landing in region n.

a. Graph the function P.
b. Give an equation for the function.

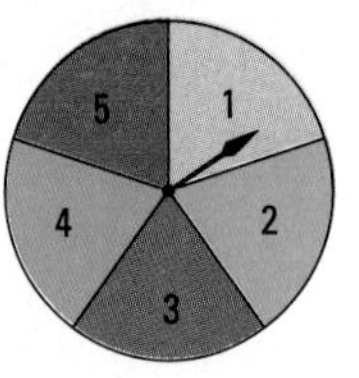

Solution

a. Since there are five regions, each with the same probability, each has probability $\frac{1}{5}$. So $P(1) = \frac{1}{5}$, $P(2) = \frac{1}{5}$, and so on. The graph is shown at the right.

b. An equation for the function is $P(n) = \frac{1}{5}$.

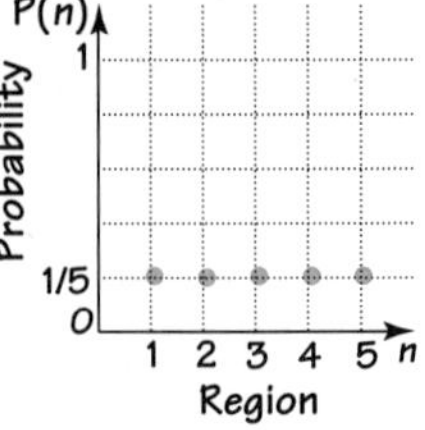

Example 2

Three fair coins—a quarter, a dime, and nickel—are tossed. Let $P(h)$ = the probability of tossing exactly h heads.

a. Make a table of values for the function P.
b. Give the domain of P.
c. Give the range of P.

Solution

a. From the Multiplication Counting Principle you know there are $2 \cdot 2 \cdot 2 = 8$ possible ways the three coins could come up. List them and count the number of heads for each outcome.

quarter	dime	nickel	no. of heads
H	H	H	3
H	H	T	2
H	T	H	2
H	T	T	1
T	H	H	2
T	H	T	1
T	T	H	1
T	T	T	0

Since the coins are fair, all of the outcomes are equally likely. There are eight outcomes, so each has probability $\frac{1}{8}$. Now make a table showing the number of heads and $P(h)$ the probability that h heads occur.

h	0	1	2	3
P(h)	$\frac{1}{8}$	$\frac{3}{8}$	$\frac{3}{8}$	$\frac{1}{8}$

Notice that getting one or two heads is more likely than getting either zero or three heads.

b. The domain is the set of h-values. domain = $\{0, 1, 2, 3\}$

c. The range is the set of probability values. range = $\{\frac{1}{8}, \frac{3}{8}\}$

Optional Activities

Activity 1 Technology Connection
Materials: Automatic graphers

You might use this activity in place of **Question 24**. In this activity, students find an equation to describe the graph on page 795. Have students **work in pairs**.

Step 1 Find an equation for the reflection image of $y = |x|$ over the x-axis. $[f(x) = -|x|]$

Step 2 Move the graph up so that $\frac{6}{36}$ is the maximum value. $[g(x) = -|x| + \frac{1}{6}]$

Step 3 Move the graph over so that the maximum value occurs when $x = 7$. $[h(x) = -|x - 7| + \frac{1}{6}]$

Step 4 Change the scale so that the sides of the angle have slopes $\frac{1}{36}$ and $-\frac{1}{36}$. This answers **Question 24d.**

Probability and Relative Frequency Functions

Relative frequencies. *Although more boys are born each year than girls, the female population eventually exceeds the male population at around age 31—and continues in that direction thereafter.*

Below are graphs for two functions related to the births of boys and girls. There were about 2,129,000 boys and 2,029,000 girls born in the U.S. in 1990. The function below left assumes that the two events—birth of a girl and birth of a boy—are equally likely. The function below right gives the relative frequencies of boy and girl births in 1990. A basic question for statisticians is: "Could you expect relative frequencies like those below right if the probabilities below at the left are true?"

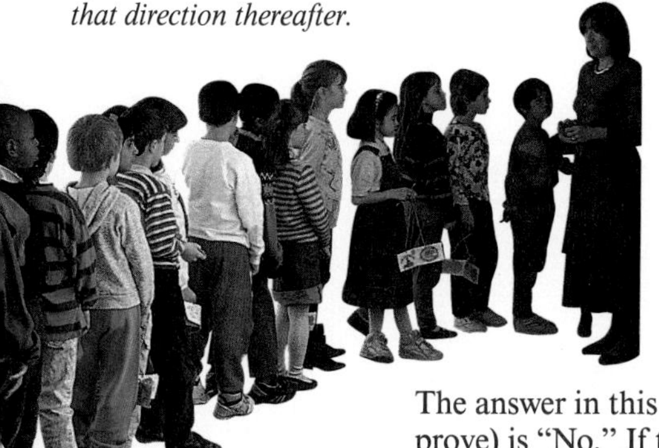

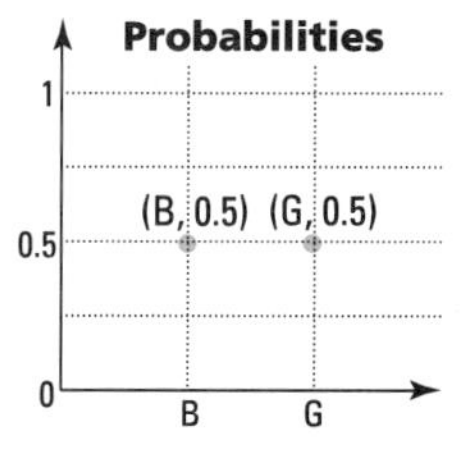

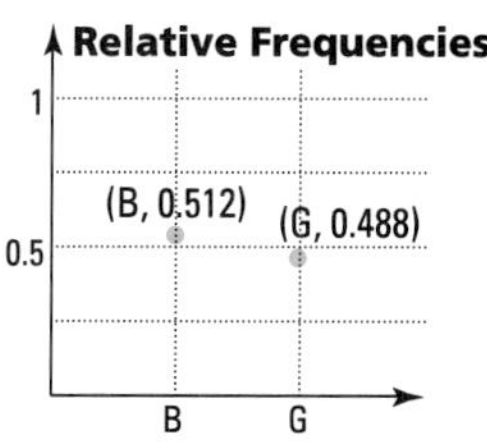

The answer in this case (which requires more advanced mathematics to prove) is "No." If the probabilities were equal, these relative frequencies would be unlikely. Thus, a baby is more likely to be a boy than a girl. However, the ratio of males to females changes with age. The U.S. population contains more women than men because women live longer.

QUESTIONS

Covering the Reading

1. Explain what a probability function is. **See left.**

In 2–5, let $P(n)$ = the probability of getting a sum of n when two fair dice are tossed.

2. Evaluate $P(3)$. $\frac{2}{36} = \frac{1}{18}$

3. If $P(n) = \frac{5}{36}$, then n = __?__ or __?__. **6; 8**

4. Describe the shape of the graph of P. **an angle**

5. Does the graph of P have an axis of symmetry? If so, what is an equation for that line? **Yes, x = 7.**

6. Consider the spinner at the right. Assume the spinner has the same probability of landing in each region. Let $P(n)$ = the probability of the spinner landing in region n.

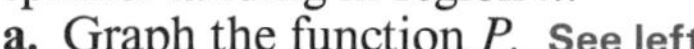

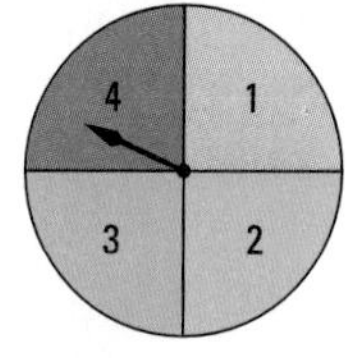

   a. Graph the function P. **See left.**
   b. Find an equation for the function. $P(n) = \frac{1}{4}$
   c. What is the domain of P? $\{1, 2, 3, 4\}$
   d. What is the range of P? $\{\frac{1}{4}\}$

1) A probability function is a function whose domain is a set of outcomes for a situation, and in which each ordered pair contains an outcome and its probability.

6a)

P(n)

1

0 1 2 3 4 n

Optional Activities

Activity 2

Materials: Eight-sided (octahedral) dice, automatic graphers

After completing the lesson, students may enjoy making graphs of other probability functions. Have them find the probabilities of different sums when rolling two eight-sided dice and graph the result.

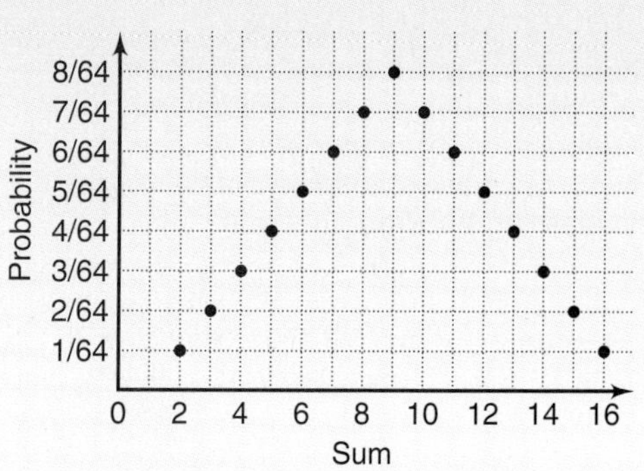

Additional Examples

These examples are also given on **Teaching Aid 138.** Additional Example 2 is suited to small-group work.

1. Assume that it is equally likely that the spinner below lands on each position.

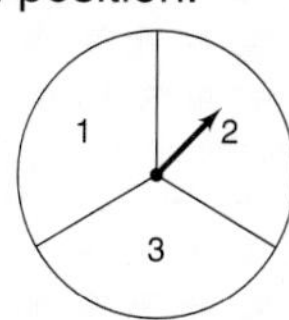

   a. Let $P(n)$ = the probability of landing in region n. Graph the function P.

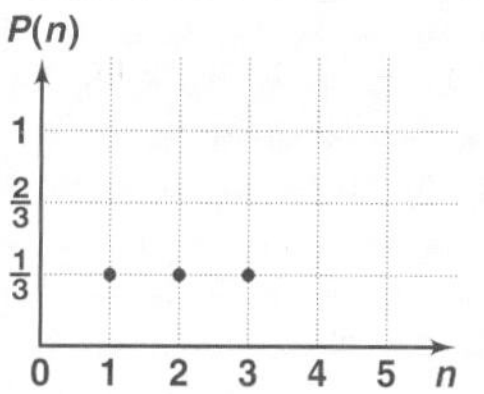

   b. Give an equation for the function. $P(n) = \frac{1}{3}$

2. Assume that four fair coins are tossed. Let $P(t)$ = the probability of tossing t tails. Give the ordered pairs of this function. Hint: First make a list of all possible outcomes. $(0, \frac{1}{16})$, $(1, \frac{4}{16})$, $(2, \frac{6}{16})$, $(3, \frac{4}{16})$, $(4, \frac{1}{16})$

Notes on Questions

Questions 2–5 These questions relate to the probability function that is graphed on page 794. You might want to use **Teaching Aid 61** with these questions.

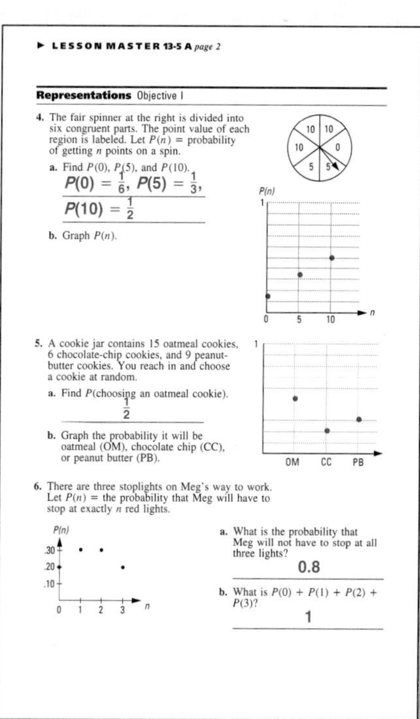

▶ **LESSON MASTER 13-5 A** *page 2*

Representations Objective I

4. The fair spinner at the right is divided into six congruent parts. The point value of each region is labeled. Let $P(n)$ = probability of getting n points on a spin.
   a. Find $P(0)$, $P(5)$, and $P(10)$. $P(0) = \frac{1}{6}$, $P(5) = \frac{1}{3}$, $P(10) = \frac{1}{2}$
   b. Graph $P(n)$.

5. A cookie jar contains 15 oatmeal cookies, 6 chocolate-chip cookies, and 9 peanut-butter cookies. You reach in and choose a cookie at random.
   a. Find P(choosing an oatmeal cookie). $\frac{1}{2}$
   b. Graph the probability it will be oatmeal (OM), chocolate chip (CC), or peanut butter (PB).

6. There are three stoplights on Meg's way to work. Let $P(n)$ = the probability that Meg will have to stop at exactly n red lights.
   a. What is the probability that Meg will not have to stop at all three lights? **0.8**
   b. What is $P(0) + P(1) + P(2) + P(3)$? **1**

Notes on Questions

Question 15a Error Alert Some students may think that there is not enough information to answer the question. Remind them that the sum of the values in the range of a probability function must be 1.

Question 19 Part a is a hint for **part b.** The product will be prime only if the value of one of the factors is 1 or −1.

Questions 22–23 These questions are an important review for Lesson 13-7.

Question 24 You might use Activity 1 in *Optional Activities* on page 796 in place of this question.

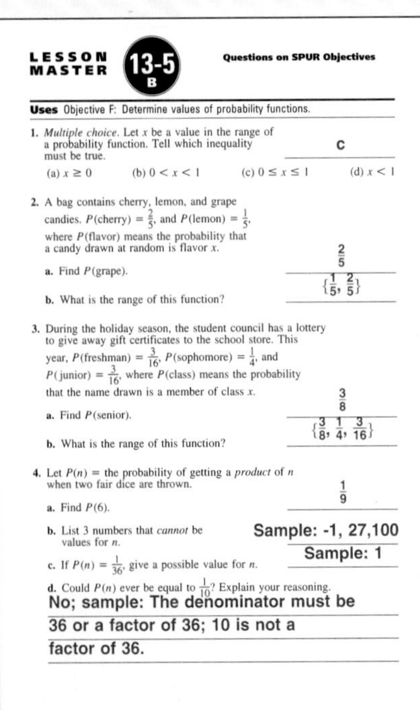

LESSON MASTER 13-5 B Questions on SPUR Objectives

Uses Objective F: Determine values of probability functions.

1. *Multiple choice.* Let x be a value in the range of a probability function. Tell which inequality must be true. **C**
   (a) $x \geq 0$ (b) $0 < x < 1$ (c) $0 \leq x \leq 1$ (d) $x < 1$

2. A bag contains cherry, lemon, and grape candies. $P(\text{cherry}) = \frac{2}{5}$, and $P(\text{lemon}) = \frac{1}{5}$, where $P(\text{flavor})$ means the probability that a candy drawn at random is flavor x.
   a. Find $P(\text{grape})$. **$\frac{2}{5}$**
   b. What is the range of this function? **$\{\frac{1}{5}, \frac{2}{5}\}$**

3. During the holiday season, the student council has a lottery to give away gift certificates to the school store. This year, $P(\text{freshman}) = \frac{3}{16}$, $P(\text{sophomore}) = \frac{1}{4}$, and $P(\text{junior}) = \frac{3}{16}$, where $P(\text{class})$ means the probability that the name drawn is a member of class x.
   a. Find $P(\text{senior})$. **$\frac{3}{8}$**
   b. What is the range of this function? **$\{\frac{3}{8}, \frac{1}{4}, \frac{3}{16}\}$**

4. Let $P(n)$ = the probability of getting a *product* of n when two fair dice are thrown.
   a. Find $P(6)$. **$\frac{1}{9}$**
   b. List 3 numbers that *cannot* be values for n. **Sample: -1, 27,100**
   c. If $P(n) = \frac{1}{36}$, give a possible value for n. **Sample: 1**
   d. Could $P(n)$ ever be equal to $\frac{1}{10}$? Explain your reasoning.
   No; sample: The denominator must be 36 or a factor of 36; 10 is not a factor of 36.

14b)

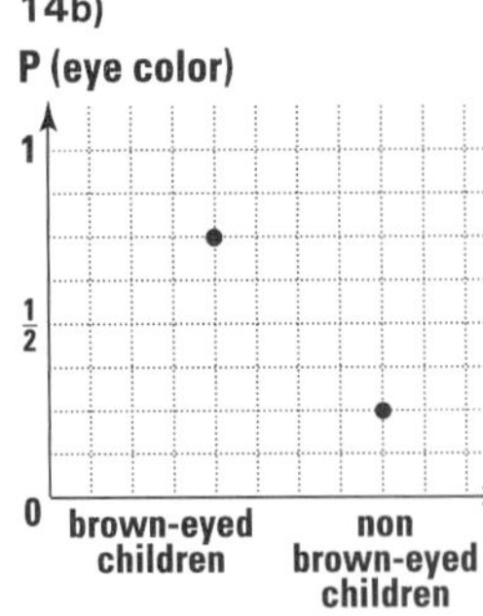

15b)

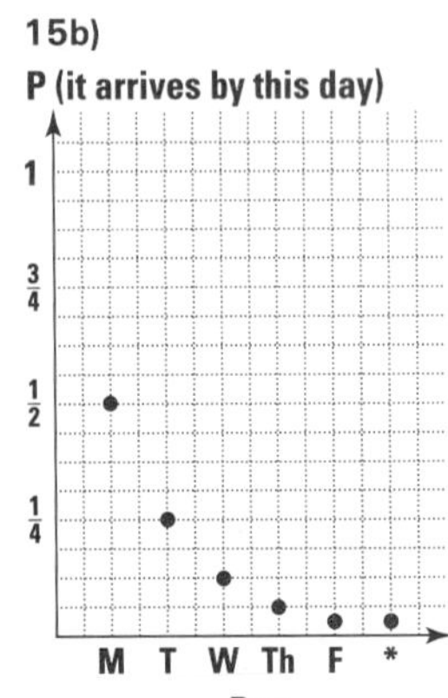

In 7 and 8, refer to Example 2.

7. *True or false.* $P(0) = P(3)$. **True**

8. Explain why $P(1) = \frac{3}{8}$. **Because in 3 of the 8 equally likely outcomes exactly one head comes up in the toss.**

In 9–11, *true or false.*

9. In 1990, more boys than girls were born in the U.S. **True**

10. In 1990, more men than women were living in the U.S. **False**

11. In births in the U.S., the relative frequency that a baby is a boy is $\frac{1}{2}$. **False**

Applying the Mathematics

12. At the right is graphed a probability function for a weighted (unfair) 6-sided die.

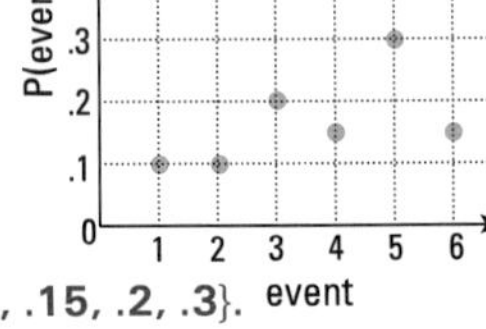

a. $P(3) =$ __?__ **0.2**
b. $P(\text{a number greater than } 3) =$ __?__ **0.6**
c. Describe how this graph differs from the probability function for a fair die. **The range in the graph for a fair die is $\{\frac{1}{6}\}$, while the range in this graph is $\{.1, .15, .2, .3\}$.**

13. Why is it impossible for the number 2 to be in the range of a probability function? **Sample: Probabilities must be from 0 to 1.**

14. Suppose that when two brown-eyed people have a child, the probability that the child is brown-eyed is $\frac{3}{4}$.
a. What is the probability that they have a child who is not brown-eyed? **$\frac{1}{4}$**
b. Graph the probability function suggested by this situation. **See left.**
c. What is the range of this function? **$\{\frac{1}{4}, \frac{3}{4}\}$**

Where do they put large packages? *This is one of the smallest post office buildings in the United States. It is located in Ochopee, Florida.*

15. A letter is mailed Saturday with the following probabilities.
$P(\text{it arrives Monday}) = \frac{1}{2}$. $P(\text{it arrives Tuesday}) = \left(\frac{1}{2}\right)^2$.
$P(\text{it arrives Wednesday}) = \left(\frac{1}{2}\right)^3$. $P(\text{it arrives Thursday}) = \left(\frac{1}{2}\right)^4$.
$P(\text{it arrives Friday}) = \left(\frac{1}{2}\right)^5$.
a. Calculate $P(\text{it does not arrive by Friday})$. **$\frac{1}{32}$**
b. Graph an appropriate probability function. **See left.**

16. Make up a probability function for which the only two elements in the range are 0 and 1.
Sample: $P(E)$ = probability that a two-headed coin toss comes up E.

Optional Activities

Activity 3 Technology Connection
You may wish to assign *Technology Sourcebook, Computer Master 32.* Students use spreadsheet software to simulate probability situations.

Adapting to Individual Needs

Extra Help Refer students to page 795. Some students may be confused by the fact that the range of the function P has only 6 values, but there are 36 possible outcomes when two dice are tossed. Point out that several sums occur more than once. For example, a sum of 5 occurs in 4 different ways. So, $P(5) = \frac{4}{36}$. Then point out that the numerators of the fractions in the column for $P(n)$ have a sum of 36.

Gone fishin'. *Pictured are fish being processed at a plant in Newport, Oregon.*

Review

17. Refer to the table and graph below showing domestic catch of fish for human food between 1960 and 1990. Let $f(x)$ = the amount (in millions of pounds) of fish caught in year x. *(Lessons 13-2, 7-1)*

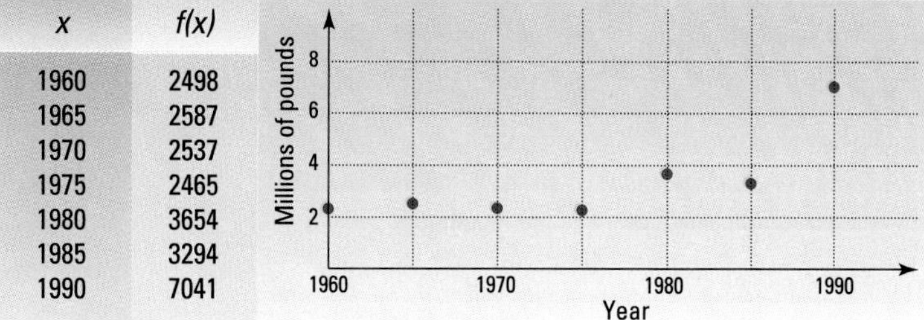

x	f(x)
1960	2498
1965	2587
1970	2537
1975	2465
1980	3654
1985	3294
1990	7041

a. What is $f(1980)$? **3654** **b) See below.**

b. Evaluate $f(1980) - f(1985)$ and tell what the answer represents.

c. What is the domain of f? **See left.**

d. Between 1980 and 1985, tell whether the rate of change of fish caught was positive, negative, or zero. **negative**

e. In which five-year period was the rate of change of fish caught the lowest? **between 1965 and 1970**

b) 360 million pounds more fish were caught in 1980 than in 1985.

17c) {1960, 1965, 1970, 1975, 1980, 1985, 1990}

18. Does $y < -3x + 1$ describe a function? Why or why not? *(Lesson 13-1)*
No, because there are infinitely many y-values for each x-value.

19. a. Factor $2x^2 + 3x - 20$. **$(2x - 5)(x + 4)$**

b. Find a value of x for which $2x^2 + 3x - 20$ is a prime number. *(Lessons 12-5, 12-1)* **$x = 3$**

20. Do this problem in your head. Since one thousand times one thousand equals one million, $1005 \cdot 995$ = _?_. *(Lesson 10-6)*
999,975

21. Simplify $\sqrt{12} + \sqrt{3}$. *(Lesson 9-7)* **$3\sqrt{3}$**

In 22 and 23, suppose the two triangles below are similar with corresponding sides parallel. *(Lessons 6-9, 4-7)*

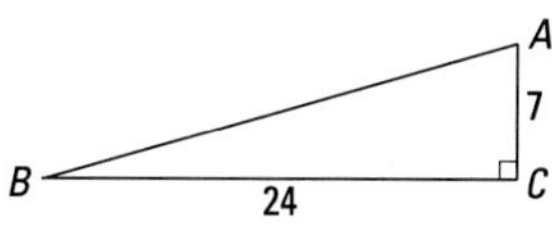

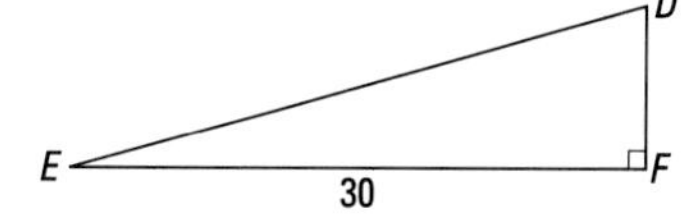

22. Find the missing lengths. **23.** If $m\angle B \approx 16°$, then $m\angle D \approx$ _?_

22) $BA = 25$, $DF = \frac{35}{4} = 8.75$, $DE = \frac{125}{4} = 31.25$ **23) 74°**

Exploration

24. Consider the function $f(x) = -\frac{1}{36}|x - 7| + \frac{1}{6}$.

a. Use an automatic grapher to plot $f(x)$ for $2 \le x \le 12$. **See left.**

b. What is $f(2)$? **$\frac{1}{36} = 0.02\overline{7}$**

c. What is $f(7)$? **$\frac{1}{6} = 0.1\overline{6}$**

d. How is this graph related to one of the probability functions in this lesson? **It contains all of the points in the probability function for the outcomes from the toss of two fair dice.**

24a)

0.5
0.4
0.3
0.2
0.1
0
3 4 5 6 7 8 9 10 11 12

Follow-up for Lesson 13-5

Practice

For more questions on SPUR Objectives, use **Lesson Master 13-5A** (shown on pages 796–797) or **Lesson Master 13-5B** (shown on pages 798–799).

Assessment

Written Communication Have each student write a paragraph describing in his or her own words the term *probability function.* [Students' explanations are clear, and they understand that the domain of a probability function is the set of all possible outcomes, and the range is the set of the probabilities that the outcomes will occur.]

Extension

Ask students to calculate the probabilities of obtaining different sums with two dice like the ones described in **Question 12**. Explain that counting the outcomes, as one does for two fair dice, will not work because the outcomes are not all equally likely to occur. However, the probability of each of the 36 outcomes can be found by multiplying.

Project Update Project 1, *Tossing Three Dice*, on page 817, relates to the content of this lesson.

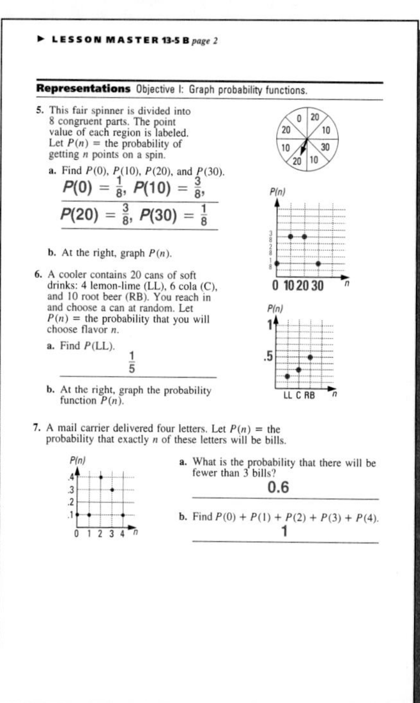

▶ **LESSON MASTER 13-5 B** *page 2*

Representations Objective I: Graph probability functions.

5. This fair spinner is divided into 8 congruent parts. The point value of each region is labeled. Let $P(n)$ = the probability of getting n points on a spin.

a. Find $P(0)$, $P(10)$, $P(20)$, and $P(30)$.
$P(0) = \frac{1}{8}$, $P(10) = \frac{3}{8}$,
$P(20) = \frac{3}{8}$, $P(30) = \frac{1}{8}$

b. At the right, graph $P(n)$.

6. A cooler contains 20 cans of soft drinks: 4 lemon-lime (LL), 6 cola (C), and 10 root beer (RB). You reach in and choose a can at random. Let $P(n)$ = the probability that you will choose flavor n.

a. Find $P(LL)$.
$\frac{1}{5}$

b. At the right, graph the probability function $P(n)$.

7. A mail carrier delivered four letters. Let $P(n)$ = the probability that exactly n of these letters will be bills.

a. What is the probability that there will be fewer than 3 bills?
0.6

b. Find $P(0) + P(1) + P(2) + P(3) + P(4)$.
1

Adapting to Individual Needs

Challenge

Tell students to consider tossing four fair coins—a quarter, a dime, a nickel, and a penny—and make a table to determine the domain and range of P if $P(h)$ = the probability of tossing exactly h heads. [Sample table is shown at the right.]

h	0	1	2	3	4
$P(h)$	$\frac{1}{16}$	$\frac{4}{16}$	$\frac{6}{16}$	$\frac{4}{16}$	$\frac{1}{16}$

Domain: {0, 1, 2, 3, 4}; range: $\{\frac{1}{16}, \frac{4}{16}, \frac{6}{16}\}$

Q	N	D	P	Number of heads
H	H	H	H	4
H	H	H	T	3
H	H	T	H	3
H	H	T	T	2
H	T	H	H	3
H	T	H	T	2
H	T	T	H	2
H	T	T	T	1

Q	N	D	P	Number of heads
T	H	H	H	3
T	H	H	T	2
T	H	T	H	2
T	H	T	T	1
T	T	H	H	2
T	T	H	T	1
T	T	T	H	1
T	T	T	T	0

Lesson 13-6

Objectives

J Graph polynomial functions.

Resources

From the ***Teacher's Resource File***

- Lesson Master 13-6A or 13-6B
- Answer Master 13-6
- Assessment Sourcebook: Quiz for Lessons 13-4 through 13-6
- Teaching Aids
  - 28 Four-Quadrant Graph Paper
  - 132 Warm-up
  - 139 Examples 1–3
- Activity Kit, Activity 29
- Technology Connection Computer Master 33

Additional Resources

- Visuals for Teaching Aids 28, 132, 139
- GraphExplorer or other automatic graphers

Teaching Lesson 13-6

Warm-up

1. Sketch the graphs of $y = x$ and $y = x^3$ on the same grid. **The graphs are given on page 801.**
2. Sketch the graphs of $y = x^2$ and $y = x^4$ on the same grid. **The graphs are given on page 801.**

(Warm-up continues on page 801.)

LESSON

13-6 *Polynomial Functions*

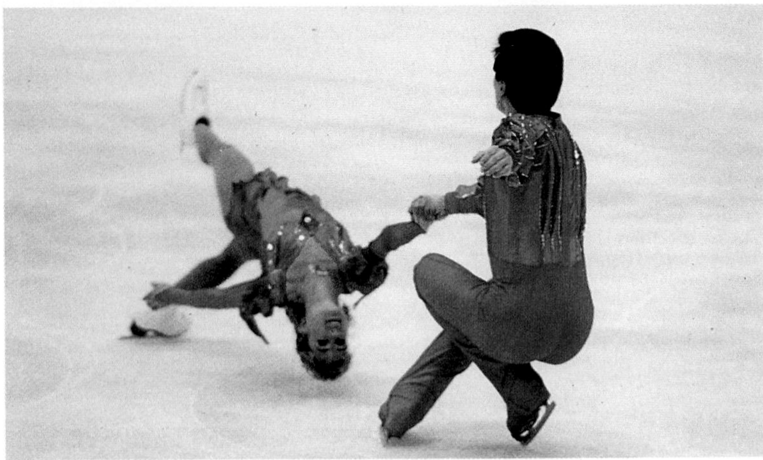

One good turn deserves another. *These pairs skaters are performing a "death spiral," in which both people rotate 360° around the fixed back toe of the male skater.*

The Cubing Function

As you know, some types of equations have graphs with predictable shapes. Graphs of all equations of the form $y = mx + b$ are lines with slope m and y-intercept b. Therefore, the function f with $f(x) = mx + b$ is called a *linear function.* Graphs of quadratic functions of the form $f(x) = ax^2 + bx + c$ are parabolas. In contrast, not all polynomial equations of degree 3 have the same shape. Consider the simplest cubic equation $y = x^3$. This equation defines the **cubing function** with equation $f(x) = x^3$.

Example 1

Graph $f(x) = x^3$ by plotting points where x goes from -2 to 2 in steps of 0.5.

Solution

Make a table of values. Connect them with a smooth curve. A table and graph are shown below.

x	y = f(x)
-2	-8
-1.5	-3.375
-1	-1
-0.5	-0.125
0	0
0.5	0.125
1	1
1.5	3.375
2	8

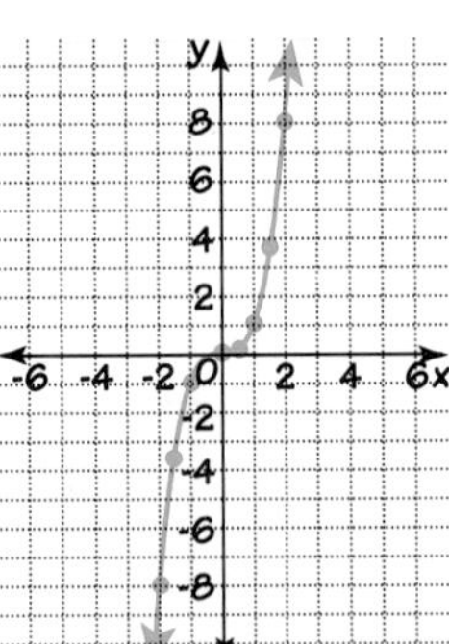

Lesson 13-6 Overview

Broad Goals The goal of this lesson is to introduce students to polynomial functions and their diverse graphs.

Perspective Until the availability of automatic graphers, the content of this lesson had to be relegated to courses in analytic geometry or calculus. It was not easy to determine the shape of a polynomial function without rather in-depth analysis and quite a bit of algebraic manipulation. The student would calculate or estimate the x-intercepts, find derivatives, and solve equations to obtain relative maximum or minimum points, and perhaps even find second derivatives.

With automatic graphers, it is much easier to obtain the graph of a polynomial function. From the graph, one can identify all of those characteristics that used to be necessary to find the graph itself.

In this lesson, students sketch the graph of a polynomial function. This requires the selection of an appropriate window—one that displays relative maxima (peaks) and minima (valleys), x-intercepts (if any), and the y-intercept. It also requires knowing whether the values of the polynomial increase or decrease as x gets larger or as x gets smaller.

To explore the shape of the graph of $y = x^3$, we can use an automatic grapher. Below are two views of this curve on different windows.

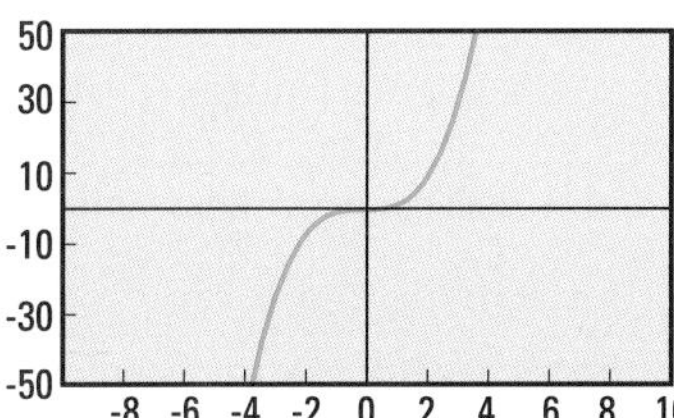
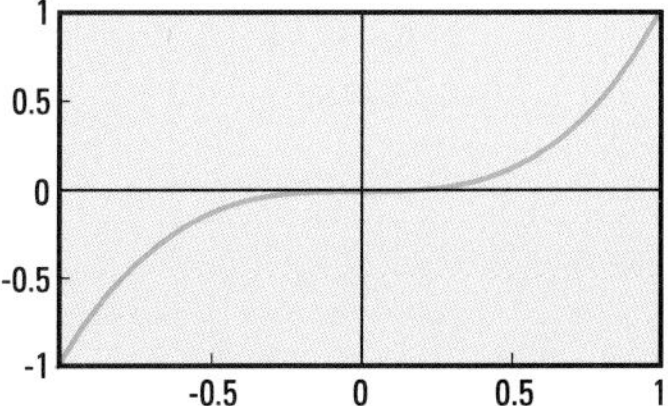

The window on the left above shows that when $|x|$ is moderately large, $|x^3|$ is very large. The one on the right shows that near the origin the graph of $y = x^3$ is fairly flat.

Activity a) Sample:

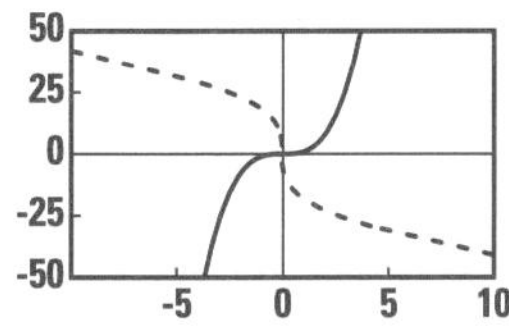

b) The 180° clockwise rotations should look just like the original graphs.

Activity

Trace one of the graphs of $y = x^3$ onto a piece of paper. Trace the axes also. Hold the paper in place on your book with a pencil point at the origin. Rotate the paper

a. 90° clockwise; **b.** 180° clockwise.

After each rotation, describe how the graph on the paper is related to the graph in your book.

You are asked for your answer to part **a** in the Questions. For part **b,** you should find that the graph on your paper coincides exactly with that in your book. Because a half-turn makes these two graphs coincide, we say that the graph of $y = x^3$ has **180° rotation symmetry** around the origin. In general, if you can turn a figure 180° around some point so that the figure coincides with itself, that point is called a **center of symmetry.** Notice that the graph does *not* have a *line* of symmetry.

Spin cycle. *Pictured are folk dancers in Mexico City. Many dances include examples of rotations and rotational symmetry.*

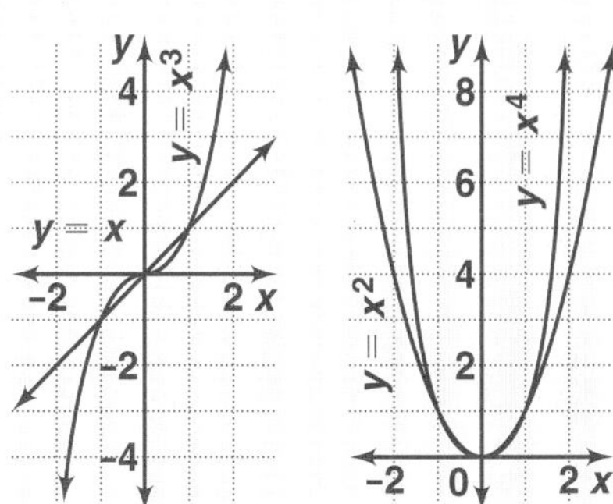

3. In which quadrants will the graph of $y = x^5$ fall? Why? **First and third quadrants; if x is positive, y is positive and if x is negative, y is negative.**
4. In which quadrants will the graph of $y = x^6$ fall? Why? **First and second quadrants; if x is positive, y is positive and if x is negative, y is positive.**

Notes on Reading

The graphs for **Examples 1–3** are given on **Teaching Aid 139.**

Cooperative Learning It is appropriate to have students review this lesson in groups. If students did Activity 1 of *Optional Activities* below, have them compare their papers to see what they drew. Point out that the rotation symmetry is a property of all cubics but not of all polynomials of any other degree greater than or equal to one. Then have students draw the graph of the function in **Example 2** and find the x-intercept. Finally, have students graph the polynomial in **Example 3**; they will see that they can obtain the two different graphs which are shown by changing the window.

Optional Activities

Activity 1

Materials: Automatic graphers

After discussing the lesson, students might **work in groups** and sketch the graphs of each of the following equations using the same window. Then ask them to predict how the graph of $y = x^6 - 2x$ would look. [Like $y = x^4 - 2x$, but steeper]

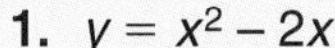

1. $y = x^2 - 2x$
2. $y = x^3 - 2x$
3. $y = x^4 - 2x$
4. $y = x^5 - 2x$

1. 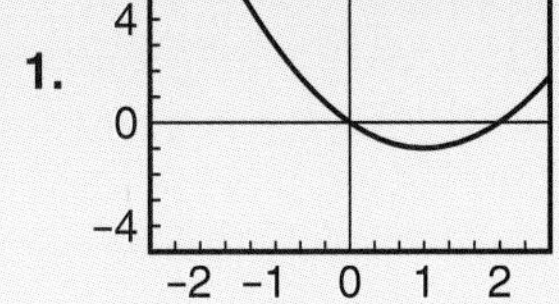

2. 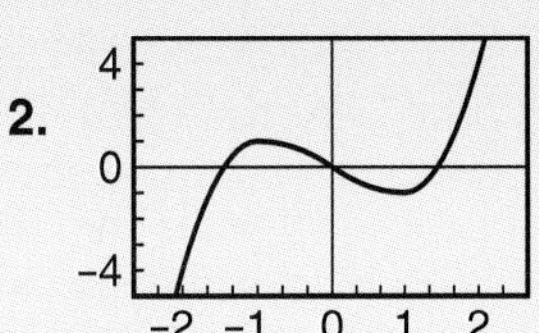

3. 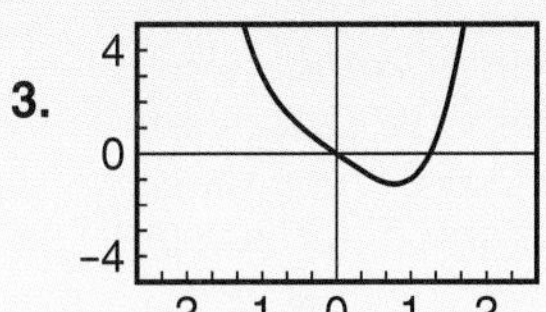

4. 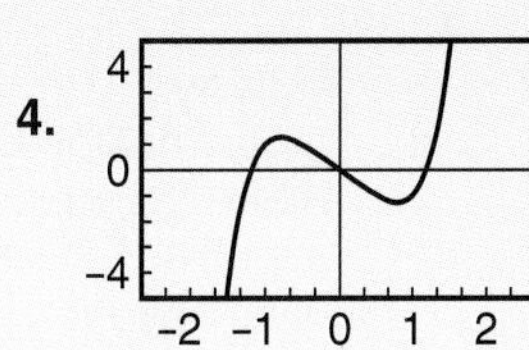

Additional Examples

1. Graph $y = x^4 - 8$ for $-2 \le x \le 2$ in increments of $\frac{1}{3}$.

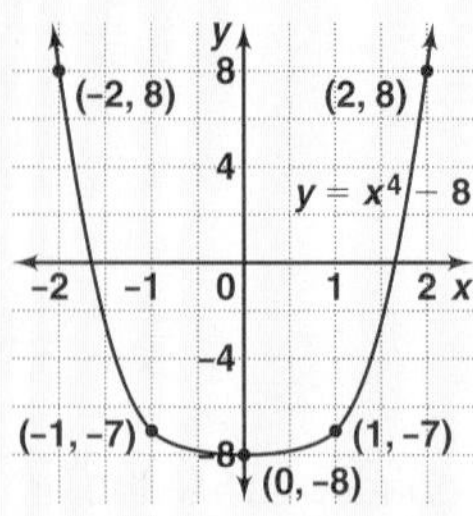

2. Graph $y = 2x^3 + 3x^2 - 10x + 1$ for values of x from -4 to 4 using an automatic grapher. Estimate the x-intercepts to the nearest tenth. **Sample: x-intercepts: -3.1, 0.1, and 1.5**

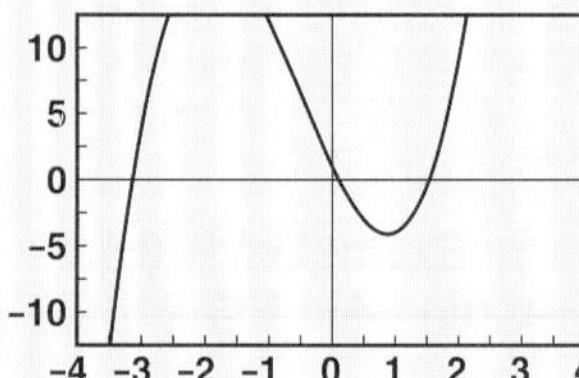

3. Graph $y = x^5 - 3x^4 - 5x^3 + 15x^2 + 4x - 12$ using an automatic grapher. Determine the number of x-intercepts. **5**

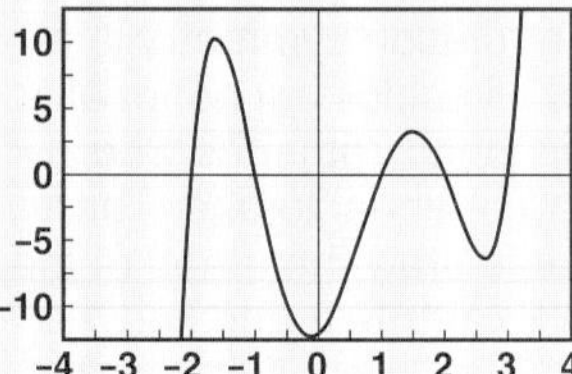

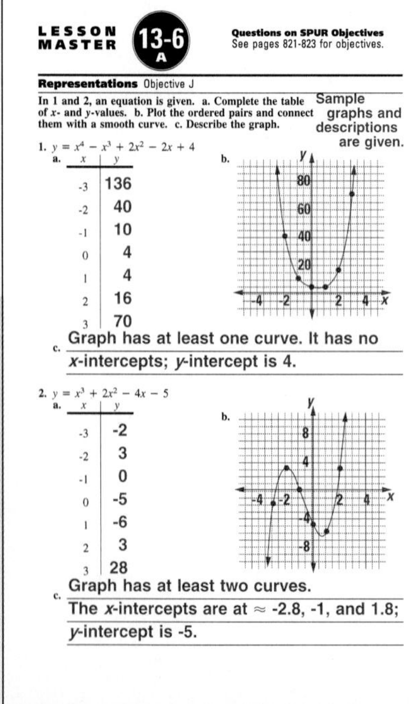

LESSON MASTER 13-6 A — Questions on SPUR Objectives. See pages 821-823 for objectives.

Representations Objective J

In 1 and 2, an equation is given. a. Complete the table of x- and y-values. b. Plot the ordered pairs and connect them with a smooth curve. c. Describe the graph. Sample graphs and descriptions are given.

1. $y = x^4 - x^3 + 2x^2 - 2x + 4$

a.

x	y
-3	136
-2	40
-1	10
0	4
1	4
2	16
3	70

c. Graph has at least one curve. It has no x-intercepts; y-intercept is 4.

2. $y = x^3 + 2x^2 - 4x - 5$

a.

x	y
-3	-2
-2	3
-1	0
0	-5
1	-6
2	3
3	28

c. Graph has at least two curves. The x-intercepts are at ≈ -2.8, -1, and 1.8; y-intercept is -5.

Another Cubic Polynomial Function

Example 2

Consider $g(x) = x^3 - x + 1$. Make a table of values satisfying the equation, using $x = -2, -1.5, -1, -0.5, 0, 0.5, 1, 1.5,$ and 2. Plot the points in the table and connect them with a smooth curve.

Solution

Be careful with the negative signs when evaluating the expression. For instance, when $x = -2$, $x^3 - x + 1 = (-2)^3 - (-2) + 1 = -8 + 2 + 1 = -5$.

Below are the table and graph.

x	$g(x) = x^3 - x + 1$
-2	-5
-1.5	-.875
-1	1
-0.5	1.375
0	1
0.5	.625
1	1
1.5	2.875
2	7

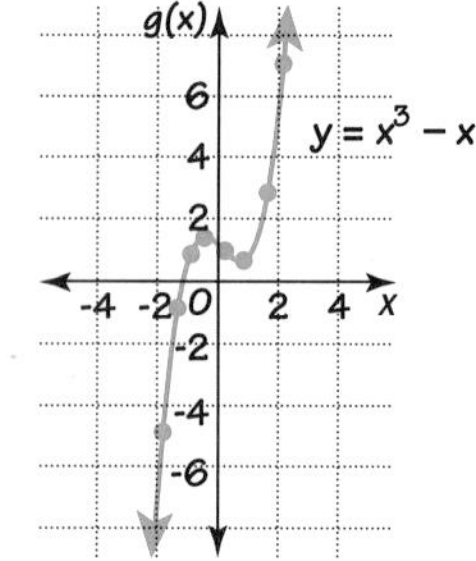

Like the graph in Example 1, the graph of $g(x) = x^3 - x + 1$ increases very quickly as x gets large, and decreases very quickly when x gets very small (that is, "very negative").

The table and graph show that there is an x-intercept between -2 and -1. To estimate the x-intercept, you can use an automatic grapher to draw the graph of $g(x) = x^3 - x + 1$. On the window $-1.5 \le x \le 1.5$, $-1 \le y \le 2$, we trace and find the x-intercept to be about -1.3.

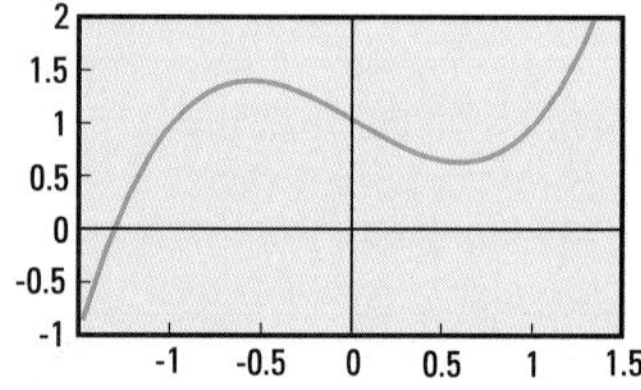

Like the graph of $y = x^3$, the graph of $y = x^3 - x + 1$ also has a point of symmetry. (In the Questions you are asked to find it.) In a later course you will see that algebra can be used to show that the graph of every polynomial function of degree 3 has 180° rotation symmetry.

Higher-Degree Polynomial Functions

Making graphs by hand of most higher-degree polynomial functions involves a great deal of computation. For all but the simplest cases, we recommend that you use an automatic grapher. Unless you are looking only for a certain point or at a certain region on the graph, it is a good idea to use a window that allows you to see all the intercepts, and all the "peaks and valleys" polynomial graphs can have.

It is helpful to know that every graph of a polynomial function of degree n has exactly one y-intercept and no more than n x-intercepts. Aside from the above, there are very few "rules" about polynomial graphs.

Optional Activities

Activity 2

You might want to use *Activity Kit, Activity 29,* after completing the lesson. This activity extends the exploration of rotation symmetry introduced on page 801.

Activity 3 Technology Connection

In *Technology Sourcebook, Computer Master 33,* students use *GraphExplorer* or similar software to graph polynomial functions.

Adapting to Individual Needs

Extra Help

The activity on page 801 refers to 180° rotation symmetry around the origin. Some students might automatically conclude that the origin is always the center of symmetry. As the text points out, the graph of every polynomial function of degree 3 has 180° rotation symmetry. However, the center of symmetry does not have to be the origin. You may have to emphasize this concept for **Question 6.**

Example 3

a. Draw a graph of $y = x^4 - 7x^3 - 9x^2 + 63x$.
b. Identify all intercepts and show the behavior of the graph between those intercepts.

Solution

a. Calculate a few values to get some idea of what window to use.

When $x = -2$, $y = (-2)^4 - 7(-2)^3 - 9(-2)^2 + 63(-2) = -90$.
When $x = 0$, $y = 0^4 - 7(0)^3 - 9(0)^2 + 63(0) = 0$.
When $x = 2$, $y = (2)^4 - 7(2)^3 - 9(2)^2 + 63(2) = 50$.

We use a window of $-15 \le x \le 15$ and $-100 \le y \le 100$. The graph is below on the left.

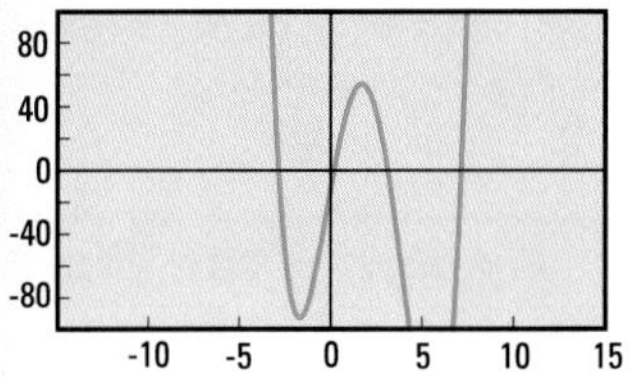

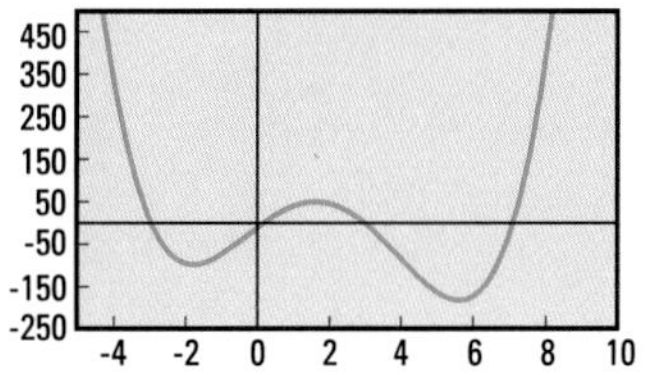

This graph shows the y-intercept and 4 x-intercepts, the maximum possible for a 4th-degree polynomial function. So there are no other intercepts. We need to increase the range of y-values to see where the peaks and valleys occur. The sketch above on the right, using the window $-5 \le x \le 10$, $-250 \le y \le 500$, captures all the essential features of the graph.

b. The x-intercepts are -3, 0, 3, and 7; the y-intercept is 0.

Check

b. The y-intercept was checked as part of the solution. To check the x-intercepts, calculate y when $x = -3, 3$, and 7.

When $x = -3$, $y = (-3)^4 - 7(-3)^3 - 9(-3)^2 + 63(-3) = 81 + 189 - 81 - 189 = 0$.

You are asked to check the other two x-intercepts in the Questions.

QUESTIONS

Covering the Reading

1. How many x-intercepts does the graph of $f(x) = x^3$ have? 1
2. *True or false.* The graph of the cubing function has a line of symmetry. False
3. What does it mean to say that a graph has 180° rotation symmetry?
4. Copy one of the graphs of $f(x) = x^3$ from the book. On the same set of axes, draw the image you got by doing part **a** of the Activity on page 801. See left.

3) You can turn the figure 180° around some point and the figure will coincide with itself.

4)

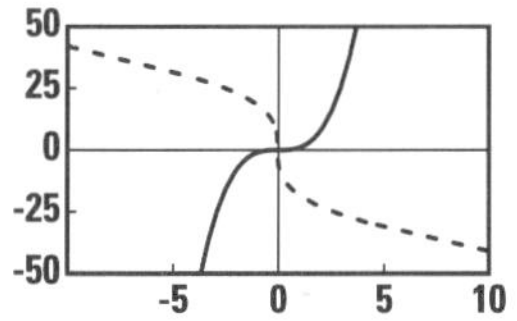

Notes on Questions

Question 3 Students may associate symmetry only with reflection symmetry. Point out that symmetry is a more general property and that it can apply to any transformation. For instance, if a figure can be *translated* onto itself, then it has *translation* symmetry. Tessellations, lines, and coordinate grids have translation symmetry.

Additional Answers, page 804, continued

13a.

x	y
-2	8
-1.5	3.375
-1	1
-0.5	0.125
0	0
0.5	-0.125
1	-1
1.5	-3.375
2	-8

b.

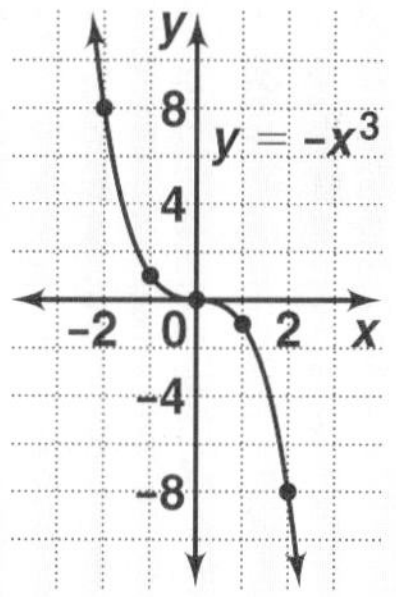

c. The graph has one intercept at (0, 0); no axis of symmetry; point of rotation symmetry: (0, 0)

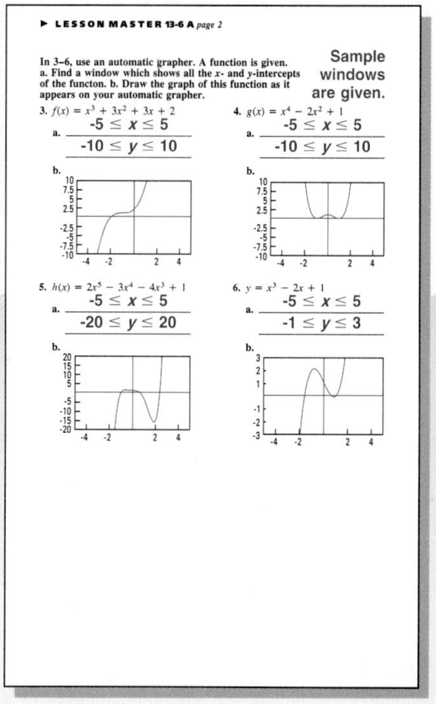
► LESSON MASTER 13-6 A *page 2*

In 3–6, use an automatic grapher. A function is given. a. Find a window which shows all the x- and y-intercepts of the functon. b. Draw the graph of this function as it appears on your automatic grapher.

Sample windows are given.

3. $f(x) = x^3 + 3x^2 + 3x + 2$
a. $-5 \le x \le 5$, $-10 \le y \le 10$
b.

4. $g(x) = x^4 - 2x^2 + 1$
a. $-5 \le x \le 5$, $-10 \le y \le 10$
b.

5. $h(x) = 2x^5 - 3x^4 - 4x^3 + 1$
a. $-5 \le x \le 5$, $-20 \le y \le 20$
b.

6. $y = x^3 - 2x + 1$
a. $-5 \le x \le 5$, $-1 \le y \le 3$
b.

Adapting to Individual Needs

Challenge

Explain that a polynomial $f(x)$ has a factor $(x - a)$ if and only if $f(a) = 0$. Then have students do the following problems.

In 1–3, determine if the expression is a factor of $f(x) = x^3 + x^2 - 9x - 9$.

1. $x - 3$ [$f(3) = 3^3 + 3^2 - 9(3) - 9 = 27 + 9 - 27 - 9 = 0$; $x - 3$ is a factor of $f(x)$].
2. $x + 3$ (If necessary, remind students that $x + 3 = x - (-3)$). [$f(-3) = (-3)^3 + (-3)^2 - 9(-3) - 9 = -27 + 9 + 27 - 9 = 0$; $x + 3$ is a factor.]
3. $x + 1$ [$f(-1) = (-1)^3 + (-1)^2 - 9(-1) - 9 = -1 + 1 + 9 - 9 = 0$; $x + 1$ is a factor.]
4. Multiply $x - 3$, $x + 3$, and $x + 1$. What do you notice? [The product is the original expression, $x^3 + x^2 - 9x - 9$.]
5. Consider $x^4 - 3x^2 + 6x - 4$.
   a. Is $x + 1$ a factor? [No]
   b. Is $x - 1$ a factor? [Yes]

Notes on Questions

Question 12 Error Alert Some automatic graphers allow $y = 2x^3$ to be entered without a multiplication symbol between the 2 and the x. Have students check to see what their graphers will allow.

Questions 17–19 Art Connection The American artist Andy Warhol is best remembered for his pictures of familiar objects, such as soup cans and soft-drink bottles. Some of his pictures involved isolating and simplifying images, enlarging them in a series of pictures, and repeating the same image many times.

Question 20 Error Alert Some students may not realize that there is a system to be solved. Graphing shows that there are two integer solutions with small absolute values which can be easily determined.

Additional Answers

11a, b. $y = x^3 + 3$

x	y
-2	-5
-1.5	-0.375
-1	2
-0.5	2.875
0	3
0.5	3.125
1	4
1.5	6.375
2	11

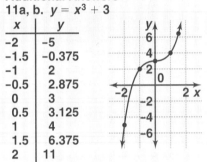

c. y-intercept: 3; x-intercept: ≈ -1.4; no axis of symmetry; point of rotation symmetry: (0, 3)

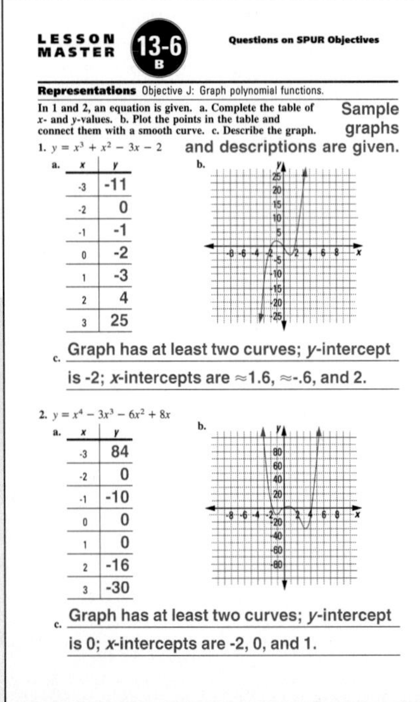

LESSON MASTER 13-6 B Questions on SPUR Objectives

Representations Objective J: Graph polynomial functions.

In 1 and 2, an equation is given. a. Complete the table of x- and y-values. b. Plot the points in the table and connect them with a smooth curve. c. Describe the graph. Sample graphs and descriptions are given.

1. $y = x^3 + x^2 - 3x - 2$

a.

x	y
-3	-11
-2	0
-1	-1
0	-2
1	-3
2	4
3	25

c. Graph has at least two curves; y-intercept is -2; x-intercepts are ≈ 1.6, $\approx -.6$, and 2.

2. $y = x^4 - 3x^3 - 6x^2 + 8x$

a.

x	y
-3	84
-2	0
-1	-10
0	0
1	0
2	-16
3	-30

c. Graph has at least two curves; y-intercept is 0; x-intercepts are -2, 0, and 1.

5) $y = (-1.3)^3 - (-1.3) + 1 = -2.197 + 2.3 = .103$

9) $(3)^4 - 7(3)^3 - 9(3)^2 + 63 \cdot 3 = 81 - 189 - 81 + 189 = 0$; $(7)^4 - 7(7)^3 - 9(7)^2 + 63 \cdot 7 = 2401 - 2401 - 441 + 441 = 0$; It checks.

14b)

$f(x) = x^3 + 3x^2 - 10x$

15b)

In 5 and 6, refer to Example 2.

5. Check by substitution that -1.3 is close to an x-intercept of $y = x^3 - x + 1$. See left.

6. Use tracing to estimate the coordinates of the point of symmetry of the graph. (0, 1)

7. *True or false.* The graph of every cubic polynomial function has a point of symmetry. True

8. The graph of a polynomial equation of degree n has __?__ y-intercept(s) and __?__ x-intercept(s). 1; no more than n

In 9 and 10, refer to Example 3.

9. Verify that $x = 3$ and $x = 7$ are x-intercepts of the graph. See left.

10. Does the graph appear to have a symmetry point? If so, what are its coordinates? No

Applying the Mathematics

In 11–13, an equation for a polynomial function is given. See margin.
 a. Make a table of x- and y-values for x-values -2, -1.5, -1, -0.5, 0, 0.5, 1, 1.5, and 2.
 b. Graph the values.
 c. Describe the graphs.

11. $y = x^3 + 3$ **12.** $y = 2x^3$ **13.** $y = -x^3$

14. A spreadsheet was used to compute the value of $f(x) = x^3 + 3x^2 - 10x$ for values of x from -6 to 4. b) See left.
 a. Copy and complete the table.
 b. Graph the function f for $-6 \le x \le 4$.
 c. Identify all x-intercepts. -5, 0, 2

	A	B
1	x	VALUE
2	-6	-48
3	-5	0
4	-4	24
5	-3	30
6	-2	24
7	-1	12
8	0	0
9	1	-6
10	2	0
11	3	24
12	4	72

15. You will need an automatic grapher. In Example 2 of Lesson 10-2 you learned that the polynomial $50x^3 + 60x^2 + 70x + 80$ represents the amount of Cole's savings after three years when invested at a scale factor x. Let $A(x) = 50x^3 + 60x^2 + 70x + 80$.
 a. What is a reasonable domain for the function A?
 b. Graph A on this domain. See left.
 c. Use the trace feature to find the amount Cole would have if he invested at a 10% annual yield. about \$298
 d. At what annual yield would Cole have to invest in order to accumulate \$400 in 3 years? 33%

a) Sample: the set of numbers x with $1 \le x \le 2$

Additional Answers, continued

12a.

x	y
-2	-16
-1.5	-6.75
-1	-2
-0.5	-0.25
0	0
0.5	0.25
1	2
1.5	6.75
2	16

b. $y = 2x^3$

c. y-intercept: 0; x-intercept: 0; no axis of symmetry; point of rotation symmetry: (0, 0)

(Answers continue on page 803.)

Review

16. Assume a coin is fair. Then if it is tossed twice, there are four possible arrangements of heads and tails: HH, HT, TT, and TH. Let $P(n)$ be the probability that there are n heads in 2 tosses of the coin.
  a. What is the domain of the function P? {0, 1, 2}
  b. Give all the ordered pairs in this function. *(Lessons 13-5, 13-4)* (0, 1/4), (1, 1/2), (2, 1/4)

In 17–19, an open soup can has radius r and height h. Its volume $V = \pi r^2 h$, and its surface area $S = \pi r^2 + 2\pi rh$.

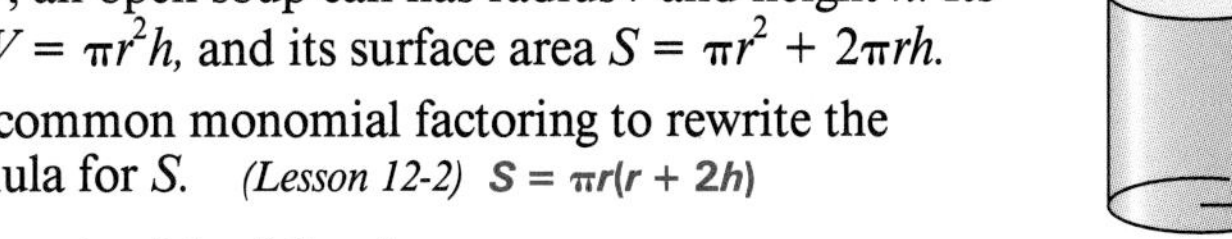

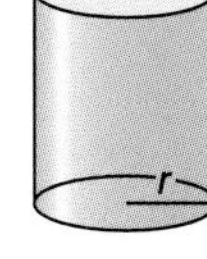

17. Use common monomial factoring to rewrite the formula for S. *(Lesson 12-2)* $S = \pi r(r + 2h)$

18. Find each of the following. *(Lesson 10-1)*
  a. the degree of V 3 **b.** the degree of S 2

19. If the can has a diameter of 8 cm and a height of 12 cm, about how many milliliters of soup can it hold? (Remember that 1 liter = 1000 cm³.) *(Lesson 1-5, Previous course)* about 603 milliliters

20. Solve by graphing. See left. $\begin{cases} y = |x| \\ y = \frac{1}{2}x^2 \end{cases}$ *(Lesson 11-1)*

In 21–26, solve. *(Lessons 9-8, 9-5, 6-8, 5-6, 3-5)*

21. $|3y - 6| = 2$ $y = \frac{4}{3}$ or $\frac{8}{3}$

22. $100B^2 + 100B - 100 = 0$ $B = \frac{-1 \pm \sqrt{5}}{2}$

23. $\frac{A}{0.2} = \frac{10}{A}$ $A = \pm\sqrt{2}$

24. $2y + 14 > 5y - 19$ $y < 11$

25. $14.7 = 7x + 21$ $x = -.9$

26. $\sqrt{C} = 400$ $C = 160{,}000$

20) $y = \frac{1}{2}x^2$, $y = |x|$, $P_2 = (-2, 2)$, $P_1 = (2, 2)$

Solutions: (0, 0), (-2, 2), (2, 2)

27b) Sample: $y = x^5 - 4x^3$

$f(x) = x^5 - 4x^3$

Exploration

27. Explore with an automatic grapher. b) See left.
  a. Find an equation for a polynomial function with a graph shaped like the following. Samples given.

i. $y = x^3 - 2x$

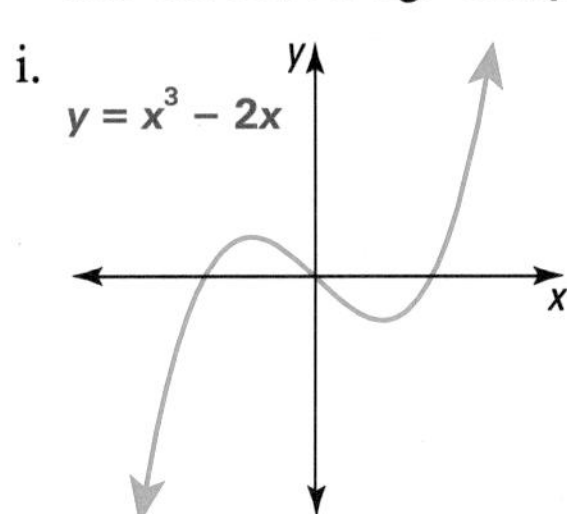

ii. $y = -x^4 + 2x^2$

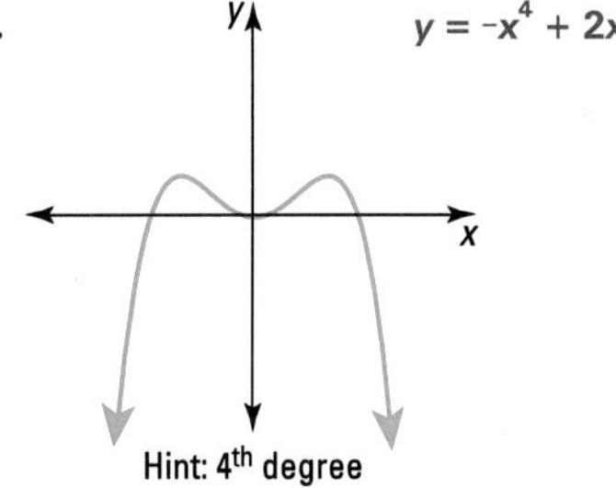

  b. Draw the graph of some 5th degree polynomial function.

Follow-up for Lesson 13-6

Practice

For more questions on SPUR Objectives, use **Lesson Master 13-6A** (shown on pages 802–803) or **Lesson Master 13-6B** (shown on pages 804–805).

Assessment

Quiz A quiz covering Lessons 13-4 through 13-6 is provided in the *Assessment Sourcebook.*

Oral Communication Ask students to explain what they can tell about the degree of a polynomial function by examining the graph of the function. [Students understand that if a graph of a polynomial function has n x-intercepts, it will be a polynomial of degree greater than or equal to n.]

Extension

Refer students to **Example 1** and the graph of $f(x) = x^3$. Have them determine how the graph of $f(x)$ is affected by replacing x^3 with $(x - a)^3$ or $(x + a)^3$. [The graph is shifted a units along the x-axis.]

Project Update Project 2, *Polynomial Functions of Degree 4*, and Project 5, *Birthday Cubics*, on pages 817–818, relate to the content of this lesson.

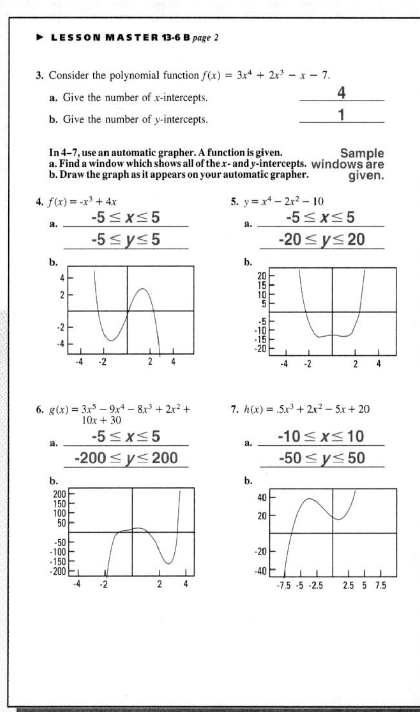

► LESSON MASTER 13-6 B *page 2*

3. Consider the polynomial function $f(x) = 3x^4 + 2x^3 - x - 7$.
  a. Give the number of x-intercepts. 4
  b. Give the number of y-intercepts. 1

In 4–7, use an automatic grapher. A function is given.
a. Find a window which shows all of the x- and y-intercepts.
b. Draw the graph as it appears on your automatic grapher.
Sample windows are given.

4. $f(x) = -x^3 + 4x$
  a. $-5 \le x \le 5$, $-5 \le y \le 5$
  b.

5. $y = x^4 - 2x^2 - 10$
  a. $-5 \le x \le 5$, $-20 \le y \le 20$
  b.

6. $g(x) = 3x^5 - 9x^4 - 8x^3 + 2x^2 + 10x + 30$
  a. $-5 \le x \le 5$, $-200 \le y \le 200$
  b.

7. $h(x) = .5x^3 + 2x^2 - 5x + 20$
  a. $-10 \le x \le 10$, $-50 \le y \le 50$
  b.

Adapting to Individual Needs

English Language Development
Recall that a *rotation* is a turn. Then draw an X on an overhead transparency. Trace the X on another transparency and rotate it 90° ($\frac{1}{4}$ turn), 180° ($\frac{1}{2}$ turn), and 360° (one full turn). Ask which rotation results in the tracing matching exactly or *coinciding* with the original X. [180° turn] Explain that X has 180° rotation symmetry. Repeat the activity with the symbols + [90°] and ☆ [72°].

Setting Up Lesson 13-7

Materials Students will need protractors and rulers marked in inches and centimeters or they can use their **Geometry Templates** for the In-class Activity on page 806.

In-class Activity

Resources

From the ***Teacher's Resource File***
- Answer Master 13-7

Additional Resources
- Protractors and rulers marked in centimeters and inches or **Geometry Templates**

Tell students to measure carefully and to use their calculators to find the ratios. Students should notice that the greater the angle, the greater the ratio of the side opposite it to the side adjacent to it. Students will use their results in **Questions 2–4** on page 810.

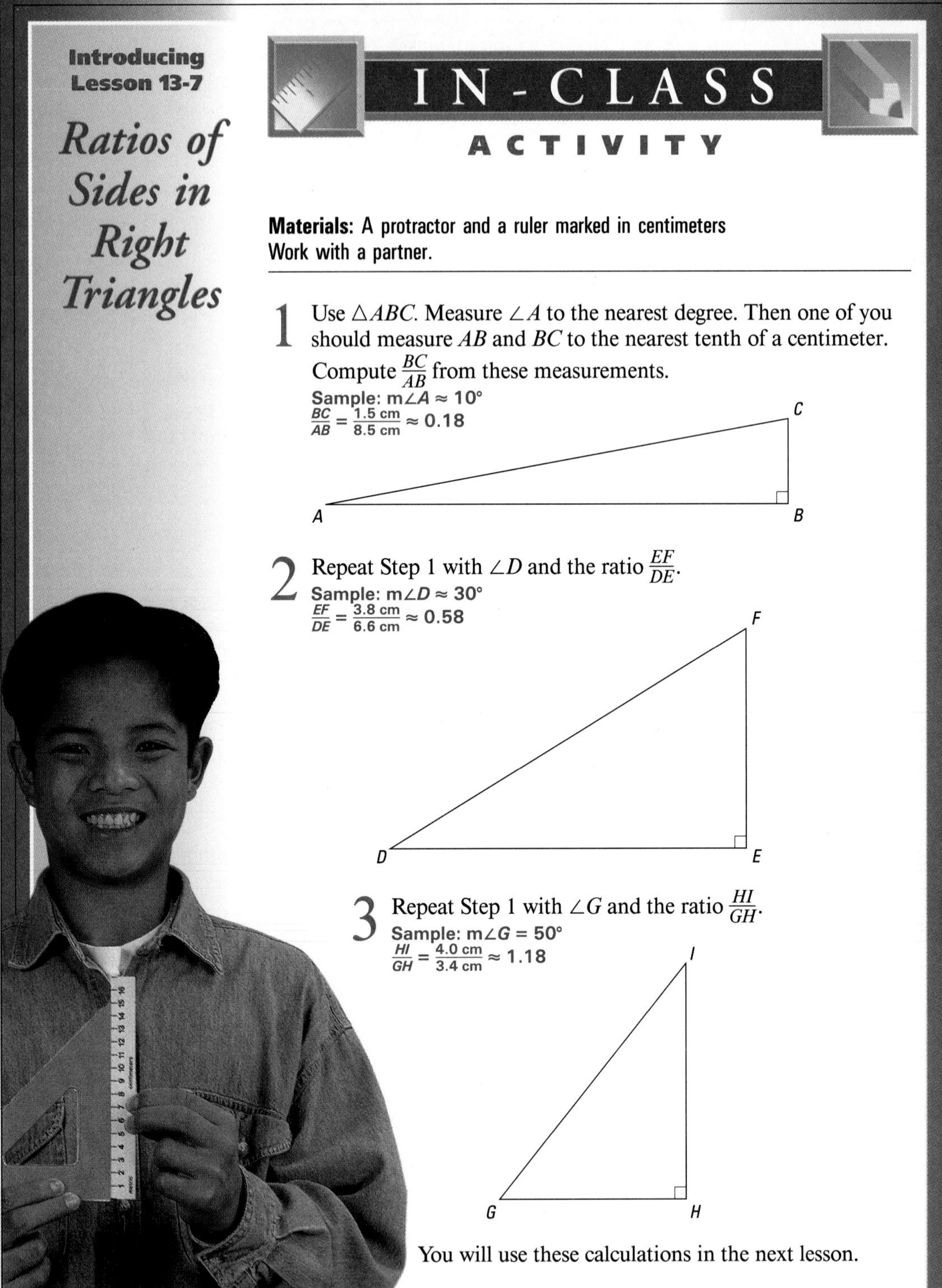

Introducing Lesson 13-7

Ratios of Sides in Right Triangles

IN-CLASS ACTIVITY

Materials: A protractor and a ruler marked in centimeters
Work with a partner.

1 Use $\triangle ABC$. Measure $\angle A$ to the nearest degree. Then one of you should measure AB and BC to the nearest tenth of a centimeter. Compute $\frac{BC}{AB}$ from these measurements.
Sample: $m\angle A \approx 10°$
$\frac{BC}{AB} = \frac{1.5 \text{ cm}}{8.5 \text{ cm}} \approx 0.18$

2 Repeat Step 1 with $\angle D$ and the ratio $\frac{EF}{DE}$.
Sample: $m\angle D \approx 30°$
$\frac{EF}{DE} = \frac{3.8 \text{ cm}}{6.6 \text{ cm}} \approx 0.58$

3 Repeat Step 1 with $\angle G$ and the ratio $\frac{HI}{GH}$.
Sample: $m\angle G = 50°$
$\frac{HI}{GH} = \frac{4.0 \text{ cm}}{3.4 \text{ cm}} \approx 1.18$

You will use these calculations in the next lesson.

LESSON

13-7 The Tangent Function

Tree top. *To determine the height of this Giant Sequoia tree, a person could use the tangent function. See the Example on pages 808–809.*

Calculators and computers have built-in functions. Sometimes these functions can be accessed by pressing buttons. At other times they may be found by highlighting an item on a menu. In this lesson, we consider the tan key. It gives the values of the **tangent function.** In the Activity preceding this lesson, you calculated three values of this function.

The Tangent of an Angle

Consider right triangle ABC below, with legs of lengths 3 and 4. $\angle C$ is a right angle and is thus a 90° angle. Since the sum of the measures of the three angles of a triangle is 180°, $m\angle A + m\angle B = 90°$.

The tangent of angle A in a right triangle is defined as a particular ratio of legs in the triangle. Here $\overline{BC}$ is the **leg opposite** $\angle A$, and $\overline{AC}$ is the **leg adjacent** to $\angle A$.

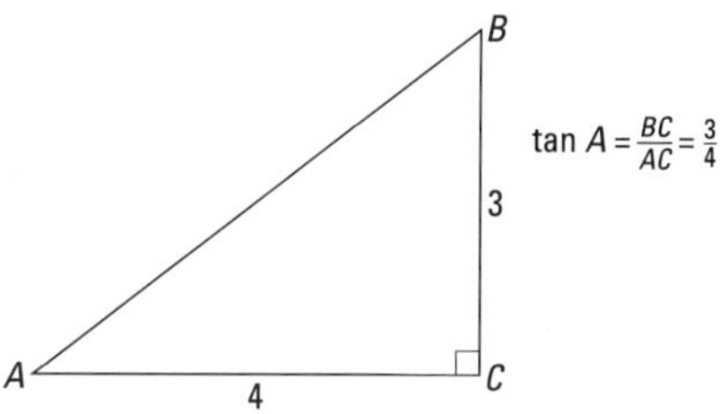

$\tan A = \frac{BC}{AC} = \frac{3}{4}$

The Tangent of an Angle
In a right triangle with acute angle A, the **tangent of $\angle A$**, abbreviated **tan A**, equals the ratio:

$$\frac{\text{length of the leg opposite angle } A}{\text{length of the leg adjacent to angle } A}.$$

Lesson 13-7

Objectives

B Use the tangent key on a calculator.

G Find lengths of sides or tangents of angles in right triangles using the tangent function.

Resources

From the ***Teacher's Resource File***
- Lesson Master 13-7A or 13-7B
- Answer Master 13-7
- Teaching Aids
  133 Warm-up
  140 Activity and Example
  141 Additional Examples

Additional Resources
- Visuals for Teaching Aids 133, 140, 141
- Protractors and rulers, or **Geometry Templates**

Lesson 13-7 Overview

Broad Goals The goal of this lesson is to introduce the tangent function and to present two of its applications: the finding of unknown distances and the relationship it has to the slope of a line.

Perspective A lesson on a trigonometric function is included for a number of reasons. (1) It ties together some of the mathematics that students have encountered in this book. (2) It explains the function of a calculator key. (3) It helps take away some of the fear that students have of trigonometric functions. (4) It makes it far easier for students when they study these functions in a later course.

The tangent of an angle is defined here as a ratio of the legs in a right triangle. If the angle measure is known, then a calculator provides a decimal approximation to the tangent of the angle. If the lengths of the legs are known, the definition can be used to obtain an equation of the form $\tan x = a$. These ideas are applied in problems that involve the finding of unknown heights. It is noted that the tangent of the angle formed by a line and the positive ray of the x-axis is the slope of the line. Thus, this trigonometric idea relates to slope, graphing, angles, ratios, triangles, equations, calculators, and functions.

(Overview continues on page 808.)

Teaching Lesson 13-7

Warm-up

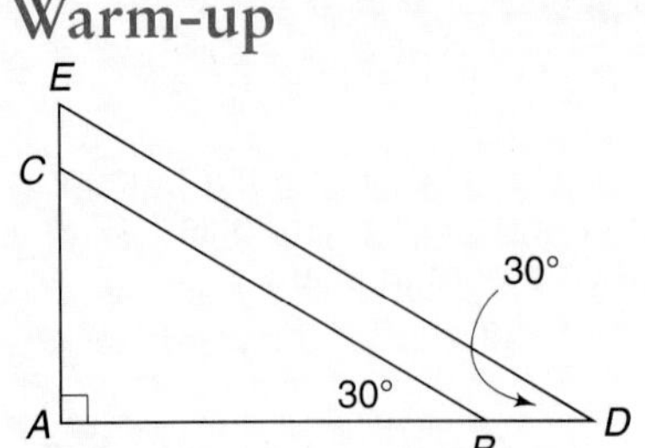

Sample answers are given for the diagram on Teaching Aid 133.

1. Measure the sides of $\triangle ABC$ in millimeters. **48 mm, 80 mm, 94 mm**
2. In $\triangle ABC$ find the ratio of these measures.
$$\frac{\text{leg opposite the } 30° \text{ angle}}{\text{leg adjacent to the } 30° \text{ angle}}$$
$\frac{48}{80} = .6$
3. Repeat Questions 1–2 for $\triangle ADE$. **60 mm, 100 mm, and 117 mm; $\frac{60}{100} = .6$**
4. Find the tangent of 30° on your calculator. Compare it with the ratios you found for $\triangle ABC$ and $\triangle ADE$. **tan 30° ≈ 0.57735 ≈ .6**

Notes on Reading

The graphs for the **Activity** and the **Example** are given on **Teaching Aid 140.**

To convince students that tan *A* depends only on the measure of $\angle A$ and not on the size of the triangle, calculate the tangent of corresponding angles in similar triangles. Relate the example that opens the lesson to triangles with sides 6, 8, and 10 or other multiples of 3, 4, and 5, to the *Warm-up,* or to the In-class Activity on page 806.

Most scientific calculators have a tan key. This key gives values of the tangent of $\angle A$ when you specify the measure of $\angle A$. Make sure your calculator is set to handle degrees when you do this lesson.

The key sequence that finds tangents varies by calculator. Here are some possible sequences for tan 30°.

30 tan,   tan 30,   30 tan =,   tan 30 =

Your calculator may use enter for =. You should see the display 0.5773503 or something close. If you see -6.4053312, then your calculator is not set to degrees, but is using another unit, the *radian.* If you see 0.5095254, then your calculator is using a third unit, called the *grad.*

Activity

Determine the key sequence that will give you tan 30° on your calculator. The values you obtained in the In-class Activity should be in agreement with this table of some values of the tangent function, rounded to three decimal places. (All the tangents in this table are irrational, so none is a finite or repeating decimal.) Below is a graph of this function for values of *x* between 0° and 90°.

x	tan *x*
10°	0.176
20°	0.364
30°	0.577
40°	0.839
50°	1.192
60°	1.732
70°	2.747
80°	5.671

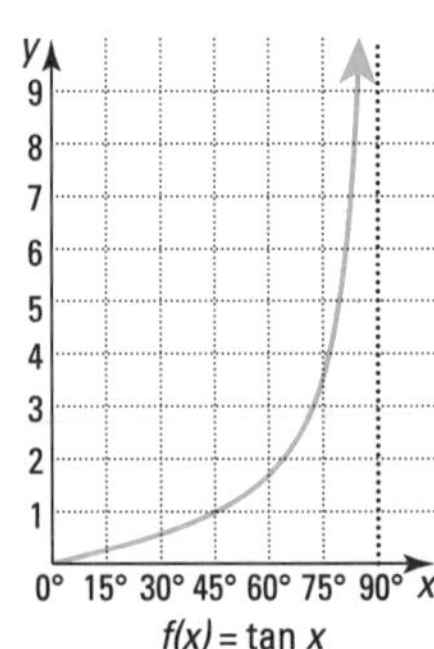

$f(x) = \tan x$

Using the Tangent Function

The tangent function can be used to estimate inaccessible heights.

Example

Nancy wants to estimate the height of a tree. As shown at the left she has to look up 50° to see the top of a tree 5 meters away. If her eyes are 1.5 meters above the ground, about how tall is the tree?

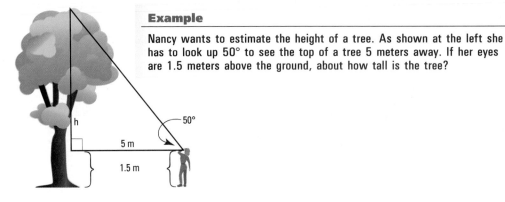

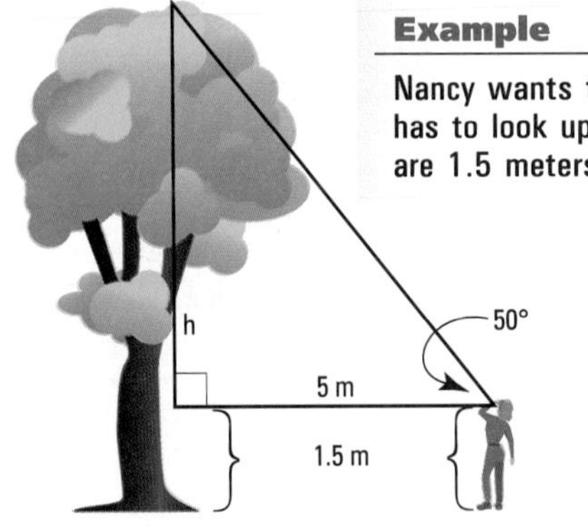

Lesson 13-7 Overview, continued

When doing the reading and the questions, students work in all four dimensions of understanding: Skills (calculating tangents and solving equations); Properties (the relation between measures of the two acute angles in a right triangle); Uses (the use of tangents to calculate unknown distances as in the Example); and Representations (the line as a representation of a linear equation).

Optional Activities

Materials: Protractors and rulers, or **Geometry Templates**

After discussing the lesson, you can use this activity which relates the tangent of an angle and the tangent to a circle. Have students do the following steps:

1. Draw circle *A* with a radius of 1 unit.
2. Draw an acute central $\angle A$. Extend the rays beyond the circle.
3. At the intersection of one ray and the circle (point *B*), draw a line perpendicular to that ray. This line is a tangent to the circle—a tangent is a line that intersects a circle in exactly one point. Label the point at which the tangent intersects the other ray point *C*.
4. Measure the length of $\overline{BC}$. This is the tangent of $\angle A$ since *AB* is 1.
5. Verify that the answer in step 4 is the tangent by measuring $\angle A$ and using a calculator to find the tangent.

Solution

If h is the height of the tree above eye level, then the height of the tree is $h + 1.5$. First use the triangle to find h and then add 1.5. To find h, note that

$$\tan 50° = \frac{h}{5}.$$

Use your calculator to evaluate tan 50°.

$$\tan 50° \approx 1.19$$

Substitute. $$\frac{h}{5} \approx 1.19$$

$$h \approx 5 \cdot 1.19$$

$$h \approx 6.0$$

So the full height of the tree is about
$h + 1.5 \approx 6.0 + 1.5 = 7.5$ m.

You may not realize it, but you have already calculated tangents on the coordinate plane. Tangents of angles are related to equations of lines.

Consider the line $y = \frac{2}{3}x - 2$. This is a line with a slope of $\frac{2}{3}$. It crosses the x-axis at (3, 0) and also passes through $\left(4, \frac{2}{3}\right)$.

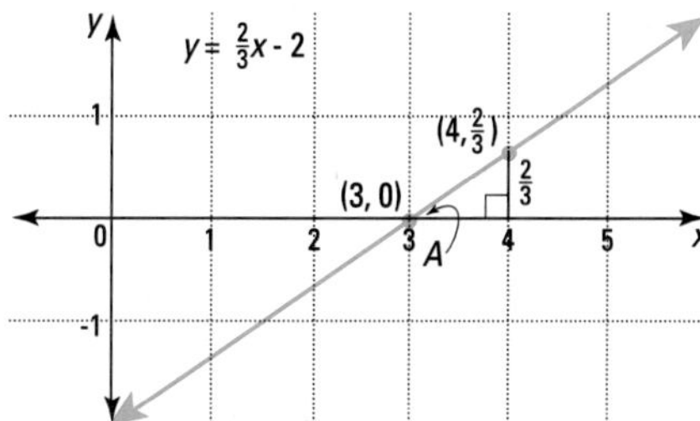

Consider the acute angle formed by the line and the positive ray of the x-axis. From the graph you can see that

$$\tan A = \frac{\frac{2}{3}}{1} = \frac{2}{3}$$

This is the slope of the line!

> If A is the acute angle formed by the upper half of the oblique line $y = mx + b$ and the positive ray of the x-axis, then
> $$\tan A = m.$$

The tangent function is quite a function; it combines the concepts of slope, graphing, angles, ratios, and triangles. In later courses you will study the tangent function with a larger domain and learn of other applications.

We do not give questions that require students to find A directly, given tan A. This would involve use of the $\tan^{-1}$ function, which we do not want to introduce at this time. However, we do ask students to approximate the value of A in **Questions 11, 13, and 15.**

Additional Examples

These examples are also given on **Teaching Aid 141.**

1. In $\triangle PQR$, find tan P. $\approx$ **0.533**

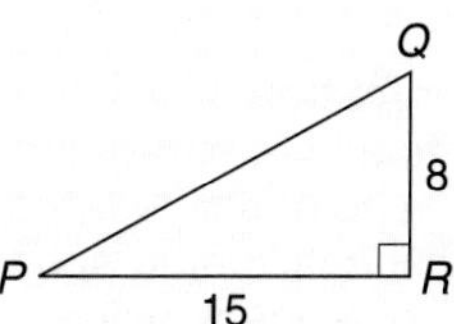

2. When Betsy and her brother Bob are standing 4 feet apart, Betsy must look up at an angle of 36° to see the top of his head. If her eyes are 40 inches above the ground, about how tall is Bob? $\approx$ **75 inches or 6 feet 3 inches**

3. Suppose an airplane takes off at a 15° angle. After it has traveled 600 feet horizontally, how far is the plane above the ground? $\approx$ **161 feet**

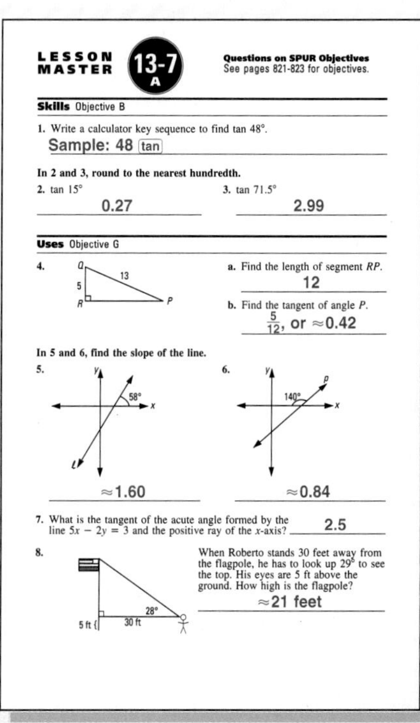

LESSON MASTER 13-7 A

Questions on SPUR Objectives
See pages 821-823 for objectives.

Skills Objective B

1. Write a calculator key sequence to find tan 48°.
Sample: 48 [tan]

In 2 and 3, round to the nearest hundredth.

2. tan 15° 0.27

3. tan 71.5° 2.99

Uses Objective G

4. a. Find the length of segment RP. 12

b. Find the tangent of angle P. $\frac{5}{12}$, or ≈ 0.42

In 5 and 6, find the slope of the line.

5. ≈ 1.60

6. ≈ 0.84

7. What is the tangent of the acute angle formed by the line $5x - 2y = 3$ and the positive ray of the x-axis? 2.5

8. When Roberto stands 30 feet away from the flagpole, he has to look up 29° to see the top. His eyes are 5 ft above the ground. How high is the flagpole? $\approx$21 feet

Sample: $m\angle A = 50°$;
$BC = 1.2$ cm;
$\tan A = 1.2$.

Adapting to Individual Needs

English Language Development

Review the terms *adjacent* and *opposite*. Name several numbers on the clock face and for each number, have students identify the two adjacent numbers and the opposite number. Then draw a right triangle XYZ on the board. Ask which sides are the legs and which side is the hypotenuse. Have students identify both the leg that is adjacent to and the leg that is opposite one of the acute angles.

Follow-up for Lesson 13-7

Practice

For more questions on SPUR Objectives, use **Lesson Master 13-7A** (shown on page 809) or **Lesson Master 13-7B** (shown on pages 810–811).

Assessment

Written Communication Have each student write a brief paragraph summarizing what he or she learned in this lesson. [Students' paragraphs include definition of tangent, using the tangent key on a calculator, and the relationship between the slope of a line and the tangent of the angle formed by the line and the positive ray of the *x*-axis.]

Extension

✎ **Writing** Tell students to refer to the graph of the tangent function on page 808. Have them explain what happens to the tangent values between 80° and 90° and why the tangent cannot be calculated for 90°. [Tan 80° ≈ 5.671, and tan 89° ≈ 57.290; between 89° and 90° the tangent becomes infinitely large—tan 89.9° ≈ 572.957, tan 89.99° ≈ 5729.578, and so on. Sample explanation: The tangent values become increasingly greater because, as the measure of the angle approaches 90°, the length of the opposite side

LESSON MASTER 13-7 B Questions on SPUR Objectives

Vocabulary

1. Use right triangle *RST* at the right.
   a. Name the leg adjacent to ∠*T*. $\overline{TS}$
   b. Name the leg opposite ∠*T*. $\overline{RS}$
   c. What ratio equals tan *T*? $\frac{RS}{TS}$
   d. Name the leg adjacent to ∠*R*. $\overline{RS}$
   e. Name the leg opposite ∠*R*. $\overline{TS}$
   f. What ratio equals tan *R*? $\frac{TS}{RS}$

R, T, S

2. Write a key sequence to find tan 72° on your calculator.
   Sample: 72 [tan]

Skills Objective B: Use the tangent key on a calculator.

In 3–6, round to the nearest hundredth.

3. tan 18° 0.32
4. tan 84° 9.51
5. tan 58° 1.60
6. tan 22.5° 0.41

Uses Objective G: Find lengths of sides or tangents of angles in right triangles using the tangent function.

7. Use right triangle *XYZ* at the right.
   a. Measure ∠*Y*. m ∠ y = 38°
   b. Measure the legs.
      xy = 3.2 cm ($1\frac{1}{4}$ in.);
      xz = 2.4 cm ($\frac{15}{16}$ in.)
   c. Compute tan *Y* by dividing the lengths given in Part **b**. Round to the nearest hundredth. 0.75
   d. On your calculator find tan *Y* to the nearest hundredth. 0.78
   e. How close are the values you found in Parts **c** and **d**? 0.03

Z, X, Y

QUESTIONS

Covering the Reading

1. Use right triangle *DEF* at the right.
   a. Name the side opposite ∠*E*. $\overline{DF}$
   b. Name the side adjacent to ∠*E*. $\overline{EF}$
   c. What ratio equals tan *E*? $\frac{DF}{EF}$

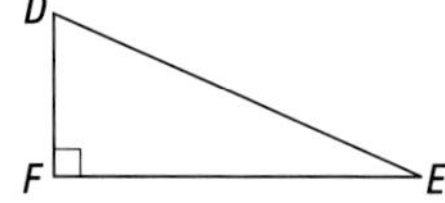

In 2–4, give the values you found from the In-class Activity on page 806.

2. m∠*A* and tan *A* ≈ 10°; ≈ .18
3. m∠*D* and tan *D* ≈ 30°; ≈ 0.58
4. m∠*G* and tan *G* ≈ 50°; ≈ 1.19

In 5 and 6, use your calculator. Give a three-place decimal approximation.

5. tan 57° 1.540
6. tan 3° 0.052
7. Refer to △*KLM* at the right. Find tan *K*. $0.5\overline{3}$

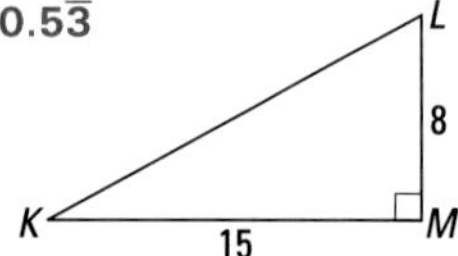

8. a. Measure angle *F* on page 806. a) m ∠*F* = 60° b) 1.74
   b. Estimate tan *F* by dividing the lengths of two sides of △*DEF*.
   c. Estimate tan *F* by using a calculator. tan 60° ≈ 1.73
   d. How close are the values you found in parts **b** and **c**? The difference is only 0.01.
9. Lester had to look up 65° to see the top of a tree 6 meters away. If his eyes are 1.7 meters above the ground, how tall is the tree? ≈ 14.6 m
10. What is the relationship between the tangent function and the slope of a line $y = mx + b$? If *A* is the angle formed by the upper half of the oblique line $y = mx + b$ and the positive ray of the x-axis, then tan *A* = *m*.
11. Consider the line with equation $y = \frac{6}{5}x + 12$.
    a. What is the tangent of the acute angle formed by this line and the positive ray of the *x*-axis? $\frac{6}{5} = 1.2$
    b. Use the table on page 808 to give the approximate measure of that angle. ≈ 50°

Applying the Mathematics

12. Refer to △*ABC* at the right.
    a. Find *AC*. 24
    b. Find tan *A*. ≈ .29
    c. Find tan *B*. ≈ 3.43

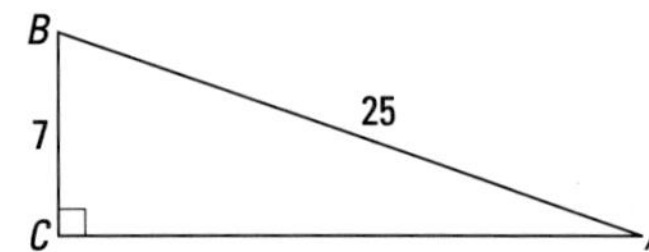

810

Adapting to Individual Needs

Extra Help

Students sometimes think that the tangent of an angle depends on the size of the triangle. Show students the triangles at the right. Point out that tan *A* for △*ABC* is $\frac{3}{4}$ or .75. For △*ADE*, tan $A = \frac{6}{8}$ or .75. Also remind students that the ratios of corresponding sides of similar triangles are equal. Since △*ABD* and △*ADE* are similar, the tangent ratios are the same.

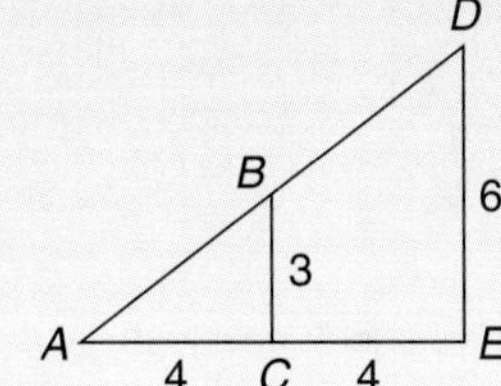

13. A meter stick casts a shadow 0.6 meter long. Use the table in this lesson to estimate the measure of the angle at which the sun appears above the horizon, to the nearest 10°. (This is called the *angle of elevation* of the sun.) **≈ 60°**

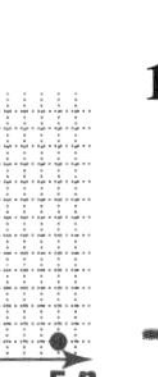

14. A line goes through the origin and the upper half makes an angle of 140° with the positive ray of the x-axis.
   a. Find the slope of this line. **≈ -0.84**
   b. Find an equation for this line. **$y = -0.84x$**

15. To the nearest 10°, find the measure of the angle formed by the upper half of the line $y = 4x - 3$ and the positive ray of the x-axis. Use the table on page 808. **80°**

16c)

Review

16. In the spinner at the left, the two diameters are perpendicular and the central angle of sector 4 has a measure of 60°.
   a. What is the measure of the angle in sector 5? **30°**
   b. Give P(landing in sector n) for n = 1, 2, 3, 4, and 5.
   c. Graph the probability function P. *(Lessons 13-5, 6-6)* **See left.**
   b) $P(1) = \frac{1}{4}$, $P(2) = \frac{1}{4}$, $P(3) = \frac{1}{4}$, $P(4) = \frac{1}{6}$, $P(5) = \frac{1}{12}$

17. What number is not in the domain of the function f if $f(x) = \frac{x - 1}{2x + 4}$? *(Lesson 13-4)* **-2**

18. **a.** If $y = 3(x + 1)^2 - 4$, what is the smallest possible value of y? **-4**
   b. If $y = 3|x + 1| - 4$, what is the smallest possible value of y? **-4**
   c. If $y = 3\sqrt{x + 1} - 4$, what is the smallest possible value of y? **-4**
   (Lessons 13-3, 13-2)

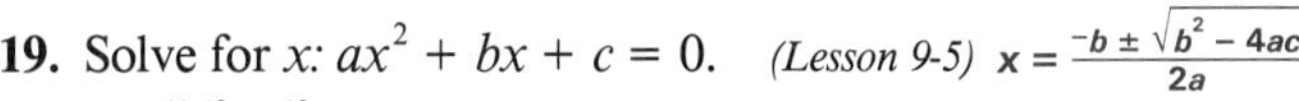

19. Solve for x: $ax^2 + bx + c = 0$. *(Lesson 9-5)* **$x = \frac{-b \pm \sqrt{b^2 - 4ac}}{2a}$**

20. If $\frac{(a^{11})^{12} \cdot a^{13}}{a^{14}} = a^t$, what is the value of t? *(Lessons 8-7, 8-5)* **131**

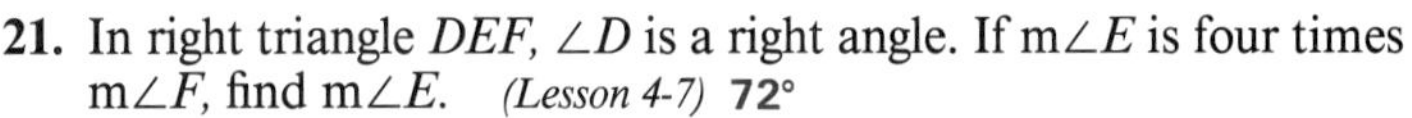

21. In right triangle DEF, $\angle D$ is a right angle. If $m\angle E$ is four times $m\angle F$, find $m\angle E$. *(Lesson 4-7)* **72°**

Las Gatas Beach at Aihuatanejo, Mexico, on the Gulf of Mexico.

Exploration

22. When $x = \frac{\pi A}{180°}$, $\tan A \approx \frac{2x^5 + 5x^3 + 15x}{15}$.
   a. Let A = 10°. How close is the polynomial approximation to the calculator value of tan A? **≈ .0000003 difference**
   b. Repeat part **a** when A = 70°. **≈ .555 difference**

23. What became of the man who sat on a beach along the Gulf of Mexico? **He became a tangent.**

becomes greater and greater, and the length of the adjacent side approaches 0. At 90°, the length of the adjacent side is 0, so the ratio of the length of the opposite side to the length of the adjacent side is $\frac{x}{0}$, which is undefined.]

Notes on Questions

Questions 2–4, 8 Answers to these questions may vary depending upon how students measure the segments and what tools they use.

Questions 5–6 Error Alert If students have made errors, have them check that their calculators are set to degrees and not to radians or grads.

Question 9 Encourage students to make drawings as the first step. As with the **Example** in the lesson, the most useful drawings are side-views of the situation.

Question 12 Students must use the Pythagorean Theorem to find the length of $\overline{AC}$.

Question 18 Encourage students to graph these equations. You might suggest that they call the functions $a(x)$, $b(x)$, and $c(x)$ to correspond to the parts of the question, or use $s(x)$, $a(x)$, and $r(x)$ for square, absolute value, and root.

Question 22 The polynomial approximation is closest when x is near zero.

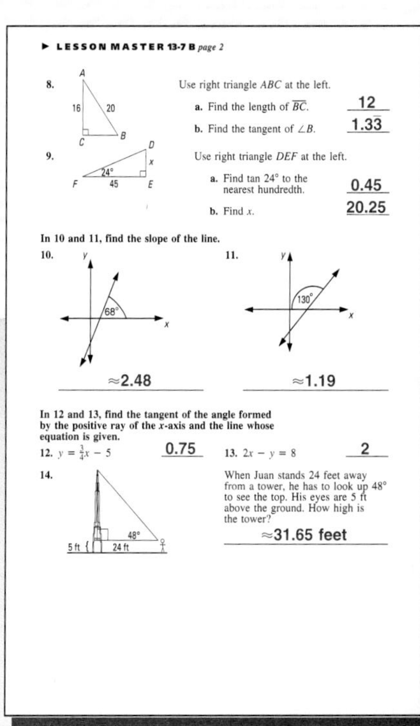

► **LESSON MASTER 13-7 B** *page 2*

8. Use right triangle ABC at the left.
   a. Find the length of $\overline{BC}$. **12**
   b. Find the tangent of $\angle B$. **$1.3\overline{3}$**

9. Use right triangle DEF at the left.
   a. Find tan 24° to the nearest hundredth. **0.45**
   b. Find x. **20.25**

In 10 and 11, find the slope of the line.

10. **≈2.48**

11. **≈1.19**

In 12 and 13, find the tangent of the angle formed by the positive ray of the x-axis and the line whose equation is given.

12. $y = \frac{3}{4}x - 5$ **0.75**

13. $2x - y = 8$ **2**

14. When Juan stands 24 feet away from a tower, he has to look up 48° to see the top. His eyes are 5 ft above the ground. How high is the tower? **≈31.65 feet**

Adapting to Individual Needs

Challenge

Give students the following problems.

1. Use a reference book to look up the meaning of the *cotangent* of an angle. [The cotangent of $\angle A$ is the reciprocal of the tangent of $\angle A$.]
2. Use the definition to find the cotangent of $\angle A$ and the cotangent of $\angle B$ in **Exercise 12**. [Cotangent $A = \frac{24}{7}$; cotangent $B = \frac{7}{24}$.]
3. Explain how you could use a calculator without a cotangent key to find the cotangent of an angle. [Use the tangent key to find the tangent. Then use the reciprocal key to find the cotangent value.]

Lesson 13-8

Objectives

B Use function keys on a calculator.

Resources

From the ***Teacher's Resource File***
- Lesson Master 13-8A or 13-8B
- Answer Master 13-8
- Teaching Aids
  133 Warm-up
  142 Graphs of $y = \sin x$ and $y = \log x$
- Activity Kit, Activity 30

Additional Resources
- Visuals for Teaching Aids 133, 142
- GraphExplorer or other automatic graphers

Teaching Lesson 13-8

Warm-up

Writing Make a list of at least 5 keys on your calculator that you have used this year other than the number keys or basic operation keys. Give an example of how each key is used. **Responses will vary.**

Notes on Reading

We suggest that you keep the following statement in mind as you teach this lesson—the intent of the lesson is to give an extremely brief look at other functions that are available on a calculator. Do not expect students to do more than evaluate these functions using calculators or graphs.

LESSON

13-8 *Functions on Calculators and Computers*

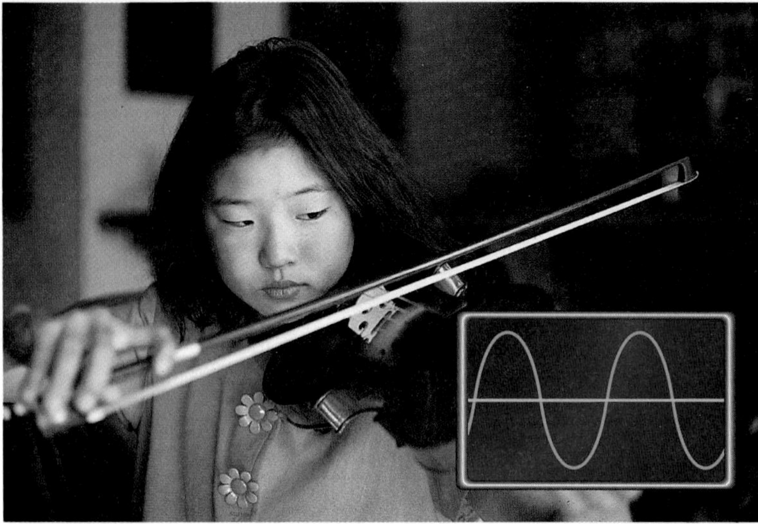

Purely musical. *The music played by a violinist or other musician is based on combinations of pure tones. The graph of a pure tone is a sine wave, like the one shown here.*

Some Familiar Calculator Functions

❶ In Lesson 13-7, you studied a function whose values are given or estimated by the tan key on a calculator. Other functions are preprogrammed into calculators. They will give you function values when you enter a value from their domains. The calculator will indicate an error message when you enter a value not in the domain. Here are some familiar keys, the functions they define, and their domains.

Key	Function	Domain	Error if:
the square root key $\sqrt{\ }$	$\text{SQR}(x) = \sqrt{x}$	set of nonnegative reals	$x < 0$
the factorial function key !	$\text{FACT}(n) = n!$	set of nonnegative integers	x is not an integer or $x < 0$
reciprocal function key 1/x	$f(x) = \frac{1}{x}$	set of nonzero reals	$x = 0$
the squaring key x^2	$s(x) = x^2$	set of all reals	none

You have learned some applications of these functions earlier in this book. In this lesson, we introduce you to the meaning of some of the other function keys on a calculator. These keys represent functions that are built into virtually every computer language as well.

Lesson 13-8 Overview

Broad Goals The goal of the last lesson in this book is to give students a glimpse of the kinds of functions that they will encounter in later courses.

Perspective Having just been introduced to a new key on their calculators, students are naturally curious about the other keys. Here we give a very brief explanation of the roles of some of those keys and emphasize the importance of the functions. As in the preceding lesson, we want to stimulate curiosity and lessen the fears that students may have of the keys that are mysterious to them.

The idea of domain is also important when using a calculator or computer. An error message will result if a person tries to use a calculator to evaluate a function for a value outside its domain. This provides a good reason for having introduced the idea.

Trigonometric Functions

Two functions related to the tangent function are the **sine** and **cosine functions,** whose values are found by the sin and cos keys on your calculator. Like the tangent function value, these values are ratios of sides in right triangles. Again, consider a right triangle *ABC*. The sine and the cosine of $\angle A$ are as follows.

$$\sin A = \frac{\text{length of the leg opposite angle } A}{\text{length of the hypotenuse}}$$

$$\cos A = \frac{\text{length of leg adjacent to angle } A}{\text{length of the hypotenuse}}$$

So in this case, $\sin A = \frac{2}{\sqrt{13}}$ and $\cos A = \frac{3}{\sqrt{13}}$. For $\angle B$, the values of sine and cosine are interchanged: $\sin B = \frac{3}{\sqrt{13}}$ and $\cos B = \frac{2}{\sqrt{13}}$.

It is possible to define these functions and the tangent function for angles of any measure. The graph of the sine or cosine function may surprise you. Below is the graph of $y = \sin x$.

❷

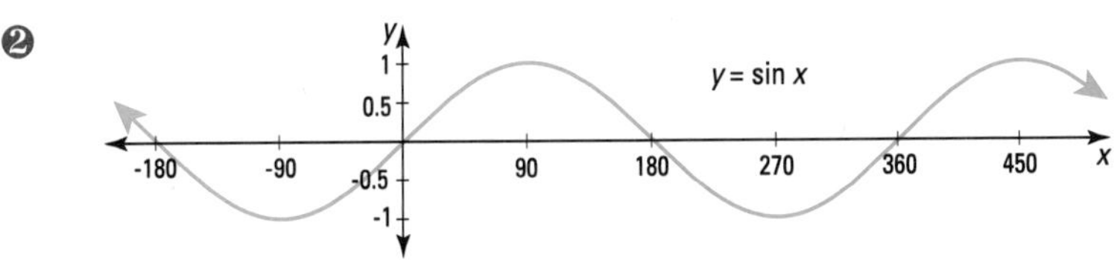

The curve is called a *sinusoidal curve,* or *sinusoid,* and it has the same shape as sound waves and radio waves. The sine, cosine, and tangent functions are part of a branch of mathematics called **trigonometry.** These functions are so important that some high schools offer a full course in trigonometry devoted to studying them and their applications. In BASIC, values of the tangent, sine, and cosine functions are denoted by TAN(x), SIN(x), and COS(x), respectively.

The Common Logarithm Function

Another function built into almost all scientific calculators is the **common logarithm function,** activated by the log key. This key defines a function $y = \log x$, read "*y* equals the common logarithm of *x*." The common logarithm of a number is the power to which 10 must be raised to equal that number. So, since 1 million $= 10^6$, $\log(1000000) = 6$. Logarithms provide a way to deal easily with very large or very small numbers. A graph of the common logarithm function is given on the next page.

❶ Remind students that error messages do not always arise from attempting to do a function calculation with a number that is not in the domain. An error message often signals that a number is too large or too small for the calculator to handle. At other times an error message means that a mistake has been made in entering, such as using the (−) key where the (+/−) key is needed.

❷ The graphs of $y = \sin x$ and $y = \log x$ are given on **Teaching Aid 142.**

Optional Activities

Activity 1

You might use this activity after students have completed the lesson. Have them use calculators to find the pairs of values to the nearest thousandth. Then have them make a generalization. [$\sin A = \cos (90 - A)$]

1. sin 30° and cos 60° [0.5, 0.5]
2. sin 28° and cos 62° [0.469, 0.469]
3. sin 10° and cos 80° [0.174, 0.174]
4. sin 1° and cos 89° [0.017, 0.017]

Activity 2

You can use *Activity Kit, Activity 30,* as a follow-up to the lesson. This activity introduces students to the unit circle as studied in future work in trigonometry.

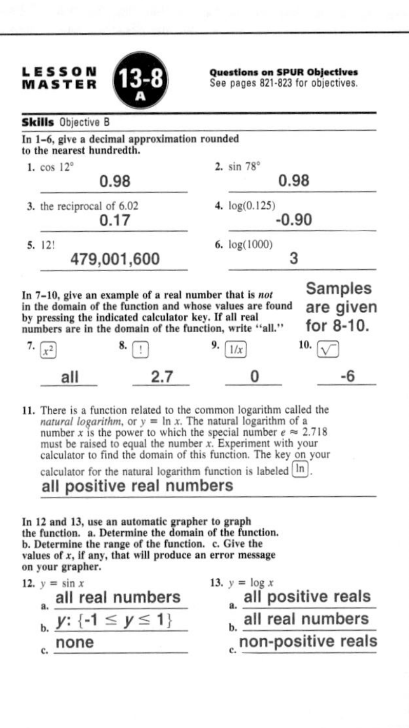
LESSON MASTER 13-8 A

Questions on SPUR Objectives
See pages 821-823 for objectives.

Skills Objective B

In 1–6, give a decimal approximation rounded to the nearest hundredth.

1. cos 12° 0.98
2. sin 78° 0.98
3. the reciprocal of 6.02 0.17
4. log(0.125) -0.90
5. 12! 479,001,600
6. log(1000) 3

In 7–10, give an example of a real number that is *not* in the domain of the function and whose values are found by pressing the indicated calculator key. If all real numbers are in the domain of the function, write "all." Samples are given for 8-10.

7. x^2 all
8. ! 2.7
9. 1/x 0
10. √ -6

11. There is a function related to the common logarithm called the *natural logarithm*, or $y = \ln x$. The natural logarithm of a number *x* is the power to which the special number $e \approx 2.718$ must be raised to equal the number *x*. Experiment with your calculator to find the domain of this function. The key on your calculator for the natural logarithm function is labeled ln. all positive real numbers

In 12 and 13, use an automatic grapher to graph the function. a. Determine the domain of the function. b. Determine the range of the function. c. Give the values of *x*, if any, that will produce an error message on your grapher.

12. $y = \sin x$
a. all real numbers
b. y: $\{-1 \le y \le 1\}$
c. none

13. $y = \log x$
a. all positive reals
b. all real numbers
c. non-positive reals

Additional Examples

In 1–3, use these function keys on your calculator:

[√], [!], [1/x], and [x^2].

1. Which of these keys produce an error messages when -9 is entered before them? [√], [!]
2. Which keys produce an error message when 14.1 is entered before them? [!]
3. Which function key gives an error message when 0 is entered before them? [1/x]

In 4–7, find each value using a calculator.

4. sin 0° 0
5. cos 1° ≈ .9998 . . .
6. cos 180° -1
7. log 100 2
8. Use an automatic grapher to graph $y = \cos x + \sin x$ for values of x between -360° and 360°. Describe the shape of the graph you see. **Sample: a wave with a range from about -1.4 to 1.4**

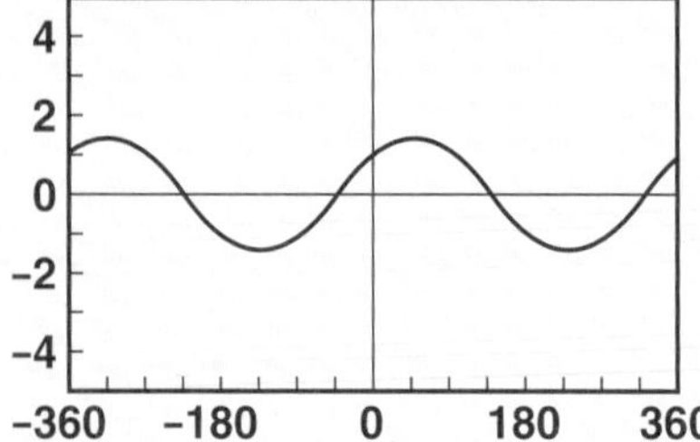

LESSON MASTER 13-8 B Questions on SPUR Objectives

Vocabulary

1. Use right triangle *RST* at the right.
   a. What ratio equals sin *T*? $\frac{RS}{RT}$
   b. What ratio equals cos *T*? $\frac{ST}{RT}$

Skills Objective B: Use function keys on a calculator.

In 2–9, a. write a key sequence to find each value. b. Write what is shown in the display. Sample key sequences are given.

2. sin 16°
   a. 16 [sin]
   b. 0.2756374
3. cos 8°
   a. 8 [cos]
   b. 0.9902681
4. $(-2.088)^2$
   a. 2.088 [+/–] [x^2]
   b. 4.359744
5. $\sqrt{0.008}$
   a. 0.008 [$\sqrt{x}$]
   b. 0.0894427
6. log(43)
   a. 43 [log]
   b. 1.6334685
7. log (10^6)
   a. 10 [y^x] 6 [=] [log]
   b. 6
8. 8!
   a. 8 [x!]
   b. 40320
9. $\frac{1}{3.44}$
   a. 3.44 [$\frac{1}{x}$]
   b. 0.2906977

The part of this graph above the x-axis pictures the kind of growth often found in learning. At first, one learns an idea quickly, so the curve increases quickly. But after a while it is more difficult to improve one's performance, so the curve increases more slowly.

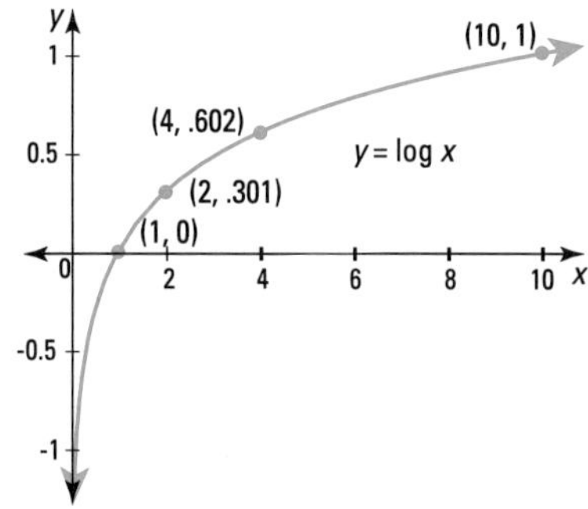

Learning curve. *When learning ballet techniques, the beginner often masters the basic movements quickly. As the dancer progresses, the techniques usually take longer to master, as suggested by the graph at the right.*

Most scientific calculators and computers have other built-in functions. These functions would not be there unless many people needed to get values of that function. Calculators and computers have made it possible for people to obtain values of these functions more easily than most people ever imagined. The algebra that you have studied this year gives you the background to understand these functions and to deal with them.

QUESTIONS

Covering the Reading

In 1–6, use a calculator to approximate each value to the nearest thousandth.

1. tan 11° **0.194**
2. sin 45° **0.707**
3. cos 47° **0.682**
4. $\log(10^7)$ **7**
5. $(-3.489)^2$ **12.173**
6. $\sqrt{0.5}$ **0.707**

In 7–9, consider the [√], [!], [1/x], and [x^2] function keys on your calculator. Which produce error messages when the given number is entered?

7. 3.5 **!**
8. -4 **√, !**
9. 0 **$\frac{1}{x}$**
10. Which function has a graph that is the shape of a sound wave? **sine**
11. Which function has a graph that is sometimes used to model learning? **common logarithm**
12. Find each number related to $\triangle ABC$ at the left.
    a. AC **29**
    b. $\sin A$ **≈ 0.690**
    c. $\cos A$ **≈ 0.724**
    d. $\tan A$ **≈ 0.952**

In 13–15, refer to the graph of $y = \sin x$ in this lesson.
a. Estimate the value from the graph.
b. Use a calculator to check your estimate.

13. sin 90° **a) 1; b) 1**
14. sin 360° **a) 0; b) 0**
15. sin(-70°) **Sample: a) -0.9; b) -0.940**

Adapting to Individual Needs

Extra Help

Some students have difficulty with the trigonometric functions because they confuse the adjacent side of an acute angle of a right triangle with the hypotenuse. You might use colored chalk for the side opposite the angle and then ask students to identify the longest side as the hypotenuse. Explain that the remaining side is the adjacent side.

English Language Development

Have students write the new terms in this lesson on index cards. Tell them to include *tangent, sine, cosine, trigonometry,* and *logarithm.*

16a)

x	cos x
0	1
15	0.97
30	0.87
45	0.71
60	0.50
75	0.26
90	0
105	-0.26
120	-0.50
135	-0.71
150	-0.87
165	-0.97
180	-1
195	-0.97
210	-0.87
225	-0.71
240	-0.50
255	-0.26
270	0
285	0.26
300	0.50
315	0.71
330	0.87
345	0.97
360	1

b)

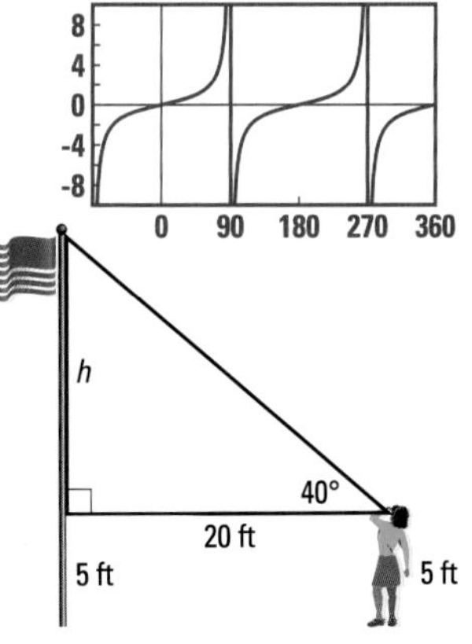

h

40°

20 ft

5 ft

5 ft

Applying the Mathematics

16. a. Make a table of values for $y = \cos x$ for values of x from 0° to 360° in increments of 15°. **See left.**

b. Carefully graph this function. **See left.**

c. What graph in this lesson does the graph of $y = \cos x$ most resemble? **y = sin x**

17. Many computers use the name LOG to refer to a logarithm function different from the one in the lesson. **See margin.**

a. Run this program or use your calculator to determine what is printed.

```
10 PRINT "X", "LOG X"
20 FOR X = 1 TO 10
30 Y = LOG(X)
40 PRINT X, Y
50 NEXT X
60 END
```

b. Graph the ordered pairs that are printed.

c. Is your graph like the one in the lesson, or is it different? If it is different, how does it differ?

18. a. Graph $y = \tan x$ on an automatic grapher with x set for degrees. Use the graph to estimate tan(-10°). **≈ -0.2; See left for graph.**

b. Verify the value of tan(-10°) on your calculator. **≈ -0.18**

19. Graph the reciprocal function $f(x) = \frac{1}{x}$. **See margin.**

20. What curve is the graph of the squaring function, $s(x) = x^2$? **a parabola**

Review

21. In $\triangle ABC$ at the right, find tan A to the nearest tenth. *(Lesson 13-7)* **≈ 0.7**

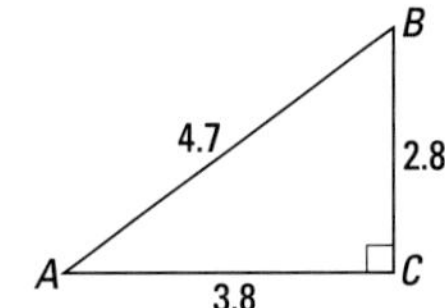

22. Joan had to look up 40° to see the top of a flagpole 20 feet away. The situation is pictured at the left. If her eyes are 5 feet above the ground, how tall is the flagpole? *(Lesson 13-7)* **21.78 ft**

23. a. In a toss of two fair dice, what sum is most likely to appear? **7**

b. What is its probability of occurring? *(Lesson 13-5)* $\frac{1}{6}$

Notes on Questions

Question 9 Explain that 0! is defined as 1 so that patterns which arise in working with factorials are consistent. One such pattern is $(n + 1)! = (n + 1) \cdot n!$. If $n = 0$, then $(0 + 1)! = (0 + 1) \cdot 0!$. So, $1! = 1 \cdot 0!$, from which $0! = 1$.

Additional Answers

17a. The table below left displays common logarithmic values. The table below right shows natural logarithmic values.

X	LOG X	LOG X
1	0	0
2	0.30103	.69315
3	0.47712	1.0986
4	0.60206	1.3863
5	0.69897	1.6094
6	0.77815	1.7918
7	0.84510	1.9459
8	0.90309	2.0794
9	0.95424	2.1972
10	1	2.3026

b. The graph below displays the right-hand data. The graph of the left-hand data is on student page 814.

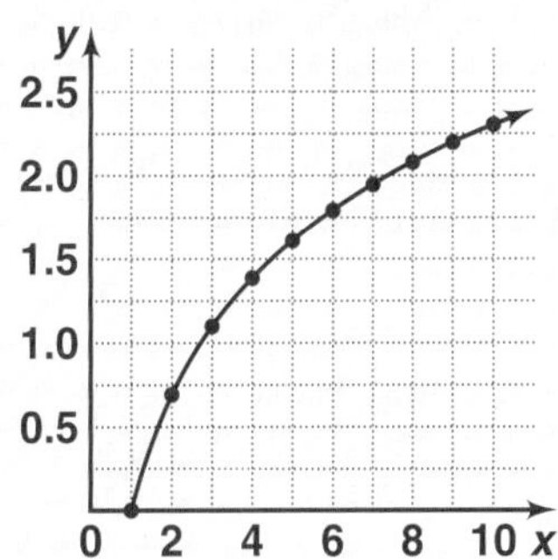

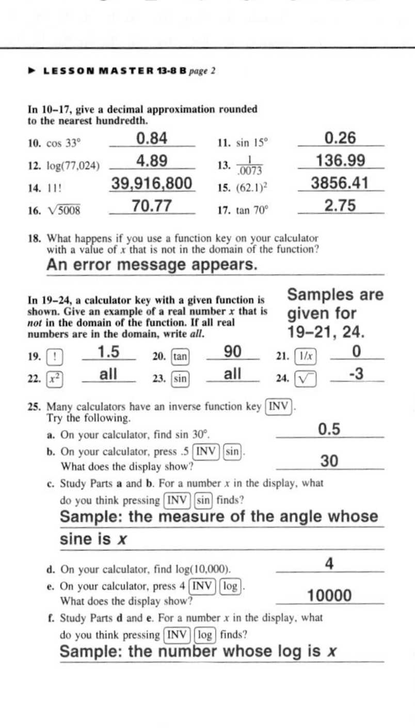

► LESSON MASTER 13-8 B *page 2*

In 10–17, give a decimal approximation rounded to the nearest hundredth.

10. cos 33°	0.84	11. sin 15°	0.26
12. log(77,024)	4.89	13. $\frac{1}{.0073}$	136.99
14. 11!	39,916,800	15. $(62.1)^2$	3856.41
16. $\sqrt{5008}$	70.77	17. tan 70°	2.75

18. What happens if you use a function key on your calculator with a value of x that is not in the domain of the function?
An error message appears.

In 19–24, a calculator key with a given function is shown. Give an example of a real number *x* that is *not* in the domain of the function. If all real numbers are in the domain, write *all*. Samples are given for 19–21, 24.

19. [!] 1.5 20. [tan] 90 21. [1/x] 0

22. [x^2] all 23. [sin] all 24. [√] -3

25. Many calculators have an inverse function key [INV]. Try the following.

a. On your calculator, find sin 30°. 0.5

b. On your calculator, press .5 [INV] [sin]. What does the display show? 30

c. Study Parts **a** and **b**. For a number x in the display, what do you think pressing [INV] [sin] finds?
Sample: the measure of the angle whose sine is x

d. On your calculator, find log(10,000). 4

e. On your calculator, press 4 [INV] [log]. What does the display show? 10000

f. Study Parts **d** and **e**. For a number x in the display, what do you think pressing [INV] [log] finds?
Sample: the number whose log is x

Addtional Answers, continued

17c. The graph of the left-hand data is the same as in the text. The graph of the right-hand data is different from that in the text. As *x* increases, the point (*x*, log *x*) is farther away from the *x*-axis than that in the graph of the left-hand data.

19.

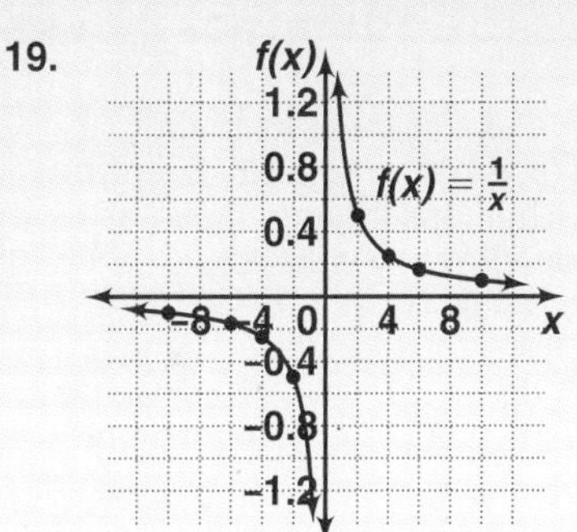

Notes on Questions

Question 28 The reference to Lesson 7-6 suggests that this question can be solved by writing an equation of a line. Another method is to calculate the speed traveled and use that speed to find the time needed to travel 13 km.

Follow-up for Lesson 13-8

Practice

For more questions on SPUR Objectives, use **Lesson Master 13-8A** (shown on page 813) or **Lesson Master 13-8B** (shown on pages 814–815).

Assessment

Written Communication Have each student write the key sequences they use on their calculators to find the square root of a number, the reciprocal of a number, the square of a number, and the factorial of a number. Ask students to include an example of each function. [Students' key sequences are correct, and their examples use numbers that are in the domain of each function.]

Extension

Writing Have students draw right triangle *ABC* with right $\angle C$, and then measure each side. Have them find the sine, cosine, and tangent of $\angle A$. Then have them show that the sine of $\angle A$ divided by the cosine of $\angle A$ equals the tangent of $\angle A$. Finally, have each student write a paragraph showing that this relationship is true for the sine, cosine, and tangent of any angle. [Sample: Suppose the lengths of the sides in a right triangle are *a*, *b*, and *c*; and that *a* is opposite $\angle A$, *b* is opposite $\angle B$, and *c* is the hypotenuse. Then, $\sin A = \frac{a}{c}$, $\cos A = \frac{b}{c}$, and $\tan A = \frac{a}{b}$.

Therefore, $\frac{\sin A}{\cos A} = \frac{\frac{a}{c}}{\frac{b}{c}} = \frac{a}{b}$.]

Project Update Project 4, *Reciprocal Functions*, on page 817, relates to the content of this lesson.

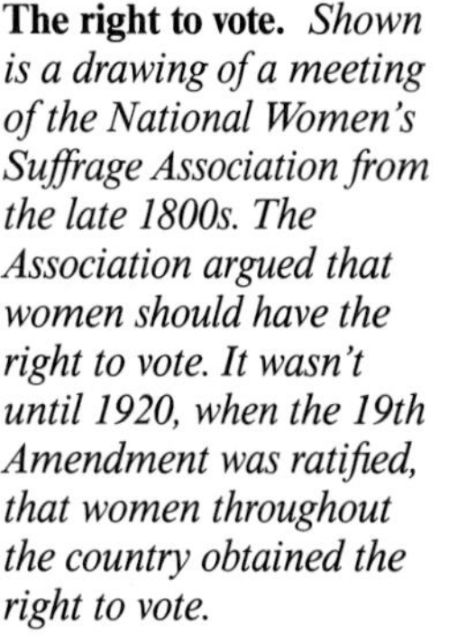

The right to vote. *Shown is a drawing of a meeting of the National Women's Suffrage Association from the late 1800s. The Association argued that women should have the right to vote. It wasn't until 1920, when the 19th Amendment was ratified, that women throughout the country obtained the right to vote.*

In 24 and 25, $S(x) = x + \frac{x^2}{20}$ gives the number of feet a car traveling at x miles per hour will take to stop. $B(x) = \frac{x^2}{20}$ gives the number of feet traveled after brakes are applied.

24. About how many feet does it take to stop at 40 mph? *(Lesson 13-2)* 120 ft

25. If skid marks in an accident are 100 feet long, about how fast was the car traveling? *(Lesson 13-2)* about 44.7 mph

26. A rectangular box has dimensions p by $p + 5$ by $p - 1$. *(Lessons 10-1, 10-5)*
a. Find its volume. $p^3 + 4p^2 - 5p$
b. What is the degree of the polynomial in part **a**? 3

27. Here are the total numbers of votes (to the nearest million) cast for all of the major candidates in the presidential elections since 1940.

Year	All Votes Cast for Major Candidates (millions)	Winner
1940	50	Franklin D. Roosevelt
1944	48	Franklin D. Roosevelt
1948	49	Harry S. Truman
1952	61	Dwight D. Eisenhower
1956	62	Dwight D. Eisenhower
1960	68	John F. Kennedy
1964	70	Lyndon B. Johnson
1968	73	Richard M. Nixon
1972	77	Richard M. Nixon
1976	81	Jimmy Carter
1980	86	Ronald Reagan
1984	92	Ronald Reagan
1988	91	George Bush
1992	104	Bill Clinton

a. Graph the ordered pairs (year, number of votes). See left.
b. Use the graph to predict how many votes will be cast for major candidates in the presidential election of 2000. *(Lessons 7-7, 3-3)* Sample: 115 million

28. The explorers were 13 km from home base at 2 P.M. and 10 km from home base at 3:30 P.M. At this rate, when will they reach home? *(Lesson 7-6)* 8:30 P.M.

29. If you read 17 pages of a 300-page novel in 45 minutes, about how long will it take you to read the entire novel? *(Lesson 6-8)* 794.12 min ≈ 13 hr 14 min

In 30–32, solve. *(Lessons 6-8, 5-6, 1-6)*

30. $\frac{m}{2} = \frac{m + 36}{11}$ $m = 8$

31. $3x + 9 > x$ $x > \frac{-9}{2}$

32. $\sqrt{v - 6} = 4$ $v = 22$

27a)

Number of votes (in millions): 30, 40, 50, 60, 70, 80, 90, 100, 110, 120

Year: 1944, 1952, 1960, 1968, 1976, 1984, 1992, 2000

Exploration

33. List all the function keys of a scientific calculator to which you have access. Separate those you have studied from those you have not. Identify at least one situation in which each function you have studied might be used. Answers will vary.

Adapting to Individual Needs

Challenge

Have students use calculators to find $(\sin A)^2$ and $(\cos A)^2$ for several different values of A. Then ask them to make a generalization. [For any value of A, $(\sin A)^2 + (\cos A)^2 = 1$.]

A project presents an opportunity for you to extend your knowledge of a topic related to the material of this chapter. You should allow more time for a project than you do for typical homework questions.

1 Tossing Three Dice

Imagine tossing three fair dice.

a. What are the possible outcomes for the sum of the three numbers that appear?

b. If each die is fair, what is the probability of each possible sum?

c. What is the most likely sum? What is the least likely sum?

d. Graph the ordered pairs (sum of x, probability of that sum).

e. Describe the graph.

f. What is the probability of an even sum? What is the probability of an odd sum?

2 Polynomial Functions of Degree 4

With an automatic grapher, explore the possible shapes of functions of the form $P(x) = ax^4 + bx^3 + cx^2 + dx + e$, for various values of a, b, c, d, and e. Find and describe as many shapes as you can.

3 Absolute Value Functions

a. Graph $f(x) = 3|x - 5| + 4$.

b. Discuss the effects of the values of a, b, and c on the graphs of the absolute value functions with equations of the form $y = a|x - b| + c$. (Hint: Let $b = 0$ and $c = 0$ to find the effects of a. Let $a = 1$ and $c = 0$ to find the effects of b. Let $a = 1$ and $b = 0$ to find the effects of c.)

4 Reciprocal Functions

When $f(x)$ is the value of a function, then $\frac{1}{f(x)}$ is the corresponding value of its *reciprocal function.*

a. On the same pair of axes, graph the function with equation $y = x$ and its reciprocal, the function with equation $y = \frac{1}{x}$.

b. On another pair of axes, graph the function $y = x^2$ and its reciprocal.

c. On a third pair of axes, graph the function $f(x) = 2^x$ and its reciprocal.

d. Pick two other functions and graph them and their reciprocals.

e. Describe any general patterns that you find.

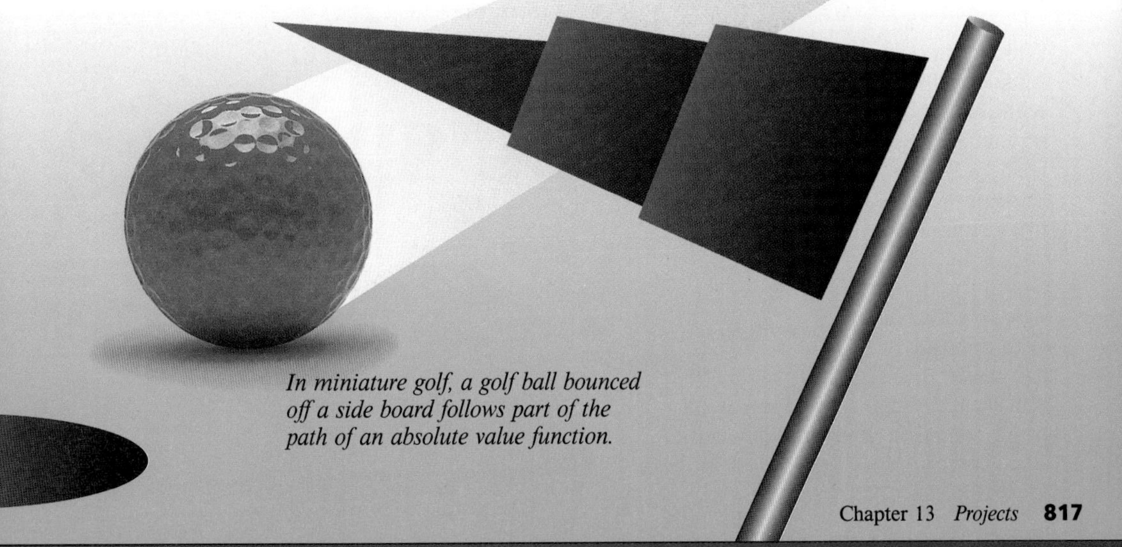

In miniature golf, a golf ball bounced off a side board follows part of the path of an absolute value function.

Possible responses

1. a. Sums of 3 through 18

b. $P(3) = P(18) = \frac{1}{216}$, $P(4) = P(17) = \frac{3}{216}$, $P(5) = P(16) = \frac{6}{216}$, $P(6) = P(15) = \frac{10}{216}$, $P(7) = P(14) = \frac{15}{216}$, $P(8) = P(13) = \frac{21}{216}$, $P(9) = P(12) = \frac{25}{216}$, $P(10) = P(11) = \frac{27}{216}$

c. 10 or 11; 3 or 18

d. Sample graph is on page 818.

e. The graph is symmetric; it raises sharply from $(3, \frac{1}{216})$ to $(10, \frac{27}{216})$, and then falls sharply from $(11, \frac{27}{216})$ to $(18, \frac{1}{216})$.

f. $\frac{1}{2}$, $\frac{1}{2}$

(Responses continue on page 818.)

Chapter 13 Projects

The projects relate chiefly to the content of the lessons of this chapter as follows:

Project	Lesson(s)
1	13-5
2	13-6
3	13-3
4	13-8
5	13-2, 13-6
6	13-4
7	13-2

1 Tossing Three Dice Students can use the Multiplication Counting Principle to find that there is a total of 216 possible outcomes. Suggest that they use the diagram which shows all the possibilities for two fair dice that is found at the beginning of Lesson 13-5. This diagram will help them make an organized list of all possible outcomes for three fair dice.

2 Polynomial Functions of Degree 4 There are many possible shapes for students to explore. Have students classify the shapes they find as opening up or down, by the number of maximum or minimum points, and by the number of times the shape intersects the *x*-axis. Students can find more information about higher-degree polynomials in advanced algebra texts.

3 Absolute Value Functions Students can complete this project without the use of an automatic grapher. Tell students to test positive and negative values for *a* when they are determining the effect of *a* on the graph. Also, have students identify the coordinates of the vertex for each equation they graph. This should lead them to see that the vertex of each graph has coordinates (*b*, *c*).

4 Reciprocal Functions This project provides an opportunity for students to graph a variety of shapes.

If students do not have access to an automatic grapher, have them make a table of several values for each equation. This will provide an opportunity to work with exponents as well as reciprocals. If students are using automatic graphers, have them use the trace feature to find the point(s) of intersection of a function and its reciprocal.

5 Birthday Cubics Check that students have written their birthday equation correctly. You might ask students which months, if any, will have the same *y*-intercept and have them explain their answers. (November and January have the same *y*-intercept because for each month, d = 1. December and February have the same *y*-intercept because for each month, $d = 2$.)

6 Functions in the Real World If students have difficulty finding ten or more functions, suggest that they look through newspapers, magazines, almanacs, or other resource materials that may contain data and graphs. Remind students that a situation is considered a function when two items are related and the first item determines the second item. It is not necessary to have a formula to describe the situation. Students may decide to find ten functions in an area of special interest to them, such as sports, music, or current affairs.

7 Functions of Two Variables You might want to have students work with a partner. They will need a computer program that can graph functions of two variables in three dimensions.

(continued)

5 Birthday Cubics

Consider the equation $y = ax^3 - bx^2 + cx - d$, and the following code. Let a and b be the last two digits of your birth year. Let c and d be your 2-digit birth month, where January = 01 and December = 12. For instance, if you were born in April 1982, your code is $a = 8$, $b = 2$, $c = 0$, $d = 4$, and your birthday cubic is $y = 8x^3 - 2x^2 - 4$.

a. Graph your birthday cubic. Describe its intercepts and its peaks and valleys.

b. What is true about the graph of all people who share your birth month?

c. All graphs of cubic polynomials have one of the following shapes.

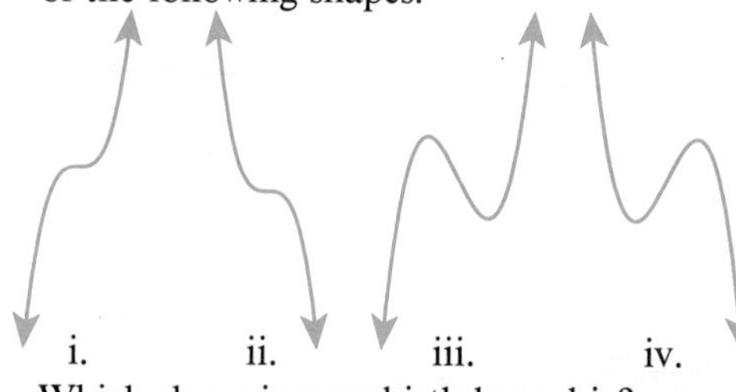

Which shape is your birthday cubic?

d. Make a collage or poster with the birthday cubics of your family or friends. Group them in some way you find interesting. Describe your results.

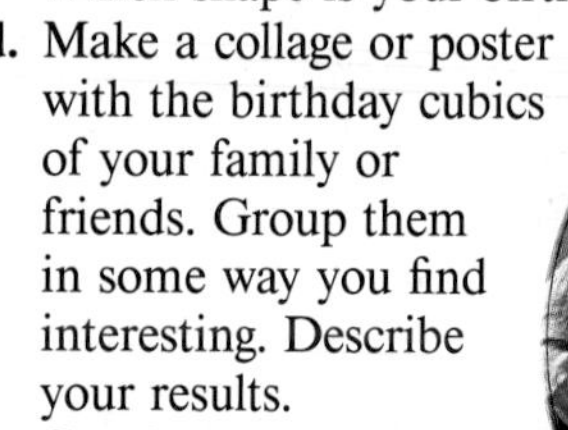

e. Graph the opposite of your birthday cubic. (The opposite of $f(x) = 8x^3 - 2x^2 - 4$ is $g(x) = -8x^3 + 2x^2 + 4$.) How is its graph related to your birthday cubic?

6 Functions in the Real World

Make a list of at least ten real-life situations in which the value of one variable determines the value of another. Give the domain and the range of each variable. Pick at least two of the situations to be of the type that you can graph, and show the graphs.

7 Functions of Two Variables

Recall the formula $p = ns$ for the perimeter of a regular polygon with n sides in which each side has length s. In this case, p is a function of the two variables n and s. We could write $p = f(n, s) = ns$. There are computer programs that can graph functions of two variables in three dimensions. Find such a program and graph this function and some other functions of your own choosing. Print out the graphs and arrange them in a nice display.

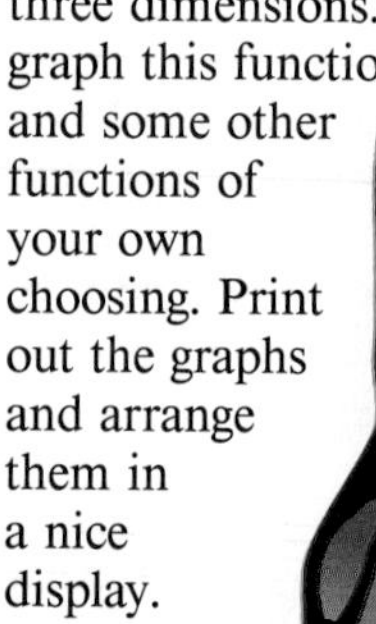

Additional responses, page 817

1.

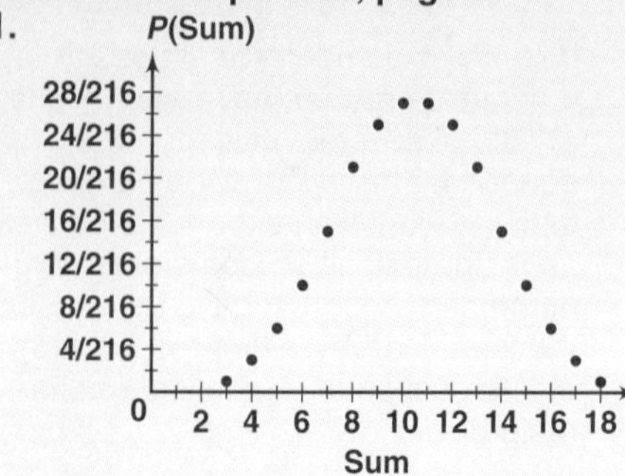

2. **Samples are given. All shapes will either open up or open down. There may be one or two maximum points or one or two minimum points.**

Diagram 1:
Solid line: $P(x) = x^4 + x^3 + x^2 + x + 1$
Dotted line: $P(x) = -x^4 + x^3 - x^2 + x - 1$

Diagram 2:
Solid line: $P(x) = x^4 - 10x^2 + 9$
Dotted line: $P(x) = -x^4 + 10x^2 - 9$

Diagram 1:

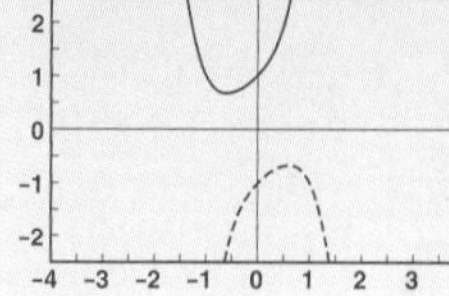

Diagram 2:

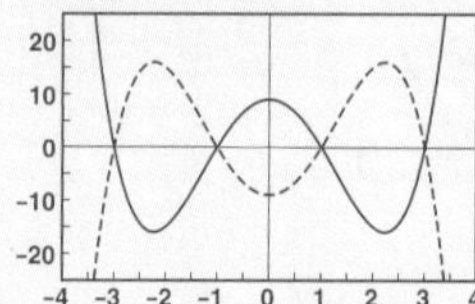

SUMMARY

A function is a set of ordered pairs in which each first coordinate appears with exactly one second coordinate. A function may be described by a graph, a written rule, a list of pairs, or an equation. The key idea in functions is that knowing the first coordinate of a pair is enough to determine the second coordinate. Thus, functions exist whenever one variable determines another.

If a function f contains the ordered pair (a, b), then we write $f(a) = b$. We say that b is the value of the function at a. The set of possible values of a is the domain of the function. The set of possible values of b is the range of the function. If a and b are real numbers, then the function can be graphed on the coordinate plane and values of the function can be approximated by reading the graph. An automatic grapher saves time when investigating complicated functions or multiple functions. Though convenient, it is not necessary to use $f(x)$ notation for functions; y is often used to stand for the second coordinate. Many of the graphs you studied in earlier chapters describe functions.

Equation	Graph	Type of Function
$f(x) = mx + b$	line	linear
$f(x) = ax^2 + bx + c$	parabola	quadratic
$f(x) = ab^x$	exponential curve	exponential
$f(x) = \|x\|$	angle	absolute value
$f(x) = ax^n + bx^{n-1} + \ldots + d$	(varied)	polynomial

In a probability function the domain is a set of outcomes in a situation and the range is the set of probabilities of these outcomes.

An important use of calculators is to evaluate functions at various values of their domains. Trigonometric functions, such as tangent, sine, and cosine, have their own keys on most calculators: tan, sin, cos. The tangent function can be used to determine lengths of sides and measures of angles in right triangles. Other functions on your calculator may include the squaring function x^2, factorial function !, and common logarithmic function log. Computer languages often build in functions such as SQR(X) and ABS(X) in BASIC.

VOCABULARY

Below are the most important terms and phrases for this chapter. You should be able to give a general description and a specific example of each.

Lesson 13-1
function
value of a function
linear function
quadratic function
exponential function
polynomial function

Lesson 13-2
$f(x)$ notation, function notation
squaring function

Lesson 13-3
absolute value function

Lesson 13-4
domain of a function
range of a function

Lesson 13-5
probability function

Lesson 13-6
cubing function
rotation symmetry, center of symmetry

Lesson 13-7
tangent function
tangent of an angle, tan A tan
leg opposite, leg adjacent

Lesson 13-8
sine function, sin A
cosine function, cos A
sinusoidal curve, sinusoid
sin, cos, log
trigonometry
common logarithm function

Summary

The Summary gives an overview of the entire chapter and provides an opportunity for students to consider the material as a whole. Thus, the Summary can be used to help students relate and unify the concepts that are presented in the chapter.

Vocabulary

Terms, symbols, and properties are listed by lesson to provide a checklist of concepts that students must know. Emphasize that students should read the vocabulary list carefully before starting the Progress Self-Test. If students do not understand the meaning of a term, they should refer back to the indicated lesson.

Additional Answers
Progress Self-Test, page 820

19.

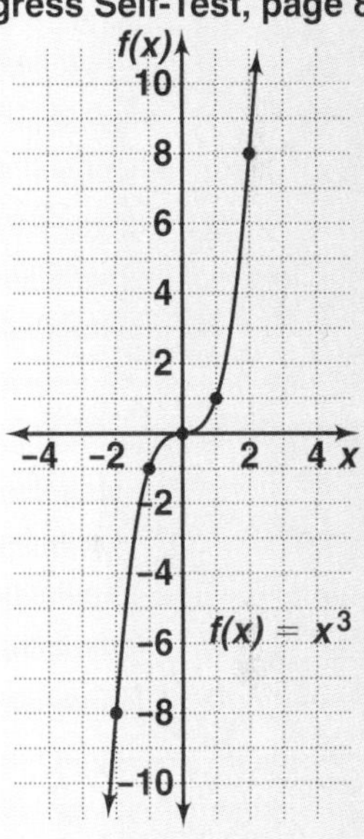

3a.

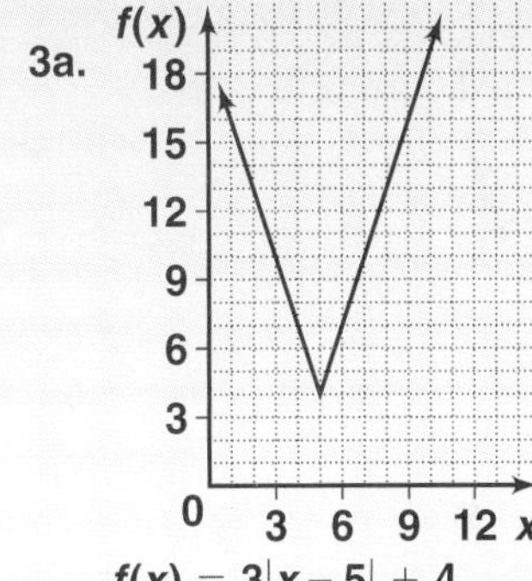

b. Sample response: The graph is always an angle. The angle has a smaller measure when $|a|$ is larger, it opens up when a is positive and down when a is negative. The line $x = b$ is a line of symmetry. The vertex of the angle is at (b, c). Changing b moves the vertex left or right; changing c moves the vertex up or down.

4. Samples:
a.

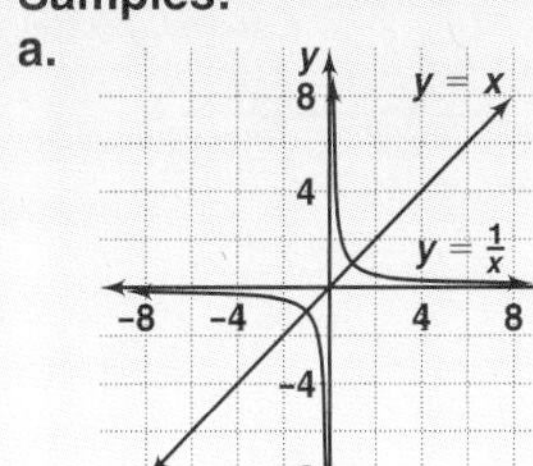

(Responses continue on page 820.)

Progress Self-Test

Assign the Progress Self-Test as a one-night assignment. Worked-out *solutions* for all questions are in the Selected Answers section of the student book. Encourage students to take the Progress Self-Test honestly, grade themselves, and then be prepared to discuss the test in class.

Additional Answers

2. **Sample: A function is a set of ordered pairs for which every first coordinate is paired with exactly one second coordinate.**

13. $P(n)$

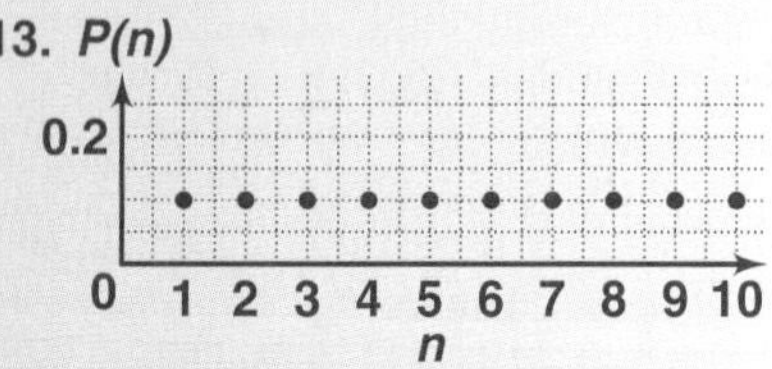

15.

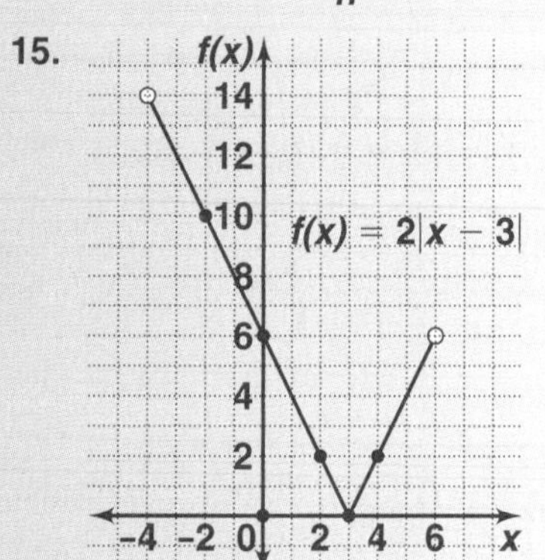

16.

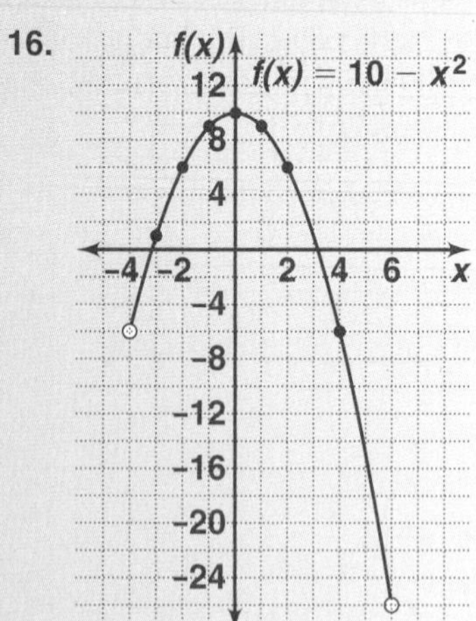

18.

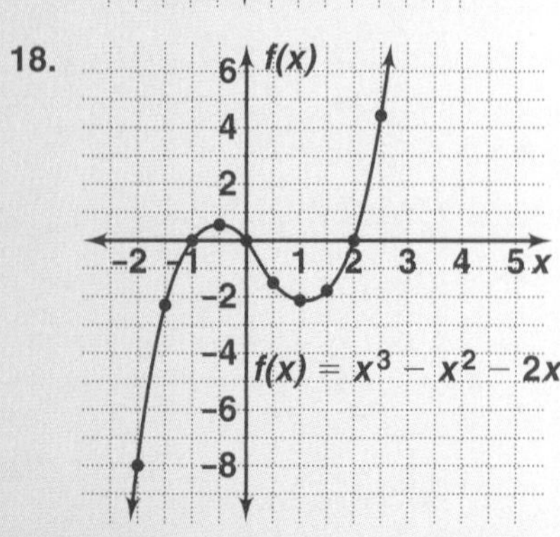

(Answers continue on page 819.)

PROGRESS SELF-TEST

2) See margin.

Take this test as you would take a test in class. You will need graph paper and a calculator. Then check your work with the solutions in the Selected Answers section in the back of the book.

1. If $f(x) = 3x + 5$, then $f(2) =$ __?__. 11
2. Explain in your own words what a function is.
3. Estimate tan 82° to the nearest thousandth. 7.115
4. Give the value of sin 30°. 0.5
5. What is the tangent of the acute angle formed by the line $y = 4x - 2$ and the x-axis? 4
6. Give an example of a quadratic function. 6) Sample: $f(x) = x^2 + 2x + 1$
7. Explain why the equation $x = |y|$ does not describe a function. See below right.
8. If the set $\{(10, 4), (x, 5), (30, 6)\}$ is a function, what values can x not have? $x \neq 30$, $x \neq 10$

In 9 and 10, tell whether the graph represents a function. If so, give its domain and range. If not, tell why not. 9) Yes; domain = {0, 1, 2, 3, 4}, range = {1, 2}.

9. **10.**

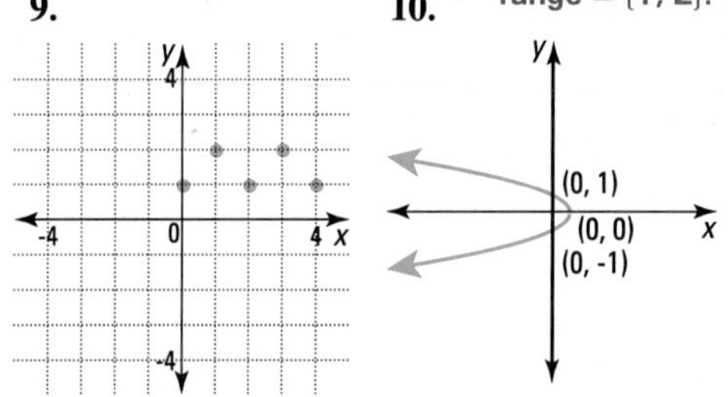

10) No; (0, 1) and (0, −1) are both in the relation.

11. In the tossing of two fair dice, what is the probability of obtaining a sum of 10? $\frac{1}{12}$

In 12 and 13, assume each of ten regions in the circle at the right has the same probability that the spinner will land on it. Let $P(n)$ = the probability of landing in region n.

12. Calculate $P(2)$. $\frac{1}{10}$
13. Graph the function P. See margin.

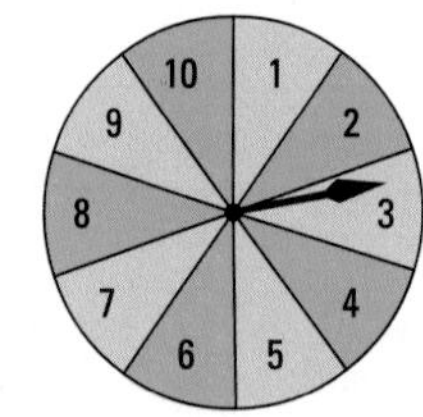

14. Suppose you are making double decker hamburger sandwiches. Let n = the number of sandwiches to be made. If $b(n)$ = (number of pieces of bread) and $h(n)$ = (number of hamburger patties), when is $b(n) > h(n)$? when $n > 0$

In 15 and 16, graph the function for values of x between -4 and 6. See margin.

15. $f(x) = 2|x - 3|$ **16.** $g(x) = 10 - x^2$

17. Kristin is having carpet installed. Fabulous Floor's price is determined by the function $f(x) = 11.75x + 300$ and Rapid Rug's price is determined by the function $r(x) = 14.50x$, where x is the number of square yards to be installed. If she needs 220 yards, where should she buy her carpet, and how much will it cost? Fabulous Floor; $2885
18. Use the spreadsheet information to the right to graph $y = x^3 - x^2 - 2x$. See margin.

	A	B
1	x	y
2	-2	-8
3	-1.5	-2.625
4	-1	0
5	-.5	.625
6	0	0
7	.5	-1.125
8	1	-2
9	1.5	-1.875
10	2	0
11	2.5	4.375

19. Graph the cubing function and give its domain and range. See margin for graph. domain = all real numbers; range = all real numbers

7) When $x > 0$, each x-value is paired with two y-values.

Additional responses, page 817

4b.

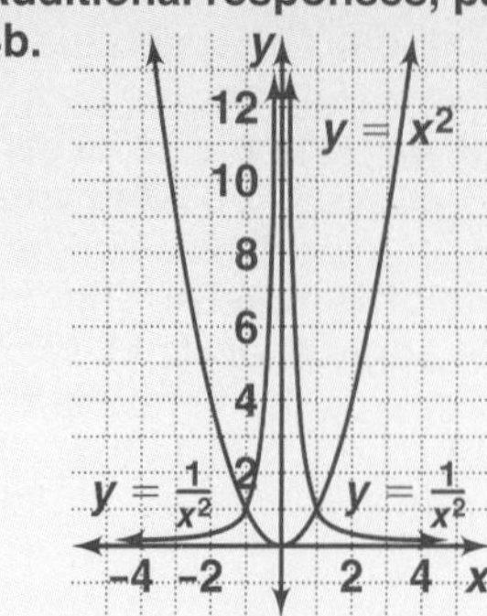

c.

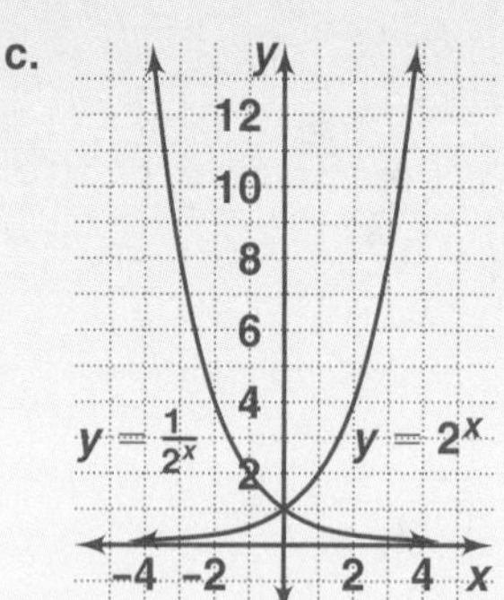

CHAPTER REVIEW

Questions on SPUR Objectives

SPUR stands for **S**kills, **P**roperties, **U**ses, and **R**epresentations. The Chapter Review questions are grouped according to the SPUR Objectives for this chapter.

SKILLS DEAL WITH THE PROCEDURES USED TO GET ANSWERS.

Objective A: *Evaluate functions and solve equations involving function notation.* *(Lessons 13-2, 13-3)*

In 1–4, $f(x) = x^2 - 3x + 8$. Calculate.

1. $f(2)$ 6 **2.** $f(3)$ 8 **3.** $f(-7)$ 78 **4.** $f(0)$ 8

5. If $A(t) = 2|t - 5|$, calculate $A(1)$. 8

6. If $g(n) = 2^n$, calculate $g(3) + g(4)$. 24

7. If $f(x) = -x$, what is $f(-1.5)$? 1.5

8. If $h(t) = 64t - 16t^2$, find $h(4)$. 0

9. If $f(x) = |x + 3|$, solve $f(x) = 5$. $x = 2$ or $x = -8$

10. If $A(r) = \pi r^2$, what is r if $A(r) = 18\pi$? $\pm 3\sqrt{2}$

Objective B: *Use function keys on a calculator.* *(Lessons 13-7, 13-8)*

In 11–20, approximate answers to the nearest thousandth.

11. $\frac{1}{12.5}$ 0.080 **12.** 10! 3,628,800

13. $\sqrt{11469}$ 107.093 **14.** 0.8^{-3} 1.953

15. tan 30° 0.577 **16.** SQR(6.5) 2.550

17. sin 82.4° 0.991 **18.** log 5 0.699

19. ABS(16 − 20) 4 **20.** tan 89.5° 114.589

PROPERTIES DEAL WITH THE PRINCIPLES BEHIND THE MATHEMATICS.

31) The domain is the largest set for which the function makes sense.

Objective C: *Determine whether a set of ordered pairs is a function.* *(Lesson 13-1)*

In 21–26, tell whether or not the sentence determines a function.

21. $y = 300(1.04)^x$ Yes **22.** $x^2 = y$ Yes

23. $3x - 5y = 7$ Yes **24.** $y = \tan x$ Yes

25. $y^2 = x$ No **26.** $y < 3$ No

In 27–30, tell whether or not the set of ordered pairs is a function.

27. {(0, 1), (1, 2), (2, 3), (3, 4)} Yes

28. {(1, 8), (1, 9), (1, 10), (1, 11)} No

29. the set of pairs (students, age) for students in your class Yes

30. the set of pairs (day of the week, date), for the days in this month No

Objective D: *Find the domain and the range of a function from its formula, graph, or rule.* *(Lesson 13-4)*

31. If the domain of a function is not given, what should you assume? See above.

32. *True or false.* a) True b) False

a. If $m \neq 0$, the range of the function $f(x) = mx + b$ is the set of all real numbers.

b. If $a \neq 0$, the range of the function $g(x) = ax^2$ is the set of all nonnegative real numbers.

33. *Multiple choice.* The domain of a function is {1, 2, 3}. The range is {4, 5, 6}. Which of these could *not* be a rule for the function? c

(a) $y = x + 3$ (b) $y = 7 - x$
(c) $y = x - 3$ (d) $y = |x| + 3$

34. What is the range of the function $A(x) = |x - 2|$? set of nonnegative real numbers

Chapter 13 Review

Resources

From the ***Teacher's Resource File***

- Answer Master for Chapter 13 Review
- Assessment Sourcebook: Chapter 13 Test, Forms A–D; Chapter 13 Test, Cumulative Form; Comprehensive Test, Chapters 1-13

Additional Resources

- Quiz and Test Writer

The main objectives for the chapter are organized in the Chapter Review under the four types of understanding this book promotes—Skills, Properties, Uses, and Representations.

Whereas end-of-chapter material may be considered optional in some texts, in *UCSMP Algebra* we have selected these objectives and questions with the expectation that they will be covered. Students should be able to answer these questions with about 85% accuracy after studying the chapter.

You may assign these questions over a single night to help students prepare for a test the next day, or you may assign the questions over a two-day period. If you work the questions over two days, then we recommend assigning the *evens* for homework the first night so that students get feedback in class the next day, then assigning the *odds* the night before the test because answers are provided to the odd-numbered questions.

It is effective to ask students which questions they still do not understand and use the day or days as a total class discussion of the material that the class finds most difficult.

d. Answers will vary.
e. Some information students may include in their responses follows. The graph of a function and its reciprocal intersect at the points where the value of the function is 1 or -1. Both graphs are above or below the *x*-axis at the same time. Students should find that the reciprocal graphs in a and b do not cross the *y*-axis because division by zero is undefined. None of the reciprocal graphs cross the *x*-axis because 1 divided by any number will never equal zero. When the absolute value of a function is greater than 1, the absolute value of its reciprocal is less than 1.

5. a. Responses will vary.
b. The graphs have the same *y*-intercept.
c. Either i or iii. The other two graphs have negative leading coefficients and no birth years can be negative.
d. Responses will vary.

(Responses continue on page 822.)

Assessment

Evaluation The *Assessment Sourcebook* provides five forms of the Chapter 13 Test. Forms A and B present parallel versions in a short-answer format. Forms C and D offer performance assessment. The fifth test is Chapter 13 Test, Cumulative Form. About 50% of this test covers Chapter 13; 25% covers Chapter 12, and 25% covers earlier chapters.

For information on grading, see *General Teaching Suggestions; Grading* in the *Professional Sourcebook*, which begins on page T20 in Part 1 of the Teacher's Edition.

Feedback After students have taken the test for Chapter 13 and you have scored the results, return the tests to students for discussion. Class discussion on the questions that caused trouble for most students can be very effective in identifying and clarifying misunderstandings. You might want to have them write down the items they missed and work either in groups or at home to correct them. It is important for students to receive feedback on every chapter test.

In 35 and 36, determine the domain and range of the function from its graph.

35.

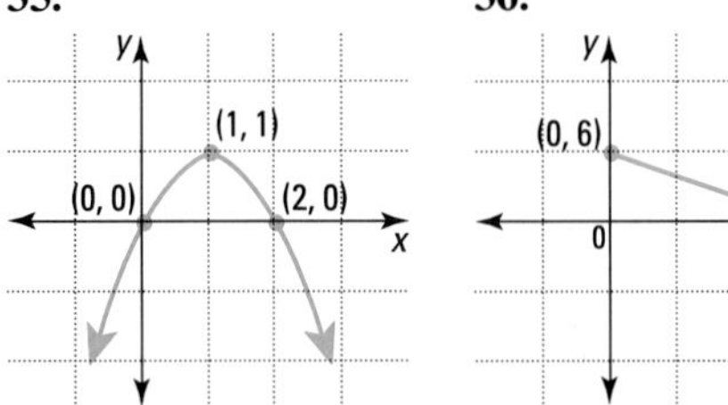

36.

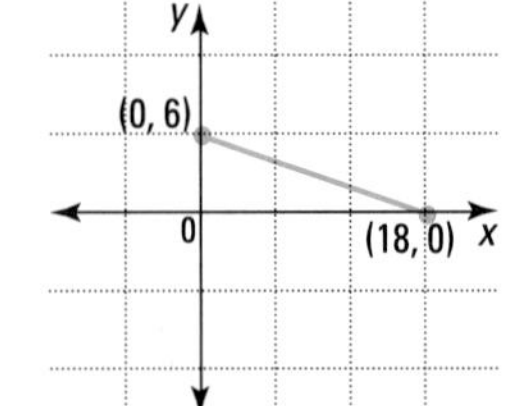

35) set of all real numbers; set of real numbers ≤ 1
36) set of all real numbers between 0 and 18, inclusive; set of all real numbers between 0 and 6, inclusive

37. What is the largest possible value in the range of the function f, where $f(x) = -x^2 + 10$? 10

38. What is the smallest possible value in the range of the function A with equation $A(n) = 5|n - 3| - 9$? -9

39. What is the domain of $d(x) = \frac{6x + 2}{2x - 6}$?

40. What is the range of $r(x) = \frac{1}{2}\sqrt{r}$?

39) set of all real numbers except 3
40) set of all nonnegative real numbers

USES DEAL WITH APPLICATIONS OF MATHEMATICS IN REAL SITUATIONS.

Objective E: *Use function notation and language in real situations.* (*Lessons 13-2, 13-3, 13-4*)

41. Let $N(t)$ = the number of chirps of a cricket in a minute at a temperature $t°$ Fahrenheit. If $N(t) = \frac{1}{4}t + 37$, for what value of t is $N(t) = 60$? $t = 92$

42. Give the domain and range of this population function for Los Angeles, California. See below.
{(1850, 1610), (1900, 102479), (1950, 1970358), (1960, 2479015), (1970, 2811801), (1980, 2966850), (1990, 3485398)}.

In 43 and 44, Bonita is comparing two health clubs. Daily Workout charges \$100 to join, and \$3.50 per visit. George's Gym charges \$250 to join and \$1 per visit. 43) $G(v) = v + 250$

43. If $D(v)$ is a function for the cost of joining Daily Workout and making v visits, and $D(v) = 3.5v + 100$, write a function $G(v)$ for joining and making v visits to George's Gym.

44. For how many visits is $G(v) < D(v)$? $v > 60$

45. S is the function that relates the time of day to the number of shoppers in MacGregor's Mart. S(time) = number of shoppers. See below right.
a. What is the domain of S?
b. What is the range of S?

46. A jar contains 437 jelly beans. You guess that there are G beans in the jar. Let $f(G)$ be the error in your guess. Find a formula for $f(G)$. $f(G) = |G - 437|$

42) domain = {1850, 1900, 1950, 1960, 1970, 1980, 1990}
range = {1610; 102,479; 1,970,358; 2,479,015; 2,811,801; 2,966,850; 3,485,398}

Objective F: *Determine values of probability functions.* (*Lesson 13-5*)

47. If it is equally likely that the spinner below will land in any direction, find each probability.
a. $P(1)$ $\frac{1}{2}$ **b.** $P(2)$ $\frac{1}{3}$ **c.** $P(3)$ $\frac{1}{6}$

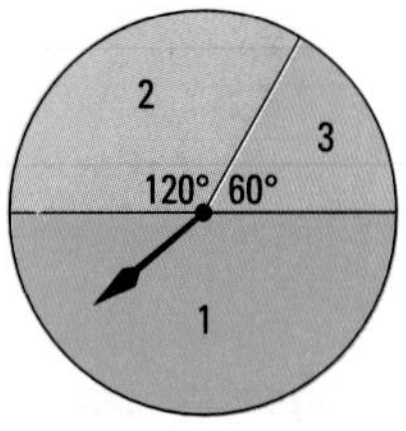

48. What is the probability of tossing a sum of 12 with two fair dice? $\frac{1}{36}$

49. If you guess on three multiple-choice questions with four choices each, the probability that you will get exactly n correct is $\frac{3!}{n!(3-n)!} \cdot \left(\frac{1}{4}\right)^n \cdot \left(\frac{3}{4}\right)^{3-n}$. Calculate the probability that you will get exactly 2 questions correct. $\frac{9}{64}$

50. A letter is mailed. Suppose P(the letter arrives the next day) = 0.75. What is P(the letter does not arrive the next day)? 0.25

45a) Set of times of the day the Mart is open
b) Sample: If the largest number of people MacGregor's Mart can hold is 2000, then the range = set of whole numbers ≤ 2000.

Additional responses, page 818

5. e. The graph of the opposite is the reflection image of the graph over the x-axis.
6. The following information is a sample of what students might include in their projects:
(month, average temperature), domain = months of year, range = real numbers
(year, population of the world), domain = integers, range = nonnegative integers
(person, person's height) domain = people, range = nonnegative real numbers
(speed of car, stopping distance), domain = nonnegative real numbers, range = nonnegative real numbers
(number of items, total cost), domain = nonnegative integers, range = nonnegative real numbers
7. Responses will vary.

Objective G: *Find lengths of sides or tangents of angles in right triangles using the tangent function.* *(Lesson 13-7)*

51. Find tan A in the triangle at the right, to the nearest thousandth. **.870**

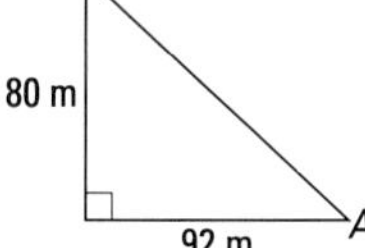

52. On a field trip, a girl whose eyes are 150 cm above the ground sights a nest high on a pole 25 meters away. If she has to raise her eyes 70° to see the nest, how high is the nest? **70.2 m**

53. What is the slope of line ℓ graphed below? **≈ 0.325**

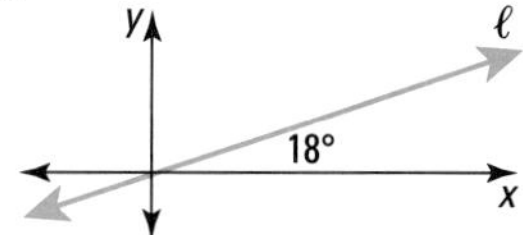

54. A line has equation $4x - 3y = 2$. What is the tangent of the acute angle this line makes with the x-axis? $\frac{4}{3}$

REPRESENTATIONS DEAL WITH PICTURES, GRAPHS, OR OBJECTS THAT ILLUSTRATE CONCEPTS.

Objective H: *Determine whether or not a graph represents a function.* *(Lesson 13-1)*

In 55–60, tell whether or not the set of ordered pairs graphed is a function.

55.

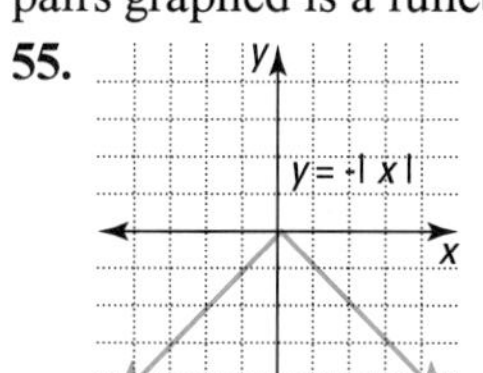

Yes

56.

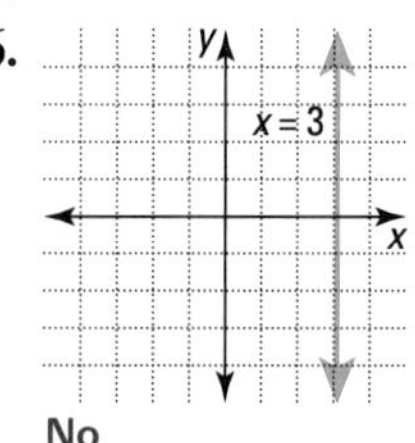

No

57.

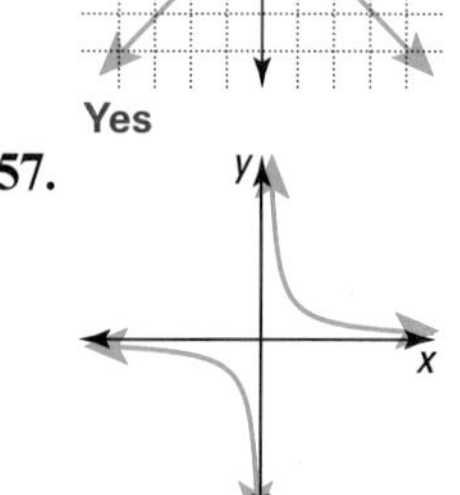

Yes

58.

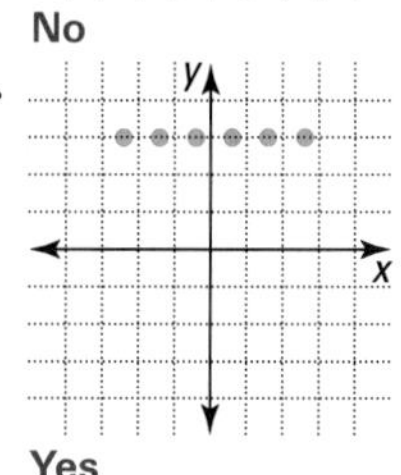

Yes

59.

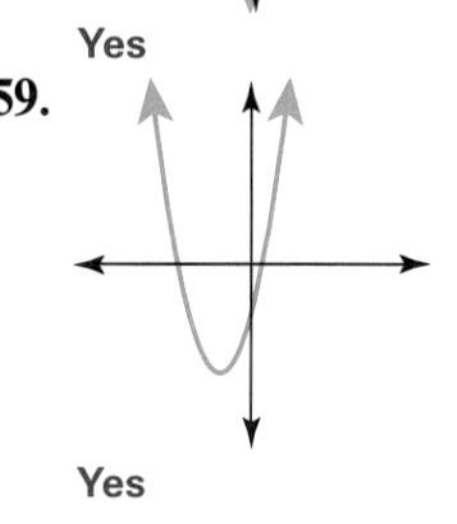

Yes

60.

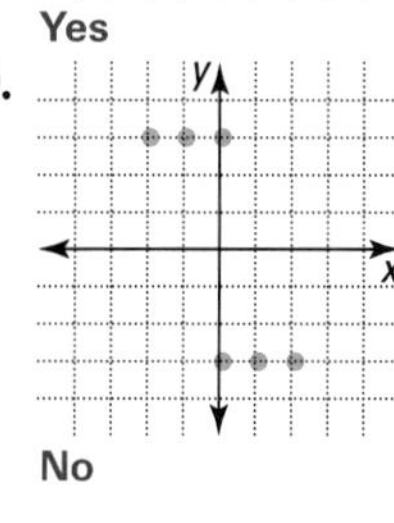

No

71a) See margin.
71b) Since the degree of $P(x)$ is 3, it has at most 3 x-intercepts.

Objective I: *Graph functions.* *(Lessons 13-1, 13-2, 13-3, 13-5)* **61–67) See margin.**

In 61–64, graph each function over the domain $-5 \le x \le 5$.

61. $f(x) = 3|2x + 1|$ **62.** $g(t) = t^2 - 10$

63. $y = \frac{1}{5}x$ **64.** $f(n) = 2^n$, n an integer

65. A weighted die has the following probabilities of landing on its sides. $P(1) = 0.12$; $P(2) = 0.19$; $P(3) = 0.09$; $P(4) = 0.21$; $P(5) = 0.15$. Find $P(6)$ and graph the probability function.

66. Graph the probability function for a fair die.

67. Graph the probability function for the number of heads that appear when two fair coins are tossed.

65) $P(6) = 0.24$; See margin for graph.

Objective J: *Graph polynomial functions.* *(Lesson 13-6)* **68, 69, 71a) See margin.**

68. Graph $y = x^3 - 4x^2 + 5$ from $x = -4$ to $x = 4$.

69. Graph $y = x^4 - 5x^2$ from $x = -3$ to $x = 3$.

70. *Multiple choice.* The graph of which equation is symmetric about the origin? **b**
(a) $y = x^2$ (b) $y = x^3$
(c) $y = x^4$ (d) $y = -x^2$

71. a. Use the table at the right to plot the graph of the polynomial function $P(x) = 2x^3 - x^2 - 6x$.

b. Why can you be sure that all the x-intercepts are listed in the table?

x	value
-3	-45
-2	-8
-1.5	0
-1	3
-.5	2.5
0	0
.5	-3
1	-5
1.5	-4.5
2	0
3	27

Additional Answers, p. 823

61. $f(x) = 3|2x + 1|$

62. $g(t) = t^2 - 10$

63. $y = \frac{1}{5}x$

64.

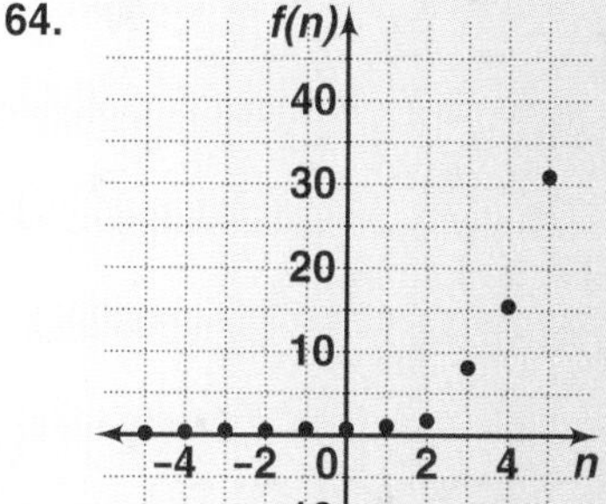

65.

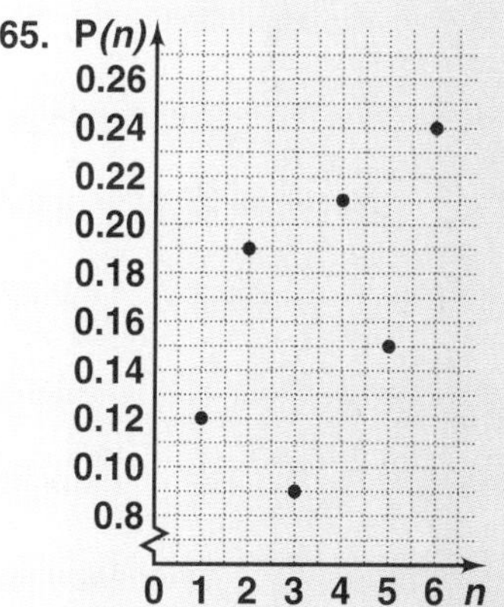

66.

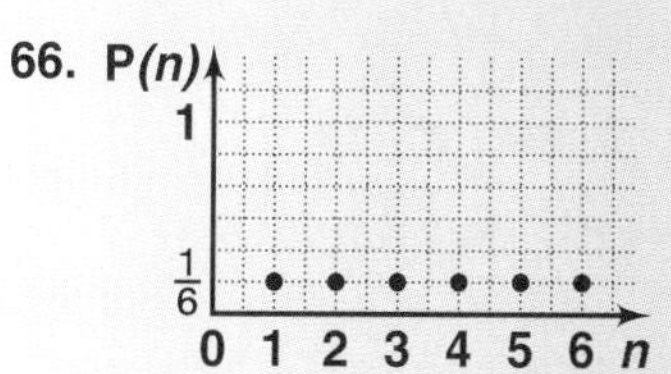

Additional Answers, p. 823, continued

67.

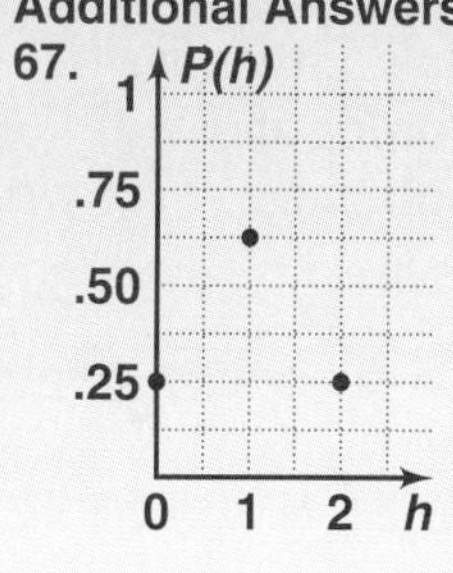

68.

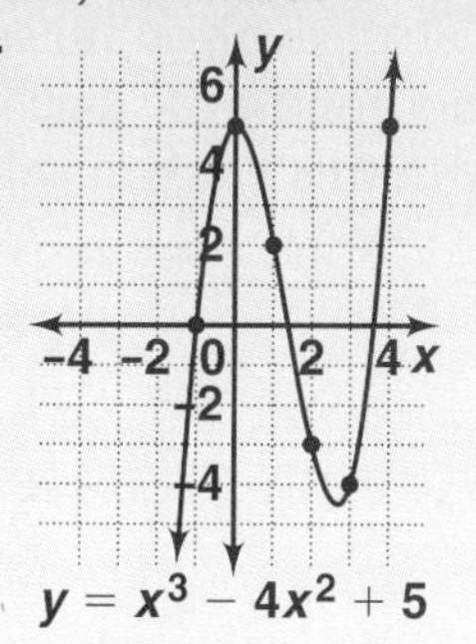

$y = x^3 - 4x^2 + 5$

69.

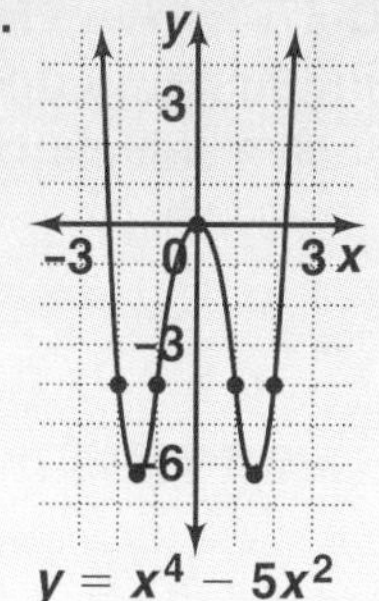

$y = x^4 - 5x^2$

71a.

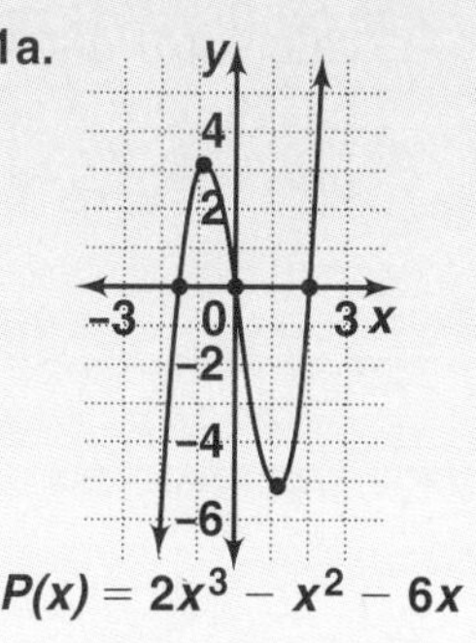

$P(x) = 2x^3 - x^2 - 6x$

APPENDIX A

Scientific Calculators

You should be using a scientific calculator throughout this book, so it is important for you to know how to use one. As you read, you should use your calculator to do all the calculations described. Some of the problems are very easy. They were selected so that you can check whether your calculator does the computations in the proper order. Your scientific calculator should follow the order of operations used in algebra.

Suppose you want to use your calculator to find 3 + 4. Here is one way to do it:

	Display shows
Press 3	3
Now press [+]	3
Now press 4	4
Now press [=]	7

Pressing calculator keys is called **entering** or **keying in.** The set of instructions in the left column is called the **key sequence** for this problem. We write the key sequence for this problem using boxes for everything pressed except the numbers.

3 [+] 4 [=]

Sometimes we put what you would see in the calculator display underneath the key presses.

Key sequence:	3	[+]	4	[=]
Display:	3	3	4	7

Some calculators do not have an equal sign, but have a key that enters the calculation [ENTER] or executes it [EXE]. Key sequences for finding 3 + 4 on these calculators are:

3 [+] 4 [ENTER]   3 [+] 4 [EXE]

Next consider $12 + 3 \cdot 5$. In the algebraic order of operations, multiplication is performed before addition. Perform the key sequence below on your calculator. See what your calculator does.

Key sequence:	12	[+]	3	[×]	5	[=]
Display:	12	12	3	3	5	27

Different calculators may give different answers even when the same buttons are pushed. If you have a calculator appropriate for algebra, the calculator displayed 27. If your calculator gave you the answer 75, then it has done the addition first and does not follow the algebraic order of operations. Using such a calculator with this book may be confusing.

824

Example 1

Evaluate $ay + bz$ when $a = 0.05$, $y = 2000$, $b = 0.06$ and $z = 9000$. (This is the total interest in a year if \$2000 is earning 5% and \$9000 is earning 6%.)

Solution

Key sequence: a [×] y [+] b [×] z [=]

Substitute in the key sequence:

Key sequence:	0.05	[×]	2000	[+]	0.06	[×]	9000	[=]
Display	0.05	0.05	2000	100	0.06	0.06	9000	640

The total interest is \$640.

Most scientific calculators have parentheses keys, [(] and [)]. To use them just enter the parentheses when they appear in the problem. You may need to use the [×] key every time you do a multiplication, even if × is not in the expression.

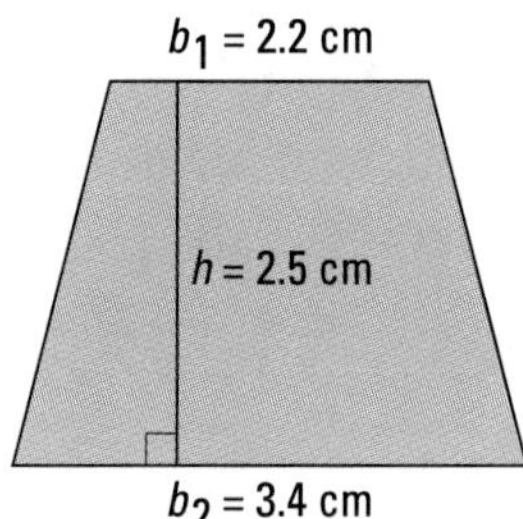

Example 2

Use the formula $A = 0.5h(b_1 + b_2)$ to calculate the area of the trapezoid at the left.

Solution

Remember that $0.5h(b_1 + b_2)$ means $0.5 \cdot h \cdot (b_1 + b_2)$.

Key sequence:	0.5	[×]	h	[×]	[(]	b_1	[+]	b_2	[)]	[=]
Substitute:	0.5	[×]	2.5	[×]	[(]	2.2	[+]	3.4	[)]	[=]
Display:	0.5	0.5	2.5	1.25	1.25	2.2	2.2	3.4	5.6	7.

The area of the trapezoid is 7 square centimeters.

Some frequently used numbers have special keys on the calculator.

Example 3

Find the circumference of the circle at the left.

Solution

The circumference is the distance around the circle, and is calculated using the formula $C = 2\pi r$, where C = circumference and r = radius. Use the π key.

Key sequence: 2 [×] [π] [×] r [=]

Substitute: 2 [×] [π] [×] 4.6 [=]

Rounding to the nearest tenth, the circumference is 28.9 miles.

As a decimal, $\pi = 3.141592653\ldots$ and the decimal is unending. Since it is impossible to list all the digits, the calculator rounds the decimal. Some calculators, like the one in Example 3, round to the nearest value that can be displayed. Some calculators truncate or round down. If the calculator in the example had truncated, it would have displayed 3.1415926 instead of 3.1415927 for π.

On some calculators you must press two keys to display π. If a small π is written above a key, two keys are probably needed. Then you should press [INV], [2nd], or [F] before pressing the key below π.

Negative numbers can be entered in your calculator. On many calculators this is done with a plus-minus key [+/–] or [±]. Enter -19.

Key sequence:	19	[+/–]
Display:	19	-19

If your scientific calculator has an opposite key [(–)], you can enter a negative number in the same order as you write it.

Key sequence:	[(–)]	19
Display:	-	-19

You will use powers of numbers throughout this book. The scientific calculator has a key [y^x] (or [x^y] or [^]) used to raise numbers to powers.

The key sequence for 3^4 is	3	[y^x]	4	[=]
You should see displayed	3	3	4	81.

This display shows that $3^4 = 81$.

Example 4

A formula for the volume of a sphere is $V = \frac{4\pi r^3}{3}$, where r is the radius. The radius of the moon is about 1080 miles. Estimate the volume of the moon.

Solution

Key sequence:	4	[×]	[π]	[×]	r	[y^x]
Substitute:	4	[×]	[π]	[×]	1080	[y^x]
Display:	4	4	3.1415927	12.566371	1080	1080 . . .

Key sequence:	3	[÷]	3	[=]
Substitute:	3	[÷]	3	[=]
Display: . . .	3	1.583 10	3	5.2767 09

The display shows the answer in scientific notation. If you do not understand scientific notation, read Appendix B.

The volume of the moon is about $5.28 \cdot 10^9$ cubic miles.

826

Note: You may be unable to use a negative number as a base on your calculator. Try the key sequence 2 [±] [y^x] 5 to evaluate $(-2)^5$. The answer should be -32. However, some calculators will give you an error message. You can, however, use negative *exponents* on scientific calculators.

QUESTIONS

Covering the Reading

1. What is meant by the phrase "keying in"? **pressing calculator keys**

2. To calculate 28.5 · 32.7 + 14.8, what key sequence can you use?
28.5 [×] 32.7 [+] 14.8 [=]

3. Consider the key sequence 13.4 [–] 15 [÷] 3 [=]. What arithmetic problem does this represent? **13.4 – 15 ÷ 3**

4. a. To evaluate $ab - c$ on a calculator, what key sequence should you use? ***a* [×] *b* [–] *c* [=]**

b. Evaluate 297 · 493 – 74,212. **72,209**

5. Estimate 26π to the nearest thousandth. **81.681**

6. What number does the key sequence 104 [±] yield? **-104**

7. a. Write a key sequence for entering -104 divided by -8 on your calculator. **104 [±] [÷] 8 [±] [=]**

b. Calculate -104 divided by -8 on your calculator. **13**

8. Calculate the area of the trapezoid below. **36.08**

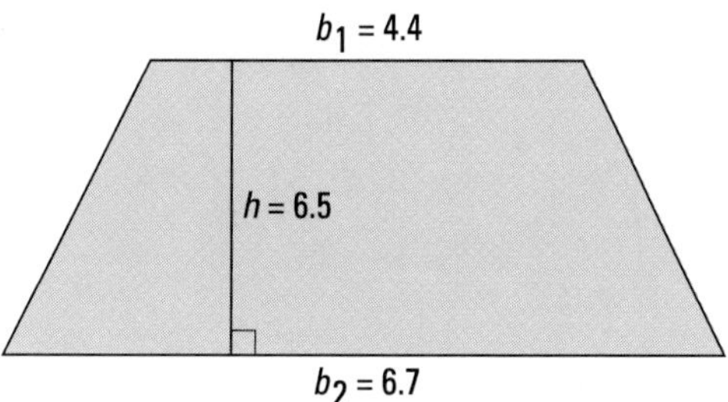

9. Find the circumference of a circle with radius 6.7 inches. **42.10 in.**

10. Which is greater, $\pi \cdot \pi$ or 10? **10**

11. What expression is evaluated by 5 [y^x] 2 [=]? **5^2**

12. A softball has a radius of about 1.92 in. What is its volume? **29.65 in^3**

13. What kinds of numbers may not be allowed as bases when you use the [y^x] key on some calculators? **negative numbers**

Applying the Mathematics

14. Use your calculator to help find the surface area $2LH + 2HW + 2LW$ of the box below. **455.3 in^2**

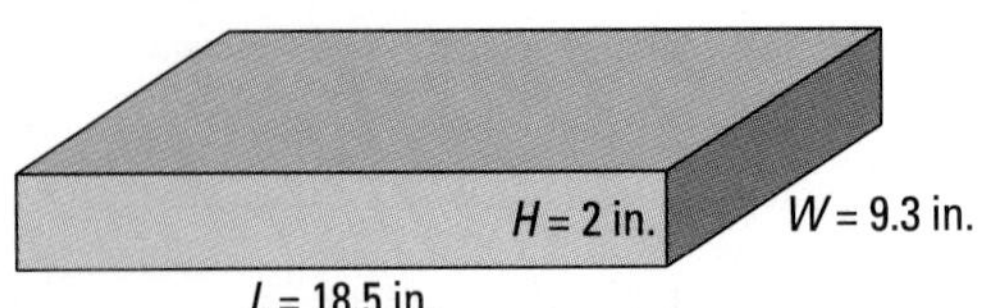

15. Remember that $\frac{2}{3} = 2 \div 3$. **Answers will vary.**

a. What decimal for $\frac{2}{3}$ is given by your calculator?

b. Does your calculator *truncate* or *round to the nearest*?

16. Order $\frac{3}{5}$, $\frac{4}{7}$, and $\frac{5}{9}$ from smallest to largest. **$\frac{5}{9}, \frac{4}{7}, \frac{3}{5}$**

17. Use the clues to find the mystery number y.

Clue 1: y will be on the display if you alternately press 2 and [×] again and again. . . .

Clue 2: $y > 20$.

Clue 3: $y < 40$. **32**

18. $A = \pi r^2$ is a formula for the area A of a circle with radius r. Find the area of the circle in Example 3. **66.5 mi^2**

19. What is the total interest in a year if $350 is earning 5% and $2000 is earning 8%? (Hint: use Example 1.) **$177.50**

20. To multiply the sum of 2.08 and 5.76 by 2.24, what key sequence can you use? **[(] 2.08 [+] 5.76 [)] [×] 2.24 [=]**

828

APPENDIX B

Scientific Notation

The first three columns in the chart below show three ways to represent integer powers of ten: in exponential notation, with word names, and as decimals. The fourth column describes a distance or length in meters. For example, the top row tells that Mercury is about ten billion meters from the sun.

Integer Powers of Ten

Exponential Notation	Word Name	Decimal	Something near this length in meters
10^{10}	ten billion	10,000,000,000	distance of Mercury from Sun
10^{9}	billion	1,000,000,000	radius of Sun
10^{8}	hundred million	100,000,000	diameter of Jupiter
10^{7}	ten million	10,000,000	radius of Earth
10^{6}	million	1,000,000	radius of Moon
10^{5}	hundred thousand	100,000	length of Lake Erie
10^{4}	ten thousand	10,000	average width of Grand Canyon
10^{3}	thousand	1,000	5 long city blocks
10^{2}	hundred	100	length of a football field
10^{1}	ten	10	height of shade tree
10^{0}	one	1	height of waist
10^{-1}	tenth	0.1	width of hand
10^{-2}	hundredth	0.01	diameter of pencil
10^{-3}	thousandth	0.001	thickness of window pane
10^{-4}	ten-thousandth	0.000 1	thickness of paper
10^{-5}	hundred-thousandth	0.000 01	diameter of red blood corpuscle
10^{-6}	millionth	0.000 001	mean distance between successive collisions of molecules in air
10^{-7}	ten-millionth	0.000 000 1	thickness of thinnest soap bubble with colors
10^{-8}	hundred-millionth	0.000 000 01	mean distance between molecules in a liquid
10^{-9}	billionth	0.000 000 001	size of air molecule
10^{-10}	ten-billionth	0.000 000 000 1	mean distance between molecules in a crystal

You probably know the quick way to multiply by 10, 100, 1000, and so on. Just move the decimal point as many places to the right as there are zeros.

$$84.3 \cdot 100 = 8430 \qquad 84.3 \cdot 10{,}000 = 843{,}000$$

It is just as quick to multiply by these numbers when they are written as powers.

$$489.76 \cdot 10^2 = 48{,}976 \qquad 489.76 \cdot 10^4 = 4{,}897{,}600$$

The general pattern is as follows.

To multiply by 10 raised to a positive power, move the decimal point to the *right* as many places as indicated by the exponent.

The patterns in the chart on the previous page also help to explain powers of 10 where the exponent is negative. Each row describes a number that is $\frac{1}{10}$ of the number in the row above it. So 10^0 is $\frac{1}{10}$ of 10^1.

$$10^0 = \frac{1}{10} \cdot 10 = 1$$

To see the meaning of 10^{-1}, think: 10^{-1} is $\frac{1}{10}$ of 10^0 (which equals 1).

$$10^{-1} = \frac{1}{10} \cdot 1 = \frac{1}{10} = .1$$

Remember that to multiply a decimal by 0.1, just move the decimal point one unit to the left. Since $10^{-1} = 0.1$, to multiply by 10^{-1}, just move the decimal point one unit to the left.

$$435.86 \cdot 10^{-1} = 43.586$$

To multiply a decimal by 0.01, or $\frac{1}{100}$, move the decimal point two units to the left. Since $10^{-2} = 0.01$, the same goes for multiplying by 10^{-2}.

$$435.86 \cdot 10^{-2} = 4.3586$$

The following pattern emerges.

To multiply by 10 raised to a negative power, move the decimal point to the *left* as many places as indicated by the exponent.

Example 1

Write $68.5 \cdot 10^{-6}$ as a decimal.

Solution

To multiply by 10^{-6}, move the decimal point six places to the left. So $68.5 \cdot 10^{-6} = 0.0000685$.

The names of the negative powers are very similar to those for the positive powers. For instance, 1 billion = 10^9 and 1 billionth = 10^{-9}.

Example 2

Write 8 billionths as a decimal.

Solution

8 billionths = $8 \cdot 10^{-9} = 0.000000008$

830

Most calculators can display only the first 8, 9, or 10 digits of a number. This presents a problem if you need to key in a large number like 455,000,000,000 or a small number like 0.00000000271. However, powers of 10 can be used to rewrite these numbers in **scientific notation.**

$$455{,}000{,}000{,}000 = 4.55 \cdot 10^{11}$$
$$0.00000000271 = 2.71 \cdot 10^{-9}$$

Definition: In scientific notation, a number is represented as $x \cdot 10^n$, where $1 \leq x < 10$ and n is an integer.

Scientific calculators can display numbers in scientific notation. The display for $4.55 \cdot 10^{11}$ will usually look like one of these shown here.

4.55 E 11 | 4.55 11 | 4.55 ×10 11

The display for $2.71 \cdot 10^{-9}$ is usually one of these

2.71 E -09 | 2.71 -09 | 2.71 ×10 -09

Numbers written in scientific notation are entered into a calculator using the EXP or EE key. For instance, to enter $6.0225 \cdot 10^{23}$ (known as Avogadro's number), key in

6.0225 EE 23.

You should see this display.

6.0225 23

In general, to enter $x \cdot 10^n$, key in x EE n.

Example 3

The total number of hands possible in the card game bridge is about 635,000,000,000. Write this number in scientific notation.

Solution

Move the decimal point to get a number between 1 and 10. In this case the number is 6.35. This tells you the answer will be:

$$6.35 \cdot 10^{\text{exponent}}.$$

The exponent of 10 is the number of places you must move the decimal point in 6.35 in order to get 635,000,000,000. You must move it 11 places to the right, so the answer is $6.35 \cdot 10^{11}$.

Example 4

The charge of an electron is 0.00000000048 electrostatic units. Put this number in scientific notation.

Solution

Move the decimal point to get a number between 1 and 10. The result is 4.8. To find the power of 10, count the number of places you must move the decimal to change 4.8 to 0.00000000048. The move is 10 places to the left, so the charge of the electron is $4.8 \cdot 10^{-10}$ *electrostatic units.*

Example 5

Enter 0.00000000123 into a calculator.

Solution

Rewrite the number in scientific notation.

$0.00000000123 = 1.23 \cdot 10^{-9}$.

Key in 1.23 EE 9 +/− or 1.23 EE 9 (−).

QUESTIONS

Covering the Reading

9a) set of real numbers greater than or equal to one and less than ten

1. Write one million as a power of ten. 10^6

2. Write 1 billionth as a power of 10. 10^{-9}

In 3–5, write as a decimal.

3. 10^{-4} 0.0001

4. $28.5 \cdot 10^7$ 285,000,000

5. 10^0 1

6. To multiply by a negative power of 10, move the decimal point to the _?_ as many places as indicated by the _?_. left, exponent

7. Write $2.46 \cdot 10^{-8}$ as a decimal. 0.0000000246

8. Why is $38.25 \cdot 10^{-2}$ not in scientific notation? 38.25 is not between one and ten.

9. Suppose $x \cdot 10^y$ is in scientific notation.
   a. What is the domain of x? See left.
   b. What is the domain of y? set of integers

In 10–14, rewrite the number in scientific notation.

10. 5,020,000,000,000,000,000,000,000,000 tons, the mass of Sirius, the brightest star $5.02 \cdot 10^{27}$

11. 0.0009 meters, the approximate width of a human hair $9 \cdot 10^{-4}$

12. 763,000 $7.63 \cdot 10^5$

13. 0.00000328 $3.28 \cdot 10^{-6}$

14. 754.9876 $7.549876 \cdot 10^2$

832

15. One computer can do an arithmetic problem in $2.4 \cdot 10^{-9}$ seconds. What key sequence can you use to display this number on your calculator? 2.4 (EE) 9 (+/−)

Applying the Mathematics

In 16 and 17, write in scientific notation.

16. 645 billion $6.45 \cdot 10^{11}$ **17.** 27.2 million $2.72 \cdot 10^{7}$

In 18–21, use the graph below. Write the estimated world population in the given year: **a.** as a decimal; **b.** in scientific notation. See left.

18. 10,000 B.C. **19.** 1 A.D. **20.** 1700 **21.** 1970

18a) 10,000,000;
b) $1.0 \cdot 10^{7}$
19a) 300,000,000;
b) $3.0 \cdot 10^{8}$
20a) 625,000,000;
b) $6.25 \cdot 10^{8}$
21a) 3,575,000,000;
b) $3.575 \cdot 10^{9}$

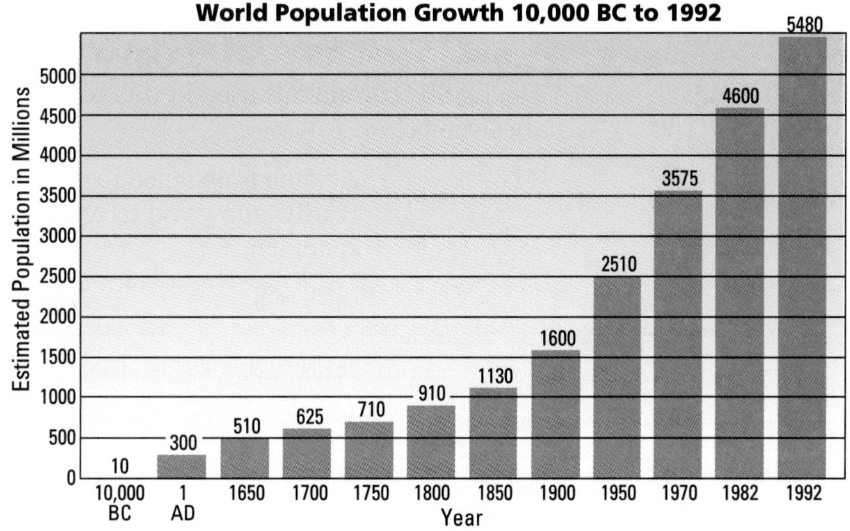

22. How can you enter the world population in 1992 into your calculator? 5.480 (EE) 9

23. How many digits are in $1.7 \cdot 10^{100}$? 101

In 24–26, write the number in scientific notation.

24. 0.00002 $2 \cdot 10^{-5}$ **25.** 0.0000000569 $5.69 \cdot 10^{-8}$ **26.** 400.007 $4.00007 \cdot 10^{2}$

In 27–29, write as a decimal.

27. $3.921 \cdot 10^{5}$ 392,100 **28.** $3.921 \cdot 10^{-5}$ 0.00003921 **29.** $8.6 \cdot 10^{-2}$ 0.086

Exploration

30. a. What is the largest number you can display on your calculator? Answers may vary.

b. What is the smallest number you can display? (Use scientific notation and consider negative numbers.) Answers may vary.

c. Find out what key sequence you could use to enter -5×10^{-7} in your calculator. Sample: 5 (+/−) (EE) 7 (+/−)

APPENDIX C

BASIC

In BASIC (Beginner's All Purpose Symbolic Instruction Code), the arithmetic symbols are: + (for addition), − (for subtraction), * (for multiplication), / (for division), and ^ (for powering). In some versions of BASIC, ↑ is used for powering. The computer evaluates expressions according to the usual order of operations. Parentheses ( ) may be used. The comparison symbols =, >, < are also used in the standard way, but BASIC uses <= instead of ≤, >= instead of ≥, and <> instead of ≠.

Variables are represented by letters or letters in combination with digits. Consult the manual for your version of BASIC for restrictions on the length or other aspects of variable names. Examples of variable names allowed in most versions are N, X1, and AREA.

COMMANDS

The BASIC commands used in this course and examples of their uses are given below.

LET . . . A value is assigned to a given variable. Some versions of BASIC allow you to omit the word LET in the assignment statement.

LET X = 5	The number 5 is stored in a memory location called X.
LET N = N + 2	The value in the memory location called N is increased by 2 and the result is stored in the location N.

PRINT . . . The computer prints on the screen what follows the PRINT command. If what follows is not in quotes, it is a constant or variable, and the computer will print the value of that constant or variable. If what follows is in quotes, the computer prints exactly what is in quotes.

PRINT X	The computer prints the number stored in memory location X.
PRINT "X-VALUES"	The computer prints the phrase X-VALUES.

INPUT . . . The computer asks the user to give a value to the variable named and stores that value.

INPUT X	When the program is run, the computer will prompt you to give X a value by printing a question mark, and then will store that value in memory location X.
INPUT "HOW OLD?";AGE	The computer prints HOW OLD? and stores your response in memory location AGE. (Note: Some computers will not print the question mark.)

834

REM . . . REM stands for *remark.* This command allows remarks to be inserted in a program. These may describe what the variables represent, what the program does, or how the program works. REM statements are often used in long complex programs or programs other people will use.

REM PYTHAGOREAN THEOREM	
	A statement that begins with REM has no effect when the program is run.

FOR . . .
NEXT . . .
STEP . . . FOR and NEXT are used when a set of instructions must be performed more than once, a process which is called a *loop.* The FOR command assigns a beginning and ending value to a variable. The first time through the loop, the variable has the beginning value in the FOR command. When the computer hits the line reading NEXT, the value of the variable is increased by the amount indicated by STEP. The commands between FOR and NEXT are then repeated. When the value of the incremented variable is larger than the ending value in the FOR command, the computer leaves the loop and executes the rest of the program. If STEP is not written, the computer increases the variable by 1 each time through the loop.

```
10 FOR N = 3 TO 6 STEP 2
20 PRINT N
30 NEXT N
40 END
```

The computer assigns 3 to N and then prints the value of N. On reaching NEXT, the computer increases N by 2 (the STEP amount) and prints 5. The next N would be 7 which is too large. The computer executes the command after NEXT, ending the program.

IF . . . THEN . . . The computer performs the consequent (the THEN part) only if the antecedent (the IF part) is true. When the antecedent is false, the computer *ignores* the consequent and goes directly to the next line of the program.

```
IF X > 100 THEN END
PRINT X
```

If the X value is less than or equal to 100, the computer ignores END, goes to the next line, and prints the value stored in X. If the X value is greater than 100, the computer stops and the value stored in X is not printed.

GOTO . . . The computer goes to whatever line of the program is indicated. GOTO statements are generally avoided because they interrupt program flow and make programs hard to interpret.

GOTO 70	The computer goes to line 70 and executes that command.

END . . . The computer stops running the program. No program should have more than one END statement.

FUNCTIONS

The following built-in functions and many others are available in most versions of BASIC. Each function name must be followed by a variable or a constant enclosed in parentheses.

ABS The absolute value of the number that follows is calculated.

LET X = ABS(-10)	The computer calculates $|-10| = 10$ and assigns the value 10 to memory location X.

SQR The square *root* of the number that follows is calculated.

C = SQR(A*A + B*B)	The computer calculates $\sqrt{A^2 + B^2}$ using the values stored in A and B and stores the result in C.

PROGRAMS

A program is a set of instructions to the computer. In most versions of BASIC, every step in the program must begin with a line number. We usually start numbering at 10 and count by ten, so intermediate steps can be added later. The computer reads and executes a BASIC program in order of the line numbers. It will not go back to a previous line unless told to do so.

To enter a new program, type NEW, and then type the lines of the program. At the end of each line press the key named RETURN or ENTER. You may enter the lines in any order. The computer will keep track of them in numerical order. If you type LIST, the program currently in the computer's memory will be printed on the screen. To change a line, retype the line number and the complete line as you now want it.

To run a new program after it has been entered, type RUN, and then press the RETURN or ENTER key.

Programs can be saved on disk. Consult your manual on how to do this for your version of BASIC. To run a program already saved on disk you must know the exact name of the program including any spaces or punctuation. To run a program called TABLE SOLVE, type RUN "TABLE SOLVE" and press the RETURN or ENTER key.

836

The following program illustrates many of the commands used in this course.

10 PRINT "A DIVIDING SEQUENCE"	The computer prints A DIVIDING SEQUENCE.
20 INPUT "NUMBER PLEASE?";X	The computer prints NUMBER PLEASE? and waits for you to enter a number. You must give a value to store in the location X. Suppose you use 20. X now contains 20.
30 LET Y = 2	2 is stored in location Y.
40 FOR Z = -5 TO 4	Z is given the value -5. Each time through the loop, the value of Z will be increased by 1.
50 IF Z = 0 THEN GOTO 70	When Z = 0 the computer goes directly to line 70. When Z ≠ 0 the computer executes line 60.
60 PRINT X" TIMES "Y " DIVIDED BY "Z" = " (X*Y)/Z	On the first pass through the loop, the computer prints -8 because (20 · 2)/(-5) = -8.
70 NEXT Z	The value in Z is increased by 1 to -4 and the computer goes back to line 50.
80 END	After going through the FOR . . . NEXT . . . loop with Z = 4, the computer stops.

The output of this program is:

```
A DIVIDING SEQUENCE
NUMBER PLEASE? 20
20  TIMES 2 DIVIDED BY -5 = -8
20  TIMES 2 DIVIDED BY -4 = -10
20  TIMES 2 DIVIDED BY -3 = -13.3333
20  TIMES 2 DIVIDED BY -2 = -20
20  TIMES 2 DIVIDED BY -1 = -40
20  TIMES 2 DIVIDED BY 1 = 40
20  TIMES 2 DIVIDED BY 2 = 20
20  TIMES 2 DIVIDED BY 3 = 13.3333
20  TIMES 2 DIVIDED BY 4 = 10
```

LESSON 7-1 (pp. 418–424)
15. 0 meters per second **17.** positive **19.** ⁻0.1 inch per year **21.** $\frac{y}{4}$ meters per minute **23. a.** $\frac{1}{a}$ **b.** $\frac{1}{3+4x}$ **c.** $x = -\frac{4}{5}$ **25. a.** $y = 5.25$ **b.** $y = \frac{3}{4}x - 3$ **c.** $y = \frac{a}{4}x - 3$ **27.** 10,000 + 1,000x **29.** $-\frac{1}{2}$

LESSON 7-2 (pp. 425–431)
13. C; D **15. a.** r **b.** q **c.** p **17.** $y = 3$
19. a.

	A	B	C
1	x	y	rate of change
2	0	1	
3	4	6	1.25
4	8	7	0.25
5	12	11	1

b. No; the rate of change is not constant. **21. a.** Zero is in the denominator; division by zero is impossible. **b.** No
23. a. 1980–1985 **b.** 1960–1965 and 1965–1970 **c.** $\frac{1}{2}$ cent per year **d.** The cost of stamps does not gradually rise over each 5-year period. For example, the cost was never 4.3 cents. **25. See below.** **27.** No. If you substitute 7 for x in the equation $y = 2x - 5$, then $y = 9$. $y \neq 6$. **29.** ⁻350 + 5x

25.
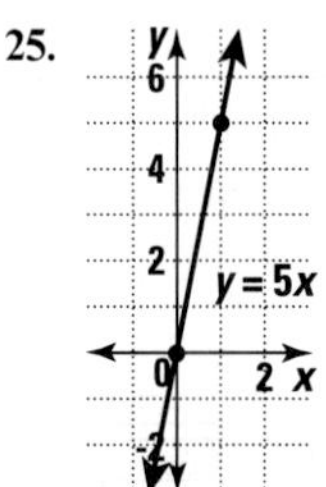

LESSON 7-3 (pp. 432–438)
17.

	A	B
1	x	y
2	6	10
3	7	2
4	8	⁻6
5	9	⁻14

19. 0.79 **21. a.,b. See below.** **c.** Sample: Lines with positive slopes slant upward from left to right. Lines with negative slopes slant downward from left to right. **23.** ⁻2 **25. a.** $y = .25x + 39$ **b. See below.** **c.** $\frac{1}{4}$ **27. a.** 1987–1988 **b.** 1987–1988 **29.** 40 **31. a.** \$1.20 **b.** 2x dollars

21. a.,b.
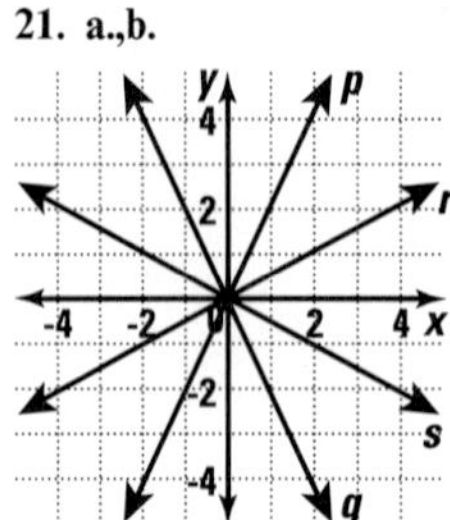

25. b.
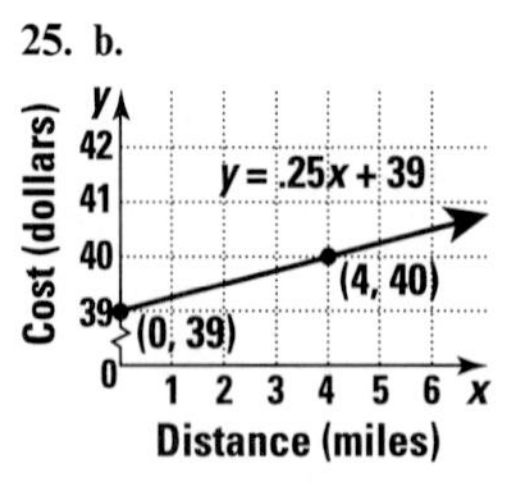

LESSON 7-4 (pp. 439–444)
13. a. ii **b.** 4 **c.** 100 **15. a.** iii **b.** 4 **c.** ⁻100 **17. a.** $y = .50x + 8$ **b. See below.** **19. a.** q **b.** p **c.** n **d.** r **21. a. See below.** **b.** a line that passes through (0, 5) with slope m **c.** Yes; **See below.** **23.** ≈.09 **25.** $\frac{-10}{3}$ **27.** $z = 2$

17. b.
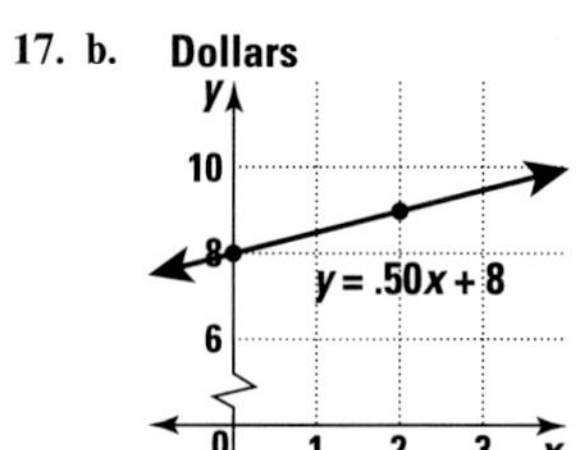

21. a.
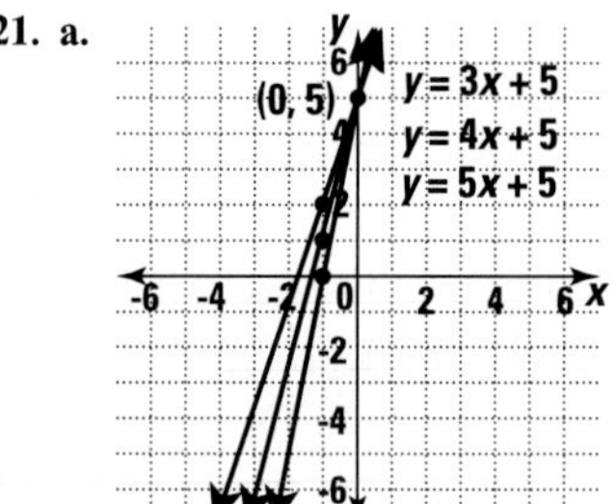

21. c.
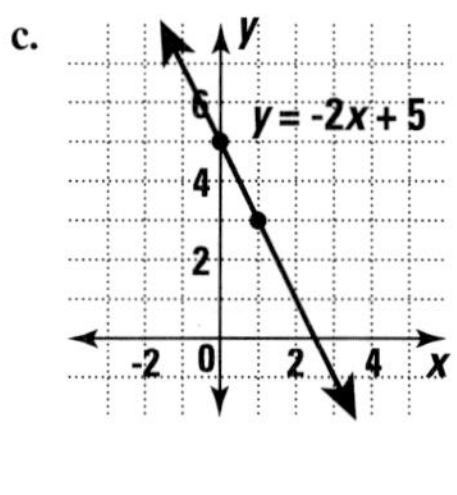

LESSON 7-5 (pp. 445–449)
9. a. $y = 5200x - 10{,}121{,}700$ **b.** 288,700 **c.** 2042
11. a. $y = 3x - 7$ **b.** (0, ⁻7) **c.** (4, 5) **d.** 5 = 3 (4) − 7? 5 = 12 − 7? 5 = 5? Yes, it checks. **13. a.** ⁻3; (14, 68) **b.** $y = {}^{-}3x + 110$ **15. a.** p **b.** n **c.** q **17. See below.** **19.** No, (1, 3) and (⁻3, ⁻5) lie on a line with slope 2, while (1, 3) and (3, 6) lie on a line with slope $\frac{3}{2}$. **21. a.** A **b.** C **23.** 455 adults **25. a.** $n = \sqrt{24}$ or $n = {}^{-}\sqrt{24}$ **b.** $n = 576$ **c.** $n = 4$

17.
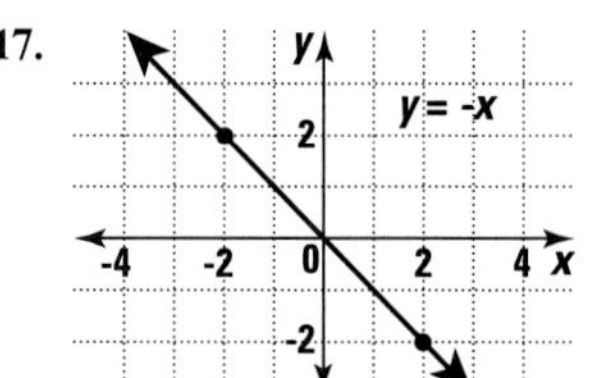

LESSON 7-6 (pp. 450–455)
11. $y = -\frac{5}{6}x + 5$ **13. a.** $C = \frac{5}{9}F - \frac{160}{9}$ **b.** ≈ 65.6°C **c.** 302°F **15.** $S = 180n - 360$ **17.** $y = 7x - 42$ **19.** $a = 10$ **21. a.** 8 or ⁻8 **b.** ⁻1 or 15 **c.** $-\frac{1}{3}$ or 5

LESSON 7-7 (pp. 458–461)
11. No. Negative values of x would result in values of y that are always greater than 107.6°F. **13.** Yes **15.** No **17.** $y = -\frac{7}{4}x + 7$ **19. a.** $-\frac{3}{5}$; $\frac{2}{5}$ **b. See right.** **21. a.** 0 **b.** ⁻2 **23.** about 91 points **25.** 2a + 4c

19. b.
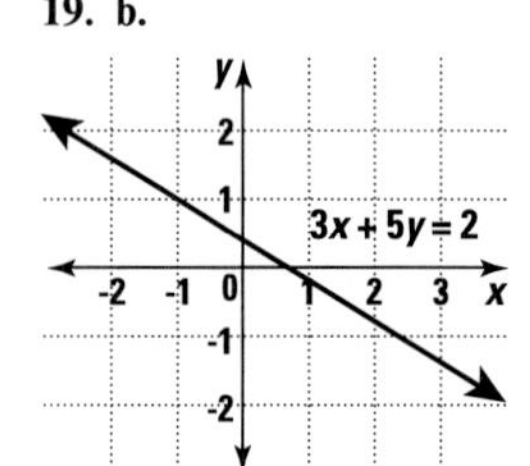

LESSON 7-8 (pp. 463–468)

15. a. $2x + 4y = 100$ **b.** Samples: (2, 24); (10, 20); (30, 10) **17. a.** $y = -\frac{A}{B}x + \frac{C}{B}$ **b.** slope $= -\frac{A}{B}$; y-intercept $= \frac{C}{B}$ **19. a. See below. b.** $y = .38x - 686$ **c.** $\approx$ 74 meters **d.** Sample: The line is not exact, or there may be a threshold distance beyond which it is physically impossible to throw. **21. a.** No; $\frac{0 - (-2)}{10 - 8} = \frac{2}{2} = 1 \neq 2$. **b.** Yes; $\frac{18 - (-2)}{18 - 8} = \frac{20}{10} = 2$. **23. See below. 25.** Yes

19. a.

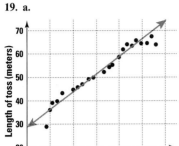

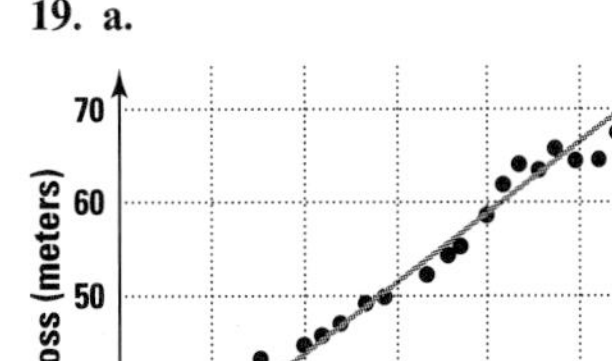

23.

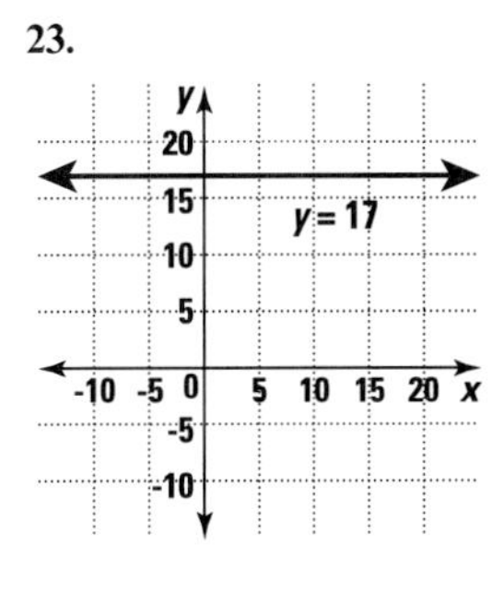

LESSON 7-9 (pp. 469–475)

13. a. $2W + T \geq 20$ **b. See below. 15. a.** 6 **b. See below.**
17. a. Sample: Average annual snowfall is linearly related to the latitude. **b.** Sample: $y = 5x - 172$ **c.** Sample: 8 inches **d.** Sample: altitude, geographic position **19.** $-\frac{1}{3}$ **21. a.** 1.44% **b.** 1.26% **c.** No **23. a.** 0.1 **b.** .03 **c.** .00001

13. b.

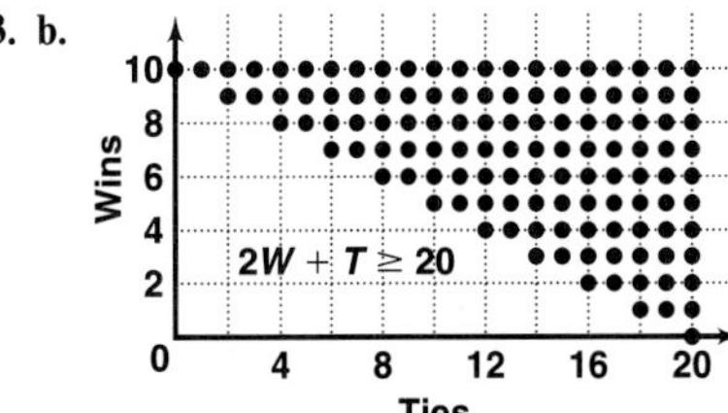

15. b.

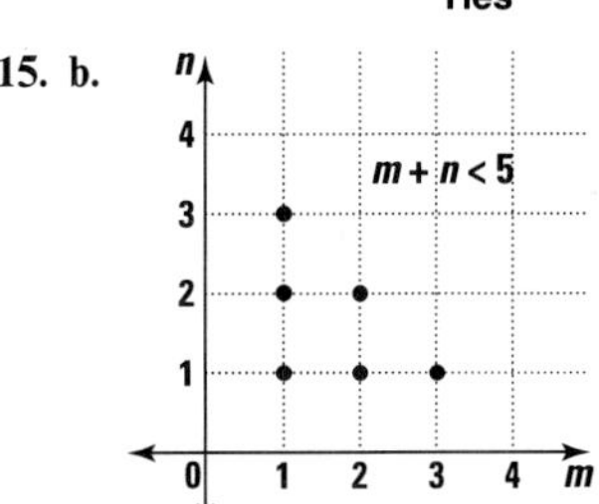

CHAPTER 7 PROGRESS SELF-TEST (p. 479)

1. y-intercept = 5 **2.** x-intercept = 2 **3.** $\frac{0 - 5}{2 - 0} = \frac{-5}{2}$ **4.** $\frac{1 - (-5)}{-2 - 4} = \frac{6}{-6} = -1$; $\frac{-20 - 1}{20 - (-2)} = \frac{-21}{22}$. No; the slope of the line between the first two points (−1) is not the same as the slope of the line between the last two points $\left(-\frac{21}{22}\right)$. **5.** slope = −4; y-intercept = 8 **6.** Rewrite as $y = \frac{-5}{2}x + \frac{1}{2}$, so the slope is $\frac{-5}{2}$ and the y-intercept is $\frac{1}{2}$. **7.** $y = \frac{3}{4}x + 13$ **8.** $-5x + y = -2$; $A = -5$, $B = 1$, $C = -2$ **9.** Slope is 0. **10.** up $\frac{3}{5}$ unit **11.** Do a quick estimate. Between 1942 and 1943 the increase was from about 3 million to 7 million, or an increase of 4 million personnel. From 1943 to 1944 the increase was a little under 1 million. From 1944 to 1945 the increase was about 270,000. From 1945 to 1946 there was a decrease. So the greatest increase was between 1942 and 1943. **12.** $\frac{1{,}889{,}690 - 3{,}074{,}184}{1946 - 1942} = -296{,}123.5$ personnel per year **13.** $2x + y = 67$ **14. a. See right. b.** Substitute $m = -2$ and $(x, y) = (-5, 6)$ in the equation $y = mx + b$ to solve for b. Since $b = -4$, the equation of the line is $y = -2x - 4$. **15.** First find the rate of increase of weight; $\frac{50 - 43}{14 - 12} = \frac{7}{2}$ kg per year. Next, substitute $m = \frac{7}{2}$ and $(x, y) = (14, 50)$ in $y = mx + b$ and solve for b; $50 = \frac{7}{2} \cdot 14 + b$, $50 = 49 + b$, $b = 1$. Therefore, the equation of the line is $y = \frac{7}{2}x + 1$. **16. See right. 17. See right. 18.** First graph $y = x + 1$. It has y-intercept 1 and slope 1. Then test (0, 0) in $y < x + 1$; $0 < 0 + 1$, $0 < 1$. So (0, 0) is a solution and the region below the line is shaded. **See right. 19.** c is the only line with negative slope and positive y-intercept. **20. a.** $\overline{AB}$, since it increases from left to right. **b.** $\overline{DE}$, since it is vertical. **21.** Sample: **a.** (60, 70); (120, 140) **See right. b.** Answers will vary. Sample: $\frac{140 - 70}{120 - 60} = \frac{7}{6}$ **c.** Answers will vary. Sample: Substitute (60, 70) in $y = \frac{7}{6}x + b$ to solve for b. Since $b = 0$, an equation of the line is $y = \frac{7}{6}x$. **d.** Answers will vary. Sample: In the equation $y = \frac{7}{6}x$, substitute 100 for x. $y = \frac{7}{6} \cdot 100 \approx 117$; about 117 feet

14. a.

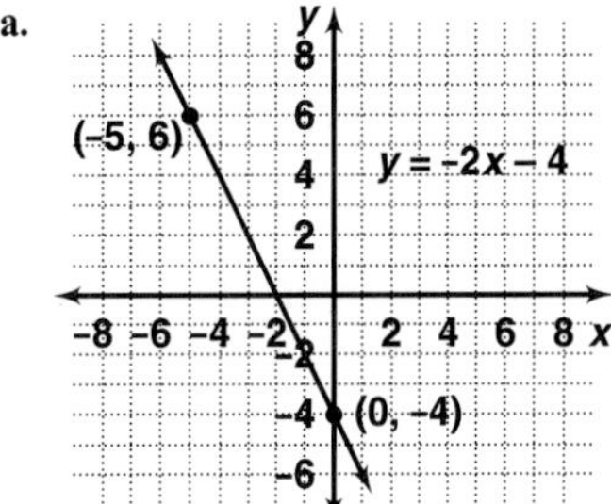

16.

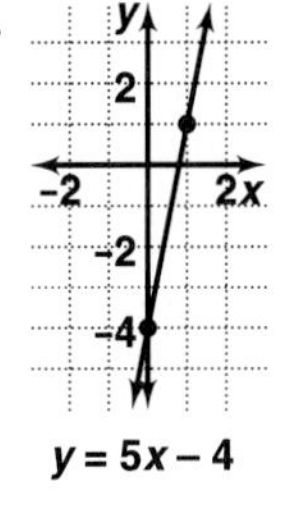

17.

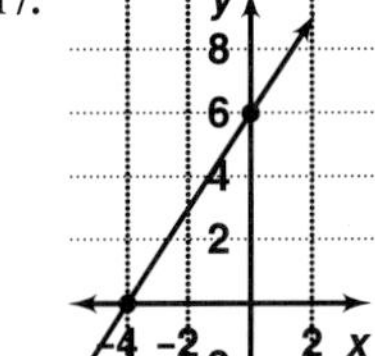

18.

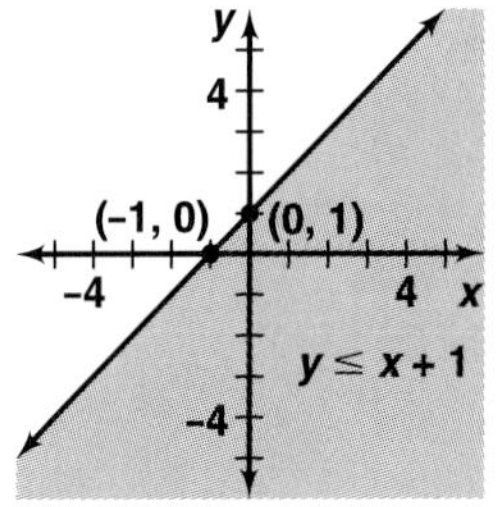

21. a.

Length and wingspan of 2- and 3-engine jets

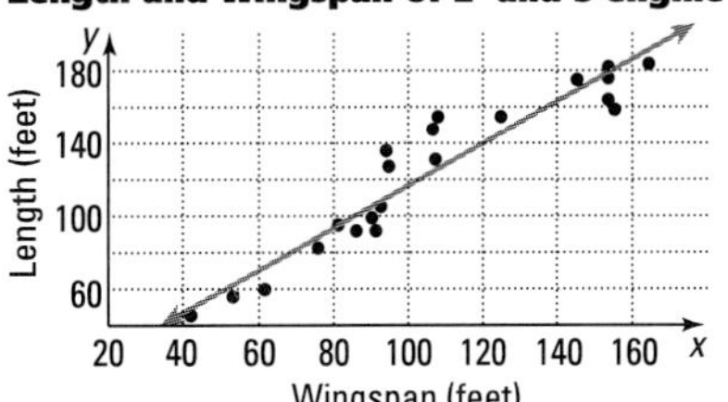

The chart below keys the **Progress Self-Test** questions to the objectives in the **Chapter Review** on pages 480–483. This will enable you to locate those **Chapter Review** questions that correspond to questions missed on the **Progress Self-Test.** The lesson where the material is covered is also indicated on the chart.

Question	**1**	**2**	**3**	**4**	**5**	**6**	**7**	**8**	**9**	**10**
Objective	Voc.	Voc.	A	D	C	C	B	C	D	D
Lesson	7-4	7-5	7-2	7-2	7-4	7-4	7-4	7-8	7-3	7-3
Question	**11**	**12**	**13**	**14**	**15**	**16**	**17**	**18**	**19**	**20**
Objective	E	E	F	B, H	F	H	H	I	H	D
Lesson	7-1	7-1	7-8	7-5	7-6	7-4	7-8	7-9	7-4	7-3
Question	**21**									
Objective	G									
Lesson	7-7									

CHAPTER 7 REVIEW (pp. 480–483)

1. $-\frac{1}{2}$ **3.** $.\overline{54}$ or $\frac{6}{11}$ **5.** $y = 2$ **7.** $y = 4x + 3$ **9.** $y = -2x - 7$ **11.** $y = 30x - \frac{359}{4}$ **13.** $y = \frac{1}{2}x - \frac{9}{2}$ **15.** $x = 6$ **17.** $x - 5y = 22$; $A = 1, B = -5, C = 22$ **19.** $y = -2x + 4$ **21.** 7; −3 **23.** −1; 0 **25.** $\frac{d-b}{c-a}$ or $\frac{b-d}{a-c}$ **27.** height; right **29.** ℓ, n **31.** It is undefined. **33. a. See below. b.** 0.46 **35.** 5.1 cm per year **37. a.** birth to 2 years **b.** 16.5 cm per year **39.** −0.2° per hour **41.** 0.25; 15 **43.** $y = 3x + 50$ **45.** $w = 0.2d + 37.2$ **47.** $2.5B + 5L = 25$ **49.** Sample: **a. See below. b.** ≈ −0.029 **c.** Olympic swimmers drop about .03 minute off their racing time each year. **d.** $y = -0.029x + 61.7$ **e.** 3.82 min or 3:49.2 min **51. See below. 53. See below. 55. See right. 57. See right. 59.** half-planes **61. See right. 63. See right. 65. See right. 67. See right.**

33. a.

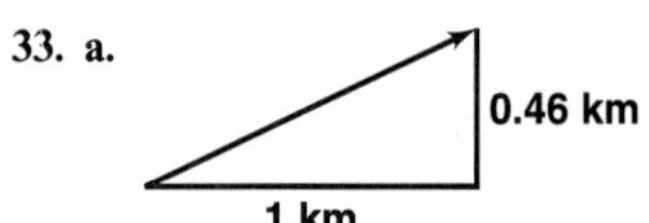

49. a.

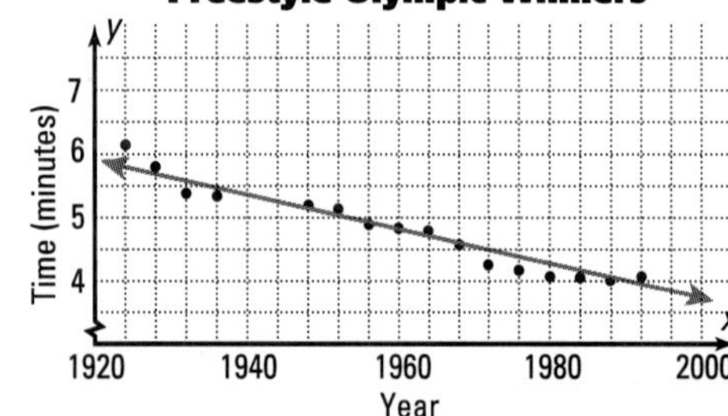

51.

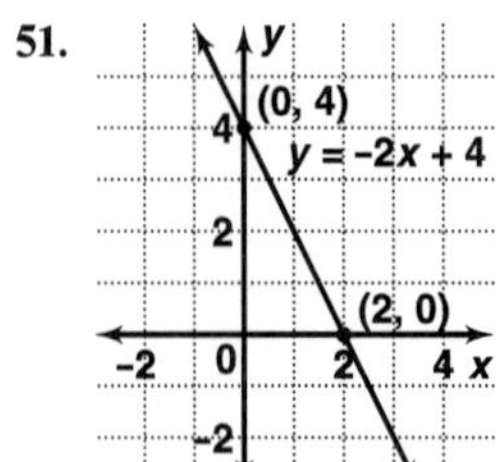

53.

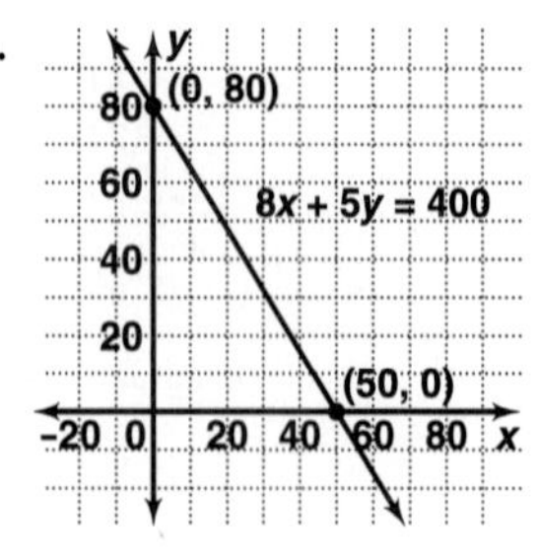

55.

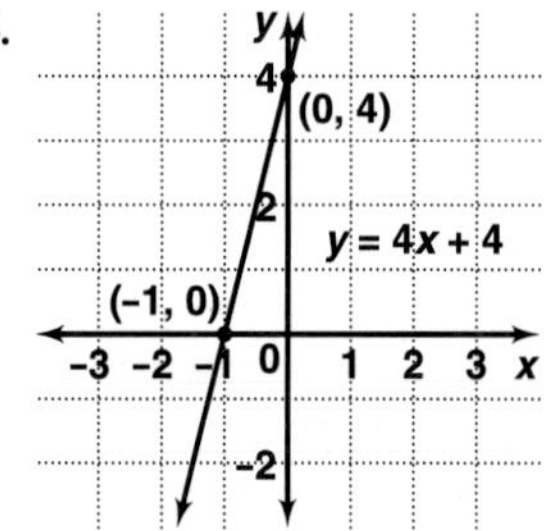

57.

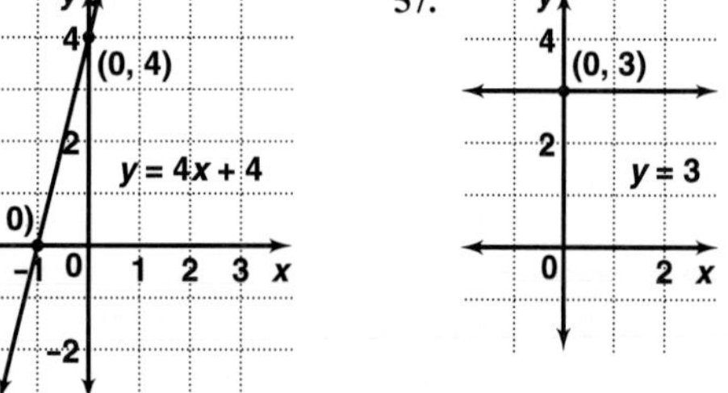

61.

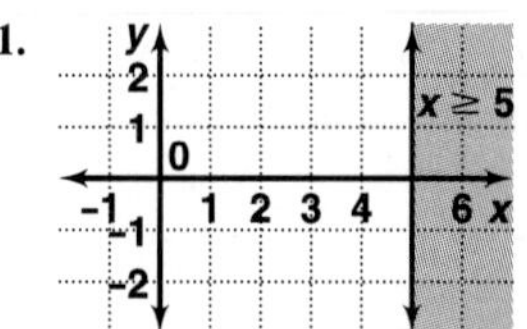

63.

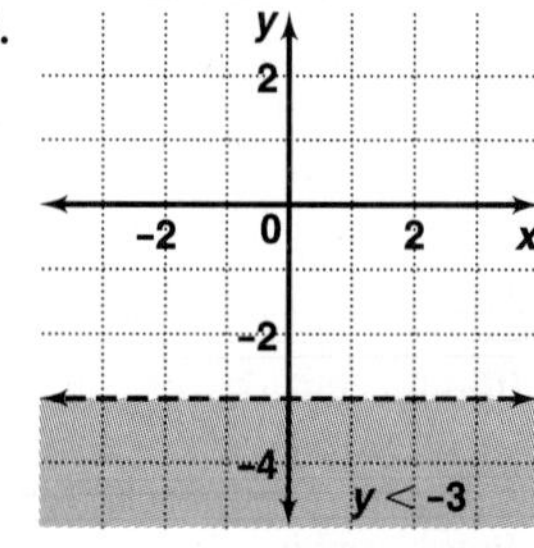

65.

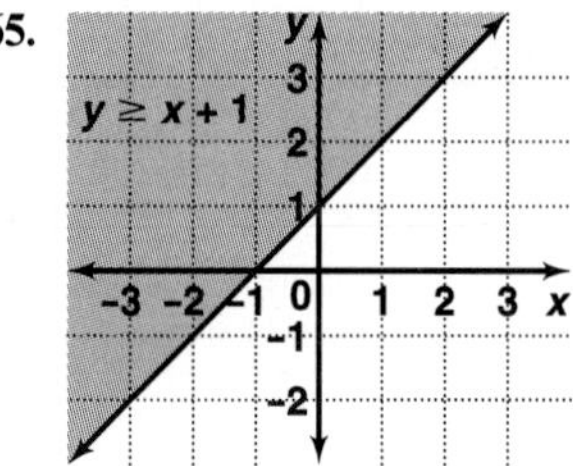

67.

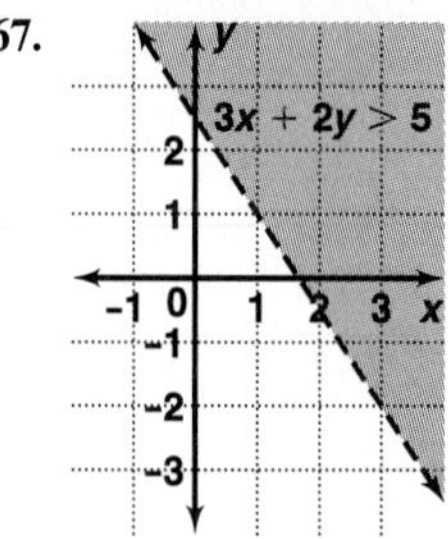

LESSON 8-1 (pp. 486–491)

15. (a); $P(1.06)^5 > P(1.1)^3$ **17. a.** = 1000*1.015^A4−1000*1.015^A3 **b.** 15.45 **19. a.** $T = 2W + 7$ **b.** **See below.** **21.** $\frac{3}{13}$ **23. a.** $36n$ **b.** $36n - 84$ **c.** $-36n + 84$ **d.** $3n^2 - 7n$ **25.** $-\frac{1}{8}$ **27.** 6

19. b.

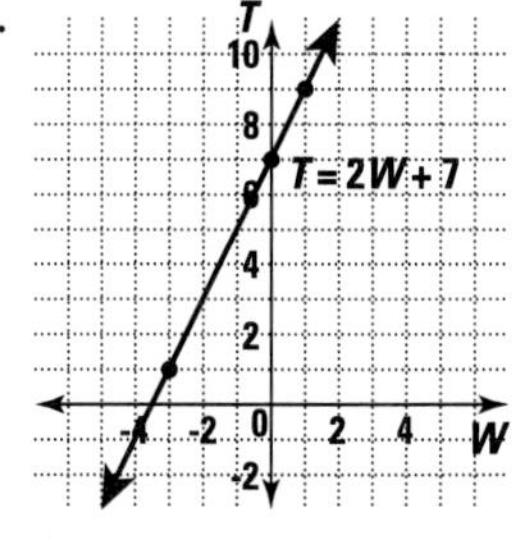

LESSON 8-2 (pp. 492–497)

15. a. 20 minutes **b.** 6 **c.** 1,458,000 **d.** 1,062,882,000 **17.** about 5.14 trillion dollars **19.** 343 **21.** $\frac{8}{9}$ **23.** $\frac{3}{26}$ **25.** $14n + 44$ **27.** $\frac{1}{128}$

LESSON 8-3 (pp. 498–504)

11. a. linear **b.** line **13. a.** exponential **b.** curve

15. a.

	A	B	C
1	years from now	constant increase	exponential
2	0	2520	2520
3	1	2640	2646
4	2	2760	2778
5	3	2880	2917
6	4	3000	3063
7	5	3120	3216

b. There are 96 more students if the growth is exponential. **c.** 4,320 students **d.** 5,239 students **17.** (c) **19.** (b) **21. a.** 100*1.06^YEAR **b.** 20 320.7135 **c.** 30 FOR YEAR = 1 TO 100 **d.** 100 33930.2084 **23. a.** **See below.** **b.** Sample: (1, 41), (6, 19) **c.** Sample: $y = -4.4x + 45.4$ **d.** about 12.4 in. **25.** $6a^3 + 12a^2 - 2a$ **27. a.** $\frac{9}{x}$ **b.** $\frac{13}{2y}$ **c.** $\frac{8z+5}{2z}$ **29.** −13,824

23. a.

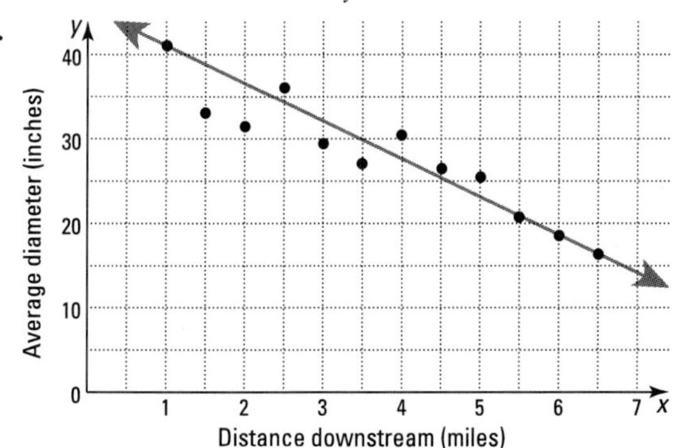

LESSON 8-4 (pp. 505–509)

13. True **15. a.** A decays exponentially; the growth factor is $\frac{1}{2}$. **b.** t grows exponentially; the growth factor is 2. **17. a.** 5832 **b.** 1.033174×10^{-2} **c.** $P = 100{,}000$; $X = 1.02$; $N = 10$ **19.** $k = 30(1.05)^n$ **21.** about 6,230,000 **23.** 1 **25. a.** $\frac{1}{2}y^3$ **b.** Sample: $\frac{1}{2}y$, y, y

LESSON 8-5 (pp. 510–514)

17. $a = 7$ **19.** Samples: x^1, x^6; x^3, x^4; x^0, x^7 **21. a.** $P \cdot 3^5$ **b.** 12 **23.** $2x^7$ **25.** x^{14} **27.** $y^8 - y^3$ **29.** **See below.** **31.** Sample: the population T of a city of 5000 that is growing at a rate of 3.5% per year, n years from now **33.** \$4.49/lb **35. a.** −1; 1; −1; 1; −1; 1; −1; 1 **b.** 1 **37.** 3.6×10^{-4}

29.

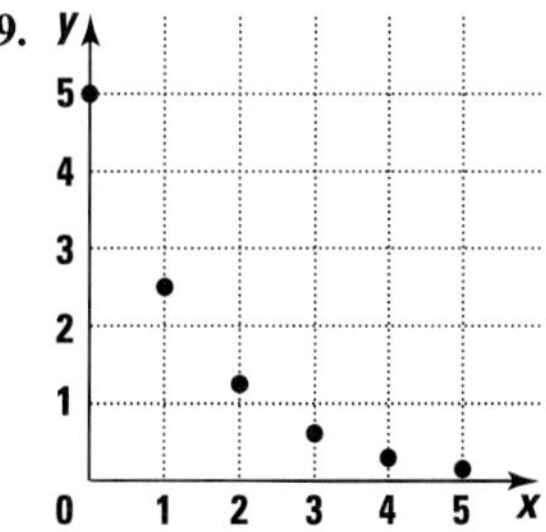

LESSON 8-6 (pp. 515–520)

19. a. **See below.** **b.** The y-coordinate approaches zero. **21.** $z = -2$; Check answers using a calculator. **23.** t^{-6} **25. a.** ≈ 5.82 billion **b.** ≈ 4.60 billion **27.** $12x^3$ **29.** $22c^7$ **31.** (c) **33.** $y = -5x + 9$ **35. a.** $28a - 8$ **b.** $13a - 11$ **c.** $a = 2$

19. a.

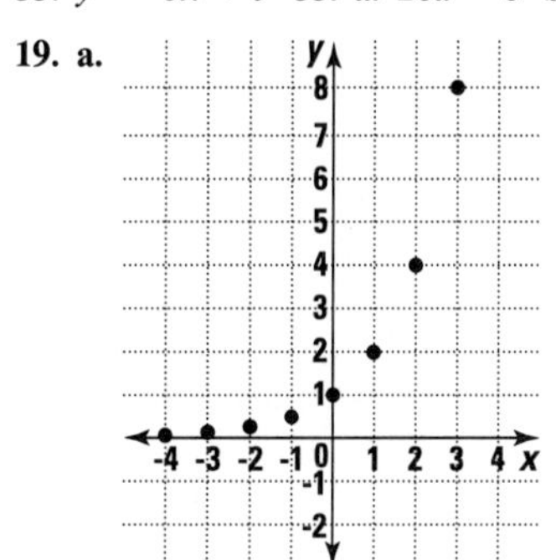

LESSON 8-7 (pp. 521–526)

15. $\frac{6 \cdot 10^9}{2.56 \cdot 10^8} \approx 2.34 \cdot 10^1 \approx 23 \frac{\text{pounds}}{\text{person}}$ **17.** 3^6 **19.** x^n **21.** 1 **23.** $5p$ **25.** about $0.40 \frac{\text{people}}{\text{km}^2}$ **27.** 4^{x+y} **29.** $-2y^{16}$ **31.** x^{15}; $3^5 \cdot 3^5 \cdot 3^5 = 243 \cdot 243 \cdot 243 = 14{,}348{,}907$; $3^{15} = 14{,}348{,}907$. **33. a.** slope = −2, y-intercept = 54; The elevator descends at a rate of 2 floors per second and started on the 54th floor. **b.** **See below.** **35.** 1000 times as much

33. b.

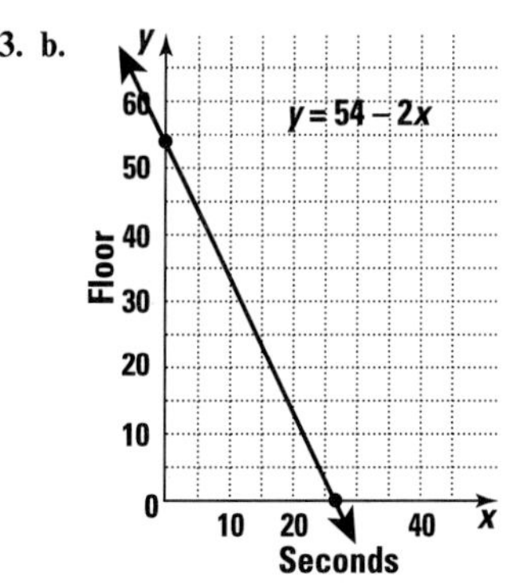

LESSON 8-8 (pp. 527–532)

17. $18x^2$ **19.** $\frac{u^t}{3^t}$ **21.** $\frac{16z^5}{2401}$ **23. a.** $\left(\frac{2}{3}\right)^5 = \frac{32}{243} \approx 0.13$ **b.** Example 4 **25.** (a) **27.** 1728 **29.** $\frac{1}{5}$ **31.** k^3 **33.** v^{-6} **35. a.** $2x^5$ **b.** x^9 **c.** x^{10} **37. a.** \$6.00 − \$.12 = \$5.88 **b.** \$60.00 + \$.20 = \$60.20 **c.** \$6.00

LESSON 8-9 (pp. 533–538)
11. Sample: Let $x = 3$. Then $(2x)^3 = (2 \cdot 3)^3 = 6^3 = 216$ and $2x^3 = 2 \cdot 3^3 = 2 \cdot 27 = 54$. **13.** Power of a Product Property **15.** Negative Exponent Property **17.** Sample: Let $a = 2$ and $b = 3$. Then $2^2 + 3^2 = 13$ and $(2 + 3)^2 = 5^2 = 25$. The counterexample shows that the pattern is not true. **19.** $\frac{36}{625x^4}$ **21.** $\frac{25a^9}{2b^3}$ **23.** $2n^4$ **25.** $a = 20$ **27.** 2^{1492} **29. a. See right. b. See right. c.** Sample: Of two accounts, each starting with \$100, which has more money in it at the end of 10 years, one that earns 6% compound interest or one that has \$6 deposited in it each year? Answer: the one that earns 6% compound interest. **31.** (a); $5m^2$

29. a.

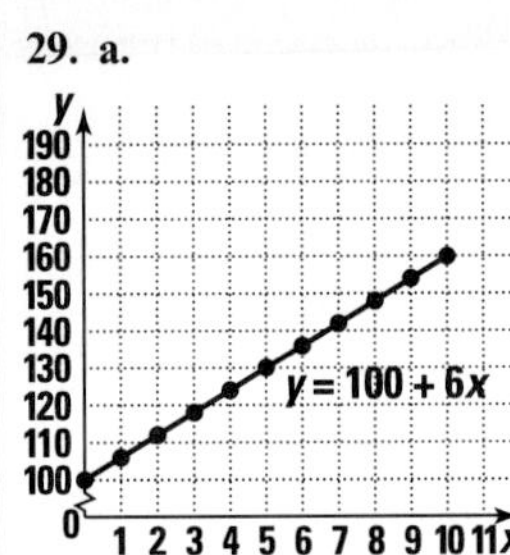

29. b.

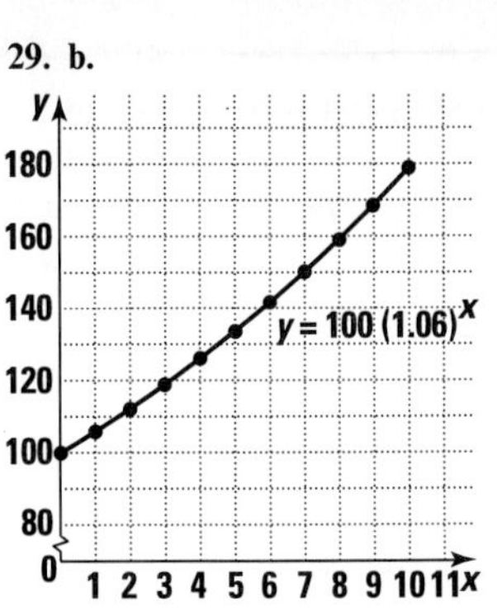

CHAPTER 8 PROGRESS SELF-TEST (p. 542)

1. a. $\frac{4^{12}}{4^6} = 4^{12-6} = 4^6 = 4096$; Apply the Quotient of Powers Property. **b.** $\frac{4^{12}}{4^6} = \frac{16777216}{4096} = 4096$ **2.** $\frac{5 \cdot 10^{20}}{5 \cdot 10^{10}} = \frac{10^{20}}{10^{10}} = 10^{20-10} = 10^{10} = 10{,}000{,}000{,}000$ **3.** $(8)^{-5} = \frac{1}{8^5} = \frac{1}{32768}$ **4.** $b^7 \cdot b^{11} = b^{7+11} = b^{18}$ **5.** $(5y^4)^3 = 5^3(y^4)^3 = 125y^{4 \cdot 3} = 125y^{12}$ **6.** $\frac{3z^6}{12z^4} = \frac{z^6}{4z^4} = \frac{1}{4}(z^{6-4}) = \frac{z^2}{4}$ **7.** $(y^{10})^4 = y^{10 \cdot 4} = y^{40}$ **8.** $\left(\frac{3}{x}\right)^2 \cdot \left(\frac{x}{3}\right)^4 = \frac{3^2}{x^2} \cdot \frac{x^4}{3^4} = \frac{x^2}{3^2} = \frac{x^2}{9}$ **9.** $\frac{48a^3b^7}{12a^4b} = \frac{4b^6}{a} = 4a^{-1}b^6$ **10.** 6 **11.** Sample: $3 \cdot 2^2 = 3 \cdot 4 = 12$ and $(3 \cdot 2)^2 = 6^2 = 36$; $12 \neq 36$. **12.** Product of Powers Property **13.** $\$6500(1.05)^5 \approx \$8295.83$ **14.** $1900(1.058)^3 \approx 2250.15$; interest earned $= 2250.15 - 1900 = \$350.15$ **15.** $135{,}000(1.03)^5 \approx 157{,}000$ **16.** $135000(1.03)^{-2} \approx 127{,}000$ **17. a.** exponential **b.** not exponential **c.** not exponential **d.** exponential **18.** See right. **19.** $(1.30)^3 = 2.197$ times as large **20.** $V = \frac{4}{3}\pi(6.96 \cdot 10^6)^3 \approx 1.41 \cdot 10^{21}$ km^3

18.

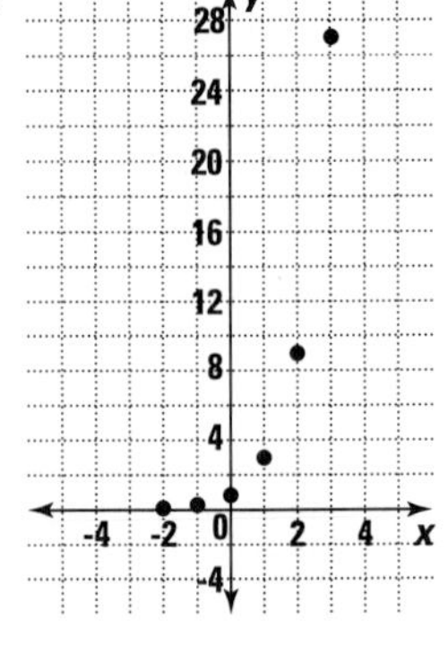

The chart below keys the **Progress Self-Test** questions to the objectives in the **Chapter Review** on pages 543–545. This will enable you to locate those **Chapter Review** questions that correspond to questions missed on the **Progress Self-Test.** The lesson where the material is covered is also indicated on the chart.

Question	**1**	**2**	**3**	**4**	**5**	**6**	**7**	**8**	**9**	**10**
Objective	A	A	A	B	C	B	B	C	B	A
Lesson	8-7	8-7	8-6	8-5	8-8	8-7	8-5	8-8	8-7	8-2
Question	**11**	**12**	**13**	**14**	**15**	**16**	**17**	**18**	**19**	**20**
Objective	D	E	F	F	G	G	I	I	H	H
Lesson	8-9	8-5	8-1	8-1	8-2	8-6	8-4	8-6	8-3	8-8

CHAPTER 8 REVIEW (pp. 543–545)

1. a. 81 **b.** −81 **c.** 81 **3.** 4 **5.** 8 **7.** $\frac{1}{125}$ **9.** $\frac{8}{343}$ **11.** 81 **13.** x^{11} **15.** x^3y^{12} **17.** n^{13} **19.** $\frac{c}{a}$ **21.** $28x^{15}$ **23.** $5^{-1}m^4$ **25.** 1000 **27.** $\frac{x}{y^2}$ **29.** $\frac{x^3}{y^3}$ **31.** $1024x^5$ **33.** $\frac{32}{n^5}$ **35.** $-27n^3$ **37.** $\frac{4k^3}{27}$ **39.** $32x^2$ **41. a.** True **b.** True **c.** False **d.** False **43.** Sample: $(1 + 1)^3 = 2^3 = 8$; $1^3 + 1^3 = 1 + 1 = 2$. **45.** Power of a Product Property **47.** Zero Exponent Property **49.** Power of a Quotient Property **51.** Negative Exponent Property **53.** Samples: $\left(\frac{x^3}{x}\right)^8 = (x^{3-1})^8$; $(x^2)^8 = x^{16}$; $\left(\frac{x^3}{x}\right)^8 = \frac{x^{3 \cdot 8}}{x^{1 \cdot 8}} = \frac{x^{24}}{x^8} = x^{24-8} = x^{16}$ **55.** \$2952.33 **57.** \$1348.32 **59.** \$9.02 per hour **61. a.** 128,000 **b.** After 4 hours there will be 128,000 bacteria. **63.** $P = 1{,}500{,}000 \cdot (0.97)^n$ **65.** 1,500,000; the population now **67. a.** ≈ 459 **b.** The death rate three years earlier (1977) was about 459 per 100,000 people. **69.** $(0.9)^x$ **71.** 5 billion cubic miles **73.** $\left(\frac{1}{3}\right)^4 = \left(\frac{1}{81}\right)$ **75.** exponential **77.** linear

79. See right.

x	y
−3	15.625
−2	6.25
−1	2.5
0	1
1	0.4
2	0.16
3	0.064

81. $y = 5 \cdot (1.04)^x$; because if the growth factor g is greater than one, exponential growth always overtakes constant increase.

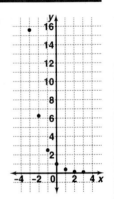

LESSON 9-1 (pp. 548–553)
15. about 1.46 seconds **17.** 11; −13 **19.** x^7y^4 **21.** $\frac{8a^3}{125}$ **23.** It is at least 5 km and at most 11 km since 8 − 3 = 5 and 8 + 3 = 11. **25.** 7500 − 1800 − k or 5700 − k

LESSON 9-2 (pp. 554–560)
15. (c) **17.** (a) **19.** (b); Sample: if $x = 1$, $y = 1 - 6 + 8 = 3$. Graph (b) contains this point. **21.** He must change the y-values in his table to their opposites, and turn his graph upside down. **23.** $\sqrt{640} \approx 25.3$ or $-\sqrt{640} \approx -25.3$ **25.** $\frac{c + d + r}{(c - 3) + d + r}$

LESSON 9-3 (pp. 562–566)
11. a. See below. **b.** Sample: All the parabolas look alike in the window. They have different vertices and are translation images of each other. **c.** It will be congruent to those in part **a.** It will be a translation image of them and have vertex (0, 6). **d.** See below. **13. a.** See below. **b.** (−1.25, 1.25), (4, 17) **15. a.** See below. **b.** 1 **17.** $x = 7$ or $x = -7$ **19. a.** 4 miles **b.** for distances less than 4 miles; $.45 + 1.20x < 1.25 + 1.00x$; $x < 4$ **21. a.** 420 **b.** 456 **c.** ≈ 21.4

11. a.

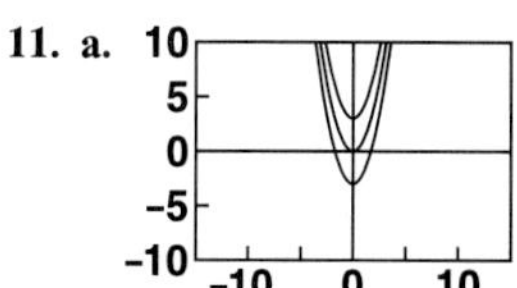

11. d.

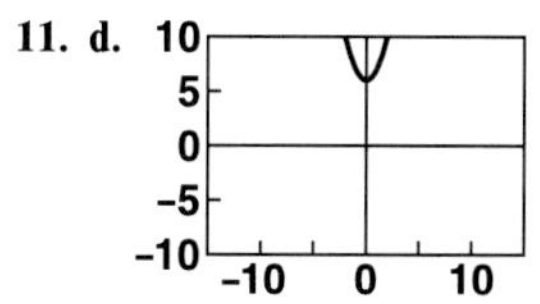

13. a.

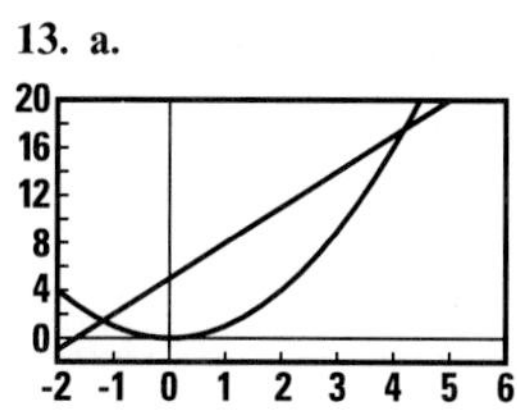

15. a.

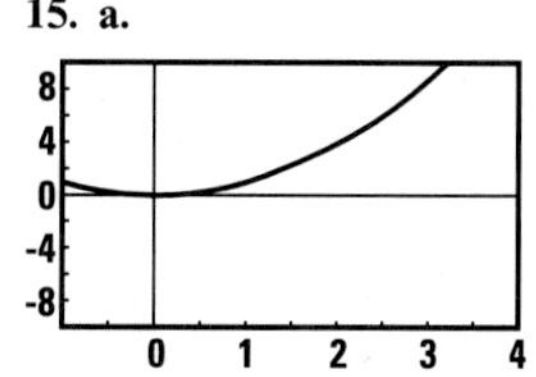

LESSON 9-4 (pp. 567–572)
13. 25 meters **15.** 35 meters **17.** 10.125 m above the surface of the water

19. a.

x	y
−3	$\frac{9}{2}$
−2	2
−1	$\frac{1}{2}$
0	0
1	$\frac{1}{2}$
2	2
3	$\frac{9}{2}$

b. See right. **21.** $A = (8, 4)$; $B = (9, 7)$ **23.** (c) **25.** (−3, −5) **27.** ≈3.4

19. b.

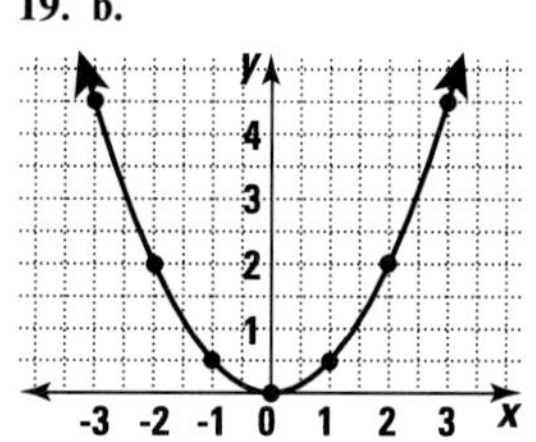

LESSON 9-5 (pp. 573–578)
15. a. 2.5; 5.0 **b.; c.** See right. **17.** 5.1 meters above the cliff **19.** 10 ft **21.** ≈ 13.5 ft **23.** (d) **25.** 540

15. b.; c.

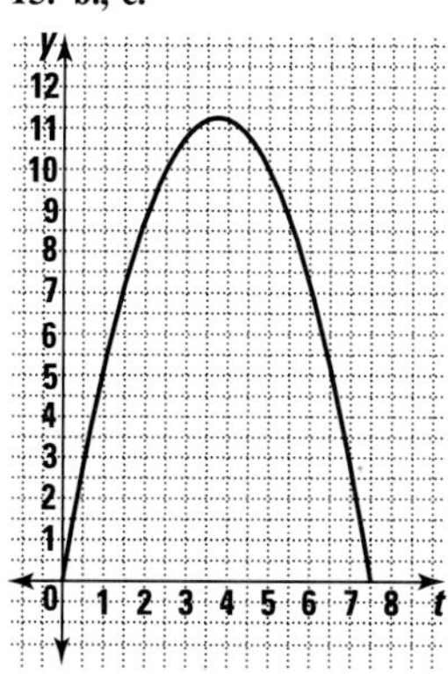

LESSON 9-6 (pp. 579–585)
11. a. 5 meters **b.** 0.4 seconds and 1.6 seconds **c.** ≈ 2.4 seconds **13.** 9 **15. a.** $0 = -16t^2 + 28t - 12$ **b.** positive **c.** 2 **17.** 12 feet **19.** down **21.** Sample: $y = -x^2$ **23. a.** $a < \frac{1}{3}$ **b.** $b > \frac{1}{3}$ **c.** $c > 0$ **25.** 6435

LESSON 9-7 (pp. 586–592)
15. $t = 3\sqrt{3}$ **17.** 6 **19.** 55 **21.** $a = \pm 2\sqrt{3}$; $(2(2\sqrt{3}))^2 = (4\sqrt{3})^2 = 16 \cdot 3 = 48$; $(2(-2\sqrt{3}))^2 = (-4\sqrt{3})^2 = 16 \cdot 3 = 48$ **23. a.** $5\sqrt{3}$ **b.** $2\sqrt{3}$ **c.** $7\sqrt{3}$ **25.** Sample: $w\sqrt{20}$, $2w\sqrt{5}$ **27.** It equals zero. **29.** 49 meters **31.** 2 **33.** 2 **35.** $0.25q + 0.10d \geq 5.20$

LESSON 9-8 (pp. 593–598)
23. 0; no real solutions **25.** 2; $q = -31$ or $q = 31$ **27.** −10; 4 **29. a.** $x - y$ **b.** $y - x$ **c.** $|y - x|$ or $|x - y|$ **31.** $\sqrt{7}$ **33.** $4 \pm 3\sqrt{2}$ **35. a.** $10\sqrt{2}$ ft **b.** 14.1 ft **c.** $10\sqrt{5} \approx 22.4$ ft **37.** $x = 4 + 3\sqrt{2}$ or $x = 4 - 3\sqrt{2}$

LESSON 9-9 (pp. 599–604)
15. 7 km **17.** $\sqrt{5} \approx 2.2$ km **19.** $\sqrt{|a|^2 + |b|^2}$ **21.** $JK = 10$, $KL = 17$, $JL = 21$ **23.** 29.732137 **25.** $x = 3.5$ or $x = -3.5$ **27. a.** $t = \pm 3$ **b.** $t = 81$ **c.** $t = 9$ or $t = -9$ **29.** 2; the line $y = 10$ intersects the graph twice. **31. a.** $\frac{-9 \pm \sqrt{101}}{2} \approx 0.52$ or -9.52 **b.** See below. **33.** $12\frac{2}{3}$

31. b.

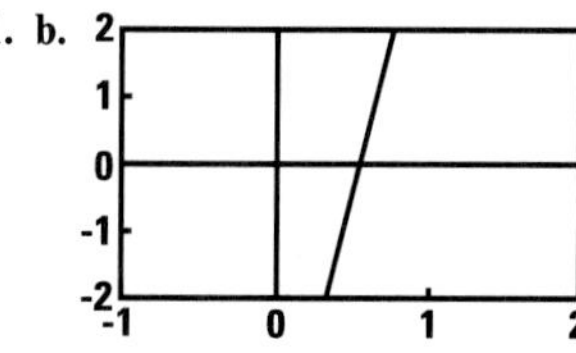

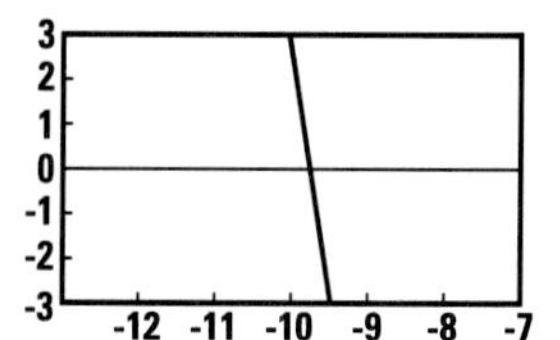

CHAPTER 9 PROGRESS SELF-TEST (p. 609)
1. $\frac{-(-9) \pm \sqrt{(-9)^2 - 4(1)(20)}}{2(1)} = \frac{9 \pm \sqrt{81 - 80}}{2} = \frac{9 \pm 1}{2} = 5$ or 4 **2.** $5y^2 - 3y - 11 = 0$, $\frac{-(-3) \pm \sqrt{(-3)^2 - 4(5)(-11)}}{2(5)} = \frac{3 \pm \sqrt{9 + 220}}{10} = \frac{3 \pm \sqrt{229}}{10}$; $y \approx 1.81$ or $y \approx -1.21$ **3.** no real solutions **4.** $z^2 - 16y + 64 = 0$, $\frac{-(-16) \pm \sqrt{(-16)^2 - 4(1)(64)}}{2(1)} = \frac{16 \pm \sqrt{256 - 256}}{2} = \frac{16}{2} = 8$ **5.** 2 **6.** (a) because as a gets larger than 1 in $y = ax^2$, the graph gets narrower.

7. a.

x	y
−3	−18
−2	−8
−1	−2
0	0
1	−2
2	−8
3	−18

b. See p. 857.

8. a.

x	y
−3	24
−2	15
−1	8
0	3
1	0
2	−1
3	0

b. See p. 857.

9. False. If $a > 0$, the graph opens up. **10.** (2, −2) **11.** (1, 0); (3, 0)

12. $x = 2$ 13. $\sqrt{500} = \sqrt{100 \cdot 5} = \sqrt{100} \cdot \sqrt{5} = 10\sqrt{5}$ 14. $\frac{\sqrt{75}}{5} = \frac{\sqrt{25 \cdot 3}}{5} = \frac{\sqrt{25} \cdot \sqrt{3}}{5} = \frac{5\sqrt{3}}{5} = \sqrt{3}$ 15. $\sqrt{5x} \cdot \sqrt{45y} = \sqrt{225xy} = \sqrt{225} \cdot \sqrt{xy} = 15\sqrt{xy}$ 16. Sample: Plot the vertex. When a is negative, the graph opens down. You can't tell how narrow or broad the parabola is without more information. **See right.** 17. $(7, -6)$ 18. True 19. $\sqrt{|2 - 7|^2 + |-6 - 4|^2} = \sqrt{|-5|^2 + |-10|^2} = \sqrt{25 + 100} = \sqrt{125} = \sqrt{25 \cdot 5} = \sqrt{25} \cdot \sqrt{5} = 5\sqrt{5}$ 20. $\sqrt{|3 - x|^2 + |-2 - y|^2}$ 21. Since $h = 0$ at ground-level, solve $0 = -16t^2 + 21t + 40$. $t \approx 2.4$ seconds 22. Solve $43 = -16t^2 + 21t + 40$. $t = 0.2$ or 1.1 seconds 23. The highest point is the vertex located at 10 yards. When $x = 10$, $h = -0.07 \cdot 10^2 + 1.4 \cdot 10 + 5 = 12$. So the vertex is (10, 12). **See right.** 24. Melody is $20 - 2 = 18$ yards from Harry. So, the ball is $-0.07(18)^2 + 1.4(18) + 5 = 7.52$ ft high when it passes Melody. 25. $\frac{1}{3}x^2 = 10$ $x^2 = 30$; $x \approx 5.48$ or -5.48 26. $x = -57$ or $x = 57$ 27. $3 - n = 0.5$ or $3 - n = -0.5$; $-n = -2.5$ or $-n = -3.5$; $n = 2.5$ or $n = 3.5$

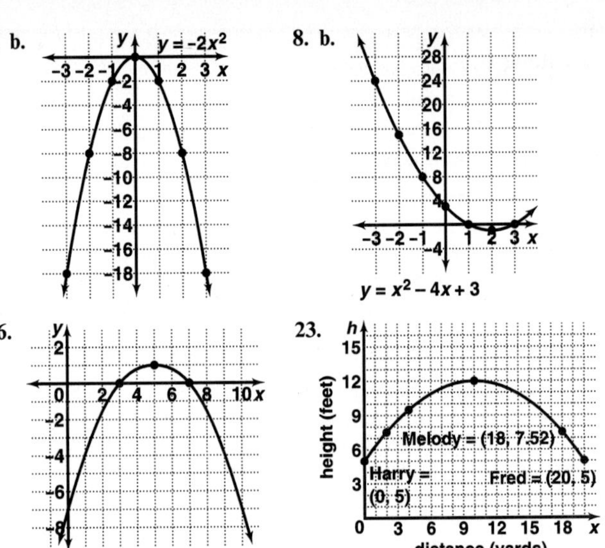

The chart below keys the **Progress Self-Test** questions to the objectives in the **Chapter Review** on pages 610–613. This will enable you to locate those **Chapter Review** questions that correspond to questions missed on the **Progress Self-Test.** The lesson where the material is covered is also indicated on the chart.

Question	**1**	**2**	**3**	**4**	**5**	**6**	**7**	**8**	**9**	**10**
Objective	A	A	A	A	D	F	F	F	F	F
Lesson	9-5	9-5	9-5	9-5	9-6	9-3	9-1	9-2	9-2	9-3
Question	**11**	**12**	**13**	**14**	**15**	**16**	**17**	**18**	**19**	**20**
Objective	F	F	B	B	B	F	G	G	G	G
Lesson	9-3	9-3	9-7	9-7	9-7	9-3	9-9	9-9	9-9	9-9
Question	**21**	**22**	**23**	**24**	**25**	**26**	**27**			
Objective	E	E	E	E	A	C	C			
Lesson	9-4	9-4	9-4	9-4	9-1	9-8	9-8			

CHAPTER 9 REVIEW (pp. 610–613)

1. $-5, 5$ 3. $-7, 7$ 5. $-\frac{5}{2}, \frac{4}{3}$ 7. 7 9. $-0.27, 7.27$ 11. no real solutions 13. $\frac{5 \pm \sqrt{13}}{2}$, ≈ 4.30 or ≈ 0.70 15. 14 17. $20\sqrt{2}$ 19. $10\sqrt{5}$ 21. $18\sqrt{2}$ 23. $3 \pm 3\sqrt{6}$ 25. $x\sqrt{5}$ 27. 17 29. 43 31. 6 33. -5 35. -16; 16 37. -7; 7 39. 5; 15 41. $\frac{-b \pm \sqrt{b^2 - 4ac}}{2a}$ 43. True 45. $D = -23$; no real solutions 47. $D = -1199$; no real solutions 49. a. 576 ft b. ≈ 11.2 seconds 51. a. See right. b. 2.5 and 10 seconds c. $\frac{-10 \pm \sqrt{10^2 - 4(-0.8)(-20)}}{2(-0.8)} = \frac{-10 \pm \sqrt{100 - 64}}{-1.6} = \frac{-10 \pm 6}{-1.6} =$ 2.5 or 10 53. a. ≈ 48 ft b. ≈ 0.65 or 3.35 seconds c. 4 d. Sample: What is the maximum height the ball will reach? 55. a. (10, 13) b. $x = 10$ c. $A = (7, 4)$, $B = (8, 9)$, $C = (9, 12)$ 57. False 59. $ax^2 + bx + c = 0$

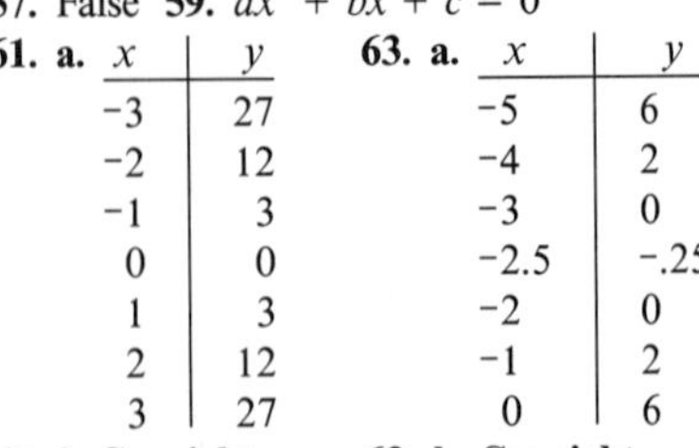

61. a.

x	y
−3	27
−2	12
−1	3
0	0
1	3
2	12
3	27

63. a.

x	y
−5	6
−4	2
−3	0
−2.5	−.25
−2	0
−1	2
0	6

61. b. See right. 63. b. See right.

65. (b) 67. a. See below. b. (20, 16) c. maximum 69. Sample: $-5 \le x \le 15$ and $-10 \le y \le 30$ 71. 31 73. -8, 2 75. 5 77. 9 79. $10\sqrt{26} \approx 50.99$ 81. 17 83. a. See below. b. $7\sqrt{2} \approx 9.9$ miles

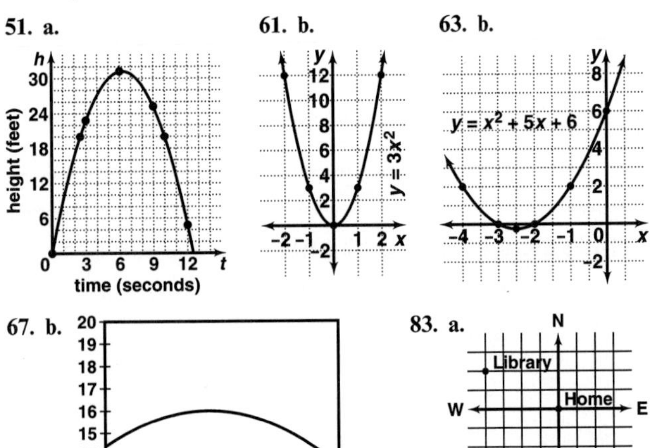

67. b.

83. a.

LESSON 10-1 (pp. 616–620)

21. a. $4x$ **b.** 1 **23. a.** $-30n^4$ **b.** 4 **25. a.** $64a^6b^6$ **b.** 12 **27. a.** See below. **b.** See below. **29. a.** $2 \cdot 10^2 + 4 \cdot 10 + 6$; $1 \cdot 10^3 + 3 \cdot 10 + 2$ **b.** The sum is $1 \cdot 10^3 + 2 \cdot 10^2 + 7 \cdot 10 + 8 = 1278$; $246 + 1032 = 1278$; yes the sums are equal. **31. a.** See below. **b.** See below. **33.** See below. **35. a.** $h = 5$ **b.** $r = 5\sqrt{10}$

27. a.

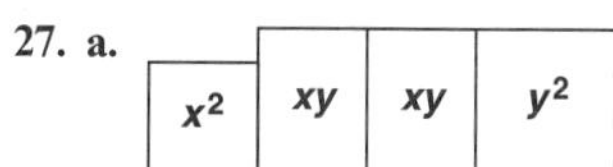

27. b.

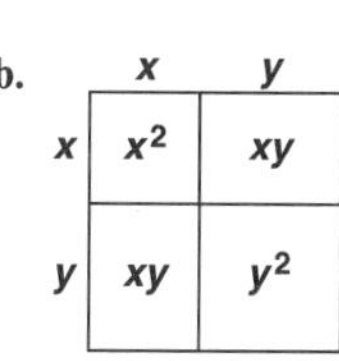

31. a.

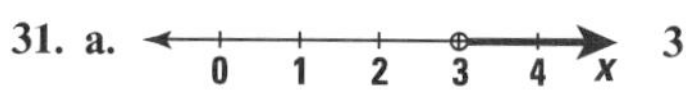

31. b.

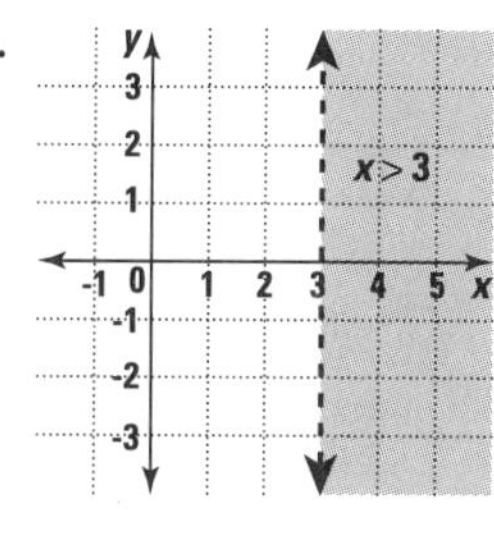

33.

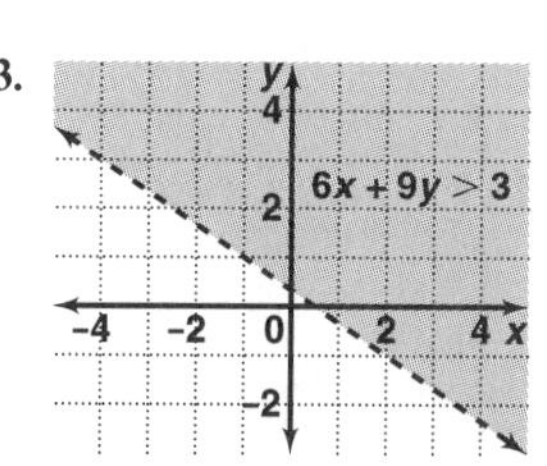

LESSON 10-2 (pp. 621–626)

9. \$100 **11.** \$480.18 **13.** $16y^2 + y - 17$ **15.** $2w^2 - w + 11$ **17.** $-\frac{3}{2}, -\frac{2}{3}$ **19.** $7y^2 - 47$ **21.** Sample: area of a circle with radius r; degree 2 **23.** Sample: amount saved if \$16 was invested at some rate t two years ago, and then \$48 was added to the account one year ago; degree 2 **25.** 61.4 ft **27.** $-\frac{b}{a}$ **29.** (c) **31.** .5 and 1.5

LESSON 10-3 (pp. 627–632)

15. $w(5w - 1)$; $5w^2 - w$ **17.** $8x^2 + 2x$ **19.** $6a^3b^3c^2$ **21.** $ay^2 + ya^2$ **23.** $6n^2$ **25. a.** No. **b.** $2x^{-3}$ cannot be written as a product of variables with nonnegative exponents. **27.** Sample: $-4x^7 + 5x^5 + 1$ **29.** $\frac{3 + 4v}{2v}$ **31.** $y = 3x + 5$ **33. a.** 2 meters per second **b.** Sample: snake

LESSON 10-4 (pp. 633–637)

11. $2n^3 + 7n^2 - 19n - 60$ **13.** $40x - 240$ **15.** $m^2 - 4n^2 - 9p^2 - 16q^2 - 12np - 16nq - 24pq$ **17.** Plan B; At the end of 10 years, Plan B is worth \$1437.86 and Plan A is worth \$1397.16. **19. a.** $10\sqrt{2}$ **b.** Sample: What is the length of the diagonal of a square with side 10? **21.** $\frac{7a^2}{3b}$ **23.** $y = 81$

LESSON 10-5 (pp. 639–645)

15. $9x^2 + 24x + 16$ **17.** $9y^2 + 12y + 4$

19. a.

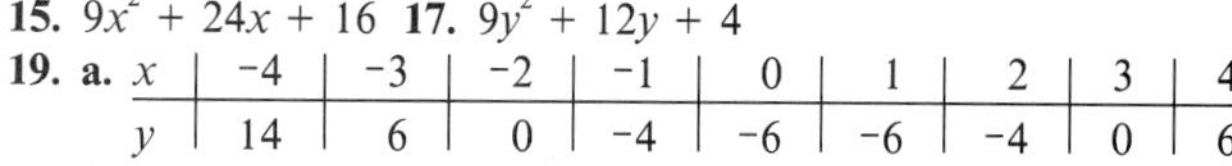

x	−4	−3	−2	−1	0	1	2	3	4
y	14	6	0	−4	−6	−6	−4	0	6

b. See right. **c.** They are the same; the equation in part **b** is the expansion of the equation in part **a.** **21.** $x^3 + 3x^2 + 2x$ **23.** 2, 3 **25.** $a + b - c + d$ **27. a.** −106 feet **b.** The rocket hit the ground before seven seconds elapsed. **29.** $y = \frac{1}{2}x + \frac{17}{2}$ **31.** $x = 512$

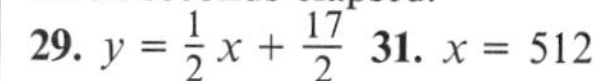

19.

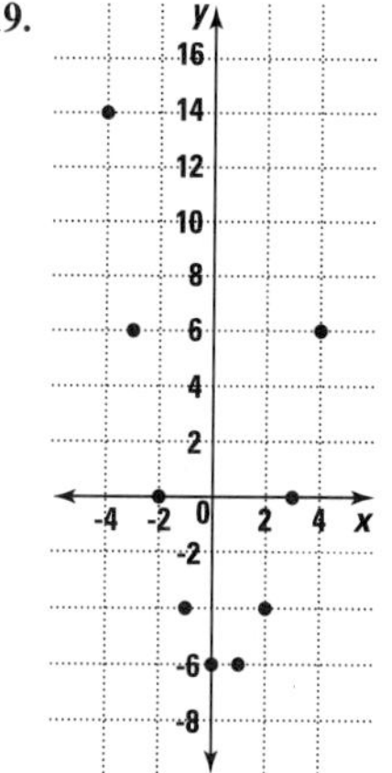

LESSON 10-6 (pp. 646–650)

19. Sample: $(1 + 2)^2 = 3^2 = 9$; $1^2 + 2^2 = 1 + 4 = 5$. So, $(1 + 2)^2 \neq 1^2 + 2^2$. **21.** $324 + 72y + 4y^2$ **23.** $x^2 - 11$ **25. a.** $s^3 + 10s^2 + 25s$ **b.** $s^3 - 2s^2$ **c.** $2s^3 + 8s^2 + 25s$ **d.** Does $2 \cdot 4^3 + 8 \cdot 4^2 + 25 \cdot 4 = 9 \cdot 9 \cdot 4 + 4 \cdot 4 \cdot 2$? Yes, $356 = 356$. **27.** $2c^2 + 3cd - 35d^2$ **29. a.** $x = 0$ **b.** $x < 0$ **31.** 61, 62, 63, 64

LESSON 10-7 (pp. 651–656)

9. Chi-square value ≈ 3.58. Such a value would occur over 10% of the time. This is not enough evidence to support the view that earthquakes occur more in certain seasons. **11.** Chi-square value ≈ 5.52. Such a value would occur over 5% of the time. This is not enough evidence to say that more runs are scored in one part of the game. **13. a.** $9a^2 - 6ab + b^2$ **b.** $9a^2 + 6ab + b^2$ **c.** $9a^2 - b^2$ **15. a.** $8p(4p + 2)$ **b.** $3p(p + 1)$ **c.** $8p(4p + 2) - 3p(p + 1) = 29p^2 + 13p$ **17.** $12y^4 + 2y^3 - 10y^2 + y + 7$ **19.** 98 teams

CHAPTER 10 PROGRESS SELF-TEST (p. 660)

1. $4x^2 - 7x + 9x^2 - 12 - 11 = 4x^2 + 9x^2 - 7x - 23 = 13x^2 - 7x - 23$ **2.** 2 **3.** trinomial **4.** $4(3v^2 - 9 + 2v) = 4 \cdot 3v^2 - 4 \cdot 9 + 4 \cdot 2v = 12v^2 - 36 + 8v = 12v^2 + 8v - 36$ **5.** $-5z(z^2 - 7z + 8) = -5z \cdot z^2 - (-5z) \cdot 7z + -5z \cdot 8 = -5z^3 + 35z^2 - 40z$ **6.** $(3x - 8)(3x + 8) = 9x^2 + 24x - 24x - 64 = 9x^2 - 64$ **7.** $(4y - 2)(3y - 16) = 12y^2 - 64y - 6y + 32 = 12y^2 - 70y + 32$ **8.** $(d - 12)^2 = (d - 12)(d - 12) = d^2 - 12d - 12d + 12^2 = d^2 - 24d + 144$ **9.** $(x - 3)(x^2 - 6x + 9) = x^3 - 6x^2 + 9x - 3x^2 + 18x - 27 = x^3 - 9x^2 + 27x - 27$ **10.** $(3x^2 - 10x) + (15x^3 - 7x^2 + x - 1) = 15x^3 + 3x^2 - 7x^2 - 10x + x - 1 = 15x^3 - 4x^2 - 9x - 1$ **11.** $8t^3 + t^2 - 7t + 1 - (5t^3 - 7t^2) = 8t^3 - 5t^3 + t^2 + 7t^2 - 7t + 1 = 3t^3 + 8t^2 - 7t + 1$ **12.** $(x + y + 5)(a + b + 2) = ax + bx + 2x + ay + by + 2y + 5a + 5b + 10$ **13.** $(4x + 1)(3x + 2) - x(x + 10) = 12x^2 + 11x + 2 - x^2 - 10x = 11x^2 + x + 2$ **14.** See right. **15.** $(30 - 1)(30 + 1) = 900 - 1 = 899$ **16.** $2 \cdot 10^4 + 6 \cdot 10^3 + 3 \cdot 10^2 + 8 \cdot 10 + 4$ **17.** $80x^2 + 60x + 90$ **18.** $80 \cdot 1.04^2 + 60 \cdot 1.04 + 90 = \$238.93$ **19.** $\frac{861 + 748 + 812 + 939}{4} = 840$ **20.** $\frac{(861 - 840)^2}{840} + \frac{(748 - 840)^2}{840} + \frac{(812 - 840)^2}{840} + \frac{(939 - 840)^2}{840} = \frac{441 + 8464 + 784 + 9801}{840} = \frac{19490}{840} \approx 23.2$ **21.** The number of events n is 4; $n - 1$ is 3. So look at the 3rd row of the Critical Chi-Square Values table. The chi-square statistic 23.2 is greater than the critical value 16.3 that would occur with probability .001. So it is very unlikely that the sophomores were being given fewer lines in the newspaper simply by chance. **22.** $(2y)(5y - 3)(y + 9) = 2y(5y^2 + 45y - 3y - 27) = 2y(5y^2 + 42y - 27) = 10y^3 + 84y^2 - 54y$

14.

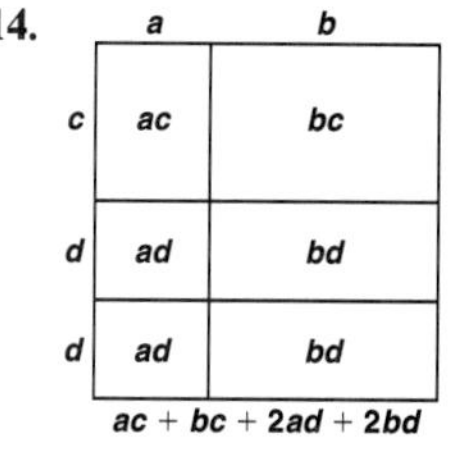

SELECTED ANSWERS

The chart below keys the **Progress Self-Test** questions to the objectives in the **Chapter Review** on pages 661–663. This will enable you to locate those **Chapter Review** questions that correspond to questions missed on the **Progress Self-Test.** The lesson where the material is covered is also indicated on the chart.

Question	**1**	**2**	**3**	**4**	**5**	**6**	**7**	**8**	**9**	**10**
Objective	Voc.	E	E	C	C	C	C	D	B	A
Lesson	10-1, 10-2	10-1	10-1	10-3	10-3	10-6	10-5	10-6	10-4	10-2
Question	**11**	**12**	**13**	**14**	**15**	**16**	**17**	**18**	**19**	**20**
Objective	A	B	I	I	G	G	C	F	H	H
Lesson	10-2	10-4	10-3, 10-5	10-5	10-2	10-2	10-6	10-1	10-7	10-7
Question	**21**	**22**								
Objective	H	I								
Lesson	10-7	10-4								

CHAPTER 10 REVIEW (pp. 661–663)

1. $7x^2 + 4x + 1$; 2 **3.** $3.9x^2 + 1.7x + 19$; 2 **5.** $-k^2 + k - 5$ **7.** $ac + ad + a + bc + bd + b + c + d + 1$ **9.** $y^3 - y^2 - y + 1$ **11.** $ax + bx + 3x + a + b + 3$ **13.** $3k^3 + 12k^2 - 3k$ **15.** $8x^2 + 10x - 36$ **17.** $y^2 - 12y - 13$ **19.** $a^2 - 225$ **21.** $-4z^2 - 5z - 1$ **23.** $d^2 - 2d + 1$ **25.** $48x^2 + 120x + 75$ **27.** $x^3 + 2x^2 + x$ **29.** (a) **31.** (a), (c), (d) **33.** Sample: z^4 **35.** 30,200,901 **37.** $9 \cdot 10^4 + 8 \cdot 10^3 + 1 \cdot 10^2 + 3$ **39. a.** $250x^4 + 250x^3 + 250x^2 + 250x + 250$ **b.** \$811.60 **41. a.** 50 **b.** 1.44 **c.** No; the chi-square value 1.44 is less than the critical value 3.84 that would occur with probability .05. **43. a.** 65° **b.** ≈ 7.11 **c.** Yes. The chi-square value 7.11 is smaller than the value 19.7 that would occur with probability .05. This is not a high enough chi-square value to support a claim that the temperatures are different throughout the year. **45. a.** $2x^2 + 8x + 8$ **b.** $(x + 2)(2x + 4)$ **47. a.** $xy + 3y + 2x + 6$ **b.** $(x + 3)(y + 2)$ **c.** Yes **49.** $10x^4 + 10x^3 + 9x^2 - 2x - 1$ **51.** $21x^2 - \pi x^2$ **53.** $x^3 - x$

LESSON 11-1 (pp. 666–671)

13. a., b. (−2, −2); (1, −.5) **See right.** **15.** Sample: Yes. Since the two lines are not parallel and the women's times are decreasing faster than the men's times, the times will be equal in 2044. **17. a.** −4 **b.** the slope of the line that passes through the points (10, −1) and (8, 7) **19.** ≈ 10,333 miles **21.** $y = 7 - 2x$ **23.** $P + XY$

13. a.

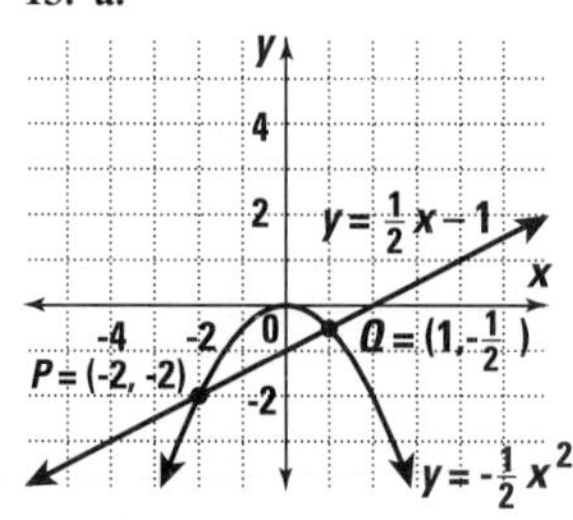

LESSON 11-2 (pp. 672–675)

11. a. (12, 1) **b.** Does $\frac{1}{2}(12) - 5 = 1$? Yes. Does $-\frac{3}{4}(12) + 10 = 1$? Yes. **13. a.** (0, 0), (4, 4) **b.** Does 0 = 0? Yes. Does $0 = \frac{1}{4}(0)^2$? Yes. Does 4 = 4? Yes. Does $4 = \frac{1}{4}(4)^2$? Yes. **15.** 1.5 hours **17. See below.** **19. a.** 2 **b.** (5, 5), (−5, 5) **21.** $2p^3 + 6p^2 + 2p$ **23. See below.** **25.** $4x + 2y$ dollars

17.

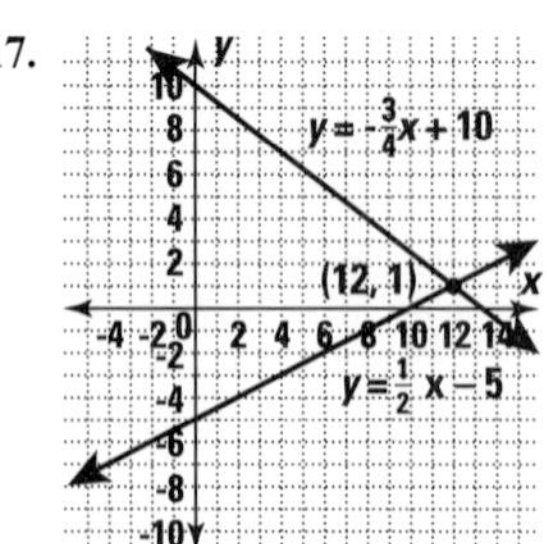

23.

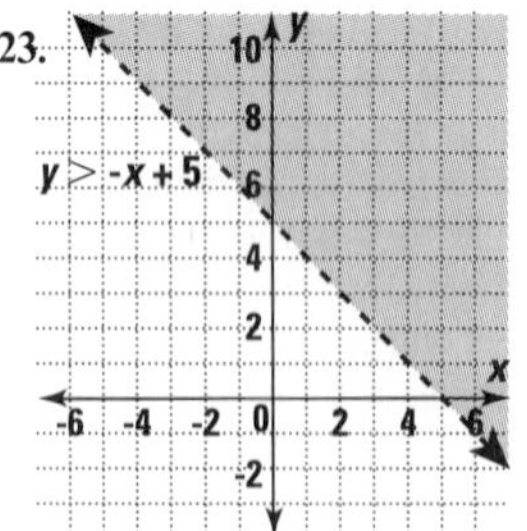

LESSON 11-3 (pp. 676–680)

9. L = \$800,000 and T = \$1,000,000 **11.** $A = 800$, $B = 600$, $K = 20$ **13.** $\left(\frac{1}{2}, -2\right)$ **15. a.** $m - 70 = v$; $m + v = 1250$ **b.** $m = 660$; $v = 590$ **17. a.** (4, 5) **See below.** **b.** Does 5 = 4 + 1? Yes. Does 5 = −2(4) + 13 = −8 + 13 = 5? Yes, it checks. **19.** $2a - 3b$ **21. See below.** **23.** 180 mph

17. a.

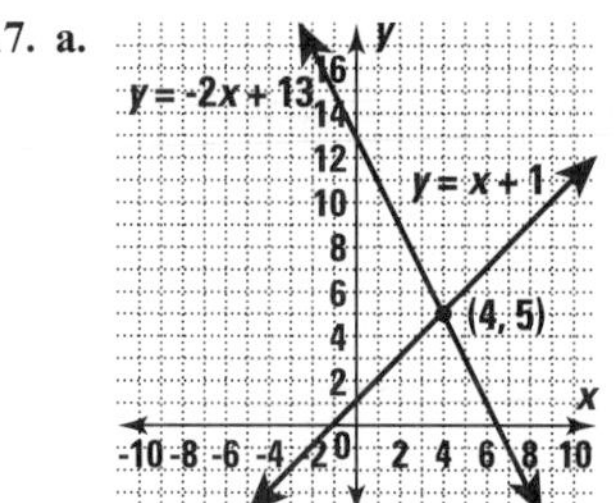

21.

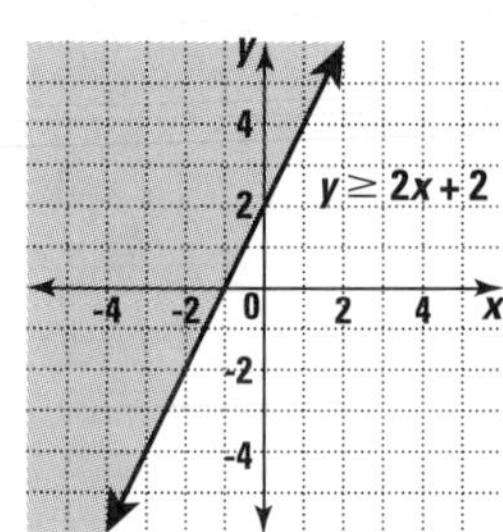

LESSON 11-4 (pp. 681–686)

13. Yes; by the Generalized Addition Property of Equality, $\frac{3}{4} - \frac{1}{5} = \frac{3}{4} + -\frac{1}{5} = 75\% + -20\% = 75\% - 20\% = 55\%$. **15.** $z = 0$; $w = -3$ **17.** regular, \$1.19 per gallon; premium, \$1.40 per gallon **19.** (−1, −3) **21. a.** $x = -7$ or $x = 2$ **b.** −7; 2 **23.** $\frac{3}{4}$ **25.** $8x + y = -15$ or $-8x - y = 15$ **27. a.** $30 + 6p$ **b.** $\frac{30 + 6p}{p + 2}$

LESSON 11-5 (pp. 687–693)

13. $\begin{cases} 3x - 4y = 2 \\ 9x - 5y = 7 \end{cases}$; $\left(\frac{6}{7}, \frac{1}{7}\right)$ **15.** (50, −5) **17.** 13 birds and 12 deer; sample reasoning: Let b = number of birds and d = number of deer. Since all animals have one head, $b + d = 25$. Since birds have two feet and deer have four feet, $2b + 4d = 74$. Solving the system gives $b = 13$ and $d = 12$. **19.** $\left(11, -\frac{78}{5}\right)$ **21. a.** $3x + y = 29$ **b.** $3x = y + 19$ **c.** (8, 5) **d.** Does 3(8) + 5 = 29? Does 24 + 5 = 29? Yes, it checks. Does 3(8) = 24 = 5 + 19? Yes, it checks.

23. a. PYTHAGOREAN TRIPLES
ENTER M
5
ENTER N
3
A = 16
B = 30
C = 34

b. PYTHAGOREAN TRIPLES
ENTER M
7
ENTER N
1
A = 48
B = 14
C = 50

c. Does $16^2 + 30^2 = 34^2$? Does $256 + 900 = 1156$? Yes. Does $48^2 + 14^2 = 50^2$? Does $2304 + 196 = 2500$? Yes. **25.** about 9 days

LESSON 11-6 (pp. 694–698)

13. (b) **15.** (a) **17.** No, with a 10% discount, you should have paid only \$9000. The full price would have been \$10,000. You paid only \$400 less, getting a 4% discount. **19. a.** $\begin{cases} t + u = 20 \\ t + 3u = 32 \end{cases}$ **b.** Sample: Substitute $t = 20 - u$ in the equation $t + 3u = 32$ to get $20 - u + 3u = 32$. Solve this equation for u to get $u = 6$. Substitute $u = 6$ in $t = 20 - u$ to get $t = 14$. **c.** Does $14 + 6 = 20$? Yes. Does $14 + 3(6) = 32$? Yes. **21. a.** $10^3 + 8 \cdot 10^2 + 7 \cdot 10 + 2$ **b.** $1.872 \cdot 10^3$ **23.** \$21.70

LESSON 11-7 (pp. 699–703)

11. a. $375m < 315m + 60m + 25$ **b.** $0 < 25$; m can be any number of months. **13.** A **15.** Sample: $y - 2 = y - 1$ **17.** two intersecting lines because the slopes, $\frac{11}{10}$ and $\frac{12}{10}$, are not equal **19.** $\left(\frac{7}{5}, 0\right)$ **21. a. See above right. b.** $-x = x + 3$; $-2x = 3$; $x = -1.5$, $y = 1.5$ **c.** $2y = 3$; $y = 1.5$, $x = -1.5$ **d.** Sample: The addition method requires the fewest steps. **23. a.** $\frac{4}{9}$ **b.** $25d^4g^2$ **c.** $\frac{c^2}{4a^2}$ **25.** about 26% **27.** $525w$

21. a.

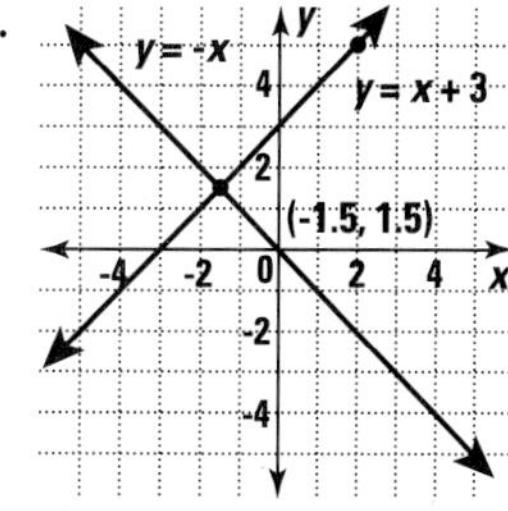

LESSON 11-8 (pp. 704–710)

13. $x \ge 0$, $y \ge 0$, $x + 2y \le 10$ **15.** Sample: $\begin{cases} x \ge -1 \\ x \le 1 \\ y \ge -2 \\ y \le 2 \end{cases}$
17. a. See right. b. 25 **19. a.** no solution **b.** y may be any real number. **21.** intersecting **23. a.** $\begin{cases} d + q = 27 \\ 0.1d + 0.25q = 5.10 \end{cases}$ **b.** 11 dimes and 16 quarters **25.** $4x^2 - 20x + 25$ **27.** \$1503.66

17.

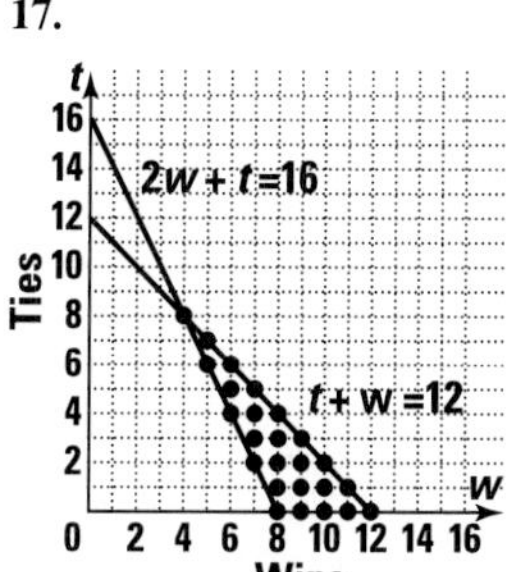

CHAPTER 11 PROGRESS SELF-TEST (p. 714)

1. Sample: Substitute $3a$ for b in the first equation and solve for a. $a - 3(3a) = a - 9a = -8a = -8a = -8$; $a = 1$, $b = 3 \cdot 1 = 3$; $a = 1$; $b = 3$ **2.** Sample: Substitute $-7r - 40$ for p in first equation. $5r + 80 = -7r - 40$; $120 = -12r$; $r = -10$; $p = 5 \cdot -10 + 80 = 30$; $p = 30$; $r = -10$ **3.** Sample: Add equations. $n = 3$; $m - 3 = -1$; $m = 2$; $m = 2$; $n = 3$ **4.** Sample: Multiply second equation by 3 and add. $3 \cdot 4x - 3 \cdot y = 3 \cdot 6$; $12x - 3y = 18$; $19x = 19$; $x = 1$; $4 \cdot 1 - y = 6$; $4 - y = 6$; $-2 = y$; $x = 1$; $y = -2$ **5.** $-5x - 15 = x - 3$; $-6x = 12$; $x = -2$; $y = -2 - 3 = -5$; $(x, y) = (-2, -5)$ **6.** $2 \cdot 5 = 2 \cdot 2A + 2 \cdot 7B$; $10 = 4A + 14B$; $0 = 0$; the lines coincide. **7.** $2x + 7 - 2x = 3$; $7 \ne 3$; the lines are parallel. **8. See right. 9. See right. 10.** Let L = Lisa's weight and S = sister's weight. $L = 4S$; $L + S = 95$; $4S + S = 95$; $5S = 95$; $S = 19$; $L = 4 \cdot 19 = 76$; Lisa weighs 76 pounds and the baby weighs 19 pounds. **11.** Let h = price of hamburgers and s = price of salads. $3h + 4s = 21.30$; $5h + 2s = 22.90$; $10h + 4s = 45.80$; $7h = 24.50$; $h = 3.50$; $5 \cdot 3.50 + 2s = 22.90$; $2s = 5.40$; $s = 2.70$. A small salad costs \$2.70. **12.** No, let r = price of a rose and d = price of a daffodil. $5(2r + 3d) = 5 \cdot 8$; $10r + 15d = 40$; $35 \ne 40$. **13.** They will never have the same population, because $6{,}016{,}000 + 30{,}000y = 4{,}375{,}000 + 30{,}000y$ has no solution. The lines are parallel. **14. See below. 15.** $12z + 8 - 12z = -3$; $8 = -3$; z has no solutions. **16.** $0 < 22 - 19p + 19p$; $0 < 22$; p may be any real number.

8.

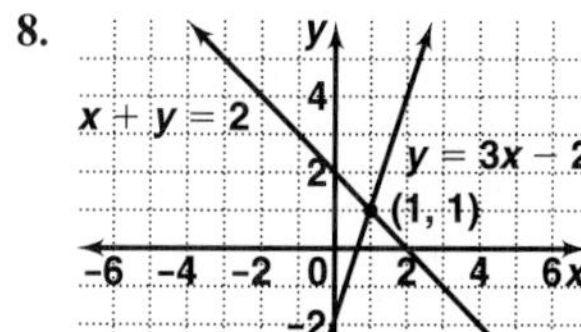

9.

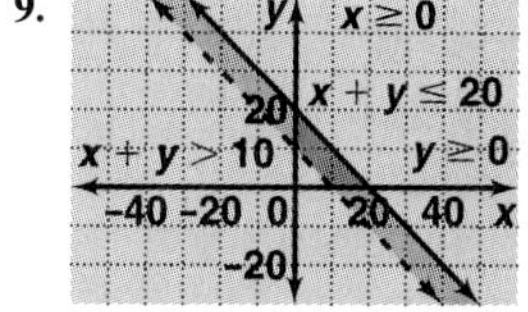

14.

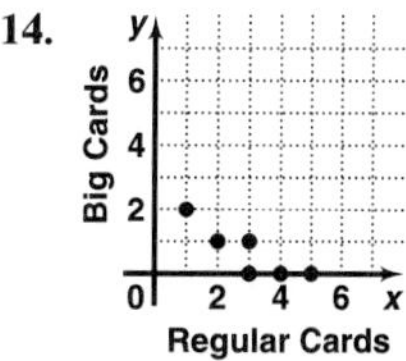

The chart below keys the **Progress Self-Test** questions to the objectives in the **Chapter Review** on pages 715–717. This will enable you to locate those **Chapter Review** questions that correspond to questions missed on the **Progress Self-Test.** The lesson where the material is covered is also indicated on the chart.

Question	**1**	**2**	**3**	**4**	**5**	**6**	**7**	**8**	**9**	**10**
Objective	A	A	B	C	A	E	E	H	I	F
Lesson	11-3	11-2	11-4	11-5	11-2	11-6	11-6	11-1	11-8	11-3
Question	**11**	**12**	**13**	**14**	**15**	**16**	**17**	**18**	**19**	**20**
Objective	F	F	F	G, I	D	D				
Lesson	11-5	11-6	11-7	11-8	11-7	11-7				

CHAPTER 11 REVIEW (pp. 715–717)

1. (15, 45) **3.** $p = 20$, $q = 30$ **5.** $(x, y) = (5, 10)$ **7.** $(a, b) = (-31, -17)$ **9.** $(-9, -43)$ **11.** $(m, b) = (-22, 77)$ **13.** $(x, y) = (-3, -7.25)$ **15.** $(f, g) = (3, 3)$ **17. a.** Sample: Multiply equation (1) by 4 to give $20x + 4y = 120$. **b.** $(x, y) = (7, -5)$ **19.** $(y, z) = (3, 3)$ **21.** $(a, b) = (4, -1)$ **23.** no solutions **25.** $x = 0$ (one solution) **27.** No. Subtract $2k$. $-7 = 0$ is never true; so the original sentence is never true. **29.** No solutions when $b \neq c$. If the slopes are equal but the y-intercepts are different, the lines are parallel and the system has no solutions. **31.** coincide **33.** coincide **35.** (c) **37.** Marty, \$52.50; Joe, \$157.50. **39.** always **41.** egg, \$0.45; muffin, \$0.90 **43.** Yes; $16p + 5e = 8$ is equivalent to $32p + 10e = 16$. There are infinitely many solutions. **45. See below. 47. a. See below. b.** 19 ft **49. See right. 51. See right. 53.** parallel, no solutions **See right. 55. See right. 57. See right. 59. See right.**

45.

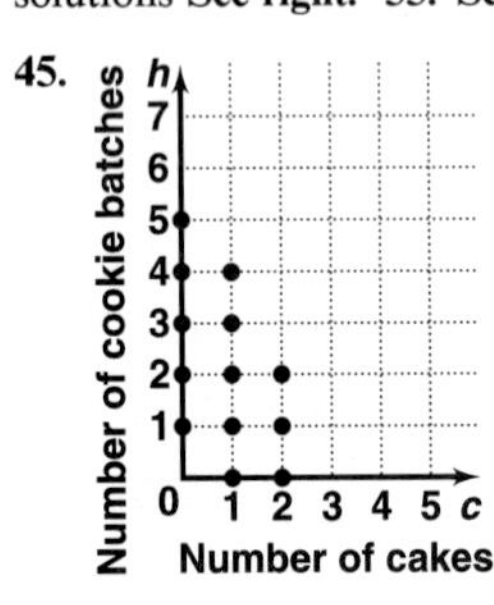

47. a.

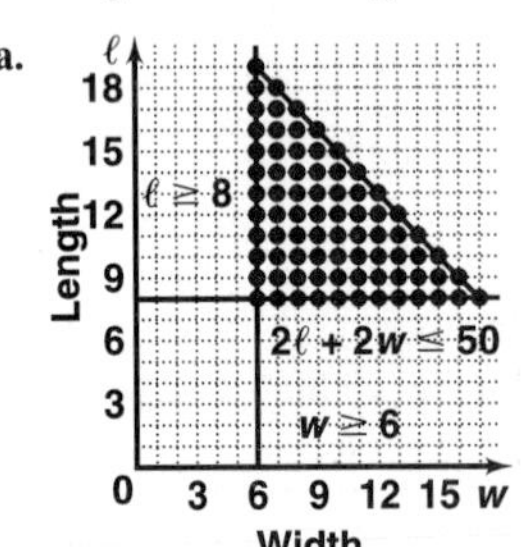

49.

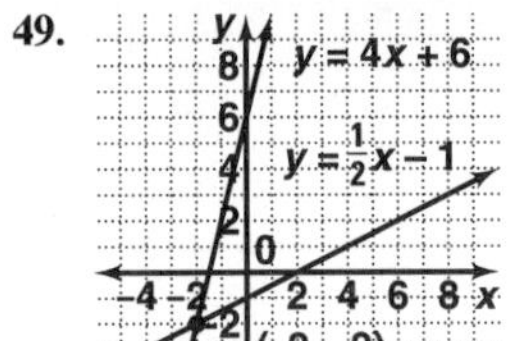

51.

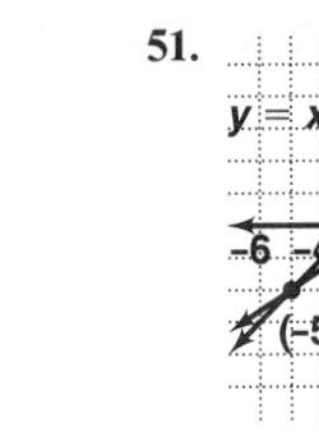

53.

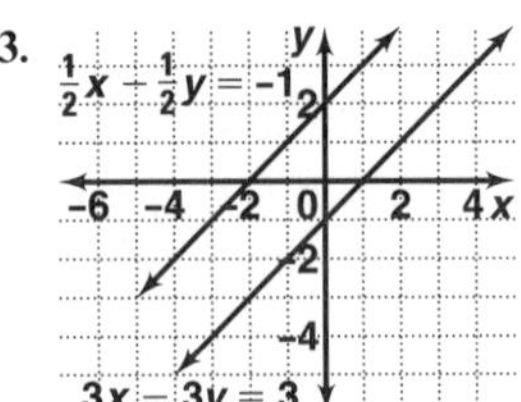

55.

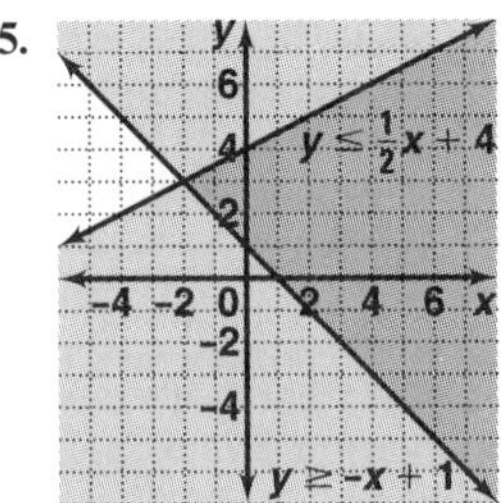

57.

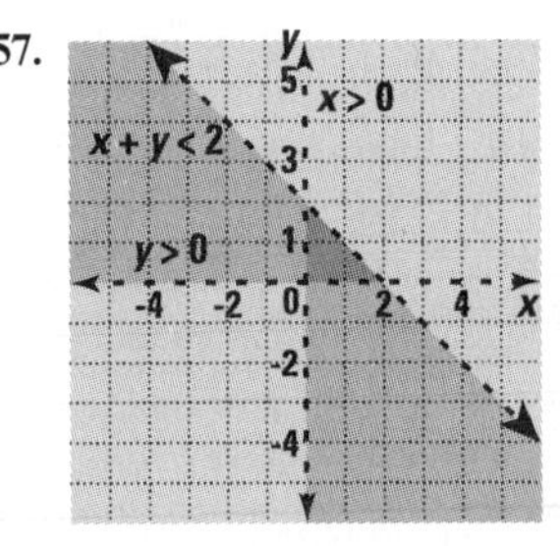

59.

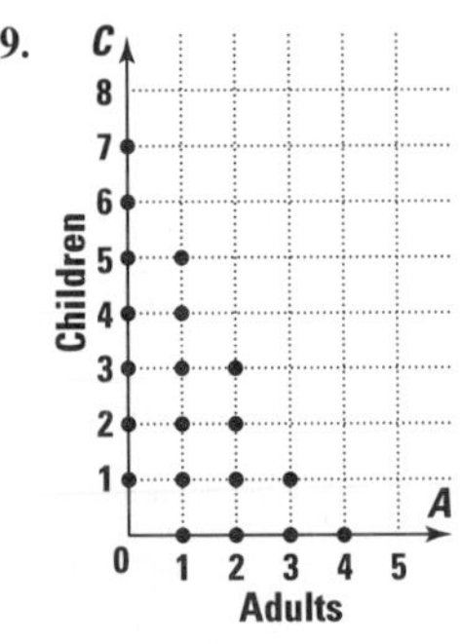

LESSON 12-1 (pp. 718–725)

15. 53, 59, 61, 67, 71, 73, 79, 83, 89, 97 **17.** 20 **19. a.** Because $2^{40} = 4 \cdot 2^{38}$ and $332 = 4 \cdot 83$, then 4 is a common factor of 2^{40} and 332. Using the Common Factor Sum Property, we know 4 is also a factor of $2^{40} + 332$. **b.** No; 332 is not divisible by 8. **21.** $d = 3; -3; 4; -4$ **23.** about 102 games **25.** (a)

LESSON 12-2 (pp. 726–732)

19. See below. 21. $-x + 2 - 3y$ **23. a.** $4r^2$ **b.** πr^2 **c.** $(4 - \pi)r^2$ **d.** $(48 - 12\pi)r^2$ **e.** 3 **25. a.** $x^2 + 3x - 154$ **b.** $x = 11$ or $x = -14$ **27.** $3a^2 + 11a + 10$ **29.** $x = 9$

19. Samples:

	x	x	x	1	1
x	x^2	x^2	x^2	x	x
x	x^2	x^2	x^2	x	x
x	x^2	x^2	x^2	x	x
x	x^2	x^2	x^2	x	x

	x	x	x	x	x	x	1	1	1	1
x	x^2	x^2	x^2	x^2	x^2	x^2	x	x	x	x
x	x^2	x^2	x^2	x^2	x^2	x^2	x	x	x	x

LESSON 12-3 (pp. 733–737)

17. $(7 - x)(5 - x)$ **19.** $10(2 + x)(2 - x)$ **21.** $1{,}032{,}060 = 10^6 + 32 \cdot 10^3 + 60 = (10^3 + 2)(10^3 + 30) = 1002 \cdot 1030 = 2 \cdot 501 \cdot 2 \cdot 515 = 2^2 \cdot 3 \cdot 167 \cdot 5 \cdot 103 = 2^2 \cdot 3 \cdot 5 \cdot 103 \cdot 167$ **23.** The chi-square value is 12.8. A chi-square value this size would be expected more than 10% of the time, so this is not an unusual distribution of digits. **25.** $64a^2 - 1$ **27.** 53.64° **29.** $x = 9$ **31.** \$8.26

LESSON 12-4 (pp. 738–743)

17. $y = 5$ or $y = -1$ **19.** $p = 3$ or $p = -4$ **21.** $v = 0$ or $v = \frac{3}{2}$ **23. a.** $5^2 + 5 = 30$; $2 + 4 + 6 + 8 + 10 = 30$ **b.** 11 **25.** (d) **27.** $13x - 14$ **29.** $16e^2 + 8e + 1$ square units **31. a. See right. b.** about 6.6 thicknesses **33.** $2x^2$

31. a.

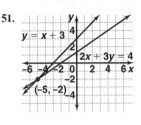

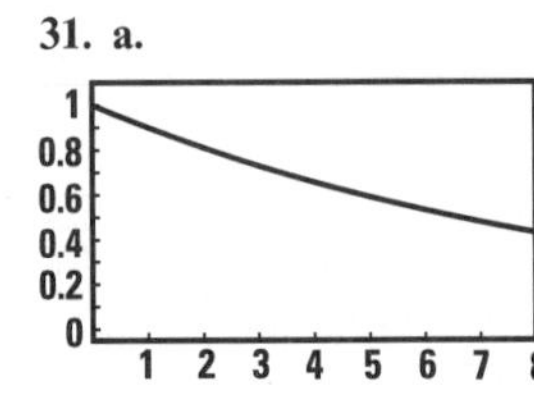

LESSON 12-5 (pp. 744–748)

13. $k = 5$ **15. a. i.** **See p. 862.** **ii.** $(x - 4)(2x + 9) = 0$; $x = 4$ or $x = \frac{-9}{2}$ **b.** Sample: Factoring, because it is more precise **17.** $p^2(3p + 2)^2$ **19.** If the product of two real numbers a and b is zero, then $a = 0$ or $b = 0$. **21.** $y \approx -8.14$ or $y \approx -0.86$ **23.** $v = 2$ or $v = -2$ **25. a.** $(x^2 + 9)(x^2 - 9)$ **b.** $(x^2 + 9)(x + 3)(x - 3)$ **27.** 37 ft

29. a. $(2x + 2y)(x - y) = 2(x^2 - y^2)$ **b.** Yes; $(2x + 2y)(x - y) = 2x^2 - 2xy + 2xy - 2y^2 = 2x^2 - 2y^2 = 2(x^2 - y^2)$
31. slope $= \frac{1}{2}$, y-intercept $= 0$
33. $\frac{13}{40}$, or .325 of a picture

15. a.

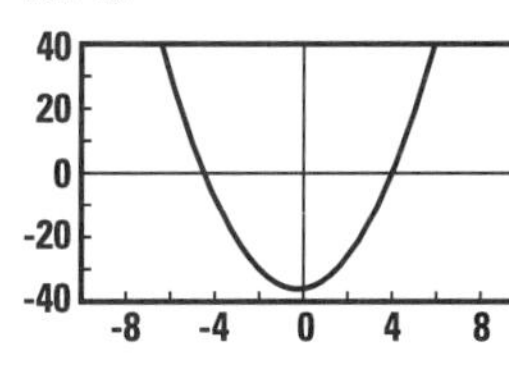

LESSON 12-6 (pp. 749–753)

13. The discriminant is negative. **15.** $4ab(2c - b + 3d)$ **17.** $(4m - 5)(5m - 4)$ **19.** Each factor gives a possibility. Here we organize by the powers of 2, and then by multiplying each by a power of 5. 1 year and \$1,000,000 (per year); 2 years and \$500,000; 4 years and \$250,000; 8 years and \$125,000; 16 years and \$62,500; 32 years and \$31,250; 64 years and \$15,625; 5 years and \$200,000; 25 years and \$40,000; 125 years and \$8,000; 625 years and \$1600; 3,125 years and \$320; 15,625 years and \$64; 10 years and \$100,000; 50 years and \$20,000; 250 years and \$4,000; 1,250 years and \$800; 6,250 years and \$160; 31,250 years and \$32; 20 years and \$50,000; 100 years and \$10,000; 500 years and \$2,000; 2,500 years and \$400; 12,500 years and \$80; 62,500 years and \$16; 40 years and \$25,000; 200 years and \$5,000; 1,000 years and \$1,000; 5,000 years and \$200; 25,000 years and \$40; 125,000 years and \$8; 80 years and \$12,500; 400 years and \$2,500; 2,000 years and \$500; 10,000 years and \$100; 50,000 years and \$20; 250,000 years and \$4; 160 years and \$6,250; 800 years and \$1,250; 4,000 years and \$250; 20,000 years and \$50; 100,000 years and \$10; 500,000 years and \$2; 320 years and \$3,125; 1,600 years and \$625; 8,000 years and \$125; 40,000 years and \$25; 200,000 years and \$5; 1,000,000 years and \$1

LESSON 12-7 (pp. 754–758)

15. Yes, because it equals a simple fraction. **17.** $30\sqrt{2}$ in.; irrational **19.** irrational **21. a.** $7y$ **b.** $a = 3$, $b = 4y$, $c = 4$, and $d = -3y$ or $a = 4$, $b = -3y$, $c = 3$, and $d = 4y$ **23.** $\left(2 + \frac{\pi}{2}\right)r$ **25.** 92.16 ft

LESSON 12-8 (pp. 759–764)

11. Sample: $k = 4$ **13.** only at 28 meters per second **15. a.** rational **b.** irrational **c.** irrational **17.** 12; 7 **19. a.** $2\pi r(r + h)$ **b.** $\approx 408.4\text{ cm}^2$; the factored form

CHAPTER 12 PROGRESS SELF-TEST (p. 768)

1. $300 = 2^2 \cdot 3 \cdot 5^2$ **2.** Both 6^{1000} and 36 are divisible by 3, so by the Common Factor Sum Property, $6^{1000} + 36$ is divisible by 3. **3.** $15a^2b^3 = 3 \cdot 5a^2b^3$, $30a^2b = 6 \cdot 5a^2b$, $25a^3b = 5 \cdot 5a^3b$. So, $5a^2b$ is the greatest common factor. **4.** $\frac{8c^2 + 4c}{c} = \frac{4c(2c + 1)}{c} = 4(2c + 1) = 8c + 4$ **5.** $12m - 2m^3 = 2m(6 - m^2)$ **6.** $500x^2y + 100xy + 50y = 50y(10x^2 + 2x + 1)$ **7.** $z^2 - 81 = (z + 9)(z - 9)$ **8.** Factors of 14 = −1, −14 and −2, −7; −2 + −7 = −9; $k^2 - 9k + 14 = (k - 2)(k - 7)$ **9.** $3y^2 - 17y - 6 = (3y + 1)(y - 6)$ **10.** $4x^2 - 20xy + 25y^2 = (2x - 5y)^2$ **11.** The discriminants are: **(a)** $7^2 - 4 \cdot 1 \cdot 12 = 1$ **(b)** $7^2 - 4 \cdot 1 \cdot (-12) = 97$ **(c)** $12^2 - 4 \cdot 1 \cdot (-7) = 172$ **(d)** $(-12)^2 - 4 \cdot 1 \cdot (-7) = 172$. Equation **(a)** is the only one with a discriminant which is a perfect square. So it is the only one which can be factored over the integers. **12.** Calculate the discriminant ($b^2 - 4ac$). If it is a perfect square, the trinomial is factorable over the integers. **13.** One side of the rectangle has length $2x + 5$, the other side has length $x + 3$. **See right.** **14.** $(q - 7)^2 = 0$; $(q - 7)^2 = (q - 7)(q - 7)$; by the Zero Product Theorem $q - 7 = 0$; $q = 7$ **15.** $d^2 - 20 - d = 0$; $(d - 5)(d + 4) = 0$; by the Zero Product Theorem $d = 5$ or $d = -4$. **16.** By the Zero Product Theorem $2a - 5 = 0$ or $3a + 1 = 0$; $a = \frac{5}{2}$ or $a = -\frac{1}{3}$ **17.** $x^3 + 6x^2 - 7x = x(x^2 + 6x - 7) = x(x + 7)(x - 1) = 0$; by the Zero Product Theorem $x = 0$ or $x = -7$ or $x = 1$. **18.** Because 3.54 can be written as the simple fraction $\frac{354}{100}$. **19.** 26 has the prime factors 13 and 2. $(26)^2$ has twice the number of prime factors, or 4. **20.** $10^2 = 100$ and $11^2 = 121$. Any integer between 100 and 121 will not be a perfect square; its square root will therefore be an irrational number between 10 and 11. Sample: $\sqrt{105}$ **21.** Because $b^2 - 4ac = 18^2 + 4 \cdot 5 \cdot 18 = 684$ is not a perfect square, both solutions are irrational. **22.** $(d - 1)(d - 2) = 12$; $d^2 - 3d + 2 = 12$; $d^2 - 3d - 10 = 0$; $(d - 5)(d + 2) = 0$; $d = 5$ or $d = -2$. The frame is 5 ft by 5 ft. **23.** When the ball returns to the ground, $d = 0$. $0 = 8t - 5t^2 = t(8 - 5t)$ when $t = 0$ and $t = \frac{8}{5}$. So, the ball returns to the ground after $\frac{8}{5} = 1.6$ seconds.
24. $\begin{cases} x + y = 8 \\ xy = 15.51 \end{cases}$; $x(8 - x) = 15.51$; $-x^2 + 8x - 15.51 = 0$
$x^2 - 8x + 15.51 = 0$;
$x = \frac{8 \pm \sqrt{8^2 - 4 \cdot 15.51}}{2} = \frac{8 \pm 1.4}{2}$.
When $x = \frac{8 + 1.4}{2} = 4.7$,
$y = 8 - x = 3.3$. When
$x = \frac{8 - 1.4}{2} = 3.3$,
$y = 8 - x = 4.7$. So, the two numbers are 3.3 and 4.7.

13.

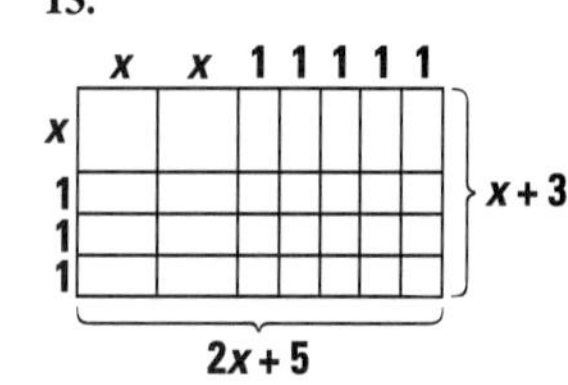

The chart below keys the **Progress Self-Test** questions to the objectives in the **Chapter Review** on pages 769–771. This will enable you to locate those **Chapter Review** questions that correspond to questions missed on the **Progress Self-Test.** The lesson where the material is covered is also indicated on the chart.

Question	**1**	**2**	**3**	**4**	**5**	**6**	**7**	**8**	**9**	**10**
Objective	A	E	B	B	B	B	C	C	C	C
Lesson	12-1	12-1	12-2	12-2	12-2	12-2	12-3	12-3	12-5	12-5
Question	**11**	**12**	**13**	**14**	**15**	**16**	**17**	**18**	**19**	**20**
Objective	G	G	J	F	D	F	D	H	E	H
Lesson	12-8	12-8	12-3	12-4	12-4	12-4	12-4	12-7	12-7	12-7
Question	**21**	**22**	**23**	**24**						
Objective	G	I	I	I						
Lesson	12-8	12-4	12-4	12-6						

CHAPTER 12 REVIEW (pp. 769–771)

1. $5^2 \cdot 7$ **3.** $3^4 \cdot 7^2$ **5.** x^3; 7 **7.** $3x^2y$ **9.** $m^2(14m^2 + 1)$ **11.** $6z^2 - 1$ **13.** $(x + 6)(x + 1)$ **15.** not factorable **17.** $4L(L + 3)(L + 4)$ **19.** (c) **21.** $(3y - 4x)(y + 2x)$ **23.** $3m(4m + 3)(m + 9)$ **25.** $(m + 8)^2$ **27.** $(a + 2)(a - 2)$ **29.** $(2x + 1)(2x - 1)$ **31.** 0 or 2 **33.** −1 or 3 **35.** −6 or 8 **37.** −4 or 4 **39.** $\frac{1}{2}$ or $-\frac{2}{3}$ **41.** No; $203 = 7 \cdot 29$ **43.** 4 **45.** For any two numbers a and b, if $ab = 0$, then $a = 0$ or $b = 0$. **47.** There is no product. **49.** 0 or $\frac{7}{2}$ **51.** $\frac{3}{2}$ or $-\frac{5}{3}$ **53.** (b) **55.** No **57.** Yes **59.** 6 and 4; it is factorable over the integers. **61.** rational **63.** irrational **65.** rational **67.** rational **69.** $\frac{428}{999}$ **71.** ≈ 7.7 in. by 11.7 in. **73.** after 2 and 3 seconds **75.** 115 cm by 90 cm **77. a.** $2x^2 + 8x + 8$ **b.** $(x + 2)(2x + 4)$ or $2(x + 2)^2$ **79.** $3a + 5b$

LESSON 13-1 (pp. 774–777)

9. (c) **11.** Yes **13.** One x value is paired with infinitely many y values. **15. a.** Sample: (−2, 5), (0, 3), (2, 5) **b.** See below. **17. a.** Let cost in dollars = y and weight in pounds = x. $y = 0.50x + 3.00$ **b.** Sample: (0, 3.00), (1, 3.50), (2, 4.00) **19.** 140 adult tickets and 380 student tickets **21.** False **23.** $100bc$

15. b.

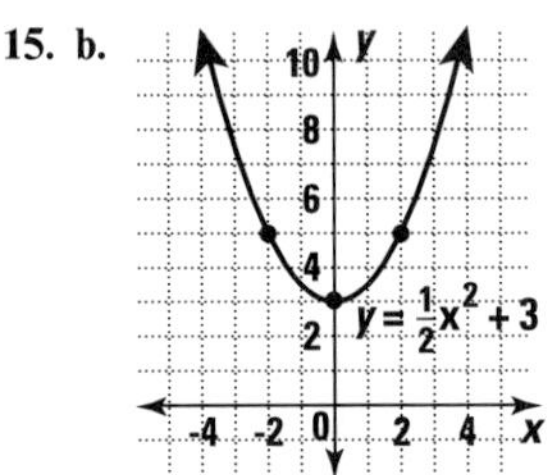

LESSON 13-2 (pp. 778–783)

15. a. After 24 months, the cost of owning the new car is greater than the cost of owning the used car. **b.** more than 120 months **17. a.** See below. **b.** 0, 5; 0 **19. a.** Answers will vary. **b.** total number of children in p's family **21. a.** 95 **b.** 44 **c.** 17 **d.** the slope of the graph of L **23. a.** See below. **b.** No **25.** $\frac{2}{3}$ **27.** $3 \pm \sqrt{6}$ **29. a.** $p = 6^m$ **b.** exponential **c.** 11 **d.** 13

17. a.

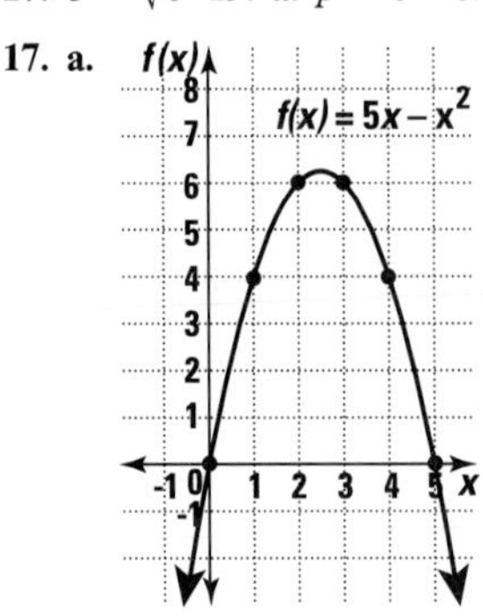

23. a.

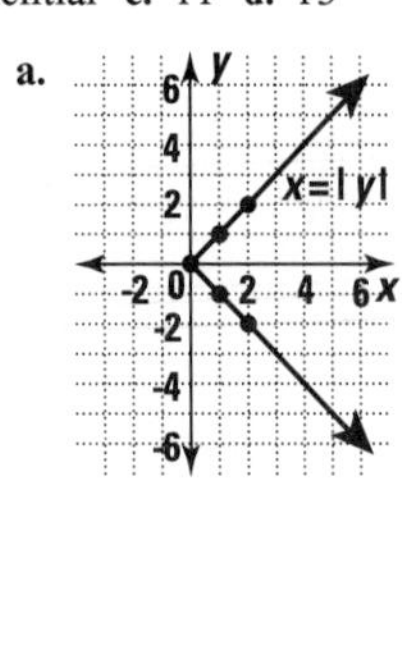

LESSON 13-3 (pp. 784–788)

15. See below. **17.** See below. **19.** 47°F **21.** 2 and 4 **23. a.** $\frac{3x}{2}$ **b.** $\frac{5x}{6}$ **c.** $\frac{2x + 3y}{6}$

15.

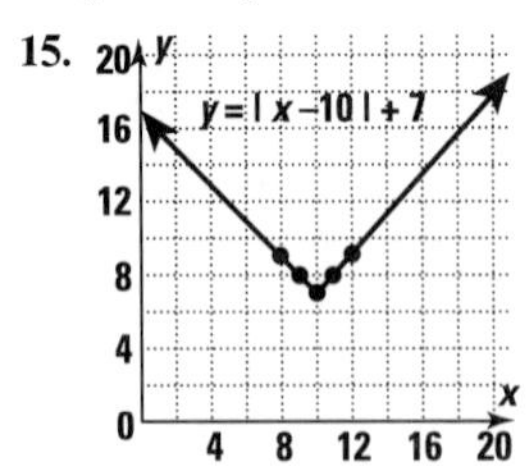

17.

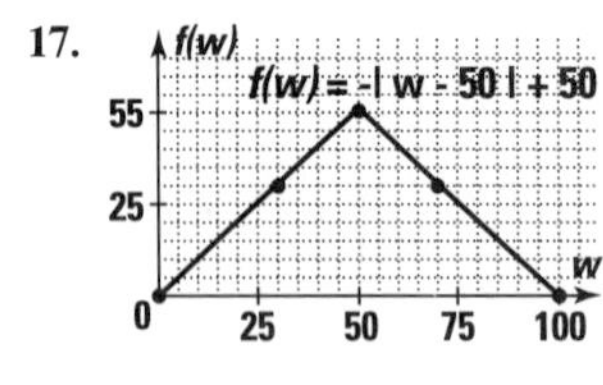

LESSON 13-4 (pp. 789–793)

15. a. See right. **b.** 6 **17.** domain: set of all real numbers; range: set of all real numbers > 0 **19.** domain: set of all real numbers ≥ 0; range: set of all real numbers ≥ 100 **21.** Yes **23. a.** $\left(1 + \frac{1}{3}\right)x = 10.00$ **b.** \$7.50 **25. a.** Cuba **b.** $\frac{10{,}700{,}000}{44{,}218} > \frac{90{,}000{,}000}{761{,}604}$ or 242 > 118

15. a.

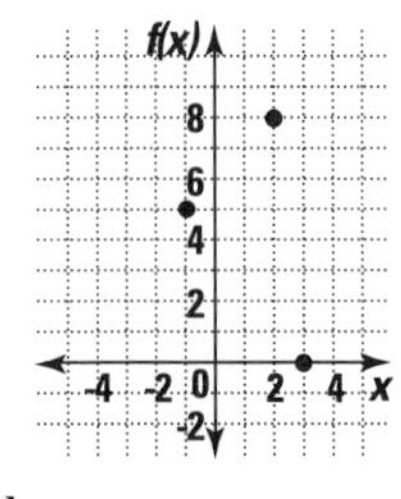

LESSON 13-5 (pp. 795–799)

13. Sample: Probabilities must be from 0 to 1. **15. a.** $\frac{1}{32}$ **b.** See right. **17. a.** 3654 **b.** 360 million pounds more fish were caught in 1980 than in 1985. **c.** {1960, 1965, 1970, 1975, 1980, 1985, 1990} **d.** negative **e.** between 1965 and 1970 **19. a.** $(2x - 5)(x + 4)$ **b.** $x = 3$ **21.** $3\sqrt{3}$ **23.** 74°

15. b.

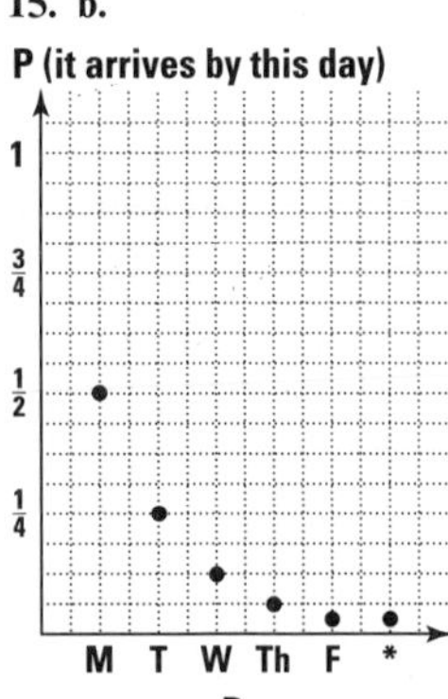

LESSON 13-6 (pp. 800–805)

11. a.

x	$y = x^3 + 3$
−2	−5
−1.5	−0.375
−1	2
−0.5	2.875
0	3
0.5	3.125
1	4
1.5	6.375
2	11

b. See page 864. **c.** This curve has a y-intercept of 3 and an x-intercept of ≈ −1.4 and does not have an axis of symmetry, but it does have a point of rotation symmetry at (0, 3).

13. a.

x	$y = -x^3$
−2	8
−1.5	3.375
−1	1
−0.5	0.125
0	0
0.5	−0.125
1	−1
1.5	−3.375
2	−8

b. See page 864. **c.** This curve has its one intercept and its point of rotation symmetry at the origin. It does not have an axis of symmetry. **15. a.** Sample: the set of numbers x with $1 \le x \le 2$. **b.** See page 864. **c.** about \$298 **d.** 33% **17.** $S = \pi r(r + 2h)$ **19.** about 603 milliliters **21.** $y = \frac{4}{3}$ or $\frac{8}{3}$ **23.** $A = \pm\sqrt{2}$ **25.** $x = -.9$

11. b.

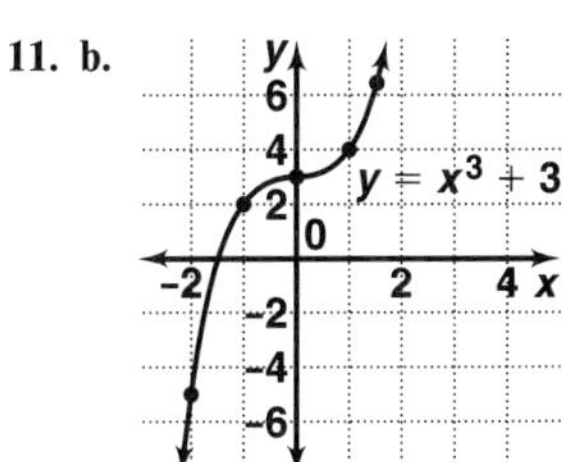

13. b.

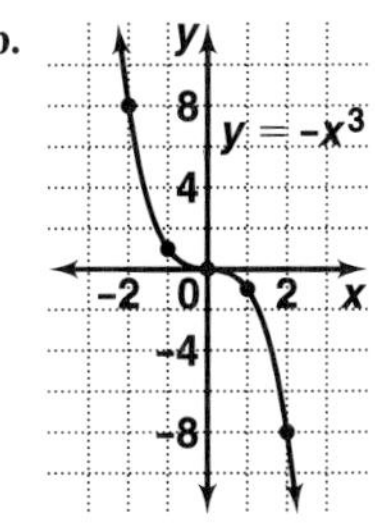

15. b.

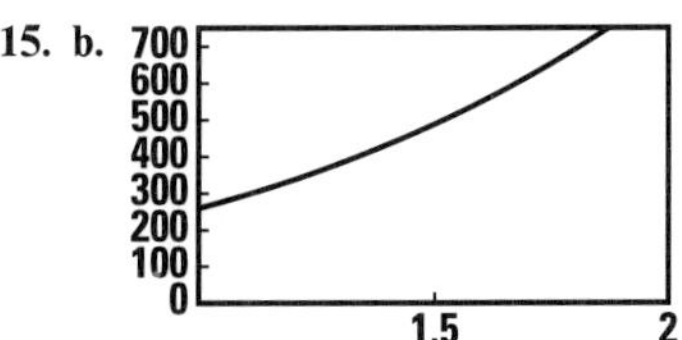

LESSON 13-7 (pp. 807–811)

13. ≈ 60° **15.** 80° **17.** −2 **19.** $x = \frac{-b \pm \sqrt{b^2 - 4ac}}{2a}$ **21.** 72°

LESSON 13-8 (pp. 812–816)

17. a. It depends on how your BASIC defines LOG(X).

X	LOG X	or	LOG X
1	0		0
2	0.30103		.69315
3	0.47712		1.0986
4	0.60206		1.3863
5	0.69897		1.6094
6	0.77815		1.7918
7	0.84510		1.9459
8	0.90309		2.0794
9	0.95424		2.1972
10	1		2.3026

b. See below. c. The graph of the left-hand data is the same as in the text. The graph of the right-hand data is different from that in the text. As x increases, the point (x, log x) is further away from the x-axis than that in the graph of the left-hand data. **19. See below.** **21.** ≈ 0.7 **23. a.** 7 **b.** $\frac{1}{6}$ **25.** about 44.7 miles per hour **b.** 3 **27. a. See below. b.** Sample: 115 million **29.** 794.12 min ≈ 13 hr 14 min **31.** $x > -\frac{9}{2}$

17. b.

19.

27. a.

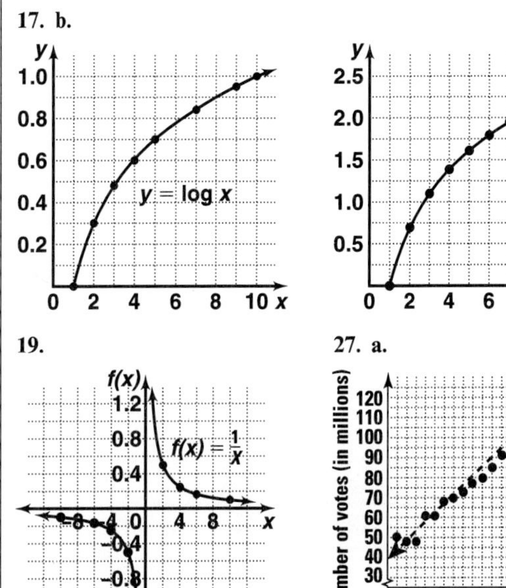

CHAPTER 13 PROGRESS SELF-TEST (p. 820)

1. $f(2) = 3 \cdot 2 + 5 = 11$ **2.** Sample: A function is a set of ordered pairs for which every first coordinate is paired with exactly one second coordinate. **3.** 7.115 **4.** 0.5 **5.** It's the slope of the line: 4. **6.** Sample: $f(x) = x^2 + 2x + 1$ **7.** $x = |y|$ is not a function because when $x > 0$, each x-value is paired with two y-values. For instance, if $x = 3$, $y = 3$ or $y = -3$. **8.** x cannot equal 10 or 30. $x \neq 30$, $x \neq 10$ **9.** Yes; domain = {0, 1, 2, 3, 4}, range = {1, 2}. **10.** No; (0, 1) and (0, −1) are both in the relation. When $x < 1$, every x-value is paired with two y-values. **11.** There are three ways of tossing a sum of 10: (6, 4), (5, 5), (4, 6). So $P(10) = \frac{3}{36} = \frac{1}{12}$. **12.** $P(2) = \frac{1}{10}$ **13. See right. 14.** $b(n) = 3n$ and $h(n) = 2n$; $3n > 2n$; $3n - 2n > 0$; $n > 0$, $b(n) > h(n)$ for any positive number of sandwiches. **15. See right. 16.** **See page 865.** **17.** $f(220) = 11.75 \cdot 220 + 300 = 2885$. $r(220) = 14.50 \cdot 220 = 3190$. Fabulous Floor's carpet costs $2885, which is cheaper than Rapid Rug's price. Therefore, she should buy her carpet from Fabulous Floor. **18.** **See page 865.** **19.** **See page 865.** The cubing function is $f(x) = x^3$. domain = the set of all real numbers; range = the set of all real numbers

13.

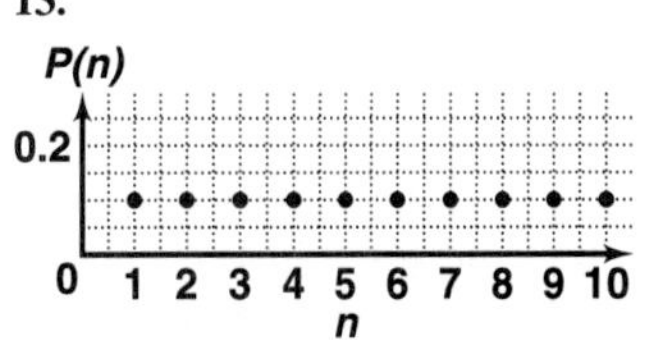

15.

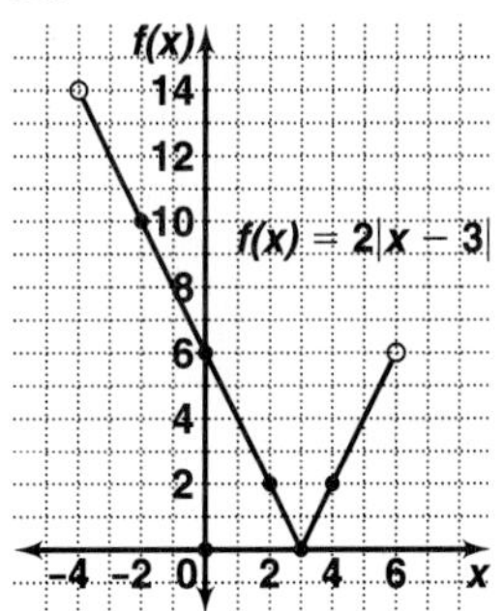

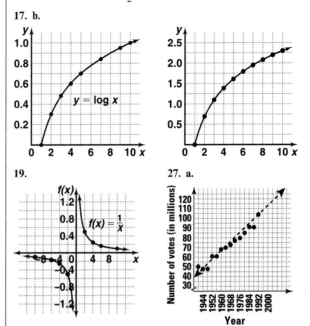

16.

$f(x) = 10 - x^2$

18.

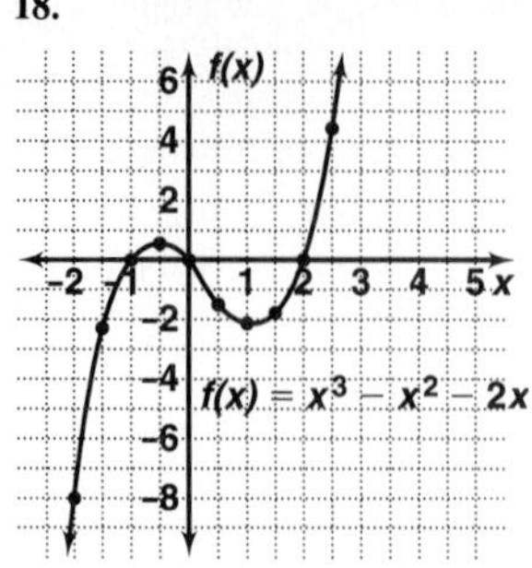

19.

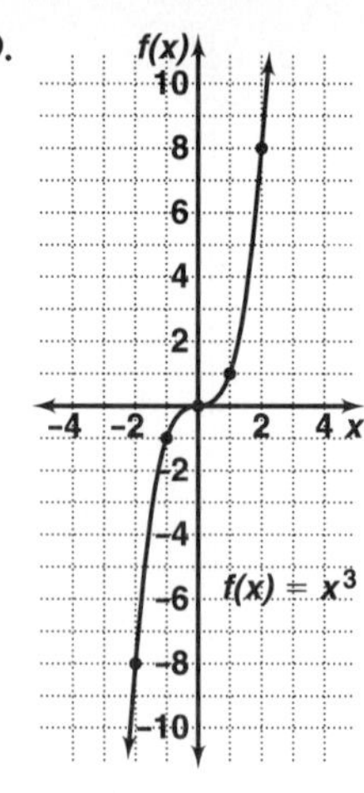

The chart below keys the **Progress Self-Test** questions to the objectives in the **Chapter Review** on pages 821–823. This will enable you to locate those **Chapter Review** questions that correspond to questions you missed on the **Progress Self-Test.** The lesson where the material is covered is also indicated on the chart.

Question	**1**	**2**	**3**	**4**	**5**	**6**	**7**	**8**	**9**	**10**
Objective	A	C	B	B	C	Voc.	C	C	D	H
Lesson	13-2	13-1	13-7	13-8	13-7	13-2	13-1	13-1	13-4	13-1
Question	**11**	**12**	**13**	**14**	**15**	**16**	**17**	**18**	**19**	
Objective	F	F	I	E	I	I	E	J	J	
Lesson	13-5	13-5	13-5	13-3	13-3	13-2	13-2	13-6	13-6	

CHAPTER 13 REVIEW (pp. 821–823)

1. 6 **3.** 78 **5.** 8 **7.** 1.5 **9.** $x = 2$ or $x = -8$ **11.** 0.080 **13.** 107.093 **15.** 0.577 **17.** 0.991 **19.** 4 **21.** Yes **23.** Yes **25.** No **27.** Yes **29.** Yes **31.** The domain is the largest set for which the function makes sense. **33.** (c) **35.** set of all real numbers; set of real numbers ≤ 1 **37.** 10 **39.** set of all real numbers except 3 **41.** $t = 92$ **43.** $G(v) = v + 250$ **45. a.** set of times of the day the Mart is open **b.** Sample: If the largest number of people MacGregor's Mart can hold is 2000, then the range = set of whole numbers ≤ 2000. **47. a.** $\frac{1}{2}$ **b.** $\frac{1}{3}$ **c.** $\frac{1}{6}$ **49.** $\frac{9}{64}$ **51.** .870 **53.** ≈ 0.325 **55.** Yes **57.** Yes **59.** Yes **61. See below.** **63. See below.** **65. See right.** **67. See right.** **69. See right.** **71. a. See right. b.** Since the degree of $P(x)$ is 3, it has at most 3 x-intercepts.

61.

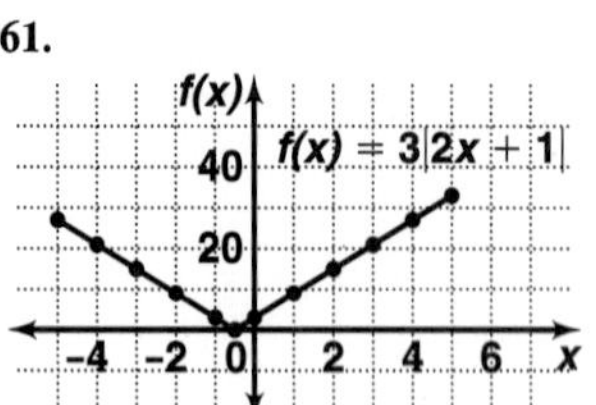

63.

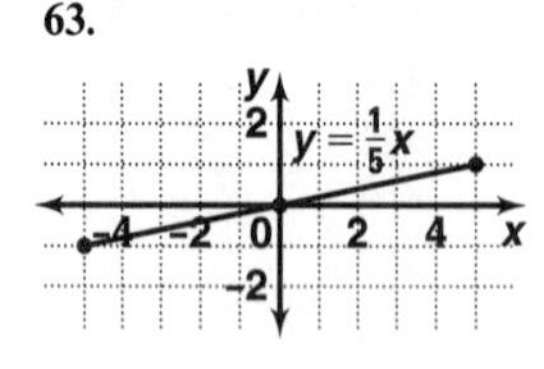

65.

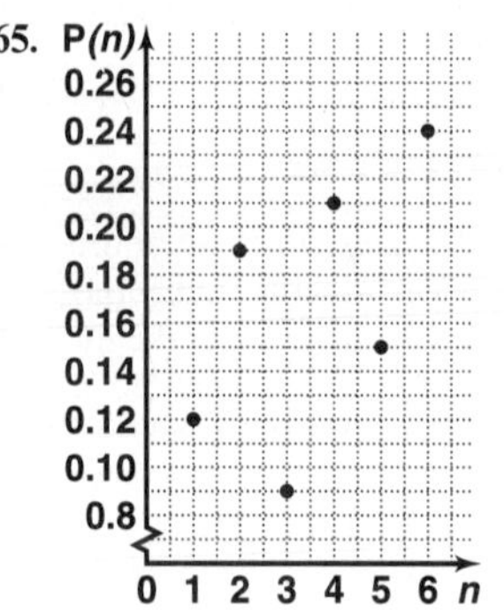

67.

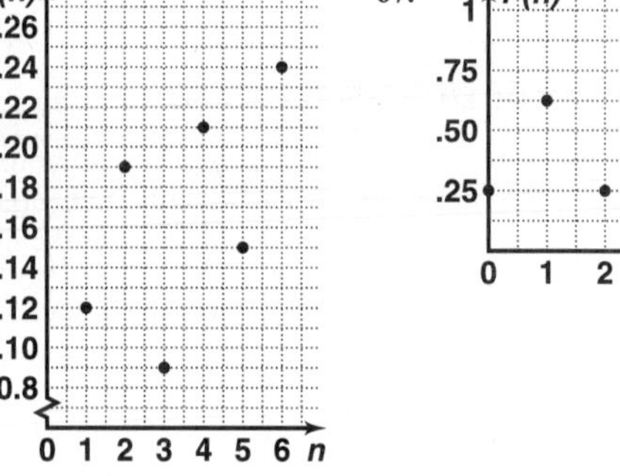

69.

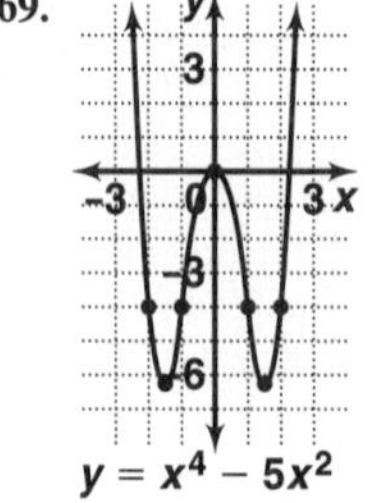

71. a.

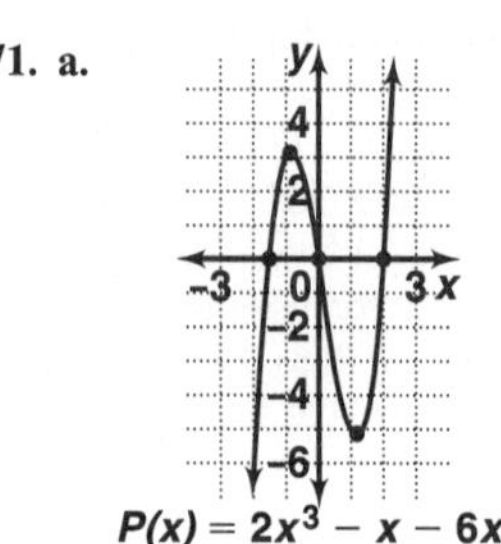

ABS(x) A BASIC function that gives the absolute value of x. (784, 836)

absolute value function A function with an equation of the form $f(x) = a|x - b| + c$. (784)

absolute value If $n < 0$, then the absolute value of n equals $-n$; if $n \geq 0$, then the absolute value of n is n. The absolute value of a number is its distance on a number line from the point with coordinate 0. (593)

Absolute Value-Square Root Property For all real numbers x, $\sqrt{x^2} = |x|$. (596)

acute angle An angle with measure between 0° and 90°.

Adding Like Terms Property See *Distributive Property: Adding or Subtracting Like Terms.*

addition method for solving a system The method of adding the sides of two equations to yield a third equation which contains solutions to the system. (682)

Addition Property of Equality For all real numbers a, b, and c: if $a = b$, then $a + c = b + c$. (151)

Addition Property of Inequality For all real numbers a, b, and c: if $a < b$, then $a + c < b + c$. (200)

additive identity The number zero. (149)

Additive Identity Property For any real number a: $a + 0 = a$. (149)

additive inverse The additive inverse of any real number x is $-x$. Also called *opposite.* (149)

Algebraic Definition of Division For any real numbers a and b, $b \neq 0$: $a \div b = a \cdot \frac{1}{b}$. (350)

Algebraic Definition of Subtraction For all real numbers a and b: $a - b = a + -b$. (216)

algebraic expression An expression that includes one or more variables. (23)

algebraic fraction A fraction which has variables in its numerator or denominator. (86)

algorithm A finite step-by-step recipe or procedure. (640)

annual yield The percent the money in an account earns per year. (486)

array See *rectangular array.* (74)

Area Model for Multiplication (discrete version) The number of elements in a rectangular array with x rows and y columns is xy. (74)

Area Model for Multiplication The area of a rectangle with length ℓ and width w is ℓw. (72)

"as the crow flies" The straight line distance between two points. (599)

Associative Property of Addition For any real numbers a, b, and c: $(a + b) + c = a + (b + c)$. (145)

Associative Property of Multiplication For any real numbers a, b, and c: $(ab)c = a(bc)$. (75)

automatic grapher A graphing calculator or a computer graphing program that enables equations to be graphed on a coordinate plane. (310)

average See *mean.*

axes The perpendicular number lines in a coordinate graph from which the coordinates of points are determined. (155)

axis of symmetry The line over which a figure coincides with its reflection image. (548)

bar graph A way of displaying data using rectangles or bars with lengths corresponding to the data. (142)

base The number x in the power x^n. (486)

BASIC A computer language, short for Beginner's All-purpose Symbolic Instruction Code. (24, 834)

binomial A polynomial with two terms. (617)

boundary point A point that separates solutions from nonsolutions on a number line. (469)

box See *rectangular solid*. (74)

cell The location or box formed by the intersection of a row and a column in a spreadsheet. (233)

Celsius scale The temperature scale in which 0° is the freezing point of water and 100° is the boiling point. Also called the centigrade scale. (322)

center of symmetry A point about which a figure with rotation symmetry can be rotated 180° to coincide with itself. (801)

centigrade scale See *Celsius scale*. (322)

changing the sense (direction) of an inequality Changing from $<$ to $>$, or from $\leq$ to $\geq$, or vice-versa. (116)

Chi-Square Statistic A number calculated from data used to determine whether the difference in two frequency distributions is greater than that expected by chance. (652)

chunking The process of grouping several bits of information into a single piece of information. In algebra, viewing an entire algebraic expression as one variable. (333)

clearing fractions Multiplying each side of an equation by a constant to get an equivalent equation without fractions as coefficients. (327)

closed interval An interval that includes its endpoints. (14)

coefficient A number multiplied by a variable or variables. In the term $-6x$, -6 is the coefficient of x. (178)

coincident lines Two lines that contain the same points. (695)

column A vertical list in a table, rectangular array, or spreadsheet. (233)

common denominator The same denominator for two or more fractions. (720)

common factor A number that is a factor of two or more given numbers. (721)

Common Factor Sum Property If a is a common factor of b and c, then it is a factor of $b + c$. (721)

common logarithm function The function with equation $y = \log x$. (813)

common monomial factoring Isolating a common factor from each term of a polynomial. (726)

GLOSSARY

Commutative Property of Addition For any real numbers a and b: $a + b = b + a$. (144)

Commutative Property of Multiplication For any real numbers a and b: $ab = ba$. (73)

Comparison Model for Subtraction The quantity $x - y$ tells how much quantity x differs from the quantity y. (223)

complementary angles Two angles whose measures have a sum of 90°. Also called *complements*. (256)

complementary events Two events which have no elements in common and whose union is the set of all possible outcomes. (372)

complete factorization A factorization into prime polynomials in which there are no common numerical factors left in the terms. (727)

complex fraction A fraction whose numerator and/or denominator contains a fraction. (351)

composite number A positive integer that has one or more positive integer factors other than 1 and itself. (721)

compound interest A form of interest payment in which the interest is placed back into the account so that it too earns interest. (487)

Compound Interest Formula $T = P(1 + i)^n$ where T is the total after n years if a principal P earns an annual yield of i. (488)

condition of the system A sentence in a system. (666)

constant decrease A situation in which a positive number is repeatedly subtracted. (425)

constant increase A situation in which a positive number is repeatedly added. (498)

constant term A term in a polynomial without a variable. (574)

constant difference A situation in which the difference of two expressions is a constant. (247)

constant sum A situation in which the sum of two expressions is a constant. (247)

continuous A situation in which numbers between any two numbers have meaning. (158)

contraction A size change in which the factor k is nonzero and between -1 and 1. (388)

coordinate graph A graph displaying points as ordered pairs of numbers. (155)

coordinates The numbers identified with a point in the coordinate plane. (156)

coordinate plane A plane in which every point can be identified by two numbers. (156)

cosine function A function defined by $y = \cos x$. (813)

cosine of an angle (cos *A*) The ratio of sides in a right triangle given by $\cos A = \frac{\text{length of leg adjacent}}{\text{length of hypotenuse}}$. (813)

counterexample An example for which a pattern is false. (39)

cubic unit The basic unit of volume. (74)

cubing function The function defined by $y = x^3$. (800)

cursor An arrow or pixel which may be moved along a graph on an automatic grapher. (312)

default window The window preprogrammed for use by an automatic grapher when no other window is specified. (311)

degree of a monomial The sum of the exponents of the variables in the monomial. (616)

degree of a polynomial The highest degree of any of its terms after the polynomial has been simplified. (617)

degrees of freedom The number of events minus one, used in the Chi-Square Statistic. (653)

deviation The absolute value of the difference between an expected number and an actual observed number. (652)

difference of squares An expression of the form $x^2 - y^2$. (648)

Difference of Two Squares Pattern $(a + b)(a - b) = a^2 - b^2$. The product of the sum and the difference of two numbers equals the difference of squares of the numbers. (648)

dimensions The number of rows and columns of an array. The lengths of sides of a rectangle or a rectangular solid. (74)

discount The percent by which the original price of an item is lowered. (222)

discrete A situation in which some numbers between given numbers do not have meaning. (13)

discrete set A set of objects that can be counted. (74)

discriminant In the quadratic equation $ax^2 + bx + c = 0$, $b^2 - 4ac$. (582)

Discriminant Property Suppose $ax^2 + bx + c = 0$ and a, b, and c are real numbers with $a \neq 0$. Let $D = b^2 - 4ac$. Then when $D > 0$, the equation has exactly two real solutions. When $D = 0$, the equation has exactly one real solution. When $D < 0$, the equation has no real solutions. (582)

Discriminant Theorem When a, b, and c are integers, with $a \neq 0$, three conditions happen at exactly the same time:
1. The solutions to $ax^2 + bx + c = 0$ are rational numbers.
2. $b^2 - 4ac$ is a perfect square.
3. $ax^2 + bx + c$ is factorable over the set of polynomials with integer coefficients. (761)

Distance Formula in the Coordinate Plane The distance AB between points $A = (x_1, y_1)$ and $B = (x_2, y_2)$ is $AB = \sqrt{(x_2 - x_1)^2 + (y_2 - y_1)^2}$. Also called *Pythagorean Distance Formula*. (601)

Distance Formula on a Number Line If two points on a line have coordinates x_1 and x_2, the distance between them is $|x_1 - x_2|$. (595)

Distributive Property: Adding Fractions For all real numbers a, b, and c, with $c \neq 0$, $\frac{a}{c} + \frac{b}{c} = \frac{a + b}{c}$. (195)

Distributive Property: Adding or Subtracting Like Terms For all real numbers a, b, and c: $ac + bc = (a + b)c$ and $ac - bc = (a - b)c$. (177)

Distributive Property: Removing Parentheses For all real numbers a, b, and c: $c(a + b) = ca + cb$ and $c(a - b) = ca - cb$. (183)

domain The values which may be meaningfully substituted for a variable. (12)

domain of a function The set of possible replacements for the first variable in a function. (789)

edge A line in a plane that separates solutions from nonsolutions. (470)

element An object in a set. Also called *member*. (11)

empty set A set which has no elements in it. (19)

END A BASIC command which stops a program. (30, 836)

endpoints The smallest and largest numbers in an interval. The points A and B in the segment $\overline{AB}$. (14)

Equal Fractions Property For all numbers a, b, and k, if $k \neq 0$ and $b \neq 0$, then $\frac{a}{b} = \frac{ak}{bk}$. (87)

equal sets Two sets that have the same elements. (11)

equally likely outcomes Outcomes in a situation where the likelihood of each outcome is assumed to be the same. (381)

equation A sentence with an equal sign. (6)

equivalent formulas Formulas in which every set of values that satisfies one formula also satisfies the other. (323)

equivalent systems Systems with exactly the same solutions. (687)

evaluating an expression Finding the numerical value of an expression. (23)

event A set of possible outcomes. (371)

expanding a power of a polynomial Writing the power of a polynomial as a single polynomial. (646)

expansion A size change in which the size change factor k is greater than 1 or less than -1. (388)

expected number The mean frequency of a given event that is predicted by probability. (651)

exponent The number n in the power x^n. (486)

exponential decay A situation in which the original amount is repeatedly multiplied by a growth factor between zero and one. (505)

exponential function A function with an equation of the form $y = ab^x$. (775)

exponential growth A situation in which the original amount is repeatedly multiplied by a growth factor greater than one. (493)

exponential curves Graphs of equations of the form $y = b \cdot g^x$, where $b \neq 0$, $g > 0$, and $g \neq 1$. (493)

Extended Distributive Property To multiply two sums, multiply each term in the first sum by each term in the second sum. (634)

extremes The numbers a and d in the proportion $\frac{a}{b} = \frac{c}{d}$. (394)

f(x) The value of the function f at x. (778)

factorial The product of the integers from 1 to a given number. $n! = 1 \cdot 2 \cdot \ldots \cdot (n - 1) \cdot n$. (126)

factoring The process of expressing a given number (or expression) as the product of two or more numbers (or expressions). (726)

factorization The result of factoring a number or expression. (727)

factors Numbers or expressions whose product is a given number or expression. If $ab = c$, then a and b are factors of c. (720)

Fahrenheit scale A temperature scale in which 32° is the freezing point and 212° is the boiling point. (322)

fitting a line to data Finding a line that closely describes data points which themselves may not all lie on a line. (458)

FOIL algorithm A method for multiplying two binomials; the sum of the product of the First terms, plus the product of the Outside terms, plus the product of the Inside terms, plus the product of the Last terms: $(a + b)(c + d) = ac + ad + bc + bd$. (640)

FOR/NEXT loop A sequence of steps in a BASIC program which enables a procedure to be repeated a certain number of times. (191, 835)

formula A sentence in which one variable is given in terms of other variables and numbers. (27)

frequency The number of times an event occurs. (370)

function A set of ordered pairs in which each first coordinate appears with exactly one second coordinate. (774)

function key on a calculator A key which produces the value of a function when a value in the domain is entered. (812)

function notation Notation to indicate a function, such as $f(x)$, and read as "f of x". (778)

general form of a linear equation An equation of the form $ax + b = cx + d$, where $a \neq 0$. (299)

general form of a quadratic equation A quadratic equation in which one side is 0 and the other side is arranged in descending order of exponents: $ax^2 + bx + c = 0$, where $a \neq 0$. (573)

Generalized Addition Property of Equality For all numbers or expressions a, b, c, and d: if $a = b$ and $c = d$, then $a + c = b + d$. (681)

GOTO A BASIC command which tells the computer to go to the indicated program line number. (836)

greatest common factor for integers The greatest integer that is a common factor of two or more integers. (726)

greatest common factor for monomials The product of the greatest common factor of their coefficients and the greatest common factor of their variables. (726)

growth factor In exponential growth or decay, the nonzero number which is repeatedly multiplied by the original amount. (493)

growth model for powering When an amount is multiplied by g, the growth factor in each of x time periods, then after the x periods, the original amount will be multiplied by g^x. (493)

half-plane In a plane, the region on either side of a line. (470)

hard copy A paper copy printed by a computer or calculator of a graph or other information on a screen. (313)

GLOSSARY

horizontal line A line with an equation of the form $y = k$, where k is a fixed real number. (285)

hypotenuse The longest side of a right triangle. (45)

IF . . . THEN A BASIC command which tells the computer to perform the THEN part only if the IF part is true. (111, 835)

image The final figure resulting from a transformation. (164)

inequality A sentence with one of the following signs: "$\neq$", "$<$", "$>$", "$\leq$", or "$\geq$". (7)

INPUT A BASIC statement that makes the computer pause and wait for a value of the variable to be entered. (30, 834)

instance An example for which a pattern is true. (37)

integers The whole numbers and their opposites. (12)

interest The money a bank pays on the principal in an account. (486)

intersection of sets The set of elements in both set A and set B, written $A \cap B$. (17)

interval The set of numbers between two numbers a and b, possibly including a or b. (14)

irrational number A real number that is not rational. A number that can be written as a nonrepeating infinite decimal. (755)

latitude A measure of the distance of a place on Earth north or south
of the equator, given in degrees. (456)

leg of a right triangle One of the sides forming the right angle of a triangle. (45)

LET A BASIC command which assigns a value to a given variable. (30, 834)

like terms Two or more terms in which the variables and corresponding exponents are the same. (178)

linear equation An equation in which the variable or variables are all to the first power and none multiply each other. (267)

linear expression An expression in which all variables are to the first power. (190)

linear function A function with an equation of the form $y = mx + b$. (775)

linear inequality A linear sentence with an inequality symbol. (317, 472)

linear polynomial A polynomial of degree one. (618)

linear sentence A sentence in which the variable or variables are all to the first power and none multiply each other. (618)

linear system A system of equations, each of degree one. (665)

log (x) The common logarithm of x. (813)

loop Repetition of a set of instructions in a computer program. (191, 835)

magnitude of a size change See *size change factor*. (388)

mark-up A percent by which the original price of an item is raised. (222)

maximum The greatest value in a set of numbers; the highest point on a graph. (223, 550)

mean The sum of the numbers in a collection divided by the number of numbers in the collection. Also called *average*. (725)

means The numbers b and c in the proportion $\frac{a}{b} = \frac{c}{d}$. (394)

Means-Extremes Property For all real numbers a, b, c, and d (b and d nonzero): if $\frac{a}{b} = \frac{c}{d}$, then $ad = bc$. (395)

median In a collection consisting of an odd number of numbers in numerical order, the middle number. In a collection of an even number of numbers arranged in numerical order, the average of the two middle terms. (725)

member An object in a set. Also called *element*. (11)

minimum The smallest value in a set of numbers; the lowest point on a graph. (223, 550)

mirror image The reflection image of a figure. (548)

mode The object(s) in a collection that appear(s) most often. (725)

model A general pattern for an operation that includes many of the uses of the operation. (72)

monomial A polynomial with one term. An expression that can be written as a real number, a variable, or a product of a real number and one or more variables with non-negative exponents. (616)

multiple of a number n A number that has n as a factor. (720)

Multiplication Counting Principle If one choice can be made in m ways and a second choice can be made in n ways, then there are mn ways of making the choices in order. (119)

Multiplication of Positive and Negative Numbers The product of an odd number of negative numbers is negative, and the product of an even number of negative numbers is positive. (98)

Multiplication Property of -1 For any real number a: $a \cdot -1 = -1 \cdot a = -a$. (97)

Multiplication Property of Equality For all real numbers a, b, and c: if $a = b$, then $ca = cb$. (102)

Multiplication Property of Inequality If $x < y$ and a is positive, then $ax < ay$. If $x < y$ and a is negative, then $ax > ay$. (114, 116)

Multiplication Property of Zero For any real number a: $a \cdot 0 = 0 \cdot a = 0$. (81)

Multiplicative Identity Property of One For any real number a: $a \cdot 1 = 1 \cdot a = a$. (79)

multiplicative identity The number 1. (79)

multiplicative inverse The multiplicative inverse of a nonzero number n is $\frac{1}{n}$. Also called *reciprocal*. (79)

Multiplying Fractions Property For all real numbers a and c, and all nonzero b and d: $\frac{a}{b} \cdot \frac{c}{d} = \frac{ac}{bd}$. (86)

Multiplying Positive and Negative Numbers If two numbers have the same sign, their product is positive. If the two numbers have different signs, their product is negative. (97)

multiplying through The process of multiplying each side of an equation by a common multiple of the denominators to result in an equation for which all coefficients are integers. (327)

n factorial ($n!$) The product of the integers from 1 to n. (126)

Negative Exponent Property For all n and all nonzero b, $b^{-n} = \frac{1}{b^n}$, the reciprocal of b^n. (516)

nth power The nth power of a number x is the number x^n. (486)

null set A set which has no elements in it. Also called *empty set.* (19)

numerical expression An expression which includes numbers and operations and no variables. (23)

oblique line A line which is neither horizontal nor vertical. (465)

obtuse angle An angle with measure greater than 90° and less than 180°.

open interval An interval that does not include its endpoints. (14)

open sentence A sentence that contains at least one variable. (7)

Opposite of a Difference Property For all real numbers a and b, $-(a - b) = -a + b$. (241)

Opposite of a Sum Property For all real numbers a and b, $-(a + b) = -a + -b = -a - b$. (240)

Opposite of Opposites Property (Op-op Property) For any real number a: $-(-a) = a$. (150)

opposite The opposite of any real number x is $-x$. Also called *additive inverse.* (97)

order of operations The correct order of evaluating numerical expressions: first, work inside parentheses, then do powers. Then do multiplications or divisions, from left to right. Then do additions or subtractions, from left to right. (23)

origin The point (0, 0) on a coordinate plane. (155)

outcome A result of an experiment. (371)

parabola The curve that is the graph of an equation of the form $y = ax^2 + bx + c$, where $a \neq 0$. (548)

pattern A general idea for which there are many instances. (37)

P(E) The probability of event E or "P of E." (371)

percent (%), times $\frac{1}{100}$, or "per 100." (376)

percent of discount The ratio of the discount to the original price. (363)

percent of tax The ratio of tax to the selling price. (363)

perfect square A number which is the square of a whole number. (32)

perfect square trinomial A trinomial which is the square of a binomial. (647)

Perfect Square Patterns $(a + b)^2 = a^2 + 2ab + b^2$ and $(a - b)^2 = a^2 - 2ab + b^2$. (647)

permutation An arrangement of letters, names, or objects. (126)

Permutation Theorem There are $n!$ possible permutations of n different objects, when each object is used exactly once. (126)

plus or minus symbol (±) A symbol which shows that a calculation should be done twice, once by adding and once by subtracting. (574)

polynomial An algebraic expression that is either a monomial or a sum of monomials. (617)

polynomial function A function whose range values are given by a polynomial. (800)

polynomial in the variable x An expression of the form $a_nx^n + a_{n-1}x^{n-1} + \ldots a_1x + a_0$, where $a_0, a_1, \ldots, a_n$ are real numbers. (617)

polynomial over the integers A polynomial with integer coefficients. (727)

population density The number of people per unit of area. (360)

power An expression written in the form x^n. (486)

Power of a Power Property For all m and n, and all nonzero b, $(b^m)^n = b^{mn}$. (512)

Power of a Product Property For all n, and all nonzero a and b, $(ab)^n = a^n \cdot b^n$. (527)

Power of a Quotient Property For all n, and all nonzero a and b, $\left(\frac{a}{b}\right)^n = \frac{a^n}{b^n}$. (529)

preimage The original figure before a transformation takes place. (164)

prime factorization The writing of a number as a product of primes. (722)

prime number An integer greater than 1 whose only integer factors are itself and 1. (721)

prime polynomial A polynomial which cannot be factored into polynomials of a lower degree. (727)

principal Money deposited in an account. (486)

PRINT A BASIC command which tells the computer to print what follows the command. (30, 834)

Probability Formula for Geometric Regions If all points occur randomly in a region, then the probability P of an event is given by $P = \frac{\text{measure of region for event}}{\text{measure of entire region}}$, where the measure may be length, area, etc. (382)

probability function A function that maps a set of outcomes onto their probabilities. (795)

probability of an event A number from 0 to 1 that measures the likelihood that an event will occur. (371)

Product of Powers Property For all m and n, and all nonzero b, $b^m \cdot b^n = b^{m+n}$. (510)

Product of Square Roots Property For all nonnegative real numbers a and b, $\sqrt{a} \cdot \sqrt{b} = \sqrt{ab}$. (587)

projectile An object that is thrown, dropped, or shot by an external force and continues to move on its own. (567)

Property of Opposites For any real number a: $a + -a = 0$. (149)

Property of Reciprocals For any nonzero real number a: $a \cdot \frac{1}{a} = \frac{1}{a} \cdot a = 1$. (79)

proportion A statement that two fractions are equal. Any equation of the form $\frac{a}{b} = \frac{c}{d}$. (394)

Putting-Together Model for Addition If a quantity x is put together with a quantity y with the same units and if there is no overlap, then the result is the quantity $x + y$. (143)

Pythagorean Distance Formula See *Distance Formula in the Coordinate Plane.*

Pythagorean Theorem In a right triangle with legs a and b and hypotenuse c, $a^2 + b^2 = c^2$. (46)

GLOSSARY

quadrant One of the four regions of the coordinate plane formed by the x-axis and y-axis. (163)

quadratic equation An equation that can be written in the form $ax^2 + bx + c = 0$. (573)

Quadratic Formula If $a \neq 0$ and $ax^2 + bx + c = 0$, then $x = \frac{-b \pm \sqrt{b^2 - 4ac}}{2a}$. (574)

quadratic function A function with an equation of the form $y = ax^2 + bx + c$ or $y = \frac{k}{x}$. (775)

quadratic polynomial A polynomial of degree two. (618)

Quotient of Powers Property For all m and n, and all nonzero b, $\frac{b^m}{b^n} = b^{m-n}$. (521)

radical sign ($\sqrt{\ }$) The symbol for square root. (31)

random outcomes Outcomes in a situation where each outcome is assumed to have the same probability. (372)

range The length of an interval. The maximum value minus the minimum value. (223)

range of a function The set of possible values of a function. (789)

Rate Factor Model for Multiplication When a rate is multiplied by another quantity, the unit of the product is the product of units. Units are multiplied as though they were fractions. The product has meaning when the units have meaning. (92)

Rate Model for Division If a and b are quantities with different units, then $\frac{a}{b}$ is the amount of quantity a per quantity b. (356)

rate of change The rate of change between points (x_1, y_1) and (x_2, y_2) is $\frac{y_2 - y_1}{x_2 - x_1}$. (419)

ratio A quotient of quantities with the same units. (363)

Ratio Model for Division Let a and b be quantities with the same units. Then the ratio $\frac{a}{b}$ compares a to b. (363)

ratio of similitude The ratio of corresponding sides of two similar figures. (402)

rational number A number that can be written as the ratio of two integers. (754)

real numbers Numbers which can be represented as finite or infinite decimals. (12)

reciprocal The reciprocal of a nonzero number n is $\frac{1}{n}$. Also called *multiplicative inverse.* (79)

Reciprocal of a Fraction Property If $a \neq 0$ and $b \neq 0$ the reciprocal of $\frac{a}{b}$ is $\frac{b}{a}$. (80)

reciprocal rates Two rates in which the quantities are compared in reverse order. (93)

rectangular array A two-dimensional display of numbers or symbols arranged in rows and columns. (74)

rectangular solid A 3-dimensional figure with 6 rectangular faces. (74)

reflection symmetry The property held by a figure that coincides with its image under a reflection over a line. Also called *symmetry with respect to a line.* (548)

relation A set of ordered pairs. (775)

relative frequency The ratio of the number of times an event occurred to the total number of possible occurrences. (370)

Relative Frequency of an Event Suppose a particular event has occurred with a frequency of f times in a total of T opportunities for it to happen. Then the relative frequency of the event is $\frac{f}{T}$. (370)

REM A BASIC statement for a remark or explanation that will be ignored by the computer. (835)

Removing Parentheses Property See *Distributive Property.* (183)

Repeated Multiplication Model for Powering When n is a positive integer, $x^n = x \cdot x \cdot \ldots \cdot x$ where there are n factors of x. (486)

replication The process of copying a formula in a spreadsheet in which the cell references in the original formula are adjusted for new positions in the spreadsheet. (235)

right angle An angle with measure 90°.

rotation symmetry A property held by some figures where a rotation of some amount other than 360° results in an image which coincides with the original image. (801)

row A horizontal list in a table, rectangular array, or spreadsheet. (233)

scatterplot A two-dimensional coordinate graph of individual points. (156)

scientific notation A number represented as $x \cdot 10^n$, where $1 \leq x < 10$ and n is an integer. (829–833)

sentence Two algebraic expressions connected by "=", "≠", "<", ">", "≤", "≥", or "≈". (6)

sequence A set of numbers or objects in a specific order. (188)

set A collection of objects called elements. (11)

similar figures Two or more figures that have the same shape. (402)

simple fraction A numerical expression of the form $\frac{a}{b}$, where a and b are integers. (754)

simplifying radicals Rewriting a radical with a smaller integer under the radical sign. (588)

sine function The function defined by $y = \sin x$. (813)

sine of an angle (sin *A*) In a right triangle, $\sin A = \frac{\text{length of the leg opposite}}{\text{length of the hypotenuse}}$. (813)

sinusoidal curve The curve that is the graph of a sine or cosine function. (813)

size change A transformation in which the image of (x,y) is (kx, ky). (387)

size change factor The number k in the transformation in which the image of (x, y) is (kx, ky). Also called *magnitude.* (387)

Size Change Model for Multiplication If a quantity x is multiplied by a size change factor k, $k \neq 0$, then the resulting quantity is kx. (388)

Slide Model for Addition If a slide x is followed by a slide y, the result is the slide $x + y$. (143)

Slope and Parallel Lines Property If two lines have the same slope, then they are parallel. (694)

slope The rate of change between points on a line. The amount of change in the height of the line as you go 1 unit to the right. The slope of the line through (x_1, y_1) and (x_2, y_2) is $\frac{y_2 - y_1}{x_2 - x_1}$. (432)

slope-intercept form An equation of a line in the form $y = mx + b$, where m is the slope and b is the y-intercept. (439)

Slope-Intercept Property The line with equation $y = mx + b$ has slope m and y-intercept b. (440)

solution A replacement of the variable(s) in a sentence that makes the sentence true. (7)

solution set of an open sentence The set of numbers from the domain that are solutions. (12)

solution set to a system The intersection of the solution sets for each of the sentences in the system. (666)

spreadsheet A computer program in which data are presented in a table and calculations upon entries in the table can be made. The table itself. (232)

SQR(X) A BASIC function that gives the square root of x. (812, 836)

Square of the Square Root Property For any nonnegative number n, $\sqrt{n} \cdot \sqrt{n} = n$. (33)

square root If $A = s^2$, then s is a square root of A. (31)

square unit The basic unit for area. (72)

squaring function A function defined by $y = x^2$. (778)

stacked bar graph A display of data using rectangles or bars stacked on top of each other. (143)

standard form for an equation of a line An equation of the form $Ax + By = C$, where not both A and B are zero. (464)

standard form of a prime factorization The form of a factorization where the factors are primes in increasing order and where exponents are used if primes are repeated. (722)

standard form of a polynomial A polynomial written with the terms in descending order of the exponents of its terms, with the largest exponent first. (618)

standard form of a quadratic equation An equation written in the form $ax^2 + bx + c = 0$, where $a \neq 0$. (576)

STEP A BASIC command that tells the computer how much to add to the counter each time through a FOR/NEXT loop. (556, 835)

stopping distance The length of time for a car to slow down from the instant the brake is applied until the car is no longer moving. (554)

substitution method for solving a system A method in which one variable is written in terms of other variables, and then this expression is used in place of the original variable in subsequent equations. (672)

supplementary angles Two angles whose measures have a sum of 180°. Also called *supplements*. (254)

symmetric Having some symmetry. See *reflection symmetry* and *rotation symmetry*. (548)

symmetry with respect to a line See *reflection symmetry*.

system A set of conditions separated by the word *and*. (666)

system of equations A system in which the conditions are equations. (666)

system of inequalities A system in which the conditions are inequalities. (704)

Take-Away Model for Subtraction If a quantity y is taken away from an original quantity x, the quantity left is $x - y$. (221)

tangent function A function defined by $y = \tan x$. (807)

tangent of an angle (tan *A*) The ratio of sides given by $\tan A = \frac{\text{length of leg opposite}}{\text{length of leg adjacent}}$. (808)

term A number, a variable, or a product of numbers and variables. (178)

testing a special case A strategy for determining whether a pattern is true by trying out specific instances. (535)

theorem A property that has been proved to be true. (46)

Third Side Property If x and y are the lengths of two sides of a triangle, and $x > y$, then the length z of the third side must satisfy the inequality $x - y < z < x + y$. (263)

tick marks Marks on the x and y axes of a graph to show distance. (312)

tolerance The specific amount that manufactured parts are allowed to vary from an accepted standard size. (598)

trace An option on an automatic grapher that allows the user to move a cursor along the graph while displaying the coordinates of the point the cursor indicates. (312)

translation A two-dimensional slide. (164)

tree-diagram A tree-like way of organizing the possibilities of choices in a situation. (120)

Triangle Inequality The sum of the lengths of two sides of any triangle is greater than the length of the third side. (261)

Triangle-Sum Theorem In any triangle with angle measures a, b, and c: $a + b + c = 180$. (255)

trigonometry The study of the trigonometric functions sine, cosine, and tangent, and their properties. (813)

trinomial A polynomial with three terms. (617)

two dimensional slide A transformation in which the image of (x, y) is $(x + h, y + k)$. (164)

undefined slope The situation regarding the slope of a vertical line, which does not exist. (435)

union of sets The set of elements in either set A or set B, written $A \cup B$. (18)

Unique Factorization Theorem Every integer can be represented as a product of primes in exactly one way, disregarding order of the factors. (722)

Unique Factorization Theorem for Polynomials Every polynomial can be represented as a product of prime polynomials in exactly one way, disregarding order and real number multiples. (727)

value of a function The value of the second variable (often called y) in a function for a given value of the first variable. (774)

variable A letter or other symbol that can be replaced by a number (or other object). (6)

Venn diagram A diagram used to show relationships among sets. (18)

vertex The intersection of a parabola with its axis of symmetry. (548)

vertical line A line with an equation of the form $x = h$, where h is a fixed real number. (286)

volume The space contained by a three-dimensional figure. The volume of a rectangular solid is the product of its dimensions. (74)

whole numbers The set of numbers $\{0, 1, 2, 3, \ldots\}$. (12)

window The part of the coordinate grid that is shown on an automatic grapher. (310)

x-axis The horizontal axis in a coordinate graph. (163)

x-coordinate The first coordinate of a point. (163)

x-intercept The x-coordinate of a point where a graph crosses the x-axis. (446)

y-axis The vertical axis in a coordinate graph. (163)

y-coordinate The second coordinate of a point. (163)

y-intercept The y-coordinate of a point where a graph crosses the y-axis. (439)

Zero Exponent Property If g is any nonzero real number, then $g^0 = 1$. (493)

Zero Product Property For any real numbers a and b, if $ab = 0$, then $a = 0$ or $b = 0$. (739)

zoom A feature on an automatic grapher that allows the user to see a graph on a window of different dimensions without having to input the dimensions. (313)

$=$	is equal to
$\neq$	is not equal to
$<$	is less than
$\leq$	is less than or equal to
$\approx$	is approximately equal to
$>$	is greater than
$\geq$	is greater than or equal to
$\pm$	plus or minus
π	Greek letter pi; $= 3.141592...$ or $\approx \frac{22}{7}$
A'	image of point A
$\overleftrightarrow{AB}$	line through A and B
$\overrightarrow{AB}$	ray starting at A and containing B
$\overline{AB}$	segment with endpoints A and B
AB	length of segment from A to B
$\angle ABC$	angle ABC
$m\angle ABC$	measure of angle ABC
$\triangle ABC$	triangle ABC
$\{\ldots\}$	the symbol used for a set
$\varnothing, \{\ \}$	the empty or null set
$A \cap B$	the intersection of sets A and B
$A \cup B$	the union of sets A and B
W	the set of whole numbers
I	the set of integers
R	the set of real numbers
⌐	symbol for 90° angle
%	percent
$\sqrt{}$	square root symbol; radical sign
$\sqrt{n}$	positive square root of n
$\|x\|$	absolute value of x
$-x$	opposite of x
$n°$	n degrees
$n!$	n factorial
$f(x)$	function notation "f of x"; the second coordinates of the points of a function
(x, y)	ordered pair x, y
$N(E)$	the number of elements in set E
$P(E)$	the probability of an event E
$P(A$ and $B)$	the probability that A and B occur
$\tan A$	tangent of $\angle A$
$\sin A$	sine of $\angle A$
$\cos A$	cosine of $\angle A$
ABS(X)	in BASIC, the absolute value of X
SQR(X)	in BASIC, the square root of X
X * X	in BASIC, $X \cdot X$
X ^ Y	in BASIC, X^Y
[1/x]	calculator reciprocal key
[y^x]	calculator powering key
[x^2]	calculator squaring function key
[$\sqrt{}$]	calculator square root function key
[x!]	calculator factorial function key
[tan]	calculator tangent function key
[sin]	calculator sine function key
[cos]	calculator cosine function key
[log]	calculator logarithm function key
[INV], [2nd], or [F]	calculator second function key
[EE] or [EXP]	calculator scientific notation key

Acknowledgments

Unless otherwise acknowledged, all photographs are the property of Scott, Foresman and Company. Page abbreviations are as follows: (T) top, (C) center, (B) bottom, (L) left, (R) right.

COVER & TITLE PAGE Steven Hunt (c)1994 **vi(L)** Stephen Studd/Tony Stone Images **vi(R)** George Hall/Check Six **vii** Uniphoto **viii** Index Stock International **ix** Nadia Mackenzie/Tony Stone Images **x(R)** Pete McArthur/Tony Stone Images **x(L)** West Light **3** AP/Wide World **4-5T** Profiles West **4C** Stephen Studd/Tony Stone Images **4BL** Backgrounds/West Light **4BR** Michael Mazzeo/The Stock Market **5** Ed Manowicz/Tony Stone Images **6** Tom Ives **8** Clive Brunskill/ALLSPORT USA **9** PhotoFest **11** B.Markel/Gamma-Liaison **13** Bob Daemmrich/The Image Works **15** Louis Psihoyos/Matrix **16** Hank Ketcham **17** Clearwater Florida Fire and Police Departments of Public Safety **20** Tony Freeman/Photo Edit **21** Paul Conklin **22** Rita Boseruf **26T** Library of Congress **26C&B** Courtesy United Air Lines **27** Michael Newman/Photo Edit **28** Robinson/ANIMALS ANIMALS **29B** Zig Leszczynski/ANIMALS ANIMALS **30** California Institute of Technology **31** Sidney Harris **35** Milt & Joan Mann/Cameramann International, Ltd. **37** Lawrence Migdale **38** Dr. Duane de Temple **39** Bob Daemmrich/Tony Stone Images **41** Photo: Bill Hogan/Copyrighted, Chicago Tribune Company, all rights reserved, **45** The Vatican/Art Resource, New York **53** L.Rorke/The Image Works **55** David Spangler **59** NASA **60B** Telegraph Color Library/FPG **60T** Eddie Adams/Leo de Wys **70-71T** Tony Hallas/SPL/Photo Researchers **70C** George Hall/Check Six **70-71B** David Lawrence/Panoramic Stock Images **71C** Steven E.Sutton/Duomo Photography Inc. **72** Robert Frerck/Odyssey Productions, Chicago **74-75** Tony Stone Images **76** NASA **79** Martha Swope **82** Milt & Joan Mann/Cameramann International, Ltd. **85** AP/Wide World **89** Christopher Morris/Black Star **91** Focus On Sports **92** Milt & Joan Mann/Cameramann International, Ltd. **93** Bob Daemmrich/Stock Boston **94** John Elk III/Stock Boston **95** David Falconer/David R. Frazier Photolibrary **96** JPL/NASA **99B** Grant Heilman/Grant Heilman Photography **101** Johnny Johnson/ANIMALS ANIMALS **104** Pasley/Stock Boston **106** Don DuBroff Photo **107T** Mary Kate Denny/Photo Edit **107B** Conte/ANIMALS ANIMALS **108** Michael Newman/Photo Edit **109** Milt & Joan Mann/Cameramann International, Ltd. **112** Robert Frerck/Tony Stone Images **113** Robert Frerck/Odyssey Productions, Chicago **117** Milt & Joan Mann/Cameramann International, Ltd. **118** Scala/Art Resource, New York **119** Alex S.MacLean/Landslides **120** Bob Daemmrich **122** Milt & Joan Mann/Cameramann International, Ltd. **125** Patrick Ward/Stock Boston **126** AP/Wide World **129** Chip Henderson/Tony Stone Images **131T** Julian Baum/SPL/Photo Researchers **131B** Ken Korsh/FPG **132T** Charly Franklin/FPG **132C** Scott Spiker/The Stock Shop **132B** Donovan Reese/Tony Stone Images **134** Ron Thomas/FPG **135** Bob Daemmrich/Stock Boston **140B** Telegraph Colour Library/FPG **140-141(TR)** Imtek Imagineering/Masterfile **140TL** Pelton & Associates/West Light **140-141C** G.Biss/Masterfile **141B** Uniphoto **143** Milt & Joan Mann/Cameramann International, Ltd. **144** Mark Burnett/Photo Edit **146** William Johnson/Stock Boston **149** Kevin Syms/David R. Frazier Photolibrary **150** Robert W.Ginn/Photo Edit **155** Jeff Greenberg/dMRp/Photo Edit **156** Jeff Greenberg/dMRp/Photo Edit **157** Rick Maiman/Sygma **160** David R. Frazier Photolibrary **165** Andy Hayt 1994 **167** Lee Boltin **169** Mark Twain National Forest/U.S.Forestry Service **172** J.C.Stevenson/ANIMALS ANIMALS **174** William Johnson/Stock Boston **177** Milt & Joan Mann/Cameramann International, Ltd. **179** Milt & Joan Mann/Cameramann International, Ltd. **182** Janice Rubin/Black Star **184** Mary Kate Denny/Photo Edit **185** Milt & Joan Mann/Cameramann International, Ltd. **187** Elk/Bruce Coleman Inc. **188** Scott Camazine/Photo Researchers **190** Art Pahlke **195** FPG **196** Reuters/Bettmann **197** Milt & Joan Mann/Cameramann International, Ltd. **198** Freeman/Grishaber/Photo Edit **199** Ralph Nelson, Jr./PhotoFest **200ALL** Carol Zacny **201** Leslye Borden/Photo Edit **205C** Ken Reid/FPG **205BL** Bruce Bishop/PhotoFile **205BR** Marc Chamberlain/Tony Stone Images **206T** John Terence Turner/FPG **206B** The name Cuisenaire and the color sequences of the rods, squares, and cubes are registered trademarks of the Cuisenaire Company of America, Inc. **214-215T** Steven Curtis **214CL** Gordon Wilts/Adventure Photo **214CR** C.Moore/West Light **214-215B** Gerry Ellis Nature Photography **215C** Perry Conway/The Stock Broker **216** Jan Kanter **219** Jim Pickerell/Stock Boston **220B** Museum of Modern Art/Film Stills Archive **221** Naoki Okamoto/The Stock Market **222** Photo Edit **223** Mike Penney/David R. Frazier Photolibrary **224** James Blank/Bruce Coleman Inc. **225** Beryl Goldberg **226** Independence National Historical Park Collection **227** JPL/NASA **228-229** Don W.Fawcett/Visuals Unlimited **230** Sygma **232** Karen Usiskin **233** David Young-Wolff/Photo Edit **234** The Stock Market **240** Reproduced from the Story of the Great American West ©1977 The Readers Digest Association, Inc. Used by permission. Artist: David K.Stone **243** Milt & Joan Mann/Cameramann International, Ltd. **245** David R. Frazier Photolibrary **246** Milt & Joan Mann/Cameramann International, Ltd. **249** Tony Freeman/Photo Edit **250** Max Gibbs/Oxford Scientific Films/ANIMALS ANIMALS **251** Tony Stone Images **253** Milt & Joan Mann/Cameramann International, Ltd. **255** Sygma **257** Focus On Sports **260** Copyright, National Geographic Society **261** Stacy Pick/Stock Boston **268** Milt & Joan Mann/Cameramann International, Ltd. **272** Rhodes/Earth Scenes **273** Milt & Joan Mann/Cameramann International, Ltd. **274** Adamsmith Productions/ West Light **275L** Telegraph Color Library/FPG **282-283T** The Stock Market **282C** Charly Franklin/FPG **282BL** Ron Watts/West Light **282-283B** Jean Miele/The Stock Market **283C** Jeff Schultz/Leo de Wys **284** M.Richards/Photo Edit **285** Phil McCarten/Photo Edit **290** Focus On Sports **291** Milt & Joan Mann/Cameramann International, Ltd. **293** Milt & Joan Mann/Cameramann International, Ltd. **294** Carol Zacny **297** Jerry Wachter/Focus On Sports **300** Bob Daemmrich/Stock Boston **302** Focus On Sports **304** Arthur Rackham Illustration **309** Courtesy General Electric Corp. **315** Michael Newman/Photo Edit **317** Stephen McBrady/Photo Edit **318T** John Stern/Earth Scenes **318B** Donald Specker/Earth Scenes **321** John Neubauer/Photo Edit **322** Scott Zapel **325** Danny Daniels/AlaskaStock Images **326T** Gail McCann/Photo Researchers **326B** Jerry Cooke **327** Focus On Sports **328** Milt & Joan Mann/Cameramann International, Ltd. **330** Joanne K.Peterson **331** Nick Sapiena/Stock Boston **337** Milt & Joan Mann/Cameramann International, Ltd. **338T** C.Brewer/H. Armstrong Roberts **338C** Dave Reede/First Light **338-339B** Alan Briere/Natural Selection **339T** Chris Springman/PhotoFile **339** Charly Franklin/FPG **348-349T** R.Gage/FPG **348CL** Robert George Young/Masterfile **348CR** Charly Franklin/FPG **348-349B** Richard Fukubara/West Light **349C** Index Stock International **350** David R. Austen/Stock Boston **353** Robert Torre/Tony Stone Images **355** Yoram Lehmann/Peter Arnold, Inc. **356** Jim Schwabel/Southern Stock Photos **357** James Blank/Southern Stock Photos **359** William Johnson/Stock Boston **360** B.P.Wolff/Photo Researchers **362** Michael Newman/Photo Edit **364** Courtesy Levi Strauss & Company, San Francisco, CA. **366** City Art Museum of St.Louis **367** Focus On Sports **370** Barbara Campbell/Gamma-Liaison **371** NASA **372B** Timothy White/ABC News **376** Stephen Ferry/Gamma-Liaison **377** Michael Newman/Photo Edit **379** Ric Patzke **380** Zig Leszczynski/Earth Scenes **385** Focus On Sports **386T** AP/Wide World **387** Everett Collection **390** Milt & Joan Mann/Cameramann International, Ltd. **391** David Spangler **393** Everett Collection **398** Joseph F.Viesti/Viesti Associates **399** Barry Iverson/Woodfin Camp & Associates **407** Milt & Joan Mann/Cameramann International, Ltd. **408** Telegraph Colour Library/FPG **409T** Tony Garcia/Tony Stone Images **409BR** Tecmap/West Light **416T** Index Stock International **416C** Richard Laird/FPG **416BL** Penny Tweedie/Tony Stone Images **416BR** R.Ian Lloyd/West Light **417T** Deuter/Zefa/H. Armstrong Roberts **418** Robert E. Daemmrich/Tony Stone Images **420** Willard Luce/ANIMALS ANIMALS **422** M.A.Chappell/ANIMALS ANIMALS **423** Charles Gupton/Stock Boston **425** Mark M. Lawrence/The Stock Market **432** James Blank/Stock Boston **434** David Spangler **437** Patricia Woeber **439** Owaki/Kulla/The Stock Market **442** Tony Freeman/Photo Edit **444** Katoomba Scenic Railway, New South Wales, Australia **447** Milt & Joan Mann/Cameramann International, Ltd. **448** Fritz Prenzel/ANIMALS ANIMALS **449** Robert Frerck/Odyssey Productions, Chicago **454** Jim Merli/Visuals Unlimited **457** Imtek Imagineering-1/Masterfile **458** Boisvieux/Photo Researchers **459** Robert Frerck/Woodfin Camp & Associates **460T** Robert Frerck/Odyssey Productions, Chicago **460B** Delip Mehta/Woodfin Camp & Associates **463** Robert E. Daemmrich/The Image Works **465** Felicia Martinez/Photo Edit **469** Robert Rathe/Stock Boston **472** Robert E. Daemmrich/The Image Works **473** Focus On Sports **476T** AP/Wide World **476BL** C.Ursillo/H. Armstrong Roberts **476R** L.Powers/H. Armstrong Roberts **477C** Gregory Heisler/The Image Bank **482** Historical Pictures/Stock Montage, Inc. **483** Mike Andrews/Earth Scenes **484T** Chris Michaels/FPG **484CL** SuperStock, Inc. **484CR** Ralph Mercer/Tony Stone Images **484B** Joe Riley/Folio **485B** Mark Tomalty/Masterfile **488ALL** Scott Zapel **491** Murray Alcosser/The Image Bank **492T** Oxford Scientific Films/ANIMALS ANIMALS **492B** Leonard Lee Rue III/ANIMALS ANIMALS **493ALL** Leonard Lee Rue III/ANIMALS ANIMALS **496** Benn Mitchell/The Image Bank **498** Everett Collection **500** PhotoFest **502** Mary Kate Denny/Photo Edit **505** Carol Zacny **506** John Elk III/Stock Boston **509** Leo Touchet/Woodfin Camp & Associates **511** Dr. Kari Lounatmaa/SPL/Photo Researchers **513** The National Archives **515** Nuridsany et Perennou/Photo Researchers **519** MGM/Photo:/The Kobal Collection **522T** Smithsonian Institution **522-523** U. S. Bureau of Printing and Engraving **525** Felicia Martinez/Photo Edit **526** Martin Rogers/Stock Boston **527** NASA **528** Copyright the British Museum **530** NASA **531** Cara Moore/The Image Bank **537** D.Woo/Stock Boston **538** Mark C. Burnett/Photo Researchers **539** Imtek Imagineering/ Masterfile **540** Telegraph Colour Library/FPG **546-547T** R.Krubner/H. Armstrong Roberts **546C** Nadia Mackenzie/Tony Stone Images **546B** Tom Sanders/Adventure Photo **547C** R.Faris/West Light **547B** SuperStock, Inc. **550-551** David Young-Wolff/Photo Edit **558** Marco Corsetti/FPG **560** Robert Pearcy/ANIMALS ANIMALS **562** Jan Kanter **567** Eric Meola/The Image Bank **568** Focus On Sports **569** Focus On Sports **570** Shaun Botterill/ALLSPORT USA **571** Focus On Sports **573** Travelpix/FPG **576** Ken Cole/Earth Scenes **579** Tom Nebbia/The Stock Market **583** Richard Hutchings/Photo Edit **584T** Tony Freeman/Photo Edit **584B** Dennis MacDonald/Photo Edit **586** Brett Froomer/The Image Bank **591** NASA **593** Oliver Strewe/Tony Stone Images **598** Kaluzny/Thatcher/Tony Stone Images **605T** Mason Morfit/FPG **605B** Erich Lessing/Art Resource, New York **606L** Dennis Hallinan/FPG **606R** Tim Davis/AllStock Inc. **611** Bruno Brokken/ALLSPORT USA **613** Tom Stewart/The Stock Market **614TL** Charles O'Rear/West Light **614-615T** Steven Hunt/The Image Bank **614-615C** Telegraph Colour Library/FPG **614B** Telegraph Colour Library/FPG **615B** Arthur Tilley/FPG **619** THE STUDIO, 1977, Jacob Lawrence, Gift of Gull Industries John H. and Ann H. Hauberg, Links, Seattle and by exchange from the estate of Mark Tobey. Photo: Paul Macapia **620** Goltzer/The Stock Market **622-623** Brett Froomer/The Image Bank **625** David R. Frazier/Tony Stone Images **631** Photo: Milbert Orlando Brown/Copyrighted, Chicago Tribune Company, all rights reserved **633** COMPOSITION WITH RED, BLUE, YELLOW & BLACK, 1921 - Piet Mondrian. Collection Haags Gemeentemuseum, The Hague. **642** Myrleen Ferguson Cate/Photo Edit **644** Therese Smith **645** Mark Antman/The Image Works **650** John Running **651** Milt & Joan Mann/Cameramann International, Ltd. **654T** Courtesy The White House Collection **654C** Courtesy The White House Collection **654B** Copyright by the White House Historical Association **655** AP/Wide World **658L** Gwendolen Cates/Sygma **658CL** Aloma/Shooting Star **658C** Julie Dennis/SS/Shooting Star **658CR** Stephen Begleiter/Shooting Star **658R** Terry O'Neill/Sygma **658BG** Pelton & Associates, Inc./West Light **664T** Chuck O'Rear/West Light **664C** The Stock Shop **664B** Mike Fizer/Check Six **665C** Mark MacLaren **665B** Gary Conner/Photo Edit **666** Focus On Sports **671** Bob Thomason/Tony Stone Images **672** John Madere/The Stock Market **673** Fay Torresyap/Stock Boston **674** Paul Steel/The Stock Market **675** Mark E. Gibson/The Stock Market **676** Spunbarg/Photo Edit **679** James J. Hill. Reference Library St. Paul, MN. **680** Pechter/The Stock Market **681** Reuters/Bettmann **682** Photo by Gail Toerpe Publishers-Washington Island Observer Newspaper **685** The Museum of the City of New York **687** Tony Freeman/Photo Edit **689** Milt & Joan Mann/Cameramann International, Ltd. **691** Milt & Joan Mann/Cameramann International, Ltd. **692** Fred Whitehead/ANIMALS ANIMALS **693** Mark Greenberg/Visions **693B** Mark Greenberg/Visions **694** Photo Edit **695** Focus On Sports **698** Tony Freeman/Photo Edit **699** Walt Disney Collection/Everett Collection **701** Bob Torrez/Tony Stone Images **707** Focus On Sports **709** The Kobal Collection **710** Paul Conklin **711T** FPG **711C** McFarland/SuperStock, Inc. **712T** L.Powers/H. Armstrong Roberts **712C** D.E.Cox/Tony Stone Images **712B** Denis Scott/FPG **714** Vince Streano/Tony Stone Images **716** Tony Freeman/Photo Edit **718T** Jook Leung/FPG **718-719C** West Light **718BL** Larry Lee/West Light **718-719BR** Ralph Mercer/Tony Stone Images **719T** West Light **720** Pete Saloutos/The Stock Market **722** U.S. Department of Defense **725** Michael Yelman **730** Roy Morsch/The Stock Market **731** Argonne National Lab **738** Jan Kanter **740** Everett Collection **741** Focus On Sports **742** David Parker/SPL/Photo Researchers **743** Fred Whitehead/Earth Scenes **746** Milt & Joan Mann/Cameramann International, Ltd. **748** Susan Copen Oken/Dot **749** Bettmann Archive **753** MISS PEACH/Creators Syndicate **754** Sidney Harris **757** Carol Zacny **758** Owen Franken/Stock Boston **764** Calspan Corporation **765L** Jon Riley/Tony Stone Images **765C** Ralph Mercer/Tony Stone Images **765R** Karageorge/H. Armstrong Roberts **766T** Roxana Villa/Stock Illustration Source, Inc. **772-773T** Index Stock International **772C** Pete McArthur/Tony Stone Images **772-773B** Kazu Studio Ltd/FPG **773C** Telegraph Colour Library/FPG **775** Peter Sidebotham/Tony Stone Images **777** Tony Freeman/Photo Edit **778** Reprinted with Special permission of King Features Syndicate **780** Claudia Parks/The Stock Market **784** Judith Canty/Stock Boston **787** Mark Richards/Photo Edit **789** Woodfin Camp & Associates **790** Jean-Claude Figenwald **793** Rob Crandall/Stock Boston **797** Frank Siteman/Stock Boston **798** Gay Bumgarner/Tony Stone Images **799** Milt & Joan Mann/Cameramann International, Ltd. **800** Focus On Sports **801** Tom & Michelle Grimm/Tony Stone Images **807** John Gerlach/Earth Scenes **811** Mark Lewis/Tony Stone Images **812** David Young-Wolff/Photo Edit **814** Paul Conklin/Photo Edit **816** Culver Pictures **818TL** Elizabeth Simpson/FPG **818TR** The Wood River Gallery **818CL** D.Degnan/H. Armstrong Roberts **818CR** Stock Illustration Source, Inc. **818B** Arthur Tilley/FPG

INDEX

NOTES